Some Acid/Base Indicators and Their Color Changes

The pH ranges shown are approximate. Specific transition ranges depend on the indicator solvent chosen.

PERIODIC TABLE OF THE ELEMENTS

Legend:
- Metals
- Nonmetals
- Metalloids

IA 1	IIA 2	IIIB 3	IVB 4	VB 5	VIB 6	VIIB 7	VIIIB 8	VIIIB 9	VIIIB 10	IB 11	IIB 12	IIIA 13	IVA 14	VA 15	VIA 16	VIIA 17	0 18
1 **H** 1.008																1 **H** 1.008	2 **He** 4.0026
3 **Li** 6.941	4 **Be** 9.0122											5 **B** 10.81	6 **C** 12.011	7 **N** 14.007	8 **O** 15.999	9 **F** 18.9984	10 **Ne** 20.1797
11 **Na** 22.9898	12 **Mg** 24.3050											13 **Al** 26.9815	14 **Si** 28.085	15 **P** 30.9738	16 **S** 32.06	17 **Cl** 35.453	18 **Ar** 39.948
19 **K** 39.0983	20 **Ca** 40.078	21 **Sc** 44.9559	22 **Ti** 47.867	23 **V** 50.9415	24 **Cr** 51.9961	25 **Mn** 54.9380	26 **Fe** 55.845	27 **Co** 58.9332	28 **Ni** 58.6934	29 **Cu** 63.546	30 **Zn** 65.38	31 **Ga** 69.723	32 **Ge** 72.63	33 **As** 74.9216	34 **Se** 78.96	35 **Br** 79.904	36 **Kr** 83.798
37 **Rb** 85.4678	38 **Sr** 87.62	39 **Y** 88.9058	40 **Zr** 91.224	41 **Nb** 92.9064	42 **Mo** 95.96	43 **Tc** (98)	44 **Ru** 101.07	45 **Rh** 102.9055	46 **Pd** 106.42	47 **Ag** 107.8682	48 **Cd** 112.411	49 **In** 114.818	50 **Sn** 118.710	51 **Sb** 121.760	52 **Te** 127.60	53 **I** 126.9045	54 **Xe** 131.293
55 **Cs** 132.9055	56 **Ba** 137.327	57 **La*** 138.9055	72 **Hf** 178.49	73 **Ta** 180.9479	74 **W** 183.84	75 **Re** 186.207	76 **Os** 190.23	77 **Ir** 192.217	78 **Pt** 195.084	79 **Au** 196.9666	80 **Hg** 200.59	81 **Tl** 204.38	82 **Pb** 207.2	83 **Bi** 208.9804	84 **Po** (209)	85 **At** (210)	86 **Rn** (222)
87 **Fr** (223)	88 **Ra** (226)	89 **Ac**** (227)	104 **Rf** (265)	105 **Db** (268)	106 **Sg** (271)	107 **Bh** (270)	108 **Hs** (277)	109 **Mt** (276)	110 **Ds** (281)	111 **Rg** (280)	112 **Cn** (285)	113 **Uut** (284)	114 **Fl** (289)	115 **Uup** (288)	116 **Lv** (293)	117 **Uus** (294)	118 **Uuo** (294)

*Lanthanide Series

58 **Ce** 140.116	59 **Pr** 140.9076	60 **Nd** 144.242	61 **Pm** (145)	62 **Sm** 150.36	63 **Eu** 151.964	64 **Gd** 157.25	65 **Tb** 158.9254	66 **Dy** 162.500	67 **Ho** 164.9303	68 **Er** 167.259	69 **Tm** 168.9342	70 **Yb** 173.054	71 **Lu** 174.9668

** Actinide Series

90 **Th** 232.0381	91 **Pa** 231.0359	92 **U** 238.0289	93 **Np** (237)	94 **Pu** (244)	95 **Am** (243)	96 **Cm** (247)	97 **Bk** (247)	98 **Cf** (251)	99 **Es** (252)	100 **Fm** (257)	101 **Md** (258)	102 **No** (259)	103 **Lr** (262)

Note: Atomic masses are 2009 IUPAC values (up to four decimal places). More accurate values for some elements are given in the table inside the back cover.

Excel Shortcut Keystrokes for the PC*
*Macintosh equivalents, if different, appear in square brackets

TO ACCOMPLISH THIS TASK	TYPE THESE KEYSTROKES
Alternate between displaying cell values and displaying cell formulas	**Ctrl+` [⌘+`]**
Calculate all sheets in all open workbooks	**F9**
Calculate the active worksheet	**Shift+F9**
Cancel an entry in a cell or formula bar	**Esc**
Complete a cell entry and move down in the selection	**Enter [Return]**
Complete a cell entry and move to the left in the selection	**Shift+Tab**
Complete a cell entry and move to the right in the selection	**Tab**
Complete a cell entry and move up in the selection	**Shift+Enter**
Copy a formula from the cell above the active cell into the cell or the formula bar	**Ctrl+'** (Apostrophe) **[⌘+']**
Copy a selection	**Ctrl+C[⌘+C]**
Copy the value from the cell above the active cell into the cell or the formula bar	**Ctrl+Shift+"** (Quotation Mark) **[⌘+Shift+"]**
Cut a selection	**Ctrl+X [⌘+X]**
Define a name	**Ctrl+F3 [⌘+F3]**
Delete the character to the left of the insertion point, or delete the selection	**Backspace [Delete]**
Delete the character to the right of the insertion point, or delete the selection	**Delete [Del]**
Displays the Insert Function dialog box	**Shift+F3**
Displays Key Tips for ribbon shortcuts	**ALT**
Edit a cell comment	**Shift+F2**
Edit the active cell	**F2 [None]**
Edit the active cell and then clear it, or delete the preceding character in the active cell as you edit the cell contents	**Backspace [Delete]**
Enter a formula as an array formula	**Ctrl+Shift+Enter**
Fill down	**Ctrl+D[⌘+D]**
Fill the selected cell range with the current entry	**Ctrl+Enter [None]**
Fill to the right	**Ctrl+R [⌘+R]**
Format cells dialog box	**Ctrl+1 [⌘+1]**
Insert the AutoSum formula	**Alt+=** (Equal Sign) **[⌘+Shift+T]**
Move one character up, down, left, or right	**Arrow Keys**
Move to the beginning of the line	**Home**
Paste a name into a formula	**F3 [None]**
Paste a selection	**Ctrl+V [⌘+V]**
Repeat the last action	**F4 Or Ctrl+Y [⌘+Y]**
Selects the entire worksheet	**Ctrl+A**
Start a formula	**=** (Equal Sign)
Start a new line in the same cell	**Alt+Enter [⌘+Option+Enter]**
Undo	**Ctrl+Z[⌘+Z]**

Home Ribbon Wide View

Home Ribbon Narrow View

Insert Tab

Formulas Tab

Data Tab

Not shown are Page Layout, Review and View Tabs

Fundamentals of
Analytical
Chemistry

NINTH EDITION

Douglas A. Skoog
Stanford University

Donald M. West
San Jose State University

F. James Holler
University of Kentucky

Stanley R. Crouch
Michigan State University

BROOKS/COLE
CENGAGE Learning·

Australia • Brazil • Japan • Korea • Mexico • Singapore • Spain • United Kingdom • United States

BROOKS/COLE
CENGAGE Learning

**Fundamentals of Analytical Chemistry,
Ninth Edition, International Edition**
Douglas A. Skoog, Donald M. West,
F. James Holler, Stanley R. Crouch

Publisher: Mary Finch

Acquisitions Editor: Chris Simpson

Developmental Editor: Sandra Kiselica

Assistant Editor: Alicia Landsberg

Media Editor: Rebecca Berardy Schwartz

Brand Manager: Nicole Hamm

Market Development Manager:
Janet del Mundo

Content Project Manager: Jennifer Risden

Art Director: Maria Epes

Manufacturing Planner: Judy Inouye

Rights Acquisitions Specialist: Don Schlotman

Production Service and Composition:
PreMediaGlobal

Photo Researcher: Bill Smith Group

Text Researcher: Isabel Saraiva

Copy Editor: PreMediaGlobal

Illustrator: Rolin Graphics, Inc.;
Nesbitt Graphics; PreMediaGlobal

Text Designer: Bill Reuter

Interior Design Image: John F. Barry

Cover Image: Charles D. Winters/Photo
Researchers, Inc.

Cover Designer: Bruce Saltzman

International Edition:
ISBN-13: 978-1-285-05624-1
ISBN-10: 1-285-05624-8

Cengage Learning International Offices

Asia
www.cengageasia.com
tel: (65) 6410 1200

Australia/New Zealand
www.cengage.com.au
tel: (61) 3 9685 4111

Brazil
www.cengage.com.br
tel: (55) 11 3665 9900

India
www.cengage.co.in
tel: (91) 11 4364 1111

Latin America
www.cengage.com.mx
tel: (52) 55 1500 6000

UK/Europe/Middle East/Africa
www.cengage.co.uk
tel: (44) 0 1264 332 424

Represented in Canada by Nelson Education, Ltd.
www.nelson.com
tel: (416) 752 9100 / (800) 668 0671

Excel® screen captures are used courtesy of Microsoft Corporation.

Cengage Learning is a leading provider of customized learning solutions with office locations around the globe, including Singapore, the United Kingdom, Australia, Mexico, Brazil, and Japan. Locate your local office at:
www.cengage.com/global

For product information and free companion resources:
www.cengage.com/international
Visit your local office: **www.cengage.com/global**
Visit our corporate website: **www.cengage.com**

Printed in the United States of America
1 2 3 4 5 6 7 16 15 14 13 12

Contents in Brief

Contents

Preface

The ninth edition of *Fundamentals of Analytical Chemistry* is an introductory textbook designed primarily for a one- or two-semester course for chemistry majors. Since the publication of the eighth edition, the scope of analytical chemistry has continued to evolve, and thus, we have included in this edition many applications to biology, medicine, materials science, ecology, forensic science, and other related fields. As in the previous edition, we have incorporated many spreadsheet applications, examples, and exercises. We have revised some older treatments to incorporate contemporary instrumentation and techniques. In response to the comments of many readers and reviewers, we have added a chapter on mass spectrometry to provide detailed instruction on this vital topic as early as possible in the chemistry curriculum. Our companion book, *Applications of Microsoft® Excel in Analytical Chemistry*, 2nd ed., provides students with a tutorial guide for using spreadsheets in analytical chemistry and introduces many additional spreadsheet operations.

We recognize that courses in analytical chemistry vary from institution to institution and depend on the available facilities and instrumentation, the time allocated to analytical chemistry in the chemistry curriculum, and the unique instructional philosophies of teachers. We have, therefore, designed the ninth edition of *Fundamentals of Analytical Chemistry* so that instructors can tailor the text to meet their needs and students can learn the concepts of analytical chemistry on several levels: in descriptions, in pictorials, in illustrations, in interesting and relevant features, and in using online learning.

Since the production of the eighth edition of this text, the duties and responsibilities for planning and writing a new edition have fallen to two of us (FJH and SRC). While making the many changes and improvements cited above and in the remainder of the preface, we have maintained the basic philosophy and organization of the eight previous editions and endeavored to preserve the same high standards that characterized those texts.

OBJECTIVES

The primary objective of this text is to provide a thorough background in the chemical principles that are particularly important to analytical chemistry. Second, we want students to develop an appreciation for the difficult task of judging the accuracy and precision of experimental data and to show how these judgments can be sharpened by applying statistical methods to analytical data. Third, we aim to introduce a broad range of modern and classic techniques that are useful in analytical chemistry. Fourth, we hope that, with the help of this book, students will develop the skills necessary to solve quantitative analytical problems and, where appropriate, use powerful spreadsheet tools to solve problems, perform calculations, and create simulations of chemical phenomena. Finally, we aim to teach laboratory skills that will give students confidence in their ability to obtain high-quality analytical data and that will highlight the importance of attention to detail in acquiring these data.

COVERAGE AND ORGANIZATION

The material in this text covers both fundamental and practical aspects of chemical analysis. We have organized the chapters into Parts that group together related topics. There are seven major Parts to the text that follow the brief introduction in Chapter 1.

- **Part I** covers the tools of analytical chemistry and comprises seven chapters. Chapter 2 discusses the chemicals and equipment used in analytical laboratories and includes many photographs of analytical operations. Chapter 3 is a tutorial introduction to the use of spreadsheets in analytical chemistry. Chapter 4 reviews the basic calculations of analytical chemistry, including expressions of chemical concentration and stoichiometric relationships. Chapters 5, 6, and 7 present topics in statistics and data analysis that are important in analytical chemistry and incorporate extensive use of spreadsheet calculations. Analysis of variance, ANOVA, is included in Chapter 7, and Chapter 8 provides details about acquiring samples, standardization, and calibration.

- **Part II** covers the principles and application of chemical equilibrium systems in quantitative analysis. Chapter 9 explores the fundamentals of chemical equilibria. Chapter 10 discusses the effect of electrolytes on equilibrium systems. The systematic approach for attacking equilibrium problems in complex systems is the subject of Chapter 11.

- **Part III** brings together several chapters dealing with classical gravimetric and volumetric analytical chemistry. Gravimetric analysis is described in Chapter 12. In Chapters 13 through 17, we consider the theory and practice of titrimetric methods of analysis, including acid/base titrations, precipitation titrations, and complexometric titrations. We take advantage of the systematic approach to equilibria and the use of spreadsheets in the calculations.

- **Part IV** is devoted to electrochemical methods. After an introduction to electrochemistry in Chapter 18, Chapter 19 describes the many uses of electrode potentials. Oxidation/reduction titrations are the subject of Chapter 20, while Chapter 21 presents the use of potentiometric methods to measure concentrations of molecular and ionic species. Chapter 22 considers the bulk electrolytic methods of electrogravimetry and coulometry, and Chapter 23 discusses voltammetric methods, including linear sweep and cyclic voltammetry, anodic stripping voltammetry, and polarography.

- **Part V** presents spectroscopic methods of analysis. The nature of light and its interaction with matter are explored in Chapter 24. Spectroscopic instruments and their components are the topics covered in Chapter 25. The various applications of molecular absorption spectrometric methods are discussed in some detail in Chapter 26, while Chapter 27 is concerned with molecular fluorescence spectroscopy. Chapter 28 covers various atomic spectrometric methods, including plasma and flame emission methods and electrothermal and flame atomic absorption spectroscopy. Chapter 29 on mass spectrometry is new to this edition and provides an introduction to ionization sources, mass analyzers, and ion detectors. Both atomic and molecular mass spectrometry are included.

- **Part VI** includes five chapters dealing with kinetics and analytical separations. We investigate kinetic methods of analysis in Chapter 30. Chapter 31 introduces analytical separations including ion exchange and the various chromatographic methods. Chapter 32 discusses gas chromatography, while high-performance liquid chromatography is covered in Chapter 33. The final chapter in this Part, Chapter 34, introduces some miscellaneous separation methods, including supercritical fluid chromatography, capillary electrophoresis, and field-flow fractionation.

- The final **Part VII** consists of four chapters dealing with the practical aspects of analytical chemistry. These chapters are published on our website at **www.cengage .com/chemistry/skoog/fac9.** We consider real samples and compare them to ideal samples in Chapter 35. Methods for preparing samples are discussed in Chapter 36, while techniques for decomposing and dissolving samples are covered in Chapter 37. The text ends with Chapter 38, which provides detailed procedures for laboratory experiments that cover many of the principles and applications discussed in previous chapters.

FLEXIBILITY

Because the text is divided into Parts, there is substantial flexibility in the use of the material. Many of the Parts can stand alone or be taken in a different order. For example, some instructors may want to cover spectroscopic methods prior to electrochemical methods or separations prior to spectroscopic methods.

HIGHLIGHTS

This edition incorporates many features and methods intended to enhance the learning experience for the student and to provide a versatile teaching tool for the instructor.

Important Equations. Equations that we feel are the most important have been highlighted with a color screen for emphasis and ease of review.

Mathematical Level. Generally the principles of chemical analysis developed here are based on college algebra. A few of the concepts presented require basic differential and integral calculus.

Worked Examples. A large number of worked examples serve as aids in understanding the concepts of analytical chemistry. In this edition, we title the examples for easier identification. As in the eighth edition, we follow the practice of including units in chemical calculations and using the factor-label method to check correctness. The examples also are models for the solution of problems found at the end of most of the chapters. Many of these use spreadsheet calculations as described next. Where appropriate, solutions to the worked examples are clearly marked with the *word* Solution for ease in identification.

Spreadsheet Calculations. Throughout the book we have introduced spreadsheets for problem solving, graphical analysis, and many other applications. Microsoft Excel® on the PC has been adopted as the standard for these calculations, but the instructions can be easily adapted to other spreadsheet programs and platforms. Many other detailed examples are presented in our companion book, *Applications of Microsoft® Excel in Analytical Chemistry*, 2nd ed. We have attempted to document each stand-alone spreadsheet with working formulas and entries.

Spreadsheet Summaries. References to our companion book *Applications of Microsoft® Excel in Analytical Chemistry*, 2nd ed., are given as Spreadsheet Summaries in the text. These are intended to direct the user to examples, tutorials, and elaborations of the text topics.

Questions and Problems. An extensive set of questions and problems is included at the end of most chapters. Answers to approximately half of the problems are given at the end of the book. Many of the problems are best solved using spreadsheets. These are identified by a spreadsheet icon ▦ placed in the margin next to the problem.

Challenge Problems. Most of the chapters have a challenge problem at the end of the regular questions and problems. Such problems are intended to be open-ended, research-type problems that are more challenging than normal. These problems may consist of multiple steps, dependent on one another, or may require library or Web searches to find information. We hope that these challenge problems stimulate discussion and extend the topics of the chapter into new areas. We encourage instructors to use them in innovative ways, such as for group projects, inquiry-driven learning assignments, and case study discussions. Because many challenge problems are open-ended and may have multiple solutions, we do not provide answers or explanations for them.

Features. A series of boxed and highlighted Features are found throughout the text. These essays contain interesting applications of analytical chemistry to the modern world, derivation of equations, explanations of more difficult theoretical points, or historical notes. Examples include, W. S. Gosset ("Student") (Chapter 7), Antioxidants (Chapter 20), Fourier Transform Spectroscopy (Chapter 25), LC/MS/MS (Chapter 33), and Capillary Electrophoresis in DNA Sequencing (Chapter 34).

Illustrations and Photos. We feel strongly that photographs, drawings, pictorials, and other visual aids greatly assist the learning process. Hence, we have included new and updated visual materials to aid the student. Most of the drawings are done in two colors to increase the information content and to highlight important aspects of the figures. Photographs and color plates taken exclusively for this book by renowned chemistry photographer Charles Winters are intended to illustrate concepts, equipment, and procedures that are difficult to illustrate with drawings.

Expanded Figure Captions. Where appropriate, we have attempted to make the figure captions quite descriptive so that reading the caption provides a second level of explanation for many of the concepts. In some cases, the figures can stand by themselves much in the manner of a *Scientific American* illustration.

Web Works. In most of the chapters we have included a brief Web Works feature at the end of the chapter. In these features, we ask the student to find information on the web, do online searches, visit the websites of equipment manufacturers, or solve analytical problems. These Web Works and the links given are intended to stimulate student interest in exploring the information available on the World Wide Web. The links will be updated regularly on our website, **www.cengage.com/chemistry/skoog/fac9.**

Glossary. At the end of the book we have placed a glossary that defines the most important terms, phrases, techniques, and operations used in the text. The glossary is intended to provide students with a means for rapidly determining a meaning without having to search through the text.

Appendixes and Endpapers. Included in the appendixes are an updated guide to the literature of analytical chemistry; tables of chemical constants, electrode potentials, and recommended compounds for the preparation of standard materials; sections on the use of logarithms and exponential notation and normality and equivalents (terms that are not used in the text itself); and a derivation of the propagation of error equations. The endpapers of this book provide a full-color chart of chemical indicators, a periodic table, a 2009 IUPAC table of atomic masses, and a table of molar masses of compounds of particular interest in analytical chemistry based on the 2009 atomic masses. In addition, included in the book is a tear-out reference card for the 2010 and 2007 versions of Microsoft Excel.

WHAT'S NEW

Readers of the eighth edition will find numerous changes in the ninth edition in content as well as in style and format.

Content. Several changes in content have been made to strengthen the book.

- Many chapters have been strengthened by adding spreadsheet examples, applications, and problems. Chapter 3 gives tutorials on the construction and use of spreadsheets. Many other tutorials are included in our supplement, *Applications of Microsoft® Excel in Analytical Chemistry*, 2nd ed., and a number of these have been corrected, updated, and augmented.

- The definitions of molar concentration have been updated in Chapter 4 to conform to current IUPAC usage, and the associated terminology including *molar concentration* and *molar analytical concentration* have been infused throughout the text.

- The chapters on statistics (5–7) have been updated and brought into conformity with the terminology of modern statistics. Analysis of variance (ANOVA) has been included in Chapter 7. ANOVA is very easy to perform with modern spreadsheet programs and quite useful in analytical problem solving. These chapters are closely linked to our Excel supplement through Examples, Features, and Summaries.

- In Chapter 8, explanations of external standard, internal standard, and standard additions methods have been clarified, expanded, and described more thoroughly. Special attention is paid to the use of least-squares methods in standardization and calibration.

- A new introduction and explanation of mass balance has been written for Chapter 11.

- An explanation and a marginal note have been added on the gravimetric factor.

- A new feature on the master equation approach was added to Chapter 14.

- Chapter 17 has been rewritten to include both complexation and precipitation titrations.

- Chapters 18, 19, 20, and 21 on electrochemical cells and cell potentials have been revised to clarify and unify the discussion. Chapter 23 has been altered to decrease the emphasis on classical polarography. The chapter now includes a discussion of cyclic voltammetry.

- In Chapter 25, the discussion on thermal IR detectors now puts more emphasis on the DTGS pyroelectric detector.

- Chapter 29 introduces both atomic and molecular mass spectrometry and covers the similarities and differences in these methods. The introduction of mass spectrometry allows the separation chapters (31–34) to place additional emphasis on combined techniques, such as chromatographic methods with mass spectrometric detection.

- The challenge problems have been updated, augmented, and replaced where appropriate.

- References to the analytical chemistry literature have been updated and corrected as necessary.

- *Digital Object Identifiers* (*DOIs*) have been added to most references to the primary literature. These universal identifiers greatly simplify the task of locating articles by a link on the website **www.doi.org.** A DOI may be typed into a form on the home page, and when the identifier is submitted, the browser transfers directly to the article on the publisher's website. For example, 10.1351/goldbook.C01222 can be typed into the form, and the browser is directed to the IUPAC article on concentration. Alternatively, DOIs may be entered directly into the URL blank of any browser as http://dx.doi.org/10.1351/goldbook.C01222. Please note that students or instructors must have authorized access to the publication of interest.

Style and Format. We have continued to make style and format changes to make the text more readable and student friendly.

- We have attempted to use shorter sentences, a more active voice, and a more conversational writing style in each chapter.
- More descriptive figure captions are used whenever appropriate to allow a student to understand the figure and its meaning without alternating between text and caption.
- Molecular models are used liberally in most chapters to stimulate interest in the beauty of molecular structures and to reinforce structural concepts and descriptive chemistry presented in general chemistry and upper-level courses.
- Several new figures have replaced obsolete figures of past editions.
- Photographs, taken specifically for this text, are used whenever appropriate to illustrate important techniques, apparatus, and operations.
- Marginal notes are used throughout to emphasize recently discussed concepts or to reinforce key information.
- Key terms are now defined in the margins throughout the book.
- All examples now delineate the question and its answer or solution.

SUPPORTING MATERIALS

Please visit **www.cengage.com/chemistry/skoog/fac9** for information about student and instructor resources for this text.

ACKNOWLEDGMENTS

We wish to acknowledge with thanks the comments and suggestions of many reviewers who critiqued the eighth edition prior to our writing or who evaluated the current manuscript in various stages.

REVIEWERS

Lane Baker,
  Indiana University
Heather Bullen,
  Northern Kentucky University
Peter de Boves Harrington,
  Ohio University
Jani Ingram,
  Northern Arizona University
R. Scott Martin,
  St. Louis University

Gary Rice,
  College of William and Mary
Kathryn Severin,
  Michigan State University
Dana Spence,
  Michigan State University
Scott Waite,
  University of Nevada, Reno

We especially acknowledge the assistance of Professor David Zellmer, California State University, Fresno, who served as the accuracy reviewer for the book. Dave's deep knowledge of analytical chemistry, his tight focus on detail, and his problem-solving and spreadsheet prowess are powerful assets for our team. We are especially indebted to the late Bryan Walker, who, while a student in Dave Zellmer's analytical chemistry course, gleefully reported a number of errors that Dave (and we) had not detected in the eighth edition. Bryan's pleasant personality, academic talent, and attention to detail inspired Dave as he worked with us on this edition. We extend a special thanks to James Edwards of St. Louis University for checking all the back-of-the-book answers to the Questions and and Problems. We also appreciate the good works of Professor Bill Vining of the State University of New York, Oneonta, who prepared many online tutorials and Charles M. Winters, who contributed many of the photos in the text and most of the color plates.

Our writing team enjoys the services of a superb technical reference librarian, Ms. Janette Carver of the University of Kentucky Science Library. She assisted us in many ways in the production of this book, including checking references, performing literature searches, and arranging for interlibrary loans. We appreciate her competence, enthusiasm, and good humor.

We are grateful to the many members of the staff of Cengage, who provided solid support during the production of this text. Acquiring Sponsoring Editor Chris Simpson has provided excellent leadership and encouragement throughout the course of this project. This is our fourth book with Senior Developmental Editor Sandi Kiselica. As always, she has done a marvelous job of overseeing and organizing the project, maintaining continuity, and making many important comments and suggestions. Simply put, she's the best in the business, and we sincerely appreciate her work. We are grateful to our copy editor, James Corrick, for his consistency and attention to detail. His keen eye and excellent editorial skills have contributed significantly to the quality of the text. Alicia Landsberg has done a fine job coordinating the various ancillary materials, and Jeremy Glover, our photo researcher, has handled the many tasks associated with acquiring new photos and securing permissions for graphics. Project Manager Erin Donahue of PreMediaGlobal kept the project moving with daily reminders and frequent schedule updates while coordinating the entire production process. Her counterpart at Cengage was Content Project Manager Jennifer Risden, who coordinated the editorial process. Finally, we thank Rebecca Berardy Schwartz, our Cengage media editor for this edition.

This is the first edition of *Fundamentals of Analytical Chemistry* written without the skill, guidance, and counsel of our senior coauthors Douglas A. Skoog and Donald M. West. Doug died in 2008, and Don followed in 2011. Doug was Don's preceptor while he was a graduate student at Stanford University, and they began writing analytical chemistry textbooks together in the 1950s. They produced twenty editions of three best-selling textbooks over a period of forty-five years. Doug's vast knowledge of analytical chemistry and consummate writing skill coupled with Don's organizational expertise and attention to detail formed an outstanding complement. We aspire to maintain the high standard of excellence of Skoog and West as we continue to build on their legacy. In honor of their manifest contributions to the philosophy, organization, and writing of this book and many others, we have chosen to list their names above the title. Since the publication of the eighth edition, the team lost another partner in Judith B. Skoog, Doug's wife who died in 2010. Judy was a world-class editorial assistant who typed and proofread twenty editions of three books (and most of the instructor's manuals), amounting to well over 100,000 pages. We miss her accuracy, speed, tenacity, good humor, and friendship in producing beautiful manuscripts.

Finally, we are deeply grateful to our wives Vicki Holler and Nicky Crouch for their counsel, patience, and support during the several years of writing this text and preparing it for production.

F. James Holler
Stanley R. Crouch

The Nature of Analytical Chemistry

Analytical chemistry is a measurement science consisting of a set of powerful ideas and methods that are useful in all fields of science, engineering, and medicine. Some exciting illustrations of the power and significance of analytical chemistry have occurred, are occurring, and will occur during NASA's rover explorations of the planet Mars. On July 4, 1997, the Pathfinder spacecraft delivered the Sojourner rover to the Martian surface. Analytical instruments returned information on the chemical composition of rocks and soil. Investigations by the lander and rover suggested that Mars was at one time in its past warm and wet with liquid water on the surface and water vapor in the atmosphere. In January 2004, the Mars rovers Spirit and Opportunity arrived on Mars for a three-month mission. A major result from Spirit's alpha particle X-ray spectrometer (APXS) and Mossbauer spectrometer was finding concentrated deposits of silica and, at a different site, high concentrations of carbonate. Spirit continued to explore and transmit data until 2010, outliving even the most optimistic predictions. Even more amazing, Opportunity continues to travel the surface of Mars and, by March, 2012, had covered more than 21 miles exploring and transmitting images of craters, small hills, and other features.

In late 2011, the Mars Science Laboratory aboard the rover Curiosity was launched. It arrived on August 6, 2012 with a host of analytical instruments on board. The Chemistry and Camera package includes a laser-induced breakdown spectrometer (LIBS, see Chapter 28) and a remote microimager. The LIBS instrument will provide determination

NASA/JPL-Caltech

NASA/JPL-Caltech

Mars Science Laboratory aboard rover Curiosity   Curiosity observing Martian landscape from Gale crater, August 2012 of many elements with no sample preparation. It can determine the identity and amounts of major, minor, and trace elements and can detect hydrated minerals. The sample analysis package contains a quadrupole mass spectrometer (Chapter 29), a gas chromatograph (Chapter 32), and a tunable laser spectrometer (Chapter 25). Its goals are to survey carbon compound sources, search for organic compounds important to life, reveal the chemical and isotopic states of several elements, determine the composition of the Martian atmosphere, and search for noble gas and light element isotopes.[1]

These examples demonstrate that both qualitative and quantitative information are required in an analysis. **Qualitative analysis** establishes the chemical identity of the species in the sample. **Quantitative analysis** determines the relative amounts of these species, or **analytes**, in numerical terms. The data from the various spectrometers on the rovers contain both types of information. As is common with many analytical instruments, the gas chromatograph and mass spectrometer incorporate a separation step as a necessary part of the analytical process. With a few analytical tools, exemplified here by the APXS and LIBS experiments, chemical separation of the various elements contained in the rocks is unnecessary since the methods provide highly selective information. In this text, we will explore quantitative methods of analysis, separation methods, and the principles behind their operation. A qualitative analysis is often an integral part of the separation step, and determining the identity of the analytes is an essential adjunct to quantitative analysis.

> **Qualitative analysis** reveals the *identity* of the elements and compounds in a sample.
>
> **Quantitative analysis** indicates the *amount* of each substance in a sample.
>
> **Analytes** are the components of a sample that are determined.

1A THE ROLE OF ANALYTICAL CHEMISTRY

Analytical chemistry is applied throughout industry, medicine, and all the sciences. To illustrate, consider a few examples. The concentrations of oxygen and of carbon dioxide are determined in millions of blood samples every day and used to diagnose and treat illnesses. Quantities of hydrocarbons, nitrogen oxides, and carbon monoxide present in automobile exhaust gases are measured to determine the effectiveness of emission-control devices. Quantitative measurements of ionized calcium in blood serum help diagnose parathyroid disease in humans. Quantitative determination of nitrogen in foods establishes their protein content and thus their nutritional value. Analysis of steel during its production permits adjustment in the concentrations of such elements as carbon, nickel, and chromium to achieve a desired strength, hardness, corrosion resistance, and ductility. The mercaptan content of household gas supplies is monitored continually to ensure that the gas has a sufficiently obnoxious odor to warn of dangerous leaks. Farmers tailor fertilization and irrigation schedules to meet changing plant needs during the growing season, gauging these needs from quantitative analyses of plants and soil.

Quantitative analytical measurements also play a vital role in many research areas in chemistry, biochemistry, biology, geology, physics, and the other sciences. For example, quantitative measurements of potassium, calcium, and sodium ions in the body fluids of animals permit physiologists to study the role these ions play in nerve-signal conduction as well as muscle contraction and relaxation. Chemists unravel the mechanisms of chemical reactions through reaction rate studies. The rate of consumption of reactants or formation of products

---

[1]For details on the Mars Science Laboratory mission and the rover Curiosity, see http://www.nasa.gov.

in a chemical reaction can be calculated from quantitative measurements made at precise time intervals. Materials scientists rely heavily on quantitative analyses of crystalline germanium and silicon in their studies of semiconductor devices whose impurities lie in the concentration range of 1×10^{-6} to 1×10^{-9} percent. Archaeologists identify the sources of volcanic glasses (obsidian) by measuring concentrations of minor elements in samples taken from various locations. This knowledge in turn makes it possible to trace prehistoric trade routes for tools and weapons fashioned from obsidian.

Many chemists, biochemists, and medicinal chemists devote much time in the laboratory gathering quantitative information about systems that are important and interesting to them. The central role of analytical chemistry in this enterprise and many others is illustrated in **Figure 1-1**. All branches of chemistry draw on the ideas and techniques of analytical chemistry. Analytical chemistry has a similar function with respect to the many other scientific fields listed in the diagram. Chemistry is often called *the central science*; its top center position and the central position of analytical chemistry in the figure

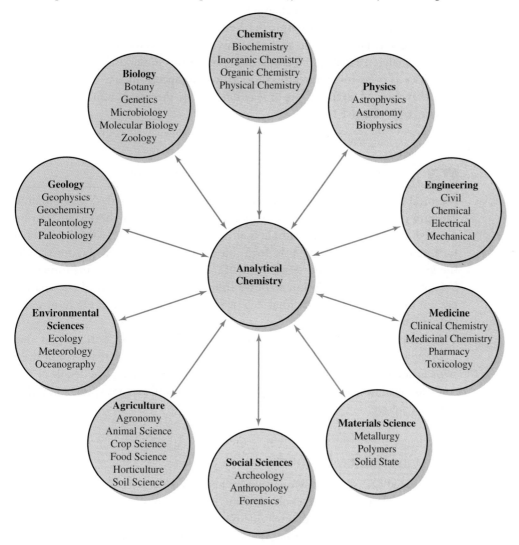

Figure 1-1 The relationship between analytical chemistry, other branches of chemistry, and the other sciences. The central location of analytical chemistry in the diagram signifies its importance and the breadth of its interactions with many other disciplines.

emphasize this importance. The interdisciplinary nature of chemical analysis makes it a vital tool in medical, industrial, government, and academic laboratories throughout the world.

1B QUANTITATIVE ANALYTICAL METHODS

We compute the results of a typical quantitative analysis from two measurements. One is the mass or the volume of sample being analyzed. The second measurement is of some quantity that is proportional to the amount of analyte in the sample such as mass, volume, intensity of light, or electrical charge. This second measurement usually completes the analysis, and we usually classify analytical methods according to the nature of this final measurement. In **gravimetric methods**, we determine the mass of the analyte or some compound chemically related to it. In a **volumetric method**, we measure the volume of a solution containing sufficient reagent to react completely with the analyte. In **electroanalytical methods**, we measure electrical properties such as potential, current, resistance, and quantity of electrical charge. In **spectroscopic methods**, we explore the interaction between electromagnetic radiation and analyte atoms or molecules or the emission of radiation by analytes. Finally, in a group of miscellaneous methods, we measure such quantities as mass-to-charge ratio of ions by mass spectrometry, rate of radioactive decay, heat of reaction, rate of reaction, sample thermal conductivity, optical activity, and refractive index.

1C A TYPICAL QUANTITATIVE ANALYSIS

A typical quantitative analysis includes the sequence of steps shown in the flow diagram of **Figure 1-2**. In some instances, one or more of these steps can be omitted. For example, if the sample is already a liquid, we can avoid the dissolution step. Chapters 1 through 34 focus on the last three steps in Figure 1-2. In the measurement step, we measure one of the physical properties mentioned in Section 1B. In the calculation step, we find the relative amount of the analyte present in the samples. In the final step, we evaluate the quality of the results and estimate their reliability.

In the paragraphs that follow, you will find a brief overview of each of the nine steps shown in Figure 1-2. We then present a case study to illustrate the use of these steps in solving an important and practical analytical problem. The details of this study foreshadow many of the methods and ideas you will explore as you study analytical chemistry.

1C-1 Choosing a Method

The essential first step in any quantitative analysis is the selection of a method as depicted in Figure 1-2. The choice is sometimes difficult and requires experience as well as intuition. One of the first questions that must be considered in the selection process is the level of accuracy required. Unfortunately, high reliability nearly always requires a large investment of time. The selected method usually represents a compromise between the accuracy required and the time and money available for the analysis.

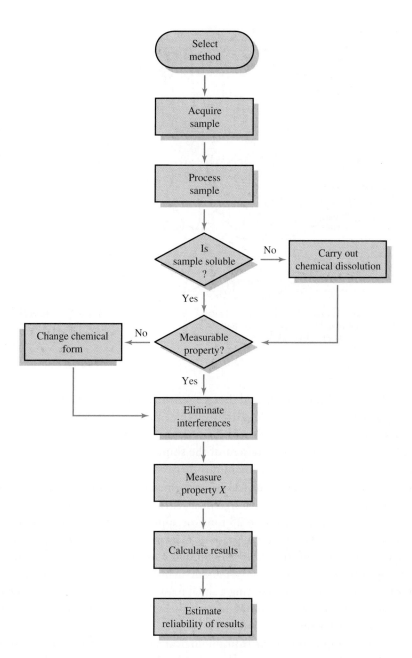

Figure 1-2 Flow diagram showing the steps in a quantitative analysis. There are a number of possible paths through these steps. In the simplest example represented by the central vertical pathway, we select a method, acquire and process the sample, dissolve the sample in a suitable solvent, measure a property of the analyte, calculate the results, and estimate the reliability of the results. Depending on the complexity of the sample and the chosen method, various other pathways may be necessary.

A second consideration related to economic factors is the number of samples that will be analyzed. If there are many samples, we can afford to spend a significant amount of time in preliminary operations such as assembling and calibrating instruments and equipment and preparing standard solutions. If we have only a single sample or a just a few samples, it may be more appropriate to select a procedure that avoids or minimizes such preliminary steps.

Finally, the complexity of the sample and the number of components in the sample always influence the choice of method to some degree.

1C-2 Acquiring the Sample

As illustrated in Figure 1-2, the second step in a quantitative analysis is to acquire the sample. To produce meaningful information, an analysis must be performed on a sample that has the same composition as the bulk of material from which it was

"TODAY EVERYONE HAS TO KNOW 'WHAT'S IN THE FOOD?', 'WHAT'S IN THE WATER?', 'WHAT'S IN THE AIR?' THIS IS TRULY THE 'GOLDEN' AGE OF ANALYTICAL CHEMISTRY."

ScienceCartoonsPlus.com

A material is **heterogeneous** if its constituent parts can be distinguished visually or with the aid of a microscope. Coal, animal tissue, and soil are heterogeneous.

An **assay** is the process of determining how much of a given sample is the material by its indicated name. For example, a zinc alloy is assayed for its zinc content, and its assay is a particular numerical value.

We *analyze* samples, and we *determine* substances. For example, a blood sample is analyzed to determine the concentrations of various substances such as blood gases and glucose. We, therefore, speak of the determination of blood gases or glucose, *not* the analysis of blood gases or glucose.

taken. When the bulk is large and **heterogeneous**, great effort is required to get a representative sample. Consider, for example, a railroad car containing 25 tons of silver ore. The buyer and seller of the ore must agree on a price, which will be based primarily on the silver content of the shipment. The ore itself is inherently heterogeneous, consisting of many lumps that vary in size as well as in silver content. The **assay** of this shipment will be performed on a sample that weighs about one gram. For the analysis to have significance, the composition of this small sample must be representative of the 25 tons (or approximately 22,700,000 g) of ore in the shipment. Isolation of one gram of material that accurately represents the average composition of the nearly 23,000,000 g of bulk sample is a difficult undertaking that requires a careful, systematic manipulation of the entire shipment. **Sampling** is the process of collecting a small mass of a material whose composition accurately represents the bulk of the material being sampled. Sampling is discussed in more detail in Chapter 8.

The collection of specimens from biological sources represents a second type of sampling problem. Sampling of human blood for the determination of blood gases illustrates the difficulty of acquiring a representative sample from a complex biological system. The concentration of oxygen and carbon dioxide in blood depends on a variety of physiological and environmental variables. For example, applying a tourniquet incorrectly or hand flexing by the patient may cause the blood oxygen concentration to fluctuate. Because physicians make life-and-death decisions based on results of blood gas analyses, strict procedures have been developed for sampling and transporting specimens to the clinical laboratory. These procedures ensure that the sample is representative of the patient at the time it is collected and that its integrity is preserved until the sample can be analyzed.

Many sampling problems are easier to solve than the two just described. Whether sampling is simple or complex, however, the analyst must be sure that the laboratory sample is representative of the whole before proceeding. Sampling is frequently the most difficult step in an analysis and the source of greatest error. The final analytical result will never be any more reliable than the reliability of the sampling step.

1C-3 Processing the Sample

As shown in Figure 1-2, the third step in an analysis is to process the sample. Under certain circumstances, no sample processing is required prior to the measurement step. For example, once a water sample is withdrawn from a stream, a lake, or an ocean, the pH of the sample can be measured directly. Under most circumstances, we must process the sample in one of several different ways. The first step in processing the sample is often the preparation of a laboratory sample.

Preparing a Laboratory Sample

A solid laboratory sample is ground to decrease particle size, mixed to ensure homogeneity, and stored for various lengths of time before analysis begins. Absorption or desorption of water may occur during each step, depending on the humidity of the environment. Because any loss or gain of water changes the chemical composition of solids, it is a good idea to dry samples just before starting an analysis. Alternatively, the moisture content of the sample can be determined at the time of the analysis in a separate analytical procedure.

Liquid samples present a slightly different but related set of problems during the preparation step. If such samples are allowed to stand in open containers, the solvent may evaporate and change the concentration of the analyte. If the analyte is a gas dissolved in a liquid, as in our blood gas example, the sample container must be kept inside a second sealed container, perhaps during the entire analytical procedure, to prevent contamination by atmospheric gases. Extraordinary measures, including sample manipulation and measurement in an inert atmosphere, may be required to preserve the integrity of the sample.

Defining Replicate Samples

Most chemical analyses are performed on **replicate samples** whose masses or volumes have been determined by careful measurements with an analytical balance or with a precise volumetric device. Replication improves the quality of the results and provides a measure of their reliability. Quantitative measurements on replicates are usually averaged, and various statistical tests are performed on the results to establish their reliability.

Replicate samples, or **replicates**, are portions of a material of approximately the same size that are carried through an analytical procedure at the same time and in the same way.

Preparing Solutions: Physical and Chemical Changes

Most analyses are performed on solutions of the sample made with a suitable solvent. Ideally, the solvent should dissolve the entire sample, including the analyte, rapidly and completely. The conditions of dissolution should be sufficiently mild that loss of the analyte cannot occur. In our flow diagram of Figure 1-2, we ask whether the sample is soluble in the solvent of choice. Unfortunately, many materials that must be analyzed are insoluble in common solvents. Examples include silicate minerals, high-molecular-mass polymers, and specimens of animal tissue. With such substances, we must follow the flow diagram to the box on the right and perform some rather harsh chemistry. Converting the analyte in such materials into a soluble form is often the most difficult and time-consuming task in the analytical process. The sample may require heating with aqueous solutions of strong acids, strong bases, oxidizing agents, reducing agents, or some combination of such reagents. It may be necessary to ignite the sample in air or oxygen or to perform a high-temperature fusion of the sample in the presence of various fluxes. Once the analyte is made soluble, we then ask whether the sample has a property that is proportional to analyte concentration and that we can measure. If it does not, other chemical steps may be necessary, as shown in Figure 1-2, to convert the analyte to a form that is suitable for the measurement step. For example, in the determination of manganese in steel, the element must be oxidized to MnO_4^- before the absorbance of the colored solution is measured (see Chapter 26). At this point in the analysis, it may be possible to proceed directly to the measurement step, but more often than not, we must eliminate interferences in the sample before making measurements, as illustrated in the flow diagram.

1C-4 Eliminating Interferences

An **interference** or **interferent** is a species that causes an error in an analysis by enhancing or attenuating (making smaller) the quantity being measured.

Once we have the sample in solution and converted the analyte to an appropriate form for measurement, the next step is to eliminate substances from the sample that may interfere with measurement (see Figure 1-2). Few chemical or physical properties of importance in chemical analysis are unique to a single chemical species. Instead, the reactions used and the properties measured are characteristic of a group of elements of compounds. Species other than the analyte that affect the final measurement are called **interferences**, or **interferents**. A scheme must be devised to isolate the analytes from interferences before the final measurement is made. No hard and fast rules can be given for eliminating interference. This problem can certainly be the most demanding aspect of an analysis. Chapters 31 through 34 describe separation methods in detail.

1C-5 Calibrating and Measuring Concentration

The **matrix**, or **sample matrix**, is the collection of all of the components in the sample containing an analyte.

Techniques or reactions that work for only one analyte are said to be **specific**. Techniques or reactions that apply to only a few analytes are **selective**.

All analytical results depend on a final measurement X of a physical or chemical property of the analyte, as shown in Figure 1-2. This property must vary in a known and reproducible way with the concentration c_A of the analyte. Ideally, the measurement of the property is directly proportional to the concentration, that is,

$$c_A = kX$$

Calibration is the process of determining the proportionality between analyte concentration and a measured quantity.

where k is a proportionality constant. With a few exceptions, analytical methods require the empirical determination of k with chemical standards for which c_A is known.[2] The process of determining k is thus an important step in most analyses; this step is called a **calibration**. Calibration methods are discussed in some detail in Chapter 8.

1C-6 Calculating Results

Computing analyte concentrations from experimental data is usually relatively easy, particularly with computers. This step is depicted in the next-to-last block of the flow diagram of Figure 1-2. These computations are based on the raw experimental data collected in the measurement step, the characteristics of the measurement instruments, and the stoichiometry of the analytical reaction. Samples of these calculations appear throughout this book.

1C-7 Evaluating Results by Estimating Reliability

As the final step in Figure 1-2 shows, analytical results are complete only when their reliability has been estimated. The experimenter must provide some measure of the uncertainties associated with computed results if the data are to have any value.

---

[2]Two exceptions are gravimetric methods, discussed in Chapter 12, and coulometric methods, considered in Chapter 22. In both these methods, k can be computed from known physical constants.

Chapters 5, 6, and 7 present detailed methods for carrying out this important final step in the analytical process.

1D AN INTEGRAL ROLE FOR CHEMICAL ANALYSIS: FEEDBACK CONTROL SYSTEMS

Analytical chemistry is usually not an end in itself but is part of a bigger picture in which the analytical results may be used to help control a patient's health, to control the amount of mercury in fish, to control the quality of a product, to determine the status of a synthesis, or to find out whether there is life on Mars. Chemical analysis is the measurement element in all of these examples and in many other cases. Consider the role of quantitative analysis in the determination and control of the concentration of glucose in blood. The system flow diagram of **Figure 1-3** illustrates the process. Patients suffering from insulin-dependent diabetes mellitus develop hyperglycemia, which manifests itself in a blood glucose concentration above the normal concentration range of 65 to 100 mg/dL. We begin our example by determining that the desired state is a blood glucose level below 100 mg/dL. Many patients must monitor their blood glucose levels by periodically submitting samples to a clinical laboratory for analysis or by measuring the levels themselves using a handheld electronic glucose monitor.

The first step in the monitoring process is to determine the actual state by collecting a blood sample from the patient and measuring the blood glucose level. The results are displayed, and then the actual state is compared to the desired state, as shown in Figure 1-3. If the measured blood glucose level is above 100 mg/dL, the patient's insulin level, which is a controllable quantity, is increased by injection or oral administration. After a delay to allow the insulin time to take effect, the glucose level is measured again to determine if the desired state has been achieved. If the level is below the threshold, the insulin level has been maintained, so no insulin is required. After a suitable delay time, the blood glucose level is measured again, and the cycle is repeated. In this way, the insulin level in the patient's blood, and thus the

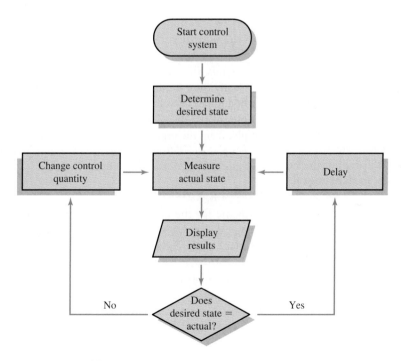

Figure 1-3 Feedback system flow diagram. The desired system state is defined, the actual state of the system is measured, and the two states are compared. The difference between the two states is used to change a controllable quantity that results in a change in the state of the system. Quantitative measurements are again performed on the system, and the comparison is repeated. The new difference between the desired state and the actual state is again used to change the state of the system if necessary. The process provides continuous monitoring and feedback to maintain the controllable quantity, and thus the actual state, at the proper level. The text describes the monitoring and control of blood glucose as an example of a feedback control system.

blood glucose level, is maintained at or below the critical threshold, which keeps the metabolism of the patient under control.

The process of continuous measurement and control is often referred to as a **feedback system**, and the cycle of measurement, comparison, and control is called a **feedback loop**. These ideas are widely applied in biological and biomedical systems, mechanical systems, and electronics. From the measurement and control of the concentration of manganese in steel to maintaining the proper level of chlorine in a swimming pool, chemical analysis plays a central role in a broad range of systems.

FEATURE 1-1

Deer Kill: A Case Study Illustrating the Use of Analytical Chemistry to Solve a Problem in Toxicology

Analytical chemistry is a powerful tool in environmental investigations. In this feature, we describe a case study in which quantitative analysis was used to determine the agent that caused deaths in a population of white-tailed deer in a wildlife area of a national recreational area in Kentucky. We begin with a description of the problem and then show how the steps illustrated in Figure 1-2 were used to solve the analytical problem. This case study also shows how chemical analysis is used in a broad context as an integral part of the feedback control system depicted in Figure 1-3.

White-tailed deer have proliferated in many parts of the country.

The Problem

The incident began when a park ranger found a dead white-tailed deer near a pond in the Land between the Lakes National Recreation Area in western Kentucky. The ranger enlisted the help of a chemist from the state veterinary diagnostic laboratory to find the cause of death so that further deer kills might be prevented.

The ranger and the chemist carefully inspected the site where the badly decomposed carcass of the deer had been found. Because of the advanced state of decomposition, no fresh organ tissue samples could be gathered. A few days after the original inquiry, the ranger found two more dead deer near the same location. The chemist was summoned to the site of the kill, where he and the ranger loaded the deer onto a truck for transport to the veterinary diagnostic laboratory. The investigators then conducted a careful examination of the surrounding area in an attempt to find clues to establish the cause of death.

The search covered about 2 acres surrounding the pond. The investigators noticed that grass surrounding nearby power line poles was wilted and discolored. They speculated that a herbicide might have been used on the grass. A common ingredient in herbicides is arsenic in any one of a variety of forms, including arsenic trioxide, sodium arsenite, monosodium methanearsenate, and disodium methanearsenate. The last compound is the disodium salt of methanearsenic acid, $CH_3AsO(OH)_2$, which is very soluble in water and thus finds use as the active ingredient in many herbicides. The herbicidal activity of disodium methanearsenate is due to its reactivity with the sulfhydryl (S—H) groups in the amino acid cysteine. When cysteine in plant enzymes reacts with arsenical compounds, the enzyme function is inhibited, and the plant eventually dies. Unfortunately, similar chemical effects occur in animals as well. The investigators, therefore, collected samples of the discolored dead grass for testing along with samples from the organs of the deer. They planned to analyze the samples to confirm the presence of arsenic and, if present, to determine its concentration in the samples.

Selecting a Method

A scheme for the quantitative determination of arsenic in biological samples is found in the published methods of the Association of Official Analytical Chemists (AOAC).[3] In this method, arsenic is distilled as arsine, AsH_3, and is then determined by colorimetric measurements.

---

[3]*Official Methods of Analysis,* 18th ed., Method 973.78 , Washington, DC: Association of Official Analytical Chemists, 2005.

Processing the Sample: Obtaining Representative Samples

Back at the laboratory, the deer were dissected, and the kidneys were removed for analysis. The kidneys were chosen because the suspected pathogen (arsenic) is rapidly eliminated from an animal through its urinary tract.

Processing the Sample: Preparing a Laboratory Sample

Each kidney was cut into pieces and homogenized in a high-speed blender. This step served to reduce the size of the pieces of tissue and to homogenize the resulting laboratory sample.

Processing the Sample: Defining Replicate Samples

Three 10-g samples of the homogenized tissue from each deer were placed in porcelain crucibles. These served as replicates for the analysis.

Doing Chemistry: Dissolving the Samples

To obtain an aqueous solution of for analysis, it was necessary to convert its organic matrix to carbon dioxide and water by the process of **dry ashing**. This process involved heating each crucible and sample cautiously over an open flame until the sample stopped smoking. The crucible was then placed in a furnace and heated at 555°C for two hours. Dry ashing served to free the analyte from organic material and convert it to arsenic pentoxide. The dry solid in each sample crucible was then dissolved in dilute HCl, which converted the As_2O_5 to soluble H_3AsO_4.

Eliminating Interferences

Arsenic can be separated from other substances that might interfere in the analysis by converting it to arsine, AsH_3, a toxic, colorless gas that is evolved when a solution of H_3AsO_3 is treated with zinc. The solutions resulting from the deer and grass samples were combined with Sn^{2+}, and a small amount of iodide ion was added to catalyze the reduction of H_3AsO_4 to H_3AsO_3 according to the following reaction:

$$H_3AsO_4 + SnCl_2 + 2HCl \rightarrow H_3AsO_3 + SnCl_4 + H_2O$$

The H_3AsO_3 was then converted to AsH_3 by the addition of zinc metal as follows:

$$H_3AsO_3 + 3Zn + 6HCl \rightarrow AsH_3(g) + 3ZnCl_2 + 3H_2O$$

Throughout this text, we will present models of molecules that are important in analytical chemistry. Here we show arsine, AsH_3. Arsine is an extremely toxic, colorless gas with a noxious garlic odor. Analytical methods involving the generation of arsine must be carried out with caution and proper ventilation.

The entire reaction was carried out in flasks equipped with a stopper and delivery tube so that the arsine could be collected in the absorber solution as shown in **Figure 1F-1**. The arrangement ensured that interferences were left in the reaction flask and that only arsine was collected in the absorber in special transparent containers called cuvettes.

Figure 1F-1 An easily constructed apparatus for generating arsine, AsH_3.

(continued)

Arsine bubbled into the solution in the cuvette reacts with silver diethyldithiocarbamate to form a colored complex compound according to the following equation:

Molecular model of diethyldithiocarbamate. This compound is an analytical reagent used in determining arsenic.

$$AsH_3 + 6Ag^+ + 3 \left[\begin{matrix} C_2H_5 \\ C_2H_5 \end{matrix} N{-}C \begin{matrix} S \\ S \end{matrix} \right]^- \longrightarrow$$

$$As \left[\begin{matrix} C_2H_5 \\ C_2H_5 \end{matrix} N{-}C \begin{matrix} S \\ S \end{matrix} \right]_3 + 6Ag + 3H^+$$

Measuring the Amount of the Analyte

The amount of arsenic in each sample was determined by measuring the intensity of the red color formed in the cuvettes with an instrument called a spectrophotometer. As shown in Chapter 26, a spectrophotometer provides a number called **absorbance** that is directly proportional to the color intensity, which is also proportional to the concentration of the species responsible for the color. To use absorbance for analytical purposes, a calibration curve must be generated by measuring the absorbance of several solutions that contain known concentrations of analyte. The upper part of **Figure 1F-2** shows that the color becomes more intense as the arsenic content of the standards increases from 0 to 25 parts per million (ppm).

Calculating the Concentration

The absorbances for the standard solutions containing known concentrations of arsenic are plotted to produce a calibration curve, shown in the lower part of Figure 1F-2. Each vertical line between the upper and lower parts of Figure 1F-2 ties a solution to its corresponding point on the plot. The color intensity of each solution is represented by its absorbance, which is plotted on the vertical axis of the calibration curve.

Note that the absorbance increases from 0 to about 0.72 as the concentration of arsenic increases from 0 to 25 parts per million. The concentration of arsenic in each standard solution corresponds to the vertical grid lines of the calibration curve as shown. This curve is then used to determine the concentration of the two unknown solutions shown on the right. We first find the absorbances of the unknowns on the absorbance axis of the plot and then read the corresponding concentrations on the concentration axis. The lines leading from the cuvettes to the calibration curve show that the concentrations of arsenic in the two deer samples were 16 ppm and 22 ppm, respectively.

Arsenic in kidney tissue of an animal is toxic at levels above about 10 ppm, so it was probable that the deer were killed by ingesting an arsenic compound. The tests also showed that the samples of grass contained about 600 ppm arsenic. This very high level of arsenic suggested that the grass had been sprayed with an arsenical herbicide. The investigators concluded that the deer had probably died as a result of eating the poisoned grass.

Estimating the Reliability of the Data

The data from these experiments were analyzed using the statistical methods described in Chapters 5- 8. For each of the standard arsenic solutions and the deer samples, the average of the three absorbance measurements was calculated. The average absorbance for the replicates is a more reliable measure of the concentration of arsenic than a single measurement. Least-squares analysis of the standard data (see Section 8D) was used to find the best straight line among the points and to calculate the concentrations of the unknown samples along with their statistical uncertainties and confidence limits.

Conclusion

In this analysis, the formation of the highly colored product of the reaction served both to confirm the probable presence of arsenic and to provide a reliable estimate of its concentration in the deer and in the grass. Based on their results, the investigators recommended that the use of arsenical herbicides be suspended in the wildlife area to protect the deer and other animals that might eat plants there.

Figure 1F-2   Constructing and using a calibration curve to determine the concentration of arsenic. The absorbances of the solutions in the cuvettes are measured using a spectrophotometer. The absorbance values are then plotted against the concentrations of the solutions in the cuvettes, as illustrated in the graph. Finally, the concentrations of the unknown solutions are read from the plot, as shown by the dark arrows.

The case study of Feature 1-1 illustrates how chemical analysis is used in the identification and determination of quantities of hazardous chemicals in the environment. Many of the methods and instruments of analytical chemistry are used routinely to provide vital information in environmental and toxicological studies of this type. The system flow diagram of Figure 1-3 may be applied to this case study. The desired state is a concentration of arsenic that is below the toxic level. Chemical analysis is used to determine the actual state, or the concentration of arsenic in the environment, and this value is compared to the desired concentration. The difference is then used to determine appropriate actions (such as decreased use of arsenical pesticides) to ensure that deer are not poisoned by excessive amounts of arsenic in the environment, which in this example is the controlled system. Many other examples are given in the text and in features throughout this book.

PART I

Tools of Analytical Chemistry

Chemicals, Apparatus, and Unit Operations of Analytical Chemistry

CHAPTER 2

At the heart of analytical chemistry is a core set of operations and equipment. This set is necessary for laboratory work in the discipline and serves as the foundation for its growth and development. In this photo, a student records titration data in a laboratory notebook for an experiment to determine nitrogen in a sample of organic matter.

In this chapter, we shall introduce the tools, techniques, and chemicals that are used by analytical chemists. The development of these tools began over two centuries ago and continues today. As the technology of analytical chemistry has improved with the advent of electronic analytical balances, automated titrators, and other computer-controlled instruments, the speed, convenience, accuracy, and precision of analytical methods have generally improved as well. For example, the determination of the mass of a sample that required 5 to 10 minutes 40 years ago is now accomplished in a few seconds. Computations that took 10 to 20 minutes using tables of logarithms may now be carried out almost instantaneously with a computer spreadsheet or calculator. Our experience with such magnificent technological innovations often elicits impatience with the sometimes tedious techniques of classical analytical chemistry. It is this impatience that drives the quest to develop better methodologies. Indeed, basic methods have often been modified in the interest of speed or convenience without sacrificing accuracy or precision.

We must emphasize, however, that many of the unit operations encountered in the analytical laboratory are timeless. These tried and true operations have gradually evolved over the past two centuries. From time to time, the directions given in this chapter may seem somewhat didactic. Although we attempt to explain why unit operations are carried out in the way that we describe, you may be tempted to modify a procedure or skip a step here or there to save time and effort. We must caution you against modifying techniques and procedures unless you have discussed your proposed modification with your instructor and have considered its consequences carefully. Such modifications may cause unanticipated results, including unacceptable levels of accuracy or precision. In a worst-case scenario, a serious accident could result. Today, the time required to prepare a carefully standardized solution of sodium hydroxide is about the same as it was 100 years ago.

Mastery of the tools of analytical chemistry will serve you well in chemistry courses and in related scientific fields. In addition, your efforts will be rewarded with the considerable satisfaction of having completed an analysis with high standards of good analytical practice and with levels of accuracy and precision consistent with the limitations of the technique.

2A SELECTING AND HANDLING REAGENTS AND OTHER CHEMICALS

The purity of reagents has an important bearing on the accuracy attained in any analysis. It is, therefore, essential that the quality of a reagent be consistent with its intended use.

2A-1 Classifying Chemicals

Reagent Grade

Reagent-grade chemicals conform to the minimum standards set forth by the Reagent Chemical Committee of the American Chemical Society (ACS)[1] and are used whenever possible in analytical work. Some suppliers label their products with the maximum limits of impurity allowed by the ACS specifications while others print actual concentrations for the various impurities.

Primary-Standard Grade

The qualities required of a **primary standard**, in addition to extraordinary purity, are discussed in Section 13A-2. Primary-standard reagents have been carefully analyzed by the supplier, and the results are printed on the container label. The National Institute of Standards and Technology (NIST) is an excellent source for primary standards. This agency also prepares and sells **reference standards**, which are complex substances that have been exhaustively analyzed.[2]

> The National Institute of Standards and Technology (NIST) is the current name of what was formerly the National Bureau of Standards.

Special-Purpose Reagent Chemicals

Chemicals that have been prepared for a specific application are also available. Included among these are solvents for spectrophotometry and high-performance liquid chromatography. Information pertinent to the intended use is supplied with these reagents. Data provided with a spectrophotometric solvent, for example, might include its absorbance at selected wavelengths and its ultraviolet cutoff wavelength.

2A-2 Rules for Handling Reagents and Solutions

A high-quality chemical analysis requires reagents and solutions of known purity. A freshly opened bottle of a reagent-grade chemical can usually be used with confidence. Whether this same confidence is justified when the bottle is half empty

---

[1]Committee on Analytical Reagents, *Reagent Chemicals,* 10th ed., Washington, DC: American Chemical Society, 2005, available on-line or hard-bound.
[2]The Standard Reference Materials Program (SRMP) of the NIST provides thousands of reference materials for sale. The NIST maintains a catalog and price list of these materials at a website that is linked to the main NIST site at www.nist.gov. Standard reference materials may be purchased online.

depends entirely on the way it has been handled after being opened. We observe the following rules to prevent the accidental contamination of reagents and solutions:

1. Select the best grade of chemical available for analytical work. Whenever possible, pick the smallest bottle that is sufficient to do the job.
2. Replace the top of every container *immediately* after removing reagent. Do not rely on someone else to do so.
3. Hold the stoppers of reagent bottles between your fingers. Never set a stopper on a desk top.
4. *Unless specifically directed otherwise, never return any excess reagent to a bottle.* The money saved by returning excesses is seldom worth the risk of contaminating the entire bottle.
5. Unless directed otherwise, never insert spatulas, spoons, or knives into a bottle that contains a solid chemical. Instead, shake the capped bottle vigorously or tap it gently against a wooden table to break up an encrustation. Then pour out the desired quantity. These measures are occasionally ineffective, and in such cases a clean porcelain spoon should be used.
6. Keep the reagent shelf and the laboratory balance clean and neat. Clean up any spills immediately.
7. Follow local regulations concerning the disposal of surplus reagents and solutions.

2B CLEANING AND MARKING OF LABORATORY WARE

A chemical analysis is usually performed in duplicate or triplicate. Each vessel that holds a sample must be marked so that its contents can be positively identified. Flasks, beakers, and some crucibles have small etched areas on which semipermanent markings can be made with a pencil.

Special marking inks are available for porcelain surfaces. The marking is baked permanently into the glaze by heating at a high temperature. A saturated solution of iron(III) chloride, although not as satisfactory as the commercial preparation, can also be used for marking.

Every beaker, flask, or crucible that will contain the sample must be thoroughly cleaned before being used. The apparatus should be washed with a hot detergent solution and then rinsed—initially with large amounts of tap water and finally with several small portions of deionized water.[3] Properly cleaned glassware will be coated with a uniform and unbroken film of water. *It is seldom necessary to dry the interior surface of glassware before use.* Drying is usually a waste of time and is always a potential source of contamination.

> Unless you are directed otherwise, do not dry the interior surfaces of glassware or porcelain ware.

An organic solvent, such as methyl ethyl ketone or acetone, may be effective in removing grease films. Chemical suppliers also market preparations for eliminating such films.

---

[3]References to deionized water in this chapter and Chapter 38 apply equally to distilled water.

Charles D. Winters

Figure 2-1 Arrangement for the evaporation of a liquid.

Bumping is the sudden, often violent boiling that tends to spatter solution out of its container.

Wet ashing is the oxidation of the organic constituents of a sample with oxidizing reagents such as nitric acid, sulfuric acid, hydrogen peroxide, aqueous bromine, or a combination of these reagents.

An **analytical balance** has a maximum capacity that ranges from 1 g to several kilograms and a precision at maximum capacity of at least 1 part in 10^5.

A **macrobalance** is the most common type of analytical balance, and it has a maximum load of 160 to 200 g and a precision of 0.1 mg.

A **semimicroanalytical balance** has a maximum load of 10 to 30 g and a precision of 0.01 mg.

A **microanalytical balance** has a maximum load of 1 to 3 g and a precision of 0.001 mg, or 1 μg.

2C EVAPORATING LIQUIDS

It is often necessary to reduce the volume of a solution that contains a nonvolatile solute. **Figure 2-1** illustrates how this procedure is accomplished. The ribbed cover glass permits vapors to escape and protects the remaining solution from accidental contamination.

Evaporation is frequently difficult to control because of the tendency of some solutions to overheat locally. The **bumping** that results can be sufficiently vigorous to cause partial loss of the solution. Careful and gentle heating will minimize the danger of such loss. Glass beads may also minimize bumping if their use is permissible.

Some unwanted substances can be eliminated during evaporation. For example, chloride and nitrate can be removed from a solution by adding sulfuric acid and evaporating until copious white fumes of sulfur trioxide are observed (this operation must be performed in a hood). Urea is effective in removing nitrate ion and nitrogen oxides from acidic solutions. Ammonium chloride is best removed by adding concentrated nitric acid and evaporating the solution to a small volume. Ammonium ion is rapidly oxidized when it is heated. The solution is then evaporated to dryness.

Organic constituents can frequently be eliminated from a solution by adding sulfuric acid and heating to the appearance of sulfur trioxide fumes (in a hood). This process is known as **wet ashing**. Nitric acid can be added toward the end of heating to hasten oxidation of the last traces of organic matter.

2D MEASURING MASS

In most analyses, an *analytical balance* must be used to measure masses with high accuracy. Less accurate *laboratory balances* are also used for mass measurements when the demands for reliability are not critical.

2D-1 Types of Analytical Balances

An **analytical balance** is an instrument for determining mass with a maximum capacity that ranges from 1 g to a few kilograms with a precision of at least 1 part in 10^5 at maximum capacity. The precision and accuracy of many modern analytical balances exceed 1 part in 10^6 at full capacity.

The most common analytical balances (**macrobalances**) have a maximum capacity ranging between 160 and 200 g. With these balances, measurements can be made with a standard deviation of ± 0.1 mg. **Semimicroanalytical balances** have a maximum loading of 10 to 30 g with a precision of ± 0.01 mg. A typical **microanalytical balance** has a capacity of 1 to 3 g and a precision of ± 0.001 mg (1 μg).

The analytical balance has evolved dramatically over the past several decades. The traditional analytical balance had two pans attached to either end of a lightweight beam that pivoted about a knife edge located in the center of the beam. The object to be weighed was placed on one pan. Standard masses were then added to the other pan to restore the beam to its original position. Weighing with such an **equal-arm balance** was tedious and time consuming.

The first **single-pan analytical balance** appeared on the market in 1946. The speed and convenience of weighing with this balance were vastly superior to what could be realized with the traditional equal-arm balance. As a result, this balance rapidly replaced the latter in most laboratories. The single-pan balance is currently being replaced by the **electronic analytical balance**, which has neither a beam nor a knife edge. This type of balance is discussed in Section 2D-2. The single-pan balance is still used in some laboratories, but the speed, ruggedness, convenience, accuracy, and capability for computer control and data logging of electronic balances ensure that the mechanical single-pan analytical balance will soon disappear from the scene. The design and operation of a single-pan balance are discussed briefly in Section 2D-3.

2D-2 The Electronic Analytical Balance[4]

Figure 2-2 shows a diagram and a photo of an electronic analytical balance. The pan rides above a hollow metal cylinder that is surrounded by a coil that fits over the inner pole of a cylindrical permanent magnet. An electric current in the coil produces a magnetic field that supports or **levitates** the cylinder, the pan and indicator arm, and whatever load is on the pan. The current is adjusted so that the level of the indicator arm is in the null position when the pan is empty. Placing an object on the pan causes the pan and indicator arm to move downward, thus increasing the amount of light striking the photocell of the null detector. The increased current from the photocell is amplified and fed into the coil, creating a larger magnetic field, which returns the pan to its original null position. A device such as this, in which a small electric current causes a mechanical system to maintain a null position, is called a **servo system**. The current required to keep the pan and object in the null position

To **levitate** means to cause an object to float in air.

A **servo system** is a device in which a small electric signal causes a mechanical system to return to a null position.

Figure 2-2 Electronic analytical balance. (a) Block diagram. (b) Photo of electronic balance. (a) *Reprinted (adapted) with permission from R. M. Schoonover,* Anal. Chem., *1982, 54, 973A. Published 1982, American Chemical Society.*

[4]For a more detailed discussion, see R. M. Schoonover, *Anal. Chem.,* **1982**, *54,* 973A, **DOI**: 10.1021/ac00245a003.

Figure 2-3   Electronic analytical balances. (a) Classical configuration with pan beneath the cell. (b) A top-loading design. Note that the mechanism is enclosed in a windowed case. *(a) Reprinted (adapted) with permission from R. M. Schoonover, Anal. Chem., 1982, 54, 973A. Published 1982, American Chemical Society.* (b) K.M. Lang. *Amer. Lab.*, **1983**, 15(3), 72. Copyright 1983 by International Scientific Communications, Inc.

is directly proportional to the mass of the object and is easily measured, digitized, and displayed. Electronic balances are calibrated by weighing a standard mass and adjusting the current so that the exact mass of the standard appears on the display.

Figure 2-3 shows the configurations for two electronic analytical balances. In each, the pan is tethered to a system of constraints known collectively as a **cell**. The cell incorporates several **flexures** that permit limited movement of the pan and prevent torsional forces (resulting from off-center loading) from disturbing the alignment of the balance mechanism. At null, the beam is parallel to the gravitational horizon, and each flexure pivot is in a relaxed position.

Figure 2-3a shows an electronic balance with the pan located below the cell. Higher precision is achieved with this arrangement than with the top-loading design shown in Figure 2-3b. Even so, top-loading electronic balances have a precision that equals or exceeds that of the best mechanical balances and additionally provides unencumbered access to the pan.

Electronic balances generally feature an automatic **taring control** that causes the display to read zero with a container (such as a boat or weighing bottle) on the pan. Most balances permit taring up to 100% of the capacity of the balance. Some electronic balances have dual capacities and dual precisions. These features permit the capacity to be decreased from that of a macrobalance to that of a semimicrobalance (30 g) with a corresponding gain in precision to 0.01 mg. These types of balances are effectively two balances in one.

A modern electronic analytical balance provides unprecedented speed and ease of use. For example, one instrument is controlled by touching a single bar at various positions along its length. One position on the bar turns the instrument on or off, another automatically calibrates the balance against a standard mass or pair of masses, and a third zeros the display, either with or without an object on the pan. Reliable mass measurements are obtainable with little or no instruction or practice.

A **tare** is the mass of an empty sample container. Taring is the process of setting a balance to read zero in the presence of the tare.

Photographs of a modern electronic balance are shown in color plates 19 and 20.

2D-3 The Single-Pan Mechanical Analytical Balance

Components

Although the single-pan mechanical balance is no longer manufactured, many of these rugged and reliable devices are still found in laboratories. We include a description of this balance for reference and historical purposes. **Figure 2-4** is a diagram of a typical single-pan mechanical balance. Fundamental to this instrument is a lightweight **beam** that is supported on a planar surface by a prism-shaped **knife edge** *(A)*. Attached to the left end of the beam is a pan for holding the object to be weighed and a full set of masses held in place by hangers. These masses can be lifted from the beam one at a time by a mechanical arrangement that is controlled by a set of knobs on the exterior of the balance case. The right end of the beam holds a counterweight of such size as to just balance the pan and masses on the left end of the beam.

A second knife edge *(B)* is located near the left end of the beam and support as a second planar surface, which is located in the inner side of a **stirrup** that couples the pan to the beam. The two knife edges and their planar surfaces are fabricated from extremely hard materials (agate or synthetic sapphire) and form two bearings that permit motion of the beam and pan with a minimum of friction. The performance of a mechanical balance is critically dependent on the perfection of these two bearings.

Single-pan balances are also equipped with a **beam arrest** and a **pan arrest**. The beam arrest is a mechanical device that raises the beam so that the central knife edge no longer touches its bearing surface and simultaneously frees the stirrup from contact with the outer knife edge. The purpose of both arrest mechanisms is to prevent damage to the bearings while objects are being placed on or removed from the pan. When engaged, the pan arrest supports most of the mass of the pan and its contents and thus prevents oscillation. Both arrests are controlled by a lever mounted on the outside of the balance case and should be engaged whenever the balance is not in use.

An **air damper** (also known as a **dashpot**) is mounted near the end of the beam opposite the pan. This device consists of a piston that moves within a concentric cylinder attached to the balance case. Air in the cylinder undergoes expansion and

The two **knife edges** in a mechanical balance are prism-shaped agate or sapphire devices that form low-friction bearings with two planar surfaces contained in **stirrups** also of agate or sapphire.

◀ To avoid damage to the knife edges and bearing surfaces, the arrest system for a mechanical balance should be engaged at all times other than during actual weighing.

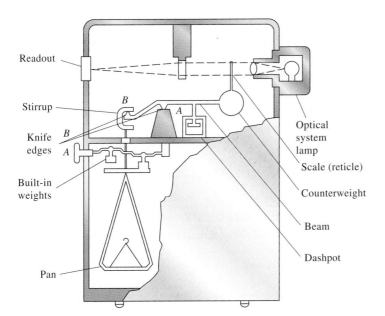

Figure 2-4 Single-pan mechanical analytical balance. [*Reprinted (adapted) with permission from R. M. Schoonover,* Anal. Chem., *1982, 54, 973A. Published 1982, American Chemical Society.*]

contraction as the beam is set in motion. The beam rapidly comes to rest as a result of this opposition to motion.

To discriminate between small differences in mass (<1 mg), analytical balances must be protected from air currents. These devices are always enclosed in cases with doors so that samples can be placed on the pan for weighing and removed when weighing is complete.

Weighing with a Single-Pan Balance

The beam of a properly adjusted balance assumes an essentially horizontal position with no object on the pan and all of the masses in place. When the pan and beam arrests are disengaged, the beam is free to rotate around the knife edge. Placing an object on the pan causes the left end of the beam to move downward. Masses are then removed systematically one by one from the beam until the imbalance is less than 100 mg. The angle of deflection of the beam with respect to its original horizontal position is directly proportional to the additional mass that must be removed to restore the beam to its original horizontal position. The optical system shown in the upper part of Figure 2-4 measures this angle of deflection and converts this angle to milligrams. A **reticle**, which is a small transparent screen mounted on the beam, is scribed with a scale that reads 0 to 100 mg. A beam of light passes through the scale to an enlarging lens, which in turn focuses a small part of the enlarged scale onto a frosted glass plate located on the front of the balance. A vernier makes it possible to read this scale to the nearest 0.1 mg.

Precautions in Using an Analytical Balance

An analytical balance is a delicate instrument that you must handle with care. Consult with your instructor for detailed instructions on weighing with your particular model of balance. Observe the following general rules for working with an analytical balance regardless of make or model:

1. Center the load on the pan as well as possible.
2. Protect the balance from corrosion. Objects to be placed on the pan should be limited to nonreactive metals, nonreactive plastics, and vitreous, or glasslike, materials.
3. Observe special precautions (see Section 2E-6) for the weighing of liquids.
4. Consult your instructor if the balance appears to need adjustment.
5. Keep the balance and its case scrupulously clean. A camel's-hair brush is useful for removing spilled material or dust.
6. Always allow an object that has been heated to return to room temperature before weighing it.
7. Use tongs, finger pads, or a glassine paper strip to handle dried objects to prevent transferring moisture to them.

2D-4 Sources of Error in Weighing[5]

Correction for Buoyancy

A **buoyancy error** will affect data if the density of the object being weighed differs significantly from that of the standard masses. This error has its origin in the

---

Glassine paper is specially treated through a process called calendering. The process begins with breaking down paper pulp fibers by beating. The beaten pulp is then squeezed into molds and dried into sheets. These sheets are then rolled through an alternating series of hot steel and fiber rollers called a supercalender. This step, which makes the pulp fibers in the sheets lie flat and in the same direction, is repeated several times. The final product is an extremely smooth paper that can be used as barrier protection from many kinds of grease, air, and liquids. Glassine is used as an interleaving paper in bookbinding, especially to protect fine illustrations against contact with facing pages. The paper can be manufactured with neutral pH and can prevent damage from spilling, exposure, or rubbing. It is used in foodservice as a barrier between layers of products: meat, baked goods, and cheese, for example. In chemistry, we use glassine as an inexpensive weighing paper for powdered or granular samples because particles have little tendency to adhere to the paper, it is quite light, and it is inexpensive. Narrow strips of glassine are nearly ideal for handling weighing bottles or any common items that must be transferred by hand to and from a balance pan.

A **buoyancy error** is the weighing error that develops when the object being weighed has a significantly different density than the masses.

[5]For further information, see R. Battino and A. G. Williamson, *J. Chem. Educ.,* **1984**, *61,* 51, **DOI**: 10.1021/ed061p51.

difference in buoyant force exerted by the medium (air) on the object and on the masses. Buoyancy corrections for electronic balances[6] may be accomplished with the equation

$$W_1 = W_2 + W_2\left(\frac{d_{air}}{d_{obj}} - \frac{d_{air}}{d_{wts}}\right) \qquad (2\text{-}1)$$

where W_1 is the corrected mass of the object, W_2 is the mass of the standard masses, d_{obj} is the density of the object, d_{wts} is the density of the masses, and d_{air} is the density of the air displaced by masses and object. The value of d_{air} is 0.0012 g/cm^3.

The consequences of Equation 2-1 are shown in **Figure 2-5** in which the relative error due to buoyancy is plotted against the density of objects weighed in air against stainless steel masses. Note that this error is less than 0.1% for objects that have a density of 2 g/cm^3 or greater. It is thus seldom necessary to correct the masses of most solids. The same cannot be said for low-density solids, liquids, or gases, however. For these, the effects of buoyancy are significant, and a correction must be applied.

The density of masses used in single-pan balances (or to calibrate electronic balances) ranges from 7.8 to 8.4 g/cm^3, depending on the manufacturer. Use of 8 g/cm^3 is close enough for most purposes. If greater accuracy is required, the manufacturer's specifications for the balance usually give the necessary density data.

Figure 2-5 Effect of buoyancy on weighing data (density of weights = 8 g/cm^3). Plot of relative error as a function of the density of the object weighed.

[6]Air buoyancy corrections for single-pan mechanical balances are somewhat different from those for electronic balances. For a thorough discussion of the differences in the corrections, see M. R. Winward et al., *Anal. Chem.,* **1977,** *49,* 2126, **DOI:** 10.1021/ac50021a062.

Temperature Effects

Attempts to weigh an object whose temperature is different from that of its surroundings will result in a significant error. Failure to allow sufficient time for a heated object to return to room temperature is the most common source of this problem. Errors due to a difference in temperature have two sources. First, convection currents within the balance case exert a buoyant effect on the pan and object. Second, warm air trapped in a closed container weighs less than the same volume at a lower temperature. Both effects cause the apparent mass of the object to be low. This error can amount to as much as 10 or 15 mg for typical porcelain filtering crucibles or weighing bottles (see **Figure 2-6**). Heated objects must always be cooled to room temperature before being weighed.

> Always allow heated objects to return to room temperature before you attempt to weigh them.

EXAMPLE 2-1

A bottle weighed 7.6500 g empty and 9.9700 g after introduction of an organic liquid with a density of 0.92 g/cm^3. The balance was equipped with stainless steel masses ($d = 8.0$ g/cm^3). Correct the mass of the sample for the effects of buoyancy.

Solution

The apparent mass of the liquid is $9.9700 - 7.6500 = 2.3200$ g. The same buoyant force acts on the container during both weighings. Thus, we need to consider only the force that acts on the 2.3200 g of liquid. By substituting 0.0012 g/cm^3 for d_{air}, 0.92 g/cm^3 for d_{obj}, and 8.0 g/cm^3 for d_{wts} in Equation 2-1, we find that the corrected mass is

$$W_1 = 2.3200 + 2.3200 \left(\frac{0.0012}{0.92} - \frac{0.0012}{8.0} \right) = 2.3227 \text{ g}$$

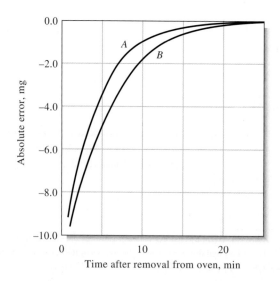

Figure 2-6 Effect of temperature on weighing data. Absolute error as a function of time after an object was removed from a 110°C drying oven. *A:* porcelain filtering crucible. *B:* weighing bottle containing about 7.5 g of KCl.

Other Sources of Error

A porcelain or glass object will occasionally acquire a static charge sufficient to cause a balance to perform erratically. This problem is particularly serious when the relative humidity is low. Spontaneous discharge frequently occurs after a short period. A low-level source of radioactivity (such as a Static-Master® photographer's brush containing a miniscule amount of polonium) in the balance case will ionize enough ions to neutralize the charge. Alternatively, the object can be wiped with a faintly damp chamois.

The optical scale of a single-pan mechanical balance should be checked regularly for accuracy, particularly under loading conditions that require the full-scale range. A standard 100-mg mass is used for this check.

2D-5 Auxiliary Balances

Balances that are less precise than analytical balances find extensive use in the analytical laboratory. These balances offer the advantages of speed, ruggedness, large capacity, and convenience. Low-precision balances should be used whenever high sensitivity is not required.

Top-loading auxiliary balances are particularly convenient. A sensitive top-loading balance will accommodate 150 to 200 g with a precision of about 1 mg—an order of magnitude less than a macroanalytical balance. Some balances of this type tolerate loads as great as 25,000 g with a precision of ±0.05 g. Most are equipped with a taring device that brings the balance reading to zero with an empty container on the pan. Some are fully automatic, require no manual dialing or mass handling, and provide a digital readout of the mass. Modern top-loading balances are electronic.

A triple-beam balance that is less sensitive than a typical top-loading auxiliary balance is also useful. This is a single-pan balance with three decades of masses that slide along individual calibrated scales. The precision of a triple-beam balance may be one or two orders of magnitude less than that of a top-loading instrument but can be adequate for many weighing operations. This type of balance is simple, durable, and inexpensive.

❰ Use auxiliary laboratory balances for determining masses that do not require great accuracy.

2E EQUIPMENT AND MANIPULATIONS ASSOCIATED WITH WEIGHING

The mass of many solids changes with humidity because they tend to absorb weighable amounts of moisture. This effect is especially pronounced when a large surface area is exposed, as with a reagent chemical or a sample that has been ground to a fine powder. In the first step in a typical analysis, then, the sample is dried so that the results will not be affected by the humidity of the surrounding atmosphere.

A sample, a precipitate, or a container is brought to constant mass by a cycle of heating (usually for one hour or more) at an appropriate temperature, cooling, and weighing. This cycle is repeated as many times as needed to obtain successive masses that agree within 0.2 to 0.3 mg of one another. The establishment of constant mass provides some assurance that the chemical or physical processes that occur during the heating (or ignition) are complete.

Drying or **ignition to constant mass** is a process in which a solid is cycled through heating, cooling, and weighing steps until its mass becomes constant to within 0.2 to 0.3 mg.

2E-1 Weighing Bottles

Weighing bottles are convenient for drying and storing solids. Two common varieties of these handy tools are shown in **Figure 2-7**. The ground-glass portion of the

Figure 2-7 Typical weighing bottles.

A **desiccator** is a device for drying substances or objects.

cap-style bottle shown on the left is on the outside and does not come into contact with the contents. This design eliminates the possibility of some of the sample becoming trapped on the ground-glass surface and subsequently being lost. Ruggedness is a principal advantage of using plastic weighing bottles rather than glass, but plastic abrades easily and is not as easily cleaned as glass.

2E-2 Desiccators and Desiccants

Oven drying is the most common way of removing moisture from solids. This approach is not appropriate for substances that decompose or for those from which water is not removed at the temperature of the oven.

To minimize the uptake of moisture, dried materials are stored in **desiccators** while they cool. **Figure 2-8** shows the components of a typical desiccator. The base section contains a chemical drying agent, such as anhydrous calcium chloride, calcium sulfate (Drierite), anhydrous magnesium perchlorate (Anhydrone or Dehydrite), or phosphorus pentoxide. The ground-glass surfaces between the top and the base are lightly coated with grease to ensure a good seal when the top is in place.

When removing or replacing the lid of a desiccator, use a sliding motion to minimize the likelihood of disturbing the sample. An airtight seal is achieved by slight rotation and downward pressure on the positioned lid.

When placing a heated object in a desiccator, the increase in pressure as the enclosed air is warmed may be sufficient to break the seal between lid and base. Conversely, if the seal is not broken, the cooling of heated objects can cause a partial vacuum to develop. Both of these conditions can cause the contents of the desiccator to be physically lost or contaminated. Although it defeats the purpose of the desiccator somewhat, allow some cooling to occur before the lid is seated. It is also helpful

(a)

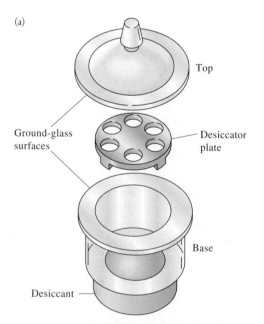

Top

Ground-glass surfaces

Desiccator plate

Base

Desiccant

(b)

Figure 2-8   (a) Components of a typical desiccator. The base contains a chemical drying agent, which is usually covered with a wire screen and a porcelain plate with holes to accommodate weighing bottles or crucibles. (b) Photo of desiccator containing weighing bottles with dry solids.

to break the seal once or twice during cooling to relieve any excessive vacuum that develops. Finally, lock the lid in place with your thumbs while moving the desiccator from one place to another.

Very hygroscopic materials should be stored in containers equipped with snug covers, such as weighing bottles. The bottles remain covered while in the desiccator. Most other solids can be safely stored uncovered.

2E-3 Manipulating Weighing Bottles

Heating at 105°C to 110°C for 1 hour is sufficient to remove the moisture from the surface of most solids. **Figure 2-9** shows the recommended way to dry a sample. The weighing bottle is contained in a labeled beaker with a cover glass. This arrangement protects the sample from accidental contamination and also allows for free access of air. Crucibles containing a precipitate that can be freed of moisture by simple drying can be treated similarly. The beaker holding the weighing bottle or crucible to be dried must be carefully marked for identification.

Avoid touching dried objects with your fingers because detectable amounts of water or oil from the skin may be transferred to the objects. Instead, use tongs, chamois finger cots, clean cotton gloves, or strips of paper to handle dried objects for weighing. **Figure 2-10** shows how a weighing bottle is manipulated with tongs and strips of paper.

2E-4 Weighing by Difference

Weighing by difference is a simple method for determining a series of sample masses. First, the bottle and its contents are weighed. One sample is then transferred from the bottle to a container. Gentle tapping of the bottle with its top and slight rotation of the bottle provide control over the amount of sample removed. Following transfer, the bottle and its residual contents are weighed. The mass of the sample is the difference between the two masses. It is essential that all the solid removed from the weighing bottle be transferred without loss to the container.

2E-5 Weighing Hygroscopic Solids

Hygroscopic substances rapidly absorb moisture from the atmosphere and, therefore, require special handling. You need a weighing bottle for each sample to be weighed. Place the approximate amount of sample needed in the individual bottles and heat for an appropriate time. When heating is complete, quickly cap the bottles and cool in a desiccator. Weigh one of the bottles after opening it momentarily to relieve any vacuum. Quickly empty the contents of the bottle into its receiving vessel, cap immediately, and weigh the bottle again along with any solid that did not get transferred. Repeat for each sample and determine the sample masses by difference.

2E-6 Weighing Liquids

The mass of a liquid is always obtained by difference. Liquids that are noncorrosive and relatively nonvolatile can be transferred to previously weighed containers with snugly fitting covers (such as weighing bottles). The mass of the container is subtracted from the total mass.

A volatile or corrosive liquid should be sealed in a weighed glass ampoule. The ampoule is heated, and the neck is then immersed in the sample. As cooling occurs,

Figure 2-9 Arrangement for the drying of samples.

Figure 2-10 Quantitative transfer of solid sample. Note the use of tongs to hold the weighing bottle and a paper strip to hold the cap to avoid contact between glass and skin.

the liquid is drawn into the bulb. The ampoule is then inverted and the neck sealed off with a small flame. The ampoule and its contents, along with any glass removed during sealing, are cooled to room temperature and weighed. The ampoule is then transferred to an appropriate container and broken. A volume correction for the glass of the ampoule may be needed if the receiving vessel is a volumetric flask.

2F FILTRATION AND IGNITION OF SOLIDS

Several techniques and experimental arrangements allow solids to be filtered and ignited with minimal contamination and error.

2F-1 Apparatus

Simple Crucibles

Simple crucibles serve only as containers. Porcelain, aluminum oxide, silica, and platinum crucibles maintain constant mass—within the limits of experimental error—and are used principally to convert a precipitate into a suitable weighing form. The solid is first collected on a filter paper. The filter and contents are then transferred to a weighed crucible, and the paper is ignited.

Simple crucibles of nickel, iron, silver, and gold are used as containers for the high-temperature fusion of samples that are not soluble in aqueous reagents. Attack by both the atmosphere and the contents may cause these crucibles to suffer mass changes. Moreover, such attack will contaminate the sample with species derived from the crucible. The crucible whose products will offer the least interference in subsequent steps of the analysis should be used.

Filtering Crucibles

Filtering crucibles serve not only as containers but also as filters. A vacuum is used to hasten the filtration. A tight seal between crucible and filtering flask is made with any of several types of rubber adaptors (see **Figure 2-11**). A complete filtration train is shown in Figure 2-16. Collection of a precipitate with a filtering crucible is frequently less time consuming than with paper.

Figure 2-11 Adaptors for filtering crucibles.

Sintered-glass (also called **fritted-glass**) crucibles are manufactured in fine, medium, and coarse porosities (marked *f, m,* and *c*). The upper temperature limit for a sintered-glass crucible is usually about 200°C. Filtering crucibles made entirely of quartz can tolerate substantially higher temperatures without damage. The same is true for crucibles with unglazed porcelain or aluminum oxide frits. The latter are not as costly as quartz.

A **Gooch crucible** has a perforated bottom that supports a fibrous mat. Asbestos was at one time the filtering medium of choice for a Gooch crucible. However, current regulations on the use of asbestos have virtually eliminated its use. Small circles of glass matting have now replaced asbestos. They are used in pairs to protect against breaking during the filtration. Glass mats can tolerate temperatures in excess of 500°C and are substantially less hygroscopic than asbestos.

Filter Paper

Paper is an important filtering medium. Ashless paper is manufactured from cellulose fibers that have been treated with hydrochloric and hydrofluoric acids to remove metallic impurities and silica; ammonia is then used to neutralize the acids. The residual ammonium salts in many filter papers may be sufficient to affect the analysis for nitrogen by the Kjeldahl method (see Section 38C-11).

All papers tend to pick up moisture from the atmosphere, and ashless paper is no exception. It is thus necessary to destroy the paper by ignition if the precipitate collected on it is to be weighed. Typically, 9- or 11-cm circles of ashless paper leave a residue that weighs less than 0.1 mg, which is negligible under most circumstances. Ashless paper can be obtained in several porosities.

Gelatinous precipitates, such as hydrous iron(III) oxide, clog the pores of any filtering medium. A coarse-porosity ashless paper is most effective for filtering such solids, but even with such paper, clogging occurs. This problem can be minimized by mixing a dispersion of ashless filter paper with the precipitate prior to filtration. Filter paper pulp is available in tablet form from chemical suppliers. If no commercial pulp is available, it can be prepared by treating a piece of ashless paper with concentrated hydrochloric acid and washing the disintegrated mass free of acid.

Table 2-1 summarizes the characteristics of common filtering media. None satisfies all requirements.

TABLE 2-1

Comparison of Filtering Media for Gravimetric Analyses

Characteristic	Paper	Gooch Crucible, Glass Mat	Glass Crucible	Porcelain Crucible	Aluminum Oxide Crucible
Speed of filtration	Slow	Rapid	Rapid	Rapid	Rapid
Convenience and ease of preparation	Troublesome, inconvenient	Convenient	Convenient	Convenient	Convenient
Maximum ignition temperature, °C	None	>500	200–500	1100	1450
Chemical reactivity	Carbon has reducing properties	Inert	Inert	Inert	Inert
Porosity	Many available	Several available	Several available	Several available	Several available
Convenience with gelatinous precipitates	Satisfactory	Unsuitable; filter tends to clog	Unsuitable; filter tends to clog	Unsuitable; filter tends to clog	Unsuitable; filter tends to clog
Cost	Low	Low	High	High	High

Heating Equipment

Many precipitates can be weighed directly after being brought to constant mass in a low-temperature drying oven. Such an oven is electrically heated and capable of maintaining a constant temperature to within 1°C (or better). The maximum attainable temperature ranges from 140°C to 260°C, depending on make and model. For many precipitates, 110°C is a satisfactory drying temperature. The efficiency of a drying oven is greatly increased by the forced circulation of air. The passage of predried air through an oven designed to operate under a partial vacuum represents an additional improvement.

Microwave laboratory ovens are currently quite popular, and where applicable, they greatly shorten drying cycles. For example, slurry samples that require 12 to 16 hours for drying in a conventional oven are reported to be dried within 5 to 6 minutes in a microwave oven.[7] The time needed to dry silver chloride, calcium oxalate, and barium sulfate precipitates for gravimetric analysis is also shortened significantly.[8]

An ordinary heat lamp can be used to dry a precipitate that has been collected on ashless paper and to char the paper as well. The process is conveniently completed by ignition at an elevated temperature in a muffle furnace.

Burners are convenient sources of intense heat. The maximum attainable temperature depends on the design of the burner and the combustion properties of the fuel. Of the three common laboratory burners, the Meker burner provides the highest temperatures, followed by the Tirrill and Bunsen types.

A heavy-duty electric furnace (**muffle furnace**) is capable of maintaining controlled temperatures of 1100°C or higher. Long-handled tongs and heat-resistant gloves are needed for protection when transferring objects to or from such a furnace.

2F-2 Filtering and Igniting Precipitates

Preparation of Crucibles

A crucible used to convert a precipitate to a form suitable for weighing must maintain—within the limits of experimental error—a constant mass throughout drying or ignition. The crucible is first cleaned thoroughly (filtering crucibles are conveniently cleaned by backwashing on a filtration train) and then subjected to the same regimen of heating and cooling as that required for the precipitate. This process is repeated until constant mass (page 25) has been achieved, that is, until consecutive weighings differ by 0.3 mg or less.

> Backwashing a filtering crucible is done by turning the crucible upside down in the adaptor (Figure 2-11) and sucking water through the inverted crucible.

Filtering and Washing Precipitates

The steps in filtering an analytical precipitate are **decantation**, **washing**, and **transfer**. In decantation, as much supernatant liquid as possible is passed through the filter while the precipitated solid is kept essentially undisturbed in the beaker where it was formed. This procedure speeds the overall filtration rate by delaying the time at which the pores of the filtering medium become clogged with precipitate. A stirring rod is used to direct the flow of the decanted liquid (**Figure 2-12a**).

When flow ceases, the drop of liquid at the end of the pouring spout is collected with the stirring rod and returned to the beaker. Wash liquid is next added to the

> **Decantation** is the process of pouring a liquid gently so as to not disturb a solid in the bottom of the container.

[7]E. S. Beary, *Anal. Chem.,* **1988,** *60,* 742, **DOI**: 10.1021/ac00159a003.
[8]R. Q. Thompson and M. Ghadradhi, *J. Chem. Educ.,* **1993,** *70,* 170, **DOI**: 10.1021/ed070p170.

(a)   (b)

Charles D. Winters

Figure 2-12 (a) Washing by decantation. (b) Transferring the precipitate.

beaker and thoroughly mixed with precipitate. The solid is allowed to settle, and then this liquid is also decanted through the filter. Several such washings may be required, depending on the precipitate. Most washing should be carried out *before* the bulk of the solid is transferred. This technique results in a more thoroughly washed precipitate and a more rapid filtration.

The transfer process is illustrated in **Figure 2-12b**. The bulk of the precipitate is moved from beaker to filter by directed streams of wash liquid. As in decantation and washing, a stirring rod provides direction for the flow of material to the filtering medium.

The last traces of precipitate that cling to the inside of the beaker are dislodged with a **rubber policeman**, which is a small section of rubber tubing that has been crimped on one end. The open end of the tubing is fitted onto the end of a stirring rod and is wetted with wash liquid before use. Any solid collected with it is combined with the main portion on the filter. Small pieces of ashless paper can be used to wipe the last traces of hydrous oxide precipitates from the wall of the beaker. These papers are ignited along with the paper that holds the bulk of the precipitate.

Many precipitates possess the exasperating property of **creeping**, or spreading over a wetted surface against the force of gravity. Filters are never filled to more than three quarters of capacity to prevent the possible loss of precipitate through creeping. The addition of a small amount of nonionic detergent, such as Triton X-100, to the supernatant liquid or wash liquid can help minimize creeping.

A gelatinous precipitate must be completely washed before it is allowed to dry. These precipitates shrink and develop cracks as they dry. Further additions of wash liquid simply pass through these cracks and accomplish little or no washing.

Creeping is a process in which a solid moves up the side of a wetted container or filter paper.

❮ Do not permit a gelatinous precipitate to dry until it has been washed completely.

2F-3 Directions for Filtering and Igniting Precipitates

Preparation of a Filter Paper

Figure 2-13 shows the sequence for folding and seating a filter paper in a 60-deg funnel. The paper is folded exactly in half (a), firmly creased, and folded again (b). A triangular piece from one of the corners is torn off parallel to the second fold (c). The

(a)   (b)

(c)   (d)

(e)   (f)

Charles D. Winters

Figure 2-13 Folding and seating a filter paper. (a) Fold the paper exactly in half and crease it firmly. (b) Fold the paper a second time. (c) Tear off one of the corners on a line parallel to the second fold. (d) Open the untorn half of the folded paper to form a cone. (e) Seat the cone firmly into the funnel. (f) Moisten the paper slightly and gently pat the paper into place.

paper is then opened so that the untorn quarter forms a cone (d). The cone is fitted into the funnel, and the second fold is creased (e). Seating is completed by dampening the cone with water from a wash bottle and *gently* patting it with a finger (f). There will be no leakage of air between the funnel and a properly seated cone. In addition, the stem of the funnel will be filled with an unbroken column of liquid.

Transferring Paper and Precipitate to a Crucible

After filtration and washing have been completed, the filter and its contents must be transferred from the funnel to a crucible that has been brought to constant mass. Ashless paper has very low wet strength and must be handled with care during the transfer. The danger of tearing is lessened considerably if the paper is allowed to dry somewhat before it is removed from the funnel.

(a)   (b)   (c)   (d)   (e)

Charles D. Winters

Figure 2-14   Transferring a filter paper and precipitate from a funnel to a crucible. (a) Pull the triple-thick portion of the cone to the opposite side of the funnel. (b) Remove the filter cone from the funnel, and flatten the cone along its upper edge. (c) Fold the corners inward. (d) Fold the top edge of the cone toward the tip to enclose the precipitate in the paper. (e) Gently ease the folded paper and its contents into the crucible.

Figure 2-14 illustrates the transfer process. The triple-thick portion of the filter paper is drawn across the funnel (a) to flatten the cone along its upper edge (b); the corners are next folded inward (c); and the top edge is then folded over (d). Finally, the paper and its contents are eased into the crucible (e) so that the bulk of the precipitate is near the bottom.

Ashing Filter Papers

If a heat lamp is used, the crucible is placed on a clean, nonreactive surface, such as a wire screen covered with aluminum foil. The lamp is then positioned about 1 cm above the rim of the crucible and turned on. Charring takes place without further attention. The process is considerably accelerated if the paper is moistened with no more than one drop of concentrated ammonium nitrate solution. The residual carbon is eliminated with a burner, as described in the next paragraph.

Considerably more attention must be paid if a burner is used to ash a filter paper because the burner produces much higher temperatures than a heat lamp. Thus, mechanical loss of precipitate may occur if moisture is expelled too rapidly in the initial stages of heating or if the paper bursts into flame. Also, partial reduction of some precipitates can occur through reaction with the hot carbon of the charring paper. This reduction is a serious problem if reoxidation following ashing is inconvenient. These difficulties can be minimized by positioning the crucible as illustrated in **Figure 2-15**. The tilted position allows for the easy access of air. A clean crucible cover should be kept handy to extinguish any flame.

Heating should begin with a small flame. The temperature is gradually increased as moisture is evolved and the paper begins to char. The amount of smoke given off indicates the intensity of heating that can be tolerated. Thin wisps are normal. A significant increase in smoke indicates that the paper is about to flash and that heating should be temporarily discontinued. Any flame should be immediately extinguished with a crucible cover. (The cover may become discolored from the condensation of carbonaceous products. These products must ultimately be removed from the cover by ignition to confirm the absence of entrained particles of precipitate.) When no further smoking can be detected, heating is increased to eliminate the residual carbon. Strong heating, as necessary, can then be undertaken. This sequence usually precedes the final ignition of a precipitate in a muffle furnace, where a reducing atmosphere is equally undesirable.

> You should have a burner for each crucible. You can tend to the ashing of several filter papers at the same time.

Figure 2-15   Ignition of a precipitate. Proper crucible position for preliminary charring is shown.

To vacuum

Trap

Figure 2-16 Train for vacuum filtration. The trap isolates the filter flask from the source of vacuum.

Using Filtering Crucibles

A vacuum filtration train (**Figure 2-16**) is used when a filtering crucible can be used instead of paper. The trap isolates the filter flask from the source of vacuum.

2F-4 Rules for Manipulating Heated Objects

Careful adherence to the following rules will minimize the possibility of accidental loss of a precipitate:

1. Practice unfamiliar manipulations before putting them to use.
2. *Never* place a heated object on the benchtop. Instead, place it on a wire gauze or a heat-resistant ceramic plate.
3. Allow a crucible that has been subjected to the full flame of a burner or to a muffle furnace to cool momentarily (on a wire gauze or ceramic plate) before transferring it to the desiccator.
4. Keep the tongs and forceps used to handle heated objects scrupulously clean. In particular, do not allow the tips to touch the benchtop.

2G | MEASURING VOLUME

The precise measurement of volume is as important to many analytical methods as the precise measurement of mass.

2G-1 Units of Volume

The **liter** is one cubic decimeter. The **milliliter** is 10^{-3} L.

The unit of volume is the **liter** (L), defined as one cubic decimeter. The **milliliter** (mL) is one one-thousandth of a liter (0.001 L) and is used when the liter represents an inconveniently large volume unit. The microliter (μL) is 10^{-6} L or 10^{-3} mL.

2G-2 The Effect of Temperature on Volume Measurements

The volume occupied by a given mass of liquid varies with temperature, as does the device that holds the liquid during measurement. Most volumetric measuring devices are made of glass, which fortunately has a small coefficient of expansion. Thus, variations in the volume of a glass container with temperature need not be considered in ordinary analytical work.

The coefficient of expansion for dilute aqueous solutions (approximately 0.025%/°C) is such that a 5°C change has a measurable effect on the reliability of ordinary volumetric measurements.

EXAMPLE 2-2

A 40.00-mL sample is taken from an aqueous solution at 5°C. What volume does it occupy at 20°C?

$$V_{20°} = V_{5°} + 0.00025(20 - 5)(40.00) = 40.00 + 0.15 = 40.15 \text{ mL}$$

Volumetric measurements must be referred to a standard temperature, often 20°C. The ambient temperature of most laboratories is usually close enough to 20°C so that there is no need for temperature corrections in volume measurements for aqueous solutions. In contrast, the coefficient of expansion for organic liquids may be large enough to require corrections for temperature differences of 1°C or less.

2G-3 Apparatus for Precisely Measuring Volume

Volume may be measured reliably with a **pipet**, a **buret**, or a **volumetric flask**.

Volumetric equipment is marked by the manufacturer to indicate not only the manner of calibration (usually TD for "to deliver" or TC for "to contain") but also the temperature at which the calibration strictly applies. Pipets and burets are usually calibrated to deliver specified volumes. Volumetric flasks, on the other hand, are calibrated to contain a specific volume.

Pipets

Pipets permit the transfer of accurately known volumes from one container to another. Common types are shown in **Figure 2-17**, and information concerning their use is given in **Table 2-2**. A **volumetric**, or **transfer**, pipet (Figure 2-17a) delivers a single,

> Glassware types include Class A and Class B. Class A glassware is manufactured to the highest tolerances from Pyrex, borosilicate, or Kimax glass (see tables on pages 36 and 37). Class B (economy ware) tolerances are about twice those of Class A.

Figure 2-17 Typical pipets: (a) volumetric pipet, (b) Mohr pipet, (c) serological pipet, (d) Eppendorf micropipet, (e) Ostwald–Folin pipet, (f) lambda pipet.

TABLE 2-2

Characteristics of Pipets

Name	Type of Calibration*	Function	Available Capacity, mL	Type of Drainage
Volumetric	TD	Delivery of fixed volume	1–200	Free
Mohr	TD	Delivery of variable volume	1–25	To lower calibration line
Serological	TD	Delivery of variable volume	0.1–10	Blow out last drop**
Serological	TD	Delivery of variable volume	0.1–10	To lower calibration line
Ostwald-Folin	TD	Delivery of fixed volume	0.5–10	Blow out last drop**
Lambda	TC	Containment of fixed volume	0.001–2	Wash out with suitable solvent
Lambda	TD	Delivery of fixed volume	0.001–2	Blow out last drop**
Eppendorf	TD	Delivery of variable or fixed volume	0.001–1	Tip emptied by air displacement

*TD, to deliver; TC, to contain.

**A frosted ring near the top of pipets indicates that the last drop is to be blown out.

Tolerances, Class A Transfer Pipets

Capacity, mL	Tolerances, mL
0.5	±0.006
1	±0.006
2	±0.006
5	±0.01
10	±0.02
20	±0.03
25	±0.03
50	±0.05
100	±0.08

Range and Precision of Typical Eppendorf Micropipets

Volume Range, μL	Standard Deviation, μL
1–20	<0.04 @ 2 μL
	<0.06 @ 20 μL
10–100	<0.10 @ 15 μL
	<0.15 @ 100 μL
20–200	<0.15 @ 25 μL
	<0.30 @ 200 μL
100–1000	<0.6 @ 250 μL
	<1.3 @ 1000 μL
500–5000	<3 @ 1.0 mL
	<8 @ 5.0 mL

fixed volume between 0.5 and 200 mL. Many such pipets are color coded by volume for convenience in identification and sorting. **Measuring pipets** (Figure 2-17b and c) are calibrated in convenient units to permit delivery of any volume up to a maximum capacity ranging from 0.1 to 25 mL.

All volumetric and measuring pipets are first filled to a calibration mark, but the manner in which the transfer is completed depends on the particular type. Because most liquids are attracted to glass, a small amount of liquid tends to remain in the tip after the pipet is emptied. This residual liquid is never blown out of a volumetric pipet or from some measuring pipets, but it is blown out of other types of pipets (see Table 2-2).

Handheld Eppendorf micropipets (see Figure 2-17d and **Figure 2-18a**) deliver adjustable microliter volumes of liquid. With these pipets, a known and adjustable volume of air is displaced from the plastic disposable tip by depressing the pushbutton on the top of the pipet to a first stop. This button operates a spring-loaded piston that forces air out of the pipet. The volume of displaced air can be varied by a locking digital micrometer adjustment located on the front or top of the device. The plastic tip is then inserted into the liquid, and the pressure on the button released, causing liquid to be drawn into the tip. The tip is then placed against the walls of the receiving vessel, and the pushbutton is again depressed to the first stop. After 1 second, the pushbutton is depressed further to a second stop, which completely empties the tip. The range of volumes and precision of typical pipets of this type are shown in the margin. The accuracy and precision of automatic pipets depend somewhat on the skill and experience of the operators and thus should be calibrated for critical work.

Numerous *automatic* pipets are available for situations that call for the repeated delivery of a particular volume. In addition, motorized, computer-controlled microliter pipets are now available (see **Figure 2-18b**). These devices are programmed to function as pipets, dispensers of multiple volumes, burets, and sample dilutors. The volume desired is entered using a joystick and buttons and is displayed on an LCD panel. A motor-driven piston dispenses the liquid. Maximum volumes range from 10 μL to 20 mL.

(a)

Digital display

Joy stick

Tip ejector button

Disposable tip

(b)

Charles D. Winters

Mettler-Toledo, Inc

Figure 2-18 (a) Variable-volume automatic pipet, 100–1000 µL. At 100 µL, accuracy is 3.0%, and precision is 0.6%. At 1000 µL, accuracy is 0.6%, and precision is 0.2%. Volume is adjusted using the thumbwheel as shown. Volume shown is 525 µL.

Burets

Burets, like measuring pipets, make it possible to deliver any volume up to the maximum capacity of the device. The precision attainable with a buret is substantially greater than the precision with a pipet.

A buret consists of a calibrated tube to hold titrant plus a valve arrangement by which the flow of titrant is controlled. This valve is the principal source of difference among burets. The simplest pinchcock valve consists of a close-fitting glass bead inside a short length of rubber tubing that connects the buret and its tip (see **Figure 2-19a**). Only when the tubing is deformed does liquid flow past the bead.

A buret equipped with a glass stopcock for a valve relies on a lubricant between the ground-glass surfaces of stopcock and barrel for a liquid-tight seal. Some solutions, notably bases, cause glass stopcocks to freeze when they are in contact with ground glass for long periods. Therefore, glass stopcocks must be thoroughly cleaned after each use. Most burets made in the last several of decades have Teflon® valves, which are unaffected by most common reagents and require no lubricant (see **Figure 2-19b**).

Volumetric Flasks

Volumetric flasks (see **Figure 2-20**) are manufactured with capacities ranging from 5 mL to 5 L and are usually calibrated *to contain* (TC) a specified volume when filled to a line etched on the neck. They are used for the preparation of standard solutions and for the dilution of samples to a fixed volume prior to taking aliquots with a pipet. Some are also calibrated on a *to-deliver* (TD) basis, and they are distinguished by two reference lines on the neck. If delivery of the stated volume is desired, the flask is filled to the upper line.

Tolerances, Class A Burets	
Volume, mL	**Tolerances, mL**
5	±0.01
10	±0.02
25	±0.03
50	±0.05
100	±0.20

Tolerances, Class A Volumetric Flasks	
Capacity, mL	**Tolerances, mL**
5	±0.02
10	±0.02
25	±0.03
50	±0.05
100	±0.08
250	±0.12
500	±0.20
1000	±0.30
2000	±0.50

(a)   (b)

Figure 2-19   Burets:
(a) glass-bead valve,
(b) Teflon valve.

Figure 2-20   Typical volumetric flasks.

Charles D. Winters

2G-4 Using Volumetric Equipment

Volume markings are blazed on clean volumetric equipment by the manufacturer. An equal degree of cleanliness is needed in the laboratory if these markings are to have their stated meanings. Only clean glass surfaces support a uniform film of liquid. Dirt or oil causes breaks in this film, so if breaks are present, the surface is almost certainly dirty.

Cleaning

A brief soaking in a warm detergent solution is usually sufficient to remove the grease and dirt responsible for water breaks. Prolonged soaking should be avoided because a rough area or ring is likely to develop at a detergent/air interface. This ring cannot be removed and causes a film break that destroys the usefulness of the equipment.

After being cleaned, the apparatus must be thoroughly rinsed with tap water and then with three or four portions of distilled water. It is seldom necessary to dry volumetric ware.

Avoiding Parallax

A **meniscus** is the curved surface of a liquid at its interface with the atmosphere.

Parallax is the apparent displacement of a liquid level or of a pointer as an observer changes position. Parallax occurs when an object is viewed from a position that is not at a right angle to the object.

The top surface of a liquid confined in a narrow tube exhibits a marked curvature, or **meniscus**. It is common practice to use the bottom of the meniscus as the point of reference in calibrating and using volumetric equipment. This minimum can be established more exactly by holding an opaque card or piece of paper behind the graduations.

In reading volumes, the eye must be at the level of the liquid surface to avoid an error due to **parallax**. Parallax is a condition that causes the volume to appear smaller than its actual value if the meniscus is viewed from above and larger if the meniscus is viewed from below (see **Figure 2-21**).

(a)

(b)

(c)

(d)

(e)

(f)

Charles D. Winters

Figure 2-21 Reading a buret.
(a) The student reads the buret from a position *above* a line perpendicular to the buret and makes a reading (b) of 12.58 mL. (c) The student reads the buret from a position *along* a line perpendicular to the buret and makes a reading (d) of 12.62 mL. (e) The student reads the buret from a position *below* a line perpendicular to the buret and makes a reading (f) of 12.67 mL. To avoid the problem of parallax, buret readings should be made consistently along a line perpendicular to the buret, as shown in (c) and (d).

2G-5 Directions for Using a Pipet

The following directions are appropriate specifically for volumetric pipets but can be modified for the use of other types as well.

Liquid is drawn into a pipet through the application of a slight vacuum. *Never pipet by mouth because there is risk of accidentally ingesting the liquid being pipetted.* Instead, use a rubber suction bulb (such as the one shown at the top of the next page) or one of a number of similar, commercially available devices.

Many devices are commercially available for filling pipets and dispensing liquids from them. The device shown here is offered by many suppliers and manufacturers. Originally called the Propipette®, it is a very handy device for the task. It consists of a rubber bulb (B) attached to three short sections of tubing. Each section of tubing contains a small chemically inert ball (A, C, and D) that functions as a valve to permit air to flow normally in the directions indicated by the arrows. The valves are opened by pinching with your thumb and forefinger. The bottom of the device fits snugly on the top of a pipet. Operation begins by opening valve A and squeezing bulb B to expel the air in the bulb. Valve A is then closed, and valve C is opened to draw liquid into the pipet to the desired level, after which C is closed. The liquid level is then adjusted in the pipet by carefully opening valve D, and finally, the liquid in the pipet is delivered by opening valve D completely.

Cleaning

Draw detergent solution to a level 2 to 3 cm above the calibration mark of the pipet. Drain this solution and then rinse the pipet with several portions of tap water. Inspect for film breaks, and repeat this portion of the cleaning cycle if necessary. Finally, fill the pipet with distilled water to perhaps one third of its capacity and carefully rotate it so that the entire interior surface is wetted. Repeat this rinsing step at least twice.

Measuring an Aliquot

An **aliquot** is a measured fraction of the volume of a liquid sample.

Draw a small volume of the liquid to be sampled into the pipet (see **Figure 2-22a**) and thoroughly wet the entire interior surface (**Figure 2-22b**). Repeat with *at least* two additional portions. Then carefully fill the pipet to a level somewhat above the graduation mark. Be sure that there are no bubbles in the bulk of the liquid or foam at the surface. Touch the tip of the pipet to the wall of a glass vessel as shown in **Figure 2-22c** (*not* the container into which the aliquot is to be transferred), and slowly allow the liquid level to drop. As the bottom of the meniscus coincides exactly with the graduation mark (**Figure 2-22d**), stop the flow. Remove the pipet from the volumetric flask, tilt it until liquid is drawn slightly up into the pipet, and wipe the tip with a lintless tissue as shown in **Figure 2-22e**. Then place the pipet tip well within the receiving vessel, and allow the liquid to drain (**Figure 2-22f**). When free flow ceases, rest the tip against the inner wall of the receiver for a full 10 seconds (**Figure 2-22g, h**). Finally, withdraw the pipet with a rotating motion to remove any liquid adhering to the tip. *The small volume remaining inside the tip of a volumetric pipet should not be blown or rinsed into the receiving vessel.* Rinse the pipet thoroughly after use.

2G-6 Directions for Using a Buret

A buret must be scrupulously clean before it is used, and its valve must be liquid-tight.

Cleaning

Thoroughly clean the tube of the buret with detergent and a long brush. Rinse thoroughly with tap water and then with distilled water. Inspect for water breaks. Repeat the treatment if necessary.

Lubricating a Glass Stopcock

Carefully remove all old grease from a glass stopcock and its barrel with a paper towel and dry both parts completely. Lightly grease the stopcock, taking care to

(a)   (b)   (c)   (d)

(e)   (f)   (g)   (h)

Charles D. Winters

Figure 2-22 Dispensing an aliquot. (a) Draw a small amount of the liquid into the pipet and (b) wet the interior surface of the glass by tilting and rotating the pipet. Repeat this procedure two more times. Then draw liquid into the pipet so that the level is a few centimeters above the line etched on the stem of the pipet. While holding the tip of the pipet against the inside surface of the volumetric flask (c), allow the liquid level to descend until the bottom of the meniscus is aligned with the line (d). Remove the pipet from the volumetric flask, tilt it (e) until liquid is drawn slightly up into the pipet, and wipe the tip with a lintless tissue as shown. Then while holding the pipet vertically, (f) allow the liquid to flow into the receiving flask until just a small amount of liquid remains in the inside of the tip and a drop remains on the outside. Tilt the flask slightly as shown in (g), and finally, touch the tip of the pipet to the inside of the flask (h). When this step is completed, a small amount of liquid will remain in the pipet. Do not remove this remaining liquid. The pipet is calibrated to reproducibly deliver its rated volume when this liquid remains in the tip.

avoid the area adjacent to the hole. Insert the stopcock into the barrel and rotate it vigorously with slight inward pressure. A proper amount of lubricant has been used when (1) the area of contact between stopcock and barrel appears nearly transparent, (2) the seal is liquid-tight, and (3) no grease has worked its way into the tip.

Notes

1. Grease films that are unaffected by cleaning solution may yield to such organic solvents as acetone or alcohols. Thorough washing with detergent should follow such treatment. Silicone lubricants are not recommended because contamination by such preparations is difficult—if not impossible—to remove.

2. So long as the flow of liquid is not impeded, fouling of a buret tip with stopcock grease is not a serious matter. Removal is best accomplished with organic solvents. A stoppage during a titration can be freed by *gentle* warming of the tip with a lighted match.

3. Before a buret is returned to service after reassembly, it is advisable to test for leakage. Simply fill the buret with water and establish that the volume reading does not change with time.

❮ Buret readings should be estimated to the nearest 0.01 mL.

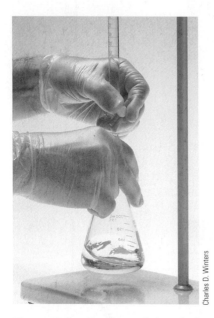

Charles D. Winters

Figure 2-23 Recommended method for manipulating a buret stopcock.

Filling

Make certain the stopcock is closed. Add 5 to 10 mL of the titrant, and carefully rotate the buret to wet the interior completely. Allow the liquid to drain through the tip. *Repeat this procedure at least two more times.* Then fill the buret well above the zero mark. Free the tip of air bubbles by rapidly rotating the stopcock and permitting small quantities of the titrant to pass. Finally, lower the level of the liquid just to or somewhat below the zero mark. Allow for drainage (~ 1 min), and then record the initial volume reading, estimating to the nearest 0.01 mL.

Titration

Figure 2-23 illustrates the preferred method for manipulating a stopcock. When you position your hand as shown, your grip on the stopcock tends to keep the stopcock firmly seated. Be sure the tip of the buret is well within the titration flask, and introduce the titrant in increments of about 1 mL. Swirl (or stir) constantly to ensure thorough mixing. Decrease the volume of the increments as the titration progresses, and add titrant drop by drop as you reach the immediate vicinity of the end point (Note 2). When it appears that only a few more drops are needed to reach the end point, rinse the walls of the container (Note 3). Allow the titrant to drain from the inner wall of the buret (at least 30 seconds) at the completion of the titration. Then record the final volume, again to the nearest 0.01 mL.

Notes

1. When unfamiliar with a particular titration, many workers prepare an extra sample. No care is taken with its titration since its functions are to reveal the nature of the end point and to provide a rough estimate of titrant requirements. This deliberate sacrifice of one sample frequently results in an overall saving of time.

2. Increments smaller than one drop can be taken by allowing a small volume of titrant to form on the tip of the buret and then touching the tip to the wall of the flask. This partial drop is then combined with the bulk of the liquid as in Note 3.

3. Instead of being rinsed toward the end of a titration, the flask can be tilted and rotated so that the bulk of the liquid picks up any drops that adhere to the inner surface.

2G-7 Directions for Using a Volumetric Flask

Before being put into use, volumetric flasks should be washed with detergent and thoroughly rinsed. Only rarely do they need to be dried. If required, however, drying is best accomplished by clamping the flask in an inverted position. Insertion of a glass tube connected to a vacuum line hastens the process.

Direct Weighing into a Volumetric Flask

The direct preparation of a standard solution requires the introduction of a known mass of solute to a volumetric flask. Use of a powder funnel minimizes the possibility of losing solid during the transfer. Rinse the funnel thoroughly, and collect the washings in the flask.

The foregoing procedure may be inappropriate if heating is needed to dissolve the solute. Instead, weigh the solid into a beaker or flask, add solvent, heat to dissolve the solute, and allow the solution to cool to room temperature. Transfer this solution quantitatively to the volumetric flask, as described in the next section.

Quantitative Transfer of Liquid to a Volumetric Flask

> The solute should be completely dissolved *before* you dilute to the mark.

Insert a funnel into the neck of the volumetric flask, and use a stirring rod to direct the flow of liquid from the beaker into the funnel. With the stirring rod, tip off the last drop of liquid on the spout of the beaker. Rinse both the stirring rod and the interior of the beaker with distilled water and transfer the washings to the volumetric flask as before. Repeat the rinsing process *at least* two more times.

Diluting to the Mark

After the solute has been transferred, fill the flask about half full and swirl the contents to hasten solution. Add more solvent and again mix well. Bring the liquid level almost to the mark, and allow time for drainage (~1 min). Then use a medicine dropper to make any necessary final additions of solvent (see Note below). Firmly stopper the flask, and invert it repeatedly to ensure thorough mixing. Transfer the contents to a storage bottle that either is dry or has been thoroughly rinsed with several small portions of the solution from the flask.

Note

If, as sometimes happens, the liquid level accidentally exceeds the calibration mark, the solution can be saved by correcting for the excess volume. Use a selfstick label to mark the location of the meniscus. After the flask has been emptied, carefully refill to the manufacturer's etched mark with water. Use a buret to determine the additional volume needed to fill the flask so that the meniscus is at the gummed-label mark. This volume must be added to the nominal volume of the flask when calculating the concentration of the solution.

2H CALIBRATING VOLUMETRIC GLASSWARE

Volumetric glassware is calibrated by measuring the mass of a liquid (usually distilled or deionized water) of known density and temperature that is contained in (or delivered by) the volumetric ware. In carrying out a calibration, a buoyancy correction must be made (Section 2D-4) since the density of water is quite different from that of the masses.

The calculations associated with calibration are a bit time consuming if done manually, but they can be automated in a spreadsheet so that they require little more time than entering the data. The raw mass data are first corrected for buoyancy with Equation 2-1. Next, the volume of the apparatus at the temperature of calibration *(T)* is obtained by dividing the density of the liquid at that temperature into the corrected mass. Finally, this volume is corrected to the standard temperature of 20°C, as in Example 2-2.

Table 2-3 is provided to help with buoyancy calculations. Corrections for buoyancy with respect to stainless steel or brass mass (the density difference between the two is small enough to be neglected) and for the volume change of water and of glass containers have been incorporated into these data. Multiplication by the appropriate factor from Table 2-3 converts the mass of water at temperature *T* to (1) the corresponding volume at that temperature or (2) the volume at 20°C.

EXAMPLE 2-3

A 25-mL pipet delivers 24.976 g of water weighed against stainless steel mass at 25°C. Use the data in Table 2-3 to calculate the volume delivered by this pipet at 25°C and the volume if the weighing were carried out at 20°C.

Solution

$$\text{At } 25°\text{C:} \quad V = 24.976 \text{ g} \times 1.0040 \text{ mL/g} = 25.08 \text{ mL}$$

$$\text{At } 20°\text{C:} \quad V = 24.976 \text{ g} \times 1.0037 \text{ mL/g} = 25.07 \text{ mL}$$

2H-1 General Directions for Calibration

All volumetric ware should be painstakingly freed of water breaks before being calibrated. Burets and pipets need not be dry, but volumetric flasks should be thoroughly drained and dried at room temperature. The water used for calibration should be

TABLE 2-3

Volume Occupied by 1.000 g of Water Weighed in Air against Stainless Steel Weights*

Temperature, T, °C	Volume, mL	
	At T	Corrected to 20°C
10	1.0013	1.0016
11	1.0014	1.0016
12	1.0015	1.0017
13	1.0016	1.0018
14	1.0018	1.0019
15	1.0019	1.0020
16	1.0021	1.0022
17	1.0022	1.0023
18	1.0024	1.0025
19	1.0026	1.0026
20	1.0028	1.0028
21	1.0030	1.0030
22	1.0033	1.0032
23	1.0035	1.0034
24	1.0037	1.0036
25	1.0040	1.0037
26	1.0043	1.0041
27	1.0045	1.0043
28	1.0048	1.0046
29	1.0051	1.0048
30	1.0054	1.0052

*Corrections for buoyancy (stainless steel weights) and change in container volume have been applied.

in thermal equilibrium with its surroundings. This condition is best established by drawing the water well in advance, noting its temperature at frequent intervals, and waiting until no further changes occur.

Although an analytical balance can be used for calibration, weighings to the nearest milligram are perfectly satisfactory for all but the very smallest volumes. Thus, a top-loading balance is more convenient to use than an analytical balance. Weighing bottles or small, well-stoppered conical flasks can serve as receivers for the calibration liquid.

Calibrating a Volumetric Pipet

Determine the empty mass of the stoppered receiver to the nearest milligram. Transfer a portion of temperature-equilibrated water to the receiver with the pipet, weigh the receiver and its contents (again, to the nearest milligram), and calculate the mass of water delivered from the difference in these masses. With the aid of Table 2-3, calculate the volume delivered. Repeat the calibration several times, and calculate the mean volume delivered and its standard deviation.

Calibrating a Buret

Fill the buret with temperature-equilibrated water and make sure that no air bubbles are trapped in the tip. Allow about 1 minute for drainage, and then lower the liquid level to bring the bottom of the meniscus to the 0.00-mL mark. Touch the tip to the wall of a beaker to remove any adhering drop. Wait 10 minutes and recheck the volume. If the stopcock is tight, there should be no perceptible change. During this interval, weigh (to the nearest milligram) a 125-mL conical flask fitted with a rubber stopper.

Once tightness of the stopcock has been established, slowly transfer (at about 10 mL/min) approximately 10 mL of water to the flask. Touch the tip to the wall of the flask. Wait 1 minute, record the volume that was apparently delivered, and refill the buret. Weigh the flask and its contents to the nearest milligram. The difference between this mass and the initial value is the mass of water delivered. Use Table 2-3 to convert this mass to the true volume. Subtract the apparent volume from the true volume. This difference is the correction that should be applied to the apparent volume to give the true volume. Repeat the calibration until agreement within ± 0.02 mL is achieved.

Starting again from the zero mark, repeat the calibration, this time delivering about 20 mL to the receiver. Test the buret at 10-mL intervals over its entire volume. Prepare a plot of the correction to be applied as a function of volume delivered. The correction associated with any interval can be determined from this plot.

Calibrating a Volumetric Flask

Weigh the clean, dry flask to the nearest milligram. Then fill to the mark with equilibrated water and reweigh. With the aid of Table 2-3, calculate the volume contained.

Calibrating a Volumetric Flask Relative to a Pipet

The calibration of a volumetric flask relative to a pipet provides an excellent method for partitioning a sample into aliquots. These directions are for a 50-mL pipet and a 500-mL volumetric flask. Other combinations of volumes are equally convenient.

Carefully transfer ten 50-mL aliquots from the pipet to a dry 500-mL volumetric flask. Mark the location of the meniscus with a gummed label. Cover with a label varnish to ensure permanence. Dilution to the label mark permits the same pipet to deliver precisely a one-tenth aliquot of the solution in the flask. Note that recalibration is necessary if another pipet is used.

2I THE LABORATORY NOTEBOOK

A laboratory notebook is needed to record measurements and observations concerning an analysis. The book should be permanently bound with consecutively numbered pages (if necessary, the pages should be hand numbered before any entries are made). Most notebooks have more than ample room, so there is no need to crowd entries.

The first few pages should be saved for a table of contents that is updated as entries are made.

2I-1 Maintaining a Laboratory Notebook

1. *Record all data and observations directly into the notebook in ink.* Neatness is desirable, but you should not achieve neatness by transcribing data from a sheet of paper to the notebook or from one notebook to another. The risk of misplacing—or incorrectly transcribing—crucial data and thereby ruining an experiment is unacceptable.

2. Supply each entry or series of entries with a heading or label. A series of weighing data for a set of empty crucibles, for example, should carry the heading "empty crucible mass" (or something similar), and the mass of each crucible should be identified by the same number or letter used to label the crucible.

3. Date each page of the notebook as it is used.

4. *Never* attempt to erase or obliterate an incorrect entry. Instead, cross it out with a single horizontal line and locate the correct entry as nearby as possible. Do not write over incorrect numbers. With time, it may become impossible to distinguish the correct entry from the incorrect one.

5. *Never* remove a page from the notebook. Draw diagonal lines across any page that is to be disregarded. Provide a brief rationale for disregarding the page.

❮ Remember that you can discard an experimental measurement *only if you have certain knowledge that you made an experimental error.* Thus, you must carefully record experimental observations in your notebook as soon as they occur.

❮ An entry in a laboratory notebook should never be erased but should be crossed out instead.

2I-2 Notebook Format

The instructor should be consulted concerning the format to be used in keeping the laboratory notebook.[9] In one convention, data and observations are recorded on consecutive pages as they occur. The completed analysis is then summarized on the next available page spread (that is, left- and right-facing pages). As shown in **Figure 2-24**, the first of these two facing pages should contain the following entries:

1. The title of the experiment ("The Gravimetric Determination of Chloride").
2. A brief statement of the principles on which the analysis is based.
3. A complete summary of the weighing, volumetric, and/or instrument response data needed to calculate the results.
4. A report of the best value for the set and a statement of its precision.

The second page should contain the following items:

1. Equations for the principal reactions in the analysis.
2. An equation showing how the results were calculated.
3. A summary of observations that appear to bear on the validity of a particular result or the analysis as a whole. *Any such entry must have been originally recorded in the notebook at the time the observation was made.*

2J   SAFETY IN THE LABORATORY

There is necessarily a degree of risk associated with any work in a chemical laboratory. Accidents can and do happen. Strict adherence to the following rules will go far toward preventing (or minimizing the effect of) accidents:

1. Before you begin work in any laboratory, learn the location of the nearest eye fountain, fire blanket, shower, and fire extinguisher. Learn the proper use of each, and do not hesitate to use this equipment if the need arises.
2. **Wear eye protection at all times.** The potential for serious and perhaps permanent eye injury makes it mandatory that adequate eye protection be worn at all times by students, instructors, and visitors. Eye protection should be donned before entering the laboratory and should be used continuously until it is time to leave. Serious eye injuries have occurred to people performing such innocuous tasks as computing or writing in a laboratory notebook. Incidents such as these usually result from someone else's loss of control over an experiment. Regular prescription glasses are not adequate substitutes for eye protection approved by the Office of Safety and Health Administration (OSHA). Contact lenses should never be used in the laboratory because laboratory fumes may react with them and have a harmful effect on the eyes.
3. Most of the chemicals in a laboratory are toxic, some are very toxic, and some—such as concentrated solutions of acids and bases—are highly corrosive. Avoid contact between these liquids and the skin. In the event of such contact, *immediately* flood the affected area with large quantities of water. If a corrosive solution is spilled on clothing, remove the garment immediately. Time is of the essence, so do not be concerned about modesty.
4. *NEVER* perform an unauthorized experiment. Unauthorized experiments are grounds for disqualification at many institutions.

[9]See also Howard M. Kanare, *Writing the Laboratory Notebook*, Washington, DC: American Chemical Society, 1985.

08

Gravimetric Determination of Chloride

The chloride in a soluble sample was precipitated as AgCl and weighed as such.

Sample masses	1	2	3
Mass bottle plus sample, g	27.6115	27.2185	26.8105
-less bottle, g	27.2185	26.8105	26.4517
mass sample, g	0.3930	0.4080	0.3588
Crucible masses, empty	~~20.7925~~	~~22.8311~~	~~21.2488~~
	20.7926	22.8311	~~21.2482~~
			21.2483
Crucible masses, with AgCl, g	~~21.4294~~	~~23.4920~~	~~21.8327~~
	~~21.4297~~	~~23.4914~~	21.8323
	21.4296	23.4915	
Mass of AgCl, g	0.6370	0.6604	0.5840
Percent Cl⁻	40.10	40.04	40.27
Average percent Cl⁻		40.12	
Relative standard deviation	3.0 parts per thousand		
Date Started		1-10-12	
Date Completed		1-16-12	

Figure 2-24 Laboratory notebook data page.

5. Never work alone in the laboratory. Always be certain that someone is within earshot.

6. Never bring food or beverages into the laboratory. *NEVER* drink from laboratory glassware. *NEVER* smoke in the laboratory.

7. Always use a bulb or other device to draw liquids into a pipet. *NEVER* pipet by mouth.

8. Wear adequate foot covering (no sandals). Confine long hair with a net. A laboratory coat or apron will provide some protection and may be required.

9. Be extremely tentative in touching objects that have been heated because hot glass looks exactly like cold glass.

10. Always fire-polish the ends of freshly cut glass tubing. *NEVER* attempt to force glass tubing through the hole of a stopper. Instead, make sure that both tubing and hole are wet with soapy water. Protect your hands with several layers of towel while inserting glass into a stopper.

11. Use fume hoods whenever toxic or noxious gases are likely to be evolved. Be cautious in testing for odors. Use your hand to waft vapors above containers toward your nose.

12. Notify your instructor immediately in the event of an injury.

13. Dispose of solutions and chemicals as instructed. It is illegal to flush solutions containing heavy metal ions or organic liquids down the drain in most localities. Alternative arrangements are required for the disposal of such liquids.

CHAPTER 3

Using Spreadsheets in Analytical Chemistry

From the ways that we deal with our finances using software applications such as Quicken to our modes of communication with friends, relatives, and colleagues using Mozilla Thunderbird, and Microsoft® Outlook, the personal computer has revolutionized nearly every aspect of our lives. Physical chemists use applications such as Gaussian, GAMESS, and MPQCC to carry out quantum calculations. Biological chemists and organic chemists use molecular mechanics programs such as Spartan to build and investigate the properties of molecules, and inorganic chemists exploit ChemDraw to visualize molecules. Certain software programs transcend specialization and are used in a broad range of fields. In analytical chemistry and many other areas of science, spreadsheet programs provide a means for storing, analyzing, and organizing numerical and textual data. Microsoft® Excel is an example of this type of program.

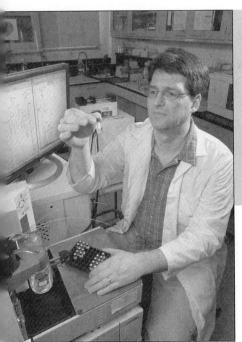

Stephen Ausmus/US Department of Agriculture

The personal computer revolution has produced many useful tools for students, chemists, biologists, and many other scientists and engineers. The spreadsheet is one of the best examples of these applications. Spreadsheets are versatile, powerful, and easy to use. They are used for record keeping, mathematical calculations, statistical analysis, curve fitting, data plotting, financial analysis, database management, and a variety of other tasks limited only by our imaginations. State-of-the-art spreadsheet programs have many built-in functions to help us accomplish the computational tasks of analytical chemistry. Throughout this text, we present examples to illustrate some of these tasks and show the actual spreadsheets for performing them. We use Microsoft® Excel 2010 or 2007 for these examples and exercises because of the program's popularity and widespread availability. We assume throughout that Excel is configured with default options as delivered from the manufacturer unless we specifically note otherwise. You will find many more examples, more elaborate explanations of spreadsheet methodology, and expanded treatments of some of the theory of analytical chemistry in the ancillary text *Applications of Microsoft® Excel in Analytical Chemistry*, 2nd ed.[1]

In this chapter, we present some introductory spreadsheet operations, including entering text and data, formatting cells, and making several useful calculations. In later chapters, we explore how to process and display large amounts of data using Excel's built-in numerical, statistical, and graphing functions.

---

[1] For more information on the use of spreadsheets in chemistry, see S. R. Crouch and F. J. Holler *Applications of Microsoft® Excel in Analytical Chemistry*, 2nd ed., Belmont, CA: Brooks/Cole, 2014.

3A KEEPING RECORDS AND MAKING CALCULATIONS

It is our feeling that we learn best by doing, not by reading about doing. Although software manufacturers have made great strides in the production of manuals for their products, it is still generally true that, when we know enough to read a software manual efficiently, we no longer need the manual. With that in mind, we have designed a series of spreadsheet exercises that evolve in the context of analytical chemistry. We introduce commands and syntax only when they are needed to accomplish a particular task, so if you need more detailed information, please consult the Excel help screens or your software documentation. Help is available at the click of a mouse button up by clicking on the Help icon in the upper-right-hand corner of the Excel screen or by pressing **F1**. In either case, a new window opens that permits you to type questions and obtain context-sensitive help.

3A-1 Getting Started

In this book, we will assume that you are familiar with *Windows*™. If you need assistance with Windows, please consult the Windows guide *Getting Started* or use the available on-line help facility. To start Excel, double click on the Excel icon, as shown in the margin, or use the **Start** button and click on **Start/All Programs/Microsoft Office/Microsoft Office Excel 2010** (or 2007 if you have that version). The window shown in **Figure 3-1** then appears on your computer screen.

Versions of Excel prior to Excel 2007 contained menus such as File, Edit, View, Insert, Format, and Tools, among others. The menus and toolbars have been

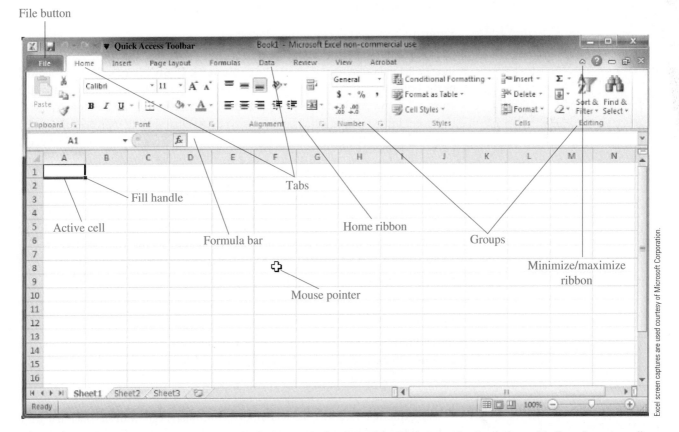

Figure 3-1   The opening window in Microsoft Excel. Note the location of the File button, the Quick Access Toolbar, the active cell, and the mouse pointer.

completely removed from Excel 2007 and 2010 and replaced by the *ribbon*, a two-dimensional layout of icons and words. Each tab, such as Home, Insert, Page Layout, Formulas, Data, Review, and View, brings up a different ribbon with its own set of icons and descriptions. Although the ribbon takes up space, it can be minimized by clicking the minimize ribbon arrow, by entering **Ctrl+F1**, or by right clicking anywhere on the ribbon and selecting Mi<u>n</u>imize the Ribbon from the list that appears. In order to maximize workspace, we suggest that you operate with the ribbon minimized. Excel 2007 does not have the File button. Instead commands such as Save, Print, Open, Close, and Send are located in the Office button to the left of the Home tab.

Below the ribbon in Figure 3-1 is the *worksheet* consisting of a grid of *cells* arranged in rows and columns. The rows are labeled 1, 2, 3, and so on, and the columns are labeled A, B, C, and so on. Each cell has a unique location specified by its address. For example, the *active cell,* which is surrounded by a dark outline as shown in Figure 3-1, has the address A1. The address of the active cell is always displayed in the box just above the first column of the displayed worksheet in the *formula bar.* You can verify this display of the active cell by clicking on various cells of the worksheet. A *workbook* is a collection of worksheets and can be comprised of multiple worksheets available by clicking the tabs at the bottom labeled Sheet1, Sheet2, and so forth. The term *spreadsheet* is a generic term and usually refers to a worksheet.

3A-2 Calculating a Molar Mass

We begin by constructing a worksheet to calculate the molar mass of sulfuric acid. Here you will learn how to enter text and numbers, how to format text and data, how to enter a formula, and how to document the worksheet.

Entering Text and Data in the Worksheet

Cells may contain text, numbers, or formulas. We start by typing some text into the worksheet. Click on cell A1, and type **Molar Mass of Sulfuric Acid** followed by the Enter key [⏎]. This is the spreadsheet title. Notice after entering the title that the active cell is now A2. In this cell, type **AM H**[⏎] as a label to indicate the atomic mass of hydrogen. In A3 type **AM S**[⏎], and in A4 type **AM O**[⏎]. In cell A6, type **Sulfuric Acid**[⏎]. In cell B2 to the right of the label **AM H**, enter the atomic mass of hydrogen, 1.00794. Likewise in cell B3, enter the atomic mass of sulfur, 32.066, and in cell B4, enter the atomic mass of oxygen, 15.9994. As you type, the data that you enter appears in the formula bar. If you make a mistake, just click the mouse in the formula bar, and make necessary corrections. Format the title of your spreadsheets in boldface font so that you can easily distinguish it from the body. This procedure can be done by selecting cell A1. In the formula bar, select the entire title by dragging the mouse over the words **Molar Mass of Sulfuric Acid**. When the title has been selected, click the Bold button (see the margin) in the Font group on the Home tab. This action will make the title appear in boldface font.

B

Entering an Equation

In cell B6 we will enter the formula that we want Excel to use to calculate the molar mass of sulfuric acid. Type the following into cell B6:

$$=2*B2+B3+4*B4 \ [⏎]$$

Excel formulas always begin with an equals sign [=].

This expression is called a *formula*. In Excel, formulas begin with an equal sign [=] followed by the desired numerical expression. This formula will calculate the molar mass of H_2SO_4 by summing twice the atomic mass of hydrogen (cell B2), the atomic mass of sulfur (cell B3), and four times the atomic mass of oxygen (cell B4). The result should be as shown in **Figure 3-2**.

Note in Figure 3-2, that Excel presents the molar mass of sulfuric acid to five digits past the decimal point. In Chapter 6, we discuss the significant figure convention, which indicates that the molar mass should only be expressed to three digits beyond the decimal point since the atomic mass of sulfur is only known to this number of digits. Hence, a more appropriate result would be 98.079 for the molar mass of H_2SO_4. To change the number of digits, display the Home ribbon, and then click on cell B6. From the Cells group, select the Format command and Format Cells . . . from the pull down menu. The Format Cells window shown in **Figure 3-3** then appears on the screen.

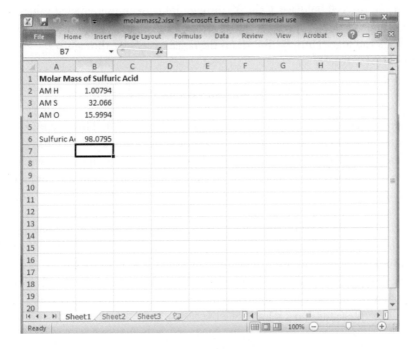

Figure 3-2 Excel spreadsheet to calculate the molar mass of sulfuric acid.

Figure 3-3 Format Cells window.

Select the Number tab and then Number from the list. In the Decimal places box select or type 3. Click the OK button. Cell B6 should now contain 98.079. Note that the effect of changing the number of decimal places can be previewed in the Sample box in the Format Cells window. You can also increase or decrease the number of decimal places by clicking the Increase or Decrease Decimal button in the Number group on the Home ribbon (see the margin).

Documenting the Worksheet

Since the spreadsheet results normally do not contain the equations entered or indicate which cells contained data, it is important to document what was done. There are several different documentation schemes, but we shall introduce a simple method for documenting. Make cell A9 the active cell and type **Documentation[↵]**. Make the font for this cell boldface. Cells B2 through B4 contain user-entered values, so in cell A10 enter **Cells B2:B4=user entries[↵]**. The colon between B2 and B4 specifies a range. Thus, B2:B4 means the range of cells B2 through B4.

In cell A11, type

$$\textbf{Cell B6=2*B2+B3+4*B4 [↵]}$$

The spreadsheet should now appear as shown in **Figure 3-4**. This documentation indicates the data entered by the user and shows the formula entered in cell B6 to calculate the molar mass of sulfuric acid. In many cases, it is apparent which cells contain user-entered data. Hence, often the documentation section will contain only formulas.

If desired, you can save your file to the hard disk by clicking on the File (Office in Excel 2007) button and choosing **Save As**. You can save as an Excel Workbook and various other formats including a format compatible with Excel 97-2003. Choose **Excel Workbook** and enter a location and a file name such as **molarmass**. Excel will automatically append the file extension **.xlsx** to the file name so that it will appear as **molarmass.xlsx**. Choosing to save in a format compatible with Excel 97-2003 appends the file extension **.xls** to the file.

> All formulas in a worksheet can be revealed by holding the control key (Ctrl) while pressing and releasing the grave accent key ` located to the left of the number 1 key on the keyboard. Release the Ctrl key. To return to the results, repeat the Ctrl + ` operation.

3B MORE COMPLEX EXAMPLES

Excel can be used for many more complex operations including numerical, statistical, and graphical functions. We illustrate a few of these operations in this chapter.

◢	A	B	C
1	**Molar Mass of Sulfuric Acid**		
2	AM H	1.00794	
3	AM S	32.066	
4	AM O	15.9994	
5			
6	Sulfuric A⁣|	98.079	
7			
8			
9	**Documentation**		
10	Cells B2:B4=user entries		
11	Cell B6=2*B2+B3+4*B4		

Figure 3-4 Final spreadsheet for calculating the molar mass of sulfuric acid including a documentation section. Note that we have omitted the Excel ribbon and formula bar for clarity in this figure.

3B-1 A Laboratory Notebook Example

For our next example, we will use Excel to carry out some functions of the laboratory notebook illustrated in Figure 2-24 for the gravimetric determination of chloride. With this example, we learn how to change the width of columns, how to fill cells with the fill handle, and how to make more complex calculations.

Entering Text in the Worksheet

Click on cell A1, and as the worksheet title, type `Gravimetric Determination of Chloride` followed by the Enter key [↵]. Continue to type text into the cells of column A as shown below.

`Mass of Bottle plus sample, g[↵]`

`Mass of bottle less sample, g[↵]`

`Mass of sample, g[↵]`

`[↵]`

`Crucible masses, with AgCl, g[↵]`

`Crucible masses, empty, g[↵]`

`Mass of AgCl, g[↵]`

`[↵]`

`%Chloride[↵]`

When you have finished entering the text, the worksheet should appear as shown in **Figure 3-5**.

Changing the Width of a Column

Notice that the labels that you typed into column A are wider than the column. You can change the width of the column by placing the mouse pointer on the boundary between column A and column B in the column head as shown in **Figure 3-6a** and dragging the boundary to the right so that all of the text shows in the column as in **Figure 3-6b**.

	A	B	C	D
1	Gravimetric Determination of Chloride			
2	Samples			
3	Mass of bottle plus sample, g			
4	Mass of bottle less sample, g			
5	Mass of sample, g			
6				
7	Crucible masses, with AgCl, g			
8	Crucible masses, empty, g			
9	Mass of AgCl, g			
10				
11	%Chloride			

Figure 3-5 Appearance of the worksheet after entering text.

Figure 3-6   Changing the column width. Left: place the mouse pointer on the boundary between column A and column B, and drag to the right to the position shown on the right.

Entering Numbers into the Spreadsheet

Now let us enter some numerical data into the spreadsheet. Click on cell B2 and type

1 [↵]

27.6115 [↵]

27.2185 [↵]

To find the mass of the sample in cell B5, we need to calculate the difference between the value in cell B3 and that in cell B4, so we type

=b3−b4 [↵]

> Note that it does not matter to Excel whether you use lower or upper case letters to refer to cells in a formula. Excel treats them all as upper case.

Continue entering data for samples 2 and 3 so that the worksheet appears as shown in **Figure 3-7**.

	A	B	C	D
1	**Gravimetric Determination of Chloride**			
2	Samples	1	2	3
3	Mass of bottle plus sample, g	27.6115	27.2185	26.8105
4	Mass of bottle less sample, g	27.2185	26.8105	26.4517
5	Mass of sample, g	0.393		
6				
7	Crucible masses, with AgCl, g			
8	Crucible masses, empty, g			
9	Mass of AgCl, g			
10				
11	%Chloride			

Figure 3-7   Sample data entry for gravimetric determination of chloride.

Filling Cells Using the Fill Handle

The formulas for cells C5 and D5 are identical to the formula in cell B5 except that the cell references for the data are different. In cell C5, we want to compute the difference between the contents of cells C3 and C4, and in cell D5, we want the difference between D3 and D4. We could type the formulas in cells C5 and D5 as we did for cell B5, but Excel provides an easy way to duplicate formulas, and it automatically changes the cell references to the appropriate values for us. To duplicate a formula in cells adjacent to an existing formula, simply click on the cell containing the formula, which is cell B5 in our example, then click on the fill handle (see Figure 3-1), and drag the corner of the rectangle to the right so that it encompasses the cells where you want the formula to be duplicated. Try it now. Click on cell B5, click on the fill handle, and drag to the right to fill cells C5 and D5. When you let up on the mouse button, the spreadsheet should look like **Figure 3-8**. Now click on cell B5, and view the formula in the formula bar. Compare the formula to those in cells C5 and D5. The cell references that change are called **relative references**.

Now enter the data into rows 7 and 8 as shown in **Figure 3-9**. Next, click on cell B9, and type the following formula:

$$=b7-b8 \; [\lrcorner]$$

> Relative cell references change when you copy a formula to another cell. By default, Excel creates relative references unless instructed not to do so.

> The fill handle permits you to copy the contents of a cell to other cells either horizontally or vertically, but not both. Just click on the fill handle, and drag from the current cell to the last cell where you want the original cell copied.

	A	B	C	D
1	**Gravimetric Determination of Chloride**			
2	Samples	1	2	3
3	Mass of bottle plus sample, g	27.6115	27.2185	26.8105
4	Mass of bottle less sample, g	27.2185	26.8105	26.4517
5	Mass of sample, g	0.393	0.408	0.3588

Figure 3-8 Use of the fill handle to copy formulas into adjacent cells of a spreadsheet. In this example, we clicked on cell B5, clicked on the fill handle, and dragged the rectangle to the right to fill cells C5 and D5. The formulas in cells B5, C5, and D5 are identical, but the cell references in the formulas refer to data in columns B, C, and D, respectively.

	A	B	C	D
1	**Gravimetric Determination of Chloride**			
2	Samples	1	2	3
3	Mass of bottle plus sample, g	27.6115	27.2185	26.8105
4	Mass of bottle less sample, g	27.2185	26.8105	26.4517
5	Mass of sample, g	0.393	0.408	0.3588
6				
7	Crucible masses, with AgCl, g	21.4296	23.4915	21.8323
8	Crucible masses, empty, g	20.7926	22.8311	21.2483
9	Mass of AgCl, g			
10				
11	%Chloride			

Figure 3-9 Entering the data into the spreadsheet in preparation for calculating the mass of dry silver chloride in the crucibles.

Again click on cell B9, click on the fill handle, and drag through columns C and D to copy the formula to cells C9 and D9. The mass of silver chloride should now be calculated for all three crucibles.

Making Complex Calculations with Excel

As we shall learn in Chapter 12, the equation for finding the %chloride in each of the samples is

$$\%\text{chloride} = \frac{\dfrac{\text{mass AgCl}}{\text{molar mass AgCl}} \times \text{molar mass Cl}}{\text{mass sample}} \times 100\%$$

$$= \frac{\dfrac{\text{mass AgCl}}{143.321 \text{ gram/mol}} \times 35.4527 \text{ grams/mol}}{\text{mass sample}} \times 100\%$$

Our task is now to translate this equation into an Excel formula and type it into cell B11 as shown below.

=B9*35.4527*100/143.321/B5[↵]

Once you have typed the formula, click on cell B11, and drag on the fill handle to copy the formula into cells C11 and D11. The %chloride for samples 2 and 3 should now appear in the worksheet as shown in **Figure 3-10**.

Documenting the Worksheet

We can now document the spreadsheet since our calculations are complete. In cell A13, type **Documention[↵]**. Cells B2 through D5 and B7 through D9 contain user-entered values. In cell A14, enter **Cells B2:D5 and B7:D9=Data entries[↵]**.

We now want to document the calculations done in cells B5:D5, B9:D9, and B11:D11. Instead of retyping the formulas in these cells from scratch as we did in the molar mass example, there is an easy way to copy them into the documentation cells. This shortcut also prevents typing errors in entering the formulas. To illustrate,

	A	B	C	D
1	**Gravimetric Determination of Chloride**			
2	Samples	1	2	3
3	Mass of bottle plus sample, g	27.6115	27.2185	26.8105
4	Mass of bottle less sample, g	27.2185	26.8105	26.4517
5	Mass of sample, g	0.393	0.408	0.3588
6				
7	Crucible masses, with AgCl, g	21.4296	23.4915	21.8323
8	Crucible masses, empty, g	20.7926	22.8311	21.2483
9	Mass of AgCl, g	0.637	0.6604	0.584
10				
11	%Chloride	40.09464	40.03929	40.26242

Figure 3-10 Completing the calculation of percent chloride. Type the formula in cell B11, click on the fill handle, and drag to the right through cell D11.

⊿	A	B	C	D
1	**Gravimetric Determination of Chloride**			
2	Samples	1	2	3
3	Mass of bottle plus sample, g	27.6115	27.2185	26.8105
4	Mass of bottle less sample, g	27.2185	26.8105	26.4517
5	Mass of sample, g	0.393	0.408	0.3588
6				
7	Crucible masses, with AgCl, g	21.4296	23.4915	21.8323
8	Crucible masses, empty, g	20.7926	22.8311	21.2483
9	Mass of AgCl, g	0.637	0.6604	0.584
10				
11	%Chloride	40.09464	40.03929	40.26242
12				
13	**Documentation**			
14	Cells B2:D4 and B7:D8=Data entries			
15	Cell B5=B3-B4			
16	Cell B9=B7-B8			
17	Cell B11=B9*35.4527*100/143.321/B5			

Figure 3-11 Completed worksheet with documentation.

select cell A15, and type `Cell B5 [↵]`. Now select cell B5, and highlight the formula displayed in the formula bar. Click on the Copy icon in the Clipboard group on the Home tab as shown in the margin. To prevent Excel from copying the formula and changing the cell references, hit the Escape key on the keyboard to cancel the operation. The text copied, however, is still in the Windows clipboard. Now select cell A15, and position the cursor after the B5 in the formula bar. Click on the Paste icon as shown in the margin. This operation will copy the formula for the mass of sample into cell A15 as a text string. In cells C5 and D5, the same formula is used except that the relative references changed to columns C and D when we used the fill handle. Because the same formula is used, the documentation does not need to include these cells. In cell A16, type `Cell B9 [↵]`. Copy the formula from cell B9 as before. In Cell A17, type `Cell B11 [↵]`, and copy the formula from this cell. When you have finished, the worksheet should appear as in **Figure 3-11**.

Copy ▾

Paste
▾

3B-2 Another Example from Gravimetric Analysis

Let's now use some of the basics we have learned to solve a problem of gravimetric analysis. In this problem, we are to compute the percentage of Fe and Fe_3O_4 in two samples of an iron ore. The samples were precipitated as $Fe_2O_3 \cdot xH_2O$, and the residue was ignited to give pure Fe_2O_3.

More Cell Formatting

First, select cell A1, and type a title such as `Gravimetric Analysis Example` in bold. You can either do as before by typing the title in regular font, selecting it, and clicking the Bold button shown in the margin, or you can click the Bold button before typing so that all subsequent typing in the active cell appears in bold. Next, in cell A2 type `Sample` and put the sample numbers in cell B2 and C2. In cell A3 type `mppt`. Now we'll learn how to make the abbreviation ppt

B

appear as a subscript as in m_{ppt}. Select cell A3. In the formula bar, use the mouse to highlight (select) the **ppt** part of **mppt**. With **ppt** highlighted, click the right mouse button, and select **Format Cells** from the list. The Format Cells window shown in **Figure 3-12** should appear.

Note that since cell A3 contained only text, the **Font** tab automatically appears in the window. Recall from the molar mass example that, when there are numbers in the cell or an entire cell is selected, the Format Cells window contains tabs for **Number**, **Alignment**, **Font**, **Border**, **Fill**, and **Protection**. Select **Subscript** in the **Effects** box so that a checkmark appears as shown. Click on the OK button, and note that cell A2 now contains m_{ppt} as a label for mass of precipitate. Similarly in cells A4, A5, and A6, type m_{samp}, M_{Fe}, and M_O as labels for sample mass, atomic mass of iron, and atomic mass of oxygen.

Entering the Data

The first sample analyzed was a 1.1324-g sample, which gave a precipitate with a mass of 0.5394 g. In cell B3, enter the number 0.5394 for the mass of the precipitate. In cell B4, type the number 1.1324 for the sample mass. In cells B5 and B6, type the atomic masses of iron (55.847) and oxygen (15.9994). Your spreadsheet should now look like **Figure 3-13**.

Calculating Molar Masses

In order to obtain the desired percentages, we need the molar masses of Fe_2O_3 and Fe_3O_4 in addition to the atomic masses of iron and oxygen. We can use Excel to calculate these molar masses. In cell A8, type M_{Fe2O3}. Because Excel cannot make sub-subscripts, Fe_2O_3 will be used as the subscript. Likewise in cell A9, type M_{Fe3O4}. We will put the calculated molar mass of Fe_2O_3 in cell B8 and the molar mass for Fe_3O_4 in cell B9. In cell B8, type

$$=2*B5+3*B6 \ [\llcorner]$$

and in B9, type

$$=3*B5+4*B6 \ [\llcorner]$$

Figure 3-12 The Format Cells window for formatting a subscript.

	A	B	C
1	**Gravimetric Analysis Example**		
2	Sample	1	2
3	m_{ppt}	0.5394	0.6893
4	m_{samp}	1.1324	1.4578
5	M_{Fe}	55.847	
6	M_O	15.9994	
7			

Figure 3-13 Data entry for sample 1 of the gravimetric analysis example.

The molar masses of Fe_2O_3 (159.692) and Fe_3O_4 (231.539) should appear in cells B8 and B9. If more than three digits beyond the decimal point are displayed, change the number format to show three digits.

Calculating the Percentages

Our next task is to use the mass of the sample, the mass of the precipitate, the molar masses, and stoichiometric information to calculate the desired percentages. Type into cells A11 and A12 the labels **%Fe** and **%Fe₃O₄**. For Fe, the following equation allows us to calculate the percentage.

$$\% \ Fe = \frac{\dfrac{m_{ppt}}{\mathcal{M}_{Fe_2O_3}} \times 2 \ \mathcal{M}_{Fe}}{m_{samp}} \times 100\%$$

Type into cell B11 the formula

$$\mathtt{=B3/B8*2*B5/B4*100 \, [\lrcorner]}$$

The calculation should return the result 33.32 for % Fe. Again adjust the number of significant figures if too many digits are displayed.

For Fe_3O_4, the equation for the percentage is

$$\% \ Fe_3O_4 = \frac{\dfrac{m_{ppt}}{\mathcal{M}_{Fe_2O_3}} \times \dfrac{2}{3} \times \mathcal{M}_{Fe_3O_4}}{m_{samp}} \times 100\%$$

Type into cell B12, the formula

$$\mathtt{=B3/B8*2/3*B9/B4*100 \, [\lrcorner]}$$

This action should return the result 46.04 for % Fe_3O_4. Note that because these calculations involve only multiplications and division, it is not necessary to tell Excel the order in which to do the calculations. This *hierarchy of operations* is necessary when there is a combination of multiplications or divisions and additions or subtractions.

Finding the Percentages for Sample 2: Using Absolute References

For Sample 2, a sample mass of 1.4578 g gave a precipitate mass of 0.6893 g. Enter these values into cells C3 and C4. We will use the same atomic mass for Fe in cell B5 and the same molar masses for Fe_2O_3 and Fe_3O_4 in cells B8 and B9 for our calculation of the percentages. Therefore, we do not want the references

Figure 3-14 Locating and replacing a relative value with an absolute value.

in cells B8 and B9 to be relative references when we copy the formulas into cells C11 and C12. We can make a reference an *absolute reference* by putting a dollar sign before the column letter and a second dollar sign before the row number. In order to change these references, we will use Excel's **Find and Replace** functions. With the cursor on cell A1, click on **Find & Select** on the Home ribbon. Choose **Replace…** from the drop-down menu. Enter B5 in the **Find** box and $B$5 in the **Replace** box as shown in **Figure 3-14**. Repeat for references B8 and B9. As a result, these cells become

$$\texttt{=B3/\$B\$8*2*\$B\$5/B4*100} \quad \text{for B11}$$

$$\texttt{=B3/\$B\$8*2/3*\$B\$9/B4*100} \quad \text{for B12}$$

We can then copy these results into cells C11 and C12 to calculate the percentages for sample 2. Click on cell C11 after copying, and note that only the relative references without the dollar signs have changed to column C values. The final worksheet after adding Documentation is shown in **Figure 3-15**. Save your worksheet to the disk with a file name such as **grav_analysis.xls**.

In this chapter, we have begun to explore the use of spreadsheets in analytical chemistry. We have examined many of the basic operations of spreadsheet use including data and text entry and formatting, basic calculations, and the use of relative and absolute cell references. In other spreadsheets in this book and in our ancillary, *Applications of Microsoft® Excel in Analytical Chemistry*,[2] we build on the techniques that we have acquired and learn much more about Excel.

Figure 3-15 Completed worksheet for gravimetric analysis example.

	A	B	C	D
1	**Gravimetric Analysis Example**			
2	Sample	1	2	
3	m_{ppt}	0.5394	0.6893	
4	m_{samp}	1.1324	1.4578	
5	M_{Fe}	55.847		
6	M_O	15.9994		
7				
8	M_{Fe2O3}	159.6922		
9	M_{Fe3O4}	231.5386		
10				
11	%Fe	33.32	33.07	
12	%Fe$_3$O$_4$	46.04	45.70	
13				
14	**Documentation**			
15	Cell B8=2*B5+3*B6			
16	Cell B9=3*B5+4*B6			
17	Cell B11=B3/$B$8*2*$B$5/B4*100			
18	Cell B12=B3/$B$8*2/3*$B$9/B4*100			

[2]S. R. Crouch and F. J. Holler, *Applications of Microsoft® Excel in Analytical Chemistry*, 2nd ed., Belmont, CA: Brooks/Cole, 2014.

WEB WORKS Use a search engine to locate Excel 2010 supported file formats. Describe files with extensions **.csv, .dbf**, and **.ods**. Which of these file formats can Excel open and which can it save? Find file formats that are not supported in Excel 2010. What is a **.wks** file format? What is a **.xlc** format? Are there any file converters that will translate a Lotus 1-2-3 file into an Excel compatible file?

QUESTIONS AND PROBLEMS

*3-1. Describe the use of the following Excel functions after reading about them in the Excel help facility.
   (a) SQRT
   (b) AVERAGE
   (c) PI
   (d) FACT
   (e) EXP
   (f) LOG

3-2. Use the Excel help facility to look up the use of the COUNT function. Use the function to determine the number of data values in each column of the worksheet of Figure 3-10. The count function is quite useful for determining the number of cells containing numbers in a given area of a worksheet.

	A	B	C
1			
2		45	
3		22	
4		36	
5		27	
6		61	
7		23	
8		33	
9		48	
10		35	
11		55	
12		31	

3-3. There are many ways to document the worksheet entries and calculations. Use a search engine to find some of these methods, and describe them in detail using a worksheet example.

3-4. Use Excel's **Find/Replace** function to replace all the values containing 27 in the worksheet of Figure 3-10 with 26.

3-5. Enter the values shown in the accompanying worksheet into a blank worksheet. Use Excel's help facility to learn about Excel's **Sort&Filter** operations. Have Excel sort the numbers from smallest to largest.

3-6. Next we will add the numbers in column B in the worksheet accompanying Problem 3-5. There are several ways to accomplish this operation in Excel. In a cell at the bottom of column B, you can invoke the SUM function by typing `=SUM(B2:B12)`. This action should return the value 416. You can also invoke the AUTOSUM function by clicking on AUTOSUM in the Editing group on the Home tab. Use the mouse to select the values to sum in this way, and show that the same result is found by Excel. In the formula bar, verify that AUTOSUM produces the exact same formula as that entered manually.

CHAPTER 4

Calculations Used in Analytical Chemistry

CSIRO Australia

Avogadro's number is one of the most important of all physical constants and is central to the study of chemistry. A worldwide effort is under way to determine this important number to 1 part in 100 million. Several spheres like the one shown in the photo have been fabricated specifically for this task, and it is claimed they are the most perfect spheres in the world. The diameter of the 10-cm sphere is uniform to within 40 nm. By measuring the diameter, the mass, the molar mass of silicon, and the spacing between silicon atoms, it is possible to calculate Avogadro's number. Once determined, this number may be used to provide a new standard mass—the silicon kilogram. For more information, see Problem 4-41 and Web Works.

In this chapter, we describe several methods used to compute the results of a quantitative analysis. We begin by presenting the SI system of units and the distinction between mass and weight. We then discuss the mole, a measure of the amount of a chemical substance. Next, we consider the various ways that concentrations of solutions are expressed. Finally, we treat chemical stoichiometry. You may have studied much of the material in this chapter in your general chemistry courses.

4A SOME IMPORTANT UNITS OF MEASUREMENT

4A-1 SI Units

> SI is the acronym for the French "Système International d'Unités."

Scientists throughout the world have adopted a standardized system of units known as the **International System of Units** (SI). This system is based on the seven fundamental base units shown in **Table 4-1**. Numerous other useful units, such as volts, hertz, coulombs, and joules, are derived from these base units.

> The **ångstrom unit Å** is a non-SI unit of length that is widely used to express the wavelength of very short radiation such as X-rays (1 Å = 0.1 nm = 10^{-10} m). Thus, typical X-radiation lies in the range of 0.1 to 10 Å.

To express small or large measured quantities in terms of a few simple digits, prefixes are used with these base units and other derived units. As shown in **Table 4-2**, these prefixes multiply the unit by various powers of 10. For example, the wavelength of yellow radiation used for determining sodium by flame photometry is about 5.9×10^{-7} m, which can be expressed more compactly as 590 nm (nanometers); the volume of a liquid injected onto a chromatographic column is often roughly 50×10^{-6} L, or 50 μL (microliters); or the amount of memory on some computer hard disks is about 20×10^9 bytes, or 20 Gbytes (gigabytes).

TABLE 4-1

SI Base Units		
Physical Quantity	**Name of Unit**	**Abbreviation**
Mass	kilogram	kg
Length	meter	m
Time	second	s
Temperature	kelvin	K
Amount of substance	mole	mol
Electric current	ampere	A
Luminous intensity	candela	cd

In analytical chemistry, we often determine the amount of chemical species from mass measurements. For such measurements, metric units of kilograms (kg), grams (g), milligrams (mg), or micrograms (µg) are used. Volumes of liquids are measured in units of liters (L), milliliters (mL), microliters (µL), and sometimes nanoliters (nL). The liter, the SI unit of volume, is defined as exactly 10^{-3} m³. The milliliter is defined as 10^{-6} m³, or 1 cm³.

4A-2 The Distinction Between Mass and Weight

It is important to understand the difference between mass and weight. **Mass** is an invariant measure of the quantity of matter in an object. **Weight** is the force of attraction between an object and its surroundings, principally the earth. Because gravitational attraction varies with geographical location, the weight of an object depends on where you weigh it. For example, a crucible *weighs* less in Denver than in Atlantic City (both cities are at approximately the same latitude) because the attractive force between the crucible and the earth is smaller at the higher altitude of Denver. Similarly, the crucible *weighs* more in Seattle than in Panama (both cities are at sea level) because the Earth is somewhat flattened at the poles, and the force of attraction

AFP/Getty Images

For more than a century, the kilogram has been defined as the mass of a single platinum-iridium standard housed in a laboratory in Sèvres, France. Unfortunately, the standard is quite imprecise relative to other standards such as the meter, which is defined to be the distance that light travels in 1/299792458 of a second. A worldwide consortium of metrologists is working on determining Avogadro's number to 1 part in 100 million, and this number may then be used to define the standard kilogram as 1000/12 of Avogadro's number of carbon atoms. For more on this project, see the chapter opening photo and Problem 4-41.

Mass *m* is an invariant measure of the quantity of matter. **Weight *w*** is the force of gravitational attraction between that matter and Earth.

TABLE 4-2

Prefixes for Units		
Prefix	**Abbreviation**	**Multiplier**
yotta-	Y	10^{24}
zetta-	Z	10^{21}
exa-	E	10^{18}
peta-	P	10^{15}
tera-	T	10^{12}
giga-	G	10^{9}
mega-	M	10^{6}
kilo-	k	10^{3}
hecto-	h	10^{2}
deca-	da	10^{1}
deci-	d	10^{-1}
centi-	c	10^{-2}
milli-	m	10^{-3}
micro-	µ	10^{-6}
nano-	n	10^{-9}
pico-	p	10^{-12}
femto-	f	10^{-15}
atto-	a	10^{-18}
zepto-	z	10^{-21}
yocto-	y	10^{-24}

Photo of Edwin "Buzz" Aldrin taken by Neil Armstrong in July 1969. Armstrong's reflection may be seen in Aldrin's visor. The suits worn by Armstrong and Aldrin during the Apollo 11 mission to the Moon in 1969 appear to be massive. But because the mass of the Moon is only 1/81 that of Earth and the acceleration due to gravity is only 1/6 that on Earth, the weight of the suits on the Moon was only 1/6 of their weight on Earth. The mass of the suits, however, was identical in both locations.

A **mole** of a chemical species is 6.022×10^{23} atoms, molecules, ions, electrons, ion pairs, or subatomic particles.

increases measurably with latitude. The *mass* of the crucible, however, remains constant regardless of where you measure it.

Weight and mass are related by the familiar expression

$$w = mg$$

where w is the weight of an object, m is its mass, and g is the acceleration due to gravity.

A chemical analysis is always based on mass so that the results will not depend on locality. A balance is used to compare the mass of an object with the mass of one or more standard masses. Because g affects both unknown and known equally, the mass of the object is identical to the standard masses with which it is compared.

The distinction between mass and weight is often lost in common usage, and the process of comparing masses is usually called *weighing*. In addition, the objects of known mass as well as the results of weighing are frequently called *weights*. Always bear in mind, however, that analytical data are based on mass rather than weight. Therefore, throughout this text, we will use mass rather than weight to describe the quantities of substances or objects. On the other hand, for lack of a better word, we will use "weigh" for the act of determining the mass of an object. Also, we will often say "weights" to mean the standard masses used in weighing.

4A-3 The Mole

The **mole** (abbreviated mol) is the SI unit for the amount of a chemical substance. It is always associated with specific microscopic entities such as atoms, molecules, ions, electrons, other particles, or specified groups of such particles as represented by a chemical formula. It is the amount of the specified substance that contains the same number of particles as the number of carbon atoms in exactly 12 grams of ^{12}C. This important number is Avogadro's number $N_A = 6.022 \times 10^{23}$. The **molar mass** $\mathcal{M}$ of a substance is the mass in grams of 1 mole of that substance. We calculate molar masses by summing the atomic masses of all the atoms appearing in a chemical formula. For example, the molar mass of formaldehyde CH_2O is

$$\mathcal{M}_{CH_2O} = \frac{1\ mol\ C}{mol\ CH_2O} \times \frac{12.0\ g}{mol\ C} + \frac{2\ mol\ H}{mol\ CH_2O} \times \frac{1.0\ g}{mol\ H}$$
$$+ \frac{1\ mol\ O}{mol\ CH_2O} \times \frac{16.0\ g}{mol\ O}$$
$$= 30.0\ g/mol\ CH_2O$$

and that of glucose, $C_6H_{12}O_6$, is

$$\mathcal{M}_{C_6H_{12}O_6} = \frac{6\ mol\ C}{mol\ C_6H_{12}O_6} \times \frac{12.0\ g}{mol\ C} + \frac{12\ mol\ H}{mol\ C_6H_{12}O_6} \times \frac{1.0\ g}{mol\ H}$$
$$+ \frac{6\ mol\ O}{mol\ C_6H_{12}O_6} \times \frac{16.0\ g}{mol\ O} = 180.0\ g/mol\ C_6H_{12}O_6$$

Thus, 1 mole of formaldehyde has a mass of 30.0 g, and 1 mole of glucose has a mass of 180.0 g.

FEATURE 4-1

Unified Atomic Mass Units and the Mole

The masses for the elements listed in the table inside the back cover of this text are *relative masses* in terms of *unified atomic mass units* (u) or *daltons* (Da). The unified atomic mass unit (often shortened to just atomic mass) is based on a relative scale in which the reference is the ^{12}C carbon isotope, which is *assigned* a mass of exactly 12 u. Thus, the u is by definition 1/12 of the mass of one neutral ^{12}C atom. The *molar mass* M of ^{12}C is then defined as the mass in *grams* of 6.022×10^{23} atoms of the carbon-12 isotope, or exactly 12 g. Likewise, the molar mass of any other element is the mass in grams of 6.022×10^{23} atoms of that element and is numerically equal to the atomic mass of the element in u units. Therefore, the atomic mass of naturally occurring oxygen is 15.999 u, and its molar mass is 15.999 g.

Approximately one mole of each of several different elements. Clockwise from the upper left we see 64 g of copper beads, 27 g of crumpled aluminum foil, 207 g of lead shot, 24 g of magnesium chips, 52 g of chromium chunks, and 32 g of sulfur powder. The beakers in the photo have a volume of 50 mL.

Charles D. Winters

❮ CHALLENGE: Show that the following interesting and useful relationship is correct: 1 mol of unified atomic mass units = 6.022×10^{23} u = 1 g.

❮ The number of moles n_X of a species X of molar mass M_X is given by

$$\text{amount X} = n_X = \frac{m_X}{M_X}$$

The units work out to

$$\text{mol X} = \frac{\text{g X}}{\text{g X/mol X}}$$

$$= \cancel{\text{g X}} \times \frac{\text{mol X}}{\cancel{\text{g X}}}$$

The number of millimoles (mmol) is given by

$$\text{mmol X} = \frac{\text{g X}}{\text{g X/mmol X}}$$

$$= \cancel{\text{g X}} \times \frac{\text{mmol X}}{\cancel{\text{g X}}}$$

When you make calculations of this kind, you should include all units as we do throughout this chapter. This practice often reveals errors in setting up equations.

4A-4 The Millimole

Sometimes it is more convenient to make calculations with millimoles (mmol) rather than moles. The millimole is 1/1000 of a mole, and the mass in grams of a millimole, the millimolar mass (mM), is likewise 1/1000 of the molar mass.

❮ 1 mmol = 10^{-3} mol, and 10^3 mmol = 1 mol

4A-5 Calculating the Amount of a Substance in Moles or Millimoles

The two examples that follow illustrate how the number of moles or millimoles of a species can be determined from its mass in grams or from the mass of a chemically related species.

Molecular model of benzoic acid, C_6H_5COOH. Benzoic acid occurs widely in nature, particularly in berries. It finds broad use as a preservative in foods, fats, and fruit juices; as a mordant for dying fabric; and as a standard in calorimetry and in acid/base analysis.

EXAMPLE 4-1

Find the number of moles and millimoles of benzoic acid ($\mathcal{M} = 122.1$ g/mol) that are contained in 2.00 g of the pure acid.

Solution

If we use HBz to represent benzoic acid, we can write that 1 mole of HBz has a mass of 122.1 g. Therefore,

$$\text{amount HBz} = n_{HBz} = 2.00 \text{ g HBz} \times \frac{1 \text{ mol HBz}}{122.1 \text{ g HBz}} \qquad (4\text{-}1)$$

$$= 0.0164 \text{ mol HBz}$$

To obtain the number of millimoles, we divide by the millimolar mass (0.1221 g/mmol), that is,

$$\text{amount HBz} = 2.00 \text{ g HBz} \times \frac{1 \text{ mmol HBz}}{0.1221 \text{ g HBz}} = 16.4 \text{ mmol HBz}$$

EXAMPLE 4-2

What is the mass in grams of Na^+ (22.99 g/mol) in 25.0 g of Na_2SO_4 (142.0 g/mol)?

Solution

The chemical formula tells us that 1 mole of Na_2SO_4 contains 2 moles of Na^+, that is,

$$\text{amount Na}^+ = n_{Na^+} = \text{mol Na}_2\text{SO}_4 \times \frac{2 \text{ mol Na}^+}{\text{mol Na}_2\text{SO}_4}$$

To find the number of moles of Na_2SO_4, we proceed as in Example 4-1:

$$\text{amount Na}_2\text{SO}_4 = n_{Na_2SO_4} = 25.0 \text{ g Na}_2\text{SO}_4 \times \frac{1 \text{ mol Na}_2\text{SO}_4}{142.0 \text{ g Na}_2\text{SO}_4}$$

Combining this equation with the first leads to

$$\text{amount Na}^+ = n_{Na^+} = 25.0 \text{ g Na}_2\text{SO}_4 \times \frac{1 \text{ mol Na}_2\text{SO}_4}{142.0 \text{ g Na}_2\text{SO}_4} \times \frac{2 \text{ mol Na}^+}{\text{mol Na}_2\text{SO}_4}$$

To obtain the mass of sodium in 25.0 g of Na_2SO_4, we multiply the number of moles of Na^+ by the molar mass of Na^+, or 22.99 g. And so,

$$\text{mass Na}^+ = \text{mol Na}^+ \times \frac{22.99 \text{ g Na}^+}{\text{mol Na}^+}$$

Substituting the previous equation gives the mass in grams of Na^+:

$$\text{mass Na}^+ = 25.0 \text{ g Na}_2\text{SO}_4 \times \frac{1 \text{ mol Na}_2\text{SO}_4}{142.0 \text{ g Na}_2\text{SO}_4} \times \frac{2 \text{ mol Na}^+}{\text{mol Na}_2\text{SO}_4} \times \frac{22.99 \text{ g Na}^+}{\text{mol Na}^+}$$

$$= 8.10 \text{ g Na}^+$$

The Factor-Label Approach to Example 4-2

Some students and instructors find it easier to write out the solution to a problem so that units in the denominator of each succeeding term eliminate the units in the numerator of the preceding one until the units of the answer are obtained. This method has been referred to as the **factor-label method**, **dimensional analysis**, or the **picket fence method**. For instance, in Example 4-2, the units of the answer are g Na^+, and the units given are g Na_2SO_4. Thus, we can write

$$25.0 \text{ g Na}_2\text{SO}_4 \times \frac{\text{mol Na}_2\text{SO}_4}{142.0 \text{ g Na}_2\text{SO}_4}$$

First eliminate moles of Na_2SO_4

$$25.0 \text{ g Na}_2\text{SO}_4 \times \frac{\text{mol Na}_2\text{SO}_4}{142.0 \text{ g Na}_2\text{SO}_4} \times \frac{2 \text{ mol Na}^+}{\text{mol Na}_2\text{SO}_4}$$

and then eliminate moles of Na^+. The result is:

$$25.0 \text{ g Na}_2\text{SO}_4 \times \frac{1 \text{ mol Na}_2\text{SO}_4}{142.0 \text{ g Na}_2\text{SO}_4} \times \frac{2 \text{ mol Na}^\pm}{\text{mol Na}_2\text{SO}_4} \times \frac{22.99 \text{ g Na}^+}{\text{mol Na}^\pm} = 8.10 \text{ g Na}^+$$

4B SOLUTIONS AND THEIR CONCENTRATIONS

Over the course of history, measurements and their corresponding units were invented at the local level. By necessity of primitive communication and local technology, standards were nearly nonexistent, and conversions among the many systems were difficult.[1] The result was many hundreds of distinct ways of expressing concentrations of solutions. Fortunately for us, the advent of rapid communications technology and the development of efficient travel have forced globalization of measurement science and, along with it, the definition of global measurement standards. No field has enjoyed more benefit in this regard than chemistry in general and analytical chemistry in particular. Even so, we use a number of methods for expressing concentration.

4B-1 Concentration of Solutions

In the pages that follow, we describe the four fundamental ways of expressing solution concentration: molar concentration, percent concentration, solution-diluent volume ratio, and p-functions.

Molar Concentration

The **molar concentration** c_x of a solution of a solute species X is the number of moles of that species that is contained in 1 liter of the solution (*not 1 L of the solvent*). In terms of the number of moles of solute, n, and the volume, V, of solution, we write

$$c_x = \frac{n_X}{V} \tag{4-2}$$

$$\text{molar concentration} = \frac{\text{no. moles solute}}{\text{volume in liters}}$$

---

[1] In a humorous (and perhaps geeky) parody of local proliferation of measurement units, Robinson Crusoe's friend Friday measured moles in units of chipmunks and volume in old goat bladders. See J. E. Bissey, *J. Chem. Educ.*, **1969**, *46* (8), 497, **DOI**: 10.1021/ed046p497.

The unit of molar concentration is **molar**, symbolized by **M**, which has the dimensions of mol/L, or mol L^{-1}. Molar concentration is also the number of millimoles of solute per milliliter of solution.

$$1\ M = 1\ mol\ L^{-1} = 1\ \frac{mol}{L} = 1\ mmol\ L^{-1} = 1\ \frac{mmol}{L}$$

EXAMPLE 4-3

Calculate the molar concentration of ethanol in an aqueous solution that contains 2.30 g of C_2H_5OH (46.07 g/mol) in 3.50 L of solution.

Solution

To calculate molar concentration, we must find both the amount of ethanol and the volume of the solution. The volume is given as 3.50 L, so all we need to do is convert the mass of ethanol to the corresponding amount of ethanol in moles.

$$\text{amount } C_2H_5OH = n_{C_2H_5OH} = 2.30\ \text{g } C_2H_5OH \times \frac{1\ \text{mol } C_2H_5OH}{46.07\ \text{g } C_2H_5OH}$$

$$= 0.04992\ \text{mol } C_2H_5OH$$

To obtain the molar concentration, $c_{C_2H_5OH}$, we divide the amount by the volume. Thus,

$$c_{C_2H_5OH} = \frac{2.30\ \text{g } C_2H_5OH \times \dfrac{1\ \text{mol } C_2H_5OH}{46.07\ \text{g } C_2H_5OH}}{3.50\ L}$$

$$= 0.0143\ \text{mol } C_2H_5OH/L = 0.0143\ M$$

We will see that there are two ways of expressing molar concentration: molar analytical concentration and molar equilibrium concentration. The distinction between these two expressions is in whether the solute undergoes chemical change in the solution process.

Molar Analytical Concentration

The **molar analytical concentration**, or for the sake of brevity, just **analytical concentration**, of a solution gives the *total* number of moles of a solute in 1 liter of the solution (or the total number of millimoles in 1 mL). In other words, the molar analytical concentration specifies a recipe by which the solution can be prepared regardless of what might happen to the solute during the solution process. Note that in Example 4-3, the molar concentration that we calculated is also the molar analytical concentration $c_{C_2H_5OH} = 0.0143$ M because the solute ethanol molecules are intact following the solution process.

In another example, a sulfuric acid solution that has an analytical concentration of $c_{H_2SO_4} = 1.0$ M can be prepared by dissolving 1.0 mole, or 98 g, of H_2SO_4 in water and diluting the acid to exactly 1.0 L. As we shall see, there are important differences between the ethanol and sulfuric acid examples.

Molar analytical concentration is the total number of moles of a solute, regardless of its chemical state, in 1 L of solution. The molar analytical concentration describes how a solution of a given concentration can be prepared.

Molar Equilibrium Concentration

The **molar equilibrium concentration**, or just **equilibrium concentration**, refers to the molar concentration of a *particular species* in a solution at equilibrium. To specify the molar equilibrium concentration of a species, it is necessary to know how the solute behaves when it is dissolved in a solvent. For example, the molar equilibrium concentration of H_2SO_4 in a solution with a molar analytical concentration $c_{H_2SO_4} = 1.0$ M is actually 0.0 M because the sulfuric acid is completely dissociated into a mixture of H^+, HSO_4^-, SO_4^- ions. There are essentially no H_2SO_4 molecules in this solution. The equilibrium concentrations of the ions are 1.01, 0.99, and 0.01 M, respectively.

Equilibrium molar concentrations are usually symbolized by placing square brackets around the chemical formula for the species. So, for our solution of H_2SO_4 with an analytical concentration of $c_{H_2SO_4} = 1.0$ M, we write

$$[H_2SO_4] = 0.00 \text{ M} \qquad [H^+] = 1.01 \text{ M}$$

$$[HSO_4^-] = 0.99 \text{ M} \qquad [SO_4^{2-}] = 0.01 \text{ M}$$

> Molar equilibrium concentration is the molar concentration of a particular species in a solution.

> ◀ In your study of chemistry, you will find that terminology constantly evolves as we refine our understanding of the processes that we study and endeavor to describe them more accurately. Molarity, which is a synonym for molar concentration, is an example of a term that is rapidly going out of fashion. Although you may find a few occurrences of molarity as a synonym for molar concentration in this textbook, we avoid it whenever possible.

> ◀ The IUPAC recommends the general term concentration to express the composition of a solution with respect to its volume, with four subterms: amount concentration, mass concentration, volume concentration, and number concentration. Molar concentration, molar analytical concentration, and molar equilibrium concentration are all amount concentrations by this definition.

> ◀ In this example, the *molar analytical concentration* of H_2SO_4 is given by
> $$c_{H_2SO_4} = [SO_4^{2-}] + [HSO_4^-]$$
> because SO_4^{2-} and HSO_4^- are the only two sulfate-containing species in the solution. The *molar equilibrium concentrations* of the ions are $[SO_4^{2-}]$ and $[HSO_4^-]$.

EXAMPLE 4-4

Calculate the analytical and equilibrium molar concentrations of the solute species in an aqueous solution that contains 285 mg of trichloroacetic acid, Cl_3CCOOH (163.4 g/mol), in 10.0 mL (the acid is 73% ionized in water).

Solution

As in Example 4-3, we calculate the number of moles of Cl_3CCOOH, which we designate as HA, and divide by the volume of the solution, 10.0 mL, or 0.0100 L. Therefore,

$$\text{amount HA} = n_{HA} = 285 \text{ mg HA} \times \frac{1 \text{ g HA}}{1000 \text{ mg HA}} \times \frac{1 \text{ mol HA}}{163.4 \text{ g HA}}$$

$$= 1.744 \times 10^{-3} \text{ mol HA}$$

The molar analytical concentration, c_{HA}, is then

$$c_{HA} = \frac{1.744 \times 10^{-3} \text{ mol HA}}{10.0 \text{ mL}} \times \frac{1000 \text{ mL}}{1 \text{ L}} = 0.174 \frac{\text{mol HA}}{\text{L}} = 0.174 \text{ M}$$

In this solution, 73% of the HA dissociates, giving H^+ and A^-:

$$HA \rightleftharpoons H^+ + A^-$$

The equilibrium concentration of HA is then 27% of c_{HA}. Thus,

$$[HA] = c_{HA} \times (100 - 73)/100 = 0.174 \times 0.27 = 0.047 \text{ mol/L}$$

$$= 0.047 \text{ M}$$

The equilibrium concentration of A^- is equal to 73% of the analytical concentration of HA, that is,

$$[A^-] = \frac{73 \text{ mol A}^-}{100 \text{ mol HA}} \times 0.174 \frac{\text{mol HA}}{\text{L}} = 0.127 \text{ M}$$

(continued)

Molecular model of trichloroacetic acid, Cl_3CCOOH. The rather strong acidity of trichloroacetic acid is usually ascribed to the inductive effect of the three chlorine atoms attached to the end of the molecule opposite the acidic proton. Electron density is withdrawn away from the carboxylate group so that the trichloroacetate anion formed when the acid dissociates is stabilized. The acid is used in protein precipitation and in dermatological preparations for the removal of undesirable skin growths.

The number of moles of the species A in a solution of A is given by

$$\text{no. mol A} = n_A = c_A \times V_A$$

$$\text{mol}_A = \frac{\text{mol}_A}{L} \times L$$

where V_A is the volume of the solution in liters.

Because 1 mole of H^+ is formed for each mole of A^-, we can also write

$$[H^+] = [A^-] = 0.127 \text{ M}$$

and

$$c_{HA} = [HA] + [A^-] = 0.047 + 0.127 = 0.174 \text{ M}$$

EXAMPLE 4-5

Describe the preparation of 2.00 L of 0.108 M $BaCl_2$ from $BaCl_2 \cdot 2H_2O$ (244.3 g/mol).

Solution

To determine the number of grams of solute to be dissolved and diluted to 2.00 L, we note that 1 mole of the dihydrate yields 1 mole of $BaCl_2$. Therefore, to produce this solution we will need

$$2.00 \text{ L} \times \frac{0.108 \text{ mol } BaCl_2 \cdot 2H_2O}{L} = 0.216 \text{ mol } BaCl_2 \cdot 2H_2O$$

The mass of $BaCl_2 \cdot 2H_2O$ is then

$$0.216 \text{ mol } BaCl_2 \cdot 2H_2O \times \frac{244.3 \text{ g } BaCl_2 \cdot 2H_2O}{\text{mol } BaCl_2 \cdot 2H_2O} = 52.8 \text{ g } BaCl_2 \cdot 2H_2O$$

Dissolve 52.8 g of $BaCl_2 \cdot 2H_2O$ in water and dilute to 2.00 L.

EXAMPLE 4-6

Describe the preparation of 500 mL of 0.0740 M Cl^- solution from solid $BaCl_2 \cdot 2H_2O$ (244.3 g/mol).

Solution

$$\text{mass } BaCl_2 \cdot 2H_2O = \frac{0.0740 \text{ mol } Cl}{L} \times 0.500 \text{ L} \times \frac{1 \text{ mol } BaCl_2 \cdot 2H_2O}{2 \text{ mol } Cl}$$

$$\times \frac{244.3 \text{ g } BaCl_2 \cdot 2H_2O}{\text{mol } BaCl_2 \cdot 2H_2O} = 4.52 \text{ g } BaCl_2 \cdot 2H_2O$$

Dissolve 4.52 g of $BaCl_2 \cdot 2H_2O$ in water and dilute to 0.500 L or 500 mL.

Percent Concentration

Chemists frequently express concentrations in terms of percent (parts per hundred). Unfortunately, this practice can be a source of ambiguity because percent composition of a solution can be expressed in several ways. Three common methods are

$$\text{weight percent (w/w)} = \frac{\text{weight solute}}{\text{weight solution}} \times 100\%$$

$$\text{volume percent (v/v)} = \frac{\text{volume solute}}{\text{volume solution}} \times 100\%$$

$$\text{weight/volume percent (w/v)} = \frac{\text{weight solute, g}}{\text{volume solution, mL}} \times 100\%$$

Note that the denominator in each of these expressions is the mass or volume of *solution* rather than mass or volume of solvent. Note also that the first two expressions do not depend on the units used for weight (mass) as long as the same units are used in the numerator and the denominator. In the third expression, units must be defined because the numerator and denominator have different units that do not cancel. Of the three expressions, only weight percent has the advantage of being temperature independent.

Weight percent is often used to express the concentration of commercial aqueous reagents. For example, nitric acid is sold as a 70% (w/w) solution, meaning that the reagent contains 70 g of HNO_3 per 100 g of solution (see Example 4-10).

Volume percent is commonly used to specify the concentration of a solution prepared by diluting a pure liquid compound with another liquid. For example, a 5% (v/v) aqueous solution of methanol *usually* describes a solution prepared by diluting 5.0 mL of pure methanol with enough water to give 100 mL.

Weight or volume percent is often used to indicate the composition of dilute aqueous solutions of solid reagents. For example, 5% (w/v) aqueous silver nitrate *often* refers to a solution prepared by dissolving 5 g of silver nitrate in sufficient water to give 100 mL of solution.

To avoid uncertainty, always specify explicitly the type of percent composition being discussed. If this information is missing, the investigator must decide intuitively which of the several types is to be used. The potential error resulting from a wrong choice is considerable. For example, commercial 50% (w/w) sodium hydroxide contains 763 g NaOH per liter, which corresponds to 76.3% (w/v) sodium hydroxide.

Parts per Million and Parts per Billion

For very dilute solutions, **parts per million** (ppm) is a convenient way to express concentration:

$$c_{\text{ppm}} = \frac{\text{mass of solute}}{\text{mass of solution}} \times 10^6 \text{ ppm}$$

where c_{ppm} is the concentration in parts per million. The units of mass in the numerator and denominator must agree so that they cancel. For even more dilute solutions, 10^9 ppb rather than 10^6 ppm is used in the previous equation to give the results in **parts per billion** (ppb). The term **parts per thousand** (ppt) is also used, especially in oceanography.

> Weight percent should more properly be called mass percent and abbreviated m/m. The term "weight percent" is so widely used in the chemical literature, however, that we will use it throughout this text. In IUPAC terminology, weight percent is mass concentration.

> In IUPAC terminology, volume percent is volume concentration.

> Always specify the type of percent when reporting concentrations in this way.

> In IUPAC terminology, parts per billion, parts per million, and parts per thousand are mass concentrations.

> A handy rule in calculating parts per million is to remember that for dilute aqueous solutions whose densities are approximately 1.00 g/mL, 1 ppm = 1.00 mg/L. That is,
>
> $$c_{\text{ppm}} = \frac{\text{mass solute (g)}}{\text{mass solution (g)}} \times 10^6 \text{ ppm}$$
>
> $$c_{\text{ppm}} = \frac{\text{mass solute (mg)}}{\text{volume solution (L)}} \text{ ppm}$$
>
> (4-3)

EXAMPLE 4-7

What is the molar concentration of K^+ in a solution that contains 63.3 ppm of $K_3Fe(CN)_6$ (329.3 g/mol)?

(continued)

In terms of the units, we have

$$\underset{\substack{\text{Density of}\\\text{solution}}}{\dfrac{g}{g}} = \dfrac{g}{g} \times \underset{\substack{\text{Conversion}\\\text{factor}}}{\dfrac{g}{mL}} \times \dfrac{10^3\,mg}{1\,g}$$

$$\times \underset{\substack{\text{Conversion}\\\text{factor}}}{\dfrac{10^3\,mL}{1\,L}} = 10^6\,\dfrac{mg}{L}$$

In other words, the mass concentration expressed in g/g is a factor of 10^6 larger than the mass concentration expressed in mg/L. Therefore, if we wish to express the mass concentration in ppm and the units are mg/L, we merely use ppm. If it is expressed in g/g, we must multiply the ratio by 10^6 ppm.

$$c_{ppb} = \dfrac{\text{mass solute (g)}}{\text{mass solution (g)}}$$
$$\times 10^9\,ppb$$

$$c_{ppb} = \dfrac{\text{mass solute } (\mu g)}{\text{volume solution (g)}}\,ppb$$

Similarly, if we wish to express the mass concentration in ppb, we convert the units to μg/L and use ppb.

The best-known p-function is pH, which is the negative logarithm of $[H^+]$. We discuss the nature of H^+, its nature in aqueous solution, and the alternative representation H_3O^+ in Section 9A-2.

> **Solution**
>
> Because the solution is so dilute, it is reasonable to assume that its density is 1.00 g/mL. Therefore, according to Equation 4-2,
>
> $$63.3 \text{ ppm } K_3Fe(CN)_6 = 63.3 \text{ mg } K_3Fe(CN)_6/L$$
>
> $$\dfrac{\text{no. mol } K_3Fe(CN)_6}{L} = \dfrac{63.3 \text{ mg } K_3Fe(CN)_6}{L} \times \dfrac{1 \text{ g } K_3Fe(CN)_6}{1000 \text{ mg } K_3Fe(CN)_6}$$
>
> $$\times \dfrac{1 \text{ mol } K_3Fe(CN)_6}{329.3 \text{ g } K_3Fe(CN)_6} = 1.922 \times 10^{-4}\,\dfrac{mol}{L}$$
>
> $$= 1.922 \times 10^{-4}\,M$$
>
> $$[K^+] = \dfrac{1.922 \times 10^{-4} \text{ mol } K_3Fe(CN)_6}{L} \times \dfrac{3 \text{ mol } K^+}{1 \text{ mol } K_3Fe(CN)_6}$$
>
> $$= 5.77 \times 10^{-4}\,\dfrac{mol\ K^+}{L} = 5.77 \times 10^{-4}\,M$$

Solution-Diluent Volume Ratios

The composition of a dilute solution is sometimes specified in terms of the volume of a more concentrated solution and the volume of solvent used in diluting it. The volume of the former is separated from that of the latter by a colon. Thus, a 1:4 HCl solution contains four volumes of water for each volume of concentrated hydrochloric acid. This method of notation is frequently ambiguous in that the concentration of the original solution is not always obvious to the reader. Moreover, under some circumstances 1:4 means dilute one volume with three volumes. Because of such uncertainties, you should avoid using solution-diluent ratios.

p-Functions

Scientists frequently express the concentration of a species in terms of its **p-function**, or **p-value**. The p-value is the negative logarithm (to the base 10) of the molar concentration of that species. Thus, for the species X,

$$pX = -\log\,[X]$$

As shown by the following examples, p-values offer the advantage of allowing concentrations that vary over ten or more orders of magnitude to be expressed in terms of small positive numbers.

> **EXAMPLE 4-8**
>
> Calculate the p-value for each ion in a solution that is 2.00×10^{-3} M in NaCl and 5.4×10^{-4} M in HCl.
>
> **Solution**
>
> $$pH = -\log\,[H^+] = -\log\,(5.4 \times 10^{-4}) = 3.27$$
>
> To obtain pNa, we write
>
> $$pNa = -\log[Na^+] = -\log\,(2.00 \times 10^{-3}) = -\log\,(2.00 \times 10^{-3}) = 2.699$$

The total Cl^- concentration is given by the sum of the concentrations of the two solutes:

$$[Cl^-] = 2.00 \times 10^{-3}\,M + 5.4 \times 10^{-4}\,M$$

$$= 2.00 \times 10^{-3}\,M + 0.54 \times 10^{-3}\,M = 2.54 \times 10^{-3}\,M$$

$$pCl = -\log[Cl^-] = -\log 2.54 \times 10^{-3} = 2.595$$

Molecular model of HCl. Hydrogen chloride is a gas consisting of hetero-nuclear diatomic molecules. The gas is extremely soluble in water; when a solution of the gas is prepared, only then do the molecules dissociate to form aqueous hydrochloric acid, which consists of H_3O^+ and Cl^- ions. See Figure 9-1 and the accompanying discussion of the nature of H_3O^+.

Note that in Example 4-8, and in the one that follows, the results are rounded according to the rules listed on page 117.

EXAMPLE 4-9

Calculate the molar concentration of Ag^+ in a solution that has a pAg of 6.372.

Solution

$$pAg = -\log[Ag^+] = 6.372$$

$$\log[Ag^+] = -6.372$$

$$[Ag^+] = 4.246 \times 10^{-7} \approx 4.25 \times 10^{-7}\,M$$

4B-2 Density and Specific Gravity of Solutions

Density and specific gravity are related terms often found in the analytical literature. The **density** of a substance is its mass per unit volume, and its **specific gravity** is the ratio of its mass to the mass of an equal volume of water at 4°C. Density has units of kilograms per liter or grams per milliliter in the metric system. Specific gravity is dimensionless and so is not tied to any particular system of units. For this reason, specific gravity is widely used in describing items of commerce (see **Figure 4-1**). Since the density of water is approximately 1.00 g/mL and since we use the metric system throughout this text, we use density and specific gravity interchangeably. The specific gravities of some concentrated acids and bases are given in **Table 4-3**.

Density expresses the mass of a substance per unit volume. In SI units, density is expressed in units of kg/L or alternatively g/mL.

Specific gravity is the ratio of the mass of a substance to the mass of an equal volume of water.

EXAMPLE 4-10

Calculate the molar concentration of HNO_3 (63.0 g/mol) in a solution that has a specific gravity of 1.42 and is 70.5% HNO_3 (w/w).

Solution

Let us first calculate the mass of acid per liter of concentrated solution

$$\frac{g\,HNO_3}{L\,reagent} = \frac{1.42\,kg\,reagent}{L\,reagent} \times \frac{10^3\,g\,reagent}{kg\,reagent} \times \frac{70.5\,g\,HNO_3}{100\,g\,reagent} = \frac{1001\,g\,HNO_3}{L\,reagent}$$

Then,

$$c_{HNO_3} = \frac{1001\,g\,HNO_3}{L\,reagent} \times \frac{1\,mol\,HNO_3}{63.0\,g\,HNO_3} = \frac{15.9\,mol\,HNO_3}{L\,reagent} \approx 16\,M$$

Figure 4-1 Label from a bottle of reagent-grade hydrochloric acid. Note that the specific gravity of the acid over the temperature range of 60° to 80°F is specified on the label. (*Label provided by Mallinckrodt Baker, Inc., Phillipsburg, NJ 08865*)

TABLE 4-3

Specific Gravities of Commercial Concentrated Acids and Bases

Reagent	Concentration, % (w/w)	Specific Gravity
Acetic acid	99.7	1.05
Ammonia	29.0	0.90
Hydrochloric acid	37.2	1.19
Hydrofluoric acid	49.5	1.15
Nitric acid	70.5	1.42
Perchloric acid	71.0	1.67
Phosphoric acid	86.0	1.71
Sulfuric acid	96.5	1.84

EXAMPLE 4-11

Describe the preparation of 100 mL of 6.0 M HCl from a concentrated solution that has a specific gravity of 1.18 and is 37% (w/w) HCl (36.5 g/mol).

Solution

Proceeding as in Example 4-10, we first calculate the molar concentration of the concentrated reagent. We then calculate the number of moles of acid that we need for the diluted solution. Finally, we divide the second figure by the first to obtain the volume of concentrated acid required. Thus, to obtain the concentration of the reagent, we write

$$c_{HCl} = \frac{1.18 \times 10^3 \text{ g reagent}}{\text{L reagent}} \times \frac{37 \text{ g HCl}}{100 \text{ g reagent}} \times \frac{1 \text{ mol HCl}}{36.5 \text{ g HCl}} = 12.0 \text{ M}$$

The number of moles HCl required is given by

$$\text{no. mol HCl} = 100 \text{ mL} \times \frac{1 \text{ L}}{1000 \text{ mL}} \times \frac{6.0 \text{ mol HCl}}{\text{L}} = 0.600 \text{ mol HCl}$$

Finally, to obtain the volume of concentrated reagent, we write

$$\text{vol concd reagent} = 0.600 \text{ mol HCl} \times \frac{1 \text{ L reagent}}{12.0 \text{ mol HCl}} = 0.0500 \text{ L or } 50.0 \text{ mL}$$

Therefore, dilute 50 mL of the concentrated reagent to 600 mL.

The solution to Example 4-11 is based on the following useful relationship, which we will be using countless times:

$$V_{concd} \times c_{concd} = V_{dil} \times c_{dil} \tag{4-4}$$

where the two terms on the left are the volume and molar concentration of a concentrated solution that is being used to prepare a diluted solution having the volume and concentration given by the corresponding terms on the right. This equation is based on the fact that the number of moles of solute in the diluted solution must equal the number of moles in the concentrated reagent. Note that the volumes can be in milliliters or liters as long as the same units are used for both solutions.

> Equation 4-4 can be used with L and mol/L or mL and mmol/mL. Thus,
>
> $$L_{concd} \times \frac{mol_{concd}}{L_{concd}} = L_{dil} \times \frac{mol_{dil}}{L_{dil}}$$
>
> $$mL_{concd} \times \frac{mmol_{concd}}{mL_{concd}} = mL_{dil}$$
>
> $$\times \frac{mmol_{dil}}{mL_{dil}}$$

4C CHEMICAL STOICHIOMETRY

Stoichiometry is the quantitative relationship among the amounts of reacting chemical species. This section provides a brief review of stoichiometry and its applications to chemical calculations.

> The **stoichiometry** of a reaction is the relationship among the number of moles of reactants and products as represented by a balanced chemical equation.

4C-1 Empirical Formulas and Molecular Formulas

An **empirical formula** gives the simplest whole number ratio of atoms in a chemical compound. In contrast, a **molecular formula** specifies the number of atoms in a molecule. Two or more substances may have the same empirical formula but different molecular formulas. For example, CH_2O is both the empirical and the molecular formula for formaldehyde; it is also the empirical formula for such diverse substances as acetic acid, $C_2H_4O_2$; glyceraldehyde, $C_3H_6O_3$; and glucose, $C_6H_{12}O_6$, as well as more than 50 other substances containing 6 or fewer carbon atoms. We may calculate the empirical formula of a compound from its percent composition. To determine the molecular formula, we must know the molar mass of the compound.

Figure 4-2   Flow diagram for making stoichiometric calculations. (1) When the mass of a reactant or product is given, the mass is first converted to the number of moles, using the molar mass. (2) The stoichiometric ratio given by the chemical equation for the reaction is then used to find the number of moles of another reactant that combines with the original substance or the number of moles of product that forms. (3) Finally, the mass of the other reactant or the product is computed from its molar mass.

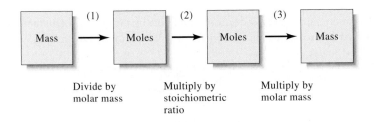

A **structural formula** provides additional information. For example, the chemically different ethanol and dimethyl ether share the same molecular formula C_2H_6O. Their structural formulas, C_2H_5OH and CH_3OCH_3, reveal structural differences between these compounds that are not shown in their common molecular formula.

4C-2   Stoichiometric Calculations

A balanced chemical equation gives the combining ratios, or stoichiometry—in units of moles—of reacting substances and their products. Therefore, the equation

$$2NaI(aq) + Pb(NO_3)_2(aq) \rightarrow PbI_2(s) + 2NaNO_3(aq)$$

> Often the physical state of substances appearing in equations is indicated by the letters (g), (l), (s), and (aq), which refer to gaseous, liquid, solid, and aqueous solution states, respectively.

indicates that 2 moles of aqueous sodium iodide combine with 1 mole of aqueous lead nitrate to produce 1 mole of solid lead iodide and 2 moles of aqueous sodium nitrate.[2]

Example 4-12 demonstrates how the mass in grams of reactants and products in a chemical reaction are related. As shown in **Figure 4-2**, a calculation of this type is a three-step process of (1) transforming the known mass of a substance in grams to a corresponding number of moles, (2) multiplying the number of moles by a factor that accounts for the stoichiometry, and (3) converting the number of moles back to the metric units called for in the answer.

EXAMPLE 4-12

(a) What mass of $AgNO_3$ (169.9 g/mol) is needed to convert 2.33 g of Na_2CO_3 (106.0 g/mol) to Ag_2CO_3? (b) What mass of Ag_2CO_3 (275.7 g/mol) will be formed?

Solution

(a) $Na_2CO_3(aq) + 2AgNO_3(aq) \rightarrow Ag_2CO_3(s) + 2NaNO_3(aq)$

Step 1.

$$\text{amount } Na_2CO_3 = n_{Na_2CO_3} = 2.33 \text{ g } Na_2CO_3 \times \frac{1 \text{ mol } Na_2CO_3}{106.0 \text{ g } Na_2CO_3}$$

$$= 0.02198 \text{ mol } Na_2CO_3$$

---

[2]In this example, it is advantageous to depict the reaction in terms of chemical compounds. If we wish to focus on reacting species, the net ionic equation is preferable:

$$2I^-(aq) + Pb^{2+}(aq) \rightarrow PbI_2(s)$$

Step 2. The balanced equation reveals that

$$\text{amount AgNO}_3 = n_{\text{AgNO}_3} = 0.02198 \text{ mol } \cancel{\text{Na}_2\text{CO}_3} \times \frac{2 \text{ mol AgNO}_3}{1 \text{ mol } \cancel{\text{Na}_2\text{CO}_3}}$$

$$= 0.04396 \text{ mol AgNO}_3$$

In this instance, the stoichiometric factor is $(2 \text{ mol AgNO}_3)/(1 \text{ mol Na}_2\text{CO}_3)$.

Step 3.

$$\text{mass AgNO}_3 = 0.04396 \text{ mol } \cancel{\text{AgNO}_3} \times \frac{169.9 \text{ g AgNO}_3}{\cancel{\text{mol AgNO}_3}} = 7.47 \text{ g AgNO}_3$$

(b) amount Ag_2CO_3 = amount Na_2CO_3 = 0.02198 mol

$$\text{mass Ag}_2\text{CO}_3 = 0.02198 \text{ mol } \cancel{\text{Ag}_2\text{CO}_3} \times \frac{275.7 \text{ g Ag}_2\text{CO}_3}{\cancel{\text{mol Ag}_2\text{CO}_3}} = 6.06 \text{ g Ag}_2\text{CO}_3$$

EXAMPLE 4-13

What mass of Ag_2CO_3 (275.7 g/mol) is formed when 25.0 mL of 0.200 M AgNO_3 are mixed with 50.0 mL of 0.0800 M Na_2CO_3?

Solution

Mixing these two solutions will result in one (and only one) of three possible outcomes:

(a) An excess of AgNO_3 will remain after the reaction is complete.
(b) An excess of Na_2CO_3 will remain after the reaction is complete.
(c) There will be no excess of either reagent (that is, the number of moles of Na_2CO_3 is exactly equal to twice the number of moles of AgNO_3).

As a first step, we must establish which of these situations applies by calculating the amounts of reactants (in moles) available before the solutions are mixed.

The initial amounts are

$$\text{amount AgNO}_3 = n_{\text{AgNO}_3} = 25.0 \text{ mL } \cancel{\text{AgNO}_3} \times \frac{1 \text{ L } \cancel{\text{AgNO}_3}}{1000 \text{ mL } \cancel{\text{AgNO}_3}}$$

$$\times \frac{0.200 \text{ mol AgNO}_3}{\cancel{\text{L AgNO}_3}} = 5.00 \times 10^{-3} \text{ mol AgNO}_3$$

$$\text{amount Na}_2\text{CO}_3 = n_{\text{Na}_2\text{CO}_3} = 50.0 \text{ mL } \cancel{\text{Na}_2\text{CO}_3} \text{ soln} \times \frac{1 \text{ L } \cancel{\text{Na}_2\text{CO}_3}}{1000 \text{ mL } \cancel{\text{Na}_2\text{CO}_3}}$$

$$\times \frac{0.0800 \text{ mol Na}_2\text{CO}_3}{\cancel{\text{L Na}_2\text{CO}_3}} = 4.00 \times 10^{-3} \text{ mol Na}_2\text{CO}_3$$

Because each CO_3^{2-} ion reacts with two Ag^+ ions, $2 \times 4.00 \times 10^{-3} = 8.00 \times 10^{-3}$ mol AgNO_3 is required to react with the Na_2CO_3. Since we have insufficient AgNO_3, situation (b) prevails, and the number of moles of Ag_2CO_3 produced will be limited by the amount of AgNO_3 available. Thus,

$$\text{mass Ag}_2\text{CO}_3 = 5.00 \times 10^{-3} \text{ mol } \cancel{\text{AgNO}_3} \times \frac{1 \text{ mol } \cancel{\text{Ag}_2\text{CO}_3}}{2 \text{ mol } \cancel{\text{AgNO}_3}} \times \frac{275.7 \text{ g Ag}_2\text{CO}_3}{\cancel{\text{mol Ag}_2\text{CO}_3}}$$

$$= 0.689 \text{ g Ag}_2\text{CO}_3$$

EXAMPLE 4-14

What will be the molar analytical concentration of Na_2CO_3 in the solution produced when 25.0 mL of 0.200 M $AgNO_3$ is mixed with 50.0 mL of 0.0800 M Na_2CO_3?

Solution

We have seen in the previous example that formation of 5.00×10^{-3} mol of $AgNO_3$ requires 2.50×10^{-3} mol of Na_2CO_3. The number of moles of unreacted Na_2CO_3 is then given by

$$n_{Na_2CO_3} = 4.00 \times 10^{-3} \text{ mol } Na_2CO_3 - 5.00 \times 10^{-3} \text{ mol } \cancel{AgNO_3} \times \frac{1 \text{ mol } Na_2CO_3}{2 \text{ mol } \cancel{AgNO_3}}$$

$$= 1.50 \times 10^{-3} \text{ mol } Na_2CO_3$$

By definition, the molar concentration is the number of moles of Na_2CO_3/L. Therefore,

$$c_{Na_2CO_3} = \frac{1.50 \times 10^{-3} \text{ mol } Na_2CO_3}{(50.0 + 25.0) \text{ } \cancel{mL}} \times \frac{1000 \text{ } \cancel{mL}}{1 \text{ L}} = 0.0200 \text{ M } Na_2CO_3$$

In this chapter, we have reviewed many of the basic chemical concepts and skills necessary for effective study of analytical chemistry. In the remaining chapters of this book, you will build on this firm foundation as you explore methods of chemical analysis.

WEB WORKS

This chapter opened with a photo of a nearly perfect silicon sphere being used to determine Avogadro's number. When this measurement is complete, the kilogram will be redefined from the mass of a Pt-Ir cylinder housed in Paris to the mass of a known multiple of Avogadro's number of silicon atoms. This will be the so-called **silicon kilogram**. Use your Web browser to connect to **www.cengage.com/chemistry/skoog/fac9**. From the Chapter Resources Menu, choose Web Works. Locate the Chapter 4 section and click on the link to the article on the Royal Society of Chemistry website by Peter Atkins that discusses the significance of the silicon kilogram and read the article. Then click on the link to the article on the consistency of Avogadro's number on the same website. How are Planck's constant, Avogadro's number, and the silicon kilogram related? Why is the kilogram being redefined? What is the uncertainty in Avogadro's number at present? What is the rate of improvement in the uncertainty of Avogadro's number?

QUESTIONS AND PROBLEMS

4-1. Define
  *(a) millimole.
  (b) molar mass.
  *(c) millimolar mass.
  (d) parts per million.

4-2. What is the difference between molar species concentration and molar analytical concentration?

*__4-3.__ Give two examples of units derived from the fundamental base SI units.

4-4. Simplify the following quantities using a unit with an appropriate prefix:
  *(a) 3.2×10^8 Hz.
  (b) 4.56×10^{-7} g.
  *(c) 8.43×10^7 μmol.
  (d) 6.5×10^{10} s.
  *(e) 8.96×10^6 nm.
  (f) 48,000 g.

*__4-5.__ Show that one gram is one mole of unified atomic mass units.

4-6. In one of the figure captions, we suggest that the standard kilogram may soon be defined as 1000/12 of Avogadro's number of carbon atoms. Prove that this statement is correct mathematically, discuss the implications of this new definition of the kilogram.

*4-7. Find the number of Na^+ ions in 2.92 g of Na_3PO_4?

4-8. Find the number of K^+ ions in 3.41 mol of K_2HPO_4?

*4-9. Find the amount of the indicated element (in moles) in
(a) 8.75 g of B_2O_3.
(b) 167.2 mg of $Na_2B_4O_7 \cdot 10H_2O$.
(c) 4.96 g of Mn_3O_4.
(d) 333 mg of CaC_2O_4.

4-10. Find the amount in millimoles of the indicated species in
(a) 850 mg of P_2O_5.
(b) 40.0 g of CO_2.
(c) 12.92 g of $NaHCO_3$.
(d) 57 mg of $MgNH_4PO_4$.

*4-11. Find the number of millimoles of solute in
(a) 2.00 L of 0.0555 M $KMnO_4$.
(b) 750 mL of 3.25×10^{-3} M KSCN.
(c) 3.50 L of a solution that contains 3.33 ppm of $CuSO_4$.
(d) 250 mL of 0.414 M KCl.

4-12. Find the number of millimoles of solute in
(a) 226 mL of 0.320 M $HClO_4$.
(b) 25.0 L of 8.05×10^{-3} M K_2CrO_4.
(c) 6.00 L of an aqueous solution that contains 6.75 ppm of $AgNO_3$.
(d) 537 mL of 0.0200 M KOH.

*4-13. What is the mass in milligrams of
(a) 0.367 mol of HNO_3?
(b) 245 mmol of MgO?
(c) 12.5 mol of NH_4NO_3?
(d) 4.95 mol of $(NH_4)_2Ce(NO_3)_6$ (548.23 g/mol)?

4-14. What is the mass in grams of
(a) 3.20 mol of KBr?
(b) 18.9 mmol of PbO?
(c) 6.02 mol of $MgSO_4$?
(d) 10.9 mmol of $Fe(NH_4)_2(SO_4)_2 \cdot 6H_2O$?

4-15. What is the mass in milligrams of solute in
*(a) 16.0 mL of 0.350 M sucrose (342 g/mol)?
*(b) 1.92 L of 3.76×10^{-3} M H_2O_2?
(c) 356 mL of a solution that contains 2.96 ppm of $Pb(NO_3)_2$?
(d) 5.75 mL of 0.0819 M KNO_3?

4-16. What is the mass in grams of solute in
*(a) 250 mL of 0.264 M H_2O_2?
*(b) 37.0 mL of 5.75×10^{-4} M benzoic acid (122 g/mol)?
(c) 4.50 L of a solution that contains 31.7 ppm of $SnCl_2$?
(d) 11.7 mL of 0.0225 M $KBrO_3$?

4-17. Calculate the p-value for each of the indicated ions in the following:
*(a) Na^+, Cl^-, and OH^- in a solution that is 0.0635 M in NaCl and 0.0403 M in NaOH.

(b) Ba^{2+}, Mn^{2+}, and Cl^- in a solution that is 4.65×10^{-3} M in $BaCl_2$ and 2.54 M in $MnCl_2$.
*(c) H^+, Cl^-, and Zn^{2+} in a solution that is 0.400 M in HCl and 0.100 M in $ZnCl_2$.
(d) Cu^{2+}, Zn^{2+}, and NO_3^- in a solution that is 5.78×10^{-2} M in $Cu(NO_3)_2$ and 0.204 M in $Zn(NO_3)_2$.
*(e) K^+, OH^-, and $Fe(CN)_6^{4-}$ in a solution that is 1.62×10^{-7} M in $K_4Fe(CN)_6$ and 5.12×10^{-7} M in KOH.
(f) H^+, Ba^{2+}, and ClO_4^- M in a solution that is 2.35×10^{-4} M in $Ba(ClO_4)_2$ and 4.75×10^{-4} M in $HClO_4$.

4-18. Calculate the molar H_3O^+ ion concentration of a solution that has a pH of
*(a) 4.31. *(c) 0.59. *(e) 7.62. *(g) −0.76.
(b) 4.48.   (d) 13.89. (f) 5.32.   (h) −0.42.

4-19. Calculate the p-functions for each ion in a solution that is
*(a) 0.0300 M in NaBr.
(b) 0.0200 M in $BaBr_2$.
*(c) 5.5×10^{-3} M in $Ba(OH)_2$.
(d) 0.020 M in HCl and 0.010 M in NaCl.
*(e) 8.7×10^{-3} M in $CaCl_2$ and 6.6×10^{-3} M in $BaCl_2$.
(f) 2.8×10^{-8} M in $Zn(NO_3)_2$ and 6.6×10^{-7} M $Cd(NO_3)_2$.

4-20. Convert the following p-functions to molar concentrations:
*(a) pH = 1.020.   *(e) pLi = 12.35.
(b) pOH = 0.0025.   (f) pNO_3 = 0.034.
*(c) pBr = 7.77.   *(g) pMn = 0.135.
(d) pCa = −0.221.   (h) pCl = 9.67.

*4-21. Sea water contains an average of 1.08×10^3 ppm of Na^+ and 270 ppm of SO_4^{2-}. Calculate
(a) the molar concentrations of Na^+ and SO_4^{2-} given that the average density of sea water is 1.02 g/mL.
(b) the pNa and pSO_4 for sea water.

4-22. Average human blood contains 300 nmoles of hemoglobin(Hb) per liter of plasma and 2.2 mmol per liter of whole blood. Calculate
(a) the molar concentration in each of these media.
(b) pHb in plasma in human serum.

*4-23. A solution was prepared by dissolving 5.76 g of $KCl \cdot MgCl_2 \cdot 6H_2O$ (277.85 g/mol) in sufficient water to give 2.000 L. Calculate
(a) the molar analytical concentration of $KCl \cdot MgCl_2$ in this solution.
(b) the molar concentration of Mg^{2+}.
(c) the molar concentration of Cl^-.
(d) the weight/volume percentage of $KCl \cdot MgCl_2 \cdot 6H_2O$.
(e) the number of millimoles of Cl^- in 25.0 mL of this solution.
(f) ppm K^+.
(g) pMg for the solution.
(h) pCl for the solution.

4-24. A solution was prepared by dissolving 1210 mg of $K_3Fe(CN)_6$ (329.2 g/mol) in sufficient water to give 775 mL. Calculate

(a) the molar analytical concentration of $K_3Fe(CN)_6$.

(b) the molar concentration of K^+.

(c) the molar concentration of $Fe(CN)_6^{3-}$.

(d) the weight/volume percentage of $K_3Fe(CN)_6$.

(e) the number of millimoles of K^+ in 50.0 mL of this solution.

(f) ppm $Fe(CN)_6^{3-}$.

(g) pK for the solution.

(h) $pFe(CN)_6$ for the solution.

*__4-25.__ A 6.42% (w/w) $Fe(NO_3)_3$ (241.86 g/mol) solution has a density of 1.059 g/mL. Calculate

(a) the molar analytical concentration of $Fe(NO_3)_3$ in this solution.

(b) the molar NO_3^- concentration in the solution.

(c) the mass in grams of $Fe(NO_3)_3$ contained in each liter of this solution.

4-26. A 12.5% (w/w) $NiCl_2$ (129.61 g/mol) solution has a density of 1.149 g/mL. Calculate

(a) the molar concentration of $NiCl_2$ in this solution.

(b) the molar Cl^- concentration of the solution.

(c) the mass in grams of $NiCl_2$ contained in each liter of this solution.

*__4-27.__ Describe the preparation of

(a) 500 mL of 4.75% (w/v) aqueous ethanol (C_2H_5OH, 46.1 g/mol).

(b) 500 g of 4.75% (w/w) aqueous ethanol.

(c) 500 mL of 4.75% (v/v) aqueous ethanol.

4-28. Describe the preparation of

(a) 2.50 L of 21.0% (w/v) aqueous glycerol ($C_3H_8O_3$, 92.1 g/mol).

(b) 2.50 kg of 21.0% (w/w) aqueous glycerol.

(c) 2.50 L of 21.0% (v/v) aqueous glycerol.

*__4-29.__ Describe the preparation of 750 mL of 6.00 M H_3PO_4 from the commercial reagent that is 86% H_3PO_4 (w/w) and has a specific gravity of 1.71.

4-30. Describe the preparation of 900 mL of 3.00 M HNO_3 from the commercial reagent that is 70.5% HNO_3 (w/w) and has a specific gravity of 1.42.

*__4-31.__ Describe the preparation of

(a) 500 mL of 0.0750 M $AgNO_3$ from the solid reagent.

(b) 1.00 L of 0.285 M HCl, starting with a 6.00 M solution of the reagent.

(c) 400 mL of a solution that is 0.0810 M in K^+, starting with solid $K_4Fe(CN)_6$.

(d) 600 mL of 3.00% (w/v) aqueous $BaCl_2$ from a 0.400 M $BaCl_2$ solution.

(e) 2.00 L of 0.120 M $HClO_4$ from the commercial reagent [71.0% $HClO_4$ (w/w), sp gr 1.67].

(f) 9.00 L of a solution that is 60.0 ppm in Na^+, starting with solid Na_2SO_4.

4-32. Describe the preparation of

(a) 5.00 L of 0.0500 M $KMnO_4$ from the solid reagent.

(b) 4.00 L of 0.250 M $HClO_4$, starting with an 8.00 M solution of the reagent.

(c) 400 mL of a solution that is 0.0250 M in I^-, starting with MgI_2.

(d) 200 mL of 1.00% (w/v) aqueous $CuSO_4$ from a 0.365 M $CuSO_4$ solution.

(e) 1.50 L of 0.215 M NaOH from the concentrated commercial reagent [50% NaOH (w/w), sp gr 1.525].

(f) 1.50 L of a solution that is 12.0 ppm in K^+, starting with solid $K_4Fe(CN)_6$.

*__4-33.__ What mass of solid $La(IO_3)_3$ (663.6 g/mol) is formed when 50.0 mL of 0.250 M La^{3+} are mixed with 75.0 mL of 0.302 M IO_3^-?

4-34. What mass of solid $PbCl_2$ (278.10 g/mol) is formed when 200 mL of 0.125 M Pb^{2+} are mixed with 400 mL of 0.175 M Cl^-?

*__4-35.__ Exactly 0.2220 g of pure Na_2CO_3 was dissolved in 100.0 mL of 0.0731 M HCl.

(a) What mass in grams of CO_2 were evolved?

(b) What was the molar concentration of the excess reactant (HCl or Na_2CO_3)?

4-36. Exactly 25.0 mL of a 0.3757 M solution of Na_3PO_4 were mixed with 100.00 mL of 0.5151 M $HgNO_3$.

(a) What mass of solid Hg_3PO_4 was formed?

(b) What is the molar concentration of the unreacted species (Na_3PO_4 or $HgNO_3$) after the reaction was complete?

*__4-37.__ Exactly 75.00 mL of a 0.3132 M solution of Na_2SO_3 were treated with 150.0 mL of 0.4025 M $HClO_4$ and boiled to remove the SO_2 formed.

(a) What was the mass in grams of SO_2 that was evolved?

(b) What was the concentration of the unreacted reagent (Na_2SO_3 or $HClO_4$) after the reaction was complete?

4-38. What mass of $MgNH_4PO_4$ precipitated when 200.0 mL of a 1.000% (w/v) solution of $MgCl_2$ were treated with 40.0 mL of 0.1753 M Na_3PO_4 and an excess of NH_4^+? What was the molar concentration of the excess reagent (Na_3PO_4 or $MgCl_2$) after the precipitation was complete?

*__4-39.__ What volume of 0.01000 M $AgNO_3$ would be required to precipitate all of the I^- in 200.0 mL of a solution that contained 24.32 ppt KI?

4-40. Exactly 750.0 mL of a solution that contained 480.4 ppm of $Ba(NO_3)_2$ were mixed with 200.0 mL of a solution that was 0.03090 M in $Al_2(SO_4)_3$.

(a) What mass of solid $BaSO_4$ was formed?

(b) What was the molar concentration of the unreacted reagent [$Al_2(SO_4)_3$ or $Ba(NO_3)_2$]?

4-41. **Challenge Problem:** According to Kenny et al.,[3] Avogadro's number N_A may be calculated from the following equation using measurements on a sphere fabricated from an ultrapure single crystal of silicon:

$$N_A = \frac{n \mathcal{M}_{Si} V}{m a^3}$$

where

N_A = Avogadro's number
n = the number of atoms per unit cell in the crystal lattice of silicon = 8
$\mathcal{M}_{Si}$ = the molar mass of silicon
V = the volume of the silicon sphere
m = the mass of the sphere
a = the crystal lattice parameter = $d(220) \sqrt{2^2 + 2^2 + 0^2}$

(a) Derive the equation for Avogadro's number.
(b) From the recent data assembled by Andreas et al.[4] on Sphere AVO28-S5 in the table below, calculate the density of silicon and its uncertainty. You may wish to delay the uncertainty calculations until you have studied Chapter 6.

Variable	Value	Relative uncertainty
Sphere volume, cm^3	431.059059	23×10^{-9}
Sphere mass, g	1000.087560	3×10^{-9}
Molar mass, g/mol	27.97697026	6×10^{-9}
Lattice spacing $d(220)$, pm	543.099624	11×10^{-9}

(c) Calculate Avogadro's number and its uncertainty.
(d) We have presented data for only one of the two silicon spheres used in these studies. Look up the data for Sphere AVO28-S8 cited in note 3 and calculate a second value for N_A. After you have studied Chapter 7, compare your two values for N_A, and decide whether the difference in the two numbers is statistically significant. If difference between the two values is not statistically significant, calculate a mean value for Avogadro's number determined from the two spheres and the uncertainty of the mean.

(e) Which of the variables in the table have the most significant influence on the value that you calculated and why?
(f) What experimental methods were used to make the measurements shown in the table?
(g) Comment on experimental variables that might contribute to the uncertainty in each measurement.
(h) Suggest ways that the determination of Avogadro's number might be improved.
(i) Use a search engine to locate the NIST website on fundamental physical constants. Look up the accepted value of Avogadro's number and its uncertainty (2010 or later) and compare them with your computed values. Discuss any differences and suggest possible causes for the discrepancies.
(j) What technological innovations of the past several decades have led to the easy availability of ultrapure silicon? What steps have been taken in recent years to minimize errors associated with impurities in the silicon used to fabricate the near-perfect spheres?[5]

[3]M. J. Kenny et al., *IEEE Trans. Instrum. Meas.,* **2001,** *50,* 587, **DOI:** 10.1109/19.918198.
[4]B. Andreas et al., *Phys. Rev. Lett.,* **2011,** *106,* 030801, **DOI:** 10.1103/PhysRevLett.106.030801.

[5]P. Becker et al, *Meas. Sci. Technol.,* **2009,** *20,* 092002, **DOI:**10.1088/0957-0233/20/9/092002.

Errors in Chemical Analyses

ND/Roger Viollet/Getty Images

The term **error** has two slightly different meanings. First, error refers to the difference between a measured value and the "true" or "known" value. Second, error often denotes the estimated uncertainty in a measurement or experiment.

Errors can sometimes be calamitous, as this picture of the famous train accident at Montparnasse station in Paris illustrates. On October 22, 1895, a train from Granville, France, crashed through the platform and the station wall because the brakes failed. The engine fell thirty feet into the street below killing a woman. Fortunately, no one on the train was seriously hurt, although the passengers were badly shaken. The story of the train derailment was featured in the children's story *The Invention of Hugo Cabret,* by Brian Selznick (2007) and part of Hugo's nightmare in the movie *Hugo* (2011), winner of 5 academy awards in 2012.

Errors in chemical analyses are seldom this dramatic, but they may have equally serious effects as described in this chapter. Among other applications, analytical results are often used in the diagnosis of disease, in the assessment of hazardous wastes and pollution, in the solving of major crimes, and in the quality control of industrial products. Errors in these results can have serious personal and societal effects. This chapter considers the various types of errors encountered in chemical analyses and the methods we can use to detect them.

Measurements invariably involve errors and uncertainties. Only a few of these are due to mistakes on the part of the experimenter. More commonly, **errors** are caused by faulty calibrations or standardizations or by random variations and uncertainties in results. Frequent calibrations, standardizations, and analyses of known samples can sometimes be used to lessen all but the random errors and uncertainties. However, measurement errors are an inherent part of the quantized world in which we live. Because of this, it is impossible to perform a chemical analysis that is totally free of errors or uncertainties. We can only hope to minimize errors and estimate their size with acceptable accuracy.[1] In this and the next two chapters, we explore the nature of experimental errors and their effects on chemical analyses.

The effect of errors in analytical data is illustrated in **Figure 5-1**, which shows results for the quantitative determination of iron. Six equal portions of an aqueous solution with a "known" concentration of 20.00 ppm of iron(III) were analyzed in exactly the same way.[2]

---

[1]Unfortunately these ideas are not widely understood. For example, when asked by defense attorney Robert Shapiro in the O. J. Simpson case what the rate of error in a blood test was, Marcia Clark the lead prosecutor replied that the state's testing laboratories had no percentage of error because "they have not committed any errors" *San Francisco Chronicle*, June 29, 1994, p. 4.

[2]Although actual concentrations can never be "known" exactly, there are many situations in which we are quite certain of the value, as, for example, when it is derived from a highly reliable reference standard.

Note that the results range from a low of 19.4 ppm to a high of 20.3 ppm of iron. The average, or **mean** value, $\bar{x}$, of the data is 19.78 ppm, which rounds to 19.8 ppm (see Section 6D-1 for rounding numbers and the significant figures convention).

Every measurement is influenced by many uncertainties, which combine to produce a scatter of results like that in Figure 5-1. Because measurement uncertainties can never be completely eliminated, *measurement data can only give us an estimate of the "true" value.* However, the probable magnitude of the error in a measurement can often be evaluated. It is then possible to define limits within which the true value of a measured quantity lies with a given level of probability.

Estimating the reliability of experimental data is extremely important whenever we collect laboratory results *because data of unknown quality are worthless.* On the other hand, results that might not seem especially accurate may be of considerable value if the limits of uncertainty are known.

Unfortunately, there is no simple and widely applicable method for determining the reliability of data with absolute certainty. Often, estimating the quality of experimental results requires as much effort as collecting the data. Reliability can be assessed in several ways. Experiments designed to reveal the presence of errors can be performed. Standards of known composition can be analyzed, and the results compared with the known composition. A few minutes consulting the chemical literature can reveal useful reliability information. Calibrating equipment usually enhances the quality of data. Finally, statistical tests can be applied to the data. Because none of these options is perfect, we must ultimately make *judgments* as to the probable accuracy of our results. These judgments tend to become harsher and less optimistic with experience. The quality assurance of analytical methods and the ways to validate and report results are discussed further in Section 8E-3.

One of the first questions to answer before beginning an analysis is "What maximum error can be tolerated in the result?" The answer to this question often determines the method chosen and the time required to complete the analysis. For example, experiments to determine whether the mercury concentration in a river water sample exceeds a certain value can often be done more rapidly than those to accurately determine the specific concentration. To increase the accuracy of a determination by a factor of ten may take hours, days, or even weeks of added labor. *No one can afford to waste time generating data that are more reliable than is necessary for the job at hand.*

The symbol **ppm** stands for parts per million, that is, 20.00 parts of iron(III) per million parts of solution. For aqueous solutions, 20 ppm = 20 mg/dL.

❮ Measurement uncertainties cause replicate results to vary.

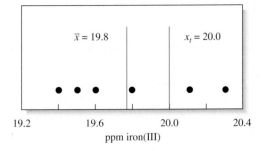

Figure 5-1 Results from six replicate determinations of iron in aqueous samples of a standard solution containing 20.0 ppm iron(III). The mean value of 19.78 has been rounded to 19.8 ppm (see Example 5-1).

5A SOME IMPORTANT TERMS

In order to improve the reliability and to obtain information about the variability of results, two to five portions (**replicates**) of a sample are usually carried through an entire analytical procedure. Individual results from a set of measurements are seldom the same (Figure 5-1), so we usually consider the "best" estimate to be the central value for the set. We justify the extra effort required to analyze replicates in two ways. First, the central value of a set should be more reliable than any of the individual results. Usually, the mean or the median is used as the central value for a set of replicate measurements. Second, an analysis of the variation in the data allows us to estimate the uncertainty associated with the central value.

Replicates are samples of about the same size that are carried through an analysis in *exactly* the same way.

5A-1 The Mean and the Median

The most widely used measure of central value is the **mean**, $\bar{x}$. The mean, also called the **arithmetic mean** or the **average**, is obtained by dividing the sum of replicate measurements by the number of measurements in the set:

The **mean** of two or more measurements is their average value.

$$\bar{x} = \frac{\displaystyle\sum_{i=1}^{N} x_i}{N} \tag{5-1}$$

The symbol $\sum x_i$ means to add all of the values x_i for the replicates.

where x_i represents the individual values of x making up the set of N replicate measurements.

The **median** is the middle result when replicate data are arranged in increasing or decreasing order. There are equal numbers of results that are larger and smaller than the median. For an odd number of results, the median can be found by arranging the results in order and locating the middle result. For an even number, the average value of the middle pair is used as shown in Example 5-1.

The **median** is the middle value in a set of data that has been arranged in numerical order. The median is used advantageously when a set of data contain an **outlier**, a result that differs significantly from others in the set. An outlier can have a significant effect on the mean of the set but has no effect on the median.

In ideal cases, the mean and median are identical. However, when the number of measurements in the set is small, the values often differ as shown in Example 5-1.

EXAMPLE 5-1

Calculate the mean and median for the data shown in Figure 5-1.

Solution

$$\text{mean} = \bar{x} = \frac{19.4 + 19.5 + 19.6 + 19.8 + 20.1 + 20.3}{6} = 19.78 \approx 19.8 \text{ ppm Fe}$$

Because the set contains an even number of measurements, the median is the average of the central pair:

$$\text{median} = \frac{19.6 + 19.8}{2} = 19.7 \text{ ppm Fe}$$

5A-2 Precision

Precision describes the reproducibility of measurements—in other words, the closeness of results that have been obtained *in exactly the same way*. Generally, the precision

Precision is the closeness of results to others obtained in exactly the same way.

of a measurement is readily determined by simply repeating the measurement on replicate samples.

Three terms are widely used to describe the precision of a set of replicate data: **standard deviation, variance,** and **coefficient of variation**. These three are functions of how much an individual result x_i differs from the mean, called the **deviation from the mean** d_i.

$$d_i = |x_i - \bar{x}| \tag{5-2}$$

The relationship between the deviation from the mean and the three precision terms is given in Section 6B.

 Spreadsheet Summary   In Chapter 2 of *Applications of Microsoft® Excel in Analytical Chemistry,* 2nd ed., the mean and the deviations from the mean are calculated with Excel.

> Note that deviations from the mean are calculated without regard to sign.

5A-3 Accuracy

Accuracy indicates the closeness of the measurement to the true or accepted value and is expressed by the *error*. **Figure 5-2** illustrates the difference between accuracy and precision. Note that accuracy measures agreement between a result and the accepted value. *Precision*, on the other hand, describes the agreement among several results obtained in the same way. We can determine precision just by measuring replicate samples. Accuracy is often more difficult to determine because the true value is usually unknown. An accepted value must be used instead. Accuracy is expressed in terms of either absolute or relative error.

Absolute Error

The **absolute error** E in the measurement of a quantity x is given by the equation

$$E = x_i - x_t \tag{5-3}$$

where x_t is the true or accepted value of the quantity. Returning to the data displayed in Figure 5-1, the absolute error of the result immediately to the left of the true value

> **Accuracy** is the closeness of a measured value to the true or accepted value.

> The term "absolute" has a different meaning in this text than it does in mathematics. An absolute value in mathematics means the magnitude of a number *ignoring its sign*. As used in this text, the absolute error is the difference between an *experimental result and an accepted value including its sign.*

The **absolute error** of a measurement is the difference between the measured value and the true value. The sign of the absolute error tells you whether the value in question is high or low. If the measurement result is low, the sign is negative; if the measurement result is high, the sign is positive.

Low accuracy, low precision

Low accuracy, high precision

High accuracy, low precision

High accuracy, high precision

Figure 5-2   Illustration of accuracy and precision using the pattern of darts on a dartboard. Note that we can have very precise results (upper right) with a mean that is not accurate and an accurate mean (lower left) with data points that are imprecise.

of 20.0 ppm is -0.2 ppm Fe; the result at 20.1 ppm is in error by $+0.1$ ppm Fe. Note that we keep the sign in stating the error. The negative sign in the first case shows that the experimental result is smaller than the accepted value, and the positive sign in the second case shows that the experimental result is larger than the accepted value.

Relative Error

The relative error of a measurement is the absolute error divided by the true value. Relative error may be expressed in percent, parts per thousand, or parts per million, depending on the magnitude of the result. As used in this chapter, relative error refers to the relative absolute error. Relative random errors (relative uncertainties) are discussed in Sections 6B and 8B.

The **relative error** E_r is often a more useful quantity than the absolute error. The percent relative error is given by the expression

$$E_r = \frac{x_i - x_t}{x_t} \times 100\% \qquad (5\text{-}4)$$

Relative error is also expressed in parts per thousand (ppt). For example, the relative error for the mean of the data in Figure 5-1 is

$$E_r = \frac{19.8 - 20.0}{20.0} \times 100\% = -1\%, \text{ or } -10 \text{ ppt}$$

5A-4 Types of Errors in Experimental Data

The precision of a measurement is readily determined by comparing data from carefully replicated experiments. Unfortunately, an estimate of the accuracy is not as easy to obtain. To determine the accuracy, we have to know the true value, which is usually what we are seeking in the analysis.

Results can be precise without being accurate and accurate without being precise. The danger of assuming that precise results are also accurate is illustrated in **Figure 5-3**, which summarizes the results for the determination of nitrogen in two pure compounds. The dots show the absolute errors of replicate results obtained by four analysts. Note that analyst 1 obtained relatively high precision and high accuracy. Analyst 2 had poor precision but good accuracy. The results of analyst 3 are surprisingly common. The precision is excellent, but there is significant error in the

benzyl isothiourea hydrochloride

nicotinic acid

Small amounts of nicotinic acid, which is often called *niacin*, occur in all living cells. Niacin is essential in the nutrition of mammals, and it is used in the prevention and treatment of pellagra.

Figure 5-3   Absolute error in the micro-Kjeldahl determination of nitrogen. Each dot represents the error associated with a single determination. Each vertical line labeled $(\bar{x}_i - x_t)$ is the absolute average deviation of the set from the true value. (Data from C. O. Willits and C. L. Ogg, *J. Assoc. Offic. Anal. Chem.*, **1949**, *32*, 561.)

numerical average for the data. Both the precision and the accuracy are poor for the results of analyst 4.

Figures 5-1 and 5-3 suggest that chemical analyses are affected by at least two types of errors. One type, called **random** (or **indeterminate**) **error**, causes data to be scattered more or less symmetrically around a mean value. Refer again to Figure 5-3, and notice that the scatter in the data, and thus the random error, for analysts 1 and 3 is significantly less than that for analysts 2 and 4. In general, then, the random error in a measurement is reflected by its precision. Random errors are discussed in detail in Chapter 6.

A second type of error, called **systematic** (or **determinate**) **error**, causes the mean of a data set to differ from the accepted value. For example, the mean of the results in Figure 5-1 has a systematic error of about -0.2 ppm Fe. The results of analysts 1 and 2 in Figure 5-3 have little systematic error, but the data of analysts 3 and 4 show systematic errors of about -0.7 and -1.2% nitrogen. In general, a systematic error in a series of replicate measurements causes all the results to be too high or too low. An example of a systematic error is the loss of a volatile analyte while heating a sample.

A third type of error is **gross error**. Gross errors differ from indeterminate and determinate errors. They usually occur only occasionally, are often large, and may cause a result to be either high or low. They are often the product of human errors. For example, if part of a precipitate is lost before weighing, analytical results will be low. Touching a weighing bottle with your fingers after its empty mass is determined will cause a high mass reading for a solid weighed in the contaminated bottle. Gross errors lead to **outliers**, results that appear to differ markedly from all other data in a set of replicate measurements. There is no evidence of a gross error in Figures 5-1 and 5-3. Had one of the results shown in Figure 5-1 occurred at, say, 21.2 ppm Fe, it might have been an outlier. Various statistical tests can be performed to determine if a result is an outlier (see Section 7D).

5B SYSTEMATIC ERRORS

Systematic errors have a definite value, an assignable cause, and are of the same magnitude for replicate measurements made in the same way. They lead to **bias** in measurement results. Note that bias affects all of the data in a set in the same way and that it bears a sign.

5B-1 Sources of Systematic Errors

There are three types of systematic errors:

- **Instrumental errors** are caused by nonideal instrument behavior, by faulty calibrations, or by use under inappropriate conditions.
- **Method errors** arise from nonideal chemical or physical behavior of analytical systems.
- **Personal errors** result from the carelessness, inattention, or personal limitations of the experimenter.

Instrumental Errors

All measuring devices are potential sources of systematic errors. For example, pipets, burets, and volumetric flasks may hold or deliver volumes slightly different from those indicated by their graduations. These differences arise from using glassware at a temperature that differs significantly from the calibration temperature, from distortions in container walls due to heating while drying, from errors in the original calibration, or from contaminants on the inner surfaces of the containers. Calibration eliminates most systematic errors of this type.

Electronic instruments are also subject to systematic errors. These can arise from several sources. For example, errors may emerge as the voltage of a battery-operated power supply decreases with use. Errors can also occur if instruments are not calibrated frequently or if they are calibrated incorrectly. The experimenter may also use an instrument under conditions where errors are large. For example, a pH meter used in strongly acidic media is prone to an acid error as discussed in Chapter 21. Temperature changes cause variation in many electronic components, which can lead to drifts and errors. Some instruments are susceptible to noise induced from the alternating current (ac) power lines, and this noise may influence precision and accuracy. In many cases, errors of these types are detectable and correctable.

Method Errors

The nonideal chemical or physical behavior of the reagents and reactions on which an analysis is based often introduce systematic method errors. Such sources of nonideality include the slowness of some reactions, the incompleteness of others, the instability of some species, the lack of specificity of most reagents, and the possible occurrence of side reactions that interfere with the measurement process. As an example, a common method error in volumetric analysis results from the small excess of reagent required to cause an indicator to change color and signal the equivalence point. The accuracy of such an analysis is thus limited by the very phenomenon that makes the titration possible.

Another example of method error is illustrated by the data in Figure 5-3 in which the results by analysts 3 and 4 show a negative bias that can be traced to the chemical nature of the sample, nicotinic acid. The analytical method used involves the decomposition of the organic samples in hot concentrated sulfuric acid, which converts the nitrogen in the samples to ammonium sulfate. Often a catalyst, such as mercuric oxide or a selenium or copper salt, is added to speed the decomposition. The amount of ammonia in the ammonium sulfate is then determined in the measurement step. Experiments have shown that compounds containing a pyridine ring such as nicotinic acid (see structural formula, page 86) are incompletely decomposed by the sulfuric acid. With such compounds, potassium sulfate is used to raise the boiling temperature. Samples containing N—O or N—N linkages must be pretreated or subjected to reducing conditions.[3] Without these precautions, low results are obtained. It is highly likely the negative errors, $(\bar{x}_3 - x_t)$ and $(\bar{x}_4 - x_t)$ in Figure 5-3 are systematic errors resulting from incomplete decomposition of the samples.

Errors inherent in a method are often difficult to detect and are thus the most serious of the three types of systematic error.

Personal Errors

Many measurements require personal judgments. Examples include estimating the position of a pointer between two scale divisions, the color of a solution at the end point in a titration, or the level of a liquid with respect to a graduation in a pipet or buret (see Figure 6-5 page 116). Judgments of this type are often subject to systematic, unidirectional errors. For example, one person may read a pointer consistently high, while another may be slightly slow in activating a timer. Yet, a third may be less sensitive to color changes, with an analyst who is insensitive to color changes tending

> Of the three types of systematic errors encountered in a chemical analysis, method errors are usually the most difficult to identify and correct.

> Color blindness is a good example of a limitation that could cause a personal error in a volumetric analysis. A famous color-blind analytical chemist enlisted his wife to come to the laboratory to help him detect color changes at end points of titrations.

---

[3]J. A. Dean, *Analytical Chemistry Handbook,* New York: McGraw-Hill, 1995, section 17, p. 17.4.

to use excess reagent in a volumetric analysis. Analytical procedures should always be adjusted so that any known physical limitations of the analyst cause negligibly small errors. Automation of analytical procedures can eliminate many errors of this type.

A universal source of personal error is *prejudice,* or *bias.* Most of us, no matter how honest, have a natural, subconscious tendency to estimate scale readings in a direction that improves the precision in a set of results. Alternatively, we may have a preconceived notion of the true value for the measurement. We then subconsciously cause the results to fall close to this value. Number bias is another source of personal error that varies considerably from person to person. The most frequent number bias encountered in estimating the position of a needle on a scale involves a preference for the digits 0 and 5. Also common is a prejudice favoring small digits over large and even numbers over odd. Again, automated and computerized instruments can eliminate this form of bias.

Digital and computer displays on pH meters, laboratory balances, and other electronic instruments eliminate number bias because no judgment is involved in taking a reading. However, many of these devices produce results with more figures than are significant. The rounding of insignificant figures can also produce bias (see Section 6D-1).

5B-2 The Effect of Systematic Errors on Analytical Results

Systematic errors may be either **constant** or **proportional**. The magnitude of a constant error stays essentially the same as the size of the quantity measured is varied. With constant errors, the absolute error is constant with sample size, but the relative error varies when the sample size is changed. Proportional errors increase or decrease according to the size of the sample taken for analysis. With proportional errors, the absolute error varies with sample size, but the relative error stays constant when the sample size is changed.

Constant errors are independent of the size of the sample being analyzed. **Proportional errors** decrease or increase in proportion to the size of the sample.

Constant Errors

The effect of a constant error becomes more serious as the size of the quantity measured decreases. The effect of solubility losses on the results of a gravimetric analysis, shown in Example 5-2, illustrates this behavior.

EXAMPLE 5-2

Suppose that 0.50 mg of precipitate is lost as a result of being washed with 200 mL of wash liquid. If the precipitate weighs 500 mg, the relative error due to solubility loss is $-(0.50/500) \times 100\% = -0.1\%$. Loss of the same quantity from 50 mg of precipitate results in a relative error of -1.0%.

The excess of reagent needed to bring about a color change during a titration is another example of constant error. This volume, usually small, remains the same regardless of the total volume of reagent required for the titration. Again, the relative error from this source becomes more serious as the total volume decreases. One way of reducing the effect of constant error is to increase the sample size until the error is acceptable.

Proportional Errors

A common cause of proportional errors is the presence of interfering contaminants in the sample. For example, a widely used method for the determination of copper is based on the reaction of copper(II) ion with potassium iodide to give iodine (see Sections 20B-2, 38H-3, and 38H-4.). The quantity of iodine is then measured and is proportional to the amount of copper. Iron(III), if present, also liberates iodine from potassium iodide. Unless steps are taken to prevent this interference, high results are observed for the percentage of copper because the iodine produced will be a measure of the copper(II) and iron(III) in the sample. The size of this error is fixed by the *fraction* of iron contamination, which is independent of the size of sample taken. If the sample size is doubled, for example, the amount of iodine liberated by both the copper and the iron contaminant is also doubled. Thus, the magnitude of the reported percentage of copper is independent of sample size.

5B-3 Detection of Systematic Instrument and Personal Errors

Some systematic instrument errors can be found and corrected by calibration. Periodic calibration of equipment is always desirable because the response of most instruments changes with time as a result of component aging, corrosion, or mistreatment. Many systematic instrument errors involve interferences where a species present in the sample affects the response of the analyte. Simple calibration does not compensate for these effects. Instead, the methods described in Section 8D-3 can be employed when such interference effects exist.

Most personal errors can be minimized by careful, disciplined laboratory work. It is a good habit to check instrument readings, notebook entries, and calculations systematically. Errors due to limitations of the experimenter can usually be avoided by carefully choosing the analytical method or using an automated procedure.

> After entering a reading into the laboratory notebook, many scientists habitually make a second reading and then check this against what has been entered to ensure the correctness of the entry.

5B-4 Detection of Systematic Method Errors

Bias in an analytical method is particularly difficult to detect. One or more of the following steps can be taken to recognize and adjust for a systematic error in an analytical method.

Analysis of Standard Samples

> **Standard reference materials (SRMs)** are substances sold by the National Institute of Standards and Technology (NIST) and certified to contain specified concentrations of one or more analytes.

The best way to estimate the bias of an analytical method is by analyzing **standard reference materials (SRMs)**, materials that contain one or more analytes at known concentration levels. Standard reference materials are obtained in several ways.

Standard materials can sometimes be prepared by synthesis. In this process, carefully measured quantities of the pure components of a material are measured out and mixed in such a way as to produce a homogeneous sample whose composition is known from the quantities taken. The overall composition of a synthetic standard material must closely approximate the composition of the samples to be analyzed. Great care must be taken to ensure that the concentration of analyte is known exactly. Unfortunately, a synthetic standard may not reveal unexpected interferences so that the accuracy of determinations may not be known. Hence, this approach is not often practical.

SRMs can be purchased from a number of governmental and industrial sources. For example, the National Institute of Standards and Technology (NIST) (formerly the National Bureau of Standards) offers over 1300 standard reference materials including rocks and minerals, gas mixtures, glasses, hydrocarbon mixtures, polymers, urban dusts, rainwaters, and river sediments.[4] The concentration of one or more of

---

[4]See U.S. Department of Commerce, *NIST Standard Reference Materials Catalog*, 2011 ed., NIST Special Publication 260, Washington, D.C.: U.S. Government Printing Office, 2011; see also http://www.nist.gov.

the components in these materials has been determined in one of three ways: (1) by analysis with a previously validated reference method, (2) by analysis by two or more independent, reliable measurement methods, or (3) by analysis by a network of co-operating laboratories that are technically competent and thoroughly knowledgeable with the material being tested. Several commercial supply houses also offer analyzed materials for method testing.[5]

Often, analysis of standard reference materials gives results that differ from the accepted value. The question then becomes one of establishing whether such a difference is due to bias or to random error. In Section 7B-1, we demonstrate a statistical test that can help answer this question.

Standard reference materials from NIST.

> In using SRMs, it is often difficult to separate bias from ordinary random error.

Independent Analysis

If standard samples are not available, a second independent and reliable analytical method can be used in parallel with the method being evaluated. The independent method should differ as much as possible from the one under study. This practice minimizes the possibility that some common factor in the sample has the same effect on both methods. Again, a statistical test must be used to determine whether any difference is a result of random errors in the two methods or due to bias in the method under study (see Section 7B-2).

Blank Determinations

A **blank** contains the reagents and solvents used in a determination, but no analyte. Often, many of the sample constituents are added to simulate the analyte environment, which is called the **sample matrix**. In a blank determination, all steps of the analysis are performed on the blank material. The results are then applied as a correction to the sample measurements. Blank determinations reveal errors due to interfering contaminants from the reagents and vessels employed in the analysis. Blanks are also used to correct titration data for the volume of reagent needed to cause an indicator to change color.

A **blank** solution contains the solvent and all of the reagents in an analysis. Whenever feasible, blanks may also contain added constituents to simulate the sample matrix.

The term **matrix** refers to the collection of all the constituents in the sample.

Variation in Sample Size

Example 5-2 on page 89 demonstrates that as the size of a measurement increases, the effect of a constant error decreases. Thus, constant errors can often be detected by varying the sample size.

WEB WORKS

Statistical methods are extremely important, not only in chemistry but in all walks of life. Newspapers, magazines, television, and the internet bombard us with confusing and often misleading statistics. Go to **www.cengage.com/chemistry/skoog/fac9**, choose Chapter 5, and go to the Web Works. There you will find a link to a website that contains an interesting presentation of statistics for writers. Use the links there to look up the definitions of mean and median. You will find some nice examples of salary data that clarify the distinction between the two measures of central tendency, show the utility of comparing the two, and point out the importance of using the appropriate measure for a particular data set. For the nine salaries given, which is larger, the mean or the median? Why are the two so different in this case?

---

[5]For example, in the clinical and biological sciences area, see Sigma-Aldrich Chemical Co., 3050 Spruce St., St. Louis, MO 63103, or Bio-Rad Laboratories, 1000 Alfred Nobel Dr., Hercules, CA 94547.

QUESTIONS AND PROBLEMS

5-1. Explain the difference between
  *(a) random and systematic error.
  (b) constant and proportional error.
  *(c) absolute and relative error.
  (d) mean and median.

*5-2. Suggest two sources of systematic error and two sources of random error in measuring the width of a 3-m table with a 1-m metal rule.

5-3. Name three types of systematic errors.

*5-4. Describe at least three systematic errors that might occur while weighing a solid on an analytical balance.

*5-5. Describe at least three ways in which a systematic error might occur while using a pipet to transfer a known volume of liquid.

5-6. How are systematic method errors detected?

*5-7. What kind of systematic errors are detected by varying the sample size?

5-8. A method of analysis yields masses of gold that are low by 0.4 mg. Calculate the percent relative error caused by this result if the mass of gold in the sample is
  *(a) 500 mg.   (b) 250 mg.
  *(c) 150 mg.   (d) 70 mg.

5-9. The method described in Problem 5-8 is to be used for the analysis of ores that assay about 1.2% gold. What minimum sample mass should be taken if the relative error resulting from a 0.4-mg loss is not to exceed
  *(a) −0.1%?   (b) −0.4%?
  *(c) −0.8%?   (d) −1.1%?

5-10. The color change of a chemical indicator requires an overtitration of 0.03 mL. Calculate the percent relative error if the total volume of titrant is
  *(a) 50.00 mL.   *(b) 10.0 mL.
  *(c) 25.0 mL.   (d) 30.0 mL.

5-11. A loss of 0.4 mg of Zn occurs in the course of an analysis for that element. Calculate the percent relative error due to this loss if the mass of Zn in the sample is
  *(a) 30 mg.   (b) 150 mg.
  *(c) 300 mg.   (d) 500 mg.

5-12. Find the mean and median of each of the following sets of data. Determine the deviation from the mean for each data point within the sets and find the mean deviation for each set. Use a spreadsheet if it is convenient.
  *(a) 0.0110   0.0104   0.0105
  (b) 24.53   24.68   24.77   24.81   24.73
  *(c) 188   190   194   187
  (d) 4.52×10^{-3} 4.47×10^{-3}
       4.63×10^{-3} 4.48×10^{-3}
       4.53×10^{-3} 4.58×10^{-3}
  *(e) 39.83   39.61   39.25   39.68
  (f) 850   862   849   869   865

5-13. **Challenge Problem:** Richards and Willard determined the molar mass of lithium and collected the following data.[6]

Experiment	Molar mass, g/mol
1	6.9391
2	6.9407
3	6.9409
4	6.9399
5	6.9407
6	6.9391
7	6.9406

(a) Find the mean molar mass determined by these workers.

(b) Find the median molar mass.

(c) Assuming that the currently accepted value for the molar mass of lithium is the true value, calculate the absolute error and the percent relative error of the mean value determined by Richards and Willard.

(d) Find in the chemical literature at least three values for the molar mass of lithium determined since 1910 and arrange them chronologically in a table or spreadsheet along with the values since 1817 given in the table on page 10 of the paper by Richards and Willard. Construct a graph of molar mass vs. year to illustrate how the molar mass of lithium has changed over the past two centuries. Suggest possible reason(s) why the value changes abruptly about 1830.

(e) The incredibly detailed experiments described by Richards and Willard suggest that it is unlikely that major changes in the molar mass of lithium will occur. Discuss this assertion in light of your calculation in part (c).

(f) What factors have led to changes in molar mass since 1910?

(g) How would you determine the accuracy of a molar mass?

[6]T. W. Richards and H. H. Willard, *J. Am. Chem. Soc.*, **1910**, *32*, 4, **DOI**: 10.1021/ja01919a002.

Random Errors in Chemical Analysis

CHAPTER 6

The probability distributions discussed in this chapter are fundamental to the use of statistics for judging the reliability of data and for testing various hypotheses. The quincunx shown in the upper part of the photo is a mechanical device that forms a normal probability distribution. Every 10 minutes, 30,000 balls drop from the center top of the machine, which contains a regular pattern of pegs to randomly deflect the balls. Each time a ball hits a peg, it has a 50:50 chance of falling to the right or to the left. After each ball passes through the array of pegs, it drops into one of the vertical "bins" of the transparent case. The height of the column of balls in each bin is proportional to the probability of a ball falling into a given bin. The smooth curve shown in the bottom half of the photo traces out the probability distribution.

Museum of Science, Boston

Random errors are present in every measurement no matter how careful the experimenter. In this chapter, we consider the sources of random errors, the determination of their magnitude, and their effects on computed results of chemical analyses. We also introduce the significant figure convention and illustrate its use in reporting analytical results.

6A THE NATURE OF RANDOM ERRORS

Random, or indeterminate, errors can never be totally eliminated and are often the major source of uncertainty in a determination. Random errors are caused by the many uncontrollable variables that accompany every measurement. Usually, most contributors to random error cannot be positively identified. Even if we can identify random error sources, it is often impossible to measure them because most are so small that they cannot be detected individually. The accumulated effect of the individual uncertainties, however, causes replicate results to fluctuate randomly around the mean of the set. For example, the scatter of data in Figures 5-1 and 5-3 is a direct result of the accumulation of small random uncertainties. We have replotted the Kjeldahl nitrogen data from Figure 5-3 as a three-dimensional plot in **Figure 6-1** to better reveal the precision and accuracy of each analyst. Note that the random error in the results of analysts 2 and 4 is much larger than that seen in the results of analysts 1 and 3. The results of analyst 3 show outstanding precision but poor accuracy. The results of analyst 1 show excellent precision and good accuracy.

Figure 6-1 A three-dimensional plot showing absolute error in Kjeldahl nitrogen determinations for four different analysts. Note that the results of analyst 1 are both precise and accurate. The results of analyst 3 are precise, but the absolute error is large. The results of analysts 2 and 4 are both imprecise and inaccurate.

6A-1 Random Error Sources

We can get a qualitative idea of the way small undetectable uncertainties produce a detectable random error in the following way. Imagine a situation in which just four small random errors combine to give an overall error. We will assume that each error has an equal probability of occurring and that each can cause the final result to be high or low by a fixed amount $\pm U$.

Table 6-1 shows all the possible ways the four errors can combine to give the indicated deviations from the mean value. Note that only one combination leads to

TABLE 6-1

Possible Combinations of Four Equal-Sized Uncertainties

Combinations of Uncertainties	Magnitude of Random Error	Number of Combinations	Relative Frequency
$+ U_1 + U_2 + U_3 + U_4$	$+ 4U$	1	$1/16 = 0.0625$
$- U_1 + U_2 + U_3 + U_4$ $+ U_1 - U_2 + U_3 + U_4$ $+ U_1 + U_2 - U_3 + U_4$ $+ U_1 + U_2 + U_3 - U_4$	$+ 2U$	4	$4/16 = 0.250$
$- U_1 - U_2 + U_3 + U_4$ $+ U_1 + U_2 - U_3 - U_4$ $+ U_1 - U_2 + U_3 - U_4$ $- U_1 + U_2 - U_3 + U_4$ $- U_1 + U_2 + U_3 - U_4$ $+ U_1 - U_2 - U_3 + U_4$	0	6	$6/16 = 0.375$
$+ U_1 - U_2 - U_3 - U_4$ $- U_1 + U_2 - U_3 - U_4$ $- U_1 - U_2 + U_3 - U_4$ $- U_1 - U_2 - U_3 + U_4$	$- 2U$	4	$4/16 = 0.250$
$- U_1 - U_2 - U_3 - U_4$	$- 4U$	1	$1/16 = 0.0625$

a deviation of $+4\ U$, four combinations give a deviation of $+2\ U$, and six give a deviation of $0\ U$. The negative errors have the same relationship. This ratio of 1:4:6:4:1 is a measure of the probability for a deviation of each magnitude. If we make a sufficiently large number of measurements, we can expect a frequency distribution like that shown in **Figure 6-2a**. Note that the y-axis in the plot is the relative frequency of occurrence of the five possible combinations.

Figure 6-2b shows the theoretical distribution for ten equal-sized uncertainties. Again we see that the most frequent occurrence is zero deviation from the mean. At the other extreme a maximum deviation of $10\ U$ occurs only about once in 500 results.

When the same procedure is applied to a very large number of individual errors, a bell-shaped curve like that shown in **Figure 6-2c** results. Such a plot is called a **Gaussian curve** or a **normal error curve**.

6A-2 Distribution of Experimental Results

From experience with many determinations, we find that the distribution of replicate data from most quantitative analytical experiments approaches that of the Gaussian curve shown in Figure 6-2c. As an example, consider the data in the spreadsheet in **Table 6-2** for the calibration of a 10-mL pipet.[1] In this experiment a small flask and stopper were weighed. Ten milliliters of water were transferred to the flask with the pipet, and the flask was stoppered. The flask, the stopper, and the water were then weighed again. The temperature of the water was also measured to determine its density. The mass of the water was then calculated by taking the difference between the two masses. The mass of water divided by its density is the volume delivered by the pipet. The experiment was repeated 50 times.

The data in Table 6-2 are typical of those obtained by an experienced worker weighing to the nearest milligram (which corresponds to 0.001 mL) on a top-loading balance and being careful to avoid systematic error. Even so, the results vary from a low of 9.969 mL to a high of 9.994 mL. This 0.025-mL **spread** of data results directly from an accumulation of all random uncertainties in the experiment.

The information in Table 6-2 is easier to visualize when the data are rearranged into frequency distribution groups, as in **Table 6-3**. In this instance, we count and tabulate the number of data points falling into a series of adjacent 0.003-mL ranges and calculate the percentage of measurements in each range. Note that 26% of the results occur in the volume range from 9.981 to 9.983 mL. This is the group containing the mean and median value of 9.982 mL. Note also that more than half the results are within ± 0.004 mL of this mean.

The frequency distribution data in Table 6-3 are plotted as a bar graph, or **histogram** in **Figure 6-3**, labeled A. We can imagine that as the number of measurements increases, the histogram approaches the shape of the continuous curve shown as plot B in Figure 6-3. This plot shows a Gaussian curve, or normal error curve, which applies to an infinitely large set of data. The Gaussian curve has the same mean (9.982 mL), the same precision, and the same area under the curve as the histogram.

---

[1] See Section 38A-4 for an experiment on calibration of a pipet.

> In our example, all the uncertainties have the same magnitude. This restriction is not necessary to derive the equation for a Gaussian curve.

(a)

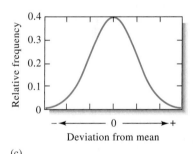

(b)

(c)

Figure 6-2 Frequency distribution for measurements containing (a) four random uncertainties, (b) ten random uncertainties, and (c) a very large number of random uncertainties.

The **spread** in a set of replicate measurements is the difference between the highest and lowest result.

A **histogram** is a bar graph such as that shown by plot A in Figure 6-3.

TABLE 6-2[†]

	A	B	C	D	E	F	G	H
1	Replicate Data for the Calibration of a 10-mL Pipet*							
2	Trial	Volume, mL		Trial	Volume, mL		Trial	Volume, mL
3	1	9.988		18	9.975		35	9.976
4	2	9.973		19	9.980		36	9.990
5	3	9.986		20	9.994		37	9.988
6	4	9.980		21	9.992		38	9.971
7	5	9.975		22	9.984		39	9.986
8	6	9.982		23	9.981		40	9.978
9	7	9.986		24	9.987		41	9.986
10	8	9.982		25	9.978		42	9.982
11	9	9.981		26	9.983		43	9.977
12	10	9.990		27	9.982		44	9.977
13	11	9.980		28	9.991		45	9.986
14	12	9.989		29	9.981		46	9.978
15	13	9.978		30	9.969		47	9.983
16	14	9.971		31	9.985		48	9.980
17	15	9.982		32	9.977		49	9.984
18	16	9.983		33	9.976		50	9.979
19	17	9.988		34	9.983			
20	*Data listed in the order obtained							
21	Mean	9.982		Maximum	9.994			
22	Median	9.982		Minimum	9.969			
23	Std. Dev.	0.0056		Spread	0.025			

[†]For Excel calculations of the statistical quantities listed at the bottom of Table 6-2, see S. R. Crouch and F. J. Holler, *Applications of Microsoft® Excel in Analytical Chemistry,* 2nd ed., Belmont, CA: Brooks/Cole, 2014, ch. 2.

A **Gaussian,** or **normal error curve**, is a curve that shows the symmetrical distribution of data around the mean of an infinite set of data such as the one in Figure 6-2c.

Variations in replicate measurements, such as those in Table 6-2, result from numerous small and individually undetectable random errors that are caused by uncontrollable variables in the experiment. Such small errors usually tend to cancel one another and thus have a minimal effect on the mean value. Occasionally, however, they occur in the same direction and produce a large positive or negative net error.

Sources of random uncertainties in the calibration of a pipet include: (1) visual judgments, such as the level of the water with respect to the marking on the pipet and the mercury level in the thermometer; (2) variations in the drainage

TABLE 6-3

Frequency Distribution of Data from Table 6-2

Volume Range, mL	Number in Range	% in Range
9.969 to 9.971	3	6
9.972 to 9.974	1	2
9.975 to 9.977	7	14
9.978 to 9.980	9	18
9.981 to 9.983	13	26
9.984 to 9.986	7	14
9.987 to 9.989	5	10
9.990 to 9.992	4	8
9.993 to 9.995	1	2
	Total = 50	Total = 100%

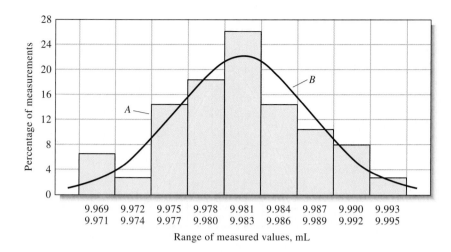

Figure 6-3 A histogram (*A*) showing distribution of the 50 results in Table 6-3 and a Gaussian curve (*B*) for data having the same mean and standard deviation as the data in the histogram.

time and in the angle of the pipet as it drains; (3) temperature fluctuations, which affect the volume of the pipet, the viscosity of the liquid, and the performance of the balance; and (4) vibrations and drafts that cause small variations in the balance readings. Undoubtedly, there are many other sources of random uncertainty in this calibration process that we have not listed. Even the simple process of calibrating a pipet is affected by many small and uncontrollable variables. The cumulative influence of these variables is responsible for the observed scatter of results around the mean.

The normal distribution of data that results from a large number of experiments is illustrated in Feature 6-1.

FEATURE 6-1

Flipping Coins: A Student Activity to Illustrate a Normal Distribution

If you flip a coin 10 times, how many heads will you get? Try it, and record your results. Repeat the experiment. Are your results the same? Ask friends or members of your class to perform the same experiment and tabulate the results. The table below contains the results obtained by several classes of analytical chemistry students over an 18-year period.

Number of Heads	0	1	2	3	4	5	6	7	8	9	10
Frequency	1	1	22	42	102	104	92	48	22	7	1

Add your results to those in the table, and plot a histogram similar to the one shown in Figure 6F-1. Find the mean and the standard deviation (see Section 6B-3) for your results and compare them to the values shown in the plot. The smooth curve in the figure is a normal error curve for an infinite number of trials with the same mean and standard deviation as the data set. Note that the mean of 5.06 is very close to the value of 5 that you would predict based on the laws of probability. As the number of trials increases, the histogram approaches the shape of the smooth curve, and the mean approaches five.

(continued)

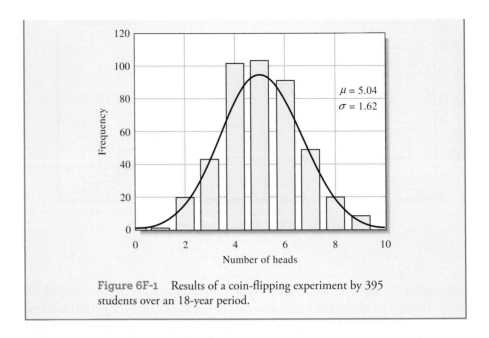

Figure 6F-1 Results of a coin-flipping experiment by 395 students over an 18-year period.

6B STATISTICAL TREATMENT OF RANDOM ERRORS

> Statistical analysis only reveals information that is already present in a data set. *No new information is created* by statistical treatments. Statistical methods, do allow us to categorize and characterize data in different ways and to make objective and intelligent decisions about data quality and interpretation.

We can use statistical methods to evaluate the random errors discussed in the preceding section. Generally, we base statistical analyses on the assumption that random errors in analytical results follow a Gaussian, or normal, distribution, such as that illustrated in Figure 6-2c, by curve *B* of Figure 6-3, or by the smooth curve in Figure 6F-1. Analytical data can follow distributions other than the Gaussian distribution. For example, experiments in which there is either a successful outcome or a failure produce data that follow the binomial distribution. Radioactive or photon-counting experiments produce results that follow the Poisson distribution. However, we often use a Gaussian distribution to approximate these distributions. The approximation becomes better in the limit of a large number of experiments. As a rule of thumb, if we have more than 30 results and the data are not heavily skewed, we can safely use a Gaussian distribution. Thus, we base this discussion entirely on normally distributed random errors.

6B-1 Samples and Populations

>
> A **population** is the collection of all measurements of interest to the experimenter, while a **sample** is a subset of measurements selected from the population.

Typically in a scientific study, we infer information about a **population** or **universe** from observations made on a subset or **sample**. The population is the collection of all measurements of interest and must be carefully defined by the experimenter. In some cases, the population is finite and real, while in others, the population is hypothetical or conceptual in nature.

As an example of a real population, consider a production run of multivitamin tablets that produces hundreds of thousands of tablets. Although the population is finite, we usually would not have the time or resources to test all the tablets for quality control purposes. Hence, we select a sample of tablets for analysis according to statistical sampling principles (see Section 8B). We then infer the characteristics of the population from those of the sample.

In many of the cases encountered in analytical chemistry, the population is conceptual. Consider, for example, the determination of calcium in a community water supply to determine water hardness. In this example, the population is the very large, nearly infinite, number of measurements that could be made if we analyzed the entire water supply. Similarly, in determining glucose in the blood of a patient, we could hypothetically make an extremely large number of measurements if we used the entire blood supply. The subset of the population analyzed in both these cases is the sample. Again, we infer characteristics of the population from those obtained with the sample. Hence, it is very important to define the population being characterized.

Statistical laws have been derived for populations, but they can be used for samples after suitable modification. Such modifications are needed for small samples because a few data points may not represent the entire population. In the discussion that follows, we first describe the Gaussian statistics of populations. Then we show how these relationships can be modified and applied to small samples of data.

6B-2 Properties of Gaussian Curves

Figure 6-4a shows two Gaussian curves in which we plot the relative frequency y of various deviations from the mean versus the deviation from the mean. As shown in the margin, curves such as these can be described by an equation that contains just two parameters, the **population mean** μ and the **population standard deviation** σ. The term **parameter** refers to quantities such as μ and σ that define a population or distribution. Data values such as x are **variables**. The term **statistic** refers to an estimate of a parameter that is made from a sample of data as discussed below. The sample mean and the sample standard deviation are examples of statistics that estimate parameters μ and σ respectively.

The Population Mean μ and the Sample Mean $\bar{x}$

Scientists find it useful to differentiate between the **sample mean** and the **population mean**. The sample mean $\bar{x}$ is the arithmetic average of a limited sample drawn from a population of data. The sample mean is defined as the sum of the measurement values divided by the number of measurements as given by Equation 5-1, page 84. In that equation, N represents the number of measurements

> Do not confuse the **statistical sample** with the **analytical sample**. Consider four water samples taken from the same water supply and analyzed in the laboratory for calcium. The four analytical samples result in four measurements selected from the population. They are thus a single statistical sample. This is an unfortunate duplication of the term sample.

> The equation for a normalized Gaussian curve has the form
>
> $$y = \frac{e^{-(x-\mu)^2/2\sigma^2}}{\sigma\sqrt{2\pi}}$$

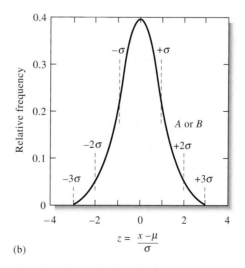

(a)   (b)

Figure 6-4   Normal error curves. The standard deviation for curve B is twice that for curve A, that is, $\sigma_B = 2\sigma_A$. In (a) the abscissa is the deviation from the mean $(x - \mu)$ in the units of measurement. In (b) the abscissa is the deviation from the mean in units of σ. For this plot, the two curves A and B are identical.

The sample mean $\bar{x}$ is found from

$$\bar{x} = \frac{\sum_{i=1}^{N} x_i}{N}$$

where N is the number of measurements in the sample set. The same equation is used to calculate the population mean μ

$$\mu = \frac{\sum_{i=1}^{N} x_i}{N}$$

where N is now the total number of measurements in the population.

In the absence of systematic error, the population mean μ is the true value of a measured quantity.

The quantity $(x_i - \mu)$ in Equation 6-1 is the deviation of data value x_i from the mean μ of a population; compare with Equation 6-4, which is for a sample of data.

in the sample set. The population mean μ, in contrast, is the true mean for the population. It is also defined by Equation 5-1 with the added provision that N represents the total number of measurements in the population. *In the absence of systematic error, the population mean is also the true value for the measured quantity.* To emphasize the difference between the two means, the sample mean is symbolized by $\bar{x}$ and the population mean by μ. More often than not, particularly when N is small, $\bar{x}$ differs from μ because a small sample of data may not exactly represent its population. In most cases we do not know μ and must infer its value from $\bar{x}$. The probable difference between $\bar{x}$ and μ decreases rapidly as the number of measurements making up the sample increases; usually by the time N reaches 20 to 30, this difference is negligible. Note that the sample mean $\bar{x}$ is a statistic that estimates the population parameter μ.

The Population Standard Deviation σ

The **population standard deviation σ**, which is a measure of the *precision* of the population, is given by the equation

$$\sigma = \sqrt{\frac{\sum_{i=1}^{N} (x_i - \mu)^2}{N}} \tag{6-1}$$

where N is the number of data points making up the population.

The two curves in Figure 6-4a are for two populations of data that differ only in their standard deviations. The standard deviation for the data set yielding the broader but lower curve B is twice that for the measurements yielding curve A. The breadth of these curves is a measure of the precision of the two sets of data. Thus, the precision of the data set leading to curve A is twice as good as that of the data set represented by curve B.

Figure 6-4b shows another type of normal error curve in which the x axis is now a new variable z, which is defined as

$$z = \frac{(x - \mu)}{\sigma} \tag{6-2}$$

The quantity z represents the deviation of a result from the population mean relative to the standard deviation. It is commonly given as a variable in statistical tables since it is a dimensionless quantity.

Note that z is the relative deviation of a data point from the mean, that is, the deviation relative to the standard deviation. Hence, when $x - \mu = \sigma$, z is equal to one; when $x - \mu = 2\sigma$, z is equal to two; and so forth. Since z is the deviation from the mean relative to the standard deviation, a plot of relative frequency versus z yields a single Gaussian curve that describes all populations of data regardless of standard deviation. Thus, Figure 6-4b is the normal error curve for both sets of data used to plot curves A and B in Figure 6-4a.

The equation for the Gaussian error curve is

$$y = \frac{e^{-(x-\mu)^2/2\sigma^2}}{\sigma\sqrt{2\pi}} = \frac{e^{-z^2/2}}{\sigma\sqrt{2\pi}} \tag{6-3}$$

Because it appears in the Gaussian error curve expression, the square of the standard deviation σ^2 is also important. This quantity is called the **variance** (see Section 6B-5).

A normal error curve has several general properties: (a) The mean occurs at the central point of maximum frequency, (b) there is a symmetrical distribution of positive and negative deviations about the maximum, and (c) there is an exponential decrease in frequency as the magnitude of the deviations increases. Thus, small uncertainties are observed much more often than very large ones.

Areas under a Gaussian Curve

Feature 6-2 shows that, regardless of its width, 68.3% of the area beneath a Gaussian curve for a population lies within one standard deviation ($\pm 1\sigma$) of the mean μ. Thus, roughly 68.3% of the results making up the population will lie within these bounds. Furthermore, approximately 95.4% of all data points are within $\pm 2\sigma$ of the mean and 99.7% within $\pm 3\sigma$. The vertical dashed lines in Figure 6-4 show the areas bounded by $\pm 1\sigma$, $\pm 2\sigma$, and $\pm 3\sigma$.

Because of area relationships such as these, the standard deviation of a population of data is a useful predictive tool. For example, we can say that the chances are 68.3 in 100 that the random uncertainty of any single measurement is no more than $\pm 1\sigma$. Similarly, the chances are 95.4 in 100 that the error is less than $\pm 2\sigma$, and so forth. The calculation of areas under the Gaussian curve is described in Feature 6-2.

FEATURE 6-2

Calculating the Areas under the Gaussian Curve

We often refer to the area under a curve. In the context of statistics, it is important to be able to determine the area under the Gaussian curve between defined limits. The area under the curve between a pair of limits gives the probability of a measured value occurring between those limits. A practical question arises: how do we determine the area under a curve? Equation 6-3 describes the Gaussian curve in terms of the population mean μ and the standard deviation σ or the variable z. Let us suppose that we want to know the area under the curve between -1σ and $+1\sigma$ of the mean. In other words, we want the area from $\mu - \sigma$ to $\mu + \sigma$.

We can perform this operation using calculus because integration of an equation gives the area under the curve described by the equation. In this case, we wish to find the definite integral from $-\sigma$ to $+\sigma$.

$$\text{area} = \int_{-\sigma}^{\sigma} \frac{e^{-(x-\mu)^2/2\sigma^2}}{\sigma\sqrt{2\pi}}\, dx$$

It is easier to use the form of Equation 6-3 with the variable z, so our equation becomes

$$\text{area} = \int_{-1}^{1} \frac{e^{-z^2/2}}{\sqrt{2\pi}}\, dz$$

Since there is no closed form solution, the integral must be evaluated numerically. The result is

$$\text{area} = \int_{-1}^{1} \frac{e^{-z^2/2}}{\sqrt{2\pi}}\, dz = 0.683$$

(continued)

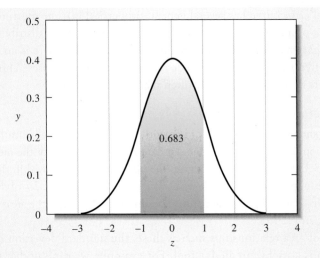

Curve showing area of 0.683.

Likewise, if we want to know the area under the Gaussian curve 2σ on either side of the mean, we evaluate the following integral.

$$\text{area} = \int_{-2}^{2} \frac{e^{-z^2/2}}{\sqrt{2\pi}} \, dz = 0.954$$

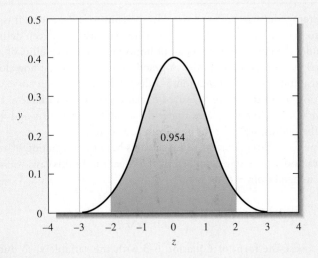

Curve showing area of 0.954.

For $\pm 3\sigma$, we have

$$\text{area} = \int_{-3}^{3} \frac{e^{-z^2/2}}{\sqrt{2\pi}} \, dz = 0.997$$

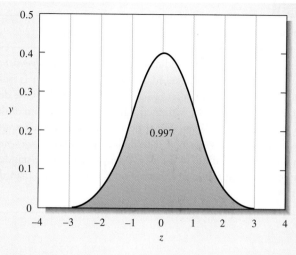

Curve showing area of 0.997.

Finally, it is important to know the area under the entire Gaussian curve, so we find the following integral.

$$\text{area} = \int_{-\infty}^{\infty} \frac{e^{-z^2/2}}{\sqrt{2\pi}} \, dz = 1$$

We can see from the integrals that the areas under the Gaussian curve for one, two, and three standard deviations from the mean are, respectively, 68.3%, 95.4%, and 99.7% of the total area under the curve.

6B-3 The Sample Standard Deviation: A Measure of Precision

Equation 6-1 must be modified when it is applied to a small sample of data. Thus, the **sample standard deviation** s is given by the equation

$$s = \sqrt{\frac{\sum_{i=1}^{N}(x_i - \bar{x})^2}{N - 1}} = \sqrt{\frac{\sum_{i=1}^{N}d_i^2}{N - 1}} \qquad (6\text{-}4)$$

where the quantity $(x_i - \bar{x})$ represents the deviation d_i of value x_i from the mean $\bar{x}$. Note that Equation 6-4 differs from Equation 6-1 in two ways. First, the sample mean, $\bar{x}$, appears in the numerator in place of the population mean, μ. Second, N in Equation 6-1 is replaced by the **number of degrees of freedom** $(N - 1)$. When $N - 1$ is used instead of N, s is said to be an unbiased estimator of the population standard deviation σ. If this substitution is not used, the calculated s will be less on average than the true standard deviation σ, that is, s will have a negative bias (see Feature 6-3).

The **sample variance** s^2 is also of importance in statistical calculations. It is an estimate of the population variance σ^2, as discussed in Section 6B-5.

Equation 6-4 applies to small sets of data. It says, "Find the deviations from the mean d_i, square them, sum them, divide the sum by $N - 1$, and take the square root." The quantity $N - 1$ is called **the number of degrees of freedom**. Scientific calculators usually have the standard deviation function built in. Many can find the population standard deviation σ as well as the sample standard deviation, s. For any small data set, you should use the sample standard deviation, s.

FEATURE 6-3

The Significance of the Number of Degrees of Freedom

The number of degrees of freedom indicates the number of *independent* results that enter into the computation of the standard deviation. When μ is unknown, two quantities must be extracted from a set of replicate data: $\bar{x}$ and s. One degree of freedom is used to establish $\bar{x}$ because, with their signs retained, the sum of the individual deviations must be zero. Thus, when $N-1$ deviations have been computed, the final one is known. Consequently, only $N-1$ deviations provide an *independent* measure of the precision of the set. Failure to use $N-1$ in calculating the standard deviation for small samples results in values of s that are on average smaller than the true standard deviation σ.

An Alternative Expression for Sample Standard Deviation

To find s with a calculator that does not have a standard deviation key, the following rearrangement is easier to use than directly applying Equation 6-4:

$$s = \sqrt{\frac{\sum\limits_{i=1}^{N} x_i^2 - \dfrac{\left(\sum\limits_{i=1}^{N} x_i\right)^2}{N}}{N-1}} \tag{6-5}$$

Example 6-1 illustrates the use of Equation 6-5 to calculate s.

EXAMPLE 6-1

The following results were obtained in the replicate determination of the lead content of a blood sample: 0.752, 0.756, 0.752, 0.751, and 0.760 ppm Pb. Find the mean and the standard deviation of this set of data.

Solution

To apply Equation 6-5, we calculate $\sum x_i^2$ and $(\sum x_i)^2/N$.

Sample	x_i	x_i^2
1	0.752	0.565504
2	0.756	0.571536
3	0.752	0.565504
4	0.751	0.564001
5	0.760	0.577600
	$\sum x_i = 3.771$	$\sum x_i^2 = 2.844145$

$$\bar{x} = \frac{\sum x_i}{N} = \frac{3.771}{5} = 0.7542 \approx 0.754 \text{ ppm Pb}$$

$$\frac{\left(\sum x_i\right)^2}{N} = \frac{(3.771)^2}{5} = \frac{14.220441}{5} = 2.8440882$$

Substituting into Equation 6-5 leads to

$$s = \sqrt{\frac{2.844145 - 2.8440882}{5 - 1}} = \sqrt{\frac{0.0000568}{4}} = 0.00377 \approx 0.004 \text{ ppm Pb}$$

Note in Example 6-1 that the difference between $\sum x_i^2$ and $(\sum x_i)^2/N$ is very small. If we had rounded these numbers before subtracting them, a serious error would have appeared in the calculated value of s. To avoid this source of error, *never round a standard deviation calculation until the very end*. Furthermore, and for the same reason, never use Equation 6-5 to calculate the standard deviation of numbers containing five or more digits. Use Equation 6-4 instead.[2] Many calculators and computers with a standard deviation function use a version of Equation 6-5 internally in the calculation. You should always be alert for roundoff errors when calculating the standard deviation of values that have five or more significant figures.

> Any time you subtract two large, approximately equal numbers, the difference will usually have a relatively large uncertainty. Hence, you should never round a standard deviation calculation until the end.

When you make statistical calculations, remember that, because of the uncertainty in $\bar{x}$, a sample standard deviation may differ significantly from the population standard deviation. As N becomes larger, $\bar{x}$ and s become better estimators of μ, and σ.

> As $N \to \infty$, $\bar{x} \to \mu$, and $s \to \sigma$.

Standard Error of the Mean

The probability figures for the Gaussian distribution calculated as areas in Feature 6-2 refer to the probable error for a *single* measurement. Thus, it is 95.4% probable that a single result from a population will lie within $\pm 2\sigma$ of the mean μ. If a series of replicate results, each containing N measurements, are taken randomly from a population of results, the mean of each set will show less and less scatter as N increases. The standard deviation of each mean is known as the **standard error of the mean** and is given the symbol s_m. The standard error is inversely proportional to the square root of the number of data points N used to calculate the mean as given by Equation 6-6.

> The **standard error** of a mean, s_m, is the standard deviation of a set of data divided by the square root of the number of data points in the set.

$$s_m = \frac{s}{\sqrt{N}} \tag{6-6}$$

Equation 6-6 tells us that the mean of four measurements is more precise by $\sqrt{4} = 2$ than individual measurements in the data set. For this reason, averaging results is often used to improve precision. However, the improvement to be gained by averaging is somewhat limited because of the square root dependence on N shown in Equation 6-6. For example, to increase the precision by a factor of 10 requires 100 times as many measurements. It is better, if possible, to decrease s than to keep averaging more results since s_m is directly proportional to s but only inversely proportional to the *square root* of N. The standard deviation can sometimes be decreased by being more precise in individual operations, by changing the procedure, and by using more precise measurement tools.

---

[2]In most cases, the first two or three digits in a set of data are identical to each other. As an alternative, then, to using Equation 6-4, these identical digits can be dropped, and the remaining digits used with Equation 6-5. For example, the standard deviation for the data in Example 6-1 could be based on 0.052, 0.056, 0.052, and so forth (or even 52, 56, 52, etc.).

Spreadsheet Summary   In Chapter 2 of *Applications of Microsoft®* *Excel in Analytical Chemistry*, 2nd ed., two different ways of calculating the sample standard deviation with Excel are shown.

6B-4 Reliability of *s* as a Measure of Precision

In Chapter 7 we describe several statistical tests that are used to test hypotheses, to produce confidence intervals for results, and to reject outlying data points. Most of these tests are based on sample standard deviations. The probability that these statistical tests provide correct results increases as the reliability of *s* becomes greater. As *N* in Equation 6-4 increases, *s* becomes a better estimator of the population standard deviation, σ. When *N* is greater than about 20, *s* is usually a good estimator of σ, and these quantities can be assumed to be identical for most purposes. For example, if the 50 measurements in Table 6-2 are divided into 10 subgroups of 5 each, the value of *s* varies widely from one subgroup to another (0.0023 to 0.0079 mL) even though the average of the computed values of *s* is that of the entire set (0.0056 mL). In contrast, the computed values of *s* for two subsets of 25 each are nearly identical (0.0054 and 0.0058 mL).

The rapid improvement in the reliability of *s* with increases in *N* makes it feasible to obtain a good approximation of σ when the method of measurement is not excessively time consuming and when an adequate supply of sample is available. For example, if the pH of numerous solutions is to be measured in the course of an investigation, it is useful to evaluate *s* in a series of preliminary experiments. This measurement is simple, requiring only that a pair of rinsed and dried electrodes be immersed in the test solution and the pH read from a scale or a display. To determine *s*, 20 to 30 portions of a buffer solution of fixed pH can be measured with all steps of the procedure being followed exactly. Normally, it is safe to assume that the random error in this test is the same as that in subsequent measurements. The value of *s* calculated from Equation 6-4 is a good estimator of the population value, σ.

CHALLENGE: Construct a spreadsheet using the data in Table 6-2, and show that *s* better estimates σ as *N* becomes larger. Also show that *s* is nearly equal to σ for *N* > 20.

Spreadsheet Summary   In Chapter 2 of *Applications of Microsoft®* *Excel in Analytical Chemistry*, 2nd ed, we introduce the use of Excel's Analysis ToolPak to compute the mean, standard deviation, and other quantities. The Descriptive Statistics option finds the standard error of the mean, the median, the range, the maximum and minimum values, and parameters that reflect the symmetry of the data set.

Pooling Data to Improve the Reliability of s

If we have several subsets of data, a better estimate of the population standard deviation can be obtained by pooling (combining) the data instead of using only one data set. Again, we must assume the same sources of random error in all the measurements. This assumption is usually valid if the samples have similar compositions and have been analyzed in exactly the same way. We must also assume that the samples are randomly drawn from the same population and thus have a common value of σ.

The pooled estimate of s, which we call s_{pooled}, is a weighted average of the individual estimates. To calculate s_{pooled}, deviations from the mean for each subset are squared; the squares of the deviations of all subsets are then summed and divided by the appropriate number of degrees of freedom. The pooled s is obtained by taking the square root of the resulting number. One degree of freedom is lost for each subset. Thus, the number of degrees of freedom for the pooled s is equal to the total number of measurements minus the number of subsets. Equation 6-7 in Feature 6-4 gives the full equation for obtaining s_{pooled} for t data sets. Example 6-2 illustrates the application of this type of computation.

FEATURE 6-4

Equation for Calculating the Pooled Standard Deviation

The equation for computing a pooled standard deviation from several sets of data takes the form

$$s_{pooled} = \sqrt{\frac{\sum_{i=1}^{N_1}(x_i - \bar{x}_1)^2 + \sum_{j=1}^{N_2}(x_j - \bar{x}_2)^2 + \sum_{k=1}^{N_3}(x_k - \bar{x}_3)^2 + \cdots}{N_1 + N_2 + N_3 + \cdots - N_t}} \tag{6-7}$$

where N_1 is the number of results in set 1, N_2 is the number in set 2, and so forth. The term N_t is the total number of data sets pooled.

EXAMPLE 6-2

Glucose levels are routinely monitored in patients suffering from diabetes. The glucose concentrations in a patient with mildly elevated glucose levels were determined in different months by a spectrophotometric analytical method. The patient was placed on a low-sugar diet to reduce the glucose levels. The following results were obtained during a study to determine the effectiveness of the diet. Calculate a pooled estimate of the standard deviation for the method.

Time	Glucose Concentration, mg/L	Mean Glucose, mg/L	Sum of Squares of Deviations from Mean	Standard Deviation
Month 1	1108, 1122, 1075, 1099, 1115, 1083, 1100	1100.3	1687.43	16.8
Month 2	992, 975, 1022, 1001, 991	996.2	1182.80	17.2
Month 3	788, 805, 779, 822, 800	798.8	1086.80	16.5
Month 4	799, 745, 750, 774, 777, 800, 758	771.9	2950.86	22.2

Total number of measurements = 24   Total sum of squares = 6907.89

Solution

For the first month, the sum of the squares in the next to last column was calculated as follows:

$$\text{Sum of squares} = (1108 - 1100.3)^2 + (1122 - 1100.3)^2$$
$$+ (1075 - 1100.3)^2 + (1099 - 1100.3)^2 + (1115 - 1100.3)^2$$
$$+ (1083 - 1100.3)^2 + (1100 - 1100.3)^2 = 1687.43$$

(continued)

The other sums of squares were obtained similarly. The pooled standard deviation is then

$$s_{pooled} = \sqrt{\frac{6907.89}{24 - 4}} = 18.58 \approx 19 \text{ mg/L}$$

Note this pooled value is a better estimate of σ than any of the individual s values in the last column. Note also that one degree of freedom is lost for each of the four sets. Because 20 degrees of freedom remain, however, the calculated value of s can be considered a good estimate of σ.

A glucose analyzer.

Spreadsheet Summary In Chapter 2 of *Applications of Microsoft® Excel in Analytical Chemistry*, 2nd ed, we develop a worksheet to calculate the pooled standard deviation of the data from Example 6-2. The Excel function DEVSQ () is introduced to find the sum of the squares of the deviations. As extensions of this exercise, you may use the worksheet to solve some of the pooled standard deviation problems at the end of this chapter. You can also expand the worksheet to accommodate more data points within data sets and larger numbers of sets.

6B-5 Variance and Other Measures of Precision

Although the sample standard deviation is usually used in reporting the precision of analytical data, we often find three other terms.

Variance (s^2)

The variance is just the square of the standard deviation. The **sample variance** s^2 is an estimate of the population variance σ^2 and is given by

> The **variance**, s^2, is equal to the square of the standard deviation.

$$s^2 = \frac{\sum_{i=1}^{N} (x_i - \bar{x})^2}{N - 1} = \frac{\sum_{i=1}^{N} (d_i)^2}{N - 1} \tag{6-8}$$

Note that the standard deviation has the same units as the data, while the variance has the units of the data squared. Scientists tend to use standard deviation rather than variance because it is easier to relate a measurement and its precision if they both have the same units. The advantage of using variance is that variances are additive in many situations, as we discuss later in this chapter.

Relative Standard Deviation (RSD) and Coefficient of Variation (CV)

Frequently standard deviations are given in relative rather than absolute terms. We calculate the relative standard deviation by dividing the standard deviation by the mean value of the data set. The relative standard deviation, RSD, is sometimes given the symbol s_r.

$$RSD = s_r = \frac{s}{\bar{x}}$$

The result is often expressed in parts per thousand (ppt) or in percent by multiplying this ratio by 1000 ppt or by 100%. For example,

$$RSD \text{ in ppt} = \frac{s}{\bar{x}} \times 1000 \text{ ppt}$$

The relative standard deviation multiplied by 100% is called the **coefficient of variation** (CV).

$$CV = RSD \text{ in percent} = \frac{s}{\bar{x}} \times 100\% \qquad (6\text{-}9)$$

> The International Union of Pure and Applied Chemistry recommends that the symbol s_r be used for relative sample standard deviation and σ_r for relative population standard deviation. In equations where it is cumbersome to use RSD, we will use s_r and σ_r.

> The **coefficient of variation**, CV, is the percent relative standard deviation.

Relative standard deviations often give a clearer picture of data quality than do absolute standard deviations. As an example, suppose that a copper determination has a standard deviation of 2 mg. If the sample has a mean value of 50 mg of copper, the CV for this sample is 4% $\left(\frac{2}{50} \times 100\%\right)$. For a sample containing only 10 mg, the CV is 20%.

Spread or Range (w)

The **spread**, or **range**, w, is another term that is sometimes used to describe the precision of a set of replicate results. It is the difference between the largest value in the set and the smallest. Thus, the spread of the data in Figure 5-1 is $(20.3 - 19.4) = 0.9$ ppm Fe. The spread in the results for month 1 in Example 6-2 is $1122 - 1075 = 47$ mg/L glucose.

EXAMPLE 6-3

For the set of data in Example 6-1, calculate (a) the variance, (b) the relative standard deviation in parts per thousand, (c) the coefficient of variation, and (d) the spread.

Solution

In Example 6-1, we found

$$\bar{x} = 0.754 \text{ ppm Pb} \quad \text{and} \quad s = 0.0038 \text{ ppm Pb}$$

(a) $s^2 = (0.0038)^2 = 1.4 \times 10^{-5}$

(b) $RSD = \frac{0.0038}{0.754} = \times 1000 \text{ ppt} = 5.0 \text{ ppt}$

(c) $CV = \frac{0.0038}{0.754} \times 100\% = 0.50\%$

(d) $w = 0.760 - 0.751 = 0.009 \text{ ppm Pb}$

6C STANDARD DEVIATION OF CALCULATED RESULTS

Often we must estimate the standard deviation of a result that has been calculated from two or more experimental data points, each of which has a known sample standard deviation. As shown in Table 6-4, the way such estimates are made depends on the types of calculations that are involved. The relationships shown in this table are derived in Appendix 9.

TABLE 6-4

Error Propagation in Arithmetic Calculations

Type of Calculation	Example*	Standard Deviation of y[†]	
Addition or subtraction	$y = a + b - c$	$s_y = \sqrt{s_a^2 + s_b^2 + s_c^2}$	(1)
Multiplication or division	$y = a \times b/c$	$\dfrac{s_y}{y} = \sqrt{\left(\dfrac{s_a}{a}\right)^2 + \left(\dfrac{s_b}{b}\right)^2 + \left(\dfrac{s_c}{c}\right)^2}$	(2)
Exponentiation	$y = a^x$	$\dfrac{s_y}{y} = x\left(\dfrac{s_a}{a}\right)$	(3)
Logarithm	$y = \log_{10} a$	$s_y = 0.434 \dfrac{s_a}{a}$	(4)
Antilogarithm	$y = \text{antilog}_{10} a$	$\dfrac{s_y}{y} = 2.303\, s_a$	(5)

*a, b, and c are experimental variables with standard deviations of s_a, s_b, and s_c, respectively
[†]These relationships are derived in Appendix 9. The values for s_y/y are absolute values if y is a negative number.

6C-1 Standard Deviation of a Sum or Difference

Consider the summation:

$$
\begin{array}{ll}
+\ 0.50 & (\pm\ 0.02) \\
+\ 4.10 & (\pm\ 0.03) \\
-\ 1.97 & (\pm\ 0.05) \\
\hline
2.63 &
\end{array}
$$

where the numbers in parentheses are absolute standard deviations. If the three individual standard deviations happen by chance to have the same sign, the standard deviation of the sum could be as large as $+0.02 + 0.03 + 0.05 = +0.10$ or $-0.02 - 0.03 - 0.05 = -0.10$. On the other hand, it is possible that the three standard deviations could combine to give an accumulated value of zero: $-0.02 - 0.03 + 0.05 = 0$ or $+0.02 + 0.03 - 0.05 = 0$. More likely, however, the standard deviation of the sum will lie between these two extremes. The variance of a sum or difference is equal to the sum of the individual variances.[3] The most probable value for a standard deviation of a sum or difference can be found by taking the square root of the sum of the squares of the individual absolute standard deviations. So, for the computation

> The variance of a sum or difference is equal to the *sum* of the variances of the numbers making up that sum or difference.

$$y = a(\pm s_a) + b(\pm s_b) - c(\pm s_c)$$

the variance of y, s_y^2, is given by

$$s_y^2 = s_a^2 + s_b^2 + s_c^2$$

---

[3]See P. R. Bevington and D. K. Robinson, *Data Reduction and Error Analysis for the Physical Sciences*, 3rd ed., New York: McGraw-Hill, 2002, ch. 3.

Hence, the standard deviation of the result s_y is

$$s_y = \sqrt{s_a^2 + s_b^2 + s_c^2} \tag{6-10}$$

where s_a, s_b, and s_c are the standard deviations of the three terms making up the result. Substituting the standard deviations from the example gives

$$s_y = \sqrt{(\pm 0.02)^2 + (\pm 0.03)^2 + (\pm 0.05)^2} = \pm 0.06$$

and the sum should be reported as 2.64 (± 0.06).

> For a sum or a difference, the *standard deviation of the answer* is the square root of the sum of the squares of the *standard deviations* of the numbers used in the calculation.

> ◀ Throughout the rest of this chapter, we highlight the uncertain digit, by showing it in a second color.

6C-2 Standard Deviation of a Product or Quotient

Consider the following computation where the numbers in parentheses are again absolute standard deviations:

$$\frac{4.10(\pm 0.02) \times 0.0050(\pm 0.0001)}{1.97(\pm 0.04)} = 0.010406(\pm ?)$$

In this situation, the standard deviations of two of the numbers in the calculation are larger than the result itself. Evidently, we need a different approach for multiplication and division. As shown in Table 6-4, the *relative standard deviation* of a product or quotient is determined by the *relative standard deviations* of the numbers forming the computed result. For example, in the case of

$$y = \frac{a \times b}{c} \tag{6-11}$$

we obtain the relative standard deviation s_y/y of the result by summing the squares of the relative standard deviations of a, b, and c and then calculating the square root of the sum:

$$\frac{s_y}{y} = \sqrt{\left(\frac{s_a}{a}\right)^2 + \left(\frac{s_b}{b}\right)^2 + \left(\frac{s_c}{c}\right)^2} \tag{6-12}$$

> For multiplication or division, the *relative standard deviation of the answer* is the square root of the sum of the squares of the *relative standard deviations* of the numbers that are multiplied or divided.

Applying this equation to the numerical example gives

$$\frac{s_y}{y} = \sqrt{\left(\frac{\pm 0.02}{4.10}\right)^2 + \left(\frac{\pm 0.0001}{0.0050}\right)^2 + \left(\frac{\pm 0.04}{1.97}\right)^2}$$

$$= \sqrt{(0.0049)^2 + (0.0200)^2 + (0.0203)^2} = \pm 0.0289$$

To complete the calculation, we must find the absolute standard deviation of the result,

$$s_y = y \times (\pm 0.0289) = 0.0104 \times (\pm 0.0289) = \pm 0.000301$$

> ◀ To find the absolute standard deviation in a product or a quotient, first find the relative standard deviation in the result and then multiply it by the result.

and we can write the answer and its uncertainty as 0.0104 (± 0.0003). Note that if y is a negative number, we should treat s_y/y as an absolute value.

Example 6-4 demonstrates the calculation of the standard deviation of the result for a more complex calculation.

EXAMPLE 6-4

Calculate the standard deviation of the result of

$$\frac{[14.3(\pm0.2) - 11.6(\pm0.2)] \times 0.050(\pm0.001)}{[820(\pm10) + 1030(\pm5)] \times 42.3(\pm0.4)} = 1.725(\pm?) \times 10^{-6}$$

Solution

First, we must calculate the standard deviation of the sum and the difference. For the difference in the numerator,

$$s_a = \sqrt{(\pm0.2)^2 + (\pm0.2)^2} = \pm0.283$$

and for the sum in the denominator,

$$s_b = \sqrt{(\pm10)^2 + (\pm5)^2} = 11.2$$

We may then rewrite the equation as

$$\frac{2.7(\pm0.283) \times 0.050(\pm0.001)}{1850(\pm11.2) \times 42.3(\pm0.4)} = 1.725 \times 10^{-6}$$

The equation now contains only products and quotients, and Equation 6-12 applies. Thus,

$$\frac{s_y}{y} = \sqrt{\left(\pm\frac{0.283}{2.7}\right)^2 + \left(\pm\frac{0.001}{0.050}\right)^2 + \left(\pm\frac{11.2}{1850}\right)^2 + \left(\pm\frac{0.4}{42.3}\right)^2} = 0.107$$

To obtain the absolute standard deviation, we write

$$s_y = y \times 0.107 = 1.725 \times 10^{-6} \times (\pm0.107) = \pm0.185 \times 10^{-6}$$

and round the answer to $1.7(\pm0.2) \times 10^{-6}$.

6C-3 Standard Deviations in Exponential Calculations

Consider the relationship

$$y = a^x$$

where the exponent x can be considered to be free of uncertainty. As shown in Table 6-4 and Appendix 9, the relative standard deviation in y resulting from the uncertainty in a is

$$\frac{s_y}{y} = x\left(\frac{s_a}{a}\right) \tag{6-13}$$

Therefore, the relative standard deviation of the square of a number is twice the relative standard deviation of the number, the relative standard deviation of the cube root of a number is one third that of the number, and so forth. Example 6-5 illustrates this type of calculation.

EXAMPLE 6-5

The solubility product K_{sp} for the silver salt AgX is 4.0 ($\pm$0.4) $\times$ 10^{-8}, and the molar solubility is

$$\text{solubility} = (K_{sp})^{1/2} = (4.0 \times 10^{-8})^{1/2} = 2.0 \times 10^{-4}M$$

What is the uncertainty in the calculated solubility of AgX?

Solution

Substituting y = solubility, $a = K_{sp}$, and x = ½ into Equation 6-13 gives

$$\frac{s_a}{a} = \frac{0.4 \times 10^{-8}}{4.0 \times 10^{-8}}$$

$$\frac{s_y}{y} = \frac{1}{2} \times \frac{0.4}{4.0} = 0.05$$

$$s_y = 2.0 \times 10^{-4} \times 0.05 = 0.1 \times 10^{-4}$$

$$\text{solubility} = 2.0\ (\pm0.1) \times 10^{-4}\ M$$

It is important to note that the error propagation in taking a number to a power is different from the error propagation in multiplication. For example, consider the uncertainty in the square of 4.0 ($\pm$0.2). The relative error in the result (16.0) is given by Equation 6-13:

$$\frac{s_y}{y} = 2\left(\frac{0.2}{4}\right) = 0.1\ \text{or}\ 10\%$$

The result is then $y = 16\ (\pm2)$.

Consider now the situation where y is the product of *two independently measured* numbers that by chance happen to have identical values of $a_1 = 4.0$ ($\pm$0.2) and $a_2 = 4.0$ ($\pm$0.2). The relative error of the product $a_1a_2 = 16.0$ is given by Equation 6-12:

$$\frac{s_y}{y} = \sqrt{\left(\frac{0.2}{4}\right)^2 + \left(\frac{0.2}{4}\right)^2} = 0.07\ \text{or}\ 7\%$$

The result is now $y = 16\ (\pm1)$. The reason for the difference between this and the previous result is that with measurements that are independent of one another, the sign associated with one error can be the same as or different from that of the other error. If they happen to be the same, the error is identical to that encountered in the first case where the signs *must* be the same. On the other hand, if one sign is positive and the other negative, the relative errors tend to cancel. Thus the probable error for the case of independent measurements lies somewhere between the maximum (10%) and zero.

> The relative standard deviation of $y = a^3$ is *not* the same as the relative standard deviation of the product of three independent measurements $y = abc$, where $a = b = c$.

6C-4 Standard Deviations of Logarithms and Antilogarithms

The last two entries in Table 6-4 show that for $y = \log a$

$$s_y = 0.434\frac{s_a}{a} \tag{6-14}$$

and for $y = \text{antilog } a$

$$\frac{s_y}{y} = 2.303s_a \tag{6-15}$$

As shown, the *absolute* standard deviation of the logarithm of a number is determined by the *relative* standard deviation of the number; conversely, the *relative* standard deviation of the antilogarithm of a number is determined by the *absolute* standard deviation of the number. Example 6-6 illustrates these calculations.

EXAMPLE 6-6

Calculate the absolute standard deviations of the results of the following calculations. The absolute standard deviation for each quantity is given in parentheses.

(a) $y = \log[2.00(\pm0.02) \times 10^{-4} = -3.6990 \pm ?$
(b) $y = \text{antilog}[1.200(\pm0.003)] = 15.849 \pm ?$
(c) $y = \text{antilog}[45.4(\pm0.3)] = 2.5119 \times 10^{45} \pm ?$

Solution

(a) Referring to Equation 6-14, we see that we must multiply the *relative* standard deviation by 0.434:

$$s_y = \pm0.434 \times \frac{0.02 \times 10^{-4}}{2.00 \times 10^{-4}} = \pm0.004$$

Thus,

$$y = \log[2.00(\pm0.02) \times 10^{-4}] = -3.699\ (\pm0.004)$$

(b) Applying Equation 6-15, we have

$$\frac{s_y}{y} = 2.303 \times (0.003) = 0.0069$$

$$s_y = 0.0069y = 0.0069 \times 15.849 = 0.11$$

Therefore,

$$y = \text{antilog}[1.200(\pm0.003)] = 15.8 \pm 0.1$$

(c) $\dfrac{s_y}{y} = 2.303 \times (0.3) = 0.69$

$$s_y = 0.69y = 0.69 \times 2.5119 \times 10^{45} = 1.7 \times 10^{45}$$

Thus,

$$y = \text{antilog}[45.4(\pm 0.3)] = 2.5(\pm 1.7) \times 10^{45} = 3\,(\pm 2) \times 10^{45}$$

Example 6-6c demonstrates that a large absolute error is associated with the anti-logarithm of a number with few digits beyond the decimal point. This large uncertainty is due to the fact that the numbers to the left of the decimal (the *characteristic*) serve only to locate the decimal point. The large error in the antilogarithm results from the relatively large uncertainty in the *mantissa* of the number (that is, 0.4 ± 0.3).

6D REPORTING COMPUTED DATA

A numerical result is worthless to users of the data unless they know something about its quality. Therefore, it is always essential to indicate your best estimate of the reliability of your data. One of the best ways of indicating reliability is to give a confidence interval at the 90% or 95% confidence level, as we describe in Section 7A-2. Another method is to report the absolute standard deviation or the coefficient of variation of the data. If one of these is reported, it is a good idea to indicate the number of data points that were used to obtain the standard deviation so that the user has some idea of the reliability of s. A much less satisfactory but more common indicator of the quality of data is the **significant figure convention**.

6D-1 Significant Figures

We often indicate the probable uncertainty associated with an experimental measurement by rounding the result so that it contains only **significant figures**. By definition, the significant figures in a number are all of the certain digits *plus the first uncertain digit*. For example, when you read the 50-mL buret section shown in **Figure 6-5**, you can easily tell that the liquid level is greater than 30.2 mL and less than 30.3 mL. You can also estimate the position of the liquid between the graduations to about 0.02 mL. So, using the significant figure convention, you should report the volume delivered as 30.24 mL, which has four significant figures. Note that the first three digits are certain, and the last digit (4) is uncertain.

A zero may or may not be significant depending on its location in a number. A zero that is surrounded by other digits is always significant (such as in 30.24 mL) because it is read directly and with certainty from a scale or instrument readout. On the other hand, zeros that only locate the decimal point for us are not. If we write 30.24 mL as 0.03024 L, the number of significant figures is the same. The only function of the zero before the 3 is to locate the decimal point, so it is not significant. Terminal, or final, zeros may or may not be significant. For example, if the volume of a beaker is expressed as 2.0 L, the presence of the zero tells us that the volume is known to a few tenths of a liter so that both the 2 and the zero are significant figures. If this same volume is reported as 2000 mL, the situation becomes confusing. The last two zeros are not significant because the uncertainty is still a few tenths of a liter or a few

The **significant figures** in a number are all of the certain digits plus the first uncertain digit.

❮ Rules for determining the number of significant figures:
1. Disregard all initial zeros.
2. Disregard all final zeros *unless they follow a decimal point*.
3. All remaining digits including zeros between nonzero digits are significant.

Figure 6-5 A buret section showing the liquid level and meniscus.

> Express data in scientific notation to avoid confusion in determining whether terminal zeros are significant.

> As a rule of thumb, for addition and subtraction, the result should contain the same number of decimal places as the number with the *smallest* number of decimal places.

> When adding and subtracting numbers in scientific notation, express the numbers to the same power of ten. For example,
>
> $\begin{aligned} 2.432 \times 10^6 = \quad & 2.432 \times 10^6 \\ +6.512 \times 10^4 = +\, & 0.06512 \times 10^6 \\ -1.227 \times 10^5 = -\, & 0.1227 \times 10^6 \\ \hline & 2.37442 \times 10^6 \\ = \;& 2.374 \times 10^6 \text{ (rounded)} \end{aligned}$

> The weak link for multiplication and division is the number of *significant figures* in the number with the smallest number of significant figures. *Use this rule of thumb with caution.*

hundred milliliters. In order to follow the significant figure convention in a case such as this, use scientific notation and report the volume as 2.0×10^3 mL.

6D-2 Significant Figures in Numerical Computations

Determining the appropriate number of significant figures in the result of an arithmetic combination of two or more numbers requires great care.[4]

Sums and Differences

For addition and subtraction, the number of significant figures can be found by visual inspection. For example, in the expression

$$3.4 + 0.020 + 7.31 = 10.730 \text{ (round to 10.7)}$$
$$= 10.7 \text{ (rounded)}$$

the second and third decimal places in the answer cannot be significant because 3.4 is uncertain in the first decimal place. Hence, the result should be rounded to 10.7. We can generalize and say that, for addition and subtraction, the result should have the same number of decimal places as the number with the *smallest* number of decimal places. Note that the result contains three significant digits even though two of the numbers involved have only two significant figures.

Products and Quotients

Sometimes it is suggested for multiplication and division that the answer should be rounded so that it contains the same number of significant digits as the original number with the smallest number of significant digits. Unfortunately, this procedure sometimes leads to incorrect rounding. For example, consider the two calculations

$$\frac{24 \times 4.52}{100.0} = 1.08 \quad \text{and} \quad \frac{24 \times 4.02}{100.0} = 0.965$$

If we follow the suggestion, the first answer would be rounded to 1.1 and the second to 0.96. A better procedure is to assume unit uncertainty in the last digit of each number. For example, in the first quotient, the relative uncertainties associated with each of these numbers are 1/24, 1/452, and 1/1000. Because the first relative uncertainty is much larger than the other two, the relative uncertainty in the result is also 1/24; the absolute uncertainty is then

$$1.08 \times \frac{1}{24} = 0.045 \approx 0.04$$

By the same argument, the absolute uncertainty of the second answer is given by

$$0.965 \times \frac{1}{24} = 0.040 \approx 0.04$$

Therefore, the first result should be rounded to three significant figures or 1.08, but the second should be rounded to only two, that is, 0.96.

---

[4]For an extensive discussion of propagation of significant figures, see L. M. Schwartz, *J. Chem. Educ.,* **1985**, *62*, 693, **DOI**: 10.1021/ed062p693.

Logarithms and Antilogarithms

Be especially careful in rounding the results of calculations involving logarithms. The following rules apply to most situations and are illustrated in Example 6-7:

1. In a logarithm of a number, keep as many digits to the right of the decimal point as there are significant figures in the original number.
2. In an antilogarithm of a number, keep as many digits as there are digits to the right of the decimal point in the original number.[5]

> The number of significant figures in the *mantissa*, or the digits to the right of the decimal point of a logarithm, is the same as the number of significant figures in the original number. Thus, log $(9.57 \times 10^4) = 4.981$. Since 9.57 has 3 significant figures, there are 3 digits to the right of the decimal point in the result.

EXAMPLE 6-7

Round the following answers so that only significant digits are retained:
(a) log $4.000 \times 10^{-5} = -4.3979400$, and (b) antilog $12.5 = 3.162277 \times 10^{12}$

Solution

(a) Following rule 1, we retain 4 digits to the right of the decimal point

$$\log 4.000 \times 10^{-5} = -4.3979$$

(b) Following rule 2, we may retain only 1 digit

$$\text{antilog } 12.5 = 3 \times 10^{12}$$

6D-3 Rounding Data

Always round the computed results of a chemical analysis in an appropriate way. For example, consider the replicate results: 41.60, 41.46, 41.55, and 41.61. The mean of this data set is 41.555, and the standard deviation is 0.069. When we round the mean, do we take 41.55 or 41.56? A good guide to follow when rounding a 5 is always to round to the nearest even number. In this way, we eliminate any tendency to round in a fixed direction. In other words, there is an equal likelihood that the nearest even number will be the higher or the lower in any given situation. Accordingly, we might choose to report the result as 41.56 ± 0.07. If we had reason to doubt the reliability of the estimated standard deviation, we might report the result as 41.6 ± 0.1.

> In rounding a number ending in 5, always round so that the result ends with an even number. Thus, 0.635 rounds to 0.64 and 0.625 rounds to 0.62.

We should note that *it is seldom justifiable to keep more than one significant figure in the standard deviation* because the standard deviation contains error as well. For certain specialized purposes, such as reporting uncertainties in physical constants in research articles, it may be useful to keep two significant figures, and there is certainly nothing wrong with including a second digit in the standard deviation. However, it is important to recognize that the uncertainty usually lies in the first digit.[6]

6D-4 Expressing Results of Chemical Calculations

Two cases are encountered when reporting the results of chemical calculations. If the standard deviations of the values making up the final calculation are known, we then apply the propagation of error methods discussed in Section 6C and round

---

[5]D. E. Jones, *J. Chem. Educ.,* **1971,** *49,* 753, **DOI**: 10.1021/ed049p753.
[6]For more details on this topic, see **http://www.chem.uky.edu/courses/che226/download/ CI_for_sigma.html.**

the results to contain significant digits. However, if we are asked to perform calculations where the precision is indicated only by the significant figure convention, common sense assumptions must be made as to the uncertainty in each number. With these assumptions, the uncertainty of the final result is then estimated using the methods presented in Section 6C. Finally, the result is rounded so that it contains only significant digits.

It is especially important to postpone rounding until the calculation is completed. At least one extra digit beyond the significant digits should be carried through all of the computations in order to avoid a *rounding error*. This extra digit is sometimes called a "guard" digit. Modern calculators generally retain several extra digits that are not significant, and the user must be careful to round final results properly so that only significant figures are included. Example 6-8 illustrates this procedure.

EXAMPLE 6-8

A 3.4842-g sample of a solid mixture containing benzoic acid, C_6H_5COOH (122.123 g/mol), was dissolved and titrated with base to a phenolphthalein end point. The acid consumed 41.36 mL of 0.2328 M NaOH. Calculate the percent benzoic acid (HBz) in the sample.

Solution

As shown in Section 13C-3, the calculation takes the following form:

$$\%HBz = \frac{41.36\,mL \times 0.2328\,\dfrac{mmol\ NaOH}{mL\ NaOH} \times \dfrac{1\ mmol\ HBz}{mmol\ NaOH} \times \dfrac{122.123\,g\ HBz}{1000\ mmol\ HBz}}{3.842\,g\ sample}$$

$$\times\ 100\%$$

$$=\ 33.749\%$$

Since all operations are either multiplication or division, the relative uncertainty of the answer is determined by the relative uncertainties of the experimental data. Let us estimate what these uncertainties are.

1. The position of the liquid level in a buret can be estimated to ± 0.02 mL (Figure 6-5). In reading the buret, two readings (initial and final) must be made so that the standard deviation of the volume s_V will be

$$s_V = \sqrt{(0.02)^2 + (0.02)^2} = 0.028\,mL$$

The relative uncertainty in volume s_V/V is then

$$\frac{s_V}{V} = \frac{0.028}{41.36} \times 1000\,ppt = 0.68\,ppt$$

2. Generally, the absolute uncertainty of a mass obtained with an analytical balance will be on the order of ± 0.0001 g. Thus the relative uncertainty of the denominator s_D/D is

$$\frac{0.0001}{3.4842} \times 1000\,ppt = 0.029\,ppt$$

3. Usually we can assume that the absolute uncertainty in the concentration of a reagent solution is ± 0.0001, and so the relative uncertainty in the concentration of NaOH, s_c/c is

$$\frac{s_c}{c} = \frac{0.0001}{0.2328} \times 1000\,\text{ppt} = 0.43\,\text{ppt}$$

4. The relative uncertainty in the molar mass of HBz is several orders of magnitude smaller than any of the three experimental values and will not be significant. Note, however, that we should retain enough digits in the calculation so that the molar mass is given to at least one more digit (the guard digit) than any of the experimental data. Thus, in the calculation, we use 122.123 for the molar mass (we are carrying two extra digits in this instance).

5. No uncertainty is associated with 100% and the 1000 mmol HBz since these are exact numbers.

Substituting the three relative uncertainties into Equation 6-12, we obtain

$$\frac{s_y}{y} = \sqrt{\left(\frac{0.028}{41.36}\right)^2 + \left(\frac{0.0001}{3.4842}\right)^2 + \left(\frac{0.0001}{0.2328}\right)^2}$$

$$= \sqrt{(0.00068)^2 + (0.000029)^2 + (0.00043)^2} = 8.02 \times 10^{-4}$$

$$s_y = 8.02 \times 10^{-4} \times y = 8.02 \times 10^{-4} \times 33.749 = 0.027$$

Therefore, the uncertainty in the calculated result is 0.03% HBz, and we should report the result as 33.75% HBz, or better 33.75 ($\pm$ 0.03)% HBz.

We must emphasize that rounding decisions are an important part of *every calculation*. These decisions *cannot* be based on the number of digits displayed on an instrument readout, on the computer screen or on a calculator display.

There is no relationship between the number of digits displayed on a computer screen or a calculator and the true number of significant figures.

WEB WORKS

The National Institute of Standards and Technology maintains Web pages of statistical data for testing software. Go to **www.cengage.com/chemistry/skoog/fac9**, choose Chapter 6, and go to the Web Works. Here you will find a link to the NIST Statistical Reference Datasets site. Browse the site to see what kinds of data are available for testing. We use two of the NIST data sets in problems 6-22 and 6-23 at the end of this chapter. Under Databases, Scientific, choose Standard Reference Data. Find the Analytical Chemistry databases. Enter the NIST Chemistry WebBook site. Find the gas chromatographic retention index data for chlorobenzene. Find the four values for the retention index (I) of chlorobenzene on an SE-30 capillary column at a temperature of 160 °C. Determine the mean retention index and its standard deviation at this temperature.

QUESTIONS AND PROBLEMS

6-1. Define
 *(a) standard error of the mean.
 (b) coefficient of variation.
 *(c) variance.
 (d) significant figures.

6-2. Differentiate between
 *(a) parameter and statistic.
 (b) population mean and sample mean.
 *(c) random and systematic error.
 (d) accuracy and precision.

6-3. Distinguish between
*(a) the sample standard deviation and the population standard deviation.
(b) the meaning of the word "sample" as it is used in a chemical and in a statistical sense.

6-4. What is the standard error of a mean? Why is the standard deviation of the mean lower than the standard deviation of the data points in a set?

*6-5. From the Gaussian (normal) error curve, what is the probability that a result from a population lies between 0 and $+1\sigma$ of the mean? What is the probability of a result occurring that is between $+1\sigma$ and $+2\sigma$ of the mean?

6-6. From the normal curve of error, find the probability that a result is outside the limits of $\pm 2\sigma$ from the mean. What is the probability that a result has a more negative deviation from the mean than -2σ?

6-7. Consider the following sets of replicate measurements:

*A	B	*C	D	*E	F
9.5	55.35	0.612	5.7	20.63	0.972
8.5	55.32	0.592	4.2	20.65	0.943
9.1	55.20	0.694	5.6	20.64	0.986
9.3		0.700	4.8	20.51	0.937
9.1			5.0		0.954

For each set, calculate the (a) mean; (b) median; (c) spread, or range; (d) standard deviation; and (e) coefficient of variation.

6-8. The accepted values for the sets of data in Problem 6-7 are: *set A, 9.0; set B, 55.33; *set C, 0.630; set D, 5.4; *set E, 20.58; set F, 0.965. For the mean of each set, calculate (a) the absolute error and (b) the relative error in parts per thousand.

6-9. Estimate the absolute deviation and the coefficient of variation for the results of the following calculations. Round each result so that it contains only significant digits. The numbers in parentheses are absolute standard deviations.
*(a) $y = 3.95(\pm 0.03) + 0.993(\pm 0.001)$
$-7.025(\pm 0.001) = -2.082$
(b) $y = 15.57(\pm 0.04) + 0.0037(\pm 0.0001)$
$+ 3.59(\pm 0.08) = 19.1637$
*(c) $y = 29.2(\pm 0.3) \times 2.034(\pm 0.02) \times 10^{-17}$
$= 5.93928 \times 10^{-16}$
(d) $y = 326(\pm 1) \times \dfrac{740(\pm 2)}{1.964(\pm 0.006)}$
$= 122,830.9572$
*(e) $y = \dfrac{187(\pm 6) - 89(\pm 3)}{1240(\pm 1) + 57(\pm 8)} = 7.5559 \times 10^{-2}$
(f) $y = \dfrac{3.56(\pm 0.01)}{522(\pm 3)} = 6.81992 \times 10^{-3}$

6-10. Estimate the absolute standard deviation and the coefficient of variation for the results of the following calculations. Round each result to include only significant

figures. The numbers in parentheses are absolute standard deviations.
*(a) $y = 1.02(\pm 0.02) \times 10^{-8} - 3.54(\pm 0.2) \times 10^{-9}$
(b) $y = 90.31(\pm 0.08) - 89.32(\pm 0.06)$
$+ 0.200(\pm 0.004)$
*(c) $y = 0.0040(\pm 0.0005) \times 10.28(\pm 0.02)$
$\times 347(\pm 1)$
(d) $y = \dfrac{223(\pm 0.03) \times 10^{-14}}{1.47(\pm 0.04) \times 10^{-16}}$
*(e) $y = \dfrac{100(\pm 1)}{2(\pm 1)}$
(f) $y =$
$\dfrac{1.49(\pm 0.02) \times 10^{-2} - 4.97(\pm 0.06) \times 10^{-3}}{27.1(\pm 0.7) + 8.99(\pm 0.08)}$

6-11. Calculate the absolute standard deviation and the co-efficient of variation for the results of the following calculations. Round each result to include only significant figures. The numbers in parentheses are absolute standard deviations.
*(a) $y = \log[2.00(\pm 0.03) \times 10^{-4}]$
(b) $y = \log[4.42(\pm 0.01) \times 10^{37}]$
*(c) $y = \text{antilog}[1.200(\pm 0.003)]$
(d) $y = \text{antilog}[49.54(\pm 0.04)]$

6-12. Calculate the absolute standard deviation and the coefficient of variation for the results of the following calculations. Round each result to include only significant figures. The numbers in parentheses are absolute standard deviations.
*(a) $y = [4.17(\pm 0.03) \times 10^{-4}]^3$
(b) $y = [2.936(\pm 0.002)]^{1/4}$

*6-13. The standard deviation in measuring the diameter d of a sphere is ± 0.02 cm. What is the standard deviation in the calculated volume V of the sphere if $d = 2.15$ cm?

6-14. The inside diameter of an open cylindrical tank was measured. The results of four replicate measurements were 5.2, 5.7, 5.3, and 5.5 m. Measurements of the height of the tank yielded 7.9, 7.8, and 7.6 m. Calculate the volume in liters of the tank and the standard deviation of the result.

*6-15. In a volumetric determination of an analyte A, the data obtained and their standard deviations are as follows:

Initial buret reading	0.19 mL	0.02 mL
Final buret reading	9.26 mL	0.03 mL
Sample mass	45.0 mg	0.2 mg

From the data, find the coefficient of variation of the final result for the % A that is obtained by using the equation that follows and assuming there is no uncertainty in the equivalent mass.

% A = titrant volume × equivalent mass
× 100%/sample mass

6-16. In Chapter 28 we discuss inductively coupled plasma atomic emission spectrometry. In that method, the number of atoms excited to a particular energy level is a strong function of temperature. For an element of

TABLE 7-3

Values of *t* for Various Levels of Probability

Degrees of Freedom	80%	90%	95%	99%	99.9%
1	3.08	6.31	12.7	63.7	637
2	1.89	2.92	4.30	9.92	31.6
3	1.64	2.35	3.18	5.84	12.9
4	1.53	2.13	2.78	4.60	8.61
5	1.48	2.02	2.57	4.03	6.87
6	1.44	1.94	2.45	3.71	5.96
7	1.42	1.90	2.36	3.50	5.41
8	1.40	1.86	2.31	3.36	5.04
9	1.38	1.83	2.26	3.25	4.78
10	1.37	1.81	2.23	3.17	4.59
15	1.34	1.75	2.13	2.95	4.07
20	1.32	1.73	2.09	2.84	3.85
40	1.30	1.68	2.02	2.70	3.55
60	1.30	1.67	2.00	2.62	3.46
∞	1.28	1.64	1.96	2.58	3.29

For the mean of N measurements,

$$t = \frac{\bar{x} - \mu}{s/\sqrt{N}} \tag{7-4}$$

Like z in Equation 7-1, t depends on the desired confidence level. However, t also depends on the number of degrees of freedom in the calculation of s. **Table 7-3** gives t values for a few degrees of freedom. More extensive tables are found in various mathematical and statistical handbooks. Note that t approaches z as the number of degrees of freedom becomes large.

The confidence interval for the mean $\bar{x}$ of N replicate measurements can be calculated from t by Equation 7-5, which is similar to Equation 7-2 using z:

$$\text{CI for } \mu = \bar{x} \pm \frac{ts}{\sqrt{N}} \tag{7-5}$$

The use of the t statistic for confidence intervals is illustrated in Example 7-3.

FEATURE 7-1

W. S. Gossett ("Student")

William Gossett, was born in England in 1876. He attended New College Oxford where he obtained first-class degrees in both chemistry and mathematics. After graduation in 1899, Gossett secured a position at the Guinness brewery in Dublin, Ireland. In 1906, he spent time at University College in London studying with the statistician Karl Pearson, who was famous for his work on the correlation coefficient. While at University College, Gossett studied the limits of the Poisson and binomial distributions, the sampling distribution of the mean and standard deviation, and several other topics. When he returned to the brewery, he began his classic studies on the statistics of small samples of data while working in quality control. Because Guinness did not allow employees to publish their work, Gossett began to publish his results under the name "Student." His most important work on the t test was developed to determine

(continued)

how closely the yeast and alcohol content of various batches of Guinness matched the standard amounts established by the brewery. He discovered the *t* distribution through mathematical and empirical studies with random numbers. The classic paper on the *t* test was published under the pseudonym Student in *Biometrika*, **1908**, *6*, 1. The *t* statistic is now often called **Student's *t*.** Gossett's work is a testimony to the interplay of practical science (quality control of beer) and theoretical research (statistics of small samples).

W. S. Gossett ("Student")

Originally Published in the Journal of Eugenics / Wikimedia Foundation

---

EXAMPLE 7-3

A clinical chemist obtained the following data for the alcohol content of a sample of blood: % C_2H_5OH: 0.084, 0.089, and 0.079. Calculate the 95% confidence interval for the mean assuming that (a) the three results obtained are the only indication of the precision of the method and that (b), from previous experience on hundreds of samples, we know that the standard deviation of the method $s = 0.005\%$ C_2H_5OH and is a good estimate of σ.

Solution

(a)　$\sum x_i = 0.084 + 0.089 + 0.079 = 0.252$

　　$\sum x_i^2 = 0.007056 + 0.007921 + 0.006241 = 0.021218$

　　$s = \sqrt{\dfrac{0.021218 - (0.252)^2/3}{3 - 1}} = 0.0050\%$ C_2H_5OH

In this instance, $\bar{x} = 0.252/4 = 0.084$. Table 7-3 indicates that $t = 4.30$ for two degrees of freedom and the 95% confidence level. Thus, using Equation 7-5,

　　$95\% \text{ CI} = \bar{x} \pm \dfrac{ts}{\sqrt{N}} = 0.084 \pm \dfrac{4.30 \times 0.0050}{\sqrt{3}}$

　　　　　　　$= 0.084 \pm 0.012\%$ C_2H_5OH

(b)　Because $s = 0.0050\%$ is a good estimate of σ, we can use z and Equation 7-2

　　$95\% \text{ CI} = \bar{x} \pm \dfrac{z\sigma}{\sqrt{N}} = 0.094 \pm \dfrac{1.96 \times 0.0050}{\sqrt{3}}$

　　　　　　　$= 0.084 \pm 0.006\%$ C_2H_5OH

Note that a sure knowledge of σ decreases the confidence interval by a significant amount even though s and σ are identical.

7B STATISTICAL AIDS TO HYPOTHESIS TESTING

Hypothesis testing is the basis for many decisions made in science and engineering. To explain an observation, a hypothetical model is advanced and tested experimentally to determine its validity. The hypothesis tests that we describe are used to determine if the results from these experiments support the model. If they do not support our model, we reject the hypothesis and seek a new one. If agreement is found, the hypothetical model serves as the basis for further experiments. When the hypothesis is supported by sufficient experimental data, it becomes recognized as a useful theory until such time as data are obtained that refute it.

Experimental results seldom agree *exactly* with those predicted from a theoretical model. As a result, scientists and engineers frequently must judge whether a numerical difference is a result of a real difference (a systematic error) or a consequence of the random errors inevitable in all measurements. Statistical tests are useful in sharpening these judgments.

Tests of this kind use a **null hypothesis**, which assumes that the numerical quantities being compared are, in fact, the same. We then use a probability distribution to calculate the probability that the observed differences are a result of random error. Usually, if the observed difference is greater than or equal to the difference that would occur 5 times in 100 by random chance (a significance level of 0.05), the null hypothesis is considered questionable, and the difference is judged to be significant. Other significance levels, such as 0.01 (1%) or 0.001 (0.1%), may also be adopted, depending on the certainty desired in the judgment. When expressed as a fraction, the significance level is often given the symbol α. The confidence level, CL, as a percentage is related to α by $CL = (1 - \alpha) \times 100\%$.

> A **null hypothesis** postulates that two or more observed quantities are the same.

Specific examples of hypothesis tests that scientists often use include the comparison of (1) the mean of an experimental data set with what is believed to be the true value, (2) the mean to a predicted or cutoff (threshold) value, and (3) the means or the standard deviations from two or more sets of data. The sections that follow consider some of the methods for making these comparisons. Section 7C treats comparisons among more than two means (analysis of variance).

7B-1 Comparing an Experimental Mean with a Known Value

There are many cases in which a scientist or engineer needs to compare the mean of a data set with a known value. In some cases, the known value is the true or accepted value based on prior knowledge or experience. An example is in comparing measured values of cholesterol to the value certified by NIST in a standard reference serum sample. In other situations, the known value might be a value predicted from theory or it might be a threshold value that we use in making decisions about the presence or absence of some constituent. An example of a decision-making value would be in comparing the measured mercury level in a bluefin tuna sample to the threshold toxicity level. In all these cases, we use a statistical **hypothesis test** to draw conclusions about the population mean μ and its nearness to the known value, which we call μ_0.

There are two contradictory outcomes that we consider in any hypothesis test. The first, the null hypothesis H_0, states that $\mu = \mu_0$. The second, the alternative hypothesis H_a can be stated in several ways. We might reject the null hypothesis in favor of H_a if μ is different than μ_0 ($\mu \neq \mu_0$). Other alternative hypotheses are $\mu > \mu_0$ or $\mu < \mu_0$. As a first example, suppose we are interested in determining whether the concentration of lead in an industrial wastewater discharge exceeds the maximum permissible amount of 0.05 ppm. Our hypothesis test would be summarized as:

$$H_0: \mu = 0.05 \text{ ppm}$$

$$H_a: \mu > 0.05 \text{ ppm}$$

As a different example, suppose instead that experiments over a several-years period have determined that the mean lead level is 0.02 ppm. Recently, changes in the industrial process have been made, and we suspect that the mean lead level is now different than 0.02 ppm. In this instance, we do not care whether it is higher or lower than 0.02 ppm. Our hypothesis test would be summarized:

$$H_0: \mu = 0.02 \text{ ppm}$$

$$H_a: \mu \neq 0.02 \text{ ppm}$$

In order to apply the statistical test, a test procedure must be implemented. The crucial elements of a test procedure are the formation of an appropriate test statistic and the identification of a rejection region. The test statistic is formulated from the data on which we will base the decision to accept or reject H_0. The rejection region consists of all the values of the test statistic for which H_0 will be rejected. The null hypothesis is rejected if the test statistic lies within the rejection region. For tests concerning one or two means, the test statistic might be the z statistic if we have a large number of measurements or if we know σ. Quite often, however, we use the t statistic for small numbers of measurements with unknown σ. When in doubt, the t statistic should be used.

Large Sample z Test

If a large number of results are available so that s is a good estimate of σ, the z test is appropriate. The procedure that is used is summarized below:

1. State the null hypothesis: $H_0: \mu = \mu_0$

2. Form the test statistic: $z = \dfrac{\bar{x} - \mu_0}{\sigma/\sqrt{N}}$

3. State the alternative hypothesis H_a and determine the rejection region

   For $H_a: \mu \neq \mu_0$, reject H_0 if $z \geq z_{\text{crit}}$ or if $z \leq -z_{\text{crit}}$ (two-tailed test)
   For $H_a: \mu > \mu_0$, reject H_0 if $z \geq z_{\text{crit}}$ (one-tailed test)
   For $H_a: \mu < \mu_0$, reject H_0 if $z \leq -z_{\text{crit}}$ (one-tailed test)

The rejection regions are illustrated in **Figure 7-2** for the 95% confidence level. Note that for $H_a: \mu \neq \mu_0$, we can reject for either a positive value of z or for a negative value of z that exceeds the critical value. This is called a **two-tailed** test since rejection can occur for results in either tail of the distribution. For the 95% confidence level, the probability that z exceeds z_{crit} is 0.025 in each tail or 0.05 total. Hence, there is only a 5% probability that random error will lead to a value of $z \geq z_{\text{crit}}$ or $z \leq -z_{\text{crit}}$. The significance level overall is $\alpha = 0.05$. From Table 7-1, the critical value of z is 1.96 for this case.

If instead our alternative hypothesis is $H_a: \mu > \mu_0$, the test is said to be a **one-tailed test**. In this case, we can reject only when $z \geq z_{\text{crit}}$. Now, for the 95% confidence level, we want the probability that z exceeds z_{crit} to be 5% or the total

(a)

(b)

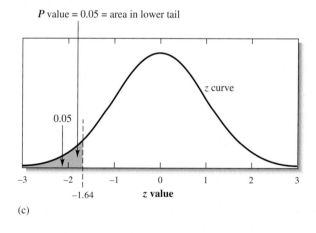

(c)

Figure 7-2 Rejection regions for the 95% confidence level. (a) Two-tailed test for $H_a: \mu \neq \mu_0$. Note the critical value of z is 1.96 as in Figure 7-1. (b) One-tailed test for $H_a: \mu > \mu_0$. The critical value of z is 1.64 so that 95% of the area is to the left of z_{crit} and 5% of the area is to the right. (c) One-tailed test for $H_a: \mu < \mu_0$. The critical value is again 1.64 so that 5% of the area lies to the left of $-z_{crit}$.

probability in both tails to be 10%. The overall significance level would be $\alpha = 0.10$, and the critical value from Table 7-1 is 1.64. Similarly, if the alternative hypothesis is $\mu < \mu_0$, we can reject only when $z \leq -z_{crit}$. The critical value of z is again 1.64 for this one-tailed test.

Example 7-4 illustrates the use of the z-test to determine whether the mean of 35 values agrees with a theoretical value.

EXAMPLE 7-4

A class of 30 students determined the activation energy of a chemical reaction to be 116 kJ mol^{-1} (mean value) with a standard deviation of 22 kJ mol^{-1}. Are the data in agreement with the literature value of 129 kJ mol^{-1} at (a) the 95% confidence level and (b) the 99% confidence level? Estimate the probability of obtaining a mean equal to the student value.

Solution

We have enough values that s should be a good estimate of σ. Accordingly, μ_0 is the literature value of 129 kJ mol^{-1} so that the null hypothesis is $\mu = 129$ kJ mol^{-1}. The alternative hypothesis is that $\mu \neq 129$ kJ mol^{-1}. This is thus a two-tailed test. From Table 7-1, $z_{crit} = 1.96$ for the 95% confidence level, and $z_{crit} = 2.58$ for the 99% confidence level. The test statistic is calculated as

$$z = \frac{\bar{x} - \mu_0}{\sigma/\sqrt{N}} = \frac{116 - 129}{22/\sqrt{30}} = -3.27$$

Since $z \leq -1.96$, we reject the null hypothesis at the 95% confidence level. Note that, since $z \leq -2.58$, we also reject H_0 at the 99% confidence level. In order to estimate the probability of obtaining a mean value $\mu = 116$ kJ mol^{-1}, we must find the probability of obtaining a z value of 3.27. From Table 7-1, the probability of obtaining a z value this large because of random error is only about 0.2%. All of these results lead us to conclude that the student mean is actually different from the literature value and not just the result of random error.

Small Sample t *Test*

For a small number of results, we use a similar procedure to the z test except that the test statistic is the t statistic. Again we test the null hypothesis H_0: $\mu = \mu_0$, where μ_0 is a specific value of μ such as an accepted value, a theoretical value, or a threshold value. The procedure is:

1. State the null hypothesis: H_0: $\mu = \mu_0$
2. Form the test statistic: $t = \dfrac{\bar{x} - \mu_0}{s/\sqrt{N}}$
3. State the alternative hypothesis H_a and determine the rejection region
   For H_a: $\mu \neq \mu_0$, reject H_0 if $t \geq t_{crit}$ or if $t \leq -t_{crit}$ (two-tailed test)
   For H_a: $\mu > \mu_0$, reject H_0 if $t \geq t_{crit}$ (one-tailed test)
   For H_a: $\mu < \mu_0$, reject H_0 if $t \leq -t_{crit}$ (one-tailed test)

As an illustration, consider the testing for systematic error in an analytical method. In this case, a sample of accurately known composition, such as a standard reference material, is analyzed. Determination of the analyte in the material gives an experimental mean that is an estimate of the population mean. If the analytical method had no systematic error, or bias, random errors would give the frequency distribution shown by curve A in **Figure 7-3**. Method B has some systematic error so that $\bar{x}_B$, which estimates μ_B, differs from the accepted value μ_0. The bias is given by

$$\text{Bias} = \mu_B - \mu_0 \tag{7-6}$$

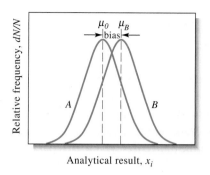

Relative frequency, dN/N

μ_0 μ_B

bias

A   B

Analytical result, x_i

Figure 7-3 Illustration of systematic error in an analytical method. Curve A is the frequency distribution for the accepted value by a method without bias. Curve B illustrates the frequency distribution of results by a method that could have a significant bias due to a systematic error.

In testing for bias, we do not know initially whether the difference between the experimental mean and the accepted value is due to random error or to an actual systematic error. The t test is used to determine the significance of the difference. Example 7-5 illustrates the use of the t test to determine whether there is bias in a method.

EXAMPLE 7-5

A new procedure for the rapid determination of sulfur in kerosenes was tested on a sample known from its method of preparation to contain 0.123% S ($\mu_0 = 0.123\%$ S). The results for % S were 0.112, 0.118, 0.115, and 0.119. Do the data indicate that there is a bias in the method at the 95% confidence level?

Solution

The null hypothesis is H_0: $\mu = 0.123\%$ S, and the alternative hypothesis is H_a: $\mu \neq 0.123\%$ S.

$$\sum x_i = 0.112 + 0.118 + 0.115 + 0.119 = 0.464$$

$$\bar{x} = 0.464/4 = 0.116\% \text{ S}$$

$$\sum x_i^2 = 0.012544 + 0.013924 + 0.013225 + 0.014161 = 0.53854$$

$$s = \sqrt{\frac{0.053854 - (0.464)^2/4}{4 - 1}} = \sqrt{\frac{0.000030}{3}} = 0.0032\% \text{ S}$$

The test statistic can now be calculated as

$$t = \frac{\bar{x} - \mu_0}{s/\sqrt{N}} = \frac{0.116 - 0.123}{0.032/\sqrt{4}} = -4.375$$

> The probability of a difference this large occurring because of only random errors can be obtained from the Excel function T.DIST.2T (x, deg_freedom) [TDIST(x,deg_freedom,tails) in Excel 2007], where x is the test value of t(4.375), deg_freedom is 3 for our case, and tails = 2 (Excel 2007). The result is T.DIST.2T (4.375,3) = 0.022. Hence, it is only 2.2% probable to get a value this large because of random errors. The critical value of t for a given confidence level can be obtained in Excel from T.INV.2T (probability,deg_freedom) [TINV(probability,deg_freedom) in Excel 2007]. In our case, T.INV.2T(0.05,3) = 3.1825.

From Table 7-3, we find that the critical value of t for 3 degrees of freedom and the 95% confidence level is 3.18. Since $t \leq -3.18$, we conclude that there is a significant difference at the 95% confidence level and thus bias in the method. Note that, if we were to do this test at the 99% confidence level, $t_{\text{crit}} = 5.84$ (Table 7-3). Since $t = -4.375$ is greater than -5.84, we would accept the null hypothesis at the 99% confidence level and conclude there is no difference between the experimental and the accepted values. Note in this case that the outcome depends on the confidence level that is used. As we will see, choice of the confidence level depends on our willingness to accept an error in the outcome. The significance level (0.05 or 0.01) is the probability of making an error by rejecting the null hypothesis (see Section 7B-3).

If it were confirmed by further experiments that the method always gave low results, we would say that the method had a **negative bias**.

7B-2 Comparison of Two Experimental Means

Frequently scientists must judge whether a difference in the means of two sets of data is real or the result of random error. In some cases, the results of chemical analyses are used to determine whether two materials are identical. In other cases, the results are used to determine whether two analytical methods give the same values or whether two analysts using the same methods obtain the same means. An extension of these procedures can be used to analyze paired data. Data are often collected in pairs to eliminate one source of variability by focusing on the differences within each pair.

The t Test for Differences in Means

We can test for differences in means with the z test, modified to take into account a comparison of two sets of data, if we have large numbers of measurements in both data sets. More often, both sets contain only a few results, and we must use the t test. To illustrate, let us assume that N_1 replicate analyses by analyst 1 yielded a mean value of $\bar{x}_1$ and that N_2 analyses by analyst 2 obtained by the same method gave $\bar{x}_2$. The null hypothesis states that the two means are identical and that any difference is the result of random errors. Thus, we can write H_0: $\mu_1 = \mu_2$. Most often when testing differences in means, the alternative hypothesis is H_a: $\mu_1 \neq \mu_2$, and the test is a two-tailed test. However, in some situations, we could test H_a: $\mu_1 > \mu_2$ or H_a: $\mu_1 < \mu_2$ and use a one-tailed test. We will assume that a two-tailed test is used.

If the data were collected in the same manner and the analysts were both careful, it is often safe to assume that the standard deviations of both data sets are similar. Thus, both s_1 and s_2 are estimates of the population standard deviation σ. To get a better estimate of σ than given by s_1 or s_2 alone, we use the pooled standard deviation (see Section 6B-4). From Equation 6-6, the standard deviation of the mean of analyst 1 is given by $s_{m1} = \dfrac{s_1}{\sqrt{N_1}}$. The variance of the mean of analyst 1 is

$$s_{m1}^2 = \frac{s_1^2}{N_1}$$

Likewise, the variance of the mean of analyst 2 is

$$s_{m2}^2 = \frac{s_2^2}{N_2}$$

In the t test, we are interested in the difference between the means or $\bar{x}_1 - \bar{x}_2$. The variance of the difference s_d^2 between the means is given by

$$s_d^2 = s_{m1}^2 + s_{m2}^2$$

The standard deviation of the difference between the means is found by taking the square root after substituting the values of s_{m1}^2 and s_{m2}^2 from above.

$$\frac{s_d}{\sqrt{N}} = \sqrt{\frac{s_1^2}{N_1} + \frac{s_2^2}{N_2}}$$

Now, if we make the further assumption that the pooled standard deviation s_{pooled} is a better estimate of σ than s_1 or s_2, we can write

$$\frac{s_d}{\sqrt{N}} = \sqrt{\frac{s_{pooled}^2}{N_1} + \frac{s_{pooled}^2}{N_2}} = s_{pooled}\sqrt{\frac{N_1 + N_2}{N_1 N_2}}$$

The test statistic t is now found from

$$t = \frac{\bar{x}_1 - \bar{x}_2}{s_{pooled}\sqrt{\dfrac{N_1 + N_2}{N_1 N_2}}} \qquad (7\text{-}7)$$

The test statistic is then compared with the critical value of t obtained from the table for the particular confidence level desired. The number of degrees of freedom for finding the critical value of t in Table 7-3 is $N_1 + N_2 - 2$. If the absolute value of the test statistic is less than the critical value, the null hypothesis is accepted, and no significant difference between the means has been demonstrated. A test value of t greater than the critical value indicates a significant difference between the means. Example 7-6 illustrates the use of the t test to determine if two barrels of wine came from the same sources.

EXAMPLE 7-6

In a forensic investigation, a glass containing red wine and an open bottle were analyzed for their alcohol content in order to determine whether the wine in the glass came from the bottle. On the basis of six analyses, the average content of the wine from the glass was established to be 12.61% ethanol. Four analyses of the wine from the bottle gave a mean of 12.53% alcohol. The 10 analyses yielded a pooled standard deviation $s_{pooled} = 0.070\%$. Do the data indicate a difference between the wines?

Solution

The null hypothesis is H_0: $\mu_1 = \mu_2$, and the alternative hypothesis is H_a: $\mu_1 \neq \mu_2$. We use Equation 7-7 to calculate the test statistic t.

$$t = \frac{\bar{x}_1 - \bar{x}_2}{s_{pooled}\sqrt{\dfrac{N_1 + N_2}{N_1 N_2}}} = \frac{12.61 - 12.53}{0.07\sqrt{\dfrac{6 + 4}{6 \times 4}}} = 1.771$$

The critical value of t at the 95% confidence level for $10 - 2 = 8$ degrees of freedom is 2.31. Since $1.771 < 2.31$, we accept the null hypothesis at the 95% confidence level and conclude that there is no difference in the alcohol content of the wines. The probability of getting a t value of 1.771 can be calculated using the Excel function T.DIST.2T() and is T.DIST.2T(1.771,8) = 0.11. Thus, there is more than a 10% chance that we could get a value this large due to random error.

In Example 7-6, no significant difference in the alcohol content of the two wines was detected at the 95% probability level. This statement is equivalent to saying that μ_1 is equal to μ_2 with a certain degree of confidence. However, the tests do not prove that the wine in the glass came from the same bottle. Indeed, it is conceivable that one wine is a merlot and the other is a cabernet sauvignon. To establish with a reasonable probability that the two wines are identical would require extensive testing of other characteristics, such as taste, color, odor, and refractive index, as well as tartaric acid, sugar, and trace element content. If no significant differences are revealed by all these tests and by others, then it might be possible to judge the glass of wine as originating in the open bottle. In contrast, the finding of *one* significant difference in any test would unmistakably show that the two wines are different. Thus, the establishment of a significant difference by a single test is much more revealing than finding that there is no significant difference in a single characteristic.

If there is good reason to believe that the standard deviations of the two data sets differ, the **two-sample *t* test** must be used.[1] However, the significance level for this *t* test is only approximate, and the number of degrees of freedom is more difficult to calculate.

 Spreadsheet Summary   In the first exercise in Chapter 3 of *Applications of Microsoft® Excel in Analytical Chemistry*, 2nd ed., we use Excel to perform the *t* test for comparing two means assuming equal variances of the two data sets. We first manually calculate the value of *t* and compare it to the critical value obtained from Excel's function T.INV.2T(). We obtain the probability from Excel's T.DIST.2T() function. Then, we use Excel's built-in function T.TEST() for the same test. Finally, we use Excel's Analysis ToolPak to automate the *t* test with equal variances.

Paired Data

Scientists and engineers often make use of pairs of measurements on the same sample in order to minimize sources of variability that are not of interest. For example, two methods for determining glucose in blood serum are to be compared. Method A could be performed on samples from five randomly chosen patients and Method B could be performed on samples from five different patients. However, there would be variability because of the different glucose levels of each patient. A better way to compare the methods would be to use both methods on the same samples and focus on the differences.

The paired *t* test uses the same type of procedure as the normal *t* test except that we analyze pairs of data and compute the differences, d_i. The standard deviation is now the standard deviation of the mean difference. Our null hypothesis is $H_0: \mu_d = \Delta_0$, where Δ_0 is a specific value of the difference to be tested, often zero. The test statistic value is

$$t = \frac{\bar{d} - \Delta_0}{s_d/\sqrt{N}}$$

where $\bar{d}$ is the average difference $= \sum d_i/N$. The alternative hypothesis could be $\mu_d \neq \Delta_0$, $\mu_d > \Delta_0$, or $\mu_d < \Delta_0$. An illustration is given in Example 7-7.

---

[1]J. L. Devore, *Probability and Statistics for Engineering and the Sciences*, 8th ed. Boston: Brooks/Cole, 2012, pp. 357–361.

EXAMPLE 7-7

A new automated procedure for determining glucose in serum (Method A) is to be compared to the established method (Method B). Both methods are performed on serum from the same six patients in order to eliminate patient-to-patient variability. Do the following results confirm a difference in the two methods at the 95% confidence level?

	Patient 1	Patient 2	Patient 3	Patient 4	Patient 5	Patient 6
Method A glucose, mg/L	1044	720	845	800	957	650
Method B glucose, mg/L	1028	711	820	795	935	639
Difference, mg/L	16	9	25	5	22	11

Solution

Let us now test the appropriate hypotheses. If μ_d is the true average difference between the methods, we want to test the null hypothesis $H_0: \mu_d = 0$ and the alternative hypothesis, $H_a: \mu_d \neq 0$. The test statistic is

$$t = \frac{\bar{d} - 0}{s_d/\sqrt{N}}$$

From the table, $N = 6$, $\sum d_i = 16 + 9 + 25 + 5 + 22 + 11 = 88$, $\sum d_i^2 = 1592$, and $\bar{d} = 88/6 = 14.67$. The standard deviation of the difference s_d is given by Equation 6-5

$$s_d = \sqrt{\frac{1592 - \frac{(88)^2}{6}}{6 - 1}} = 7.76$$

and the t statistic is

$$t = \frac{14.67}{7.76/\sqrt{6}} = 4.628$$

From Table 7-3, the critical value of t is 2.57 for the 95% confidence level and 5 degrees of freedom. Since $t > t_{crit}$, we reject the null hypothesis and conclude that the two methods give different results.

Note that, if we merely average the results of Method A ($\bar{x}_A = 836.0$ mg/L) and the results of Method B ($\bar{x}_B = 821.3$ mg/L), the large patient-to-patient variation in glucose level gives us large values for s_A (146.5) and s_B (142.7). A comparison of means gives us a test t value of 0.176, and we would accept the null hypothesis. Hence, the large patient-to-patient variability masks the method differences that are of interest. Pairing allows us to focus on the differences.

Structural formula of glucose, $C_6H_{12}O_6$.

Molecular model of glucose.

 Spreadsheet Summary In Chapter 3 of *Applications of Microsoft® Excel in Analytical Chemistry*, 2nd ed., we use Excel's Analysis ToolPak to perform the paired t test on the data of Example 7-7. We compare the results obtained to those found without pairing.

7B-3 Errors in Hypothesis Testing

The choice of a rejection region for the null hypothesis is made so that we can readily understand the errors involved. At the 95% confidence level, for example, there is a 5% chance that we will reject the null hypothesis even though it is true. This could happen if an unusual result occurred that put our test statistic z or t into the rejection region. The error that results from rejecting H_0 when it is true is called a **type I error**. The significance level α gives the frequency of rejecting H_0 when it is true.

The other type of error is that we accept H_0 when it is false. This is termed a **type II error**. The probability of a type II error is given the symbol β. No test procedure can guarantee that we will not commit one error or the other. The error probabilities are the result of using a sample of data to make inferences about the population. At first thought, making α smaller (0.01 instead of 0.05) would appear to make sense in order to minimize the type I error rate. However, decreasing the type I error rate increases the type II error rate because they are inversely related to each other.

It is important when thinking about errors in hypothesis testing to determine the consequences of making a type I or a type II error. If a type I error is much more likely to have serious consequences than a type II error, it is reasonable to choose a small value of α. On the other hand, in some situations a type II error would be quite serious, and so a larger value of α is used to keep the type II error rate under control. As a general rule of thumb, the largest α that is tolerable for the situation should be used. This ensures the smallest type II error while keeping the type I error within acceptable limits. For many cases in analytical chemistry, an α value of 0.05 (95% confidence level) provides an acceptable compromise.

> A **type I error** occurs when H_0 is rejected although it is actually true. In some sciences, a type I error is called a **false negative**. A **type II error** occurs when H_0 is accepted and it is actually false. This is sometimes termed a **false positive**.

> The consequences of making errors in hypothesis testing are often compared to the errors made in judicial procedures. Therefore, convicting an innocent person is usually considered a more serious error than setting a guilty person free. If we make it less likely that an innocent person gets convicted, we make it more likely that a guilty person goes free.

7B-4 Comparison of Variances

At times, there is a need to compare the variances (or standard deviations) of two data sets. For example, the normal t test requires that the standard deviations of the data sets being compared are equal. A simple statistical test, called the F test, can be used to test this assumption under the provision that the populations follow the normal (Gaussian) distribution. The F test is also used in comparing more than two means (see Section 7C) and in linear regression analysis (see Section 8D-2).

The F test is based on the null hypothesis that the two population variances under consideration are equal, $H_0: \sigma_1^2 = \sigma_2^2$. The test statistic F, which is defined as the ratio of the two sample variances ($F = s_1^2 / s_2^2$), is calculated and compared with the critical value of F at the desired significance level. The null hypothesis is rejected if the test statistic differs too much from unity.

Critical values of F at the 0.05 significance level are shown in **Table 7-4**. Note that two degrees of freedom are given, one associated with the numerator and the other with the denominator. Most mathematical handbooks give more extensive tables of F values at various significance levels.

The F test can be used in either a one-tailed mode or in a two-tailed mode. For a one-tailed test we test the alternative hypothesis that one variance is greater than the other. Hence, the variance of the supposedly more precise procedure is placed in the denominator and that of the less precise procedure is placed in the numerator. The alternative hypothesis is $H_a: \sigma_1^2 > \sigma_2^2$. The critical values of F for the 95% confidence level are given in Table 7-4. For a two-tailed test, we are testing whether the variances are different, $H_a: \sigma_1^2 \neq \sigma_2^2$. For this application, the larger variance always appears in the numerator. This arbitrary placement of the larger variance in the numerator makes the outcome of the test less certain; thus, the uncertainty level of the F values in Table 7-4 is doubled from 5% to 10%. Example 7-8 illustrates the use of the F test for comparing measurement precision.

TABLE 7-4

Critical Values of F at the 5% Probability Level (95 % confidence level)

Degrees of Freedom (Denominator)	Degrees of Freedom (Numerator)								
	2	**3**	**4**	**5**	**6**	**10**	**12**	**20**	**∞**
2	19.00	19.16	19.25	19.30	19.33	19.40	19.41	19.45	19.50
3	9.55	9.28	9.12	9.01	8.94	8.79	8.74	8.66	8.53
4	6.94	6.59	6.39	6.26	6.16	5.96	5.91	5.80	5.63
5	5.79	5.41	5.19	5.05	4.95	4.74	4.68	4.56	4.36
6	5.14	4.76	4.53	4.39	4.28	4.06	4.00	3.87	3.67
10	4.10	3.71	3.48	3.33	3.22	2.98	2.91	2.77	2.54
12	3.89	3.49	3.26	3.11	3.00	2.75	2.69	2.54	2.30
20	3.49	3.10	2.87	2.71	2.60	2.35	2.28	2.12	1.84
∞	3.00	2.60	2.37	2.21	2.10	1.83	1.75	1.57	1.00

EXAMPLE 7-8

A standard method for the determination of the carbon monoxide (CO) level in gaseous mixtures is known from many hundreds of measurements to have a standard deviation of 0.21 ppm CO. A modification of the method yields a value for s of 0.15 ppm CO for a pooled data set with 12 degrees of freedom. A second modification, also based on 12 degrees of freedom, has a standard deviation of 0.12 ppm CO. Is either modification significantly more precise than the original?

Solution

We test the null hypothesis $H_0: \sigma_{std}^2 = \sigma_1^2$, where σ_{std}^2 is the variance of the standard method and σ_1^2 the variance of the modified method. The alternative hypothesis is one-tailed, $H_a: \sigma_1^2 < \sigma_{std}^2$. Because an improvement is claimed, the variances of the modifications are placed in the denominator. For the first modification,

$$F_1 = \frac{s_{std}^2}{s_1^2} = \frac{(0.21)^2}{(0.15)^2} = 1.96$$

and, for the second,

$$F_2 = \frac{(0.21)^2}{(0.12)^2} = 3.06$$

For the standard procedure, s_{std} is a good estimate of σ, and the number of degrees of freedom from the numerator can be taken as infinite. From Table 7-4, the critical value of F at the 95% confidence level is $F_{crit} = 2.30$.

Since F_1 is less than 2.30, we cannot reject the null hypothesis for the first modification and conclude that there is no improvement in precision. For the second modification, however, $F_2 > 2.30$. Hence, we reject the null hypothesis and conclude that the second modification does appear to give better precision at the 95% confidence level.

It is interesting to note that if we ask whether the precision of the second modification is significantly better than that of the first, the F test dictates that we must accept the null hypothesis, that is,

$$F = \frac{s_1^2}{s_2^2} = \frac{(0.15)^2}{(0.12)^2} = 1.56$$

In this case, $F_{crit} = 2.69$. Since $F < 2.69$, we must accept H_0 and conclude that the two methods give equivalent precision.

> **Spreadsheet Summary**   In Chapter 3 of *Applications of Microsoft® Excel in Analytical Chemistry*, 2nd ed., we use two Excel functions for performing the F test. First, we use the built-in function F.TEST(), which returns the probability that the variances in two data arrays are not significantly different. Second, we use the Analysis ToolPak for the same comparison of variances.

7C ANALYSIS OF VARIANCE

In Section 7B, we introduced methods to compare two sample means or one sample mean to a known value. In this section, we extend these principles to allow comparisons among more than two population means. The methods used for multiple comparisons fall under the general category of analysis of variance, often known by the acronym **ANOVA**. These methods use a single test to determine whether there is or is not a difference among the population means rather than pairwise comparisons as is done with the t test. After ANOVA indicates a potential difference, **multiple comparison** procedures can be used to identify which specific population means differ from the others. **Experimental design methods** take advantage of ANOVA in planning and performing experiments.

7C-1 ANOVA Concepts

In ANOVA procedures, we detect difference in several population means by comparing the *variances*. For comparing I population means, $\mu_1, \mu_2, \mu_3, \cdots \mu_I$, the null hypothesis H_0 is of the form

$$H_0: \mu_1 = \mu_2 = \mu_3 = \cdots = \mu_I$$

and the alternative hypothesis H_a is

$$H_a: \text{at least two of the } \mu_i\text{'s are different.}$$

The following are typical applications of ANOVA:

1. Is there a difference in the results of five analysts determining calcium by a volumetric method?
2. Will four different solvent compositions have differing influences on the yield of a chemical synthesis?
3. Are the results of manganese determinations by three different analytical methods different?
4. Is there any difference in the fluorescence of a complex ion at six different values of pH?

In each of these situations, the populations have differing values of a common characteristic called a **factor** or sometimes a **treatment**. In the case of determining calcium by a volumetric method, the factor of interest is the analyst. The different values of the factor of interest are called **levels**. For the calcium example, there are five levels corresponding to analyst 1, analyst 2, analyst 3, analyst 4, and analyst 5. The comparisons among the various populations are made by measuring a **response** for each item sampled. In the case of the calcium determination, the response is the amount of Ca (in millimoles) determined by each analyst. For the four examples given above, the factors, levels and responses are:

Factor	Levels	Response
Analyst	Analyst 1, analyst 2, analyst 3, analyst 4, analyst 5	Amount Ca, mmol
Solvent	Composition 1, composition 2, composition 3, composition 4	Synthesis yield, %
Analytical methods	Method 1, method 2, method 3	Concentration Mn, ppm
pH	pH 1, pH 2, pH 3, pH 4, pH 5, pH 6	Fluorescence intensity

The factor can be considered the independent variable, while the response is the dependent variable. Figure 7-4 illustrates how to visualize ANOVA data for the five analysts determining Ca in triplicate.

The type of ANOVA shown in **Figure 7-4** is known as a single-factor, or one-way, ANOVA. Often, several factors may be involved, such as in an experiment to determine whether pH and temperature influence the rate of a chemical reaction. In such a case, the type of ANOVA is known as a two-way ANOVA. Procedures for dealing with multiple factors are given in statistics books.[2] We consider only single-factor ANOVA.

Take the triplicate results for each analyst in Figure 7-4 to be random samples. In ANOVA, the factor levels are often called groups. The basic principle of ANOVA is to compare the between-groups variation to the within-groups variation. In our specific case, the groups (factor levels) are the different analysts, and this case is a comparison of the variation between analysts to the within-analyst variation. **Figure 7-5**

The basic principle of ANOVA is to compare the variations between the different factor levels (groups) to that within factor levels.

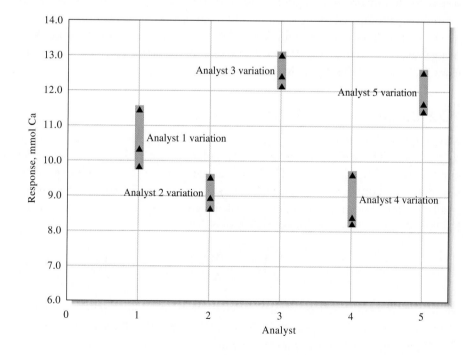

Figure 7-4 Pictorial of the results from the ANOVA study of the determination of calcium by five analysts. Each analyst does the determination in triplicate. Analyst is considered a factor, while analyst 1, analyst 2, analyst 3, analyst 4, and analyst 5 are levels of the factor.

[2]See, for example, J. L. Devore, *Probability and Statistics for Engineering and the Sciences*, 8th ed. Boston: Brooks/Cole, 2012, ch.11.

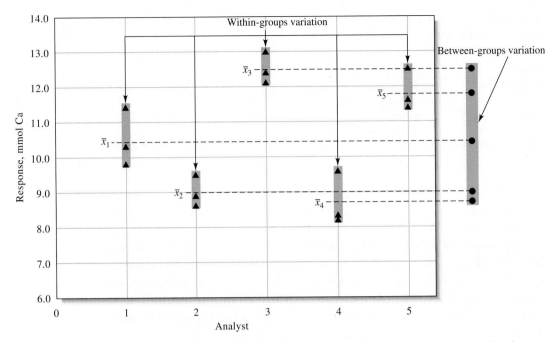

Figure 7-5 Pictorial representation of the ANOVA principle. The results of each analyst are considered a group. The triangles (▲) represent individual results, and the circles (●) represent the means. The variation between the group means is compared to that within groups.

illustrates this comparison. When H_0 is true, the variation between the group means is close to the variation within groups. When H_0 is false, the variation between group means is large compared to the variation within groups.

The basic statistical test used for ANOVA is the F test described in Section 7B-4. A large value of F compared to the critical value from the tables may give us reason to reject H_0 in favor of the alternative hypothesis.

7C-2 Single-Factor ANOVA

Several quantities are important for testing the null hypothesis H_0: $\mu_1 = \mu_2 = \mu_3 = \cdots \mu_I$. The sample means of the I populations are $\bar{x}_1, \bar{x}_2, \bar{x}_3, \cdots \bar{x}_I$ and the sample variances $s_1^2, s_2^2, s_3^2, \cdots s_I^2$. These are estimates of the corresponding population values. In addition, we can calculate the grand average $\bar{\bar{x}}$, which is the average of all the data. The grand mean can be calculated as the weighted average of the individual group means as shown in Equation 7-8

$$\bar{\bar{x}} = \left(\frac{N_1}{N}\right)\bar{x}_1 + \left(\frac{N_2}{N}\right)\bar{x}_2 + \left(\frac{N_3}{N}\right)\bar{x}_3 + \cdots + \left(\frac{N_I}{N}\right)\bar{x}_I \qquad (7\text{-}8)$$

where N_1 is the number of measurements in group 1, N_2 is the number in group 2, and so on. The grand average can also be found by summing all the data values and dividing by the total number of measurements N.

To calculate the variance ratio needed in the F test, it is necessary to obtain several other quantities called sums of squares:

1. The sum of the squares due to the factor SSF is

$$\text{SSF} = N_1(\bar{x}_1 - \bar{\bar{x}})^2 + N_2(\bar{x}_2 - \bar{\bar{x}})^2 + N_3(\bar{x}_3 - \bar{\bar{x}})^2 + \cdots + N_I(\bar{x}_I - \bar{\bar{x}})^2 \qquad (7\text{-}9)$$

2. The sum of the squares due to error SSE is

$$SSE = \sum_{j=1}^{N_1} (x_{1j} - \bar{x}_1)^2 + \sum_{j=1}^{N_2} (x_{2j} - \bar{x}_2)^2 + \sum_{j=1}^{N_3} (x_{3j} - \bar{x}_3)^2 + \cdots + \sum_{j=1}^{N_I} (x_{ij} - \bar{x}_I)^2$$

(7-10)

These two sums of squares are used to obtain the between-groups variation and the within-groups variation. The error sum of the squares is related to the individual group variances by

$$SSE = (N_1 - 1)s_1^2 + (N_2 - 1)s_2^2 + (N_3 - 1)s_3^2 + \cdots + (N_I - 1)s_I^2 \quad (7\text{-}11)$$

3. The total sum of the squares SST is obtained as the sum of SSF and SSE:

$$SST = SSF + SSE \qquad (7\text{-}12)$$

The total sum of the squares can also be obtained from $(N-1)s^2$, where s^2 is the sample variance of all the data points.

   To apply ANOVA methods, we need to make a few assumptions concerning the populations under study. First, the usual ANOVA methods are based on an equal variance assumption. That is, the variances of the I populations are assumed to be identical. This assumption is sometimes tested (Hartley test) by comparing the maximum and minimum variances in the set with an F test (see Section 7B-4). However, the Hartley test is quite susceptible to departures from the normal distribution. As a rough rule of thumb, the largest s should not be much more than twice the smallest s for equal variances to be assumed.[3] Transforming the data by working with a new variable such as $\sqrt{x}$, or log x, can also be used to give populations with more equal variances. Second, each of the I populations is assumed to follow a Gaussian distribution. For cases in which this last assumption is not true, there are distribution-free ANOVA procedures that can be applied.

4. The number of degrees of freedom for each of the sum of squares must be obtained. The total sum of the squares SST has $N - 1$ degrees of freedom. Just as SST is the sum of SSF and SSE, the total number degrees of freedom $N - 1$ can be decomposed into degrees of freedom associated with SSF and SSE. Since there are I groups being compared, SSF has $I - 1$ degrees of freedom. This leaves $N - I$ degrees of freedom for SSE. Or,

$$SST = SSF + SSE$$
$$(N - 1) = (I - 1) + (N - I)$$

5. By dividing the sums of squares by their corresponding degrees of freedom, we can obtain quantities that are estimates of the between-groups and within-groups variations. These quantities are called **mean square values** and are defined as:

$$\text{Mean square due to factor levels} = MSF = \frac{SSF}{I - 1} \qquad (7\text{-}13)$$

$$\text{Mean square due to error} = MSE = \frac{SSE}{N - I} \qquad (7\text{-}14)$$

---

[3] J. L. Devore, *Probability and Statistics for Engineering and the Sciences*, 8th ed. Boston: Brooks/Cole, 2012, p. 395.

The quantity MSE is an estimate of the variance due to error (σ_E^2), while MSF is an estimate of the error variance plus the between-groups variance ($\sigma_E^2 + \sigma_F^2$). If the factor has little effect, the between-groups variance should be small compared to the error variance. Thus, the two mean squares should be nearly identical under these circumstances. If the factor effect is significant, MSF is greater than MSE. The test statistic is the F value, calculated as

$$F = \frac{MSF}{MSE} \tag{7-15}$$

To complete the hypothesis test, we compare the value of F calculated from Equation 7-15 with the critical value from the table at a significance level of α. We reject H_0 if F exceeds the critical value. It is common practice to summarize the results of ANOVA in an **ANOVA table**, as follows:

Source of Variation	Sum of Squares (SS)	Degrees of Freedom (df)	Mean Square (MS)	Mean Square Estimates	F
Between groups (factor effect)	SSF	$I - 1$	$MSF = \dfrac{SSF}{I - 1}$	$\sigma_E^2 + \sigma_F^2$	$\dfrac{MSF}{MSE}$
Within groups (error)	SSE	$N - I$	$MSE = \dfrac{SSE}{N - I}$	σ_E^2	
Total	SST	$N - 1$			

Example 7-9 shows an application of ANOVA to the determination of calcium by five analysts. The data are those used to construct Figures 7-4 and 7-5.

Example 7-9

Five analysts determined calcium by a volumetric method and obtained the amounts (in mmol Ca) shown in the table below. Do the means differ significantly at the 95% confidence level?

Trial No.	Analyst 1	Analyst 2	Analyst 3	Analyst 4	Analyst 5
1	10.3	9.5	12.1	9.6	11.6
2	9.8	8.6	13.0	8.3	12.5
3	11.4	8.9	12.4	8.2	11.4

Solution

First, we obtain the means and standard deviations for each analyst. The mean for analyst 1 is $\bar{x}_1 = (10.3 + 9.8 + 11.4)/3 = 10.5$ mmol Ca. The remaining means are obtained in the same manner: $\bar{x}_2 = 9.0$ mmol Ca, $\bar{x}_3 = 12.5$ mmol Ca, $\bar{x}_4 = 8.7$ mmol Ca, $\bar{x}_5 = 11.833$ mmol Ca. The standard deviations are obtained as described in Section 6B-3. These results are summarized, as follows:

	Analyst 1	Analyst 2	Analyst 3	Analyst 4	Analyst 5
Mean	10.5	9.0	12.5	8.7	11.833
Standard Dev.	0.818535	0.458258	0.458258	0.781025	0.585947

The grand mean is found from Equation 7-8, where $N_1 = N_2 = N_3 = N_4 = N_5 = 3$ and $N = 15$:

$$\bar{\bar{x}} = \frac{3}{15}(\bar{x}_1 + \bar{x}_2 + \bar{x}_3 + \bar{x}_4 + \bar{x}_5) = 10.507 \text{ mmol Ca}$$

The between-groups sum of the squares is found from Equation 7-9:

$$\begin{aligned} \text{SSF} &= 3(10.5 - 10.507)^2 + 3(9.0 - 10.507)^2 + 3(12.5 - 10.507)^2 \\ &\quad + 3(8.7 - 10.507)^2 + 3(11.833 - 10.507)^2 \\ &= 33.80267 \end{aligned}$$

Note that SSF has associated with it $(5 - 1) = 4$ degrees of freedom. The error sum of squares is easiest to find from the standard deviations and Equation 7-11:

$$\begin{aligned} \text{SSE} &= 2(0.818535)^2 + 2(0.458258)^2 + 2(0.458258)^2 + \\ &\quad 2(0.781025)^2 + 2(0.585947)^2 \\ &= 4.086667 \end{aligned}$$

The error sum of the squares has $(15 - 5) = 10$ degrees of freedom.

We can now calculate the mean square values, MSF and MSE, from Equations 7-13 and 7-14:

$$\text{MSF} = \frac{33.80267}{4} = 8.450667$$

$$\text{MSE} = \frac{4.086667}{10} = 0.408667$$

The F value obtained from Equation 7-15 is

$$F = \frac{8.450667}{0.408667} = 20.68$$

From Table 7-4, the F table, the critical value of F at the 95% confidence level for 4 and 10 degrees of freedom is 3.48. Since F exceeds 3.48, we reject H_0 at the 95% confidence level and conclude that there is a significant difference among the analysts. The ANOVA table is:

Source of Variation	Sum of Squares (SS)	Degrees of Freedom (df)	Mean Square (MS)	F
Between groups	33.80267	4	8.450667	20.68
Within groups	4.086667	10	0.408667	
Total	37.88933	14		

 Spreadsheet Summary In Chapter 3 of *Applications of Microsoft® Excel in Analytical Chemistry*, 2nd ed., the use of Excel to perform ANOVA procedures is described. There are several ways to do ANOVA with Excel. First, the equations from this section are entered manually into a worksheet, and Excel is invoked to do the calculations. Second, the Analysis ToolPak is used to perform the entire ANOVA procedure automatically. The results of the five analysts from Example 7-9 are analyzed by both of these methods.

7C-3 Determining Which Results Differ

If significant differences are indicated in ANOVA, we are often interested in the cause. Is one mean different from the others? Are all the means different? Are there two distinct groups that the means fall into? There are several methods to determine which means are significantly different. One of the simplest is the **least significant difference** method. In this method, a difference is calculated that is judged to be the smallest difference that is significant. The difference between each pair of means is then compared to the least significant difference to determine which means are different.

For an equal number of replicates N_g in each group, the least significant difference LSD is calculated as follows:

$$LSD = t\sqrt{\frac{2 \times MSE}{N_g}} \tag{7-16}$$

where MSE is the mean square for error and the value of t has $N - I$ degrees of freedom. Example 7-10 illustrates the procedure.

EXAMPLE 7-10

For the results of Example 7-9, determine which analysts differ from each other at the 95% confidence level.

Solution

First, we arrange the means in increasing order: 8.7, 9.0, 10.5, 11.833, and 12.5. Each analyst did three repetitions, and so we can use Equation 7-16. From Table 7-3, we obtain a t value of 2.23 for the 95% confidence level and 10 degrees of freedom. Application of Equation 7-16 gives us

$$LSD = 2.23\sqrt{\frac{2 \times 0.408667}{3}} = 1.16$$

We now calculate the differences in means and compare them to 1.16. For the various pairs:

$$\bar{x}_{largest} - \bar{x}_{smallest} = 12.5 - 8.7 = 3.8 \quad \text{(a significant difference)}.$$
$$\bar{x}_{2ndlargest} - \bar{x}_{smallest} = 11.833 - 8.7 = 3.133 \quad \text{(significant)}.$$
$$\bar{x}_{3rdlargest} - \bar{x}_{smallest} = 10.5 - 8.7 = 1.8 \quad \text{(significant)}.$$
$$\bar{x}_{4thlargest} - \bar{x}_{smallest} = 9.0 - 8.7 = 0.3 \quad \text{(no significant difference)}.$$

We then continue to test each pair to determine which are different. From these calculations, we conclude that analysts 3, 5, and 1 differ from analyst 4; analysts 3, 5, and 1 differ from analyst 2; analysts 3 and 5 differ from analyst 1; and analyst 3 differs from analyst 5.

7D DETECTION OF GROSS ERRORS

There are times when a set of data contains an outlying result that appears to be outside the range that the random errors in the procedure would give. It is generally considered inappropriate and in some cases unethical to discard data without a reason. However, the questionable result, called an outlier, could be the result of an undetected gross error. Hence, it is important to develop a criterion to decide whether to retain or reject the outlying data point. The choice of criterion

An **outlier** is a result that is quite different from the others in the data set and might be due to a gross error.

for the rejection of a suspected result has its perils. If our standard is too strict so that it is quite difficult to reject a questionable result, we run the risk of retaining a spurious value that has an inordinate effect on the mean. If we set a lenient limit and make the rejection of a result easy, we are likely to discard a value that rightfully belongs in the set, thus introducing bias to the data. While there is no universal rule to settle the question of retention or rejection, the Q test is generally acknowledged to be an appropriate method for making the decision.[4]

7D-1 The Q Test

The Q test is a simple, widely used statistical test for deciding whether a suspected result should be retained or rejected.[5] In this test, the absolute value of the difference between the questionable result x_q and its nearest neighbor x_n is divided by the spread w of the entire set to give the quantity Q:

$$Q = \frac{|x_q - x_n|}{w} \tag{7-17}$$

This ratio is then compared with critical values Q_{crit} found in **Table 7-5**. If Q is greater than Q_{crit}, the questionable result can be rejected with the indicated degree of confidence (See **Figure 7-6**).

TABLE 7-5

Critical Values for the Rejection Quotient, Q^*			
	Q_{crit} (Reject if $Q > Q_{crit}$)		
Number of Observations	90% Confidence	95% Confidence	99% Confidence
3	0.941	0.970	0.994
4	0.765	0.829	0.926
5	0.642	0.710	0.821
6	0.560	0.625	0.740
7	0.507	0.568	0.680
8	0.468	0.526	0.634
9	0.437	0.493	0.598
10	0.412	0.466	0.568

*Reprinted (adapted) with permission from D. B. Rorabacher, *Anal. Chem.,* **1991**, *63*, 139, **DOI**: 10.1021/ac00002a010. Copyright 1991 American Chemical Society.

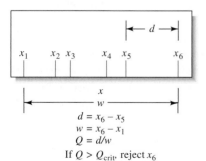

$$d = x_6 - x_5$$
$$w = x_6 - x_1$$
$$Q = d/w$$
If $Q > Q_{crit}$, reject x_6

Figure 7-6 The Q test for outliers.

[4]J. Mandel, in *Treatise on Analytical Chemistry*, 2nd ed., I. M. Kolthoff and P. J. Elving, eds., New York: Wiley, 1978, pt. I, vol. 1., pp. 282–289.

[5]R. B. Dean and W. J. Dixon, *Anal. Chem.*, **1951**, *23*, 636, **DOI**: 10.1021/ac60052a025.

EXAMPLE 7-11

The analysis of a city drinking water for arsenic yielded values of 5.60. 5.64, 5.70, 5.69, and 5.81 ppm. The last value appears anomalous; should it be rejected at the 95% confidence level?

Solution

The difference between 5.81 and 5.70 is 0.11 ppm. The spread $(5.81 - 5.60)$ is 0.21 ppm. Thus,

$$Q = \frac{0.11}{0.21} = 0.52$$

For five measurements, Q_{crit} at the 95% confidence level is 0.71. Because $0.52 < 0.71$, we must retain the outlier at the 95% confidence level.

7D-2 Other Statistical Tests

Several other statistical tests have been developed to provide criteria for rejection or retention of outliers. Such tests, like the Q test, assume that the distribution of the population data is normal, or Gaussian. Unfortunately, this condition cannot be proved or disproved for samples that have considerably fewer than 50 results. Consequently, statistical rules, which are perfectly reliable for normal distributions of data, should be *used with extreme caution* when applied to samples containing only a few values. J. Mandel, in discussing treatment of small sets of data, writes, "Those who believe that they can discard observations with statistical sanction by using statistical rules for the rejection of outliers are simply deluding themselves."[6] Thus, for small samples, statistical tests for rejection should be used only as aids to common sense.

> Use extreme caution when rejecting data for any reason.

The blind application of statistical tests to retain or reject a suspect measurement in a small set of data is not likely to be much more fruitful than an arbitrary decision. The application of good judgment based on broad experience with an analytical method is usually a sounder approach. In the end, the only valid reason for rejecting a result from a small set of data is the sure knowledge that a mistake was made in the measurement process. Without this knowledge, *a cautious approach to rejection of an outlier is wise.*

7D-3 Recommendations for Treating Outliers

There are a number of recommendations for the treatment of a small set of results that contains a suspect value:

1. Reexamine carefully all data relating to the outlying result to see if a gross error could have affected its value. This recommendation demands *a properly kept laboratory notebook containing careful notations of all observations* (see Section 2I).
2. If possible, estimate the precision that can be reasonably expected from the procedure to be sure that the outlying result actually is questionable.
3. Repeat the analysis if sufficient sample and time are available. Agreement between the newly acquired data and those of the original set that appear to be valid will lend weight to the notion that the outlying result should be rejected. Furthermore,

---

[6]J. Mandel, in *Treatise on Analytical Chemistry*, 2nd ed., I. M. Kolthoff and P. J. Elving, eds., New York: Wiley, 1978, pt. I, vol. 1., p. 282.

if retention is still indicated, the questionable result will have a small effect on the mean of the larger set of data.

4. If more data cannot be secured, apply the Q test to the existing set to see if the doubtful result should be retained or rejected on statistical grounds.

5. If the Q test indicates retention, consider reporting the median of the set rather than the mean. The median has the great virtue of allowing inclusion of all data in a set without undue influence from an outlying value. In addition, the median of a normally distributed set containing three measurements provides a better estimate of the correct value than the mean of the set after the outlying value has been discarded.

 WEB WORKS   Go to **www.cengage.com/chemistry/skoog/fac9**, choose Chapter 7, and go to the Web-Works. Click on the link and then on Statistics Textbook. Then click on the ANOVA/ MANOVA button. Now, read about the partitioning of the sum of squares in ANOVA procedures. Click on the F-distribution link in this section. Look at the tail areas for an F-distribution with both degrees of freedom equal to 10. Determine the value of F for a significance level of 0.10 with both degrees of freedom equal to 10.

QUESTIONS AND PROBLEMS

*7-1. Describe in your own words why the confidence interval for the mean of five measurements is smaller than that for a single result.

7-2. Assuming a large number of measurements so that s is a good estimate of σ, determine what confidence level was used for each of the following confidence intervals.

(a) $\bar{x} \pm \dfrac{2.58s}{\sqrt{N}}$ (b) $\bar{x} \pm \dfrac{1.96s}{\sqrt{N}}$

(c) $\bar{x} \pm \dfrac{3.29s}{\sqrt{N}}$ (d) $\bar{x} \pm \dfrac{s}{\sqrt{N}}$

7-3. Discuss how the size of the confidence interval for the mean is influenced by the following (all the other factors are constant):
  (a) the standard deviation σ.
  (b) the sample size N.
  (c) the confidence level.

7-4. Consider the following sets of replicate measurements:

*A	B	*C	D	*E	F
2.7	0.514	70.24	3.5	0.812	70.65
3.0	0.503	70.22	3.1	0.792	70.63
2.6	0.486	70.10	3.1	0.794	70.64
2.8	0.497		3.3	0.900	70.21
3.2	0.472		2.5		

Calculate the mean and the standard deviation for each of these six data sets. Calculate the 95% confidence interval for each set of data. What does this interval mean?

7-5. Calculate the 95% confidence interval for each set of data in Problem 7-4 if s is a good estimate of σ and has a value of *set A, 0.30; set B, 0.015; *set C, 0.070; set D, 0.20; *set E, 0.0090; and set F, 0.15.

7-6. The last result in each set of data in Problem 7-4 may be an outlier. Apply the Q test (95% confidence level) to determine whether or not there is a statistical basis to reject the result.

*7-7. An atomic absorption method for the determination of the amount of iron present in used jet engine oil was found from pooling 30 triplicate analyses to have a standard deviation $s = 3.6\ \mu g$ Fe/mL. If s is a good estimate of σ, calculate the 95 and 99% confidence intervals for the result 18.5 μg Fe/mL if it was based on (a) a single analysis, (b) the mean of two analyses, and (c) the mean of four analyses.

7-8. An atomic absorption method for determination of copper in fuel samples yielded a pooled standard deviation of $s_{pooled} = 0.27\ \mu g$ Cu/mL ($s \rightarrow \sigma$). The analysis of an oil from a reciprocating aircraft engine showed a copper content of 7.91 μg Cu/mL. Calculate the 95 and 99% confidence intervals for the result if it was based on (a) a single analysis, (b) the mean of 4 analyses, and (c) the mean of 16 analyses.

*7-9. How many replicate measurements are needed to decrease the 95 and 99% confidence limits for the analysis described in Problem 7-7 to ± 2.2 μg Fe/mL?

7-10. How many replicate measurements are necessary to decrease the 95 and 99% confidence limits for the analysis described in Problem 7-8 to ± 0.20 μg Cu/mL?

*7-11. A volumetric calcium analysis on triplicate samples of the blood serum of a patient believed to be suffering from a hyperparathyroid condition produced the following data: mmol Ca/L = 3.15, 3.25, 3.26. What is the 95% confidence interval for the mean of the data, assuming

  (a) no prior information about the precision of the analysis?

  (b) $s \rightarrow \sigma = 0.056$ mmol Ca/L?

7-12. A chemist obtained the following data for percent lindane in the triplicate analysis of an insecticide preparation: 7.23, 6.95, and 7.53. Calculate the 90% confidence interval for the mean of the three data, assuming that

  (a) the only information about the precision of the method is the precision for the three data.

  (b) on the basis of long experience with the method, it is believed that $s \rightarrow \sigma = 0.28\%$ lindane.

7-13. A standard method for the determination of glucose in serum is reported to have a standard deviation of 0.38 mg/dL. If $s = 0.38$ is a good estimate of σ, how many replicate determinations should be made in order for the mean for the analysis of a sample to be within

  *(a) 0.3 mg/dL of the true mean 99% of the time?

  (b) 0.3 mg/dL of the true mean 95% of the time?

  (c) 0.2 mg/dL of the true mean 90% of the time?

7-14. To test the quality of the work of a commercial laboratory, duplicate analyses of a purified benzoic acid (68.8% C, 4.953% H) sample were requested. It is assumed that the relative standard deviation of the method is $s_r \rightarrow \sigma = 4$ ppt for carbon and 6 ppt for hydrogen. The means of the reported results are 68.5% C and 4.882% H. At the 95% confidence level, is there any indication of systematic error in either analysis?

*7-15. A prosecuting attorney in a criminal case presented as principal evidence small fragments of glass found imbedded in the coat of the accused. The attorney claimed that the fragments were identical in composition to a rare Belgian stained glass window broken during the crime. The average of triplicate analyses for five elements in the glass are in the table. On the basis of these data, does the defendant have grounds for claiming reasonable doubt as to guilt? Use the 99% confidence level as a criterion for doubt.

| Element | Concentration, ppm | | Standard Deviation |
	From Clothes	From Window	$s \rightarrow \sigma$
As	129	119	9.5
Co	0.53	0.60	0.025
La	3.92	3.52	0.20
Sb	2.75	2.71	0.25
Th	0.61	0.73	0.043

7-16. Sewage and industrial pollutants dumped into a body of water can reduce the dissolved oxygen concentration and adversely affect aquatic species. In one study, weekly readings are taken from the same location in a river over a two-month period.

Week Number	Dissolved O_2, ppm
1	4.9
2	5.1
3	5.6
4	4.3
5	4.7
6	4.9
7	4.5
8	5.1

Some scientists think that 5.0 ppm is a dissolved O_2 level that is marginal for fish to live. Conduct a statistical test to determine whether the mean dissolved O_2 concentration is less than 5.0 ppm at the 95% confidence level. State clearly the null and alternative hypotheses.

*7-17. The week 3 measurement in the data set of Problem 7-16 is suspected of being an outlier. Use the Q test to determine if the value can be rejected at the 95% confidence level.

7-18. Before agreeing to the purchase of a large order of solvent, a company wants to see conclusive evidence that the mean value of a particular impurity is less than 1.0 ppb. What hypotheses should be tested? What are the type I and type II errors in this situation?

*7-19. The level of a pollutant in a river adjacent to a chemical plant is regularly monitored. Over a period of years, the normal level of the pollutant has been established by chemical analyses. Recently, the company has made several changes to the plant that appear to have increased the level of the pollutant. The Environmental Protection Agency (EPA) wants conclusive proof that the pollutant level has not increased. State the relevant null and alternative hypotheses and describe the type I and type II errors that might occur in this situation.

7-20. State quantitatively the null hypothesis H_0 and the alternative hypothesis H_a for the following situations and describe the type I and type II errors. If these hypotheses were to be tested statistically, comment on whether a one- or two-tailed test would be involved for each case.

  *(a) The mean values for Ca determinations by an ion-selective electrode method and by an EDTA titration differ substantially.

  (b) Since this sample gave a concentration lower than the 7.03 ppm level certified by the National Institute of Standards and Technology (NIST), a systematic error must have occurred.

  *(c) Results show that the batch-to-batch variation in the impurity content of Brand X acetonitrile is lower than Brand Y acetonitrile.

  (d) The atomic absorption results obtained for Cd are less precise than the electrochemical results.

*7-21. The homogeneity of the chloride level in a water sample from a lake was tested by analyzing portions drawn from the top and from near the bottom of the lake, with the following results in ppm Cl:

Top	Bottom
26.30	26.22
26.43	26.32
26.28	26.20
26.19	26.11
26.49	26.42

(a) Apply the t test at the 95% confidence level to determine if the chloride level from the top of the lake is different from that at the bottom.

(b) Now use the paired t test and determine whether there is a significant difference between the top and bottom values at the 95% confidence level.

(c) Why is a different conclusion drawn from using the paired t test than from just pooling the data and using the normal t test for differences in means?

7-22. Two different analytical methods were used to determine residual chlorine in sewage effluents. Both methods were used on the same samples, but each sample came from various locations with differing amounts of contact time with the effluent. Two methods were used to determine the concentration of Cl in mg/L, and the results are shown in the following table:

Sample	Method A	Method B
1	0.39	0.36
2	0.84	1.35
3	1.76	2.56
4	3.35	3.92
5	4.69	5.35
6	7.70	8.33
7	10.52	10.70
8	10.92	10.91

(a) What type of t test should be used to compare the two methods and why?

(b) Do the two methods give different results? State and test the appropriate hypotheses.

(c) Does the conclusion depend on whether the 90%, 95%, or 99% confidence levels are used?

*7-23. Sir William Ramsey, Lord Rayleigh, prepared nitrogen samples by several different methods. The density of each sample was measured as the mass of gas required to fill a particular flask at a certain temperature and pressure. Masses of nitrogen samples prepared by decomposition of various nitrogen compounds were 2.29280, 2.29940, 2.29849, and 2.30054 g. Masses of "nitrogen" prepared by removing oxygen from air in various ways were 2.31001, 2.31163, and 2.31028 g. Is the density of nitrogen prepared from nitrogen compounds significantly different from that prepared from air? What are the chances of the conclusion being in error? (Study of this difference led to the discovery of the inert gases by Lord Rayleigh).

7-24. The phosphorous content was measured for three different soil locations. Five replicate determinations were made on each soil sample. A partial ANOVA table follows:

Variation Source	SS	df	MS	F
Between soils	——	——	——	——
Within soils	——	——	0.0081	
Total	0.374	——		

(a) Fill in the missing entries in the ANOVA table.

(b) State the null and alternative hypotheses.

(c) Do the three soils differ in phosphorous content at the 95% confidence level?

*7-25. The ascorbic acid concentration of five different brands of orange juice was measured. Six replicate samples of each brand were analyzed. The following partial ANOVA table was obtained.

Variation Source	SS	df	MS	F
Between juices	——	——	——	8.45
Within juices	——	——	0.913	
Total	——	——		

(a) Fill in the missing entries in the table.

(b) State the null and alternative hypothesis.

(c) Is there a difference in the ascorbic acid content of the five juices at the 95% confidence level?

7-26. Five different laboratories participated in an interlaboratory study involving determinations of the iron level in water samples. The results below are replicate determinations of Fe in ppm for laboratories A through E

Result No.	Lab A	Lab B	Lab C	Lab D	Lab E
1	10.3	9.5	10.1	8.6	10.6
2	11.4	9.9	10.0	9.3	10.5
3	9.8	9.6	10.4	9.2	11.1

(a) State the appropriate hypotheses.

(b) Do the laboratories differ at the 95% confidence level? At the 99% confidence level ($F_{crit} = 5.99$)? At the 99.9% confidence level ($F_{crit} = 11.28$)?

(c) Which laboratories are different from each other at the 95% confidence level?

*7-27. Four analysts perform replicate sets of Hg determinations on the same analytical sample. The results in ppb Hg are shown in the following table:

Determination	Analyst 1	Analyst 2	Analyst 3	Analyst 4
1	10.24	10.14	10.19	10.19
2	10.26	10.12	10.11	10.15
3	10.29	10.04	10.15	10.16
4	10.23	10.07	10.12	10.10

(a) State the appropriate hypotheses.

(b) Do the analysts differ at the 95% confidence level? At the 99% confidence level ($F_{crit} = 5.95$)? At the 99.9% confidence level ($F_{crit} = 10.80$)?

(c) Which analysts differ from each other at the 95% confidence level?

7-28. Four different fluorescence flow cell designs were compared to see if they were significantly different. The following results represented relative fluorescence intensities for four replicate measurements:

Measurement No.	Design 1	Design 2	Design 3	Design 4
1	72	93	96	100
2	93	88	95	84
3	76	97	79	91
4	90	74	82	94

(a) State the appropriate hypotheses.
(b) Do the flow cell designs differ at the 95% confidence level?
(c) If a difference was detected in part (b), which designs differ from each other at the 95% confidence level?

*7-29. Three different analytical methods are compared for determining Ca in a biological sample. The laboratory is interested in knowing whether the methods differ. The results shown below represent Ca results in ppm determined by an ion-selective electrode (ISE) method, by EDTA titration, and by atomic absorption spectrometry:

Repetition No.	ISE	EDTA Titration	Atomic Absorption
1	39.2	29.9	44.0
2	32.8	28.7	49.2
3	41.8	21.7	35.1
4	35.3	34.0	39.7
5	33.5	39.2	45.9

(a) State the null and alternative hypotheses.
(b) Determine whether there are differences in the three methods at the 95% and 99% confidence levels.
(c) If a difference was found at the 95% confidence level, determine which methods differ from each other.

7-30. Apply the Q test to the following data sets to determine whether the outlying result should be retained or rejected at the 95% confidence level.
(a) 41.27, 41.61, 41.84, 41.70
(b) 7.295, 7.284, 7.388, 7.292

*7-31 Apply the Q test to the following data sets to determine whether the outlying result should be retained or rejected at the 95% confidence level.
(a) 85.10, 84.62, 84.70
(b) 85.10, 84.62, 84.65, 84.70

7-32. Determination of phosphorous in blood serum gave results of 4.40, 4.42, 4.60, 4.48, and 4.50 ppm P. Determine whether the 4.60 ppm result is an outlier or should be retained at the 95% confidence level.

7-33. **Challenge Problem:** The following are three sets of data for the atomic mass of antimony from the work of Willard and McAlpine[7]:

Set 1	Set 2	Set 3
121.771	121.784	121.752
121.787	121.758	121.784
121.803	121.765	121.765
121.781	121.794	

(a) Determine the mean and the standard deviation for each data set.
(b) Determine the 95% confidence interval for each data set.
(c) Determine whether the 121.803 value in the first data set is an outlier for that set at the 95% confidence level.
(d) Use the t test to determine whether the mean of data set 3 is identical to that of data set 1 at the 95% confidence level.
(e) The means of all three data sets are to be compared by ANOVA. State the null hypothesis. Determine whether the means differ at the 95% confidence level.
(f) Pool the data and determine the overall mean and the pooled standard deviation.
(g) Compare the overall mean of the 11 data points to the currently accepted value. Report the absolute error and the relative error in percent assuming the currently accepted value is the true value.

[7]H. H. Willard and R. K. McAlpine, *J. Am. Chem. Soc.*, **1921**, 43, 797, **DOI:** 10.1021/ja01437a010.

Sampling, Standardization, and Calibration

CHAPTER 8

Because a chemical analysis uses only a small fraction of the available sample, the process of sampling is a very important operation. The fractions of the sandy and loam soil samples shown in the photo that are collected for analyses must be representative of the bulk materials. Knowing how much sample to collect and how to further subdivide the collected sample to obtain a laboratory sample is vital in the analytical process. Sampling, standardization, and calibration are the focal points of this chapter. Statistical methods are an integral part of all three of these operations.

© Bob Rowan; Progressive Image/CORBIS

In Chapter 1, we described a typical real-world analytical procedure consisting of several important steps. In any such procedure, the specific analytical method selected depends on how much sample is available and how much analyte is present. We discuss here a general classification of the types of determinations based on these factors. After selecting the particular method to be used, a representative sample must be acquired. In the sampling process, we make every effort to select a small amount of material that accurately represents the bulk of the material being analyzed. We use statistical methods to aid in the selection of a representative sample. Once the analytical sample has been acquired, it must be processed in a dependable manner that maintains sample integrity without losing sample or introducing contaminants. Many laboratories use the automated sample handling methods discussed here because they are reliable and cost effective. Because analytical methods are not absolute, results must be compared with those obtained on standard materials of accurately known composition. Some methods require direct comparison with standards while others involve an indirect calibration procedure. Much of our discussion focuses on the details of standardization and calibration including the use of statistical procedures to construct calibration models. We conclude this chapter with a discussion of the methods used to compare analytical methods by using various performance criteria, called **figures of merit**.

8A ANALYTICAL SAMPLES AND METHODS

Many factors are involved in the choice of a specific analytical method as discussed in Section 1C-1. Among the most important factors are the amount of sample and the concentration of the analyte.

8A-1 Types of Samples and Methods

Often, we distinguish a method of identifying chemical species, a **qualitative analysis**, from one to determine the amount of a constituent, a **quantitative analysis**. Quantitative methods, as discussed in Section 1B are traditionally classified as gravimetric methods, volumetric methods, or instrumental methods. Another way to distinguish methods is based on the size of the sample and the level of the constituents.

Sample Size

As shown in **Figure 8-1**, the term **macro analysis** is used for samples whose masses are greater than 0.1 g. A **semimicro analysis** is performed on samples in the range of 0.01 to 0.1 g, and samples for a **micro analysis** are in the range 10^{-4} to 10^{-2} g. For samples whose mass is less than 10^{-4} g, the term **ultramicro analysis** is sometimes used.

From the classification in Figure 8-1, we can see that the analysis of a 1-g sample of soil for a suspected pollutant would be called a macro analysis whereas that of a 5-mg sample of a powder suspected to be an illicit drug would be a micro analysis. A typical analytical laboratory processes samples ranging from the macro to the micro and even to the ultramicro range. Techniques for handling very small samples are quite different from those for treating macro samples.

Constituent Types

The constituents determined in an analytical procedure can cover a huge range in concentration. In some cases, analytical methods are used to determine **major constituents**, which are those present in the range of 1 to 100% by mass. Many of the gravimetric and some of the volumetric procedures discussed in Part III are examples of major constituent determinations. As shown in **Figure 8-2**, species present in the range of 0.01 to 1% are usually termed **minor constituents**, while those present

Sample Size	Type of Analysis
> 0.1 g	Macro
0.01 to 0.1 g	Semimicro
0.0001 to 0.01 g	Micro
< 10^{-4} g	Ultramicro

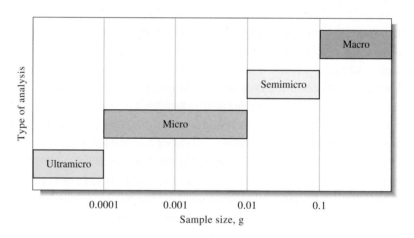

Figure 8-1   Classification of analyses by sample size.

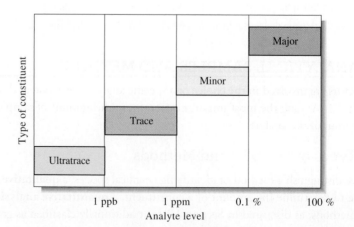

Figure 8-2   Classification of constituent types by analyte level.

in amounts between 100 ppm (0.01%) and 1 ppb are called **trace constituents**. Components present in amounts lower than 1 ppb are usually considered to be **ultratrace constituents**.

Determining Hg in the ppb to ppm range in a 1-µL ($\approx$1 mg) sample of river water would be a micro analysis of a trace constituent. Determinations of trace and ultratrace constituents are particularly demanding because of potential interferences and contaminations. In extreme cases, determinations must be performed in special rooms that are kept meticulously clean, free from dust and other contaminants. A general problem in trace procedures is that the reliability of results usually decreases dramatically with a decrease in analyte level. **Figure 8-3** shows how the relative standard deviation between laboratories increases as the level of analyte decreases. At the ultratrace level of 1 ppb, interlaboratory error (%RSD) is nearly 50%. At lower levels, the error approaches 100%.

Analyte Level	Type of Constituent
1 to 100%	Major
0.01 (100 ppm) to 1%	Minor
1 ppb to 100 ppm	Trace
< 1 ppb	Ultratrace

8A-2 Real Samples

The analysis of real samples is complicated by the presence of the sample matrix. The matrix can contain species with chemical properties similar to the analyte. Matrix components can react with the same reagents as the analyte, or they can cause an instrument response that is not easily distinguished from the analyte. These effects interfere with the determination of the analyte. If the interferences are caused by extraneous species in the matrix, they are often called **matrix effects**. Such effects can be induced not only by the sample itself but also by the reagents and solvents used to prepare the samples for the determination. The composition of the matrix containing the analyte may vary with time as is the case when materials lose water by dehydration or undergo photochemical reactions during storage. We discuss matrix effects and other interferences in the context of standardization and calibration methods in Section 8D-3.

As discussed in Section 1C, samples are *analyzed*, but species or concentrations are *determined*. Hence, we can correctly discuss the *determination* of glucose in blood serum or the *analysis* of blood serum for glucose.

 Samples are analyzed, but constituents or concentrations are determined.

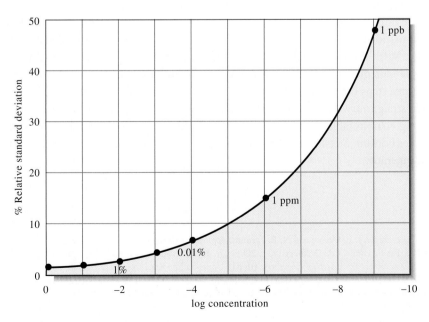

Figure 8-3 Interlaboratory error as a function of analyte concentration. Note that the relative standard deviation dramatically increases as the analyte concentration decreases. In the ultratrace range, the relative standard deviation approaches 100%. (Reprinted (adapted) with permission from W. Horowitz, *Anal. Chem.*, **1982**, *54*, 67A–76A., **DOI**: 10.1021/ac00238a002. Copyright 1982 American Chemical Society.)

8B   SAMPLING

A chemical analysis is most often performed on only a small fraction of the material of interest, for example a few milliliters of water from a polluted lake. The composition of this fraction must reflect as closely as possible the average composition of the bulk of the material if the results are to be meaningful. The process by which a *representative* fraction is acquired is termed **sampling**. Often, sampling is the most difficult step in the entire analytical process and the step that limits the accuracy of the procedure. This statement is especially true when the material to be analyzed is a large and inhomogeneous liquid, such as a lake, or an inhomogeneous solid, such as an ore, a soil, or a piece of animal tissue.

Sampling for a chemical analysis necessarily requires the use of statistics because conclusions will be drawn about a much larger amount of material from the analysis of a small laboratory sample. This is the same process that we discussed in Chapters 6 and 7 for examining a finite number of items drawn from a population. From the observation of the sample, we use statistics, such as the mean and standard deviation, to draw conclusions about the population. The literature on sampling is extensive[1]; we provide only a brief introduction in this section.

8B-1   Obtaining a Representative Sample

The sampling process must ensure that the items chosen are representative of the bulk of material or population. The items chosen for analysis are often called **sampling units** or **sampling increments**. For example, our population might be 100 coins, and we might wish to know the average concentration of lead in the collection of coins. Our sample is to be composed of 5 coins. Each coin is a sampling unit or increment. In the statistical sense, the sample corresponds to several small parts taken from different parts of the bulk material. To avoid confusion, chemists usually call the collection of sampling units or increments, the **gross sample**.

For analysis in the laboratory, the gross sample is usually reduced in size and homogenized to create the **laboratory sample**. In some cases, such as sampling powders, liquids, and gases, we do not have obvious discrete items. Such materials may not be homogeneous because they may consist of microscopic particles of different compositions or, in the case of fluids, zones where concentrations of the analyte differ. With these materials, we can prepare a representative sample by taking our sample increments from different regions of the bulk material. **Figure 8-4** illustrates the three steps that are usually involved in obtaining the laboratory sample. Step 1 is often straightforward with the population being as diverse as a carton of bottles containing vitamin tablets, a field of wheat, the brain of a rat, or the mud from a stretch of river bottom. Steps 2 and 3 are seldom simple and may require tremendous effort and ingenuity.

Sampling is often the most difficult aspect of an analysis.

The composition of the **gross sample** and the **laboratory sample** must closely resemble the average composition of the total mass of material to be analyzed.

In *sampling*, a sample population is reduced in size to an amount of homogeneous material that can be conveniently handled in the laboratory and whose composition is representative of the population.

Figure 8-4   Steps in obtaining a laboratory sample. The laboratory sample consists of a few grams to at most a few hundred grams. It may constitute as little as 1 part in 10^7 or 10^8 of the bulk material.

[1]See, for example, J. L. Devore and N. R. Farnum, *Applied Statistics for Engineers and Scientists*, 2nd ed. Pacific Grove, CA: Duxbury Press, 2005, Ch. 4; J. C. Miller and J. N. Miller, *Statistics and Chemometrics for Analytical Chemistry*, 4th ed., Upper Saddle River, NJ: Prentice Hall, 2000; B. W. Woodget and D. Cooper, *Samples and Standards*, London: Wiley, 1987; F. F. Pitard, *Pierre Gy's Sampling Theory and Sampling Practice*, Boca Raton, Fl: CRC Press, 1989.

Statistically, the goals of the sampling process are:

1. To obtain a mean analyte concentration that is an unbiased estimate of the population mean. This goal can be realized only if all members of the population have an equal probability of being included in the sample.
2. To obtain a variance in the measured analyte concentration that is an unbiased estimate of the population variance so that valid confidence limits can be found for the mean, and various hypothesis tests can be applied. This goal can be reached only if every possible sample is equally likely to be drawn.

Both goals require obtaining a **random sample**. Here the term random sample does not imply that the samples are chosen in a haphazard manner. Instead a randomization procedure is applied to obtain such a sample. For example, suppose our sample is to consist of 10 pharmaceutical tablets to be drawn from 1000 tablets off a production line. One way to ensure the sample is random is to choose the tablets to be tested from a table of random numbers. These can be conveniently generated from a random number table or from a spreadsheet as is shown in **Figure 8-5**. Here, we would assign each of the tablets a number from 1 to 1000 and use the sorted random numbers in column C of the spreadsheet to pick tablet 16, 33, 97, etc. for analysis.

8B-2 Sampling Uncertainties

In Chapter 5, we concluded that both systematic and random errors in analytical data can be traced to instrument, method, and personal causes. Most systematic errors can be eliminated by exercising care, by calibration, and by the proper use

	A	B	C	D	E
1	**Spreadsheet to generate random numbers between 1 and 1000**				
2		Random Numbers	Sorted Numbers		
3		97	16		
4		382	33		
5		507	97		
6		33	268		
7		511	382		
8		16	507		
9		268	511		
10		810	810		
11		934	821		
12		821	934		
13					
14	**Spreadsheet Documentation**				
15	Cell B3=RAND()*(1000-1)+1				

Figure 8-5 Generating 10 random numbers from 1 to 1000 by a spreadsheet. The random number function in Excel [**=RAND()**] generates random numbers between 0 and 1. The multiplier shown in the documentation ensures that the numbers generated in column B will be between 1 and 1000. In order to obtain integer numbers, we right click on the selected cells and choose Format Cells… from the drop down menu. We then choose Number and then 0 decimal places. So that the numbers do not change with every recalculation, the random numbers in column B were copied and then pasted as values into column C using the Paste Special… command on the Home ribbon. In column C the numbers were sorted in ascending order using Excel's Data Sort… command on the Data ribbon.

of standards, blanks, and reference materials. Random errors, which are reflected in the precision of data, can generally be kept at an acceptable level by close control of the variables that influence the measurements. Errors due to invalid sampling are unique in the sense that they are not controllable by the use of blanks and standards or by closer control of experimental variables. For this reason, sampling errors are ordinarily treated separately from the other uncertainties associated with an analysis.

For random and independent uncertainties, the overall standard deviation s_o for an analytical measurement is related to the standard deviation of the sampling process s_s and to the standard deviation of the method s_m by the relationship

$$s_o^2 = s_s^2 + s_m^2 \tag{8-1}$$

In many cases, the method variance will be known from replicate measurements of a single laboratory sample. Under this circumstance, s_s can be computed from measurements of s_0 for a series of laboratory samples, each of which is obtained from several gross samples. An analysis of variance (see Section 7C) can reveal whether the between samples variation (sampling plus measurement variance) is significantly greater than the within samples variation (measurement variance).

Youden has shown that, once the measurement uncertainty has been reduced to one third or less of the sampling uncertainty (that is, $s_m \leq s_s/3$), further improvement in the measurement uncertainty is fruitless.[2] This result suggests that, if the sampling uncertainty is large and cannot be improved, it is often a good idea to switch to a less precise but faster method of analysis so that more samples can be analyzed in a given length of time. Since the standard deviation of the mean is lower by a factor of $\sqrt{N}$, taking more samples can improve precision.

> When $s_m \leq s_s/3$, there is no point in trying to improve the measurement precision. Equation 8-1 shows that s_o will be determined predominately by the sampling uncertainty under these conditions.

8B-3 The Gross Sample

> The gross sample is the collection of individual sampling units. It must be representative of the whole in composition and in particle-size distribution.

Ideally, the gross sample is a miniature replica of the entire mass of material to be analyzed. It should correspond to the bulk material in chemical composition and in particle-size distribution if the sample is composed of particles.

Size of the Gross Sample

For convenience and economy, the gross sample should be no larger than absolutely necessary. Basically, gross sample size is determined by (1) the uncertainty that can be tolerated between the composition of the gross sample and that of the whole, (2) the degree of heterogeneity of the whole, and (3) the level of particle size at which heterogeneity begins.[3]

The last point warrants amplification. A well-mixed, homogeneous solution of a gas or liquid is heterogeneous only on the molecular scale, and the mass of the molecules themselves governs the minimum mass of the gross sample. A particulate solid, such as an ore or a soil, represents the opposite situation. In such materials, the individual pieces of solid differ from each other in composition. Heterogeneity develops in particles that may have dimensions on the order of a centimeter or more and may be several grams in mass. Intermediate between these extremes are colloidal materials and solidified metals. With colloidal materials, heterogeneity is first encountered in the range of 10^{-5} cm or less. In an alloy, heterogeneity first occurs in the crystal grains.

---

[2]W. J. Youden, *J. Assoc. Off. Anal. Chem.*, **1981**, *50*, 1007.
[3]For a paper on sample mass as a function of particle size, see G. H. Fricke, P. G. Mischler, F. P. Staffieri, and C. L. Housmyer, *Anal. Chem.*, **1987**, *59*, 1213, **DOI**: 10.1021/ac00135a030.

To obtain a truly representative gross sample, a certain number N of particles must be taken. The magnitude of this number depends on the uncertainty that can be tolerated (point 1 above) and how heterogeneous the material is (point 2 above). The number may range from a few particles to as many as 10^{12} particles. The need for large numbers of particles is of no great concern for homogeneous gases and liquids since heterogeneity among particles first occurs at the molecular level. Thus, even a very small amount of sample will contain more than the requisite number of particles. However, the individual particles of a particulate solid may have a mass of a gram or more, which sometimes leads to a gross sample of several tons. Sampling of such material is a costly, time-consuming procedure at best. To minimize cost, it is important to determine the smallest amount of material required to provide the desired information.

> The number of particles required in a gross sample ranges from a few particles to 10^{12} particles.

The laws of probability govern the composition of a gross sample removed randomly from a bulk of material. Because of this principle, it is possible to predict the likelihood that a selected fraction is similar to the whole. As an idealized example, let us presume that a pharmaceutical mixture contains just two types of particles: type A particles containing the active ingredient and type B particles containing only an inactive filler material. All particles are the same size. We wish to collect a gross sample that will allow us to determine the percentage of particles containing the active ingredient in the bulk material.

Assume that the probability of randomly drawing an A type particle is p and that of randomly drawing a B type particle is $(1 - p)$. If N particles of the mixture are taken, the most probable value for the number of A type particles is pN, while the most probable number of B type particles is $(1 - p)N$. For such a binary population, the Bernoulli equation[4] can be used to calculate the standard deviation of the number of A particles drawn, σ_A.

$$\sigma_A = \sqrt{Np(1 - p)} \tag{8-2}$$

The relative standard deviation σ_r of drawing A type particles[5] is σ_A/Np.

$$\sigma_r = \frac{\sigma_A}{Np} = \sqrt{\frac{1 - p}{Np}} \tag{8-3}$$

> We use the symbol σ_r to indicate relative standard deviation in accordance with the International Union of Pure and Applied Chemistry (IUPAC) recommendations (see footnote 5). You should bear in mind that σ_r is a ratio.

From Equation 8-3, we can obtain the number of particles needed to achieve a given relative standard deviation as shown in Equation 8-4.

$$N = \frac{1 - p}{p\sigma_r^2} \tag{8-4}$$

Thus, for example, if 80% of the particles are type A ($p = 0.8$) and the desired relative standard deviation is 1% ($\sigma_r = 0.01$), the number of particles making up the gross sample should be

$$N = \frac{1 - 0.8}{0.8(0.01)^2} = 2500$$

[4]A. A. Beneditti Pichler, in *Physical Methods in Chemical Analysis*, W. G. Berl, ed., New York: Academic Press, 1956, vol. 3, pp. 183–194; A. A. Beneditti-Pichler, *Essentials of Quantitative Analysis*, New York, Ronald Press, 1956, ch. 19.
[5]*Compendium of Analytical Nomenclature: Definitive Rules, 1997*, International Union of Pure and Applied Chemistry, prepared by J. Inczedy, T. Lengyel, and A. M. Ure, Malden, MA: Blackwell Science, 1998, pp. 2–8.

In this example, a random sample containing 2500 particles should be collected. A relative standard deviation of 0.1% would require 250,000 particles. Such a large number of particles would, of course, be determined by measuring the mass of the particles, not by counting.

We can now make the problem more realistic and assume that both of the components in the mixture contain the active ingredient (analyte), although in differing percentages. The type A particles contain a higher percentage of analyte, P_A and the type B particles a lesser amount, P_B. Furthermore, the average density d of the particles differs from the densities d_A and d_B of these components. We must now decide what number of particles and thus what mass we should take to ensure that we have a sample with the overall average percent of active ingredient P with a sampling relative standard deviation of σ_r. Equation 8-4 can be extended to include these conditions:

$$N = p(1 - p)\left(\frac{d_A d_B}{d^2}\right)^2 \left(\frac{P_A - P_B}{\sigma_r P}\right)^2 \tag{8-5}$$

From this equation, we see that the demands of precision are costly in terms of the sample size required because of the inverse-square relationship between the allowable relative standard deviation and the number of particles taken. Also, we can see that a greater number of particles must be taken as the average percentage P of the active ingredient becomes smaller.

The degree of heterogeneity as measured by $P_A - P_B$ has a large influence on the number of particles required since N increases with the square of the difference in composition of the two components of the mixture.

We can rearrange Equation 8-5 to calculate the relative standard deviation of sampling, σ_r.

$$\sigma_r = \frac{|P_A - P_B|}{P} \times \frac{d_A d_B}{d^2} \sqrt{\frac{p(1 - p)}{N}} \tag{8-6}$$

If we make the assumption that the sample mass m is proportional to the number of particles and the other quantities in Equation 8-6 are constant, the product of m and σ_r should be a constant. This constant K_s is called the Ingamells sampling constant.[6] Thus,

$$K_s = m \times (\sigma_r \times 100)^2 \tag{8-7}$$

where the term $\sigma_r \times 100\%$ is the percent relative standard deviation. Hence, when $\sigma_r = 0.01$, $\sigma_r \times 100\% = 1\%$, and K_s is just equal to m. We can thus interpret the sampling constant K_s to be the minimum sample mass required to reduce the sampling uncertainty to 1%.

The problem of deciding on the mass of the gross sample for a solid substance is usually even more difficult than this example because most materials not only contain more than two components, but they also consist of a range of particle sizes. In most instances, the problem of multiple components can be met by dividing the sample into an imaginary two-component system. Thus, with an actual complex mixture of substances, one component selected might be all the various analyte-containing particles and the other all the residual components containing little or no analyte.

> To simplify the problem of defining the mass of a gross sample of a multicomponent mixture, assume that the sample is a hypothetical two-component mixture.

[6]C. O. Ingamells and P. Switzer, *Talanta*, **1973**, *20*, 547–568, **DOI:** 10.1016/0039-9140(73)80135-3.

After average densities and percentages of analyte are assigned to each part, the system is treated as if it has only two components.

The problem of variable particle size can be handled by calculating the number of particles that would be needed if the sample consisted of particles of a single size. The gross sample mass is then determined by taking into account the particle-size distribution. One approach is to calculate the necessary mass by assuming that all particles are the size of the largest. Unfortunately, this procedure is not very efficient because it usually calls for removal of a larger mass of material than necessary. Benedetti-Pichler gives alternative methods for computing the mass of gross sample to be chosen.[7]

An interesting conclusion from Equation 8-5 is that the number of particles in the gross sample is independent of particle size. The mass of the sample, of course, increases directly as the volume (or as the cube of the particle diameter) so that reduction in the particle size of a given material has a large effect on the mass required for the gross sample.

A great deal of information must be known about a substance to use Equation 8-5. Fortunately, reasonable estimates of the various parameters in the equation can often be made. These estimates can be based on a qualitative analysis of the substance, visual inspection, and information from the literature on substances of similar origin. Crude measurements of the density of the various sample components may also be necessary.

EXAMPLE 8-1

A column-packing material for chromatography consists of a mixture of two types of particles. Assume that the average particle in the batch being sampled is approximately spherical with a radius of about 0.5 mm. Roughly 20% of the particles appear to be pink in color and are known to have about 30% by mass of a polymeric stationary phase attached (analyte). The pink particles have a density of 0.48 g/cm³. The remaining particles have a density of about 0.24 g/cm³ and contain little or no polymeric stationary phase. What mass of the material should the gross sample contain if the sampling uncertainty is to be kept below 0.5% relative?

Solution

We first compute values for the average density and percent polymer:

$$d = 0.20 \times 0.48 + 0.80 \times 0.24 = 0.288 \text{ g/cm}^3$$

$$P = \frac{(0.20 \times 0.48 \times 0.30) \text{ g polymer/cm}^3}{0.288 \text{ g sample/cm}^3} \times 100\% = 0.10\%$$

Then, substituting into Equation 8-5 gives

$$N = 0.20(1 - 0.20)\left[\frac{0.48 \times 0.24}{(0.288)^2}\right]^2\left(\frac{30 - 0}{0.005 \times 10.0}\right)^2$$

$$= 1.11 \times 10^5 \text{ particles required}$$

$$\text{mass of sample} = 1.11 \times 10^5 \text{ particles} \times \frac{4}{3}\pi(0.05)^3 \frac{\text{cm}^3}{\text{particle}} \times \frac{0.288 \text{ g}}{\text{cm}^3}$$

$$= 16.7 \text{ g}$$

---

[7]A. A. Beneditti-Pichler, in *Physical Methods in Chemical Analysis*, W. G. Berl, ed., New York: Academic Press, 1956, vol. 3, p. 192.

Well-mixed solutions of liquids and gases require only a very small sample because they are homogeneous down to the molecular level.

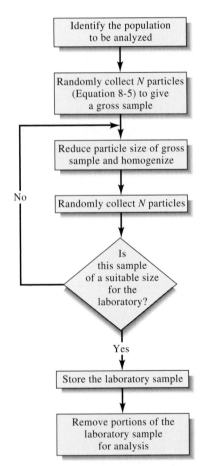

Figure 8-6 Steps in sampling a particulate solid.

The laboratory sample should have the same number of particles as the gross sample.

Sampling Homogeneous Solutions of Liquids and Gases

For solutions of liquids or gases, the gross sample can be relatively small because they are homogeneous down to the molecular level. Therefore, even small volumes of sample contain many more particles than the number computed from Equation 8-5. Whenever possible, the liquid or gas to be analyzed should be stirred well prior to sampling to make sure that the gross sample is homogeneous. With large volumes of solutions, mixing may be impossible; it is then best to sample several portions of the container with a "sample thief," a bottle that can be opened and filled at any desired location in the solution. This type of sampling is important, for example, in determining the constituents of liquids exposed to the atmosphere. For instance, the oxygen content of lake water may vary by a factor of 1000 or more over a depth difference of a few meters.

With the advent of portable sensors, it has become common in recent years to bring the laboratory to the sample instead of bringing the sample back to the laboratory. Most sensors, however, measure only local concentrations and do not average or sense remote concentrations.

In process control and other applications, samples of liquids are collected from flowing streams. Care must be taken to ensure that the sample collected represents a constant fraction of the total flow and that all portions of the stream are sampled.

Gases can be sampled by several methods. In some cases, a sampling bag is simply opened and filled with the gas. In others, gases can be *trapped* in a liquid or adsorbed onto the surface of a solid.

Sampling Particulate Solids

It is often difficult to obtain a random sample from a bulky particulate material. Random sampling can best be accomplished while the material is being transferred. Mechanical devices have been developed for handling many types of particulate matter. Details regarding sampling of these materials are beyond the scope of this book.

Sampling Metals and Alloys

Samples of metals and alloys are obtained by sawing, milling, or drilling. In general, it is not safe to assume that chips of the metal removed from the surface are representative of the entire bulk, so solid from the interior must be sampled as well. With some materials, a representative sample can be obtained by sawing across the piece at random intervals and collecting the "sawdust" as the sample. Alternatively, the specimen may be drilled, again at various randomly spaced intervals, and the drillings collected as the sample; the drill should pass entirely through the block or halfway through from opposite sides. The drillings can then be broken up and mixed or melted together in a special graphite crucible. A granular sample can often then be produced by pouring the melt into distilled water.

8B-4 Preparing a Laboratory Sample

For heterogeneous solids, the mass of the gross sample may range from hundreds of grams to kilograms or more, and so reduction of the gross sample to a finely ground and homogeneous laboratory sample, of at most a few hundred grams, is necessary. As shown in **Figure 8-6**, this process involves a cycle of operations that includes crushing and grinding, sieving, mixing, and dividing the sample (often into halves) to reduce its mass. During each division, the mass of sample that contains the number of particles computed from Equation 8-5 is retained.

EXAMPLE 8-2

A carload of lead ore containing galena ($\approx$ 70% Pb) and other particles with little or no lead is to be sampled. From the densities (galena = 7.6 g/cm³, other particles = 3.5 g/cm³, average density = 3.7 g/cm³) and rough percentage of lead, Equation 8-5 indicates that 8.45×10^5 particles are required to keep the sampling error below 0.5% relative. The particles appear spherical with a radius of 5 mm. A calculation of the sample mass required, similar to that in Example 8-1, shows that the gross sample mass should be about 1.6×10^6 g (1.8 ton). The gross sample needs to be reduced to a laboratory sample of about 100 g. How can this be done?

Solution

The laboratory sample should contain the same number of particles as the gross sample, or 8.45×10^5. The average mass of each particle, m_{avg}, is then

$$m_{avg} = \frac{100 \text{ g}}{8.45 \times 10^5 \text{ particles}} = 1.18 \times 10^{-4} \text{ g/particle}$$

The average mass of a particle is related to its radius in cm by the equation

$$m_{avg} = \frac{4}{3} \pi r^3 \times \frac{3.7 \text{ g}}{\text{cm}^3}$$

Since $m_{avg} = 1.18 \times 10^{-4}$ g / particle, we can solve for the average particle radius r:

$$r = \left(1.18 \times 10^{-4} \text{ g} \times \frac{3}{4\pi} \times \frac{\text{cm}^3}{3.7 \text{g}} \right)^{1/3} = 1.97 \times 10^{-2} \text{ cm or 0.2 mm}$$

Thus, the sample should be repeatedly ground, mixed, and divided until the particles are about 0.2 mm in diameter.

Additional information on details of preparing the laboratory sample can be found in Chapter 35 and in the literature.[8]

8B-5 Number of Laboratory Samples

Once the laboratory samples have been prepared, the question that remains is how many samples should be taken for the analysis? If we have reduced the measurement uncertainty to less than 1/3 the sampling uncertainty, the sampling uncertainty will limit the precision of the analysis. The number, of course, depends on what confidence interval we want to report for the mean value and the desired relative standard deviation of the method. If the sampling standard deviation σ_s is known from previous experience, we can use values of z from tables (see Section 7A-1).

$$\text{CI for } \mu = \bar{x} \pm \frac{z\sigma_s}{\sqrt{N}}$$

[8] *Standard Methods of Chemical Analysis*, F. J. Welcher, ed., Princeton, NJ: Van Nostrand, 1963, vol. 2, pt. A, pp. 21–55. An extensive bibliography of specific sampling information has been compiled by C. A. Bicking, in *Treatise on Analytical Chemistry*, 2nd ed., I. M. Kolthoff and P. J. Elving, eds., New York: Wiley, 1978, vol. 1, p. 299.

Often, we use an estimate of σ_s and so must use t instead of z (Section 7A-2)

$$\text{CI for } \mu = \bar{x} \pm \frac{ts_s}{\sqrt{N}}$$

The last term in this equation represents the absolute uncertainty that we can tolerate at a particular confidence level. If we divide this term by the mean value $\bar{x}$, we can calculate the relative uncertainty σ_r that is tolerable at a given confidence level:

$$\sigma_r = \frac{ts_s}{\bar{x}\sqrt{N}} \tag{8-8}$$

If we solve Equation 8-8 for the number of samples N, we obtain

$$N = \frac{t^2 s_s^2}{\bar{x}^2 \sigma_r^2} \tag{8-9}$$

Using t instead of z in Equation 8-9 does lead to the complication that the value of t itself depends on N. Usually, however, we can solve the equation by iteration as shown in Example 8-3 and obtain the desired number of samples.

EXAMPLE 8-3

The determination of copper in a seawater sample gave a mean value of 77.81 µg/L and a standard deviation s_s of 1.74 µg/L. (Note: the insignificant figures were retained here because these results are used below in another calculation.) How many samples must be analyzed to obtain a relative standard deviation of 1.7% in the results at the 95% confidence level?

Solution

We begin by assuming that we have an infinite number of samples, which corresponds to a t value of 1.96 at the 95% confidence level. Since $\sigma_r = 0.017$, $s_s = 1.74$, and $\bar{x} = 77.81$, Equation 8-9 gives

$$N = \frac{(1.96)^2 \times (1.74)^2}{(0.017)^2 \times (77.81)^2} = 6.65$$

We round this result to 7 samples and find the value of t for 6 degrees of freedom is 2.45. Using this t value, we then calculate a second value for N which is 10.38. Now if we use 9 degrees of freedom and $t = 2.26$, the next value is $N = 8.84$. The iterations converge with an N value of approximately 9. Note that it would be good strategy to reduce the sampling uncertainty so that fewer samples would be needed.

8C AUTOMATED SAMPLE HANDLING

Automated sample handling can lead to higher throughput (more analyses per unit time), higher reliability, and lower costs than manual sample handling.

Once sampling has been accomplished and the number of samples and replicates chosen, sample processing begins (recall Figure 1-2). Many laboratories are using automated sample handling methods because they are reliable and cost-effective. In some cases, automated sample handling is used for only a few specific operations, such as dissolving the sample and removing interferences. In others, all the remaining steps in the analytical procedure are automated. We describe two different methods for automated sample handling: the **batch** or **discrete** approach and the **continuous flow** approach.

Discrete Methods

Automated instruments that process samples in a discrete manner often mimic the operations that would be performed manually. Laboratory robots are used to process samples in cases where it might be dangerous for humans to be involved or where a large number of routine steps might be required. Small laboratory robots suitable for these purposes have been available commercially since the mid-1980s.[9] The robotic system is controlled by a computer so that it can be programmed by the user. Robots can be used to dilute, filter, partition, grind, centrifuge, homogenize, extract, and treat samples with reagents. They can also be trained to heat and shake samples, dispense measured volumes of liquids, inject samples into chromatographic columns, weigh samples, and transport samples to an appropriate instrument for measurement.

Some discrete sample processors automate only the measurement step of the procedure or a few chemical steps and the measurement step. Discrete analyzers have long been used in clinical chemistry, and today a wide variety of these analyzers are available. Some of these analyzers are general purpose and capable of performing several different determinations, often on a random access basis. Others are intended for one application or a few specific methods, such as blood glucose or blood electrolyte determinations.[10]

Continuous Flow Methods

In continuous flow methods, the sample is inserted into a flowing stream where a number of operations can be performed prior to transporting it to a flow-through detector. Hence, these systems behave as automated analyzers in that they can perform not only sample processing operations but also the final measurement step. Such sample-processing operations as reagent addition, dilution, incubation, mixing, dialysis, extraction, and many others can be implemented between the point of sample introduction and detection. There are two different types of continuous flow systems: segmented flow analyzers and flow injection analyzers.

The segmented flow analyzer divides the sample into discrete segments separated by gas bubbles as shown in **Figure 8-7a**. As shown in **Figure 8-7b**, the gas bubbles provide barriers to prevent the sample from spreading out along the tube due to dispersion processes. The bubbles thus confine the sample and minimize cross-contamination between different samples. They also enhance mixing between the samples and the reagents. The concentration profiles of the analyte are shown in **Figure 8-7c**. Samples are introduced at the sampler as plugs (left). Some broadening due to dispersion occurs by the time the samples reach the detector. Hence, the type of signal shown on the right is typically used to obtain quantitative information about the analyte. Samples can be analyzed at a rate of 30 to 120 samples per hour.

Two types of continuous flow analyzers are the segmented flow analyzer and the flow injection analyzer.

Dispersion is a band-spreading or mixing phenomenon that results from the coupling of fluid flow with molecular diffusion. **Diffusion** is mass transport due to a concentration gradient.

[9]For a description of laboratory robots, see *Handbook of Clinical Automation, Robotics and Optimization*, G. J. Kost, ed. New York: Wiley, 1996; J. R. Strimaitis, *J. Chem. Educ.*, **1989**, *66*, A8, **DOI**: 10.1021/ed066pA8, and **1990**, *67*, A20, **DOI**: 10.1021/ed067pA20; W. J. Hurst and J. W. Mortimer, *Laboratory Robotics*, New York: VCH Publishers, 1987.

[10]For a more extensive discussion of discrete clinical analyzers, see D. A. Skoog, F. J. Holler, and S. R. Crouch, *Principles of Instrumental Analysis*, 6th ed., Belmont, CA: Brooks/Cole, 2007, pp. 942–947.

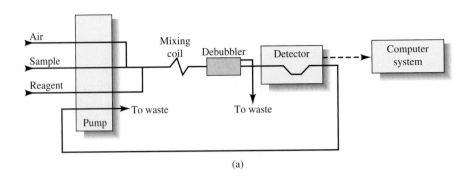

(a)

Figure 8-7 Segmented continuous flow analyzer. Samples are aspirated from sample cups in the sampler and pumped into the manifold where they are mixed with one or more reagents. Air is also injected to segment the samples with bubbles. The bubbles are usually removed by a debubbler before the stream reaches the detector. The segmented sample is shown in more detail in (b). The bubbles minimize dispersion of the sample that can cause broadening of the zones and cross-contamination from different samples. The analyte concentration profiles at the sampler and at the detector are shown in (c). Normally the height of a sample peak is related to the concentration of the analyte.

(b)

(c)

Flow injection analysis (FIA) is a more recent development.[11] With FIA, samples are injected from a sample loop into a flowing stream containing one or more reagents, as shown in **Figure 8-8a**. The sample plug is allowed to disperse in a controlled manner before it reaches the detector, as illustrated in **Figure 8-8b**. Injection of the sample into a reagent stream yields the type of responses shown on the right. In merging zones FIA, the sample and reagent are both injected into carrier streams and merged at a tee mixer. In either normal or merging zones FIA, sample dispersion is controlled by the sample size, the flow rate, and the length and diameter of the tubing. It is also possible to stop the flow when the sample reaches the detector to allow concentration-time profiles to be measured for kinetic methods (see Chapter 30).

Flow injection systems can also incorporate several sample processing units, such as solvent extraction modules, dialysis modules, heating modules, and others. Samples can be processed with FIA at rates varying from 60 to 300 samples per hour. Since the introduction of FIA in the mid-1970s, several variations of normal FIA have appeared. These include flow reversal FIA, sequential injection analysis, and

---

[11]For more information on FIA, see J. Ruzicka and E. H. Hansen, *Flow Injection Analysis*, 2nd ed. New York: Wiley, 1988; M. Valcarcel and M. D. Luque de Castro, *Flow Injection Analysis: Principles and Applications*, Chichester, England: Ellis Horwood, 1987; B. Karlberg and G. E. Pacey, *Flow Injection Analysis: A Practical Guide*, New York: Elsevier, 1989; M. Trojanowicz, *Flow Injection Analysis: Instrumentation and Applications*, River Edge, NJ: World Scientific Publication, 2000; E. A. G. Zagatto, C.C. Olivera, A. Townshend and P. J. Worsfold, *Flow Analysis with Spectrophotometric and Luminometric Detection*, Waltham MA: Elsevier, 2012.

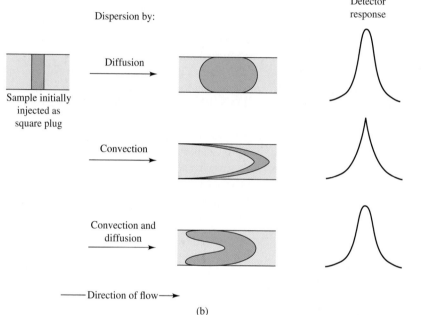

Figure 8-8 Flow injection analyzer. In (a) the sample is loaded from a sampler into the sample loop of a sampling valve. The valve, shown in the load position, also has a second inject position shown by the dotted lines. When switched to the inject position, the stream containing the reagent flows through the sample loop. Sample and reagent are allowed to mix and react in the mixing coil before reaching the detector. In this case, the sample plug is allowed to disperse prior to reaching the detector (b). The resulting concentration profile (detector response) depends on the degree of dispersion.

lab-on-a-valve technology.[12] Miniaturized FIA systems using microfluidics, often called lab-on-a-chip technology, have also been reported (see Feature 8-1).

8D STANDARDIZATION AND CALIBRATION

A very important part of all analytical procedures is the calibration and standardization process. **Calibration** determines the relationship between the analytical response and the analyte concentration. This relationship is usually determined by the use of **chemical standards**. The standards used can be prepared from purified reagents, if available, or standardized by classical quantitative methods (see Chapters 12–17). Most commonly, the

---

[12]For more information on variations of FIA, see D. A. Skoog, F. J. Holler, and S. R. Crouch, *Principles of Instrumental Analysis*, 6th ed., Belmont, CA: Brooks/Cole, 2007, pp. 939–940.

FEATURE 8-1

Lab-on-a-Chip

The development of microfluidic systems in which operations are miniaturized to the scale of an integrated circuit has enabled the fabrication of a complete **laboratory-on-a-chip** or **micro total analysis system** (μTAS).[13] Miniaturization of laboratory operations to the chip scale promises to reduce analysis costs by lowering reagent consumption and waste production by automating the procedures and by increasing the numbers of analyses per day. There have been several approaches to implementing the lab-on-a-chip concept. The most successful use the same photolithography technology as is used for preparing electronic integrated circuits. This technology is used to produce the valves, propulsion systems, and reaction chambers needed for performing chemical analyses. The development of microfluidic devices is an active research area involving scientists and engineers from academic and industrial laboratories.[14]

At first, microfluidic flow channels and mixers were coupled with traditional macroscale fluid propulsion systems and valves. The downsizing of the fluid flow channels showed great promise, but the advantages of low reagent consumption and complete automation were not realized. However, in more recent developments, monolithic systems have been used in which the propulsion systems, mixers, flow channels, and valves are integrated into a single structure.[15]

Several different fluid propulsion systems have been investigated for microfluidic systems, including electroosmosis (see Chapter 34), microfabricated mechanical pumps, and hydrogels that emulate human muscles. Flow injection techniques as well as such separation methods as liquid chromatography (Chapter 33), capillary electrophoresis, and capillary electrokinetic chromatography (Chapter 34) have been implemented. Figure 8F-1 shows the layout of a microstructure used for FIA. The monolithic unit is made of two polydimethyl siloxane (PDMS) layers that are permanently bonded together. The fluidic channels are 100 μm wide and 10 μm high. The entire device is only 2.0 cm by 2.0 cm. A glass cover allows for optical imaging of the channels by fluorescence excited with an Ar ion laser.

Lab-on-a chip analyzers are now available from several instrument companies. One commercial analyzer allows the analysis of DNA, RNA, proteins, and cells. Another commercial microfluidics device is used for nanoflow liquid chromatography and provides an interface to an electrospray mass spectrometry detector (see Chapter 29). Lab-on-a chip analyzers are envisioned for drug screening, for DNA sequencing, and for detecting life forms on Earth, Mars, and other planets. These devices should become more important as the technology matures.

Figure 8F-1 Layout of a microfabricated structure for FIA. Microfluidic channels are shown in blue, while control channels (pumps and valves) are shown in black. The components are (a) peristaltic pump, (b), injection valve, (c), mixing/reaction chamber, and (d), sample selector. Blue circles represent fluid reservoirs. Numbers (1) and (2) are samples, (3) is the carrier, (4) the reagent, and (5) and (6) are waste reservoirs. The entire structure is 2.0 cm by 2.0 cm. (Reprinted (adapted) with permission from A. M. Leach, A. R. Wheeler, and R. N. Zare, *Anal. Chem.*, **2003**, *75*, 967. Copyright 2003 American Chemical Society.)

[13]For reviews of these systems, see P. S. Dittrich, K. Tachikawa, and A. Manz, *Anal. Chem.*, **2006**. *78*, 3887, **DOI**: 10.1021/ac0605602; T. Vilkner, D. Janasek, and A. Manz, *Anal. Chem.*, **2004**, *76*, 3373, **DOI**: 10.1021/ac040063q; D. R. Reyes, D. Iossifidis, P. A. Auroux, and A. Manz, *Anal. Chem.*, **2002**, *74*, 2623, **DOI**: 10.1021/ac0202435; P. A. Auroux, D. Iossifidis, D. R. Reyes, and A. Manz, *Anal. Chem.* **2002**, *74*, 2637, **DOI**: 10.1021/ac020239t.

[14]See N. A. Polson and M. A. Hayes, *Anal. Chem.*, **2001** *73*, 313A, **DOI**: 10.1021/ac0124585.

[15]A. M. Leach, A. R. Wheeler, and R. N. Zare, *Anal. Chem.* **2003**, *75*, 967, **DOI**: 10.1021/ac026112l.

standards used are prepared externally to the analyte solutions (external standard methods). In the deer kill case study of Feature 1-1, the arsenic concentration was determined by calibration of the absorbance scale of a spectrophotometer with external standard solutions of known arsenic concentration. In some cases, an attempt is made to reduce interferences from other constituents in the sample matrix, called **concomitants**, by using standards added to the analyte solution (internal standard methods or standard addition methods) or by matrix matching or modification. Almost all analytical methods require some type of calibration with chemical standards. Gravimetric methods (Chapter 12) and some coulometric methods (Chapter 22) are among the few **absolute** methods that do not rely on calibration with chemical standards. The most common types of calibration procedures are described in this section.

8D-1 Comparison with Standards

We now describe two types of comparison methods: direct comparison techniques and titration procedures.

Direct Comparison

Some analytical procedures involve comparing a property of the analyte (or the product of a reaction with the analyte) with standards such that the property being tested matches or nearly matches that of the standard. For example, in early colorimeters, the color produced as the result of a chemical reaction of the analyte was compared with the color produced by using standards in place of the analyte in the same reaction. If the concentration of the standard was varied by dilution, for example, it was possible to match colors fairly precisely. The concentration of the analyte was then equal to the concentration of the diluted standard. Such procedures are called **null comparison** or **isomation methods**.[16]

With some modern instruments, a variation of this procedure is used to determine if an analyte concentration exceeds or is less than some threshold level. Feature 8-2 gives an example of how such a **comparator** can be used to determine whether the level of aflatoxin in a sample exceeds the level that would be toxic. The exact concentration of aflatoxin is not needed. The comparator only needs to indicate that the threshold has been exceeded. Alternatively, the approximate concentration of the analyte can be determined by comparing the color of the unknown solution with those of several standards.

FEATURE 8-2

A Comparison Method for Aflatoxins[17]

Aflatoxins are potential carcinogens produced by certain molds that may be found in corn, peanuts, and other food items. They are colorless, odorless, and tasteless. The toxicity of aflatoxins was revealed in the aftermath of a "turkey kill" involving over one hundred thousand birds in England in 1960.

One method to detect aflatoxins is by means of a competitive binding immunoassay (see Feature 11-2).

In the comparison method, antibodies specific to the aflatoxin are coated on the base of a plastic compartment or microtiter well in an array on a plate such as that shown in

(continued)

---

[16]See, for example, H. V. Malmstadt and J. D. Winefordner, *Anal. Chim. Acta*, **1960**, *20*, 283, **DOI**: 10.1016/0003-2670(59)80066-0; L. Ramaley and C. G. Enke, *Anal. Chem.*, **1965**, *37*,1073, **DOI**: 10.1021/ac60227a041.

[17]P. R. Kraus, A. P. Wade, S. R. Crouch, J. F. Holland, and B. M. Miller, *Anal. Chem.*, **1988**, *60*, 1387, **DOI**: 10.1021/ac00165a007.

Figure 8F-2. The aflatoxin behaves as the antigen. During the analysis, an enzyme reaction causes a blue product to be formed. As the amount of aflatoxin in the sample increases, the blue color decreases in intensity. The color-measuring instrument is the basic fiber optic comparator shown in Figure 8F-3. In the mode shown, the instrument compares the color intensity of the sample to that of a reference solution and indicates whether the aflatoxin level exceeds the threshold level. In another mode, a series of increasingly concentrated standards is placed in the reference well holder. The sample aflatoxin concentration is then between the two standards that are slightly more and slightly less concentrated than the analyte as indicated by the green and red indicator light-emitting diodes (LEDs).

Figure 8F-2 Microtiter plates. Several different sizes and configurations are available commercially. Most are arrays of 24 or 96 wells. Some are strips or can be cut into strips.

Figure 8F-3 Optical comparator. (a) An optical fiber that splits into two branches carries light from a light-emitting diode (LED) through sample and reference wells in a microtiter plate holder. In the comparison mode, a standard containing the threshold level of analyte (aflatoxin) is placed in one of the reference well holders. The samples containing unknown amounts of the analyte are placed in the sample well holder. If the sample contains more aflatoxin than the standard (b), the sample well absorbs less light at 650 nm than the reference well. An electronic circuit lights a red LED to indicate a dangerous amount of aflatoxin. If the sample contains less aflatoxin than the standard (c), a green LED is lit (recall that more aflatoxin means less intense color).

Titrations

Titrations are among the most accurate of all analytical procedures. In a titration, the analyte reacts with a standardized reagent (the titrant) in a known stoichiometric manner. Usually the amount of titrant is varied until chemical equivalence is reached as indicated by the color change of a chemical indicator or by the change in an instrument response. The amount of the standardized reagent needed to achieve chemical equivalence can then be related to the amount of analyte present by means of the stoichiometry. Titration is thus a type of chemical comparison.

For example, in the titration of the strong acid HCl with the strong base NaOH, a standardized solution of NaOH is used to determine the amount of HCl present. The reaction is

$$HCl + NaOH \rightarrow NaCl + H_2O$$

The standardized solution of NaOH is added from a buret until an indicator like phenolphthalein changes color. At this point, called the **end point**, the number of moles of NaOH added is approximately equal to the number of moles of HCl initially present.

The titration procedure is very general and used for a broad range of determinations. Chapters 13 through 17 treat the details of acid-base titrations, complexation titrations, and precipitation titrations. Titrations based on oxidation/reduction reactions are the subject of Chapter 19.

8D-2 External Standard Calibration

In **external standard calibration**, a series of standard solutions is prepared separately from the sample. The standards are used to establish the instrument **calibration function**, which is obtained from analysis of the instrument response as a function of the known analyte concentration. Ideally, three or more standard solutions are used in the calibration process, although in some routine determinations, two-point calibrations can be reliable.

The calibration function can be obtained graphically or in mathematical form. Generally, a plot of instrument response versus known analyte concentrations is used to produce a **calibration curve**, sometimes called a **working curve**. It is often desirable that the calibration curve be linear in at least the range of the analyte concentrations. A linear calibration curve of absorbance versus analyte concentration is shown in **Figure 8-9**. For graphical methods, a straight line is drawn through the data points (shown as circles). The linear relationship is then used to *predict* the concentration of an unknown analyte solution shown here with an absorbance of 0.505. Graphically, this prediction is done by locating the absorbance on the line and then finding the concentration corresponding to that absorbance (0.0044 M). The concentration found is then related back to the analyte concentration in the original sample by applying appropriate dilution factors from the sample preparation steps.

Computerized numerical data analysis has largely replaced graphical calibration methods, which are now seldom used except for visual confirmation of results. Statistical methods, such as the method of least squares, are routinely used to find the mathematical equation describing the calibration function. The concentration of the unknown is then found from the calibration function.

The Least-Squares Method

The calibration curve shown in Figure 8-9 is for the determination of Ni(II) by reaction with excess thiocyanate to form an absorbing complex ion [Ni(SCN)⁺]. The

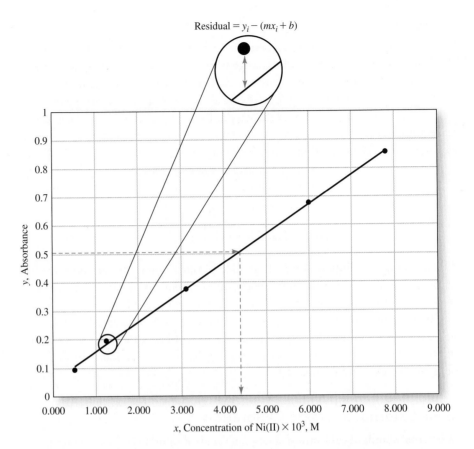

Figure 8-9 Calibration curve of absorbance versus analyte concentration for a series of standards. Data for standards shown as solid circles. The calibration curve is used in an inverse fashion to obtain the concentration of an unknown with an absorbance of 0.505. The absorbance is located on the line, and then the concentration corresponding to that absorbance is obtained by extrapolating to the x-axis (dashed lines). Residuals are distances on the y-axis between the data points and the predicted line as shown in the inset.

ordinate is the dependent variable, absorbance, while the abcissa is the independent variable, concentration of Ni(II). As is typical and usually desirable, the plot approximates a straight line. Note that, because of indeterminate errors in the measurement process, not all the data fall exactly on the line. Thus, the investigator must try to draw the "best" straight line among the data points. **Regression analysis** provides the means for objectively obtaining such a line and also for specifying the uncertainties associated with its subsequent use. We consider here only the basic **method of least squares** for two-dimensional data.

Assumptions of the Least-Squares Method. Two assumptions are made in using the method of least squares. The first is that there is actually a linear relationship between the measured response y (absorbance in Figure 8-9) and the standard analyte concentration x. The mathematical relationship that describes this assumption is called the **regression model**, which may be represented as

$$y = mx + b$$

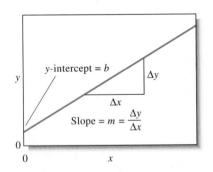

Figure 8-10 The slope-intercept form of a straight line.

The linear least-squares method assumes an actual linear relationship between the response y and the independent variable x. In addition, it is assumed that there is no error in the x values.

where b is the y intercept (the value of y when x is zero), and m is the slope of the line (see **Figure 8-10**). We also assume that any deviation of the individual points from the straight line arises from error in the *measurement*. That is, we assume there is no error in x values of the points (concentrations). Both of these assumptions are appropriate for many analytical methods, but bear in mind that, whenever there is significant uncertainty in the x data, basic linear least-squares analysis may not give the best straight line. In such a case, a more complex **correlation analysis** may be necessary. In addition, simple least-squares analysis may not be appropriate when the uncertainties in the y values vary significantly with x. In that instance, it may be necessary to apply different weighting factors to the points and perform a **weighted least-squares analysis**.

Finding the Least-Squares Line. The least-squares procedure can be illustrated with the aid of the calibration curve for the determination of Ni(II) shown in Figure 8-9. Thiocyanate was added to the Ni(II) standards, and the absorbances measured as a function of the Ni(II) concentration. The vertical deviation of each point from the straight line is called a **residual** as shown in the inset. The line generated by the least-squares method is the one that minimizes the sum of the squares of the residuals for all the points. In addition to providing the best fit between the experimental points and the straight line, the method gives the standard deviations for m and b.

The least-squares method finds the sum of the squares of the residuals SS_{resid} and minimizes the sum using calculus.[18] The value of SS_{resid} is found from

$$SS_{resid} = \sum_{i=1}^{N} [y_i - (b + mx_i)]^2$$

where N is the number of points used. The calculation of slope and intercept is simplified when three quantities are defined, S_{xx}, S_{yy}, and S_{xy} as follows:

$$S_{xx} = \sum (x_i - \bar{x})^2 = \sum x_i^2 - \frac{(\sum x_i)^2}{N} \tag{8-10}$$

$$S_{yy} = \sum (y_i - \bar{y})^2 = \sum y_i^2 - \frac{(\sum y_i)^2}{N} \tag{8-11}$$

$$S_{xy} = \sum (x_i - \bar{x})(y_i - \bar{y}) = \sum x_i y_i - \frac{\sum x_i \sum y_i}{N} \tag{8-12}$$

where x_i and y_i are individual pairs of data for x and y, N is the number of pairs, and $\bar{x}$ and $\bar{y}$ are the average values for x and y, that is, $\bar{x} = \dfrac{\sum x_i}{N}$, and $\bar{y} = \dfrac{\sum y_i}{N}$.

Note that S_{xx} and S_{yy} are the sum of the squares of the deviations from the mean for individual values of x and y. The expressions shown on the far right in Equations 8-10 through 8-12 are more convenient when a calculator without a built-in regression function is used.

Six useful quantities can be derived from S_{xx}, S_{yy}, and S_{xy}:

1. The slope of the line, m:

$$m = \frac{S_{xy}}{S_{xx}} \tag{8-13}$$

2. The intercept, b:

$$b = \bar{y} - m\bar{x} \tag{8-14}$$

> When there is uncertainty in the x values, basic least-squares analysis may not give the best straight line. Instead, a correlation analysis should be used.

> The equations for S_{xx} and S_{yy} are the numerators in the equations for the variance in x and the variance in y. Likewise, S_{xy} is the numerator in the covariance of x and y.

[18]The procedure involves differentiating SS_{resid} with respect to first m and then b and setting the derivatives equal to zero. This operation yields two equations, called normal equations, in the two unknowns m and b. These equations are then solved to give the least-squares best estimates of these parameters.

3. The standard deviation about regression, s_r:

$$s_r = \sqrt{\frac{S_{yy} - m^2 S_{xx}}{N - 2}} \tag{8-15}$$

4. The standard deviation of the slope, s_m:

$$s_m = \sqrt{\frac{s_r^2}{S_{xx}}} \tag{8-16}$$

5. The standard deviation of the intercept, s_b:

$$s_b = s_r\sqrt{\frac{\sum x_i^2}{N\sum x_i^2 - (\sum x_i)^2}} = s_r\sqrt{\frac{1}{N - (\sum x_i)^2/\sum x_i^2}} \tag{8-17}$$

6. The standard deviation for results obtained from the calibration curve, s_c:

$$s_c = \frac{s_r}{m}\sqrt{\frac{1}{M} + \frac{1}{N} + \frac{(\bar{y}_c - \bar{y})^2}{m^2 S_{xx}}} \tag{8-18}$$

Equation 8-18 gives us a way to calculate the standard deviation from the mean $\bar{y}_c$ of a set of M replicate analyses of unknowns when a calibration curve that contains N points is used; recall that $\bar{y}$ is the mean value of y for the N calibration points. This equation is only approximate and assumes that slope and intercept are independent parameters, which is not strictly true.

The standard deviation about regression s_r (Equation 8-15) is the standard deviation for y when the deviations are measured not from the mean of y (as is the usual case) but from the straight line that results from the least-squares prediction. The value of s_r is related to SS_{resid} by

$$s_r = \sqrt{\frac{\sum_{i=1}^{N}[y_i - (b + mx_i)]^2}{N - 2}} = \sqrt{\frac{SS_{resid}}{N - 2}}$$

In this equation the number of degrees of freedom is $N - 2$ since one degree of freedom is lost in calculating m and one in determining b. The standard deviation about regression is often called the **standard error of the estimate**. It roughly corresponds to the size of a typical deviation from the estimated regression line. Examples 8-4 and 8-5 illustrate how these quantities are calculated and used. With computers, the calculations are typically done using a spreadsheet program, such as Microsoft® Excel.[19]

The **standard deviation about regression**, also called the **standard error of the estimate** or just the **standard error**, is a rough measure of the magnitude of a typical deviation from the regression line.

EXAMPLE 8-4

Carry out a least-squares analysis of the calibration data for the determination of isooctane in a hydrocarbon mixture provided in the first two columns of **Table 8-1**.

---

[19]See S. R. Crouch and F. J. Holler, *Applications of Microsoft® Excel in Analytical Chemistry*, 2nd ed., Belmont CA: Brooks-Cole, 2014, ch. 4.

TABLE 8-1

Calibration Data for the Chromatographic Determination of Isooctane in a Hydrocarbon Mixture

Mole Percent Isooctane, x_i	Peak Area y_i	x_i^2	y_i^2	$x_i y_i$
0.352	1.09	0.12390	1.1881	0.38368
0.803	1.78	0.64481	3.1684	1.42934
1.08	2.60	1.16640	6.7600	2.80800
1.38	3.03	1.90440	9.1809	4.18140
1.75	4.01	3.06250	16.0801	7.01750
5.365	12.51	6.90201	36.3775	15.81992

Columns 3, 4, and 5 of the table contain computed values for x_i^2, y_i^2, and $x_i y_i$, with their sums appearing as the last entry in each column. Note that the number of digits carried in the computed values should be the *maximum allowed by the calculator or computer*, that is, *rounding should not be performed until the calculation is complete.*

> Do *not* round until calculations are complete.

Solution

We now substitute into Equations 8-10, 8-11, and 8-12 and obtain

$$S_{xx} = \sum x_i^2 - \frac{(\sum x_i)^2}{N} = 6.9021 - \frac{(5.365)^2}{5} = 1.14537$$

$$S_{yy} = \sum y_i^2 - \frac{(\sum y_i)^2}{N} = 36.3775 - \frac{(12.51)^2}{5} = 5.07748$$

$$S_{xy} = \sum x_i y_i - \frac{\sum x_i \sum y_i}{N} = 15.81992 - \frac{5.365 \times 12.51}{5} = 2.39669$$

Substitution of these quantities into Equations 8-13 and 8-14 yields

$$m = \frac{2.39669}{1.14537} = 2.0925 \approx 2.09$$

$$b = \frac{12.51}{5} - 2.0925 \times \frac{5.365}{5} = 0.2567 \approx 0.26$$

Thus, the equation for the least-squares line is

$$y = 2.09x + 0.26$$

Substitution into Equation 8-15 yields the standard deviation about regression,

$$s_r = \sqrt{\frac{S_{yy} - m^2 S_{xx}}{N-2}} = \sqrt{\frac{5.07748 - (2.0925)^2 \times 1.14537}{5-2}} = 0.1442 \approx 0.14$$

and substitution into Equation 8-16 gives the standard deviation of the slope,

$$s_m = \sqrt{\frac{s_r^2}{S_{xx}}} = \sqrt{\frac{(0.1442)^2}{1.14537}} = 0.13$$

Finally, we find the standard deviation of the intercept from Equation 8-17:

$$s_b = 0.1442\sqrt{\frac{1}{5 - (5.365)^2/6.9021}} = 0.16$$

EXAMPLE 8-5

The calibration curve found in Example 8-4 was used for the chromatographic determination of isooctane in a hydrocarbon mixture. A peak area of 2.65 was obtained. Calculate the mole percent of isooctane in the mixture and the standard deviation if the area was (a) the result of a single measurement and (b) the mean of four measurements.

Solution

In either case, the unknown concentration is found from rearranging the least-squares equation for the line, which gives

$$x = \frac{y - b}{m} = \frac{y - 0.2567}{2.0925} = \frac{2.65 - 0.2567}{2.0925} = 1.144 \text{ mol \%}$$

(a) Substituting into Equation 8-18, we obtain

$$s_c = \frac{0.1442}{2.0925} \sqrt{\frac{1}{1} + \frac{1}{5} + \frac{(2.65 - 12.51/5)^2}{(2.0925)^2 \times 1.145}} = 0.076 \text{ mole \%}$$

(b) For the mean of four measurements,

$$s_c = \frac{0.1442}{2.0925} \sqrt{\frac{1}{4} + \frac{1}{5} + \frac{(2.65 - 12.51/5)^2}{(2.0925)^2 \times 1.145}} = 0.046 \text{ mole \%}$$

Interpretation of Least-Squares Results.   The closer the data points are to the line predicted by a least-squares analysis, the smaller are the residuals. The sum of the squares of the residuals, SS_{resid}, measures the variation in the observed values of the dependent variable (y values) that are not explained by the presumed linear relationship between x and y.

$$SS_{resid} = \sum_{i=1}^{N} [y_i - (b + mx_i)]^2 \tag{8-19}$$

We can also define a total sum of the squares SS_{tot} as

$$SS_{tot} = S_{yy} = \sum (y_i - \bar{y})^2 = \sum y_i^2 - \frac{\left(\sum y_i\right)^2}{N} \tag{8-20}$$

The total sum of the squares is a measure of the total variation in the observed values of y since the deviations are measured from the mean value of y.

An important quantity called the **coefficient of determination** (R^2) measures the fraction of the observed variation in y that is explained by the linear relationship and is given by

$$R^2 = 1 - \frac{SS_{resid}}{SS_{tot}} \tag{8-21}$$

The closer R^2 is to unity, the better the linear model explains the y variations, as shown in Example 8-6. The difference between SS_{tot} and SS_{resid} is the sum of the squares due to regression, SS_{regr}. In contrast to SS_{resid}, SS_{regr} is a measure of the explained variation. We can write

$$SS_{regr} = SS_{tot} - SS_{resid} \quad \text{and} \quad R^2 = \frac{SS_{regr}}{SS_{tot}}$$

By dividing the sum of squares by the appropriate number of degrees of freedom, we can obtain the mean square values for regression and for the residuals (error) and then the F value. The F value gives us an indication of the significance of the regression. The F value is used to test the null hypothesis that the total variance in y is equal to the variance due to error. A value of F smaller than the value from the tables at the chosen confidence level indicates that the null hypothesis should be accepted and that the regression is not significant. A large value of F indicates that the null hypothesis should be rejected and that the regression is significant.

> A significant regression is one in which the variation in the y values due to the presumed linear relationship is large compared to that due to error (residuals). When the regression is significant, a large value of F occurs.

EXAMPLE 8-6

Find the coefficient of determination for the chromatographic data of Example 8-4.

Solution

For each value of x_i, we can find a predicted value of y_i from the linear relationship. Let us call the predicted values of y_i, $\hat{y}_i$. We can write $\hat{y}_i = b + mx_i$ and make a table of the observed y_i values, the predicted values $\hat{y}_i$, the residuals $y_i - \hat{y}_i$, and the squares of the residuals $(y_i - \hat{y}_i)^2$. By summing the latter values, we obtain SS_{resid} as shown in Table 8-2.

TABLE 8-2

Finding the Sum of the Squares of the Residuals

	x_i	y_i	$\hat{y}_i$	$y_i - \hat{y}_i$	$(y_i - \hat{y}_i)^2$
	0.352	1.09	0.99326	0.09674	0.00936
	0.803	1.78	1.93698	−0.15698	0.02464
	1.08	2.60	2.51660	0.08340	0.00696
	1.38	3.03	3.14435	−0.11435	0.01308
	1.75	4.01	3.91857	0.09143	0.00836
Sums	5.365	12.51			0.06240

From Example 8-4, the value of $S_{yy} = 5.07748$. Hence,

$$R^2 = 1 - \frac{SS_{resid}}{SS_{tot}} = 1 - \frac{0.0624}{5.07748} = 0.9877$$

This calculation shows that over 98% of the variation in peak area can be explained by the linear model.

We can also calculate SS_{regr} as

$$SS_{regr} = SS_{tot} - SS_{resid} = 5.07748 - 0.06240 = 5.01508$$

Let us now calculate the F value. There were five xy pairs used for the analysis. The total sum of the squares has 4 degrees of freedom associated with it since one is lost in

(continued)

calculating the mean of the y values. The sum of the squares due to the residuals has 3 degrees of freedom because two parameters m and b are estimated. Hence SS_{regr} has only 1 degree of freedom since it is the difference between SS_{tot} and SS_{resid}. In our case, we can find F from

$$F = \frac{SS_{regr}/1}{SS_{resid}/3} = \frac{5.01508/1}{0.0624/3} = 241.11$$

This very large value of F has a very small chance of occurring by random chance, and therefore, we conclude that this is a significant regression.

Transformed Variables.   Sometimes an alternative to a simple linear model is suggested by a theoretical relationship or by examining residuals from a linear regression. In some cases linear least-squares analysis can be used after one of the simple transformations shown in **Table 8-3**.

Although transforming variables is quite common, beware of pitfalls inherent in this process. Linear least squares gives best estimates of the transformed variables, but these may not be optimal when transformed back to obtain estimates of the original parameters. For the original parameters, **nonlinear regression methods**[20] may give better estimates. Sometimes, the relationship between the analytical response and concentration is inherently nonlinear. In other cases, nonlinearities arise because solutions do not behave ideally. Transforming variables does not give good estimates if the errors are not normally distributed. The statistics produced by ANOVA after transformation always refer to the transformed variables.

> **Spreadsheet Summary**   Chapter 4 of *Applications of Microsoft® Excel in Analytical Chemistry*, 2nd ed., introduces several ways to perform least-squares analysis. The built-in SLOPE and INTERCEPT functions of Excel are used with the data of Example 8-4. Then, the Excel function LINEST is used with the same data. The Analysis ToolPak Regression tool has the advantage of producing a complete ANOVA table for the results. A chart of the fit and the residuals can be produced directly from the Regression window. An unknown concentration is found with the calibration curve, and a statistical analysis is used to find the standard deviation of the concentration.

TABLE 8-3

Transformations to Linearize Functions

Function	Transformation to Linearize	Resulting Equation
Exponential: $y = be^{mx}$	$y' = \ln(y)$	$y' = \ln(b) + mx$
Power: $y = bx^m$	$y' = \log(y), x' = \log(x)$	$y' = \log(b) + mx'$
Reciprocal $y = b + m\left(\dfrac{1}{x}\right)$	$x' = \dfrac{1}{x}$	$y = b + mx'$

[20]See D. M. Bates and D. G. Watts, *Nonlinear Regression Analysis and Its Applications*, New York: Wiley, 1988.

Errors in External Standard Calibration

When external standards are used, it is assumed that, when the same analyte concentration is present in the sample and in the standard, the same response will be obtained. Thus, the calibration functional relationship between the response and the analyte concentration must apply to the sample as well. Usually in a determination, the raw response from the instrument is not used. Instead, the raw analytical response is corrected by measuring a **blank** (see Section 5B-4). The **ideal blank** is identical to the sample but without the analyte. In practice, with complex samples, it is too time consuming or impossible to prepare an ideal blank, and so a compromise must be made. Most often a real blank is either a **solvent blank**, containing the same solvent in which the sample is dissolved, or a **reagent blank**, containing the solvent plus all the reagents used in sample preparation.

Even with blank corrections, several factors can cause the basic assumption of the external standard method to break down. Matrix effects, due to extraneous species in the sample that are not present in the standards or blank, can cause the same analyte concentrations in the sample and standards to give different responses. Differences in experimental variables at the times at which blank, sample, and standard are measured can also invalidate the established calibration function. Even when the basic assumption is valid, errors can still occur due to contamination during the sampling or sample preparation steps.

Systematic errors can also occur during the calibration process. For example, if the standards are prepared incorrectly, an error will occur. The accuracy with which the standards are prepared depends on the accuracy of the gravimetric and volumetric techniques and of the equipment used. The chemical form of the standards must be identical to that of the analyte in the sample; the state of oxidation, isomerization, or complexation of the analyte can alter the response. Once prepared, the concentration of the standards can change due to decomposition, volatilization, or adsorption onto container walls. Contamination of the standards can also result in higher analyte concentrations than expected. A systematic error can occur if there is some bias in the calibration model. For example, errors can occur if the calibration function is obtained without using enough standards to obtain good statistical estimates of the parameters.

The accuracy of a determination can sometimes by checked by analyzing real samples of a similar matrix but with known analyte concentrations. The National Institute of Standards and Technology (NIST) and other organizations provide biological, geological, forensic, and other sample types with certified concentrations of several species (see Sections 5B-4 and 35B-4).

Random errors can also influence the accuracy of results obtained from calibration curves. From Equation 8-18, it can be seen that the standard deviation in the concentration of analyte s_c obtained from a calibration curve is lowest when the response $\bar{y}_c$ is close to the mean value $\bar{y}$. The point $\bar{x}, \bar{y}$ represents the centroid of the regression line. Points close to this value are determined with more certainty than those far away from the centroid. **Figure 8-11** shows a calibration curve with confidence limits. Note that measurements made near the center of the curve will give less uncertainty in analyte concentration than those made at the extremes.

❮ To avoid systematic errors in calibration, standards must be accurately prepared, and their chemical state must be identical to that of the analyte in the sample. The standards should be stable in concentration, at least during the calibration process.

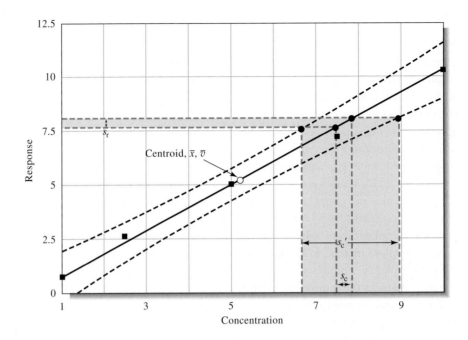

Figure 8-11 Effect of calibration curve uncertainty. The dashed lines show confidence limits for concentrations determined by the regression line. Note that uncertainties increase at the extremities of the plot. Usually we estimate the uncertainty in analyte concentration only from the standard deviation of the response. Calibration curve uncertainty can significantly increase the uncertainty in the analyte concentration from s_c to $s_c{'}$ as shown.

FEATURE 8-3

Multivariate Calibration

The least-squares procedure just described is an example of a univariate calibration procedure because only one response is used per sample. The process of relating multiple instrument responses to an analyte or a mixture of analytes is known as **multivariate calibration**. Multivariate calibration methods[21] have become quite popular in recent years as new instruments are now available that produce multidimensional responses (absorbance of several samples at multiple wavelengths, mass spectrum of chromatographically separated components, and so on). Multivariate calibration methods are very powerful. They can be used to determine multiple components in mixtures simultaneously and can provide redundancy in measurements to improve precision. Recall that repeating a measurement N times provides a $\sqrt{N}$ improvement in the precision of the mean value. These methods can also be used to detect the presence of interferences that would not be identified in a univariate calibration.

Multivariate techniques are **inverse calibration methods**. In normal least-squares methods, often called **classical least-squares methods**, the system response is modeled as a function of analyte concentration. In inverse methods, the concentrations are treated as functions of the responses. This latter approach can lead to some advantages in that concentrations can be accurately predicted even in the presence of chemical and physical sources of interference. In classical methods, all components in the system must be considered in the mathematical model produced (regression equation).

[21]For a more extensive discussion, see K. R. Beebe, R. J. Pell, and M. B. Seasholtz, *Chemometrics: A Practical Guide*, New York: Wiley, 1998, ch. 5; H. Martens and T. Naes, *Multivariate Calibration*, New York: Wiley, 1989; K. Varmuza and P. Filzmoser, *Introduction to Multivariate Statistical Analyis in Chemometrics*, Boca Raton, FL: CRC Press, 2009.

The common multivariate calibration methods are **multiple linear regression**, **partial least-squares regression**, and **principal components regression**. These differ in the details of the ways in which variations in the data (responses) are used to predict the concentration. Software for accomplishing multivariate calibration is available from several companies. The use of multivariate statistical methods for quantitative analysis is part of the subdiscipline of chemistry called **chemometrics**.

The multicomponent determination of Ni(II) and Ga(III) in mixtures is an example of the use of multivariate calibration.[22] Both metals react with 4-(2-pyridylazo)-resorcinol (PAR) to form colored products. The absorption spectra of the products are slightly different, and they form at slightly different rates. Advantage can be taken of these small differences to perform simultaneous determinations of the metals in mixtures. In the study cited, 16 standard mixtures containing the two metals were used to determine the calibration model. A multichannel (multiwavelength) diode array spectrometer (Section 25B-3) collected data for 26 time intervals at 26 wavelengths. Concentrations of the metals in the μM range were determined with relative errors of less than 10% in unknown mixtures at pH 8.5 by partial least squares and principal components regression.

Structural formula of 4-(2-pyridylazo)-resorcinol.

Molecular model of PAR.

8D-3 Minimizing Errors in Analytical Procedures

There are several steps that can be taken to ensure accuracy in analytical procedures.[23] Most of these depend on minimizing or correcting errors that might occur in the measurement step. We should note, however, that the overall accuracy and precision of an analysis might not be limited by the measurement step and might instead be limited by factors such as sampling, sample preparation, and calibration as discussed earlier in this chapter.

Separations

Sample cleanup by separation methods is an important way to minimize errors from possible interferences in the sample matrix. Techniques such as filtration, precipitation, dialysis, solvent extraction, volatilization, ion exchange, and chromatography are all very useful in ridding the sample of potential interfering constituents. Most separation methods are, however, time consuming and may increase the chances that some of the analyte will be lost or that the sample can be contaminated. In

[22]T. F. Cullen and S. R. Crouch, *Anal. Chim. Acta*, **2000**, *407*,135, **DOI**: 10.1016/S0003-2670(99)00836-3.

[23]For a more extensive discussion of error minimization, see J. D. Ingle, Jr., and S. R. Crouch, *Spectrochemical Analysis*, Upper Saddle River, NJ: Prentice-Hall, 1988, pp. 176–183.

many cases, though, separations are the only way to eliminate an interfering species. Some modern instruments include an automated front-end sample delivery system that includes a separation step (flow injection or chromatography).

Saturation, Matrix Modification, and Masking

The **saturation method** involves adding the interfering species to all the samples, standards, and blanks so that the interference effect becomes independent of the original concentration of the interfering species in the sample. This approach can, however, degrade the sensitivity and detectability of the analyte.

A **matrix modifier** is a species, not itself an interfering species, added to samples, standards, and blanks in sufficient amounts to make the analytical response independent of the concentration of the interfering species. For example, a buffer might be added to keep the pH within limits regardless of the sample pH. Sometimes, a **masking agent** is added that reacts selectively with the interfering species to form a complex that does not interfere. In both these methods, care must be taken that the added reagents do not contain significant quantities of the analyte or other interfering species.

Dilution and Matrix Matching

The **dilution method** can sometimes be used if the interfering species produces no significant effect below a certain concentration level. So, the interference effect is minimized simply by diluting the sample. Dilution may influence our ability to detect the analyte or to measure its response with accuracy and precision, and therefore, care is necessary in using this method.

> Errors in procedures can be minimized by saturating with interfering species, by adding matrix modifiers or masking agents, by diluting the sample, or by matching the matrix of the sample.

The **matrix-matching method** attempts to duplicate the sample matrix by adding the major matrix constituents to the standard and blank solutions. For example in the analysis of seawater samples for a trace metal, the standards can be prepared in a synthetic seawater containing Na^+, K^+, Cl^-, Ca^{2+}, Mg^{2+}, and other components. The concentrations of these species are well known and fairly constant in seawater. In some cases, the analyte can be removed from the original sample matrix, and the remaining components used to prepare standards and blanks. Again, we must be careful that added reagents do not contain the analyte or cause extra interference effects.

Internal Standard Methods

In the **internal standard method**, a known amount of a reference species is added to all the samples, standards, and blanks. The response signal is then not the analyte signal itself but the *ratio* of the analyte signal to the reference species signal. A calibration curve is prepared where the y-axis is the ratio of responses and the x-axis is the analyte concentration in the standards as usual. **Figure 8-12** illustrates the use of the internal standard method for peak-shaped responses.

> An **internal standard** is a reference species, chemically and physically similar to the analyte, that is added to samples, standards, and blanks. The ratio of the response of the analyte to that of the internal standard is plotted versus the concentration of analyte.

The internal standard method can compensate for certain types of errors if these influence both the analyte and the reference species to the same proportional extent. For example, if temperature influences both the analyte and reference species to the same extent, taking the ratio can compensate for variations in temperature. For compensation to occur, the reference species is chosen to have very similar chemical and physical properties to the analyte. The use of an internal standard in flame spectrometry is illustrated in Example 8-7.

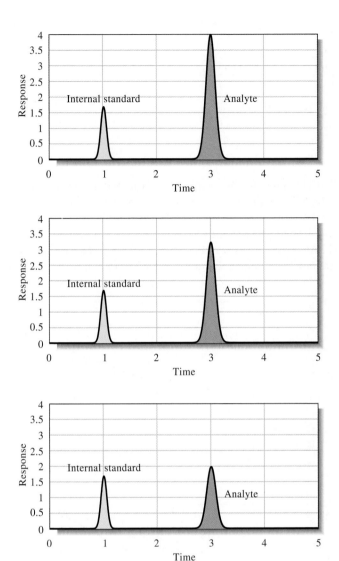

Figure 8-12 Illustration of the internal standard method. A fixed amount of the internal standard species is added to all samples, standards, and blanks. The calibration curve plots the ratio of the analyte signal to the internal standard signal against the concentration of the analyte.

EXAMPLE 8-7

The intensities of flame emission lines can be influenced by a variety of instrumental factors, including flame temperature, flow rate of solution, and nebulizer efficiency. We can compensate for variations in these factors by using the internal standard method. Thus, we add the same amount of internal standard to mixtures containing known amounts of the analyte and to the samples of unknown analyte concentration. We then take the ratio of the intensity of the analyte line to that of the internal standard. The internal standard should be absent in the sample to be analyzed.

In the flame emission determination of sodium, lithium is often added as an internal standard. The following emission intensity data were obtained for solutions containing Na and 1000 ppm Li.

(*continued*)

c_{Na}, ppm	Na intensity, I_{Na}	Li intensity, I_{Li}	I_{Na}/I_{Li}
0.10	0.11	86	0.00128
0.50	0.52	80	0.0065
1.00	1.8	128	0.0141
5.00	5.9	91	0.0648
10.00	9.5	73	0.1301
Unknown	4.4	95	0.0463

A plot of the Na emission intensity versus the Na concentration is shown in **Figure 8-13a**. Note that there is some scatter in the data and the R^2 value is 0.9816. In **Figure 8-13b**, the ratio of the Na to Li emission intensities is against the Na concentration. Note that the linearity is improved as indicated by the R^2 value of 0.9999. The unknown intensity ratio (0.0463) is then located on the curve, and the concentration of Na corresponding to this ratio is found to be 3.55 ± 0.05 ppm.

Figure 8-13   In (a) the Na flame emission intensity is plotted versus the Na concentration in ppm. The internal standard calibration curve is shown in (b), where the ratio of the Na to Li intensities is plotted versus the Na concentration.

Spreadsheet Summary   In Chapter 4 of *Applications of Microsoft® Excel in Analytical Chemistry*, 2nd ed., the data of Example 8-7 are used to construct a spreadsheet and plot the results. The unknown concentration is determined, and the statistics are presented.

A suitable reference species must be available for the internal standard method to compensate for errors. The reference species must not have unique interferences different from the analyte. There must be no analyte contamination in the materials used to prepare the internal standard. Also, both species must be present in concentrations that are in the linear portions of their calibration curves. Because of the difficulty in finding an appropriate internal standard species, the internal standard method is not as commonly used as some other error-compensating methods.

Standard Addition Methods

We use the **method of standard additions** when it is difficult or impossible to duplicate the sample matrix. In general, the sample is "spiked" with a known amount or known amounts of a standard solution of the analyte. In the single-point standard addition method, two portions of the sample are taken. One portion is measured as usual, but a known amount of standard analyte solution is added to the second portion. The responses for the two portions are then used to calculate the unknown concentration assuming a linear relationship between response and analyte concentration (see Example 8-8). In the **multiple additions method**, additions of known amounts of standard analyte solution are made to several portions of the sample, and a multiple additions calibration curve is obtained. The multiple additions method verifies to some extent that the linear relationship between response and analyte concentration holds. We discuss the multiple additions method further in Chapter 26 where it is used in conjunction with molecular absorption spectroscopy (Fig. 26-8).

The method of standard additions is a quite powerful method when used properly. First, there must be a good blank measurement so that extraneous species do not contribute to the analytical response. Second, the calibration curve for the analyte must be linear in the sample matrix. The multiple additions method provides a check on this assumption. A significant disadvantage of the multiple additions method is the extra time required for making the additions and measurements. The major benefit is the potential compensation for complex interference effects that may be unknown to the user.

> In the **method of standard additions**, a known amount of a standard solution of analyte is added to one portion of the sample. The responses before and after the addition are measured and used to obtain the analyte concentration. Alternatively multiple additions are made to several portions of the sample. The standard additions method assumes a linear response. Linearity should always be confirmed, or the **multiple additions method** used to check linearity.

EXAMPLE 8-8

The single-point standard addition method was used in the determination of phosphate by the molybdenum blue method. A 2.00-mL urine sample was treated with molybdenum blue reagents to produce a species absorbing at 820 nm, after which the sample was diluted to 100.00 mL. A 25.00-mL aliquot gave an instrument reading (absorbance) of 0.428 (solution 1). Addition of 1.00 mL of a solution containing 0.0500 mg of phosphate to a second 25.0-mL aliquot gave an absorbance of 0.517 (solution 2). Use these data to calculate the concentration of phosphate in milligrams per mL of the sample. Assume that there is a linear relationship between absorbance and concentration and that a blank measurement has been made.

Molecular model of phosphate ion (PO_4^{3-}).

(continued)

Solution

The absorbance of the first solution is given by

$$A_1 = kc_u$$

where c_u is the unknown concentration of phosphate in the first solution and k is a proportionality constant. The absorbance of the second solution is given by

$$A_2 = \frac{kV_u c_u}{V_t} + \frac{kV_s c_s}{V_t}$$

where V_u is the volume of the solution of unknown phosphate concentration (25.00 mL), V_s is the volume of the standard solution of phosphate added (1.00 mL), V_t is the total volume after the addition (26.00 mL), and c_s is the concentration of the standard solution (0.500 mg mL^{-1}). If we solve the first equation for k, substitute the result into the second equation, and solve for c_u, we obtain

$$c_u = \frac{A_1 c_s V_s}{A_2 V_t - A_1 V_u} =$$

$$= \frac{0.428 \times 0.0500 \text{ mg mL}^{-1} \times 1.00 \text{ mL}}{0.517 \times 26.00 \text{ mL} - 0.428 \times 25.00 \text{ mL}} = 0.0780 \text{ mg mL}^{-1}$$

This is the concentration of the diluted sample. To obtain the concentration of the original urine sample, we need to multiply by 100.00/2.00. Thus,

$$\text{concentration of phosphate} = 0.00780 \text{ mg mL}^{-1} \times 100.00 \text{ mL}/2.00 \text{ mL}$$
$$= 0.390 \text{ mg mL}^{-1}$$

Spreadsheet Summary   In Chapter 4 of *Applications of Microsoft® Excel in Analytical Chemistry*, 2nd ed., a multiple standard additions procedure is illustrated. The determination of strontium in seawater by inductively coupled plasma atomic emission spectrometry is used as an example. The worksheet is prepared, and the standard additions plot is made. Multiple linear regression and polynomial regression are also discussed.

8E   FIGURES OF MERIT FOR ANALYTICAL METHODS

Analytical procedures are characterized by a number of figures of merit such as accuracy, precision, sensitivity, detection limit, and dynamic range. We discussed in Chapter 5 the general concepts of accuracy and precision. Now, we describe those additional figures of merit that are commonly used and discuss the validation and reporting of analytical results.

8E-1 Sensitivity and Detection Limit

The term **sensitivity** is often used in describing an analytical method. Unfortunately, it is occasionally used indiscriminately and incorrectly. The definition of sensitivity most often used is the **calibration sensitivity**, or the change in the response signal per unit change in analyte concentration. The calibration sensitivity is thus the slope of the calibration curve, as shown in **Figure 8-14**. If the calibration curve

Figure 8-14 Calibration curve of response R versus concentration c. The slope of the calibration curve is called the calibration sensitivity m. The detection limit, DL, designates the lowest concentration that can be measured at a specified confidence level.

is linear, the sensitivity is constant and independent of concentration. If nonlinear, sensitivity changes with concentration and is not a single value.

The calibration sensitivity does not indicate what concentration differences can be detected. Noise in the response signals must be taken into account in order to be quantitative about what differences can be detected. For this reason, the term **analytical sensitivity** is sometimes used. The analytical sensitivity is the ratio of the calibration curve slope to the standard deviation of the analytical signal at a given analyte concentration. The analytical sensitivity is usually a strong function of concentration.

The **detection limit**, DL, is the smallest concentration that can be reported with a certain level of confidence. Every analytical technique has a detection limit. For methods that require a calibration curve, the detection limit is defined in a practical sense by Equation 8-22. It is the analyte concentration that produces a response equal to k times the standard deviation of the blank s_b:

$$\text{DL} = \frac{ks_b}{m} \tag{8-22}$$

where k is called the confidence factor and m is the calibration sensitivity. The factor k is usually chosen to be 2 or 3. A k value of 2 corresponds to a confidence level of 92.1%, while a k value of 3 corresponds to a 98.3% confidence level.[24]

Detection limits reported by researchers or instrument companies may not apply to real samples. The values reported are usually measured on ideal standards with optimized instruments. These limits are useful, however, in comparing methods or instruments.

8E-2 Linear Dynamic Range

The **linear dynamic range** of an analytical method most often refers to the concentration range over which the analyte can be determined using a linear calibration curve (see Figure 8-14). The lower limit of the dynamic range is generally considered to be the detection limit. The upper end is usually taken as the concentration at which the analytical signal or the slope of the calibration curve deviates by a specified amount.

---

[24]See J. D. Ingle, Jr., and S. R. Crouch, *Spectrochemical Analysis*, Upper Saddle River, NJ: Prentice Hall, 1988, p. 174.

Usually a deviation of 5% from linearity is considered the upper limit. Deviations from linearity are common at high concentrations because of nonideal detector responses or chemical effects. Some analytical techniques, such as absorption spectrophotometry, are linear over only one to two orders of magnitude. Other methods, such as mass spectrometry, may exhibit linearity over four to five orders of magnitude.

A linear calibration curve is preferred because of its mathematical simplicity and because it makes it easy to detect an abnormal response. With linear calibration curves, fewer standards and a linear regression procedure can be used. Nonlinear calibration curves are often useful, but more standards are required to establish the calibration function than with linear cases. A large linear dynamic range is desirable because a wide range of concentrations can be determined without dilution of samples, which is time consuming and a potential source of error. In some determinations, only a small dynamic range is required. For example, in the determination of sodium in blood serum, only a small range is needed because variations of the sodium level in humans is quite limited.

8E-3 Quality Assurance of Analytical Results

When analytical methods are applied to real-world problems, the quality of results as well as the performance quality of the tools and instruments used must be evaluated constantly. The major activities involved are quality control, validation of results, and reporting.[25] We briefly describe each of these here.

Control Charts

A **control chart** is a sequential plot of some characteristic that is a criterion of quality.

A control chart is a sequential plot of some quality characteristic that is important in quality assurance. The chart also shows the statistical limits of variation that are permissible for the characteristic being measured.

As an example, we will consider monitoring the performance of an analytical balance. Both the accuracy and the precision of the balance can be monitored by periodically determining the mass of a standard. We can then determine whether the measurements on consecutive days are within certain limits of the standard mass. These limits are called the **upper control limit** (UCL) and the **lower control limit** (LCL). They are defined as

$$\text{UCL} = \mu + \frac{3\sigma}{\sqrt{N}}$$

$$\text{LCL} = \mu - \frac{3\sigma}{\sqrt{N}}$$

where μ is the population mean for the mass measurement, σ is the population standard deviation for the measurement, and N is the number of replicates that are obtained for each sample. The population mean and standard deviation for the standard mass must be estimated from preliminary studies. Note that the UCL and the LCL are three standard deviations on either side of the population mean and form a range within which a measured mass is expected to lie 99.7% of the time.

Figure 8-15 is a typical instrument control chart for an analytical balance. Mass data were collected on twenty-four consecutive days for a 20.000-g standard mass

---

[25]For more information, see J. K. Taylor, *Quality Assurance of Chemical Measurements*, Chelsea, MI: Lewis Publishers, 1987.

Figure 8-15 A control chart for a modern analytical balance. The results appear to fluctuate normally about the mean except for those obtained on day 17. Investigation led to the conclusion that the questionable value resulted from a dirty balance pan. UCL = upper control limit; LCL = lower control limit.

certified by the National Institute of Standards and Technology. On each day, five replicate determinations were made. From independent experiments, estimates of the population mean and standard deviation were found to be $\mu = 20.000$ g and $\sigma = 0.00012$ g, respectively. For the mean of five measurements, $3 \times \dfrac{0.00012}{\sqrt{5}} = 0.00016$.

Hence, the UCL value = 20.00016 g, and the LCL value = 19.99984 g. With these values and the mean masses for each day, the control chart shown in **Figure 8-15** can be constructed. As long as the mean mass remains between the LCL and the UCL, the balance is said to be in **statistical control**. On day 17, the balance went out of control, and an investigation was launched to find the cause for this condition. In this example, the balance was not properly cleaned on day 17 so that there was dust on the balance pan. Systematic deviations from the mean are relatively easy to spot on a control chart.

In another example, a control chart was used to monitor the production of medications containing benzoyl peroxide used for treating acne. Benzoyl peroxide is a bactericide that is effective when applied to the skin as a cream or gel containing 10% of the active ingredient. These substances are regulated by the Food and Drug Administration (FDA). Concentrations of benzoyl peroxide must, therefore, be monitored and maintained in statistical control. Benzoyl peroxide is an oxidizing agent that can be combined with an excess of iodide to produce iodine that is titrated with standard sodium thiosulfate to provide a measure of the benzoyl peroxide in the sample.

Structural formula of benzoyl
peroxide.

Molecular model of benzoyl peroxide.

The control chart of **Figure 8-16** shows the results of 89 production runs of a cream containing a nominal 10% benzoyl peroxide measured on consecutive days. Each sample is represented by the mean percent benzoyl peroxide determined from the results of five titrations of different analytical samples of the cream.

The chart shows that, until day 83, the manufacturing process was in statistical control with normal random fluctuations in the amount of benzoyl peroxide. On day 83, the system went out of control with a dramatic systematic increase above the UCL. This increase caused considerable concern at the manufacturing facility

Figure 8-16 A control chart for monitoring the concentration of benzoyl peroxide in a commercial acne preparation. The manufacturing process became out of statistical control with sample 83 and exhibited a systematic change in the mean concentration.

until its source was discovered and corrected. These examples show how control charts are effective for presenting quality control data in a variety of situations.

Validation

Validation determines the suitability of an analysis for providing the sought-for information and can apply to samples, to methodologies, and to data. Validation is often done by the analyst, but it can also be done by supervisory personnel.

Validation of samples is often used to accept samples as members of the population being studied, to admit samples for measurement, to establish the authenticity of samples, and to allow for resampling if necessary. In the validation process, samples can be rejected because of questions about the sample identity, questions about sample handling, or knowledge that the method of sample collection was not appropriate or in doubt. For example, contamination of blood samples during collection as evidence in a forensic examination would be reason to reject the samples.

There are several different ways to validate analytical methods. Some of these were discussed in Section 5B-4. The most common methods include analysis of standard reference materials when available, analysis by a different analytical method, analysis of "spiked" samples, and analysis of synthetic samples approximating the chemical composition of the test samples. Individual analysts and laboratories often must periodically demonstrate the validity of the methods and techniques used.

Data validation is the final step before release of the results. This process starts with validating the samples and methods used. Then, the data are reported with statistically valid limits of uncertainty after a thorough check has been made to eliminate blunders in sampling and sample handling, mistakes in performing the analysis, errors in identifying samples, and mistakes in the calculations used.

Reporting Analytical Results

Specific reporting formats and procedures vary from laboratory to laboratory. However, a few general guidelines can be mentioned here. Whenever appropriate, reports should follow the procedure of a good laboratory practice (GLP).[26]

Generally, analytical results should be reported as the mean value and the standard deviation. Sometimes the standard deviation of the mean is reported instead of that of the data set. Either of these is acceptable as long as it is clear what is being reported. A confidence interval for the mean should also be reported. Usually the 95% confidence level is a reasonable compromise between being too inclusive or too restrictive.

---

[26]J. K. Taylor, *Quality Assurance of Chemical Measurements*, Chelsea, MI: Lewis Publishers, 1987, pp. 113–114.

Again, the interval and its confidence level should be explicitly reported. The results of various statistical tests on the data should also be reported when appropriate, as should the rejection of any outlying results along with the rejection criterion.

Significant figures are quite important when reporting results and should be based on statistical evaluation of the data. Whenever possible the significant figure convention stated in Section 6D-1 should be followed. Rounding of the data should be done with careful attention to the guidelines.

Whenever possible graphical presentation should include error bars on the data points to indicate uncertainty. Some graphical software allows the user to choose different error bar limits of $\pm 1s$, $\pm 2s$, and so forth, while other software packages automatically choose the size of the error bars. Whenever appropriate the regression equation and its statistics should also be reported.

Validating and reporting analytical results are not the most glamorous parts of an analysis, but they are among the most important because validation gives us confidence in the conclusions drawn. The report is often the "public" part of the procedure and may be brought to light during hearings, trials, patent applications, and other events.

WEB WORKS   Use a search engine to find the **method of standard additions**. Locate five different instrumental techniques (e.g., atomic absorption spectrometry and gas chromatography) that use the method of standard additions and provide references to a website or a journal article for each technique. Describe one method in detail. Include the instrumental technique, the analyte, the sample matrix, and any data treatment (single or multiple additions) procedures.

QUESTIONS AND PROBLEMS

*8-1. A 0.005-g sample of a rock is to be analyzed, and iron is to be determined at the ppm level. Determine the type of analysis and type of constituent.

8-2. What is the object of the sampling step in an analysis?

*8-3. Describe the steps in a sampling operation.

8-4. What factors determine the mass of a gross sample?

*8-5. The following results were obtained for the determination of calcium in a NIST limestone sample: %CaO = 50.33, 50.22, 50.36, 50.21, and 50.44. Five gross samples were then obtained for a carload of limestone. The average percent CaO values for the gross samples were found to be 49.53, 50.12, 49.60, 49.87, and 50.49. Calculate the relative standard deviation associated with the sampling step.

8-6. A coating that weighs at least 3.00 mg is needed to impart adequate shelf life to a pharmaceutical tablet. A random sampling of 250 tablets revealed that 14 failed to meet this requirement.

   (a) Use this information to estimate the relative standard deviation for the measurement.

   (b) What is the 95% confidence interval for the number of unsatisfactory tablets.

   (c) Assuming that the fraction of rejects remains unchanged, how many tablets should be taken to ensure a relative standard deviation of 5% in the measurement?

*8-7. Changes in the method used to coat the tablets in Problem 8-6 lowered the percentage of rejects from 5.6% to 2.0%. How many tablets should be taken for inspection if the permissible relative standard deviation in the measurement is to be

   (a) 20%?   (b) 12%?   (c) 7%?   (d) 2%?

8-8. The mishandling of a shipping container loaded with 750 cases of wine caused some of the bottles to break. An insurance adjuster proposed to settle the claim at 20.8% of the value of the shipment, based on a random 250-bottle sample in which 52 were cracked or broken. Calculate

   (a) the relative standard deviation of the adjuster's evaluation.

   (b) the absolute standard deviation for the 750 cases (12 bottles per case).

   (c) the 90% confidence interval for the total number of bottles.

   (d) the size of a random sampling needed for a relative standard deviation of 5.0%, assuming a breakage rate of about 21%.

*8-9. Approximately 15% of the particles in a shipment of silver-bearing ore are judged to be argentite, Ag_2S

$(d = 7.3$ g cm^{-3}, 87% Ag); the remainder are siliceous $(d = 2.6$ g cm^{-3}) and contain essentially no silver.

(a) Calculate the number of particles that should be taken for the gross sample if the relative standard deviation due to sampling is to be 2% or less.

(b) Estimate the mass of the gross sample, assuming that the particles are spherical and have an average diameter of 3.5 mm.

(c) The sample taken for analysis is to weigh 0.500 g and contain the same number of particles as the gross sample. To what diameter must the particles be ground to satisfy these criteria?

8-10. In the determination of lead in a paint sample, it is known that the sampling variance is 10 ppm while the measurement variance is 4 ppm. Two different sampling schemes are under consideration:

Scheme a: Take five sample increments and blend them. Perform a duplicate analysis of the blended sample.

Scheme b: Take three sample increments and perform a duplicate analysis on each.

Which sampling scheme, if any, should have the lower variance of the mean?

***8-11.** The data in the accompanying table represent the concentration of glucose in the blood serum of an adult patient. On four consecutive days, a blood sample was drawn from the patient and analyzed in triplicate. The variance for a given sample is an estimate of the measurement variance while the day-to-day variance reflects both the measurement variance and the sampling variance.

Day	Glucose Concentration, mg/100 mL		
1	62	60	63
2	58	57	57
3	51	47	48
4	54	59	57

(a) Perform an analysis of variance, and see whether the mean concentrations vary significantly from day to day.

(b) Estimate the sampling variance.

(c) What is the best way to lower the overall variance?

8-12. The seller of a mining claim took a random ore sample that weighed approximately 5 lb and had an average particle diameter of 5.0 mm. Inspection revealed that about 1% of the sample was argentite (see Problem 8-9), and the remainder had a density of about 2.6 g/cm^3 and contained no silver. The prospective buyer insisted on knowing the silver content of the claim with a relative error no greater than 5%. Determine whether the seller provided a sufficiently large sample to permit such an evaluation. Give the details of your analysis.

***8-13.** A method for the determination of the corticosteroid methylprednisolone acetate in solutions obtained from pharmaceutical preparations yielded a mean value of 3.7 mg mL^{-1} with a standard deviation of 0.3 mg mL^{-1}. For quality control purposes, the relative uncertainty in

the concentration should be no more than 3%. How many samples of each batch should be analyzed to ensure that the relative standard deviation does not exceed 7% at the 95% confidence level?

8-14. The sulfate ion concentration in natural water can be determined by measuring the turbidity that results when an excess of $BaCl_2$ is added to a measured quantity of the sample. A turbidimeter, the instrument used for this analysis, was calibrated with a series of standard Na_2SO_4 solutions. The following data were obtained in the calibration for sulfate concentrations, c_x:

c_x, mg SO_4^{2-}/L	Turbidimeter Reading, R
0.00	0.06
5.00	1.48
10.00	2.28
15.0	3.98
20.0	4.61

Assume that there is a linear relationship between the instrument reading and concentration.

(a) Plot the data, and draw a straight line through the points by eye.

(b) Compute the least-squares slope and intercept for the best straight line among the points.

(c) Compare the straight line from the relationship determined in (b) with that in (a).

(d) Use ANOVA to find the R^2 value, the adjusted R^2 value, and the significance of the regression. Comment on the interpretation of these values.

(e) Compute the concentration of sulfate in a sample yielding a turbidimeter reading of 2.84. Find the absolute standard deviation and the coefficient of variation.

(f) Repeat the calculations in (e) assuming that the 2.84 was the mean of six turbidimeter readings.

8-15. The following data were obtained in calibrating a calcium ion electrode for the determination of pCa. A linear relationship between the potential and pCa is known to exist.

pCa $= -\log [Ca^{2+}]$	E, mV
5.00	-53.8
4.00	-27.7
3.00	$+2.7$
2.00	$+31.9$
1.00	$+65.1$

(a) Plot the data and draw a line through the points by eye.

*(b) Find the least-squares expression for the best straight line among the points. Plot this line.

(c) Do ANOVA and report the statistics given in the ANOVA table. Comment on the meaning of the ANOVA statistics.

*(d) Calculate the pCa of a serum solution in which the electrode potential was 15.3 mV. Find the absolute and relative standard deviations for pCa if the result was from a single voltage measurement.

(e) Find the absolute and relative standard deviations for pCa if the potential reading in (d) was the mean of two replicate measurements. Repeat the calculation based on the mean of eight measurements.

8-16. The following are relative peak areas for chromatograms of standard solutions of methyl vinyl ketone (MVK).

MVK concentration, mmol/L	Relative Peak Area
0.500	3.76
1.50	9.16
2.50	15.03
3.50	20.42
4.50	25.33
5.50	31.97

(a) Determine the coefficients of the best fit line using the least-squares method.
(b) Construct an ANOVA table.
(c) Plot the least-squares line as well as the experimental points.
(d) A sample containing MVK yielded relative peak area of 12.9. Calculate the concentration of MVK in the solution.
(e) Assume that the result in (d) represents a single measurement as well as the mean of four measurements. Calculate the respective absolute and relative standard deviations for the two cases.
(f) Repeat the calculations in (d) and (e) for a sample that gave a peak area of 21.3.

*8-17. The data in the table below were obtained during a colorimetric determination of glucose in blood serum.

Glucose concentration, mM	Absorbance, A
0.0	0.002
2.0	0.150
4.0	0.294
6.0	0.434
8.0	0.570
10.0	0.704

(a) Assuming a linear relationship between the variables, find the least-squares estimates of the slope and intercept.
(b) What are the standard deviations of the slope and intercept? What is the standard error of the estimate?
(c) Determine the 95% confidence intervals for the slope and intercept.
(d) A serum sample gave an absorbance of 0.413. Find the 95% confidence interval for glucose in the sample.

8-18. The data in the table below represent electrode potential E vs. concentration c

E, mV	c, mol L^{-1}	E, mV	c, mol L^{-1}
106	0.20000	174	0.00794
115	0.07940	182	0.00631
121	0.06310	187	0.00398
139	0.03160	211	0.00200
153	0.02000	220	0.00126
158	0.01260	226	0.00100

(a) Transform the data to E vs. $-\log c$ values.
(b) Plot E vs. $-\log c$, and find the least-squares estimate of the slope and intercept. Write the least-squares equation.
(c) Find the 95 % confidence intervals for the slope and intercept.
(d) Use the F test to comment on the significance of regression.
(e) Find the standard error of the estimate, the correlation coefficient, and the multiple correlation coefficient squared.

8-19. A study was made to determine the activation energy E_A for a chemical reaction. The rate constant k was determined as a function of temperature T, and the data in the table below obtained.

T, K	k, s^{-1}
599	0.00054
629	0.0025
647	0.0052
666	0.014
683	0.025
700	0.064

The data should fit a linear model of the form $\log k = \log A - E_A/(2.303RT)$, where A is the preexponential factor, and R is the gas constant.

(a) Fit the data to a straight line of the form $\log k = a - 1000b/T$.
*(b) Find the slope, intercept, and standard error of the estimate.
*(c) Noting that $E_A = -b \times 2.303R \times 1000$, find the activation energy and its standard deviation (Use $R = 1.987$ cal mol^{-1} K^{-1}).
*(d) A theoretical prediction gave $E_A = 41.00$ kcal mol^{-1}K^{-1}. Test the null hypothesis that E_A is this value at the 95% confidence level.

8-20. Water can be determined in solid samples by infrared spectroscopy. The water content of calcium sulfate hydrates is to be measured using calcium carbonate as an internal standard to compensate for some systematic errors in the procedure. A series of standard solutions containing calcium sulfate dihydrate and a constant known amount of the internal standard is prepared. The solution of unknown water content is also prepared with the same amount of internal standard. The absorbance of the dihydrate is measured at one wavelength (A_{sample}) along with that of the internal standard at another wavelength (A_{std}). The following results were obtained.

A_{sample}	A_{std}	% water
0.15	0.75	4.0
0.23	0.60	8.0
0.19	0.31	12.0
0.57	0.70	16.0
0.43	0.45	20.0
0.37	0.47	Unknown

(a) Plot the absorbance of the sample (A_{sample}) vs. the % water and determine whether the plot is linear from the regression statistics.

(b) Plot the ratio A_{sample}/A_{std} vs. % water, and comment on whether using the internal standard improves the linearity from that in part (a). If it improves the linearity, why?

(c) Calculate % water in the unknown using the internal standard data.

8-21. Potassium can be determined by flame emission spectrometry (flame photometry) using a lithium internal standard. The following data were obtained for standard solutions of KCl and an unknown containing a constant, known amount of LiCl as the internal standard. All the intensities were corrected for background by subtracting the intensity of a blank.

c_K, ppm	Intensity of K Emission	Intensity of Li Emission
1.0	10.0	10.0
2.0	15.3	7.5
5.0	34.7	6.8
7.5	65.2	8.5
10.0	95.8	10.0
20.0	110.2	5.8
Unknown	47.3	9.1

(a) Plot the K emission intensity vs. the concentration of K, and determine the linearity from the regression statistics.

(b) Plot the ratio of the K intensity to the Li intensity vs. the concentration of K, and compare the resulting linearity to that in part (a). Why does the internal standard improve linearity?

*(c) Calculate the concentration of K in the unknown.

8-22. Copper was determined in a river water sample by atomic absorption spectrometry and the method of standard additions. For the addition, 100.0 μL of a 1000.0-μg/mL Cu standard was added to 100.0 mL of solution. The following data were obtained:

Absorbance of reagent blank = 0.020
Absorbance of sample = 0.520
Absorbance of sample plus addition − blank = 1.020

(a) Calculate the copper concentration in the sample.

(b) Later studies showed that the reagent blank used to obtain the above data was inadequate and that the actual blank absorbance was 0.100. Find the copper concentration with the appropriate blank, and determine the error caused by using an improper blank.

*8-23. The method of standard additions was used to determine nitrite in a soil sample. A 1.00-mL portion of the sample was mixed with 24.00 mL of a colorimetric reagent, and the nitrite was converted to a colored product that produced a blank-corrected absorbance of 0.300. To 50.00 mL of the original sample, 1.00 mL of a standard

solution of 1.00×10^{-3} M nitrite was added. The same color-forming procedure was followed, and the new absorbance was 0.530. What was the concentration of nitrite in the original undiluted sample?

8-24. The following atomic absorption results were obtained for determinations of Zn in multivitamin tablets. All absorbance values are corrected for the appropriate reagent blank ($c_{Zn} = 0.0$ ng/mL). The mean value for the blank was 0.0000 with a standard deviation of 0.0047 absorbance units.

c_{Zn}, ng/mL	A
5.0	0.0519
5.0	0.0463
5.0	0.0485
10.0	0.0980
10.0	0.1033
10.0	0.0925
Tablet sample	0.0672
Tablet sample	0.0614
Tablet sample	0.0661

(a) Find the mean absorbance values for the 5.0- and 10.0-ng/mL standards and for the tablet sample. Find the standard deviations of these values.

(b) Find the least-squares best line through the points at $c_{Zn} = 0.0$, 5.0, and 10.0 ng/mL. Find the calibration sensitivity and the analytical sensitivity.

(c) Find the detection limit for a k value of 3. To what level of confidence does this correspond?

(d) Find the concentration of Zn in the tablet sample and the standard deviation in the concentration.

8-25. Atomic emission measurements were made to determine sodium in a blood serum sample. The following emission intensities were obtained for standards of 5.0 and 10.0 ng/mL and for the serum sample. All emission intensities were corrected for any blank emission. The mean value for the blank intensity ($c_{Na} = 0.0$) was 0.000 with a standard deviation of 0.0071 (arbitrary units).

c_{Na}, ng/mL	Emission Intensity
5.0	0.51
5.0	0.49
5.0	0.48
10.0	1.02
10.0	1.00
10.0	0.99
Serum	0.71
Serum	0.77
Serum	0.78

(a) Find the mean emission intensity values for the 5.0- and 10.0-ng/mL standards and for the serum sample. Find the standard deviations of these values.

(b) Find the least squares best line through the points at $c_{Na} = 0.0$, 5.0, and 10.0 ng/mL. Find the calibration sensitivity and the analytical sensitivity.

*(c) Find the detection limit for k values of 2 and 3. To what level of confidence do these correspond?

(d) Find the concentration of Na in the serum sample and the standard deviation of the concentration.

8-26. The following data represent measurements made on a process for 30 days. One measurement was made each day. Assuming that 30 measurements are enough that $\bar{x} \rightarrow \mu$ and $s \rightarrow \sigma$, find the mean of the values, the standard deviation, and the upper and lower control limits. Plot the data points along with the statistical quantities on a chart, and determine whether the process was always in statistical control.

Day	Value	Day	Value	Day	Value
1	49.8	11	49.5	21	58.8
2	48.4	12	50.5	22	51.3
3	49.8	13	48.9	23	50.6
4	50.8	14	49.7	24	48.8
5	49.6	15	48.9	25	52.6
6	50.2	16	48.8	26	54.2
7	51.7	17	48.6	27	49.3
8	50.5	18	48.1	28	47.9
9	47.7	19	53.8	29	51.3
10	50.3	20	49.6	30	49.3

*8-27. The following table gives the sample means and standard deviations for six measurements each day of the purity of a polymer in a process. The purity is monitored for 24 days. Determine the overall mean and standard deviation of the measurements and construct a control chart with upper and lower control limits. Do any of the means indicate a loss of statistical control?

Day	Mean	SD	Day	Mean	SD
1	96.50	0.80	13	96.64	1.59
2	97.38	0.88	14	96.87	1.52
3	96.85	1.43	15	95.52	1.27
4	96.64	1.59	16	96.08	1.16
5	96.87	1.52	17	96.48	0.79
6	95.52	1.27	18	96.63	1.48
7	96.08	1.16	19	95.47	1.30
8	96.48	0.79	20	96.43	0.75
9	96.63	1.48	21	97.06	1.34
10	95.47	1.30	22	98.34	1.60
11	97.38	0.88	23	96.42	1.22
12	96.85	1.43	24	95.99	1.18

8-28. Challenge Problem: Zwanziger and Sârbu[27] conducted a study to validate analytical methods and instruments. The following data are results obtained in the determination of mercury in solid wastes by atomic absorption spectroscopy using two different sample preparation methods: a microwave digestion method and a traditional digestion method.

x, Mercury Concentration, ppm (Traditional)	y, Mercury Concentration, ppm (Microwave)
7.32	5.48
15.80	13.00
4.60	3.29
9.04	6.84
7.16	6.00
6.80	5.84
9.90	14.30
28.70	18.80

(a) Perform a least-squares analysis on the data in the table assuming that the traditional method (x) is the independent variable. Determine the slope, the intercept, the R^2 value, the standard error, and any other relevant statistics.

(b) Plot the results obtained in part (a), and give the equation of the regression line.

(c) Now assume that the microwave digestion method (y) is the independent variable, once again perform a regression analysis, and determine the relevant statistics.

(d) Plot the data in part (c), and determine the regression equation.

(e) Compare the regression equation obtained in (b) with the equation from (d). Why are the equations different?

(f) Is there any conflict between the procedure that you have just performed and the assumptions of the least-squares method? What type of statistical analysis would be more appropriate than linear least-squares in dealing with data sets of this type?

(g) Look up the paper in Footnote 27, and compare your results to those presented for Example 4 in Table 2. You will note that your results from (d) differ from the authors' results. What is the most probable explanation for this discrepancy?

(h) Download the test data found in Table 1 of Footnote 27 from Chapter 8 in **www.cengage.com/chemistry/skoog/fac9**, and perform the same type of analysis for Example 1 and Example 3. Compare your results to those in Table 2 of the paper. Note that in Example 3, you must include all 37 data pairs.

(i) What other methods for dealing with method comparison data are suggested in the paper?

(j) What is implied when we compare two methods by linear regression and the slope is not equal to one? What is implied when the intercept is not zero?

[27]H. W. Zwanziger and C. Sârbu, *Anal. Chem.*, **1998**, *70*, 1277, **DOI**: 10.1021/ac970926y.

PART II

Chemical Equilibria

Aqueous Solutions and Chemical Equilibria

CHAPTER 9

Most analytical techniques require the state of chemical equilibrium. At equilibrium, the rate of a forward process or reaction and that of the reverse process are equal. The photo at right shows the beautiful natural formation called "Frozen Niagara" in Mammoth Cave National Park in Kentucky. As water slowly seeps over the limestone surface of the cave, calcium carbonate dissolves in the water according to the chemical equilibrium

$$CaCO_3(s) + CO_2(g) + H_2O(l) \rightleftharpoons Ca^{2+}(aq) + 2HCO_3^-(aq)$$

The flowing water becomes saturated with calcium carbonate. As carbon dioxide is swept away, the reverse reaction becomes favored, and limestone is deposited in formations whose shapes are governed by the path of the flowing water. Stalactites and stalagmites are examples of similar formations found where water saturated with calcium carbonate drips from the ceiling to the floor of caves over eons.

© Jim Roshan

This chapter presents a fundamental approach to chemical equilibrium, including calculations of chemical composition and of equilibrium concentrations for monoprotic acid/base systems. We also discuss buffer solutions, which are extremely important in many areas of science, and describe the properties of these solutions.

9A THE CHEMICAL COMPOSITION OF AQUEOUS SOLUTIONS

Water is the most plentiful solvent on Earth, is easily purified, and is not toxic. It is, therefore, widely used as a medium for chemical analyses.

9A-1 Classifying Solutions of Electrolytes

Most of the solutes we will discuss are **electrolytes**, which form ions when dissolved in water (or certain other solvents) and thus produce solutions that conduct electricity. **Strong electrolytes** ionize essentially completely in a solvent, but **weak electrolytes** ionize only partially. These characteristics mean that a solution of a weak electrolyte will not conduct electricity as well as a solution containing an equal

TABLE 9-1

Classification of Electrolytes

Strong	Weak
1. Inorganic acids such as HNO_3, $HClO_4$, $H_2SO_4^*$, HCl, HI, HBr, $HClO_3$, $HBrO_3$	1. Many inorganic acids, including H_2CO_3, H_3BO_3, H_3PO_4, H_2S, H_2SO_3
2. Alkali and alkaline-earth hydroxides	2. Most organic acids
3. Most salts	3. Ammonia and most organic bases
	4. Halides, cyanides, and thiocyanates of Hg, Zn, and Cd

*H_2SO_4 is completely dissociated into HSO_4^- and H_3O^+ ions and for this reason is classified as a strong electrolyte. Note, however, that the HSO_4^- ion is a weak electrolyte and is only partially dissociated into SO_4^{2-} and H_3O^+.

concentration of a strong electrolyte. **Table 9-1** shows various solutes that act as strong and weak electrolytes in water. Among the strong electrolytes listed are acids, bases, and **salts**.

9A-2 Acids and Bases

> A **salt** is produced in the reaction of an acid with a base. Examples include $NaCl$, Na_2SO_4, and $NaOOCCH_3$ (sodium acetate).

In 1923, J. N. Brønsted in Denmark and J. M. Lowry in England proposed independently a theory of acid/base behavior that is especially useful in analytical chemistry. According to the Brønsted-Lowry theory, an **acid** is a proton donor, and a **base** is a proton acceptor. For a molecule to behave as an acid, it must encounter a proton acceptor (or base). Likewise, a molecule that can accept a proton behaves as a base if it encounters an acid.

> An **acid** donates protons. A **base** accepts protons.

Conjugate Acids and Bases

> An acid donates protons only in the presence of a proton acceptor (a base). Likewise, a base accepts protons only in the presence of a proton donor (an acid).

An important feature of the Brønsted-Lowry concept is the idea that the product formed when an acid gives up a proton is a potential proton acceptor and is called the **conjugate base** of the parent acid. For example, when the species $acid_1$ gives up a proton, the species $base_1$ is formed, as shown by the reaction

$$acid_1 \rightleftharpoons base_1 + proton$$

> A **conjugate base** is formed when an acid loses a proton. For example, acetate ion is the conjugate base of acetic acid. Similarly, ammonium ion is the conjugate acid of the base ammonia.

We refer to $acid_1$ and $base_1$ as a **conjugate acid/base pair**, or just a **conjugate pair**.

Similarly, every base accepts a proton to produce a **conjugate acid**. That is,

$$base_2 + proton \rightleftharpoons acid_2$$

> A **conjugate acid** is formed when a base accepts a proton.

When these two processes are combined, the result is an acid/base, or **neutralization**, reaction:

$$acid_1 + base_2 \rightleftharpoons base_1 + acid_2$$

> A substance acts as an acid only in the presence of a base and vice versa.

This reaction proceeds to an extent that depends on the relative tendencies of the two bases to accept a proton (or the two acids to donate a proton). Examples of conjugate acid/base relationships are shown in Equations 9-1 through 9-4.

Many solvents are proton donors or proton acceptors and can thus induce basic or acidic behavior in solutes dissolved in them. For example, in an aqueous solution of ammonia, water can donate a proton and acts as an acid with respect to the solute NH_3:

$$NH_3 + H_2O \rightleftharpoons NH_4^+ + OH^- \qquad (9\text{-}1)$$

In this reaction, ammonia (base$_1$) reacts with water, which is labeled acid$_2$, to give the conjugate acid ammonium ion (acid$_1$) and hydroxide ion, which is the conjugate base (base$_2$) of the acid water.

On the other hand, water acts as a proton acceptor, or base, in an aqueous solution of nitrous acid:

$$H_2O + HNO_2 \rightleftharpoons H_3O^+ + NO_2^- \qquad (9\text{-}2)$$

The conjugate base of the acid HNO_2 is nitrite ion. The conjugate acid of water is the hydrated proton written as H_3O^+. This species is called the **hydronium ion**, and it consists of a proton covalently bonded to a single water molecule. Higher hydrates such as $H_5O_2^+$, $H_9O_4^+$, and the dodecahedral cage structure shown in **Figure 9-1** may also appear in aqueous solutions of protons. For convenience, however, we generally use the notation H_3O^+, or more simply H^+, when we write chemical equations containing the hydrated proton.

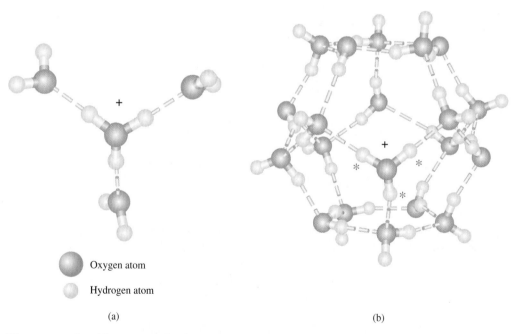

● Oxygen atom

○ Hydrogen atom

(a)   (b)

Figure 9-1 Possible structures for the hydronium ion. (a) The species $H_9O_4^+$ has been observed in the solid state and may be an important contributor in aqueous solution. (b) The species $(H_2O)_{20}H^+$ exhibits a dodecahedral caged structure. The extra proton in the structure, which may be any one of the three marked with an asterisk, is free to move around the surface of the dodecahedron by being transferred to an adjacent water molecule.

Svante Arrhenius (1859–1927), Swedish chemist, formulated many of the early ideas regarding ionic dissociation in solution. His ideas were not accepted at first. In fact, he was given the lowest possible passing grade for his Ph.D. examination in 1884. In 1903, Arrhenius was awarded the Nobel Prize in chemistry for his revolutionary ideas. He was one of the first scientists to suggest the relationship between the amount of carbon dioxide in the atmosphere and global temperature, a phenomenon that has come to be known as the **greenhouse effect**. You may like to read Arrhenius's original paper "On the Influence of Carbonic Acid in the Air upon the Temperature of the Ground," *London Edinburgh Dublin Philos. Mag. J. Sci.*, **1896**, *41*, 237–276.

A **zwitterion** is an ion that has both a positive and a negative charge.

Water can act as either an acid or a base.

Amphiprotic solvents behave as acids in the presence of basic solutes and bases in the presence of acidic solutes.

An acid that has donated a proton becomes a conjugate base capable of accepting a proton to reform the original acid. Similarly, a base that has accepted a proton becomes a conjugate acid that can donate a proton to form the original base. Thus, nitrite ion, the species produced by the loss of a proton from nitrous acid, is a potential acceptor of a proton from a suitable donor. It is this reaction that causes an aqueous solution of sodium nitrite to be slightly basic:

$$\underset{\text{base}_1}{NO_2^-} + \underset{\text{acid}_2}{H_2O} \rightleftharpoons \underset{\substack{\text{conjugate} \\ \text{acid}_1}}{HNO_2} + \underset{\substack{\text{conjugate} \\ \text{base}_2}}{OH^-}$$

9A-3 Amphiprotic Species

Species that have both acidic and basic properties are **amphiprotic**. An example is dihydrogen phosphate ion, $H_2PO_4^-$, which behaves as a base in the presence of a proton donor such as H_3O^+.

$$\underset{\text{base}_1}{H_2PO_4^-} + \underset{\text{acid}_2}{H_3O^+} \rightleftharpoons \underset{\text{acid}_1}{H_3PO_4} + \underset{\text{base}_2}{H_2O}$$

Here, H_3PO_4 is the conjugate acid of the original base. In the presence of a proton acceptor, such as hydroxide ion, however, $H_2PO_4^-$ behaves as an acid and donates a proton to form the conjugate base HPO_4^{2-}.

$$\underset{\text{acid}_1}{H_2PO_4^-} + \underset{\text{base}_2}{OH^-} \rightleftharpoons \underset{\text{base}_1}{HPO_4^{2-}} + \underset{\text{acid}_2}{H_2O}$$

The simple amino acids are an important class of amphiprotic compounds that contain both a weak acid and a weak base functional group. When dissolved in water, an amino acid, such as glycine, undergoes a kind of internal acid/base reaction to produce a **zwitterion**—a species that has both a positive and a negative charge. Thus,

$$\underset{\text{glycine}}{NH_2CH_2COOH} \rightleftharpoons \underset{\text{zwitterion}}{NH_3^+CH_2COO^-}$$

This reaction is analogous to the acid/base reaction between a carboxylic acid and an amine:

$$\underset{\text{acid}_1}{R'COOH} + \underset{\text{base}_2}{R''NH_2} \rightleftharpoons \underset{\text{base}_1}{R'COO^-} + \underset{\text{acid}_2}{R''NH_3^+}$$

Water is the classic example of an **amphiprotic solvent**, that is, a solvent that can act either as an acid (Equation 9-1) or as a base (Equation 9-2), depending on the solute. Other common amphiprotic solvents are methanol, ethanol, and anhydrous acetic acid. In methanol, for example, the equilibria analogous to the water equilibria shown in Equations 9-1 and 9-2 are

$$\underset{\text{base}_1}{NH_3} + \underset{\text{acid}_2}{CH_3OH} \rightleftharpoons \underset{\substack{\text{conjugate} \\ \text{acid}_1}}{NH_4^+} + \underset{\text{base}_2}{CH_3O^-} \qquad (9\text{-}3)$$

$$\underset{\text{base}_1}{CH_3OH} + \underset{\text{acid}_2}{HNO_2} \rightleftharpoons \underset{\substack{\text{conjugate} \\ \text{acid}_1}}{CH_3OH_2^+} + \underset{\substack{\text{conjugate} \\ \text{base}_2}}{NO_2^-} \qquad (9\text{-}4)$$

9A-4 Autoprotolysis

Amphiprotic solvents undergo self-ionization, or **autoprotolysis**, to form a pair of ionic species. Autoprotolysis is yet another example of acid/base behavior, as illustrated by the following equations:

$$
\begin{array}{llll}
\text{base}_1 & + \text{acid}_2 & \rightleftharpoons \text{acid}_1 & + \text{base}_2 \\
H_2O & + H_2O & \rightleftharpoons H_3O^+ & + OH^- \\
CH_3OH & + CH_3OH & \rightleftharpoons CH_3OH_2^+ & + CH_3O^- \\
HCOOH & + HCOOH & \rightleftharpoons HCOOH_2^+ & + HCOO^- \\
NH_3 & + NH_3 & \rightleftharpoons NH_4^+ & + NH_2^-
\end{array}
$$

The extent to which water undergoes autoprotolysis at room temperature is slight. Thus, the hydronium and hydroxide ion concentrations in pure water are only about 10^{-7} M. Despite the small values of these concentrations, this dissociation reaction is of utmost importance in understanding the behavior of aqueous solutions.

9A-5 Strengths of Acids and Bases

Figure 9-2 shows the dissociation reactions of a few common acids in water. The first two are **strong acids** because reaction with the solvent is sufficiently complete that no undissociated solute molecules are left in aqueous solution. The rest are **weak acids**, which react incompletely with water to give solutions containing significant quantities of both the parent acid and its conjugate base. Note that acids can be cationic, anionic, or electrically neutral. The same holds for bases.

The acids in Figure 9-2 become progressively weaker from top to bottom. Perchloric acid and hydrochloric acid are completely dissociated, but only about 1% of acetic acid ($HC_2H_3O_2$) is dissociated. Ammonium ion is an even weaker acid with only about 0.01% of this ion being dissociated into hydronium ions and ammonia molecules. Another generality illustrated in Figure 9-2 is that the weakest acid forms the strongest conjugate base, that is, ammonia has a much stronger affinity for protons than any base above it. Perchlorate and chloride ions have no affinity for protons.

The tendency of a solvent to accept or donate protons determines the strength of a solute acid or base dissolved in it. For example, perchloric and hydrochloric acids are strong acids in water. If anhydrous acetic acid, a weaker proton acceptor than water, is substituted *as the solvent*, neither of these acids undergoes complete dissociation. Instead, equilibria such as the following are established:

$$
\underset{\text{base}_1}{CH_3COOH} + \underset{\text{acid}_2}{HClO_4} \rightleftharpoons \underset{\text{acid}_1}{CH_3COOH_2^+} + \underset{\text{base}_2}{ClO_2^-}
$$

Perchloric acid is, however, about 5000 times stronger than hydrochloric acid in this solvent. Acetic acid thus acts as a **differentiating solvent** toward the two acids by

Autoprotolysis (also called autoionization) is the spontaneous reaction of molecules of a substance to give a pair of ions.

The **hydronium ion** is the hydrated proton formed when water reacts with an acid. It is usually formulated as H_3O^+, although there are several possible higher hydrates, as shown in Figure 9-1.

◄ In this text, we use the symbol H_3O^+ in those chapters that deal with acid/base equilibria and acid/base equilibrium calculations. In the remaining chapters, we simplify to the more convenient H^+, with the understanding that this symbol represents the hydrated proton.

◄ The common strong bases include NaOH, KOH, Ba(OH)$_2$, and the quaternary ammonium hydroxide R$_4$NOH, where R is an alkyl group such as CH_3 or C_2H_5.

◄ The common strong acids include HCl, HBr, HI, HClO$_4$, HNO$_3$, the first proton in H_2SO_4, and the organic sulfonic acid RSO$_3$H.

$$
\begin{array}{l}
\boxed{\text{Strongest acid}} \quad HClO_4 + H_2O \rightarrow H_3O^+ + ClO_4^- \qquad \boxed{\text{Weakest base}} \\
\qquad\qquad HCl + H_2O \rightarrow H_3O^+ + Cl^- \\
\qquad\qquad H_3PO_4 + H_2O \rightleftharpoons H_3O^+ + H_2PO_4^- \\
\qquad\qquad Al(H_2O)_6^{3+} + H_2O \rightleftharpoons H_3O^+ + AlOH(H_2O)_5^{2+} \\
\qquad\qquad HC_2H_3O_2 + H_2O \rightleftharpoons H_3O^+ + C_2H_3O_2^- \\
\boxed{\text{Weakest acid}} \quad H_2PO_4^- + H_2O \rightleftharpoons H_3O^+ + HPO_3^{2-} \qquad \boxed{\text{Strongest base}} \\
\qquad\qquad NH_4^+ + H_2O \rightleftharpoons H_3O^+ + NH_3
\end{array}
$$

Figure 9-2 Dissociation reactions and relative strengths of some common acids and their conjugate bases. Note that HCl and HClO$_4$ are completely dissociated in water.

In a **differentiating solvent**, various acids dissociate to different degrees and have different strengths. In a **leveling solvent**, several acids are completely dissociated and show the same strength.

Of all the acids listed in the marginal note on page 200 and in Figure 9-2, only perchloric acid is a strong acid in methanol and ethanol. Therefore, these two alcohols are also differentiating solvents.

revealing the inherent differences in their acidities. Water, on the other hand, is a **leveling solvent** for perchloric, hydrochloric, and nitric acids because all three are completely ionized in this solvent and show no differences in strength. There are differentiating and leveling solvents for bases as well.

9B CHEMICAL EQUILIBRIUM

Many reactions used in analytical chemistry never result in complete conversion of reactants to products. Instead, they proceed to a state of **chemical equilibrium** in which the ratio of concentrations of reactants and products is constant. **Equilibrium-constant expressions** are *algebraic* equations that describe the concentration relationships among reactants and products at equilibrium. Among other things, equilibrium-constant expressions permit calculation of the error in an analysis resulting from the quantity of unreacted analyte that remains when equilibrium has been reached.

In the discussion that follows, we cover the use of equilibrium-constant expressions to gain information about analytical systems in which no more than one or two equilibria are present. Chapter 11 extends these methods to systems containing several simultaneous equilibria. Such complex systems are often found in analytical chemistry.

9B-1 The Equilibrium State

Consider the chemical reaction

$$H_3AsO_4 + 3I^- + 2H^+ \rightleftharpoons H_3AsO_3 + I_3^- + H_2O \tag{9-5}$$

We can follow the rate of this reaction and the extent to which it proceeds to the right by monitoring the appearance of the orange-red color of the triiodide ion I_3^-. (The other participants in the reaction are colorless.) For example, if 1 mmol of arsenic acid, H_3AsO_4, is added to 100 mL of a solution containing 3 mmol of potassium iodide, the red color of the triiodide ion appears almost immediately. Within a few seconds, the intensity of the color becomes constant, showing that the triiodide concentration has become constant (see color plates 1b and 2b).

A solution of identical color intensity (and hence identical triiodide concentration) can also be produced by adding 1 mmol of arsenous acid, H_3AsO_3, to 100 mL of a solution containing 1 mmol of triiodide ion (see color plate 1a). Here, the color intensity is initially greater than in the first solution but rapidly decreases as a result of the reaction

$$H_3AsO_3 + I_3^- + H_2O \rightleftharpoons H_3AsO_4 + 3I^- + 2H^+$$

Ultimately, the color of the two solutions is identical. Many other combinations of the four reactants can be used to yield solutions that are indistinguishable from the two just described.

The final position of a chemical equilibrium is independent of the route to the equilibrium state.

The results of the experiments shown in color plates 1–3 illustrate that the concentration relationship at chemical equilibrium (that is, the *position of equilibrium*) is independent of the route to the equilibrium state. This relationship is altered by applying stress to the system, however. Such stresses include changes in temperature, in pressure (if one of the reactants or products is a gas), or in total concentration of a reactant or a product. These effects can be predicted qualitatively from the **Le Châtelier's principle**. This principle states that the position of chemical equilibrium always shifts in a direction that tends to relieve the effect of an applied stress. For example, an increase in temperature of a system alters the concentration relationship in the direction that tends to absorb heat, and an increase in pressure favors those participants that occupy a smaller total volume.

In an analysis, the effect of introducing an additional amount of a reactant or product to the reaction mixture is particularly important. The resulting stress is relieved by a shift in equilibrium in the direction that tends to use up the added substance. Thus, for the equilibrium we have been considering (Equation 9-5), the addition of arsenic acid (H_3AsO_4) or hydrogen ions causes an increase in color as more triiodide ion and arsenous acid are formed. Adding arsenous acid has the reverse effect. An equilibrium shift brought about by changing the amount of one of the participating reactants or products is called a **mass-action effect**.

Theoretical and experimental studies of reacting systems on the molecular level show that reactions among the participating species continue even after equilibrium is achieved. The concentration ratio of reactants and products is constant because the rates of the forward and reverse reactions are precisely equal. In other words, chemical equilibrium is a dynamic state in which the rates of the forward and reverse reactions are identical.

9B-2 Equilibrium-Constant Expressions

The influence of concentration or pressure (if the participants are gases) on the position of a chemical equilibrium is conveniently described in quantitative terms by means of an equilibrium-constant expression. These expressions are derived from thermodynamics. They are important because they allow us to predict the direction and completeness of chemical reactions. An equilibrium-constant expression, however, yields no information concerning the rate of a reaction. In fact, we sometimes find reactions that have highly favorable equilibrium constants but are of little analytical use because they are so slow. This limitation can often be overcome by the use of a catalyst, which speeds the approach to equilibrium without changing its position.

Consider a generalized equation for a chemical equilibrium

$$wW + xX \rightleftharpoons yY + zZ \tag{9-6}$$

where the capital letters represent the formulas of participating chemical reactants and products, and the lowercase italic letters are the small whole numbers required to balance the equation. Thus, the equation says that w moles of W react with x moles of X to form y moles of Y and z moles of Z. The equilibrium-constant expression for this reaction is

$$K = \frac{[Y]^y[Z]^z}{[W]^w[X]^x} \tag{9-7}$$

where the square-bracketed terms are:

1. molar concentrations if they represent dissolved solutes.
2. partial pressures in atmospheres if they are gas-phase reactants or products. In such an instance, we will often replace the square bracketed term (say [Z] in Equation 9-7) with the symbol p_z, which stands for the partial pressure of the gas Z in atmospheres.

Le Châtelier's principle states that the position of an equilibrium always shifts in such a direction as to relieve a stress that is applied to the system.

The **mass-action effect** is a shift in the position of an equilibrium caused by adding one of the reactants or products to a system.

❮ Equilibrium is a dynamic process. Although chemical reactions appear to stop at equilibrium, in fact, the amounts of reactants and products are constant because the rates of the forward and reverse processes are exactly the same.

Chemical thermodynamics is a branch of chemistry that concerns the flow of heat and energy in chemical reactions. The position of a chemical equilibrium is related to these energy changes.

❮ Equilibrium-constant expressions provide *no* information about whether a chemical reaction is fast enough to be useful in an analytical procedure.

Cato Guldberg (1836–1902) and Peter Waage (1833–1900) were Norwegian chemists whose primary interests were in the field of thermodynamics. In 1864, these workers were the first to propose the law of mass action, which is expressed in Equation 9-7. If you would like to learn more about Guldberg and Waage and read a translation of their original paper on the law of mass action, go to www.cengage.com/chemistry/skoog/fac9, choose Chapter 9, and go to the Web Works.

$[Z]^z$ in Equation 9-7 is replaced with p_z in atmospheres if Z is a gas. No term for Z is included in the equation if this species is a pure solid, a pure liquid, or the solvent of a dilute solution.

Remember: Equation 9-7 is only an approximate form of an equilibrium-constant expression. The exact expression takes the form

$$K = \frac{a_Y^y a_Z^z}{a_W^w a_X^x} \qquad (9\text{-}8)$$

where a_Y, a_Z, a_W, and a_X are the *activities* of species Y, Z, W, and X (see Section 10B).

If a reactant or product in Equation 9-7 is a pure liquid, a pure solid, or the solvent present in excess, no term for this species appears in the equilibrium-constant expression. For example, if Z in Equation 9-6 is the solvent H_2O, the equilibrium-constant expression simplifies to

$$K = \frac{[Y]^y}{[W]^w [X]^x}$$

We discuss the basis for this simplification in the sections that follow.

The constant K in Equation 9-7 is a temperature-dependent numerical quantity called the *equilibrium constant*. By convention, the concentrations of the products, *as the equation is written*, are always placed in the numerator and the concentrations of the reactants are always in the denominator.

Equation 9-7 is only an approximate form of a thermodynamic equilibrium-constant expression. The exact form is given by Equation 9-8 (in the margin). Generally, we use the approximate form of this equation because it is less tedious and time consuming. In Section 10B, we show when the use of Equation 9-7 is likely to lead to serious errors in equilibrium calculations and how Equation 9-8 can be modified in these cases.

9B-3 Types of Equilibrium Constants in Analytical Chemistry

Table 9-2 summarizes the types of chemical equilibria and equilibrium constants that are of importance in analytical chemistry. Basic applications of some of these constants are illustrated in the three sections that follow.

TABLE 9-2

Equilibria and Equilibrium Constants Important in Analytical Chemistry

Type of Equilibrium	Name and Symbol of Equilibrium-Constant	Typical Example	Equilibrium-Constant Expression
Dissociation of water	Ion-product constant, K_w	$2H_2O \rightleftharpoons H_3O^+ + OH^-$	$K_w = [H_3O^+][OH^-]$
Heterogeneous equilibrium between a slightly soluble substance and its ions in a saturated solution	Solubility product, K_{sp}	$BaSO_4(s) \rightleftharpoons Ba^{2+} + SO_4^{2-}$	$K_{sp} = [Ba^{2+}][SO_4^{2-}]$
Dissociation of a weak acid or base	Dissociation constant, K_a or K_b	$CH_3COOH + H_2O \rightleftharpoons H_3O^+ + CH_3COO^-$ $CH_3COO^- + H_2O \rightleftharpoons OH^- + CH_3COOH$	$K_a = \dfrac{[H_3O^+][CH_3COO^-]}{[CH_3COOH]}$ $K_b = \dfrac{[OH^-][CH_3COOH]}{[CH_3COO^-]}$
Formation of a complex ion	Formation constant, β_n	$Ni^{2+} + 4CN^- \rightleftharpoons Ni(CN)_4^{2-}$	$\beta_4 = \dfrac{[Ni(CN)_4^{2-}]}{[Ni^{2+}][CN^-]^4}$
Oxidation/reduction equilibrium	K_{redox}	$MnO_4^- + 5Fe^{2+} + 8H^+ \rightleftharpoons Mn^{2+} + 5Fe^{3+} + 4H_2O$	$K_{redox} = \dfrac{[Mn^{2+}][Fe^{3+}]^5}{[MnO_4^-][Fe^{2+}]^5[H^+]^8}$
Distribution equilibrium for a solute between immiscible solvents	K_d	$I_2(aq) \rightleftharpoons I_2(org)$	$K_d = \dfrac{[I_2]_{org}}{[I_2]_{aq}}$

FEATURE 9-1

Stepwise and Overall Formation Constants for Complex Ions

The formation of $Ni(CN)_4^{2-}$ (Table 9-2) is typical in that it occurs in steps as shown. Note that **stepwise formation constants** are symbolized by K_1, K_2, and so forth.

$$Ni^{2+} + CN^- \rightleftharpoons Ni(CN)^+ \qquad K_1 = \frac{[Ni(CN)^+]}{[Ni^{2+}][CN^-]}$$

$$Ni(CN)^+ + CN^- \rightleftharpoons Ni(CN)_2 \qquad K_2 = \frac{[Ni(CN)_2]}{[Ni(CN)^+][CN^-]}$$

$$Ni(CN)_2 + CN^- \rightleftharpoons Ni(CN)_3^- \qquad K_3 = \frac{[Ni(CN)_3^-]}{[Ni(CN)_2][CN^-]}$$

$$Ni(CN)_3^- + CN^- \rightleftharpoons Ni(CN)_4^{2-} \qquad K_4 = \frac{[Ni(CN)_4^{2-}]}{[Ni(CN)_3^-][CN^-]}$$

Overall constants are designated by the symbol β_n. Thus,

$$Ni^{2+} + 2CN^- \rightleftharpoons Ni(CN)_2 \qquad \beta_2 = K_1K_2 = \frac{[Ni(CN)_2]}{[Ni^{2+}][CN^-]^2}$$

$$Ni^{2+} + 3CN^- \rightleftharpoons Ni(CN)_3^- \qquad \beta_3 = K_1K_2K_3 = \frac{[Ni(CN)_3^-]}{[Ni^{2+}][CN^-]^3}$$

$$Ni^{2+} + 4CN^- \rightleftharpoons Ni(CN)_4^{2-} \qquad \beta_4 = K_1K_2K_3K_4 = \frac{[Ni(CN)_4^{2-}]}{[Ni^{2+}][CN^-]^4}$$

9B-4 Applying the Ion-Product Constant for Water

Aqueous solutions contain small concentrations of hydronium and hydroxide ions as a result of the dissociation reaction

$$2H_2O \rightleftharpoons H_3O^+ + OH^- \tag{9-9}$$

An equilibrium constant for this reaction can be written as shown in Equation 9-7:

$$K = \frac{[H_3O^+][OH^-]}{[H_2O]^2} \tag{9-10}$$

The concentration of water in dilute aqueous solutions is enormous, however, when compared with the concentration of hydronium and hydroxide ions. As a result, $[H_2O]^2$ in Equation 9-10 can be taken as constant, and we write

$$K[H_2O]^2 = K_w = [H_3O^+][OH^-] \tag{9-11}$$

where the new constant K_w is given a special name, the **ion-product constant for water**.

> If we take the negative logarithm of Equation 9-11, we discover a very useful relationship.
>
> $$-\log K_w = -\log[H_3O^+] - \log[OH^-]$$
>
> By the definition of p-function, (see Section 4B-1)
>
> $$pK_w = pH + pOH \tag{9-12}$$
>
> At 25°C, $pK_w = 14.00$.

FEATURE 9-2

Why [H_2O] Does Not Appear in Equilibrium-Constant Expressions for Aqueous Solutions

In a dilute aqueous solution, the molar concentration of water is

$$[H_2O] = \frac{1000 \text{ g } H_2O}{L \text{ } H_2O} \times \frac{1 \text{ mol } H_2O}{18.0 \text{ g } H_2O} = 55.6 \text{ M}$$

Suppose we have 0.1 mol of HCl in 1 L of water. The presence of this acid will shift the equilibrium shown in Equation 9-9 to the left. Originally, however, there was only 10^{-7} mol/L OH^- to consume the added protons. Therefore, even if all the OH^- ions are converted to H_2O, the water concentration will increase to only

$$[H_2O] = 55.6 \frac{\text{mol } H_2O}{L \text{ } H_2O} + 1 \times 10^{-7} \frac{\text{mol } OH^-}{L \text{ } H_2O} \times \frac{1 \text{ mol } H_2O}{\text{mol } OH^-} \approx 55.6 \text{ M}$$

The percent change in water concentration is

$$\frac{10^{-7} \text{ M}}{55.6 \text{ M}} \times 100\% = 2 \times 10^{-7}\%$$

which is insignificant. Thus, $K[H_2O]^2$ in Equation 9-10 is for all practical purposes a constant, that is,

$$K(55.6)^2 = K_w = 1.00 \times 10^{-14} \text{ at } 25°C$$

TABLE 9-3

Variation of K_w with Temperature

Temperature, °C	K_w
0	0.114×10^{-14}
25	1.01×10^{-14}
50	5.47×10^{-14}
75	19.9×10^{-14}
100	49×10^{-14}

At 25°C, the ion-product constant for water is 1.008×10^{-14}. For convenience, we use the approximation that at room temperature $K_w \approx 1.00 \times 10^{-14}$. **Table 9-3** shows how K_w depends on temperature. The ion-product constant for water permits us to easily find the hydronium and hydroxide ion concentrations of aqueous solutions.

EXAMPLE 9-1

Calculate the hydronium and hydroxide ion concentrations of pure water at 25°C and 100°C.

Solution

Because OH^- and H_3O^+ are formed only from the dissociation of water, their concentrations must be equal:

$$[H_3O^+] = [OH^-]$$

We substitute this equality into Equation 9-11 to give

$$[H_3O^+]^2 = [OH^-]^2 = K_w$$
$$[H_3O^+] = [OH^-] = \sqrt{K_w}$$

At 25°C,

$$[H_3O^+] = [OH^-] = \sqrt{1.00 \times 10^{-14}} = 1.00 \times 10^{-7}\ M$$

At 100°C, from Table 9-3,

$$[H_3O^+] = [OH^-] = \sqrt{49 \times 10^{-14}} = 7.0 \times 10^{-7}\ M$$

EXAMPLE 9-2

Calculate the hydronium and hydroxide ion concentrations and the pH and pOH of 0.200 M aqueous NaOH at 25°C.

Solution

Sodium hydroxide is a strong electrolyte, and its contribution to the hydroxide ion concentration in this solution is 0.200 mol/L. As in Example 9-1, hydroxide ions and hydronium ions are formed in equal amounts from the dissociation of water. Therefore, we write

$$[OH^-] = 0.200 + [H_3O^+]$$

where $[H_3O^+]$ is equal to the hydroxide ion concentration from the dissociation of water. The concentration of OH^- from the water is insignificant, however, compared with 0.200, so we can write

$$[OH^-] \approx 0.200$$

$$pOH = -\log 0.200 = 0.699$$

Equation 9-11 is then used to calculate the hydronium ion concentration:

$$[H_3O^+] = \frac{K_w}{[OH^-]} = \frac{1.00 \times 10^{-14}}{0.200} = 5.00 \times 10^{-14}\ M$$

$$pH = -\log 0.500 \times 10^{-14} = 13.301$$

Note that the approximation

$$[OH^-] = 0.200 + 5.00 \times 10^{-14} \approx 0.200\ M$$

causes no significant error in our answer.

9B-5 Using Solubility-Product Constants

Most, but not all, sparingly soluble salts are essentially completely dissociated in saturated aqueous solution. For example, when an excess of barium iodate is equilibrated with water, the dissociation process is adequately described by the equation

$$Ba(IO_3)_2(s) \rightleftharpoons Ba^{2+}(aq) + 2IO_3^-(aq)$$

> When we say that a sparingly soluble salt is completely dissociated, *we do not imply* that all of the salt dissolves. What we mean is that the very small amount that *does* go into solution dissociates completely.

What does it mean to say that "an excess of barium iodate is equilibrated with water"? It means that more solid barium iodate is added to a portion of water than would dissolve at the temperature of the experiment. Some solid $BaIO_3$ is in contact with the solution.

Using Equation 9-7, we write

$$K = \frac{[Ba^{2+}][IO_3^-]^2}{[Ba(IO_3)_2(s)]}$$

The denominator represents the molar concentration of $Ba(IO_3)_2$ *in the solid*, which is a phase that is separate from but in contact with the saturated solution. The concentration of a compound in its solid state is, however, constant. In other words, the number of moles of $Ba(IO_3)_2$ divided by the *volume* of the solid $Ba(IO_3)_2$ is constant no matter how much excess solid is present. Therefore, the previous equation can be rewritten in the form

For Equation 9-13 to be valid, it is necessary only that *some solid be present. You should always keep in mind that if there is no $Ba(IO_3)(s)$ in contact with the solution, Equation 9-13 is not applicable.*

$$K[Ba(IO_3)_2(s)] = K_{sp} = [Ba^{2+}][IO_3^-]^2 \qquad (9\text{-}13)$$

where the new constant is called the **solubility-product constant** or the **solubility product**. It is important to appreciate that Equation 9-13 shows that the position of this equilibrium is independent of the *amount* of $Ba(IO_3)_2$ as long as some solid is present. In other words, it does not matter whether the amount is a few milligrams or several grams.

A table of solubility-product constants for numerous inorganic salts is found in Appendix 2. The examples that follow demonstrate some typical uses of solubility-product expressions. Further applications are considered in later chapters.

The Solubility of a Precipitate in Pure Water

With the solubility-product expression, we can calculate the solubility of a sparingly soluble substance that ionizes completely in water.

EXAMPLE 9-3

What mass (in grams) of $Ba(IO_3)_2$ (487 g/mol) can be dissolved in 500 mL of water at 25°C?

Solution

The solubility-product constant for $Ba(IO_3)_2$ is 1.57×10^{-9} (see Appendix 2). The equilibrium between the solid and its ions in solution is described by the equation

$$Ba(IO_3)_2(s) \rightleftharpoons Ba^{2+} + 2IO_3^-$$

and so

$$K_{sp} = [Ba^{2+}][IO_3^-]^2 = 1.57 \times 10^{-9}$$

The equation describing the equilibrium reveals that 1 mol of Ba^{2+} is formed for each mole of $Ba(IO_3)_2$ that dissolves. Therefore,

$$\text{molar solubility of } Ba(IO_3)_2 = [Ba^{2+}]$$

Since two moles of iodate are produced for each mole of barium ion, the iodate concentration is twice the barium ion concentration:

$$[IO_3^-] = 2[Ba^{2+}]$$

Substituting this last equation into the equilibrium-constant expression gives

$$[Ba^{2+}](2[Ba^{2+}])^2 = 4[Ba^{2+}]^3 = 1.57 \times 10^{-9}$$

$$[Ba^{2+}] = \left(\frac{1.57 \times 10^{-9}}{4}\right)^{1/3} = 7.32 \times 10^{-4} \text{ M}$$

Since 1 mol Ba^{2+} is produced for every mole of $Ba(IO_3)_2$,

$$\text{solubility} = 7.32 \times 10^{-4} \text{ M}$$

To compute the number of millimoles of $Ba(IO_3)_2$ dissolved in 500 mL of solution, we write

$$\text{no. mmol } Ba(IO_3)_2 = 7.32 \times 10^{-4} \frac{\text{mmol } Ba(IO_3)_2}{\text{mL}} \times 500 \text{ mL}$$

The mass of $Ba(IO_3)_2$ in 500 mL is given by

$$\text{mass } Ba(IO_3)_2 =$$

$$(7.32 \times 10^{-4} \times 500) \text{ mmol } Ba(IO_3)_2 \times 0.487 \frac{\text{g } Ba(IO_3)_2}{\text{mmol } Ba(IO_3)_2}$$

$$= 0.178 \text{ g}$$

> Notice that the molar solubility is equal to $[Ba^{2+}]$ or to $\frac{1}{2}[IO_3^-]$.

The Effect of a Common Ion on the Solubility of a Precipitate

The **common-ion effect** is a mass-action effect predicted from Le Châtelier's principle and is demonstrated by the following examples.

> The solubility of an ionic precipitate decreases when a soluble compound containing one of the ions of the precipitate is added to the solution (see color plate 4). This behavior is called the **common-ion effect**.

EXAMPLE 9-4

Calculate the molar solubility of $Ba(IO_3)_2$ in a solution that is 0.0200 M in $Ba(NO_3)_2$.

Solution

The solubility is not equal to $[Ba^{2+}]$ in this case because $Ba(NO_3)_2$ is also a source of barium ions. We know, however, that the solubility is related to $[IO_3^-]$:

$$\text{molar solubility of } Ba(IO_3)_2 = \tfrac{1}{2}[IO_3^-]$$

There are two sources of barium ions: $Ba(NO_3)_2$ and $Ba(IO_3)_2$. The contribution from the nitrate is 0.0200 M, and that from the iodate is equal to the molar solubility, or $\tfrac{1}{2}[IO_3^-]$. Thus,

$$[Ba^{2+}] = 0.0200 + \tfrac{1}{2}[IO_3^-]$$

By substituting these quantities into the solubility-product expression, we find that

$$\left(0.0200 + \tfrac{1}{2}[IO_3^-]\right)[IO_3^-]^2 = 1.57 \times 10^{-9}$$

(continued)

Since this is a cubic equation, we would like to make an assumption that would simplify the algebra required to find $[IO_3^-]$. The small numerical value of K_{sp} suggests that the solubility of $Ba(IO_3)_2$ is quite small, and this finding is confirmed by the result obtained in Example 9-3. Also, barium ion from $Ba(NO_3)_2$ will further suppress the limited solubility of $Ba(IO_3)_2$. Therefore, it seems reasonable to assume that 0.0200 is large with respect to $\frac{1}{2}[IO_3^-]$ in order to find a provisional answer to the problem. That is, we assume that $\frac{1}{2}[IO_3^-] \ll 0.0200$, so

$$[Ba^{2+}] = 0.0200 + \tfrac{1}{2}[IO_3^-] \approx 0.0200 \text{ M}$$

The original equation then simplifies to

$$0.0200\,[IO_3^-]^2 = 1.57 \times 10^{-9}$$
$$[IO_3^-] = \sqrt{1.57 \times 10^{-9}/0.0200} = \sqrt{7.85 \times 10^{-8}} = 2.80 \times 10^{-4} \text{ M}$$

The assumption that $(0.0200 + \frac{1}{2} \times 2.80 \times 10^{-4}) \approx 0.0200$ causes minimal error because the second term, representing the amount of Ba^{2+} arising from the dissociation of $Ba(IO_3)_2$, is only about 0.7% of 0.0200. Usually, we consider an assumption of this type to be satisfactory if the discrepancy is less than 10%.[1] Finally, then,

$$\text{solubility of } Ba(IO_3)_2 = \tfrac{1}{2}[IO_3^-] = \tfrac{1}{2} \times 2.80 \times 10^{-4} = 1.40 \times 10^{-4} \text{ M}$$

If we compare this result with the solubility of barium iodate in pure water (Example 9-3), we see that the presence of a small concentration of the common ion has decreased the molar solubility of $Ba(IO_3)_2$ by a factor of about 5.

EXAMPLE 9-5

Calculate the solubility of $Ba(IO_3)_2$ in a solution prepared by mixing 200 mL of 0.0100 M $Ba(NO_3)_2$ with 100 mL of 0.100 M $NaIO_3$.

Solution

First, establish whether either reactant is present in excess at equilibrium. The amounts taken are

$$\text{no. mmol } Ba^{2+} = 200 \text{ mL} \times 0.0100 \text{ mmol/mL} = 2.00$$

$$\text{no. mmol } IO_3^- = 100 \text{ mL} \times 0.100 \text{ mmol/mL} = 10.0$$

If the formation of $Ba(IO_3)_2$ is complete,

$$\text{no. mmol excess } NaIO_3 = 10.0 - 2 \times 2.00 = 6.0$$

---

[1]Ten percent error is a somewhat arbitrary cutoff, but since we do not consider activity coefficients in our calculations, which often create errors of at least 10%, our choice is reasonable. Many general chemistry and analytical chemistry texts suggest that 5% error is appropriate, but such decisions should be based on the goal of the calculation. If you require an exact answer, the method of successive approximations presented in Feature 9-4 may be used. A spreadsheet solution may be appropriate for complex examples.

Thus,

$$[\text{IO}_3^-] = \frac{6.0\ \text{mmol}}{200\ \text{mL} + 100\ \text{mL}} = \frac{6.0\ \text{mmol}}{300\ \text{mL}} = 0.0200\ \text{M}$$

The uncertainty in $[\text{IO}_3^-]$ is 0.1 part in 6.0 or 1 part in 60. Thus, $0.0200\ (1/60) = 0.0003$, and we round to 0.0200 M.

As in Example 9-3,

$$\text{molar solubility of Ba(IO}_3)_2 = [\text{Ba}^{2+}]$$

In this case, however,

$$[\text{IO}_3^-] = 0.0200 + 2[\text{Ba}^{2+}]$$

where $2[\text{Ba}^{2+}]$ represents the iodate contributed by the sparingly soluble $\text{Ba(IO}_3)_2$. We find a provisional answer after making the assumption that $[\text{IO}_3^-] \approx 0.0200$. Therefore,

$$\text{solubility of Ba(IO}_3)_2 = [\text{Ba}^{2+}] = \frac{K_{sp}}{[\text{IO}_3^-]^2} = \frac{1.57 \times 10^{-9}}{(0.0200)^2} = 3.93 \times 10^{-6}\ \text{M}$$

Since the provisional answer is nearly four orders of magnitude less than 0.0200 M, our approximation is justified, and the solution does not need further refinement.

Notice that the results from the last two examples demonstrate that an excess of iodate ions is more effective in decreasing the solubility of $\text{Ba(IO}_3)_2$ than is the same excess of barium ions.

A 0.02 M excess of Ba^{2+} decreases the solubility of $\text{Ba(IO}_3)_2$ by a factor of about 5; this same excess of IO_3^- lowers the solubility by a factor of about 200.

9B-6 Using Acid/Base Dissociation Constants

When a weak acid or a weak base is dissolved in water, partial dissociation occurs. Thus, for nitrous acid, we can write

$$\text{HNO}_2 + \text{H}_2\text{O} \rightleftharpoons \text{H}_3\text{O}^+ + \text{NO}_2^- \qquad K_a = \frac{[\text{H}_3\text{O}^+][\text{NO}_2^-]}{[\text{HNO}_2]}$$

where K_a is the **acid dissociation constant** for nitrous acid. In an analogous way, the **base dissociation constant** for ammonia is

$$\text{NH}_3 + \text{H}_2\text{O} \rightleftharpoons \text{NH}_4^+ + \text{OH}^- \qquad K_b = \frac{[\text{NH}_4^+][\text{OH}^-]}{[\text{NH}_3]}$$

Notice that $[\text{H}_2\text{O}]$ does not appear in the denominator of either equation because the concentration of water is so large relative to the concentration of the weak acid or base that the dissociation does not alter $[\text{H}_2\text{O}]$ appreciably (see Feature 9-2). Just as in the derivation of the ion-product constant for water, $[\text{H}_2\text{O}]$ is incorporated into the equilibrium constants K_a and K_b. Dissociation constants for weak acids are found in Appendix 3.

Dissociation Constants for Conjugate Acid/Base Pairs

Consider the base dissociation-constant expression for ammonia and the acid dissociation-constant expression for its conjugate acid, ammonium ion:

$$NH_3 + H_2O \rightleftharpoons NH_4^+ + OH^- \quad K_b = \frac{[NH_4^+][OH^-]}{[NH_3]}$$

$$NH_4^+ + H_2O \rightleftharpoons NH_3 + H_3O^+ \quad K_a = \frac{[NH_3][H_3O^+]}{[NH_4^+]}$$

By multiplying one equilibrium-constant expression by the other, we have

$$K_a K_b = \frac{[NH_3][H_3O^+]}{[NH_4^+]} \times \frac{[NH_4^+][OH^-]}{[NH_3]} = [H_3O^+][OH^-]$$

but

$$K_w = [H_3O^+][OH^-]$$

and, therefore,

$$K_w = K_a K_b \qquad\qquad (9\text{-}14)$$

This relationship is general for all conjugate acid/base pairs. Many compilations of equilibrium-constant data list only acid dissociation constants because it is so easy to calculate dissociation constants for bases by using Equation 9-14. For example, in Appendix 3, we find no data on the basic dissociation of ammonia (nor for any other bases). Instead, we find the acid dissociation constant for the conjugate acid, ammonium ion. That is,

$$NH_4^+ + H_2O \rightleftharpoons H_3O^+ + NH_3 \quad K_a = \frac{[H_3O^+][NH_3]}{[NH_4^+]} = 5.70 \times 10^{-10}$$

and we can write

$$NH_3 + H_2O \rightleftharpoons NH_4^+ + OH^-$$

> To find a dissociation constant for a base at 25°C in water, we look up the dissociation constant for its conjugate acid and then divide 1.00×10^{-14} by the K_a.

$$K_b = \frac{[NH_4^+][OH^-]}{[NH_3]} = \frac{K_w}{K_a} = \frac{1.00 \times 10^{-14}}{5.00 \times 10^{-10}} = 1.75 \times 10^{-5}$$

FEATURE 9-3

Relative Strengths of Conjugate Acid/Base Pairs

Equation 9-14 confirms the observation in Figure 9-2 that as the acid of a conjugate acid/base pair becomes weaker, its conjugate base becomes stronger and vice versa. Thus, the conjugate base of an acid with a dissociation constant of 10^{-2} will have a basic dissociation constant of 10^{-12}, and an acid with a dissociation constant of 10^{-9} has a conjugate base with a dissociation constant of 10^{-5}.

EXAMPLE 9-6

What is K_b for the equilibrium

$$CN^- + H_2O \rightleftharpoons HCN + OH^-$$

Solution

Appendix 3 lists a K_a value of 6.2×10^{-10} for HCN. Thus,

$$K_b = \frac{K_w}{K_a} = \frac{[HCN][OH^-]}{[CN^-]}$$

$$K_b = \frac{1.00 \times 10^{-14}}{6.2 \times 10^{-10}} = 1.61 \times 10^{-5}$$

Hydronium Ion Concentration of Solutions of Weak Acids

When the weak acid HA is dissolved in water, two equilibria produce hydronium ions:

$$HA + H_2O \rightleftharpoons H_3O^+ + A^- \qquad K_a = \frac{[H_3O^+][A^-]}{[HA]}$$

$$2H_2O \rightleftharpoons H_3O^+ + OH^- \qquad K_w = [H_3O^+][OH^-]$$

Normally, the hydronium ions produced from the first reaction suppress the dissociation of water to such an extent that the contribution of hydronium ions from the second equilibrium is negligible. Under these circumstances, one H_3O^+ ion is formed for each A^- ion, and we write

$$[A^-] \approx [H_3O^+] \tag{9-15}$$

Furthermore, the sum of the molar concentrations of the weak acid and its conjugate base must equal the analytical concentration of the acid c_{HA} because the solution contains no other source of A^- ions. Therefore,

$$c_{HA} = [A^-] + [HA] \tag{9-16}$$

In Chapter 11, we will learn that Equation 9-16 is called a **mass-balance equation**.

Substituting $[H_3O^+]$ for $[A^-]$ (see Equation 9-15) in Equation 9-16 yields

$$c_{HA} = [H_3O^+] + [HA]$$

which rearranges to

$$[HA] = c_{HA} - [H_3O^+] \tag{9-17}$$

When $[A^-]$ and $[HA]$ are replaced by their equivalent terms from Equations 9-15 and 9-17, the equilibrium-constant expression becomes

$$K_a = \frac{[H_3O^+]^2}{c_{HA} - [H_3O^+]} \tag{9-18}$$

which rearranges to

$$[H_3O^+]^2 + K_a[H_3O^+] - K_a c_{HA} = 0 \tag{9-19}$$

The positive solution to this quadratic equation is

$$[H_3O^+] = \frac{-K_a + \sqrt{K_a^2 + 4K_ac_{HA}}}{2} \tag{9-20}$$

As an alternative to using Equation 9-20, Equation 9-19 may be solved by successive approximations, as shown in Feature 9-4.

Equation 9-17 can frequently be simplified by making the additional assumption that dissociation does not appreciably decrease the molar concentration of HA. Thus, if $[H_3O^+] \ll c_{HA}$, $c_{HA} - [H_3O^+] \approx c_{HA}$, and Equation 9-18 reduces to

$$K_a = \frac{[H_3O^+]^2}{c_{HA}} \tag{9-21}$$

and

$$[H_3O^+] = \sqrt{K_ac_{HA}} \tag{9-22}$$

Table 9-4 shows that the error introduced by the assumption that $[H_3O^+] \ll c_{HA}$ increases as the molar concentration of acid becomes smaller and its dissociation constant becomes larger. Notice that the error introduced by the assumption is about 0.5% when the ratio c_{HA}/K_a is 10^4. The error increases to about 1.6% when the ratio is 10^3, to about 5% when it is 10^2, and to about 17% when it is 10. Figure 9-3 illustrates the effect graphically. Notice that the hydronium ion concentration computed with the approximation becomes greater than or equal to the molar concentration of the acid when the ratio is less than or equal to 1, which is not meaningful.

In general, it is a good idea to make the simplifying assumption and calculate a trial value for $[H_3O^+]$ that can be compared with c_{HA} in Equation 9-17. If the trial value alters [HA] by an amount smaller than the allowable error in the calculation, we consider the solution satisfactory. Otherwise, the quadratic equation must be solved to find a better value for $[H_3O^+]$. Alternatively, the method of successive approximations (see Feature 9-4) may be used.

TABLE 9-4

Error Introduced by Assuming H_3O^+ Concentration Is Small Relative to c_{HA} in Equation 9-16

K_a	c_{HA}	$[H_3O^+]$ Using Assumption	$\dfrac{c_{HA}}{K_a}$	$[H_3O^+]$ Using More Exact Equation	Percent Error
1.00×10^{-2}	1.00×10^{-3}	3.16×10^{-3}	10^{-1}	0.92×10^{-3}	244
	1.00×10^{-2}	1.00×10^{-2}	10^{0}	0.62×10^{-2}	61
	1.00×10^{-1}	3.16×10^{-2}	10^{1}	2.70×10^{-2}	17
1.00×10^{-4}	1.00×10^{-4}	1.00×10^{-4}	10^{0}	0.62×10^{-4}	61
	1.00×10^{-3}	3.16×10^{-4}	10^{1}	2.70×10^{-4}	17
	1.00×10^{-2}	1.00×10^{-3}	10^{2}	0.95×10^{-3}	5.3
	1.00×10^{-1}	3.16×10^{-3}	10^{3}	3.11×10^{-3}	1.6
1.00×10^{-6}	1.00×10^{-5}	3.16×10^{-6}	10^{1}	2.70×10^{-6}	17
	1.00×10^{-4}	1.00×10^{-5}	10^{2}	0.95×10^{-5}	5.3
	1.00×10^{-3}	3.16×10^{-5}	10^{3}	3.11×10^{-5}	1.6
	1.00×10^{-2}	1.00×10^{-4}	10^{4}	9.95×10^{-5}	0.5
	1.00×10^{-1}	3.16×10^{-4}	10^{5}	3.16×10^{-4}	0.0

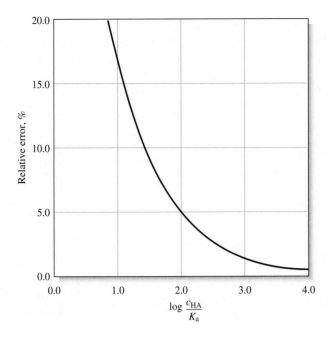

Figure 9-3 Relative error resulting from the assumption that $[H_3O^+] \ll c_{HA}$ in Equation 9-18.

EXAMPLE 9-7

Calculate the hydronium ion concentration in 0.120 M nitrous acid.

Solution

The principal equilibrium is

$$HNO_2 + H_2O \rightleftharpoons H_3O^+ + NO_2^-$$

for which (see Appendix 2)

$$K_a = 7.1 \times 10^{-4} = \frac{[H_3O^+][NO_2^-]}{[HNO_2]}$$

Substitution into Equations 9-15 and 9-17 gives

$$[NO_2^-] = [H_3O^+]$$
$$[HNO_2] = 0.120 - [H_3O^+]$$

When these relationships are introduced into the expression for K_a, we obtain

$$K_a = \frac{[H_3O^+]^2}{0.120 - [H_3O^+]} = 7.1 \times 10^{-4}$$

If we now assume that $[H_3O^+] \ll 0.120$, we find

$$\frac{[H_3O^+]^2}{0.120} = 7.1 \times 10^{-4}$$

$$[H_3O^+] = \sqrt{0.120 \times 7.1 \times 10^{-4}} = 9.2 \times 10^{-3} \text{ M}$$

(continued)

We now examine the assumption that $0.120 - 0.0092 \approx 0.120$ and see that the error is about 8%. The relative error in $[H_3O^+]$ is actually smaller than this figure, however, as we can see by calculating $\log(c_{HA}/K_a) = 2.2$, which from Figure 9-3, suggests an error of about 4%. If a more accurate figure is needed, the quadratic equation gives an answer of 8.9×10^{-3} M for the hydronium ion concentration.

EXAMPLE 9-8

Calculate the hydronium ion concentration in a solution that is 2.0×10^{-4} M in aniline hydrochloride, $C_6H_5NH_3Cl$.

Solution

In aqueous solution, dissociation of the salt to Cl^- and $C_6H_5NH_3^+$ is complete. The weak acid $C_6H_5NH_3^+$ dissociates as follows:

$$C_6H_5NH_3^+ + H_2O \rightleftharpoons C_6H_5NH_2 + H_3O^+ \quad K_a = \frac{[H_3O^+][C_6H_5NH_2]}{[C_6H_5NH_3^+]}$$

If we look in Appendix 3, we find that the K_a for $C_6H_5NH_3^+$ is 2.51×10^{-5}. Proceeding as in Example 9-7, we have

$$[H_3O^+] = [C_6H_5NH_2]$$

$$[C_6H_5NH_3^+] = 2.0 \times 10^{-4} - [H_3O^+]$$

Assume that $[H_3O^+] \ll 2.0 \times 10^{-4}$, and substitute the simplified value for $[C_6H_5NH_3^+]$ into the dissociation-constant expression to obtain (see Equation 9-21)

$$\frac{[H_3O^+]^2}{2.0 \times 10^{-4}} = 2.51 \times 10^{-5}$$

$$[H_3O^+] = \sqrt{5.02 \times 10^{-9}} = 7.09 \times 10^{-5} \text{ M}$$

If we compare 7.09×10^{-5} with 2.0×10^{-4}, we see that a significant error has been introduced by the assumption that $[H_3O^+] \ll c_{C_6H_5NH_3^+}$ (Figure 9-3 indicates that this error is about 20%.) Thus, unless only an approximate value for $[H_3O^+]$ is needed, it is necessary to use the more accurate expression (Equation 9-19)

$$\frac{[H_3O^+]^2}{2.0 \times 10^{-4} - [H_3O^+]} = 2.51 \times 10^{-5}$$

which rearranges to

$$[H_3O^+]^2 + 2.51 \times 10^{-5}[H_3O^+] - 5.02 \times 10^{-9} = 0$$

$$[H_3O^+] = \frac{-2.51 \times 10^{-5} + \sqrt{(2.54 \times 10^{-5})^2 + 4 \times 5.02 \times 10^{-9}}}{2}$$

$$= 5.94 \times 10^{-5} \text{ M}$$

The quadratic equation can also be solved by the iterative method shown in Feature 9-4.

FEATURE 9-4

The Method of Successive Approximations

For convenience, let us write the quadratic equation in Example 9-8 in the form

$$x^2 + 2.51 \times 10^{-5}x - 5.02 \times 10^{-9} = 0$$

where $x = [H_3O^+]$.

As a first step, rearrange the equation to the form

$$x = \sqrt{5.02 \times 10^{-9} - 2.51 \times 10^{-5}x}$$

We then assume that x on the right-hand side of the equation is zero and calculate a provisional solution, x_1.

$$x_1 = \sqrt{5.02 \times 10^{-9} - 2.51 \times 10^{-5} \times 0} = 7.09 \times 10^{-5}$$

We then substitute this value into the original equation and calculate a second value, x_2.

$$x_2 = \sqrt{5.02 \times 10^{-9} - 2.51 \times 10^{-5} \times 7.09 \times 10^{-5}} = 5.69 \times 10^{-5}$$

Repeating this calculation gives

$$x_3 = \sqrt{5.02 \times 10^{-9} - 2.51 \times 10^{-5} \times 5.69 \times 10^{-5}} = 5.99 \times 10^{-5}$$

Continuing in the same way, we find

$$x_4 = 5.93 \times 10^{-5}$$
$$x_5 = 5.94 \times 10^{-5}$$
$$x_6 = 5.94 \times 10^{-5}$$

Note that after three iterations, x_3 is 5.99×10^{-5}, which is within about 0.8% of the final value of 5.94×10^{-5} M.

The method of successive approximations is particularly useful when cubic or higher-power equations must be solved.

As shown in Chapter 5 of *Applications of Microsoft® Excel in Analytical Chemistry*, 2nd ed., iterative solutions can be found quite conveniently using a spreadsheet.

Hydronium Ion Concentration of Solutions of Weak Bases

We can adapt the techniques of the previous sections to calculate the hydroxide or hydronium ion concentration in solutions of weak bases.

Aqueous ammonia is basic as a result of the reaction

$$NH_3 + H_2O \rightleftharpoons NH_4^+ + OH^-$$

The predominant species in this solution is certainly NH_3. Nevertheless, solutions of ammonia are still called ammonium hydroxide occasionally because at one time chemists thought that NH_4OH rather than NH_3 was the undissociated form of the base. We write the equilibrium constant for the reaction as

$$K_b = \frac{[NH_4^+][OH^-]}{[NH_3]}$$

EXAMPLE 9-9

Calculate the hydroxide ion concentration of a 0.0750 M NH_3 solution.

Solution

The predominant equilibrium is

$$NH_3 + H_2O \rightleftharpoons NH_4^+ + OH^-$$

As we showed on page 211,

$$K_b = \frac{[NH_4^+][OH^-]}{[NH_3]} = \frac{1.00 \times 10^{-14}}{5.70 \times 10^{-10}} = 1.75 \times 10^{-5}$$

The chemical equation shows that

$$[NH_4^+] = [OH^-]$$

Both NH_4^+ and NH_3 come from the 0.0750 M solution. Thus,

$$[NH_4^+] + [NH_3] = c_{NH_3} = 0.0750 \text{ M}$$

If we substitute $[OH^-]$ for $[NH_4^+]$ in the second of these equations and rearrange, we find that

$$[NH_3] = 0.0750 - [OH^-]$$

By substituting these quantities into the dissociation-constant, we have

$$\frac{[OH^-]^2}{7.50 \times 10^{-2} - [OH^-]} = 1.75 \times 10^{-5}$$

which is analogous to Equation 9-17 for weak acids. If we assume that $[OH^-] \ll 7.50 \times 10^{-2}$, this equation simplifies to

$$[OH^-]^2 \approx 7.50 \times 10^{-2} \times 1.75 \times 10^{-5}$$

$$[OH^-] = 1.15 \times 10^{-3} \text{ M}$$

Comparing the calculated value for $[OH^-]$ with 7.50×10^{-2}, we see that the error in $[OH^-]$ is less than 2%. If necessary, a better value for $[OH^-]$ can be obtained by solving the quadratic equation.

EXAMPLE 9-10

Calculate the hydroxide ion concentration in a 0.0100 M sodium hypochlorite solution.

Solution

The equilibrium between OCl^- and water is

$$OCl^- + H_2O \rightleftharpoons HOCl + OH^-$$

for which

$$K_b = \frac{[\text{HOCl}][\text{OH}^-]}{[\text{OCl}^-]}$$

The acid dissociation constant for HOCl from Appendix 3 is 3.0×10^{-8}. Therefore, we rearrange Equation 9-14 and write

$$K_b = \frac{K_w}{K_a} = \frac{1.00 \times 10^{-14}}{3.0 \times 10^{-8}} = 3.33 \times 10^{-7}$$

Proceeding as in Example 9-9, we have

$$[\text{OH}^-] = [\text{HOCl}]$$
$$[\text{OCl}^-] + [\text{HOCl}] = 0.0100$$
$$[\text{OCl}^-] = 0.0100 - [\text{OH}^-] \approx 0.0100$$

In this case, we have assumed that $[\text{OH}^-] \ll 0.0100$. We substitute this value into the equilibrium constant expression and calculate

$$\frac{[\text{OH}^-]^2}{0.0100} = 3.33 \times 10^{-7}$$
$$[\text{OH}^-] = 5.8 \times 10^{-5}\,\text{M}$$

Verify for yourself that the error resulting from the approximation is small.

Spreadsheet Summary In the first three exercises in Chapter 5 of *Applications of Microsoft® Excel in Analytical Chemistry*, 2nd ed., we explore the solution to the types of equations found in chemical equilibria. A general purpose quadratic equation solver is developed and used for equilibrium problems. Then, Excel is used to find iterative solutions by successive approximations. Excel's Solver is next employed to solve quadratic, cubic, and quartic equations of the type encountered in equilibrium calculations.

9C BUFFER SOLUTIONS

A **buffer solution** resists changes in pH when it is diluted or when acids or bases are added to it. Generally, buffer solutions are prepared from a conjugate acid/base pair, such as acetic acid/sodium acetate or ammonium chloride/ammonia. Scientists and technologists in most areas of science and in many industries use buffers to maintain the pH of solutions at a relatively constant and predetermined level. You will find many references to buffers throughout this text.

Buffers are used in all types of chemical applications whenever it is important to maintain the pH of a solution at a constant and predetermined level.

9C-1 Calculating the pH of Buffer Solutions

A solution containing a weak acid, HA, and its conjugate base, A^-, may be acidic, neutral, or basic, depending on the positions of two competitive equilibria:

$$\text{HA} + \text{H}_2\text{O} \rightleftharpoons \text{H}_3\text{O}^+ + \text{A}^- \qquad K_a = \frac{[\text{H}_3\text{O}^+][\text{A}^-]}{[\text{HA}]} \qquad (9\text{-}23)$$

Buffered aspirin contains buffers to help prevent stomach irritation from the acidity of the carboxylic acid group in aspirin.

Molecular model and structure of aspirin. The analgesic action is thought to arise because aspirin interferes with the synthesis of prostaglandins, which are hormones involved in the transmission of pain signals.

$$A^- + H_2O \rightleftharpoons OH^- + HA \quad K_b = \frac{[OH^-][HA]}{[A^-]} = \frac{K_w}{K_a} \qquad (9\text{-}24)$$

If the first equilibrium lies farther to the right than the second, the solution is acidic. If the second equilibrium is more favorable, the solution is basic. These two equilibrium-constant expressions show that the relative concentrations of the hydronium and hydroxide ions depend not only on the magnitudes of K_a and K_b but also on the ratio between the concentrations of the acid and its conjugate base.

To find the pH of a solution containing both an acid, HA, and its conjugate base, NaA, we need to express the equilibrium concentrations of HA and NaA in terms of their analytical concentrations, c_{HA} and c_{NaA}. If we look closely at the two equilibria, we find that the first reaction decreases the concentration of HA by an amount equal to $[H_3O^+]$, while the second increases the HA concentration by an amount equal to $[OH^-]$. Thus, the species concentration of HA is related to its analytical concentration by the equation

$$[HA] = c_{HA} - [H_3O^+] + [OH^-] \qquad (9\text{-}25)$$

Similarly, the first equilibrium will increase the concentration of A^- by an amount equal to $[H_3O^+]$, and the second will decrease this concentration by the amount $[OH^-]$. Therefore, the equilibrium concentration is given by a second equation that looks a lot like Equation 9-25.

$$[A^-] = c_{NaA} + [H_3O^+] - [OH^-] \qquad (9\text{-}26)$$

Because of the inverse relationship between $[H_3O^+]$ and $[OH^-]$, it is *always* possible to eliminate one or the other from Equations 9-25 and 9-26. Additionally, the *difference* in concentration between $[H_3O^+]$ and $[OH^-]$ is usually so small relative to the molar concentrations of acid and conjugate base that Equations 9-25 and 9-26 simplify to

$$[HA] \approx c_{HA} \qquad (9\text{-}27)$$

$$[A^-] \approx c_{NaA} \qquad (9\text{-}28)$$

If we then substitute Equations 9-27 and 9-28 into the dissociation-constant expression and rearrange the result, we have

$$[H_3O^+] = K_a \frac{c_{HA}}{c_{NaA}} \qquad (9\text{-}29)$$

The assumption leading to Equations 9-27 and 9-28 sometimes breaks down with acids or bases that have dissociation constants greater than about 10^{-3} or when the molar concentration of either the acid or its conjugate base (or both) is very small. In these circumstances, either $[OH^-]$ or $[H_3O^+]$ must be retained in Equations 9-25 and 9-26 depending on whether the solution is acidic or basic. In any case, Equations 9-27 and 9-28 should always be used initially. Provisional values for $[H_3O^+]$ and $[OH^-]$ can then be used to test the assumptions.

Within the limits imposed by the assumptions made in deriving Equation 9-29, it says that the hydronium ion concentration of a solution containing a weak acid and its conjugate base depends *only on the ratio* of the molar concentrations of these two solutes. Furthermore, this ratio is *independent of dilution* because the concentration of each component changes proportionally when the volume changes.

FEATURE 9-5

The Henderson-Hasselbalch Equation

The Henderson-Hasselbalch equation, which is used to calculate the pH of buffer solutions, is frequently encountered in the biological literature and biochemical texts. It is obtained by expressing each term in Equation 9-29 in the form of its negative logarithm and inverting the concentration ratio to keep all signs positive:

$$-\log\left[H_3O^+\right] = -\log K_a + \log\frac{c_{NaA}}{c_{HA}}$$

Therefore,

$$pH = pK_a + \log\frac{c_{NaA}}{c_{HA}} \qquad (9\text{-}30)$$

If the assumptions leading to Equation 9-28 are not valid, the values for [HA] and [A$^-$] are given by Equations 9-24 and 9-25, respectively. If we take the negative logarithms of these expressions, we derive extended Henderson-Hasselbalch equations.

EXAMPLE 9-11

What is the pH of a solution that is 0.400 M in formic acid and 1.00 M in sodium formate?

Solution

The pH of this solution is affected by the K_w of formic acid and the K_b of formate ion.

$$HCOOH + H_2O \rightleftharpoons H_3O^+ + HCOO^- \quad K_a = 1.80 \times 10^{-4}$$

$$HCOO^- + H_2O \rightleftharpoons HCOOH + OH^- \quad K_b = \frac{K_w}{K_a} = 5.56 \times 10^{-11}$$

Because the K_a for formic acid is orders of magnitude larger than the K_b for formate, the solution is acidic, and K_a determines the H_3O^+ concentration. We can thus write

$$K_a = \frac{[H_3O^+][HCOO^-]}{[HCOOH]} = 1.80 \times 10^{-4}$$

$$[HCOO^-] \approx c_{HCOO^-} = 1.00 \text{ M}$$

$$[HCOOH] \approx c_{HCOOH} = 0.400 \text{ M}$$

By substituting these expressions into Equation 9-29 and rearranging, we have

$$[H_3O^+] = 1.80 \times 10^{-4} \times \frac{0.400}{1.00} = 7.20 \times 10^{-5} \text{ M}$$

Notice that our assumptions that $[H_3O^+] \ll c_{HCOOH}$ and that $[H_3O^+] \ll c_{HCOO^-}$ are valid. Therefore,

$$pH = -\log(7.20 \times 10^{-5}) = 4.14$$

As shown in Example 9-12, Equations 9-25 and 9-26 also apply to buffer systems consisting of a weak base and its conjugate acid. Furthermore, in most cases it is possible to simplify these equations so that Equation 9-29 can be used.

EXAMPLE 9-12

Calculate the pH of a solution that is 0.200 M in NH_3 and 0.300 M in NH_4Cl.

Solution

In Appendix 3, we find that the acid dissociation constant K_a for NH_4^+ is 5.70×10^{-10}. The equilibria we must consider are

$$NH_4^+ + H_2O \rightleftharpoons NH_3 + H_3O^+ \qquad K_a = 5.70 \times 10^{-10}$$

$$NH_3 + H_2O \rightleftharpoons NH_4^+ + OH^- \qquad K_b = \frac{K_w}{K_a} = \frac{1.00 \times 10^{-14}}{5.70 \times 10^{-10}} = 1.75 \times 10^{-5}$$

Using the arguments that led to Equations 9-25 and 9-26, we find that

$$[NH_4^+] = c_{NH_4Cl} + [OH^-] - [H_3O^+] \approx c_{NH_4Cl} + [OH^-]$$

$$[NH_3] = c_{NH_3} + [H_3O^+] - [OH^-] \approx c_{NH_3} - [OH^-]$$

Because K_b is several orders of magnitude larger than K_a, we have assumed that the solution is basic and that $[OH^-]$ is much larger than $[H_3O^+]$. Thus, we have neglected the concentration of H_3O^+ in these approximations.

We also assume that $[OH^-]$ is much smaller than c_{NH_4Cl} and c_{NH_3}, so that

$$[NH_4^+] \approx c_{NH_4Cl} = 0.300 \text{ M}$$

$$[NH_3] \approx c_{NH_3} = 0.200 \text{ M}$$

When we substitute these expressions into the acid dissociation constant for NH_4^+, we have a relationship similar to Equation 9-29. That is,

$$[H_3O^+] = \frac{K_a \times [NH_4^+]}{[NH_3]} = \frac{5.70 \times 10^{-10} \times c_{NH_4Cl}}{c_{NH_3}}$$

$$= \frac{5.70 \times 10^{-10} \times 0.300}{0.200} = 8.55 \times 10^{-10} \text{ M}$$

To check the validity of our approximations, we calculate $[OH^-]$. Thus,

$$[OH^-] = \frac{1.00 \times 10^{-14}}{8.55 \times 10^{-10}} = 1.17 \times 10^{-5} \text{ M}$$

which is far smaller than c_{NH_4Cl} or c_{NH_3}. Finally, we write

$$pH = -\log(8.55 \times 10^{-10}) = 9.07$$

9C-2 Properties of Buffer Solutions

In this section, we demonstrate the resistance of buffers to changes of pH brought about by dilution or addition of strong acids or bases.

The Effect of Dilution

The pH of a buffer solution remains essentially independent of dilution until the concentrations of the species it contains are decreased to the point where the approximations used to develop Equations 9-27 and 9-28 become invalid. **Figure 9-4** contrasts the behavior of buffered and unbuffered solutions with dilution. For each, the initial solute concentration is 1.00 M. The resistance of the buffered solution to changes in pH during dilution is clearly shown.

The Effect of Added Acids and Bases

Example 9-13 illustrates a second property of buffer solutions, their resistance to pH change after addition of small amounts of strong acids or bases.

Figure 9-4 The effect of dilution of the pH of buffered and unbuffered solutions. The dissociation constant for HA is 1.00×10^{-4}. Initial solute concentrations are 1.00 M.

EXAMPLE 9-13

Calculate the pH change that takes place when a 100-mL portion of (a) 0.0500 M NaOH and (b) 0.0500 M HCl is added to 400 mL of the buffer solution that was described in Example 9-12.

Solution

(a) Adding NaOH converts part of the NH_4^+ in the buffer to NH_3:

$$NH_4^+ + OH^- \rightleftharpoons NH_3 + H_2O$$

The analytical concentrations of NH_3 and NH_4Cl then become

$$c_{NH_3} = \frac{400 \times 0.200 + 100 \times 0.0500}{500} = \frac{85.0}{500} = 0.170 \text{ M}$$

$$c_{NH_4Cl} = \frac{400 \times 0.300 - 100 \times 0.0500}{500} = \frac{115}{500} = 0.230 \text{ M}$$

When substituted into the acid dissociation-constant expression for NH_4^+, these values yield

$$[H_3O^+] = 5.70 \times 10^{-10} \times \frac{0.230}{0.170} = 7.71 \times 10^{-10} \text{ M}$$

$$pH = -\log 7.71 \times 10^{-10} = 9.11$$

and the change in pH is

$$\Delta pH = 9.11 - 9.07 = 0.04$$

(b) Adding HCl converts part of the NH_3 to NH_4^+. Thus,

$$NH_3 + H_3O^+ \rightleftharpoons NH_4^+ + H_2O$$

$$c_{NH_3} = \frac{400 \times 0.200 - 100 \times 0.0500}{500} = \frac{75}{500} = 0.150 \text{ M}$$

$$c_{NH_4^+} = \frac{400 \times 0.300 + 100 \times 0.0500}{500} = \frac{125}{500} = 0.250 \text{ M}$$

$$[H_3O^+] = 5.70 \times 10^{-10} \times \frac{0.250}{0.150} = 9.50 \times 10^{-10}$$

$$pH = -\log 9.50 \times 10^{-10} = 9.02$$

$$\Delta pH = 9.02 - 9.07 = -0.05$$

◀ Buffers do not maintain pH at an absolutely constant value, but changes in pH are relatively small when small amounts of acid or base are added.

It is interesting to contrast the behavior of an unbuffered solution with a pH of 9.07 to that of the buffer in Example 9-13. We can show that adding the same quantity of base to the unbuffered solution would increase the pH to 12.00—a pH change of 2.93 units. Adding the acid would decrease the pH by slightly more than 7 units.

The Composition of Buffer Solutions as a Function of pH: Alpha Values

The composition of buffer solutions can be visualized by plotting the *relative* equilibrium concentrations of the two components of a conjugate acid/base as a function of the pH of the solution. These relative concentrations are called **alpha values**. For example, if we let c_T be the sum of the analytical concentrations of acetic acid and sodium acetate in a typical buffer solution, we may write

$$c_T = c_{HOAc} + c_{NaOAc} \tag{9-31}$$

We then define α_0, the fraction of the total concentration of acid that is undissociated, as

$$\alpha_0 = \frac{[HOAc]}{c_T} \tag{9-32}$$

and α_1, the fraction dissociated, as

$$\alpha_1 = \frac{[OAc^-]}{c_T} \tag{9-33}$$

Alpha values are unitless ratios whose sum must equal unity. That is,

$$\alpha_0 + \alpha_1 = 1$$

> Alpha values do not depend on c_T.

Alpha values depend *only* on $[H_3O^+]$ and K_a and are independent of c_T. To derive expressions for α_0, we rearrange the dissociation-constant expression to

$$[OAc^-] = \frac{K_a[HOAc]}{[H_3O^+]} \tag{9-34}$$

The total concentration of acetic acid, c_T, is in the form of either HOAc or OAc$^-$. Thus,

$$c_T = [HOAc] + [OAc^-] \tag{9-35}$$

Substituting Equation 9-34 into Equation 9-35 gives

$$c_T = [HOAc] + \frac{K_a[HOAc]}{[H_3O^+]} = [HOAc]\left(\frac{[H_3O^+] + K_a}{[H_3O^+]}\right)$$

When rearranged, this equation becomes

$$\frac{[HOAc]}{c_T} = \frac{[H_3O^+]}{[H_3O^+] + K_a}$$

But according to Equation 9-32, $[HOAc]/c_T = \alpha_0$, so

$$\alpha_0 = \frac{[HOAc]}{c_T} = \frac{[H_3O^+]}{[H_3O^+] + K_a} \tag{9-36}$$

To derive a similar expression for α_1, we rearrange the dissociation-constant expression to

$$[HOAc] = \frac{[H_3O^+][OAc^-]}{K_a}$$

and substitute into Equation 9-36

$$c_T = \frac{[H_3O^+][OAc^-]}{K_a} + [OAc^-] = [OAc^-]\left(\frac{[H_3O^+] + K_a}{K_a}\right)$$

Rearranging this expression gives α_1 as defined by Equation 9-33.

$$\alpha_1 = \frac{[OAc^-]}{c_T} = \frac{K_a}{[H_3O^+] + K_a} \tag{9-37}$$

Note that the denominator is the same in Equations 9-36 and 9-37.

Figure 9-5 illustrates how α_0 and α_1 vary as a function of pH. The data for these plots were calculated from Equations 9-36 and 9-37.

You can see that the two curves cross at the point where pH = pK_{HOAc} = 4.74. At this point, the concentrations of acetic acid and acetate ion are equal, and the fractions of the total analytical concentration of acid both equal one half.

Buffer Capacity

Figure 9-4 and Example 9-13 demonstrate that a solution containing a conjugate acid/base pair is remarkably resistant to changes in pH. For example, the pH of a 400-mL portion of a buffer formed by diluting the solution described in Example 9-13 by

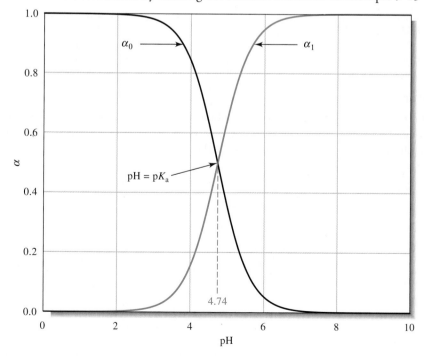

Figure 9-5 Variation in α with pH. Note that most of the transition between α_0 and α_1 occurs within ± 1 pH unit of the crossover point of the two curves. The crossover point where $\alpha_0 = \alpha_1 = 0.5$ occurs when pH = pK_{HOAc} = 4.74.

10 would change by about 0.4 to 0.5 unit when treated with 100 mL of 0.0500 M sodium hydroxide or 0.0500 M hydrochloric acid. We showed in Example 9-13 that the change is only about 0.04 to 0.05 unit for the concentrated buffer.

The **buffer capacity of a buffer is the number of moles of strong acid or strong base that 1 L of the buffer can absorb without changing pH by more than 1.**

The **buffer capacity**, β, of a solution is defined as the number of moles of a strong acid or a strong base that causes 1.00 L of the buffer to undergo a 1.00-unit change in pH. Mathematically, buffer capacity is given by

$$\beta = \frac{dc_b}{dpH} = -\frac{dc_a}{dpH}$$

where dc_b is the number of moles per liter of strong base, and dc_a is the number of moles per liter of strong acid added to the buffer. Since adding strong acid to a buffer causes the pH to decrease, dc_a/dpH is negative, and *buffer capacity is always positive.*

Buffer capacity depends not only on the total concentration of the two buffer components but also on their concentration ratio. As **Figure 9-6** shows buffer capacity decreases fairly rapidly as the concentration ratio of acid to conjugate base becomes larger or smaller than 1 (the logarithm of the ratio increases above or decreases below zero). For this reason, the pK_a of the acid chosen for a given application should lie within ± 1 unit of the desired pH for the buffer to have a reasonable capacity.

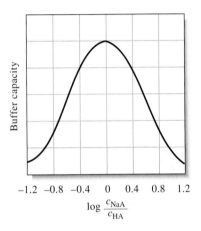

Figure 9-6 Buffer capacity as a function of the logarithm of the ratio c_{NaA}/c_{HA}. The maximum buffer capacity occurs when the concentration of acid and conjugate base are equal, that is, when $\alpha_0 = \alpha_1 = 0.5$.

Preparation of Buffers

In principle, a buffer solution of any desired pH can be prepared by combining calculated quantities of a suitable conjugate acid/base pair. In practice, however, the pH values of buffers prepared from recipes calculated from theory differ from the predicted values because of uncertainties in the numerical values of many dissociation constants and from the simplifications used in calculations. Because of these discrepancies, we prepare buffers by making up a solution of approximately the desired pH (see Example 9-14) and then adjusting it by adding strong acid or strong base until the required pH is indicated by a pH meter. Alternatively, empirical recipes for preparing buffer solutions of known pH are available in chemical handbooks and reference works.[2]

EXAMPLE 9-14

Describe how you might prepare approximately 500.0 mL of a pH 4.5 buffer solution from 1.0 M acetic acid (HOAc) and sodium acetate (NaOAc).

Solution

It is reasonable to assume there is little volume change if we add solid sodium acetate to the acetic acid solution. We then calculate the mass of NaOAc to add to 500.0 mL of 1.0 M HOAc. The H_3O^+ concentration should be

$$[H_3O^+] = 10^{-4.5} = 3.16 \times 10^{-5}\ M$$

$$K_a = \frac{[H_3O^+][OAc^-]}{[HOAc]} = 1.75 \times 10^{-5}$$

$$\frac{[OAc^-]}{[HOAc]} = \frac{1.75 \times 10^{-5}}{[H_3O^+]} = \frac{1.75 \times 10^{-5}}{3.16 \times 10^{-5}} = 0.5534$$

---

[2]See, for example, J. A. Dean, *Analytical Chemistry Handbook,* New York: McGraw-Hill, 1995, pp. 14–29 through 14–34.

The acetate concentration should be

$$[OAc^-] = 0.5534 \times 1.0\ M = 0.5534\ M$$

The mass of NaOAc needed is then

$$mass\ NaOAc = \frac{0.5534\ mol\ \cancel{NaOAc}}{\cancel{L}} \times 0.500\ \cancel{L} \times \frac{82.034\ g\ NaOAc}{mol\ \cancel{NaOAc}}$$

$$= 22.7\ g\ NaOAc$$

After dissolving this quantity of NaOAc in the acetic acid solution, we would check the pH with a pH meter and, if necessary, adjust it slightly by adding a small amount of acid or base.

Buffers are tremendously important in biological and biochemical studies in which a low but constant concentration of hydronium ions (10^{-6} to 10^{-10} M) must be maintained throughout experiments. Chemical and biological supply houses offer a variety of such buffers.

FEATURE 9-6

Acid Rain and the Buffer Capacity of Lakes

Acid rain has been the subject of considerable controversy over the past few decades. Acid rain forms when the gaseous oxides of nitrogen and sulfur dissolve in water droplets in the air. These gases form at high temperatures in power plants, automobiles, and other combustion sources. The combustion products pass into the atmosphere, where they react with water to form nitric acid and sulfuric acid as shown by the equations

$$4NO_2(g) + 2H_2O(l) + O_2(g) \rightarrow 4HNO_3(aq)$$

$$SO_3(g) + H_2O(l) \rightarrow H_2SO_4(aq)$$

Eventually, the droplets coalesce with other droplets to form acid rain. The profound effects of acid rain have been highly publicized. Stone buildings and monuments literally dissolve as acid rain flows over their surfaces. Forests are slowly being killed off in some locations. To illustrate the effects on aquatic life, consider the changes in pH that have occurred in the lakes of the Adirondack Mountains area of New York, illustrated in the bar graphs of **Figure 9F-1**. The graphs show the distribution of pH in these lakes, which were studied first in the 1930s and then again in 1975.[3] The shift in pH of the lakes over a 40-year period is dramatic. The average pH of the lakes changed from 6.4 to about 5.1, which represents a 20-fold change in the hydronium ion concentration. Such changes in pH have a profound effect on aquatic life, as shown by a study of the fish population in lakes in the same area.[4] In the graph of **Figure 9F-2**, the number of lakes is plotted as a function of pH. The darker bars represent lakes containing fish, and lakes having no fish are lighter in color. There is a distinct correlation between pH changes in the lakes and diminished fish population.

(continued)

[3]R. F. Wright and E. T. Gjessing, *Ambio,* **1976**, *5,* 219.
[4]C. L. Schofield, *Ambio,* **1976**, *5,* 228.

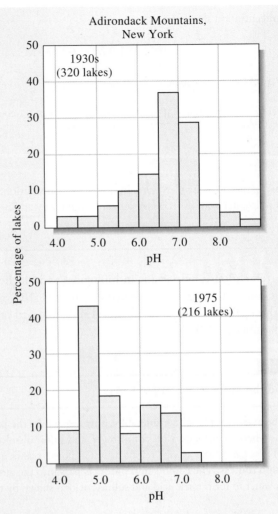

Figure 9F-1   Changes in pH of lakes between 1930 and 1975.

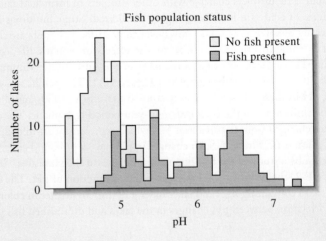

Figure 9F-2   Effect of pH of lakes on fish population.

Many factors contribute to pH changes in groundwater and lakes in a given geographical area. These include the prevailing wind patterns and weather, types of soils, water sources, nature of the terrain, characteristics of plant life, human activity, and geological characteristics. The susceptibility of natural water to acidification is largely determined by its buffer capacity, and the principal buffer of natural water is a mixture of bicarbonate ion and carbonic acid. Recall that the buffer capacity of a solution is proportional to the concentration of the buffering agent. So the higher the concentration of dissolved bicarbonate, the greater is the capacity of the water to neutralize acid from acid rain. The most important source of bicarbonate ion in natural water is limestone, or calcium carbonate, which reacts with hydronium ion as shown in the following equation:

$$CaCO_3(s) + H_3O^+(aq) \rightleftharpoons HCO_3^-(aq) + Ca^{2+}(aq) + H_2O(l)$$

Limestone-rich areas have lakes with relatively high concentrations of dissolved bicarbonate and thus low susceptibility to acidification. Granite, sandstone, shale, and other rock containing little or no calcium carbonate are associated with lakes having high susceptibility to acidification.

The map of the United States shown in **Figure 9F-3** vividly illustrates the correlation between the absence of limestone-bearing rocks and the acidification of groundwater.[5] Areas containing little limestone are shaded in blue, while areas rich in limestone are white. Contour lines (isopleths) of equal pH for groundwater during the period 1978–1979 are superimposed on the map. The Adirondack Mountains area, located in northeastern New York, contains little limestone and exhibits pH in the range of 4.2 to 4.4. The low buffer capacity of the lakes in this region combined

Figure 9F-3   Effect of the presence of limestone on pH of lakes in the United States. Shaded areas contain little limestone.

(continued)

[5]J. Root et al., cited in *The Effects of Air Pollution and Acid Rain on Fish, Wildlife, and Their Habitats—Introduction.* U.S. Fish and Wildlife Service, Biological Services Program, Eastern Energy and Land Use Team, U.S. Government Publication FWS/OBS-80/40.3, 1982 M. A. Peterson, ed., p. 63.

with the low pH of precipitation appears to have caused the decline of fish populations. Similar correlations among acid rain, buffer capacity of lakes, and wildlife decline occur throughout the industrialized world.

Although natural sources such as volcanos produce sulfur trioxide and lightning discharges in the atmosphere generate nitrogen dioxide, large quantities of these compounds come from the burning of high-sulfur coal and from automobile emissions. To minimize emissions of these pollutants, some states have enacted legislation imposing strict standards on automobiles sold and operated within their borders. Some states have required the installation of scrubbers to remove oxides of sulfur from the emissions of coal-fired power plants. To minimize the effects of acid rain on lakes, powdered limestone is dumped on their surfaces to increase the buffer capacity of the water. Solutions to these problems require the expenditure of much time, energy, and money. We must sometimes make difficult economic decisions to preserve the quality of our environment and to reverse trends that have operated for many decades.

The 1990 Clean Air Act Amendments provided a dramatic new way to regulate sulfur dioxide. Congress issued specific emission limits to power plant operators, as shown in **Figure 9F-4**, but no specific methods were proposed for meeting the standards. In addition, Congress established an emissions trading system by which power plants could buy, sell, and trade rights to pollute. Although detailed scientific and economic analysis of the effects of these congressional measures is still under way, it is clear from the results so far that the Clean Air Act Amendments have had a profound positive effect on the causes and effects of acid rain.[6]

Figure 9F-4 shows that sulfur dioxide emissions have decreased dramatically since 1990 and are well below levels forecasted by the EPA and within the limits

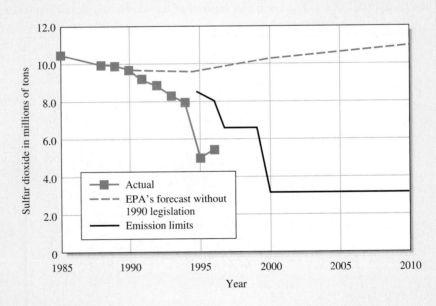

Figure 9F-4   Sulfur dioxide emissions from selected plants in the United States have dropped below the levels required by law. (R.A. Kerr, *Science*, **1998**, *282*, 1024. Copyright 1998 American Association of the Advancement of Science. Reprinted with permission of AAAS.)

[6]R. A. Kerr, *Science*, **1998**, *282*(5391), 1024, **DOI**: 10.1126/science.282.5391.1024.

set by Congress. The effects of these measures on acid rain are depicted in the map in **Figure 9F-5**, which shows the percent change in acidity in various regions of the eastern United States from 1983 to 1994. The significant improvement in acid rain shown on the map has been attributed tentatively to the flexibility of the regulatory statutes imposed in 1990. Another surprising result of the statutes is that their implementation has apparently been much less costly than originally projected. Initial estimates of the cost of meeting the emission standards were as high as $10 billion per year, but recent surveys indicate that actual costs may be as low as $1 billion per year.[7]

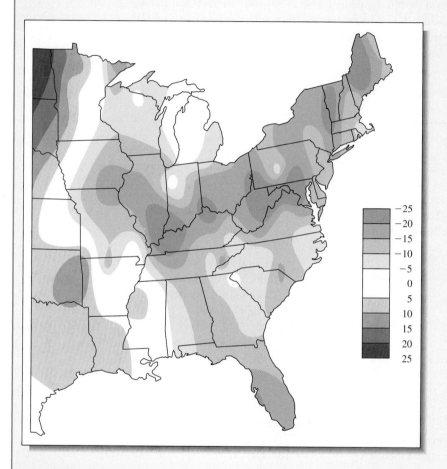

Figure 9F-5 Precipitation over much of the eastern United States has become less acidic, as shown by the percent change from 1983 to 1994. (R.A. Kerr, *Science*, **1998**, *282*, 1024. Copyright 1998 American Association of the Advancement of Science. Reprinted by permission of AAAS.)

[7]C. C. Park, *Acid Rain*, New York: Methuen, 1987.

WEB WORKS

Use a search engine to find the website for the Swedish Environmental Protection Agency. Perform a search on the site for keywords *acidification* and *liming*. Find the web page on these topics, read the article on the page, and answer the following questions. According to the article, what is the source of most of Sweden's acid rain pollution? Roughly how much has soil pH changed in Sweden over the last few decades? What is liming? Find the link to an article on sulfur dioxide emissions that is in Swedish. Use the Google translator to translate the article. How much sulfur dioxide was emitted in Sweden in 1990? How much in 2009? How was this improvement achieved?

Browse to the *Scientific American* website (www.sciam.com) and do a search using the keywords, "acid rain." One of the hits should be a short 2010 article entitled "Sour Showers." The article suggests that acid rain may be returning. What is the cause of its return? What measures are suggested to reduce this new rise of acid rain?

QUESTIONS AND PROBLEMS

9-1. Briefly describe or define and give an example of
*(a) a weak electrolyte.
(b) a Brønsted-Lowry acid.
*(c) the conjugate acid of a Brønsted-Lowry base.
(d) neutralization, in terms of the Brønsted-Lowry concept.
*(e) an amphiprotic solvent.
(f) a zwitterion.
*(g) autoprotolysis.
(h) a strong acid.
*(i) Le Châtelier's principle.
(j) the common-ion effect.

9-2. Briefly describe or define and give an example of
*(a) an amphiprotic solute.
(b) a differentiating solvent.
*(c) a leveling solvent.
(d) a mass-action effect.

*9-3. Briefly explain why there is no term in an equilibrium-constant expression for water or for a pure solid, even though one (or both) appears in the balanced net ionic equation for the equilibrium.

9-4. Identify the acid on the left and its conjugate base on the right in the following equations:
*(a) $HOCl + H_2O \rightleftharpoons H_3O^+ + OCl$
(b) $HONH_2 + H_2O \rightleftharpoons HONH_3^+ + OH^-$
*(c) $NH_4^+ + H_2O \rightleftharpoons NH_3 + H_3O^+$
(d) $2HCO_3^- \rightleftharpoons H_2CO_3 + CO_3^{2-}$
*(e) $PO_4^{3-} + H_2PO_4^- \rightleftharpoons 2HPO_4^{2-}$

9-5. Identify the base on the left and its conjugate acid on the right in the equations for Problem 9-4.

9-6. Write expressions for the autoprotolysis of
*(a) H_2O.
(b) CH_3COOH.
*(c) CH_3NH_2.
(d) CH_3OH.

9-7. Write the equilibrium-constant expressions and obtain numerical values for each constant in
*(a) the basic dissociation of aniline, $C_6H_5NH_2$.
(b) the acidic dissociation of hypochlorous acid, $HClO$.
*(c) the acidic dissociation of methyl ammonium hydrochloride, CH_3NH_3Cl.
(d) the basic dissociation of $NaNO_2$.
*(e) the dissociation of H_3AsO_3 to H_3O^+ and AsO_3^{3-}.
(f) the reaction of $C_2O_4^{2-}$ with H_2O to give $H_2C_2O_4$ and OH^-.

9-8. Generate the solubility-product expression for
*(a) $CuBr$.   (d) $La(IO_3)_3$.
*(b) $HgClI$.   (e) Ag_3AsO_4.
*(c) $PbCl_2$.

9-9. Express the solubility-product constant for each substance in Problem 9-8 in terms of its molar solubility S.

9-10. Calculate the solubility-product constant for each of the following substances, given that the molar concentrations of their saturated solutions are as indicated:
(a) $AgSeCN$ (2.0×10^{-8} M; products are Ag^+ and $SeCN^-$).
*(b) $RaSO_4$ (6.6×10^{-6} M).
(c) $Pb(BrO_3)_2$ (1.7×10^{-1} M).
*(d) $Ce(IO_3)_3$ (1.9×10^{-3} M).

9-11. Calculate the solubility of the solutes in Problem 9-10 for solutions in which the cation concentration is 0.030 M.

9-12. Calculate the solubility of the solutes in Problem 9-10 for solutions in which the anion concentration is 0.030 M.

*9-13. What CrO_4^{2-} concentration is required to
(a) initiate precipitation of Ag_2CrO_4 from a solution that is 4.13×10^{-3} M in Ag^+?
(b) lower the concentration of Ag^+ in a solution to 9.00×10^{-7} M?

9-14. What hydroxide concentration is required to
(a) initiate precipitation of Al^{3+} from a 4.60×10^{-2} M solution of $Al_2(SO_4)_3$?
(b) lower the Al^{3+} concentration in the foregoing solution to 3.50×10^{-7} M?

*9-15. The solubility-product constant for $Ce(IO_3)_3$ is 3.2×10^{-10}. What is the Ce^{3+} concentration in a solution prepared by mixing 50.00 mL of 0.0450 M Ce^{3+} with 50.00 mL of
(a) water?
(b) 0.0450 M IO_3^-?
(c) 0.250 M IO_3^-?
(d) 0.0500 M IO_3^-?

9-16. The solubility-product constant for K_2PdCl_6 is 6.0×10^{-6} ($K_2PdCl_6 \rightleftharpoons 2K^+ + PdCl_6^{2-}$). What is the K^+ concentration of a solution prepared by mixing 50.0 mL of 0.200 M KCl with 50.0 mL of
(a) 0.0800 M $PdCl_6^{2-}$?
(b) 0.160 M $PdCl_6^{2-}$?
(c) 0.240 M $PdCl_6^{2-}$?

*9-17. The solubility products for a series of iodides are

CuI	$K_{sp} = 1 \times 10^{-12}$
AgI	$K_{sp} = 8.3 \times 10^{-17}$
PbI_2	$K_{sp} = 7.1 \times 10^{-9}$
BiI_3	$K_{sp} = 8.1 \times 10^{-19}$

List these four compounds in order of decreasing molar solubility in
(a) water.
(b) 0.20 M NaI.
(c) a 0.020 M solution of the solute cation.

9-18. The solubility products for a series of hydroxides are

BiOOH	$K_{sp} = 4.0 \times 10^{-10} = [BiO^+][OH^-]$
$Be(OH)_2$	$K_{sp} = 7.0 \times 10^{-22}$
$Tm(OH)_3$	$K_{sp} = 3.0 \times 10^{-24}$
$Hf(OH)_4$	$K_{sp} = 4.0 \times 10^{-26}$

Which hydroxide has
(a) the lowest molar solubility in H_2O?
(b) the lowest molar solubility in a solution that is 0.30 M in NaOH?

9-19. Calculate the pH of water at 25°C and 75°C. The values for pK_w at these temperatures are 13.99 and 12.70, respectively.[8]

9-20. At 25°C, what are the molar H_3O^+ and OH^- concentrations in
*(a) 0.0300 M C_6H_5COOH?
(b) 0.0600 M HN_3?
*(c) 0.100 M ethylamine?
(d) 0.200 M trimethylamine?
*(e) 0.200 M C_6H_5COONa (sodium benzoate)?
(f) 0.0860 M CH_3CH_2COONa?
*(g) 0.250 M hydroxylamine hydrochloride?
(h) 0.0250 M ethyl ammonium chloride?

9-21. At 25°C, what is the hydronium ion concentration in
*(a) 0.200 M chloroacetic acid?
*(b) 0.200 M sodium chloroacetate?
(c) 0.0200 M methylamine?
(d) 0.0200 M methylamine hydrochloride?
*(e) 2.00×10^{-3} M aniline hydrochloride?
(f) 0.300 M HIO_3?

9-22. What is a buffer solution, and what are its properties?

*9-23. Define buffer capacity.

9-24. Which has the greater buffer capacity: (a) a mixture containing 0.100 mol of NH_3 and 0.200 mol of NH_4Cl or (b) a mixture containing 0.0500 mol of NH_3 and 0.100 mol of NH_4Cl?

*9-25. Consider solutions prepared by
(a) dissolving 8.00 mmol of NaOAc in 200 mL of 0.100 M HOAc.
(b) adding 100 mL of 0.0500 M NaOH to 100 mL of 0.175 M HOAc.
(c) adding 40.0 mL of 0.1200 M HCl to 160.0 mL of 0.0420 M NaOAc.

In what respects do these solutions resemble one another? How do they differ?

9-26. Consult Appendix 3, and pick out a suitable acid/base pair to prepare a buffer with a pH of
*(a) 4.5. (b) 8.1. ***(c)** 10.3. (d) 6.1.

*9-27. What mass of sodium formate must be added to 500.0 mL of 1.00 M formic acid to produce a buffer solution that has a pH of 3.50?

9-28. What mass of sodium glycolate should be added to 400.0 mL of 1.00 M glycolic acid to produce a buffer solution with a pH of 4.00?

---

*9-29. What volume of 0.200 M HCl must be added to 500.0 mL of 0.300 M sodium mandelate to produce a buffer solution with a pH of 3.37?

9-30. What volume of 2.00 M NaOH must be added to 200.0 mL of 1.00 M glycolic acid to produce a buffer solution having a pH of 4.00?

9-31. Is the following statement true or false or both? Define your answer with equations, examples, or graphs. "A buffer maintains the pH of a solution constant."

9-32. **Challenge Problem:** It can be shown[9] that the buffer capacity is

$$\beta = 2.303\left(\frac{K_w}{[H_3O^+]} + [H_3O^+] + \frac{c_T K_a [H_3O^+]}{(K_a + [H_3O^+])^2}\right)$$

where c_T is the molar analytical concentration of the buffer.

(a) Show that

$$\beta = 2.303([OH^-] + [H_3O^+] + c_T \alpha_0 \alpha_1)$$

(b) Use the equation in (a) to explain the shape of Figure 9-6.

(c) Differentiate the equation presented at the beginning of the problem and show that the buffer capacity is at a maximum when $\alpha_0 = \alpha_1 = 0.5$.

(d) Describe the conditions under which these relationships apply.

(e) Buffer capacity is sometime called *inverse slope*. Explain the origin of this term.

---

[9]J. N. Butler, *Ionic Equilibrium: Solubility and pH Calculations,* New York: Wiley-Interscience, 1998, p. 134.

Effect of Electrolytes on Chemical Equilibria

CHAPTER 10

This calotype of a leaf was taken in 1844 by the inventor of the process, William Henry Fox Talbot. In its earliest form, the photosensitive paper was created by coating the paper with a sodium chloride solution, allowing the paper to dry, and then applying a second coat of silver nitrate, which produced a film of silver chloride. The leaf was then placed on the paper and exposed to light. The silver chloride in the paper was produced by the chemical equilibrium $Ag^+ + Cl^- \rightleftharpoons AgCl(s)$, which is driven by the activities of reactants and products.

In this chapter, we explore the detailed effects of electrolytes on chemical equilibria. The equilibrium constants for chemical reactions should be, strictly speaking, written in terms of the activities of the participating species. The **activity** of a species is related to its concentration by a factor called the **activity coefficient**. In some cases, the activity of a reactant is essentially equal to its concentration, and we can write the equilibrium constant in terms of the concentrations of the participating species. In the case of ionic equilibria, however, activities and concentrations can be substantially different. Such equilibria are also affected by the concentrations of electrolytes in solution that may not participate directly in the reaction.

Concentration-based equilibrium constants, such as those represented by Equation 9-7 on page 203, provide a reasonable estimate, but they do not approach the accuracy of real laboratory measurements. In this chapter, we show how concentration-based equilibrium constants often lead to significant error. We explore the difference between the activity of a solute and its concentration as well as calculate activity coefficients and use them to modify concentration-based expressions to compute species concentrations that more closely match real laboratory systems at chemical equilibrium.

10A THE EFFECT OF ELECTROLYTES ON CHEMICAL EQUILIBRIA

Experimentally, we find that the position of most solution equilibria depends on the electrolyte concentration of the medium, even when the added electrolyte contains no ion in common with those participating in the equilibrium. For example, consider again the oxidation of iodide ion by arsenic acid that we described in Section 9B-1:

$$H_3AsO_4 + 3I^- + 2H^+ \rightleftharpoons H_3AsO_3 + I_3^- + H_2O$$

If an electrolyte, such as barium nitrate, potassium sulfate, or sodium perchlorate, is added to this solution, the color of the triiodide ion becomes less intense. This decrease in color intensity indicates that the concentration of I_3^- has decreased and that the equilibrium has been shifted to the left by the added electrolyte.

Figure 10-1 further illustrates the effect of electrolytes. Curve A is a plot of the product of the molar hydronium and hydroxide ion *concentrations* ($\times 10^{14}$) as a function of the concentration of sodium chloride. This *concentration*-based ion product is designated K_w'. At low sodium chloride concentrations, K_w' becomes independent of the electrolyte concentration and is equal to 1.00×10^{-14}, which is the *thermodynamic* ion-product constant for water, K_w (curve A, dashed line). A relationship that approaches a constant value as some variable (in this instance, the electrolyte concentration) approaches zero is called a **limiting law**. The constant numerical value observed at this limit is referred to as a **limiting value**.

The vertical axis for curve B in Figure 10-1 is the product of the molar concentrations of barium and sulfate ions ($\times 10^{10}$) in saturated solutions of barium sulfate. This concentration-based solubility product is designated as K_{sp}'. At low electrolyte concentrations, K_{sp}' has a limiting value of 1.1×10^{-10}, which is the accepted thermodynamic value of K_{sp} for barium sulfate.

Curve C is a plot of K_a' ($\times 10^5$), the concentration-based equilibrium constant for the acetic acid dissociation as a function of electrolyte concentration. We see once again that the ordinate function approaches a limiting value $K_a = 1.75 \times 10^{-5}$, which is the thermodynamic acid dissociation constant for acetic acid.

The dashed lines in Figure 10-1 represent ideal behavior of the solutes. Note that departures from ideality can be significant. For example, the product of the molar concentrations of hydrogen and hydroxide ion increases from 1.0×10^{-14} in pure water to about 1.7×10^{-14} in a solution that is 0.1 M in sodium chloride, a 70% increase. The effect is even more pronounced with barium sulfate. In 0.1 M sodium chloride, the K_{sp}' is more than double that of its limiting value.

> Concentration-based equilibrium constants are often indicated by adding a prime mark, for example, K_w', K_{sp}', K_a'.

> As the electrolyte concentration becomes very small, concentration-based equilibrium constants approach their thermodynamic values: K_w, K_{sp}, K_a.

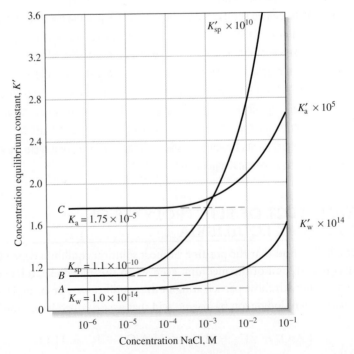

Figure 10-1   Effect of electrolyte concentration on concentration-based equilibrium constants.

The electrolyte effect shown in Figure 10-1 is not unique to sodium chloride. In fact, we would see nearly identical curves if potassium nitrate or sodium perchlorate were substituted for sodium chloride. In each case, the origin of the effect is the electrostatic attraction between the ions of the electrolyte and the ions of reacting species of opposite charge. Since the electrostatic forces associated with all singly charged ions are approximately the same, the three salts exhibit essentially identical effects on equilibria.

Next, we consider how we can take the electrolyte effect into account when we wish to make more accurate equilibrium calculations than those that you may have made in your previous work.

10A-1 The Effect of Ionic Charges on Equilibria

Extensive studies have revealed that the magnitude of the electrolyte effect is highly dependent on the charges of the participants in an equilibrium. When only neutral species are involved, the position of equilibrium is essentially independent of electrolyte concentration. With ionic participants, the magnitude of the electrolyte effect increases with charge. This generality is demonstrated by the three solubility curves in **Figure 10-2**. Note, for example, that, in a 0.02 M solution of potassium nitrate, the solubility of barium sulfate with its pair of doubly charged ions is larger than it is in pure water by a factor of 2. This same change in electrolyte concentration increases the solubility of barium iodate by a factor of only 1.25 and that of silver chloride by 1.2. The enhanced effect due to doubly charged ions is also reflected in the greater slope of curve B in Figure 10-1.

10A-2 The Effect of Ionic Strength

Systematic studies have shown that the effect of added electrolyte on equilibria is *independent* of the chemical nature of the electrolyte but depends on a property of the solution called the **ionic strength**. This quantity is defined as

$$\text{ionic strength} = \mu = \frac{1}{2}([A]Z_A^2 + [B]Z_B^2 + [C]Z_C^2 + \cdots) \qquad (10\text{-}1)$$

where [A], [B], [C], … represent the species molar concentrations of ions A, B, C, … and Z_A, Z_B, Z_C, … are their charges.

Figure 10-2 Effect of electrolyte concentration on the solubility of some salts for compounds containing ions of different charge.

EXAMPLE 10-1

Calculate the ionic strength of (a) a 0.1 M solution of KNO_3 and (b) a 0.1 M solution of Na_2SO_4.

Solution

(a) For the KNO_3 solution, $[K^+]$ and $[NO_3^-]$ are 0.1 M and

$$\mu = \frac{1}{2}(0.1 \text{ M} \times 1^2 + 0.1 \text{ M} \times 1^2) = 0.1 \text{ M}$$

(b) For the Na_2SO_4 solution, $[Na^+] = 0.2$ M and $[SO_4^{2-}]$ 0.1 M. Therefore,

$$\mu = \frac{1}{2}(0.2 \text{ M} \times 1^2 + 0.1 \text{ M} \times 2^2) = 0.3 \text{ M}$$

EXAMPLE 10-2

What is the ionic strength of a solution that is 0.05 M in KNO_3 and 0.1 M in Na_2SO_4?

Solution

$$\mu = \frac{1}{2}(0.05 \text{ M} \times 1^2 + 0.05 \text{ M} \times 1^2 + 0.2 \text{ M} \times 1^2 + 0.1 \text{ M} \times 2^2) = 0.35 \text{ M}$$

These examples show that the ionic strength of a solution of a strong electrolyte consisting solely of singly charged ions is identical to its total molar salt concentration. The ionic strength is greater than the molar concentration, however, if the solution contains ions with multiple charges (see **Table 10-1**).

For solutions with ionic strengths of 0.1 M or less, the electrolyte effect is *independent of the kind of ions* and *dependent only on the ionic strength*. Thus, the solubility of barium sulfate is the same in aqueous sodium iodide, potassium nitrate, or aluminum chloride provided the concentrations of these species are such that the ionic strengths are identical. Note that this independence with respect to electrolyte species disappears at high ionic strengths.

TABLE 10-1

Effect of Charge on Ionic Strength

Type Electrolyte	Example	Ionic Strength*
1:1	NaCl	c
1:2	$Ba(NO_3)_2$, Na_2SO_4	$3c$
1:3	$Al(NO_3)_3$, Na_3PO_4	$6c$
2:2	$MgSO_4$	$4c$

*c = molar concentration of the salt.

10A-3 The Salt Effect

The electrolyte effect (also called the **salt effect**), which we have just described, results from the electrostatic attractive and repulsive forces between the ions of an electrolyte and the ions involved in an equilibrium. These forces cause each ion from the dissociated reactant to be surrounded by a sheath of solution that contains a slight excess of electrolyte ions of opposite charge. For example, when a barium sulfate precipitate is equilibrated with a sodium chloride solution, each dissolved barium ion tends to attract Cl^- and repel Na^+, therefore creating a slightly negative ionic atmosphere around the barium ion. Similarly, each sulfate ion is surrounded by an ionic atmosphere that tends to be slightly positive. These charged layers make the barium ions appear to be somewhat less positive and the sulfate ions somewhat less negative than in the absence of sodium chloride. The result of this shielding effect is a decrease in overall attraction between barium and sulfate ions and a corresponding increase in the solubility of $BaSO_4$. The solubility becomes greater as the number of electrolyte ions in the solution becomes larger. In other words, the *effective concentration* of barium ions and of sulfate ions becomes less as the ionic strength of the medium becomes greater.

10B ACTIVITY COEFFICIENTS

Chemists use a term called activity, a, to account for the effects of electrolytes on chemical equilibria. The activity, or effective concentration, of species X depends on the ionic strength of the medium and is defined by

$$a_X = [X]\gamma_X \tag{10-2}$$

where a_X is the activity of the species X, $[X]$ is its molar concentration, and γ_X is a dimensionless quantity called the **activity coefficient**. The activity coefficient and thus the activity of X vary with ionic strength. If we substitute a_X for $[X]$ in any equilibrium-constant expression, we find that the equilibrium constant is then independent of the ionic strength. To illustrate this point, if X_mY_n is a precipitate, the thermodynamic solubility product expression is defined by the equation

$$K_{sp} = a_X^m \cdot a_Y^n \tag{10-3}$$

Applying Equation 10-2 gives

$$K_{sp} = [X]^m[Y]^n \cdot \gamma_X^m\gamma_Y^n = K'_{sp} \cdot \gamma_X^m\gamma_Y^n \tag{10-4}$$

In this equation, K'_{sp} is the **concentration solubility product constant**, and K_{sp} is the thermodynamic equilibrium constant.[1] The activity coefficients γ_X and γ_Y vary with

> The activity of a species is a measure of its effective concentration as determined by colligative properties such as increasing the boiling point or decreasing the freezing point of water, by electrical conductivity, and by the mass-action effect.

[1]In the chapters that follow, we use the prime notation only when it is necessary to distinguish between thermodynamic and concentration equilibrium constants.

ionic strength in such a way as to keep K_{sp} numerically constant and independent of ionic strength (in contrast to the concentration constant, K'_{sp}).

10B-1 Properties of Activity Coefficients

Activity coefficients have the following properties:

> Although we use only molar concentrations, activities can also be based on molality, mole fraction, and so on. The activity coefficients will be different depending on the concentration scale used.

1. The activity coefficient of a species is a measure of the effectiveness with which that species influences an equilibrium in which it is a participant. In very dilute solutions in which the ionic strength is minimal, this effectiveness becomes constant, and the activity coefficient is unity. Under these circumstances, the activity and the molar concentration are identical (as are thermodynamic and concentration equilibrium constants). As the ionic strength increases, however, an ion loses some of its effectiveness, and its activity coefficient decreases. We may summarize this behavior in terms of Equations 10-2 and 10-3. At moderate ionic strengths, $\gamma_X < 1$. As the solution approaches infinite dilution, however, $\gamma_X \rightarrow 1$, and thus, $a_X \rightarrow [X]$ while $K'_{sp} \rightarrow K_{sp}$. At high ionic strengths ($\mu > 0.1$ M), activity coefficients often increase and may even become greater than unity. Because interpretation of the behavior of solutions in this region is difficult, we confine our discussion to regions of low or moderate ionic strength (that is, where $\mu \leq 0.1$ M). The variation of typical activity coefficients as a function of ionic strength is shown in **Figure 10-3**.

> As $\mu \rightarrow 0$, $\gamma_X \rightarrow 1$, $a_X \rightarrow [X]$, and $K'_{sp} \rightarrow K_{sp}$.

2. In solutions that are not too concentrated, the activity coefficient for a given species is independent of the nature of the electrolyte and dependent only on the ionic strength.

3. For a given ionic strength, the activity coefficient of an ion decreases more dramatically from unity as the charge on the species increases. This effect is shown in Figure 10-3.

4. The activity coefficient of an uncharged molecule is approximately unity, no matter what the level of ionic strength.

5. At any given ionic strength, the activity coefficients of ions of the same charge are approximately equal. The small variations among ions of the same charge can be correlated with the effective diameter of the hydrated ions.

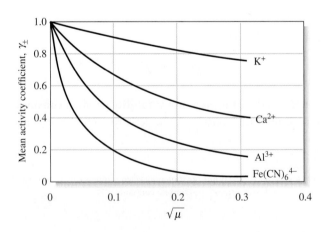

Figure 10-3 Effect of ionic strength on activity coefficients.

6. The activity coefficient of a given ion describes its effective behavior in all equilibria in which it participates. For example, at a given ionic strength, a single activity coefficient for cyanide ion describes the influence of that species on any of the following equilibria:

$$HCN + H_2O \rightleftharpoons H_3O^+ + CN^-$$

$$Ag^+ + CN^- \rightleftharpoons AgCN(s)$$

$$Ni^{2+} + 4CN^- \rightleftharpoons Ni(CN)_4{}^{2-}$$

10B-2 The Debye-Hückel Equation

In 1923, P. Debye and E. Hückel used the ionic atmosphere model, described in Section 10A-3, to derive an equation that permits the calculation of activity coefficients of ions from their charge and their average size.[2] This equation, which has become known as the **Debye-Hückel equation**, takes the form

$$-\log \gamma_X = \frac{0.51 Z_X^2 \sqrt{\mu}}{1 + 3.3\alpha_X \sqrt{\mu}} \tag{10-5}$$

where

γ_X = activity coefficient of the species X
Z_X = charge on the species X
μ = ionic strength of the solution
α_X = effective diameter of the hydrated ion X in nanometers (10^{-9} m)

The constants 0.51 and 3.3 are applicable to aqueous solutions at 25°C. Other values must be used at other temperatures. Unfortunately, there is considerable uncertainty in the magnitude of α_X in Equation 10-5. Its value appears to be approximately 0.3 nm for most singly charged ions. For these species, then, the denominator of the Debye-Hückel equation simplifies to approximately $1 + \sqrt{\mu}$. For ions with higher charge, α_X may be as large as 1.0 nm. This increase in size with increase in charge makes good chemical sense. The larger the charge on an ion, the larger the number of polar water molecules that will be held in the solvation shell around the ion. The second term of the denominator is small with respect to the first when the ionic strength is less than 0.01 M. At these ionic strengths, uncertainties in α_X have little effect on calculating activity coefficients.

Kielland[3] has estimated values of α_X for numerous ions from a variety of experimental data. His best values for effective diameters are given in **Table 10-2**. Also presented are activity coefficients calculated from Equation 10-5 using these values for the size parameter. It is unfortunately impossible to determine experimentally single-ion activity coefficients such as those shown in Table 10-2 because experimental methods give only a mean activity coefficient for the positively and negatively

Peter Debye (1884–1966) was born and educated in Europe but became professor of chemistry at Cornell University in 1940. He was noted for his work in several distinct areas of chemistry, including electrolyte solutions, X-ray diffraction, and the properties of polar molecules. He received the 1936 Nobel Prize in Chemistry.

❮ When μ is less than 0.01 M, $1 + \sqrt{\mu} \approx 1$, and Equation 10-5 becomes

$$-\log \gamma_X = 0.51 Z_X^2 \sqrt{\mu}.$$

This equation is referred to as the Debye-Hückel Limiting Law (DHLL). Thus, in solutions of very low ionic strength ($\mu < 0.01$ M), the DHLL can be used to calculate approximate activity coefficients.

[2]P. Debye and E. Hückel, *Physik. Z.*, **1923,** *24*, 185 (See the Web Works).

[3]J. Kielland, *J. Amer. Chem. Soc.*, **1937,** *59*, 1675, **DOI:** 10.1021/ja01288a032.

TABLE 10-2

Activity Coefficients for Ions at 25°C

Ion	α_X, nm	Activity Coefficient at Indicated Ionic Strength				
		0.001	0.005	0.01	0.05	0.1
H_3O^+	0.9	0.967	0.934	0.913	0.85	0.83
$Li^+, C_6H_5COO^-$	0.6	0.966	0.930	0.907	0.83	0.80
$Na^+, IO_3^-, HSO_3^-, HCO_3^-, H_2PO_4^-, H_2AsO_4^-, OAc^-$	0.4–0.45	0.965	0.927	0.902	0.82	0.77
$OH^-, F^-, SCN^-, HS^-, ClO_3^-, ClO_4^-, BrO_3^-, IO_3^-, MnO_4^-$	0.35	0.965	0.926	0.900	0.81	0.76
$K^+, Cl^-, Br^-, I^-, CN^-, NO_2^-, NO_3^-, HCOO^-$	0.3	0.965	0.925	0.899	0.81	0.75
$Rb^+, Cs^+, Tl^+, Ag^+, NH_4^+$	0.25	0.965	0.925	0.897	0.80	0.75
Mg^{2+}, Be^{2+}	0.8	0.872	0.756	0.690	0.52	0.44
$Ca^{2+}, Cu^{2+}, Zn^{2+}, Sn^{2+}, Mn^{2+}, Fe^{2+}, Ni^{2+}, Co^{2+}, Phthalate^{2-}$	0.6	0.870	0.748	0.676	0.48	0.40
$Sr^{2+}, Ba^{2+}, Cd^{2+}, Hg^{2+}, S^{2-}$	0.5	0.869	0.743	0.668	0.46	0.38
$Pb^{2+}, CO_3^{2-}, SO_3^{2-}, C_2O_4^{2-}$	0.45	0.868	0.741	0.665	0.45	0.36
$Hg_2^{2+}, SO_4^{2-}, S_2O_3^{2-}, Cr_4^{2-}, HPO_4^{2-}$	0.40	0.867	0.738	0.661	0.44	0.35
$Al^{3+}, Fe^{3+}, Cr^{3+}, La^{3+}, Ce^{3+}$	0.9	0.737	0.540	0.443	0.24	0.18
$PO_4^{3-}, Fe(CN)_6^{3-}$	0.4	0.726	0.505	0.394	0.16	0.095
$Th^{4+}, Zr^{4+}, Ce^{4+}, Sn^{4+}$	1.1	0.587	0.348	0.252	0.10	0.063
$Fe(CN)_6^{4-}$	0.5	0.569	0.305	0.200	0.047	0.020

Source: Reprinted (adapted) with permission from J. Kielland, *J. Am. Chem. Soc.*, **1937**, *59*, 1675, **DOI**: 10.1021/ja01288a032. Copyright 1937 American Chemical Society.

charged ions in a solution. In other words, it is impossible to measure the properties of individual ions in the presence of counter-ions of opposite charge and solvent molecules. We should point out, however, that mean activity coefficients calculated from the data in Table 10-2 agree satisfactorily with the experimental values.

FEATURE 10-1

Mean Activity Coefficients

The mean activity coefficient of the electrolyte A_mB_n is defined as

$$\gamma_\pm = \text{mean activity coefficient} = (\gamma_A^m \gamma_B^n)^{1/(m+n)}$$

The mean activity coefficient can be measured in any of several ways, but it is impossible experimentally to resolve this term into the individual activity coefficients for γ_A and γ_B. For example, if

$$K_{sp} = [A]^m[B]^n \cdot \gamma_A^m \gamma_B^n = [A]^m[B]^n \gamma_\pm^{(m+n)}$$

we can obtain K_{sp} by measuring the solubility of A_mB_n in a solution in which the electrolyte concentration approaches zero (that is, where both γ_A and $\gamma_B \rightarrow 1$). A second solubility measurement at some ionic strength μ_1 gives values for [A] and [B]. These data then permit the calculation of $\gamma_A^m \gamma_B^n = \gamma_\pm^{(m+n)}$ for ionic strength μ_1. It is important to understand that this procedure does not provide enough experimental data to permit the calculation of the *individual* quantities γ_A and γ_B and that there appears to be no additional experimental information that would permit evaluation of these quantities. This situation is general, and the *experimental* determination of an individual activity coefficient is impossible.

EXAMPLE 10-3

(a) Use Equation 10-5 to calculate the activity coefficient for Hg^{2+} in a solution that has an ionic strength of 0.085 M. Use 0.5 nm for the effective diameter of the ion. (b) Compare the value obtained in (a) with the activity coefficient obtained by linear interpolation of the data in Table 10-2 for coefficients of the ion at ionic strengths of 0.1 M and 0.05 M.

Solution

(a)
$$-\log \gamma_{Hg^{2+}} = \frac{(0.51)(2)^2 \sqrt{0.085}}{1 + (3.3)(0.5)\sqrt{0.085}} \approx 0.4016$$

$$\gamma_{Hg^{2+}} = 10^{-0.4016} = 0.397 \approx 0.40$$

> Values for activity coefficients at ionic strengths not listed in Table 10-2 can be approximated by interpolation, as shown in Example 10-3(b).

(b) From Table 10-1

μ	$\gamma_{Hg^{2+}}$
0.1M	0.38
0.05M	0.46

Thus, when $\Delta\mu = (0.10 \text{ M} - 0.05 \text{ M}) = 0.05 \text{ M}$, $\Delta\gamma_{Hg^{2+}} = 0.46 - 0.38 = 0.08$. At an ionic strength of 0.085 M,

$$\Delta\mu = (0.100 \text{ M} - 0.085 \text{ M}) = 0.015 \text{ M}$$

and

$$\Delta\gamma_{Hg^{2+}} = \frac{0.015}{0.05} \times 0.08 = 0.024$$

Thus,

$$\Delta\gamma_{Hg^{2+}} = 0.38 + 0.024 = 0.404 \approx 0.40$$

Based on agreement between calculated and experimental values of mean ionic activity coefficients, we can infer that the Debye-Hückel relationship and the data in Table 10-2 give satisfactory activity coefficients for ionic strengths up to about 0.1 M. Beyond this value, the equation fails, and we must determine mean activity coefficients experimentally.

> The Debye-Hückel limiting law is usually assumed to be accurate at values of μ up to about 0.01 for singly charged ions.

10B-3 Equilibrium Calculations Using Activity Coefficients

Equilibrium calculations using activities produce results that agree with experimental data more closely than those obtained with molar concentrations. Unless otherwise specified, equilibrium constants found in tables are usually based on activities and are thus thermodynamic constants. The examples that follow illustrate how activity coefficients from Table 10-2 are used with thermodynamic equilibrium constants.

EXAMPLE 10-4

Find the relative error introduced by neglecting activities in calculating the solubility of $Ba(IO_3)_2$ in a 0.033 M solution of $Mg(IO_3)_2$. The thermodynamic solubility product for $Ba(IO_3)_2$ is 1.57×10^{-9} (see Appendix 2).

Solution

First, we write the solubility-product expression in terms of activities:

$$K_{sp} = a_{Ba^{2+}} \cdot a_{IO_3^-}^2 = 1.57 \times 10^{-9}$$

where $a_{Ba^{2+}}$ and $a_{IO_3^-}$ are the activities of barium and iodate ions. Replacing activities in this equation with activity coefficients and concentrations from Equation 10-2 yields

$$K_{sp} = \gamma_{Ba^{2+}}[Ba^{2+}] \cdot \gamma_{IO_3^-}^2[IO_3^-]^2 \qquad (10\text{-}6)$$

where $\gamma_{Ba^{2+}}$ and $\gamma_{IO_3^-}$ are the activity coefficients for the two ions. Rearranging this expression gives

$$K_{sp}' = \frac{K_{sp}}{\gamma_{Ba^{2+}}\gamma_{IO_3^-}^2} = [Ba^{2+}][IO_3^-]^2$$

where K_{sp}' is the *concentration-based solubility product.*

The ionic strength of the solution is obtained by substituting into Equation 10-1:

$$\mu = \frac{1}{2}([Mg^{2+}] \times 2^2 + [IO_3^-] \times 1^2)$$

$$= \frac{1}{2}(0.033 \text{ M} \times 4 + 0.066 \text{ M} \times 1) = 0.099 \text{ M} \approx 0.1 \text{ M}$$

In calculating μ, we have assumed that the Ba^{2+} and IO_3^- ions from the precipitate do not contribute significantly to the ionic strength of the solution. This simplification seems justified considering the low solubility of barium iodate and the relatively high concentration of $Mg(IO_3)_2$. In situations in which it is not possible to make such an assumption, the concentrations of the two ions can be approximated by solubility calculation in which activities and concentrations are assumed to be identical (as in Examples 9-3, 9-4, and 9-5). These concentrations can then be introduced to give a better value for μ (see spreadsheet summary).

Turning now to Table 10-2, we find that at an ionic strength of 0.1 M,

$$\gamma_{Ba^{2+}} = 0.38 \qquad \gamma_{IO_3^-} = 0.77$$

If the calculated ionic strength did not match that of one of the columns in the table, $\gamma_{Ba^{2+}}$ and $\gamma_{IO_3^-}$ could be calculated from Equation 10-5.

Substituting into the thermodynamic solubility-product expression gives

$$K_{sp}' = \frac{1.57 \times 10^{-9}}{(0.38)(0.77)^2} = 6.97 \times 10^{-9}$$

$$[Ba^{2+}][IO_3^-]^2 = 6.97 \times 10^{-9}$$

Proceeding now as in earlier solubility calculations,

$$\text{solubility} = [Ba^{2+}]$$

$$[IO_3^-] = 2 \times 0.033\ M + 2[Ba^{2+}] \approx 0.066\ M$$

$$[Ba^{2+}](0.066)^2 = 6.97 \times 10^{-9}$$

$$[Ba^{2+}] = \text{solubility} = 1.60 \times 10^{-6}\ M$$

If we neglect activities, we find the solubility as follows:

$$[Ba^{2+}](0.066)^2 = 1.57 \times 10^{-9}$$

$$[Ba^{2+}] = \text{solubility} = 3.60 \times 10^{-7}\ M$$

$$\text{relative error} = \frac{3.60 \times 10^{-7} - 1.60 \times 10^{-6}}{1.60 \times 10^{-6}} \times 100\% = -77\%$$

EXAMPLE 10-5

Use activities to calculate the hydronium ion concentration in a 0.120 M solution of HNO_2 that is also 0.050 M in NaCl. What is the relative percent error incurred by neglecting activity corrections?

Solution

The ionic strength of this solution is

$$\mu = \frac{1}{2}(0.0500\ M \times 1^2 + 0.0500\ M \times 1^2) = 0.0500\ M$$

In Table 10-2, at ionic strength 0.050 M, we find

$$\gamma_{H_3O^+} = 0.85 \qquad \gamma_{NO_2^-} = 0.81$$

Also, from Rule 4 (page 240), we can write

$$\gamma_{HNO_2} = 1.0$$

These three values for γ permit us to calculate the concentration-based dissociation constant from the thermodynamic constant of 7.1×10^{-4} (see Appendix 3).

$$K'_a = \frac{[H_3O^+][NO_2^-]}{[HNO_2]} = \frac{K_a \cdot \gamma_{HNO_2}}{\gamma_{H_3O^+}\gamma_{NO_2^-}} = \frac{7.1 \times 10^{-4} \times 1.0}{0.85 \times 0.81} = 1.03 \times 10^{-3}$$

Proceeding as in Example 9-7, we write

$$[H_3O^+] = \sqrt{K_a \times c_a} = \sqrt{1.03 \times 10^{-3} \times 0.120} = 1.11 \times 10^{-2}\ M$$

(continued)

Note that assuming unit activity coefficients gives $[H_3O^+] = 9.2 \times 10^{-3}$ M.

$$\text{relative error} = \frac{9.2 \times 10^{-3} - 1.11 \times 10^{-2}}{1.11 \times 10^{-2}} \times 100\% = -17\%$$

In this example, we assumed that the contribution of the acid dissociation to the ionic strength was negligible. In addition, we used the approximate solution for calculating the hydronium ion concentration. See Problem 10-19 for a discussion of these approximations.

10B-4 Omitting Activity Coefficients in Equilibrium Calculations

We normally neglect activity coefficients and simply use molar concentrations in applications of the equilibrium law. This approximation simplifies the calculations and greatly decreases the amount of data needed. For most purposes, the error introduced by the assumption of unity for the activity coefficient is not large enough to lead to false conclusions. The preceding examples illustrate, however, that disregarding activity coefficients may introduce significant numerical errors in calculations of this kind. Note, for example, that neglecting activities in Example 10-4 resulted in an error of about -77%. Be alert to situations in which the substitution of concentration for activity is likely to lead to maximum error. Significant discrepancies occur when the ionic strength is large (0.01 M or larger) or when the participating ions have multiple charges (see Table 10-2). With dilute solutions ($\mu < 0.01$ M) of nonelectrolytes or of singly charged ions, mass-law calculations using concentrations are often reasonably accurate. When, as is often the case, solutions have ionic strengths greater than 0.01 M, activity corrections must be made. Computer applications such as Excel greatly reduce the time and effort required to make these calculations. It is also important to note that the decrease in solubility resulting from the presence of an ion common to the precipitate (the common-ion effect) is in part counteracted by the larger electrolyte concentration of the salt containing the common ion.

Spreadsheet Summary   In Chapter 5 of *Applications of Microsoft® Excel in Analytical Chemistry*, 2nd ed., we explore the solubility of a salt in the presence of an electrolyte that changes the ionic strength of the solution. The solubility also changes the ionic strength. An iterative solution is first found in which the solubility is determined by assuming that activity coefficients are unity. The ionic strength is then calculated and used to find the activity coefficients, which in turn are used to obtain a new value for the solubility. The iteration process continues until the results reach a steady value. Excel's Solver is then used to find the solubility directly from an equation containing all the variables.

It is often interesting and instructive to read the original papers describing important discoveries in your field of interest. Two websites, *Selected Classic Papers from the History of Chemistry* and *Classic Papers from the History of Chemistry (and Some Physics Too)*, present many original papers (or their translations) for those who wish to explore pioneering work in chemistry. Go to **www.cengage.com/chemistry/skoog/fac9**, choose Chapter 10 and go to the Web Works. Click on the link to one of the websites just listed. Find and click on the link to the famous 1923 paper by Debye and Hückel on the theory of electrolytic solutions. Read the paper and compare the notation in the paper to the notation in this chapter. What symbol do the authors use for the activity coefficient? What important phenomena do the authors relate to their theory? Note that the mathematical details are missing from the translation of the paper.

QUESTIONS AND PROBLEMS

*10-1. Make a distinction between
  (a) activity and activity coefficient.
  (b) thermodynamic and concentration equilibrium constants.

10-2. List general properties of activity coefficients.

*10-3. Neglecting any effects caused by volume changes, would you expect the ionic strength to (1) increase, (2) decrease, or (3) remain essentially unchanged when NaOH is added to a dilute solution of
  (a) magnesium chloride [$Mg(OH)_2(s)$ forms]?
  (b) hydrochloric acid?
  (c) acetic acid?

10-4. Neglecting any effects caused by volume changes, would you expect the ionic strength to (1) increase, (2) decrease, or (3) remain essentially unchanged by the addition of iron(III) chloride to
  (a) HCl?
  (b) NaOH?
  (c) $AgNO_3$?

*10-5. Explain why the activity coefficient for dissolved ions in water is usually less than that for water itself.

10-6. Explain why the activity coefficient for a neutral molecule is usually 1.

*10-7. Explain why the initial slope for Ca^{2+} in Figure 10-3 is steeper than that for K^+?

10-8. What is the numerical value of the activity coefficient of aqueous ammonia (NH_3) at an ionic strength of 0.2?

10-9. Calculate the ionic strength of a solution that is
  *(a) 0.030 M in $FeSO_4$.
  (b) 0.30 M in $(NH_4)_2CrO_4$.
  *(c) 0.30 M in $FeCl_3$ and 0.20 M in $FeCl_2$.
  (d) 0.030 M in $La(NO_3)_3$ and 0.060 M in $Fe(NO_3)_2$.

10-10. Use Equation 10-5 to calculate the activity coefficient of
  *(a) Fe^{3+} at $\mu = 0.062$.
  (b) Pb^{2+} at $\mu = 0.042$.
  *(c) Ce^{4+} at $\mu = 0.070$.
  (d) Sn^{4+} at $\mu = 0.045$.

10-11. Calculate activity coefficients for the species in Problem 10-10 by linear interpolation of the data in Table 10-2.

10-12. For a solution in which $\mu = 8.0 \times 10^{-2}$, calculate K'_{sp} for
  *(a) AgSCN.
  (b) PbI_2.
  *(c) $La(IO_3)_3$.
  (d) $MgNH_4PO_4$.

*10-13. Use activities to calculate the molar solubility of $Zn(OH)_2$ in
  (a) 0.0200 M KCl.
  (b) 0.0300 M K_2SO_4.
  (c) the solution that results when you mix 40.0 mL of 0.250 M KOH with 60.0 mL of 0.0250 M $ZnCl_2$.
  (d) the solution that results when you mix 20.0 mL of 0.100 M KOH with 80.0 mL of 0.0250 M $ZnCl_2$.

*10-14. Calculate the solubilities of the following compounds in a 0.0333 M solution of $Mg(ClO_4)_2$ using (1) activities and (2) molar concentrations:
  (a) AgSCN.
  (b) PbI_2.
  (c) $BaSO_4$.
  (d) $Cd_2Fe(CN)_6$.

$$Cd_2Fe(CN)_6(s) \rightleftharpoons 2Cd^{2+} + Fe(CN)_6^{4-}$$
$$K_{sp} = 3.2 \times 10^{-17}$$

*10-15. Calculate the solubilities of the following compounds in a 0.0167 M solution of $Ba(NO_3)_2$ using (1) activities and (2) molar concentrations:
   (a) $AgIO_3$.
   (b) $Mg(OH)_2$.
   (c) $BaSO_4$.
   (d) $La(IO_3)_3$.

10-16. Calculate the % relative error in solubility by using concentrations instead of activities for the following compounds in 0.0500 M KNO_3 using the thermodynamic solubility products listed in Appendix 2.
   *(a) $CuCl$ (α_{Cu^+} = 0.3 nm)
   (b) $Fe(OH)_2$
   *(c) $Fe(OH)_3$
   (d) $La(IO_3)_3$
   *(e) Ag_3AsO_4 ($\alpha_{AsO_4^{3-}}$ = 0.4 nm)

10-17. Calculate the % relative error in hydronium ion concentration by using concentrations instead of activities in calculating the pH of the following buffer solutions using the thermodynamic constants found in Appendix 3.
   *(a) 0.150 M HOAc and 0.250 M NaOAc
   (b) 0.0400 M NH_3 and 0.100 M NH_4Cl
   (c) 0.0200 M $ClCH_2COOH$ and 0.0500 M $ClCH_2COONa$ ($a_{ClCH2COO^-}$ = 0.35)

10-18. Design and construct a spreadsheet to calculate activity coefficients in a format similar to Table 10-2. Enter values of a_X in cells A3, A4, A5, and so forth and enter ionic charges in cells B3, B4, B5, and so forth. Enter in cells C2:G2 the same set of values for ionic strength listed in Table 10-2. Enter the formula for the activity coefficients in cells C3:G3. Be sure to use absolute cell references for ionic strength in your formulas for the activity coefficients. Finally, copy the formulas for the activity coefficients into the rows below row C by highlighting C3:G3 and dragging the fill handle downward. Compare the activity coefficients that you calculate to those in Table 10-2. Do you find any discrepancies? If so, explain how they arise.

10-19. **Challenge Problem:** In Example 10-5, we neglected the contribution of nitrous acid to the ionic strength.

We also used the simplified solution for the hydronium ion concentration,

$$[H_3O^+] = \sqrt{K_a c_a}$$

   (a) Perform an iterative solution to the problem in which you actually calculate the ionic strength, first without taking into account the dissociation of the acid. Then, calculate corresponding activity coefficients for the ions using the Debye-Hückel equation, compute a new K_a, and find a new value for $[H_3O^+]$. Repeat the process, but use the concentrations of H_3O^+ and NO_2^- along with the 0.05 M NaCl to calculate a new ionic strength. Once again, find the activity coefficients, K_a and a new value for $[H_3O^+]$. Iterate until you obtain two consecutive values of $[H_3O^+]$ that are equal to within 0.1%. How many iterations did you need? What is the relative error between your final value and the value obtained in Example 10-5 with no activity correction? What is the relative error between the first value that you calculated and the last one? You may want to use a spreadsheet to assist you in these calculations.

   (b) Now perform the same calculation, except this time calculate the hydronium ion concentration using the quadratic equation or the method of successive approximations each time you compute a new ionic strength. How much improvement do you observe over the results that you found in (a)?

   (c) When are activity corrections like those that you carried out in (a) necessary? What variables must be considered in deciding whether to make such corrections?

   (d) When are corrections such as those in (b) necessary? What criteria do you use to decide whether these corrections should be made?

   (e) Suppose that you are attempting to determine ion concentrations in a complex matrix such as blood serum or urine. Is it possible to make activity corrections in such a system? Explain your answer.

Solving Equilibrium Problems for Complex Systems

CHAPTER 11

Complex equilibria are extremely important in many areas of science. Such equilibria play important roles in the environment. Rivers and lakes are subject to many sources of pollution that may make the water unsuitable for drinking, swimming, or fishing. One of the most common problems with lakes is the nutrient overload caused by the increased flow of plant nutrients, such as phosphates and nitrates, from sewage treatment plants, fertilizers, detergents, animal wastes, and soil erosion. These nutrients are involved in complex equilibria that cause rooted plants like water hyacinths (photo) and algae to undergo population explosions. When the plants die and fall to the bottom of the lake, the decomposing bacteria deplete the lake's lower layers of dissolved oxygen, which may lead food fish to die of oxygen starvation. The calculations involved in complex equilibria are the major subject of this chapter. The systematic approach to solving multiple-equilibrium problems is described. The calculation of solubility, when the equilibrium is influenced by pH and the formation of complexes, is also discussed.

Wolfgang Kaehler/CORBIS

Aqueous solutions often contain species that interact with one another and water to yield two or more simultaneous equilibria. For example, when we dissolve a sparingly soluble salt in water, there are three equilibria:

$$BaSO_4(s) \rightleftharpoons Ba^{2+} + SO_4^{2-} \tag{11-1}$$

$$SO_4^{2-} + H_3O^+ \rightleftharpoons HSO_4^- + H_2O \tag{11-2}$$

$$2H_2O \rightleftharpoons H_3O^+ + OH^- \tag{11-3}$$

> ◀ The introduction of a new equilibrium system into a solution does not change the equilibrium constants for any existing equilibria.

If hydronium ions are added to this system, the second equilibrium is shifted to the right by the common-ion effect. The resulting decrease in sulfate concentration causes the first equilibrium to shift to the right as well, which increases the solubility of the barium sulfate.

The solubility of barium sulfate is also increased when acetate ions are added to an aqueous suspension of barium sulfate because acetate ions tend to form a soluble complex with barium ions, as shown by the reaction

$$Ba^{2+} + OAc^- \rightleftharpoons BaOAc^+ \tag{11-4}$$

The common-ion effect again causes both this equilibrium and the solubility equilibrium of Equation 11-1 to shift to the right. The solubility of barium sulfate thus increases.

If we wish to calculate the solubility of barium sulfate in a system containing hydronium and acetate ions, we must take into account not only the solubility equilibrium but also the other three equilibria. We find, however, that using four equilibrium-constant expressions to calculate solubility is much more difficult and complex than the simple procedure illustrated in Examples 9-3, 9-4, and 9-5. To solve complex problems of this type, a systematic approach is essential. We use this approach to illustrate the effect of pH and complex formation on the solubility of typical analytical precipitates. In later chapters, we use the principles of this same systematic method to solve problems with multiple equilibria of several types.

11A SOLVING MULTIPLE-EQUILIBRIUM PROBLEMS USING A SYSTEMATIC METHOD

To solve a multiple-equilibrium problem, we must write as many independent equations as there are chemical species in the system being studied. For example, if our task is to compute the solubility of barium sulfate in an acidic solution, we must calculate the concentrations of all the species in the solution. In this example, there are five species: $[Ba^{2+}]$, $[SO_4^{2-}]$, $[HSO_4^-]$, $[H_3O^+]$, and $[OH^-]$. To calculate the solubility of barium sulfate in this solution rigorously, it is necessary to create five independent algebraic equations that can be solved simultaneously to give the five concentrations.

We use three types of algebraic equations to solve multiple-equilibrium problems: (1) equilibrium-constant expressions, (2) *mass-balance* equations, and (3) a single *charge-balance* equation. We showed in Section 9B how equilibrium-constant expressions are written. Let us now focus on the other two types of equations.

11A-1 Mass-Balance Equations

Mass-balance equations relate the *equilibrium* concentrations of various species in a solution to one another and to the *analytical* concentrations of the various solutes. To illustrate, suppose that we want to prepare an aqueous solution of a salt NaA in which A^- is the conjugate base of a weak acid HA. For every mole of NaA that dissolves, at equilibrium, the solution must contain one mole of A in all of its forms. Some of the A^- from the dissociation of the salt will abstract protons from water to become HA, and the rest will remain in solution as A^-. Regardless of any chemistry that occurs as the solution is formed, however, we must have the same number of moles of A (and thus the same mass of A) before we dissolve the solute as we find after the solution has reached equilibrium. Because all the solutes are contained in the same volume of solution, the concentration of A before dissolving is equal to the total concentration of all forms of A in the equilibrium solution. In essence, mass-balance equations are a direct result of the conservation of mass, moles, and in this application, concentration. To write mass-balance expression(s), we must know the properties and amounts of all solutes in the solution, how the solution was prepared, and the equilibria in the solution.

As our first example of mass balance, let us explore the details of what happens when a weak acid HA is dissolved in water with a molar analytical concentration of c_{HA}. Our goal is to write the mass-balance expression(s) for this system. Our first type of mass-balance expression is based on precise knowledge of the value of c_{HA} from the description of the solution.

As shown on page 213, there are two equilibria at play in this solution:

$$HA + H_2O \rightleftharpoons H_3O^+ + A^-$$

$$2H_2O \rightleftharpoons H_3O^+ + OH^-$$

The widely used term "mass-balance equation" is somewhat misleading because such equations are really based on balancing *concentrations* rather than *masses*. We introduce no problems by equating masses to concentrations because all solute species are in the same volume of solution.

The only source of the two A-containing species, HA and A^-, is the original solute, HA, whose *analytical* concentration is c_{HA}. Because all of the A^- and the HA in the solution comes from the measured quantity of solute HA, we can write our first mass-balance equation.

$$c_{HA} = [HA] + [A^-]$$

The second type of mass-balance expression relies on our detailed knowledge of the equilibria in the solution. Hydronium ions in the solution come from two sources: the dissociation of HA and the dissociation of water. The overall concentration of H_3O^+ is then the sum of the two concentrations from these sources, or

$$[H_3O^+] = [H_3O^+]_{HA} + [H_3O^+]_{H_2O}$$

But, from the equilibria above, the concentration of hydronium from the dissociation of the acid $[H_3O^+]_{HA}$ is equal to $[A^-]$, and the hydronium concentration from water $[H_3O^+]_{H_2O}$ is equal to $[OH^-]$. Thus, we have

$$[H_3O^+] = [A^-] + [OH^-]$$

This type of mass-balance expression is often referred to as the **proton balance equation** because it accounts for all sources of protons. As we will see, this last equation is very interesting and useful because it demonstrates the conservation of another quantity, charge.

In Examples 11-1 and 11-2, we consider mass-balance expressions for two nearly insoluble salts in the presence of other solutes that affect the solubility of the salts.

> Writing mass balance equations may be as straightforward as the case of a weak acid discussed here. In complex solutions containing many solutes participating in numerous equilibria, the task may be quite difficult.

EXAMPLE 11-1

Write mass-balance expressions for a 0.0100 M solution of HCl that is in equilibrium with an excess of solid $BaSO_4$.

Solution

As shown by Equations 11-1, 11-2, and 11-3, there are three equilibria in this solution:

$$BaSO_4(s) \rightleftharpoons Ba^{2+} + SO_4^{2-}$$
$$SO_4^{2-} + H_3O^+ \rightleftharpoons HSO_4^- + H_2O$$
$$2H_2O \rightleftharpoons H_3O^+ + OH^-$$

Because the only source for the two sulfate species is the dissolved $BaSO_4$, the barium ion concentration must equal the total concentration of sulfate-containing species, so we can write our first mass-balance equation.

$$[Ba^{2+}] = [SO_4^{2-}] + [HSO_4^-]$$

According to the second reaction above, hydronium ions in the solution are either free H_3O^+ or they react with SO_4^{2-} to form HSO_4^-. We may express this as

$$[H_3O^+]_{tot} = [H_3O^+] + [HSO_4^-]$$

(continued)

> For a slightly soluble salt with a 1:1 stoichiometry, the equilibrium concentration of the cation is equal to the equilibrium concentration of the anion. This equality is the mass-balance expression. For anions that can be protonated, the equilibrium concentration of the cation is equal to the sum of the concentrations of the various forms of the anion.

where $[H_3O^+]_{tot}$ is the hydronium concentration from all sources and $[H_3O^+]$ the free equilibrium concentration of hydronium. The protons that contribute to $[H_3O^+]_{tot}$ have two sources: aqueous HCl and the dissociation of water. In this example, we refer to the hydronium ion concentration from the complete dissociation of HCl as $[H_3O^+]_{HCl}$ and the concentration from the autoprotolysis of water as $[H_3O^+]_{H_2O}$. The total concentration of hydronium ion is then

$$[H_3O^+]_{tot} = [H_3O^+]_{HCl} + [H_3O^+]_{H_2O}$$

And from above,

$$[H_3O^+]_{tot} = [H_3O^+] + [HSO_4^-] = [H_3O^+]_{HCl} + [H_3O^+]_{H_2O}$$

But $[H_3O^+]_{HCl} = c_{HCl}$, and since the only source of hydroxide is the dissociation of water, we may also write that $[H_3O^+]_{H_2O} = [OH^-]$. By substituting these two quantities into the equation above,

$$[H_3O^+]_{tot} = [H_3O^+] + [HSO_4^-] = c_{HCl} + [OH^-]$$

and the mass-balance equation is then

$$[H_3O^+] + [HSO_4^-] = 0.0100 + [OH^-]$$

EXAMPLE 11-2

Write mass-balance expressions for the system formed when a 0.010 M NH_3 solution is saturated with the slightly soluble AgBr.

Solution

In this example, the equations for the equilibria in the solution are

$$AgBr(s) \rightleftharpoons Ag^+ + Br^-$$
$$Ag^+ + NH_3 \rightleftharpoons Ag(NH_3)^+$$
$$Ag(NH_3)^+ + NH_3 \rightleftharpoons Ag(NH_3)_2^+$$
$$NH_3 + H_2O \rightleftharpoons NH_4^+ + OH^-$$
$$2H_2O \rightleftharpoons H_3O^+ + OH^-$$

In this solution, AgBr is the only source of Br^-, Ag^+, $Ag(NH_3)^+$, and $Ag(NH_3)_2^+$. As AgBr dissolves, silver and bromide ions appear in a 1:1 ratio. While Ag^+ reacts with ammonia to form $Ag(NH_3)^+$ and $Ag(NH_3)_2^+$, bromide appears only as Br^-, so our first mass-balance equation is

$$[Ag^+] + [Ag(NH_3)^+] + [Ag(NH_3)_2^+] = [Br^-]$$

where the bracketed terms are molar species concentrations. Also, we know that the only source of ammonia-containing species is the 0.010 M NH_3. Therefore,

$$c_{NH_3} = [NH_3] + [NH_4^+] + [Ag(NH_3)^+] + 2[Ag(NH_3)_2^+] = 0.010 \text{ M}$$

For slightly soluble salts with stoichiometries other than 1:1, the mass-balance expression is obtained by multiplying the concentration of one of the ions by the stoichiometric ratio. For example, in a solution saturated with PbI_2, the iodide ion concentration is twice that of the Pb^{2+}. That is,

$$[I^-] = 2[Pb^{2+}]$$

This result seems counterintuitive to most people because two iodide ions appear in the solution for every lead (II) ion that appears. Remember that it is for precisely this reason that we must multiply $[Pb^{2+}]$ by two in order to set the equality.

The coefficient 2 in this equation arises because $Ag(NH_3)_2^+$ contains two ammonia molecules. From the last two equilibria, we see that one hydroxide ion is formed for each NH_4^+ and each hydronium ion. Therefore,

$$[OH^-] = [NH_4^+] + [H_3O^+]$$

11A-2 Charge-Balance Equation

Electrolyte solutions are electrically neutral even though they may contain up to several moles per liter of charged ions. Solutions are neutral because the *molar concentration of positive charge* in an electrolyte solution always equals the *molar concentration of negative charge*. In other words, for any solution containing electrolytes, we may write

no. moles/L positive charge = no. moles/L negative charge

This equation represents the charge-balance condition and is called the **charge-balance equation**. To be useful for equilibrium calculations, the equality must be expressed in terms of the molar concentrations of the species that carry a charge in the solution.

How much charge is contributed to a solution by 1 mol of Na^+? How about 1 mol of Mg^{2+} or 1 mol of PO_4^{3-}? The concentration of charge contributed to a solution by an ion is equal to the molar concentration of that ion multiplied by its charge. Thus, the molar concentration of positive charge in a solution due to the presence of sodium ions is the molar sodium ion concentration:

$$\frac{\text{mol positive charge}}{L} = \frac{1 \text{ mol positive charge}}{\cancel{\text{mol Na}^+}} \times \frac{\cancel{\text{mol Na}^+}}{L}$$

$$= 1 \times [Na^+]$$

> Always remember that a charge-balance equation is based on the equality in *molar charge concentrations* and that to obtain the charge concentration of an ion, you must multiply the molar concentration of the ion by its charge.

The concentration of positive charge due to magnesium ions is

$$\frac{\text{mol positive charge}}{L} = \frac{2 \text{ mol positive charge}}{\cancel{\text{mol Mg}^{2+}}} \times \frac{\cancel{\text{mol Mg}^{2+}}}{L}$$

$$= 2 \times [Mg^{2+}]$$

since each mole of magnesium ion contributes 2 mol of positive charge to the solution. Similarly, we may write for phosphate ion

$$\frac{\text{mol negative charge}}{L} = \frac{3 \text{ mol negative charge}}{\cancel{\text{mol PO}_4^{3-}}} \times \frac{\cancel{\text{mol PO}_4^{3-}}}{L}$$

$$= 3 \times [PO_4^{3-}]$$

> In some systems, a useful charge-balance equation cannot be written because not enough information is available or because the charge-balance equation is identical to one of the mass-balance equations.

Now, consider how we would write a charge-balance equation for a 0.100 M solution of sodium chloride. Positive charges in this solution are supplied by Na^+ and H_3O^+ (from dissociation of water). Negative charges come from Cl^- and OH^-. The concentrations of positive and negative charges are

$$\text{mol/L positive charge} = [Na^+] + [H_3O^+] = 0.100 + 1 \times 10^{-7}$$

$$\text{mol/L negative charge} = [Cl^-] + [OH^-] = 0.100 + 1 \times 10^{-7}$$

The equilibrium concentrations $[OH^-]$ and $[H_3O^+]$ are near 1×10^{-7} M in these examples, but these concentrations can change if other equilibria come into play.

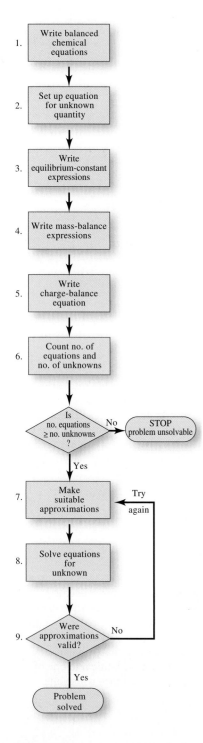

Figure 11-1 A systematic method for solving multiple-equilibrium problems.

The flowchart shows:
1. Write balanced chemical equations
2. Set up equation for unknown quantity
3. Write equilibrium-constant expressions
4. Write mass-balance expressions
5. Write charge-balance equation
6. Count no. of equations and no. of unknowns
 Is no. equations ≥ no. unknowns? — No → STOP problem unsolvable
 Yes
7. Make suitable approximations ← Try again
8. Solve equations for unknown
9. Were approximations valid? — No → (back to 7)
 Yes
 Problem solved

We write the charge-balance equation by equating the concentrations of positive and negative charges:

$$[Na^+] + [H_3O^+] = [Cl^-] + [OH^-] = 0.100 + [OH^-]$$

Now consider a solution that has an analytical concentration of magnesium chloride of 0.100 M. In this example, the concentrations of positive and negative charge are given by

$$\text{mol/L positive charge} = 2[Mg^{2+}] + [H_3O^+] = 2 \times 0.100 + [H_3O^+]$$
$$\text{mol/L negative charge} = [Cl^-] + [OH^-] = 2 \times 0.100 + [OH^-]$$

In the first equation, the molar concentration of magnesium ion is multiplied by two (2×0.100) because 1 mol of that ion contributes 2 mol of positive charge to the solution. In the second equation, the molar chloride ion concentration is twice that of the magnesium chloride concentration, or 2×0.100. To obtain the charge-balance equation, we equate the concentration of positive charge with the concentration of negative charge to obtain

$$2[Mg^{2+}] + [H_3O^+] = [Cl^-] + [OH^-] = 0.200 + [OH^-]$$

For a neutral solution, $[H_3O^+]$ and $[OH^-]$ are very small ($\approx 1 \times 10^{-7}$ M) and equal so that we can usually simplify the charge-balance equation to

$$2[Mg^{2+}] \approx [Cl^-] = 0.200 \text{ M}$$

EXAMPLE 11-3

Write a charge-balance equation for the system in Example 11-2.

Solution

$$[Ag^+] + [Ag(NH_3)^+] + [Ag(NH_3)_2^+] + [H_3O^+] + [NH_4^+] = [OH^-] + [Br^-]$$

EXAMPLE 11-4

Write a charge-balance equation for an aqueous solution that contains NaCl, $Ba(ClO_4)_2$, and $Al_2(SO_4)_3$.

Solution

$$[Na^+] + [H_3O^+] + 2[Ba^{2+}] + 3[Al^{3+}] =$$
$$[ClO_4^-] + [NO_3^-] + 2[SO_4^{2-}] + [HSO_4^-] + [OH^-]$$

11A-3 Steps for Solving Problems with Several Equilibria

Step 1. Write a set of balanced chemical equations for all pertinent equilibria.

Step 2. State the quantity being sought in terms of equilibrium concentrations.

Step 3. Write equilibrium-constant expressions for all equilibria developed in Step 1, and find numerical values for the constants in tables of equilibrium constants.

Step 4. Write mass-balance expressions for the system.

Step 5. If possible, write a charge-balance expression for the system.

Step 6. Count the number of unknown concentrations in the equations developed in Steps 3, 4, and 5, and compare this number with the number of independent equations. Step 6 is critical because it shows whether an exact solution to the problem is possible. If the number of unknowns is identical to the number of equations, the problem has been reduced to one of *algebra* alone. In other words, answers can be obtained with sufficient perseverance. On the other hand, if there are not enough equations even after approximations are made, the problem should be abandoned. If a sufficient number of equations have been developed, proceed to either Step 7a or Step 7b.

Step 7a. Make suitable approximations to reduce the number of unknown equilibrium concentrations and thus the number of equations needed to provide an answer, as defined in Step 2. Proceed to Steps 8 and 9.

Step 7b. Solve the simultaneous equations exactly for the concentrations required by Step 2 by means of a computer program.

Step 8. Solve manually the simplified algebraic equations to give provisional concentrations for the species in the solution.

Step 9. Check the validity of the approximations.

These steps are illustrated in **Figure 11-1**.

11A-4 Using Approximations to Solve Equilibrium Calculations

When Step 6 of the systematic approach is complete, we have a *mathematical* problem of solving several nonlinear simultaneous equations. This job requires that a suitable computer program be available or that approximations be made that decrease the number of unknowns and equations. In this section, we consider in general terms how equations describing equilibrium relationships can be simplified by suitable approximations.

Bear in mind that *only* the mass-balance and charge-balance equations can be simplified because only in these equations do the concentration terms appear as sums or differences rather than as products or quotients. It is always possible to assume that one (or more) of the terms in a sum or difference is so much smaller than the others that it can be ignored without significantly affecting the equality. The assumption that a concentration term in an equilibrium-constant expression is zero makes the expression meaningless.

The assumption that a given term in a mass- or charge-balance equation is sufficiently small that it can be neglected is generally based on a knowledge of the chemistry of the system. For example, in a solution containing a reasonable concentration of an acid, the hydroxide concentration will often be negligible with respect to the other species in the solution. Therefore, the term for the hydroxide concentration can usually be neglected in a mass- or charge-balance expression without introducing significant error to such a calculation.

Do not worry that invalid approximations in Step 7a will lead to serious errors in your computed results. Experienced workers are often as puzzled as beginners when making approximations to simplify equilibrium calculations. Nonetheless, they make such approximations without fear because they know that the effects of an invalid assumption will be revealed by the time a computation is completed (see Example 11-6). It is a good idea to try questionable assumptions early during the solution of a

❮ Do not waste time starting the algebra in an equilibrium calculation until you are absolutely sure that you have enough independent equations to make the solution feasible.

❮ Approximations can be made only in charge-balance and mass-balance equations, never in equilibrium-constant expressions.

❮ Never be afraid to make an assumption while attempting to solve an equilibrium problem. If the assumption is not valid, you will know it as soon as you have an approximate answer.

problem. If the assumption leads to an intolerable error (which is generally easy to recognize), recalculate without the faulty approximation to arrive at a tentative answer. It is usually more efficient to try a questionable assumption at the beginning of a problem than to make a more laborious and time-consuming calculation without the assumption.

11A-5 Use of Computer Programs to Solve Multiple-Equilibrium Problems

So far, we have learned that, if we know all of the chemical equilibria in a system, we can write a corresponding system of equations that allows us to solve for the concentrations of all species in the system. Although the systematic method gives us the means to solve equilibrium problems of great complexity, it can be tedious and time consuming, particularly when a system must be solved for various sets of experimental conditions. For example, if we wish to find the solubility of silver chloride as a function of the concentration of added chloride, the system of five equations and five unknowns must be solved repetitively for each different concentration of chloride (see Example 11-9).

> Several software packages are available for solving multiple nonlinear simultaneous equations rigorously. Three such programs are Mathcad, Mathematica, and Excel.

A number of powerful general-purpose software applications are available for solving equations. Such so-called **solvers** include Mathcad, Mathematica, MATLAB, TK Solver, and Excel, among many others. Once a system of equations has been set up, they may be solved repetitively for many different sets of conditions. Furthermore, the accuracy of the solutions to the equations can be controlled by choosing appropriate tolerances within the programs. The equation-solving features of these applications coupled with their graphical capabilities enable you to solve complex systems of equations and present the results in graphical form. In this way, you can explore many different types of systems quickly and efficiently and develop your chemical intuition based on the results.

A word of caution is in order, however. Solvers often require initial estimates of the solutions to solve systems of equations. To provide these estimates, you must think about the chemistry a bit before beginning to solve the equations, and you should check the solutions that you find to be sure that they make good chemical sense.

Also, computers *do not know chemistry*. They will dutifully find solutions to the equations that you write based on the initial estimates that you give. If you make errors in the equations, software applications can sometimes flag errors based on certain mathematical constraints, but they will not find errors in the chemistry. If a program does not find a solution to a set of equations, it is often because of faulty initial estimates. Always be skeptical of computer results and respectful of software limitations. Used wisely, computer applications can be a marvelous aid in your study of chemical equilibria. For examples of the use of Excel in solving systems of equations such as those found in this chapter, see Chapter 6 of *Applications of Microsoft® Excel in Analytical Chemistry*, 2nd ed.

11B CALCULATING SOLUBILITIES BY THE SYSTEMATIC METHOD

In these sections, we illustrate the systematic method with examples of precipitate solubility under various conditions. In later chapters, we apply the method to other types of equilibria.

11B-1 The Solubility of Metal Hydroxides

In Examples 11-5 and 11-6, we calculate the solubilities of two metal hydroxides. These examples illustrate how to make approximations and check their validity.

EXAMPLE 11-5

Calculate the molar solubility of $Mg(OH)_2$ in water.

Solution

Step 1. Write Equations for the Pertinent Equilibria Two equilibria need to be considered:

$$Mg(OH)_2(s) \rightleftharpoons Mg^{2+} + 2OH^-$$

$$2H_2O \rightleftharpoons H_3O^+ + OH^-$$

Step 2. Define the Unknown Since 1 mol of Mg^{2+} is formed for each mole of $Mg(OH)_2$ dissolved,

$$\text{solubility } Mg(OH)_2 = [Mg^{2+}]$$

Step 3. Write All Equilibrium-Constant Expressions

$$K_{sp} = [Mg^{2+}][OH^-]^2 = 7.1 \times 10^{-12} \tag{11-5}$$

$$K_w = [H_3O^+][OH^-] = 1.00 \times 10^{-14} \tag{11-6}$$

Step 4. Write Mass-Balance Expressions As shown by the two equilibrium equations, there are two sources of hydroxide ions: $Mg(OH)_2$ and H_2O. The hydroxide ion concentration resulting from dissociation of $Mg(OH)_2$ is twice the magnesium ion concentration, and the hydroxide ion concentration from the dissociation of water is equal to the hydronium ion concentration. Thus,

$$[OH^-] = 2[Mg^{2+}] + [H_3O^+] \tag{11-7}$$

Step 5. Write the Charge-Balance Expression

$$[OH^-] = 2[Mg^{2+}] + [H_3O^+] \tag{11-8}$$

Note that this equation is identical to Equation 11-7. Often, a mass-balance equation for a system is identical to the charge-balance equation.

Step 6. Count the Number of Independent Equations and Unknowns We have developed three independent algebraic equations (Equations 11-5, 11-6, and 11-7) and have three unknowns ($[Mg^{2+}]$, $[OH^-]$, and $[H_3O^+]$). Therefore, the problem can be solved rigorously.

Step 7a. Make Approximations We can make approximations only in Equation 11-7. Since the solubility-product constant for $Mg(OH)_2$ is relatively large, the solution will be somewhat basic. Therefore, it is reasonable to assume that $[H_3O^+] \ll [OH^-]$. Equation 11-7 then simplifies to

$$2[Mg^{2+}] \approx [OH^-]$$

(continued)

> To arrive at Equation 11-7, we reasoned that if $[OH^-]_{H_2O}$ and $[OH^-]_{Mg(OH)_2}$ are the concentrations of OH^- produced from H_2O and $Mg(OH)_2$, respectively, then
>
> $$[OH^-]_{H_2O} = [H_3O^+]$$
> $$[OH^-]_{Mg(OH)_2} = 2[Mg^{2+}]$$
> $$[OH^-]_{total} = [OH^-]_{H_2O} + [OH^-]_{Mg(OH)_2}$$
> $$= [H_3O^+] + 2[Mg^{2+}]$$

Step 8. Solve the Equations Substitution of Equation 11-8 into Equation 11-5 gives

$$[Mg^{2+}](2[Mg^{2+}])^2 = 7.1 \times 10^{-12}$$

$$[Mg^{2+}]^3 = \frac{7.1 \times 10^{-12}}{4} = 1.78 \times 10^{-12}$$

$$[Mg^{2+}] = \text{solubility} = (1.78 \times 10^{-12})^{1/3} = 1.21 \times 10^{-4} \text{ or } 1.2 \times 10^{-4} \text{ M}$$

Step 9. Check the Assumptions Substitution into Equation 11-8 yields

$$[OH^-] = 2 \times 1.21 \times 10^{-4} = 2.42 \times 10^{-4} \text{ M}$$

and, from Equation 11-6,

$$[H_3O^+] = \frac{1.00 \times 10^{-14}}{2.42 \times 10^{-4}} = 4.1 \times 10^{-11} \text{ M}$$

Thus, our assumption that $[H_3O^+] \ll [OH^-]$ is certainly valid.

EXAMPLE 11-6

Calculate the solubility of $Fe(OH)_3$ in water.

Solution

Proceeding by the systematic approach used in Example 11-5, we write.

Step 1. Write the Equations for the Pertinent Equilibria

$$Fe(OH)_3(s) \rightleftharpoons Fe^{3+} + 3OH^-$$

$$2H_2O \rightleftharpoons H_3O^+ + OH^-$$

Step 2. Define the Unknown

$$\text{solubility} = [Fe^{3+}]$$

Step 3. Write All the Equilibrium-Constant Expressions

$$K_{sp} = [Fe^{3+}][OH^-]^3 = 2 \times 10^{-39}$$

$$K_w = [H_3O^+][OH^-] = 1.00 \times 10^{-14}$$

Step 4 and 5. Write Mass-Balance and Charge-Balance Equations As in Example 11-5, the mass-balance equation and the charge-balance equations are identical, that is,

$$[OH^-] = 3[Fe^{3+}] + [H_3O^+]$$

Step 6. Count the Number of Independent Equations and Unknown We see that we have enough equations to calculate the three unknowns.

Step 7a. Make Approximations As in Example 11-5, assume that $[H_3O^+]$ is very small so that $[H_3O^+] \ll 3[Fe^{3+}]$ and

$$3[Fe^{3+}] \approx [OH^-]$$

Step 8. Solve the Equations Substituting $[OH^-] = 3[Fe^{3+}]$ into the solubility-product expression gives

$$[Fe^{3+}](3[Fe^{3+}])^3 = 2 \times 10^{-39}$$

$$[Fe^{3+}] = \left(\frac{2 \times 10^{-39}}{27}\right)^{1/4} = 9 \times 10^{-11}$$

$$\text{solubility} = [Fe^{3+}] = 9 \times 10^{-11} \text{ M}$$

Step 9. Check the Assumption From the assumption made in Step 7, we can calculate a provisional value of $[OH^-]$:

$$[OH^-] \approx 3[Fe^{3+}] = 3 \times 9 \times 10^{-11} = 3 \times 10^{-10} \text{ M}$$

Using this value of $[OH^-]$ to compute a *provisional* value for $[H_3O^+]$, we have

$$[H_3O^+] = \frac{1.00 \times 10^{-14}}{3 \times 10^{-10}} = 3 \times 10^{-5} \text{ M}$$

But 3×10^{-5} is not much smaller than three times our provisional value of $[Fe^{3+}]$. This discrepancy means that our assumption was invalid and the provisional values for $[Fe^{3+}]$, $[OH^-]$, and $[H_3O^+]$ are all significantly in error. Therefore, go back to Step 7a and assume that

$$3[Fe^{3+}] \ll [H_3O^+]$$

Now, the mass-balance expression becomes

$$[H_3O^+] = [OH^-]$$

Substituting this equality into the expression for K_w gives

$$[H_3O^+] = [OH^-] = 1.00 \times 10^{-7} \text{ M}$$

Substituting this number into the solubility-product expression developed in Step 3 gives

$$[Fe^{3+}] = \frac{2 \times 10^{-39}}{(1.00 \times 10^{-7})^3} = 2 \times 10^{-18} \text{ M}$$

Since $[H_3O^+] = [OH^-]$, our assumption is that $3[Fe^{3+}] \ll [OH^-]$ or $3 \times 2 \times 10^{-18} \ll 10^{-7}$. Thus, our assumption is valid, and we may write

$$\text{solubility} = 2 \times 10^{-18} \text{ M}$$

Note the very large error (~8 orders of magnitude!) introduced by the invalid assumption.

11B-2 The Effect of pH on Solubility

> All precipitates containing an anion that is the conjugate base of a weak acid are more soluble at low than at high pH.

The solubility of precipitates containing an anion with basic properties, a cation with acidic properties, or both will depend on pH.

Solubility Calculations When the pH Is Constant

Analytical precipitations are usually performed in buffered solutions in which the pH is fixed at some predetermined and known value. The calculation of solubility under this circumstance is illustrated by the following example.

EXAMPLE 11-7

Calculate the molar solubility of calcium oxalate in a solution that has been buffered so that its pH is constant and equal to 4.00.

Solution

Step 1. Write Pertinent Equilibria

$$CaC_2O_4(s) \rightleftharpoons Ca^{2+} + C_2O_4^{2-} \tag{11-9}$$

Oxalate ions react with water to form $HC_2O_4^-$ and $H_2C_2O_4$. Thus, there are three other equilibria present in this solution:

$$H_2C_2O_4 + H_2O \rightleftharpoons H_3O^+ + HC_2O_4^- \tag{11-10}$$

$$HC_2O_4^- + H_2O \rightleftharpoons H_3O^+ + C_2O_4^{2-} \tag{11-11}$$

$$2H_2O \rightleftharpoons H_3O^+ + OH^-$$

Step 2. Define the Unknown Calcium oxalate is a strong electrolyte so that its molar analytical concentration is equal to the equilibrium calcium ion concentration, that is,

$$\text{solubility} = [Ca^{2+}] \tag{11-12}$$

Step 3. Write All the Equilibrium-Constant Expressions

$$[Ca^{2+}][C_2O_4^{2-}] = K_{sp} = 1.7 \times 10^{-9} \tag{11-13}$$

$$\frac{[H_3O^+][HC_2O_4^-]}{[H_2C_2O_4]} = K_1 = 5.60 \times 10^{-2} \tag{11-14}$$

$$\frac{[H_3O^+][C_2O_4^{2-}]}{[HC_2O_4^-]} = K_2 = 5.42 \times 10^{-5} \tag{11-15}$$

$$[H_3O^+][OH^-] = K_w = 1.0 \times 10^{-14}$$

Step 4. Mass-Balance Expressions Because CaC_2O_4 is the only source of Ca^{2+} and the three oxalate species.

$$[Ca^{2+}] = [C_2O_4^{2-}] + [HC_2O_4^-] + [H_2C_2O_4] = \text{solubility} \tag{11-16}$$

The molecular structure of oxalic acid. Oxalic acid occurs naturally in many plants as the potassium or sodium salt, and molds produce oxalic acid as the calcium salt. The sodium salt is used as a primary standard in redox titrimetry (see Chapter 20). The acid is widely used in the dye industry as a cleaning agent in a variety of applications, including the cleaning and restoration of wood surfaces; in the ceramics industry; in metallurgy; in the paper industry; and in photography. It is poisonous if ingested and may cause severe gastroenteritis or kidney damage. It can be prepared by passing carbon monoxide into concentrated sodium hydroxide.

Since the problem states that the pH is 4.00, we can also write that

$$[H_3O^+] = 1.00 \times 10^{-4} \text{ and } [OH^-] = K_w/[H_3O^+] = 1.00 \times 10^{-10}$$

Step 5. Write Charge-Balance Expression A **buffer** is required to maintain the pH at 4.00. The buffer most likely consists of some weak acid HA and its conjugate base, A^-. The nature of the three species and their concentrations have not been specified, however, so we do not have enough information to write a charge-balance equation.

A **buffer** keeps the pH of a solution nearly constant (see Chapter 9).

Step 6. Count the Number of Independent Equations and Unknowns We have four unknowns ($[Ca^{2+}]$, $[C_2O_4^{2-}]$, $[HC_2O_4^-]$, and $[H_2C_2O_4]$) as well as four independent algebraic relationships (Equations 11-13, 11-14, 11-15, and 11-16). Therefore, an exact solution can be obtained, and the problem becomes one of algebra.

Step 7a. Make Approximations It is relatively easy to solve the system of equations exactly in this case, so we will not bother with approximations.

Step 8. Solve the Equations A convenient way to solve the problem is to substitute Equations 11-14 and 11-15 into 11-16 in such a way as to develop a relationship between $[Ca^{2+}]$, $[C_2O_4^{2-}]$, and $[H_3O^+]$. Thus, we rearrange Equation 11-15 to give

$$[HC_2O_4^-] = \frac{[H_3O^+][C_2O_4^{2-}]}{K_2}$$

Substituting numerical values for $[H_3O^+]$ and K_2 gives

$$[HC_2O_4^-] = \frac{1.00 \times 10^{-4}[C_2O_4^{2-}]}{5.42 \times 10^{-5}} = 1.85[C_2O_4^{2-}]$$

Substituting this relationship into Equation 11-14 and rearranging gives

$$[H_2C_2O_4] = \frac{[H_3O^+][C_2O_4^{2-}] \times 1.85}{K_1}$$

Substituting numerical values for $[H_3O^+]$ and K_1 yields

$$[H_2C_2O_4] = \frac{1.85 \times 10^{-4}[C_2O_4^{2-}]}{5.60 \times 10^{-2}} = 3.30 \times 10^{-3}[C_2O_4^{2-}]$$

Substituting these expressions for $[HC_2O_4^-]$ and $[H_2C_2O_4]$ into Equation 11-16 gives

$$[Ca^{2+}] = [C_2O_4^{2-}] + 1.85[C_2O_4^{2-}] + 3.30 \times 10^{-3}[C_2O_4^{2-}]$$
$$= 2.85[C_2O_4^{2-}]$$

or $\quad [C_2O_4^{2-}] = [Ca^{2+}]/2.85$

(continued)

Substituting into Equation 11-13 gives

$$\frac{[Ca^{2+}][Ca^{2+}]}{2.85} = 1.7 \times 10^{-9}$$

$$[Ca^{2+}] = \text{solubility} = \sqrt{2.85 \times 1.7 \times 10^{-9}} = 7.0 \times 10^{-5} \text{ M}$$

Solubility Calculations When the pH Is Variable

Computing the solubility of a precipitate such as calcium oxalate in a solution in which the pH is not fixed and known is considerably more complicated than in the example that we just explored. Thus, to determine the solubility of CaC_2O_4 in pure water, we must take into account the change in OH^- and H_3O^+ that accompanies the solution process. In this example, there are four equilibria to consider.

$$CaC_2O_4(s) \rightleftharpoons Ca^{2+} + C_2O_4^{2-}$$

$$C_2O_4^{2-} + H_2O \rightleftharpoons HC_2O_2^- + OH^-$$

$$HC_2O_4^- + H_2O \rightleftharpoons H_2C_2O_4 + OH^-$$

$$2H_2O \rightleftharpoons H_3O^+ + OH^-$$

In contrast to Example 11-7, the hydroxide ion concentration now becomes an unknown, and an additional algebraic equation must therefore be developed to calculate the solubility of calcium oxalate.

It is not difficult to write the six algebraic equations needed to calculate the solubility of calcium oxalate (see Feature 11-1). Solving the six equations manually, however, is somewhat tedious and time consuming.

FEATURE 11-1

Algebraic Expressions Needed to Calculate the Solubility of CaC₂O₄ in Water

As in Example 11-7, the solubility is equal to the cation concentration, $[Ca^{2+}]$.

$$\text{solubility} = [Ca^{2+}] = [C_2O_4^{2-}] + [HC_2O_4^-] + [H_2C_2O_4]$$

In this case, however, we must take into account one additional equilibrium—the dissociation of water. The equilibrium-constant expressions for the four equilibria are then

$$K_{sp} = [Ca^{2+}][C_2O_4^{2-}] = 1.7 \times 10^{-9} \qquad (11\text{-}17)$$

$$K_2 = \frac{[H_3O^+][C_2O_4^{2-}]}{[HC_2O_4^-]} = 5.42 \times 10^{-5} \qquad (11\text{-}18)$$

$$K_1 = \frac{[H_3O^+][HC_2O_4^{2-}]}{[H_2C_2O_4]} = 5.60 \times 10^{-2} \qquad (11\text{-}19)$$

$$K_w = [H_3O^+][OH^-] = 1.00 \times 10^{-14} \qquad (11\text{-}20)$$

The mass-balance equation is

$$[Ca^{2+}] = [C_2O_4^{2-}] + [HC_2O_4^-] + [H_2C_2O_4] \qquad (11\text{-}21)$$

The charge-balance equation is

$$2[Ca^{2+}] + [H_3O^+] = 2[C_2O_4^{2-}] + [HC_2O_4^-] + [OH^-] \tag{11-22}$$

We now have six unknowns ($[Ca^{2+}]$, $[C_2O_4^{2-}]$, $[HC_2O_4^-]$, $[H_2C_2O_4]$, $[H_3O^+]$, and $[OH^-]$) and six equations (11-17 through 11-22). Thus, in principle, the problem can be solved exactly.

11B-3 The Effect of Undissociated Solutes on Precipitation Calculations

So far, we have considered only solutes that dissociate completely when dissolved in aqueous media. However, there are some inorganic substances, such as calcium sulfate and the silver halides, that act as weak electrolytes and only partially dissociate in water. For example, a saturated solution of silver chloride contains significant amounts of undissociated silver chloride molecules as well as silver and chloride ions. In this case, two equilibria are required to describe the system:

$$AgCl(s) \rightleftharpoons AgCl(aq) \tag{11-23}$$

$$AgCl(aq) \rightleftharpoons Ag^+ + Cl^- \tag{11-24}$$

The equilibrium constant for the first reaction takes the form

$$\frac{[AgCl(aq)]}{[AgCl(s)]} = K$$

where the numerator is the concentration of the undissociated species in *the solution* and the denominator is the concentration of silver chloride in *the solid phase*. The latter term is a constant, however (page 208), and so the equation can be written

$$[AgCl(aq)] = K[AgCl(s)] = K_s = 3.6 \times 10^{-7} \tag{11-25}$$

where K is the constant for the equilibrium shown in Equation 11-23. It is evident from this equation that at a given temperature, the concentration of the undissociated silver chloride is constant and *independent* of the chloride and silver ion concentrations.

The equilibrium constant K_d for the dissociation reaction (Equation 11-24) is

$$\frac{[Ag^+][Cl^-]}{[AgCl(aq)]} = K_d = 5.0 \times 10^{-4} \tag{11-26}$$

The product of these two constants is equal to the solubility product:

$$[Ag^+][Cl^-] = K_dK_s = K_{sp}$$

As shown by Example 11-8, both Reaction 11-23 and Reaction 11-24 contribute to the solubility of silver chloride in water.

EXAMPLE 11-8

Calculate the solubility of AgCl in distilled water.

Solution

$$\text{Solubility} = S = [\text{AgCl}(aq)] + [\text{Ag}^+]$$

$$[\text{Ag}^+] = [\text{Cl}^-]$$

$$[\text{Ag}^+][\text{Cl}^-] = K_{sp} = 1.82 \times 10^{-10}$$

$$[\text{Ag}^+] = \sqrt{1.82 \times 10^{-10}} = 1.35 \times 10^{-5}$$

Substituting this value and K_s from Equation 11-25 gives

$$S = 1.35 \times 10^{-5} + 3.6 \times 10^{-7} = 1.38 \times 10^{-5} \text{ M}$$

Note that neglecting [AgCl(aq)] leads to a 2% error in this example.

11B-4 The Solubility of Precipitates in the Presence of Complexing Agents

> The solubility of a precipitate always increases in the presence of a complexing agent that reacts with the cation of the precipitate.

The solubility of a precipitate may increase dramatically in the presence of reagents that form complexes with the anion or the cation of the precipitate. For example, fluoride ions prevent the quantitative precipitation of aluminum hydroxide even though the solubility product of this precipitate is remarkably small (2×10^{-32}). The cause of the increase in solubility is shown by the equations

$$\text{Al(OH)}_3(s) \rightleftharpoons \text{Al}^{3+} + 3\text{OH}^-$$
$$+$$
$$6\text{F}^-$$
$$\updownarrow$$
$$\text{AlF}_6^{3-}$$

The fluoride complex is sufficiently stable to permit fluoride ions to compete successfully with hydroxide ions for aluminum ions.

Many precipitates react with excesses of the precipitating reagent to form soluble complexes. In a gravimetric analysis, this tendency may have the undesirable effect of reducing the recovery of analytes if too large an excess of reagent is used. For example, silver is often determined by precipitation of silver ion by adding an excess of a potassium chloride solution. The effect of excess reagent is complex, as revealed by the following set of equations that describe the system:

$$\text{AgCl}(s) \rightleftharpoons \text{AgCl}(aq) \tag{11-27}$$

$$\text{AgCl}(aq) \rightleftharpoons \text{Ag}^+ + \text{Cl}^- \tag{11-28}$$

$$\text{AgCl}(s) + \text{Cl}^- \rightleftharpoons \text{AgCl}_2^- \tag{11-29}$$

$$\text{AgCl}_2^- + \text{Cl}^- \rightleftharpoons \text{AgCl}_3^{2-} \tag{11-30}$$

Note that Equilibrium 11-28 and thus Equilibrium 11-27 shift to the left with added chloride ion, but Equilibria 11-29 and 11-30 shift to the right under the same circumstance. The consequence of these opposing effects is that a plot of silver chloride solubility as a function of concentration of added chloride exhibits a minimum. Example 11-9 illustrates how this behavior can be described in quantitative terms.

EXAMPLE 11-9

Derive an equation that describes the effect of the analytical concentration of KCl on the solubility of AgCl in an aqueous solution. Calculate the concentration of KCl at which the solubility is a minimum.

Solution

Step 1. Pertinent Equilibria Equations 11-27 through 11-30 describe the pertinent equilibria.

Step 2. Definition of Unknown The molar solubility S of AgCl is equal to the sum of the concentrations of the silver-containing species:

$$\text{solubility} = S = [\text{AgCl}(aq)] + [\text{Ag}^+] + [\text{AgCl}_2^-] + [\text{AgCl}_3^{2-}] \qquad (11\text{-}31)$$

Step 3. Equilibrium-Constant Expressions Equilibrium constants available in the literature include

$$[\text{Ag}^+][\text{Cl}^-] = K_{sp} = 1.82 \times 10^{-10} \qquad (11\text{-}32)$$

$$\frac{[\text{Ag}^+][\text{Cl}^-]}{[\text{AgCl}(aq)]} = K_d = 3.9 \times 10^{-4} \qquad (11\text{-}33)$$

$$\frac{[\text{AgCl}_2^-]}{[\text{AgCl}(aq)][\text{Cl}^-]} = K_2 = 2.0 \times 10^{-5} \qquad (11\text{-}34)$$

$$\frac{[\text{AgCl}_3^{2-}]}{[\text{AgCl}_2^-][\text{Cl}^-]} = K_3 = 1 \qquad (11\text{-}35)$$

Step 4. Mass-Balance Equation

$$[\text{Cl}^-] = c_{\text{KCl}} + [\text{Ag}^+] - [\text{AgCl}_2^-] - 2[\text{AgCl}_3^{2-}] \qquad (11\text{-}36)$$

The second term on the right-hand side of this equation gives the chloride ion concentration produced by the dissolution of the precipitate, and the next two terms correspond to the *decrease* in chloride ion concentration resulting from the formation of the two chloro complexes from AgCl.

Step 5. Charge-Balance Equation As in some of the earlier examples, the charge-balance equation is identical to the mass-balance equation. We begin with the basic charge balance equation,

$$[\text{K}^+] + [\text{Ag}^+] = [\text{Cl}^-] + [\text{AgCl}_2^-] + 2[\text{AgCl}_3^{2-}]$$

If we substitute $c_{\text{KCl}} = [\text{K}^+]$ into this equation, we find that

$$c_{\text{KCl}} + [\text{Ag}^+] = [\text{Cl}^-] + [\text{AgCl}_2^-] + 2[\text{AgCl}_3^{2-}]$$

and

$$[\text{Cl}^-] = c_{\text{KCl}} + [\text{Ag}^+] - [\text{AgCl}_2^-] - 2[\text{AgCl}_3^{2-}]$$

(continued)

This last expression is identical to the mass balance expression in Step 4.

Step 6. Number of Equations and Unknowns We have five equations (11-32 through 11-36) and five unknowns ($[Ag^+]$, $[AgCl(aq)]$, $[AgCl_2^-]$, $[AgCl_3^{2-}]$, and $[Cl^-]$).

Step 7a. Assumptions We assume that, over a considerable range of chloride ion concentrations, the solubility of AgCl is so small that Equation 11-36 can be greatly simplified by the assumption that

$$[Ag^+] - [AgCl_2^-] - 2[AgCl_3^{2-}] \ll c_{KCl}$$

It is not certain that this is a valid assumption, but it is worth trying because it greatly simplifies the problem. With this assumption, then, Equation 11-36 reduces to

$$[Cl^-] = c_{KCl} \tag{11-37}$$

Step 8. Solution of Equations For convenience, we multiply Equations 11-34 and 11-35 to give

$$\frac{[AgCl_3^{2-}]}{[Cl^-]^2} = K_2 K_3 = 2.0 \times 10^{-5} \times 1 = 2.0 \times 10^{-5} \tag{11-38}$$

To calculate $[AgCl(aq)]$, we divide Equation 11-32 by Equation 11-33 and rearrange:

$$[AgCl(aq)] = \frac{K_{sp}}{K_d} = \frac{1.82 \times 10^{-10}}{3.9 \times 10^{-4}} = 4.7 \times 10^{-7} \tag{11-39}$$

Note that the concentration of this species is *constant and independent of the chloride concentration.*

Substitution of Equations 11-39, 11-32, 11-33, and 11-38 into Equation 11-31 permits us to express the solubility in terms of the chloride ion concentration and the several constants.

$$S = \frac{K_{sp}}{K_d} + \frac{K_{sp}}{[Cl^-]} + K_2[Cl^-] + K_2 K_3 [Cl^-]^2 \tag{11-40}$$

By substituting Equation 11-37 into Equation 11-40, we find the desired relationship between the solubility and the analytical concentration of KCl:

$$S = \frac{K_{sp}}{K_d} + \frac{K_{sp}}{c_{KCl}} + K_2 c_{KCl} + K_2 K_3 c_{KCl}^2 \tag{11-41}$$

To find the minimum in S, we set the derivative of S with respect to c_{KCl} equal to zero:

$$\frac{dS}{dc_{KCl}} = 0 = \frac{K_{sp}}{c_{KCl}^2} + K_2 + 2K_2 K_3 c_{KCl}$$

$$2K_2 K_3 c_{KCl}^3 + c_{KCl}^2 K_2 - K_{sp} = 0$$

Substituting numerical values gives

$$(4.0 \times 10^{-5})c_{KCl}^3 + (2.0 \times 10^{-5})c_{KCl}^2 - 1.82 \times 10^{-10} = 0$$

Following the procedure shown in Feature 9-4, we can solve this equation by successive approximations to obtain

$$c_{KCl} = 0.0030 = [Cl^-]$$

To check the assumption made earlier, we calculate the concentrations of the various species. Substitutions into Equations 11-32, 11-34, and 11-36 yield

$$[Ag^+] = (1.82 \times 10^{-10})/0.0030 = 6.1 \times 10^{-8} \text{ M}$$

$$[AgCl_2^-] = 2.0 \times 10^{-5} \times 0.0030 = 6.0 \times 10^{-8} \text{ M}$$

$$[AgCl_3^{2-}] = 2.0 \times 10^{-5} \times (0.0030)^2 = 1.8 \times 10^{-10} \text{ M}$$

Thus, our assumption that c_{KCl} is much larger than the concentrations of the silver-containing ions is reasonable. The minimum solubility is obtained by substitution of these concentrations and [AgCl(aq)] into Equation 11-31:

$$S = 4.7 \times 10^{-7} + 6.1 \times 10^{-8} + 6.0 \times 10^{-8} + 1.8 \times 10^{-10}$$

$$= 5.9 \times 10^{-7} \text{ M}$$

The solid curve in Figure 11-2 illustrates the effect of chloride ion concentration on the solubility of silver chloride; data for the curve were obtained by substituting various chloride concentrations into Equation 11-41. Note that at high concentrations of the common ion, the solubility becomes greater than that in pure water. The broken lines represent the equilibrium concentrations of the various silver-containing species as a function of c_{KCl}. Note that at the solubility minimum, undissociated silver chloride, AgCl(aq), is the major silver species in the solution, representing about 80% of the total dissolved silver. Its concentration is invariant, as has been demonstrated.

Unfortunately, there are few reliable equilibrium data regarding undissociated species such as AgCl(aq) and complex species such as $AgCl_2^-$. Because of this lack of data, solubility calculations are often, of necessity, based on solubility-product

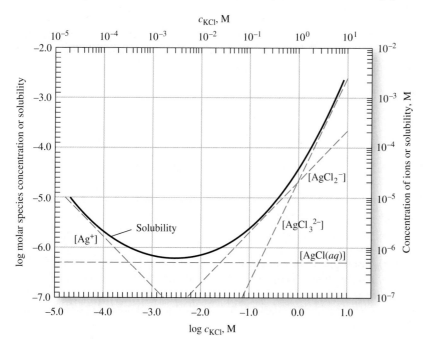

Figure 11-2 The effect of chloride ion concentration on the solubility of AgCl. The solid curve shows the total concentration of dissolved AgCl. The broken lines show the concentrations of the various silver-containing species.

equilibria alone. Example 11-9 shows that, under some circumstances, such neglect of other equilibria can lead to serious error. In addition, in solutions containing high concentrations of diverse ions and thus high ionic strength, it may be necessary to apply activity corrections as discussed in Chapter 10.

> **Spreadsheet Summary**   In the first exercise in Chapter 6 of *Applications of Microsoft® Excel in Analytical Chemistry*, 2nd ed., we explore the use of Excel's Solver to find the concentrations of Mg^{2+}, OH^-, and H_3O^+ in the $Mg(OH)_2$ system of Example 11-5. Solver finds the concentrations from the mass-balance expression, the solubility product of $Mg(OH)_2$, and the ion product of water. Then Excel's built-in tool Goal Seek is used to solve a cubic equation for the same system. The final exercise in Chapter 6 uses Solver to find the solubility of calcium oxalate at a known pH (see Example 11-7) and when the pH is unknown (see Feature 11-1).

11C SEPARATION OF IONS BY CONTROL OF THE CONCENTRATION OF THE PRECIPITATING AGENT

Several precipitating agents permit separation of ions based on solubility differences. Such separations require close control of the active reagent concentration at a suitable and predetermined level. Most often, such control is achieved by controlling the pH of the solution with suitable buffers. This technique is applicable to anionic reagents in which the anion is the conjugate base of a weak acid. Examples include sulfide ion (the conjugate base of hydrogen sulfide), hydroxide ion (the conjugate base of water), and the anions of several organic weak acids.

11C-1 Calculation of the Feasibility of Separations

The following example illustrates how solubility-product calculations are used to determine the feasibility of separations based on solubility differences.

EXAMPLE 11-10

Can Fe^{3+} and Mg^{2+} be separated quantitatively as hydroxides from a solution that is 0.10 M in each cation? If the separation is possible, what range of OH^- concentrations is permissible?

Solution

Solubility-product constants for the two precipitates are

$$K_{sp} = [Fe^{3+}][OH^-]^3 = 2 \times 10^{-39}$$

$$K_{sp} = [Mg^{2+}][OH^-]^2 = 7.1 \times 10^{-12}$$

The K_{sp} for $Fe(OH)_3$ is so much smaller than that for $Mg(OH)_2$ that it appears likely that the former will precipitate at a lower OH^- concentration. We can answer the questions posed in this problem by (1) calculating the OH^- concentration required to achieve quantitative precipitation of Fe^{3+} and (2) computing the OH^- concentration at which $Mg(OH)_2$ just begins to precipitate. If (1) is smaller than (2), a separation is feasible in principle, and the range of permissible OH^- concentrations is defined by the two values.

To determine (1), we must first specify what constitutes a quantitative removal of Fe^{3+} from the solution. The decision here is arbitrary and depends on the purpose of the separation. In this example and the next, we consider a precipitation to be quantitative when all but 1 part in 1000 of the ion has been removed from the solution, that is, when $[Fe^{3+}] < 1 \times 10^{-4}$ M.

We can calculate the OH^- concentration in equilibrium with 1×10^{-1} M Fe^{3+} by substituting directly into the solubility-product expression:

$$K_{sp} = (1.0 \times 10^{-4})[OH^-]^3 = 2 \times 10^{-39}$$

$$[OH^-] = [(2 \times 10^{-39})/(1.0 \times 10^{-4})]^{1/3} = 3 \times 10^{-12} \text{ M}$$

Thus, if we maintain the OH^- concentration at about 3×10^{-12} M, the Fe^{3+} concentration will be lowered to 1×10^{-4} M. Note that quantitative precipitation of $Fe(OH)_3$ is achieved in a distinctly acidic medium (pH $\approx$ 2.5).

To determine what maximum OH^- concentration can exist in the solution without causing formation of $Mg(OH)_2$, we note that precipitation cannot occur until the product $[Mg^{2+}][OH^-]^2$ exceeds the solubility product, 7.1×10^{-12}. Substitution of 0.1 (the molar Mg^{2+} concentration of the solution) into the solubility-product expression permits the calculation of the maximum OH^- concentration that can be tolerated:

$$K_{sp} = 0.10[OH^-]^2 = 7.1 \times 10^{-12}$$

$$[OH^-] = 8.4 \times 10^{-6} \text{ M}$$

When the OH^- concentration exceeds this level, the solution will be supersaturated with respect to $Mg(OH)_2$, and precipitation should begin.

From these calculations, we conclude that quantitative separation of $Fe(OH)_3$ can be achieved if the OH^- concentration is greater than 3×10^{-12} M and that $Mg(OH)_2$ will not precipitate until an OH^- concentration of 8.4×10^{-6} M is reached. Therefore, it is possible, in principle, to separate Fe^{3+} from Mg^{2+} by maintaining the OH^- concentration between these levels. In practice, the concentration of OH^- is kept as low as practical—often about 10^{-10} M. Note that these calculations neglect activity effects.

11C-2 Sulfide Separations

Sulfide ion forms precipitates with heavy metal cations that have solubility products that vary from 10^{-10} to 10^{-90} or smaller. In addition, the concentration of S^{2-} can be varied over a range of about 0.1 M to 10^{-22} M by controlling the pH of a saturated solution of hydrogen sulfide. These two properties make possible a number of useful cation separations. To illustrate the use of hydrogen sulfide to separate cations based on pH control, consider the precipitation of the divalent cation M^{2+} from a solution that is kept saturated with hydrogen sulfide by bubbling the gas continuously through the solution. The important equilibria in this solution are:

$$MS(s) \rightleftharpoons M^{2+} + S^{2-} \qquad K_{sp} = [M^{2+}][S^{2-}]$$

$$H_2S + H_2O \rightleftharpoons H_3O^+ + HS^- \qquad K_1 = \frac{[H_3O^+][HS^-]}{[H_2S]} = 9.6 \times 10^{-8}$$

$$HS^- + H_2O \rightleftharpoons H_3O^+ + S^{2-} \qquad K_2 = \frac{[H_3O^+][S^{2-}]}{[HS^-]} = 1.3 \times 10^{-14}$$

We may also write

$$\text{solubility} = [M^{2+}]$$

The concentration of hydrogen sulfide in a saturated solution of the gas is approximately 0.1 M. Thus, we may write as a mass-balance expression

$$[S^{2-}] + [HS^-] + [H_2S] = 0.1$$

Because we know the hydronium ion concentration, we have four unknowns, the concentration of the metal ion and the three sulfide species.

We can simplify the calculation greatly by assuming that $([S^{2-}] + [HS^-]) \ll [H_2S]$, so that

$$[H_2S] \approx 0.10 \text{ M}$$

The two dissociation-constant expressions for hydrogen sulfide can be multiplied to give an expression for the overall dissociation of hydrogen sulfide to sulfide ion:

$$H_2S + 2H_2O \rightleftharpoons 2H_2O^+ + S^{2-} \qquad K_1K_2 = \frac{[H_3O^+]^2[S^{2-}]}{[H_2S]} = 1.2 \times 10^{-21}$$

The constant for this overall reaction is simply the product of K_1 and K_2.

Substituting the numerical value for $[H_2S]$ into this equation gives

$$\frac{[H_3O^+]^2[S^{2-}]}{0.10} = 1.2 \times 10^{-21}$$

On rearranging this equation, we obtain

$$[S^{2-}] = \frac{1.2 \times 10^{-22}}{[H_3O^+]^2} \tag{11-42}$$

Therefore, we see that the sulfide ion concentration of a saturated hydrogen sulfide solution varies inversely with the square of the hydrogen ion concentration. **Figure 11-3**, which was obtained with this equation, reveals that the sulfide ion

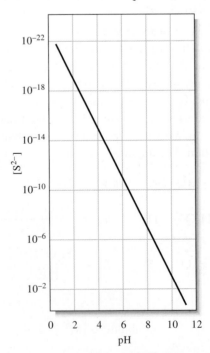

Figure 11-3   Sulfide ion concentration as a function of pH in a saturated H_2S solution.

concentration of an aqueous solution can be changed by more than 20 orders of magnitude by varying the pH from 1 to 11.

Substituting Equation 11-42 into the solubility-product expression gives

$$K_{sp} = \frac{[M^{2+}] \times 1.2 \times 10^{-22}}{[H_3O^+]^2}$$

$$[M^{2+}] = \text{solubility} = \frac{[H_3O^+]^2 K_{sp}}{1.2 \times 10^{-22}}$$

Thus, the solubility of a divalent metal sulfide increases with the square of the hydronium ion concentration.

Hydrogen sulfide is a colorless, flammable gas with important chemical and toxicological properties. It is the product of a number of natural processes, including the decay of sulfur-containing material. Its noxious odor of rotten eggs permits its detection at extremely low concentration (0.02 ppm). Because the olfactory sense is dulled by its action, however, higher concentrations may be tolerated, and the lethal concentration of 100 ppm may be exceeded. Aqueous solutions of the gas were used traditionally as a source of sulfide for the precipitation of metals, but because of the toxicity of H_2S, this role has been taken over by other sulfur-containing compounds such as thioacetamide.

EXAMPLE 11-11

Cadmium sulfide is less soluble than thallium(I) sulfide. Find the conditions under which Cd^{2+} and Tl^+ can, in theory, be separated quantitatively with H_2S from a solution that is 0.1 M in each cation.

Solution

The constants for the two solubility equilibria are:

$$CdS(s) \rightleftharpoons Cd^{2+} + S^{2-} \quad K_{sp} = [Cd^{2+}][S^{2-}] = 1 \times 10^{-27}$$

$$Tl_2S(s) \rightleftharpoons 2Tl^+ + S^{2-} \quad K_{sp} = [Tl^{2+}]^2[S^{2-}] = 6 \times 10^{-22}$$

Since CdS precipitates at a lower $[S^{2-}]$ than does Tl_2S, we first compute the sulfide ion concentration necessary for quantitative removal of Cd^{2+} from solution. As in Example 11-10, we arbitrarily specify that separation is quantitative when all but 1 part in 1000 of the Cd^{2+} has been removed, that is, the concentration of the cation has been lowered to 1.00×10^{-4} M. Substituting this value into the solubility-product expression gives

$$K_{sp} = 10^{-4}[S^{2-}] = 1 \times 10^{-27}$$

$$[S^{2-}] = 1 \times 10^{-23} \text{ M}$$

If we maintain the sulfide concentration at this level or greater, we may assume that quantitative removal of the cadmium will take place. Next, we compute the $[S^{2-}]$ needed to initiate precipitation of Tl_2S from a 0.1 M solution. Precipitation will begin when the solubility product is just exceeded. Since the solution is 0.1 M in Tl^+,

$$(0.1)^2[S^{2-}] = 6 \times 10^{-22}$$

$$[S^{2-}] = 6 \times 10^{-20} \text{ M}$$

These two calculations show that quantitative precipitation of Cd^{2+} takes place if $[S^{2-}]$ is made greater than 1×10^{-23} M. No precipitation of Tl^+ occurs, however, until $[S^{2-}]$ becomes greater than 6×10^{-20} M.

Substituting these two values for $[S^{2-}]$ into Equation 11-42 permits us to calculate the $[H_3O^+]$ range required for the separation.

$$[H_3O^+]^2 = \frac{1.2 \times 10^{-22}}{1 \times 10^{-23}} = 12$$

$$[H_3O^+] = 3.5 \text{ M}$$

(continued)

and

$$[H_3O^+]^2 = \frac{1.2 \times 10^{-22}}{6 \times 10^{-20}} = 2.0 \times 10^{-3}$$

$$[H_3O^+] = 0.045 \text{ M}$$

By maintaining $[H_3O^+]$ between approximately 0.045 and 3.5 M, we should be able to separate Cd^{2+} quantitatively from Tl^+. Because of the high ionic strength of such acidic solutions, it may be necessary to correct for activity effects.

FEATURE 11-2

Immunoassay: Equilibria in the Specific Determination of Drugs

The determination of drugs in the human body is a matter of great importance in drug therapy and in the detection and prevention of drug abuse. The diversity of drugs and their typical low concentrations in body fluids make them difficult to identify and measure. Fortunately, it is possible to harness one of nature's own mechanisms—the immune response—to determine quantitatively several therapeutic and illicit drugs.

When a foreign substance, or antigen (Ag), shown schematically in **Figure 11F-1a**, is introduced into the body of a mammal, the immune system synthesizes protein-based molecules (**Figure 11F-1b**) called antibodies (Ab). Antibodies specifically bind to the antigen molecules via electrostatic interactions, hydrogen bonding, and other noncovalent short-range forces. These massive molecules (molar mass $\approx$ 150,000) form a complex with antigens, as shown in the following reaction and in **Figure 11F-1c**.

$$Ag + Ab \rightleftharpoons AgAb \quad K = \frac{[AgAb]}{[Ag][Ab]}$$

The immune system does not recognize relatively small molecules, so we must use a trick to prepare antibodies with binding sites that are specific for a particular drug. As shown in **Figure 11F-1d**, we attach the drug covalently to an antigenic carrier molecule such as bovine serum albumin (BSA), a protein that is obtained from the blood of cattle:

$$D + Ag \rightarrow D - Ag$$

When the resulting drug-antigen conjugate (D-Ag) is injected into the bloodstream of a rabbit, the immune system of the rabbit synthesizes antibodies with binding sites that are specific for the drug, as illustrated in **Figure 11F-1e**. Approximately three weeks after injection of the antigen, blood is drawn from the rabbit, the serum is separated from the blood, and the antibodies of interest are separated from the serum and other antibodies, usually by chromatographic methods (see Chapters 32 and 33). It is important to note that once the drug-specific antibody has been synthesized by the immune system of the rabbit, the drug can bind directly to the antibody without the aid of the carrier molecule, as

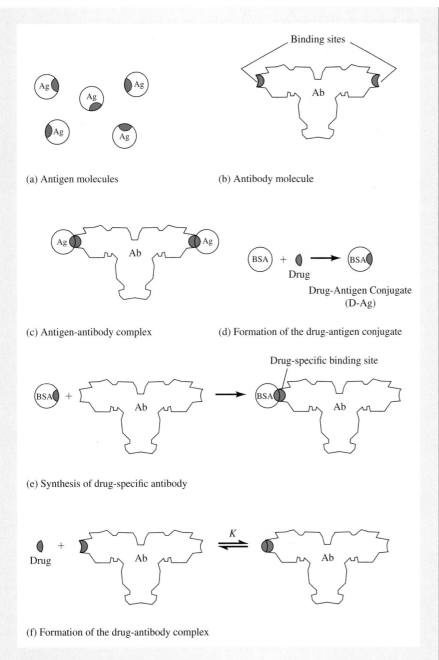

(a) Antigen molecules

(b) Antibody molecule

(c) Antigen-antibody complex

(d) Formation of the drug-antigen conjugate

(e) Synthesis of drug-specific antibody

(f) Formation of the drug-antibody complex

Figure 11F-1   Antibody-antigen interaction.

shown in **Figure 11F-1f**. This direct drug-antibody forms the basis for the specific determination of the drug.

The measurement step of the immunoassay is accomplished by mixing the sample containing the drug with a measured amount of drug-specific antibody. At this point, the quantity of Ab-D must be determined by adding a standard sample of the drug that has been chemically altered to contain a detectable label. Typical labels are enzymes, fluorescent or chemiluminescent molecules, or radioactive atoms. For our example, we assume that a fluorescent molecule has been attached to the drug to produce the labeled drug D*.[1]

(continued)

---

[1] For a discussion of molecular fluorescence, see Chapter 27.

If the amount of the antibody is somewhat less than the sum of the amounts of D and D*, then D and D* compete for the antibody, as shown in the following equilibria.

$$D^* + Ab \rightleftharpoons Ab - D^* \quad K^* = \frac{[Ab - D^*]}{[D^*][Ab]}$$

$$D + Ab \rightleftharpoons Ab - D \quad K = \frac{[Ab - D]}{[D][Ab]}$$

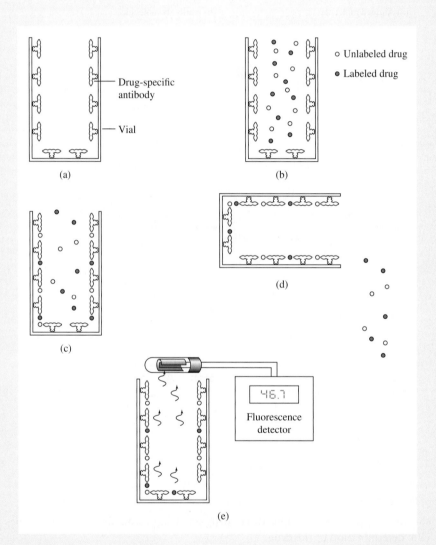

Figure 11F-2   Procedure for determining drugs by immunoassay with fluorescence labeling. (a) Vial is lined with drug-specific antibodies; (b) vial is filled with solution containing both labeled and unlabeled drug; (c) labeled and unlabeled drug binds to antibodies; (d) solution is discarded leaving bound drug behind; (e) fluorescence of the bound, labeled drug is measured. The concentration of drug is determined by using the dose-response curve of Figure 11F-3.

It is important to select a label that does not substantially alter the affinity of the drug for the antibody so that the labeled and unlabeled drugs bind with the antibody equally well. If the binding affinities are equal, then $K = K^*$. Typical values for equilibrium constants of this type, called **binding constants**, range from 10^7 to 10^{12}. The larger the concentration of the unknown, unlabeled drug, the smaller the concentration of Ab-D* and vice versa. This inverse relationship between D and Ab-D* forms the basis for the quantitative determination of the drug. We can find the amount of D if we measure *either* Ab-D* or D*.

To differentiate between bound drug and unbound labeled drug, it is necessary to separate them before measurement. The amount of Ab-D* can then be found by using a fluorescence detector to measure the intensity of the fluorescence resulting from the Ab-D*. A determination of this type using a fluorescent drug and radiation detection is called a **fluorescence immunoassay**. Determinations of this type are very sensitive and selective.

One convenient way to separate D* and Ag-D* is to prepare polystyrene vials that are coated on the inside with antibody molecules, as illustrated in **Figure 11F-2a**. A sample of blood serum, urine, or other body fluid containing an unknown concentration of D along with a volume of solution containing labeled drug D* is added to the vial, as depicted in **Figure 11F-2b**. After equilibrium is achieved in the vial (**Figure 11F-2c**), the solution containing residual D and D* is decanted, and the vial is rinsed. An amount of D* is then left bound to the antibody that is inversely proportional to the concentration of D in the sample (**Figure 11F-2d**). Finally, the fluorescence intensity of the bound D* is determined using a fluorometer, as shown in **Figure 11F-2e**.

This procedure is repeated for several standard solutions of D to produce a nonlinear working curve called a **dose-response curve** similar to the curve of **Figure 11F-3**. The fluorescence intensity for an unknown solution of D is located on the calibration curve, and the concentration is read from the concentration axis.

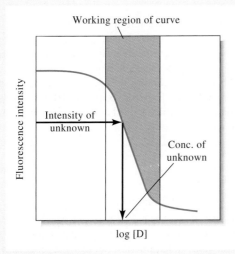

Figure 11F-3   Dose-response curve for determining drugs by fluorescence-based immunoassay.

(continued)

Immunoassay is a powerful tool in the clinical laboratory and is one of the most widely used of all analytical techniques. Reagent kits for many different immunoassays are available commercially, as are automated instruments for carrying out fluorescent immunoassays and immunoassays of other types. In addition to concentrations of drugs, vitamins, proteins, growth hormones, allergens, pregnancy hormones, cancer and other disease indicators, and pesticide residues in natural waters and food are determined by immunoassay. The structure of an antigen-antibody complex is shown in **Figure 11F-4**.

Figure 11F-4 The molecular structure of an antigen-antibody complex. Pictured are two representations of the complex formed between a digestion fragment of intact mouse antibody A6 and genetically engineered human interferon-gamma receptor alpha chain. (a) A space-filling model of the molecular structure of the complex. (b) A ribbon diagram showing the protein chains in the complex. (From the Protein Data Bank, Rutgers University, Structure 1JRH, S. Sogabe, F. Stuart, C. Henke, A. Bridges, G. Williams, A. Birch, F. K. Winkler, and J. A. Robinson, 1997, http://www.rcsb.org)

WEB WORKS   The Centers for Disease Control and Prevention (CDC) maintains a website to provide information related to AIDS and HIV. Locate this site on the web, and use its search facility or another search engine to find pages containing information on FDA-approved rapid HIV screening tests. What is the basis for most of the tests? What physical or chemical properties are used in the tests for detection? What are the chemical principles behind these methods?

QUESTIONS AND PROBLEMS

11-1. Demonstrate how the sulfide ion concentration is related to the hydronium ion concentration of a solution that is kept saturated with hydrogen sulfide.

*11-2. Why are simplifying assumptions restricted to relationships that are sums or differences?

11-3. In the margin note on page 250, we suggest that the term mass-balance equation may be something of a misnomer. Using a specific chemical system, discuss mass balance and show that mass balance and concentration balance are equivalent.

*11-4. Why do molar concentrations of some species appear as multiples in charge-balance equations?

11-5. Write the mass-balance expressions for a solution that is
*(a) 0.2 M in HF.
(b) 0.35 M in NH_3.
*(c) 0.10 M in H_3PO_4.
(d) 0.20 M in Na_2HPO_4.
*(e) 0.0500 M in $HClO_2$ and 0.100 M in $NaClO_2$.
(f) 0.12 M in NaF and saturated with CaF_2.
*(g) 0.100 M in NaOH and saturated with $Zn(OH)_2$, which undergoes the reaction $Zn(OH)_2 + 2OH^- \rightleftharpoons Zn(OH)_4^{2-}$.
(h) saturated with $Ag_2C_2O_4$.
*(i) saturated with $PbCl_2$.

11-6. Write the charge-balance equations for the solutions in Problem 11-5.

11-7. Calculate the molar solubility of SrC_2O_4 in a solution that has a fixed H_3O^+ concentration of
*(a) 1.0×10^{-6} M.
(b) 1.0×10^{-7} M.
*(c) 1.0×10^{-9} M.
(d) 1.0×10^{-11} M.

11-8. Calculate the molar solubility of $BaSO_4$ in a solution in which $[H_3O^+]$ is
*(a) 3.5 M.
(b) 0.5 M.
*(c) 0.080 M.
(d) 0.100 M.

*11-9. Calculate the molar solubility of PbS in a solution in which $[H_3O^+]$ is held constant at (a) 3.0×10^{-1} M and (b) 3.0×10^{-4} M.

11-10. Calculate the concentration of CuS in a solution in which $[H_3O^+]$ is held constant at (a) 2.0×10^{-1} M and (b) 2.0×10^{-4} M.

11-11. Calculate the molar solubility of MnS (pink) in a solution with a constant $[H_3O^+]$ of *(a) 3.00×10^{-5} and (b) 3.00×10^{-7}.

*11-12. Calculate the molar solubility of $ZnCO_3$ in a solution buffered to a pH of 7.00.

11-13. Calculate the molar solubility of Ag_2CO_3 in a solution buffered to a pH of 7.50.

*11-14. Dilute NaOH is introduced into a solution that is 0.050 M in Cu^{2+} and 0.040 M in Mn^{2+}.
(a) Which hydroxide precipitates first?
(b) What OH^- concentration is needed to initiate precipitation of the first hydroxide?
(c) What is the concentration of the cation forming the less soluble hydroxide when the more soluble hydroxide begins to form?

11-15. A solution is 0.040 M in Na_2SO_4 and 0.050 M in $NaIO_3$. To this is added a solution containing Ba^{2+}. Assuming that no HSO_4^- is present in the original solution,
(a) which barium salt will precipitate first?
(b) what is the Ba^{2+} concentration as the first precipitate forms?
(c) what is the concentration of the anion that forms the less soluble barium salt when the more soluble precipitate begins to form?

*11-16. Silver ion is being considered for separating I^- from SCN^- in a solution that is 0.040 M in KI and 0.080 M in NaSCN.
(a) What Ag^+ concentration is needed to lower the I^- concentration to 1.0×10^{-6} M?
(b) What is the Ag^+ concentration of the solution when AgSCN begins to precipitate?
(c) What is the ratio of SCN^- to I^- when AgSCN begins to precipitate?
(d) What is the ratio of SCN^- to I^- when the Ag^+ concentration is 1.0×10^{-3} M?

11-17. Using 1.0×10^{-6} M as the criterion for quantitative removal, determine whether it is feasible to use
(a) SO_4^{2-} to separate Ba^{2+} and Sr^{2+} in a solution that is initially 0.040 M in Sr^{2+} and 0.20 M in Ba^{2+}.
(b) SO_4^{2-} to separate Ba^{2+} and Ag^+ in a solution that is initially 0.030 M in each cation. For Ag_2SO_4, $K_{sp} = 1.6 \times 10^{-5}$.
(c) OH^- to separate Be^{2+} and Hf^{4+} in a solution that is initially 0.030 M in Be^{2+} and 0.020 M in Hf^{4+}. For $Be(OH)_2$, $K_{sp} = 7.0 \times 10^{-22}$, and for $Hf(OH)_4$, $K_{sp} = 4.0 \times 10^{-26}$.
(d) IO_3^- to separate In^{3+} and Tl^+ in a solution that is initially 0.30 M in In^{3+} and 0.10 M in Tl^+. For $In(IO_3)_3$, $K_{sp} = 3.3 \times 10^{-11}$, and for $TlIO_3$, $K_{sp} = 3.1 \times 10^{-6}$.

*11-18. What mass of AgBr dissolves in 200 mL of 0.200 M NaCN?

$$Ag^+ + 2CN^- \rightleftharpoons Ag(CN)_2^- \quad \beta_2 = 1.3 \times 10^{21}$$

11-19. The equilibrium constant for formation of $CuCl_2^-$ is given by

$$Cu^+ + 2Cl^- \rightleftharpoons CuCl_2^-$$

$$\beta_2 = \frac{[CuCl_2^-]}{[Cu^+][Cl^-]^2} = 7.9 \times 10^4$$

What is the solubility of CuCl in solutions having the following analytical NaCl concentrations:
(a) 5.0 M?
(b) 5.0×10^{-1} M?
(c) 5.0×10^{-2} M?
(d) 5.0×10^{-3} M?
(e) 5.0×10^{-4} M?

*11-20. In contrast to many salts, calcium sulfate is only partially dissociated in aqueous solution:

$$CaSO_4(aq) \rightleftharpoons Ca^{2+} + SO_4^{2-}$$

$$K_d = 5.2 \times 10^{-3}$$

The solubility-product constant for $CaSO_4$ is 2.6×10^{-5}. Calculate the solubility of $CaSO_4$ in (a) water and (b) 0.0100 M Na_2SO_4. In addition, calculate the percent of undissociated $CaSO_4$ in each solution.

11-21. Calculate the molar solubility of Tl_2S as a function of pH over the range of pH 10 to pH 1. Find values at every 0.5 pH unit and use the charting function of Excel to plot solubility versus pH.

11-22. **Challenge Problem:** (a) The solubility of CdS is normally very low, but it can be increased by lowering the solution pH. Calculate the molar solubility of CdS as a function of pH from pH 11 to pH 1. Find values at every 0.5 pH unit and plot solubility versus pH.

(b) A solution contains 1×10^{-4} M of both Fe^{2+} and Cd^{2+}. Sulfide ions are slowly added to this solution to precipitate either FeS or CdS. Determine which ion precipitates first and the range of S^{2-} concentration that will allow a clean separation of the two ions.

(c) The analytical concentration of H_2S in a solution saturated with $H_2S(g)$ is 0.10 M. What pH range is necessary for the clean separation described in part (b)?

(d) If there is no pH control from a buffer, what is the pH of a saturated H_2S solution?

(e) Plot the α_0 and α_1 values for H_2S over the pH range of 10 to 1.

(f) A solution contains H_2S and NH_3. Four Cd^{2+} complexes form with NH_3 in a stepwise fashion: $Cd(NH_3)^{2+}$, $Cd(NH_3)_2^{2+}$, $Cd(NH_3)_3^{2+}$, and $Cd(NH_3)_4^{2+}$. Find the molar solubility of CdS in a solution of 0.1 M NH_3.

(g) For the same solution components as in part (f), buffers are prepared with a total concentration of $NH_3 + NH_4Cl = 0.10$ M. The pH values are 8.0, 8.5, 9.0, 9.5, 10.0, 10.5, and 11.0. Find the molar solubility of CdS in these solutions.

(h) For the solutions in part (g), how could you determine whether the solubility increase with pH is due to complex formation or to an activity effect?

Classical Methods of Analysis

PART III

CHAPTER 12

Gravimetric Methods of Analysis

The formation and growth of precipitates and crystals are very important in analytical chemistry and in other areas of science. Shown in the photo is the growth of sodium acetate crystals from a supersaturated solution. Because supersaturation leads to small particles that are difficult to filter, it is desirable in gravimetric analysis to minimize the supersaturation and thus increase the particle size of the solid that is formed. The properties of precipitates that are used in chemical analysis are described in this chapter. The techniques for obtaining easily filterable precipitates that are free from contaminants are major topics. Such precipitates are used in gravimetric analysis and in the separation of interferences for other analytical procedures.

Charlies D. Winters

Gravimetric methods are quantitative methods that are based on determining the mass of a pure compound to which the analyte is chemically related.

Gravimetric methods of analysis are based on mass measurements with an analytical balance, an instrument that yields highly accurate and precise data. In fact, if you perform a gravimetric determination in your laboratory, you may make some of the most accurate and precise measurements of your life.

Several analytical methods are based on mass measurements. In **precipitation gravimetry**, the analyte is separated from a solution of the sample as a precipitate and is converted to a compound of known composition that can be weighed. In **volatilization gravimetry**, the analyte is separated from other constituents of a sample by converting it to a gas of known chemical composition. The mass of the gas then serves as a measure of the analyte concentration. These two types of gravimetry are considered in this chapter.[1] In **electrogravimetry**, the analyte is separated by deposition on an electrode by an electrical current. The mass of this product then provides a measure of the analyte concentration. Electrogravimetry is described in Section 22C.

Two other types of analytical methods are based on mass. In **gravimetric titrimetry**, which is described in Section 13D, the mass of a reagent of known concentration required to react completely with the analyte provides the information needed to determine the analyte concentration. **Atomic mass spectrometry** uses a mass spectrometer to separate the gaseous ions formed from the elements making up a sample of matter. The concentration of the resulting ions is then determined by measuring the electrical current produced when they fall on the surface of an ion detector. This technique is described briefly in Chapter 29.

12A PRECIPITATION GRAVIMETRY

In precipitation gravimetry, the analyte is converted to a sparingly soluble precipitate. This precipitate is then filtered, washed free of impurities, converted to a product of known composition by suitable heat treatment, and weighed. For

[1]For an extensive treatment of gravimetric methods, see C. L. Rulfs, in *Treatise on Analytical Chemistry*, I. M. Kolthoff and P. J. Elving, eds., Part I, Vol. 11, Chap. 13, New York: Wiley, 1975.

example, a precipitation method for determining calcium in water is one of the official methods of the Association of Official Analytical Chemists.[2] In this technique, an excess of oxalic acid, $H_2C_2O_4$, is added to an aqueous solution of the sample. Ammonia is then added, which neutralizes the acid and causes essentially all of the calcium in the sample to precipitate as calcium oxalate. The reactions are

$$2NH_3 + H_2C_2O_4 \rightarrow 2NH_4^+ + C_2O_4^{2-}$$

$$Ca^{2+}(aq) + C_2O_4^{2-}(aq) \rightarrow CaC_2O_4(s)$$

The CaC_2O_4 precipitate is filtered using a weighed filtering crucible, then dried and ignited. This process converts the precipitate entirely to calcium oxide. The reaction is

$$CaC_2O_4(s) \xrightarrow{\Delta} CaO(s) + CO(g) + CO_2(g)$$

After cooling, the crucible and precipitate are weighed, and the mass of calcium oxide is determined by subtracting the known mass of the crucible. The calcium content of the sample is then computed as shown in Example 12-1, Section 12B.

12A-1 Properties of Precipitates and Precipitating Reagents

Ideally, a gravimetric precipitating agent should react *specifically* or at least *selectively* with the analyte. Specific reagents, which are rare, react only with a single chemical species. Selective reagents, which are more common, react with a limited number of species. In addition to specificity and selectivity, the ideal precipitating reagent would react with the analyte to give a product that is

1. easily filtered and washed free of contaminants;
2. of sufficiently low solubility that no significant loss of the analyte occurs during filtration and washing;
3. unreactive with constituents of the atmosphere;
4. of known chemical composition after it is dried or, if necessary, ignited (Section 12A-7).

> An example of a selective reagent is $AgNO_3$. The only common ions that it precipitates from acidic solution are Cl^-, Br^-, I^-, and SCN^-. Dimethylglyoxime, which is discussed in Section 12C-3, is a specific reagent that precipitates only Ni^{2+} from alkaline solutions.

Few, if any, reagents produce precipitates that have all these desirable properties.

The variables that influence solubility (the second property in our list) are discussed in Section 11B. In the sections that follow, we are concerned with methods that allow us to obtain easily filtered and pure solids of known composition.[3]

12A-2 Particle Size and Filterability of Precipitates

Precipitates consisting of large particles are generally desirable for gravimetric work because these particles are easy to filter and wash free of impurities. In addition, precipitates of this type are usually purer than are precipitates made up of fine particles.

---

[2]W. Horwitz and G. Latimer, eds., *Official Methods of Analysis*, 18th ed., Official Method 920.199, Gaithersburg, MD: Association of Official Analytical Chemists International, 2005.

[3]For a more detailed treatment of precipitates, see H. A. Laitinen and W. E. Harris, *Chemical Analysis*, 2nd ed., Chaps. 8 and 9, New York: McGraw-Hill, 1975; A. E. Nielsen, in *Treatise on Analytical Chemistry*, 2nd ed., I. M. Kolthoff and P. J. Elving, eds., Part I, Vol. 3, Chap. 27, New York: Wiley, 1983.

A **colloid** consists of solid particles with diameters that are less than 10^{-4} cm.

In diffuse light, **colloidal suspensions** may be perfectly clear and appear to contain no solid. The presence of the second phase can be detected, however, by shining the beam of a flashlight into the solution. Because particles of colloidal dimensions scatter visible radiation, the path of the beam through the solution can be seen by the eye. This phenomenon is called the **Tyndall effect** (see color plate 6).

It is very difficult to filter the particles of a colloidal suspension. To trap these particles, the pore size of the filtering medium must be so small that filtrations take a very long time. With suitable treatment, however, the individual colloidal particles can be made to stick together, or coagulate, to produce large particles that are easy to filter.

Equation 12-1 is known as the Von Weimarn equation in recognition of the scientist who proposed it in 1925.

A **supersaturated solution** is an unstable solution that contains a higher solute concentration than a saturated solution. As excess solute precipitates with time, supersaturation decreases to zero (see color plate 5).

To increase the particle size of a precipitate, minimize the relative supersaturation during precipitate formation.

Nucleation is a process in which a minimum number of atoms, ions, or molecules join together to give a stable solid.

Precipitates form by nucleation and by particle growth. If nucleation predominates, a large number of very fine particles is produced. If particle growth predominates, a smaller number of larger particles is obtained.

Factors That Determine the Particle Size of Precipitates

The particle size of solids formed by precipitation varies enormously. At one extreme are **colloidal suspensions**, whose tiny particles are invisible to the naked eye (10^{-7} to 10^{-4} cm in diameter). Colloidal particles show no tendency to settle from solution and are difficult to filter. At the other extreme are particles with dimensions on the order of tenths of a millimeter or greater. The temporary dispersion of such particles in the liquid phase is called a **crystalline suspension**. The particles of a crystalline suspension tend to settle spontaneously and are easily filtered.

Precipitate formation has been studied for many years, but the mechanism of the process is still not fully understood. What is certain, however, is that the particle size of a precipitate is influenced by precipitate solubility, temperature, reactant concentrations, and the rate at which reactants are mixed. The net effect of these variables can be accounted for, at least qualitatively, by assuming that the particle size is related to a single property of the system called **relative supersaturation**, where

$$\text{relative supersaturation} = \frac{Q - S}{S} \tag{12-1}$$

In this equation, Q is the concentration of the solute at any instant, and S is its equilibrium solubility.

Generally, precipitation reactions are slow so that, even when a precipitating reagent is added drop by drop to a solution of an analyte, some supersaturation is likely. Experimental evidence indicates that the particle size of a precipitate varies inversely with the average relative supersaturation during the time when the reagent is being introduced. Thus, when $(Q - S)/S$ is large, the precipitate tends to be colloidal, and when $(Q - S)/S$ is small, a crystalline solid is more likely.

Mechanism of Precipitate Formation

The effect of relative supersaturation on particle size can be explained if we assume that precipitates form in two ways: by **nucleation** and by **particle growth**. The particle size of a freshly formed precipitate is determined by the mechanism that predominates.

In nucleation, a few ions, atoms, or molecules (perhaps as few as four or five) come together to form a stable solid. Often, these nuclei form on the surface of suspended solid contaminants, such as dust particles. Further precipitation then is governed by the competition between additional nucleation and growth of existing nuclei (particle growth). If nucleation predominates, a precipitate containing a large number of small particles results, and if growth predominates, a smaller number of larger particles is produced.

The rate of nucleation is believed to increase enormously with increasing relative supersaturation. In contrast, the rate of particle growth is only moderately enhanced by high relative supersaturations. Therefore, when a precipitate is formed at high relative supersaturation, nucleation is the major precipitation mechanism, and a large number of small particles is formed. At low relative supersaturations, on the other hand, the rate of particle growth tends to predominate, and deposition of solid on existing particles occurs rather than further nucleation. Low relative supersaturation produces crystalline suspensions.

Experimental Control of Particle Size

Experimental variables that minimize supersaturation and thus produce crystalline precipitates include elevated temperatures to increase the solubility of the precipitate (*S* in Equation 12-1), dilute solutions (to minimize *Q*), and slow addition of the precipitating agent with good stirring. The last two measures also minimize the concentration of the solute (*Q*) at any given instant.

If the solubility of the precipitate depends on pH, larger particles can also be produced by controlling pH. For example, large, easily filtered crystals of calcium oxalate are obtained by forming the bulk of the precipitate in a mildly acidic environment in which the salt is moderately soluble. The precipitation is then completed by slowly adding aqueous ammonia until the acidity is sufficiently low for removal of substantially all of the calcium oxalate. The additional precipitate produced during this step deposits on the solid particles formed in the first step.

Unfortunately, many precipitates cannot be formed as crystals under practical laboratory conditions. A colloidal solid is generally formed when a precipitate has such a low solubility that *S* in Equation 12-1 always remains negligible relative to *Q*. The relative supersaturation thus remains enormous throughout precipitate formation, and a colloidal suspension results. For example, under conditions feasible for an analysis, the hydrous oxides of iron(III), aluminum, and chromium(III) and the sulfides of most heavy-metal ions form only as colloids because of their very low solubilities.[4]

> ❮ Precipitates that have very low solubilities, such as many sulfides and hydrous oxides, generally form as colloids.

12A-3 Colloidal Precipitates

Individual colloidal particles are so small that they are not retained by ordinary filters. Moreover, Brownian motion prevents their settling out of solution under the influence of gravity. Fortunately, however, we can coagulate, or agglomerate, the individual particles of most colloids to give a filterable, amorphous mass that will settle out of solution.

Coagulation of Colloids

Coagulation can be hastened by heating, by stirring, and by adding an electrolyte to the medium. To understand the effectiveness of these measures, we need to look into why colloidal suspensions are stable and do not coagulate spontaneously.

Colloidal suspensions are stable because all of the particles of the colloid are either positively or negatively charged and thus repel one another. The charge results from cations or anions that are bound to the surface of the particles. We can show that colloidal particles are charged by placing them between charged plates where some of the particles migrate toward one electrode while others move toward the electrode of the opposite charge. The process by which ions are retained *on the surface of a solid* is known as **adsorption**.

The adsorption of ions on an ionic solid originates from the normal bonding forces that are responsible for crystal growth. For example, a silver ion at the surface of a silver chloride particle has a partially unsatisfied bonding capacity for anions because of its surface location. Negative ions are attracted to this site by the same forces

> **Adsorption** is a process in which a substance (gas, liquid, or solid) is held *on the surface* of a solid. In contrast, **absorption** is retention of a substance *within the pores* of a solid.

---

[4]Silver chloride illustrates that the relative supersaturation concept is imperfect. This compound forms as a colloid, yet its molar solubility is not significantly different from that of other compounds, such as $BaSO_4$, which generally form as crystals.

that hold chloride ions in the silver chloride lattice. Chloride ions at the surface of the solid exert an analogous attraction for cations dissolved in the solvent.

The kind of ions retained on the surface of a colloidal particle and their number depend in a complex way on several variables. For a suspension produced in a gravimetric analysis, however, the species adsorbed, and hence the charge on the particles, can be easily predicted because lattice ions are generally more strongly held than others. For example, when silver nitrate is first added to a solution containing chloride ion, the colloidal particles of the precipitate are negatively charged as a result of adsorption of some of the excess chloride ions. This charge, though, becomes positive when enough silver nitrate has been added to provide an excess of silver ions. The surface charge is at a minimum when the supernatant liquid does not contain an excess of either ion.

The extent of adsorption and thus the charge on a given particle increase rapidly as the concentration of a common ion becomes greater. Eventually, however, the surface of the particles becomes covered with the adsorbed ions, and the charge becomes constant and independent of concentration.

Figure 12-1 shows a colloidal silver chloride particle in a solution that contains an excess of silver nitrate. Attached directly to the solid surface is the **primary adsorption layer**, which consists mainly of adsorbed silver ions. Surrounding the charged particle is a layer of solution, called the **counter-ion layer**, which contains sufficient excess of negative ions (principally nitrate) to just balance the charge on the surface of the particle. The primarily adsorbed silver ions and the negative counter-ion layer constitute an **electric double layer** that imparts stability to the colloidal suspension. As colloidal particles approach one another, this double layer exerts an electrostatic repulsive force that prevents particles from colliding and adhering.

Figure 12-2a shows the effective charge on two silver chloride particles. The upper curve represents a particle in a solution that contains a reasonably large excess

> The charge on a colloidal particle formed in a gravimetric analysis is determined by the charge of the lattice ion that is in excess when the precipitation is complete.

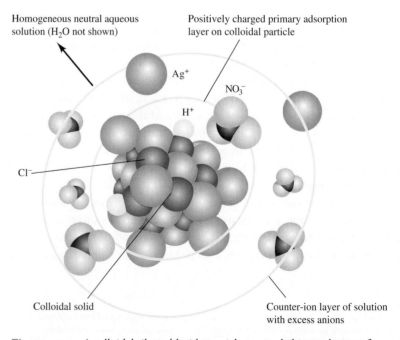

Figure 12-1   A colloidal silver chloride particle suspended in a solution of silver nitrate.

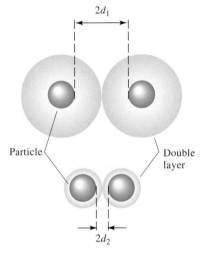

Figure 12-2 Effect of AgNO$_3$ and electrolyte concentration on the thickness of the double layer surrounding a colloidal AgCl particle in a solution containing excess AgNO$_3$.

Figure 12-3 The electrical double layer of a colloid consists of a layer of charge adsorbed on the surface of the particle (the primary adsorption layer) and a layer of opposite charge (the counter-ion layer) in the solution surrounding the particle. Increasing the electrolyte concentration has the effect of decreasing the volume of the counter-ion layer, thereby increasing the chances for coagulation.

of silver nitrate, and the lower curve depicts a particle in a solution that has a much lower silver nitrate content. The effective charge can be thought of as a measure of the repulsive force that the particle exerts on like particles in the solution. Note that the effective charge falls off rapidly as the distance from the surface increases, and it approaches zero at the points d_1 or d_2. These decreases in effective charge (in both cases positive) are caused by the negative charge of the excess counter-ions in the double layer surrounding each particle. At points d_1 and d_2, the number of counter-ions in the layer is approximately equal to the number of primarily adsorbed ions on the surfaces of the particles; therefore, the effective charge of the particles approaches zero at this point.

The upper portion of **Figure 12-3** depicts two silver chloride particles and their counter-ion layers as they approach each other in the concentrated silver nitrate just considered. Note that the effective charge on the particles prevents them from approaching one another more closely than about $2d_1$—a distance that is too great for coagulation to occur. As shown in the lower part of Figure 12-3, in the more dilute silver nitrate solution, the two particles can approach within $2d_2$ of one another. Ultimately, as the concentration of silver nitrate is further decreased, the distance between particles becomes small enough for the forces of agglomeration to take effect and a coagulated precipitate to appear.

Coagulation of a colloidal suspension can often be brought about by a short period of heating, particularly if accompanied by stirring. Heating decreases the number of adsorbed ions and thus the thickness, d_i, of the double layer. The particles may also gain enough kinetic energy at the higher temperature to overcome the barrier to close approach imposed by the double layer.

An even more effective way to coagulate a colloid is to increase the electrolyte concentration of the solution. If we add a suitable ionic compound to a colloidal suspension, the concentration of counter-ions increases in the vicinity of each particle. As a result, the volume of solution that contains sufficient counter-ions to balance the charge of the primary adsorption layer decreases. The net effect of adding an electrolyte is thus a shrinkage of the counter-ion layer, as shown in **Figure 12-2b**. The particles can then approach one another more closely and agglomerate.

◀ Colloidal suspensions can often be coagulated by heating, stirring, and adding an electrolyte.

Peptization of Colloids

Peptization is the process by which a coagulated colloid reverts to its original dispersed state. When a coagulated colloid is washed, some of the electrolyte responsible for its coagulation is leached from the internal liquid in contact with the solid

Peptization is a process by which a coagulated colloid returns to its dispersed state.

particles. Removal of this electrolyte has the effect of increasing the volume of the counter-ion layer. The repulsive forces responsible for the original colloidal state are then reestablished, and particles detach themselves from the coagulated mass. The washings become cloudy as the freshly dispersed particles pass through the filter.

We are thus faced with a dilemma in working with coagulated colloids. On the one hand, washing is needed to minimize contamination, but on the other, there is a risk of losses resulting from peptization if pure water is used. The problem is usually solved by washing the precipitate with a solution containing an electrolyte that volatilizes when the precipitate is dried or ignited. For example, silver chloride is usually washed with a dilute solution of nitric acid. While the precipitate no doubt becomes contaminated with acid, no harm is done, since the nitric acid is lost during the drying step.

Practical Treatment of Colloidal Precipitates

Colloids are best precipitated from hot, stirred solutions containing sufficient electrolyte to ensure coagulation. The filterability of a coagulated colloid often improves if it is allowed to stand for an hour or more in contact with the hot solution from which it was formed. During this process, which is known as **digestion**, weakly bound water appears to be lost from the precipitate. The result is a denser mass that is easier to filter.

12A-4 Crystalline Precipitates

Crystalline precipitates are generally more easily filtered and purified than are coagulated colloids. In addition, the size of individual crystalline particles, and thus their filterability, can be controlled to some extent.

Methods of Improving Particle Size and Filterability

The particle size of crystalline solids can often be improved significantly by minimizing Q or maximizing S, or both, in Equation 12-1. The value of Q is can often be minimized by using dilute solutions and adding the precipitating reagent slowly, with good mixing. Often, S is increased by precipitating from hot solution or by adjusting the pH of the precipitation medium.

Digestion of crystalline precipitates (without stirring) for some time after formation often yields a purer, more filterable product. The improvement in filterability undoubtedly results from the dissolution and recrystallization that occur continuously and at an enhanced rate at elevated temperatures. Recrystallization apparently results in bridging between adjacent particles, a process that yields larger and more easily filtered crystalline aggregates. This view is supported by the observation that little improvement in filtering characteristics occurs if the mixture is stirred during digestion.

12A-5 Coprecipitation

When *otherwise soluble* compounds are removed from solution during precipitate formation, we refer to the process as **coprecipitation**. Contamination of a precipitate by a second substance whose solubility product has been exceeded *is not coprecipitation*.

There are four types of coprecipitation: **surface adsorption, mixed-crystal formation, occlusion,** and **mechanical entrapment**.[5] Surface adsorption and mixed-crystal formation are equilibrium processes, and occlusion and mechanical entrapment arise from the kinetics of crystal growth.

Digestion is a process in which a precipitate is heated in the solution from which it was formed (the mother liquor) and allowed to stand in contact with the solution.

Mother liquor is the solution from which a precipitate was formed.

Digestion improves the purity and filterability of both colloidal and crystalline precipitates.

Coprecipitation is a process in which *normally soluble* compounds are carried out of solution by a precipitate.

---

[5]We follow the simple system of classification of coprecipitation phenomena proposed by A. E. Nielsen, in *Treatise on Analytical Chemistry*, 2nd ed., I. M. Kolthoff and P. J. Elving, eds., Part I, Vol. 3, p. 333, New York: Wiley, 1983.

Surface Adsorption

Adsorption is a common source of coprecipitation and is likely to cause significant contamination of precipitates with large specific surface areas, that is, coagulated colloids (see Feature 12-1 for the definition of specific area). Although adsorption does occur in crystalline solids, its effects on purity are usually undetectable because of the relatively small specific surface area of these solids.

Coagulation of a colloid does not significantly decrease the amount of adsorption because the coagulated solid still contains large internal surface areas that remain exposed to the solvent (**Figure 12-4**). The coprecipitated contaminant on the coagulated colloid consists of the lattice ion originally adsorbed on the surface before coagulation plus the counter-ion of opposite charge held in the film of solution immediately adjacent to the particle. *The net effect of surface adsorption is, therefore, the carrying down of an otherwise soluble compound as a surface contaminant.* For example, the coagulated silver chloride formed in the gravimetric determination of chloride ion is contaminated with primarily adsorbed silver ions with nitrate or other anions in the counter-ion layer. The result is that silver nitrate, a normally soluble compound, is coprecipitated with the silver chloride.

Minimizing Adsorbed Impurities on Colloids   The purity of many coagulated colloids is improved by digestion. During this process, water is expelled from the solid to give a denser mass that has a smaller specific surface area for adsorption.

Washing a coagulated colloid with a solution containing a volatile electrolyte may also be helpful because any nonvolatile electrolyte added earlier to cause coagulation is displaced by the volatile species. Washing generally does not remove much of the primarily adsorbed ions because the attraction between these ions and the surface of the solid is too strong. Exchange occurs, however, between existing *counter-ions* and ions in the wash liquid. For example, in the determination of silver by precipitation with chloride ion, the primarily adsorbed species is chloride. Washing with an acidic solution converts the counter-ion layer largely to hydrogen ions so that both chloride and hydrogen ions are retained by the solid. Volatile HCl is then given off when the precipitate is dried.

> Adsorption is often the major source of contamination in coagulated colloids but of no significance in crystalline precipitates.

> In adsorption, a normally soluble compound is carried out of solution on the surface of a coagulated colloid. This compound consists of the primarily adsorbed ion and an ion of opposite charge from the counter-ion layer.

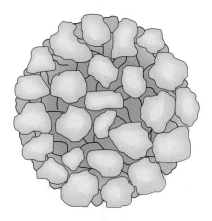

Figure 12-4   A coagulated colloid. This figure suggests that a coagulated colloid continues to expose a large surface area to the solution from which it was formed.

FEATURE 12-1

Specific Surface Area of Colloids

Specific surface area is defined as the surface area per unit mass of solid and usually has the units of square centimeters per gram. For a given mass of solid, the specific surface area increases dramatically as particle size decreases, and it becomes enormous for colloids. For example, the solid cube shown in **Figure 12F-1**, which has dimensions of 1 cm on an edge, has a surface area of 6 cm^2. If this cube weighs 2 g, its specific surface area is 6 cm^2/2 g = 3 cm^2/g. Now, let us divide the cube into 1000 cubes, each having an edge length of 0.1 cm. The surface area of each face of these cubes is now 0.1 cm $\times$ 0.1 cm = 0.01 cm^2, and the total area for the six faces of the cube is 0.06 cm^2. Because there are 1000 of these cubes, the total surface area for the 2 g of solid is now 60 cm^2; the specific surface area is 30 cm^2/g. Continuing in this way, we find that the specific surface area becomes 300 cm^2/g when we have 10^6 cubes that are 0.01 cm on a side. The particle size of a typical crystalline suspension lies in the region of 0.01 to 0.1 cm so that a typical crystalline precipitate has a specific surface area between 30 cm^2/g and 300 cm^2/g. Contrast these figures with those for 2 g of a colloid made up of 10^{18} particles, each having an edge of 10^{-6} cm. In this case, the specific area is 3×10^6 cm^2/g, which is over 3000 ft^2/g. Based on these calculations, 1 g of a typical colloidal suspension has a surface area that is equal to the average floor area of a fairly large family home.

(continued)

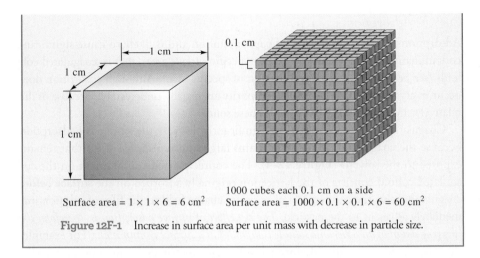

Figure 12F-1   Increase in surface area per unit mass with decrease in particle size.

Regardless of the method of treatment, a coagulated colloid is always contaminated to some degree, even after extensive washing. The error introduced into the analysis from this source can be as low as 1 to 2 ppt, as in the coprecipitation of silver nitrate on silver chloride. In contrast, coprecipitation of heavy-metal hydroxides on the hydrous oxides of trivalent iron or aluminum can result in errors as large as several percent, which is intolerable.

Reprecipitation   A drastic but effective way to minimize the effects of adsorption is **reprecipitation**. In this process, the filtered solid is redissolved and reprecipitated. The first precipitate usually carries down only a fraction of the contaminant present in the original solvent. Thus, the solution containing the redissolved precipitate has a significantly lower contaminant concentration than the original, and even less adsorption occurs during the second precipitation. Reprecipitation adds substantially to the time required for an analysis. However, it is often necessary for such precipitates as the hydrous oxides of iron(III) and aluminum, which have extraordinary tendencies to adsorb the hydroxides of heavy-metal cations such as zinc, cadmium, and manganese.

Mixed-Crystal Formation

In mixed-crystal formation, one of the ions in the crystal lattice of a solid is replaced by an ion of another element. For this exchange to occur, it is necessary that the two ions have the same charge and that their sizes differ by no more than about 5%. Furthermore, the two salts must belong to the same crystal class. For example, barium sulfate formed by adding barium chloride to a solution containing sulfate, lead, and acetate ions is found to be severely contaminated by lead sulfate. This contamination occurs even though acetate ions normally prevent precipitation of lead sulfate by complexing the lead. In this case, lead ions replace some of the barium ions in the barium sulfate crystals. Other examples of coprecipitation by mixed-crystal formation include $MgKPO_4$ in $MgNH_4PO_4$, $SrSO_4$ in $BaSO_4$, and MnS in CdS.

The extent of mixed-crystal contamination is governed by the law of mass action and increases as the ratio of contaminant to analyte concentration increases. Mixed-crystal formation is a particularly troublesome type of coprecipitation because little can be done about it when certain combinations of ions are present in a sample matrix. This problem occurs with both colloidal suspensions and crystalline precipitates. When mixed-crystal formation occurs, the interfering ion may have to be separated before the final precipitation step. Alternatively, a different precipitating reagent that does not give mixed crystals with the ions in question may be used.

Mixed-crystal formation is a type of coprecipitation in which a contaminant ion replaces an ion in the lattice of a crystal.

Occlusion and Mechanical Entrapment

When a crystal is growing rapidly during precipitate formation, foreign ions in the counter-ion layer may become trapped, or *occluded*, within the growing crystal. Because supersaturation and thus growth rate decrease as precipitation progresses, the amount of occluded material is greatest in that part of a crystal that forms first.

Mechanical entrapment occurs when crystals lie close together during growth. Several crystals grow together and in so doing trap a portion of the solution in a tiny pocket.

Both occlusion and mechanical entrapment are at a minimum when the rate of precipitate formation is low, that is, under conditions of low supersaturation. In addition, digestion often reduces the effects of these types of coprecipitation. Undoubtedly, the rapid dissolving and reprecipitation that occur at the elevated temperature of digestion open up the pockets and allow the impurities to escape into the solution.

> **Occlusion** is a type of coprecipitation in which a compound is trapped within a pocket formed during rapid crystal growth.

> ❮ Mixed-crystal formation may occur in both colloidal and crystalline precipitates, but occlusion and mechanical entrapment are confined to crystalline precipitates.

Coprecipitation Errors

Coprecipitated impurities may cause either negative or positive errors in an analysis. If the contaminant is not a compound of the ion being determined, a positive error will always result. Therefore, a positive error is observed whenever colloidal silver chloride adsorbs silver nitrate during a chloride analysis. In contrast, when the contaminant does contain the ion being determined, either positive or negative errors may occur. For example, in the determination of barium by precipitation as barium sulfate, occlusion of other barium salts occurs. If the occluded contaminant is barium nitrate, a positive error is observed because this compound has a larger molar mass than the barium sulfate that would have formed had no coprecipitation occurred. If barium chloride is the contaminant, the error is negative because its molar mass is less than that of the sulfate salt.

> ❮ Coprecipitation can cause either negative or positive errors.

12A-6 Precipitation from Homogeneous Solution

Precipitation from homogeneous solution is a technique in which a precipitating agent is generated in a solution of the analyte by a slow chemical reaction.[6] Local reagent excesses do not occur because the precipitating agent appears gradually and homogeneously throughout the solution and reacts immediately with the analyte. As a result, the relative supersaturation is kept low during the entire precipitation. In general, homogeneously formed precipitates, both colloidal and crystalline, are better suited for analysis than a solid formed by direct addition of a precipitating reagent.

Urea is often used for the homogeneous generation of hydroxide ion. The reaction can be expressed by the equation

$$(H_2N)_2CO + 3H_2O \rightarrow CO_2 + 2NH_4^+ + 2OH^-$$

This hydrolysis proceeds slowly at temperatures just below 100°C, with 1 to 2 hours needed to complete a typical precipitation. Urea is particularly valuable for the precipitation of hydrous oxides or basic salts. For example, hydrous oxides of iron(III) and aluminum, formed by direct addition of base, are bulky and gelatinous masses that are heavily contaminated and difficult to filter. In contrast, when these same products are produced by homogeneous generation of hydroxide ion, they are dense, are easily filtered, and have considerably higher purity. **Figure 12-5** shows hydrous oxide precipitates of aluminum formed by direct addition of base and by homogeneous precipitation with urea. Homogeneous precipitation of crystalline precipitates also results in marked increases in crystal size as well as improvements in purity.

> **Homogeneous precipitation** is a process in which a precipitate is formed by slow generation of a precipitating reagent homogeneously throughout a solution.

> ❮ Solids formed by homogeneous precipitation are generally purer and more easily filtered than precipitates generated by direct addition of a reagent to the analyte solution.

Figure 12-5 Aluminum hydroxide formed by the direct addition of ammonia (left) and the homogeneous production of hydroxide (right).

---

[6]For a general reference on this technique, see L. Gordon, M. L. Salutsky, and H. H. Willard, *Precipitation from Homogeneous Solution,* New York: Wiley, 1959.

Representative methods based on precipitation by homogeneously generated reagents are given in **Table 12-1**.

TABLE 12-1

Methods for Homogeneous Generation of Precipitating Agents

Precipitating Agent	Reagent	Generation Reaction	Elements Precipitated
OH^-	Urea	$(NH_2)_2CO + 3H_2O \rightarrow CO_2 + 2NH_4^+ + 2OH^-$	Al, Ga, Th, Bi, Fe, Sn
PO_4^{3-}	Trimethyl phosphate	$(CH_3O)_3PO + 3H_2O \rightarrow 3CH_3OH + H_3PO_4$	Zr, Hf
$C_2O_4^{2-}$	Ethyl oxalate	$(C_2H_5)_2C_2O_4 + 2H_2O \rightarrow 2C_2H_5OH + H_2C_2O_4$	Mg, Zn, Ca
SO_4^{2-}	Dimethyl sulfate	$(CH_3O)_2SO_2 + 4H_2O \rightarrow 2CH_3OH + SO_4^{2-} + 2H_3O^+$	Ba, Ca, Sr, Pb
CO_3^{2-}	Trichloroacetic acid	$Cl_3CCOOH + 2OH^- \rightarrow CHCl_3 + CO_3^{2-} + H_2O$	La, Ba, Ra
H_2S	Thioacetamide*	$CH_3CSNH_2 + H_2O \rightarrow CH_3CONH_2 + H_2S$	Sb, Mo, Cu, Cd
DMG†	Biacetyl + hydroxylamine	$CH_3COCOCH_3 + 2H_2NOH \rightarrow DMG + 2H_2O$	Ni
HOQ‡	8-Acetoxyquinoline§	$CH_3COOQ + H_2O \rightarrow CH_3COOH + HOQ$	Al, U, Mg, Zn

$$*CH_3-\overset{\overset{\displaystyle S}{\|}}{C}-NH_2$$

$$†DMG = Dimethylglyoxime = CH_3-\overset{\overset{\displaystyle OH}{\|}}{\underset{}{C}}\ \overset{\overset{\displaystyle OH}{\|}}{\underset{}{C}}-CH_3$$

‡HOQ = 8-Hydroxyquinoline =

$$§CH_3-\overset{\overset{\displaystyle O}{\|}}{C}-O$$

12A-7 Drying and Ignition of Precipitates

After filtration, a gravimetric precipitate is heated until its mass becomes constant. Heating removes the solvent and any volatile species carried down with the precipitate. Some precipitates are also ignited to decompose the solid and form a compound of known composition. This new compound is often called the *weighing form*.

The temperature required to produce a suitable weighing form varies from precipitate to precipitate. **Figure 12-6** shows mass loss as a function of temperature for several common analytical precipitates. These data were obtained with an automatic thermobalance,[7] an instrument that records the mass of a substance continuously as its temperature is increased at a constant rate (**Figure 12-7**). Heating three of the precipitates—silver chloride, barium sulfate, and aluminum oxide—simply causes removal of water and perhaps volatile electrolytes. Note the vastly different temperatures required to produce an anhydrous precipitate of constant mass. Moisture is completely removed from silver chloride at temperatures higher than 110°C, but dehydration of aluminum oxide is not complete until a temperature greater than 1000°C is achieved. Aluminum oxide formed homogeneously with urea can be completely dehydrated at about 650°C.

Figure 12-6 Effect of temperature on precipitate mass.

---

[7]For descriptions of thermobalances, see D. A. Skoog, F. J. Holler, and S. R. Crouch, *Principles of Instrumental Analysis*, 6th ed., Chap. 31, Belmont, CA: Brooks/Cole, 2007; P. Gabbot, ed. *Principles and Applications of Thermal Analysis*, Chap. 3, Ames, IA: Blackwell, 2008; W. W. Wendlandt, *Thermal Methods of Analysis*, 3rd ed., New York: Wiley, 1985; A. J. Paszto, in *Handbook of Instrumental Techniques for Analytical Chemistry*, F. Settle, ed. , Chap. 50, Upper Saddle River, NJ: Prentice Hall, 1997.

Figure 12-7 Schematic of a thermobalance: *A:* beam; *B:* sample cup and holder; *C:* counterweight; *D:* lamp and photodiodes; *E:* coil; *F:* magnet; *G:* control amplifier; *H:* tare calculator; *I:* amplifier; and *J:* recorder. (Reprinted by permission of Mettler Toledo, Inc., Columbus, OH.)

The thermal curve for calcium oxalate is considerably more complex than the others shown in Figure 12-6. Below about 135°C, unbound water is eliminated to give the monohydrate $CaC_2O_4 \cdot H_2O$. This compound is then converted to the anhydrous oxalate CaC_2O_4 at 225°C. The abrupt change in mass at about 450°C signals the decomposition of calcium oxalate to calcium carbonate and carbon monoxide. The final step in the curve depicts the conversion of the carbonate to calcium oxide and carbon dioxide. As can be seen, the compound finally weighed in a gravimetric calcium determination based on oxalate precipitation is highly dependent on the ignition temperature.

> The temperature required to dehydrate a precipitate completely may be as low as 100°C or as high as 1000°C.

Recording thermal decomposition curves is called **thermogravimetric analysis**, and the mass versus temperature curves are termed **thermograms**.

12B CALCULATION OF RESULTS FROM GRAVIMETRIC DATA

The results of a gravimetric analysis are generally computed from two experimental measurements: the mass of sample and the mass of a product of known composition. The examples that follow illustrate how such computations are carried out.

EXAMPLE 12-1

The calcium in a 200.0-mL sample of a natural water was determined by precipitating the cation as CaC_2O_4. The precipitate was filtered, washed, and ignited in a crucible with an empty mass of 26.6002 g. The mass of the crucible plus CaO (56.077 g/mol) was 26.7134 g. Calculate the concentration of Ca (40.078 g/mol) in water in units of grams per 100 mL of the water.

Solution

The mass of CaO is

$$26.7134g - 26.6002 = 0.1132g$$

The number of moles of Ca in the sample is equal to the number of moles of CaO, or

$$\text{amount of Ca} = 0.1132\,g\,CaO \times \frac{1\,mol\,CaO}{56.077\,g\,CaO} \times \frac{1\,mol\,Ca}{mol\,CaO}$$

$$= 2.0186 \times 10^{-3}\,mol\,Ca$$

$$\text{conc. Ca} = \frac{2.0186 \times 10^{-3}\,mol\,Ca \times 40.078\ g\ Ca/mol\,Ca}{200\ mL\ sample} \times \frac{100}{100}$$

$$= 0.04045\ g/100\ mL\ sample$$

EXAMPLE 12-2

An iron ore was analyzed by dissolving a 1.1324-g sample in concentrated HCl. The resulting solution was diluted with water, and the iron(III) was precipitated as the hydrous oxide $Fe_2O_3 \cdot xH_2O$ by the addition of NH_3. After filtration and washing, the residue was ignited at a high temperature to give 0.5394 g of pure Fe_2O_3 (159.69 g/mol). Calculate (a) the % Fe (55.847 g/mol) and (b) the % Fe_3O_4 (231.54 g/mol) in the sample.

Solution

For both parts of this problem, we need to calculate the number of moles of Fe_2O_3. Thus,

$$\text{amount } Fe_2O_3 = 0.5394 \text{ g } Fe_2O_3 \times \frac{1 \text{ mol } Fe_2O_3}{159.69 \text{ g } Fe_2O_3}$$

$$= 3.3778 \times 10^{-3} \text{ mol } Fe_2O_3$$

(a) The number of moles of Fe is twice the number of moles of Fe_2O_3, and

$$\text{mass Fe} = 3.3778 \times 10^{-3} \text{ mol } Fe_2O_3 \times \frac{2 \text{ mol Fe}}{\text{mol } Fe_2O_3} \times \frac{55.847 \text{ g Fe}}{\text{mol Fe}}$$

$$= 0.37728 \text{ g Fe}$$

$$\% \text{ Fe} = \frac{0.37728 \text{ g Fe}}{1.1324 \text{ g sample}} \times 100\% = 33.32\%$$

(b) As shown by the following balanced equation, 3 mol of Fe_2O_3 is chemically equivalent to 2 mol of Fe_3O_4, that is,

$$3Fe_2O_3 \rightarrow 2Fe_3O_4 + \frac{1}{2}O_2$$

$$\text{mass } Fe_3O_4 = 3.3778 \times 10^{-3} \text{ mol } Fe_2O_3 \times \frac{2 \text{ mol } Fe_3O_4}{3 \text{ mol } Fe_2O_3} \times \frac{231.54 \text{ g } Fe_3O_4}{\text{mol } Fe_3O_4}$$

$$= 0.52140 \text{ g } Fe_3O_4$$

$$\% \text{ } Fe_3O_4 = \frac{0.5140 \text{ g } Fe_3O_4}{1.1324 \text{ g sample}} \times 100\% = 46.04\%$$

Notice that all of the constant factors in each part of this example, such as the molar masses and the stoichiometric ratio, can be combined into a single factor called the **gravimetric factor**. For part (a), we have

The combined constant factors in a gravimetric calculation are referred to as the **gravimetric factor**. When the gravimetric factor is multiplied by the mass of the weighed substance, the result is the mass of the sought for substance.

$$\text{gravimetric factor} = \frac{1 \text{ mol } Fe_2O_3}{159.69 \text{ g } Fe_2O_3} \times \frac{2 \text{ mol Fe}}{\text{mol } Fe_2O_3} \times \frac{55.847 \text{ g Fe}}{\text{mol Fe}} = 0.69944 \frac{\text{g Fe}}{\text{g } Fe_2O_3}$$

For part (b), the gravimetric factor is

$$\text{gravimetric factor} = \frac{1 \text{ mol } Fe_2O_3}{159.69 \text{ g } Fe_2O_3} \times \frac{2 \text{ mol } Fe_3O_4}{3 \text{ mol } Fe_2O_3} \times \frac{231.54 \text{ g } Fe_3O_4}{\text{mol } Fe_3O_4}$$

$$= 0.96662 \frac{\text{g } Fe_3O_4}{\text{g } Fe_2O_3}$$

EXAMPLE 12-3

A 0.2356-g sample containing *only* NaCl (58.44 g/mol) and $BaCl_2$ (208.23 g/mol) yielded 0.4637 g of dried AgCl (143.32 g/mol). Calculate the percent of each halogen compound in the sample.

Solution

If we let x be the mass of NaCl in grams and y be the mass of $BaCl_2$ in grams, we can write as a first equation

$$x + y = 0.2356 \text{ g sample}$$

To obtain the mass of AgCl from the NaCl, we write an expression for the number of moles of AgCl formed from the NaCl, that is,

$$\text{amount AgCl from NaCl} = x \text{ g NaCl} \times \frac{1 \text{ mol NaCl}}{58.44 \text{ g NaCl}} \times \frac{1 \text{ mol AgCl}}{\text{mol NaCl}}$$

$$= 0.017111x \text{ mol AgCl}$$

The mass of AgCl from this source is

$$\text{mass AgCl from NaCl} = 0.017111x \text{ mol AgCl} \times 143.32 \frac{\text{g AgCl}}{\text{mol AgCl}}$$

$$= 2.4524x \text{ g AgCl}$$

Proceeding in the same way, we can write that the number of moles of AgCl from the $BaCl_2$ is given by

$$\text{amount AgCl from } BaCl_2 = y \text{ g } BaCl_2 \times \frac{1 \text{ mol } BaCl_2}{208.23 \text{ g } BaCl_2} \times \frac{2 \text{ mol AgCl}}{\text{mol } BaCl_2}$$

$$= 9.605 \times 10^{-3}y \text{ mol AgCl}$$

$$\text{mass AgCl from } BaCl_2 = 9.605 \times 10^{-3}y \text{ mol AgCl} \times 143.32 \frac{\text{g AgCl}}{\text{mol AgCl}}$$

$$= 1.3766y \text{ g AgCl}$$

Because 0.4637 g of AgCl comes from the two compounds, we can write

$$2.4524x \text{ g AgCl} + 1.3766y \text{ g AgCl} = 0.4637 \text{ g AgCl, or to simplify,}$$

$$2.4524x + 1.3766y = 0.4637$$

Our first equation can then be rewritten as

$$y = 0.2356 - x$$

Substituting into the previous equation gives

$$2.4524x + 1.3766(0.2356 - x) = 0.4637$$

(*continued*)

which rearranges to

$$1.0758x = 0.13942$$

$$x = \text{mass NaCl} = 0.12960 \text{ g NaCl}$$

$$\% \text{NaCl} = \frac{0.12960 \text{ g NaCl}}{0.2356 \text{ g sample}} \times 100\% = 55.01\%$$

$$\% \text{BaCl}_2 = 100.00\% - 55.01\% = 44.99\%$$

Spreadsheet Summary   In some chemical problems, two or more simultaneous equations must be solved to obtain the desired result. Example 12-3 is such a problem. In Chapter 6 of *Applications of Microsoft® Excel in Analytical Chemistry*, 2nd ed., the method of determinants and the matrix inversion method are explored for solving such equations. The matrix method is extended to solve a system of four equations in four unknowns. The matrix method is used to confirm the results of Example 12-3.

> Gravimetric methods do not require a calibration or standardization step (as do all other analytical procedures except coulometry) because the results are calculated directly from the experimental data and atomic masses. Thus, when only one or two samples are to be analyzed, a gravimetric procedure may be the method of choice because it requires less time and effort than a procedure that requires preparation of standards and calibration.

12C APPLICATIONS OF GRAVIMETRIC METHODS

Gravimetric methods have been developed for most inorganic anions and cations, as well as for such neutral species as water, sulfur dioxide, carbon dioxide, and iodine. A variety of organic substances can also be determined gravimetrically. Examples include lactose in milk products, salicylates in drug preparations, phenolphthalein in laxatives, nicotine in pesticides, cholesterol in cereals, and benzaldehyde in almond extracts. Indeed, gravimetric methods are among the most widely applicable of all analytical procedures.

12C-1 Inorganic Precipitating Agents

Table 12-2 lists common inorganic precipitating agents. These reagents typically form slightly soluble salts or hydrous oxides with the analyte. As you can see from the many entries for each reagent, few inorganic reagents are selective.

TABLE 12-2

Some Inorganic Precipitating Agents

Precipitating Agent	Element Precipitated*
$NH_3(aq)$	**Be** (BeO), **Al** (Al_2O_3), **Sc** (Sc_2O_3), Cr (Cr_2O_3)†, **Fe** (Fe_2O_3), Ga (Ga_2O_3), Zr (ZrO_2), **In** (In_2O_3), Sn (SnO_2), U (U_3O_8)
H_2S	Cu (CuO)†, **Zn** (ZnO or $ZnSO_4$), **Ge** (GeO_2), As (As_2O_3 or As_2O_5), Mo (MoO_3), Sn (SnO_2)†, Sb (Sb_2O_3), or Sb_2O_5), Bi (Bi_2S_3)
$(NH_4)_2S$	Hg (HgS), Co (Co_3O_4)
$(NH_4)_2HPO_4$	**Mg** ($Mg_2P_2O_7$), Al ($AlPO_4$), Mn ($Mn_2P_2O_7$), Zn ($Zn_2P_2O_7$), Zr ($Zr_2P_2O_7$), Cd ($Cd_2P_2O_7$), Bi ($BiPO_4$)
H_2SO_4	Li, Mn, **Sr**, **Cd**, **Pb**, **Ba** (all as sulfates)
H_2PtCl_6	K (K_2PtCl_6 or Pt), Rb (Rb_2PtCl_6), Cs (Cs_2PtCl_6)
$H_2C_2O_4$	Ca (CaO), Sr (SrO), **Th** (ThO_2)
$(NH_4)_2MoO_4$	Cd ($CdMoO_4$)†, Pb ($PbMoO_4$)

HCl	**Ag** (AgCl), Hg (Hg$_2$Cl$_2$), Na (as NaCl from butyl alcohol), Si (SiO$_2$)
AgNO$_3$	**Cl** (AgCl), Br (AgBr), I(AgI)
(NH$_4$)$_2$CO$_3$	**Bi** (Bi$_2$O$_3$)
NH$_4$SCN	Cu [Cu$_2$(SCN)$_2$]
NaHCO$_3$	Ru, Os, Ir (precipitated as hydrous oxides, reduced with H$_2$ to metallic state)
HNO$_3$	Sn (SnO$_2$)
H$_5$IO$_6$	Hg [Hg$_5$(IO$_6$)$_2$]
NaCl, Pb(NO$_3$)$_2$	F (PbClF)
BaCl$_2$	**SO$_4$$^{2-}$** (BaSO$_4$)
MgCl$_2$, NH$_4$Cl	PO$_4$$^{3-}$ (Mg$_2$P$_2$O$_7$)

*Boldface type indicates that gravimetric analysis is the preferred method for the element or ion. The weighed form is indicated in parentheses.

†A dagger indicates that the gravimetric method is seldom used. An underscore indicates the most reliable gravimetric method.

Source: From W. F. Hillebrand, G. E. F. Lundell, H. A. Bright, and J. I. Hoffman, Applied Inorganic Analysis, New York: Wiley, 1953. Reprinted by permission of author Lundell's estate.

12C-2 Reducing Agents

Table 12-3 lists several reagents that convert an analyte to its elemental form for weighing.

12C-3 Organic Precipitating Agents

Numerous organic reagents have been developed for the gravimetric determination of inorganic species. Some of these reagents are significantly more selective in their reactions than are most of the inorganic reagents listed in Table 12-2. There are two types of organic reagents: one forms slightly soluble nonionic products called **coordination compounds**, and the other forms products in which the bonding between the inorganic species and the reagent is largely ionic.

Organic reagents that yield sparingly soluble coordination compounds typically contain at least two functional groups. Each of these groups is capable of bonding with a cation by donating a pair of electrons. The functional groups are located in the molecule such that a five- or six-membered ring results from the reaction. Reagents that form compounds of this type are called **chelating agents**, and their products are called **chelates** (see Chapter 17).

Metal chelates are relatively nonpolar and, as a consequence, have solubilities that are low in water but high in organic liquids. Usually, these compounds possess low densities and are often intensely colored. Because they are not wetted by water, coordination compounds are easily freed of moisture at low temperatures. Two widely used chelating reagents are described in the paragraphs that follow.

8-Hydroxyquinoline (oxine)

Approximately two dozen cations form sparingly soluble chelates with 8-hydroxyquinoline. The structure of magnesium 8-hydroxyquinolate is typical of these chelates.

Chelates are cyclical metal-organic compounds in which the metal is a part of one or more five- or six-membered rings. The chelate pictured here is heme, which is a part of hemoglobin, the oxygen-carrying molecule in human blood. Notice the four six-membered rings that are formed with Fe^{2+}.

TABLE 12-3

Some Reducing Agents Used in Gravimetric Methods

Reducing Agent	Analyte
SO$_2$	Se, Au
SO$_2$ + H$_2$NOH	Te
H$_2$NOH	Se
H$_2$C$_2$O$_4$	Au
H$_2$	Re, Ir
HCOOH	Pt
NaNO$_2$	Au
SnCl$_2$	Hg
Electrolytic reduction	Co, Ni, Cu, Zn Ag, In, Sn, Sb, Cd, Re, Bi

© Cengage Learning.

Magnesium complex with 8-hydroxyquinoline.

Nickel dimethylglyoxime is spectacular in appearance. As shown in color plate 7, it has a beautiful vivid red color.

Sodium tetraphenylborate.

Molecular model for sodium tetraphenylborate.

The solubilities of metal 8-hydroxyquinolates vary widely from cation to cation and are pH dependent because 8-hydroxyquinoline is always deprotonated during a chelation reaction. Therefore, we can achieve a considerable degree of selectivity in the use of 8-hydroxyquinoline by controlling pH.

Dimethylglyoxime

Dimethylglyoxime is an organic precipitating agent of unparalleled specificity. Only nickel(II) is precipitated from a weakly alkaline solution. The reaction is

This precipitate is so bulky that only small amounts of nickel can be handled conveniently. It also has an exasperating tendency to creep up the sides of the container as it is filtered and washed. The solid is conveniently dried at 110°C and has the composition $C_8H_{14}N_4NiO_4$.

Sodium Tetraphenylborate

Sodium tetraphenylborate, $(C_6H_5)_4B^-Na^+$, is an important example of an organic precipitating reagent that forms salt-like precipitates. In cold mineral acid solutions, it is a near-specific precipitating agent for potassium and ammonium ions. The precipitates have stoichiometric composition and contain one mole of potassium or ammonium ion for each mole of tetraphenylborate ion. These ionic compounds are easily filtered and can be brought to constant mass at 105°C to 120°C. Only mercury(II), rubidium, and cesium interfere and must be removed by prior treatment.

12C-4 Organic Functional Group Analysis

Several reagents react selectively with certain organic functional groups and thus can be used for the determination of most compounds containing these groups. A list of gravimetric functional group reagents is given in **Table 12-4**. Many of the reactions shown can also be used for volumetric and spectrophotometric determinations.

TABLE 12-4

Gravimetric Methods for Organic Functional Groups

Functional Group	Basis for Method	Reaction and Product Weighed*
Carbonyl	Mass of precipitate with 2,4-dinitrophenylhydrazine	$RCHO + H_2NNHC_6H_3(NO_2)_2 \rightarrow$ $R{-}CH = NNHC_6H_3(NO_2)_2(s) + H_2O$ (RCOR′ reacts similarly)
Aromatic carbonyl	Mass of CO_2 formed at 230°C in quinoline; CO_2 distilled, absorbed, and weighed	$ArCHO \xrightarrow[CuCO_3]{230°C} Ar + \underline{CO_2(g)}$
Methoxyl and ethoxyl	Mass of AgI formed after distillation and decomposition of CH_3I or C_2H_5I	$\left.\begin{array}{l} ROCH_3 + HI \rightarrow ROH + CH_3I \\ RCOOH_3 + HI \rightarrow RCOOH + CH_3I \\ ROC_2H_5 + HI \rightarrow ROH + C_2H_5I \end{array}\right\}$ $\begin{array}{l}CH_3I + Ag^+ + H_2O \rightarrow \\ \underline{AgI(s)} + CH_3OH\end{array}$
Aromatic nitro	Mass loss of Sn	$RNO_2 + \frac{3}{2}Sn(s) + 6H^+ \rightarrow RNH_2 + \frac{3}{2}Sn^{4+} + 2H_2O$
Azo	Mass loss of Cu	$RN = NR' + 2Cu(s) + 4H^+ \rightarrow RNH_2 + R'NH_2 + 2Cu^{2+}$
Phosphate	Mass of Ba salt	$\underset{\substack{\parallel \\ }}{O}$ $\quad\quad\quad\quad\quad$ $\underset{\substack{\parallel \\ }}{O}$ $ROP(OH)_2 + Ba^{2+} \rightarrow \underline{ROPO_2Ba(s)} + 2H^+$
Sulfamic acid	Mass of $BaSO_4$ after oxidation with HNO_2	$RNHSO_3H + HNO_2 + Ba^{2+} \rightarrow ROH + \underline{BaSO_4(s)} + N_2 + 2H^+$
Sulfinic acid	Mass of Fe_2O_3 after ignition of Fe(III) sulfinate	$3ROSOH + Fe^{3+} \rightarrow (ROSO)_3Fe(s) + 3H^+$ $(ROSO)_3Fe \xrightarrow[O_2]{} CO_2 + H_2O + SO_2 + \underline{Fe_2O_3(s)}$

*The substance weighed is underlined.

12C-5 Volatilization Gravimetry

The two most common gravimetric methods based on volatilization are those for determining water and carbon dioxide. Water is quantitatively distilled from many materials by heating. In direct determination, water vapor is collected on any of several solid desiccants, and its mass is determined from the mass gain of the desiccant. The indirect method in which the amount of water is determined by the loss of mass of the sample during heating is less satisfactory because it must be assumed that water is the only component that is volatilized. This assumption can present problems, however, if any component of the precipitate is volatile. Nevertheless, the indirect method is widely used to determine water in items of commerce. For example, a semiautomated instrument for the determination of moisture in cereal grains can be purchased. It consists of a platform balance on which a 10-g sample is heated with an infrared lamp. The percent moisture is read directly.

> Automatic instruments for the routine determination of water in various products of agriculture and commerce are marketed by several instrument manufacturers.

An example of a gravimetric procedure involving volatilization of carbon dioxide is the determination of the sodium hydrogen carbonate content of antacid tablets. A weighed sample of the finely ground tablets is treated with dilute sulfuric acid to convert the sodium hydrogen carbonate to carbon dioxide:

$$NaHCO_3(aq) + H_2SO_4(aq) \rightarrow CO_2(g) + H_2O(l) + NaHSO_4(aq)$$

As shown in **Figure 12-8**, this reaction is carried out in a flask connected first to a tube containing $CaSO_4$ that removes water vapor from the initial reaction stream to produce a stream of pure CO_2 in nitrogen. These gases then pass through a weighed absorption tube containing the absorbent Ascarite II,[8] which consists of sodium hydroxide absorbed on a nonfibrous silicate. This material retains carbon dioxide by the reaction

$$2NaOH + CO_2 \rightarrow Na_2CO_3 + H_2O$$

---

[8]Thomas Scientific, Swedesboro, NJ.

Nitrogen

Water vapor and CO_2

CO_2

NaOH on nonfibrous
silicate (Ascarite)

$CaSO_4$

$CO_2 + 2NaOH \rightarrow Na_2CO_3 + H_2O$

Reaction mixture containing
bicarbonate and sulfuric acid

Drying tube containing
$CaSO_4$ (Drierite)

Figure 12-8 Apparatus for determining the sodium hydrogen carbonate content of antacid tablets by a gravimetric volatilization procedure.

The absorption tube must also contain a desiccant such as $CaSO_4$ to prevent loss of the water produced by this last reaction.

Sulfides and sulfites can also be determined by volatilization. Hydrogen sulfide or sulfur dioxide evolved from the sample after treatment with acid is collected in a suitable absorbent.

Finally, the classical method for the determination of carbon and hydrogen in organic compounds is a gravimetric volatilization procedure in which the combustion products (H_2O and CO_2) are collected selectively on weighed absorbents. The increase in mass serves as the analytical variable.

WEB WORKS

If you have online access to American Chemical Society journals through your campus information system, locate one of the articles on classical analysis by C. M. Beck.[9] You may locate these articles using their digital object identifiers (DOI) at the DOI website: http:// **www.doi.org/**. Beck makes a strong case for the revival of classical analysis. What is Beck's definition of classical analysis? Why does Beck maintain that classical analysis should be cultivated in this age of automated, computerized instrumentation? What solution does he propose for the problem of dwindling numbers of qualified classical analysts? List three reasons why, in Beck's view, a supply of classical analysts must be maintained.

[9]C. M. Beck, *Anal. Chem.*, **1994,** *66*(4), 224A–239A, **DOI**: 10.1021/ac00076a001; C. M. Beck, *Anal. Chem.,* **1991,** *63*(20), 993A–1003A, **DOI**: 10.1021/ac00020a002; C. M. Beck, *Metrologia,* **1997,** *34*(1), 19–30, **DOI**: 10.1088/0026-1394/34/1/4.

QUESTIONS AND PROBLEMS

12-1. Explain the difference between
    *(a) a colloidal and a crystalline precipitate.
    (b) a gravimetric precipitation method and a gravimetric volatilization method.
    *(c) precipitation and coprecipitation.
    (d) peptization and coagulation of a colloid.
    *(e) occlusion and mixed-crystal formation.
    (f) nucleation and particle growth.

12-2. Define
    *(a) digestion.
    (b) adsorption.
    *(c) reprecipitation.
    (d) precipitation from homogeneous solution.
    *(e) counter-ion layer.
    (f) mother liquor.
    *(g) supersaturation.

12-3. What are the structural characteristics of a chelating agent?

12-4. How can the relative supersaturation be varied during precipitate formation?

12-5. An aqueous solution contains $NaNO_3$ and KBr. The bromide ion is precipitated as $AgBr$ by addition of $AgNO_3$. After an excess of the precipitating reagent has been added,
    (a) what is the charge on the surface of the coagulated colloidal particles?
    (b) what is the source of the charge?
    (c) what ions make up the counter-ion layer?

12-6. Suggest a method by which Ni^{2+} can be precipitated homogeneously as NiS.

12-7. What is peptization and how is it avoided?

12-8. Suggest a precipitation method for separation of K^+ from Na^+ and Li^+.

12-9. Write an equation showing how the mass of the substance sought can be converted to the mass of the weighed substance on the right.

Sought	Weighed	Sought	Weighed
*(a) SO_2	$BaSO_4$	(f) $MnCl_2$	Mn_3O_4
(b) Mg	$Mg_2P_2O_7$	(g) Pb_3O_4	PbO_2
*(c) In	In_2O_3	(h) $U_2P_2O_{11}$	P_2O_5
(d) K	K_2PtCl_6	*(i) $Na_2B_4O_7 \cdot 10H_2O$	B_2O_3
*(e) CuO	$Cu_2(SCN)_2$	(j) Na_2O	†

†$NaZn(UO_2)_3(C_2H_3O_2)_9 \cdot 6H_2O$

12-10. Treatment of a 0.2500-g sample of impure potassium chloride with an excess of $AgNO_3$ resulted in the formation of 0.2912 g of $AgCl$. Calculate the percentage of KCl in the sample.

12-11. The aluminum in a 1.200-g sample of impure ammonium aluminum sulfate was precipitated with aqueous ammonia as the hydrous $Al_2O_3 \cdot xH_2O$. The precipitate was filtered and ignited at 1000°C to give anhydrous Al_2O_3, which weighed 0.2001 g. Express the result of this analysis in terms of
    (a) % $NH_4Al(SO_4)_2$.
    (b) % Al_2O_3.
    (c) % Al.

12-12. What mass of $Cu(IO_3)_2$ can be formed from 0.650 g of $CuSO_4 \cdot 5H_2O$?

12-13. What mass of KIO_3 is needed to convert the copper in 0.2750 g of $CuSO_4 \cdot 5H_2O$ to $Cu(IO_3)_2$?

12-14. What mass of AgI can be produced from a 0.512-g sample that assays 20.1% AlI_3?

12-15. Precipitates used in the gravimetric determination of uranium include $Na_2U_2O_7$ (634.0 g/mol), $(UO_2)_2P_2O_7$ (714.0 g/mol), and $V_2O_5 \cdot 2UO_3$ (753.9 g/mol). Which of these weighing forms provides the greatest mass of precipitate from a given quantity of uranium?

12-16. A 0.8102-g sample of impure $Al_2(CO_3)_3$ decomposed with HCl; the liberated CO_2 was collected on calcium oxide and found to weigh 0.0515 g. Calculate the percentage of aluminum in the sample.

12-17. The hydrogen sulfide in a 80.0-g sample of crude petroleum was removed by distillation and uncollected in a solution of $CdCl_2$. The precipitated CdS was then filtered, washed, and ignited to $CdSO_4$. Calculate the percentage of H_2S in the sample if 0.125 g of $CdSO_4$ was recovered.

12-18. A 0.2121-g sample of an organic compound was burned in a stream of oxygen, and the CO_2 produced was collected in a solution of barium hydroxide. Calculate the percentage of carbon in the sample if 0.6006 g of $BaCO_3$ was formed.

12-19. A 7.000-g sample of a pesticide was decomposed with metallic sodium in alcohol, and the liberated chloride ion was precipitated as $AgCl$. Express the results of this analysis in terms of percent DDT ($C_{14}H_9Cl_5$) based on the recovery of 0.2513 g of $AgCl$.

12-20. The mercury in a 1.0451-g sample was precipitated with an excess of paraperiodic acid, H_5IO_6:

$$5Hg^{2+} + 2H_5IO_6 \rightarrow Hg_5(IO_6)_2 + 10H^+$$

The precipitate was filtered, washed free of precipitating agent, dried, and weighed, and 0.5718 g was recovered. Calculate the percentage of Hg_2Cl_2 in the sample.

12-21. The iodide in a sample that also contained chloride was converted to iodate by treatment with an excess of bromine:

$$3H_2O + 3Br_2 + I^- \rightarrow 6Br^- + IO_3^- + 6H^+$$

The unused bromine was removed by boiling; an excess of barium ion was then added to precipitate the iodate:

$$Ba^{2+} + 2IO_3^- \rightarrow Ba(IO_3)_2$$

In the analysis of a 1.59-g sample, 0.0538 g of barium iodate was recovered. Express the results of this analysis as percent potassium iodide.

*12-22. Ammoniacal nitrogen can be determined by treatment of the sample with chloroplatinic acid; the product is slightly soluble ammonium chloroplatinate:

$$H_2PtCl_6 + 2NH_4^+ \rightarrow (NH_4)_2PtCl_6 + 2H^+$$

The precipitate decomposes on ignition, yielding metallic platinum and gaseous products:

$$(NH_4)_2PtCl_6 \rightarrow Pt(s) + 2Cl_2(g) + 2NH_3(g) + 2HCl(g)$$

Calculate the percentage of ammonia in a sample if 0.2115 g gave rise to 0.4693 g of platinum.

12-23. A 0.6447-g portion of manganese dioxide was added to an acidic solution in which 1.1402 g of a chloride-containing sample was dissolved. Evolution of chlorine took place as a consequence of the following reaction:

$$MnO_2(s) + 2Cl^- + 4H^+ \rightarrow Mn^{2+} + Cl_2(g) + 2H_2O$$

After the reaction was complete, the excess MnO_2 was collected by filtration, washed, and weighed, and 0.3521 g was recovered. Express the results of this analysis in terms of percent aluminum chloride.

*12-24. A series of sulfate samples is to be analyzed by precipitation as $BaSO_4$. If it is known that the sulfate content in these samples ranges between 20% and 55%, what minimum sample mass should be taken to ensure that a precipitate mass no smaller than 0.200 g is produced? What is the maximum precipitate mass to be expected if this quantity of sample is taken?

12-25. The addition of dimethylglyoxime, $H_2C_4H_6O_2N_2$, to a solution containing nickel(II) ion gives rise to a precipitate:

$$Ni^{2+} + 2H_2C_4H_6O_2N_2 \rightarrow 2H^+ + Ni(HC_4H_6O_2N_2)_2$$

Nickel dimethylglyoxime is a bulky precipitate that is inconvenient to manipulate in amounts greater than 175 mg. The amount of nickel in a type of permanent-magnet alloy ranges between 24% and 35%. Calculate the sample size that should not be exceeded when analyzing these alloys for nickel.

*12-26. The efficiency of a particular catalyst is highly dependent on its zirconium content. The starting material for this preparation is received in batches that assay between 68% and 84% $ZrCl_4$. Routine analysis based on precipitation of AgCl is feasible, it having been established that there are no sources of chloride ion other than the $ZrCl_4$ in the sample.
   (a) What sample mass should be taken to ensure an AgCl precipitate that weighs at least 0.400 g?
   (b) If this sample mass is used, what is the maximum mass of AgCl that can be expected in this analysis?
   (c) To simplify calculations, what sample mass should be taken to have the percentage of $ZrCl_4$ exceed the mass of AgCl produced by a factor of 100?

12-27. A 0.8720-g sample of a mixture consisting solely of sodium bromide and potassium bromide yields 1.505 g of silver bromide. What are the percentages of the two salts in the sample?

*12-28. A 0.6407-g sample containing chloride and iodide ions gave a silver halide precipitate weighing 0.4430 g. This precipitate was then strongly heated in a stream of Cl_2 gas to convert the AgI to AgCl; on completion of this treatment, the precipitate weighed 0.3181 g. Calculate the percentage of chloride and iodide in the sample.

12-29. The phosphorus in a 0.2091-g sample was precipitated as the slightly soluble $(NH_4)_3PO_4 \cdot 12MoO_3$. This precipitate was filtered, washed, and then redissolved in acid. Treatment of the resulting solution with an excess of Pb^{2+} resulted in the formation of 0.2922 g of $PbMoO_4$. Express the results of this analysis in terms of percent P_2O_5.

*12-30. What mass in grams of CO_2 is evolved in the complete decomposition of a 2.300-g sample that is 38.0% $MgCO_3$ and 42.0% K_2CO_3 by mass?

12-31. A 6.881-g sample containing magnesium chloride and sodium chloride was dissolved in sufficient water to give 500 mL of solution. Analysis for the chloride content of a 50.0-mL aliquot resulted in the formation of 0.5923 g of AgCl. The magnesium in a second 50.0-mL aliquot was precipitated as $MgNH_4PO_4$; on ignition, 0.1796 g of $Mg_2P_2O_7$ was found. Calculate the percentage of $MgCl_2 \cdot 6H_2O$ and of NaCl in the sample.

*12-32. A 50.0-mL portion of a solution containing 0.200 g of $BaCl_2 \cdot 2H_2O$ is mixed with 50.0 mL of a solution containing 0.300 g of $NaIO_3$. Assume that the solubility of $Ba(IO_3)_2$ in water is negligibly small and calculate
   (a) the mass of the precipitated $Ba(IO_3)_2$.
   (b) the mass of the unreacted compound that remains in solution.

12-33. When a 100.0-mL portion of a solution containing 0.500 g of $AgNO_3$ is mixed with 100.0 mL of a solution containing 0.300 g of K_2CrO_4, a bright red precipitate of Ag_2CrO_4 forms.

    (a) Assuming that the solubility of Ag_2CrO_4 is negligible, calculate the mass of the precipitate.

    (b) Calculate the mass of the unreacted component that remains in solution.

12-34. Challenge Problem: Stones form in the urinary tract when certain chemicals become too concentrated in urine. By far the most common kidney stones are those formed from calcium and oxalate. Magnesium is known to inhibit the formation of kidney stones.

    (a) The solubility of calcium oxalate (CaC_2O_4) in urine is 9×10^{-5} M. What is the solubility product, K_{sp}, of CaC_2O_4 in urine?

    (b) The solubility of magnesium oxalate (MgC_2O_4) in urine is 0.0093 M. What is the solubility product, K_{sp}, of MgC_2O_4 in urine?

    (c) The concentration of calcium in urine is approximately 5 mM. What is the maximum concentration of oxalate that can be tolerated and not precipitate CaC_2O_4?

    (d) The pH of Subject A's urine was 5.9. What fraction of total oxalate, c_T, is present as oxalate ion, $C_2O_4^{2-}$, at pH 5.9? The K_a values for oxalic acid in urine are the same as in water. *Hint:* Find the ratio $[C_2O_4^{2-}]/c_T$ at pH 5.9.

    (e) If the total oxalate concentration in Subject A's urine was 15.0 mM, should a calcium oxalate precipitate form?

    (f) In actuality, Subject A does not show the presence of calcium oxalate crystals in urine. Give a plausible reason for this observation.

    (g) Why would magnesium inhibit the formation of CaC_2O_4 crystals?

    (h) Why are patients with CaC_2O_4 kidney stones often advised to drink large amounts of water?

    (i) The calcium and magnesium in a urine sample were precipitated as oxalates. A mixed precipitate of CaC_2O_4 and MgC_2O_4 resulted and was analyzed by a thermogravimetric procedure. The precipitate mixture was heated to form $CaCO_3$ and MgO. This second mixture weighed 0.0433 g. After ignition to form CaO and MgO, the resulting solid weighed 0.0285 g. What was the mass of Ca in the original sample?

CHAPTER 13

Titrations in Analytical Chemistry

Titrations are widely used in analytical chemistry to determine acids, bases, oxidants, reductants, metal ions, proteins, and many other species. Titrations are based on a reaction between the analyte and a standard reagent known as the titrant. The reaction is of known and reproducible stoichiometry. The volume, or the mass, of the titrant needed to react completely with the analyte is determined and used to calculate the quantity of analyte. A volume-based titration is shown in this figure in which the standard solution is added from a buret, and the reaction occurs in the Erlenmeyer flask. In some titrations, known as coulometric titrations, the quantity of charge required to completely consume the analyte is obtained. In any titration, the point of chemical equivalence, called the end point when determined experimentally, is signaled by an indicator color change or a change in an instrumental response. This chapter introduces the titration principle and the calculations that determine the amount of the unknown. Titration curves, which show the progress of the titration, are also introduced. Such curves will be used in several of the following chapters.

Charles D. Winters

Titration methods are based on determining the quantity of a reagent of known concentration that is required to react completely with the analyte. The reagent may be a standard solution of a chemical or an electric current of known magnitude.

In **volumetric titrations**, the volume of a standard reagent is the measured quantity.

In **coulometric titrations,** the quantity of charge required to complete a reaction with the analyte is the measured quantity.

Titration methods, often called titrimetric methods, include a large and powerful group of quantitative procedures based on measuring the amount of a reagent of known concentration that is consumed by an analyte in a chemical or electrochemical reaction. **Volumetric titrations** involve measuring the volume of a solution of known concentration that is needed to react completely with the analyte. In **Gravimetric titrations,** the mass of the reagent is measured instead of its volume. In **coulometric titrations,** the "reagent" is a constant direct electrical current of known magnitude that consumes the analyte. For this titration, the time required (and thus the total charge) to complete the electrochemical reaction is measured (see Section 22D-5).

This chapter provides introductory material that applies to all the different types of titrations. Chapters 14, 15, and 16 are devoted to the various types of neutralization titrations in which the analyte and titrants undergo acid/base reactions. Chapter 17 provides information about titrations in which the analytical reactions involve complex formation or formation of a precipitate. These methods are particularly important for determining a variety of cations. Finally, Chapters 18 and 19 are devoted to volumetric methods in which the analytical reactions involve electron transfer. These methods are often called **redox titrations.** Some additional titration methods are explored in later chapters. These methods include **amperometric titrations** in Section 23B-4 and **spectrophotometric titrations** in Section 26A-4.

13A SOME TERMS USED IN VOLUMETRIC TITRATIONS[1]

A **standard solution** (or a **standard titrant**) is a reagent of known concentration that is used to carry out a volumetric titration. The **titration** is performed by slowly adding a standard solution from a buret or other liquid-dispensing device to a solution of the analyte until the reaction between the two is judged complete. The volume or mass of reagent needed to complete the titration is determined from the difference between the initial and final readings. A volumetric titration process is depicted in **Figure 13-1**.

It is sometimes necessary to add an excess of the standard titrant and then determine the excess amount by **back-titration** with a second standard titrant. For example, the amount of phosphate in a sample can be determined by adding a measured excess of standard silver nitrate to a solution of the sample, which leads to the formation of insoluble silver phosphate:

$$3Ag^+ + PO_4^{3-} \rightarrow Ag_3PO_4(s)$$

The excess silver nitrate is then back-titrated with a standard solution of potassium thiocyanate:

$$Ag^+ + SCN^- \rightarrow AgSCN(s)$$

The amount of silver nitrate is chemically equivalent to the amount of phosphate ion plus the amount of thiocyanate used for the back-titration. The amount of phosphate is then the difference between the amount of silver nitrate and the amount of thiocyanate.

13A-1 Equivalence Points and End Points

The **equivalence point** in a titration is a theoretical point reached when the amount of added titrant is chemically equivalent to the amount of analyte in the sample. For example, the equivalence point in the titration of sodium chloride with silver nitrate occurs after exactly one mole of silver ion has been added for each mole of chloride ion in the sample. The equivalence point in the titration of sulfuric acid with sodium hydroxide is reached after introducing 2 moles of base for each mole of acid.

We cannot determine the equivalence point of a titration experimentally. Instead, we can only estimate its position by observing some physical change associated with the condition of chemical equivalence. The position of this change is called the **end point** for the titration. We try very hard to ensure that any volume or mass difference between the equivalence point and the end point is small. Such differences do exist, however, as a result of inadequacies in the physical changes and in our ability to observe them. The difference in volume or mass between the equivalence point and the end point is the **titration error**.

Indicators are often added to the analyte solution to produce an observable physical change (signaling the end point) at or near the equivalence point. Large changes in the relative concentration of analyte or titrant occur in the equivalence-point region. These concentration changes cause the indicator to change in appearance. Typical

A **standard solution** is a reagent of known concentration. Standard solutions are used in titrations and in many other chemical analyses.

Back-titration is a process in which the excess of a standard solution used to consume an analyte is determined by titration with a second standard solution. Back-titrations are often required when the rate of reaction between the analyte and reagent is slow or when the standard solution lacks stability.

The **equivalence point** is the point in a titration when the amount of added standard reagent is equivalent to the amount of analyte.

The **end point** is the point in a titration when a physical change occurs that is associated with the condition of chemical equivalence.

In volumetric methods, the **titration error**, E_t, is given by

$$E_t = V_{ep} - V_{eq}$$

where V_{ep} is the actual volume of reagent required to reach the end point and V_{eq} is the theoretical volume necessary to reach the equivalence point.

[1]For a detailed discussion of volumetric methods, see J. I. Watters, in *Treatise on Analytical Chemistry*, I. M. Kolthoff and P. J. Elving, Eds., Part I, Vol. 11, Chap. 114. New York: Wiley, 1975.

Typical setup for carrying out a titration. The apparatus consists of a buret, a buret stand and clamp with a white porcelain base to provide an appropriate background for viewing indicator changes, and a wide-mouth Erlenmeyer flask containing a precisely known volume of the solution to be titrated. The solution is normally delivered into the flask using a pipet, as shown in Figure 2-22.

Detail of the buret graduations. Normally, the buret is filled with titrant solution to within 1 or 2 mL of the zero position at the top. The initial volume of the buret is read to the nearest 0.01 mL. The reference point on the meniscus and the proper position of the eye for reading are depicted in Figure 2-21.

Before the titration begins. The solution to be titrated, an acid in this example, is placed in the flask, and the indicator is added as shown in the photo. The indicator in this case is phenolphthalein, which turns pink in basic solution.

During titration. The titrant is added to the flask with swirling until the color of the indicator persists. In the initial region of the titration, titrant may be added rather rapidly, but as the end point is approached, increasingly smaller portions are added; at the end point, less than half a drop of titrant should cause the indicator to change color.

Titration end point. The end point is achieved when the barely perceptible pink color of phenolphthalein persists. The flask on the left shows the titration less than half a drop prior to the end point; the middle flask shows the end point. The final reading of the buret is made at this point, and the volume of base delivered in the titration is calculated from the difference between the initial and final buret readings. The flask on the right shows what happens when a slight excess of base is added to the titration mixture. The solution turns a deep pink color, and the end point has been exceeded. In color plate 9, the color change at the end point is much easier to see than in this black-and-white version.

Figure 13-1   The titration process.

indicator changes include the appearance or disappearance of a color, a change in color, or the appearance or disappearance of turbidity. As an example, the indicator used in the neutralization titration of hydrochloric acid with sodium hydroxide is phenolphthalein, which causes the solution to change from colorless to a pink color once excess sodium hydroxide has been added.

We often use instruments to detect end points. These instruments respond to properties of the solution that change in a characteristic way during the titration. Among such instruments are colorimeters, turbidimeters, spectrophotometers, temperature monitors, refractometers, voltmeters, current meters, and conductivity meters.

13A-2 Primary Standards

A **primary standard** is a highly purified compound that serves as a reference material in titrations and in other analytical methods. The accuracy of a method critically depends on the properties of the primary standard. Important requirements for a primary standard are the following:

1. High purity. Established methods for confirming purity should be available.
2. Atmospheric stability.
3. Absence of hydrate water so that the composition of the solid does not change with variations in humidity.
4. Modest cost.
5. Reasonable solubility in the titration medium.
6. Reasonably large molar mass so that the relative error associated with weighing the standard is minimized.

Very few compounds meet or even approach these criteria, and only a limited number of primary-standard substances are available commercially. As a consequence, less pure compounds must sometimes be used in place of a primary standard. The purity of such a **secondary standard** must be established by careful analysis.

> A **primary standard** is an ultrapure compound that serves as the reference material for a titration or for another type of quantitative analysis.

> A **secondary standard** is a compound whose purity has been determined by chemical analysis. The secondary standard serves as the working standard material for titrations and for many other analyses.

13B STANDARD SOLUTIONS

Standard solutions play a central role in all titrations. Therefore, we must consider the desirable properties for such solutions, how they are prepared, and how their concentrations are expressed. The *ideal* standard solution for a titrimetric method will

1. be sufficiently stable so that it is necessary to determine its concentration only once;
2. react rapidly with the analyte so that the time required between additions of reagent is minimized;
3. react more or less completely with the analyte so that satisfactory end points are realized;
4. undergo a selective reaction with the analyte that can be described by a balanced equation.

Few reagents completely meet these ideals.

The accuracy of a titration can be no better than the accuracy of the concentration of the standard solution used. Two basic methods are used to establish the concentration of such solutions. The first is the **direct method** in which a carefully determined mass of a primary standard is dissolved in a suitable solvent and diluted to a known volume in a volumetric flask. The second is by **standardization** in which the titrant to be standardized is used to titrate (1) a known mass of a primary standard, (2) a known mass of a secondary standard, or (3) a measured volume of another standard solution. A titrant that is standardized is sometimes referred to as a **secondary-standard solution**. The concentration of a secondary-standard solution is subject to a larger uncertainty than is the concentration of a primary-standard solution. If there is a choice, then, solutions are best prepared by the direct method. Many reagents, however, lack the properties required for a primary standard and, therefore, require standardization.

In a **standardization,** the concentration of a volumetric solution is determined by titrating it against a carefully measured quantity of a primary or secondary standard or an exactly known volume of another standard solution.

13C VOLUMETRIC CALCULATIONS

As we indicated in Section 4B-1, we can express the concentration of solutions in several ways. For the standard solutions used in most titrations, either **molar concentration**, c, or **normal concentration**, c_N, is usually used. Molar concentration is the number of moles of reagent contained in one liter of solution, and normal concentration is the number of **equivalents** of reagent in the same volume.

Throughout this text, we base volumetric calculations exclusively on molar concentration and molar masses. We have also included in Appendix 7 a discussion of how volumetric calculations are performed based on normal concentration and equivalent masses because you may encounter these terms and their uses in the industrial and health science literature.

13C-1 Some Useful Relationships

$$n_A = \frac{m_A}{\mathscr{M}_A}$$

where n_A is the amount of A, m_A is the mass of A, and $\mathscr{M}_A$ is the molar mass of A.

Most volumetric calculations are based on two pairs of simple equations that are derived from definitions of the mole, the millimole, and the molar concentration. For the chemical species A, we can write

$$\text{amount A (mol)} = \frac{\text{mass A (g)}}{\text{molar mass A (g/mol)}} \tag{13-1}$$

$$\text{amount A (mmol)} = \frac{\text{mass A (g)}}{\text{millimolar mass A (g/mmol)}} \tag{13-2}$$

$$c_A = \frac{n_A}{V} \quad \text{or} \quad n_A = V \times c_A$$

The second pair of equations is derived from the definition of molar concentration, that is,

Any combination of grams, moles, and liters can be expressed in milligrams, millimoles, and milliliters. For example, a 0.1 M solution contains 0.1 mol of a species per liter or 0.1 mmol per milliliter. Similarly, the number of moles of a compound is equal to the mass in grams of that compound divided by its molar mass in grams or the mass in milligrams divided by its millimolar mass in milligrams.

$$\text{amount A (mol)} = V(L) \times c_A\left(\frac{\text{mol A}}{L}\right) \tag{13-3}$$

$$\text{amount A (mmol)} = V(mL) \times c_A\left(\frac{\text{mmol A}}{L}\right) \tag{13-4}$$

where V is the volume of the solution.

Equations 13-1 and 13-3 are used when volumes are measured in liters, and Equations 13-2 and Equations 13-4 when the units are milliliters.

13C-2 Calculating the Molar Concentration of Standard Solutions

The following three examples illustrate how the concentrations of volumetric reagents are calculated.

EXAMPLE 13-1

Describe the preparation of 2.000 L of 0.0500 M $AgNO_3$ (169.87 g/mol) from the primary-standard-grade solid.

Solution

$$\text{amount } AgNO_3 = V_{soln}(L) \times c_{AgNO_3}(\text{mol/L})$$

$$= 2.00 \text{ L} \times \frac{0.0500 \text{ mol } AgNO_3}{\text{L}} = 0.100 \text{ mol } AgNO_3$$

To obtain the mass of $AgNO_3$, we rearrange Equation 13-2 to give

$$\text{mass } AgNO_3 = 0.1000 \text{ mol } AgNO_3 \times \frac{169.87 \text{ g } AgNO_3}{\text{mol } AgNO_3}$$

$$= 16.987 \text{ g } AgNO_3$$

Therefore, the solution should be prepared by dissolving 16.987 g of $AgNO_3$ in water and diluting to the mark in a 2.000 L volumetric flask.

EXAMPLE 13-2

A standard 0.0100 M solution of Na^+ is required to calibrate an ion-selective electrode method to determine sodium. Describe how 500 mL of this solution can be prepared from primary standard Na_2CO_3 (105.99 g/mL).

Solution

We wish to compute the mass of reagent required to produce a species concentration of 0.0100 M. In this instance, we will use millimoles since the volume is in milliliters. Because Na_2CO_3 dissociates to give two Na^+ ions, we can write that the number of millimoles of Na_2CO_3 needed is

$$\text{amount } Na_2CO_3 = 500 \text{ mL} \times \frac{0.0100 \text{ mmol } Na^+}{\text{mL}} \times \frac{1 \text{ mmol } Na_2CO_3}{2 \text{ mmol } Na^+}$$

$$= 2.50 \text{ mmol}$$

From the definition of millimole, we write

$$\text{mass } Na_2CO_3 = 2.50 \text{ mmol } Na_2CO_3 \times 105.99 \frac{\text{mg } Na_2CO_3}{\text{mmol } Na_2CO_3}$$

$$= 264.975 \text{ mg } Na_2CO_3$$

Since there are 1000 mg/g, or 0.001 g/mg, the solution should be prepared by dissolving 0.265 g of Na_2CO_3 in water and diluting to 500 mL.

EXAMPLE 13-3

How would you prepare 50.0-mL portions of standard solutions that are 0.00500 M, 0.00200 M, and 0.00100 M in Na^+ from the solution in Example 13-2?

Solution

The number of millimoles of Na^+ taken from the concentrated solution must equal the number in the dilute solutions. Thus,

$$\text{amount } Na^+ \text{ from concd soln} = \text{amount } Na^+ \text{ in dil soln}$$

Recall that the number of millimoles is equal to the number of millimoles per milliliter times the number of milliliters, that is,

$$V_{concd} \times c_{concd} = V_{dil} \times c_{dil}$$

where V_{concd} and V_{dil} are the volumes in milliliters of the concentrated and diluted solutions, respectively, and c_{concd} and c_{dil} are their molar Na^+ concentrations. For the 0.00500-M solution, this equation can be rearranged to

$$V_{concd} = \frac{V_{dil} \times c_{dil}}{c_{concd}} = \frac{50.0 \text{ mL} \times 0.005 \text{ mmol } Na^+/mL}{0.0100 \text{ mmol } Na^+/mL} = 25.0 \text{ mL}$$

Therefore, to produce 50.0 mL of 0.00500 M Na^+, 25.0 mL of the concentrated solution should be diluted to exactly 50.0 mL.

Repeat the calculation for the other two molarities to confirm that diluting 10.0 and 5.00 mL of the concentrated solution to 50.0 mL produces the desired concentrations.

13C-3 Working with Titration Data

Two types of volumetric calculations are discussed here. In the first, we compute concentrations of solutions that have been standardized against either a primary-standard or another standard solution. In the second, we calculate the amount of analyte in a sample from titration data. Both types of calculation are based on three algebraic relationships. Two of these are Equations 13-2 and 13-4, which are based on millimoles and milliliters. The third relationship is the stoichiometric ratio of the number of millimoles of the analyte to the number of millimoles of titrant.

Calculating Molar Concentrations from Standardization Data

Examples 13-4 and 13-5 illustrate how standardization data are treated.

EXAMPLE 13-4

A 50.00-mL portion of an HCl solution required 29.71 mL of 0.01963 M $Ba(OH)_2$ to reach an end point with bromocresol green indicator. Calculate the molar concentration of the HCl.

Solution

In the titration, 1 mmol of $Ba(OH)_2$ reacts with 2 mmol of HCl:

$$Ba(OH)_2 + 2HCl \rightarrow BaCl_2 + 2H_2O$$

Thus, the stoichiometric ratio is

$$\text{stoichiometric ratio} = \frac{2 \text{ mmol HCl}}{1 \text{ mmol Ba(OH)}_2}$$

The number of millimoles of the standard is calculated by substituting into Equation 13-4:

$$\text{amount Ba(OH)}_2 = 29.71 \text{ mL Ba(OH)}_2 \times 0.01963 \frac{\text{mmol Ba(OH)}_2}{\text{mL Ba(OH)}_2}$$

To find the number of millimoles of HCl, we multiply this result by the stoichiometric ratio determined from the titration reaction:

$$\text{amount HCl} = (29.71 \times 0.01963) \text{ mmol Ba(OH)}_2 \times \frac{2 \text{ mmol HCl}}{1 \text{ mmol Ba(OH)}_2}$$

To obtain the number of millimoles of HCl per mL, we divide by the volume of the acid. Therefore,

$$c_{\text{HCl}} = \frac{(29.71 \times 0.01963 \times 2) \text{ mmol HCl}}{50.0 \text{ mL HCl}}$$

$$= 0.023328 \frac{\text{mmol HCl}}{\text{mL HCl}} = 0.02333 \text{ M}$$

> In determining the number of significant figures to retain in volumetric calculations, the stoichiometric ratio is assumed to be known exactly without uncertainty.

EXAMPLE 13-5

Titration of 0.2121 g of pure $Na_2C_2O_4$ (134.00 g/mol) required 43.31 mL of $KMnO_4$. What is the molar concentration of the $KMnO_4$ solution? The chemical reaction is

$$2MnO_4^- + 5C_2O_4^{2-} + 16H^+ \rightarrow 2Mn^{2+} + 10CO_2 + 8H_2O$$

Solution

From this equation we see that

$$\text{stoichiometric ratio} = \frac{2 \text{ mmol KMnO}_4}{5 \text{ mmol Na}_2\text{C}_2\text{O}_4}$$

The amount of primary-standard $Na_2C_2O_4$ is given by Equation 13-2

$$\text{amount Na}_2\text{C}_2\text{O}_4 = 0.2121 \text{ g Na}_2\text{C}_2\text{O}_4 \times \frac{1 \text{ mmol Na}_2\text{C}_2\text{O}_4}{0.13400 \text{ g Na}_2\text{C}_2\text{O}_4}$$

(continued)

To obtain the number of millimoles of $KMnO_4$, we multiply this result by the stoichiometric ratio:

$$\text{amount } KMnO_4 = \frac{0.2121}{0.1340} \text{ mmol } \cancel{Na_2C_2O_4} \times \frac{2 \text{ mmol } KMnO_4}{5 \text{ mmol } \cancel{Na_2C_2O_4}}$$

The concentration of $KMnO_4$ is then obtained by dividing by the volume consumed. Thus,

$$c_{KMnO_4} = \frac{\left(\dfrac{0.2121}{0.13400} \times \dfrac{2}{5}\right) \text{ mmol } KMnO_4}{43.31 \text{ mL } KMnO_4} = 0.01462 \text{ M}$$

Note that units are carried through all calculations as a check on the correctness of the relationships used in Examples 13-4 and 13-5.

Calculating the Quantity of Analyte from Titration Data

As shown by the examples that follow, the systematic approach just described is also used to compute analyte concentrations from titration data.

EXAMPLE 13-6

A 0.8040-g sample of an iron ore is dissolved in acid. The iron is then reduced to Fe^{2+} and titrated with 47.22 mL of 0.02242 M $KMnO_4$ solution. Calculate the results of this analysis in terms of (a) % Fe (55.847 g/mol) and (b) % Fe_3O_4 (231.54 g/mol).

Solution

The reaction of the analyte with the reagent is described by the equation

$$MnO_4^- + 5Fe^{2+} + 8H^+ \rightarrow Mn^{2+} + 5Fe^{3+} + 4H_2O$$

(a) $$\text{stoichiometric ratio} = \frac{5 \text{ mmol } Fe^{2+}}{1 \text{ mmol } KMnO_4}$$

$$\text{amount } KMnO_4 = 47.22 \text{ mL } \cancel{KMnO_4} \times \frac{0.02242 \text{ mmol } KMnO_4}{\text{mL } \cancel{KMnO_4}}$$

$$\text{amount } Fe^{2+} = (47.22 \times 0.02242) \text{ mmol } \cancel{KMnO_4} \times \frac{5 \text{ mmol } Fe^{2+}}{1 \text{ mmol } \cancel{KMnO_4}}$$

The mass of Fe^{2+} is then given by

$$\text{mass } Fe^{2+} = (47.22 \times 0.02242 \times 5) \text{ mmol } \cancel{Fe^{2+}} \times 0.055847 \frac{\text{g } Fe^{2+}}{\text{mmol } \cancel{Fe^{2+}}}$$

The percent Fe^{2+} is

$$\% \ Fe^{2+} = \frac{(47.22 \times 0.02242 \times 5 \times 0.055847) \ g \ Fe^{2+}}{0.8040 \ g \ sample} \times 100\% = 36.77\%$$

(b) To determine the correct stoichiometric ratio, we note that

$$5 \ Fe^{2+} \equiv 1 \ MnO_4^-$$

Therefore,

$$5 \ Fe_3O_4 \equiv 15 \ Fe^{2+} \equiv 3 \ MnO_4^-$$

and

$$\text{stoichiometric ratio} = \frac{5 \ \text{mmol} \ Fe_3O_4}{3 \ \text{mmol} \ KMnO_4}$$

As in part (a),

$$\text{amount} \ KMnO_4 = \frac{47.22 \ \text{mL} \ \cancel{KMnO_4^-} \times 0.02242 \ \text{mmol} \ KMnO_4}{\cancel{\text{mL} \ KMnO_4^-}}$$

$$\text{amount} \ Fe_3O_4 = (47.22 \times 0.02242) \ \cancel{\text{mmol} \ KMnO_4^-} \times \frac{5 \ \text{mmol} \ Fe_3O_4}{3 \ \cancel{\text{mmol} \ KMnO_4^-}}$$

$$\text{mass} \ Fe_3O_4 = \left(47.22 \times 0.02242 \times \frac{5}{3}\right) \cancel{\text{mmol} \ Fe_3O_4^-} \times 0.23154 \ \frac{g \ Fe_3O_4}{\cancel{\text{mmol} \ Fe_3O_4^-}}$$

$$\% \ Fe_3O_4 = \frac{\left(47.22 \times 0.02242 \times \dfrac{5}{3}\right) \times 0.23154 \ g \ Fe_3O_4}{0.8040 \ g \ sample} \times 100\% = 50.81\%$$

FEATURE 13-1

Another Approach to Example 13-6(a)

Some people find it easier to write out the solution to a problem in such a way that the units in the denominator of each succeeding term eliminate the units in the numerator of the preceding one until the units of the answer are obtained.[2] For example, the solution to part (a) of Example 13-6 can be written

$$47.22 \ \text{mL} \ \cancel{KMnO_4^-} \times \frac{0.02242 \ \cancel{\text{mmol} \ KMnO_4^-}}{\cancel{\text{mL} \ KMnO_4^-}} \times \frac{5 \ \cancel{\text{mmol} \ Fe}}{1 \ \cancel{\text{mmol} \ KMnO_4^-}} \times \frac{0.055847 \ g \ Fe}{\cancel{\text{mmol} \ Fe}}$$

$$\times \frac{1}{0.8040 \ g \ sample} \times 100\% = 36.77\% \ Fe$$

---

[2]This process is often referred to as the factor-label method. It is sometimes erroneously called dimensional analysis. For an explanation of dimensional analysis, perform a search at the Wikipedia website. In earlier texts, the factor-label method was sometimes called the "picket fence" method.

EXAMPLE 13-7

A 100.0-mL sample of brackish water was made ammoniacal, and the sulfide it contained was titrated with 16.47 mL of 0.02310 M $AgNO_3$. The analytical reaction is

$$2Ag^+ + S^{2-} \rightarrow Ag_2S(s)$$

Calculate the concentration of H_2S in the water in parts per million, c_{ppm}.

Solution

At the end point,

$$\text{stoichiometric ratio} = \frac{1 \text{ mmol } H_2S}{2 \text{ mmol } AgNO_3}$$

$$\text{amount } AgNO_3 = 16.47 \text{ mL } \cancel{AgNO_3} \times 0.02310 \frac{\text{mmol } AgNO_3}{\text{mL } \cancel{AgNO_3}}$$

$$\text{amount } H_2S = (16.47 \times 0.02310) \cancel{\text{mmol } AgNO_3} \times \frac{1 \text{ mmol } H_2S}{2 \cancel{\text{mmol } AgNO_3}}$$

$$\text{mass } H_2S = \left(16.47 \times 0.02310 \times \frac{1}{2} \right) \cancel{\text{mmol } H_2S} \times 0.034081 \frac{\text{g } H_2S}{\cancel{\text{mmol } H_2S}}$$

$$= 6.483 \times 10^{-3} \text{ g } H_2S$$

$$c_{ppm} = \frac{6.483 \times 10^{-3} \text{ g } H_2S}{100.0 \text{ mL } \cancel{\text{sample}} \times 1.00 \text{ g sample/mL } \cancel{\text{sample}}} \times 10^6 \text{ ppm}$$

$$= 64.8 \text{ ppm}$$

FEATURE 13-2

Rounding the Answer to Example 13-7

Note that the input data for Example 13-7 all contained four or more significant figures, but the answer was rounded to three. Why?

We can make the rounding decision by doing a couple of rough calculations in our heads. Assume that the input data are uncertain to 1 part in the last significant figure. The largest *relative* error will then be associated with the sample size. In Example 13-7, the relative uncertainty is 0.1/100.0. Thus, the uncertainty is about 1 part in 1000 (compared with about 1 part in 1647 for the volume of $AgNO_3$ and 1 part in 2300 for the reagent concentration). We then assume that the calculated result is uncertain to about the same amount as the least precise measurement, or 1 part in 1000. The absolute uncertainty of the final result is then 64.8 ppm $\times$ 1/1000 = 0.065, or about 0.1 ppm, and we round to the first figure to the right of the decimal point. Thus, we report 64.8 ppm.

Practice making this rough type of rounding decision whenever you make a computation.

EXAMPLE 13-8

The phosphorus in a 4.258-g sample of a plant food was converted to PO_4^{3-} and precipitated as Ag_3PO_4 by adding 50.00 mL of 0.0820 M $AgNO_3$. The excess $AgNO_3$ was back-titrated with 4.06 mL of 0.0625 M KSCN. Express the results of this analysis in terms of % P_2O_5.

Solution

The chemical reactions are

$$P_2O_5 + 9H_2O \rightarrow 2PO_4^{3-} + 6H_3O^+$$

$$2PO_4^{3-} + \underset{\text{excess}}{6\ Ag^+} \rightarrow 2Ag_3PO_4(s)$$

$$Ag^+ + SCN^- \rightarrow AgSCN(s)$$

The stoichiometric ratios are

$$\frac{1\ \text{mmol}\ P_2O_5}{6\ \text{mmol}\ AgNO_3} \quad \text{and} \quad \frac{1\ \text{mmol}\ KSCN}{1\ \text{mmol}\ AgNO_3}$$

$$\text{total amount}\ AgNO_3 = 50.00\ \cancel{mL} \times 0.0820\ \frac{\text{mmol}\ AgNO_3}{\cancel{mL}} = 4.100\ \text{mmol}$$

$$\text{amount}\ AgNO_3\ \text{consumed by KSCN} = 4.06\ \cancel{mL} \times 0.0625\ \frac{\text{mmol}\ \cancel{KSCN}}{\cancel{mL}}$$

$$\times \frac{1\ \text{mmol}\ AgNO_3}{\text{mmol}\ \cancel{KSCN}}$$

$$= 0.2538\ \text{mmol}$$

$$\text{amount}\ P_2O_5 = (4.100 - 0.254)\ \text{mmol}\ \cancel{AgNO_3} \times \frac{1\ \text{mmol}\ P_2O_5}{6\ \text{mmol}\ \cancel{AgNO_3}}$$

$$= 0.6410\ \text{mmol}\ P_2O_5$$

$$\%\ P_2O_5 = \frac{0.6410\ \cancel{mmol} \times \dfrac{0.1419\ \text{g}\ P_2O_5}{\cancel{mmol}}}{4.258\ \text{g sample}} \times 100\% = 2.14\%$$

EXAMPLE 13-9

The CO in a 20.3-L sample of gas was converted to CO_2 by passing the sample over iodine pentoxide heated to 150°C:

$$I_2O_5(s) + 5CO(g) \rightarrow 5CO_2(g) + I_2(g)$$

The iodine was distilled at this temperature and was collected in an absorber containing 8.25 mL of 0.01101 M $Na_2S_2O_3$.

$$I_2(g) + 2S_2O_3^{2-}(aq) \rightarrow 2I^-(aq) + S_4O_6^{2-}(aq)$$

The excess $Na_2S_2O_3$ was back-titrated with 2.16 mL of 0.00947 M I_2 solution. Calculate the concentration of CO (28.01 g/mol) in mg per liter of sample.

(*continued*)

Solution

Based on the two reactions, the stoichiometric ratios are

$$\frac{5 \text{ mmol CO}}{1 \text{ mmol I}_2} \quad \text{and} \quad \frac{2 \text{ mmol Na}_2S_2O_3}{1 \text{ mmol I}_2}$$

We divide the first ratio by the second to get a third useful ratio

$$\frac{5 \text{ mmol CO}}{2 \text{ mmol Na}_2S_2O_3}$$

This relationship reveals that 5 mmol of CO are responsible for the consumption of 2 mmol of $Na_2S_2O_3$. The total amount of $Na_2S_2O_3$ is

$$\text{amount Na}_2S_2O_3 = 8.25 \text{ mL Na}_2S_2O_3 \times 0.01101 \frac{\text{mmol Na}_2S_2O_3}{\text{mL Na}_2S_2O_3}$$

$$= 0.09083 \text{ mmol Na}_2S_2O_3$$

The amount of $Na_2S_2O_3$ consumed in the back-titration is

$$\text{amount Na}_2S_2O_3 = 2.16 \text{ mL I}_2 \times 0.00947 \frac{\text{mmol I}_2}{\text{mL I}_2} \times \frac{2 \text{ mmol Na}_2S_2O_3}{\text{mmol I}_2}$$

$$= 0.04091 \text{ mmol Na}_2S_2O_3$$

The number of millimoles of CO can then be calculated by using the third stoichiometric ratio:

$$\text{amount CO} = (0.09083 - 0.04091) \text{ mmol Na}_2S_2O_3 \times \frac{5 \text{ mmol CO}}{2 \text{ mmol Na}_2S_2O_3}$$

$$= 0.1248 \text{ mmol CO}$$

$$\text{mass CO} = 0.1248 \text{ mmol CO} \times \frac{28.01 \text{ mg CO}}{\text{mmol CO}} = 3.4956 \text{ mg}$$

$$\frac{\text{mass CO}}{\text{vol sample}} = \frac{3.4956 \text{ mg CO}}{20.3 \text{ L sample}} = 0.172 \frac{\text{mg CO}}{\text{L sample}}$$

13D GRAVIMETRIC TITRATIONS

Mass (weight) or **gravimetric titrations** differ from their volumetric counterparts in that the mass of titrant is measured rather than the volume. Therefore, in a mass titration, a balance and a weighable solution dispenser are substituted for a buret and its markings. Gravimetric titrations actually predate volumetric titrations by more than 50 years. With the advent of reliable burets, however, mass titrations were largely supplanted by volumetric methods because the former required relatively elaborate equipment and were tedious and time consuming. The availability of sensitive, low-cost, top-loading digital analytical balances and convenient plastic solution dispensers has changed this situation completely, and mass titrations can now be performed as easily and rapidly as volumetric titrations.

Remember that for historical reasons we often refer to *weight* or *weighing*, but we really mean mass, although most of us cannot bring ourselves to say *massing*.

13D-1 Calculations Associated with Mass Titrations

The most common way to express concentration for mass titrations is the **weight concentration**, c_w, in weight molar concentration units, M_w, which is the number of moles of a reagent in one kilogram of solution or the number of millimoles in one gram of solution. Thus, aqueous 0.1 M_w NaCl contains 0.1 mol of the salt in 1 kg of solution or 0.1 mmol in 1 g of the solution.

The weight molar concentration $c_w(A)$ of a solution of a solute A is computed using either one of two equations that are analogous to Equation 4-2:

$$\text{weight molar concentration} = \frac{\text{no. mol A}}{\text{no. kg solution}} = \frac{\text{no. mmol A}}{\text{no. g solution}} \quad (13\text{-}5)$$

$$c_w(A) = \frac{n_A}{m_{\text{soln}}}$$

where n_A is the number of moles of species A and m_{soln} is the mass of the solution. Gravimetric titration data can then be treated by using the methods illustrated in Sections 13C-2 and 13C-3 after substitution of weight concentration for molar concentration and grams and kilograms for milliliters and liters.

13D-2 Advantages of Gravimetric Titrations

In addition to greater speed and convenience, mass titrations offer certain other advantages over their volumetric counterparts:

1. Calibration of glassware and tedious cleaning to ensure proper drainage are completely eliminated.
2. Temperature corrections are unnecessary because the mass (weight) molar concentration does not change with temperature, in contrast to the volume molar concentration. This advantage is particularly important in nonaqueous titrations because of the high coefficients of expansion of most organic liquids (about 10 times that of water).
3. Mass measurements can be made with considerably greater precision and accuracy than can volume measurements. For example, 50 g or 100 g of an aqueous solution can be readily measured to ± 1 mg, which corresponds to ± 0.001 mL. This greater sensitivity makes it possible to choose sample sizes that lead to significantly smaller consumption of standard reagents.
4. Gravimetric titrations are more easily automated than are volumetric titrations.

13E TITRATION CURVES

As noted in Section 13A-1, an end point is signaled by an observable physical change near the equivalence point of a titration. The two most widely used signals involve (1) changes in color due to the reagent (titrant), the analyte, or an indicator and (2) a change in potential of an electrode that responds to the titrant concentration or the analyte concentration.

To understand the theoretical basis of end point determinations and the sources of titration errors, we calculate the data points necessary to construct **titration curves** for the systems under consideration. A titration curve is a plot of some function of the analyte or titrant concentration on the y axis versus titrant volume on the x axis.

Titration curves are plots of a concentration-related variable versus titrant volume.

13E-1 Types of Titration Curves

> The vertical axis in a sigmoidal titration curve is either the p-function of the analyte or titrant or the potential of an analyte- or titrant-sensitive electrode.

> The vertical axis of a linear-segment titration curve is an instrument signal that is proportional to the concentration of the analyte or titrant.

Two general types of titration curves (and thus two general types of end points) occur in titrimetric methods. In the first type, called a sigmoidal curve, important observations are confined to a small region (typically ±0.1 to ±0.5 mL) surrounding the equivalence point. A **sigmoidal curve** in which the p-function of analyte (or sometimes the titrant) is plotted as a function of titrant volume is shown in **Figure 13-2a**.

In the second type of curve, called a **linear segment curve**, measurements are made on both sides of, but well away from, the equivalence point. Measurements near equivalence are avoided. In this type of curve, the vertical axis represents an instrument reading that is directly proportional to the concentration of the analyte or the titrant. A typical linear segment curve is found in **Figure 13-2b**.

The sigmoidal type offers the advantages of speed and convenience. The linear segment type is advantageous for reactions that are complete only in the presence of a considerable excess of the reagent or analyte.

In this chapter and several that follow, we deal exclusively with sigmoidal titration curves. We explore linear segment curves in Chapters 23 and 26.

(a) Sigmoidal curve

(b) Linear segment curve

Figure 13-2 Two types of titration curves.

13E-2 Concentration Changes During Titrations

The equivalence point in a titration is characterized by major changes in the *relative* concentrations of reagent and analyte. **Table 13-1** illustrates this phenomenon. The data in the second column of the table show the changes in the hydronium ion concentration as a 50.00-mL aliquot of a 0.1000 M solution of hydrochloric acid is titrated with 0.1000 M sodium hydroxide. The neutralization reaction is described by the equation

$$H_3O^+ + OH^- \rightarrow 2H_2O \tag{13-6}$$

To emphasize the changes in *relative* concentration that occur in the equivalence point region, the volume increments computed are those required to cause tenfold decreases in the concentration of H_3O^+ (or tenfold increases in hydroxide ion concentration). Thus, we see in the third column that an addition of 40.91 mL of base is needed to decrease the concentration of H_3O^+ by one order of magnitude from 0.100 M to 0.0100 M. An addition of only 8.11 mL is required to lower the concentration by another factor of 10 to 0.00100 M; 0.89 mL causes yet another tenfold decrease. Corresponding increases in OH^- concentration occur at the same time. End-point detection then depends on this large change in the *relative* concentration of the analyte (or titrant) that occurs at the equivalence point for every type of titration. Feature 13-3 describes how the volumes in the first column of Table 13-1 are calculated.

The large changes in relative concentration that occur in the region of chemical equivalence are shown by plotting the negative logarithm of the analyte or the titrant concentration (the p-function) against reagent volume, as seen in **Figure 13-3**. The data for these plots are found in the fourth and fifth columns of Table 13-1. Titration curves for reactions involving complex formation, precipitation, and oxidation/reduction all exhibit the same sharp increase or decrease in p-function in the equivalence-point region as those shown in Figure 13-3. Titration curves define the properties required of an indicator or instrument and allow us to estimate the error associated with titration methods.

TABLE 13-1

Concentration Changes During a Titration of 50.00 mL of 0.1000 M HCl

Volume of 0.1000 M NaOH, mL	$[H_3O^+]$, mol/L	Volume of 0.1000 M NaOH to Cause a Tenfold Decrease in $[H_3O^+]$, mL	pH	pOH
0.00	0.1000		1.00	13.00
40.91	0.0100	40.91	2.00	12.00
49.01	1.000×10^{-3}	8.11	3.00	11.00
49.90	1.000×10^{-4}	0.89	4.00	10.00
49.99	1.000×10^{-5}	0.09	5.00	9.00
49.999	1.000×10^{-6}	0.009	6.00	8.00
50.00	1.000×10^{-7}	0.001	7.00	7.00
50.001	1.000×10^{-8}	0.001	8.00	6.00
50.01	1.000×10^{-9}	0.009	9.00	5.00
50.10	1.000×10^{-10}	0.09	10.00	4.00
51.10	1.000×10^{-11}	0.91	11.00	3.00
61.11	1.000×10^{-12}	10.10	12.00	2.00

Figure 13-3 Titration curves of pH and pOH versus volume of base for the titration of 0.1000 M HCl with 0.1000 M NaOH.

FEATURE 13-3

Calculating the NaOH Volumes Shown in the First Column of Table 13-1

Prior to the equivalence point, $[H_3O^+]$ equals the concentration of unreacted HCl (c_{HCl}). The concentration of HCl is equal to the original number of millimoles of HCl (50.00 mL $\times$ 0.1000 M) minus the number of millimoles of NaOH added ($V_{NaOH} \times 0.1000$ M) divided by the total volume of the solution:

$$c_{HCl} = [H_3O^+] = \frac{50.00 \times 0.1000 - V_{NaOH} \times 0.1000}{50.00 + V_{NaOH}}$$

where V_{NaOH} is the volume of 0.1000 M NaOH added. This equation reduces to

$$50.00[H_3O^+] + V_{NaOH}[H_3O^+] = 5.000 - 0.1000V_{NaOH}$$

Collecting the terms containing V_{NaOH} gives

$$V_{NaOH}(0.1000 + [H_3O^+]) = 5.000 - 50.00[H_3O^+]$$

(continued)

or

$$V_{NaOH} = \frac{5.000 - 50.00[H_3O^+]}{0.1000 + [H_3O^+]}$$

Thus to obtain $[H_3O^+] = 0.0100$ M, we find

$$V_{NaOH} = \frac{5.000 - 50.00 \times 0.0100}{0.1000 + 0.0100} = 40.91 \text{ mL}$$

Challenge: Use the same reasoning to show that beyond the equivalence point,

$$V_{NaOH} = \frac{50.000[OH^-] + 5.000}{0.1000 - [OH^-]}$$

 Spreadsheet Summary Chapter 7 of *Applications of Microsoft® Excel in Analytical Chemistry,* 2nd ed., deals with plotting titration curves. Several types of titrations are presented and ordinary titration curves are plotted along with derivative plots and Gran plots. The stoichiometric approach developed in this chapter is used and a master equation approach is explored.

WEB WORKS Look up *titration* in Wikipedia, the online encyclopedia. Give the definition of titration found there. Is a chemical reaction necessary for a quantitative procedure to be called a titration? From what Latin word does titration derive? Who developed the first buret and in what year? List five different methods to determine the end point of a titration. Define the term *acid number*, also called the *acid value*. How are titrations applied to biodiesel fuels?

QUESTIONS AND PROBLEMS

13-1. Define
  *(a) millimole.
  (b) titration.
  *(c) stoichiometric ratio.
  (d) titration error.
13-2. Write two equations that—along with the stoichiometric factor—form the basis for the calculations of volumetric titrations.
13-3. Distinguish between
  *(a) the equivalence point and the end point of a titration.
  (b) a primary standard and a secondary standard.
13-4. Briefly explain why the concentration units of milligrams of solute per liter and parts per million can be used interchangeably for a dilute aqueous solution.
13-5. Calculations of volumetric analysis ordinarily consist of transforming the quantity of titrant used (in chemical units) to a chemically equivalent quantity of analyte (also in chemical units) through use of a stoichiometric factor. Use chemical formulas (NO CALCULATIONS

REQUIRED) to express this ratio for calculation of the percentage of
  *(a) hydrazine in rocket fuel by titration with standard iodine. Reaction:

$$H_2NNH_2 + 2I_2 \rightarrow N_2(g) + 4I^- + 4H^+$$

  (b) hydrogen peroxide in a cosmetic preparation by titration with standard permanganate. Reaction:

$$5H_2O_2 + 2MnO_4^- + 6H^+ \rightarrow$$
$$2Mn^{2+} + 5O_2(g) + 8H_2O$$

  *(c) boron in a sample of borax, $Na_2B_4O_7 \cdot 10H_2O$, by titration with standard acid. Reaction:

$$B_4O_7^{2-} + 2H^+ + 5H_2O \rightarrow 4H_3BO_3$$

(d) sulfur in an agricultural spray that was converted to thiocyanate with an unmeasured excess of cyanide. Reaction:

$$S(s) + CN^- \rightarrow SCN^-$$

After removal of the excess cyanide, the thiocyanate was titrated with a standard potassium iodate solution in strong HCl. Reaction:

$$2SCN^- + 3IO_3^- + 2H^+ + 6Cl^- \rightarrow$$
$$2SO_4^{2-} + 2CN^- + 3ICl_2^- + H_2O$$

13-6. How many millimoles of solute are contained in
(a) 2.00 L of 2.76×10^{-3} M $KMnO_4$?
(b) 250.0 mL of 0.0423 M KSCN?
(c) 500.0 mL of a solution containing 2.97 ppm $CuSO_4$?
(d) 2.50 L of 0.352 M KCl?

*13-7. How many millimoles of solute are contained in
(a) 2.95 mL of 0.0789 M KH_2PO_4?
(b) 0.2011 L of 0.0564 M $HgCl_2$?
(c) 2.56 L of a 47.5 ppm solution of $Mg(NO_3)_2$?
(d) 79.8 mL of 0.1379 M NH_4VO_3 (116.98 g/mol)?

13-8. What mass of solute in milligrams is contained in
(a) 26.0 mL of 0.250 M sucrose (342 g/mol)?
(b) 2.92 L of 5.23×10^{-4} M H_2O_2?
(c) 673 mL of a solution that contains 5.76 ppm $Pb(NO_3)_2$ (331.20 g/mol)?
(d) 6.75 mL of 0.0426 M KNO_3?

*13-9. What mass of solute in grams is contained in
(a) 450.0 mL of 0.0986 M H_2O_2?
(b) 26.4 mL of 9.36×10^{-4} M benzoic acid (122.1 g/mol)?
(c) 2.50 L of a solution that contains 23.4 ppm $SnCl_2$?
(d) 21.7 mL of 0.0214 M $KBrO_3$?

13-10. Calculate the molar concentration of a solution that is 50.0% NaOH (w/w) and has a specific gravity of 1.52.

*13-11. Calculate the molar concentration of a 20.0% solution (w/w) of KCl that has a specific gravity of 1.13.

13-12. Describe the preparation of
(a) 500 mL of 0.0750 M $AgNO_3$ from the solid reagent.
(b) 2.00 L of 0.325 M HCl, starting with a 6.00 M solution of the reagent.
(c) 750 mL of a solution that is 0.0900 M in K^+, starting with solid $K_4Fe(CN)_6$.
(d) 600 mL of 2.00% (w/v) aqueous $BaCl_2$ from a 0.500 M $BaCl_2$ solution.
(e) 2.00 L of 0.120 M $HClO_4$ from the commercial reagent [60% $HClO_4$ (w/w), sp gr 1.60].
(f) 9.00 L of a solution that is 60.0 ppm in Na^+, starting with solid Na_2SO_4.

*13-13. Describe the preparation of
(a) 1.00 L of 0.150 M $KMnO_4$ from the solid reagent.
(b) 2.50 L of 0.500 M $HClO_4$, starting with a 9.00 M solution of the reagent.
(c) 400 mL of a solution that is 0.0500 M in I^-, starting with MgI_2.
(d) 200 mL of 1.00% (w/v) aqueous $CuSO_4$ from a 0.218 M $CuSO_4$ solution.
(e) 1.50 L of 0.215 M NaOH from the concentrated commercial reagent [50% NaOH (w/w), sp gr 1.525].
(f) 1.50 L of a solution that is 12.0 ppm in K^+, starting with solid $K_4Fe(CN)_6$.

13-14. A solution of $HClO_4$ was standardized by dissolving 0.4008 g of primary-standard-grade HgO in a solution of KBr:

$$HgO(s) + 4Br^- + H_2O \rightarrow HgBr_4^{2-} + 2OH^-$$

The liberated OH^- consumed 43.75 mL of the acid. Calculate the molar concentration of the $HClO_4$.

*13-15. A 0.4723-g sample of primary-standard-grade Na_2CO_3 required 34.78 mL of an H_2SO_4 solution to reach the end point in the reaction

$$CO_3^{2-} + 2H^+ \rightarrow H_2O + CO_2(g)$$

What is the molar concentration of the H_2SO_4?

13-16. A 0.5002-g sample that assayed 96.4% Na_2SO_4 required 48.63 mL of a barium chloride solution. Reaction:

$$Ba^{2+} + SO_4^{2-} \rightarrow BaSO_4(s)$$

Calculate the analytical molar concentration of $BaCl_2$ in the solution.

*13-17. A 0.4126-g sample of primary-standard Na_2CO_3 was treated with 40.00 mL of dilute perchloric acid. The solution was boiled to remove CO_2, following which the excess $HClO_4$ was back-titrated with 9.20 mL of dilute NaOH. In a separate experiment, it was established that 26.93 mL of the $HClO_4$ neutralized the NaOH in a 25.00-mL portion. Calculate the molarities of the $HClO_4$ and NaOH.

13-18. Titration of 50.00 mL of 0.04715 M $Na_2C_2O_4$ required 39.25 mL of a potassium permanganate solution.

$$2MnO_4^- + 5H_2C_2O_4 + 6H^+ \rightarrow$$
$$2Mn^{2+} + 10CO_2(g) + 8H_2O$$

Calculate the molar concentration of the $KMnO_4$ solution.

*13-19. Titration of the I_2 produced from 0.1142 g of primary-standard KIO_3 required 27.95 mL of sodium thiosulfate.

$$IO_3^- + 5I^- + 6H^+ \rightarrow 3I_2 + 3H_2O$$

$$I_2 + 2S_2O_3^{2-} \rightarrow 2I^- + S_4O_6^{2-}$$

Calculate the concentration of the $Na_2S_2O_3$.

13-20. A 4.912-g sample of a petroleum product was burned in a tube furnace, and the SO_2 produced was collected in 3% H_2O_2. Reaction:

$$SO_2(g) + H_2O_2 \rightarrow H_2SO_4$$

A 25.00-mL portion of 0.00873 M NaOH was introduced into the solution of H_2SO_4, following which the excess base was back-titrated with 15.17 mL of 0.01102 M HCl. Calculate the sulfur concentration in the sample in parts per million.

*13-21. A 100.0-mL sample of spring water was treated to convert any iron present to Fe^{2+}. Addition of 25.00-mL of 0.002517 M $K_2Cr_2O_7$ resulted in the reaction

$$6Fe^{2+} + Cr_2O_7^{2-} + 14H^+ \rightarrow$$
$$6Fe^{3+} + 2Cr^{3+} + 7H_2O$$

The excess $K_2Cr_2O_7$ was back-titrated with 8.53 mL of 0.00949 M Fe^{2+} solution. Calculate the concentration of iron in the sample in parts per million.

13-22. The arsenic in a 1.203-g sample of a pesticide was converted to H_3AsO_4 by suitable treatment. The acid was then neutralized, and 40.00 mL of 0.05871 M $AgNO_3$ was added to precipitate the arsenic quantitatively as Ag_3AsO_4. The excess Ag^+ in the filtrate and in the washings from the precipitate was titrated with 9.63 mL of 0.1000 M KSCN, and the reaction was

$$Ag^+ + SCN^- \rightarrow AgSCN(s)$$

Find the percentage of As_2O_3 in the sample.

*13-23. The thiourea in a 1.455-g sample of organic material was extracted into a dilute H_2SO_4 solution and titrated with 37.31 mL of 0.009372 M Hg^{2+} via the reaction

$$4(NH_2)_2CS + Hg^{2+} \rightarrow [(NH_2)_2CS]_4Hg^{2+}$$

Find the percentage of $(NH_2)_2CS$ (76.12 g/mol) in the sample.

13-24. A solution of $Ba(OH)_2$ was standardized against 0.1215 g of primary-standard-grade benzoic acid,

C_6H_5COOH (122.12 g/mol). An end point was observed after addition of 43.25 mL of base.

(a) Calculate the molar concentration of the base.

(b) Calculate the standard deviation of the molar concentration if the standard deviation for the mass measurement was ± 0.3 mg and that for the volume measurement was ± 0.02 mL.

(c) Assuming an error of -0.3 mg in the mass measurement, calculate the absolute and relative systematic error in the molar concentration.

*13-25. The ethyl acetate concentration in an alcoholic solution was determined by diluting a 10.00-mL sample to 100.00 mL. A 20.00-mL portion of the diluted solution was refluxed with 40.00 mL of 0.04672 M KOH:

$$CH_3COOC_2H_5 + OH^- \rightarrow$$
$$CH_3COO^- + C_2H_5OH$$

After cooling, the excess OH^- was back-titrated with 3.41 mL of 0.05042 M H_2SO_4. Calculate the amount of ethyl acetate (88.11 g/mol) in the original sample in grams.

13-26. A 0.1475-M solution of $Ba(OH)_2$ was used to titrate the acetic acid (60.05 g/mol) in a dilute aqueous solution. The following results were obtained.

Sample	Sample Volume, mL	$Ba(OH)_2$ Volume, mL
1	50.00	43.17
2	49.50	42.68
3	25.00	21.47
4	50.00	43.33

(a) Calculate the mean w/v percentage of acetic acid in the sample.

(b) Calculate the standard deviation for the results.

(c) Calculate the 90% confidence interval for the mean.

(d) At the 90% confidence level, could any of the results be discarded?

*13-27. (a) A 0.3147-g sample of primary-standard-grade $Na_2C_2O_4$ was dissolved in H_2SO_4 and titrated with 31.67 mL of dilute $KMnO_4$:

$$2MnO_4^- + 5C_2O_4^{2-} + 16H^+ \rightarrow$$
$$2Mn^{2+} + 10CO_2(g) + 8H_2O$$

Calculate the molar concentration of the $KMnO_4$ solution.

(b) The iron in a 0.6656-g ore sample was reduced quantitatively to the +2 state and then titrated with 26.75 mL of the $KMnO_4$ solution from part (a). Calculate the percent Fe_2O_3 in the sample.

13-28. (a) A 0.1527-g sample of primary-standard $AgNO_3$ was dissolved in 502.3 g of distilled water.

Calculate the weight molar concentration of Ag^+ in this solution.

(b) The standard solution described in part (a) was used to titrate a 25.171-g sample of a KSCN solution. An end point was obtained after adding 24.615 g of the $AgNO_3$ solution. Calculate the weight molar concentration of the KSCN solution.

(c) The solutions described in parts (a) and (b) were used to determine the $BaCl_2 \cdot 2H_2O$ in a 0.7120-g sample. A 20.102-g sample of the $AgNO_3$ was added to a solution of the sample, and the excess $AgNO_3$ was back-titrated with 7.543 g of the KSCN solution. Calculate the percent $BaCl_2 \cdot 2H_2O$ in the sample.

*13-29. A solution was prepared by dissolving 7.48 g of $KCl \cdot MgCl_2 \cdot 6H_2O$ (277.85 g/mol) in sufficient water to give 2.000 L. Calculate

(a) the molar analytical concentration of $KCl \cdot MgCl_2$ in this solution.

(b) the molar concentration of Mg^{2+}.

(c) the molar concentration of Cl^-.

(d) the weight/volume percentage of $KCl \cdot MgCl_2 \cdot 6H_2O$.

(e) the number of millimoles of Cl^- in 25.0 mL of this solution.

(f) the concentration of K^+ in ppm.

13-30. A solution was prepared by dissolving 367 mg of $K_3Fe(CN)_6$ (329.2 g/mol) in sufficient water to give 750.0 mL. Calculate

(a) the molar analytical concentration of $K_3Fe(CN)_6$.

(b) the molar concentration of K^+.

(c) the molar concentration of $Fe(CN)_6^{3-}$.

(d) the weight/volume percentage of $K_3Fe(CN)_6$.

(e) the number of millimoles of K^+ in 50.0 mL of this solution.

(f) the concentration of $Fe(CN)_6^{3-}$ in ppm.

13-31. **Challenge Problem:** For each of the following acid/base titrations, calculate the H_3O^+ and OH^- concentrations at equivalence and at titrant volumes corresponding to ±20.00 mL, ±10.00 mL, and ±1.00 mL of equivalence. Construct a titration curve from the data, plotting the p-function versus titrant volume.

(a) 25.00 mL of 0.05000 M HCl with 0.02500 M NaOH.

(b) 20.00 mL of 0.06000 M HCl with 0.03000 M NaOH.

(c) 30.00 mL of 0.07500 M H_2SO_4 with 0.1000 M NaOH.

(d) 40.00 mL of 0.02500 M NaOH with 0.05000 M HCl.

(e) 35.00 mL of 0.2000 M Na_2CO_3 with 0.2000 M HCl.

CHAPTER 14

Principles of Neutralization Titrations

Courtesy of HANNA Instruments

Neutralization titrations are widely used to determine the amounts of acids and bases. In addition, neutralization titrations can be used to monitor the progress of reactions that produce or consume hydrogen ions. In clinical chemistry, for example, pancreatitis can be diagnosed by measuring the activity of serum lipase. Lipases hydrolyze the long-chain fatty acid triglyceride. The reaction liberates two moles of fatty acid and one mole of β-monoglyceride for each mole of triglyceride present according to

$$\text{triglyceride} \xrightarrow{\text{lipase}} \text{monoglyceride} + 2 \text{ fatty acid}$$

The reaction is allowed to proceed for a certain amount of time, and then the liberated fatty acid is titrated with NaOH using a phenolphthalein indicator or a pH meter. The amount of fatty acid produced in a fixed time is related to the lipase activity (see Chapter 30). The entire procedure can be automated with an automatic titrator such as that shown here.

A cid/base equilibria are ubiquitous in chemistry and science in general. For example, you will find that the material in this chapter and in Chapter 15 is directly relevant to the acid/base reactions that are so important in biochemistry and the other biological sciences.

Standard solutions of strong acids and strong bases are used extensively for determining analytes that are themselves acids or bases or analytes that can be converted to such species. This chapter explores the principles of acid/base titrations. In addition, we investigate titration curves that are plots of pH vs. volume of titrant, and present several examples of pH calculations.

14A SOLUTIONS AND INDICATORS FOR ACID/BASE TITRATIONS

Like all titrations, neutralization titrations depend on a chemical reaction of the analyte with a standard reagent. There are several different types of acid/base titrations. One of the most common is the titration of a strong acid, such as hydrochloric or sulfuric acid, with a strong base, such as sodium hydroxide. Another common type is the titration of a weak acid, such as acetic or lactic acid, with a strong base. Weak bases, such as sodium cyanide or sodium salicylate, can also be titrated with strong acids.

In all titrations, we must have a method of determining the point of chemical equivalence. Typically, a chemical indicator or an instrumental method is used to locate the end point, which we hope is very close to the equivalence point. Our discussion focuses on the types of standard solutions and the chemical indicators that are used for neutralization titrations.

14A-1 Standard Solutions

The standard solutions used in neutralization titrations are strong acids or strong bases because these substances react more completely with an analyte than do weak acids and bases, and as a result, they produce sharper end points. Standard solutions of acids are prepared by diluting concentrated hydrochloric, perchloric, or sulfuric acid. Nitric acid is seldom used because its oxidizing properties offer the potential for undesirable side reactions. *Hot concentrated perchloric and sulfuric acids are potent oxidizing agents and are very hazardous.* Fortunately, cold dilute solutions of these reagents are safe to use in the analytical laboratory without any special precautions other than eye protection.

Standard solutions of bases are usually prepared from solid sodium, potassium, and occasionally barium hydroxides. Again, always use eye protection when handling dilute solutions of these reagents.

> The standard reagents used in acid/base titrations are always strong acids or strong bases, most commonly HCl, $HClO_4$, H_2SO_4, NaOH, and KOH. Weak acids and bases are never used as standard reagents because they react incompletely with analytes.

14A-2 Acid/Base Indicators

Many naturally occurring and synthetic compounds exhibit colors that depend on the pH of the solutions in which they are dissolved. Some of these substances, which have been used for centuries to indicate the acidity or alkalinity of water, are still applied today as acid/base indicators.

An acid/base indicator is a weak organic acid or a weak organic base whose undissociated form differs in color from its conjugate base or its conjugate acid form. For example, the behavior of an acid-type indicator, HIn, is described by the equilibrium

$$\underset{\text{acid color}}{\text{HIn}} + H_2O \rightleftharpoons \underset{\text{base color}}{\text{In}^-} + H_3O^+$$

In this reaction, internal structural changes accompany dissociation and cause the color change (for example, see **Figure 14-1**). The equilibrium for a base-type indicator, In, is

$$\underset{\text{base color}}{\text{In}} + H_2O \rightleftharpoons \underset{\text{acid color}}{\text{InH}^+} + OH^-$$

In the paragraphs that follow, we focus on the behavior of acid-type indicators. The principles can be easily extended to base-type indicators as well.

The equilibrium-constant expression for the dissociation of an acid-type indicator takes the form

$$K_a = \frac{[H_3O^+][In^-]}{[HIn]} \tag{14-1}$$

Rearranging leads to

$$[H_3O^+] = K_a\frac{[HIn]}{[In^-]} \tag{14-2}$$

> For a list of common acid/base indicators and their colors, look inside the front cover of this book. See also color plate 8 for photographs showing the colors and transition ranges of 12 common indicators.

We see then that the hydronium ion is proportional to the ratio of the concentration of the acid form to the concentration of the base form of the indicator, which in turn controls the color of the solution.

colorless
(a)

red
(b)

+

H_3O^+

Figure 14-1   Color change and molecular modes for phenolphthalein. (a) Acidic form after hydrolysis of the lactone form. (b) Basic form.

The human eye is not very sensitive to color differences in a solution containing a mixture of HIn and In⁻, particularly when the ratio [HIn]/[In⁻] is greater than about 10 or smaller than about 0.1. Because of this restriction, the color change detected by an average observer occurs within a limited range of concentration ratios from about 10 to about 0.1. At greater or smaller ratios, the color appears essentially constant to the eye and is independent of the ratio. As a result, we can write that the average indicator, HIn, exhibits its pure acid color when

$$\frac{[\text{HIn}]}{[\text{In}^-]} \geq \frac{10}{1}$$

and its base color when

$$\frac{[\text{HIn}]}{[\text{In}^-]} \leq \frac{1}{10}$$

The color appears to be intermediate for ratios between these two values. These ratios vary considerably from indicator to indicator. Furthermore, people differ significantly in their ability to distinguish between colors.

If we substitute the two concentration ratios into Equation 14-2, the range of hydronium ion concentrations needed for the indicator to change color can be estimated. For full acid color,

$$[\text{H}_3\text{O}^+] = 10K_a$$

and for the full base color,

$$[H_3O^+] = 0.1\, K_a$$

To obtain the indicator pH range, we take the negative logarithms of the two expressions:

$$pH(\text{acid color}) = -\log(10K_a) = pK_a + 1$$

$$pH(\text{basic color}) = -\log(0.1K_a) = pK_a - 1$$

$$\text{indicator pH range} = pK_a \pm 1 \qquad (14\text{-}3)$$

> The pH transition range of most acid type indicators is roughly $pK_a \pm 1$.

This expression shows that an indicator with an acid dissociation constant of 1×10^{-5} ($pK_a = 5$) typically shows a complete color change when the pH of the solution in which it is dissolved changes from 4 to 6 (see **Figure 14-2**). We can derive a similar relationship for a basic-type indicator.

Titration Errors with Acid/Base Indicators

We find two types of titration error in acid/base titrations. The first is a determinate error that occurs when the pH at which the indicator changes color differs from the pH at the equivalence point. This type of error can usually be minimized by choosing the indicator carefully or by making a blank correction.

The second type is an indeterminate error that originates from the limited ability of the human eye to distinguish reproducibly the intermediate color of the indicator. The magnitude of this error depends on the change in pH per milliliter of reagent at the equivalence point, on the concentration of the indicator, and on the sensitivity of the eye to the two indicator colors. On average, the visual uncertainty with an acid/base indicator is in the range of ± 0.5 to ± 1 pH unit. This uncertainty can often be decreased to as little as ± 0.1 pH unit by matching the color of the solution being titrated with that of a reference standard containing a similar amount of indicator at the appropriate pH. These uncertainties are approximations that vary considerably from indicator to indicator as well as from person to person.

Variables That Influence the Behavior of Indicators

The pH interval over which a given indicator exhibits a color change is influenced by temperature, by the ionic strength of the medium, and by the presence of organic solvents and colloidal particles. Some of these effects, particularly the last two, can cause the transition range to shift by one or more pH units.[1]

The Common Acid/Base Indicators

The list of acid/base indicators is large and includes a number of organic compounds. Indicators are available for almost any desired pH range. A few common indicators and their properties are listed in **Table 14-1**. Note that the transition ranges vary from 1.1 to 2.2, with the average being about 1.6 units. These indicators and several more are shown along with their transition ranges in the colored chart inside the front cover of this book.

[1]For a discussion of these effects, see H.A. Laitinen and W. E. Harris, *Chemical Analysis*, 2nd ed., pp. 48–51. New York: McGraw-Hill, 1975.

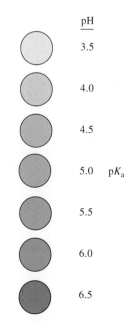

pH
3.5
4.0
4.5
5.0 pK_a
5.5
6.0
6.5

Figure 14-2 Indicator color as a function of pH ($pK_a = 5.0$).

TABLE 14-1

Some Important Acid/Base Indicators

Common Name	Transition Range, pH	pK_a*	Color Change[†]	Indicator Type[‡]
Thymol blue	1.2–2.8	1.65§	R – Y	1
	8.0–9.6	8.96§	Y – B	
Methyl yellow	2.9–4.0		R – Y	2
Methyl orange	3.1–4.4	3.46§	R – O	2
Bromocresol green	3.8–5.4	4.66§	Y – B	1
Methyl red	4.2–6.3	5.00§	R – Y	2
Bromocresol purple	5.2–6.8	6.12§	Y – P	1
Bromothymol blue	6.2–7.6	7.10§	Y – B	1
Phenol red	6.8–8.4	7.81§	Y – R	1
Cresol purple	7.6–9.2		Y – P	1
Phenolphthalein	8.3–10.0		C – R	1
Thymolphthalein	9.3–10.5		C – B	1
Alizarin yellow GG	10–12		C – Y	2

*At ionic strength of 0.1.

[†]B = blue; C = colorless; O = orange; P = purple; R = red; Y = yellow.

[‡](1) Acid type: $HIn + H_2O \rightleftharpoons H_3O^+ + In^-$; (2) Base type: $In + H_2O \rightleftharpoons InH^+ + OH^-$

§For the reaction $InH^+ + H_2O \rightleftharpoons H_3O^+ + In$

14B TITRATION OF STRONG ACIDS AND BASES

The hydronium ions in an aqueous solution of a strong acid have two sources: (1) the reaction of the acid with water and (2) the dissociation of water itself. In all but the most dilute solutions, however, the contribution from the strong acid far exceeds that from the solvent. Thus, for a solution of HCl with a concentration greater than about 10^{-6} M, we can write

$$[H_3O^+] = c_{HCl} + [OH^-] \approx c_{HCl}$$

where $[OH^-]$ represents the contribution of hydronium ions from the dissociation of water. An analogous relationship applies for a solution of a strong base, such as sodium hydroxide. That is,

$$[OH^-] = c_{NaOH} + [H_3O^+] \approx c_{NaOH}$$

14B-1 Titrating a Strong Acid with a Strong Base

We will be interested in this chapter and in the next several, in calculating *hypothetical* titration curves of pH vs. volume of titrant. We must make a clear distinction between the curves constructed by computing the values of pH and the *experimental* titration curves that we observe in the laboratory. Three types of calculations must be done in order to construct the hypothetical curve for titrating a solution of a strong acid with a strong base. Each of these types corresponds to a distinct stage in the titration: (1) preequivalence, (2) equivalence, and (3) postequivalence. In the preequivalence stage, we compute the concentration of the acid from its starting concentration and the amount of base added. At the equivalence point, the hydronium and hydroxide ions are present in equal concentrations, and the hydronium ion concentration can be calculated directly from the ion-product constant for water, K_w. In the postequivalence stage, the analytical concentration of the excess base is computed, and the hydroxide ion concentration is assumed to be equal to or a multiple of the analytical concentration.

> In solutions of a strong acid that are more concentrated than about 1×10^{-6} M, we can assume that the equilibrium concentration of H_3O^+ is equal to the analytical concentration of the acid. The same is true for $[OH^-]$ in solutions of strong bases.

> Before the equivalence point, we calculate the pH from the molar concentration of unreacted acid.

A convenient way of converting hydroxide concentration to pH can be developed by taking the negative logarithm of both sides of the ion-product constant expression for water. Thus,

$$K_w = [H_3O^+][OH^-]$$

$$-\log K_w = -\log[H_3O^+][OH^-] = -\log[H_3O^+] -\log[OH^-]$$

$$pK_w = pH + pOH$$

And, at 25°C,

$$-\log 10^{-14} = 14.00 = pH + pOH$$

◀ At the equivalence point, the solution is neutral, and pH = pOH. Both pH and pOH = 7.00, at 25°C.

◀ Beyond the equivalence point, we first calculate pOH and then pH. Remember that pH = pK_w − pOH. At 25°C, pH = 14.00 − pOH.

EXAMPLE 14-1

Generate the hypothetical titration curve for the titration of 50.00 mL of 0.0500 M HCl with 0.1000 M NaOH at 25°C.

Initial Point

Before any base is added, the solution is 0.0500 M in H_3O^+, and

$$pH = -\log[H_3O^+] = -\log 0.0500 = 1.30$$

After Addition of 10.00 mL of Reagent

The hydronium ion concentration is decreased as a result of both reaction with the base and dilution. So, the remaining HCl concentration, c_{HCl}, is

$$c_{HCl} = \frac{\text{no. mmol HCl remaining after addition of NaOH}}{\text{total volume soln}}$$

$$= \frac{\text{original no. mmol HCl} - \text{no. mmol NaOH added}}{\text{total volume soln}}$$

$$= \frac{(50.00 \text{ mL} \times 0.0500 \text{ M}) - (10.00 \text{ mL} \times 0.1000 \text{ M})}{50.00 \text{ mL} + 10.00 \text{ mL}}$$

$$= \frac{(2.500 \text{ mmol} - 1.00 \text{ mmol})}{60.00 \text{ mL}} = 2.50 \times 10^{-2} \text{ M}$$

$$[H_3O^+] = 2.50 \times 10^{-2} \text{ M}$$

$$pH = -\log[H_3O^+] = -\log(2.50 \times 10^{-2}) = 1.602 \approx 1.60$$

Note that we usually compute pH to two decimal places in titration curve calculations. We calculate additional points defining the curve in the region before the equivalence point in the same way. The results of these calculations are shown in the second column of Table 14-2.

(continued)

TABLE 14-2

Changes in pH during the Titration of a Strong Acid with a Strong Base

Volume of NaOH, mL	pH	
	50.00 mL of 0.0500 M HCl with 0.100 M NaOH	50.00 mL of 0.000500 M HCl with 0.00100 M NaOH
0.00	1.30	3.30
10.00	1.60	3.60
20.00	2.15	4.15
24.00	2.87	4.87
24.90	3.87	5.87
25.00	7.00	7.00
25.10	10.12	8.12
26.00	11.12	9.12
30.00	11.80	9.80

After Addition of 25.00 mL of Reagent: The Equivalence Point

At the equivalence point, neither HCl nor NaOH is in excess, and so, the concentrations of hydronium and hydroxide ions must be equal. Substituting this equality into the ion-product constant for water yields

$$[H_3O^+] = [OH^-] = \sqrt{K_w} = \sqrt{1.00 \times 10^{-14}} = 1.00 \times 10^{-7} \text{ M}$$

$$pH = -\log[H_3O^+] = -\log(1.00 \times 10^{-7}) = 7.00$$

After Addition of 25.10 mL of Reagent

The solution now contains an excess of NaOH, and we can write

$$c_{NaOH} = \frac{\text{no. mmol NaOH added } - \text{ original no. of mmoles HCl}}{\text{total volume soln}}$$

$$= \frac{25.10 \times 0.1000 - 50.00 \times 0.0500}{75.10} = 1.33 \times 10^{-4} \text{ M}$$

The equilibrium concentration of hydroxide ion is

$$[OH^-] = c_{NaOH} = 1.33 \times 10^{-4} \text{ M}$$

$$pOH = -\log[OH^-] = -\log(1.33 \times 10^{-4}) = 3.88$$

$$pH = 14.00 - pOH = 14.00 - 3.88 = 10.12$$

Additional values beyond the equivalence point are calculated in the same way. The results of these computations are shown in the last three rows of Table 14-2.

FEATURE 14-1

Using the Charge-balance Equation to Construct Titration Curves

In Example 14-1, we generated an acid/base titration curve from the reaction stoichiometry. We can show that all points on the curve can also be calculated from the charge-balance equation.

For the system treated in Example 14-1, the charge-balance equation is given by

$$[H_3O^+] + [Na^+] = [OH^-] + [Cl^-]$$

The sodium and chloride ion concentrations are given by

$$[Na^+] = \frac{c^0_{NaOH} V_{NaOH}}{V_{NaOH} + V_{HCl}}$$

$$[Cl^-] = \frac{c^0_{HCl} V_{HCl}}{V_{NaOH} + V_{HCl}}$$

where c^0_{NaOH} and c^0_{HCl} are the initial concentrations of base and acid, respectively. We can rewrite the first equation in the form

$$[H_3O^+] = [OH^-] + [Cl^-] - [Na^+]$$

For volumes of NaOH short of the equivalence point, $[OH^-] \ll [Cl^-]$ so that

$$[H_3O^+] \approx [Cl^-] - [Na^+] \approx c_{HCl}$$

and

$$[H_3O^+] = \frac{c^0_{HCl} V_{HCl}}{V_{HCl} + V_{NaOH}} - \frac{c^0_{NaOH} V_{NaOH}}{V_{HCl} + V_{NaOH}} = \frac{c^0_{HCl} V_{HCl} - c^0_{NaOH} V_{NaOH}}{V_{HCl} + V_{NaOH}}$$

At the equivalence point, $[Na^+] = [Cl^-]$, and

$$[H_3O^+] = [OH^-]$$
$$[H_3O^+] = \sqrt{K_w}$$

Beyond the equivalence point, $[H_3O^+] \ll [Na^+]$, and the original equation rearranges to

$$[OH^-] \approx [Na^+] - [Cl^-] \approx c_{NaOH}$$

$$= \frac{c^0_{NaOH} V_{NaOH}}{V_{NaOH} + V_{HCl}} - \frac{c^0_{HCl} V_{HCl}}{V_{NaOH} + V_{HCl}} = \frac{c^0_{NaOH} V_{NaOH} - c^0_{HCl} V_{HCl}}{V_{NaOH} + V_{HCl}}$$

The Effect of Concentration

The effects of reagent and analyte concentration on the neutralization titration curves for strong acids are shown by the two sets of data in Table 14-2 and the plots in **Figure 14-3**. Note that with 0.1 M NaOH as the titrant, the change in pH in the equivalence-point region is large. With 0.001 M NaOH, the change is much smaller, but still pronounced.

Choosing an Indicator

Figure 14-3 shows that the selection of an indicator is not critical when the reagent concentration is approximately 0.1 M. In that case, the volume differences in titrations with the three indicators shown are of the same magnitude as the uncertainties associated with reading the buret and so are negligible. Note, however, that bromocresol green is unsuited for a titration involving the 0.001 M reagent because the color change occurs over a 5-mL range well before the equivalence point. The use of phenolphthalein is subject to similar objections. Of the three indicators, then, only bromothymol blue provides a satisfactory end point with a minimal systematic error in the titration of 0.001 M NaOH.

Figure 14-3 Titration curves for HCl with NaOH. Curve *A*: 50.00 mL of 0.0500 M HCl with 0.1000 M NaOH. Curve *B*: 50.00 mL of 0.000500 M HCl with 0.00100 M NaOH.

14B-2 Titrating a Strong Base with a Strong Acid

Titration curves for strong bases are calculated in a similar way to those for strong acids. Short of the equivalence point, the solution is basic, and the hydroxide ion concentration is numerically related to the analytical concentration of the base. The solution is neutral at the equivalence point and becomes acidic in the region beyond the equivalence point. After the equivalence point, the hydronium ion concentration is equal to the analytical concentration of the excess strong acid.

EXAMPLE 14-2

Calculate the pH during the titration of 50.00 mL of 0.0500 M NaOH with 0.1000 M HCl at 25°C after the addition of the following volumes of reagent: (a) 24.50 mL, (b) 25.00 mL, (c) 25.50 mL.

Solution

(a) At 24.50 mL added, $[H_3O^+]$ is very small and cannot be computed from stoichiometric considerations but can be obtained from $[OH^-]$:

$$[OH^-] = c_{NaOH} = \frac{\text{original no. mmol NaOH} - \text{no. mmol HCl added}}{\text{total volume of solution}}$$

$$= \frac{50.00 \times 0.0500 - 24.50 \times 0.1000}{50.00 + 24.50} = 6.71 \times 10^{-4} \text{ M}$$

$$[H_3O^+] = K_w/(6.71 \times 10^{-4}) = 1.00 \times 10^{-14}/(6.71 \times 10^{-4})$$

$$= 1.49 \times 10^{-11} \text{ M}$$

$$pH = -\log(1.49 \times 10^{-11}) = 10.83$$

(b) 25.00 mL added is the equivalence point where $[H_3O^+] = [OH^-]$:

$$[H_3O^+] = \sqrt{K_w} = \sqrt{1.00 \times 10^{-14}} = 1.00 \times 10^{-7} \text{ M}$$

$$pH = -\log(1.00 \times 10^{-7}) = 7.00$$

(c) At 25.50 mL added

$$[H_3O^+] = c_{HCl} = \frac{25.50 \times 0.1000 - 50.00 \times 0.0500}{75.50}$$

$$= 6.62 \times 10^{-4} \text{ M}$$

$$pH = -\log(6.62 \times 10^{-4}) = 3.18$$

Curves for the titration of 0.0500 M and 0.00500 M NaOH with 0.1000 M and 0.0100 M HCl are shown in **Figure 14-4**. We use the same criteria described for the titration of a strong acid with a strong base to select an indicator.

Figure 14-4 Titration curves for NaOH with HCl. Curve *A*: 50.00 mL of 0.0500 M NaOH with 0.1000 M HCl. Curve *B*: 50.00 mL of 0.00500 M NaOH with 0.0100 M HCl.

FEATURE 14-2

Significant Figures in Titration Curve Calculations

Concentrations calculated in the equivalence-point region of titration curves are generally of low precision because they are based on small differences between large numbers. For example, in the calculation of c_{NaOH} after introduction of 25.10 mL of NaOH in Example 14-1, the numerator (2.510 − 2.500 = 0.010) is known to only two significant figures. To minimize rounding error, however, three digits were retained in c_{NaOH} (1.33 × 10⁻⁴), and rounding was postponed until pOH and pH were computed.

To round the calculated values of p-functions, remember (see Section 6D-2) that it is *the mantissa of a logarithm*

(that is, the number to the right of the decimal point) *that should be rounded to include only significant figures* because the characteristic (the number to the left of the decimal point) merely locates the decimal point. Fortunately, the large changes in p-functions characteristic of most equivalence points are not obscured by the limited precision of the calculated data. Generally, in calculating values for titration curves, we will round p-functions to two places to the right of the decimal point whether or not such rounding is called for.

A 1 **Spreadsheet Summary** In Chapter 7 of *Applications of Microsoft® Excel in Analytical Chemistry*, 2nd ed., strong acid/strong base titrations are considered first. The stoichiometric approach and the charge-balance-equation approach are used to calculate pH at various points in these titrations. Excel's charting functions are then used to prepare titration curves for these systems.

14C   TITRATION CURVES FOR WEAK ACIDS

Four distinctly different types of calculations are needed to compute values for a weak acid (or a weak base) titration curve:

1. At the beginning, the solution contains only a weak acid or a weak base, and the pH is calculated from the concentration of that solute and its dissociation constant.
2. After various increments of titrant have been added (up to, but not including, the equivalence point), the solution consists of a series of buffers. The pH of each buffer can be calculated from the analytical concentrations of the conjugate base or acid and the concentrations of the weak acid or base that remains.
3. At the equivalence point, the solution contains only the conjugate of the weak acid or base being titrated (that is, a salt), and the pH is calculated from the concentration of this product.
4. Beyond the equivalence point, the excess of strong acid or base titrant suppresses the acidic or basic character of the reaction product to such an extent that the pH is governed largely by the concentration of the excess titrant.

> Titration curves for strong and weak acids are identical just slightly beyond the equivalence point. The same is true for strong and weak bases.

EXAMPLE 14-3

Generate a curve for the titration of 50.00 mL of 0.1000 M acetic acid (HOAc) with 0.1000 M sodium hydroxide at 25°C.

Initial pH

First, we must calculate the pH of a 0.1000 M solution of HOAc using Equation 9-22.

$$[H_3O^+] = \sqrt{K_a c_{HOAc}} = \sqrt{1.75 \times 10^{-5} \times 0.1000} = 1.32 \times 10^{-3}\ M$$

$$pH = -\log(1.32 \times 10^{-3}) = 2.88$$

pH after Addition of 10.00 mL of Reagent

A buffer solution consisting of NaOAc and HOAc has now been produced. The analytical concentrations of the two constituents are

$$c_{HOAc} = \frac{50.00\ mL \times 0.1000\ M\ -\ 10.00\ mL \times 0.1000\ M}{60.00\ mL} = \frac{4.000}{60.00}\ M$$

$$c_{NaOAc} = \frac{10.00\ mL \times 0.1000\ M}{60.00\ mL} = \frac{1.000}{60.00}\ M$$

Now, for the 10.00 mL volume, we substitute the concentrations of HOAc and OAc⁻ into the dissociation-constant expression for acetic acid and obtain

$$K_a = \frac{[H_3O^+](1.000/60.00)}{4.00/60.00} = 1.75 \times 10^{-5}$$

$$[H_3O^+] = 7.00 \times 10^{-5}$$

$$pH = 4.15$$

Note that the total volume of solution is present in both numerator and denominator and thus cancels in the expression for $[H_3O^+]$. Calculations similar to this provide points on the curve throughout the buffer region. Data from these calculations are presented in column 2 of **Table 14-3**.

TABLE 14-3

Changes in pH during the Titration of a Weak Acid with a Strong Base

Volume of NaOH, mL	pH	
	50.00 mL of 0.1000 M HOAc with 0.1000 M NaOH	**50.00 mL of 0.001000 M HOAc with 0.001000 M NaOH**
0.00	2.88	3.91
10.00	4.15	4.30
25.00	4.76	4.80
40.00	5.36	5.38
49.00	6.45	6.46
49.90	7.46	7.47
50.00	8.73	7.73
50.10	10.00	8.09
51.00	11.00	9.00
60.00	11.96	9.96
70.00	12.22	10.25

pH after Addition of 25.00 mL of Reagent

As in the previous calculation, the analytical concentrations of the two constituents are

$$c_{HOAc} = \frac{50.00 \text{ mL} \times 0.1000 \text{ M} - 25.00 \text{ mL} \times 0.1000 \text{ M}}{75.00 \text{ mL}} = \frac{2.500}{75.00} \text{ M}$$

$$c_{NaOAc} = \frac{25.00 \text{ mL} \times 0.1000 \text{ M}}{75.00 \text{ mL}} = \frac{2.500}{75.00} \text{ M}$$

Now, for the 25.00 mL volume, we substitute the concentrations of HOAc and OAc$^-$ into the dissociation-constant expression for acetic acid and obtain

$$K_a = \frac{[H_3O^+](2.500/75.00)}{2.500/75.00} = 1.75 \times 10^{-5}$$

$$pH = pK_a = -\log(1.75 \times 10^{-5}) = 4.76$$

At this half-titration point, the analytical concentrations of the acid and conjugate base cancel in the expression for $[H_3O^+]$.

Equivalence-point pH

At the equivalence point, all of the acetic acid has been converted to sodium acetate. The solution is, therefore, similar to one formed by dissolving NaOAc in water, and the pH calculation is identical to that shown in Example 9-10 (page 218) for a weak base. In the present example, the NaOAc concentration is

$$c_{NaOAc} = \frac{50.00 \text{ mL} \times 0.1000 \text{ M}}{100.00 \text{ mL}} = 0.0500 \text{ M}$$

Thus,

$$OAc^- + H_2O \rightleftharpoons HOAc + OH^-$$

$$[OH^-] = [HOAc]$$

$$[OAc^-] = 0.0500 - [OH^-] \approx 0.0500$$

(continued)

Substituting these quantities into the base dissociation-constant expression for OAc^- gives

$$\frac{[OH^-]^2}{0.0500} = \frac{K_w}{K_a} = \frac{1.00 \times 10^{-14}}{1.75 \times 10^{-5}} = 5.71 \times 10^{-10}$$

$$[OH^-] = \sqrt{0.0500 \times 5.71 \times 10^{-10}} = 5.34 \times 10^{-6} \text{ M}$$

$$pH = 14.00 - [-\log(5.34 \times 10^{-6})] = 8.73$$

> Note that the pH at the equivalence point in this titration is greater than 7. The solution is basic. A solution of the salt of a weak acid is always basic.

pH After Addition of 50.10 mL of Base

After the addition of 50.10 mL of NaOH, the excess base and the acetate ion are both sources of hydroxide ion. The contribution from the acetate ion is small, however, because the excess of strong base suppresses the reaction of acetate with water. This fact becomes evident when we consider that the hydroxide ion concentration is only 5.34×10^{-6} M at the equivalence point; once a tiny excess of strong base is added, the contribution from the reaction of the acetate is even smaller. We have then

$$[OH^-] = c_{NaOH} = \frac{50.10 \text{ mL} \times 0.1000 \text{ M} - 50.00 \text{ mL} \times 0.1000 \text{ M}}{100.10 \text{ mL}}$$

$$= 9.99 \times 10^{-5} \text{ M}$$

$$pH = 14.00 - [-\log(9.99 \times 10^{-5})] = 10.00$$

Note that the titration curve for a weak acid with a strong base is identical with that for a strong acid with a strong base in the region slightly beyond the equivalence point.

Table 14-3 and Figure 14-5 compares the pH values calculated in this example with a more dilute titration. In a dilute solution, some of the assumptions made in this example do not hold. The effect of concentration is discussed further in Section 14C-1.

Note from Example 14-3 that the analytical concentrations of acid and conjugate base are identical when an acid has been half neutralized (after the addition of exactly 25.00 mL of base in this case). Thus, these terms cancel in the equilibrium-constant expression, and the hydronium ion concentration is numerically equal to the dissociation constant. Likewise, in the titration of a weak base, the hydroxide ion concentration is numerically equal to the dissociation constant of the base at the midpoint in the titration curve. In addition, the buffer capacities of each of the solutions are at a maximum at this point. These points, often called the **half-titration points**, are used to determine the dissociation constants as discussed in Feature 14-3.

> At the half-titration point in a weak-acid titration, $[H_3O^+] = K_a$, and $pH = pK_a$.

> At the half-titration point in a weak-base titration, $[OH^-] = K_b$, and $pOH = pK_b$ (recall $K_b = K_w/K_a$).

FEATURE 14-3

Determining Dissociation Constants of Weak Acids and Bases

The dissociation constants of weak acids or weak bases are often determined by monitoring the pH of the solution while the acid or base is being titrated. A pH meter with a glass pH electrode (see Section 21D-3) is used for the measurements. The titration is recorded from the initial pH until after the end point. The pH at one-half the end point volume is then obtained and used to obtain the dissociation constant. For an acid, the measured pH when the acid is half neutralized is numerically equal to pK_a. For a weak base, the pH at half titration must be converted to pOH, which is then equal to pK_b.

Figure 14-5 Curve for the titration of acetic acid with sodium hydroxide. Curve A: 0.1000 M acid with 0.1000 M base. Curve B: 0.001000 M acid with 0.001000 M base.

14C-1 The Effect of Concentration

The second and third columns of Table 14-3 contain pH data for the titration of 0.1000 M and of 0.001000 M acetic acid with sodium hydroxide solutions of the same two concentrations. In calculating the values for the more dilute acid, none of the approximations shown in Example 14-3 were valid, and solution of a quadratic equation was necessary for each point on the curve until after the equivalence point. In the postequivalence point region, the excess OH^- predominates, and the simple calculation works nicely.

Figure 14-5 is a plot of the data in Table 14-3. Note that the initial pH values are higher and the equivalence-point pH is lower for the more dilute solution (Curve *B*). At intermediate titrant volumes, however, the pH values differ only slightly because of the buffering action of the acetic acid/sodium acetate system that is present in this region. Figure 14-5 is graphical confirmation that the pH of buffers is largely independent of dilution. Note that the change in $[OH^-]$ in the vicinity of the equivalence point becomes smaller with lower analyte and reagent concentrations. This effect is analogous to the effect for the titration of a strong acid with a strong base (see Figure 14-3).

> CHALLENGE: Show that the pH values in the third column of Table 14-3 are correct.

14C-2 The Effect of Reaction Completeness

Titration curves for 0.1000 M solutions of acids with different dissociation constants are shown in **Figure 14-6**. Note that the pH change in the equivalence-point region becomes smaller as the acid becomes weaker—that is, as the reaction between the acid and the base becomes less complete.

14C-3 Choosing an Indicator: The Feasibility of Titration

Figures 14-5 and 14-6 show that the choice of indicator is more limited for the titration of a weak acid than for the titration of a strong acid. For example, Figure 14-5 illustrates that bromocresol green is totally unsuited for titration of 0.1000 M acetic

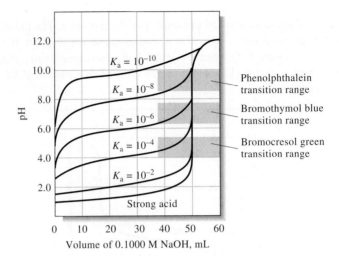

Figure 14-6 The effect of acid strength (dissociation constant) on titration curves. Each curve represents the titration of 50.00 mL of 0.1000 M weak acid with 0.1000 M strong base.

acid. Bromothymol blue does not work either because its full color change occurs over a range of titrant volume from about 47 mL to 50 mL of 0.1000 M base. On the other hand, an indicator exhibiting a color change in the basic region, such as phenolphthalein, provides a sharp end point with a minimal titration error.

The end-point pH change associated with the titration of 0.001000 M acetic acid (curve *B*, Figure 14-5) is so small that there is likely to be a significant titration error regardless of indicator. However, using an indicator with a transition range between that of phenolphthalein and that of bromothymol blue in conjunction with a suitable color comparison standard makes it possible to establish the end point in this titration with decent precision (a few percent relative standard deviation).

Figure 14-6 illustrates that similar problems occur as the strength of the acid being titrated decreases. A precision on the order of ± 2 ppt can be achieved by titrating a 0.1000 M solution of an acid with a dissociation constant of 10^{-8} if a suitable color comparison standard is available. With more concentrated solutions, weaker acids can be titrated with reasonable precision.

FEATURE 14-4

A Master Equation Approach to Weak Acid/Strong Base Titrations

With a weak acid and strong base titrations, a single master equation is used to find the H_3O^+ concentration throughout the titration. As an example let us take the titration of a hypothetical weak acid, HA (dissociation constant, K_a), with strong base, NaOH. Consider V_{HA} mL of c^0_{HA} M HA being titrated with c^0_{NaOH} M NaOH. At any point in the titration, we can write the charge balance equation as

$$[Na^+] + [H_3O^+] = [A^-] + [OH^-]$$

We now substitute to obtain an equation for H_3O^+ as a function of the volume of NaOH added, V_{NaOH}. We can express the sodium ion concentration as the number of millimoles of NaOH added divided by the total solution volume. Or

$$[Na^+] = \frac{c^0_{NaOH} V_{NaOH}}{V_{NaOH} + V_{HA}}$$

Mass balance gives the total concentration of A-containing species, c_T, as

$$c_T = [HA] + [A^-] = \frac{[A^-][H_3O^+]}{K_a} + [A^-]$$

Solving for $[A^-]$ yields

$$[A^-] = \left(\frac{K_a}{[H_3O^+] + K_a}\right)c_T$$

If we substitute these latter two equations into the charge-balance equation, we get

$$[Na^+] + [H_3O^+] = \frac{c_T K_a}{[H_3O^+] + K_a} + \frac{K_w}{[H_3O^+]}$$

By rearranging this equation, we obtain the master system equation for the entire titration:

$$[H_3O^+]^3 + (K_a + [Na^+])[H_3O^+]^2 + (K_a[Na^+] - c_T K_a - K_w)[H_3O^+] - K_w K_a = 0$$

We must solve this cubic equation for each volume of NaOH added. Mathematical or spreadsheet software simplify this task. The concentrations of H_3O^+ found are then converted to pH values in the usual way to generate a titration curve of pH vs. volume of NaOH.

> ◀ Note that a master equation can also be generated by calculating $[Na^+]$ for the range of pH values desired. The $[Na^+]$ is directly related to the volume added by the second equation of this feature.

Spreadsheet Summary   In the weak acid/strong base titrations section of Chapter 7 of *Applications of Microsoft® Excel in Analytical Chemistry*, 2nd ed., the stoichiometric method and a master equation approach are used to carry out the calculations and plot a titration curve for the titration of a weak acid with a strong base. Excel's Goal Seek is used to solve the charge-balance expression for the H_3O^+ concentration and the pH.

14D TITRATION CURVES FOR WEAK BASES

The calculations needed to draw the titration curve for a weak base are analogous to those of a weak acid, as shown in Example 14-4.

EXAMPLE 14-4

A 50.00-mL aliquot of 0.0500 M NaCN (K_a for HCN = 6.2×10^{-10}) is titrated with 0.1000 M HCl. The reaction is

$$CN^- + H_3O^+ \rightleftharpoons HCN + H_2O$$

Calculate the pH after the addition of (a) 0.00, (b) 10.00, (c) 25.00, and (d) 26.00 mL of acid.

Solution

(a) **0.00 mL of Reagent**
The pH of a solution of NaCN can be calculated by the method in Example 9-10, page 218:

$$CN^- + H_2O \rightleftharpoons HCN + OH^-$$

$$K_b = \frac{[OH^-][HCN]}{[CN^-]} = \frac{K_w}{K_a} = \frac{1.00 \times 10^{-14}}{6.2 \times 10^{-10}} = 1.61 \times 10^{-5}$$

$$[OH^-] = [HCN]$$

$$[CN^-] = c_{NaCN} - [OH^-] \approx c_{NaCN} = 0.0500 \text{ M}$$

> ◀ Note that for calculation purposes, equilibrium constants are considered exact so that the number of significant figures in the equilibrium constant does not affect the number of significant figures in the result.

(continued)

Substituting into the dissociation-constant expression gives, after rearranging,

$$[OH^-] = \sqrt{K_b c_{NaCN}} = \sqrt{1.61 \times 10^{-5} \times 0.0500} = 8.97 \times 10^{-4} \text{ M}$$

$$pH = 14.00 - [-\log(8.97 \times 10^{-4})] = 10.95$$

(b) 10.00 mL of Reagent

Addition of acid produces a buffer with a composition given by

$$c_{NaCN} = \frac{50.00 \times 0.0500 - 10.00 \times 0.1000}{60.00} = \frac{1.500}{60.00} \text{ M}$$

$$c_{HCN} = \frac{10.00 \times 0.1000}{60.00} = \frac{1.000}{60.00} \text{ M}$$

These values are then substituted into the expression for the acid dissociation constant of HCN to give $[H_3O^+]$ directly (see margin note):

CHALLENGE: Show that the pH of the buffer can be calculated with K_a for HCN, as was done here or equally well with K_b. We used K_a because it gives $[H_3O^+]$ directly; K_b gives $[OH^-]$.

$$[H_3O^+] = \frac{6.2 \times 10^{-10} \times (1.000/60.00)}{1.500/60.00} = 4.13 \times 10^{-10} \text{ M}$$

$$pH = -\log(4.13 \times 10^{-10}) = 9.38$$

(c) 25.00 mL of Reagent

This volume corresponds to the equivalence point, where the principal solute species is the weak acid HCN. Thus,

$$c_{HCN} = \frac{25.00 \times 0.1000}{75.00} = 0.03333 \text{ M}$$

Applying Equation 9-22 Gives

Since the principal solute species at the equivalence point is HCN, the pH is acidic.

$$[H_3O^+] = \sqrt{K_a c_{HCN}} = \sqrt{6.2 \times 10^{-10} \times 0.03333} = 4.55 \times 10^{-6} \text{ M}$$

$$pH = -\log(4.55 \times 10^{-6}) = 5.34$$

(d) 26.00 mL of Reagent

The excess of strong acid now present suppresses the dissociation of the HCN to the point where its contribution to the pH is negligible. Thus,

$$[H_3O^+] = c_{HCl} = \frac{26.00 \times 0.1000 - 50.00 \times 0.0500}{76.00} = 1.32 \times 10^{-3} \text{ M}$$

$$pH = -\log(1.32 \times 10^{-3}) = 2.88$$

When you titrate a weak base, use an indicator with a mostly acidic transition range. When titrating a weak acid, use an indicator with a mostly basic transition range.

Figure 14-7 shows hypothetical titration curves for a series of weak bases of different strengths. The curves show that indicators with mostly *acidic* transition ranges must be used for weak bases.

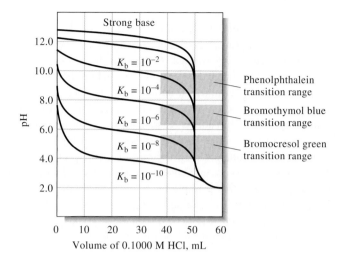

Figure 14-7 The effect of base strength (K_b) on titration curves. Each curve represents the titration of 50.00 mL of 0.1000 M base with 0.1000 M HCl.

FEATURE 14-5

Determining the pK Values for Amino Acids

Amino acids contain both an acidic and a basic group. For example, the structure of alanine is represented in **Figure 14F-1**.

Figure 14F-1 Structure and molecular model of alanine, an amino acid. Alanine can exist in two mirror image forms, the left-handed (L) form and the right-handed (D) form. All naturally occurring amino acids are left handed.

The amine group behaves as a base, and at the same time the carboxyl group acts as an acid. In aqueous solution, the amino acid is an internally ionized molecule, or "zwitterion," in which the amine group acquires a proton and becomes positively charged while the carboxyl group, having lost a proton becomes negatively charged.

(continued)

The pK values for amino acids can conveniently be determined by the general procedure described in Feature 14-3. Since the zwitterion has both acidic and basic character, two pKs can be determined. The pK for deprotonation of the protonated amine group can be determined by adding base, while the pK for protonating the carboxyl group can be determined by adding acid. In practice, a solution is prepared containing a known concentration of the amino acid. Hence, the investigator knows the amount of base or acid to add to reach halfway to the equivalence point. A curve of pH versus volume of acid or base added is shown in **Figure 14F-2**. In this type of experiment, the titration starts in the middle of the plot (0.00 mL added) and, for determining pK values, is only taken to a point that is half the volume required for equivalence. Note in this example for alanine, a volume of 20.00 mL of HCl is needed to completely protonate the carboxyl group. By adding acid to the zwitterion, the curve to the left of 0.00 volume is obtained. At a volume of 10.00 mL of HCl added, the pH is equal to the pK_a for the carboxyl group, 2.35.

Figure 14F-2   Curves for the titration of 20.00 mL of 0.1000 M alanine with 0.1000 M NaOH and 0.1000 M HCl. Note that the zwitterion is present before any acid or base has been added. Adding acid protonates the carboxylate group with a pK_a of 2.35. Adding base causes deprotonation of the protonated amine group with a pK_a of 9.89.

By adding NaOH to the zwitterion, the pK for deprotonating the NH_3^+ group can be determined. Now, 20.00 mL of base is required for complete deprotonation. At a volume of 10.00 mL of NaOH added, the pH is equal to the pK_a for the amine group, or 9.89. The pK_a values for other amino acids and more complicated biomolecules such as peptides and proteins can often be obtained in a similar manner. Some amino acids have more than one carboxyl or amine group. Aspartic acid is an example (see **Figure 14F-3**).

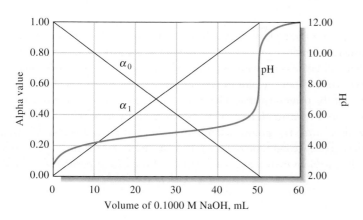

Figure 14F-3 Aspartic acid is an amino acid with two carboxyl groups. It can be combined with phenylalanine to make the artificial sweetener aspartame, which is sweeter and less fattening than ordinary sugar (sucrose).

It is important to note that in general amino acids cannot be quantitatively determined by direct titration because end points for completely protonating or deprotonating the zwitterion are often indistinct. Amino acids are normally determined by high performance liquid chromatography (see Chapter 33) or spectroscopic methods (see Part V).

14E | THE COMPOSITION OF SOLUTIONS DURING ACID/BASE TITRATIONS

We are often interested in the changes in composition that occur while a solution of a weak acid or a weak base is being titrated. These changes can be visualized by plotting the *relative* equilibrium concentration α_0 of the weak acid as well as the relative equilibrium concentration of the conjugate base α_1 as functions of the pH of the solution.

The solid straight lines labeled α_0 and α_1 in **Figure 14-8** were calculated with Equations 9-35 and 9-36 using values for $[H_3O^+]$ shown in column 2 of Table 14-3. The actual titration curve is shown as the curved line in Figure 14-8. Note that at the beginning of the titration α_0 is nearly 1 (0.987) meaning that 98.7% of the acetate containing species is present as HOAc and only 1.3% is present as OAc⁻. At the equivalence point, α_0 decreases to 1.1×10^{-4}, and α_1 approaches 1. Thus, only about 0.011% of the acetate containing species is HOAc. Notice that, at the half-titration point (25.00 mL), α_0 and α_1 are both 0.5. For polyprotic acids (see Chapter 15), the alpha values are very useful in illustrating the changes in solution composition during titrations.

Figure 14-8 Plots of relative amounts of acetic acid and acetate ion during a titration. The straight lines show the change in relative amounts of HOAc (α_0) and OAc⁻ (α_1) during the titration of 50.00 mL of 0.1000 M acetic acid. The curved line is the titration curve for the system.

FEATURE 14-6

Locating Titration End Points from pH Measurements

Although indicators are still widely used in acid/base titrations, the glass pH electrode and pH meter allow the direct measurement of pH as a function of titrant volume. The glass pH electrode is discussed in detail in Chapter 21. The titration curve for the titration of 50.00 mL of 0.1000 M weak acid ($K_a = 1.0 \times 10^{-5}$) with 0.1000 M NaOH is shown in **Figure 14F-4a**. The end point can be located in several ways from the pH versus volume data.

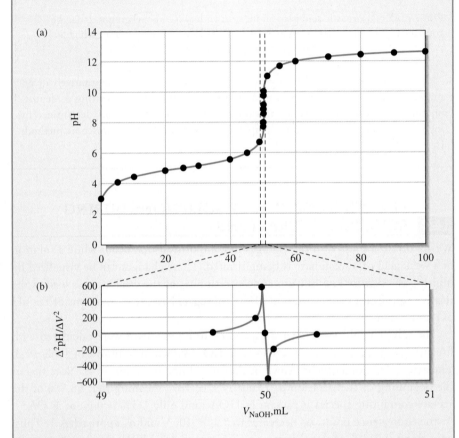

Figure 14F-4 In (a), the titration curve of 50.00 mL of 0.1000 M weak acid with 0.1000 M NaOH is shown as collected by a pH meter. In (b), the second derivative is shown on an expanded scale. Note that the second derivative crosses zero at the end point. This can be used to locate the end point very precisely.

The end point can be taken as the **inflection point** of the titration curve. With a sigmoid-shaped titration curve, the inflection point is the steepest part of the titration curve where the pH change with respect to volume is a maximum. This point can be estimated visually from the plot or we can use calculus to find the first and second derivatives of the titration curve. The first derivative, which is approximately $\Delta pH/\Delta V$, is the slope of the titration curve. It goes from nearly zero far before the end point to a maximum at the end point back to zero far beyond the end point. We can differentiate the curve a second time to locate the maximum of the first derivative since the slope of the first derivative changes dramatically from a large positive value to a large negative value as we pass through the maximum in the first derivative curve. This is the basis for locating the end point by taking the second derivative. The estimated second derivative, $\Delta^2 pH/\Delta V^2$, is zero at the end point as shown in **Figure 14F-4b**. Note that we have expanded the scale to make it easier to locate the zero crossing of the second derivative. The details of calculating the derivatives are given in Section 21G. The spreadsheet approach for obtaining these derivatives and making the plots is developed in Chapter 7 of *Applications of Microsoft® Excel in Analytical Chemistry*, 2nd ed.

The Gran plot is an alternative method for locating the end point in a titration. In this method, a linear plot is produced that can reveal both the acid dissociation constant and the volume of base required to reach the end point. Unlike the normal titration curve and derivative curves, which find the end point only from data located in the end point region, the Gran plot uses data far away from the end point. This method can decrease the tedium of making many measurements after dispensing tiny volumes of titrant in the end point region.

Prior to the equivalence point of the titration of a weak acid with a strong base, the concentration of acid remaining, c_{HA}, is given by

$$c_{HA} = \frac{\text{no. mmoles of HA initially present}}{\text{total volume of solution}} - \frac{\text{no. mmoles NaOH added}}{\text{total volume of solution}}$$

or

$$c_{HA} = \frac{c_{HA}^0 V_{HA}}{V_{HA} + V_{NaOH}} - \frac{c_{NaOH}^0 V_{NaOH}}{V_{HA} + V_{NaOH}}$$

where c_{HA}^0 is the initial analytical concentration of HA and c_{NaOH}^0 the initial concentration of base. The equivalence point volume of NaOH, V_{eq}, can be found from the stoichiometry, which for a 1:1 reaction is given by

$$c_{HA}^0 V_{HA} = c_{NaOH}^0 V_{eq}$$

Substituting for $c_{HA}^0 V_{HA}$ in the equation for c_{HA} and rearranging yield

$$c_{HA} = \frac{c_{NaOH}^0}{V_{HA} + V_{NaOH}} (V_{eq} - V_{NaOH})$$

If K_a is not too large, the equilibrium concentration of acid in the preequivalence point region is approximately equal to the analytical concentration (see Equation 9-27). That is

$$[HA] \approx c_{HA} \approx \frac{c_{NaOH}^0}{V_{HA} + V_{NaOH}} (V_{eq} - V_{NaOH})$$

With moderate dissociation of the acid, the equilibrium concentration of A^- at any point is approximately the number of millimoles of base added divided by the total solution volume.

$$[A^-] \approx \frac{c_{NaOH}^0 V_{NaOH}}{V_{HA} + V_{NaOH}}$$

(continued)

The concentration of H_3O^+ can be found from the equilibrium constant as

$$[H_3O^+] = \frac{K_a[HA]}{[A^-]} = \frac{K_a(V_{eq} - V_{NaOH})}{V_{NaOH}}$$

Multiplying both sides by V_{NaOH} gives,

$$[H_3O^+]V_{NaOH} = K_a V_{eq} - K_a V_{NaOH}$$

> When we rearrange this equation slightly, we have the slope-intercept form of a straight line,
>
> $$\underbrace{[H_3O^+]V_{NaOH}}_{y} = \underbrace{-K_a}_{m}\underbrace{V_{NaOH}}_{x} + \underbrace{K_a V_{eq}}_{b}$$
>
> or
>
> $$y = mx + b$$
>
> In which
>
> $y = [H_3O^+]V_{NaOH}$,
> $m = \text{slope} = -K_a$,
> $x = V_{NaOH}$, and
> $b = \text{intercept} = K_a V_{eq}$

A plot of the left-hand side of this equation versus the volume of titrant, V_{NaOH}, should yield a straight line with a slope of $-K_a$ and an intercept of $K_a V_{eq}$. In **Figure 14F-5**, a Gran plot of the titration of 50.00 mL of 0.1000 M weak acid ($K_a = 1.0 \times 10^{-5}$) with 0.1000 M NaOH is shown along with the least-squares equation. From the intercept value of 0.0005, we calculate an end point volume of 50.00 mL by dividing by the value for K_a. Usually, points in the middle stages of the titration are plotted and used to obtain the slope and intercept values. The Gran plot can exhibit curvature in the early stages if K_a is too large, and it can curve near the equivalence point.

Figure 14F-5 Gran plot for the titration of 50.00 mL of 0.1000 M weak acid ($K_a = 1.00 \times 10^{-5}$) with 0.1000 M NaOH. The least-squares equation for the line is given in the figure.

Spreadsheet Summary In the exercises in Chapter 7 of *Applications of Microsoft® Excel in Analytical Chemistry*, 2nd ed., we first use Excel to plot a simple distribution of species diagram (α plot) for a weak acid. Then, the first- and second-derivatives of the titration curve are plotted in order to better locate the titration end point. A combination plot is produced that simultaneously displays the pH versus volume curve and the second-derivative curve. Finally, a Gran plot is explored for locating the end point by a linear regression procedure.

QUESTIONS AND PROBLEMS

*14-1. Why does the typical acid/base indicator exhibit its color change over a range of about 2 pH units?

14-2. What factors affect end-point sharpness in an acid/base titration?

*14-3. Consider curves for the titration of 0.10 M NaOH and 0.010 M NH_3 with 0.10 M HCl.
 (a) Briefly account for the differences between curves for the two titrations.
 (b) In what respect will the two curves be indistinguishable?

14-4. Why are the standard reagents used in neutralization titrations generally strong acids and bases rather than weak acids and bases?

*14-5. What variables can cause the pH range of an indicator to shift?

14-6. Which solute would provide the sharper end point in a titration with 0.10 M HCl:
 *(a) 0.10 M NaOCl or 0.10 M hydroxylamine?
 (b) 0.10 M NH_3 or 0.10 M sodium phenolate?
 *(c) 0.10 M methylamine or 0.10 M hydroxylamine?
 (d) 0.10 M hydrazine or 0.10 M NaCN?

14-7. Which solute would provide the sharper end point in a titration with 0.10 M NaOH:
 *(a) 0.10 M nitrous acid or 0.10 M iodic acid?
 (b) 0.10 M anilinium hydrochloride ($C_6H_5NH_3Cl$) or 0.10 M benzoic acid?
 *(c) 0.10 M hypochlorous acid or 0.10 M pyruvic acid?
 (d) 0.10 M salicylic acid or 0.10 M acetic acid?

14-8. Before glass electrodes and pH meters became widely used, pH was often determined by measuring the concentration of the acid and base forms of the indicator colorimetrically (see Chapter 26 for details). If bromothymol blue is introduced into a solution and the concentration ratio of acid to base form is found to be 1.29, what is the pH of the solution?

*14-9. The procedure described in Problem 14-8 was used to determine pH with methyl orange as the indicator. The concentration ratio of the acid to base form of the indicator was 1.84. Calculate the pH of the solution.

14-10. Values for K_w at 0, 50, and 100°C are 1.14×10^{-15}, 5.47×10^{-14}, and 4.9×10^{-13}, respectively. Calculate the pH for a neutral solution at each of these temperatures.

14-11. Using the data in Problem 14-10, calculate pK_w at
 (a) 0°C.
 *(b) 50°C.
 (c) 100°C.

14-12. Using the data in Problem 14-10, calculate the pH of a 1.00×10^{-2} M NaOH solution at
 (a) 0°C.
 *(b) 50°C.
 (c) 100°C.

*14-13. What is the pH of an aqueous solution that is 3.00% HCl by mass and has a density of 1.015 g/mL?

14-14. Calculate the pH of a solution that contains 2.00% (w/w) NaOH and has a density of 1.022 g/mL.

*14-15. What is the pH of a solution that is 2.00×10^{-8} M in NaOH? (Hint: In such a dilute solution you must take into account the contribution of H_2O to the hydroxide ion concentration.)

14-16. What is the pH of a 2.00×10^{-8} M HCl solution (see hint in Problem 14-15)?

*14-17. What is the pH of the solution that results when 0.093 g of $Mg(OH)_2$ is mixed with
 (a) 75.0 mL of 0.0500 M HCl?
 (b) 100.0 mL of 0.0500 M HCl?
 (c) 15.0 mL of 0.0500 M HCl?
 (d) 30.0 mL of 0.0500 M $MgCl_2$?

14-18. Calculate the pH of the solution that results when mixing 20.0 mL of 0.1750 M HCl with 25.0 mL of
 (a) distilled water.
 (b) 0.132 M $AgNO_3$.
 (c) 0.132 M NaOH.
 (d) 0.132 M NH_3.
 (e) 0.232 M NaOH.

*14-19. Calculate the hydronium ion concentration and pH of a solution that is 0.0500 M in HCl
 (a) neglecting activities.
 (b) using activities (see Chapter 10).

14-20. Calculate the hydroxide ion concentration and the pH of a 0.0167 M $Ba(OH)_2$ solution
 (a) neglecting activities.
 (b) using activities (see Chapter 10).

*14-21. Calculate the pH of an aqueous solution that is
 (a) 1.00×10^{-1} M in HOCl.
 (b) 1.00×10^{-2} M in HOCl.
 (c) 1.00×10^{-4} M HOCl.

14-22. Calculate the pH of a solution that is
    (a) 1.00×10^{-1} M NaOCl.
    (b) 1.00×10^{-2} M NaOCl.
    (c) 1.00×10^{-4} M NaOCl.

*14-23.** Calculate the pH of an ammonia solution that is
    (a) 1.00×10^{-1} M NH_3.
    (b) 1.00×10^{-2} M NH_3.
    (c) 1.00×10^{-4} M NH_3.

14-24. Calculate the pH of a solution that is
    (a) 1.00×10^{-1} M NH_4Cl.
    (b) 1.00×10^{-2} M NH_4Cl.
    (c) 1.00×10^{-4} M NH_4Cl.

*14-25.** Calculate the pH of a solution in which the concentration of piperdine is
    (a) 1.00×10^{-1} M.
    (b) 1.00×10^{-2} M.
    (c) 1.00×10^{-4} M.

14-26. Calculate the pH of a solution that is
    (a) 1.00×10^{-1} M sulfamic acid.
    (b) 1.00×10^{-2} M sulfamic acid.
    (c) 1.00×10^{-4} M sulfamic acid.

*14-27.** Calculate the pH of a solution prepared by
    (a) dissolving 36.5 g of lactic acid in water and diluting to 500 mL.
    (b) diluting 25.0 mL of the solution in (a) to 250 mL.
    (c) diluting 10.0 mL of the solution in (b) to 1.00 L.

14-28. Calculate the pH of a solution prepared by
    (a) dissolving 2.13 g of picric acid, $(NO_2)_3C_6H_2OH$ (229.11 g/mol), in 100 mL of water.
    (b) diluting 10.0 mL of the solution in (a) to 100 mL.
    (c) diluting 10.0 mL of the solution in (b) to 1.00 L.

*14-29.** Calculate the pH of the solution that results when 20.0 mL of 0.1750 M formic acid is
    (a) diluted to 45.0 mL with distilled water.
    (b) mixed with 25.0 mL of 0.140 M NaOH solution.
    (c) mixed with 25.0 mL of 0.200 M NaOH solution.
    (d) mixed with 25.0 mL of 0.200 sodium formate solution.

14-30. Calculate the pH of the solution that results when 40.0 mL of 0.1250 M NH_3 is
    (a) diluted to 20.0 mL with distilled water.
    (b) mixed with 20.0 mL of 0.250 M HCl solution.
    (c) mixed with 20.0 mL of 0.300 M HCl solution.
    (d) mixed with 20.0 mL of 0.200 M NH_4Cl solution.
    (e) mixed with 20.0 mL of 0.100 M HCl solution.

*14-31.** A solution is 0.0500 M in NH_4Cl and 0.0300 M in NH_3. Calculate its OH^- concentration and its pH
    (a) neglecting activities.
    (b) taking activities into account.

14-32. What is the pH of a solution that is
    (a) prepared by dissolving 7.85 g of lactic acid (90.08 g/mol) and 10.09 g of sodium lactate (112.06 g/mol) in water and diluting to 1.00 L?
    (b) 0.0630 M in acetic acid and 0.0210 M in sodium acetate?

    (c) prepared by dissolving 3.00 g of salicylic acid, $C_6H_4(OH)COOH$ (138.12 g/mol), in 50.0 mL of 0.1130 M NaOH and diluting to 500.0 mL?
    (d) 0.0100 M in picric acid and 0.100 M in sodium picrate?

*14-33.** What is the pH of a solution that is
    (a) prepared by dissolving 3.30 g of $(NH_4)_2SO_4$ in water, adding 125.0 mL of 0.1011 M NaOH, and diluting to 500.0 mL?
    (b) 0.120 M in piperidine and 0.010 M in its chloride salt?
    (c) 0.050 M in ethylamine and 0.167 M in its chloride salt?
    (d) prepared by dissolving 2.32 g of aniline (93.13 g/mol) in 100 mL of 0.0200 M HCl and diluting to 250.0 mL?

14-34. Calculate the change in pH that occurs in each of the solutions listed below as a result of a tenfold dilution with water. Round calculated values for pH to three figures to the right of the decimal point.
    *(a) H_2O.
    (b) 0.0500 M HCl.
    *(c) 0.0500 M NaOH.
    (d) 0.0500 M CH_3COOH.
    *(e) 0.0500 M CH_3COONa.
    (f) 0.0500 M CH_3COOH + 0.0500 M CH_3COONa.
    *(g) 0.500 M CH_3COOH + 0.500 M CH_3COONa.

*14-35.** Calculate the change in pH that occurs when 1.00 mmol of a strong acid is added to 100 mL of the solutions listed in Problem 14-34.

14-36. Calculate the change in pH that occurs when 1.00 mmol of a strong base is added to 100 mL of the solutions listed in Problem 14-34. Calculate values to three decimal places.

14-37. Calculate the change in pH to three decimal places that occurs when 0.50 mmol of a strong acid is added to 100 mL of
    (a) 0.0200 M lactic acid + 0.0800 M sodium lactate.
    *(b) 0.0800 M lactic acid + 0.0200 M sodium lactate.
    (c) 0.0500 M lactic acid + 0.0500 M sodium lactate.

14-38. A 50.00-mL aliquot of 0.1000 M NaOH is titrated with 0.1000 M HCl. Calculate the pH of the solution after the addition of 0.00, 10.00, 25.00, 40.00, 45.00, 49.00, 50.00, 51.00, 55.00, and 60.00 mL of acid and prepare a titration curve from the data.

*14-39.** In a titration of 50.00 mL of 0.05000 M formic acid with 0.1000 M KOH, the titration error must be smaller than 0.05 mL. What indicator can be chosen to realize this goal?

14-40. In a titration of 50.00 mL of 0.1000 M ethylamine with 0.1000 M $HClO_4$, the titration error must be no more than 0.05 mL. What indicator can be chosen to realize this goal?

14-41. Calculate the pH after addition of 0.00, 5.00, 15.00, 25.00, 40.00, 45.00, 49.00, 50.00, 51.00, 55.00, and 60.00 mL of 0.1000 M NaOH in the titration of 50.00 mL of

  *(a) 0.1000 M HNO_2.

  (b) 0.1000 M pyridinium chloride.

  *(c) 0.1000 M lactic acid.

14-42. Calculate the pH after addition of 0.00, 5.00, 15.00, 25.00, 40.00, 45.00, 49.00, 50.00, 51.00, 55.00, and 60.00 mL of 0.1000 M HCl in the titration of 50.00 mL of

  (a) 0.1000 M ammonia.

  (b) 0.1000 M hydrazine.

  (c) 0.1000 M sodium cyanide.

14-43. Calculate the pH after addition of 0.00, 5.00, 15.00, 25.00, 40.00, 49.00, 50.00, 51.00, 55.00, and 60.00 mL of reagent in the titration of 50.0 mL of

  *(a) 0.01000 M chloroacetic acid with 0.01000 M NaOH.

  (b) 0.1000 M anilinium chloride with 0.1000 M NaOH.

  *(c) 0.1000 M hypochlorous acid with 0.1000 M NaOH.

  (d) 0.1000 M hydroxylamine with 0.1000 M HCl. Construct titration curves from the data.

14-44. Calculate α_0 and α_1 for

  *(a) acetic acid in a solution with a pH of 5.320.

  (b) picric acid in a solution with a pH of 1.250.

  *(c) hypochlorous acid in a solution with a pH of 7.00.

  (d) hydroxylamine acid in a solution with a pH of 5.12.

  *(e) piperidine in a solution with a pH of 10.08.

*14-45. Calculate the equilibrium concentration of methyl ammonia in a solution that has a molar analytical CH_3NH_2 concentration of 0.120 and a pH of 11.471.

14-46. Calculate the equilibrium concentration of undissociated HCOOH in a formic acid solution with an analytical formic acid concentration of 0.0850 and a pH of 3.200.

14-47. Supply the missing data in the table below.

Acid	Molar Analytical Concentration, c_T ($c_T = c_{HA} + c_{A^-}$)	pH	[HA]	[A$^-$]	α_0	α_1
*Lactic	0.120	——	——	——	0.640	——
Iodic	0.200	——	——	——	——	0.765
*Butanoic	——	5.00	0.644	——	——	——
Hypochlorous	0.280	7.00	——	——	——	——
Nitrous	——	——	——	0.105	0.413	0.587
Hydrogen cyanide	——	——	0.145	0.221	——	——
*Sulfamic	0.250	1.20	——	——	——	——

14-48. Challenge Problem: This photo shows a buret that has at least two defects on the scale that were created during its fabrication.

A mislabeled buret.

Answer the following questions about the buret, its origin, and its use.

  (a) Under what conditions is the buret usable?

  (b) Assuming that the user does not notice the defects in the buret, what type of error would occur if the liquid level was between the second 43-mL mark and the 48-mL mark?

  (c) Assume that the initial reading in a titration is 0.00 mL (very unlikely) and calculate the relative error in the volume if the final reading is 43.00 (upper mark). What is the relative error if the same reading is made on the lower mark? Perform the same calculation for a final reading made at the 48.00-mL mark. What do these calculations demonstrate about the type of error caused by the defect in the buret?

  (d) Speculate on the age of the buret. How would you suspect that the markings were made on the glass? Is it likely that the same type of defect would appear on a buret manufactured today? Explain the rationale for your answer.

  (e) Modern electronic chemical instruments such as pH meters, balances, titrators, and spectrophotometers are normally assumed to be free of manufacturing defects analogous to the one illustrated in the photo. Comment on the wisdom of making such an assumption.

  (f) Burets in automated titrators contain a motor connected to a screw-driven plunger that delivers titrant in much the same way that a hypodermic syringe delivers liquids. The distance of travel of the plunger is proportional to the volume of liquid delivered. What kinds of manufacturer's defects would lead to inaccuracy or imprecision in the volume dispensed by these devices?

  (g) What steps can you take to avoid measurement errors while using modern chemical instruments?

CHAPTER 15

Complex Acid/Base Systems

Professors Pietro M. Motta & Silvia Correr / Photo
Researchers, Inc.

Polyfunctional acids and bases play important roles in many chemical and biological systems.
The human body contains a complicated system of buffers within cells and within bodily fluids,
such as human blood. Shown here is a scanning electron micrograph of red blood cells traveling through an artery. The pH of human blood is controlled to be within the range of 7.35 to
7.45, primarily by the carbonic acid-bicarbonate buffer system:

$$CO_2(g) + H_2O(l) \rightleftharpoons H_2CO_3(aq)$$
$$H_2CO_3(aq) + H_2O(l) \rightleftharpoons H_3O^+(aq) + HCO_3^-(aq)$$

This chapter describes polyfunctional acid and base systems including buffer solutions. Calculations of pH and of titration curves are also described.

In this chapter, we describe methods for treating complex acid/base systems, including
the calculation of titration curves. We define complex systems as solutions made
up of (1) two acids or two bases of different strengths, (2) an acid or a base that has
two or more acidic or basic functional groups, or (3) an amphiprotic substance, which
is capable of acting as both an acid and a base. For more than one equilibrium, chemical
reactions and algebraic equations are required to describe the characteristics of any
of these systems.

15A MIXTURES OF STRONG AND WEAK ACIDS OR STRONG AND WEAK BASES

Each of the components in a mixture containing a strong acid and a weak acid (or
a strong base and a weak base) can be determined provided that the concentrations
of the two are of the same order of magnitude and that the dissociation constant for
the weak acid or base is somewhat less than about 10^{-4}. To demonstrate that this
statement is true, Example 15-1 shows how a titration curve can be constructed for
a solution containing roughly equal concentrations of HCl and HA, where HA is a
weak acid with a dissociation constant of 10^{-4}.

EXAMPLE 15-1

Calculate the pH of a mixture that is 0.1200 M in hydrochloric acid and 0.0800 M in the weak acid HA (K_a = 1.00 × 10^{-4}) during its titration with 0.1000 M KOH. Compute results for additions of the following volumes of base: (a) 0.00 mL and (b) 5.00 mL.

Solution

(a) 0.00 mL KOH

The molar hydronium ion concentration in this mixture is equal to the concentration of HCl plus the concentration of hydronium ions that results from dissociation of HA and H_2O. In the presence of the two acids, however, we can be certain that the concentration of hydronium ions from the dissociation of water is extremely small. We, therefore, need to take into account only the other two sources of protons. Thus, we may write

$$[H_3O^+] = c_{HCl}^0 + [A^-] = 0.1200 + [A^-]$$

Note that $[A^-]$ is equal to the concentration of hydronium ions from the dissociation of HA.

Now, assume that the presence of the strong acid so represses the dissociation of HA that $[A^-] \ll$ 0.1200 M; then,

$$[H_3O^+] \approx 0.1200 \text{ M, and the pH is 0.92}$$

To check this assumption, the provisional value for $[H_3O^+]$ is substituted into the dissociation-constant expression for HA. When this expression is rearranged, we obtain

$$\frac{[A^-]}{[HA]} = \frac{K_a}{[H_3O^+]} = \frac{1.00 \times 10^{-4}}{0.1200} = 8.33 \times 10^{-4}$$

This expression can be rearranged to

$$[HA] = [A^-]/(8.33 \times 10^{-4})$$

From the concentration of the weak acid, we can write the mass-balance expression

$$c_{HA}^0 = [HA] + [A^-] = 0.0800 \text{ M}$$

Substituting the value of [HA] from the previous equation gives

$$[A^-]/(8.33 \times 10^{-4}) + [A^-] \approx (1.20 \times 10^3) [A^-] = 0.0800 \text{ M}$$
$$[A^-] = 6.7 \times 10^{-5} \text{ M}$$

We see that $[A^-]$ is indeed much smaller than 0.1200 M, as assumed.

(b) 5.00 mL KOH

$$c_{HCl} = \frac{25.00 \times 0.1200 - 5.00 \times 0.100}{25.00 + 5.00} = 0.0833 \text{ M}$$

and we may write

$$[H_3O^+] = 0.0833 + [A^-] \approx 0.0833 \text{ M}$$
$$pH = 1.08$$

(continued)

To determine whether our assumption is still valid, we compute $[A^-]$ as we did in part (a), knowing that the concentration of HA is now $0.0800 \times 25.00/30.00 = 0.0667$, and find

$$[A^-] = 8.0 \times 10^{-5} \text{ M}$$

which is still much smaller than 0.0833 M.

Example 15-1 demonstrates that hydrochloric acid suppresses the dissociation of the weak acid in the early stages of the titration to such an extent that we can assume that $[A^-] \ll c_{HCl}$ and $[H_3O^+] = c_{HCl}$. In other words, the hydronium ion concentration is simply the molar concentration of the strong acid.

The approximation used in Example 15-1 can be shown to apply until most of the hydrochloric acid has been neutralized by the titrant. Therefore, the curve in the early stages of the titration *is identical to that for a 0.1200 M solution of a strong acid by itself*. As shown by Example 15-2, however, the presence of HA must be taken into account as the first end point in the titration is approached.

EXAMPLE 15-2

Calculate the pH of the resulting solution after the addition of 29.00 mL of 0.1000 M NaOH to 25.00 mL of the solution described in Example 15-1.

Solution

In this case,

$$c_{HCl} = \frac{25.00 \times 0.1200 - 29.00 \times 0.1000}{25.00 + 29.00} = 1.85 \times 10^{-3} \text{ M}$$

$$c_{HA} = \frac{25.00 \times 0.0800}{54.00} = 3.70 \times 10^{-2} \text{ M}$$

As in the previous example, a provisional result based on the assumption that $[H_3O^+] = 1.85 \times 10^{-3}$ M yields a value of 1.90×10^{-3} M for $[A^-]$. We see that $[A^-]$ is no longer much smaller than $[H_3O^+]$, and we must write

$$[H_3O^+] = c_{HCl} + [A^-] = 1.85 \times 10^{-3} + [A^-] \tag{15-1}$$

In addition, from mass-balance considerations, we know that

$$[HA] + [A^-] = c_{HA} = 3.70 \times 10^{-2} \tag{15-2}$$

We rearrange the acid dissociation-constant expression for HA and obtain

$$[HA] = \frac{[H_3O^+][A^-]}{1.00 \times 10^{-4}}$$

Substitution of this expression into Equation 15-2 yields

$$\frac{[H_3O^+][A^-]}{1.00 \times 10^{-4}} + [A^-] = 3.70 \times 10^{-2}$$

$$[A^-] = \frac{3.70 \times 10^{-6}}{[H_3O^+] + 1.00 \times 10^{-4}}$$

Substitution for $[A^-]$ and c_{HCl} in Equation 15-1 yields

$$[H_3O^+] = 1.85 \times 10^{-3} + \frac{3.70 \times 10^{-6}}{[H_3O^+] + 1.00 \times 10^{-4}}$$

Multiplying through to clear the denominator and collecting terms gives

$$[H_3O^+]^2 - (1.75 \times 10^{-3})[H_3O^+] - 3.885 \times 10^{-6} = 0$$

Solving the quadratic equation gives

$$[H_3O^+] = 3.03 \times 10^{-3} \text{ M}$$

$$pH = 2.52$$

Note that the contributions to the hydronium ion concentration from HCl (1.85×10^{-3} M) and HA (3.03×10^{-3} M $- 1.85 \times 10^{-3}$ M) are of comparable magnitude. Hence, we cannot make the assumption that we made in Example 15-1.

When the amount of base added is equivalent to the amount of hydrochloric acid originally present, the solution is identical in all respects to one prepared by dissolving appropriate quantities of the weak acid and sodium chloride in a suitable volume of water. The sodium chloride, however, has no effect on the pH (neglecting the increased ionic strength); thus, the remainder of the titration curve is identical to that for a dilute solution of HA.

The shape of the curve for a mixture of weak and strong acids, and hence the information that may be derived from it, depends in large measure on the strength of the weak acid. **Figure 15-1** illustrates the pH changes that occur during the titration of mixtures containing hydrochloric acid and several weak acids with different dissociation constants. Note that the rise in pH at the first equivalence point is small or essentially nonexistent when the weak acid has a relatively large dissociation constant (curves A and B). For titrations such as these, only the total number of millimoles of weak and strong acid can be determined accurately. Conversely, when the weak acid has a very small dissociation constant, only the strong acid content can be determined. For weak acids of intermediate strength (K_a somewhat less than 10^{-4} but greater than 10^{-8}), there are usually two useful end points.

It is also possible to determine the amount of each component in a mixture that contains a strong base and a weak base, subject to the constraints just described for the strong acid/weak acid system. The construction of titration curves for mixtures of bases is analogous to that for mixtures of acids.

> The composition of a mixture of a strong acid and a weak acid can be determined by titration with suitable indicators if the weak acid has a dissociation constant that lies between 10^{-4} and 10^{-8} and the concentrations of the two acids are of the same order of magnitude.

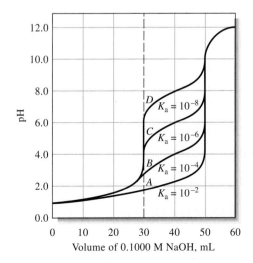

Figure 15-1 Curves for the titration of strong/weak acid mixtures with 0.1000 M NaOH. Each titration curve is for 25.00 mL of a solution that is 0.1200 M in HCl and 0.0800 M in the weak acid HA.

15B POLYFUNCTIONAL ACIDS AND BASES

There are several species of interest in analytical chemistry that have two or more acidic or basic functional groups. These species are said to exhibit polyfunctional acidic or basic behavior. Generally, with a polyfunctional acid such as phosphoric acid (H_3PO_4), the protonated species (H_3PO_4, $H_2PO_4^-$, HPO_4^{2-}) differ enough in their dissociation constants that they exhibit multiple end points in a neutralization titration.

15B-1 The Phosphoric Acid System

Phosphoric acid is a typical polyfunctional acid. In aqueous solution, it undergoes the following three dissociation reactions:

> Throughout the remainder of this chapter, we use K_{a1}, K_{a2} to represent the first and second dissociation constants of acids and K_{b1}, K_{b2} to represent the stepwise constants of bases.

$$H_3PO_4 + H_2O \rightleftharpoons H_2PO_4^- + H_3O^+ \qquad K_{a1} = \frac{[H_3O^+][H_2PO_4^-]}{[H_3PO_4]}$$

$$= 7.11 \times 10^{-3}$$

> Generally, $K_{a1} > K_{a2}$, often by a factor of 10^4 to 10^5 because of electrostatic forces. That is, the first dissociation involves separating a single positively charged hydronium ion from a singly charged anion. In the second step, the hydronium ion must be separated from a doubly charged anion, a process that requires considerably more energy.

$$H_2PO_4^- + H_2O \rightleftharpoons HPO_4^{2-} + H_3O^+ \qquad K_{a2} = \frac{[H_3O^+][HPO_4^{2-}]}{[H_2PO_4^-]}$$

$$= 6.32 \times 10^{-8}$$

$$HPO_4^{2-} + H_2O \rightleftharpoons PO_4^{3-} + H_3O^+ \qquad K_{a3} = \frac{[H_3O^+][PO_4^{3-}]}{[HPO_4^{2-}]}$$

$$= 4.5 \times 10^{-13}$$

> A second reason that $K_{a1} > K_{a2}$ is a statistical one. In the first step, a proton can be removed from more locations than in the second and third steps.

When we add two adjacent stepwise equilibria, we multiply the two equilibrium constants to obtain the equilibrium constant for the resulting overall reaction. Thus, for the first two dissociation equilibria for H_3PO_4, we write

$$H_3PO_4 + 2H_2O \rightleftharpoons HPO_4^{2-} + 2H_3O^+ \qquad K_{a1}K_{a2} = \frac{[H_3O^+]^2[HPO_4^{2-}]}{[H_3PO_4]}$$

$$= 7.11 \times 10^{-3} \times 6.32 \times 10^{-8} = 4.49 \times 10^{-10}$$

Similarly, for the reaction

$$H_3PO_4 + 3H_2O \rightleftharpoons 3H_3O^+ + PO_4^{3-}$$

we may write

$$K_{a1}K_{a2}K_{a3} = \frac{[H_3O^+]^3[PO_4^{3-}]}{[H_3PO_4]}$$

$$= 7.11 \times 10^{-3} \times 6.32 \times 10^{-8} \times 4.5 \times 10^{-13} = 2.0 \times 10^{-22}$$

15B-2 The Carbon Dioxide/Carbonic Acid System

When carbon dioxide is dissolved in water, a dibasic acid system is formed by the following reactions:

$$CO_2(aq) + H_2O \rightleftharpoons H_2CO_3 \qquad K_{hyd} = \frac{[H_2CO_3]}{[CO_2(aq)]} = 2.8 \times 10^{-3} \qquad (15\text{-}3)$$

$$H_2CO_3 + H_2O \rightleftharpoons H_3O^+ + HCO_3^-$$

$$K_1 = \frac{[H_3O^+][HCO_3^-]}{[H_2CO_3]} = 1.5 \times 10^{-4} \tag{15-4}$$

$$HCO_3^- + H_2O \rightleftharpoons H_3O^+ + CO_3^{2-}$$

$$K_2 = \frac{[H_3O^+][CO_3^{2-}]}{[HCO_3^-]} = 4.69 \times 10^{-11} \tag{15-5}$$

The first reaction describes the hydration of aqueous CO_2 to form carbonic acid. Note that the magnitude of K_{hyd} indicates that the concentration of $CO_2(aq)$ is much larger than the concentration of H_2CO_3 (that is, $[H_2CO_3]$ is only about 0.3% that of $[CO_2(aq)]$). Thus, a more useful way of discussing the acidity of solutions of carbon dioxide is to combine Equation 15-3 and 15-4 to give

$$CO_2(aq) + 2H_2O \rightleftharpoons H_3O^+ + HCO_3^- \quad K_{a1} = \frac{[H_3O^+][HCO_3^-]}{[CO_2(aq)]} \tag{15-6}$$

$$= 2.8 \times 10^{-3} \times 1.5 \times 10^{-4}$$

$$= 4.2 \times 10^{-7}$$

$$HCO_3^- + H_2O \rightleftharpoons H_3O^+ + CO_3^{2-} \quad K_{a2} = 4.69 \times 10^{-11} \tag{15-7}$$

EXAMPLE 15-3

Calculate the pH of a solution that is 0.02500 M CO_2.

Solution

The mass-balance expression for CO_2-containing species is

$$c_{CO_2}^0 = 0.02500 = [CO_2(aq)] + [H_2CO_3] + [HCO_3^-] + [CO_3^{2-}]$$

The small magnitude of K_{hyd}, K_1, and K_2 (see Equations 15-3, 15-4, and 15-5) suggests that

$$([H_2CO_3] + [HCO_3^-] + [CO_3^{2-}]) \ll [CO_2(aq)]$$

and we may write

$$[CO_2(aq)] \approx c_{CO_2}^0 = 0.02500 \text{ M}$$

The charge-balance equation is

$$[H_3O^+] = [HCO_3^-] + 2[CO_3^{2-}] + [OH^-]$$

We will then assume that

$$2[CO_3^{2-}] + [OH^-] \ll [HCO_3^-]$$

Therefore,

$$[H_3O^+] \approx [HCO_3^-]$$

(continued)

Substituting these approximations in Equation 15-6 leads to

$$\frac{[H_3O^+]^2}{0.02500} = K_{a1} = 4.2 \times 10^{-7}$$

$$[H_3O^+] = \sqrt{0.02500 \times 4.2 \times 10^{-7}} = 1.02 \times 10^{-4}\,M$$

$$pH = -\log(1.02 \times 10^{-4}) = 3.99$$

Calculating values for $[H_2CO_3]$, $[CO_3^{2-}]$, and $[OH^-]$ indicates that the assumptions were valid.

CHALLENGE: Write a sufficient number of equations to make it possible to calculate the concentrations of all species in a solution containing known molar analytical concentrations of Na_2CO_3 and $NaHCO_3$.

The pH of polyfunctional systems, such as phosphoric acid or sodium carbonate, can be computed rigorously through use of the systematic approach to multiequilibrium problems described in Chapter 11. Manually solving the several simultaneous equations that are involved can be difficult and time consuming, but a computer can simplify the work dramatically.[1] In many cases, simplifying assumptions can be made when the successive equilibrium constants for the acid (or base) differ by a factor of about 10^3 or more. These assumptions can make it possible to compute pH data for titration curves by the techniques discussed in earlier chapters.

BUFFER SOLUTIONS INVOLVING
15C **POLYPROTIC ACIDS**

Two buffer systems can be prepared from a weak dibasic acid and its salts. The first consists of free acid H_2A and its conjugate base NaHA, and the second makes use of the acid NaHA and its conjugate base Na_2A. The pH of the NaHA/Na_2A system is higher than that of the H_2A/NaHA system because the acid dissociation constant for HA^- is always less than that for H_2A.

We can write enough independent equations to permit a rigorous calculation of the hydronium ion concentration for either of these systems. Ordinarily, however, it is permissible to introduce the simplifying assumption that only one of the equilibria is important in determining the hydronium ion concentration of the solution. Thus, for a buffer prepared from H_2A and NaHA, the dissociation of HA^- to yield A^{2-} can usually be neglected so that the calculation is based only on the first dissociation. With this simplification, the hydronium ion concentration is calculated by the method described in Section 9C-1 for a simple buffer solution. As shown in Example 15-4, the validity of the assumption can be checked by calculating an approximate concentration of A^{2-} and comparing this value with the concentrations of H_2A and HA^-.

---

[1]See S. R. Crouch and F. J. Holler, *Applications of Microsoft® Excel in Analytical Chemistry*, 2nd ed., Ch. 6, Belmont, CA: Brooks/Cole, 2014.

EXAMPLE 15-4

Calculate the hydronium ion concentration for a buffer solution that is 2.00 M in phosphoric acid and 1.50 M in potassium dihydrogen phosphate.

Solution

The principal equilibrium in this solution is the dissociation of H_3PO_4.

$$H_3PO_4 + H_2O \rightleftharpoons H_3O^+ + H_2PO_4^- \qquad K_{a1} = \frac{[H_3O^+][H_2PO_4^-]}{[H_3PO_4]}$$

$$= 7.11 \times 10^{-3}$$

We assume that the dissociation of $H_2PO_4^-$ is negligible, that is, $[HPO_4^{2-}]$ and $[PO_4^{3-}] \ll [H_2PO_4^-]$ and $[H_3PO_4]$. Then,

$$[H_3PO_4] \approx c^0_{H_3PO_4} = 2.00 \text{ M}$$

$$[H_2PO_4^-] \approx c^0_{KH_2PO_4} = 1.50 \text{ M}$$

$$[H_3O^+] = \frac{7.11 \times 10^{-3} \times 2.00}{1.50} = 9.49 \times 10^{-3} \text{ M}$$

We now use the equilibrium constant expression for K_{a2} to see if our assumption was valid.

$$K_{a2} = 6.34 \times 10^{-8} = \frac{[H_3O^+][HPO_4^{2-}]}{[H_2PO_4^-]} = \frac{9.48 \times 10^{-3}[HPO_4^{2-}]}{1.50}$$

Solving this equation yields

$$[HPO_4^{2-}] = 1.00 \times 10^{-5} \text{ M}$$

Since this concentration is much smaller than the concentrations of the major species, H_3PO_4 and $H_2PO_4^-$, our assumption is valid. Note that $[PO_4^{3-}]$ is even smaller than $[HPO_4^{2-}]$.

For a buffer prepared from NaHA and Na_2A, the second dissociation usually predominates, and the equilibrium

$$HA^- + H_2O \rightleftharpoons H_2A + OH^-$$

can be neglected. The concentration of H_2A is negligible compared with that of HA^- or A^{2-}. The hydronium ion concentration can then be calculated from the second dissociation constant by the techniques for a simple buffer solution. To test the assumption, we compare an estimate of the H_2A concentration with the concentrations of HA^- and A^{2-}, as in Example 15-5.

EXAMPLE 15-5

Calculate the hydronium ion concentration of a buffer that is 0.0500 M in potassium hydrogen phthalate (KHP) and 0.150 M in potassium phthalate (K_2P).

$$HP^- + H_2O \rightleftharpoons H_3O^+ + P^{2-} \qquad K_{a2} = \frac{[H_3O^+][P^{2-}]}{[HP^-]} = 3.91 \times 10^{-6}$$

Solution

We will make the assumption that the concentration of H_2P is negligible in this solution.
Therefore,

$$[HP^-] \approx c_{KHP}^0 = 0.0500 \text{ M}$$

$$[P^{2-}] \approx c_{K_2P} = 0.150 \text{ M}$$

$$[H_3O^+] = \frac{3.91 \times 10^{-6} \times 0.0500}{0.150} = 1.30 \times 10^{-6} \text{ M}$$

To check the first assumption, an approximate value for $[H_2P]$ is calculated by substituting numerical values for $[H_3O^+]$ and $[HP^-]$ into the K_{a1} expression:

$$K_{a1} = \frac{[H_3O^+][HP^-]}{[H_2P]} = 1.12 \times 10^{-3} = \frac{(1.30 \times 10^{-6})(0.0500)}{[H_2P]}$$

$$[H_2P] = 6 \times 10^{-5} \text{ M}$$

Since $[H_2P] \ll [HP]$ and $[P^{2-}]$, our assumption that the reaction of HP^- to form OH^- is negligible is justified.

In all but a few situations, the assumption of a single principal equilibrium, as illustrated in Examples 15-4 and 15-5, provides a satisfactory estimate of the pH of buffer mixtures derived from polybasic acids. Appreciable errors occur, however, when the concentration of the acid or the salt is very low or when the two dissociation constants are numerically close. In these cases, a more rigorous calculation is needed.

15D CALCULATION OF THE pH OF SOLUTIONS OF NaHA

We have not yet considered how to calculate the pH of solutions of salts that have both acidic and basic properties, that is, salts that are amphiprotic. Such salts are formed during neutralization titrations of polyfunctional acids and bases. For example, when 1 mol of NaOH is added to a solution containing 1 mol of the acid H_2A, 1 mol of NaHA is formed. The pH of this solution is determined by two equilibria established between HA^- and water:

$$HA^- + H_2O \rightleftharpoons A^{2-} + H_3O^+$$

and

$$HA^- + H_2O \rightleftharpoons H_2A + OH^-$$

If the first reaction predominates, the solution will be acidic. If the second predominates, the solution will be basic. The relative magnitudes of the equilibrium constants for these processes determine whether a solution of NaHA is acidic or basic.

$$K_{a2} = \frac{[H_3O^+][A^{2-}]}{[HA^-]} \tag{15-8}$$

$$K_{b2} = \frac{K_w}{K_{a1}} = \frac{[H_2A][OH^-]}{[HA^-]} \tag{15-9}$$

where K_{a1} and K_{a2} are the acid dissociation constants for H_2A and K_{b2} is the basic dissociation constant for HA^-. If K_{b2} is greater than K_{a2}, the solution is basic. It is acidic if K_{a2} exceeds K_{b2}.

To derive an expression for the hydronium ion concentration of a solution of HA^-, we use the systematic approach described in Section 11A. We first write the mass-balance expression.

$$c_{NaHA} = [HA^-] + [H_2A] + [A^{2-}] \tag{15-10}$$

The charge-balance equation is

$$[Na^+] + [H_3O^+] = [HA^-] + 2[A^{2-}] + [OH^-]$$

Since the sodium ion concentration is equal to the molar analytical concentration of NaHA, the last equation can be rewritten as

$$c_{NaHA} + [H_3O^+] = [HA^-] + 2[A^{2-}] + [OH^-] \tag{15-11}$$

We now have four algebraic equations (Equations 15-10 and 15-11 and the two dissociation constant expressions for H_2A) and need one additional expression to solve for the five unknowns. The ion-product constant for water serves this purpose:

$$K_w = [H_3O^+][OH^-]$$

The rigorous solution of these five equations in five unknowns is somewhat difficult, but computer methods have made the task less formidable than previously.[2] A reasonable approximation, applicable to solutions of most acid salts, can be used, however, to simplify the problem. We first subtract the mass-balance equation from the charge-balance equation.

$$c_{NaHA} + [H_3O^+] = [HA^-] + 2[A^{2-}] + [OH^-] \quad \text{charge balance}$$

$$c_{NaHA} = [H_2A] + [HA^-] + [A^{2-}] \quad \text{mass balance}$$

$$[H_3O^+] = [A^{2-}] + [OH^-] - [H_2A] \tag{15-12}$$

---

[2]See S.R. Crouch and F. J. Holler, *Applications of Microsoft® Excel in Analytical Chemistry*, 2nd ed., Ch. 6, Belmont, CA: Brooks/Cole 2014.

We then rearrange the acid-dissociation constant expressions for H_2A and HA^- to give

$$[H_2A] = \frac{[H_3O^+][HA^-]}{K_{a1}}$$

$$[A^{2-}] = \frac{K_{a2}[HA^-]}{[H_3O^+]}$$

Substituting these expressions and that for K_w into Equation 15-12 yields

$$[H_3O^+] = \frac{K_{a2}[HA^-]}{[H_3O^+]} + \frac{K_w}{[H_3O^+]} - \frac{[H_3O^+][HA^-]}{K_{a1}}$$

Multiplying through by $[H_3O^+]$ gives

$$[H_3O^+]^2 = K_{a2}[HA^-] + K_w - \frac{[H_3O^+]^2[HA^-]}{K_{a1}}$$

We collect terms to obtain

$$[H_3O^+]^2\left(\frac{[HA^-]}{K_{a1}} + 1\right) = K_{a2}[HA^-] + K_w$$

This equation rearranges to

$$[H_3O^+] = \sqrt{\frac{K_{a2}[HA^-] + K_w}{1 + [HA^-]/K_{a1}}} \tag{15-13}$$

Under most circumstances, we can make the approximation that

$$[HA^-] \approx c_{NaHA} \tag{15-14}$$

Substituting this relationship into Equation 15-13 gives

$$[H_3O^+] = \sqrt{\frac{K_{a2}c_{NaHA} + K_w}{1 + c_{NaHA}/K_{a1}}} \tag{15-15}$$

The approximation shown as Equation 15-14 requires that $[HA^-]$ be much larger than any of the other equilibrium concentrations in Equations 15-10 and 15-11. This assumption is not valid for very dilute solutions of NaHA or in situations where K_{a2} or K_w/K_{a1} is relatively large.

Frequently, the ratio c_{NaHA}/K_{a1} is much larger than unity in the denominator of Equation 15-15, and $K_{a2}c_{NaHA}$ is considerably greater than K_w in the numerator. In this case, Equation 15-15 simplifies to

Always check the assumptions that are inherent in Equation 15-16.

$$[H_3O^+] = \sqrt{K_{a1}K_{a2}} \tag{15-16}$$

Note that Equation 15-16 does not contain c_{NaHA}, which implies that the pH of solutions of this type remains constant over a considerable range of solution concentrations where the assumptions are valid.

EXAMPLE 15-6

Calculate the hydronium ion concentration of a 1.00×10^{-3} M Na_2HPO_4 solution.

Solution

The pertinent dissociation constants are K_{a2} and K_{a3}, which both contain $[HPO_4^{2-}]$. Their values are $K_{a2} = 6.32 \times 10^{-8}$ and $K_{a3} = 4.5 \times 10^{-13}$. In the case of a Na_2HPO_4 solution, Equation 15-15 can be written

$$[H_3O^+] = \sqrt{\frac{K_{a3}c_{NaHA} + K_w}{1 + c_{NaHA}/K_{a2}}}$$

Note that we have used K_{a3} in place of K_{a2} in Equation 15-15 and K_{a2} in place of K_{a1} since these are the appropriate dissociation constants when Na_2HPO_4 is the salt.

   If we consider again the assumptions that led to Equation 15-16, we find that the term $c_{NaHA}/K_{a2} = (1.0 \times 10^{-3})/(6.32 \times 10^{-8})$ is much larger than 1 so that the denominator can be simplified. In the numerator, however, $K_{a3}c_{NaHA} = 4.5 \times 10^{-13} \times 1.00 \times 10^{-3}$ is comparable to K_w so that no simplification can be made there. We, therefore, use a partially simplified version of Equation 15-15:

$$[H_3O^+] = \sqrt{\frac{K_{a3}c_{NaHA} + K_w}{c_{NaHA}/K_{a2}}}$$

$$= \sqrt{\frac{(4.5 \times 10^{-13})(1.00 \times 10^{-3}) + 1.00 \times 10^{-14}}{(1.00 \times 10^{-3})/(6.32 \times 10^{-8})}} = 8.1 \times 10^{-10} \text{ M}$$

The simplified Equation 15-15 gave 1.7×10^{-10} M, which is in error by a large amount.

EXAMPLE 15-7

Find the hydronium ion concentration of a 0.0100 M NaH_2PO_4 solution.

Solution

The two dissociation constants of importance (those containing $[H_2PO_4^{2-}]$ are $K_{a1} = 7.11 \times 10^{-3}$ and $K_{a2} = 6.32 \times 10^{-8}$). A test shows that the denominator of Equation 15-15 cannot be simplified, but the numerator reduces to $K_{a2}c_{NaH_2PO_4}$. Thus, Equation 15-15 becomes,

$$[H_3O^+] = \sqrt{\frac{(6.32 \times 10^{-8})(1.00 \times 10^{-2})}{1.00 + (1.00 \times 10^{-2})/(7.11 \times 10^{-3})}} = 1.62 \times 10^{-5} \text{ M}$$

EXAMPLE 15-8

Calculate the hydronium ion concentration of a 0.1000 M $NaHCO_3$ solution.

Solution

We assume, as we did earlier (page 353), that $[H_2CO_3] \ll [CO_2(aq)]$ and that the following equilibria describe the system:

$$CO_2(aq) + 2H_2O \rightleftharpoons H_3O^+ + HCO_3^- \qquad K_{a1} = \frac{[H_3O^+][HCO_3^-]}{[CO_2(aq)]}$$

$$= 4.2 \times 10^{-7}$$

(continued)

$$HCO_3^- + H_2O \rightleftharpoons H_3O^+ + CO_3^{2-} \qquad K_{a2} = \frac{[H_3O^+][CO_3^{2-}]}{[HCO_3^-]}$$

$$= 4.69 \times 10^{-11}$$

We note that $c_{NaHA}/K_{a1} \gg 1$ so that the denominator of Equation 15-15 can be simplified. In addition, $K_{a2}c_{NaHA}$ has a value of 4.69×10^{-12}, which is substantially greater than K_w. Thus, Equation 15-16 applies, and

$$[H_3O^+] = \sqrt{4.2 \times 10^{-7} \times 4.69 \times 10^{-11}} = 4.4 \times 10^{-9} \, M$$

15E TITRATION CURVES FOR POLYFUNCTIONAL ACIDS

Compounds with two or more acidic functional groups yield multiple end points in a titration if the functional groups differ sufficiently in strength as acids. The computational techniques described in Chapter 14 permit construction of reasonably accurate theoretical titration curves for polyprotic acids if the ratio K_{a1}/K_{a2} is somewhat greater than 10^3. If this ratio is smaller, the error becomes excessive, particularly in the region of the first equivalence point, and a more rigorous treatment of the equilibrium relationships is required.

Figure 15-2 shows the titration curve for a diprotic acid H_2A with dissociation constants of $K_{a1} = 1.00 \times 10^{-3}$ and $K_{a2} = 1.00 \times 10^{-7}$. Because the K_{a1}/K_{a2} ratio is significantly greater than 10^3, we can calculate this curve (except for the first equivalence point) using the techniques developed in Chapter 14 for simple monoprotic weak acids. Thus, to calculate the initial pH (point A), we treat the system as if it contained a single monoprotic acid with a dissociation constant of $K_{a1} = 1.00 \times 10^{-3}$. In region B, we have the equivalent of a simple buffer solution consisting of the weak acid H_2A and its conjugate base NaHA. That is, we assume that the concentration of A^{2-} is negligible

Figure 15-2 Titration of 20.00 mL of 0.1000 M H_2A with 0.1000 M NaOH. For H_2A, $K_{a1} = 1.00 \times 10^{-3}$, and $K_{a2} = 1.00 \times 10^{-7}$. The method of pH calculation is shown for several points and regions on the titration curve.

with respect to the other two A-containing species and use Equation 9-29 (page 220) to find $[H_3O^+]$. At the first equivalence point (point C), we have a solution of an acid salt and use Equation 15-15 or one of its simplifications to compute the hydronium ion concentration. In the region labeled D, we have a second buffer consisting of a weak acid HA^- and its conjugate base Na_2A, and we calculate the pH using the second dissociation constant, $K_{a2} = 1.00 \times 10^{-7}$. At point E, the solution contains the conjugate base of a weak acid with a dissociation constant of 1.00×10^{-7}. That is, we assume that the hydroxide concentration of the solution is determined solely by the reaction of A^{2-} with water to form HA^- and OH^-. Finally, in the region labeled F, we have excess NaOH and compute the hydroxide concentration from the molar concentration of the NaOH. The pH is then found from this quantity and the ion-product of water.

Example 15-9 illustrates a somewhat more complicated example, that of titrating the diprotic maleic acid (H_2M) with NaOH. Although the ratio of K_{a1}/K_{a2} is large enough to use the techniques just described, the value of K_{a1} is so large that some of the simplifications made in previous discussions do not apply, particularly in regions just prior to and just beyond the equivalence points.

EXAMPLE 15-9

Construct a curve for the titration of 25.00 mL of 0.1000 M maleic acid, HOOC—CH=CH—COOH, with 0.1000 M NaOH.

We can write the two dissociation equilibria as

$$H_2M + H_2O \rightleftharpoons H_3O^+ + HM^- \qquad K_{a1} = 1.3 \times 10^{-2}$$

$$HM^- + H_2O \rightleftharpoons H_3O^+ + M^{2-} \qquad K_{a2} = 5.9 \times 10^{-7}$$

Because the ratio K_{a1}/K_{a2} is large (2×10^{-4}), we can proceed using the techniques just described.

Solution

Initial pH

Initially, the solution is 0.1000 M H_2M. At this point, only the first dissociation makes an appreciable contribution to $[H_3O^+]$; thus,

$$[H_3O^+] \approx [HM^-]$$

Mass balance requires that

$$c^0_{H_2M} = [H_2M] + [HM^-] + [M^{2-}] = 0.1000 \text{ M}$$

Since the second dissociation is negligible, $[M^{2-}]$ is very small so that

$$c^0_{H_2M} \approx [H_2M] + [HM^-] = 0.1000 \text{ M}$$

or

$$[H_2M] = 0.1000 - [HM^-] = 0.1000 - [H_3O^+]$$

Substituting these relationships into the expression for K_{a1} gives

$$K_{a1} = 1.3 \times 10^{-2} = \frac{[H_3O^+][HM^-]}{[H_2M]} = \frac{[H_3O^+]^2}{0.1000 - [H_3O^+]}$$

(continued)

Molecular models of maleic acid, or (Z)-butenedioic acid (top), and fumaric acid, or (E)-butenedioic acid (bottom). These geometric isomers exhibit striking differences in both their physical and their chemical properties. Because the *cis* isomer (maleic acid) has both carboxyl groups on the same side of the molecule, the compound eliminates water to form cyclic maleic anhydride, which is a very reactive precursor widely used in plastics, dyes, pharmaceuticals, and agrichemicals. Fumaric acid, which is essential to animal and vegetable respiration, is used industrially as an antioxidant to synthesize resins and to fix colors in dyeing. It is interesting to compare the pK_a values for the two acids; for fumaric acid, $pK_{a1} = 3.05$, and $pK_{a2} = 4.49$; for maleic acid, $pK_{a1} = 1.89$, and $pK_{a2} = 6.23$. CHALLENGE: Explain the differences in the pK_a values based on the differences in the molecular structures.

Rearranging yields

$$[H_3O^+]^2 + 1.3 \times 10^{-2}[H_3O^+] - 1.3 \times 10^{-3} = 0$$

Because K_{a1} for maleic acid is relatively large, we must solve the quadratic equation or find $[H_3O^+]$ by successive approximations. When we do so, we obtain

$$[H_3O^+] = 3.01 \times 10^{-2}\,M$$

$$pH = 2 - \log 3.01 = 1.52$$

First Buffer Region

The addition of base, for example 5.00 mL, results in the formation of a buffer consisting of the weak acid H_2M and its conjugate base HM^-. To the extent that dissociation of HM^- to give M^{2-} is negligible, the solution can be treated as a simple buffer system. Thus, applying Equations 9-27 and 9-28 (page 220) gives

$$c_{NaHM} \approx [HM^-] = \frac{5.00 \times 0.1000}{30.00} = 1.67 \times 10^{-2}\,M$$

$$c_{H_2M} \approx [H_2M] = \frac{25.00 \times 0.1000 - 5.00 \times 0.1000}{30.00} = 6.67 \times 10^{-2}\,M$$

Substitution of these values into the equilibrium-constant expression for K_{a1} yields a tentative value of $5.2 \times 10^{-2}\,M$ for $[H_3O^+]$. It is clear, however, that the approximation $[H_3O^+] \ll c_{H_2M}$ or c_{HM^-} is not valid; therefore, Equations 9-25 and 9-26 must be used, and

$$[HM^-] = 1.67 \times 10^{-2} + [H_3O^+] - [OH^-]$$

$$[H_2M] = 6.67 \times 10^{-2} - [H_3O^+] - [OH^-]$$

Because the solution is quite acidic, the approximation that $[OH^-]$ is very small is surely justified. Substitution of these expressions into the dissociation-constant relationship gives

$$K_{a1} = \frac{[H_3O^+](1.67 \times 10^{-2} + [H_3O^+])}{6.67 \times 10^{-2} - [H_3O^+]} = 1.3 \times 10^{-2}$$

$$[H_3O^+]^2 + (2.97 \times 10^{-2})[H_3O^+] - 8.67 \times 10^{-4} = 0$$

$$[H_3O^+] = 1.81 \times 10^{-2}\,M$$

$$pH = -\log(1.81 \times 10^{-2}) = 1.74$$

Additional points in the first buffer region are computed in a similar way until just prior to the first equivalence point.

Just Prior to First Equivalence Point

Just prior to the first equivalence point, the concentration of H_2M is so small that it becomes comparable to the concentration of M^{2-}, and the second equilibrium must also be considered. Within approximately 0.1 mL of the first equivalence point, we have a solution of primarily HM^- with a small amount of H_2M

remaining and a small amount of M^{2-} formed. For example, at 24.90 mL of NaOH added,

$$[HM^-] \approx c_{NaHM} = \frac{24.90 \times 0.1000}{49.90} = 4.99 \times 10^{-2} \, M$$

$$c_{H_2M} = \frac{25.00 \times 0.1000}{49.90} - \frac{24.90 \times 0.1000}{49.90} = 2.00 \times 10^{-4} \, M$$

Mass balance gives

$$c_{H_2M} + c_{NaHM} = [H_2M] + [HM^-] + [M^{2-}]$$

Charge balance gives

$$[H_3O^+] + [Na^+] = [HM^-] + 2[M^{2-}] + [OH^-]$$

Since the solution consists primarily of the acid HM^- at the first equivalence point, we can safely neglect $[OH^-]$ in the previous equation and replace $[Na^+]$ with c_{NaHM}. After rearranging, we obtain

$$c_{NaHM} = [HM^-] + 2[M^{2-}] - [H_3O^+]$$

Substituting this equation into the mass-balance expression and solving for $[H_3O^+]$ give

$$[H_3O^+] = c_{H_2M} + [M^{2-}] - [H_2M]$$

If we express $[M^{2-}]$ and $[H_2M]$ in terms of $[HM^-]$ and $[H_3O^+]$, the result is

$$[H_3O^+] = c_{H_2M} + \frac{K_{a2}[HM^-]}{[H_3O^+]} - \frac{[H_3O^+][HM^-]}{K_{a1}}$$

Multiplying through by $[H_3O^+]$ gives, after rearrangement,

$$[H_3O^+]^2\left(1 + \frac{[HM^-]}{K_{a1}}\right) - c_{H_2M}[H_3O^+] - K_{a2}[HM^-] = 0$$

Substituting $[HM^-] = 4.99 \times 10^{-2}$, $c_{H_2M} = 2.00 \times 10^{-4}$, and the values for K_{a1} and K_{a2} leads to

$$4.838 \, [H_3O^+]^2 - 2.00 \times 10^{-4} \, [H_3O^+] - 2.94 \times 10^{-8} = 0$$

The solution to this equation is

$$[H_3O^+] = 1.014 \times 10^{-4} \, M$$

$$pH = 3.99$$

The same reasoning applies at 24.99 mL of titrant, where we find

$$[H_3O^+] = 8.01 \times 10^{-5} \, M$$

$$pH = 4.10$$

(continued)

First Equivalence Point

At the first equivalence point,

$$[HM^-] \approx c_{NaHM} = \frac{25.00 \times 0.1000}{50.00} = 5.00 \times 10^{-2}\, M$$

Our simplification of the numerator in Equation 15-15 is certainly justified. On the other hand, the second term in the denominator is not $\ll 1$. Hence,

$$[H_3O^+] = \sqrt{\frac{K_{a2}c_{NaHM}}{1 + c_{NaHM}/K_{a1}}} = \sqrt{\frac{5.9 \times 10^{-7} \times 5.00 \times 10^{-2}}{1 + (5.00 \times 10^{-2})/(1.3 \times 10^{-2})}}$$

$$= 7.80 \times 10^{-5}\, M$$

$$pH = -\log(7.80 \times 10^{-5}\, M) = 4.11$$

Just after the First Equivalence Point

Prior to the second equivalence point, we can obtain the analytical concentrations of NaHM and Na$_2$M from the titration stoichiometry. At 25.01 mL, for example, the values are

$$c_{NaHM} = \frac{\text{mmol NaHM formed} - (\text{mmol NaOH added} - \text{mmol NaHM formed})}{\text{total volume of solution}}$$

$$= \frac{25.00 \times 0.1000 - (25.01 - 25.00) \times 0.1000}{50.01} = 0.04997\, M$$

$$c_{Na_2M} = \frac{(\text{mmol NaOH added} - \text{mmol NaHM formed})}{\text{total volume of solution}} = 1.9996 \times 10^{-5}\, M$$

In the region a few tenths of a milliliter beyond the first equivalence point, the solution is primarily HM$^-$ with some M^{2-} formed as a result of the titration. The mass balance at 25.01 mL added is

$$c_{Na_2M} + c_{NaHM} = [H_2M] + [HM^-] + [M^{2-}] = 0.04997 + 1.9996 \times 10^{-5}$$

$$= 0.04999\, M$$

and the charge balance is

$$[H_3O^+] + [Na^+] = [HM^-] + 2[M^{2-}] + [OH^-]$$

Again, the solution should be acidic, and so, we can neglect OH$^-$ as an important species. The Na$^+$ concentration equals the number of millimoles of NaOH added divided by the total volume, or

$$[Na^+] = \frac{25.01 \times 0.1000}{50.01} = 0.05001\, M$$

Subtracting the mass balance from the charge balance and solving for $[H_3O^+]$ gives

$$[H_3O^+] = [M^{2-}] - [H_2M] + (c_{Na_2M} + c_{NaHM}) - [Na^+]$$

Expressing the $[M^{2-}]$ and $[H_2M]$ in terms of the predominant species HM^-, we have

$$[H_3O^+] = \frac{K_{a2}[HM^-]}{[H_3O^+]} - \frac{[H_3O^+][HM^-]}{K_{a1}} + (c_{Na_2M} + c_{NaHM}) - [Na^+]$$

Since $[HM^-] \approx c_{NaHM} = 0.04997$. Therefore, if we substitute this value and numerical values for $c_{Na_2M} + c_{NaHM}$ and $[Na^+]$ into the previous equation, we have, after rearranging, the following quadratic equation:

$$[H_3O^+] = \frac{K_{a2}(0.04997)}{[H_3O^+]} - \frac{[H_3O^+](0.04997)}{K_{a1}} - 1.9996 \times 10^{-5}$$

$$K_{a1}[H_3O^+]^2 = 0.04997\,K_{a1}K_{a2} - 0.04997[H_3O^+]^2 - 1.9996 \times 10^{-5}K_{a1}[H_3O^+]$$

$$(K_{a1} + 0.04997)[H_3O^+]^2 + 1.9996 \times 10^{-5}K_{a1}[H_3O^+] - 0.04997\,K_{a1}K_{a2} = 0$$

This equation can then be solved for $[H_3O^+]$.

$$[H_3O^+] = 7.60 \times 10^{-5}\,M$$

$$pH = 4.12$$

Second Buffer Region

Further additions of base to the solution create a new buffer system consisting of HM^- and M^{2-}. When enough base has been added so that the reaction of HM^- with water to give OH^- can be neglected (a few tenths of a milliliter beyond the first equivalence point), the pH of the mixture may be calculated from K_{a2}. With the introduction of 25.50 mL of NaOH, for example,

$$[M^{2-}] \approx c_{Na_2M} = \frac{(25.50 - 25.00)(0.1000)}{50.50} = \frac{0.050}{50.50}\,M$$

and the molar concentration of NaHM is

$$[HM^-] \approx c_{NaHM} = \frac{(25.00 \times 0.1000) - (25.50 - 25.00)(0.1000)}{50.50} = \frac{2.45}{50.50}\,M$$

Substituting these values into the expression for K_{a2} gives

$$K_{a2} = \frac{[H_3O^+][M^{2-}]}{[HM^-]} = \frac{[H_3O^+](0.050/50.50)}{2.45/50.50} = 5.9 \times 10^{-7}$$

$$[H_3O^+] = 2.89 \times 10^{-5}\,M$$

The assumption that $[H_3O^+]$ is small relative to c_{HM^-} and $c_{M^{2-}}$ is valid, and pH = 4.54. The other values in the second buffer region are calculated in a similar manner.

Just Prior to Second Equivalence Point

Just prior to the second equivalence point (49.90 mL and more), the ratio $[M^{2-}]/[HM^-]$ becomes large, and the simple buffer equation no longer applies. At 49.90 mL, $c_{HM^-} = 1.335 \times 10^{-4}\,M$, and $c_{M^{2-}} = 0.03324$. The primary equilibrium is now

$$M^{2-} + H_2O \rightleftharpoons HM^- + OH^-$$

(continued)

We can write the equilibrium constant as

$$K_{b1} = \frac{K_w}{K_{a2}} = \frac{[OH^-][HM^-]}{[M^{2-}]} = \frac{[OH^-](1.335 \times 10^{-4} + [OH^-])}{(0.03324 - [OH^-])}$$

$$= \frac{1.00 \times 10^{-14}}{5.9 \times 10^{-7}} = 1.69 \times 10^{-8}$$

In this case, it is easier to solve for $[OH^-]$ than for $[H_3O^+]$. Solving the resulting quadratic equation gives

$$[OH^-] = 4.10 \times 10^{-6} \text{ M}$$

$$pOH = 5.39$$

$$pH = 14.00 - pOH = 8.61$$

The same reasoning for 49.99 mL leads to $[OH^-] = 1.80 \times 10^{-5}$ M, and pH = 9.26.

Second Equivalence Point

After the addition of 50.00 mL of 0.1000 M sodium hydroxide, the solution is 0.0333 M in Na_2M (2.5 mmol/75.00 mL). Reaction of the base M^{2-} with water is the predominant equilibrium in the system and the only one that we need to take into account. Thus,

$$M^{2-} + H_2O \rightleftharpoons OH^- + HM^-$$

$$K_{b1} = \frac{K_w}{K_{a2}} = \frac{[OH^-][HM^-]}{[M^{2-}]} = 1.69 \times 10^{-8}$$

$$[OH^-] \cong [HM^-]$$

$$[M^{2-}] = 0.0333 - [OH^-] \cong 0.0333$$

$$\frac{[OH^-]^2}{0.0333} = 1.69 \times 10^{-8}$$

$$[OH^-] = 2.37 \times 10^{-5} \text{ M, and } pOH = -\log(2.37 \times 10^{-5}) = 4.62$$

$$pH = 14.00 - pOH = 9.38$$

pH Just beyond Second Equivalence Point

In the region just beyond the second equivalence point (50.01 mL, for example), we still need to take into account the reaction of M^{2-} with water to give OH^- since not enough OH^- has been added in excess to suppress this reaction. The analytical concentration of M^{2-} is the number of millimoles of M^{2-} produced divided by the total solution volume:

$$c_{M^{2-}} = \frac{25.00 \times 0.1000}{75.01} = 0.03333 \text{ M}$$

The OH^- now comes from the reaction of M^{2-} with water and from the excess OH^- added as titrant. The number of millimoles of excess OH^- is then the number of millimoles of NaOH added minus the number required to reach the second equivalence point. The concentration of this excess is the number of millimoles of excess OH^- divided by the total solution volume, or

$$[OH^-]_{excess} = \frac{(50.01 - 50.00) \times 0.1000}{75.01} = 1.333 \times 10^{-5} \text{ M}$$

The concentration of HM^- can now be found from K_{b1}.

$$[M^{2-}] = c_{M^{2-}} - [HM^-] = 0.03333 - [HM^-]$$

$$[OH^-] = 1.3333 \times 10^{-5} + [HM^-]$$

$$K_{b1} = \frac{[HM^-][OH^-]}{[M^{2-}]} = \frac{[HM^-](1.3333 \times 10^{-5} + [HM^-])}{0.03333 - [HM^-]} = 1.69 \times 10^{-8}$$

Solving the quadratic equation for $[HM^-]$ gives

$$[HM^-] = 1.807 \times 10^{-5} \text{ M}$$

and

$$[OH^-] = 1.3333 \times 10^{-5} + [HM^-] = 1.33 \times 10^{-5} + 1.807 \times 10^{-5} = 3.14 \times 10^{-5} \text{ M}$$

$$pOH = 4.50 \text{ and } pH = 14.00 - pOH = 9.50$$

The same reasoning applies to 50.10 mL where the calculations give pH = 10.14

pH beyond the Second Equivalence Point

Addition of more than a few tenths of a milliliter of NaOH beyond the second equivalence point gives enough excess OH^- to repress the basic dissociation of M^{2-}. The pH is then calculated from the concentration of NaOH added in excess of that required for the complete neutralization of H_2M. Thus, when 51.00 mL of NaOH have been added, we have 1.00-mL excess of 0.1000 M NaOH, and

$$[OH^-] = \frac{1.00 \times 0.100}{76.00} = 1.32 \times 10^{-3} \text{ M}$$

$$pOH = -\log(1.32 \times 10^{-3}) = 2.88$$

$$pH = 14.00 - pOH = 11.12$$

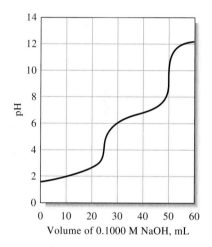

Figure 15-3 Titration curve for 25.00 mL of 0.1000 M maleic acid, H_2M, titrated with 0.1000 M NaOH.

Figure 15-3 is the titration curve for 0.1000 M maleic acid generated as shown in Example 15-9. Two end points are apparent, either of which could in principle be used as a measure of the concentration of the acid. The second end point is more satisfactory, however, because the pH change is more pronounced than in the first.

Figure 15-4 shows titration curves for three other polyprotic acids. These curves illustrate that a well-defined end point corresponding to the first equivalence point is observed only when the degree of dissociation of the two acids is sufficiently different. The ratio K_{a1}/K_{a2} for oxalic acid (curve *B*) is approximately 1000. The curve for this titration shows an inflection corresponding to the first equivalence point. The magnitude of the pH change is too small to permit precise location of the end point with an indicator. The second end point, however, can be used to accurately determine oxalic acid.

Curve *A* in Figure 15-4 is the theoretical titration curve for triprotic phosphoric acid. For this acid, the ratio K_{a1}/K_{a2} is approximately 10^5, as is K_{a2}/K_{a3}. This results in two well-defined end points, either of which is satisfactory for analytical purposes.

❮ In titrating a polyprotic acid or base, two usable end points appear if the ratio of dissociation constants is greater than 10^4 and if the weaker acid or base has a dissociation constant greater than 10^{-8}.

Figure 15-4 Curves for the titration of polyprotic acids. A 0.1000 M NaOH solution is used to titrate 25.00 mL of 0.1000 M H_3PO_4 (curve A), 0.1000 M oxalic acid (curve B), and 0.1000 M H_2SO_4 (curve C).

An acid-range indicator will provide a color change when 1 mol of base has been introduced for each mole of acid, and a base-range indicator will require 2 mol of base per mole of acid. The third hydrogen of phosphoric acid is so slightly dissociated ($K_{a3} = 4.5 \times 10^{-13}$) that no practical end point is associated with its neutralization. The buffering effect of the third dissociation is noticeable, however, and causes the pH for curve A to be lower than the pH for the other two curves in the region beyond the second equivalence point.

Curve C is the titration curve for sulfuric acid, a substance that has one fully dissociated proton and one that is dissociated to a relatively large extent ($K_{a2} = 1.02 \times 10^{-2}$). Because of the similarity in strengths of the two acids, only a single end point, corresponding to the titration of both protons, is observed. Calculation of the pH in sulfuric acid solutions is illustrated in Feature 15-1.

In general, the titration of acids or bases that have two reactive groups yields individual end points that are of practical value only when the ratio between the two dissociation constants is at least 10^4. If the ratio is much smaller than this 10^4, the pH change at the first equivalence point will prove less satisfactory for an analysis.

CHALLENGE: Construct a titration curve for 50.00 mL of 0.0500 M H_2SO_4 with 0.1000 M NaOH.

FEATURE 15-1

The Dissociation of Sulfuric Acid

Sulfuric acid is unusual in that one of its protons behaves as a strong acid in water and the other as a weak acid ($K_{a2} = 1.02 \times 10^{-2}$). Let us consider how the hydronium ion concentration of sulfuric acid solutions is computed using a 0.0400 M solution as an example.

We will first assume that the dissociation of HSO_4 is negligible because of the large excess of H_3O^+ resulting from the complete dissociation of H_2SO_4. Therefore,

$$[H_3O^+] \approx [HSO_4^-] \approx 0.0400 \text{ M}$$

An estimate of $[SO_4^{2-}]$ based on this approximation and the expression for K_{a2} reveals that

$$\frac{0.0400 \, [SO_4^{2-}]}{0.0400} = 1.02 \times 10^{-2}$$

We see that $[SO_4^{2-}]$ is *not* small relative to $[HSO_4^-]$, and a more rigorous solution is required.

From stoichiometric considerations, it is necessary that

$$[H_3O^+] = 0.0400 + [SO_4^{2-}]$$

The first term on the right is the concentration of H_3O^+ resulting from dissociation of the H_2SO_4 to HSO_4^-. The second term is the contribution of the dissociation of HSO_4^-. Rearrangement yields

$$[SO_4^{2-}] = [H_3O^+] - 0.0400$$

Mass-balance considerations require that

$$c_{H_2SO_4} = 0.0400 = [HSO_4^-] + [SO_4^{2-}]$$

Combining the last two equations and rearranging yield

$$[HSO_4^-] = 0.0800 - [H_3O^+]$$

By introducing these equations for $[SO_4^{2-}]$ and HSO_4^- into the expression for K_{a2}, we find that

$$\frac{[H_3O^+]([H_3O^+] - 0.0400)}{0.0800 - [H_3O^+]} = 1.02 \times 10^{-2}$$

Solving the quadratic equation for $[H_3O^+]$ yields

$$[H_3O^+] = 0.0471 \text{ M}$$

Spreadsheet Summary In Chapter 8 of *Applications of Microsoft® Excel in Analytical Chemistry*, 2nd ed., we extend the treatment of neutralization titration curves to polyfunctional acids. Both a stoichiometric approach and a master equation approach are used for the titration of maleic acid with sodium hydroxide.

15F TITRATION CURVES FOR POLYFUNCTIONAL BASES

The same principles just described for constructing titration curves for polyfunctional acids can be applied to titration curves for polyfunctional bases. To illustrate, consider the titration of a sodium carbonate solution with standard hydrochloric acid. The important equilibrium constants are

$$CO_3^{2-} + H_2O \rightleftharpoons OH^- + HCO_3^- \quad K_{b1} = \frac{K_w}{K_{a2}} = \frac{1.00 \times 10^{-14}}{4.69 \times 10^{-11}} = 2.13 \times 10^{-4}$$

$$HCO_3^- + H_2O \rightleftharpoons OH^- + CO_2(aq) \quad K_{b2} = \frac{K_w}{K_{a1}} = \frac{1.00 \times 10^{-14}}{4.2 \times 10^{-7}} = 2.4 \times 10^{-8}$$

The reaction of carbonate ion with water governs the initial pH of the solution, which can be found by the method shown for the second equivalence point in Example 15-9. With the first additions of acid, a carbonate/hydrogen carbonate buffer is established. In this region, the pH can be determined from *either* the hydroxide ion concentration calculated from K_{b1} *or* the hydronium ion concentration calculated from K_{a2}. Because we are usually interested in calculating $[H_3O^+]$ and pH, the expression for K_{a2} is easier to use.

Sodium hydrogen carbonate is the principal solute species at the first equivalence point, and Equation 15-16 is used to compute the hydronium ion concentration (see Example 15-8). With the addition of more acid, a new buffer consisting of sodium hydrogen carbonate and carbonic acid (from $CO_2(aq)$ as shown in Equation 15-3) is formed. The pH of this buffer is easily calculated from either K_{b2} or K_{a1}.

At the second equivalence point, the solution consists of $CO_2(aq)$ (carbonic acid) and sodium chloride. The $CO_2(aq)$ can be treated as a simple weak acid having a dissociation constant K_{a1}. Finally, after excess hydrochloric acid has been introduced, the dissociation of the weak acid is repressed to a point where the hydronium ion concentration is essentially that of the molar concentration of the strong acid.

Figure 15-5 illustrates that two end points appear in the titration of sodium carbonate, the second being appreciably sharper than the first. This suggests that the individual components in mixtures of sodium carbonate and sodium hydrogen carbonate can be determined by neutralization methods.

> CHALLENGE: Show that either K_{b2} or K_{a1} can be used to calculate the pH of a buffer that is 0.100 M in Na_2CO_3 and 0.100 M in $NaHCO_3$.

Spreadsheet Summary The titration curve for a difunctional base being titrated with strong acid is developed in Chapter 8 of *Applications of Microsoft® Excel in Analytical Chemistry*, 2nd ed. In the example studied, ethylene diamine is titrated with hydrochloric acid. A master equation approach is explored, and the spreadsheet is used to plot pH versus fraction titrated.

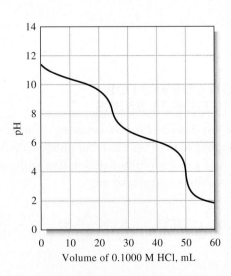

Figure 15-5 Curve for the titration of 25.00 mL of 0.1000 M Na_2CO_3 with 0.1000 M HCl.

15G TITRATION CURVES FOR AMPHIPROTIC SPECIES

An amphiprotic substance when dissolved in a suitable solvent behaves both as a weak acid and as a weak base. If either of its acidic or basic characters predominates, titration of the substance with a strong base or a strong acid may be feasible. For example, in a sodium dihydrogen phosphate solution, the principal equilibria are:

$$H_2PO_4^- + H_2O \rightleftharpoons H_3O^+ + HPO_4^{2-} \qquad K_{a2} = 6.32 \times 10^{-8}$$

$$H_2PO_4^- + H_2O \rightleftharpoons OH^- + H_3PO_4 \qquad K_{b3} = \frac{K_w}{K_{a1}} = \frac{1.00 \times 10^{-14}}{7.11 \times 10^{-3}}$$
$$= 1.41 \times 10^{-12}$$

Note that K_{b3} is much too small to permit titration of $H_2PO_4^-$ with an acid, but K_{a2} is large enough for a successful titration of dihydrogen phosphate with a standard base solution.

A different situation prevails in solutions containing disodium hydrogen phosphate for which the pertinent equilibria are

$$HPO_4^{2-} + H_2O \rightleftharpoons H_3O^+ + PO_4^{3-} \qquad K_{a3} = 4.5 \times 10^{-13}$$

$$HPO_4^{2-} + H_2O \rightleftharpoons OH^- + H_2PO_4^- \qquad K_{b2} = \frac{K_w}{K_{a2}} = \frac{1.00 \times 10^{-14}}{6.32 \times 10^{-8}}$$
$$= 1.58 \times 10^{-7}$$

The magnitude of the constants indicates that HPO_4^{2-} can be titrated with standard acid but not with standard base.

FEATURE 15-2

Acid/Base Behavior of Amino Acids

> Amino acids are amphiprotic.

The simple amino acids are an important class of amphiprotic compounds that contain both a weak acid and a weak base functional group. In an aqueous solution of a typical amino acid, such as glycine, three important equilibria operate:

$$NH_2CH_2COOH \rightleftharpoons NH_3^+CH_2COO^- \qquad (15\text{-}17)$$

$$NH_3^+CH_2COO^- + H_2O \rightleftharpoons$$
$$NH_2CH_2COO^- + H_3O^+ \qquad K_a = 2 \times 10^{-10} \qquad (15\text{-}18)$$

$$NH_3^+CH_2COO^- + H_2O \rightleftharpoons$$
$$NH_3^+CH_2COOH + OH^- \qquad K_b = 2 \times 10^{-12} \qquad (15\text{-}19)$$

The first equilibrium constitutes a kind of internal acid/base reaction and is analogous to the reaction one would observe between a carboxylic acid and an amine:

$$R_1NH_2 + R_2COOH \rightleftharpoons R_1NH_3^+ + R_2COO^- \qquad (15\text{-}20)$$

The typical aliphatic amine has a base dissociation constant of 10^{-4} to 10^{-5} (see Appendix 3), while many carboxylic acids have acid dissociation constants of about the same magnitude. As a result, both Reactions 15-18 and 15-19 proceed far to the right, with the product or products being the predominant species in the solution.

(continued)

A **zwitterion** is an ionic species that has both a positive and a negative charge.

The **isoelectric** point of a species is the pH at which no net migration occurs in an electric field.

The molecular structure of the glycine zwitterion, $NH_3^+CH_2COO^-$. Glycine is one of the so-called nonessential amino acids; it is nonessential in the sense that it is synthesized in the bodies of mammals and so is not generally essential in the diet. Because of its compact structure, glycine acts as a versatile building block in protein synthesis and in the biosynthesis of hemoglobin. A significant fraction of the collagen—or the fibrous protein constituent of bone, cartilage, tendon, and other connective tissue in the human body—is made up of glycine. Glycine is also an inhibitory *neurotransmitter* and, as a result, has been suggested as a possible therapeutic agent for diseases of the central nervous system such as multiple sclerosis and epilepsy. Glycine is also used in treating schizophrenia, stroke, and benign prostatic hyperplasia.

The amino acid species in Equation 15-17, which bears both a positive and a negative charge, is called a **zwitterion**. As shown by Equations 15-18 and 15-19, the zwitterion of glycine is stronger as an acid than as a base. Thus, an aqueous solution of glycine is somewhat acidic.

The zwitterion of an amino acid, which contains both a positive and a negative charge, has no tendency to migrate in an electric field, although the singly charged anionic and cationic species are attracted to electrodes of opposite polarity. No net migration of the amino acid occurs in an electric field when the pH of the solvent is such that the concentrations of the anionic and cationic forms are identical. The pH at which no net migration occurs is called the **isoelectric point** and is an important physical constant for characterizing amino acids. The isoelectric point is readily related to the ionization constants for the species. Thus, for glycine,

$$K_a = \frac{[NH_2CH_2COO^-][H_3O^+]}{[NH_3^+CH_2COO^-]}$$

$$K_b = \frac{[NH_3^+CH_2COOH][OH^-]}{[NH_3^+CH_2COO^-]}$$

At the isoelectric point,

$$[NH_2CH_2COO^-] = [NH_3^+CH_2COOH]$$

Therefore, if we divide K_a by K_b and substitute this relationship, we obtain for the isoelectric point

$$\frac{K_a}{K_b} = \frac{[H_3O^+][NH_2CH_2COO^-]}{[OH^-][NH_3^+CH_2COOH]} = \frac{[H_3O^+]}{[OH^-]}$$

If we substitute $K_w/[H_3O^+]$ for $[OH^-]$ and rearrange, we get

$$[H_3O^+] = \sqrt{\frac{K_a K_w}{K_b}}$$

The isoelectric point for glycine occurs at a pH of 6.0, that is

$$[H_3O^+] = \sqrt{\frac{(2 \times 10^{-10})(1 \times 10^{-14})}{2 \times 10^{-12}}} = 1 \times 10^{-6} \text{ M}$$

For simple amino acids, K_a and K_b are generally so small that their determination by direct neutralization is impossible. Addition of formaldehyde removes the amine functional group, however, and leaves the carboxylic acid available for titration with a standard base. For example, with glycine,

$$NH_3^+CH_2COO^- + CH_2O \rightarrow CH_2C{=}NCH_2COOH + H_2O$$

The titration curve for the product is that of a typical carboxylic acid.

Spreadsheet Summary   The final exercise in Chapter 8 of *Applications of Microsoft® Excel in Analytical Chemistry*, 2nd ed., considers the titration of an amphiprotic species, phenylalanine. A spreadsheet is developed to plot the titration curve of this amino acid, and the isoelectric pH is calculated.

15H COMPOSITION OF POLYPROTIC ACID SOLUTIONS AS A FUNCTION OF pH

In Section 14E, we showed how alpha values are useful in visualizing the changes in the concentration of various species that occur in a titration of a monoprotic weak acid. Alpha values provide an excellent way of thinking about the properties of polyfunctional acids and bases. For example, if we let c_T be the sum of the molar concentrations of the maleate-containing species in the solution throughout the titration described in Example 15-9, the alpha value for the free acid α_0 is defined as

$$\alpha_0 = \frac{[H_2M]}{c_T}$$

where

$$c_T = [H_2M] + [HM^-] + [M^{2-}] \tag{15-21}$$

The alpha values for HM^- and M^{2-} are given by similar equations

$$\alpha_1 = \frac{[HM^-]}{c_T}$$

$$\alpha_2 = \frac{[M^{2-}]}{c_T}$$

As noted in Section 9C-2, the sum of the alpha values for a system must equal one:

$$\alpha_0 + \alpha_1 + \alpha_2 = 1$$

We may express the alpha values for the maleic acid system very neatly in terms of $[H_3O^+]$, K_{a1}, and K_{a2}. To find the appropriate expressions, we follow the method used to derive Equations 9-35 and 9-36 in Section 9C-2 and obtain the following equations:

$$\alpha_0 = \frac{[H_3O^+]^2}{[H_3O^+]^2 + K_{a1}[H_3O^+] + K_{a1}K_{a2}} \tag{15-22}$$

$$\alpha_1 = \frac{K_{a1}[H_3O^+]}{[H_3O^+]^2 + K_{a1}[H_3O^+] + K_{a1}K_{a2}} \tag{15-23}$$

$$\alpha_2 = \frac{K_{a1}K_{a2}}{[H_3O^+]^2 + K_{a1}[H_3O^+] + K_{a1}K_{a2}} \tag{15-24}$$

◀ CHALLENGE: Derive Equations 15-22, 15-23, and 15-24.

Notice that the denominator is the same for each expression. A somewhat surprising result is that the fractional amount of each species is fixed at a given pH and is *absolutely independent* of the total concentration, c_T. A general expression for the alpha values is given in Feature 15-3.

A General Expression for Alpha Values

For the weak acid H_nA, the denominator D in all alpha-value expressions takes the form:

$$D = [H_3O^+]^n + K_{a1}[H_3O^+]^{(n-1)} + K_{a1}K_{a2}[H_3O^+]^{(n-2)} + \cdots K_{a1}K_{a2} \cdots K_{an}$$

The numerator for α_0 is the first term in the denominator, and for α_1, it is the second term, and so forth. Thus, $\alpha_0 = [H_3O^+]^n/D$, and $\alpha_1 = K_{a1}[H_3O^+]^{(n-1)}/D$.

Alpha values for polyfunctional bases are generated in an analogous way, with the equations being written in terms of base dissociation constants and $[OH^-]$.

The three curves plotted in **Figure 15-6** show the alpha values for each maleate-containing species as a function of pH. The solid curves in **Figure 15-7** depict the same alpha values but now plotted as a function of volume of sodium hydroxide as the acid is titrated. The titration curve is also shown by the dashed line in Figure 15-7. These curves give a comprehensive picture of all concentration changes that occur during the titration. For example, Figure 15-7 reveals that before the addition of any base, α_0 for H_2M is roughly 0.7, and α_1 for HM^- is approximately 0.3. For all practical purposes, α_2 is zero. Thus, initially, approximately 70% of the maleic acid exists as H_2M and 30% as HM^-. With addition of base, the pH rises, as does the fraction of HM^-. At the first equivalence point (pH = 4.11), essentially all of the maleate is present as HM^- ($\alpha_1 \to 1$). When we add more base, beyond the first equivalence point, HM^- decreases and M^{2-} increases. At the second equivalence point (pH = 9.38) and beyond, essentially all of the maleate is in the M^{2-} form.

Another way to visualize polyfunctional acid and base systems is by using logarithmic concentration diagrams, as illustrated in Feature 15-4.

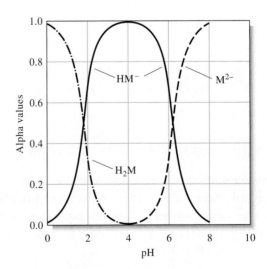

Figure 15-6 Composition of H_2M solutions as a function of pH.

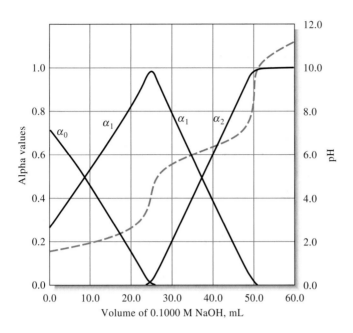

Figure 15-7 Titration of 25.00 mL of 0.1000 M maleic acid with 0.1000 M NaOH. The solid curves are plots of alpha values as a function of titrant volume. The broken curve is the titration curve of pH as a function of volume.

FEATURE 15-4

Logarithmic Concentration Diagrams

A logarithmic concentration diagram is a plot of log concentration versus a master variable such as pH. Such diagrams are useful because they express the concentrations of all species in a polyprotic acid solution as a function of pH. This type of diagram allows us to observe at a glance the species that are important at a particular pH. The logarithmic scale is used because the concentrations can vary over many orders of magnitude.

The logarithmic concentration diagram only applies for a specific acid and for a particular initial concentration of acid. We may calculate results to construct logarithmic concentration diagrams from the distribution diagrams previously discussed. The details of constructing logarithmic concentration diagrams are given in Chapter 8 of *Applications of Microsoft® Excel in Analytical Chemistry*, 2nd ed.

Logarithmic concentration diagrams can be computed from the concentration of acid and the dissociation constants. We use as an example the maleic acid system discussed previously. The diagram shown in **Figure 15F-1** is a logarithmic concentration diagram for a maleic acid concentration of 0.10M ($c_T =$ 0.10 M maleic acid). The diagram expresses the concentrations of all forms of maleic acid, H_2M, HM^-, and M^{2-} as a function of pH. We usually include the H_3O^+ and OH^- concentrations as well. The diagram is based on the mass-balance condition and the acid-dissociation constants. The changes in slope in the diagram for the maleic acid species occur near what are termed **system points**. These are defined by the total acid concentration, 0.10 M in our case, and the pK_a values. For maleic acid, the first system point occurs at log $c_T = -1$ and pH = pK_{a1} = $-\log (1.30 \times 10^{-2}) = 1.89$, while the second system point

is at pH = pK_{a2} = $-\log (5.90 \times 10^{-7}) = 6.23$ and log $c_T = -1$. Note that when pH = pK_{a1}, the concentrations of H_2M and HM^- are equal as shown by the crossing of the lines indicating these concentrations. Also, note that at this first system point $[M^{2-}] \ll [HM^-]$ and $[M^{2-}] \ll [H_2M]$. Near this first system point we can thus neglect the unprotonated maleate ion and express the mass balance as $c_T \approx [H_2M] + [HM^-]$.

To the left of this first system point, $[H_2M] \gg [HM^-]$, and so $c_T \approx [H_2M]$. This is indicated on the diagram by the slope of 0 for the H_2M line between pH values of 0 to about 1. In this same region, the HM^- concentration is steadily increasing with increasing pH since protons are removed from H_2M as the pH increases. From the K_{a1} expression we can write,

$$[HM^-] = \frac{[H_2M]K_{a1}}{[H_3O^+]} \approx \frac{c_T K_{a1}}{[H_3O^+]}$$

Taking the logarithms of both sides of this equation gives

$$\log [HM^-] = \log c_T + \log K_{a1} - \log [H_3O^+]$$
$$= \log c_T + \log K_{a1} + pH$$

Hence, to the left of the first system point (region A), a plot of log $[HM^-]$ versus pH is a straight line of slope +1.

Using similar reasoning we conclude that to the right of the first system point, $c_T \approx [HM^-]$, and

$$[H_2M] \approx \frac{c_T[H_3O^+]}{K_{a1}}$$

(continued)

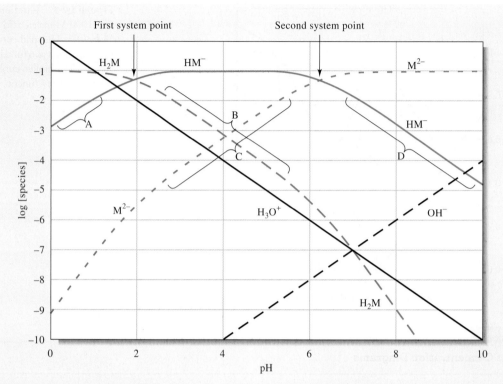

Figure 15F-1   Logarithmic concentration diagram for 0.10 M maleic acid.

Taking the logarithms of both sides of this equation reveals that a plot of log [H$_2$M] versus pH (region B) should be linear with a slope of -1. This relationship holds until we get near the second system point that occurs at pH = pK_{a2} = $-$log (5.90×10^{-7}) = 6.23 and log c_T = -1.

At the second system point, the HM$^-$ and M^{2-} concentrations are equal. Note that to the left of the second system point, [HM$^-$] $\approx c_T$, and log [M^{2-}] increases with increasing pH with a slope of $+1$ (region C). To the right of the second system point, [M^{2-}] $\approx c_T$, and log [HM$^-$] decreases with increasing pH with a slope of -1 (region D). The H$_3$O$^+$ lines and the OH$^-$ lines are easy to draw since

$$\log [H_3O^+] = - pH, \text{ and } \log [OH^-] = pH - 14.$$

We can draw a logarithmic concentration diagram easily by noting the relationships given above. An easier method is to modify the distribution diagram so that it produces the logarithmic concentration diagram. This is the method illustrated in *Applications of Microsoft® Excel in Analytical Chemistry*, 2nd ed., Chapter 8. Note that the plot is specific for a total analytical concentration of 0.10 M and for maleic acid since the acid dissociation constants are included.

Estimating Concentrations at a Given pH Value

The log concentration diagram can be very useful in making more exact calculations and in determining which species are important at a given pH. For example, if we are interested in calculating concentrations at pH 5.7, we can use the diagram in

Figure 15F-1 to tell us which species to include in the calculation. At pH 5.7, the concentrations of the maleate containing species are [H$_2$M] $\approx 10^{-5}$ M, [HM$^-$] ≈ 0.07 M, and [M^{2-}] $\approx$ 0.02 M. Hence, the only maleate species of importance at this pH are HM$^-$ and M^{2-}. Since [OH$^-$] is four orders of magnitude lower than [H$_3$O$^+$], we could carry out a more accurate calculation than the above estimates by considering only three species. If we do so, we find the following concentrations: [H$_2$M] $\approx 1.18 \times 10^{-5}$ M, [HM$^-$] ≈ 0.077 M, and [M^{2-}] = 0.023 M.

Finding pH Values

If we do not know the pH, the logarithmic concentration diagram can also be used to give us an approximate pH value. For example, find the pH of a 0.10 M maleic acid solution. Since the log concentration diagram expresses mass balance and the equilibrium constants, we need only one additional equation such as charge balance to solve the problem exactly. The charge-balance equation for this system is

$$[H_3O^+] = [HM^-] + 2[M^{2-}] + [OH^-]$$

The pH is found by graphically superimposing the charge-balance equation on the log concentration diagram. Beginning with a pH of 0, move from left to right along the H$_3$O$^+$ line until it intersects a line representing one of the species on the right hand side of the charge-balance equation. We see that the H$_3$O$^+$ line first intersects the HM$^-$ line at a pH of approximately 1.5. At this point, [H$_3$O$^+$] = [HM$^-$]. We also see that the concentrations of the other negatively charged species M^{2-} and OH^- are negligible compared to the HM^- concentration. Hence, the pH of a 0.10 M solution of maleic acid is approximately 1.5. A more accurate calculation using the quadratic formula gives pH = 1.52.

We can ask another question: "What is the pH of a 0.10 M solution of NaHM?" In this case, the charge-balance equation is

$$[H_3O^+] + [Na^+] = [HM^-] + 2[M^{2-}] + [OH^-]$$

The Na^+ concentration is just the total concentration of maleate-containing species:

$$[Na^+] = c_T = [H_2M] + [HM^-] + [M^{2-}]$$

Substituting this latter equation into the charge-balance equation gives

$$[H_3O^+] + [H_2M] = [M^{2-}] + [OH^-]$$

Now, we superimpose this equation on the log concentration diagram. If we again begin on the left at pH 0 and move along either the H_3O^+ line or the H_2M line, we see that, at pH values greater than about 2, the concentration of H_2M exceeds the H_3O^+ concentration by about an order of magnitude. Hence, we move along the H_2M line until it intersects either the M^{2-} line or the OH^- line. We see that it intersects the M^{2-} line first at pH ≈ 4.1. Thus, $[H_2M] ≈ [M^{2-}]$, and the

concentrations of the $[H_3O^+]$ and $[OH^-]$ are relatively small compared to H_2M and M^{2-}. Therefore, we conclude that the pH of a 0.10 M NaHM solution is approximately 4.1. A more exact calculation using the quadratic formula reveals that the pH of this solution is 4.08.

Finally, we will find the pH of a 0.10 M solution of Na_2M. The charge-balance equation is the same as before:

$$[H_3O^+] + [Na^+] = [HM^-] + 2[M^{2-}] + [OH^-]$$

Now, however, the Na^+ concentration is given by

$$[Na^+] = 2c_T = 2[H_2M] + 2[HM^-] + 2[M^{2-}]$$

Substituting this equation into the charge-balance equation gives

$$[H_3O^+] + 2[H_2M] + [HM^-] = [OH^-]$$

In this case, it is easier to find the OH^- concentration. This time we move down the OH^- line from right to left until it intersects the HM^- line at a pH of approximately 9.7. Since $[H_3O^+]$ and $[H_2M]$ are negligibly small at this intersection, $[HM^-] ≈ [OH^-]$, and we conclude that pH 9.7 is the approximate pH of a 0.10 M solution of Na_2M. A more exact calculation using the quadratic formula gives the pH as 9.61.

Spreadsheet Summary   In the first exercise in Chapter 8 of *Applications of Microsoft® Excel in Analytical Chemistry*, 2nd ed., we investigate the calculation of distribution diagrams for polyfunctional acids and bases. The alpha values are plotted as a function of pH. The plots are used to find concentrations at a given pH and to infer which species can be neglected in more extensive calculations. A logarithmic concentration diagram is constructed. The diagram is used to estimate concentrations at a given pH and to find the pH for various starting conditions with a weak acid system.

WEB WORKS   Go to **www.cengage.com/chemistry/skoog/fac9**, choose Chapter 15 and go to the Web Works. Click on the link to the Virtual Titrator. Click on the indicated frame to invoke the Virtual Titrator Java applet and display two windows: the Menu Panel and the Virtual Titrator main window. To begin, click on Acids on the main window menu bar and select the diprotic acid *o*-phthalic acid. Examine the titration curve that results. Then, click on Graphs/Alpha Plot vs. pH and observe the result. Click on Graphs/Alpha Plot vs. mL base. Repeat the process for several monoprotic and polyprotic acids and note the results.

QUESTIONS AND PROBLEMS

*15-1. As its name implies, NaHA is an "acid salt" because it has a proton available to donate to a base. Briefly explain why a pH calculation for a solution of NaHA differs from that for a weak acid of the type HA.

15-2. Explain the origin and significance of each of the terms on the right side of Equation 15-12. Does the equation make intuitive sense? Why or why not?

15-3. Briefly explain why Equation 15-15 can only be used to calculate the hydronium ion concentration of solutions in which NaHA is the only solute that determines the pH.

*15-4. Why is it impossible to titrate all three protons of phosphoric acid in aqueous solution?

15-5. Indicate whether an aqueous solution of the following compounds is acidic, neutral, or basic. Explain your answer.
  *(a) NH_4OAc
  (b) $NaNO_2$
  *(c) $NaNO_3$
  (d) $NaHC_2O_4$
  *(e) $Na_2C_2O_4$
  (f) Na_2HPO_4
  *(g) NaH_2PO_4
  (h) Na_3PO_4

*15-6. Suggest an indicator that could be used to provide an end point for the titration of the first proton in H_3AsO_4.

15-7. Suggest an indicator that would give an end point for the titration of the first two protons in H_3AsO_4.

*15-8. Suggest a method for determining the amounts of H_3PO_4 and NaH_2PO_4 in an aqueous solution.

15-9. Suggest a suitable indicator for a titration based on each of the following reactions. Use 0.05 M if an equivalence point concentration is needed.
  *(a) $H_2CO_3 + NaOH \rightarrow NaHCO_3 + H_2O$
  (b) $H_2P + 2NaOH \rightarrow Na_2P + 2H_2O$ (H_2P = o-phthalic acid)
  *(c) $H_2T + 2NaOH \rightarrow Na_2T + 2H_2O$ (H_2T = tartaric acid)
  (d) $NH_2C_2H_4NH_2 + HCl \rightarrow NH_2C_2H_4NH_3Cl$
  *(e) $NH_2C_2H_4NH_2 + 2HCl \rightarrow ClNH_3C_2H_4NH_3Cl$
  (f) $H_2SO_3 + NaOH \rightarrow NaHSO_3 + H_2O$
  *(g) $H_2SO_3 + 2NaOH \rightarrow Na_2SO_3 + 2H_2O$

15-10. Calculate the pH of a solution that is 0.0400 M in
  *(a) H_3PO_4.
  (b) $H_2C_2O_4$.
  *(c) H_3PO_3.
  (d) H_2SO_3.
  *(e) H_2S.
  (f) $H_2NC_2H_4NH_2$.

15-11. Calculate the pH of a solution that is 0.0400 M in
  *(a) NaH_2PO_4.
  (b) $NaHC_2O_4$.
  *(c) NaH_2PO_3.
  (d) $NaHSO_3$.
  *(e) $NaHS$.
  (f) $H_2NC_2H_4NH_3^+Cl^-$.

15-12. Calculate the pH of a solution that is 0.0400 M in
  *(a) Na_3PO_4.
  (b) $Na_2C_2O_4$.
  *(c) Na_2HPO_3.
  (d) Na_2SO_3.
  *(e) Na_2S.
  (f) $C_2H_4(NH_3^+Cl^-)_2$.

15-13. Calculate the pH of a solution that contains the following analytical concentrations:
  (a) 0.0500 M in H_3PO_4 and 0.0200 M in NaH_2PO_4.
  (b) 0.0300 M in NaH_2AsO_4 and 0.0500 M in Na_2HAsO_4.
  (c) 0.0600 M in Na_2CO_3 and 0.0300 M in $NaHCO_3$.
  (d) 0.0400 M in H_3PO_4 and 0.0200 M in Na_2HPO_4.
  (e) 0.0500 M in $NaHSO_4$ and 0.0400 M in Na_2SO_4.

*15-14. Calculate the pH of a solution that contains the following analytical concentrations:
  (a) 0.225 M in H_3PO_4 and 0.414 M in NaH_2PO_4.
  (b) 0.0670 M in Na_2SO_3 and 0.0315 M in $NaHSO_3$.
  (c) 0.640 M in $HOC_2H_4NH_2$ and 0.750 M in $HOC_2H_4NH_3Cl$.
  (d) 0.0240 in $H_2C_2O_4$ (oxalic acid) and 0.0360 M in $Na_2C_2O_4$.
  (e) 0.0100 M in $Na_2C_2O_4$ and 0.0400 M in $NaHC_2O_4$.

15-15. Calculate the pH of a solution that is
  (a) 0.0100 M in HCl and 0.0200 M in picric acid.
  (b) 0.0100 M in HCl and 0.0200 M in benzoic acid.
  (c) 0.0100 M in NaOH and 0.100 M in Na_2CO_3.
  (d) 0.0100 M in NaOH and 0.100 M in NH_3.

*15-16. Calculate the pH of a solution that is
  (a) 0.0100 M in $HClO_4$ and 0.0300 M in mono-chloroacetic acid.
  (b) 0.0100 M in HCl and 0.0150 M in H_2SO_4.
  (c) 0.0100 M in NaOH and 0.0300 M in Na_2S.
  (d) 0.0100 M in NaOH and 0.0300 M in sodium acetate.

15-17. Identify the principal conjugate acid/base pair and calculate the ratio between them in a solution that is buffered to pH 6.00 and contains
  (a) H_2SO_3.
  (b) citric acid.
  (c) malonic acid.
  (d) tartaric acid.

*15-18. Identify the principal conjugate acid/base pair and calculate the ratio between them in a solution that is buffered to pH 9.00 and contains
  (a) H_2S.
  (b) ethylenediamine dihydrochloride.
  (c) H_3AsO_4.
  (d) H_2CO_3.

15-19. What mass (g) of $Na_2HPO_4 \cdot 2H_2O$ must be added to 500 mL of 0.160 M H_3PO_4 to give a buffer of pH 7.30?

*15-20. What mass (g) of dipotassium phthalate must be added to 750 mL of 0.0500 M phthalic acid to give a buffer of pH 5.75?

15-21. What is the pH of the buffer formed by mixing 40.0 mL of 0.200 M NaH_2PO_4 with
  (a) 60.0 mL of 0.100 M HCl?
  (b) 60.0 mL of 0.100 M NaOH?

*15-22. What is the pH of the buffer formed by adding 100 mL of 0.150 M potassium hydrogen phthalate to
  (a) 100.0 mL of 0.0800 M NaOH?
  (b) 100.0 mL of 0.0800 M HCl?

15-23. How would you prepare 1.00 L of a buffer with a pH of 9.45 from 0.300 M Na_2CO_3 and 0.200 M HCl?

*15-24. How would you prepare 1.00 L of a buffer with a pH of 7.00 from 0.200 M H_3PO_4 and 0.160 M NaOH?

15-25. How would you prepare 1.00 L of a buffer with a pH of 6.00 from 0.500 M Na_3AsO_4 and 0.400 M HCl?

15-26. Identify by letter the curve you would expect in the titration of a solution containing
  (a) disodium maleate, Na_2M, with standard acid.
  (b) pyruvic acid, HP, with standard base.
  (c) sodium carbonate, Na_2CO_3, with standard acid.

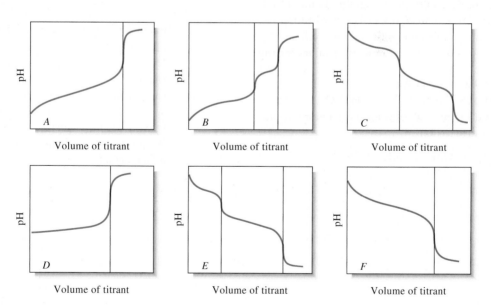

Titration curves for Problem 15-26.

15-27. Describe the composition of a solution that would be expected to yield a curve resembling (see Problem 15-26)
 (a) curve B.
 (b) curve A.
 (c) curve E.

*15-28. Briefly explain why curve B *cannot* describe the titration of a mixture consisting of H_3PO_4 and NaH_2PO_4.

15-29. Construct a curve for the titration of 50.00 mL of a 0.1000 M solution of compound A with a 0.2000 M solution of compound B in the following table. For each titration, calculate the pH after the addition of 0.00, 12.50, 20.00, 24.00, 25.00, 26.00, 37.50, 45.00, 49.00, 50.00, 51.00, and 60.00 mL of compound B.

A	B
(a) H_2SO_3	NaOH
(b) ethylenediamine	HCl
(c) H_2SO_4	NaOH

15-30. Generate a curve for the titration of 50.00 mL of a solution in which the analytical concentration of NaOH is 0.1000 M and that for hydrazine is 0.0800 M. Calculate the pH after addition of 0.00, 10.00, 20.00, 24.00, 25.00, 26.00, 35.00, 44.00, 45.00, 46.00, and 50.00 mL of 0.2000 M $HClO_4$.

15-31. Generate a curve for the titration of 50.00 mL of a solution in which the analytical concentration of $HClO_4$ is 0.1000 M and that for formic acid is 0.0800 M. Calculate the pH after addition of 0.00, 10.00, 20.00, 24.00, 25.00, 26.00, 35.00, 44.00, 45.00, 46.00, and 50.00 mL of 0.2000 M KOH.

*15-32. Formulate equilibrium constants for the following equilibria, and determine numerical values for the constants:
 (a) $2H_2AsO_4^- \rightleftharpoons H_3AsO_4 + HAsO_4^{2-}$
 (b) $2HAsO_4^{2-} \rightleftharpoons AsO_4^{3-} + H_2AsO_4^-$

15-33. Calculate a numerical value for the equilibrium constant for the reaction

$$NH_4^+ + OAc^- \rightleftharpoons NH_3 + HOAc$$

15-34. For pH values of 2.00, 6.00, and 10.00, calculate the alpha values for each species in an aqueous solution of
 *(a) phthalic acid.
 (b) phosphoric acid.
 *(c) citric acid.
 (d) arsenic acid.
 *(e) phosphorous acid.
 (f) oxalic acid.

15-35. Derive equations that define α_0, α_1, α_2, and α_3 for the acid H_3AsO_4.

15-36. Calculate alpha values for the following diprotic acids every 0.5 pH units from pH 0.0 to 10.0. Plot the distribution diagram for each of the acids and label the curve for each species.
 (a) phthalic acid.
 (b) succinic acid.
 (c) tartaric acid.

15-37. Calculate alpha values for the following triprotic acids every 0.5 pH units from pH 0.0 to 14.0. Plot the distribution diagram for each of the acids and label the curve for each species.
 (a) citric acid
 (b) arsenic acid

15-38. **Challenge Problem:**
 (a) Plot logarithmic concentration diagrams for 0.1000 M solutions of each of the acids in Problem 15-36 above.
 (b) For phthalic acid, find the concentrations of all species at pH 4.8.
 (c) For tartaric acid, find the concentrations of all species at pH 4.3.
 (d) From the log concentration diagram, find the pH of a 0.1000 M solution of phthalic acid, H_2P. Find the pH of a 0.100 M solution of HP^-.
 (e) Discuss how you might modify the log concentration diagram for phthalic acid so that it shows the pH in terms of the hydrogen ion activity, a_{H^+}, instead of the hydrogen ion concentration (pH $= -\log a_{H^+}$. instead of pH $= -\log c_{H^+}$). Be specific in your discussion and show what the difficulties might be.

Applications of Neutralization Titrations

CHAPTER 16

Acids and bases are very important in the environment, in our bodies, and in many other systems. In the environment, acid rain falling on the surface waters of lakes and rivers can cause these waters to become acidic. In the eastern United States, the number of acidic lakes increased dramatically over the period of 1930 to 1970, and this increase has been attributed to acid rain. On the other hand, many lakes in the Midwest have no problem with acidification even though the industrial Midwest is presumed to be a major source of the acids found in acid rain. In the Midwest, the surface rock is mostly limestone (calcium carbonate), which reacts with CO_2 and H_2O to form bicarbonate. Bicarbonate in turn neutralizes acids to maintain the pH relatively constant. This effect is characterized by the **acid neutralizing capacity** of the lake, which is usually quite large in limestone-rich areas. In contrast, many eastern lakes and streams are surrounded by granite, which is a much less reactive rock. These bodies of water have little neutralizing capacity and so are more susceptible to acidification. To combat this problem, limestone is often imported from limestone-rich states into the eastern states and applied to lakes and streams. The photo shows workers dumping pulverized limestone into Cedar Creek, Shenandoah County, Virginia, to neutralize acidic waters that had previously killed many stocked rainbow trout. Acid neutralizing capacity is frequently determined by titration with a standard solution of acid.

© Wally McNamee/CORBIS

Neutralization titrations are widely used to determine the concentration of acidic or basic analytes or analytes that can be converted to acids or bases by suitable treatment.[1] Water is the usual solvent for neutralization titrations because it is convenient, inexpensive, and nontoxic. An added advantage is water's low temperature coefficient of expansion. Some analytes, however, cannot be titrated in aqueous media because of low solubilities or because their strengths as acids or bases are not large enough to give satisfactory end points. Such substances can often be titrated by nonaqueous solvents.[2] We shall restrict our discussions to aqueous systems.

---

[1]For a review of applications of neutralization titrations, see J. A. Dean. *Analytical Chemistry Handbook,* Section 3.2, p. 3.28, New York: McGraw-Hill, 1995; D. Rosenthal and P. Zuman, in *Treatise on Analytical Chemistry*, 2nd ed., I. M. Kolthoff and P. J. Elving, eds., Part I, Vol. 2, Chap. 18, New York: Wiley, 1979.

[2]For a review of nonaqueous acid/base titrimetry, see J. A. Dean. *Analytical Chemistry* Handbook, Section 3.3, p. 3.48, New York: McGraw-Hill, 1995; I. M. Kolthoff and P. J. Elving, eds., *Treatise on Analytical Chemistry*, 2nd ed., Part I, Vol. 2, Chaps. 19A–19E, New York: Wiley, 1979.

16A REAGENTS FOR NEUTRALIZATION TITRATIONS

In Chapter 14, we noted that strong acids and strong bases produce the largest change in pH at the equivalence point. For this reason, standard solutions for neutralization titrations are always prepared from these reagents.

16A-1 Preparation of Standard Acid Solutions

Hydrochloric acid solutions are widely used as standard solutions for titrating bases. Dilute solutions of HCl are stable indefinitely, and many chloride salts are soluble in aqueous solution. Solutions of 0.1 M HCl can be boiled for as long as an hour without loss of acid, provided that the water lost by evaporation is periodically replaced; 0.5 M solutions can be boiled for at least ten minutes without significant loss.

> Solutions of HCl, $HClO_4$, and H_2SO_4 are stable indefinitely. Restandardization is not required unless evaporation occurs.

Solutions of perchloric acid and sulfuric acid are also stable and are useful for titrations where chloride ion interferes by forming precipitates. Standard solutions of nitric acid are seldom used because of their oxidizing properties.

To obtain most standard acid solutions, a solution of an approximate concentration is first prepared by diluting the concentrated reagent. The diluted acid solution is then standardized against a primary-standard base. Occasionally, the composition of the concentrated acid is obtained by careful density measurements. A weighed quantity of the concentrated acid is then diluted to a known volume. Most chemistry and chemical engineering handbooks contain tables relating reagent density to composition. A stock solution with a known hydrochloric acid concentration can also be prepared by diluting a quantity of the concentrated reagent with an equal volume of water followed by distillation. Under controlled conditions, the final quarter of the distillate, known as **constant-boiling** HCl, has a constant and known composition. The acid content of constant-boiling HCl depends only on atmospheric pressure. For a pressure P between 670 and 780 torr, the mass in air of the distillate that contains exactly one mole of H_3O^+ is[3]

$$\frac{\text{mass of constant-boiling HCl in g}}{\text{mol } H_3O^+} = 164.673 + 0.02039P \qquad (16\text{-}1)$$

Standard solutions are prepared by diluting weighed quantities of this acid to accurately known volumes.

16A-2 The Standardization of Acids

Sodium carbonate is the most frequently used reagent for standardizing acids. Several other reagents are also used.

Sodium Carbonate

> Sodium carbonate occurs naturally in large deposits as *washing soda*, $Na_2CO_3 \cdot 10H_2O$, and as *trona*, $Na_2CO_3 \cdot NaHCO_3 \cdot 2H_2O$. These minerals are used in the glass industry as well as in many others. Primary standard sodium carbonate is manufactured by extensively purifying these minerals.

Primary-standard-grade sodium carbonate is available commercially or can be prepared by heating purified sodium hydrogen carbonate between 270 to 300°C for 1 hr.

$$2HCO_3(s) \rightarrow Na_2CO_3 + H_2O + CO_2(g)$$

An accurately determined mass of the primary-standard material is then taken to standardize the acid.

---

[3]See *Official Methods of Analysis of the AOAC*, 18th ed. online (subscription required), Appendix A.1.06, Official Method 936.15. Washington, D.C.: Association of Official Analytical Chemists, 2005.

As shown in **Figure 16-1**, there are two end points in the titration of sodium carbonate. The first, corresponding to the conversion of carbonate to hydrogen carbonate, occurs at about pH 8.3; the second, involving the formation of carbonic acid and carbon dioxide, appears at about pH 3.8. The second end point is always used for standardization because the change in pH is greater than that at the first. An even sharper end point can be achieved by boiling the solution briefly to eliminate the reaction product, carbonic acid and carbon dioxide. The sample is titrated to the first appearance of the acid color of the indicator (such as bromcresol green or methyl orange). At this point, the solution contains a large amount of dissolved carbon dioxide and small amounts of carbonic acid and unreacted hydrogen carbonate. Boiling effectively destroys this buffer by eliminating the carbonic acid:

$$H_2CO_3(aq) \rightarrow CO_2(g) + H_2O(l)$$

The solution then becomes alkaline again due to the residual hydrogen carbonate ion. The titration is completed after the solution has cooled, resulting in a substantially larger decrease in pH during the final additions of acid. A more abrupt color change (see Figure 16-1) is the result.

Alternatively, an amount of acid slightly in excess of that needed to convert the sodium carbonate to carbonic acid can be introduced. The solution is boiled as before to remove carbon dioxide and cooled. The excess acid is then back-titrated with a dilute solution of standard base. Any indicator suitable for a strong acid/strong base titration is satisfactory. An independent titration is used to establish the volume ratio of acid to base.

Other Primary Standards for Acids

Tris-(hydroxymethyl)aminomethane, $(HOCH_2)_3CNH_2$, known also as TRIS or THAM, is available in primary-standard purity from commercial sources. The main advantage of TRIS is its much greater mass per mole of protons consumed (121.1 g/mol) than sodium carbonate (53.0 g/mol). Example 16-1 illustrates this advantage. The reaction of TRIS with acids is

$$(HOCH_2)_3CNH_2 + H_3O^+ \rightleftharpoons (HOCH_2)_3CNH_3^+ + H_2O$$

Sodium tetraborate decahydrate and mercury(II) oxide have also been recommended as primary standards. The reaction of an acid with the tetraborate is

$$B_4O_7^- + 2H_3O^+ + 3H_2O \rightarrow 4H_3BO_3$$

Figure 16-1 Titration of 25.00 mL of 0.1000 M Na_2CO_3 with 0.1000 M HCl. After about 49 mL of HCl have been added, the solution is boiled, causing the increase in pH shown. The change in pH when more HCl is added is much larger after boiling.

❮ A high mass per proton consumed is desirable in a primary standard because a larger mass of reagent must be used, thus decreasing the relative weighing error.

❮ Borax, $Na_2B_4O_7 \cdot 10H_2O$, is a mineral that is mined in the desert and is widely used in cleaning preparations. A highly purified form of borax is used as a primary standard for bases.

EXAMPLE 16-1

Use a spreadsheet to compare the masses of (a) TRIS (121 g/mol), (b) Na_2CO_3 (106 g/mol), and (c) $Na_2B_4O_7 \cdot 10H_2O$ (381 g/mol) required to standardize an approximately 0.020 M solution of HCl for the following volumes of HCl: 20.00, 30.00, 40.00, and 50.00 mL. Assume that the standard deviation of the mass of each of the primary standard bases is 0.1 mg and use the spreadsheet to calculate the percent relative standard deviation that this uncertainty would introduce into each of the calculated concentrations.

(continued)

$$CH_2OH$$
$$|$$
$$H_2N-C-CH_2OH$$
$$|$$
$$CH_2OH$$

Molecular model and structure of TRIS.

Solution

The spreadsheet is shown in **Figure 16-2**. The molar concentration of HCl is entered in cell B2 and the molar masses of the three primary standards in cells B3, B4, and B5. The volumes of HCl for which the calculations are desired are entered in cells A8 through A11. We will do a sample calculation for 20.00 mL of HCl and show the spreadsheet entry. In each case, the number of mmoles of HCl is calculated from

$$\text{mmol HCl} = \text{mL HCl} \times 0.020 \, \frac{\text{mmol HCl}}{\text{mL HCl}}$$

(a) TRIS

$$\text{mass TRIS} = \text{mmol HCl} \times \frac{1 \text{ mmol TRIS}}{\text{mmol HCl}} \times \frac{121 \text{ g TRIS/mol}}{1000 \text{ mmol TRIS/mol}}$$

The appropriate formula for this equation is entered into cell B8 and then copied into cells B9 through B11. The relative uncertainty in the molar concentration due to the mass measurement is equal to the relative uncertainty in the mass measurement process. For the first TRIS amount (0.048 g in cell B8), the percent relative standard deviation (%RSD) is (0.0001/B8) × 100% as shown in the documentation in Figure 16-2. This formula is then copied into cells C9:C11.

(b) Na$_2$CO$_3$

$$\text{mass Na}_2\text{CO}_3 = \text{mmol HCl} \times \frac{1 \text{ mmol Na}_2\text{CO}_3}{2 \text{ mmol HCl}} \times \frac{106 \text{ g Na}_2\text{CO}_3/\text{mol}}{1000 \text{ mmol Na}_2\text{CO}_3/\text{mol}}$$

This formula is entered into cell D8 and copied into D9:D11. The relative standard deviation in cell E8 is calculated as (0.0001/D8) × 100%.

(c) Na$_2$B$_4$O$_7 \cdot$ 10H$_2$O

	A	B	C	D	E	F	G
1	Spreadsheet to compare masses required for various bases in the standardization of 0.020 M HCl						
2	M HCl	0.020					
3	Molar mass TRIS	121	g/mol	Note:All mass measurements have standard deviations of 0.1 mg			
4	Molar mass Na$_2$CO$_3$	106	g/mol				
5	Molar mass Na$_2$B$_4$O$_7$.H$_2$O	381	g/mol				
6							
7	mL HCl	g TRIS	%RSD TRIS	g Na$_2$CO$_3$	%RSD Na$_2$CO$_3$	g Na$_2$B$_4$O$_7$.H$_2$O	%RSD Na$_2$B$_4$O$_7$.H$_2$O
8	20.00	0.048	0.21	0.021	0.47	0.08	0.13
9	30.00	0.073	0.14	0.032	0.31	0.11	0.09
10	40.00	0.097	0.10	0.042	0.24	0.15	0.07
11	50.00	0.121	0.08	0.053	0.19	0.19	0.05
12							
13	Documentation						
14	Cell B8=$B$2*A8*1*$B$3/1000						
15	Cell C8=(0.0001/B8)*100						
16	Cell D8=$B$2*A8*1/2*$B$4/1000						
17	Cell E8=(0.0001/D8)*100						
18	Cell F8=$B$2*A8*1/2*$B$5/1000						
19	Cell G8=(0.0001/F8)*100						

Figure 16-2 Spreadsheet to compare masses and relative errors associated with using different primary-standard bases to standardize HCl solutions.

The same formula as that for Na_2CO_3 is used except that the molecular mass of borax (381 g/mol) is substituted for that of Na_2CO_3. The other formulas are shown in the documentation in Figure 16-2.

Note in Figure 16-2 that the relative standard deviation in concentration using TRIS is 0.10% or less if the volume of HCl taken is more than 40.00 mL. For Na_2CO_3, more than 50.00 mL of HCl would be required for this level of uncertainty. For borax, any volume above about 26.00 mL would suffice.

16A-3 Preparation of Standard Solutions of Base

Sodium hydroxide is the most common base for preparing standard solutions, although potassium hydroxide and barium hydroxide are also used. These bases cannot be obtained in primary-standard purity, and so, all must be standardized after they are prepared.

The Effect of Carbon Dioxide on Standard Base Solutions

In solution as well as in the solid state, the hydroxides of sodium, potassium, and barium react rapidly with atmospheric carbon dioxide to produce the corresponding carbonate:

$$CO_2(g) + 2OH^- \rightarrow CO_3^{2-} + H_2O$$

Although production of each carbonate ion consumes two hydroxide ions, the uptake of carbon dioxide by a solution of base does not necessarily alter its combining capacity for hydronium ions. Thus, at the end point of a titration that requires an acid-range indicator (such as bromocresol green), each carbonate ion produced from sodium or potassium hydroxide will have reacted with two hydronium ions of the acid (see Figure 16-1):

$$CO_3^{2-} + 2H_3O^+ \rightarrow H_2CO_3 + 2H_2O$$

Because the amount of hydronium ion consumed by this reaction is identical to the amount of hydroxide lost during formation of the carbonate ion, no error results from the reaction of the hydroxide with CO_2.

Unfortunately, most applications of standard base require an indicator with a basic transition range (phenolphthalein, for example). In this case, carbonate ion has reacted with only one hydronium ion when the color change of the indicator is observed:

$$CO_3^{2-} + H_3O^+ \rightarrow HCO_3^- + H_2O$$

The effective concentration of the base is thus diminished by absorption of carbon dioxide, and a systematic error (called the **carbonate error**) results as illustrated in Example 16-2.

> ❮ Absorption of carbon dioxide by a standardized solution of sodium or potassium hydroxide leads to a negative systematic error in analyses in which an indicator with a basic range is used; there is no systematic error when an indicator with an acidic range is used.

EXAMPLE 16-2

The hydroxide concentration in a carbonate-free NaOH solution was found to be 0.05118 M immediately after preparation. If exactly 1.000 L of this solution was exposed to air for some time and absorbed 0.1962 g CO_2, calculate the relative carbonate error that would arise in the determination of acetic acid with the contaminated solution if phenolphthalein were used as an indicator.

(continued)

Solution

$$2NaOH + CO_2 \rightarrow Na_2CO_3 + H_2O$$

$$c_{Na_2CO_3} = \frac{0.1962 \text{ g } CO_2}{1.000 \text{ L}} \times \frac{1 \text{ mol } CO_2}{44.01 \text{ g } CO_2} \times \frac{1 \text{ mol } Na_2CO_3}{\text{mol } CO_2} = 4.458 \times 10^{-3} \text{ M}$$

The effective concentration c_{NaOH} of NaOH for acetic acid is then

$$c_{NaOH} = \frac{0.05118 \text{ mol NaOH}}{L} - \left(\frac{4.456 \times 10^{-3} \text{ mol } Na_2CO_3}{L} \right.$$

$$\left. \times \frac{1 \text{ mol } HOAc}{\text{mol } Na_2CO_3} \times \frac{1 \text{ mol NaOH}}{\text{mol } HOAc} \right)$$

$$= 0.04672 \text{ M}$$

$$\text{rel error} = \frac{0.04672 - 0.05118}{0.05118} \times 100\% = -8.7\%$$

> Carbonate ion in standard base solutions decreases the sharpness of end points and is usually removed before standardization.

> **WARNING**: Concentrated solutions of NaOH (and KOH) are extremely corrosive to the skin. A face shield, rubber gloves, and protective clothing **must be worn at all times** while working with these solutions.

> Water that is in equilibrium with atmospheric constituents contains only about 1.5×10^{-5} mol CO_2/L, an amount that has a negligible effect on the strength of most standard bases. As an alternative to boiling to remove CO_2 from supersaturated solutions of CO_2, the excess gas can be removed by bubbling air through the water for several hours. This process is called sparging and produces a solution that contains the equilibrium concentration of CO_2.

> **Sparging** is the process of removing a gas from a solution by bubbling an inert gas through the solution.

The solid reagents used to prepare standard solutions of base are always contaminated by significant amounts of carbonate ion. The presence of this contaminant does not cause a carbonate error provided the same indicator is used for both standardization and analysis. Carbonate does, however, decrease the sharpness of end points. For this reason, carbonate ion is usually removed before a solution of a base is standardized.

The best method for preparing carbonate-free sodium hydroxide solutions takes advantage of the very low solubility of sodium carbonate in concentrated solutions of the base. An approximately 50% aqueous solution of sodium hydroxide is prepared or purchased from commercial sources. The solid sodium carbonate is allowed to settle to give a clear liquid that is decanted and diluted to give the desired concentration. Alternatively, the solid can be removed by vacuum filtration.

Water that is used to prepare carbonate-free solutions of base must also be free of carbon dioxide. Distilled water, which is sometimes supersaturated with carbon dioxide, should be boiled briefly to eliminate CO_2. The water is then allowed to cool to room temperature before base is introduced because hot alkali solutions rapidly absorb carbon dioxide. Deionized water usually does not contain significant amounts of carbon dioxide.

A tightly capped low-density polyethylene bottle usually provides adequate short-term protection against the uptake of atmospheric carbon dioxide. Before capping, the bottle is squeezed to minimize the interior air space. Care should also be taken to keep the bottle closed except during the brief periods when the contents are being transferred to a buret. Over time, sodium hydroxide solutions cause polyethylene bottles to become brittle.

The concentration of a sodium hydroxide solution will decrease slowly (0.1% to 0.3% per week) if the base is stored in glass bottles. The loss in strength is caused by the reaction of the base with the glass to form sodium silicates. For this reason, standard solutions of base should not be stored for extended periods (longer than 1 or 2 weeks) in glass containers. In addition, bases should never be kept in glass-stoppered containers because the reaction between the base and the stopper may cause the latter to "freeze" after a brief period. Finally, to avoid the same type of freezing, burets with glass stopcocks should be promptly drained and thoroughly rinsed with water after use with standard base solutions. Most modern burets are equipped with Teflon stopcocks and thus do not have this problem.

16A-4 The Standardization of Bases

Several excellent primary standards are available for standardizing bases. Most are weak organic acids that require the use of an indicator with a basic transition range.

Potassium Hydrogen Phthalate

Potassium hydrogen phthalate, $KHC_8H_4O_4$ is a nearly ideal primary standard. It is a nonhygroscopic crystalline solid with a relatively large molar mass (204.2 g/mol). For most purposes, the commercial analytical-grade salt can be used without further purification. For the most exacting work, potassium hydrogen phthalate (KHP) of certified purity is available from the National Institute of Standards and Technology.

Other Primary Standards for Bases

Benzoic acid can be obtained in primary-standard purity and used for the standardization of bases. Benzoic acid has limited solubility in water, so it is usually dissolved in ethanol prior to dilution with water and titration. A blank should always be carried through this standardization because commercial alcohol is sometimes slightly acidic.

Potassium hydrogen iodate, $KH(IO_3)_2$, is an excellent primary standard with a high molar mass per mole of protons. It is also a strong acid that can be titrated using virtually any indicator with a transition range between pH 4 and 10.

16B TYPICAL APPLICATIONS OF NEUTRALIZATION TITRATIONS

Neutralization titrations are used to determine the many inorganic, organic, and biological species that possess acidic or basic properties. In addition, however, there are nearly as many applications in which the analyte is converted to an acid or base by suitable chemical treatment and then titrated with a standard strong base or acid.

There are two major types of end points that are widely used in neutralization titrations. The first is a visual end point based on indicators such as those described in Section 14A. The second is a *potentiometric* end point in which the potential of a glass/calomel electrode system is determined with a pH meter or another voltage-measuring device. The measured voltage is directly proportional to pH. We describe potentiometric end points in Section 21G.

16B-1 Elemental Analysis

Several important elements that occur in organic and biological systems are conveniently determined by methods that have an acid/base titration as the final step. Generally, the elements susceptible to this type of analysis are nonmetals such as carbon, nitrogen, chlorine, bromine, fluorine, as well as a few other less common species. Pretreatment converts the element to an inorganic acid or base that is then titrated. A few examples follow.

> Solutions of bases should be stored in polyethylene bottles rather than glass because of the reaction between bases and glass. Such solutions should never be stored in glass-stoppered bottles; after standing for a period, a stopper is often impossible to remove.

> Standard solutions of strong bases cannot be prepared directly by mass and must always be standardized against a primary-standard acid.

> In contrast to all other primary standards for bases, $KH(IO_3)_2$ has the advantage of being a strong acid, which makes it easy to choose an indicator.

Molecular model and structure of potassium hydrogen phthalate.

> Neutralization titrations are still among the most widely used analytical methods.

Nitrogen

Nitrogen occurs in a wide variety of substances of interest in the life sciences, in industry, and in agriculture. Examples include amino acids, proteins, synthetic drugs, fertilizers, explosives, soils, potable water supplies, and dyes. Thus, analytical methods for the determination of nitrogen, particularly in organic substrates, are extremely important.

The most common method for determining organic nitrogen is the **Kjeldahl method**, which is based on a neutralization titration (see Feature 16-1). The procedure is straightforward, requires no special equipment, and is readily adapted to the routine analysis of large numbers of samples. The Kjeldahl method, or one of its modifications, is the standard process for determining the protein content of grains, meats, and biological materials (see Feature 16-2 for other methods). Since most proteins contain approximately the same percentage of nitrogen, multiplication of this percentage by a suitable factor (6.25 for meats, 6.38 for dairy products, and 5.70 for cereals) gives the percentage of protein in a sample.

> Kjeldahl is pronounced *Kyell'dahl*. Hundreds of thousands of Kjeldahl nitrogen determinations are performed each year, primarily to provide a measure of the protein content of meats, grains, and animal feeds.

The Kjeldahl method was developed by the Danish chemist Johan Kjeldahl, who first described it in 1883 (J. Kjeldahl, *Z. Anal. Chem.*, **1883**, *22*, 366). Working at the Carlsberg Laboratory, Kjeldahl developed the method to determine the protein content of various grains to be used in brewing beer.

FEATURE 16-1

Determining Total Serum Protein

The determination of total serum protein is an important clinical measurement and used in diagnosing liver malfunctions. Although the Kjeldahl method is capable of high precision and accuracy, it is too slow and cumbersome to be routinely used in determining total serum protein. However, the Kjeldahl procedure has historically been the *reference method* against which other methods are compared. Methods that are commonly used include the **biuret method** and the **Lowry method**.[4] In the biuret method, a reagent containing copper(II) ions is used, and a violet colored complex is formed between the Cu^{2+} ions and peptide bonds. The increase in the absorption of visible radiation is used to measure serum protein. The biuret method is readily automated. In the Lowry procedure, the serum sample is pretreated with an alkaline copper solution followed by a phenolic reagent. A color develops because of reduction of phosphotungstic acid and phosphomolybdic acid to molybdenum blue. Both the biuret and Lowry methods use spectrophotometry (see Chapter 26) for quantitative measurements.

FEATURE 16-2

Other Methods for Determining Organic Nitrogen

Several other methods are used to determine the nitrogen content of organic materials. In the **Dumas method**, the sample is mixed with powdered copper(II) oxide and ignited in a combustion tube to give carbon dioxide, water, nitrogen, and small amounts of nitrogen oxides. A stream of carbon dioxide carries these products through a packing of hot copper, which reduces any oxides of nitrogen to elemental nitrogen. The mixture then is passed into a gas buret filled with concentrated potassium hydroxide. The only component not absorbed by the base is nitrogen, and its volume is measured directly.

A relatively recent method for determining organic nitrogen begins with the combustion of the sample at 1100°C for a few minutes to convert the nitrogen to nitric oxide, NO. Ozone is then introduced into the gaseous mixture, which oxidizes the nitric oxide to nitrogen dioxide. This reaction gives off visible radiation (*chemiluminescence*). The intensity of chemiluminescence is measured and is proportional to the nitrogen content of the sample. Commercial instruments are available for this procedure. Chemiluminescence is discussed further in Chapter 27.

---

[4]O. H. Lowry, et al., J. Biol. Chem., **1951**, *193*, 265.

In the Kjeldahl method, the sample is decomposed in hot, concentrated sulfuric acid to convert the bound nitrogen to ammonium ion. The resulting solution is then cooled, diluted, and made basic, a process that converts the ammonium ions to ammonia. The ammonia is distilled from the basic solution, collected in an acidic solution, and determined by a neutralization titration.

The critical step in the Kjeldahl method is the decomposition with sulfuric acid, which oxidizes the carbon and hydrogen in the sample to carbon dioxide and water. The fate of the nitrogen, however, depends on its state of combination in the original sample. Amine and amide nitrogens are quantitatively converted to ammonium ion. In contrast, nitro, azo, and azoxy groups are likely to yield the element or its various oxides, all of which are lost from the hot acidic medium. This loss can be avoided by pretreating the sample with a reducing agent to form amides or amines. In one such prereduction scheme, salicylic acid and sodium thiosulfate are added to the concentrated sulfuric acid solution containing the sample. After a brief period, the digestion is performed in the usual way.

Pyridine, pyridine derivatives, and some other aromatic heterocyclic compounds are particularly resistant to complete decomposition by sulfuric acid. Such compounds yield low results as a consequence (see Figure 5-3) unless special precautions are taken.

The decomposition step is frequently the most time-consuming aspect of a Kjeldahl determination. Some samples may require heating periods in excess of 1 hr. Numerous modifications of the original procedure have been proposed with the aim of shortening the digestion time. In the most widely used modification, a neutral salt, such as potassium sulfate, is added to increase the boiling point of the sulfuric acid solution and thus the temperature at which the decomposition occurs. In another modification, a solution of hydrogen peroxide is added to the mixture after the digestion has decomposed most of the organic matrix.

Many substances catalyze the decomposition of organic compounds by sulfuric acid. Mercury, copper, and selenium, either combined or in the elemental state, are effective. Mercury(II), if present, must be precipitated with hydrogen sulfide prior to distillation to prevent retention of ammonia as a mercury(II) ammine complex.

Example 16-3 illustrates the calculations used in the Kjeldahl method.

◀ $-NO_2$ $-N{=}N-$ $\begin{array}{c} -N^+{=}N- \\ | \\ O^- \end{array}$

nitro group   azo group   azoxy group

EXAMPLE 16-3

A 0.7121 g sample of wheat flour was analyzed by the Kjeldahl method. The ammonia formed by addition of concentrated base after digestion with H_2SO_4 was distilled into 25.00 mL of 0.04977 M HCl. The excess HCl was then back-titrated with 3.97 mL of 0.04012 M NaOH. Calculate the percent protein in the flour, using the 5.70 factor for cereal.

Solution

$$\text{amount HCl} = 25.00 \text{ mL HCl} \times 0.04977 \frac{\text{mmol}}{\text{mL HCl}} = 1.2443 \text{ mmol}$$

$$\text{amount NaOH} = 3.97 \text{ mL NaOH} \times 0.04012 \frac{\text{mmol}}{\text{mL NaOH}} = 0.1593 \text{ mmol}$$

$$\text{amount N} = \text{amount HCl} - \text{amount NaOH} = 1.2443 \text{ mmol}$$

$$- 0.1593 \text{ mmol} = 1.0850 \text{ mmol}$$

(continued)

$$\%N = \frac{1.0850 \; \cancel{\text{mmol N}} \times \dfrac{0.014007 \; \text{g N}}{\cancel{\text{mmol N}}}}{0.7121 \; \text{g sample}} \times 100\% = 2.1342$$

$$\% \text{ protein} = 2.1342 \; \cancel{\%\text{ N}} \times \frac{5.70\% \text{ protein}}{\cancel{\%\text{ N}}} = 12.16$$

Sulfur

Sulfur in organic and biological materials is conveniently determined by burning the sample in a stream of oxygen. The sulfur dioxide (as well as the sulfur trioxide) formed during the oxidation is collected by distillation into a dilute solution of hydrogen peroxide:

$$SO_2(g) + H_2O_2 \rightarrow H_2SO_4$$

> Sulfur dioxide in the atmosphere is often determined by drawing a sample through a hydrogen peroxide solution and then titrating the sulfuric acid that is produced.

The sulfuric acid is then titrated with standard base.

Other Elements

Table 16-1 lists other elements that can be determined by neutralization methods.

16B-2 The Determination of Inorganic Substances

Various inorganic species can be determined by titration with strong acids or bases. A few examples follow.

Ammonium Salts

Ammonium salts are conveniently determined by conversion to ammonia with strong base followed by distillation. The ammonia is collected and titrated as in the Kjeldahl method.

Nitrates and Nitrites

The method just described for ammonium salts can be extended to the determination of inorganic nitrate or nitrite. These ions are first reduced to ammonium ion by reaction with an alloy of 50% Cu, 45% Al, and 5% Zn(Devarda's alloy). Granules of the alloy are introduced into a strongly alkaline solution of the sample in a Kjeldahl flask. The ammonia is distilled after reaction is complete. An alloy of 60% Cu and 40% Mg (Arnd's alloy) has also been used as the reducing agent.

TABLE 16-1

Elemental Analyses Based on Neutralization Titrations

Element	Converted to	Adsorption or Precipitation Products	Titration
N	NH_3	$NH_3(g) + H_3O^+ \rightarrow NH_4^+ + H_2O$	Excess HCl with NaOH
S	SO_2	$SO_2(g) + H_2O_2 \rightarrow H_2SO_4$	NaOH
C	CO_2	$CO_2(g) + Ba(OH)_2 \rightarrow BaCO_3(s) + H_2O$	Excess $Ba(OH)_2$ with HCl
Cl(Br)	HCl	$HCl(g) + H_2O \rightarrow Cl^- + H_3O^+$	NaOH
F	SiF_4	$3SiF_4(g) + 2H_2O \rightarrow 2H_2SiF_6 + SiO_2$	NaOH
P	H_3PO_4	$12H_2MoO_4 + 3NH_4^+ + H_3PO_4 \rightarrow$ $(NH_4)_3PO_4 \cdot 12MoO_3(s) + 12H_2O + 3H^+$ $(NH_4)_3PO_4 \cdot 12MoO_3(s) + 26OH^- \rightarrow$ $HPO_4^{2-} + 12MoO_4^{2-} + 14H_2O + 3NH_3(g)$	Excess NaOH with HCl

Carbonate and Carbonate Mixtures

The qualitative and quantitative determination of the constituents in a solution containing sodium carbonate, sodium hydrogen carbonate, and sodium hydroxide, either alone or as various mixtures, provides interesting examples of how neutralization titrations can be applied to analyze mixtures. No more than two of these three constituents can exist in appreciable amount in any solution because reaction eliminates the third. For example, mixing sodium hydroxide with sodium hydrogen carbonate results in the formation of sodium carbonate until one or the other (or both) of the original reactants is exhausted. If the sodium hydroxide is used up, the solution will contain sodium carbonate and sodium hydrogen carbonate. If sodium hydrogen carbonate is depleted, sodium carbonate and sodium hydroxide will remain. If equimolar amounts of sodium hydrogen carbonate and sodium hydroxide are mixed, the principal solute species will be sodium carbonate.

The analysis of such mixtures requires two titrations with a strong acid: one using an alkaline-range indicator, such as phenolphthalein, and the other with an acid-range indicator, such as bromocresol green. The composition of the solution can then be deduced from the relative volumes of acid needed to titrate equal volumes of the sample (see Table 16-2 and Figure 16-3). Once the composition of the solution has been established, the volume data can be used to determine the concentration of each component in the sample. Example 16-4 illustrates the calculations needed to analyze a carbonate mixture.

The method described in Example 16-4 is not entirely satisfactory because the pH change corresponding to the hydrogen carbonate end point is not sufficient to give a sharp color change with a chemical indicator (see Figure 15-5). Because of this lack of sharpness, relative errors of 1% or more are common.

The accuracy of methods for analyzing solutions containing mixtures of carbonate and hydrogen carbonate ions or carbonate and hydroxide ions can be greatly improved by taking advantage of the limited solubility of barium carbonate in neutral and basic solutions. For example, in the **Winkler method** for the analysis of carbonate/hydroxide mixtures, both components are titrated with a standard acid to the end point with an acid-range indicator, such as bromocresol green (the end point being established after the solution is boiled to remove carbon dioxide). An unmeasured excess of neutral barium chloride is then added to a second aliquot of the sample solution to precipitate the carbonate ion, following which the hydroxide ion is titrated to a phenolphthalein end point. The presence of the sparingly soluble barium carbonate does not interfere as long as the concentration of barium ion is greater than 0.1 M.

TABLE 16-2

Volume Relationships in the Analysis of Mixures Containing Hydroxide, Carbonate, and Hydrogen Carbonate Ions

Constituents in Sample	Relationship between V_{phth} and V_{bcg} in the Titration of an Equal Volume of Sample*
NaOH	$V_{phth} = V_{bcg}$
Na_2CO_3	$V_{phth} = \frac{1}{2}V_{bcg}$
$NaHCO_3$	$V_{phth} = 0; V_{bcg} > 0$
NaOH, Na_2CO_3	$V_{phth} > \frac{1}{2}V_{bcg}$
Na_2CO_3, $NaHCO_3$	$V_{phth} < \frac{1}{2}V_{bcg}$

*V_{phth} = volume of acid needed for a phenolphthalein end point; V_{bcg} = volume of acid needed for a bromocresol green end point

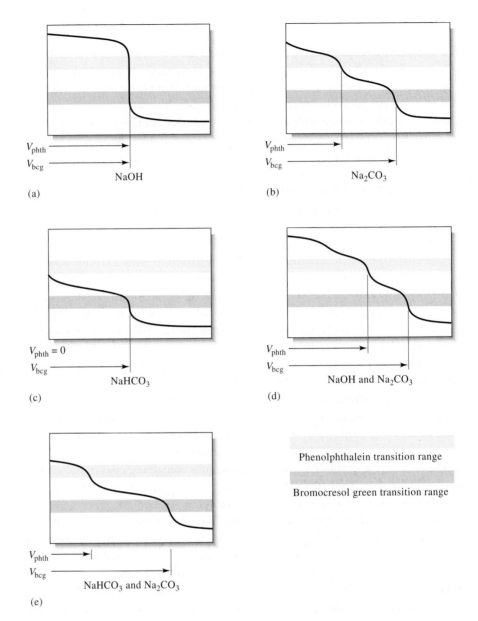

Figure 16-3 Titration curves and indicator transition ranges for the analysis of mixtures containing hydroxide, carbonate, and hydrogen carbonate ions using a strong-acid titrant.

Compatible mixtures containing two of the following can also be analyzed in a similar way: HCl, H_3PO_4, NaH_2PO_4, Na_2HPO_4, Na_3PO_4, and NaOH.

How could you analyze a mixture of HCl and H_3PO_4? A mixture of Na_3PO_4 and Na_2HPO_4? See Figure 15-4, curve A.

EXAMPLE 16-4

A solution contains $NaHCO_3$, Na_2CO_3, and NaOH, either alone or in a permissible combination. Titration of a 50.0-mL portion to a phenolphthalein end point requires 22.1 mL of 0.100 M HCl. A second 50.0-mL aliquot requires 48.4 mL of the HCl when titrated to a bromocresol green end point. Determine the composition and the molar solute concentrations of the original solution.

Solution

If the solution contained only NaOH, the volume of acid required would be the same regardless of indicator (see Figure 16-3a). Similarly, we can rule out the presence of Na_2CO_3 alone because titration of this compound to a bromocresol green end point would consume just twice the volume of acid required to reach the phenolphthalein end point (see Figure 16-3b). Actually, the second titration requires 48.4 mL. Because less than half of this amount is used in the first titration, the solution must contain some $NaHCO_3$ in addition to Na_2CO_3 (see Figure 16-3e). We can then calculate the concentration of the two constituents.

When the phenolphthalein end point is reached, the CO_3^{2-} originally present is converted to HCO_3^-. Thus,

$$\text{amount } Na_2CO_3 = 22.1 \text{ mL} \times 0.100 \frac{\text{mmol}}{\text{mL}} = 2.21 \text{ mmol}$$

The titration from the phenolphthalein to the bromocresol green end point (48.4 mL − 22.1 mL = 26.3 mL) includes both the hydrogen carbonate originally present and that formed by titration of the carbonate. Therefore,

$$\text{amount } NaHCO_3 + \text{amount } Na_2CO_3 = 26.3 \text{ mL} \times 0.100 \frac{\text{mmol}}{\text{mL}} = 2.63 \text{ mmol}$$

Hence,

$$\text{amount } NaHCO_3 = 2.63 \text{ mmol} - 2.21 \text{ mmol} = 0.42 \text{ mmol}$$

The molar concentrations can then be calculated from these results as follows:

$$c_{Na_2CO_3} = \frac{2.21 \text{ mmol}}{50.0 \text{ mL}} = 0.0442 \text{ M}$$

$$c_{NaHCO_3} = \frac{0.42 \text{ mmol}}{50.0 \text{ mL}} = 0.084 \text{ M}$$

Carbonate and hydrogen carbonate ions can be accurately determined in mixtures by first titrating both ions with standard acid to an end point with an acid-range indicator (with boiling to eliminate carbon dioxide). The hydrogen carbonate in a second aliquot is converted to carbonate by the addition of a known excess of standard base. After a large excess of barium chloride has been introduced, the excess base is titrated with standard acid to a phenolphthalein end point. The presence of solid barium carbonate does not influence end-point detection in either of these methods.

16B-3 The Determination of Organic Functional Groups

Neutralization titrations provide convenient methods for the direct or indirect determination of several organic functional groups. Brief descriptions of methods for the more common groups follow.

Carboxylic and Sulfonic Acid Groups

Carboxylic and sulfonic acids are very common organic acids. Most carboxylic acids have dissociation constants that range between 10^{-4} and 10^{-6}, and thus, these compounds are readily titrated. An indicator that changes color in a basic range, such as phenolphthalein, is required.

Many carboxylic acids are not sufficiently soluble in water to permit direct titration in aqueous solution. When this problem exists, the acid can be dissolved in ethanol and titrated with aqueous base. Alternatively, the acid can be dissolved in an excess of standard base followed by back-titration with standard acid.

Sulfonic acids are generally strong acids that easily dissolve in water. Titration with standard base can be used for the determination.

Neutralization titrations are often used to determine the equivalent masses of purified organic acids (see Feature 16-3). Equivalent masses serve as an aid in the qualitative identification of organic acids.

The **equivalent mass** of an acid or base is the mass of the compound that reacts with or contains one mole of protons. Thus, the equivalent mass of KOH (56.11 g/mol) is equal to its molar mass.

$$56.11 \frac{\text{g}}{\text{mol KOH}}$$

$$\times \frac{1 \text{ mol KOH}}{\text{mol protons reacted}}$$

$$= 56.11 \frac{\text{g}}{\text{mol protons reacted}}$$

For $Ba(OH)_2$, it is the molar mass divided by 2.

$$171.3 \frac{\text{g}}{\text{mol Ba(OH)}_2}$$

$$\times \frac{1 \text{ mol Ba(OH)}_2}{2 \text{ mol protons reacted}}$$

$$= 85.6 \frac{\text{g}}{\text{mol protons reacted}}$$

Equivalent Masses of Acids and Bases

The equivalent mass (called *equivalent weight* in the older literature) of a participant in a neutralization reaction is the mass that reacts with or supplies one mole of protons in a *particular reaction*. For example, the equivalent mass of H_2SO_4 is one half of its molar mass. The equivalent mass of Na_2CO_3 is usually one half of its molar mass because in most applications its reaction is

$$Na_2CO_3 + 2H_3O^+ \rightarrow 3H_2O + CO_2 + 2Na^+$$

When titrated with some indicators, however, Na_2CO_3 consumes only a single proton:

$$Na_2CO_3 + H_3O^+ \rightarrow NaHCO_3 + Na^+$$

In this instance, the equivalent mass and the molar mass of Na_2CO_3 are the same. These observations show that the equivalent mass of a compound cannot be defined without having a particular reaction in mind (see Appendix 7).

Amine Groups

Aliphatic amines generally have base dissociation constants on the order of 10^{-5} and can be titrated directly with a solution of strong acid. Aromatic amines such as aniline and its derivatives, however, are usually too weak for titration in aqueous solutions ($K_b \approx 10^{-10}$). Likewise, cyclic amines with aromatic character, such as pyridine and its derivatives, are usually too weak for titration in aqueous solutions. Many saturated cyclic amines, such as piperdine, tend to resemble aliphatic amines in their acid/base behavior and can thus be titrated in aqueous media. Many amines that are too weak to be titrated as bases in water are easily titrated in nonaqueous solvents, such as anhydrous acetic acid, which enhance their basicity.

Ester Groups

Saponification is the process by which an ester is hydrolyzed in alkaline solution to give an alcohol and a conjugate base. For example,

$$\begin{array}{c} O \\ \parallel \\ CH_3COCH_3 + OH^- \longrightarrow \end{array}$$

$$\begin{array}{c} O \\ \parallel \\ CH_3C-O^- + CH_3OH \end{array}$$

Esters are commonly determined by **saponification** with a measured quantity of standard base:

$$R_1COOR_2 + OH^- \rightarrow R_1COO^- + HOR_2$$

The excess base is then titrated with standard acid.

Esters vary widely in their rates of saponification. Some require several hours of heating with a base to complete the process, while a few react rapidly enough that direct titration with standard base is feasible. Typically, the ester is refluxed with standard 0.5 M KOH for 1 to 2 hours. After cooling, the excess base is titrated with standard acid.

Hydroxyl Groups

Hydroxyl groups in organic compounds can be determined by esterification with various carboxylic acid anhydrides or chlorides. The two most common reagents are acetic anhydride and phthalic anhydride. With acetic anhydride, the reaction is

$$(CH_3CO)_2O + ROH \rightarrow CH_3COOR + CH_3COOH$$

Usually, the sample is mixed with a carefully measured volume of acetic anhydride in pyridine. After heating, water is added to hydrolyze the unreacted anhydride according to

$$(CH_3CO_2)O + H_2O \rightarrow 2CH_3COOH$$

The acetic acid is then titrated with a standard solution of alcoholic sodium or potassium hydroxide. A blank is carried through the analysis to establish the original amount of anhydride.

Amines, if present, are converted quantitatively to amides by acetic anhydride. A correction for this potential interference is frequently made by direct titration of another portion of the sample with standard acid.

Carbonyl Groups

Many aldehydes and ketones can be determined with a solution of hydroxylamine hydrochloride. The reaction, which produces an oxime, is

$$\begin{array}{c} R_1 \\ \diagdown \\ C = O \\ \diagup \\ R_2 \end{array} + NH_2OH \cdot HCl \longrightarrow \begin{array}{c} R_1 \\ \diagdown \\ C = NOH \\ \diagup \\ R_2 \end{array} + HCl + H_2O$$

where R_2 may be hydrogen. The liberated HCl is titrated with base. Once again, the conditions necessary for quantitative reaction vary. Typically, 30 minutes suffices for aldehydes, while many ketones require refluxing with the reagent for 1 hour or more.

16B-4 The Determination of Salts

The total salt content of a solution can be determined accurately by acid/base titration. The salt is converted to an equivalent amount of an acid or base by passing a solution containing the salt through a column packed with an ion-exchange resin. (This application is considered in more detail in Section 31D.)

Standard acid or base solutions can also be prepared with ion-exchange resins. A solution containing a known mass of a pure compound, such as sodium chloride, is washed through the resin column and diluted to a known volume. The salt releases an equivalent amount of acid or base from the resin. The concentration of the acid or base can then be calculated from the known mass of the original salt.

WEB WORKS

Go to **www.cengage.com/chemistry/skoog/fac9**, choose Chapter 16 and go to the Web Works. Click on the link to the executive summary of the Lake Champlain Basin Agricultural Watersheds Project. You can also download the final report for this project from the second link on the Chapter 16 Web Works section. The report and summary describe a project to improve water quality in Lake Champlain in Vermont and New York. Based on your reading of these documents, what appears to be the primary general cause of the eutrophication of Lake Champlain? What types of industry are the sources of the pollution? What measures have been taken to reduce the pollution? Briefly describe the experimental design used to determine whether these measures have been effective. One of the quantities measured in the study was total Kjeldahl nitrogen (TKN). Name three other measured quantities. Explain how TKN measurements relate to pollution in the lake. Based on the TKN measurements and other data in the report, have the pollution abatement measures been effective? What are the final recommendations of the report?

QUESTIONS AND PROBLEMS

*16-1. Why is nitric acid seldom used to prepare standard acid solutions?

16-2. Describe how Na_2CO_3 of primary-standard grade can be prepared from primary-standard $NaHCO_3$.

*16-3. The boiling points of HCl and CO_2 are nearly the same ($-85°C$ and $-78°C$). Explain why CO_2 can be removed from an aqueous solution by boiling briefly while essentially no HCl is lost even after boiling for 1 h or more.

16-4. Why is it common practice to boil the solution near the equivalence point in the standardization of Na_2CO_3 with acid?

*16-5. Give two reasons why $KH(IO_3)_2$ is preferred over benzoic acid as a primary standard for a 0.010 M NaOH solution.

16-6. What types of organic nitrogen-containing compounds tend to yield low results with the Kjeldahl method unless special precautions are taken?

*16-7. Briefly describe the circumstance where the concentration of a sodium hydroxide solution will apparently be unaffected by the absorption of carbon dioxide.

16-8. How would you prepare 500 mL of
(a) 0.200 M H_2SO_4 from a reagent that has a density of 1.1539 g/mL and is 21.8% H_2SO_4 (w/w)?
(b) 0.250 M NaOH from the solid?
(c) 0.07500 M Na_2CO_3 from the pure solid?

*16-9. How would you prepare 2.00 L of
(a) 0.10 M KOH from the solid?
(b) 0.010 M $Ba(OH)_2 \cdot 8H_2O$ from the solid?
(c) 0.150 M HCl from a reagent that has a density of 1.0579 g/mL and is 11.50% HCl (w/w)?

16-10. Standardization of a sodium hydroxide solution against potassium hydrogen phthalate (KHP) yielded the results in the following table.

Mass KHP, g	0.7987	0.8365	0.8104	0.8039
Volume NaOH, mL	38.29	39.96	38.51	38.29

Calculate
(a) the average molar concentration of the base.
(b) the standard deviation and the coefficient of variation for the data.
(c) the spread of the data.

*16-11. The concentration of a perchloric acid solution was established by titration against primary standard sodium carbonate (product: CO_2); the following data were obtained.

Mass Na_2CO_3, g	0.2068	0.1997	0.2245	0.2137
Volume $HClO_4$, mL	36.31	35.11	39.00	37.54

(a) Calculate the average molar concentration of the acid.
(b) Calculate the standard deviation for the data and the coefficient of variation for the data.
(c) Use statistics to decide whether the outlier should be retained or rejected.

16-12. If 1.000 L of 0.1500 M NaOH was unprotected from the air after standardization and absorbed 11.2 mmol of CO_2, what is its new molar concentration when it is standardized against a standard solution of HCl using
(a) phenolphthalein?
(b) bromocresol green?

*16-13. A NaOH solution was 0.1019 M immediately after standardization. Exactly 500.0 mL of the reagent was left exposed to air for several days and absorbed 0.652 g of CO_2. Calculate the relative carbonate error in the determination of acetic acid with this solution if the titrations were performed with phenolphthalein.

16-14. Calculate the molar concentration of a dilute HCl solution if
(a) a 50.00-mL aliquot yielded 0.5902 g of AgCl.
(b) the titration of 25.00 mL of 0.03970 M $Ba(OH)_2$ required 17.93 mL of the acid.
(c) the titration of 0.2459 g of primary standard Na_2CO_3 required 36.52 mL of the acid (products: CO_2 and H_2O).

*16-15. Calculate the molar concentration of a dilute $Ba(OH)_2$ solution if
(a) 50.00 mL yielded 0.1791 g of $BaSO_4$.
(b) titration of 0.4512 g of primary standard potassium hydrogen phthalate (KHP) required 26.46 mL of the base.
(c) addition of 50.00 mL of the base to 0.3912 g of benzoic acid required a 4.67-mL back-titration with 0.05317 M HCl.

16-16. Suggest a range of sample masses for the indicated primary standard if it is desired to use between 35 and 45 mL of titrant:
(a) 0.175 M $HClO_4$ titrated against Na_2CO_3 (CO_2 product).
(b) 0.085 M HCl titrated against $Na_2C_2O_4$.

$$Na_2C_2O_4 \rightarrow Na_2CO_3 + CO$$
$$CO_3^{2-} + 2H^+ \rightarrow H_2O + CO_2$$

(c) 0.150 M NaOH titrated against benzoic acid.
(d) 0.050 M $Ba(OH)_2$ titrated against $KH(IO_3)_2$.
(e) 0.075 M $HClO_4$ titrated against TRIS.
(f) 0.050 M H_2SO_4 titrated against $Na_2B_4O_7 \cdot 10H_2O$. Reaction:

$$B_4O_7^{2-} + 2H_3O^+ + 3H_2O \rightarrow 4H_3BO_3$$

*16-17. Calculate the relative standard deviation in the computed molar concentration of 0.0200 M HCl if this acid was standardized against the masses found in Example 16-1 for (a) TRIS, (b) Na_2CO_3, and (c) $Na_2B_4O_7 \cdot 10H_2O$. Assume that the absolute standard deviation in the mass measurement is 0.0001 g and that this measurement limits the precision of the computed concentration.

16-18. (a) Compare the masses of potassium hydrogen phthalate (204.22 g/mol), potassium hydrogen iodate (389.91 g/mol), and benzoic acid (122.12 g/mol) needed for a 30.00-mL standardization of 0.0400 M NaOH.

(b) What would be the relative standard deviation in the molar concentration of the base if the standard deviation in the measurement of mass in (a) is 0.002 g and this uncertainty limits the precision of the calculation?

*16-19. A 50.00-mL sample of a white dinner wine required 24.57 mL of 0.03291 M NaOH to achieve a phenolphthalein end point. Express the acidity of the wine in grams of tartaric acid ($H_2C_4H_4O_6$; 150.09 g/mol) per 100 mL. (Assume that two hydrogens of the acid are titrated.)

16-20. A 25.0-mL aliquot of vinegar was diluted to 250 mL in a volumetric flask. Titration of 50.0-mL aliquots of the diluted solution required an average of 35.23 mL of 0.08960 M NaOH. Express the acidity of the vinegar in terms of the percentage (w/v) of acetic acid.

*16-21. Titration of a 0.7513-g sample of impure $Na_2B_4O_7$ required 30.79 mL of 0.1129 M HCl (see Problem 16-16(f) for reaction). Express the results of this analysis in terms of percent

(a) $Na_2B_4O_7$.

(b) $Na_2B_4O_7 \cdot 10H_2O$.

(c) B_2O_3.

(d) B.

16-22. A 0.6915-g sample of impure mercury(II) oxide was dissolved in an unmeasured excess of potassium iodide. Reaction:

$$HgO(s) + 4I^- + H_2O \rightarrow HgI_4^{2-} + 2OH^-$$

Calculate the percentage of HgO in the sample if titration of the liberated hydroxide required 40.39 mL of 0.1092 M HCl.

*16-23. The formaldehyde content of a pesticide preparation was determined by weighing 0.2985 g of the liquid sample into a flask containing 50.0 mL of 0.0959 M NaOH and 50 mL of 3% H_2O_2. Upon heating, the following reaction took place:

$$OH^- + HCHO + H_2O_2 \rightarrow HCOO^- + 2H_2O$$

After cooling, the excess base was titrated with 22.71 mL of 0.053700 M H_2SO_4. Calculate the percentage of HCHO (30.026 g/mol) in the sample.

16-24. The benzoic acid extracted from 97.2 g of catsup required a 12.91-mL titration with 0.0501 M NaOH. Express the results of this analysis in terms of percent sodium benzoate (144.10 g/mol).

*16-25. The active ingredient in Antabuse, a drug used for the treatment of chronic alcoholism, is tetraethylthiuram disulfide,

$$\begin{array}{cc} S & S \\ \parallel & \parallel \\ (C_2H_5)_2NCSSCN(C_2H_5)_2 \end{array}$$

(296.54 g/mol). The sulfur in a 0.4169-g sample of an Antabuse preparation was oxidized to SO_2, which was absorbed in H_2O_2 to give H_2SO_4. The acid was titrated with 19.25 mL of 0.04216 M base. Calculate the percentage of active ingredient in the preparation.

16-26. A 25.00-mL sample of a household cleaning solution was diluted to 250.0 mL in a volumetric flask. A 50.00-mL aliquot of this solution required 41.27 mL of 0.1943 M HCl to reach a bromocresol green end point. Calculate the mass/volume percentage of NH_3 in the sample. (Assume that all the alkalinity results from the ammonia.)

*16-27. A 0.1401-g sample of a purified carbonate was dissolved in 50.00 mL of 0.1140 M HCl and boiled to eliminate CO_2. Back-titration of the excess HCl required 24.21 mL of 0.09802 M NaOH. Identify the carbonate.

16-28. A dilute solution of an unknown weak acid required a 28.62-mL titration with 0.1084 M NaOH to reach a phenolphthalein end point. The titrated solution was evaporated to dryness. Calculate the equivalent mass of the acid if the sodium salt was found to weigh 0.2110 g.

*16-29. A 3.00-L sample of urban air was bubbled through a solution containing 50.0 mL of 0.0116 M $Ba(OH)_2$, which caused the CO_2 in the sample to precipitate as $BaCO_3$. The excess base was back-titrated to a phenolphthalein end point with 23.6 mL of 0.0108 M HCl. Calculate concentration of CO_2 in the air in parts per million (that is, mL $CO_2/10^6$ mL air); use 1.98 g/L for the density of CO_2.

16-30. Air was bubbled at a rate of 30.0 L/min through a trap containing 75 mL of 1% H_2O_2 ($H_2O_2 + SO_2 \rightarrow H_2SO_4$). After 10.0 min, the H_2SO_4 was titrated with 11.70 mL of 0.00197 M NaOH. Calculate concentration of SO_2 in parts per million (that is, mL $SO_2/10^6$ mL air) if the density of SO_2 is 0.00285 g/mL.

*16-31. The digestion of a 0.1417-g sample of a phosphorus-containing compound in a mixture of HNO_3 and H_2SO_4 resulted in the formation of CO_2, H_2O, and H_3PO_4. Addition of ammonium molybdate yielded a solid having the composition $(NH_4)_3PO_4 \cdot 12MoO_3$ (1876.3 g/mol). This precipitate was filtered, washed, and dissolved in 50.00 mL of 0.2000 M NaOH:

$$(NH_4)_3PO_4 \cdot 12MoO_3(s) + 26OH^- \rightarrow HPO_4^{2-} + 12MoO_4^{2-} + 14H_2O + 3NH_3(g)$$

After the solution was boiled to remove the NH_3, the excess NaOH was titrated with 14.17 mL of 0.1741 M HCl to a phenolphthalein end point. Calculate the percentage of phosphorus in the sample.

16-32. A 0.9471-g sample containing dimethylphthalate, $C_6H_4(COOCH_3)_2$ (194.19 g/mol), and unreactive species was refluxed with 50.00 mL of 0.1215 M NaOH to hydrolyze the ester groups (this process is called saponification).

$$C_6H_4(COOCH_3)_2 + 2OH^- \rightarrow C_6H_4(COO)_2^{2-} + 2CH_3OH$$

After the reaction was complete, the excess NaOH was back-titrated with 24.27 mL of 0.1644 M HCl. Calculate the percentage of dimethylphthalate in the sample.

*16-33. Neohetramine, $C_{16}H_{21}ON_4$ (285.37 g/mol), is a common antihistamine. A 0.1247-g sample containing this compound was analyzed by the Kjeldahl method. The ammonia produced was collected in H_3BO_3; the resulting $H_2BO_3^-$ was titrated with 26.13 mL of 0.01477 M HCl. Calculate the percentage of neohetramine in the sample.

16-34. The *Merck Index* indicates that 10 mg of guanidine, CH_5N_3, may be administered for each kilogram of body mass in the treatment of myasthenia gravis. The nitrogen in a 4-tablet sample that had a total mass of 7.50 g was converted to ammonia by a Kjeldahl digestion, followed by distillation into 100.0 mL of 0.1750 M HCl. The analysis was completed by titrating the excess acid with 11.37 mL of 0.1080 M NaOH. How many of these tablets represent a proper dose for a patient who weighs (a) 100 lb, (b) 150 lb, and (c) 275 lb?

*16-35. A 0.917-g sample of canned tuna was analyzed by the Kjeldahl method. A volume of 20.59 mL of 0.1249 M HCl was required to titrate the liberated ammonia. Calculate the percentage of nitrogen in the sample.

16-36. Calculate the mass in grams of protein in a 6.50-oz can of tuna in Problem 16-35.

*16-37. A 0.5843-g sample of a plant food preparation was analyzed for its N content by the Kjeldahl method, the liberated NH_3 being collected in 50.00 mL of 0.1062 M HCl. The excess acid required an 11.89 mL back-titration with 0.0925 M NaOH. Express the results of this analysis in terms of

(a) %N.   (c) %$(NH_4)_2SO_4$.
(b) %urea, H_2NCONH_2.   (d) %$(NH_4)_3PO_4$.

16-38. A 0.9325-g sample of a wheat flour was analyzed by the Kjeldahl procedure. The ammonia formed was distilled into 50.00 mL of 0.05063 M HCl; a 7.73-mL back-titration with 0.04829 M NaOH was required. Calculate the percentage of protein in the flour.

*16-39. A 1.219-g sample containing $(NH_4)_2SO_4$, NH_4NO_3, and nonreactive substances was diluted to 200 mL in a volumetric flask. A 50.00-mL aliquot was made

basic with strong alkali, and the liberated NH_3 was distilled into 30.00 mL of 0.08421 M HCl. The excess HCl required 10.17 mL of 0.08802 M NaOH for neutralization. A 25.00-mL aliquot of the sample was made alkaline after the addition of Devarda's alloy, and the NO_3^- was reduced to NH_3. The NH_3 from both NH_4^+ and NO_3^- was then distilled into 30.00 mL of the standard acid and back-titrated with 14.16 mL of the base. Calculate the percentage of $(NH_4)_2SO_4$ and NH_4NO_3 in the sample.

16-40. A 1.217-g sample of commercial KOH contaminated by K_2CO_3 was dissolved in water, and the resulting solution was diluted to 500.0 mL. A 50.00-mL aliquot of this solution was treated with 40.00 mL of 0.05304 M HCl and boiled to remove CO_2. The excess acid consumed 4.74 mL of 0.04983 M NaOH (phenolphthalein indicator). An excess of neutral $BaCl_2$ was added to another 50.00-mL aliquot to precipitate the carbonate as $BaCO_3$. The solution was then titrated with 28.56 mL of the acid to a phenolphthalein end point. Calculate the percentage KOH, K_2CO_3, and H_2O in the sample, assuming that these are the only compounds present.

*16-41. A 0.5000-g sample containing $NaHCO_3$, Na_2CO_3, and H_2O was dissolved and diluted to 250.0 mL. A 25.00-mL aliquot was then boiled with 50.00 mL of 0.01255 M HCl. After cooling, the excess acid in the solution required 2.34 mL of 0.01063 M NaOH when titrated to a phenolphthalein end point. A second 25.00-mL aliquot was then treated with an excess of $BaCl_2$ and 25.00 mL of the base. All the carbonate precipitated, and 7.63 mL of the HCl was required to titrate the excess base. Determine the composition of the mixture.

16-42. Calculate the volume of 0.06122 M HCl needed to titrate

(a) 20.00 mL of 0.05555 M Na_3PO_4 to a thymolphthalein end point.
(b) 25.00 mL of 0.05555 M Na_3PO_4 to a bromocresol green end point.
(c) 40.00 mL of a solution that is 0.02102 M in Na_3PO_4 and 0.01655 M in Na_2HPO_4 to a bromocresol green end point.
(d) 20.00 mL of a solution that is 0.02102 M in Na_3PO_4 and 0.01655 M in NaOH to a thymolphthalein end point.

*16-43. Calculate the volume of 0.07731 M NaOH needed to titrate

(a) 25.00 mL of a solution that is 0.03000 M in HCl and 0.01000 M in H_3PO_4 to a bromocresol green end point.
(b) the solution in (a) to a thymolphthalein end point.
(c) 30.00 mL of 0.06407 M NaH_2PO_4 to a thymolphthalein end point.

(d) 25.00 mL of a solution that is 0.02000 M in H_3PO_4 and 0.03000 M in NaH_2PO_4 to a thymolphthalein end point.

16-44. A series of solutions containing $NaOH$, Na_3AsO_4, and Na_2HAsO_4, alone or in compatible combination, was titrated with 0.08601 M HCl. Tabulated below are the volumes of acid needed to titrate 25.00-mL portions of each solution to (1) a phenolphthalein and (2) a bromocresol green end point. Use this information to deduce the composition of the solutions. In addition, calculate the mass in milligrams of each solute per milliliter of solution.

	(1)	(2)
(a)	0.00	18.15
(b)	21.00	28.15
(c)	19.80	39.61
(d)	18.04	18.03
(e)	16.00	37.37

*16-45. A series of solutions containing $NaOH$, Na_2CO_3, and $NaHCO_3$, alone or in compatible combination, was titrated with 0.1202 M HCl. Tabulated below are the volumes of acid needed to titrate 25.00-mL portions of each solution to (1) a phenolphthalein and (2) a bromocresol green end point. Use this information to deduce the composition of the solutions. In addition, calculate the mass in milligrams of each solute per milliliter of solution.

	(1)	(2)
(a)	22.42	22.44
(b)	15.67	42.13
(c)	29.64	36.42
(d)	16.12	32.23
(e)	0.00	33.333

16-46. Define the equivalent mass of (a) an acid and (b) a base.

*16-47. Calculate the equivalent mass of oxalic acid dehydrate ($H_2C_2O_4 \cdot 2H_2O$, 126.066 g/mol) when it is titrated to (a) a bromocresol green end point and (b) a phenolphthalein end point.

16-48. A 10.00-mL sample of vinegar (acetic acid, CH_3COOH) was pipetted into a flask, two drops of phenolphthalein indicator were added, and the acid was titrated with 0.1008 M NaOH.

(a) If 45.62 mL of the base was required for the titration, what was the molar concentration of acetic acid in the sample?

(b) If the density of the pipetted acetic acid solution was 1.004 g/mL, what was the percentage of acetic acid in the sample?

16-49. Challenge Problem:

(a) Why are indicators used only in the form of dilute solutions?

(b) Suppose that 0.1% methyl red (molar mass 269 g/mol) is used as the indicator in a titration to determine the acid neutralizing capacity of an Ohio lake. Five drops (0.25 mL) of methyl red solution are added to a 100-mL sample of water, and 4.74 mL of 0.01072 M hydrochloric acid is required to change the indicator to the midpoint in its transition range. Assuming that there is no indicator error, what is the acid neutralizing capacity of the lake expressed as milligrams of calcium bicarbonate per liter in the sample?

(c) If the indicator was initially in its acid form, what is the indicator error expressed as a percentage of the acid neutralizing capacity?

(d) What is the correct value for the acid neutralizing capacity?

(e) List four species other than carbonate or bicarbonate that may contribute to acid neutralizing capacity.

(f) It is normally assumed that species other than carbonate or bicarbonate do not contribute appreciably to acid neutralizing capacity. Suggest circumstances under which this assumption may not be valid.

(g) Particulate matter may make a significant contribution to acid neutralizing capacity. Explain how you would deal with this problem.

(h) Explain how you would determine separately the contribution to the acid neutralizing capacity from particulate matter and the contribution from soluble species.

Complexation and Precipitation Reactions and Titrations

CHAPTER 17

© American Chemical Society. Courtesy of R. N. Zare, Stanford University, Chemistry Dept.

© American Chemical Society. Courtesy of R. N. Zare, Stanford University, Chemistry Dept.

Complexation and precipitation reactions are important in many areas of science and everyday life as discussed in this chapter. Black-and-white photography is one such area. Although digital photography has come to dominate consumer areas, film photography is still important in many applications. Shown here are photomicrographs of a capillary chromatography column at ×1300 (top) and ×4900 (bottom) magnification. Black-and-white film consists of an emulsion of finely divided AgBr coated on a polymer strip. Exposure to light from the scanning electron microscope causes reduction of some of the Ag^+ ions to Ag atoms and corresponding oxidation of Br^- to Br atoms. These atoms remain in the crystal lattice of AgBr as invisible defects, or the so-called latent image. Developing reduces many more Ag^+ ions to Ag atoms in the granules of AgBr containing Ag atoms from the original latent image. Development produces a visible negative image where dark regions of Ag atoms represent areas where light has exposed the film. The fixing step removes the unexposed AgBr by forming the highly stable silver thiosulfate complex $[Ag(S_2O_3)_2]^{2-}$. The black metallic silver of the negative remains.

$$AgBr(s) + 2S_2O_3^{2-}(aq) \rightarrow [Ag(S_2O_3)_2]^{3-}(aq) + Br^-(aq)$$

After the negative has been fixed, a positive image is produced by projecting light through the negative onto photographic paper. (M. T. Dulay, R. P. Kulkarni, and R. N. Zare, *Anal. Chem.*, **1998**, 70, 5103, **DOI**: 10.1021/ac9806456. ©American Chemical Society. Courtesy of R. N. Zare, Stanford University.)

Complexation reactions are widely used in analytical chemistry. One of the earliest uses of these reactions was for titrating cations, a major topic of this chapter. In addition, many complexes are colored or absorb ultraviolet radiation; the formation of these complexes is often the basis for spectrophotometric determinations (see Chapter 26). Some complexes are sparingly soluble and can be used in gravimetric analysis (see Chapter 12) or for precipitation titrations as discussed in this chapter. Complexes are also widely used for extracting cations from one solvent to another and for dissolving insoluble precipitates. The most useful complex forming reagents are organic compounds containing several electron-donor groups that form multiple covalent bonds with metal ions. Inorganic complexing agents are also used to control solubility, form colored species, or form precipitates.

17A THE FORMATION OF COMPLEXES

Most metal ions react with electron-pair donors to form coordination compounds or complexes. The donor species, or **ligand**, must have at least one pair of unshared electrons available for bond formation. Water, ammonia, and halide ions are common inorganic ligands. In fact most metal ions in aqueous solution actually exist as aquo complexes. Copper(II), for example, in aqueous solution is readily complexed by water molecules to form species such as $Cu(H_2O)_4^{2+}$. We often simplify such complexes in chemical equations by writing the metal ion as if it were uncomplexed Cu^{2+}. We should remember, however, that most metal ions are actually aquo complexes in aqueous solution.

The number of covalent bonds that a cation tends to form with electron donors is its **coordination number**. Typical values for coordination numbers are two, four, and six. The species formed as a result of coordination can be electrically positive, neutral, or negative. For example, copper(II), which has a coordination number of four, forms a cationic ammine complex, $Cu(NH_3)_4^{2+}$; a neutral complex with glycine, $Cu(NH_2CH_2COO)_2$; and an anionic complex with chloride ion, $CuCl_4^{2-}$.

Titrations based on complex formation, sometimes called **complexometric titrations**, have been used for more than a century. The truly remarkable growth in their analytical application, based on a particular class of coordination compounds called **chelates**, began in the 1940s. A chelate is produced when a metal ion coordinates with two or more donor groups of a single ligand to form a five- or six-membered heterocyclic ring. The copper complex of glycine, mentioned in the previous paragraph, is an example. In this complex, copper bonds to both the oxygen of the carboxyl group and the nitrogen of the amine group:

$$Cu^{2+} + 2H\!-\!\overset{\displaystyle NH_2}{\underset{\displaystyle H}{C}}\!-\!\overset{\displaystyle O}{C}\!-\!OH \longrightarrow$$

Glycine

$$\begin{array}{c} O\!=\!C\!-\!O \quad\quad O\!-\!C\!=\!O \\ \diagdown \quad\quad \diagup \\ Cu \\ \diagup \quad\quad \diagdown \\ H_2C\!-\!NH \quad NH\!-\!CH_2 \end{array} + 2H^+$$

Cu^{2+} complex with glycine.

A ligand that has a single donor group, such as ammonia, is called **unidentate** (single-toothed), whereas one such as glycine, which has two groups available for covalent bonding, is called **bidentate**. Tridentate, tetradentate, pentadentate, and hexadentate chelating agents are also known.

18-crown-6

dibenzo-18-crown-6

cryptand 2,2,2

Crown ethers and cryptands.

Another important type of complex is formed between metal ions and cyclic organic compounds, known as **macrocycles**. These molecules contain nine or more atoms in the cycle and include at least three heteroatoms, usually oxygen, nitrogen or sulfur. Crown ethers, such as 18-crown-6 and dibenzo-18-crown-6 are examples of organic macrocycles. Some macrocyclic compounds form three dimensional cavities that can just accommodate appropriately sized metal ions. Ligands known as **cryptands** are examples. Selectivity occurs to a large extent

A **ligand** is an ion or a molecule that forms a covalent bond with a cation or a neutral metal atom by donating a pair of electrons, which are then shared by the two.

Chelate is pronounced *kee'late* and is derived from the Greek word for claw.

Dentate comes from the Latin word *dentatus* and means having toothlike projections.

Molecular model of 18-crown-6. This crown ether can form strong complexes with alkali metal ions. The formation constants of the Na^+, K^+, and Rb^+ complexes with 18-crown-6 are in the 10^5 to 10^6 range.

because of the size and shape of the cycle or cavity relative to that of the metal ion, although the nature of the heteroatoms and their electron densities, the compatibility of the donor atoms with the metal ion, and several other factors also play important roles.

17A-1 Complexation Equilibria

> The **selectivity** of a ligand for one metal ion over another refers to the stability of the complexes formed. The higher the formation constant of the metal-ligand complex, the better the selectivity of the ligand for the metal relative to similar complexes formed with other metals.

Complexation reactions involve a metal-ion M reacting with a ligand L to form a complex ML, as shown in Equation 17-1:

$$M + L \rightleftharpoons ML \tag{17-1}$$

where we have omitted the charges on the ions in order to be general. Complexation reactions occur in a stepwise fashion and the reaction above is often followed by additional reactions:

$$ML + L \rightleftharpoons ML_2 \tag{17-2}$$

$$ML_2 + L \rightleftharpoons ML_3 \tag{17-3}$$

$$\vdots \qquad\qquad \vdots$$

$$ML_{n-1} + L \rightleftharpoons ML_n \tag{17-4}$$

Unidentate ligands invariably add in a series of steps as shown above. With multidentate ligands, the maximum coordination number of the cation may be satisfied with only one or a few added ligands. For example, Cu(II), with a maximum coordination number of 4, can form complexes with ammonia that have the formulas $Cu(NH_3)^{2+}$, $Cu(NH_3)_2^{2+}$, $Cu(NH_3)_3^{2+}$, and $Cu(NH_3)_4^{2+}$. With the bidentate ligand glycine (gly), the only complexes that form are $Cu(gly)^{2+}$ and $Cu(gly)_2^{2+}$.

The equilibrium constants for complex formation reactions are generally written as formation constants, as discussed in Chapter 9. Thus, each of the reactions 17-1 through 17-4 is associated with a stepwise formation constant K_1 through K_4. For example, $K_1 = [ML]/[M][L]$, $K_2 = [ML_2]/[ML][L]$, and so on. We can also write the equilibria as the sum of individual steps. These have overall formation constants designated by the symbol β_n. Therefore,

$$M + L \rightleftharpoons ML \qquad \beta_1 = \frac{[ML]}{[M][L]} = K_1 \tag{17-5}$$

$$M + 2L \rightleftharpoons ML_2 \qquad \beta_2 = \frac{[ML_2]}{[M][L]^2} = K_1 K_2 \tag{17-6}$$

$$M + 3L \rightleftharpoons ML_3 \qquad \beta_3 = \frac{[ML_3]}{[M][L]^3} = K_1 K_2 K_3 \tag{17-7}$$

$$\vdots \qquad\qquad \vdots$$

$$M + nL \rightleftharpoons ML_n \qquad \beta_n = \frac{[ML_n]}{[M][L]^n} = K_1 K_2 \cdots K_n \tag{17-8}$$

Except for the first step, the overall formation constants are products of the stepwise formation constants for the individual steps leading to the product.

For a given species like the free metal M, we can calculate an alpha value, which is the fraction of the total metal concentration in that form. Thus, α_M is the fraction of the total metal present at equilibrium in the free metal form, α_{ML} is the fraction in the ML form, and so on. As derived in Feature 17-1, the alpha values are given by

$$\alpha_M = \frac{1}{1 + \beta_1[L] + \beta_2[L]^2 + \beta_3[L]^3 + \cdots + \beta_n[L]^n} \quad (17\text{-}9)$$

$$\alpha_{ML} = \frac{\beta_1[L]}{1 + \beta_1[L] + \beta_2[L]^2 + \beta_3[L]^3 + \cdots + \beta_n[L]^n} \quad (17\text{-}10)$$

$$\alpha_{ML_2} = \frac{\beta_2[L]^2}{1 + \beta_1[L] + \beta_2[L]^2 + \beta_3[L]^3 + \cdots + \beta_n[L]^n} \quad (17\text{-}11)$$

$$\alpha_{ML_n} = \frac{\beta_n[L]^n}{1 + \beta_1[L] + \beta_2[L]^2 + \beta_3[L]^3 + \cdots + \beta_n[L]^n} \quad (17\text{-}12)$$

FEATURE 17-1

Calculation of Alpha Values for Metal Complexes

The alpha values for metal-ligand complexes can be derived as we did for polyfunctional acids in Section 15H. The alphas are defined as

$$\alpha_M = \frac{[M]}{c_M}; \quad \alpha_{ML} = \frac{[ML]}{c_M};$$

$$\alpha_{ML_2} = \frac{[ML_2]}{c_M}; \quad \alpha_{ML_n} = \frac{[ML_n]}{c_M}$$

The total metal concentration c_M can be written

$$c_M = [M] + [ML] + [ML_2] + \cdots + [ML_n]$$

From the overall formation constants (Equations 17-5 through 17-8), the concentrations of the complexes can be expressed in terms of the free metal concentration [M] to give

$$c_M = [M] + \beta_1[M][L] + \beta_2[M][L]^2 + \cdots + \beta_n[M][L]^n$$
$$= [M]\{1 + \beta_1[L] + \beta_2[L]^2 + \cdots + \beta_n[L]^n\}$$

Now, α_M can be found as

$$\alpha_M = \frac{[M]}{c_M} = \frac{[M]}{[M] + \beta_1[M][L] + \beta_2[M][L]^2 + \cdots + \beta_n[M][L]^n}$$

$$= \frac{1}{1 + \beta_1[L] + \beta_2[L]^2 + \beta_3[L]^3 + \cdots + \beta_n[L]^n}$$

Note that the form on the right is Equation 17-9. We can find α_{ML} from

$$\alpha_{ML} = \frac{[ML]}{c_M} = \frac{\beta_1[M][L]}{[M] + \beta_1[M][L] + \beta_2[M][L]^2 + \cdots + \beta_n[M][L]^n}$$

$$= \frac{\beta_1[L]}{1 + \beta_1[L] + \beta_2[L]^2 + \beta_3[L]^3 + \cdots + \beta_n[L]^n}$$

The rightmost form of this equation is identical to Equation 17-10. The other alpha values in Equations 17-11 and 17-12 can be found in a similar manner.

Note that these expressions are analogous to the α expressions we wrote for polyfunctional acids and bases except that the equations here are written in terms of formation equilibria while those for acids or bases are written in terms of dissociation equilibria. Also, the master variable is the ligand concentration [L] instead of the hydronium ion concentration. The denominators are the same for each α value. Plots of the α values versus p[L] are known as **distribution diagrams**.

> **Spreadsheet Summary**   In the first exercise in Chapter 9 of *Applications of Microsoft® Excel in Analytical Chemistry*, 2nd ed., α values for the Cu(II)/ NH_3 complexes are calculated and used to plot distribution diagrams. The α values for the Cd(II)/Cl$^-$ system are also calculated.

17A-2 The Formation of Insoluble Species

In the cases discussed in the previous section, the complexes formed are soluble in solution. The addition of ligands to a metal ion, however, may result in insoluble species, such as the familiar nickel-dimethylglyoxime precipitate. In many cases, the intermediate uncharged complexes in the stepwise formation scheme may be sparingly soluble, whereas the addition of more ligand molecules may result in soluble species. For example, adding Cl$^-$ to Ag$^+$ results in the insoluble AgCl precipitate. Addition of a large excess of Cl$^-$ produces soluble species AgCl$_2^-$, AgCl$_3^{2-}$, and AgCl$_4^{3-}$.

In contrast to complexation equilibria, which are most often treated as formation reactions, solubility equilibria are normally treated as dissociation reactions, as discussed in Chapter 9. In general, for a sparingly soluble salt M_xA_y in a saturated solution, we can write

$$M_xA_y(s) \rightleftharpoons xM^{y+}(aq) + yA^{x-}(aq) \qquad K_{sp} = [M^{y+}]^x[A^{x-}]^y \qquad (17\text{-}13)$$

where K_{sp} is the solubility product. Hence, for BiI$_3$, the solubility product is written, $K_{sp} = [Bi^{3+}][I^-]^3$.

The formation of soluble complexes can be used to control the concentration of free metal ions in solution and thus control their reactivity. For example, we can prevent a metal ion from precipitating or taking part in another reaction by forming a stable complex, which decreases the free metal-ion concentration. The control of solubility by complex formation is also used to achieve the separation of one metal ion from another. If the ligand is capable of protonation, as discussed in the next section, even more control can be accomplished by a combination of complexation and pH adjustment.

17A-3 Ligands That Can Protonate

Complexation equilibria can be complicated by side reactions involving the metal or the ligand. Such side reactions make it possible to exert some additional control over the complexes that form. Metals can form complexes with ligands other than the one of interest. If these complexes are strong, we can effectively prevent complexation with the ligand of interest. Ligands can also undergo side reactions. One of the most common side reactions is that of a ligand that can protonate, that is, the ligand is a weak acid or the conjugate base of a weak acid.

Complexation with Protonating Ligands

Consider the case of the formation of soluble complexes between the metal M and the ligand L, where the ligand L is the conjugate base of a polyprotic acid and forms HL, $H_2L, \ldots H_nL$ for which again the charges have been omitted for generality. Adding acid to a solution containing M and L reduces the concentration of free L available to complex with M and thus decreases the effectiveness of L as a complexing agent (Le Chatelier's principle). For example, ferric ions (Fe^{3+}) form complexes with oxalate ($C_2O_4^{2-}$, which we abbreviate as ox^{2-}) with formulas $[Fe(ox)]^+$, $[Fe(ox)_2]^-$, and $[Fe(ox)_3]^{3-}$. Oxalate can protonate to form Hox^- and H_2ox. In basic solution, where most of the oxalate is present as ox^{2-} before complexation with Fe^{3+}, the ferric/oxalate complexes are very stable. Adding acid, however, protonates the oxalate ion, which in turn causes dissociation of the ferric complexes.

For a diprotic acid, like oxalic acid, the fraction of the total oxalate-containing species in any given form, ox^{2-}, Hox^-, and H_2ox, is given by an alpha value (recall Section 15H). Since

$$c_T = [H_2ox] + [Hox^-] + [ox^{2-}] \tag{17-14}$$

we can write the alpha values, α_0, α_1, and α_2, as

$$\alpha_0 = \frac{[H_2ox]}{c_T} = \frac{[H^+]^2}{[H^+]^2 + K_{a1}[H^+] + K_{a1}K_{a2}} \tag{17-15}$$

$$\alpha_1 = \frac{[Hox^-]}{c_T} = \frac{K_{a1}[H^+]}{[H^+]^2 + K_{a1}[H^+] + K_{a1}K_{a2}} \tag{17-16}$$

$$\alpha_2 = \frac{[ox^{2-}]}{c_T} = \frac{K_{a1}K_{a2}}{[H^+]^2 + K_{a1}[H^+] + K_{a1}K_{a2}} \tag{17-17}$$

Since we are interested in the free oxalate concentration, we will be most concerned with the highest α value, here α_2. From Equation 17-17, we can write

$$[ox^{2-}] = c_T\alpha_2 \tag{17-18}$$

Note that, as the solution gets more acidic, the first two terms in the denominator of Equation 17-17 dominate, and α_2 and the free oxalate concentration decrease. When the solution is very basic, the last term dominates, α_2 becomes nearly unity, and $[ox^{2-}] \approx c_T$, indicating that nearly all the oxalate is in the ox^{2-} form in basic solution.

Conditional Formation Constants

To take into account the effect of pH on the free ligand concentration in a complexation reaction, it is useful to introduce a **conditional formation constant**, or **effective formation constant**. Such constants are pH-dependent equilibrium constants that apply at a single pH only. For the reaction of Fe^{3+} with oxalate, for example, we can write the formation constant K_1 for the first complex as

$$K_1 = \frac{[Fe(ox)^+]}{[Fe^{3+}][ox^{2-}]} = \frac{[Fe(ox)^+]}{[Fe^{3+}]\alpha_2 c_T} \tag{17-19}$$

At a particular pH value, α_2 is constant, and we can combine K_1 and α_2 to yield a new conditional constant K_1':

$$K_1' = \alpha_2 K_1 = \frac{[Fe(ox)^+]}{[Fe^{3+}]c_T} \tag{17-20}$$

The use of conditional constants greatly simplifies calculations because c_T is often known or is easily computed, but the free ligand concentration is not as easily determined. The overall formation constants, β values, for the higher complexes, $[Fe(ox)_2]^-$ and $[Fe(ox)_3]^{3-}$, can also be written as conditional constants.

17B TITRATIONS WITH INORGANIC COMPLEXING AGENTS

Complexation reactions have many uses in analytical chemistry. One of the earliest uses, which is still widespread, is in **complexometric titrations**. In these titrations, a metal ion reacts with a suitable ligand to form a complex, and the equivalence point is determined by an indicator or an appropriate instrumental method. The formation of soluble inorganic complexes is not widely used for titrations, but the formation of precipitates, particularly with silver nitrate as the titrant, is the basis for many important determinations, as discussed in Section 17B-2.

17B-1 Complexation Titrations

Complexometric titration curves are usually a plot of pM = −log [M] as a function of the volume of titrant added. Usually in complexometric titrations, the ligand is the titrant, and the metal ion is the analyte, although occasionally the roles are reversed. As we shall see later, many precipitation titrations use the metal ion as the titrant. Most simple inorganic ligands are unidentate, which can lead to low complex stability and indistinct titration end points. As titrants, multidentate ligands, particularly those having four or six donor groups, have two advantages over their unidentate counterparts. First, they generally react more completely with cations and thus provide sharper end points. Second, they ordinarily react with metal ions in a single-step process, whereas complex formation with unidentate ligands usually involves two or more intermediate species (recall Equations 17-1 through 17-4).

The advantage of a single-step reaction is illustrated by the titration curves shown in **Figure 17-1**. Each of the titrations shown involves a reaction that has an overall equilibrium constant of 10^{20}. Curve *A* is computed for a reaction in which a metal-ion M having a coordination number of four reacts with a tetradentate ligand D to form the complex of MD (we have again omitted the charges on the two reactants for convenience). Curve *B* is for the reaction of M with a hypothetical bidentate ligand B to give MB_2 in two steps. The formation constant for the first step is 10^{12} and for the second

> Tetradentate or hexadentate ligands are more satisfactory as titrants than ligands with fewer donor groups because their reactions with cations are more complete and because they tend to form 1:1 complexes.

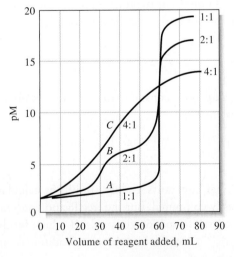

Figure 17-1 Titration curves for complexometric titrations. Titration of 60.0 mL of a solution that is 0.020 M in metal M with (A) a 0.020 M solution of the tetradentate ligand D to give MD as the product; (B) a 0.040 M solution of the bidentate ligand B to give MB_2; and (C) a 0.080 M solution of the unidentate ligand A to give MA_4. The overall formation constant for each product is 10^{20}.

Determination of Hydrogen Cyanide in Acrylonitrile Plant Streams

Acrylonitrile, $CH_2{=}CH{-}C{\equiv}N$, is an important chemical in the production of poly-acrylonitrile. This thermoplastic was drawn into fine threads and woven into synthetic fabrics such as Orlon, Acrilan, and Creslan. Although acrylic fibers are no longer produced in the US, they are still made in many countries. Hydrogen cyanide is an impurity in the plant streams that carry aqueous acrylonitrile. The cyanide is commonly determined by titration with $AgNO_3$. The titration reaction is

$$Ag^+ + 2CN^- \rightarrow Ag(CN)_2^-$$

In order to determine the end point of the titration, the aqueous sample is mixed with a basic solution of potassium iodide before the titration. Before the equivalence point, cyanide is in excess, and all the Ag^+ is complexed. As soon as all the cyanide has been reacted, the first excess of Ag^+ causes a permanent turbidity to appear in the solution because of the formation of the AgI precipitate according to

$$Ag^+ + I^- \rightarrow AgI(s)$$

10^8. Curve *C* involves a unidentate ligand, A, that forms MA_4 in four steps with successive formation constants of 10^8, 10^6, 10^4, and 10^2. These curves demonstrate that a much sharper end point is obtained with a reaction that takes place in a single step. For this reason, multidentate ligands are usually preferred for complexometric titrations.

The most widely used complexometric titration with a unidentate ligand is the titration of cyanide with silver nitrate, a method introduced by Liebig in the 1850s. This method involves the formation of soluble $Ag(CN)_2^-$, as discussed in Feature 17-2. Other common inorganic complexing agents and their applications are listed in **Table 17-1**.

 Spreadsheet Summary   The complexometric titration of Cd(II) with Cl^- is considered in Chapter 9 of *Applications of Microsoft® Excel in Analytical Chemistry*, 2nd ed. A master equation approach is used.

17B-2 Precipitation Titrations

Precipitation titrations are based on reactions that yield ionic compounds of limited solubility. Precipitation titrimetry is one of the oldest analytical techniques, dating back to the mid-1800s. The slow rate at which most precipitates form, however, limits the number of precipitating agents that can be used in titrations to a handful. We limit our discussion here to the most widely used and important precipitating reagent, silver nitrate, which is used for the determination of the halogens, the

TABLE 17-1

Typical Inorganic Complex-Forming Titrations		
Titrant	**Analyte**	**Remarks**
$Hg(NO_3)_2$	Br^-, Cl^-, SCN^-, CN^-, thiourea	Products are neutral Hg(II) complexes; various indicators used
$AgNO_3$	CN^-	Product is $Ag(CN)_2^-$; indicator is I^-; titrate to first turbidity of AgI
$NiSO_4$	CN^-	Product is $Ni(CN)_4^{2-}$; indicator is AgI; titrate to first turbidity of AgI
KCN	Cu^{2+}, Hg^{2+}, Ni^{2+}	Products are $Cu(CN)_4^{2-}$, $Hg(CN)_2$, and $Ni(CN)_4^{2-}$; various indicators used

halogenlike anions, mercaptans, fatty acids, and several divalent inorganic anions. Titrations with silver nitrate are sometimes called **argentometric titrations**.

The Shapes of Titration Curves

Titration curves for precipitation reactions are calculated in a completely analogous way to the methods described in Section 14B for titrations involving strong acids and strong bases. The only difference is that the solubility product of the precipitate is substituted for the ion-product constant for water. Most indicators for argentometric titrations respond to changes in the concentrations of silver ions. Because of this response, titration curves for precipitation reactions usually consist of a plot of pAg versus volume of the silver reagent (usually $AgNO_3$). Example 17-1 illustrates how p-functions are obtained for the preequivalence-point region, the postequivalence-point region, and the equivalence point for a typical precipitation titration.

EXAMPLE 17-1

Calculate the silver ion concentration in terms of pAg during the titration of 50.00 mL of 0.05000 M NaCl with 0.1000 M $AgNO_3$ after the addition of the following volumes of reagent: (a) in the preequivalence point region at 10.00 mL, (b) at the equivalence point (25.00 mL), (c) after the equivalence point at 26.00 mL. For AgCl, $K_{sp} = 1.82 \times 10^{-10}$.

Solution

(a) Preequivalence-Point Data

At 10.00 mL, $[Ag^+]$ is very small and cannot be computed from stoichiometric considerations, but the molar concentration of chloride, c_{NaCl} can be obtained readily. The equilibrium concentration of chloride is essentially equal to c_{NaCl}.

$$[Cl^-] \approx c_{NaCl} = \frac{\text{original no. mmol } Cl^- - \text{no. mol } AgNO_3 \text{ added}}{\text{total volume of solution}}$$

$$= \frac{(50.00 \times 0.05000 - 10.00 \times 0.1000)}{50.00 + 10.00} = 0.02500 \text{ M}$$

$$[Ag^+] = \frac{K_{sp}}{[Cl^-]} = \frac{1.82 \times 10^{-10}}{0.02500} = 7.28 \times 10^{-9} \text{ M}$$

$$pAg = -\log(7.28 \times 10^{-9}) = 8.14$$

Additional points in the preequivalence-point region can be obtained in the same way. Results of calculations of this kind are shown in the second column of **Table 17-2**.

TABLE 17-2

Changes in pAg in Titration of Cl^- with Standard $AgNO_3$

	pAg	
Volume of $AgNO_3$	50.00 mL of 0.0500 M NaCl with 0.1000 M $AgNO_3$	50.00 mL of 0.005 M NaCl with 0.0100 M $AgNO_3$
10.00	8.14	7.14
20.00	7.59	6.59
24.00	6.87	5.87
25.00	4.87	4.87
26.00	2.88	3.88
30.00	2.20	3.20
40.00	1.78	2.78

(b) Equivalence Point pAg

At the equivalence point, $[Ag^+] = [Cl^-]$, and $[Ag^+][Cl^-] = K_{sp} = 1.82 \times 10^{-10} = [Ag^+]^2$

$$[Ag^+] = \sqrt{K_{sp}} = \sqrt{1.82 \times 10^{-10}} = 1.35 \times 10^{-5}$$

$$pAg = -\log(1.35 \times 10^{-5}) = 4.87$$

(c) Postequivalence-Point Region

At 26.00 mL of $AgNO_3$, Ag^+ is in excess so

$$[Ag^+] = c_{AgNO_3} = \frac{(26.00 \times 0.1000 - 50.00 \times 0.05000)}{76.00} = 1.32 \times 10^{-3} \text{ M}$$

$$pAg = -\log(1.32 \times 10^{-3}) = 2.88$$

Additional results in the postequivalence-point region are obtained in the same way and are shown in Table 17-2. The titration curve can also be derived from the charge-balance equation as shown for an acid/base titration in Feature 14-1.

The Effect of Concentration on Titration Curves

The effect of reagent and analyte concentration on titration curves can be seen in the data in Table 17-2 and the two curves shown in **Figure 17-2**. With 0.1000 M $AgNO_3$ (Curve *A*), the change in pAg in the equivalence-point region is large, about 2 pAg units. With the 0.01000 M reagent, the change is about 1 pAg unit, but still pronounced. An indicator that produces a signal in the 4.0 to 6.0 pAg region should give a minimal error for the stronger solution. For the more dilute chloride solution (Curve *B*), the change in pAg in the equivalence-point region would be drawn out over a fairly large volume of reagent (~3 mL as shown by the dashed lines in the

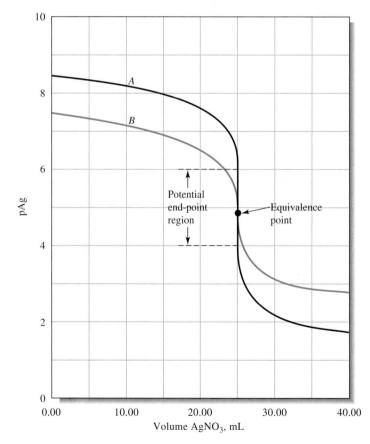

Figure 17-2 Titration curve for (*A*), 50.00 mL of 0.05000 M NaCl titrated with 0.1000 M $AgNO_3$, and (*B*), 50.00 mL of 0.00500 M NaCl titrated with 0.01000 M $AgNO_3$. Note the increased sharpness of the break at the end point with the more concentrated solution.

figure) so that to determine the end point accurately would be impossible. The effect here is analogous to that illustrated for acid/base titrations in Figure 14-4.

The Effect of Reaction Completeness on Titration Curves

Figure 17-3 illustrates the effect of solubility product on the sharpness of the end point for titrations with 0.1 M silver nitrate. Note that the change in pAg at the equivalence point becomes greater as the solubility products become smaller, that is, as the reaction between the analyte and silver nitrate becomes more complete. By choosing an indicator that changes color in the pAg region of 4 to 6, titration of chloride ions should be possible with a minimal titration error. Note that ions forming precipitates with solubility products much larger than about 10^{-10} do not yield satisfactory end points.

Titration Curves for Mixtures of Anions

The methods developed in Example 17-1 for constructing precipitation titration curves can be extended to mixtures that form precipitates of different solubilities. To illustrate, consider 50.00 mL of a solution that is 0.0500 M in iodide ion and 0.0800 M in chloride ion titrated with 0.1000 M silver nitrate. The curve for the initial stages of this titration is identical to the curve shown for iodide in Figure 17-3 because silver chloride, with its much larger solubility product, does not begin to precipitate until well into the titration.

It is interesting to determine how much iodide is precipitated before appreciable amounts of silver chloride form. With the appearance of the smallest amount of solid silver chloride, the solubility-product expressions for both precipitates apply, and division of one by the other provides the useful relationship

$$\frac{K_{sp}(AgI)}{K_{sp}(AgCl)} = \frac{[Ag^+][I^-]}{[Ag^+][Cl^-]} = \frac{8.3 \times 10^{-17}}{1.82 \times 10^{-10}} = 4.56 \times 10^{-7}$$

$$[I^-] = (4.56 \times 10^{-7})[Cl^-]$$

From this relationship, we see that the iodide concentration decreases to a tiny fraction of the chloride ion concentration before silver chloride begins to precipitate.

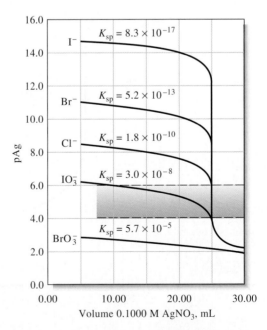

Figure 17-3 Effect of reaction completeness on precipitation titration curves. For each curve, 50.00 mL of a 0.0500 M solution of the anion was titrated with 0.1000 M $AgNO_3$. Note that smaller values of K_{sp} give much sharper breaks at the end point.

So, for all practical purposes, silver chloride forms only after 25.00 mL of titrant have been added in this titration. At this point, the chloride ion concentration is approximately

$$c_{Cl^-} \approx [Cl^-] = \frac{50.00 \times 0.0800}{50.00 + 25.00} = 0.0533 \text{ M}$$

Substituting into the previous equation yields

$$[I^-] = 4.56 \times 10^{-7}[Cl^-] = 4.56 \times 10^{-7} \times 0.0533 = 2.43 \times 10^{-8} \text{ M}$$

The percentage of iodide unprecipitated at this point can be calculated as follows:

$$\text{amount I}^- \text{ unprecipitated} = (75.00 \text{ mL})(2.43 \times 10^{-8} \text{ mmol I}^-/\text{mL}) = 1.82 \times 10^{-6} \text{ mmol}$$

$$\text{original amount I}^- = (50.00 \text{ mL})(0.0500 \text{ mmol/mL}) = 2.50 \text{ mmol}$$

$$\text{percentage I}^- \text{ unprecipitated} = \frac{1.82 \times 10^{-6}}{2.50} \times 100\% = 7.3 \times 10^{-5}\%$$

Thus, to within about 7.3×10^{-5} percent of the equivalence point for iodide, no silver chloride forms. Up to this point, the titration curve is indistinguishable from that for iodide alone, as shown in **Figure 17-4**. The data points for the first part of the titration curve, shown by the solid line, were computed on this basis.

As chloride ion begins to precipitate, however, the rapid decrease in pAg ends abruptly at a level that can be calculated from the solubility product for silver chloride and the computed chloride concentration (0.0533 M):

$$[Ag^+] = \frac{K_{sp}(AgCl)}{[Cl^-]} = \frac{1.82 \times 10^{-10}}{0.0533} = 3.41 \times 10^{-9} \text{ M}$$

$$pAg = -\log(3.41 \times 10^{-9}) = 8.47$$

The sudden end to the sharp decrease in $[Ag^+]$ can be clearly seen in Figure 17-4 at pAg = 8.47. Further additions of silver nitrate decrease the chloride ion concentration, and the curve then becomes that for the titration of chloride by itself.

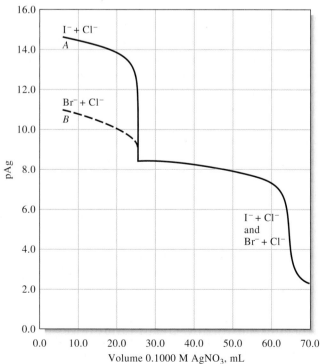

Figure 17-4 Titration curves for 50.00 mL of a solution 0.0800 M in Cl^- and 0.0500 M in I^- or Br^-.

For example, after 30.00 mL of titrant have been added,

$$c_{Cl^-} = [Cl^-] = \frac{50.00 \times 0.0800 + 50.00 \times 0.0500 - 30.00 \times 0.100}{50.00 + 30.00} = 0.0438\,M$$

In this expression, the first two terms in the numerator give the number of millimoles of chloride and iodide, respectively, and the third term is the number of millimoles of titrant. Therefore,

$$[Ag^+] = \frac{1.82 \times 10^{-10}}{0.0438} = 4.16 \times 10^{-9}\,M$$

$$pAg = 8.38$$

The remainder of the data points for this curve can be computed in the same way as for a curve of chloride by itself.

Curve A in Figure 17-4, which is the titration curve for the chloride/iodide mixture just considered, is a composite of the individual curves for the two anionic species. Two equivalence points are evident. Curve B is the titration curve for a mixture of bromide and chloride ions. Note that the change associated with the first equivalence point becomes less distinct as the solubilities of the two precipitates approach one another. In the bromide/chloride titration, the initial pAg values are lower than they are in the iodide/chloride titration because the solubility of silver bromide exceeds that of silver iodide. Beyond the first equivalence point, however, where chloride ion is being titrated, the two titration curves are identical.

Titration curves similar to those in Figure 17-4 can be obtained experimentally by measuring the potential of a silver electrode immersed in the analyte solution (see Section 21C). These curves can then be use to determine the concentration of each of the ions in mixtures of two halide ions.

End Points for Argentometric Titrations

Chemical, potentiometric, and amperometric end points are used in titrations with silver nitrate. In this section, we describe one of the chemical indicator methods. In potentiometric titrations, the potential difference between a silver electrode and a reference electrode is measured as a function of titrant volume. Titration curves similar to those shown in Figures 17-2, 17-3, and 17-4 are obtained. Potentiometric titrations are discussed in Section 21C. In amperometric titrations, the current generated between a pair of silver electrodes is measured and plotted as a function of titrant volume. Amperometric methods are considered in Section 23B-4.

Chemical indicators produce a color change or occasionally the appearance or disappearance of turbidity in the solution being titrated. The requirements for an indicator for a precipitation titration are that (1) the color change should occur over a limited range in p-function of the titrant or the analyte and (2) the color change should take place within the steep portion of the titration curve for the analyte. For example, in Figure 17-3, we see that the titration of iodide with any indicator providing a signal in the pAg range of about 4.0 to 12.0 should give a satisfactory end point. Note that, in contrast, the end-point signal for the titration of chloride would be limited to a pAg of about 4.0 to 6.0.

The Volhard Method.   The Volhard method is one of the most common argentometric methods. In this method, silver ions are titrated with a standard solution of thiocyanate ion:

$$Ag^+ + SCN^- \rightleftharpoons AgSCN(s)$$

Iron(III) serves as the indicator. The solution turns red with the first slight excess of thiocyanate ion due to the formation of $Fe(SCN)^{2+}$.

The most important application of the Volhard method is the indirect determination of halide ions. A measured excess of standard silver nitrate solution is added to the sample, and the excess silver is determined by back-titration with a standard thiocyanate solution. The strongly acidic environment of the Volhard titration is a distinct advantage over other titrations of halide ions because such ions as carbonate, oxalate, and arsenate do not interfere. The silver salts of these ions are soluble in acidic media but only slightly soluble in neutral media.

Silver chloride is more soluble than silver thiocyanate. As a result, in chloride determinations using the Volhard method, the reaction

$$AgCl(s) + SCN^- \rightleftharpoons AgSCN(s) + Cl^-$$

occurs to a significant extent near the end of the back-titration. This reaction causes the end point to fade and results in overconsumption of thiocyanate ion. The resulting low results for chloride can be overcome by filtering the silver chloride before undertaking the back-titration. Filtration is not required for other halides because they form silver salts that are less soluble than silver thiocyanate.

Other Argentometric Methods. In the **Mohr method**, sodium chromate serves as the indicator for the argentometric titration of chloride, bromide, and cyanide ions. Silver ions react with chromate to form the brick-red silver chromate (Ag_2CrO_4) precipitate in the equivalence-point region. The Mohr method is now rarely used because $Cr(VI)$ is a carcinogen.

The **Fajans method** uses an **adsorption indicator**, an organic compound that adsorbs onto or desorbs from the surface of the solid in a precipitation titration. Ideally, the adsorption or desorption occurs near the equivalence point and results not only in a color change but also in the transfer of color from the solution to the solid or vice versa.

> Adsorption indicators were first described by K. Fajans, a Polish chemist in 1926. Titrations involving adsorption indicators are rapid, accurate, and reliable, but their application is limited to the few precipitation titrations that form colloidal precipitates rapidly.

> **Spreadsheet Summary** In Chapter 9 of *Applications of Microsoft® Excel in Analytical Chemistry*, 2nd ed., we plot a curve for the titration of NaCl with $AgNO_3$. A stoichiometric approach is first used and then a master equation approach is explored. Finally, the problem is inverted, and the volume needed to achieve a given pAg value is computed.

17C ORGANIC COMPLEXING AGENTS

Several different organic complexing agents have become important in analytical chemistry because of their inherent sensitivity and potential selectivity in reacting with metal ions. Organic reagents are particularly useful in precipitating metals, in binding metals so as to prevent interferences, in extracting metals from one solvent to another, and in forming complexes that absorb light for spectrophotometric determinations. The most useful organic reagents form chelate complexes with metal ions.

Many organic reagents are useful in converting metal ions into forms that can be readily extracted from water into an immiscible organic phase. Extractions are widely used to separate metals of interest from potential interfering ions and for achieving a concentrating effect by transfer of the metal into a phase of smaller volume. Extractions are applicable to much smaller amounts of metals than precipitations, and they avoid problems associated with coprecipitation. Separations by extraction are considered in Section 31C.

Several of the most widely used organic complexing agents for extractions are listed in **Table 17-3**. Some of these same reagents normally form insoluble species

TABLE 17-3

Organic Reagents for Extracting Metals

Reagent	Metal Ions Extracted	Solvents
8-Hydroxyquinoline	Zn^{2+}, Cu^{2+}, Ni^{2+}, Al^{3+}, many others	Water $\rightarrow$ Chloroform ($CHCl_3$)
Diphenylthiocarbazone (dithizone)	Cd^{2+}, Co^{2+}, Cu^{2+}, Pb^{2+}, many others	Water $\rightarrow$ $CHCl_3$ or CCl_4
Acetylacetone	Fe^{3+}, Cu^{2+}, Zn^{2+}, U(VI), many others	Water $\rightarrow$ $CHCl_3$, CCl_4, or C_6H_6
Ammonium pyrrolidine dithiocarbamate	Transition metals	Water $\rightarrow$ Methyl isobutyl ketone
Tenoyltrifluoroacetone	Ca^{2+}, Sr^{2+}, La^{3+}, Pr^{3+} other rare earths	Water $\rightarrow$ Benzene
Dibenzo-18-crown-6	Alkali metals, some alkaline earths	Water $\rightarrow$ Benzene

with metal ions in aqueous solution. However, in extraction applications, the solubility of the metal chelate in the organic phase keeps the complex from precipitating in the aqueous phase. In many cases, the pH of the aqueous phase is used to achieve some control over the extraction process since most of the reactions are pH dependent, as shown in Equation 17-21.

$$n\text{HX}(org) + \text{M}^{n+}(aq) \rightleftharpoons \text{MX}_n(org) + n\text{H}^+(aq) \tag{17-21}$$

Another important application of organic complexing agents is in forming stable complexes that bind a metal and prevent it from interfering in a determination. Such complexing agents are called **masking agents** and are discussed in Section 17D-8. Organic complexing agents are also widely used in spectrophotometric determinations of metal ions (see Chapter 26). In this instance, the metal-ligand complex is either colored or absorbs ultraviolet radiation. Organic complexing agents are also commonly used in electrochemical determinations and in molecular fluorescence spectrometry.

17D   AMINOCARBOXYLIC ACID TITRATIONS

Tertiary amines that also contain carboxylic acid groups form remarkably stable chelates with many metal ions.[1] Gerold Schwarzenbach, a Swiss chemist, first recognized their potential as analytical reagents in 1945. Since his original work, investigators throughout the world have described applications of these compounds to the volumetric determination of most of the metals in the periodic table.

17D-1 Ethylenediaminetetraacetic Acid (EDTA)

Ethylenediaminetetraacetic acid, which is also called (ethylenedinitrilo)tetraacetic acid and which is commonly shortened to EDTA, is the most widely used complexometric titrant. EDTA has the structural formula

Structural formula of EDTA.

---

[1]See for example, R. Pribil, *Applied Complexometry*, New York: Pergamon, 1982; A. Ringbom and E. Wanninen, in *Treatise on Analytical Chemistry*, 2nd ed., I. M. Kolthoff and P. J. Elving, eds., Part I, Vol. 2, Chap, 11, New York: Wiley, 1979.

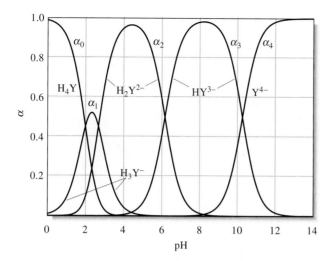

Figure 17-5 Composition of EDTA solutions as a function of pH. Note that the fully protonated form, H_4Y is only a major component in very acidic solutions (pH $<$ 3). Throughout the pH range of 3 to 10, the species H_2Y^{2-} and HY^{3-} are predominant. The fully unprotonated form Y^{4-} is a significant component only in very basic solutions (pH $>$ 10).

The EDTA molecule has six potential sites for bonding a metal ion: the four carboxyl groups and the two amino groups, each of the latter with an unshared pair of electrons. Thus, EDTA is a hexadentate ligand.

> EDTA, a hexadentate ligand, is among the most important and widely used reagents in titrimetry.

Acidic Properties of EDTA

The dissociation constants for the acidic groups in EDTA are $K_1 = 1.02 \times 10^{-2}$, $K_2 = 2.14 \times 10^{-3}$, $K_3 = 6.92 \times 10^{-7}$, and $K_4 = 5.50 \times 10^{-11}$. Note that the first two constants are of the same order of magnitude. This similarity suggests that the two protons involved dissociate from opposite ends of the rather long molecule. Because the protons are several atoms apart, the negative charge resulting from the first dissociation does not greatly influence the removal of the second proton. Note, however, that the dissociation constants of the other two protons are much smaller and different from one another. These protons are closer to the negatively charged carboxylate ions resulting from the dissociations of the first two protons, and they are more difficult to remove from the ion because of electrostatic attraction.

The various EDTA species are often abbreviated H_4Y, H_3Y^-, H_2Y^{2-}, HY^{3-}, and Y^{4-}. Feature 17-3 describes the EDTA species and shows their structural formulas. **Figure 17-5** illustrates how the relative amounts of these five species vary as a function of pH. Note that the species H_2Y^{2-} predominates from pH 3 to 6.

FEATURE 17-3

Species Present in a Solution of EDTA

When it is dissolved in water, EDTA behaves like an amino acid, such as glycine (see Features 14-5 and 15-2). With EDTA, however, a double zwitterion forms, which has the structure shown in Figure 17F-1a. Note that the net charge on this species is zero and that it contains four acidic protons, two associated with two of the carboxyl groups and the other two with the two amine groups. For simplicity, we usually abbreviate the double zwitterion as H_4Y, where Y^{4-} is the fully deprotonated form of Figure 17F-1e. The first and second steps in the dissociation process involve successive loss of protons from the two carboxylic acid groups; the third and fourth steps involve dissociation of the protonated amine groups. The structural formulas of H_3Y^-, H_2Y^{2-}, and HY^{3-} are shown in **Figure 17F-1b, c,** and **d.**

(continued)

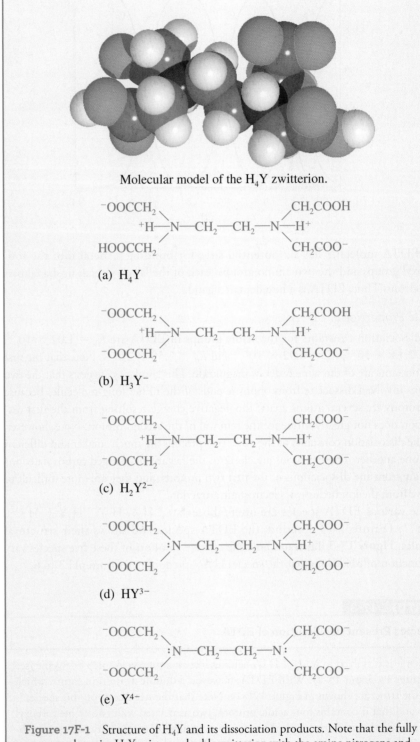

Molecular model of the H₄Y zwitterion.

(a) H_4Y

(b) H_3Y^-

(c) H_2Y^{2-}

(d) HY^{3-}

(e) Y^{4-}

Figure 17F-1 Structure of H_4Y and its dissociation products. Note that the fully protonated species H_4Y exist as a double zwitterion with the amine nitrogens and two of the carboxylic acid groups protonated. The first two protons dissociate from the carboxyl groups, while the last two come from the amine groups.

Reagents for EDTA Titrations

The free acid H_4Y and the dihydrate of the sodium salt, $Na_2H_2Y \cdot 2H_2O$, are commercially available in reagent quality. The free acid can serve as a primary standard after it has been dried for several hours at 130°C to 145°C. However, the free acid is not very soluble in water and must be dissolved in a small amount of base for complete solution.

More commonly, the dihydrate, $Na_2H_2Y \cdot 2H_2O$, is used to prepare standard solutions. Under normal atmospheric conditions, the dihydrate contains 0.3% moisture in excess of the stoichiometric water of hydration. For all but the most exacting work, this excess is sufficiently reproducible to permit use of a corrected mass of the salt in the direct preparation of a standard solution. If necessary, the pure dihydrate can be prepared by drying at 80°C for several days in an atmosphere of 50% relative humidity. Alternatively, an approximate concentration can be prepared and then standardized against primary standard $CaCO_3$.

Several compounds that are chemically related to EDTA have also been investigated. Since these do not seem to offer significant advantages, we shall limit our discussion here to the properties and applications of EDTA.

17D-2 Complexes of EDTA and Metal Ions

Solutions of EDTA are particularly valuable as titrants because the EDTA *combines with metal ions in a 1:1 ratio regardless of the charge on the cation.* For example, the silver and aluminum complexes are formed by the reactions

$$Ag^+ + Y^{4-} \rightleftharpoons AgY^{3-}$$

$$Al^{3+} + Y^{4-} \rightleftharpoons AlY^-$$

EDTA is a remarkable reagent not only because it forms chelates with all cations but also because most of these chelates are sufficiently stable for titrations. This great stability undoubtedly results from the several complexing sites within the molecule that give rise to a cagelike structure in which the cation is effectively surrounded and isolated from solvent molecules. One of the common structures for metal/EDTA complexes is shown in **Figure 17-6**. The ability of EDTA to form complexes with metals is responsible for its widespread use as a preservative in foods and in biological samples as discussed in Feature 17-4.

> Standard EDTA solutions can be prepared by dissolving weighed quantities of $Na_2H_2Y \cdot 2H_2O$ and diluting to the mark in a volumetric flask.

> Nitrilotriacetic acid (NTA) is the second most common aminopolycarboxylic acid used for titrations. It is a tetradentate chelating agent and has the structure

$$COOH-CH_2 \quad \overset{H_2}{C}-COOH$$
$$N$$
$$CH_2-COOH$$

Structural formula of NTA.

> In general, we can write the reaction of the EDTA anion with a metal ion M^{n+} as $M^{n+} + Y^{4-} \rightleftharpoons MY^{(n-4)+}$.

Figure 17-6 Structure of a metal/EDTA complex. Note that EDTA behaves here as a hexadentate ligand in that six donor atoms are involved in bonding the divalent metal cation.

TABLE 17-4

Formation Constants for EDTA Complexes

Cation	K_{MY}*	log K_{MY}	Cation	K_{MY}	log K_{MY}
Ag^+	2.1×10^7	7.32	Cu^{2+}	6.3×10^{18}	18.80
Mg^{2+}	4.9×10^8	8.69	Zn^{2+}	3.2×10^{16}	16.50
Ca^{2+}	5.0×10^{10}	10.70	Cd^{2+}	2.9×10^{16}	16.46
Sr^{2+}	4.3×10^8	8.63	Hg^{2+}	6.3×10^{21}	21.80
Ba^{2+}	5.8×10^7	7.76	Pb^{2+}	1.1×10^{18}	18.04
Mn^{2+}	6.2×10^{13}	13.79	Al^{3+}	1.3×10^{16}	16.13
Fe^{2+}	2.1×10^{14}	14.33	Fe^{3+}	1.3×10^{25}	25.1
Co^{2+}	2.0×10^{16}	16.31	V^{3+}	7.9×10^{25}	25.9
Ni^{2+}	4.2×10^{18}	18.62	Th^{4+}	1.6×10^{23}	23.2

*Constants are valid at 20°C and ionic strength of 0.1.
Source: G. Schwarzenbach, Complexometric Titrations, London: Chapman and Hall, 1957, p. 8.

Table 17-4 lists formation constants K_{MY} for common EDTA complexes. Note that the constant refers to the equilibrium involving the fully unprotonated species Y^{4-} with the metal ion:

$$M^{n+} + Y^{4-} \rightleftharpoons MY^{(n-4)+} \qquad K_{MY} = \frac{[MY^{(n-4)+}]}{[M^{n+}][Y^{4-}]} \qquad (17\text{-}22)$$

17D-3 Equilibrium Calculations Involving EDTA

A titration curve for the reaction of a cation M^{n+} with EDTA consists of a plot of pM (pM $= -\log[M^{n+}]$) versus reagent volume. In the early stage of a titration, values for pM are readily computed by assuming that the equilibrium concentration of M^{n+} is equal to its analytical concentration, which is found from stoichiometric data.

FEATURE 17-4

EDTA as a Preservative

Trace quantities of metal ions can efficiently catalyze the air oxidation of many of the compounds present in foods and biological samples (for example, proteins in blood). To prevent such oxidation reactions, it is important to inactivate or remove even trace amounts of metal ions. Processed foods can readily pick up trace quantities of metal ions while in contact with various metallic containers (kettles and vats) during the processing stages. EDTA is an excellent preservative for foods and is a common ingredient of such commercial food products as mayonnaise, salad dressings, and oils. When EDTA is added to foods, it so tightly binds most metal ions that they are unable to catalyze the air oxidation reaction. EDTA and other similar chelating agents are often called **sequestering agents** because of their ability to remove or inactivate metal ions. In addition to EDTA, some other common sequestering agents are salts of citric and phosphoric acid. These agents can protect the unsaturated side chains of triglycerides and other components against air oxidation. Such oxidation reactions are responsible for making fats and oils turn rancid. Sequestering

agents are also added to prevent oxidation of easily oxidized compounds, such as ascorbic acid.

It is important to add EDTA to preserve biological samples that are to be stored for long periods. As in foods, EDTA forms very stable complexes with metal ions and prevents them from catalyzing air oxidation reactions that can lead to decomposition of proteins and other compounds. During the murder trial of celebrity and former football player O. J. Simpson, the use of EDTA as a preservative became an important point of evidence. The prosecution team contended that if blood evidence had been planted on the back fence at his former wife's home, EDTA should be present, but if the blood were from the murderer, no preservative should be seen. Analytical evidence, obtained by using a sophisticated instrumental system (liquid chromatography combined with tandem mass spectrometry), did show traces of EDTA, but the amounts were very small and subject to differing interpretations.[2]

[2]D. Margolick, "FBI Disputes Simpson Defense on Tainted Blood," *New York Times*, July 26, 1995, p. A12.

Calculation of $[M^{n+}]$ at and beyond the equivalence point requires the use of Equation 17-22. In this region of the titration curve, it is difficult and time consuming to apply Equation 17-22 if the pH is unknown and variable because both $[MY^{(n-4)+}]$ and $[M^{n+}]$ are pH dependent. Fortunately, EDTA titrations are always performed in solutions that are buffered to a known pH to avoid interferences by other cations or to ensure satisfactory indicator behavior. Calculating $[M^{n+}]$ in a buffered solution containing EDTA is a relatively straightforward procedure provided the pH is known. In this computation, we use the alpha value for H_4Y, α_4 (see Section 15H).

$$\alpha_4 = \frac{[Y^{4-}]}{c_T} \tag{17-23}$$

where c_T is the total molar concentration of *uncomplexed* EDTA.

$$c_T = [Y^{4-}] + [HY^{3-}] + [H_2Y^{2-}] + [H_3Y^{3-}] + [H_4Y]$$

Note that, at a given pH, α_4, the fraction of total EDTA in the unprotonated form, is constant.

Conditional Formation Constants

To obtain the conditional formation constant for the equilibrium shown in Equation 17-22, we substitute $\alpha_4 c_T$ from Equation 17-23 for $[Y^{4-}]$ in the formation constant expression (right side of Equation 17-22):

$$M^{n+} + Y^{4-} \rightleftharpoons MY^{(n-4)+} \quad K_{MY} = \frac{[MY^{(n-4)+}]}{[M^{n+}]\alpha_4 c_T} \tag{17-24}$$

Combining the two constants α_4 and K_{MY} yields the conditional formation constant K'_{MY}

$$K'_{MY} = \alpha_4 K_{MY} = \frac{[MY^{(n-4)+}]}{[M^{n+}]c_T} \tag{17-25}$$

where K'_{MY} is a constant *only at the pH for which α_4 is applicable.*

❮ Conditional formation constants are pH dependent.

Conditional constants are easily computed once the pH is known. They may be used to calculate the equilibrium concentration of the metal ion and the complex at the equivalence point and where there is an excess of reactant. Note that replacement of $[Y^{4-}]$ with c_T in the equilibrium-constant expression greatly simplifies calculations because c_T is easily determined from the reaction stoichiometry whereas $[Y^{4-}]$ is not.

Computing α_4 Values for EDTA Solutions

An expression for calculating α_4 at a given hydrogen ion concentration is obtained by the method given in Section 15-H (see Feature 15-3). Thus, α_4 for EDTA is

❮ The alpha values for the other EDTA species are calculated in a similar manner and are found to be

$\alpha_0 = [H^+]^4/D$
$\alpha_1 = K_1[H^+]^3/D$
$\alpha_2 = K_1K_2[H^+]^2/D$
$\alpha_3 = K_1K_2K_3[H^+]/D$

Only α_4 is needed in calculating titration curves.

$$\alpha_4 = \frac{K_1K_2K_3K_4}{[H^+]^4 + K_1[H^+]^3 + K_1K_2[H^+]^2 + K_1K_2K_3[H^+] + K_1K_2K_3K_4} \tag{17-26}$$

$$\alpha_4 = \frac{K_1K_2K_3K_4}{D} \tag{17-27}$$

where K_1, K_2, K_3, and K_4 are the four dissociation constants for H_4Y, and D is the denominator of Equation 17-26.

Figure 17-7 Spreadsheet to calculate α_4 for EDTA at selected pH values. Note that the acid dissociation constants for EDTA are entered in column B (labels in column A). Next the pH values for which the calculations are to be done are entered in column C. The formula for calculating the denominator D in Equations 17-26 and 17-27 is placed into cell D3 and copied into D4 through D16. The final column E contains the equation for calculating the α_4 values as given in Equation 17-27. The graph shows a plot of α_4 versus pH over the pH range of 6 to 14.

Figure 17-7 shows an Excel spreadsheet for calculating α_4 at selected pH values according to Equations 17-26 and 17-27. Note the wide variation of α_4 with pH. This variation allows the effective complexing ability of EDTA to be dramatically changed by varying the pH. Example 17-2 illustrates how the concentration of Y^{4-} is calculated for a solution of known pH.

EXAMPLE 17-2

Calculate the molar Y^{4-} concentration in a 0.0200 M EDTA solution buffered to a pH of 10.00.

Solution

At pH 10.00, α_4 is 0.35 (see Figure 17-7). Thus,

$$[Y^{4-}] = \alpha_4 c_T = 0.35 \times 0.0200 \text{ M} = 7.00 \times 10^{-3} \text{ M}$$

Calculating the Cation Concentration in EDTA Solutions

In an EDTA titration, we are interested in finding the cation concentration as a function of the amount of titrant (EDTA) added. Prior to the equivalence point, the cation is in excess, and its concentration can be found from the reaction stoichiometry. At the equivalence point and in the postequivalence-point region, however, the conditional formation constant of the complex must be used to calculate the cation concentration. Example 17-3 demonstrates how the cation concentration can be found in a solution of an EDTA complex. Example 17-4 illustrates this calculation when excess EDTA is present.

EXAMPLE 17-3

Calculate the equilibrium concentration of Ni^{2+} in a solution with an analytical NiY^{2-} concentration of 0.0150 M at pH (a) 3.0 and (b) 8.0.

Solution

From Table 17-4,

$$Ni^{2+} + Y^{4-} \rightleftharpoons NiY^{2-} \qquad K_{NiY} = \frac{[NiY^{2-}]}{[Ni^{2+}][Y^{4-}]} = 4.2 \times 10^{18}$$

The equilibrium concentration of NiY^{2-} is equal to the analytical concentration of the complex minus the concentration lost by dissociation. The concentration lost by dissociation is equal to the equilibrium Ni^{2+} concentration. Thus,

$$[NiY^{2-}] = 0.0150 - [Ni^{2+}]$$

If we assume that $[Ni^{2+}] \ll 0.0150$, an assumption that is almost certainly valid in light of the large formation constant of the complex, this equation simplifies to

$$[NiY^{2-}] \cong 0.0150$$

Since the complex is the only source of both Ni^{2+} and the EDTA species,

$$[Ni^{2+}] = [Y^{4-}] + [HY^{3-}] + [H_2Y^{2-}] + [H_3Y^-] + [H_4Y] = c_T$$

Substitution of this equality into Equation 17-25 gives

$$K'_{NiY} = \frac{[NiY^{2-}]}{[Ni^{2+}]c_T} = \frac{[NiY^{2-}]}{[Ni^{2+}]^2} = \alpha_4 K_{NiY}$$

(a) The spreadsheet in Figure 17-7 indicates that α_4 is 2.51×10^{-11} at pH 3.0. If we substitute this value and the concentration of NiY^{2-} into the equation for K'_{MY}, we get

$$\frac{0.0150}{[Ni^{2+}]^2} = 2.51 \times 10^{-11} \times 4.2 \times 10^{18} = 1.05 \times 10^8$$

$$[Ni^{2+}] = \sqrt{1.43 \times 10^{-10}} = 1.2 \times 10^{-5} M$$

(b) At pH 8.0, α_4, and thus the conditional constant, is much larger. Therefore,

$$K'_{NiY} = 5.39 \times 10^{-3} \times 4.2 \times 10^{18} = 2.27 \times 10^{16}$$

and, after we substitute this into the equation for K'_{NiY}, we find that

$$[Ni^{2+}] = \sqrt{\frac{0.0150}{2.27 \times 10^{16}}} = 8.1 \times 10^{-10} M$$

Molecular model of NiY^{2-}. This complex is typical of the strong complexes that EDTA forms with metal ions. The formation constant of the Ni^{2+} complex is 4.2×10^{18}.

Note that for both pH 3.0 and pH 8.0, our assumption that $[Ni^{2+}] \ll 0.0150$ M is valid.

EXAMPLE 17-4

Calculate the concentration of Ni^{2+} in a solution that was prepared by mixing 50.0 mL of 0.0300 M Ni^{2+} with 50.00 mL of 0.0500 M EDTA. The mixture was buffered to a pH of 3.0.

Solution

The solution has an excess of EDTA, and the analytical concentration of the complex is determined by the amount of Ni^{2+} originally present. Thus,

$$c_{NiY^{2-}} = 50.00 \text{ mL} \times \frac{0.0300 \text{ M}}{100 \text{ mL}} = 0.0150 \text{ M}$$

$$c_{EDTA} = \frac{(50.00 \times 0.0500) \text{ mmol} - (50.0 \times 0.0300) \text{ mmol}}{100.0 \text{ mL}} = 0.0100 \text{ M}$$

Again, we will assume that $[Ni^{2+}] \ll [NiY^{2-}]$ so that

$$[NiY^{2-}] = 0.0150 - [Ni^{2+}] \approx 0.0150 \text{ M}$$

At this point, the total concentration of uncomplexed EDTA is given by its concentration, c_{EDTA}:

$$c_T = c_{EDTA} = 0.0100 \text{ M}$$

If we substitute this value in Equation 17-25, we get

$$K'_{NiY} = \frac{0.0150}{[Ni^{2+}] \times 0.0100} = \alpha_4 K_{NiY}$$

Using the value of α_4 at pH 3.0 from Figure 17-7, we obtain

$$[Ni^{2+}] = \frac{0.0150}{0.0100 \times 2.51 \times 10^{-11} \times 4.2 \times 10^{18}} = 1.4 \times 10^{-8} \text{ M}$$

Note again that our assumption that $[Ni^{2+}] \ll [NiY^{2-}]$ is valid.

17D-4 EDTA Titration Curves

The principles illustrated in Examples 17-3 and 17-4 can be used to generate the titration curve for a metal ion with EDTA in a solution of fixed pH. Example 17-5 demonstrates how a spreadsheet can be used to construct the titration curve.

EXAMPLE 17-5

Use a spreadsheet to construct the titration curve of pCa versus volume of EDTA for 50.0 mL of 0.00500 M Ca^{2+} titrated with 0.0100 M EDTA in a solution buffered to pH 10.0.

Solution

Initial Entries

The spreadsheet is shown in **Figure 17-8**. We enter the initial volume of Ca^{2+} in cell B3 and the initial Ca^{2+} concentration in E2. The EDTA concentration is entered into cell E3. The volumes for which pCa values are to be calculated are entered into cells A5 through A19. We also need the conditional formation constant for the CaY complex. This constant is obtained from the formation constant of the complex (Table 17-4) and the α_4 value for EDTA at pH 10 (see Figure 17-7). If we substitute into Equation 17-25, we get

$$K'_{CaY} = \frac{[CaY^{2-}]}{[Ca^{2+}]c_T} = \alpha_4 K_{CaY}$$

$$= 0.35 \times 5.0 \times 10^{10} = 1.75 \times 10^{10}$$

This value is entered into cell B2. Since the conditional constant is to be used in further calculations, we do not round off to keep only significant figures at this point.

Preequivalence-Point Values for pCa

The initial $[Ca^{2+}]$ at 0.00 mL titrant is just the value in cell E2. Hence, **=E2** is entered into cell B5. The initial pCa is calculated from the initial $[Ca^{2+}]$ by taking the negative logarithm as shown in the documentation for cell E5. This formula is copied into cells E6 through E19. For the other entries prior to the equivalence point, the equilibrium concentration of Ca^{2+} is equal to the untitrated excess of the cation plus any Ca^{2+} resulting from dissociation of the complex. The latter concentration is equal to c_T. Usually, c_T is small relative to the analytical concentration of the uncomplexed calcium ion. For example, after 5.00 mL of EDTA has been added,

Figure 17-8 Spreadsheet for the titration of 50.00 mL of 0.00500 M Ca^{2+} with 0.0100 M EDTA in a solution buffered at pH 10.0.

$$[Ca^{2+}] = \frac{50.0 \text{ mL} \times 0.00500 \text{ M} - 5.00 \text{ mL} \times 0.0100 \text{ M}}{(50 + 5.00) \text{ mL}} + c_T$$

$$\approx \frac{50.0 \text{ mL} \times 0.00500 \text{ M} - 5.00 \text{ mL} \times 0.0100 \text{ M}}{55.00 \text{ mL}}$$

We thus enter into cell B6 the formula shown in the documentation section of the spreadsheet. The reader should verify that the spreadsheet formula is equivalent to the expression for $[Ca^{2+}]$ given above. The volume of titrant (A6) is the only value that changes in this preequivalence-point region. The other preequivalence-point values of pCa are calculated by copying the formula in cell B6 into cells B7 through B10.

The Equivalence-Point pCa

At the equivalence point (25.00 mL of EDTA), we follow the method shown in Example 17-3 and first compute the analytical concentration of CaY^{2-}:

$$c_{CaY^{2-}} = \frac{(50.0 \times 0.00500) \text{ mmol}}{(50.0 + 25.0) \text{ mL}}$$

The only source of Ca^{2+} ions is the dissociation of the complex. It also follows that the Ca^{2+} concentration must be equal to the sum of the concentrations of the uncomplexed EDTA, c_T. Therefore,

$$[Ca^{2+}] = c_T, \text{ and } [CaY^{2-}] = c_{CaY^{2-}} - [Ca^{2+}] \approx c_{CaY^{2-}}$$

The formula for $[CaY^{2-}]$ is entered into cell C11. Be sure to verify this formula for yourself. To obtain $[Ca^{2+}]$, we substitute into the expression for K'_{CaY},

$$K'_{CaY} = \frac{[CaY^{2-}]}{[Ca^{2+}] c_T} \cong \frac{c_{CaY^{2-}}}{[Ca^{2+}]^2}$$

$$[Ca^{2+}] = \sqrt{\frac{c_{CaY^{2-}}}{K'_{CaY}}}$$

We enter into cell B11 the formula corresponding to this expression.

Postequivalence-Point pCa

Beyond the equivalence point, analytical concentrations of CaY^{2-} and EDTA are obtained directly from the stoichiometry. Since there is excess EDTA, a calculation similar to that in Example 17-4 is then performed. For example, after the addition of 26.0 mL of EDTA, we can write

$$c_{CaY^{2-}} = \frac{(50.0 \times 0.00500) \text{ mmol}}{(50.0 + 26.0) \text{ mL}}$$

$$c_{EDTA} = \frac{(26.0 \times 0.0100) \text{ mL} - (50.0 \times 0.00500) \text{ mL}}{76.0 \text{ mL}}$$

As an approximation,

$$[CaY^{2-}] = c_{CaY^{2-}} - [Ca^{2+}] \approx c_{CaY^{2-}} \approx \frac{(50.0 \times 0.00500) \text{ mmol}}{(50.0 + 26.0) \text{ mL}}$$

We note that this expression is the same as that previously entered into cell C11. Therefore, we copy that equation into cell C12. We also note that $[CaY^{2-}]$ will be given by this same expression (with the volume varied) throughout the remainder of the titration. Hence, the formula in cell C12 is copied into cells C13 through C19. Also, we approximate

$$c_T = c_{EDTA} + [Ca^{2+}] \approx c_{EDTA} = \frac{(26.0 \times 0.0100)\ mL - (50.0 \times 0.00500)\ mL}{76.0\ mL}$$

We enter this formula into cell D12 and copy it into cells D13 through D16.

To calculate $[Ca^{2+}]$, we then substitute this approximation for c_T in the conditional formation-constant expression, and obtain

$$K'_{CaY} = \frac{[CaY^{2-}]}{[Ca^{2+}] \times c_T} \cong \frac{c_{CaY^{2-}}}{[Ca^{2+}] \times c_{EDTA}}$$

$$[Ca^{2+}] = \frac{c_{CaY^{2-}}}{c_{EDTA} \times K'_{CaY}}$$

Hence, the $[Ca^{2+}]$ in cell B12 is computed from the values in cells C12 and D12. We copy this formula into cells B13 through B19, and plot the titration curve shown in Figure 17-8.

Spreadsheet Summary   The alpha values for EDTA are calculated and used to plot a distribution diagram in Chapter 9 of *Applications of Microsoft® Excel in Analytical Chemistry*, 2nd ed. The titration of the tetraprotic acid EDTA with base is also considered.

Curve *A* in **Figure 17-9** is a plot of data for the titration in Example 17-5. Curve *B* is the titration curve for a solution of magnesium ion under identical conditions. The formation constant for the EDTA complex of magnesium is smaller than that

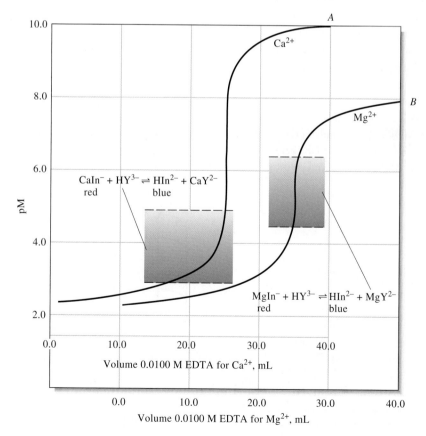

Figure 17-9   EDTA titration curves for 50.0 mL of 0.00500 M Ca^{2+} ($K'_{CaY} = 1.75 \times 10^{10}$) and Mg^{2+} ($K'_{MgY} = 1.72 \times 10^{8}$) at pH 10.0. Note that because of the larger formation constant, the reaction of calcium ion with EDTA is more complete, and a larger change occurs in the equivalence-point region. The shaded areas show the transition range for the indicator Eriochrome Black T.

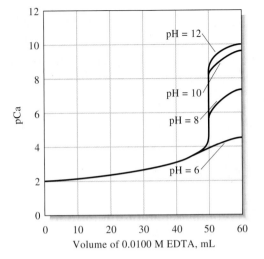

Figure 17-10 Influence of pH on the titration of 0.0100 M Ca^{2+} with 0.0100 M EDTA. Note that the end point becomes less sharp as the pH decreases because the complex-formation reaction is less complete under these circumstances.

of the calcium complex and this produces a smaller change in the p-function in the equivalence-point region.

Figure 17-10 shows titration curves for calcium ion in solutions buffered to various pH levels. Recall that α_4, and hence K'_{CaY}, becomes smaller as the pH decreases. As the conditional formation constant becomes less favorable, there is a smaller change in pCa in the equivalence-point region. Figure 17-10 shows that an adequate end point in the titration of calcium requires that the pH be greater than about 8.0. As shown in **Figure 17-11**, however, cations with larger formation constants provide sharp end points even in acidic media. If we assume that the conditional constant should be at least 10^6 to obtain a satisfactory end point with a 0.01 M solution of the metal ion, we can calculate the minimum pH needed.[3] **Figure 17-12** shows this minimum pH for a satisfactory end point in the titration of various metal ions in the absence of competing complexing agents. Note that a moderately acidic environment is satisfactory for many divalent heavy-metal cations and that a strongly acidic medium can be tolerated in the titration of such ions as iron(III) and indium(III).

> **Spreadsheet Summary**   We construct the titration curve for the titration of Ca^{2+} with EDTA by both a stoichiometric approach and a master equation approach in Chapter 9 of *Applications of Microsoft® Excel in Analytical Chemistry*, 2nd ed. The effect of pH on the shape and end point of the titration curve is examined.

Figure 17-11 Titration curves for 50.0 mL of 0.0100 M solutions of various cations at pH 6.0.

[3]C. N. Reilley and R. W. Schmid, *Anal. Chem.*, **1958**, *30*, 947, **DOI**: 10.1021/ac60137a022

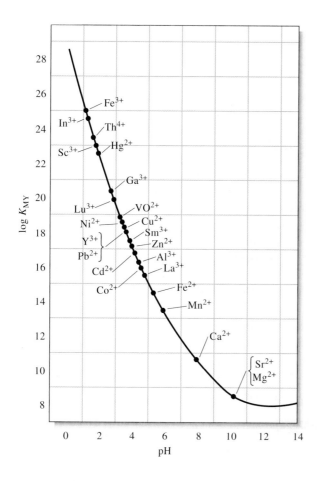

Figure 17-12 Minimum pH needed for satisfactory titration of various cations with EDTA. (Reprinted (adapted) with permission from C. N. Reilley and R. W. Schmid, *Anal. Chem.*, **1958**, *30*, 947, **DOI**: 10.1021/ac60137a022. Copyright 1958 American Chemical Society.)

17D-5 The Effect of Other Complexing Agents on EDTA Titration Curves

Many cations form hydrous oxide precipitates (hydroxides, oxides, or oxyhydroxides) when the pH is raised to the level required for their successful titration with EDTA. When we encounter this problem, an auxiliary complexing agent is needed to keep the cation in solution. For example, zinc(II) is usually titrated in a medium that has fairly high concentrations of ammonia and ammonium chloride. These species buffer the solution to a pH that ensures complete reaction between cation and titrant. In addition, ammonia forms ammine complexes with zinc(II) and prevents formation of the sparingly soluble zinc hydroxide, particularly in the early stages of the titration. A somewhat more realistic description of the reaction is then

$$Zn(NH_3)_4{}^{2+} + HY^{3-} \rightarrow ZnY^{2-} + 3NH_3 + NH_4{}^+$$

The solution also contains such other zinc/ammonia species as $Zn(NH_3)_3{}^{2+}$, $Zn(NH_3)_2{}^{2+}$ and $Zn(NH_3)^{2+}$. Calculation of pZn in a solution that contains ammonia must take these species into account as shown in Feature 17-5. Qualitatively, complexation of a cation by an auxiliary complexing reagent causes preequivalence pM values to be larger than in a comparable solution without the reagent.

Figure 17-13 shows two theoretical curves for the titration of zinc(II) with EDTA at pH 9.00. The equilibrium concentration of ammonia was 0.100 M for one titration and 0.0100 M for the other. Note that, when the ammonia concentration is higher, the change in pZn near the equivalence point decreases. For this reason, the

❮ Often, auxiliary complexing agents must be used in EDTA titrations to prevent precipitation of the analyte as a hydrous oxide. Such reagents cause the end points to be less sharp.

Figure 17-13 Influence of ammonia concentration on the end point for the titration of 50.0 mL of 0.00500 M Zn^{2+}. Solutions buffered to pH 9.00. The shaded region shows the transition range for Eriochrome Black T. Note that ammonia decreases the change in pZn in the equivalence-point region.

concentration of auxiliary complexing reagents should always be kept to the minimum required to prevent precipitation of the analyte. Note that the auxiliary complexing agent does not affect pZn beyond the equivalence point. On the other hand, keep in mind that α_4, and thus pH, plays an important role in defining this part of the titration curve (see Figure 17-10).

FEATURE 17-5

EDTA Titration Curves When a Complexing Agent Is Present

We can describe the effects of an auxiliary complexing reagent by a procedure similar to that used to determine the influence of pH on EDTA titration curves. In this case, we define a quantity α_M that is analogous to α_4:

$$\alpha_M = \frac{[M^{n+}]}{c_M} \tag{17-28}$$

where c_M is the sum of the concentrations of all species containing the metal ion that are *not* combined with EDTA. For solutions containing zinc(II) and ammonia, then

$$c_M = [Zn^{2+}] + [Zn(NH_3)^{2+}] + [Zn(NH_3)_2^{2+}]$$
$$+ [Zn(NH_3)_3^{2+}] + [Zn(NH_3)_4^{2+}] \tag{17-29}$$

The value of α_M can be expressed in terms of the ammonia concentration and the formation constants for the various ammine complexes as we describe for a general metal-ligand reaction in Feature 17-1. The result is an equation analogous to Equation 17-9:

$$\alpha_M = \frac{1}{1 + \beta_1[NH_3] + \beta_2[NH_3]^2 + \beta_3[NH_3]^3 + \beta_4[NH_3]^4} \tag{17-30}$$

Finally, we obtain a conditional constant for the equilibrium between EDTA and zinc(II) in an ammonia/ammonium chloride buffer by substituting Equation 17-28 into Equation 17-25 and rearranging

$$K''_{ZnY} = \alpha_4 \alpha_M K_{ZnY} = \frac{[ZnY^{2-}]}{c_M c_T} \tag{17-31}$$

The new conditional constant K''_{ZnY} applies at a single concentration of ammonia as well as at a single pH.

To show how Equations 17-28 to 17-31 can be used to construct a titration curve, we can calculate the pZn of solutions prepared by adding 20.0, 25.0, and 30.0 mL of 0.0100 M EDTA to 50.0 mL of 0.00500 M Zn^{2+}. Assume that both the Zn^{2+} and EDTA solutions are 0.100 M in NH_3 and 0.175 M in NH_4Cl to provide a constant pH of 9.0.

In Appendix 4, we find that the logarithms of the stepwise formation constants for the four zinc complexes with ammonia are 2.21, 2.29, 2.36, and 2.03. Thus,

$$\beta_1 = \text{antilog } 2.21 = 1.62 \times 10^2$$

$$\beta_2 = \text{antilog } (2.21 + 2.29) = 3.16 \times 10^4$$

$$\beta_3 = \text{antilog } (2.21 + 2.29 + 2.36) = 7.24 \times 10^6$$

$$\beta_4 = \text{antilog } (2.21 + 2.29 + 2.36 + 2.03) = 7.76 \times 10^8$$

Calculating the Conditional Constant

A value for α_M can be calculated from Equation 17-30 by assuming that the molar and analytical concentrations of ammonia are the same; thus, for $[NH_3] \approx c_{NH_3} = 0.100$ M,

$$\alpha_M = \frac{1}{1 + 162 \times 0.100 + 3.16 \times 10^4 \times (0.100)^2 + 7.24 \times 10^6 \times (0.100)^3 + 7.76 \times 10^8 \times (0.100)^4}$$

$$= 1.17 \times 10^{-5}$$

A value for K_{ZnY} is found in Table 17-4, and α_4 for pH 9.0 is given in Figure 17-7. Substituting into Equation 17-31, we find

$$K''_{\mathrm{ZnY}} = 5.21 \times 10^{-2} \times 1.17 \times 10^{-5} \times 3.12 \times 10^{16} = 1.9 \times 10^{10}$$

Calculating pZn after Adding 20.0 mL of EDTA

At this point, only part of the zinc has been complexed by EDTA. The remainder is present as Zn^{2+} and the four ammine complexes. By definition, the sum of the concentrations of these five species is c_M. Therefore,

$$c_M = \frac{50.00 \text{ mL} \times 0.00500 \text{ M} - 20.0 \text{ mL} \times 0.0100 \text{ M}}{70.00 \text{ mL}} = 7.14 \times 10^{-4} \text{ M}$$

Substitution of this value into Equation 17-28 gives

$$[Zn^{2+}] = c_M \alpha_M = (7.14 \times 10^{-4})(1.17 \times 10^{-5}) = 8.35 \times 10^{-9} \text{ M}$$

$$pZn = 8.08$$

Calculating pZn after Adding 25.0 mL of EDTA

Twenty-five milliliters is the equivalence point, and the analytical concentration of ZnY^{2-} is

$$c_{ZnY^{2-}} = \frac{50.00 \times 0.00500}{50.0 + 25.0} = 3.33 \times 10^{-3} \text{ M}$$

(continued)

The sum of the concentrations of the various zinc species not combined with EDTA equals the sum of the concentrations of the uncomplexed EDTA species:

$$c_M = c_T$$

And

$$[ZnY^{2-}] = 3.33 \times 10^{-3} - c_M \approx 3.33 \times 10^{-3} \text{ M}$$

Substituting this value into Equation 17-31, we have

$$K''_{ZnY} = \frac{3.33 \times 10^{-3}}{(c_M)^2} = 1.9 \times 10^{10}$$

$$c_M = 4.19 \times 10^{-7} \text{ M}$$

With Equation 17-28, we find that

$$[Zn^{2+}] = c_M \alpha_M = (4.19 \times 10^{-7})(1.17 \times 10^{-5}) = 4.90 \times 10^{-12} \text{ M}$$

$$pZn = 11.31$$

Calculating pZn after Adding 30.0 mL of EDTA

Because the solution now contains excess EDTA,

$$c_{EDTA} = c_T = \frac{30.0 \times 0.0100 - 50.0 \times 0.00500}{80.0} = 6.25 \times 10^{-4} \text{ M}$$

and since essentially all of the original Zn^{2+} is now complexed,

$$c_{ZnY^{2-}} = [ZnY^{2-}] = \frac{50.00 \times 0.00500}{80.0} = 3.12 \times 10^{-3} \text{ M}$$

Rearranging Equation 17-31 gives

$$c_M = \frac{[ZnY^{2-}]}{c_T K''_{ZnY}} = \frac{3.12 \times 10^{-3}}{(6.25 \times 10^{-4})(1.9 \times 10^{10})} = 2.63 \times 10^{-10} \text{ M}$$

and, from Equation 17-28,

$$[Zn^{2+}] = c_M \alpha_M = (2.63 \times 10^{-10})(1.17 \times 10^{-5}) = 3.08 \times 10^{-15} \text{ M}$$

$$pZn = 14.51$$

17D-6 Indicators for EDTA Titrations

Nearly 200 organic compounds have been investigated as indicators for metal ions in EDTA titrations. The most common indicators are given by Dean.[4] In general, these indicators are organic dyes that form colored chelates with metal ions in a pM range that is characteristic of the particular cation and dye. The complexes are often intensely colored and can be detected visually at concentrations in the range of 10^{-6} to 10^{-7} M.

[4]J. A. Dean, *Analytical Chemistry Handbook*, New York: McGraw-Hill, 1995, p. 3.95.

Figure 17-14 Structure and molecular model of Eriochrome Black T. The compound contains a sulfonic acid group that completely dissociates in water and two phenolic groups that only partially dissociate.

Eriochrome Black T is a typical metal-ion indicator that is used in the titration of several common cations. The structural formula of Eriochrome Black T is shown in **Figure 17-14**. Its behavior as a weak acid is described by the equations

$$H_2O + \underset{\text{red}}{H_2In^-} \rightleftharpoons \underset{\text{blue}}{HIn^{2-}} + H_3O^+ \quad K_1 = 5 \times 10^{-7}$$

$$H_2O + \underset{\text{blue}}{HIn^{2-}} \rightleftharpoons \underset{\text{orange}}{In^{3-}} + H_3O^+ \quad K_2 = 2.8 \times 10^{-12}$$

Note that the acids and their conjugate bases have different colors. Thus, Eriochrome Black T behaves as an acid/base indicator as well as a metal-ion indicator.

The metal complexes of Eriochrome Black T are generally red, as is H_2In^-. Thus, for metal-ion detection, it is necessary to adjust the pH to 7 or above so that the blue form of the species, HIn^{2-}, predominates in the absence of a metal ion. Until the equivalence point in a titration, the indicator complexes the excess metal ion so that the solution is red. With the first slight excess of EDTA, the solution turns blue as a result of the reaction

$$\underset{\text{red}}{MIn^-} + HY^{3-} \rightleftharpoons \underset{\text{blue}}{HIn^{2-}} + MY^{2-}$$

Eriochrome Black T forms red complexes with more than two dozen metal ions, but the formation constants of only a few are appropriate for end-point detection. As shown in Example 17-6, the applicability of a given indicator for an EDTA titration can be determined from the change in pM in the equivalence-point region, provided the formation constant for the metal-indicator complex is known.[5]

EXAMPLE 17-6

Determine the transition ranges for Eriochrome Black T in titrations of Mg^{2+} and Ca^{2+} at pH 10.0, given (a) that the second acid dissociation constant for the indicator is

$$HIn^{2-} + H_2O \rightleftharpoons In^{3-} + H_3O^+ \qquad K_2 = \frac{[H_3O^+][In^{3-}]}{[HIn^{2-}]} = 2.8 \times 10^{-12}$$

(continued)

[5]C. N. Reilley and R. W. Schmid, *Anal. Chem.*, **1959**, *31*, 887, **DOI:** 10.1021/ac60137a022.

(b) that the formation constant for $MgIn^-$ is

$$Mg^{2+} + In^{3-} \rightleftharpoons MgIn^- \qquad K_f = \frac{[MgIn^-]}{[Mg^{2+}][In^{3-}]} = 1.0 \times 10^7$$

and (c) that the analogous formation constant for Ca^{2+} is 2.5×10^5.

Solution

We assume, as we did earlier (see Section 14A-1), that a detectable color change requires a tenfold excess of one or the other of the colored species, that is, a detectable color change is observed when the ratio $[MgIn^-]/[HIn^{2-}]$ changes from 10 to 0.10. The product of K_2 for the indicator and K_f for $MgIn^-$ contains this ratio:

$$\frac{[MgIn^-][H_3O^+]}{[HIn^{2-}][Mg^{2+}]} = 2.8 \times 10^{-12} \times 1.0 \times 10^7 = 2.8 \times 10^{-5}$$

Substituting 1.0×10^{-10} for $[H_3O^+]$ and 10 and 0.10 for the ratio yields, the range of $[Mg^{2+}]$ over which the color change occurs is

$$[Mg^{2+}] = 3.6 \times 10^{-5} \quad \text{to} \quad 3.6 \times 10^{-7} \text{ M}$$

$$pMg = 5.4 \pm 1.0$$

Proceeding in the same way, we find the range for pCa to be 3.8 ± 1.0.

Transition ranges for magnesium and calcium are indicated on the titration curves in Figure 17-9. The curves show that, Eriochrome Black T is ideal for the titration of magnesium, but it is unsatisfactory for calcium. Note that the formation constant for $CaIn^-$ is only about 1/40 that for $MgIn^-$. Because of the lower formation constant, significant conversion of $CaIn^-$ to HIn^{2-} occurs well before equivalence. A similar calculation shows that Eriochrome Black T is also well suited for the titration of zinc with EDTA (see Figure 17-13).

A limitation of Eriochrome Black T is that its solutions decompose slowly with standing. Solutions of Calmagite (see **Figure 17-15**), an indicator that for all practical purposes is identical in behavior to Eriochrome Black T, do not appear to suffer this disadvantage. Many other metal indicators have been developed for EDTA

Figure 17-15 Structural formula and molecular model of Calmagite. Note the similarity to Eriochrome Black T (see Figure 17-14).

titrations.[6] In contrast to Eriochrome Black T, some of these indicators can be used in strongly acidic media.

17D-7 Titration Methods Involving EDTA

Next, we describe several different types of titration methods that can be used with EDTA.

Direct Titration

Many of the metals in the periodic table can be determined by titration with standard EDTA solutions. Some methods are based on indicators that respond to the analyte itself, while others are based on an added metal ion.

Methods Based on Indicators for the Analyte. Dean[7] lists nearly 40 metal ions that can be determined by direct titration with EDTA using metal-ion indicators. Indicators that respond to the metal directly cannot be used in all cases either because an indicator with an appropriate transition range is not available or because the reaction between the metal ion and EDTA is so slow as to make titration impractical.

Methods Based on Indicators for an Added Metal Ion. In cases where a good, direct indicator for the analyte is unavailable, a small amount of a metal ion for which a good indicator is available can be added. The metal ion must form a complex that is less stable than the analyte complex. For example, indicators for calcium ion are generally less satisfactory than those we have described for magnesium ion. Consequently, a small amount of magnesium chloride is often added to an EDTA solution that is to be used for the determination of calcium. In this case, Eriochrome Black T can be used as indicator. In the initial stages of the titration, magnesium ions are displaced from the EDTA complex by calcium ions and are free to combine with the Eriochrome Black T, therefore imparting a red color to the solution. When all of the calcium ions have been complexed, however, the liberated magnesium ions again combine with the EDTA until the end point is observed. This procedure requires standardization of the EDTA solution against primary-standard calcium carbonate.

Potentiometric Methods. Potential measurements can be used for end-point detection in the EDTA titration of those metal ions for which specific ion electrodes are available. Electrodes of this type are described in Section 21D-1.

Spectrophotometric Methods. Measurement of UV/visible absorption can also be used to determine the end points of titrations (see Section 26A-4). In these cases, a spectrophotometer responds to the color change in the titration rather than relying on a visual determination of the end point.

Back-Titration Methods

Back-titrations are useful for the determination of cations that form stable EDTA complexes and for which a satisfactory indicator is not available. The method is also useful for cations such as Cr(III) and Co(III) that react slowly with EDTA. A measured excess of standard EDTA solution is added to the analyte solution. After the reaction is judged complete, the excess EDTA is back-titrated with a standard

◄ Direct titration procedures with a metal-ion indicator that responds to the analyte are the easiest and most convenient to use. Methods that incorporate an added metal ion are also used.

---

[6]See, for example, J. A. Dean, *Analytical Chemistry Handbook*, New York: McGraw-Hill, 1995, pp. **3.94**–**3.96**.

[7]J. A. Dean, ibid, pp. 3.104–3.109.

Back-titration procedures are
used when no suitable indicator
is available, when the reaction
between analyte and EDTA is
slow, or when the analyte forms
precipitates at the pH required
for its titration.

magnesium or zinc ion solution to an Eriochrome Black T or Calmagite end point.[8]
For this procedure to be successful, it is necessary that the magnesium or zinc ions
form an EDTA complex that is less stable than the corresponding analyte complex.

Back-titration is also useful for analyzing samples that contain anions that could
form precipitates with the analyte under the analytical conditions. The excess EDTA
complexes the analyte and prevents precipitate formation.

Displacement Methods

In displacement titrations, an unmeasured excess of a solution containing the magnesium or zinc complex of EDTA is introduced into the analyte solution. If the analyte
forms a more stable complex than that of magnesium or zinc, the following displacement reaction occurs:

$$MgY^{2-} + M^{2+} \rightarrow MY^{2-} + Mg^{2+}$$

where M^{2+} represents the analyte cation. The liberated Mg^{2+} or, in some cases Zn^{2+},
is then titrated with a standard EDTA solution.

17D-8 The Scope of EDTA Titrations

Complexometric titrations with EDTA have been applied to the determination
of virtually every metal cation with the exception of the alkali metal ions. Because EDTA complexes most cations, the reagent might appear at first glance
to be totally lacking in selectivity. In fact, however, considerable control over
interferences can be realized by pH regulation. For example, trivalent cations
can usually be titrated without interference from divalent species by maintaining
the solution at a pH of about 1 (see Figure 17-12). At this pH, the less stable
divalent chelates do not form to any significant extent, but trivalent ions are
quantitatively complexed.

Similarly, ions such as cadmium and zinc, which form more stable EDTA
chelates than does magnesium, can be determined in the presence of the magnesium by buffering the mixture to pH 7 before titration. Eriochrome Black T
serves as an indicator for the cadmium or zinc end points without interference
from magnesium because the indicator chelate with magnesium is not formed at
this pH.

Finally, interference from a particular cation can sometimes be eliminated by
adding a suitable **masking agent**, an auxiliary ligand that preferentially forms
highly stable complexes with the potential interfering ion.[9] Thus, cyanide ion is
often used as a masking agent to permit the titration of magnesium and calcium
ions in the presence of ions such as cadmium, cobalt, copper, nickel, zinc, and
palladium. All of these ions form sufficiently stable cyanide complexes to prevent
reaction with EDTA. Feature 17-6 illustrates how masking and demasking reagents are used to improve the selectivity of EDTA reactions.

A **masking agent** is a
complexing agent that reacts
selectively with a component
in a solution to prevent that
component from interfering in a
determination.

---

[8]For a discussion of the back-titration procedure, see C. Macca and M. Fiorana, *J. Chem. Educ.*,
1986, *63*, 121, **DOI**: 10.1021/ed063p121.

[9]For further information, see D. D. Perrin, *Masking and Demasking of Chemical Reactions*, New York:
Wiley-Interscience, 1970; J. A. Dean, *Analytical Chemistry Handbook*, New York: McGraw-Hill,
1995, pp. 3.92–3.111.

FEATURE 17-6

Enhancing the Selectivity of EDTA Titrations with Masking and Demasking Agents

Lead, magnesium, and zinc can be determined in a single sample by two titrations with standard EDTA and one titration with standard Mg^{2+}. The sample is first treated with an excess of NaCN, which masks Zn^{2+} and prevents it from reacting with EDTA:

$$Zn^{2+} + 4CN^- \rightleftharpoons Zn(CN)_4^{2-}$$

The Pb^{2+} and Mg^{2+} are then titrated with standard EDTA. After the equivalence point has been reached, a solution of the complexing agent BAL (2-3-dimercapto-1-propanol, $CH_2SHCHSHCH_2OH$), which we will write as $R(SH)_2$, is added to the solution. This bidentate ligand reacts selectively to form a complex with Pb^{2+} that is much more stable than PbY^{2-}:

$$PbY^{2-} + 2R(SH)_2 \rightarrow Pb(RS)_2 + 2H^+ + Y^{4-}$$

The liberated Y^{4-} is then titrated with a standard solution of Mg^{2+}. Finally, the zinc is demasked by adding formaldehyde:

$$Zn(CN)_4^{2-} + 4HCHO + 4H_2O \rightarrow Zn^{2+} + 4HOCH_2CN + 4OH^-$$

The liberated Zn^{2+} is then titrated with the standard EDTA solution.

Molecular model of BAL (2,3-dimercapto-1-propanol, $CH_2SHCHSHCH_2OH$).

Suppose the initial titration of Mg^{2+} and Pb^{2+} required 42.22 mL of 0.02064 M EDTA. Titration of the Y^{4-} liberated by the BAL consumed 19.35 mL of 0.007657 M Mg^{2+}. After addition of formaldehyde, the liberated Zn^{2+} was titrated with 28.63 mL of the EDTA solution. Calculate the percent of the three elements if a 0.4085-g sample was used.

$$\text{amount } (Pb^{2+} + Mg^{2+}) \text{ in mmol} = 42.22 \times 0.02064 = 0.87142$$

The second titration gives the amount of Pb^{2+}. Thus,

$$\text{amount } Pb^{2+} \text{ in mmol} = 19.35 \times 0.007657 = 0.14816$$

$$\text{amount } Mg^{2+} \text{ in mmol} = 0.87142 - 0.14816 = 0.72326$$

Finally, from the third titration, we obtain

$$\text{amount } Zn^{2+} \text{ in mmol} = 28.63 \times 0.02064 = 0.59092$$

(continued)

To obtain the percentages, we write

$$\frac{0.14816 \text{ mmol Pb} \times 0.2072 \text{ g Pb/mmol Pb}}{0.4085 \text{ g sample}} \times 100\% = 7.515\% \text{ Pb}$$

$$\frac{0.72326 \text{ mmol Mg} \times 0.024305 \text{ g Mg/mmol Mg}}{0.4085 \text{ g sample}} \times 100\% = 4.303\% \text{ Mg}$$

$$\frac{0.59095 \text{ mmol Zn} \times 0.06538 \text{ g Zn/mmol Zn}}{0.4085 \text{ g sample}} \times 100\% = 9.459\% \text{ Zn}$$

17D-9 Determination of Water Hardness

> Hard water contains calcium, magnesium, and heavy metal ions that form precipitates with soap (but not detergents).

Historically, water "hardness" was defined in terms of the capacity of cations in the water to replace the sodium or potassium ions in soaps and form sparingly soluble products that cause "scum" in the sink or bathtub. Most multiply charged cations share this undesirable property. In natural waters, however, the concentrations of calcium and magnesium ions generally far exceed those of any other metal ion. Consequently, hardness is now expressed in terms of the concentration of calcium carbonate that is equivalent to the total concentration of all the multivalent cations in the sample.

The determination of hardness is a useful analytical test that provides a measure of the quality of water for household and industrial uses. The test is important to industry because hard water, on being heated, precipitates calcium carbonate, which clogs boilers and pipes.

Water hardness is usually determined by an EDTA titration after the sample has been buffered to pH 10. Magnesium, which forms the least stable EDTA complex of all of the common multivalent cations in typical water samples, is not titrated until enough reagent has been added to complex all of the other cations in the sample. Therefore, a magnesium-ion indicator, such as Calmagite or Eriochrome Black T, can serve as indicator in water-hardness titrations. Often, a small concentration of the magnesium-EDTA chelate is incorporated in the buffer or in the titrant to ensure the presence of sufficient magnesium ions for satisfactory indicator action. Feature 17-7 gives an example of a kit for testing household water for hardness.

FEATURE 17-7

Test Kits for Water Hardness

Test kits for determining the hardness of household water are available at stores selling water softeners and plumbing supplies. They usually consist of a vessel calibrated to contain a known volume of water, a packet containing an appropriate amount of a solid buffer mixture, an indicator solution, and a bottle of standard EDTA, which is equipped with a medicine dropper. A typical kit is shown in **Figure 17F-2**. The number of drops of standard reagent needed to cause a color change is counted. The EDTA solution is usually prepared with a concentration such that one drop corresponds to one grain (about 0.065 g) of calcium carbonate per gallon of water. Home water softeners that use ion-exchange processes to remove hardness are discussed in Feature 31-2.

Figure 17F-2 Typical kit for testing household water for hardness.

WEB WORKS

The disodium salt of EDTA ($Na_2H_2Y \cdot 2H_2O$) is widely used to prepare standard EDTA solutions. The free acid is also used, but it is not very soluble in water. Use a search engine to locate the Materials Safety Data Sheets for these reagents. What are the solubilities of the two reagents in water in g/100mL? What, if any, are the health effects of these chemicals? What is the J. T. Baker Safe-T-Data™ Rating for the disodium salt. What precautions are recommended when working with these reagents in the laboratory? How should the reagents or solutions containing them be disposed?

QUESTIONS AND PROBLEMS

17-1. Define
  *(a) ligand.
  (b) chelate.
  *(c) tetradentate chelating agent.
  (d) adsorption indicator.
  *(e) argentometric titration.
  (f) conditional formation constant.
  *(g) EDTA displacement titration.
  (h) water hardness.

17-2. Why are multidentate ligands preferable to unidentate ligands for complexometric titrations?

*17-3. Describe three general methods for performing EDTA titrations. What are the advantages of each?

17-4. Write chemical equations and equilibrium-constant expressions for the stepwise formation of
  *(a) $Ag(S_2O_3)_2^{3-}$.
  (b) $Ni(CN)_4^{2-}$.
  (c) $Cd(SCN)_3^{-}$.

*17-5. Explain how stepwise and overall formation constants are related.

17-6. Write chemical formulas for the following complex ions:
  (a) hexamminezinc(II)
  (b) dichloroargentate
  (c) disulfatocuprate(II)
  (d) trioxalotoferrate(III)
  (e) hexacyanoferrate(II)

*17-7. In what respect is the Fajans method superior to the Volhard method for the titration of chloride ion?

17-8. Briefly explain why the sparingly soluble product must be removed by filtration before you back-titrate the excess silver ion in the Volhard determination of
  (a) chloride ion.
  (b) cyanide ion.
  (c) carbonate ion.

*17-9. Why does the charge on the surface of precipitate particles change sign at the equivalence point of a titration?

17-10. Outline a method for the determination of K^+ based on argentometry. Write balanced equations for the chemical reactions.

*17-11. Write equations in terms of the acid dissociation constants and $[H^+]$ for the highest alpha value for each of the following weak acid ligands:
 (a) acetate (α_1).
 (b) tartrate (α_2).
 (c) phosphate (α_3).

17-12. Write conditional formation constants for 1:1 complexes of Fe(III) with each of the ligands in Problem 17-11. Express these constants in terms of the α value and the formation constant and in terms of concentrations as in Equation 17-20.

*17-13. Write a conditional overall formation constant for $[Fe(ox)_3]^{3-}$ in terms of α_2 for oxalic acid and the β value for the complex. Also express the conditional constant in terms of concentrations as in Equation 17-20.

17-14. Propose a complexometric method for the determination of the individual components in a solution containing In^{3+}, Zn^{2+}, and Mg^{2+}.

*17-15. Given an overall complex formation reaction of $M + nL \rightleftharpoons ML_n$, with an overall formation constant of β_n, show that the following relationship holds:

$$\log \beta_n = pM + npL - pML_n$$

17-16. Why is a small amount of MgY^{2-} often added to a water specimen that is to be titrated for hardness?

*17-17. An EDTA solution was prepared by dissolving 3.426 g of purified and dried $Na_2H_2Y_2 \cdot 2H_2O$ in sufficient water to give 1.000 L. Calculate the molar concentration, given that the solute contained 0.3% excess moisture (see Section 17D-1).

17-18. A solution was prepared by dissolving about 3.0 g of $NaH_2Y_2 \cdot 2H_2O$ in approximately 1 L of water and standardizing against 50.00-mL aliquots of 0.004423 M Mg^{2+}. An average titration of 30.27 mL was required. Calculate the molar concentration of the EDTA.

*17-19. A solution contains 1.569 mg of $CoSO_4$ (155.0 g/mol) per milliliter. Calculate
 (a) the volume of 0.007840 M EDTA needed to titrate a 25.00-mL aliquot of this solution.
 (b) the volume of 0.009275 M Zn^{2+} needed to titrate the excess reagent after addition of 50.00 mL of 0.007840 M EDTA to a 25.00-mL aliquot of this solution.
 (c) the volume of 0.007840 M EDTA needed to titrate the Zn^{2+} displaced by Co^{2+} following addition of an unmeasured excess of ZnY^{2-} to a 25.00-mL aliquot of the $CoSO_4$ solution. The reaction is

$$Co^{2+} + ZnY^{2-} \rightarrow CoY^{2-} + Zn^{2+}$$

17-20. Calculate the volume of 0.0500 M EDTA needed to titrate
 *(a) 29.13 mL of 0.0598 M $Mg(NO_3)_2$.
 (b) the Ca in 0.1598 g of $CaCO_3$.

*(c) the Ca in a 0.4861-g mineral specimen that is 81.4% brushite, $CaHPO_4 \cdot 2H_2O$ (172.09 g/mol).
 (d) the Mg in a 0.1795-g sample of the mineral hydromagnesite, $3MgCO_3Mg(OH)_2 \cdot 3H_2O$ (365.3 g/mol).
 *(e) the Ca and Mg in a 0.1612-g sample that is 92.5% dolomite, $CaCO_3 \cdot MgCO_3$ (184.4 g/mol).

*17-21. The Zn in a 0.7457-g sample of foot powder was titrated with 22.57 mL of 0.01639 M EDTA. Calculate the percent Zn in this sample.

17-22. The Cr plating on a surface that measured 3.00×4.00 cm was dissolved in HCl. The pH was suitably adjusted, following which 15.00 mL of 0.01768 M EDTA were introduced. The excess reagent required a 4.30-mL back-titration with 0.008120 M Cu^{2+}. Calculate the average weight of Cr on each square centimeter of surface.

17-23. A silver nitrate solution contains 14.77 g of primary-standard $AgNO_3$ in 1.00 L. What volume of this solution will be needed to react with
 *(a) 0.2631 g of NaCl?
 (b) 0.1799 g of Na_2CrO_4?
 *(c) 64.13 mg of Na_3AsO_4?
 (d) 381.1 mg of $BaCl_2 \cdot 2H_2O$?
 *(e) 25.00 mL of 0.05361 M Na_3PO_4?
 (f) 50.00 mL of 0.01808 M H_2S?

17-24. What is the molar analytical concentration of a silver nitrate solution if a 25.00-mL aliquot reacts with each amount of solute listed in Problem 17-23?

17-25. What minimum volume of 0.09621 M $AgNO_3$ will be needed to assure an excess of silver ion in the titration of
 *(a) an impure NaCl sample that weighs 0.2513 g?
 (b) a 0.3462-g sample that is 74.52% (w/w) $ZnCl_2$?
 *(c) 25.00 mL of 0.01907 M $AlCl_3$?

17-26. A Fajans titration of a 0.7908-g sample required 45.32 mL of 0.1046 M $AgNO_3$. Express the results of this analysis in terms of the percentage of
 (a) Cl^-.
 (b) $BaCl_2 \cdot H_2O$.
 (c) $ZnCl_2 \cdot 2NH_4Cl$ (243.28 g/mol).

*17-27. The Tl in a 9.57-g sample of rodenticide was oxidized to the trivalent state and treated with an unmeasured excess of Mg/EDTA solution. The reaction is

$$Tl^{3+} + MgY^{2-} \rightarrow TlY^- + Mg^{2+}$$

Titration of the liberated Mg^{2+} required 12.77 mL of 0.03610 M EDTA. Calculate the percent Tl_2SO_4 (504.8 g/mol) in the sample.

17-28. An EDTA solution was prepared by dissolving approximately 4 g of the disodium salt in approximately 1 L of water. An average of 42.35 mL of this solution was required to titrate 50.00-mL aliquots of a standard that contained 0.7682 g of $MgCO_3$ per liter. Titration of a 25.00-mL sample of mineral water at pH 10 required 18.81 mL of the EDTA solution. A 50.00-mL aliquot of the mineral water was rendered strongly alkaline to precipitate the magnesium at $Mg(OH)_2$. Titration with a calcium-specific indicator required 31.54 mL of the EDTA solution. Calculate

(a) the molar concentration of the EDTA solution.

(b) the concentration of $CaCO_3$ in the mineral water in ppm.

(c) the concentration of $MgCO_3$ in the mineral water in ppm.

*17-29. A 50.00-mL aliquot of a solution containing iron(II) and iron(III) required 10.98 mL of 0.01500 M EDTA when titrated at pH 2.0 and 23.70 mL when titrated at pH 6.0. Express the concentration of each solute in parts per million.

17-30. A 24-hr urine specimen was diluted to 2.000 L. After the solution was buffered to pH 10, a 10.00-mL aliquot was titrated with 23.57 mL of 0.004590 M EDTA. The calcium in a second 10.00-mL aliquot was isolated as $CaC_2O_4(s)$, redissolved in acid, and titrated with 10.53 mL of the EDTA solution. Assuming that 15 to 300 mg of magnesium and 50 to 400 mg of calcium per day are normal, did this specimen fall within these ranges?

*17-31. A 1.509-g sample of a Pb/Cd alloy was dissolved in acid and diluted to exactly 250.0 mL in a volumetric flask. A 50.00-mL aliquot of the diluted solution was brought to a pH of 10.0 with a NH_4^+/NH_3 buffer; the subsequent titration involved both cations and required 28.89 mL of 0.06950 M EDTA. A second 50.00-mL aliquot was brought to a pH of 10.0 with an HCN/NaCN buffer, which also served to mask the Cd^{2+}; 11.56 mL of the EDTA solution were needed to titrate the Pb^{2+}. Calculate the percent Pb and Cd in the sample.

17-32. A 0.6004-g sample of Ni/Cu condenser tubing was dissolved in acid and diluted to 100.0 mL in a volumetric flask. Titration of both cations in a 25.00-mL aliquot of this solution required 45.81 mL of 0.05285 M EDTA. Mercaptoacetic acid and NH_3 were then introduced; production of the Cu complex with the former resulted in the release of an equivalent amount of EDTA, which required a 22.85-mL titration with 0.07238 M Mg^{2+}. Calculate the percent Cu and Ni in the alloy.

*17-33. Calamine, which is used for relief of skin irritations, is a mixture of zinc and iron oxides. A 1.056-g sample of dried calamine was dissolved in acid and diluted to 250.0 mL. Potassium fluoride was added to a 10.00-mL aliquot of the diluted solution to mask the iron; after suitable adjustment of the pH, Zn^{2+} consumed 38.37 mL of 0.01133 M EDTA. A second 50.00-mL

aliquot was suitably buffered and titrated with 2.30 mL of 0.002647 M ZnY^{2-} solution:

$$Fe^{3+} + ZnY^{2-} \rightarrow FeY^- + Zn^{2+}$$

Calculate the percentages of ZnO and Fe_2O_3 in the sample.

*17-34. A 3.650-g sample containing bromate and bromide was dissolved in sufficient water to give 250.0 mL. After acidification, silver nitrate was introduced to a 25.00-mL aliquot to precipitate AgBr, which was filtered, washed, and then redissolved in an ammoniacal solution of potassium tetracyanonickelate(II):

$$Ni(CN)_4^{2-} + 2AgBr(s) \rightarrow 2Ag(CN)_2^- + Ni^{2+} + 2Br^-$$

The liberated nickel ion required 26.73 mL of 0.02089 M EDTA. The bromate in a 10.00-mL aliquot was reduced to bromide with arsenic(III) prior to the addition of silver nitrate. The same procedure was followed, and the released nickel ion was titrated with 21.94 mL of the EDTA solution. Calculate the percentages of NaBr and $NaBrO_3$ in the sample.

17-35. The potassium ion in a 250.0-mL sample of mineral water was precipitated with sodium tetraphenylborate:

$$K^+ + B(C_6H_5)_4^- \rightarrow KB(C_6H_5)(s)$$

The precipitate was filtered, washed, and redissolved in an organic solvent. An excess of the mercury(II)/EDTA chelate was added:

$$4HgY^{2-} + B(C_6H_4)_4^- + 4H_2O \rightarrow$$
$$H_3BO_3 + 4C_6H_5Hg^+ + 4HY^{3-} + OH^-$$

The liberated EDTA was titrated with 29.64 mL of 0.05581 M Mg^{2+}. Calculate the potassium ion concentration in parts per million.

*17-36. Chromel is an alloy composed of nickel, iron, and chromium. A 0.6553-g sample was dissolved and diluted to 250.0 mL. When a 50.00-mL aliquot of 0.05173 M EDTA was mixed with an equal volume of the diluted sample, all three ions were chelated, and a 5.34-mL back-titration with 0.06139 M copper(II) was required. The chromium in a second 50.0-mL aliquot was masked through the addition of hexamethylenetetramine; titration of the Fe and Ni required 36.98 mL of 0.05173 M EDTA. Iron and chromium were masked with pyrophosphate in a third 50.0-mL aliquot, and the nickel was titrated with 24.53 mL of the EDTA solution. Calculate the percentages of nickel, chromium, and iron in the alloy.

17-37. A 0.3304-g sample of brass (containing lead, zinc, copper, and tin) was dissolved in nitric acid. The sparingly soluble $SnO_2 \cdot 4H_2O$ was removed by filtration, and the combined filtrate and washings were then diluted to 500.0 mL. A 10.00-mL aliquot was suitably buffered; titration of the lead, zinc, and copper in this aliquot required 34.78 mL of 0.002700 M EDTA. The copper in a 25.00-mL aliquot was masked with thiosulfate; the lead and zinc were then titrated with 25.62 mL of the EDTA solution. Cyanide ion was used to mask the copper and zinc in a 100-mL aliquot; 10.00 mL of the EDTA solution was needed to titrate the lead ion. Determine the composition of the brass sample; evaluate the percentage of tin by difference.

*17-38. Calculate conditional constants for the formation of the EDTA complex of Fe^{2+} at a pH of (a) 6.0, (b) 8.0, and (c) 10.0.

17-39. Calculate conditional constants for the formation of the EDTA complex of Ba^{2+} at a pH of (a) 5.0, (b) 7.0, (c) 9.0, and (d) 11.0.

17-40. Construct a titration curve for 50.00 mL of 0.01000 M Sr^{2+} with 0.02000 M EDTA in a solution buffered to pH 11.0. Calculate pSr values after the addition of 0.00, 10.00, 24.00, 24.90, 25.00, 25.10, 26.00, and 30.00 mL of titrant.

17-41. Construct a titration curve for 50.00 mL of 0.0150 M Fe^{2+} with 0.0300 M EDTA in a solution buffered to pH 7.0. Calculate pFe values after the addition of 0.00, 10.00, 24.00, 24.90, 25.00, 25.10, 26.00, and 30.00 mL of titrant.

*17-42. Titration of Ca^{2+} and Mg^{2+} in a 50.00-mL sample of hard water required 23.65 mL of 0.01205 M EDTA. A second 50.00-mL aliquot was made strongly basic with NaOH to precipitate Mg^{2+} as $Mg(OH)_2(s)$. The supernatant liquid was titrated with 14.53 mL of the EDTA solution. Calculate

(a) the total hardness of the water sample, expressed as ppm $CaCO_3$.

(b) the concentration of $CaCO_3$ in the sample in ppm.

(c) the concentration of $MgCO_3$ in the sample in ppm.

17-43. **Challenge Problem:** Zinc sulfide, ZnS, is sparingly soluble in most situations. With ammonia, Zn^{2+} forms four complexes, $Zn(NH_3)^{2+}$, $Zn(NH_3)_2^{2+}$, $Zn(NH_3)_3^{2+}$, and $Zn(NH_3)_4^{2+}$. Ammonia is, of course, a base, and S^{2-} is the anion of the weak diprotic acid, H_2S. Find the molar solubility of zinc sulfide in

(a) pH-7.0 water.

(b) a solution containing 0.100 M NH_3.

(c) a pH-9.00 ammonia/ammonium ion buffer with a total NH_3/NH_4^+ concentration of 0.100 M.

(d) the same solution as in part (c) except that it also contains 0.100 M EDTA.

(e) Use a search engine and locate a Materials Safety Data Sheet (MSDS) for ZnS. Determine what health hazards ZnS poses.

(f) Determine if there is a phosphorescent pigment containing ZnS. What activates the pigment to "glow in the dark"?

(g) Determine what uses ZnS has in making optical components. Why is ZnS useful for these components?

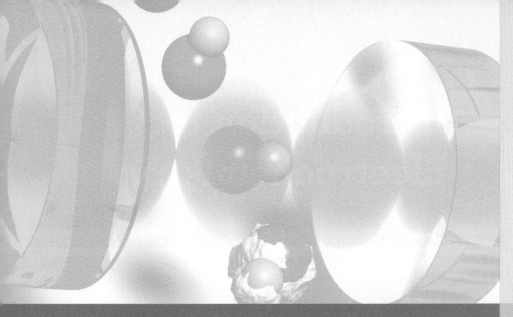

Electrochemical Methods

PART IV

CHAPTER 18

Introduction to Electrochemistry

From the earliest days of experimental science, workers such as Galvani, Volta, and Cavendish realized that electricity interacts in interesting and important ways with animal tissues. Electrical charge causes muscles to contract, for example. Perhaps more surprising is that a few animals such as the torpedo (shown in the photo) produce charge by physiological means. More than 50 billion nerve terminals in the torpedo's flat "wings" on its left and right sides rapidly emit acetylcholine on the bottom side of membranes housed in the wings. The acetylcholine causes sodium ions to surge through the membranes, producing a rapid separation of charge and a corresponding potential difference, or voltage, across the membrane.[1] The potential difference then generates an electric current of several amperes in the surrounding seawater that may be used to stun or kill prey, detect and ward off enemies, or navigate. Natural devices for separating charge and creating electrical potential difference are relatively rare, but humans have learned to separate charge mechanically, metallurgically, and chemically to create cells, batteries, and other useful charge storage devices.

© Norbert Wu/Minden Pictures/Corbis

We now turn our attention to several analytical methods that are based on oxidation/reduction reactions. These methods, which are described in Chapters 18 through 23, include oxidation/reduction titrimetry, potentiometry, coulometry, electrogravimetry, and voltammetry. In this chapter, we present the fundamentals of electrochemistry that are necessary for understanding the principles of these procedures.

18A CHARACTERIZING OXIDATION/REDUCTION REACTIONS

Oxidation/reduction reactions are sometimes called **redox** reactions.

In an **oxidation/reduction reaction** electrons are transferred from one reactant to another. An example is the oxidation of iron(II) ions by cerium(IV) ions. The reaction is described by the equation

$$Ce^{4+} + Fe^{2+} \rightleftharpoons Ce^{3+} + Fe^{3+} \tag{18-1}$$

A **reducing agent** is an electron donor. An **oxidizing agent** is an electron acceptor.

In this reaction, an electron is transferred from Fe^{2+} to Ce^{4+} to form Ce^{3+} and Fe^{3+} ions. A substance that has a strong affinity for electrons, such as Ce^{4+}, is called an **oxidizing agent**, or an **oxidant**. A **reducing agent**, or **reductant**, is a species, such

[1]Y. Dunant and M. Israel, *Sci. Am.* **1985**, *252*, 58, **DOI**: 10.1038/scientificamerican0485-58.

as Fe^{2+}, that donates electrons to another species. To describe the chemical behavior represented by Equation 18-1, we say that Fe^{2+} is oxidized by Ce^{4+}; similarly, Ce^{4+} is reduced by Fe^{2+}.

We can split any oxidation/reduction equation into two half-reactions that show which species gains electrons and which loses them. For example, Equation 18-1 is the sum of the two half-reactions

$$Ce^{4+} + e^- \rightleftharpoons Ce^{3+} \qquad \text{(reduction of } Ce^{4+}\text{)}$$

$$Fe^{2+} \rightleftharpoons Fe^{3+} + e^- \qquad \text{(oxidation of } Fe^{2+}\text{)}$$

The rules for balancing half-reactions (see Feature 18-1) are the same as those for other reaction types, that is, the number of atoms of each element as well as the net charge on each side of the equation must be the same. Thus, for the oxidation of Fe^{2+} by MnO_4^-, the half-reactions are

$$MnO_4^- + 5e^- + 8H^+ \rightleftharpoons Mn^{2+} + 4H_2O$$

$$5Fe^{2+} \rightleftharpoons 5Fe^{3+} + 5e^-$$

In the first half-reaction, the net charge on the left side is $(-1 -5 + 8) = +2$, which is the same as the charge on the right. Note also that we have multiplied the second half-reaction by 5 so that the number of electrons lost by Fe^{2+} equals the number gained by MnO_4^-. We can then write a balanced net ionic equation for the overall reaction by adding the two half-reactions

$$MnO_4^- + 5Fe^{2+} + 8H^+ \rightleftharpoons Mn^{2+} + 5Fe^{3+} + 4H_2O$$

> It is important to understand that while we can write an equation for a half-reaction in which electrons are consumed or generated, we cannot observe an isolated half-reaction experimentally because there must always be a second half-reaction that serves as a source of electrons or a recipient of electrons. In other words, an individual half-reaction is a theoretical concept.

18A-1 Comparing Redox Reactions to Acid/Base Reactions

Oxidation/reduction reactions can be viewed in a way that is analogous to the Brønsted-Lowry concept of acid/base reactions (see Section 9A-2). In both, one or more charged particles are transferred from a donor to an acceptor—the particles

> Recall that in the Brønsted/ Lowry concept an acid/base reaction is described by the equation
>
> $$acid_1 + base_2 \rightleftharpoons base_1 + acid_2$$

B.C.   **by johnny hart**

Copyright 1993 by permission of Johnny Hart and Creator's Syndicate, Inc.

FEATURE 18-1

Balancing Redox Equations

Knowing how to balance oxidation/reduction reactions is essential to understanding all the concepts covered in this chapter. Although you probably remember this technique from your general chemistry course, we present a quick review to remind you of how the process works. For practice, we will complete and balance the following equation after adding H^+, OH^-, or H_2O as needed.

$$MnO_4^- + NO_2^- \rightleftharpoons Mn^{2+} + NO_3^-$$

First, we write and balance the two half-reactions. For MnO_4^-, we write

$$MnO_4^- \rightleftharpoons Mn^{2+}$$

To account for the 4 oxygen atoms on the left-hand side of the equation, we add $4H_2O$ on the right-hand side. Then, to balance the hydrogen atoms, we must provide $8H^+$ on the left:

$$MnO_4^- + 8H^+ \rightleftharpoons Mn^{2+} + 4H_2O$$

To balance the charge, we need to add 5 electrons to the left side of the equation. Thus,

$$MnO_4^- + 8H^+ + 5e^- \rightleftharpoons Mn^{2+} + 4H_2O$$

For the other half-reaction,

$$NO_2^- \rightleftharpoons NO_3^-$$

we add one H_2O to the left side of the equation to supply the needed oxygen and $2H^+$ on the right to balance hydrogen:

$$NO_2^- + H_2O \rightleftharpoons NO_3^- + 2H^+$$

Then, we add two electrons to the right-hand side to balance the charge:

$$NO_2^- + H_2O \rightleftharpoons NO_3^- + 2H^+ + 2e^-$$

Before combining the two equations, we must multiply the first by 2 and the second by 5 so that the number of electrons lost will be equal to the number of electrons gained. We then add the two half reactions to obtain

$$2MnO_4^- + 16H^+ + 10e^- + 5NO_2^- + 5H_2O \rightleftharpoons$$
$$2Mn^{2+} + 8H_2O + 5NO_3^- + 10H^+ + 10e^-$$

This equation rearranges to the balanced equation

$$2MnO_4^- + 6H^+ + 5NO_2^- \rightleftharpoons 2Mn^{2+} + 5NO_3^- + 3H_2O$$

being electrons in oxidation/reduction and protons in neutralization. When an acid donates a proton, it becomes a conjugate base that is capable of accepting a proton. By analogy, when a reducing agent donates an electron, it becomes an oxidizing agent that can then accept an electron. This product could be called a conjugate oxidant, but that terminology is seldom, if ever, used. With this idea in mind, we can write a generalized equation for a redox reaction as

$$A_{red} + B_{ox} \rightleftharpoons A_{ox} + B_{red} \tag{18-2}$$

In this equation, B_{ox}, the oxidized form of species B, accepts electrons from A_{red} to form the new reductant, B_{red}. At the same time, reductant A_{red}, having given up electrons, becomes an oxidizing agent, A_{ox}. If we know from chemical evidence that the equilibrium in Equation 18-2 lies to the right, we can state that B_{ox} is a better electron acceptor (stronger oxidant) than A_{ox}. Likewise, A_{red} is a more effective electron donor (better reductant) than B_{red}.

EXAMPLE 18-1

The following reactions are spontaneous and thus proceed to the right, as written:

$$2H^+ + Cd(s) \rightleftharpoons H_2 + Cd^{2+}$$

$$2Ag^+ + H_2(g) \rightleftharpoons 2Ag(s) + 2H^+$$

$$Cd^{2+} + Zn(s) \rightleftharpoons Cd(s) + Zn^{2+}$$

What can we deduce regarding the strengths of H^+, Ag^+, Cd^{2+}, and Zn^{2+} as electron acceptors (or oxidizing agents)?

Solution

The second reaction establishes that Ag^+ is a more effective electron acceptor than H^+; the first reaction demonstrates that H^+ is more effective than Cd^{2+}. Finally, the third equation shows that Cd^{2+} is more effective than Zn^{2+}. Thus, the order of oxidizing strength is $Ag^+ > H^+ > Cd^{2+} > Zn^{2+}$.

18A-2 Oxidation/Reduction Reactions in Electrochemical Cells

Many oxidation/reduction reactions can be carried out in either of two ways that are physically quite different. In one, the reaction is performed by bringing the oxidant and the reductant into direct contact in a suitable container. In the second, the reaction is carried out in an electrochemical cell in which the reactants do not come in direct contact with one another. A spectacular example of direct contact is the famous "silver tree" experiment in which a piece of copper is immersed in a silver nitrate solution (see **Figure 18-1**). Silver ions migrate to the metal and are reduced:

$$Ag^+ + e^- \rightleftharpoons Ag(s)$$

At the same time, an equivalent quantity of copper is oxidized:

$$Cu(s) \rightleftharpoons Cu^{2+} + 2e^-$$

Charles D. Winters

Figure 18-1 Photograph of a "silver tree" created by immersing a coil of copper wire in a solution of silver nitrate.

For an interesting illustration of this reaction, immerse a piece of copper in a solution of silver nitrate. The result is the deposition of silver on the copper in the form of a "silver tree." See Figure 18-1 and color plate 10.

By multiplying the silver half-reaction by two and adding the reactions, we obtain a net ionic equation for the overall process:

$$2Ag^+ + Cu(s) \rightleftharpoons 2Ag(s) + Cu^{2+} \tag{18-3}$$

A unique aspect of oxidation/reduction reactions is that the transfer of electrons—and thus an identical net reaction—can often be brought about in an **electrochemical cell** in which the oxidizing agent and the reducing agent are physically separated from one another. **Figure 18-2a** shows such an arrangement. Note that a **salt bridge** isolates the reactants but maintains electrical contact between the two halves of the cell. When a voltmeter of high internal resistance is connected as shown or the electrodes are not connected externally, the cell is said to be at **open circuit** and delivers the full cell potential. When the circuit is open, no net reaction occurs in the cell, although we shall show that the cell has the **potential** for doing work. The voltmeter measures the potential difference, or **voltage**, between the two electrodes at any instant. This voltage is a measure of the tendency of the cell reaction to proceed toward equilibrium.

In **Figure 18-2b**, the cell is connected so that electrons can pass through a low-resistance external circuit. The potential energy of the cell is now converted to electrical energy to light a lamp, run a motor, or do some other type of electrical work. In the cell in Figure 18-2b, metallic copper is oxidized at the left-hand electrode, silver ions are reduced at the right-hand electrode, and electrons flow through the external circuit to the silver electrode. As the reaction goes on, the cell potential, initially 0.412 V when the circuit is open, decreases continuously and approaches zero as the overall reaction approaches equilibrium. When the cell is at equilibrium, the forward reaction (left-to-right) occurs at the same rate as the reverse reaction (right-to-left), and the cell voltage is zero. A cell with zero voltage does not perform work, as anyone who has found a "dead" battery in a flashlight or in a laptop computer can attest.

When zero voltage is reached in the cell of Figure 18-2b, the concentrations of Cu(II) and Ag(I) ions will have values that satisfy the equilibrium-constant expression shown in Equation 18-4. At this point, no further net flow of electrons will occur. *It is important to recognize that the overall reaction and its position of equilibrium are totally independent of the way the reaction is carried out*, whether it is by direct reaction in a solution or by indirect reaction in an electrochemical cell.

18B ELECTROCHEMICAL CELLS

We can study oxidation/reduction equilibria conveniently by measuring the potentials of electrochemical cells in which the two half-reactions making up the equilibrium are participants. For this reason, we must consider some characteristics of electrochemical cells.

An electrochemical cell consists of two conductors called **electrodes**, each of which is immersed in an electrolyte solution. In most of the cells that will be of interest to us, the solutions surrounding the two electrodes are different and must be separated to avoid direct reaction between the reactants. The most common way of avoiding mixing is to insert a salt bridge, such as that shown in Figure 18-2, between the solutions. Conduction of electricity from one electrolyte solution to the other then occurs by migration of potassium ions in the bridge in one direction and chloride ions in the other. However, direct contact between copper metal and silver ions is prevented.

Salt bridges are widely used in electrochemistry to prevent mixing of the contents of the two electrolyte solutions making up electrochemical cells. Normally, the two ends of the bridge are fitted with sintered glass disks or other porous materials to prevent liquid from siphoning from one part of the cell to the other.

When the $CuSO_4$ and $AgNO_3$ solutions are 0.0200 M, the cell has a potential of 0.412 V, as shown in Figure 18-2a.

The equilibrium-constant expression for the reaction shown in Equation 18-3 is

$$K_{eq} = \frac{[Cu^{2+}]}{[Ag^+]} = 4.1 \times 10^{15} \tag{18-4}$$

This expression applies whether the reaction occurs directly between reactants or within an electrochemical cell.

At equilibrium, the two half reactions in a cell continue, but their rates are equal.

The electrodes in some cells share a common electrolyte; these are known as **cells without liquid junction**. For an example of such a cell, see Figure 19-2 and Example 19-7.

(a) $[Cu^{2+}] = 0.0200$ M $[Ag^+] = 0.0200$ M

(b) $[Cu^{2+}] = 0.0200$ M $[Ag^+] = 0.0200$ M
$Cu(s) \rightleftharpoons Cu^{2+}(aq) + 2e^-$ $Ag(aq) + e^- \rightleftharpoons Ag(s)$
Anode   Cathode

(c) $[Cu^{2+}] = 0.0200$ M $[Ag^+] = 0.0200$ M
$Cu^{2+}(aq) + 2e^- \rightleftharpoons Cu(s)$ $Ag(s) \rightleftharpoons Ag^+(aq) + e^-$
Cathode   Anode

Figure 18-2   (a) A galvanic cell at open circuit. (b) A galvanic cell doing work. (c) An electrolytic cell.

18B-1 Cathodes and Anodes

A **cathode** is an electrode where reduction occurs. An **anode** is an electrode where oxidation occurs.

The **cathode** in an electrochemical cell is the electrode at which reduction occurs. The **anode** is the electrode at which an oxidation takes place.

Examples of typical cathodic reactions include

$$Ag^+ + e^- \rightleftharpoons Ag(s)$$

$$Fe^{3+} + e^- \rightleftharpoons Fe^{2+}$$

$$NO_3^- + 10H^+ + 8e^- \rightleftharpoons NH_4^+ + 3H_2O$$

The reaction $2H^+ + 2e^- \rightleftharpoons H_2(g)$ occurs at a cathode when an aqueous solution contains no other species that are more easily reduced than H^+.

We can force a desired reaction to occur by applying a suitable potential to an electrode made of an unreactive material such as platinum. Note that the reduction of NO_3^- in the third reaction reveals that anions can migrate to a cathode and be reduced.

Typical anodic reactions include

$$Cu(s) \rightleftharpoons Cu^{2+} + 2e^-$$

$$2Cl^- \rightleftharpoons Cl_2(g) + 2e^-$$

$$Fe^{2+} \rightleftharpoons Fe^{3+} + e^-$$

The Fe^{2+}/Fe^{3+} half-reaction may seem somewhat unusual because a cation rather than an anion migrates to the anode and gives up an electron. Oxidation of a cation at an anode or reduction of an anion at a cathode is a relatively common process.

The first reaction requires a copper anode, but the other two can be carried out at the surface of an inert platinum electrode.

18B-2 Types of Electrochemical Cells

Galvanic cells store electrical energy; **electrolytic cells** consume electricity.

Electrochemical cells are either galvanic or electrolytic. They can also be classified as reversible or irreversible.

Galvanic, or **voltaic**, **cells** store electrical energy. **Batteries** are usually made from several such cells connected in series to produce higher voltages than a single cell can produce. The reactions at the two electrodes in such cells tend to proceed spontaneously and produce a flow of electrons from the anode to the cathode via an external conductor. The cell shown in Figure 18-2a shows a galvanic cell that exhibits a potential of about 0.412 V when no current is being drawn from it. The silver electrode is positive with respect to the copper electrode in this cell. The copper electrode, which is negative with respect to the silver electrode, is a potential source of electrons to the external circuit when the cell is discharged. The cell in Figure 18-2b is the same galvanic cell, but now it is under discharge so that electrons move through the external circuit from the copper electrode to the silver electrode. While being discharged, the silver electrode is the *cathode* since the reduction of Ag^+ occurs here. The copper electrode is the *anode* since the oxidation of $Cu(s)$ occurs at this electrode. Galvanic cells operate spontaneously, and the net reaction during discharge is called the **spontaneous cell reaction**. For the cell of Figure 18-2b, the spontaneous cell reaction is that given by equation 18-3, that is, $2Ag^+ + Cu(s) \rightleftharpoons 2Ag(s) + Cu^{2+}$.

The reaction $2H_2O \rightleftharpoons O_2(g) + 4H^+ + 4e^-$ occurs at an anode when an aqueous solution contains no other species that are more easily oxidized than H_2O.

For both galvanic and electrolytic cells, remember that (1) reduction always takes place at the cathode, and (2) oxidation always takes place at the anode. The cathode in a galvanic cell becomes the anode, however, when the cell is operated as an electrolytic cell.

An **electrolytic cell**, in contrast to a voltaic cell, requires an external source of electrical energy for operation. The cell in Figure 18-2 can be operated as an electrolytic cell by connecting the positive terminal of an external voltage source with

Alessandro Volta (1745–1827), Italian physicist, was the inventor of the first battery, the so-called voltaic pile (shown on the right). It consisted of alternating disks of copper and zinc separated by disks of cardboard soaked with salt solution. In honor of his many contributions to electrical science, the unit of potential difference, the volt, is named for Volta. In fact, in modern usage, we often call the quantity the voltage instead of potential difference.

a potential somewhat greater than 0.412 V to the silver electrode and the negative terminal of the source to the copper electrode, as shown in **Figure 18-2c**. Since the negative terminal of the external voltage source is electron rich, electrons flow from this terminal to the copper electrode, where reduction of Cu^{2+} to $Cu(s)$ occurs. The current is sustained by the oxidation of $Ag(s)$ to Ag^+ at the right-hand electrode, producing electrons that flow to the positive terminal of the voltage source. Note that in the electrolytic cell, the direction of the current is the reverse of that in the galvanic cell in Figure 18-2b, and the reactions at the electrodes are reversed as well. The silver electrode is forced to become the *anode*, while the copper electrode is forced to become the *cathode*. The net reaction that occurs when a voltage higher than the galvanic cell voltage is applied is the opposite of the spontaneous cell reaction. That is,

$$2Ag(s) + Cu^{2+} \rightleftharpoons 2Ag^+ + Cu(s)$$

The cell in Figure 18-2 is an example of a reversible cell, in which the direction of the electrochemical reaction is reversed when the direction of electron flow is changed. In an irreversible cell, changing the direction of current causes entirely different half-reactions to occur at one or both electrodes. The lead-acid storage battery in an automobile is a common example of a series of reversible cells. When an external charger or the generator charges the battery, its cells are electrolytic. When it is used to operate the headlights, the radio, or the ignition, its cells are galvanic.

In a **reversible cell**, reversing the current reverses the cell reaction. In an **irreversible cell**, reversing the current causes a different half-reaction to occur at one or both of the electrodes.

FEATURE 18-2

The Daniell Gravity Cell

The Daniell gravity cell was one of the earliest galvanic cells to find widespread practical application. It was used in the mid-1800s to power telegraphic communication systems. As shown in **Figure 18F-1** (also see color plate 11), the cathode was a piece of copper immersed in a saturated solution of copper sulfate. A much less dense solution of dilute zinc sulfate was layered on top of the copper sulfate, and a massive zinc electrode was located in this solution. The electrode reactions were

$$Zn(s) \rightleftharpoons Zn^{2+} + 2e^-$$

$$Cu^{2+} + 2e^- \rightleftharpoons Cu(s)$$

This cell develops an initial voltage of 1.18 V, which gradually decreases as the cell discharges.

Figure 18F-1   A Daniell gravity cell.

18B-3 Representing Cells Schematically

Chemists frequently use a shorthand notation to describe electrochemical cells. The cell in Figure 18-2a, for example, is described by

$$Cu \,|\, Cu^{2+}(0.0200\ M) \,\|\, Ag^+(0.0200\ M) \,|\, Ag \tag{18-5}$$

By convention, a single vertical line indicates a phase boundary, or interface, at which a potential develops. For example, the first vertical line in this schematic indicates that a potential develops at the phase boundary between the copper electrode and the copper sulfate solution. The double vertical lines represent two-phase boundaries, one at each end of the salt bridge. There is a **liquid-junction potential** at each of these interfaces. The junction potential results from differences in the rates at which the ions in the cell compartments and the salt bridge migrate across the interfaces. A liquid-junction potential can amount to as much as several hundredths of a volt but can be negligibly small if the electrolyte in the salt bridge has an anion and a cation that migrate at nearly the same rate. A saturated solution of potassium chloride, KCl, is the electrolyte that is most widely used. This electrolyte can reduce the junction potential to a few millivolts or less. For our purposes, we will neglect the contribution of liquid-junction potentials to the total potential of the cell. There are also several examples of cells that are without liquid junction and therefore do not require a salt bridge.

An alternative way of writing the cell shown in Figure 18-2a is

$$\text{Cu} \mid \text{CuSO}_4(0.0200 \text{ M}) \parallel \text{AgNO}_3(0.0200 \text{ M}) \mid \text{Ag}$$

In this description, the compounds used to prepare the cell are indicated rather than the active participants in the cell half-reactions.

18B-4 Currents in Electrochemical Cells

Figure 18-3 shows the movement of various charge carriers in a galvanic cell during discharge. The electrodes are connected with a wire so that the spontaneous cell reaction occurs. Charge is transported through such an electrochemical cell by three mechanisms:

1. Electrons carry the charge within the electrodes as well as the external conductor. Notice that by convention, current, which is normally indicated by the symbol I, is opposite in direction to electron flow.
2. Anions and cations are the charge carriers within the cell. At the left-hand electrode, copper is oxidized to copper ions, giving up electrons to the electrode. As shown in Figure 18-3, the copper ions formed move away from the copper electrode into the bulk of solution, while anions, such as sulfate and hydrogen sulfate ions, migrate toward the copper anode. Within the salt bridge, chloride ions migrate toward and into the copper compartment, and potassium ions move in the opposite direction. In the right-hand compartment, silver ions move toward the silver electrode where they are reduced to silver metal, and the nitrate ions move away from the electrode into the bulk of solution.
3. The ionic conduction of the solution is coupled to the electronic conduction in the electrodes by the reduction reaction at the cathode and the oxidation reaction at the anode.

> In a cell, electricity is carried by the movement of ions. Both anions and cations contribute.

> The phase boundary between an electrode and its solution is called an **interface**.

18C ELECTRODE POTENTIALS

The potential difference between the electrodes of the cell in **Figure 18-4a** is a measure of the tendency for the reaction

$$2\text{Ag}(s) + \text{Cu}^{2+} \rightleftharpoons 2\text{Ag}^+ + \text{Cu}(s)$$

to proceed from a nonequilibrium state to the condition of equilibrium. The cell potential E_{cell} is related to the free energy of the reaction ΔG by

$$\Delta G = -nFE_{\text{cell}} \tag{18-6}$$

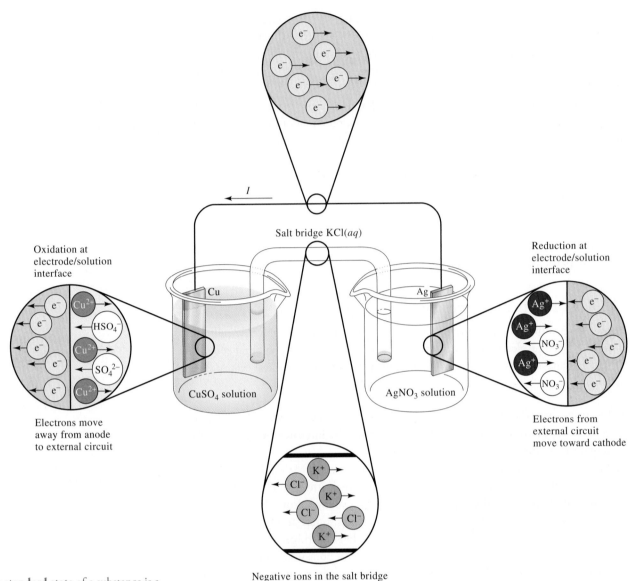

Negative ions in the salt bridge
move toward the anode; positive
ions move toward the cathode

Figure 18-3   Movement of charge in a galvanic cell.

The **standard state** of a substance is a reference state that allows us to obtain relative values of such thermodynamic quantities as free energy, activity, enthalpy, and entropy. All substances are assigned unit activity in their standard states. For gases, the standard state has the properties of an ideal gas but at one atmosphere pressure. It is thus said to be a *hypothetical* state. For pure liquids and solvents, the standard states are *real* states and are the pure substances at a specified temperature and pressure. For solutes in dilute solution, the standard state is a hypothetical state that has the properties of an infinitely dilute solute but at unit concentration (molar or molal concentration, or mole fraction). The standard state of a solid is a real state and is the pure solid in its most stable crystalline form.

If the reactants and products are in their **standard states**, the resulting cell potential is called the **standard cell potential**. This latter quantity is related to the standard free energy change for the reaction and thus to the equilibrium constant by

$$\Delta G^0 = -nFE_{cell}^0 = -RT \ln K_{eq} \qquad (18\text{-}7)$$

where R is the gas constant and T is the absolute temperature.

18C-1 Sign Convention for Cell Potentials

When we consider a normal chemical reaction, we speak of the reaction occurring from reactants on the left side of the arrow to products on the right side. By the International Union of Pure and Applied Chemistry (IUPAC) sign convention, when

Figure 18-4 Change in cell potential after passage of current until equilibrium is reached. In (a), the high-resistance voltmeter prevents any significant electron flow, and the full open-circuit cell potential is measured. For the concentrations shown, this potential is +0.412 V. In (b), the voltmeter is replaced with a low-resistance current meter, and the cell discharges with time until eventually equilibrium is reached. In (c), after equilibrium is reached, the cell potential is again measured with a voltmeter and found to be 0.000 V. The concentrations in the cell are now those at equilibrium as shown.

we consider an electrochemical cell and its resulting potential, we consider the cell reaction to occur in a certain direction as well. The convention for cells is called the **plus right rule**. This rule implies that we always measure the cell potential by connecting the positive lead of the voltmeter to the right-hand electrode in the schematic or cell drawing (Ag electrode in Figure 18-4) and the common, or ground, lead of the voltmeter to the left-hand electrode (Cu electrode in Figure 18-4). If we always follow this convention, the value of E_{cell} is a measure of the tendency of the cell reaction to occur spontaneously in the direction written below from left to right.

$$Cu\,|\,Cu^{2+}(0.0200\ M)\,||\,Ag^+(0.0200\ M)\,|\,Ag$$

That is, the direction of the overall process has Cu metal being oxidized to Cu^{2+} in the left-hand compartment and Ag^+ being reduced to Ag metal in the right-hand compartment. In other words, the reaction being considered is

$$Cu(s) + 2Ag^+ \rightleftharpoons Cu^{2+} + 2Ag(s).$$

Implications of the IUPAC Convention

The leads of voltmeters are color coded. The positive lead is red, and the common, or ground, lead is black.

There are several implications of the sign convention that may not be obvious. First, if the measured value of E_{cell} is positive, the right-hand electrode is positive with respect to the left-hand electrode, and the free energy change for the reaction in the direction being considered is negative according to Equation 18-6. Hence, the reaction in the direction being considered would occur spontaneously if the cell were short-circuited or connected to some device to perform work (e.g., light a lamp, power a radio, or start a car). On the other hand, if E_{cell} is negative, the right-hand electrode is negative with respect to the left-hand electrode, the free energy change is positive, and the reaction in the direction considered (oxidation on the left, reduction on the right) is *not* the spontaneous cell reaction. For our cell of Figure 18-4a, $E_{cell} = +0.412$ V, and the oxidation of Cu and reduction of Ag^+ occur spontaneously when the cell is connected to a device and allowed to do so.

The IUPAC convention is consistent with the signs that the electrodes actually develop in a galvanic cell. That is, in the Cu/Ag cell shown in Figure 18-4, the Cu electrode becomes electron rich (negative) because of the tendency of Cu to be oxidized to Cu^{2+}, and the Ag electrode is electron deficient (positive) because of the tendency for Ag^+ to be reduced to Ag. As the galvanic cell discharges spontaneously, the silver electrode is the cathode, while the copper electrode is the anode. Note that for the same cell written in the opposite direction

$$Ag\,|\,AgNO_3\,(0.0200\ M)\,||\,CuSO_4\,(0.0200\ M)\,|\,Cu$$

the measured cell potential would be $E_{cell} = -0.412$ V, and the reaction considered is

$$2Ag(s) + Cu^{2+} \rightleftharpoons 2Ag^+ + Cu(s)$$

This reaction is *not* the spontaneous cell reaction because E_{cell} is negative, and ΔG is thus positive. It does not matter to the cell which electrode is written in the schematic on the right and which is written on the left. The spontaneous cell reaction is *always*

$$Cu(s) + 2Ag^+ \rightleftharpoons Cu^{2+} + 2Ag(s)$$

By convention, we just measure the cell in a standard manner and consider the cell reaction in a standard direction. Finally, we must emphasize that, no matter how we may write the cell schematic or arrange the cell in the laboratory, if we connect a wire or a low-resistance circuit to the cell, *the spontaneous cell reaction will occur*. The only way to achieve the reverse reaction is to connect an external voltage source and force the electrolytic reaction $2Ag(s) + Cu^{2+} \rightleftharpoons 2Ag^+ + Cu(s)$ to occur.

Half-Cell Potentials

The potential of a cell such as that shown in Figure 18-4a is the difference between two half-cell or single-electrode potentials, one associated with the half-reaction at the right-hand electrode (E_{right}) and the other associated with the half-reaction at the left-hand electrode (E_{left}). According to the IUPAC sign convention, as long as the liquid-junction potential is negligible or there is no liquid junction, we may write the cell potential E_{cell} as

$$E_{cell} = E_{right} - E_{left} \qquad (18\text{-}8)$$

Although we cannot determine absolute potentials of electrodes such as these (see Feature 18-3), we can easily determine relative electrode potentials. For example, if we replace the copper electrode in the cell in Figure 18-2 with a cadmium electrode immersed in a cadmium sulfate solution, the voltmeter reads about 0.7 V more positive than the original cell. Since the right-hand compartment remains unaltered, we conclude that the half-cell potential for cadmium is about 0.7 V less than that for copper (that is, cadmium is a stronger reductant than is copper). Substituting other electrodes while keeping one of the electrodes unchanged allows us to construct a table of relative electrode potentials, as discussed in Section 18C-3.

Discharging a Galvanic Cell

The galvanic cell of Figure 18-4a is in a nonequilibrium state because the very high resistance of the voltmeter prevents the cell from discharging significantly. So when we measure the cell potential, no reaction occurs, and what we measure is the tendency of the reaction to occur *if* we allowed it to proceed. For the Cu/Ag cell with the concentrations shown, the cell potential measured under open circuit conditions is $+0.412$ V, as previously noted. If we now allow the cell to discharge by replacing the voltmeter with a low-resistance current meter, as shown in **Figure 18-4b**, the spontaneous cell reaction occurs. The current, initially high, decreases exponentially with time (see **Figure 18-5**). As shown in **Figure 18-4c**, when equilibrium is reached, there is no net current in the cell, and the cell potential is 0.000 V. The copper ion concentration at equilibrium is then 0.0300 M, while the silver ion concentration falls to 2.7×10^{-9} M.

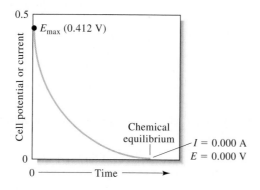

Figure 18-5 Cell potential in the galvanic cell of Figure 18-4b as a function of time. The cell current, which is directly related to the cell potential, also decreases with the same time behavior.

FEATURE 18-3

Why We Cannot Measure Absolute Electrode Potentials

Although it is not difficult to measure *relative* half-cell potentials, it is impossible to determine absolute half-cell potentials because all voltage-measuring devices measure only *differences* in potential. To measure the potential of an electrode, one contact of a voltmeter is connected to the electrode in question. The other contact from the meter must then be brought into electrical contact with the solution in the electrode compartment via another conductor. This second contact, however, inevitably creates a solid/solution interface that acts as a second half-cell when the potential is measured. Thus, an absolute half-cell potential is not obtained. What we do obtain is the difference between the half-cell potential of interest and a half-cell made up of the second contact and the solution.

Our inability to measure absolute half-cell potentials presents no real obstacle because relative half-cell potentials are just as useful provided they are all measured against the same reference half-cell. Relative potentials can be combined to give cell potentials. We can also use them to calculate equilibrium constants and generate titration curves.

The standard hydrogen electrode is sometimes called the **normal hydrogen electrode (NHE)**.

SHE is the abbreviation for standard hydrogen electrode.

Platinum black is a layer of finely divided platinum that is formed on the surface of a smooth platinum electrode by electrolytic deposition of the metal from a solution of chloroplatinic acid, H_2PtCl_6. The platinum black provides a large specific surface area of platinum at which the H^+/H_2 reaction can occur. Platinum black catalyzes the reaction shown in Equation 18-9. Remember that catalysts do not change the position of equilibrium but simply shorten the time it takes to reach equilibrium.

The reaction shown as Equation 18-9 combines two equilibria:

$$2H^+ + 2e^- \rightleftharpoons H_2(aq)$$
$$H_2(aq) \rightleftharpoons H_2(g)$$

The continuous stream of gas at constant pressure provides the solution with a constant molecular hydrogen concentration.

18C-2 The Standard Hydrogen Reference Electrode

For relative electrode potential data to be widely applicable and useful, we must have a generally agreed-upon reference half-cell against which all others are compared. Such an electrode must be easy to construct, reversible, and highly reproducible in its behavior. The **standard hydrogen electrode (SHE)** meets these specifications and has been used throughout the world for many years as a universal reference electrode. It is a typical **gas electrode**.

Figure 18-6 shows the physical arrangement of a hydrogen electrode. The metal conductor is a piece of platinum that has been coated, or **platinized**, with finely divided platinum (platinum black) to increase its specific surface area. This electrode is immersed in an aqueous acid solution of known, constant hydrogen ion activity. The solution is kept saturated with hydrogen by bubbling the gas at constant pressure over the surface of the electrode. The platinum does not take part in the electrochemical reaction and serves only as the site where electrons are transferred. The half-reaction responsible for the potential that develops at this electrode is

$$2H^+(aq) + 2e^- \rightleftharpoons H_2(g) \tag{18-9}$$

The hydrogen electrode shown in Figure 18-6 can be represented schematically as

$$Pt, H_2(p = 1.00 \text{ atm}) \mid (H^+ = x \text{ M}) \parallel$$

In Figure 18-6, the hydrogen is specified as having a partial pressure of one atmosphere and the concentration of hydrogen ions in the solution is x M. The hydrogen electrode is reversible.

The potential of a hydrogen electrode depends on temperature and the activities of hydrogen ion and molecular hydrogen in the solution. The latter, in turn, is proportional to the pressure of the gas that is used to keep the solution saturated in hydrogen. For the SHE, the activity of hydrogen ions is specified as unity, and the partial pressure of the gas is specified as one atmosphere. *By convention, the potential*

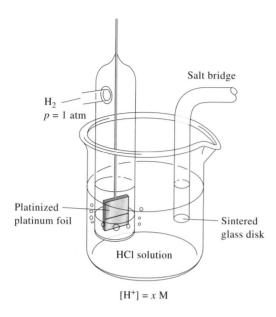

H₂
$p = 1$ atm

Salt bridge

Platinized
platinum foil

Sintered
glass disk

HCl solution

$[H^+] = x$ M

Figure 18-6 The hydrogen gas
electrode.

of the standard hydrogen electrode is assigned a value of 0.000 V at all temperatures. As a consequence of this definition, any potential developed in a galvanic cell consisting of a standard hydrogen electrode and some other electrode is attributed entirely to the other electrode.

> At $p_{H_2} = 1.00$ and $a_{H^+} = 1.00$, the potential of the hydrogen electrode is assigned a value of exactly 0.000 V at all temperatures.

Several other reference electrodes that are more convenient for routine measurements have been developed. Some of these are described in Section 21B.

18C-3 Electrode Potential and Standard Electrode Potential

An **electrode potential** is defined as the potential of a cell in which the electrode in question is the right-hand electrode and the standard hydrogen electrode is the left-hand electrode. So if we want to obtain the potential of a silver electrode in contact with a solution of Ag^+, we would construct a cell as shown in **Figure 18-7**. In this cell, the half-cell on the right consists of a strip of pure silver in contact with a solution containing silver ions; the electrode on the left is the standard hydrogen electrode. The cell potential is defined as in Equation 18-8. Because the left-hand electrode is the standard hydrogen electrode with a potential that has been assigned a value of 0.000 V, we can write

> An electrode potential is the potential of a cell that has a standard hydrogen electrode as the left electrode (reference).

$$E_{cell} = E_{right} - E_{left} = E_{Ag} - E_{SHE} = E_{Ag} - 0.000 = E_{Ag}$$

where E_{Ag} is the potential of the silver electrode. Despite its name, an electrode potential is in fact the potential of an electrochemical cell which has a carefully defined reference electrode. Often, the potential of an electrode, such as the silver electrode in Figure 18-7, is referred to as E_{Ag} versus SHE to emphasize that it is the potential of a complete cell measured against the standard hydrogen electrode as a reference.

The **standard electrode potential**, E^0, of a half-reaction is defined as its electrode potential when the activities of the reactants and products are all unity. For the cell in Figure 18-7, the E^0 value for the half reaction

$$Ag^+ + e^- \rightleftharpoons Ag(s)$$

Figure 18-7 Measurement of the electrode potential for an Ag electrode. If the silver ion activity in the right-hand compartment is 1.00, the cell potential is the standard electrode potential of the Ag^+/Ag half-reaction.

can be obtained by measuring E_{cell} with the activity of Ag^+ equal to 1.00. In this case, the cell shown in Figure 18-7 can be represented schematically as

$$Pt, H_2(p = 1.00 \text{ atm}) \mid H^+(a_{H^+} = 1.00) \parallel Ag^+(a_{Ag^+} = 1.00) \mid Ag$$

or alternatively as

$$SHE \parallel Ag^+(a_{Ag^+} = 1.00) \mid Ag$$

This galvanic cell develops a potential of +0.799 V with the silver electrode on the right, that is, the spontaneous cell reaction is oxidation in the left-hand compartment and reduction in the right-hand compartment:

$$2Ag^+ + H_2(g) \rightleftharpoons 2Ag(s) + 2H^+$$

Because the silver electrode is on the right and the reactants and products are in their standard states, the measured potential is by definition the standard electrode potential for the silver half-reaction, or the **silver couple**. Note that the silver electrode is positive with respect to the standard hydrogen electrode. Therefore, the standard electrode potential is given a positive sign, and we write

A metal ion/metal half-cell is sometimes called a **couple**.

$$Ag^+ + e^- \rightleftharpoons Ag(s) \qquad E^0_{Ag^+/Ag} = +0.799 \text{ V}$$

Figure 18-8 illustrates a cell used to measure the standard electrode potential for the half-reaction

$$Cd^{2+} + 2e^- \rightleftharpoons Cd(s)$$

In contrast to the silver electrode, the cadmium electrode is negative with respect to the standard hydrogen electrode. Therefore, the standard electrode potential of

Figure 18-8 Measurement of the standard electrode potential for $Cd^{2+} + 2e^- \rightleftharpoons Cd(s)$.

the Cd/Cd^{2+} couple is *by convention* given a negative sign, and $E^0_{Cd^{2+}/Cd} = -0.403$ V. Because the cell potential is negative, the spontaneous cell reaction is not the reaction as written (that is, oxidation on the left and reduction on the right). Rather, the spontaneous reaction is in the opposite direction.

$$Cd(s) + 2H^+ \rightleftharpoons Cd^{2+} + H_2(g)$$

A zinc electrode immersed in a solution having a zinc ion activity of unity develops a potential of -0.763 V when it is the right-hand electrode paired with a standard hydrogen electrode on the left. Thus, we can write $E^0_{Zn^{2+}/Zn} = -0.763$ V.

The standard electrode potentials for the four half-cells just described can be arranged in the following order:

Half-Reaction	Standard Electrode Potential, V
$Ag^+ + e^- \rightleftharpoons Ag(s)$	+0.799
$2H^+ + 2e^- \rightleftharpoons H_2(g)$	0.000
$Cd^{2+} + 2e^- \rightleftharpoons Cd(s)$	−0.403
$Zn^{2+} + 2e^- \rightleftharpoons Zn(s)$	−0.763

The magnitudes of these electrode potentials indicate the relative strength of the four ionic species as electron acceptors (oxidizing agents), that is, in decreasing strength, $Ag^+ > H^+ > Cd^{2+} > Zn^{2+}$.

18C-4 Additional Implications of the IUPAC Sign Convention

The sign convention described in the previous section was adopted at the IUPAC meeting in Stockholm in 1953 and is now accepted internationally. Prior to this

An **electrode potential** is by definition a reduction potential. An oxidation potential is the potential for the half-reaction written in the opposite way. The sign of an oxidation potential is, therefore, opposite that for a reduction potential, but the magnitude is the same.

The IUPAC sign convention is based on the actual sign of the half-cell of interest when it is part of a cell containing the standard hydrogen electrode as the other half-cell.

agreement, chemists did not always use the same convention, and this inconsistency was the cause of controversy and confusion in the development and routine use of electrochemistry.

Any sign convention must be based on expressing half-cell processes in a single way—either as oxidations or as reductions. According to the IUPAC convention, the term "*electrode potential*" (or, more exactly, "relative electrode potential") *is reserved exclusively to describe half-reactions written as reductions.* There is no objection to the use of the term "oxidation potential" to indicate a process written in the opposite sense, but it is not proper to refer to such a potential as an electrode potential.

The sign of an electrode potential is determined by the sign of the half-cell in question when it is coupled to a standard hydrogen electrode. When the half-cell of interest exhibits a positive potential versus the SHE (see Figure 18-7), it will behave spontaneously as the cathode when the cell is discharging. When the half-cell of interest is negative versus the SHE (see Figure 18-8), it will behave spontaneously as the anode when the cell is discharging.

18C-5 Effect of Concentration on Electrode Potentials: The Nernst Equation

An electrode potential is a measure of the extent to which the concentrations of the species in a half-cell differ from their equilibrium values. For example, there is a greater tendency for the process

$$Ag^+ + e^- \rightleftharpoons Ag(s)$$

to occur in a concentrated solution of silver(I) than in a dilute solution of that ion. It follows that the magnitude of the electrode potential for this process must also become larger (more positive) as the silver ion concentration of a solution is increased. We now examine the quantitative relationship between concentration and electrode potential.

Consider the reversible half-reaction

$$a\mathrm{A} + b\mathrm{B} + \cdots + ne^- \rightleftharpoons c\mathrm{C} + d\mathrm{D} + \cdots \tag{18-10}$$

where the capital letters represent formulas for the participating species (atoms, molecules, or ions), e^- represents the electrons, and the lower case italic letters indicate the number of moles of each species appearing in the half-reaction as it has been written. The electrode potential for this process is given by the equation

The meanings of the bracketed terms in Equations 18-11 and 18-12 are,

for a solute A [A] = molar concentration and
for a gas B [B] = p_B = partial pressure in atmospheres.
If one or more of the species appearing in Equation 18-11 is a pure liquid, pure solid, or the solvent present in excess, then no bracketed term for this species appears in the quotient because the activities of these are unity.

$$E = E^0 - \frac{RT}{nF} \ln \frac{[\mathrm{C}]^c[\mathrm{D}]^d \cdots}{[\mathrm{A}]^a[\mathrm{B}]^b \cdots} \tag{18-11}$$

where

E^0 = the *standard electrode potential*, which is characteristic for each half-reaction
R = the ideal gas constant, 8.314 J K^{-1} mol^{-1}
T = temperature, K
n = number of moles of electrons that appears in the half-reaction for the electrode process as written
F = the faraday = 96,485 C (coulombs) per mole of electrons
ln = natural logarithm = 2.303 log

Walther Nernst (1864–1941) received the 1920 Nobel Prize in chemistry for his numerous contributions to the field of chemical thermodynamics. Nernst (right) is seen here in his laboratory in 1921.

© Bettmann/CORBIS

If we substitute numerical values for the constants, convert to base 10 logarithms, and specify 25°C for the temperature, we get

$$E = E^0 - \frac{0.0592}{n} \log \frac{[C]^c[D]^d \cdots}{[A]^a[B]^b \cdots} \tag{18-12}$$

Strictly speaking, the letters in brackets represent activities, but we will usually follow the practice of substituting molar concentrations for activities in most calculations. Thus, if some participating species A is a solute, [A] is the concentration of A in moles per liter. If A is a gas, [A] in Equation 18-12 is replaced by p_A, the partial pressure of A in atmospheres. If A is a pure liquid, a pure solid, or the solvent, its activity is unity, and no term for A is included in the equation. The rationale for these assumptions is the same as that described in Section 9B-2, which deals with equilibrium-constant expressions. Equation 18-12 is known as the Nernst equation in honor of the German chemist Walther Nernst, who was responsible for its development.

EXAMPLE 18-2

Typical half-cell reactions and their corresponding Nernst expressions follow.

(1) $Zn^{2+} + 2e^- \rightleftharpoons Zn(s)$ $\qquad E = E^0 - \frac{0.0592}{2} \log \frac{1}{[Zn^{2+}]}$

No term for elemental zinc is included in the logarithmic term because it is a pure second phase (solid). Thus, the electrode potential varies linearly with the logarithm of the reciprocal of the zinc ion concentration.

(2) $Fe^{3+} + e^- \rightleftharpoons Fe^{2+}(s)$ $\qquad E = E^0 - \frac{0.0592}{1} \log \frac{[Fe^{2+}]}{[Fe^{3+}]}$

The potential for this couple can be measured with an inert metallic electrode immersed in a solution containing both iron species. The potential depends on the logarithm of the ratio between the molar concentrations of these ions.

(3) $2H^+ + 2e^- \rightleftharpoons H_2(g)$ $E = E^0 - \dfrac{0.0592}{2} \log \dfrac{p_{H_2}}{[H^+]^2}$

In this example, p_{H_2} is the partial pressure of hydrogen (in atmospheres) at the surface of the electrode. Usually, its value will be the same as atmospheric pressure.

(4) $MnO_4^- + 5e^- + 8H^+ \rightleftharpoons Mn^{2+} + 4H_2O$

$$E = E^0 - \dfrac{0.0592}{5} \log \dfrac{[Mn^{2+}]}{[MnO_4^-][H^+]^8}$$

In this situation, the potential depends not only on the concentrations of the manganese species but also on the pH of the solution.

(5) $AgCl(s) + e^- \rightleftharpoons Ag(s) + Cl^-$ $E = E^0 - \dfrac{0.0592}{1} \log [Cl^-]$

This half-reaction describes the behavior of a silver electrode immersed in a chloride solution that is *saturated* with AgCl. To ensure this condition, an excess of the solid AgCl must always be present. Note that this electrode reaction is the sum of the following two reactions:

$$AgCl(s) \rightleftharpoons Ag^+ + Cl^-$$

$$Ag^+ + e^- \rightleftharpoons Ag(s)$$

Note also that the electrode potential is independent of the amount of AgCl present as long as there is at least some present to keep the solution saturated.

> The Nernst expression in part (5) of Example 18-2 requires an excess of solid AgCl so that the solution is saturated with the compound at all times.

18C-6 The Standard Electrode Potential, E^0

When we look carefully at Equations 18-11 and 18-12, we see that the constant E^0 is the electrode potential whenever the concentration quotient (actually, the activity quotient) has a value of 1. This constant is by definition the standard electrode potential for the half-reaction. Note that the quotient is always equal to 1 when the activities of the reactants and products of a half-reaction are unity.

The standard electrode potential is an important physical constant that provides quantitative information regarding the driving force for a half-cell reaction.[2] The important characteristics of these constants are the following:

> The **standard electrode potential** for a half-reaction, E^0, is defined as the electrode potential when all reactants and products of a half-reaction are at unit activity.

1. The standard electrode potential is a relative quantity in the sense that it is the potential of an electrochemical cell in which the reference electrode (left-hand electrode) is the standard hydrogen electrode, whose potential has been assigned a value of 0.000 V.
2. The standard electrode potential for a half-reaction refers exclusively to a reduction reaction, that is, it is a relative reduction potential.
3. The standard electrode potential measures the relative force tending to drive the half-reaction from a state in which the reactants and products are at unit activity to a state in which the reactants and products are at their equilibrium activities relative to the standard hydrogen electrode.

---

[2] For further reading on standard electrode potentials, see R. G. Bates, in *Treatise on Analytical Chemistry*, 2nd ed., I. M. Kolthoff and P. J. Elving, eds., Part I, Vol. 1, Ch. 13, New York: Wiley, 1978.

4. The standard electrode potential is independent of the number of moles of reactant and product shown in the balanced half-reaction. Thus, the standard electrode potential for the half-reaction

$$Fe^{3+} + e^- \rightleftharpoons Fe^{2+} \qquad E^0 = +0.771 \text{ V}$$

does not change if we choose to write the reaction as

$$5Fe^{3+} + 5e^- \rightleftharpoons 5Fe^{2+} \qquad E^0 = +0.771 \text{ V}$$

Note, however, that the Nernst equation must be consistent with the half-reaction as written. For the first case, it will be

$$E = 0.771 - \frac{0.0592}{1} \log \frac{[Fe^{2+}]}{[Fe^{3+}]}$$

and for the second

$$E = 0.771 - \frac{0.0592}{5} \log \frac{[Fe^{2+}]^5}{[Fe^{3+}]^5} = 0.771 - \frac{0.0592}{5} \log \left(\frac{[Fe^{2+}]}{[Fe^{3+}]} \right)^5$$

$$= 0.771 - \frac{5 \times 0.0592}{5} \log \frac{[Fe^{2+}]}{[Fe^{3+}]}$$

> Note that the two log terms have identical values, that is,
>
> $$\frac{0.0592}{1} \log \frac{[Fe^{2+}]}{[Fe^{3+}]}$$
>
> $$= \frac{0.0592}{5} \log \frac{[Fe^{2+}]^5}{[Fe^{3+}]^5}$$
>
> $$= \frac{0.0592}{5} \log \left(\frac{[Fe^{2+}]}{[Fe^{3+}]} \right)^5$$

5. A positive electrode potential indicates that the half-reaction in question is spontaneous with respect to the standard hydrogen electrode half-reaction. In other words, the oxidant in the half-reaction is a stronger oxidant than is hydrogen ion. A negative sign indicates just the opposite.
6. The standard electrode potential for a half-reaction is temperature dependent.

Standard electrode potential data are available for an enormous number of half-reactions. Many have been determined directly from electrochemical measurements. Others have been computed from equilibrium studies of oxidation/reduction systems and from thermochemical data associated with such reactions. **Table 18-1** contains standard electrode potential data for several half-reactions that we will be considering in the pages that follow. A more extensive listing is found in Appendix 5.[3]

Table 18-1 and Appendix 5 illustrate the two common ways for tabulating standard potential data. In Table 18-1, potentials are listed in decreasing numerical order. As a consequence, the species in the upper left part are the most effective electron acceptors, as evidenced by their large positive values. They are therefore the strongest oxidizing agents. As we proceed down the left side of such a table, each succeeding species is less effective as an electron acceptor than the one above it. The half-cell reactions at the bottom of the table have little or no tendency to take place as they are written. On the other hand, they do tend to occur in the opposite sense. The most effective reducing agents, then, are those species that appear in the lower right portion of the table.

---

[3] Comprehensive sources for standard electrode potentials include A. J. Bard, R. Parsons, and J. Jordan, eds., *Standard Electrode Potentials in Aqueous Solution*, New York: Dekker, 1985; G. Milazzo, S. Caroli, and V. K. Sharma, *Tables of Standard Electrode Potentials*, New York: Wiley-Interscience, 1978; M. S. Antelman and F. J. Harris, *Chemical Electrode Potentials*, New York: Plenum Press, 1982. Some compilations are arranged alphabetically by element; others are tabulated according to the value of E^0.

Based on the E^0 values in Table 18-1 for Fe^{3+} and I_3^-, which species would you expect to predominate in a solution produced by mixing iron(III) and iodide ions? See color plate 12.

TABLE 18-1

Standard Electrode Potentials*

Reaction	E^0 at 25°C, V
$Cl_2(g) + 2e^- \rightleftharpoons 2Cl^-$	+1.359
$O_2(g) + 4H^+ + 4e^- \rightleftharpoons 2H_2O$	+1.229
$Br_2(aq) + 2e^- \rightleftharpoons 2Br^-$	+1.087
$Br_2(l) + 2e^- \rightleftharpoons 2Br^-$	+1.065
$Ag^+ + e^- \rightleftharpoons Ag(s)$	+0.799
$Fe^{3+} + e^- \rightleftharpoons Fe^{2+}$	+0.771
$I_3^- + 2e^- \rightleftharpoons 3I^-$	+0.536
$Cu^{2+} + 2e^- \rightleftharpoons Cu(s)$	+0.337
$UO_2^{2+} + 4H^+ + 2e^- \rightleftharpoons U^{4+} + 2H_2O$	+0.334
$Hg_2Cl_2(s) + 2e^- \rightleftharpoons 2Hg(l) + 2Cl^-$	+0.268
$AgCl(s) + e^- \rightleftharpoons Ag(s) + Cl^-$	+0.222
$Ag(S_2O_3)_2^{3-} + e^- \rightleftharpoons Ag(s) + 2S_2O_3^{2-}$	+0.017
$2H^+ + 2e^- \rightleftharpoons H_2(g)$	**0.000**
$AgI(s) + e^- \rightleftharpoons Ag(s) + I^-$	−0.151
$PbSO_4 + 2e^- \rightleftharpoons Pb(s) + SO_4^{2-}$	−0.350
$Cd^{2+} + 2e^- \rightleftharpoons Cd(s)$	−0.403
$Zn^{2+} + 2e^- \rightleftharpoons Zn(s)$	−0.763

*See Appendix 5 for a more extensive list.

FEATURE 18-4

Sign Conventions in the Older Literature

Reference works, particularly those published before 1953, often contain tabulations of electrode potentials that are not in accord with the IUPAC recommendations. For example, in a classic source of standard-potential data compiled by Latimer,[4] one finds

$$Zn(s) \rightleftharpoons Zn^{2+} + 2e^- \qquad E = +0.76\ V$$

$$Cu(s) \rightleftharpoons Cu^{2+} + 2e^- \qquad E = +0.34\ V$$

To convert these oxidation potentials to electrode potentials as defined by the IUPAC convention, we must mentally (1) express the half-reactions as reductions and (2) change the signs of the potentials.

The sign convention used in a tabulation of electrode potentials may not be explicitly stated. This information can be deduced, however, by noting the direction and sign of the potential for a familiar half-reaction. If the sign agrees with the IUPAC convention, the table can be used as is. If not, the signs of all of the data must be reversed. For example, the reaction

$$O_2(g) + 4H^+ + 4e^- \rightleftharpoons 2H_2O \qquad E = +1.229\ V$$

occurs spontaneously with respect to the standard hydrogen electrode and thus carries a positive sign. If the potential for this half-reaction is negative in a table, it and all the other potentials should be multiplied by −1.

---

[4]W. M. Latimer, *The Oxidation States of the Elements and Their Potentials in Aqueous Solutions*, 2nd ed. Englewood Cliffs, NJ: Prentice-Hall, 1952.

Compilations of electrode-potential data, such as that shown in Table 18-1, provide chemists with qualitative insights into the extent and direction of electron-transfer reactions. For example, the standard potential for silver(I) (+0.799 V) is more positive than that for copper(II) (+0.337 V). We therefore conclude that a piece of copper immersed in a silver(I) solution will cause the reduction of that ion and the oxidation of the copper. On the other hand, we would expect no reaction if we place a piece of silver in a copper(II) solution.

In contrast to the data in Table 18-1, standard potentials in Appendix 5 are arranged alphabetically by element to make it easier to locate data for a given electrode reaction.

Systems Involving Precipitates or Complex Ions

In Table 18-1, we find several entries involving Ag(I) including

$$Ag^+ + e^- \rightleftharpoons Ag(s) \qquad\qquad E^0_{Ag^+/Ag} = +0.799 \text{ V}$$

$$AgCl(s) + e^- \rightleftharpoons Ag(s) + Cl^- \qquad\qquad E^0_{AgCl/Ag} = +0.222 \text{ V}$$

$$Ag(S_2O_3)_2^{3-} + e^- \rightleftharpoons Ag(s) + 2S_2O_3^{2-} \qquad\qquad E^0_{Ag(S_2O_3)_2^{3-}/Ag} = +0.017 \text{ V}$$

Each gives the potential of a silver electrode in a different environment. Let us see how the three potentials are related.

The Nernst expression for the first half-reaction is

$$E = E^0_{Ag^+/Ag} - \frac{0.0592}{1} \log \frac{1}{[Ag^+]}$$

If we replace $[Ag^+]$ with $K_{sp}/[Cl^-]$, we obtain

$$E = E^0_{Ag^+/Ag} - \frac{0.0592}{1} \log \frac{[Cl^-]}{K_{sp}} = E^0_{Ag^+/Ag} + 0.0592 \log K_{sp} - 0.0592 \log [Cl^-]$$

By definition, the standard potential for the second half-reaction is the potential where $[Cl^-] = 1.00$. That is, when $[Cl^-] = 1.00$, $E = E^0_{AgCl/Ag}$. Substituting these values gives

$$E^0_{AgCl/Ag} = E^0_{Ag^+/Ag} - 0.0592 \log 1.82 \times 10^{-10} - 0.0592 \log (1.00)$$

$$= 0.799 + (-0.577) - 0.000 = 0.222 \text{ V}$$

Figure 18-9 shows the measurement of the standard electrode potential for the Ag/AgCl electrode.

If we proceed in the same way, we can obtain an expression for the standard electrode potential for the reduction of the thiosulfate complex of silver ion depicted in the third equilibrium shown at the start of this section. In this case, the standard potential is given by

$$E^0_{Ag(S_2O_3)_2^{3-}/Ag} = E^0_{Ag^+/Ag} - 0.0592 \log \beta_2 \qquad (18-13)$$

◀ CHALLENGE: Derive Equation 18-13.

where β_2 is the formation constant for the complex. That is,

$$\beta_2 = \frac{[Ag(S_2O_3)_2^{3-}]}{[Ag^+][S_2O_3^{2-}]^2}$$

Figure 18-9 Measurement of the standard electrode potential for an Ag/AgCl electrode.

EXAMPLE 18-3

Calculate the electrode potential of a silver electrode immersed in a 0.0500 M solution of NaCl using (a) $E^\circ_{Ag^+/Ag} = 0.799$ V and (b) $E^\circ_{AgCl/Ag} = 0.222$ V.

Solution

(a) $Ag^+ + e^- \rightleftharpoons Ag(s)$ $E^0_{Ag^+/Ag} = +0.799$ V

The Ag^+ concentration of this solution is given by

$$[Ag^+] = \frac{K_{sp}}{[Cl^-]} = \frac{1.82 \times 10^{-10}}{0.0500} = 3.64 \times 10^{-9} \text{ M}$$

Substituting into the Nernst expression gives

$$E = 0.799 - 0.0592 \log \frac{1}{3.64 \times 10^{-9}} = 0.299 \text{ V}$$

(b) We may write this last equation as

$$E = 0.222 - 0.0592 \log [Cl^-] = 0.222 - 0.0592 \log 0.0500$$
$$= 0.299$$

FEATURE 18-5

Why Are There Two Electrode Potentials for Br$_2$ in Table 18-1?

In Table 18-1, we find the following data for Br_2:

$$Br_2(aq) + 2e^- \rightleftharpoons 2Br^- \qquad E^0 = +1.087 \text{ V}$$
$$Br_2(l) + 2e^- \rightleftharpoons 2Br^- \qquad E^0 = +1.065 \text{ V}$$

The second standard potential applies only to a solution that is saturated with Br_2 and not to undersaturated solutions. You should use 1.065 V to calculate the electrode potential of a 0.0100 M solution of KBr that is saturated with Br_2 and in contact with an excess of the liquid. In such a case,

$$E = 1.065 - \frac{0.0592}{2} \log [Br^-]^2 = 1.065 - \frac{0.0592}{2} \log (0.0100)^2$$

$$= 1.065 - \frac{0.0592}{2} \times (-4.00) = 1.183 \text{ V}$$

In this calculation, no term for Br_2 appears in the logarithmic term because it is a pure liquid present in excess (unit activity). The standard electrode potential shown in the first entry for $Br_2(aq)$ is hypothetical because the solubility of Br_2 at 25°C is only about 0.18 M. Thus, the recorded value of 1.087 V is based on a system that—in terms of our definition of E^0—cannot be realized experimentally. Nevertheless, the hypothetical potential does permit us to calculate electrode potentials for solutions that are undersaturated in Br_2. For example, if we wish to calculate the electrode potential for a solution that was 0.0100 M in KBr and 0.00100 M in Br_2, we would write

$$E = 1.087 - \frac{0.0592}{2} \log \frac{[Br^-]^2}{[Br_2(aq)]} = 1.087 - \frac{0.0592}{2} \log \frac{(0.0100)^2}{0.00100}$$

$$= 1.087 - \frac{0.0592}{2} \log 0.100 = 1.117 \text{ V}$$

18C-7 Limitations to the Use of Standard Electrode Potentials

We will use standard electrode potentials throughout the rest of this text to calculate cell potentials and equilibrium constants for redox reactions as well as to calculate data for redox titration curves. You should be aware that such calculations sometimes lead to results that are significantly different from those you would obtain in the laboratory. There are two main sources of these differences: (1) the necessity of using concentrations in place of activities in the Nernst equation and (2) failure to take into account other equilibria such as dissociation, association, complex formation, and solvolysis. Measurement of electrode potentials can allow us to investigate these equilibria and determine their equilibrium constants, however.

Use of Concentrations Instead of Activities

Most analytical oxidation/reduction reactions are carried out in solutions that have such high ionic strengths that activity coefficients cannot be obtained via the Debye-Hückel equation (see Equation 10-5, Section 10B-2). Significant errors may result, however, if concentrations are used in the Nernst equation rather than activities. For example, the standard potential for the half-reaction

$$Fe^{3+} + e^- \rightleftharpoons Fe^{2+} \qquad E^0 = +0.771 \text{ V}$$

is +0.771 V. When the potential of a platinum electrode immersed in a solution that is 10^{-4} M in iron(III) ion, iron(II) ion, and perchloric acid is measured against a standard hydrogen electrode, a reading of close to +0.77 V is obtained, as predicted by theory. If, however, perchloric acid is added to this mixture until the acid concentration is 0.1 M, the potential is found to decrease to about +0.75 V. This difference is attributable to the fact that the activity coefficient of iron(III) is considerably smaller than that of iron(II) (0.4 versus 0.18) at the high ionic strength of the 0.1 M perchloric acid medium (see Table 10-2 page 242). As a consequence, the ratio of activities of the two species ($[Fe^{2+}]/[Fe^{3+}]$) in the Nernst equation is greater than unity, a condition that leads to a decrease in the electrode potential. In 1 M $HClO_4$, the electrode potential is even smaller (≈ 0.73 V).

Effect of Other Equilibria

The following further complicate application of standard electrode potential data to many systems of interest in analytical chemistry: association, dissociation, complex formation, and solvolysis equilibria of the species that appear in the Nernst equation. These phenomena can be taken into account only if their existence is known and appropriate equilibrium constants are available. More often than not, neither of these requirements is met and significant discrepancies arise. For example, the presence of 1 M hydrochloric acid in the iron(II)/iron(III) mixture we have just discussed leads to a measured potential of +0.70 V, while in 1 M sulfuric acid, a potential of +0.68 V is observed, and in a 2 M phosphoric acid, the potential is +0.46 V. In each of these cases, the iron(II)/iron(III) activity ratio is larger because the complexes of iron(III) with chloride, sulfate, and phosphate ions are more stable than those of iron(II). In these cases, the ratio of the species concentrations, $[Fe^{2+}]/[Fe^{3+}]$, in the Nernst equation is greater than unity, and the measured potential is less than the standard potential. If formation constants for these complexes were available, it would be possible to make appropriate corrections. Unfortunately, such data are often not available, or if they are, they are not very reliable.

Formal Potentials

A **formal potential** is the electrode potential when the ratio of **analytical concentrations** of reactants and products of a half-reaction are exactly 1.00 and the molar concentrations of any other solutes are specified. To distinguish the formal potential from the standard electrode potential a prime symbol is added to E^0.

Formal potentials are empirical potentials that compensate for the types of activity and competing equilibria effects that we have just described. The formal potential $E^{0'}$ of a system is the potential of the half-cell with respect to the standard hydrogen electrode measured under conditions such that the ratio of analytical concentrations of reactants and products as they appear in the Nernst equation is exactly unity and the concentrations of other species in the system are all carefully specified. For example, the formal potential for the half-reaction

$$Ag^+ + e^- \rightleftharpoons Ag(s) \qquad E^{0'} = 0.792 \text{ V in 1 M } HClO_4$$

could be obtained by measuring the potential of the cell shown in **Figure 18-10**. Here, the right-hand electrode is a silver electrode immersed in a solution that is 1.00 M in $AgNO_3$ and 1.00 M in $HClO_4$. The reference electrode on the left is a standard hydrogen electrode. This cell has a potential of +0.792 V, which is the formal potential of the Ag^+/Ag couple in 1.00 M $HClO_4$. Note that the standard potential for this couple is +0.799 V.

Formal potentials for many half-reactions are listed in Appendix 5. Note that there are large differences between the formal and standard potentials for some half-reactions. For example, the formal potential for

$$Fe(CN)_6^{3-} + e^- \rightleftharpoons Fe(CN)_6^{4-} \qquad E^0 = +0.36 \text{ V}$$

is 0.72 V in 1 M perchloric or sulfuric acids, which is 0.36 V greater than the standard electrode potential for the half-reaction. The reason for this difference is that in the presence of high concentrations of hydrogen ion, hexacyanoferrate(II) ions ($Fe(CN)_6^{4-}$), and hexacyanoferrate(III) ions ($Fe(CN)_6^{3-}$) combine with one or more protons to form hydrogen hexacyanoferrate(II) and hydrogen hexacyanoferrate(III)

$a_{H^+} = 1.00$

$c_{AgNO_3} = 1.00\ M$
$c_{HClO_4} = 1.00\ M$

Figure 18-10 Measurement of the formal potential of the Ag^+/Ag couple in 1 M $HClO_4$.

acid species. Because $H_4Fe(CN)_6$ is a weaker acid than $H_3Fe(CN)_6$, the ratio of the species concentrations, $[Fe(CN)_6^{4-}]/[Fe(CN)_6^{3-}]$, in the Nernst equation is less than 1, and the observed potentials are greater.

Substitution of formal potentials for standard electrode potentials in the Nernst equation yields better agreement between calculated and experimental results—provided, of course, that the electrolyte concentration of the solution approximates that for which the formal potential is applicable. Not surprisingly, attempts to apply formal potentials to systems that differ substantially in type and in concentration of electrolyte can result in errors that are larger than those associated with the use of standard electrode potentials. In this text, we use whichever is the more appropriate.

Spreadsheet Summary In the first exercise in Chapter 10 of *Applications of Microsoft® Excel in Analytical Chemistry*, 2nd ed., a spreadsheet is developed to calculate electrode potentials as a function of the ratio of reductant-to-oxidant concentration ([R]/[O]) for the case of two soluble species. Plots of E versus [R]/[O] and E versus log([R]/[O]) are made, and the slopes and intercepts determined. The spreadsheet is modified for metal/metal ion systems.

WEB WORKS Fuel cells have been used to provide electrical power for spacecraft since the 1960s. In recent years, fuel cell technology has begun to mature, and batteries made up of fuel cells will soon be or are now available for small-scale power generation and electric automobiles. Use a search engine to find the Fuel Cells 2000 website. Locate an article that explains the operation of the hydrogen fuel cell. Describe the proton-exchange membrane and explain its role in the hydrogen fuel cell. Discuss the advantages of the hydrogen fuel cell over other electrical energy storage devices such as lead-acid batteries, lithium-hydride batteries, and so forth. What are its disadvantages? What are the some of the reasons why this technology has not rapidly replaced current energy technologies?

QUESTIONS AND PROBLEMS

NOTE: Numerical data are molar analytical concentrations where the full formula of a species is provided. Molar equilibrium concentrations are supplied for species displayed as ions.

18-1. Briefly describe or define
 *(a) oxidation.
  (b) reducing agent.
 *(c) salt bridge.
  (d) liquid junction.
 *(e) Nernst equation.

18-2. Briefly describe or define
 *(a) electrode potential.
  (b) formal potential.
 *(c) standard electrode potential.
  (d) liquid-junction potential.
  (e) oxidation potential.

18-3. Make a clear distinction between
 *(a) oxidation and oxidizing agent.
  (b) an electrolytic cell and a galvanic cell.
 *(c) the cathode of an electrochemical cell and the right-hand electrode.
  (d) a reversible electrochemical cell and an irreversible electrochemical cell.
 *(e) the standard electrode potential and formal potential.

*18-4. The following entries are found in a table of standard electrode potentials:

$$I_2(s) + 2e^- \rightleftharpoons 2I^- \qquad E^0 = 0.5355 \text{ V}$$
$$I_2(aq) + 2e^- \rightleftharpoons 2I^- \qquad E^0 = 0.615 \text{ V}$$

What is the significance of the difference between these two standard potentials?

*18-5. Why is it necessary to bubble hydrogen through the electrolyte in a hydrogen electrode?

18-6. The standard electrode potential for the reduction of Ni^{2+} to Ni is -0.25 V. Would the potential of a nickel electrode immersed in a 1.00 M NaOH solution saturated with $Ni(OH)_2$ be more negative than $E^0_{Ni^{2+}/Ni}$ or less? Explain.

18-7. Write balanced net ionic equations for the following reactions. Supply H^+ and/or H_2O as needed to obtain balance.
 *(a) $Fe^{3+} + Sn^{2+} \rightarrow Fe^{2+} + Sn^{4+}$
  (b) $Cr(s) + Ag^+ \rightarrow Cr^{3+} + Ag(s)$
 *(c) $NO_3^- + Cu(s) \rightarrow NO_2(g) + Cu^{2+}$
  (d) $MnO_4^- + H_2SO_3 \rightarrow Mn^{2+} + SO_4^{2-}$
 *(e) $Ti^{3+} + Fe(CN)_6^{3-} \rightarrow TiO^{2+} + Fe(CN)_6^{4-}$
  (f) $H_2O_2 + Ce^{4+} \rightarrow O_2(g) + Ce^{3+}$
 *(g) $Ag(s) + I^- + Sn^{4+} \rightarrow AgI(s) + Sn^{2+}$
  (h) $UO_2^{2+} + Zn(s) \rightarrow U^{4+} + Zn^{2+}$
 *(i) $HNO_2 + MnO_4^- \rightarrow NO_3^- + Mn^{2+}$
  (j) $HN_2NNH_2 + IO_3^- + Cl^- \rightarrow N_2(g) + ICl_2^-$

*18-8. Identify the oxidizing agent and the reducing agent on the left side of each equation in Problem 18-7; write a balanced equation for each half-reaction.

18-9. Write balanced net ionic equations for the following reactions. Supply H^+ and/or H_2O as needed to obtain balance.
 *(a) $MnO_4^- + VO^{2+} \rightarrow Mn^{2+} + V(OH)_4^+$
  (b) $I_2 + H_2S(g) \rightarrow I^- + S(s)$
 *(c) $Cr_2O_7^{2-} + U^{4+} \rightarrow Cr^{3+} + UO_2^{2+}$
  (d) $Cl^- + MnO_2(s) \rightarrow Cl_2(g) + Mn^{2+}$
 *(e) $IO_3^- + I^- \rightarrow I_2(aq)$
  (f) $IO_3^- + I^- + Cl^- \rightarrow ICl_2^-$
 *(g) $HPO_3^{2-} + MnO_4^- + OH^- \rightarrow PO_4^{3-} + MnO_4^{2-}$
  (h) $SCN^- + BrO_3^- \rightarrow Br^- + SO_4^{2-} + HCN$
 *(i) $V^{2+} + V(OH)_4^+ \rightarrow VO^{2+}$
  (j) $MnO_4^- + Mn^{2+} + OH^- \rightarrow MnO_2(s)$

18-10. Identify the oxidizing agent and the reducing agent on the left side of each equation in Problem 18-9; write a balanced equation for each half-reaction.

*18-11. Consider the following oxidation/reduction reactions:

$$AgBr(s) + V^{2+} \rightarrow Ag(s) + V^{3+} + Br^-$$
$$Tl^{3+} + 2Fe(CN)_6^{4-} \rightarrow Tl^+ + 2Fe(CN)_6^{3-}$$
$$2V^{3+} + Zn(s) \rightarrow 2V^{2+} + Zn^{2+}$$
$$Fe(CN)_6^{3-} + Ag(s) + Br^- \rightarrow Fe(CN)_6^{4-} + AgBr(s)$$
$$S_2O_8^{2-} + Tl^+ \rightarrow 2SO_4^{2-} + Tl^{3+}$$

 (a) Write each net process in terms of two balanced half-reactions.
 (b) Express each half-reaction as a reduction.
 (c) Arrange the half-reactions in (b) in order of decreasing effectiveness as electron acceptors.

18-12. Consider the following oxidation/reduction reactions:

$$2H^+ + Sn(s) \rightarrow H_2(g) + Sn^{2+}$$
$$Ag^+ + Fe^{2+} \rightarrow Ag(s) + Fe^{3+}$$
$$Sn^{4+} + H_2(g) \rightarrow Sn^{2+} + 2H^+$$
$$2Fe^{3+} + Sn^{2+} \rightarrow 2Fe^{2+} + Sn^{4+}$$
$$Sn^{2+} + Co(s) \rightarrow Sn(s) + Co^{2+}$$

 (a) Write each net process in terms of two balanced half-reactions.
 (b) Express each half-reaction as a reduction.
 (c) Arrange the half-reactions in (b) in order of decreasing effectiveness as electron acceptors.

*18-13. Calculate the potential of a copper electrode immersed in
 (a) 0.0380 M $Cu(NO_3)_2$.
 (b) 0.0650 M in NaCl and saturated with CuCl.
 (c) 0.0350 M in NaOH and saturated with $Cu(OH)_2$.
 (d) 0.0375 M in $Cu(NH_3)_4^{2+}$ and 0.108 M in NH_3 (β_4 for $Cu(NH_3)_4^{2+}$ is 5.62×10^{11}).

(e) a solution in which the molar analytical concentration of $Cu(NO_3)_2$ is 3.90×10^{-3} M, that for H_2Y^{2-} is 3.90×10^{-2} M (Y = EDTA), and the pH is fixed at 4.00.

18-14. Calculate the potential of a zinc electrode immersed in
(a) 0.0500 M $Zn(NO_3)_2$.
(b) 0.0200 M in NaOH and saturated with $Zn(OH)_2$.
(c) 0.0150 M in $Zn(NH_3)_4^{2+}$ and 0.350 M in NH_3—β_4 for $Zn(NH_3)_4^{2+}$ is 7.76×10^8.
(d) a solution in which the molar analytical concentration of $Zn(NO_3)_2$ is 4.00×10^{-3}, that for H_2Y^{2-} is 0.0550 M, and the pH is fixed at 9.00.

18-15. Use activities to calculate the electrode potential of a hydrogen electrode in which the electrolyte is 0.0100 M HCl and the activity of H_2 is 1.00 atm.

*18-16. Calculate the potential of a platinum electrode immersed in a solution that is
(a) 0.0160 M in K_2PtCl_4 and 0.2450 M in KCl.
(b) 0.0650 M in $Sn(SO_4)_2$ and 3.5×10^{-3} M in $SnSO_4$.
(c) buffered to a pH of 6.50 and saturated with $H_2(g)$ at 1.00 atm.
(d) 0.0255 M in $VOSO_4$, 0.0686 M in $V_2(SO_4)_3$, and 0.100 M in $HClO_4$.
(e) prepared by mixing 25.00 mL of 0.0918 M $SnCl_2$ with an equal volume of 0.1568 M $FeCl_3$.
(f) prepared by mixing 25.00 mL of 0.0832 M $V(OH)_4^+$ with 50.00 mL of 0.01087 M $V_2(SO_4)_3$ and has a pH of 1.00.

18-17. Calculate the potential of a platinum electrode immersed in a solution that is
(a) 0.0613 M in $K_4Fe(CN)_6$ and 0.00669 M in $K_3Fe(CN)_6$.
(b) 0.0400 M in $FeSO_4$ and 0.00915 M in $Fe_2(SO_4)_3$.
(c) buffered to a pH of 5.55 and saturated with H_2 at 1.00 atm.
(d) 0.1015 M in $V(OH)_4^+$, 0.0799 M in VO^{2+}, and 0.0800 M in $HClO_4$.
(e) prepared by mixing 50.00 mL of 0.0607 M $Ce(SO_4)_2$ with an equal volume of 0.100 M $FeCl_2$ (assume solutions were 1.00 M in H_2SO_4 and use formal potentials).
(f) prepared by mixing 25.00 mL of 0.0832 M $V_2(SO_4)_3$ with 50.00 mL of 0.00628 M $V(OH)_4^+$ and has a pH of 1.00.

*18-18. If the following half-cells are the right-hand electrode in a galvanic cell with a standard hydrogen electrode on the left, calculate the cell potential. If the cell were shorted, indicate whether the electrodes shown would act as an anode or a cathode.
(a) $Ni|Ni^{2+}(0.0883 \text{ M})$
(b) $Ag|AgI(\text{sat'd}), KI(0.0898 \text{ M})$
(c) $Pt|O_2(780 \text{ torr}), HCl(2.50 \times 10^{-4} \text{ M})$
(d) $Pt|Sn^{2+}(0.0893 \text{ M}), Sn^{4+}(0.215 \text{ M})$
(e) $Ag|Ag(S_2O_3)_2^{3-}(0.00891 \text{ M}), Na_2S_2O_3(0.1035 \text{ M})$

18-19. The following half-cells are on the left and coupled with the standard hydrogen electrode on the right to form a galvanic cell. Calculate the cell potential. Indicate which electrode would be the cathode if each cell were short circuited.
(a) $Cu|Cu^{2+}(0.0805 \text{ M})$
(b) $Cu|CuI(\text{sat'd}), KI(0.0993 \text{ M})$
(c) $Pt, H_2(0.914 \text{ atm})|HCl(1.00 \times 10^{-4} \text{ M})$
(d) $Pt|Fe^{3+}(0.0886 \text{ M}), Fe^{2+}(0.1420 \text{ M})$
(e) $Ag|Ag(CN)_2^-(0.0778 \text{ M}), KCN(0.0651 \text{ M})$

*18-20. The solubility-product constant for Ag_2SO_3 is 1.5×10^{-14}. Calculate E^0 for the process

$$Ag_2SO_3(s) + 2e^- \rightleftharpoons 2Ag + SO_3^{2-}$$

18-21. The solubility-product constant for $Ni_2P_2O_7$ is 1.7×10^{-13}. Calculate E^0 for the process

$$Ni_2P_2O_7(s) + 4e^- \rightleftharpoons 2Ni(s) + P_2O_7^{4-}$$

*18-22. The solubility-product constant for Tl_2S is 6×10^{-22}. Calculate E^0 for the reaction

$$Tl_2S(s) + 2e^- \rightleftharpoons 2Tl(s) + S^{2-}$$

18-23. The solubility product for $Pb_3(AsO_4)_2$ is 4.1×10^{-36}. Calculate E^0 for the reaction

$$Pb_2(AsO_4)_2(s) + 6e^- \rightleftharpoons 3Pb(s) + 2AsO_4^{2-}$$

*18-24. Compute E^0 for the process

$$ZnY^{2-} + 2e^- \rightleftharpoons Zn(s) + Y^{4-}$$

where Y^{4-} is the completely deprotonated anion of EDTA. The formation constant for ZnY^{2-} is 3.2×10^{16}.

*18-25. Given the formation constants

$$Fe^{3+} + Y^{4-} \rightleftharpoons FeY^- \quad K_f = 1.3 \times 10^{25}$$
$$Fe^{2+} + Y^{4-} \rightleftharpoons FeY^{2-} \quad K_f = 2.1 \times 10^{14}$$

calculate E^0 for the process

$$FeY^- + e^- \rightleftharpoons FeY^{2-}$$

18-26. Calculate E^0 for the process

$$Cu(NH_3)_4^{2+} + e^- \rightleftharpoons Cu(NH_3)_2^+ + 2\,NH_3$$

given that

$$Cu^+ + 2NH_3 \rightleftharpoons Cu(NH_3)_2^+ \quad \beta_2 = 7.2 \times 10^{10}$$
$$Cu^{2+} + 4NH_3 \rightleftharpoons Cu(NH_3)_4^{2+} \quad \beta_4 = 5.62 \times 10^{11}$$

18-27. For a $Pt|Fe^{3+}, Fe^{2+}$ half-cell, find the potential for the following ratios of $[Fe^{3+}]/[Fe^{2+}]$: 0.001, 0.0025, 0.005, 0.0075, 0.010, 0.025, 0.050, 0.075, 0.100, 0.250, 0.500, 0.750, 1.00, 1.250, 1.50, 1.75, 2.50, 5.00, 10.00, 25.00, 75.00, and 100.00.

18-28. For a $Pt|Ce^{4+}, Ce^{3+}$ half-cell, find the potential for the same ratios of $[Ce^{4+}]/[Ce^{3+}]$ as given in Problem 18-27 for $[Fe^{3+}]/[Fe^{2+}]$.

18-29. Plot the half-cell potential versus concentration ratio for the half-cells of Problems 18-27 and 18-28. How would the plot look if potential were plotted against log(concentration ratio)?

18-30.   **Challenge Problem**: At one time, the standard hydrogen electrode was used for measuring pH.

   (a) Sketch a diagram of an electrochemical cell that could be used to measure pH and label all parts of the diagram. Use the SHE for both half-cells.

   (b) Derive an equation that gives the potential of the cell in terms of the hydronium ion concentration $[H_3O^+]$ in both half-cells.

   (c) One half-cell should contain a solution of known hydronium ion concentration, and the other should contain the unknown solution. Solve the equation in (b) for the pH of the solution in the unknown half-cell.

   (d) Modify your resulting equation to account for activity coefficients and express the result in terms of $pa_H = -\log a_H$, the negative logarithm of the hydronium ion activity.

   (e) Describe the circumstances under which you would expect the cell to provide accurate measurements of pa_H.

   (f) Could your cell be used to make practical absolute measurements of pa_H or would you have to calibrate your cell with solutions of known pa_H? Explain your answer in detail.

   (g) How (or where) could you obtain solutions of known pa_H?

   (h) Discuss the practical problems that you might encounter in using your cell for making pH measurements.

   (i) Klopsteg[5] discusses how to make hydrogen electrode measurements. In Figure 2 of his paper, he suggests using a slide rule, a segment of which is shown here, to convert hydronium ion concentrations to pH and vice versa.

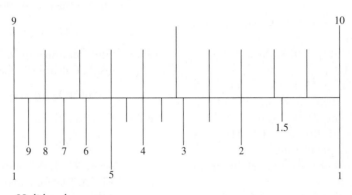

pH slide rule.

Explain the principles of operation of this slide rule and describe how it works. What reading would you obtain from the slide rule for a hydronium ion concentration of 3.56×10^{-4} M? How many significant figures are there in the resulting pH? What is the hydronium ion concentration of a solution of pH = 9.85?

[5] P. E. Klopsteg, *Ind. Eng. Chem.*, **1922**, *14*(5), 399, **DOI:** 10.1021/ie50149a011.

Applications of Standard Electrode Potentials

CHAPTER 19

This composite satellite image displays areas on the surface of the Earth where chlorophyll-bearing plants are located. Chlorophyll, which is one of nature's most important biomolecules, is a member of a class of compounds called porphyrins. This class also includes hemoglobin and cytochrome *c*, which is discussed in Feature 19-1. Many analytical techniques have been used to measure the chemical and physical properties of chlorophyll to explore its role in photosynthesis. The redox titration of chlorophyll with other standard redox couples reveals the oxidation/ reduction properties of the molecule that help explain the photophysics of the complex process that green plants use to oxidize water to molecular oxygen.

NASA/Jesse Allen, Earth Observatory/SeaWiFS/NASA/GSFC/ORBIMAGE

In this chapter, we show how standard electrode potentials can be used for (1) calculating thermodynamic cell potentials, (2) calculating equilibrium constants for redox reactions, and (3) constructing redox titration curves.

19A CALCULATING POTENTIALS OF ELECTROCHEMICAL CELLS

We can use standard electrode potentials and the Nernst equation to calculate the potential obtainable from a galvanic cell or the potential required to operate an electrolytic cell. The calculated potentials (sometimes called thermodynamic potentials) are theoretical in the sense that they refer to cells in which there is no current. As we show in Chapter 22, additional factors must be taken into account if there is current in the cell.

The thermodynamic potential of an electrochemical cell is the difference between the electrode potential of the right-hand electrode and the electrode potential of the left-hand electrode, that is,

$$E_{cell} = E_{right} - E_{left} \tag{19-1}$$

where E_{right} and E_{left} are the electrode potentials of the right-hand and left-hand electrodes, respectively. Equation 19-1 is valid when the liquid junction potential is absent or minimal. Throughout this chapter, we will assume that liquid junction potentials are negligible.

> It is important to note that E_{right} and E_{left} in Equation 19-1 are both *electrode potentials* as defined at the beginning of Section 18C-3.

Emilio Segrè Visual Archives/AIP

Gustav Robert Kirchhoff (1824–1877) was a German physicist who made many important contributions to physics and chemistry. In addition to his work in spectroscopy, he is known for Kirchhoff's laws of current and voltage in electrical circuits. These laws can be summarized by the following equations: $\Sigma I = 0$, and $\Sigma E = 0$. These equations state that the sum of the currents into any circuit point (node) is zero and the sum of the potential differences around any circuit loop is zero.

EXAMPLE 19-1

Calculate the thermodynamic potential of the following cell and the free energy change associated with the cell reaction:

$$Cu|Cu^{2+}(0.0200\ M) \parallel Ag^+(0.0200\ M)|Ag$$

Note that this cell is the galvanic cell shown in Figure 18-2a.

Solution

The two half-reactions and standard potentials are

$$Ag^+ + e^- \rightleftharpoons Ag(s) \qquad E^0 = 0.799\ V \qquad (19\text{-}2)$$

$$Cu^{2+} + 2e^- \rightleftharpoons Cu(s) \qquad E^0 = 0.337\ V \qquad (19\text{-}3)$$

The electrode potentials are

$$E_{Ag^+/Ag} = 0.799 - 0.0592 \log \frac{1}{0.0200} = 0.6984\ V$$

$$E_{Cu^{2+}/Cu} = 0.337 - \frac{0.0592}{2} \log \frac{1}{0.0200} = 0.2867\ V$$

We see from the cell diagram that the silver electrode is the right-hand electrode and the copper electrode is the left-hand electrode. Therefore, application of Equation 19-1 gives

$$E_{cell} = E_{right} - E_{left} = E_{Ag^+/Ag} - E_{Cu^{2+}/Cu} = 0.6984 - 0.2867 = +0.412\ V$$

The free energy change ΔG for the reaction $Cu(s) + 2Ag^+ \rightleftharpoons Cu^{2+} + Ag(s)$ is found from

$$\Delta G = -nFE_{cell} = -2 \times 96485\ C \times 0.412\ V = -79{,}503\ J\ (18.99\ kcal)$$

EXAMPLE 19-2

Calculate the potential for the cell

$$Ag|Ag^+(0.0200\ M) \parallel Cu^{2+}(0.0200\ M)|Cu$$

Solution

The electrode potentials for the two half-reactions are identical to the electrode potentials calculated in Example 19-1, that is,

$$E_{Ag^+/Ag} = 0.6984\ V \qquad \text{and} \qquad E_{Cu^{2+}/Cu} = 0.2867\ V$$

In contrast to the previous example, however, the silver electrode is on the left, and the copper electrode is on the right. Substituting these electrode potentials into Equation 19-1 gives

$$E_{cell} = E_{right} - E_{left} = E_{Cu^{2+}/Cu} - E_{Ag^+/Ag} = 0.2867 - 0.6984 = -0.412\ V$$

Examples 19-1 and 19-2 illustrate an important fact. The magnitude of the potential difference between the two electrodes is 0.412 V independent of which electrode is considered the left or reference electrode. If the Ag electrode is the left electrode as in Example 19-2, the cell potential has a negative sign, but if the Cu electrode is the reference as in Example 19-2, the cell potential has a positive sign. However, no matter how the cell is arranged, the spontaneous cell reaction is oxidation of Cu and reduction of Ag^+, and the free energy change is 79,503 J. Examples 19-3 and 19-4 illustrate other types of electrode reactions.

EXAMPLE 19-3

Calculate the potential of the following cell and indicate the reaction that would occur spontaneously if the cell were short-circuited (see Figure 19-1).

$$Pt|U^{4+}(0.200\ M),\ UO_2^{2+}(0.0150\ M),\ H^+(0.0300\ M)\ \|$$
$$Fe^{2+}(0.0100\ M),\ Fe^{3+}(0.0250\ M)|Pt$$

Solution

The two half-reactions are

$$Fe^{3+} + e^- \rightleftharpoons Fe^{2+} \qquad\qquad E^0 = +0.771\ V$$

$$UO_2^{2+} + 4H^+ + 2e^- \rightleftharpoons U^{4+} + 2H_2O \qquad E^0 = +0.334\ V$$

Figure 19-1 Cell for Example 19-3.

(continued)

The electrode potential for the right-hand electrode is

$$E_{right} = 0.771 - 0.0592 \log \frac{[Fe^{2+}]}{[Fe^{3+}]}$$

$$= 0.771 - 0.0592 \log \frac{0.0100}{0.0250} = 0.771 - (-0.0236)$$

$$= 0.7946 \text{ V}$$

The electrode potential for the left-hand electrode is

$$E_{left} = 0.334 - \frac{0.0592}{2} \log \frac{[U^{4+}]}{[UO_2^{2+}][H^+]^4}$$

$$= 0.334 - \frac{0.0592}{2} \log \frac{0.200}{(0.0150)(0.0300)^4}$$

$$= 0.334 - 0.2136 = 0.1204 \text{ V}$$

and

$$E_{cell} = E_{right} - E_{left} = 0.7946 - 0.1204 = 0.6742 \text{ V}$$

The positive sign means that the spontaneous reaction is the oxidation of U^{4+} on the left and the reduction of Fe^{3+} on the right, or

$$U^{4+} + 2Fe^{3+} + 2H_2O \rightarrow UO_2^{2+} + 2Fe^{2+} + 4H^+$$

EXAMPLE 19-4

Calculate the cell potential for

$$Ag \,|\, AgCl(sat'd), HCl(0.0200 \text{ M}) \,|\, H_2(0.800 \text{ atm}), Pt$$

Note that this cell does not require two compartments (nor a salt bridge) because molecular H_2 has little tendency to react directly with the low concentration of Ag^+ in the electrolyte solution. This is an example of a **cell without liquid junction** (see **Figure 19-2**).

Solution

The two half-reactions and their corresponding standard electrode potentials are (see Table 18-1).

$$2H^+ + 2e^- \rightleftharpoons H_2(g) \qquad\qquad E^0_{H^+/H_2} = 0.000 \text{ V}$$

$$AgCl(s) + e^- \rightleftharpoons Ag(s) + Cl^- \qquad E^0_{AgCl/Ag} = 0.222 \text{ V}$$

[H$^+$] = 0.0200 M
[Cl$^-$] = 0.0200 M

Figure 19-2 Cell without liquid junction for Example 19-4.

The two electrode potentials are

$$E_{\text{right}} = 0.000 - \frac{0.0592}{2} \log \frac{p_{\text{H}_2}}{[\text{H}^+]^2} = -\frac{0.0592}{2} \log \frac{0.800}{(0.0200)^2}$$

$$= -0.0977 \text{ V}$$

$$E_{\text{left}} = 0.222 - 0.0592 \log[\text{Cl}^-] = 0.222 - 0.0592 \log 0.0200$$

$$= 0.3226 \text{ V}$$

The cell potential is thus

$$E_{\text{cell}} = E_{\text{right}} - E_{\text{left}} = -0.0977 - 0.3226 = -0.420 \text{ V}$$

The negative sign indicates that the cell reaction as considered

$$2\text{H}^+ + 2\text{Ag}(s) \rightarrow \text{H}_2(g) + 2\text{AgCl}(s)$$

is nonspontaneous. In order to get this reaction to occur, we would have to apply an external voltage and construct an electrolytic cell.

EXAMPLE 19-5

Calculate the potential for the following cell using (a) concentrations and (b) activities:

$$Zn \,|\, ZnSO_4(x \text{ M}), PbSO_4(\text{sat'd}) \,|\, Pb$$

where $x = 5.00 \times 10^{-4}, 2.00 \times 10^{-3}, 1.00 \times 10^{-2}, 5.00 \times 10^{-2}$.

Solution

(a) In a neutral solution, little HSO_4^- is formed, and we can assume that

$$[SO_4^{2-}] = c_{ZnSO_4} = x = 5.00 \times 10^{-4} \text{ M}$$

The half-reactions and standard electrode potentials are (see Table 18-1).

$$PbSO_4(s) + 2e^- \rightleftharpoons Pb(s) + SO_4^{2-} \qquad E^0_{PbSO_4/Pb} = -0.350 \text{ V}$$

$$Zn^{2+} + 2e^- \rightleftharpoons Zn(s) \qquad E^0_{Zn^{2+}/Zn} = -0.763 \text{ V}$$

The lead electrode potential is

$$E_{PbSO_4/Pb} = E^0_{PbSO_4/Pb} - \frac{0.0592}{2} \log [SO_4^{2-}]$$

$$= -0.350 - \frac{0.0592}{2} \log(5.00 \times 10^{-4}) = -0.252 \text{ V}$$

The zinc electrode potential is

$$E_{Zn^{2+}/Zn} = E^0_{Zn^{2+}/Zn} - \frac{0.0592}{2} \log \frac{1}{[Zn^{2+}]}$$

$$= -0.763 - \frac{0.0592}{2} \log \frac{1}{5.00 \times 10^{-4}} = -0.860 \text{ V}$$

The cell potential is thus

$$E_{cell} = E_{right} - E_{left} = E_{PbSO_4/Pb} - E_{Zn^{2+}/Zn} = -0.252 - (-0.860) = 0.608 \text{ V}$$

Cell potentials at the other concentrations can be calculated in the same way. Their values are given in **Table 19-1**.

(b) To calculate activity coefficients for Zn^{2+} and $[SO_4^{2-}]$, we must first find the ionic strength of the solution using Equation 10-1:

$$\mu = \frac{1}{2}[5.00 \times 10^{-4} \times (2)^2 + 5.00 \times 10^{-4} \times (2)^2] = 2.00 \times 10^{-3}$$

In Table 10-2, we find and $\alpha_{SO_4^{2-}} = 0.4$ nm and $\alpha_{Zn^{2+}} = 0.6$ nm. If we substitute these values into Equation 10-5, we find that

$$-\log \gamma_{SO_4^{2-}} = \frac{0.51 \times (2)^2 \sqrt{2.00 \times 10^{-3}}}{1 + 3.3 \times 0.4 \sqrt{2.00 \times 10^{-3}}} = 8.61 \times 10^{-2}$$

$$\gamma_{SO_4^{2-}} = 0.820$$

Repeating the calculations for Zn^{2+}, we find that

$$\gamma_{Zn^{2+}} = 0.825$$

The Nernst equation for the lead electrode is now

$$E_{PbSO_4/Pb} = E^0_{PbSO_4/Pb} - \frac{0.0592}{2} \log \gamma_{SO_4^{2-}} c_{SO_4^{2-}}$$

$$= -0.350 - \frac{0.0592}{2} \log(0.820 \times 5.00 \times 10^{-4}) = -0.250 \text{ V}$$

and for the zinc electrode, we have

$$E_{Zn^{2+}/Zn} = E^0_{Zn^{2+}/Zn} - \frac{0.0592}{2} \log \frac{1}{\gamma_{Zn^{2+}} c_{Zn^{2+}}}$$

$$= -0.763 - \frac{0.0592}{2} \log \frac{1}{0.825 \times 5.00 \times 10^{-4}} = -0.863 \text{ V}$$

Finally, we find the cell potential from

$$E_{cell} = E_{right} - E_{left} = E_{PbSO_4/Pb} - E_{Zn^{2+}/Zn} = -0.250 - (-0.863) = 0.613 \text{ V}$$

Values for other concentrations and experimentally determined potentials for the cell are found in Table 19-1.

Table 19-1 shows that cell potentials calculated without activity coefficient corrections exhibit significant error. It is also clear from the data in the fifth column of the table that potentials computed with activities agree reasonably well with experiment.

TABLE 19-1

Effect of Ionic Strength on the Potential of a Galvanic Cell*

Concentration $ZnSO_4$, M	Ionic Strength, μ	(a) E, Based on Concentrations	(b) E, Based on Activities	E, Experimental Values[†]
5.00×10^{-4}	2.00×10^{-3}	0.608	0.613	0.611
2.00×10^{-3}	8.00×10^{-3}	0.573	0.582	0.583
1.00×10^{-2}	4.00×10^{-2}	0.531	0.550	0.553
2.00×10^{-2}	8.00×10^{-2}	0.513	0.537	0.542
5.00×10^{-2}	2.00×10^{-1}	0.490	0.521	0.529

*Cell described in Example 19-5. All potentials E are in volts.
[†]Experimental data from I. A. Cowperthwaite and V. K. LaMer, *J. Amer. Chem. Soc.*, **1931**, *53*, 4333, **DOI**: 10.1021/ja01363a010.

EXAMPLE 19-6

Calculate the potential required to initiate deposition of copper from a solution that is 0.010 M in $CuSO_4$ and contains sufficient H_2SO_4 to give a pH of 4.00.

Solution

The deposition of copper necessarily occurs at the cathode, which according to IUPAC convention is the right-hand-electrode. Since there is no more easily oxidizable species than water in the system, O_2 will evolve at the anode. The two half-reactions and their corresponding standard electrode potentials are (see Table 18-1):

$$Cu^{2+} + 2e^- \rightleftharpoons Cu(s) \qquad E^0_{AgCl/Ag} = +0.337 \text{ V (right)}$$

$$O_2(g) + 4H^+ + 4e^- \rightleftharpoons 2H_2O \qquad E^0_{O_2/H_2O} = +1.229 \text{ V (left)}$$

The electrode potential for the Cu electrode is

$$E_{Cu^{2+}/Cu} = +0.337 - \frac{0.0592}{2} \log \frac{1}{0.010} = +0.278 \text{ V}$$

If O_2 is evolved at 1.00 atm, the electrode potential for the oxygen electrode is

$$E_{O_2/H_2O} = +1.229 - \frac{0.0592}{4} \log \frac{1}{p_{O_2}[H^+]^4}$$

$$= +1.229 - \frac{0.0592}{4} \log \frac{1}{(1 \text{ atm})(1.00 \times 10^{-4})^4} = +0.992 \text{ V}$$

and the cell potential is thus

$$E_{cell} = E_{right} - E_{left} = E_{Cu^{2+}/Cu} - E_{O_2/H_2O} = +0.278 - 0.992 = -0.714 \text{ V}$$

The negative sign shows that the cell reaction

$$2Cu^{2+} + 2H_2O \rightarrow O_2(g) + 4H^+ + 2Cu(s)$$

is nonspontaneous and that, to cause copper to be deposited according to the following reaction, we must apply a negative potential slightly greater than -0.714 V.

 Spreadsheet Summary   In the first exercise in Chapter 10 of *Applications of Microsoft® Excel in Analytical Chemistry*, 2nd ed., a spreadsheet is developed for calculating electrode potentials for simple half-reactions. Plots are made of the potential versus the ratio of the reduced species to the oxidized species and of the potential versus the logarithm of this ratio.

DETERMINING STANDARD POTENTIALS
19B EXPERIMENTALLY

Although it is easy to look up standard electrode potentials for hundreds of half-reactions in compilations of electrochemical data, it is important to realize that none of these potentials, including the potential of the standard hydrogen electrode, can be measured directly in the laboratory. The SHE is a hypothetical electrode, as is any electrode system in which the reactants and products are at unit activity or pressure. Such electrode systems cannot be prepared in the lab because there is no way to prepare solutions containing ions whose activities are exactly 1. In other words, no theory is available that permits the calculation of the concentration of solute that must be dissolved in order to produce a solution of exactly unit activity. At high ionic strengths, the Debye Hückel relationships (see Section 10B-2), as well as other extended forms of the equation, do a relatively poor job of calculating activity coefficients, and there is no independent experimental method for determining activity coefficients in such solutions. So, for example, it is impossible to calculate the concentration of HCl or other acids that will produce a solution in which $a_{H^+} = 1$, and it is impossible to determine the activity experimentally. In spite of this difficulty, data collected in solutions of low ionic strength can be extrapolated to give valid estimates of theoretically defined standard electrode potentials. The following example shows how such hypothetical electrode potentials may be determined experimentally.

EXAMPLE 19-7

D. A. MacInnes[1] found that a cell similar to that shown in Figure 19-2 had a potential of 0.52053 V. The cell is described by the following notation:

$$Pt, H_2(1.00\ atm)\,|\,HCl(3.215 \times 10^{-3}\ M), AgCl(sat'd)\,|\,Ag$$

Calculate the standard electrode potential for the half-reaction

$$AgCl(s) + e^- \rightleftharpoons Ag(s) + Cl^-$$

Solution

In this example, the electrode potential for the right-hand electrode is

$$E_{right} = E^0_{AgCl} - 0.0592 \log (\gamma_{Cl^-})(c_{HCl})$$

where γ_{Cl^-} is the activity coefficient of Cl^-. The second half-cell reaction is

$$H^+ + e^- \rightleftharpoons \frac{1}{2} H_2(g)$$

and

$$E_{left} = E^0_{H^+/H_2} - \frac{0.0592}{1} \log \frac{p_{H_2}^{1/2}}{(\gamma_{H^+})(c_{HCl})}$$

The cell potential is then the difference between these two potentials

$$E_{cell} = E_{right} - E_{left}$$

$$= [E^0_{AgCl} - 0.0592 \log (\gamma_{Cl^-})(c_{HCl})] - \left[E^0_{H^+/H_2} - 0.0592 \log \frac{p_{H_2}^{1/2}}{(\gamma_{H^+})(c_{HCl})} \right]$$

$$= E^0_{AgCl} - 0.0592 \log (\gamma_{Cl^-})(c_{HCl}) - 0.000 - 0.0592 \log \frac{(\gamma_{H^+})(c_{HCl})}{p_{H_2}^{1/2}}$$

(continued)

---

[1]D. A. MacInnes, *The Principles of Electrochemistry*, New York: Reinhold, 1939, p. 187.

Notice that we have inverted the terms in the second logarithmic term. We now combine the two logarithmic terms to find that

$$E_{cell} = 0.52053 = E_{AgCl}^0 - 0.0592 \log \frac{(\gamma_{H^+})(\gamma_{Cl^-})(c_{HCl})^2}{p_{H_2}^{1/2}}$$

The activity coefficients for H^+ and Cl^- can be calculated from Equation 10-5 using 3.215×10^{-3} M for the ionic strength μ. These values are 0.945 and 0.939, respectively. If we substitute these values of the activity coefficients and the experimental data into the equation above and rearrange the equation, we obtain

$$E_{AgCl}^0 = 0.52053 + 0.0592 \log \frac{(0.945)(0.939)(3.215 \times 10^{-3})^2}{1.00^{1/2}}$$

$$= 0.2223 \approx 0.222 \text{ V}$$

MacInnes found the mean for this and similar measurements at other concentrations to be 0.222 V.

FEATURE 19-1

Biological Redox Systems

There are many redox systems of importance in biology and biochemistry. The cytochromes are excellent examples of such systems. Cytochromes are iron-heme proteins in which a porphyrin ring is coordinated through nitrogen atoms to an iron atom. These undergo one-electron redox reactions. The physiological functions of cytochromes are to facilitate electron transport. In the respiratory chain, the cytochromes are intimate participants in the formation of water from H_2. Reduced pyridine nucleotides deliver hydrogen to flavoproteins. The reduced flavoproteins are reoxidized by the Fe^{3+} of cytochrome b or c. The result is the formation of H^+ and the transport of electrons. The chain is completed when cytochrome oxidase transfers electrons to oxygen. The resulting oxide ion (O^{2-}) is unstable and immediately picks up two H^+ ions to produce H_2O. The scheme is illustrated in **Figure 19F-1**.

Most biological redox systems are pH dependent. It has become standard practice to list the electrode potentials of these systems at pH 7.0 in order to make comparisons of oxidizing or reducing powers. The values listed are typically formal potentials at pH 7.0 and sometimes symbolized $E_7^{0\prime}$.

Other redox systems of importance in biochemistry include the NADH/NAD system, the flavins, the pyruvate/lactate system, the oxalacetate/malate system, and the quinone/hydroquinone system.

Molecular model of cytochrome c.

19C CALCULATING REDOX EQUILIBRIUM CONSTANTS

Let us again consider the equilibrium that is established when a piece of copper is immersed in a solution containing a dilute solution of silver nitrate:

$$Cu(s) + 2Ag^+ \rightleftharpoons Cu^{2+} + 2Ag(s) \tag{19-4}$$

Figure 19F-1 Redox systems in the respiratory chain. P = phosphate ion. (From P. Karlson, *Introduction to Modern Biochemistry*, New York: Academic Press, 1963. With permission.)

The equilibrium constant for this reaction is

$$K_{eq} = \frac{[Cu^{2+}]}{[Ag^+]^2} \tag{19-5}$$

As we showed in Example 19-1, this reaction can be carried out in the galvanic cell

$$Cu|Cu^{2+}(xM) \| Ag^+(yM)|Ag$$

A sketch of a cell similar to this one is shown in Figure 18-2a. Its cell potential at any instant is given by Equation 19-1:

$$E_{cell} = E_{right} - E_{left} = E_{Ag^+/Ag} - E_{Cu^{2+}/Cu}$$

As the reaction proceeds, the concentration of Cu(II) ions increases, and the concentration of Ag(I) ions decreases. These changes make the potential of the copper electrode more positive and that of the silver electrode less positive. As shown in Figure 18-5, the net effect of these changes is a continuous decrease in the potential of the cell as it discharges. Ultimately, the concentrations of Cu(II) and Ag(I) attain their equilibrium values as determined by Equation 19-5, and the current ceases. Under these conditions, *the potential of the cell becomes zero.* Thus, *at chemical equilibrium,* we may write

$$E_{cell} = 0 = E_{right} - E_{left} = E_{Ag} - E_{Cu}$$

or

$$E_{right} = E_{left} = E_{Ag} = E_{Cu} \qquad (19\text{-}6)$$

> Remember *that, when redox systems are at equilibrium, the electrode potentials of all redox couples that are present are identical.* This generality applies whether the reactions take place directly in solution or indirectly in a galvanic cell

We can generalize Equation 19-6 by stating that, *at equilibrium, the electrode potentials for all half-reactions in an oxidation/reduction system are equal.* This generalization applies regardless of the number of half-reactions present in the system because interactions among all must take place until the electrode potentials are identical. For example, if we have four oxidation/reduction systems in a solution, interaction among all four takes place until the potentials of all four redox couples are equal.

Returning to the reaction shown in Equation 19-4, let us substitute Nernst expressions for the two electrode potentials in Equation 19-6, giving

$$E_{Ag}^0 - \frac{0.0592}{2}\log\frac{1}{[Ag^+]^2} = E_{Cu}^0 - \frac{0.0592}{2}\log\frac{1}{[Cu^{2+}]} \qquad (19\text{-}7)$$

Note that we apply the Nernst equation to the silver half-reaction as it appears in the balanced equation (Equation 19-4):

$$2Ag^+ + 2e^- \rightleftharpoons 2Ag(s) \qquad E^0 = 0.799 \text{ V}$$

Rearrangement of Equation 19-7 gives

$$E_{Ag}^0 - E_{Cu}^0 = \frac{0.0592}{2}\log\frac{1}{[Ag^+]^2} - \frac{0.0592}{2}\log\frac{1}{[Cu^{2+}]}$$

If we invert the ratio in the second log term, we must change the sign of the term. This inversion gives

$$E_{Ag}^0 - E_{Cu}^0 = \frac{0.0592}{2}\log\frac{1}{[Ag^+]^2} + \frac{0.0592}{2}\log\frac{[Cu^{2+}]}{1}$$

Finally, combining the log terms and rearranging gives

$$\frac{2(E_{Ag}^0 - E_{Cu}^0)}{0.0592} = \log\frac{[Cu^{2+}]}{[Ag^+]^2} = \log K_{eq} \qquad (19\text{-}8)$$

The concentration terms in Equation 19-8 are *equilibrium concentrations,* and the ratio $[Cu^{2+}]/[Ag^+]^2$ in the logarithmic term is, therefore, *the equilibrium constant for the reaction.* Note that the term in parenthesis in Equation 19-8 is the standard cell potential E_{cell}^0, which in general is given by

$$E_{cell}^0 = E_{right}^0 - E_{left}^0$$

We can also obtain Equation 19-8 from the free energy change for the reaction as was given in Equation 18-7. Rearrangement of this equation gives

$$\ln K_{eq} = -\frac{\Delta G^0}{RT} = \frac{nFE_{cell}^0}{RT} \tag{19-9}$$

At 25°C after conversion to base 10 logarithms, we can write

$$\log K_{eq} = \frac{nE_{cell}^0}{0.0592} = \frac{n(E_{right}^0 - E_{left}^0)}{0.0592}$$

For the reaction given in Equation 19-4, substituting E_{Ag}^0 for E_{right}^0 and E_{Cu}^0 for E_{left}^0 gives Equation 19-8.

EXAMPLE 19-8

Calculate the equilibrium constant for the reaction shown in Equation 19-4.

Solution

Substituting numerical values into Equation 19-8 yields

$$\log K_{eq} = \log\frac{[Cu^{2+}]}{[Ag^+]^2} = \frac{2(0.799 - 0.337)}{0.0592} = 15.61$$

$$K_{eq} = \text{antilog } 15.61 = 4.1 \times 10^{15}$$

> In making calculations of the sort shown in Example 19-8, you should follow the rounding rule for antilogs that is given on page 117.

EXAMPLE 19-9

Calculate the equilibrium constant for the reaction

$$2Fe^{3+} + 3I^- \rightleftharpoons 2Fe^{2+} + I_3^-$$

Solution

In Appendix 5, we find

$$2Fe^{3+} + 2e^- \rightleftharpoons 2Fe^{2+} \qquad E^0 = 0.771 \text{ V}$$
$$I_3^- + 2e^- \rightleftharpoons 3I^- \qquad E^0 = 0.536 \text{ V}$$

We have multiplied the first half-reaction by 2 so that the number of moles of Fe^{3+} and Fe^{2+} will be the same as in the balanced overall equation. We write the

(continued)

Nernst equation for Fe^{3+} based on the half-reaction for a 2-electron transfer, that is,

$$E_{Fe^{3+}/Fe^{2+}} = E^0_{Fe^{3+}/Fe^{2+}} - \frac{0.0592}{2} \log \frac{[Fe^{2+}]^2}{[Fe^{3+}]^2}$$

and

$$E_{I_3^-/I^-} = E^0_{I_3^-/I^-} - \frac{0.0592}{2} \log \frac{[I^-]^3}{[I_3^-]}$$

At equilibrium, the electrode potentials are equal, and

$$E_{Fe^{3+}/Fe^{2+}} = E_{I_3^-/I^-}$$

$$E^0_{Fe^{3+}/Fe^{2+}} - \frac{0.0592}{2} \log \frac{[Fe^{2+}]^2}{[Fe^{3+}]^2} = E^0_{I_3^-/I^-} - \frac{0.0592}{2} \log \frac{[I^-]^3}{[I_3^-]}$$

This equation rearranges to

$$\frac{2(E^0_{Fe^{3+}/Fe^{2+}} - E^0_{I_3^-/I^-})}{0.0592} = \log \frac{[Fe^{2+}]^2}{[Fe^{3+}]^2} - \log \frac{[I^-]^3}{[I_3^-]}$$

$$= \log \frac{[Fe^{2+}]^2}{[Fe^{3+}]^2} + \log \frac{[I_3^-]}{[I^-]^3}$$

$$= \log \frac{[Fe^{2+}]^2[I_3^-]}{[Fe^{3+}]^2[I^-]^3}$$

Notice that we have changed the sign of the second logarithmic term by inverting the fraction. Further arrangement gives

$$\log \frac{[Fe^{2+}]^2[I_3^-]}{[Fe^{3+}]^2[I^-]^3} = \frac{2(E^0_{Fe^{3+}/Fe^{2+}} - E^0_{I_3^-/I^-})}{0.0592}$$

Recall, however, that in this instance the concentration terms are *equilibrium concentrations*, and

$$\log K_{eq} = \frac{2(E^0_{Fe^{3+}/Fe^{2+}} - E^0_{I_3^-/I^-})}{0.0592} = \frac{2(0.771 - 0.536)}{0.0592} = 7.94$$

$$K_{eq} = \text{antilog } 7.94 = 8.7 \times 10^7$$

We round the answer to two figures because $\log K_{eq}$ contains only two significant figures (the two to the right of the decimal point).

FEATURE 19-2

A General Expression for Calculating Equilibrium Constants from Standard Potentials

To derive a general relationship for computing equilibrium constants from standard-potential data, let us consider a reaction in which a species A_{red} reacts with a species B_{ox} to yield A_{ox} and B_{red}. The two electrode reactions are

$$A_{ox} + ae^- \rightleftharpoons A_{red}$$
$$B_{ox} + be^- \rightleftharpoons B_{red}$$

We obtain a balanced equation for the desired reaction by multiplying the first equation by b and the second by a to give

$$bA_{ox} + bae^- \rightleftharpoons bA_{red}$$
$$aB_{ox} + bae^- \rightleftharpoons aB_{red}$$

We then subtract the first equation from the second to obtain a balanced equation for the redox reaction

$$bA_{red} + aB_{ox} \rightleftharpoons bA_{ox} + aB_{red}$$

When this system is at equilibrium, the two electrode potentials E_A and E_B are equal, that is,

$$E_A = E_B$$

If we substitute the Nernst expression for each couple into this equation, we find that *at equilibrium*

$$E_A^0 - \frac{0.0592}{ab} \log \frac{[A_{red}]^b}{[A_{ox}]^b} = E_B^0 - \frac{0.0592}{ab} \log \frac{[B_{red}]^a}{[B_{ox}]^a}$$

which rearranges to

$$E_B^0 - E_A^0 = \frac{0.0592}{ab} \log \frac{[A_{ox}]^b [B_{red}]^a}{[A_{red}]^b [B_{ox}]^a} = \frac{0.0592}{ab} \log K_{eq}$$

Finally, then,

$$\log K_{eq} = \frac{ab(E_B^0 - E_A^0)}{0.0592} \qquad (19\text{-}10)$$

Note that the product ab is the total number of electrons gained in the reduction (and lost in the oxidation) represented by the balanced redox equation. Thus, if $a = b$, it is not necessary to multiply the half-reactions by a and b. If $a = b = n$, the equilibrium constant is determined from

$$\log K_{eq} = \frac{n(E_B^0 - E_A^0)}{0.0592}$$

EXAMPLE 19-10

Calculate the equilibrium constant for the reaction

$$2MnO_4^- + 3Mn^{2+} + 2H_2O \rightleftharpoons 5MnO_2(s) + 4H^+$$

Solution

In Appendix 5, we find

$$2MnO_4^- + 8H^+ + 6e^- \rightleftharpoons 2MnO_2(s) + 4H_2O \qquad E^0 = +1.695 \text{ V}$$

$$3MnO_2(s) + 12H^+ + 6e^- \rightleftharpoons 3Mn^{2+} + 6H_2O \qquad E^0 = +1.23 \text{ V}$$

Again, we have multiplied both equations by integers so that the numbers of electrons are equal. When this system is at equilibrium,

$$E^0_{MnO_4^-/MnO_2} = E^0_{MnO_2/Mn^{2+}}$$

$$1.695 - \frac{0.0592}{6} \log \frac{1}{[MnO_4^-]^2[H^+]^8} = 1.23 - \frac{0.0592}{6} \log \frac{[Mn^{2+}]^3}{[H^+]^{12}}$$

If we invert the log term on the right and rearrange, we obtain

$$\frac{6(1.695 - 1.23)}{0.0592} = \log \frac{1}{[MnO_4^-]^2[H^+]^8} + \log \frac{[H^+]^{12}}{[Mn^{2+}]^3}$$

Adding the two log terms gives

$$\frac{6(1.695 - 1.23)}{0.0592} = \log \frac{[H^+]^{12}}{[MnO_4^-]^2[Mn^{2+}]^3[H^+]^8}$$

$$47.1 = \log \frac{[H^+]^4}{[MnO_4^-]^2[Mn^{2+}]^3} = \log K_{eq}$$

$$K_{eq} = \text{antilog } 47.1 = 1 \times 10^{47}$$

Note that the final result has only one significant figure.

> **Spreadsheet Summary**   In the second exercise in Chapter 10 of *Applications of Microsoft® Excel in Analytical Chemistry*, 2nd ed., cell potentials and equilibrium constants are calculated. A spreadsheet is developed for simple reactions to calculate complete cell potentials and equilibrium constants. The spreadsheet calculates E_{left}, E_{right}, E_{cell}, E^0_{cell}, $\log K_{eq}$, and K_{eq}.

19D CONSTRUCTING REDOX TITRATION CURVES

Because most redox indicators respond to changes in electrode potential, the vertical axis in oxidation/reduction titration curves is generally an electrode potential instead of the logarithmic p-functions that were used for complex-formation and neutralization titration curves. We saw in Chapter 18 that there is a logarithmic relationship between electrode potential and concentration of the analyte or titrant. Because of this relationship, redox titration curves are similar in appearance to those for other types of titrations in which a p-function is plotted as the ordinate.

19D-1 Electrode Potentials during Redox Titrations

Consider the redox titration of iron(II) with a standard solution of cerium(IV). This reaction is widely used for the determination of iron in various kinds of samples. The titration reaction is

$$Fe^{2+} + Ce^{4+} \rightleftharpoons Fe^{3+} + Ce^{3+}$$

This reaction is rapid and reversible so that the system is at equilibrium at all times throughout the titration. Consequently, the electrode potentials for the two half-reactions are always identical (Equation 19-6), that is,

$$E_{Ce^{4+}/Ce^{3+}} = E_{Fe^{3+}/Fe^{2+}} = E_{system}$$

where we have termed E_{system} as **the potential of the system**. If a redox indicator has been added to this solution, the ratio of the concentrations of its oxidized and reduced forms must adjust so that the electrode potential for the indicator, E_{In}, is also equal to the system potential. Therefore, using Equation 19-6, we may write

$$E_{In} = E_{Ce^{4+}/Ce^{3+}} = E_{Fe^{3+}/Fe^{2+}} = E_{system}$$

We can calculate the electrode potential of a system from standard potential data. Thus, for the reaction under consideration, the titration mixture is treated as if it were part of the hypothetical cell

$$SHE \| Ce^{4+}, Ce^{3+}, Fe^{3+}, Fe^{2+} | Pt$$

where SHE symbolizes the standard hydrogen electrode. The potential of the platinum electrode with respect to the standard hydrogen electrode is determined by the tendencies of iron(III) and cerium(IV) to accept electrons, that is, by the tendencies of the following half-reactions to occur:

$$Fe^{3+} + e^- \rightleftharpoons Fe^{2+}$$

$$Ce^{4+} + e^- \rightleftharpoons Ce^{3+}$$

At equilibrium, the concentration ratios of the oxidized and reduced forms of the two species are such that their attraction for electrons (and thus their electrode potentials) are identical. Note that these concentration ratios vary continuously throughout the titration, as must E_{system}. End points are determined from the characteristic variation in E_{system} that occurs during the titration.

Because $E_{Ce^{4+}/Ce^{3+}} = E_{Fe^{3+}/Fe^{2+}} = E_{system}$, data for a titration curve can be obtained by applying the Nernst equation for *either* the cerium(IV) half-reaction *or* the iron(III) half-reaction. It turns out, however, that one or the other will be more convenient, depending on the stage of the titration. Prior to the equivalence point, the analytical concentrations of Fe(II), Fe(III), and Ce(III) are immediately available

> Remember that, when redox systems are at equilibrium, *the electrode potentials of all half-reactions are identical.* This generality applies whether the reactions take place directly in solution or indirectly in a galvanic cell.

> Most end points in oxidation/reduction titrations are based on the rapid changes in E_{system} that occur at or near chemical equivalence.

> Before the equivalence point, E_{system} calculations are easiest to make using the Nernst equation for the analyte. After the equivalence point, the Nernst equation for the titrant is used.

from the volumetric data and reaction stoichiometry, while the very small amount of Ce(IV) can only be obtained by calculations based on the equilibrium constant. Beyond the equivalence point, a different situation predominates. In this region, we can evaluate concentrations of Ce(III), Ce(IV), and Fe(III) directly from the volumetric data, while the Fe(II) concentration is small and more difficult to calculate. In this region, then, the Nernst equation for the cerium couple becomes the more convenient to use. At the equivalence point, we can also evaluate the concentrations for Fe(III) and Ce(III) from the stoichiometry, but the concentrations of both Fe(II) and Ce(IV) will necessarily be quite small. A method for calculating the equivalence-point potential is given in the next section.

Equivalence-Point Potentials

At the equivalence point, the concentration of cerium(IV) and iron(II) are minute and cannot be obtained from the stoichiometry of the reaction. Fortunately, equivalence-point potentials are easily obtained by taking advantage of the fact that the two reactant species and the two product species have known concentration ratios at chemical equivalence.

At the equivalence point in the titration of iron(II) with cerium(IV), the potential of the system is given by both

$$E_{eq} = E^0_{Ce^{4+}/Ce^{3+}} - \frac{0.0592}{1} \log \frac{[Ce^{3+}]}{[Ce^{4+}]}$$

and

$$E_{eq} = E^0_{Fe^{3+}/Fe^{2+}} - \frac{0.0592}{1} \log \frac{[Fe^{2+}]}{[Fe^{3+}]}$$

Adding these two expressions gives

> The concentration quotient, $\frac{[Ce^{3+}][Fe^{2+}]}{[Ce^{4+}][Fe^{3+}]}$, in Equation 19-11 is *not* the usual ratio of product concentrations and reactant concentrations that appears in equilibrium-constant expressions.

$$2E_{eq} = E^0_{Fe^{3+}/Fe^{2+}} + E^0_{Ce^{4+}/Ce^{3+}} - \frac{0.0592}{1} \log \frac{[Ce^{3+}][Fe^{2+}]}{[Ce^{4+}][Fe^{3+}]} \tag{19-11}$$

The definition of equivalence point requires that

$$[Fe^{3+}] = [Ce^{3+}]$$
$$[Fe^{2+}] = [Ce^{4+}]$$

Substitution of these equalities into Equation 19-11 results in the concentration quotient becoming unity and the logarithmic term becoming zero:

$$2E_{eq} = E^0_{Fe^{3+}/Fe^{2+}} + E^0_{Ce^{4+}/Ce^{3+}} - \frac{0.0592}{1} \log \frac{\cancel{[Ce^{3+}]}\,\cancel{[Ce^{4+}]}}{\cancel{[Ce^{4+}]}\,\cancel{[Ce^{3+}]}} = E^0_{Fe^{3+}/Fe^{2+}} + E^0_{Ce^{4+}/Ce^{3+}}$$

$$E_{eq} = \frac{E^0_{Fe^{3+}/Fe^{2+}} + E^0_{Ce^{4+}/Ce^{3+}}}{2} \tag{19-12}$$

Example 19-11 illustrates how we calculate equivalence-point potential for a more complex reaction.

EXAMPLE 19-11

Derive an expression for the equivalence-point potential in the titration of 0.0500 M U^{4+} with 0.1000 M Ce^{4+}. Assume both solutions are 1.0 M in H_2SO_4.

$$U^{4+} + 2Ce^{4+} + 2H_2O \rightleftharpoons UO_2^{2+} + 2Ce^{3+} + 4H^+$$

Solution

In Appendix 5, we find

$$UO_2^{2+} + 4H^+ + 2e^- \rightleftharpoons U^{4+} + 2H_2O \qquad E^0 = 0.334 \text{ V}$$

$$Ce^{4+} + e^- \rightleftharpoons Ce^{3+} \qquad\qquad\qquad E^{0'} = 1.44 \text{ V}$$

Now, we use the formal potential for Ce^{4+} in 1.0 M H_2SO_4.

Proceeding as in the cerium(IV)/iron(II) equivalence-point calculation, we write

$$E_{eq} = E^0_{UO_2^{2+}/U^{4+}} - \frac{0.0592}{2} \log \frac{[U^{4+}]}{[UO_2^{2+}][H^+]^4}$$

$$E_{eq} = E^{0'}_{Ce^{4+}/Ce^{3+}} - \frac{0.0592}{1} \log \frac{[Ce^{3+}]}{[Ce^{4+}]}$$

In order to combine the log terms, we must multiply the first equation by 2 to give

$$2E_{eq} = 2E^0_{UO_2^{2+}/U^{4+}} - 0.0592 \log \frac{[U^{4+}]}{[UO_2^{2+}][H^+]^4}$$

Adding this equation to the previous equation leads to

$$3E_{eq} = 2E^0_{UO_2^{2+}/U^{4+}} + E^{0'}_{Ce^{4+}/Ce^{3+}} - 0.0592 \log \frac{[U^{4+}][Ce^{3+}]}{[UO_2^{2+}][Ce^{4+}][H^+]^4}$$

But, at equivalence,

$$[U^{4+}] = [Ce^{4+}]/2$$

and

$$[UO_2^{2+}] = [Ce^{3+}]/2$$

Substituting these equations gives on rearranging

$$E_{eq} = \frac{2E^0_{UO_2^{2+}/U^{4+}} + E^{0'}_{Ce^{4+}/Ce^{3+}}}{3} - \frac{0.0592}{3} \log \frac{2[\cancel{Ce^{4+}}][\cancel{Ce^{3+}}]}{2[\cancel{Ce^{3+}}][\cancel{Ce^{4+}}][H^+]^4}$$

$$= \frac{2E^0_{UO_2^{2+}/U^{4+}} + E^{0'}_{Ce^{4+}/Ce^{3+}}}{3} - \frac{0.0592}{3} \log \frac{1}{[H^+]^4}$$

We see that, in this titration, the equivalence-point potential is pH dependent.

19D-2 The Titration Curve

Let us first consider the titration of 50.00 mL of 0.0500 M Fe^{2+} with 0.1000 M Ce^{4+} in a medium that is 1.0 M in H_2SO_4 at all times. Formal potential data for both half-cell processes are available in Appendix 5 and are used for these calculations. Thus,

$$Ce^{4+} + e^- \rightleftharpoons Ce^{3+} \qquad E^{0'} = 1.44 \text{ V (1 M } H_2SO_4)$$
$$Fe^{3+} + e^- \rightleftharpoons Fe^{2} \qquad E^{0'} = 0.68 \text{ V (1 M } H_2SO_4)$$

Initial Potential

The solution contains no cerium species before we add titrant. It is more than likely that there is a small but unknown amount of Fe^{3+} present due to air oxidation of Fe^{2+}. In any case, we don't have enough information to calculate an initial potential.

Potential after the Addition of 5.00 mL of Cerium(IV)

> Remember, the equation for this reaction is
>
> $Fe^{2+} + Ce^{4+} \rightleftharpoons Fe^{3+} + Ce^{3+}$

When oxidant is added, Ce^{3+} and Fe^{3+} are formed, and the solution contains appreciable and easily calculated concentrations of three of the participants, while the concentration of the fourth, Ce^{4+}, is vanishingly small. Therefore, it is more convenient to use the concentrations of the two iron species to calculate the electrode potential of the system.

The equilibrium concentration of Fe(III) is equal to its molar analytical concentration minus the molar equilibrium concentration of the unreacted Ce(IV):

$$[Fe^{3+}] = \frac{5.00 \text{ mL} \times 0.1000 \text{ M}}{50.00 \text{ mL} + 5.00 \text{ mL}} - [Ce^{4+}] = \frac{0.500 \text{ mmol}}{55.00 \text{ mL}} - [Ce^{4+}]$$

$$= \left(\frac{0.500}{55.00}\right) \text{M} - [Ce^{4+}]$$

Similarly, the Fe^{2+} concentration is given by its molar analytical concentration plus the molar equilibrium concentration of unreacted $[Ce^{4+}]$:

$$[Fe^{2+}] = \frac{50.00 \text{ mL} \times 0.0500 \text{ M} - 5.00 \text{ mL} \times 0.1000 \text{ M}}{55.00 \text{ mL}} + [Ce^{4+}]$$

$$= \left(\frac{2.00}{55.00}\right) \text{M} + [Ce^{4+}]$$

Generally, redox reactions used in titrimetry are sufficiently complete that the equilibrium concentration of one of the species (in this case $[Ce^{4+}]$) is minuscule with respect to the other species present in the solution. Thus, the foregoing two equations can be simplified to

> Strictly speaking, the concentrations of Fe^{2+} and Fe^{3+} should be corrected for the concentration of unreacted Ce^{4+}. This correction would increase $[Fe^{2+}]$ and decrease $[Fe^{3+}]$. The amount of unreacted Ce^{4+} is usually so small that we can neglect the correction in both cases.

$$[Fe^{3+}] = \frac{0.500}{55.00} \text{ M} \qquad \text{and} \qquad [Fe^{2+}] = \frac{2.00}{55.00} \text{ M}$$

Substitution for $[Fe^{2+}]$ and $[Fe^{3+}]$ in the Nernst equation gives

$$E_{system} = +0.68 - \frac{0.0592}{1} \log \frac{2.00 \text{ / } 55.00}{0.50 \text{ / } 55.00} = 0.64 \text{ V}$$

TABLE 19-2

	Potential, V, vs. SHE*		
Reagent Volume, mL	50.00 mL of 0.0500 M Fe²⁺		50.00 mL of 0.02500 M U⁴⁺
5.00	0.64		0.316
15.00	0.69		0.339
20.00	0.72		0.352
24.00	0.76		0.375
24.90	0.82		0.405
25.00	1.06	← Equivalence → Point	0.703
25.10	1.30		1.30
26.00	1.36		1.36
30.00	1.40		1.40

Electrode Potential versus SHE in Titrations with 0.100 M Ce^{4+}

*H_2SO_4 concentration is such that $[H^+] = 1.0$ throughout.

Note that the volumes in the numerator and denominator cancel, indicating that the potential is independent of dilution. This independence persists until the solution becomes so dilute that the two assumptions made in the calculation become invalid.

It is worth emphasizing again that the use of the Nernst equation for the Ce(IV)/Ce(III) system would yield the same value for E_{system}, but to do so would require computing $[Ce^{4+}]$ by means of the equilibrium constant for the reaction.

Additional potentials needed to define the titration curve short of the equivalence point can be obtained similarly. Such data are given in **Table 19-2**. You may want to confirm a few of these values.

Equivalence-Point Potential

Substitution of the two formal potentials into Equation 19-12 yields

$$E_{eq} = \frac{E^{0'}_{Ce^{4+}/Ce^{3+}} + E^{0'}_{Fe^{3+}/Fe^{2+}}}{2} = \frac{1.44 + 0.68}{2} = 1.06 \text{ V}$$

Potential after Adding 25.10 mL of Cerium(IV)

The molar concentrations of Ce(III), Ce(IV), and Fe(III) are easily computed at this point, but that for Fe(II) is not. Therefore, E_{system} computations based on the cerium half-reaction are more convenient. The concentrations of the two cerium ion species are

$$[Ce^{3+}] = \frac{25.00 \times 0.1000}{75.10} - [Fe^{2+}] \approx \frac{2.500}{75.10} \text{ M}$$

$$[Ce^{4+}] = \frac{25.10 \times 0.1000 - 50.00 \times 0.0500}{75.10} + [Fe^{2+}] \approx \frac{0.010}{75.10} \text{ M}$$

In the equations for the cerium ion species, we assume the iron(II) concentration is negligible with respect to the analytical concentrations. Substitution into the Nernst equation for the cerium couple gives

$$E = +1.44 - \frac{0.0592}{1} \log \frac{[Ce^{3+}]}{[Ce^{4+}]} = +1.44 - \frac{0.0592}{1} \log \frac{2.500/75.10}{0.010/75.10}$$

$$= +1.30 \text{ V}$$

In contrast to other titration curves we have seen, oxidation/reduction curves are *independent* of reactant concentration except for very dilute solutions.

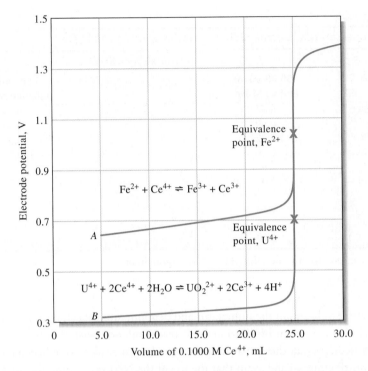

Figure 19-3 Titration curves for 0.1000 M Ce^{4+} titration. *A*: Titration of 50.00 mL of 0.05000 M Fe^{2+}. *B*: Titration of 50.00 mL of 0.02500 M U^{4+}.

The additional postequivalence potentials in Table 19-2 were calculated in a similar fashion.

The titration curve of iron(II) with cerium(IV) appears as *A* in **Figure 19-3**. This plot resembles closely the curves in neutralization, precipitation, and complex-formation titrations, with the equivalence point being signaled by a rapid change in the variable on the vertical axis. A titration involving 0.00500 M iron(II) and 0.01000 M cerium(IV) yields a curve that, for all practical purposes, is identical to the one we have computed since the electrode potential of the system is independent of dilution. A spreadsheet to calculate E_{system} as a function of the volume of Ce(IV) added is shown in **Figure 19-4**.

The data in the third column of Table 19-2 are plotted as curve *B* in Figure 19-3 to compare the two titrations. The two curves are identical for volumes greater than 25.10 mL because the concentrations of the two cerium species are identical in this region. It is also interesting that the curve for iron(II) is symmetric around the equivalence point but that the curve for uranium(IV) is not symmetric. In general, redox titration curves are symmetric when the analyte and titrant react in a 1:1 molar ratio.

❯ Why is it impossible to calculate the potential of the system before titrant is added?

❯ Redox titration curves are symmetric when the reactants combine in a 1:1 ratio. Otherwise, they are asymmetric.

EXAMPLE 19-12

Calculate data and construct a titration curve for the reaction of 50.00 mL of 0.02500 M U^{4+} with 0.1000 M Ce^{4+}. The solution is 1.0 M in H_2SO_4 throughout the titration (for the sake of simplicity, assume that [H^+] for this solution is also about 1.0 M).

Solution

The analytical reaction is

$$U^{4+} + 2Ce^{4+} + 2H_2O \rightleftharpoons UO_2^{2+} + 2Ce^{3+} + 4H^+$$

	A	B	C	D	E	F	G	H	I	J	K
1	Spreadsheet for titration of 50.00 mL of 0.0500 M Fe²⁺ with 0.1000 M Ce⁴⁺										
2	Initial Conc. Fe²⁺, M	0.0500	$E^{0'}_{Fe, V}$	0.68							
3	Vol. Fe²⁺, mL	50.00	$E^{0'}_{Ce, V}$	1.44							
4	Conc. Ce⁴⁺, M	0.1000									
5											
6	Volume Ce⁴⁺, mL	[Fe³⁺]	[Fe²⁺]	[Ce³⁺]	[Ce⁴⁺]	$E_{system, V}$					
7	5.00	0.009091	0.036364			0.64					
8	10.00	0.016667	0.025000			0.67					
9	15.00	0.023077	0.015385			0.69					
10	20.00	0.028571	0.007143			0.72					
11	24.00	0.032432	0.001351			0.76					
12	24.90	0.033244	0.000134			0.82					
13	25.00					1.06					
14	25.10			0.033289	0.000133	1.30					
15	26.00			0.032895	0.001316	1.36					
16	30.00			0.031250	0.006250	1.40					
17	35.00			0.029412	0.011765	1.42					
18	40.00			0.027778	0.016667	1.43					
19											
20	**Spreadsheet Documentation**										
21	Cell B7=A7*$B$4/($B$3+A7)			Cell D14=$B$2*$B$3/($B$3+A14)							
22	Cell C7=($B$2*$B$3-$B$4*A7)/($B$3+A7)			Cell E14=(A14*$B$4-$B$2*$B$3)/($B$3+A14)							
23	Cell F7=$D$2-0.0592*LOG10(C7/B7)			Cell F14=$D$3-0.0592*LOG10(D14/E14)							
24	Cell F13=($D$2+$D$3)/2										

Figure 19-4 Spreadsheet and plot for titration of 50.00 mL of 0.0500 M Fe²⁺ with 0.1000 M Ce⁴⁺. Prior to the equivalence point, the system potential is calculated from the Fe³⁺ and Fe²⁺ concentrations. After the equivalence point, the Ce⁴⁺ and Ce³⁺ concentrations are used in the Nernst equation. The Fe³⁺ concentration in cell B7 is calculated from the number of millimoles of Ce⁴⁺ added, divided by the total volume of solution. The formula used for the first volume is shown in documentation cell A21. In cell C7, [Fe²⁺] is calculated as the initial number of millimoles of Fe²⁺ present minus the number of millimoles of Fe³⁺ formed and divided by the total solution volume. Documentation cell A22 gives the formula for the 5.00-mL volume. The system potential prior to the equivalence point is calculated in cells F7:F12 by using the Nernst equation, expressed for the first volume by the formula shown in documentation cell A23. In cell F13, the equivalence-point potential is found from the average of the two formal potentials, as shown in documentation cell A24. After the equivalence point, the Ce(III) concentration (cell D14) is found from the number of millimoles of Fe²⁺ initially present divided by the total solution volume, as shown for the 25.10-mL volume by the formula in documentation cell D21. The Ce(IV) concentration (E14) is found from the total number of millimoles of Ce(IV) added minus the number of millimoles of Fe²⁺ initially present and divided by the total solution volume, as shown in documentation cell D22. The system potential in cell F14 is found from the Nernst equation, as shown in documentation cell D23. The chart is then the resulting titration curve.

And, in Appendix 5, we find

$$UO_2^{2+} + 4H^+ + 2e^- \rightleftharpoons U^{4+} + 2H_2O \qquad E^0 = 0.334 \text{ V}$$

$$Ce^{4+} + e^- \rightleftharpoons Ce^{3+} \qquad E^{0'} = 1.44 \text{ V}$$

Potential after Adding 5.00 mL of Ce⁴⁺

$$\text{original amount U}^{4+} = 50.00 \text{ mL U}^{4+} \times 0.02500 \frac{\text{mmol U}^{4+}}{\text{mL U}^{4+}}$$

$$= 1.250 \text{ mmol U}^{4+}$$

(continued)

$$\text{amount } Ce^{4+} \text{ added} = 5.00 \text{ mL } \cancel{Ce^{4+}} \times 0.1000 \frac{\text{mmol } Ce^{4+}}{\text{mL } \cancel{Ce^{4+}}}$$

$$= 0.5000 \text{ mmol } Ce^{4+}$$

$$\text{amount } U^{4+} \text{ remaining} = 1.250 \text{ mmol } U^{4+} - 0.2500 \cancel{\text{mmol } UO_2^{2+}}$$

$$\times \frac{1 \text{ mmol } U^{4+}}{1 \cancel{\text{mmol } UO_2^{2+}}}$$

$$= 1.000 \text{ mmol } U^{4+}$$

$$\text{total volume of solution} = (50.00 + 5.00)\text{mL} = 55.00 \text{ mL}$$

$$\text{concentration } U^{4+} \text{ remaining} = \frac{1.000 \text{ mmol } U^{4+}}{55.00 \text{ mL}}$$

$$\text{concentration } UO_2^{2+} \text{ formed} = \frac{0.5000 \cancel{\text{mmol } Ce^{4+}} \times \dfrac{1 \text{ mmol } UO_2^{2+}}{2 \cancel{\text{mmol } Ce^{4+}}}}{55.00 \text{ mL}}$$

$$= \frac{0.2500 \text{ mmol } UO_2^{2+}}{55.00 \text{ mL}}$$

Applying the Nernst equation for UO_2^{2+}, we obtain

$$E = 0.334 - \frac{0.0592}{2} \log \frac{[U^{4+}]}{[UO_2^{2+}][H^+]^4}$$

$$= 0.334 - \frac{0.0592}{2} \log \frac{[U^{4+}]}{[UO_2^{2+}](1.00)^4}$$

Substituting concentrations of the two uranium species gives

$$E = 0.334 - \frac{0.0592}{2} \log \frac{1.000 \text{ mmol } U^{4+}/55.00 \text{ mL}}{0.2500 \text{ mmol } UO_2^{2+}/55.00 \text{ mL}}$$

$$= 0.316 \text{ V}$$

Other preequivalence-point data, calculated in the same way, are given in the third column in Table 19-2.

Equivalence-Point Potential

Following the procedure shown in Example 19-11, we obtain

$$E_{eq} = \frac{(2E^0_{UO_2^{2+}/U^{4+}} + E^{0'}_{Ce^{4+}/Ce^{3+}})}{3} - \frac{0.0592}{3} \log \frac{1}{[H^+]^4}$$

Substituting gives

$$E_{eq} = \frac{2 \times 0.334 + 1.44}{3} - \frac{0.0592}{3} \log \frac{1}{(1.00)^4}$$

$$= \frac{2 \times 0.334 + 1.44}{3} = 0.703 \text{ V}$$

Potential after Adding 25.10 mL of Ce⁴⁺

$$\text{total volume of solution} = 75.10 \text{ mL}$$

$$\text{original amount U}^{4+} = 50.00 \text{ mL } \cancel{\text{U}^{4+}} \times 0.02500 \frac{\text{mmol U}^{4+}}{\text{mL } \cancel{\text{U}^{4+}}}$$

$$= 1.250 \text{ mmol U}^{4+}$$

$$\text{amount Ce}^{4+} \text{ added} = 25.10 \text{ mL } \cancel{\text{Ce}^{4+}} \times 0.1000 \frac{\text{mmol Ce}^{4+}}{\text{mL } \cancel{\text{Ce}^{4+}}}$$

$$= 2.510 \text{ mmol Ce}^{4+}$$

$$\text{concentration of Ce}^{3+} \text{ formed} = \frac{1.250 \text{ mmol } \cancel{\text{U}^{4+}} \times \dfrac{2 \text{ mmol Ce}^{3+}}{\text{mmol } \cancel{\text{U}^{4+}}}}{75.10 \text{ mL}}$$

concentration of Ce⁴⁺ remaining

$$= \frac{2.510 \text{ mmol Ce}^{4+} - 2.500 \text{ mmol } \cancel{\text{Ce}^{3+}} \times \dfrac{1 \text{ mmol Ce}^{4+}}{\text{mmol } \cancel{\text{Ce}^{3+}}}}{75.10 \text{ mL}}$$

Substituting into the expression for the formal potential gives

$$E = 1.44 - 0.0592 \log \frac{2.500/\cancel{75.10}}{0.010/\cancel{75.10}} = 1.30 \text{ V}$$

Table 19-2 contains other postequivalence-point data obtained in this same way.

FEATURE 19-3

The Inverse Master Equation Approach for Redox Titration Curves

α Values for Redox Species
The α values that we used for acid/base and complexation equilibria are also useful in studying redox equilibria. To calculate redox α values, we must solve the Nernst equation for the ratio of the concentration of the reduced species to the oxidized species. We use an approach similar to that of de Levie.[2] Since

$$E = E^0 - \frac{2.303RT}{nF} \log \frac{[\text{R}]}{[\text{O}]}$$

(continued)

---

[2]R. de Levie, *J. Electroanal. Chem.*, **1992**, *323*, 347–55. **DOI**: 10.1016/0022-0728(92)80022-V.

we can write

$$\frac{[R]}{[O]} = 10^{-\frac{nF(E-E^0)}{2.303RT}} = 10^{-nf(E-E^0)}$$

Where, at 25° C,

$$f = \frac{F}{2.303RT} = \frac{1}{0.0592}$$

Now, we can find the fractions α of the total $[R] + [O]$ as follows:

$$\alpha_R = \frac{[R]}{[R]+[O]} = \frac{[R]/[O]}{[R]/[O]+1} = \frac{10^{-nf(E-E^0)}}{10^{-nf(E-E^0)}+1}$$

As an exercise, you can show that

$$\alpha_R = \frac{1}{10^{-nf(E^0-E)}+1}$$

and that

$$\alpha_O = 1 - \alpha_R = \frac{1}{10^{-nf(E-E^0)}+1}$$

Furthermore, you can rearrange the equations as follows:

$$\alpha_R = \frac{10^{-nfE}}{10^{-nfE}+10^{-nfE^0}} \qquad \alpha_O = \frac{10^{-nfE^0}}{10^{-nfE}+10^{-nfE^0}}$$

We express α values in this way is so that they are in a form similar to the α values for a weak monoprotic acid presented in Chapter 14.

$$\alpha_0 = \frac{[H_3O^+]}{[H_3O^+]+K_a} \qquad \alpha_1 = \frac{K_a}{[H_3O^+]+K_a}$$

or, alternatively,

$$\alpha_0 = \frac{10^{-pH}}{10^{-pH}+10^{-pK_a}} \qquad \alpha_1 = \frac{10^{-pK_a}}{10^{-pH}+10^{-pK_a}}$$

Notice the very similar forms of the α values for redox species and those for the weak monoprotic acid. The term 10^{-nfE} in the redox expression is analogous to 10^{-pH} in the acid/base case, and the term 10^{-nfE^0} is analogous to 10^{-pK_a}. These analogies will become more apparent when we plot α_O and α_R versus E in the same way that we plotted α_0 and α_1 versus pH. It is important to recognize that we obtain these relatively straightforward expressions for the redox alphas only for redox half-reactions that have 1:1 stoichiometry. For other stoichiometries, which we will not consider in this feature, the expressions become considerably more complex. For simple cases, these equations provide us with a nice way to visualize redox chemistry and to calculate the data for redox titration curves. If we have formal potential data in a constant ionic strength medium, we can use the $E^{0'}$ values in place of the E^0 values in the α expressions.

Now, let us examine graphically the dependence of the redox α values on the potential E. We shall determine this dependence for both the Fe^{3+}/Fe^{2+} and the Ce^{4+}/Ce^{3+} couples in 1 M H_2SO_4, where the formal potentials are known. For these two couples, the α expressions are given by

$$\alpha_{Fe^{2+}} = \frac{10^{-fE}}{10^{-fE} + 10^{-fE_{Fe}^{0'}}} \qquad \alpha_{Fe^{3+}} = \frac{10^{-fE_{Fe}^{0'}}}{10^{-fE} + 10^{-fE_{Fe}^{0'}}}$$

$$\alpha_{Ce^{3+}} = \frac{10^{-fE}}{10^{-fE} + 10^{-fE_{Ce}^{0'}}} \qquad \alpha_{Ce^{4+}} = \frac{10^{-fE_{Ce}^{0'}}}{10^{-fE} + 10^{-fE_{Ce}^{0'}}}$$

Note that the *only* difference in the expressions for the two sets of α values is the two different formal potentials $E_{Fe}^{0'} = 0.68$ V and $E_{Ce}^{0'} = 1.44$ V in 1 M H_2SO_4. The effect of this difference will be apparent in the α plots. Since $n = 1$ for both couples, it does not appear in these equations for α.

The plot of α values is shown in Figure 19F-2. We have calculated the α values every 0.05 V from 0.50 V to 1.75 V. The shapes of the α plots are identical to those for acid/base systems (treated in Chapters 14 and 15) as you might expect from the form of the analogous expressions that were mentioned previously.

It is worth mentioning that we normally think of calculating the potential of an electrode for a redox system in terms of concentration rather than the other way around. But, just as pH is the independent variable in our α calculations with acid/base systems, potential is the independent variable in redox calculations. It is far simpler to calculate α for a series of potential values than to solve the expressions for potential given various values of α.

Inverse Master Equation Approach

At all points during the titration, the concentrations of Fe^{3+} and Ce^{3+} are equal from the stoichiometry. Or,

$$[Fe^{3+}] = [Ce^{3+}]$$

From the α values and the concentrations and volumes of the reagents, we can write

$$\alpha_{Fe^{3+}}\left(\frac{V_{Fe}c_{Fe}}{V_{Fe} + V_{Ce}}\right) = \alpha_{Ce^{3+}}\left(\frac{V_{Ce}c_{Ce}}{V_{Fe} + V_{Ce}}\right)$$

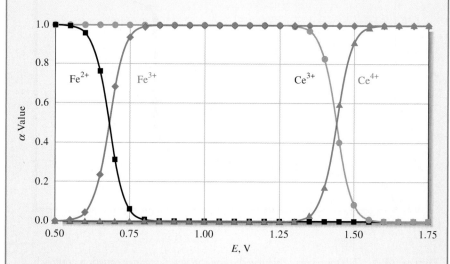

Figure 19F-2   Alpha plot for the Fe^{2+}/Ce^{4+} system.

(continued)

where V_{Fe} and c_{Fe} are the initial volume and concentration of Fe^{2+} present and V_{Ce} and c_{Ce} are the volume and concentration of the titrant. By multiplying both sides of the equation by $V_{Fe} + V_{Ce}$ and dividing both sides by $V_{Fe}c_{Fe}\,\alpha_{Ce^{3+}}$, we find that

$$\alpha_{Fe^{3+}}\left(\frac{V_{Fe}c_{Fe}}{V_{Fe}+V_{Ce}}\right)\left(\frac{V_{Fe}+V_{Ce}}{V_{Fe}c_{Fe}\,\alpha_{Ce^{3+}}}\right) = \alpha_{Ce^{3+}}\left(\frac{V_{Ce}c_{Ce}}{V_{Fe}+V_{Ce}}\right)\left(\frac{V_{Fe}+V_{Ce}}{V_{Fe}c_{Fe}\,\alpha_{Ce^{3+}}}\right)$$

and

$$\phi = \frac{V_{Ce}c_{Ce}}{V_{Fe}c_{Fe}} = \frac{\alpha_{Fe^{3+}}}{\alpha_{Ce^{3+}}}$$

where ϕ is the extent of the titration (fraction titrated). We then substitute the expressions previously derived for the α values and obtain

$$\phi = \frac{\alpha_{Fe^{3+}}}{\alpha_{Ce^{3+}}} = \frac{1 + 10^{-f(E_{Ce}^{0'} - E)}}{1 + 10^{-f(E - E_{Fe}^{0'})}}$$

where E is now the system potential. We then substitute values of E in 0.5 V increments from 0.5 to 1.40 V into this equation to calculate ϕ and plot the resulting data, as shown in **Figure 19F-3**. An additional point at 1.42 V was added since 1.45 V gave a ϕ value of more than 2. Compare this graph to Figure 19-4, which was generated using the traditional stoichiometric approach.

At this point, we should mention that some redox titration expressions are more complex than those presented here for a basic 1:1 situation. If you are interested in exploring the master equation approach for pH-dependent redox titrations or other situations, consult the paper by de Levie.[2] You can find the details of the calculations for the two plots in this feature in Chapter 10 of *Applications of Microsoft® Excel in Analytical Chemistry*, 2nd ed.

Figure 19F-3 Titration curve calculated using the inverse master equation approach. The extent of titration ϕ is calculated for various values of the system potential, E_{system}, but the graph is plotted as E_{system} versus ϕ.

19D-3 Effect of Variables on Redox Titration Curves

In earlier chapters, we considered the effects of reactant concentrations and completeness of the reaction on titration curves. Next, we describe the effects of these variables on oxidation/reduction titration curves.

Reactant Concentration

As we have just seen, E_{system} for an oxidation/reduction titration is usually independent of dilution. Consequently, titration curves for oxidation/reduction reactions are usually independent of analyte and reagent concentrations. This characteristic is in distinct contrast to that observed in the other types of titration curves we have encountered.

Completeness of the Reaction

The change in potential in the equivalence-point region of an oxidation/reduction titration becomes larger as the reaction becomes more complete. This effect is demonstrated by the two curves in Figure 19-3. The equilibrium constant for the reaction of cerium(IV) with iron(II) is 7×10^{12} while that for U(IV) is 2×10^{37}. The effect of reaction completeness is further demonstrated in **Figure 19-5**. This figure shows curves for the titration of a hypothetical reductant that has a standard electrode potential of 0.20 V with several hypothetical oxidants with standard potentials ranging from 0.40 to 1.20 V. The equilibrium constants corresponding with each of the curves lie between about 2×10^3 and 8×10^{16}. Curve A shows that the greatest change in potential of the system is associated with the reaction that is most complete, and curve E illustrates the opposite extreme. In this respect, oxidation/reduction titration curves are similar to those involving other types of reactions.

	$E_T^0 - E_A^0$	K_{eq}
A	1.00 V	8×10^{16}
B	0.80 V	3×10^{13}
C	0.60 V	1×10^{10}
D	0.40 V	6×10^6
E	0.20 V	2×10^3

Figure 19-5 Effect of titrant electrode potential on reaction completeness. The standard electrode potential for the analyte (E_A^0) is 0.200 V; starting with curve A, standard electrode potentials for the titrant (E_T^0) are 1.20, 1.00, 0.80, 0.60, and 0.40, respectively. Both analyte and titrant undergo a one-electron change.

FEATURE 19-4

Reaction Rates and Electrode Potentials

Standard potentials reveal whether or not a reaction proceeds far enough toward completion to be useful in a particular analytical problem, but they provide no information about the rate at which the equilibrium state is approached. Consequently, a reaction that appears extremely favorable thermodynamically may be totally unacceptable from the kinetic standpoint. The oxidation of arsenic(III) with cerium(IV) in dilute sulfuric acid is a typical example. The reaction is

$$H_3AsO_3 + 2Ce^{4+} + H_2O \rightleftharpoons H_3AsO_4 + 2Ce^{3+} + 2H^+$$

The formal potentials, $E^{0'}$s, for these two systems are

$$Ce^{4+} + e^- \rightleftharpoons Ce^{3+} \qquad\qquad E^{0'} = +1.44 \text{ V}$$

$$H_3AsO_4 + 2H^+ + 2e^- \rightleftharpoons H_3AsO_3 + H_2O \qquad\qquad E^{0'} = +0.577 \text{ V}$$

And an equilibrium constant of about 10^{29} can be calculated from these data. Even though this equilibrium lies far to the right, titration of arsenic(III) with cerium(IV) is impossible without a catalyst because several hours are required to achieve equilibrium. Fortunately, several substances catalyze the reaction and thus make the titration feasible.

Spreadsheet Summary   In Chapter 10 of *Applications of Microsoft®Excel in Analytical Chemistry*, 2nd ed., Excel is used to obtain α values for redox species. These values show how the equilibrium concentrations change throughout a redox titration. Redox titration curves are constructed by both a stoichiometric and a master equation approach. The stoichiometric approach is also used for a system that is pH dependent.

19E   OXIDATION/REDUCTION INDICATORS

Two types of chemical indicators are used for obtaining end points for oxidation/reduction titrations: general redox indicators and specific indicators.

19E-1 General Redox Indicators

> Color changes for general redox indicators depend only on the potential of the system.

General oxidation/reduction indicators are substances that change color on being oxidized or reduced. In contrast to specific indicators, the color changes of true redox indicators are largely independent of the chemical nature of the analyte and titrant and depend instead on the changes in the electrode potential of the system that occur as the titration progresses.

The half-reaction responsible for color change in a typical general oxidation/reduction indicator can be written as

$$In_{ox} + ne^- \rightleftharpoons In_{red}$$

If the indicator reaction is reversible, we can write

$$E = E^0_{\mathrm{In_{ox}/In_{red}}} - \frac{0.0592}{n} \log \frac{[\mathrm{In_{red}}]}{[\mathrm{In_{ox}}]} \qquad (19\text{-}13)$$

Typically, a change from the color of the oxidized form of the indicator to the color of the reduced form requires a change of about 100 in the ratio of reactant concentrations, that is, a color change appears when

$$\frac{[\mathrm{In_{red}}]}{[\mathrm{In_{ox}}]} \le \frac{1}{10}$$

changes to

$$\frac{[\mathrm{In_{red}}]}{[\mathrm{In_{ox}}]} \ge 10$$

The potential change required to produce the full color change of a typical general indicator can be found by substituting these two values into Equation 19-13, giving

$$E = E^0_{\mathrm{In}} \pm \frac{0.0592}{n}$$

This equation shows that a typical general indicator exhibits a detectable color change when a titrant causes the system potential to shift from $E^0_{\mathrm{In}} + 0.0592/n$ to $E^0_{\mathrm{In}} - 0.0592/n$ or about $(0.118/n)$ V. For many indicators, $n = 2$, and a change of 0.059 V is thus sufficient.

Table 19-3 lists transition potentials for several redox indicators. Note that indicators functioning in any desired potential range up to about $+1.25$ V are available. Structures for and reactions of a few of the indicators listed in the table are considered in the paragraphs that follow.

> Protons participate in the reduction of many indicators. Thus, the range of potentials over which a color change occurs (the *transition potential*) is often pH dependent.

The compound 1,10-phenanthroline is an excellent complexing agent for Fe(II).

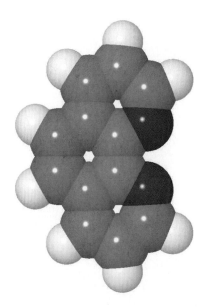

TABLE 19-3

Selected Oxidation/Reduction Indicators*

| Indicator | Color | | Transition Potential, V | Conditions |
	Oxidized	Reduced		
5-Nitro-1,10-phenanthroline iron(II) complex	Pale blue	Red-violet	$+1.25$	1 M H_2SO_4
2,3'-Diphenylamine dicarboxylic acid	Blue-violet	Colorless	$+1.12$	7-10 M H_2SO_4
1,10- Phenanthroline iron(II) complex	Pale blue	Red	$+1.11$	1 M H_2SO_4
5-Methyl 1,10-phenanthroline iron(II) complex	Pale blue	Red	$+1.02$	1 M H_2SO_4
Erioglaucin A	Blue-red	Yellow-green	$+0.98$	0.5 M H_2SO_4
Diphenylamine sulfonic acid	Red-violet	Colorless	$+0.85$	Dilute acid
Diphenylamine	Violet	Colorless	$+0.76$	Dilute acid
p-Ethoxychrysoidine	Yellow	Red	$+0.76$	Dilute acid
Methylene blue	Blue	Colorless	$+0.53$	1 M acid
Indigo tetrasulfonate	Blue	Colorless	$+0.36$	1 M acid
Phenosafranine	Red	Colorless	$+0.28$	1 M acid

*Data in part from I. M. Kolthoff and V. A. Stenger, *Volumetric Analysis,* 2nd ed., Vol. 1, p. 140, New York: Interscience, 1942.

Iron(II) Complexes of Orthophenanthrolines

A class of organic compounds known as 1,10-phenanthrolines, or orthophenanthrolines, form stable complexes with iron(II) and certain other ions. The parent compound has a pair of nitrogen atoms located in such positions that each can form a covalent bond with the iron(II) ion.

Three orthophenanthroline molecules combine with each iron ion to yield a complex with the structure shown in the margin. This complex, which is sometimes called "ferroin," is conveniently formulated as $(phen)_3Fe^{2+}$. The complexed iron in the ferroin undergoes a reversible oxidation/reduction reaction that can be written

$$(phen)_3Fe^{3+} + e^- \rightleftharpoons (phen)_3Fe^{2+}$$
$$\text{pale blue} \qquad\qquad\qquad \text{red}$$

In practice, the color of the oxidized form is so slight as to go undetected, and the color change associated with this reduction is thus from nearly colorless to red. Because of the difference in color intensity, the end point is usually taken when only about 10% of the indicator is in the iron(II) form. The transition potential is thus approximately $+1.11$ V in 1 M sulfuric acid.

Of all the oxidation/reduction indicators, ferroin approaches most closely the ideal substance. It reacts rapidly and reversibly, its color change is pronounced, and its solutions are stable and easily prepared. In contrast to many indicators, the oxidized form of ferroin is remarkably inert toward strong oxidizing agents. At temperatures above 60°C, ferroin decomposes.

A number of substituted phenanthrolines have been investigated for their indicator properties, and some have proved to be as useful as the parent compound. Among these, the 5-nitro and 5-methyl derivatives are noteworthy, with transition potentials of $+1.25$ V and $+1.02$ V, respectively.

ferroin $(phen)_3Fe^{2+}$

5-nitro-1,
10-phenanthroline

5-methyl-1,
10-phenanthroline

Starch/Iodine Solutions

Starch, which forms a blue complex with triiodide ion, is a widely used specific indicator in oxidation/reduction reactions involving iodine as an oxidant or iodide ion as a reductant. A starch solution containing a little triiodide or iodide ion can also function as a true redox indicator, however. In the presence of excess oxidizing agent, the concentration ratio of iodine to iodide is high, giving a blue color to the solution. With excess reducing agent, on the other hand, iodide ion predominates, and the blue color is absent. Thus, the indicator system changes from colorless to blue in the titration of many reducing agents with various oxidizing agents. This color change is quite independent of the chemical composition of the reactants, depending only on the potential of the system at the equivalence point.

The Choice of Redox Indicator

Figure 19-5 demonstrates that all the indicators in Table 19-3 except for the first and the last could be used with titrant A. In contrast, with titrant D, only indigo tetrasulfonate could be used. The change in potential with titrant E is too small to be satisfactorily detected by an indicator.

19E-2 Specific Indicators

Perhaps the best-known specific indicator is starch, which forms a dark blue complex with triiodide ion. This complex signals the end point in titrations in which iodine is either produced or consumed.

Another specific indicator is potassium thiocyanate, which may be used, for example, in the titration of iron(III) with solutions of titanium(III) sulfate. The end point occurs when the red color of the iron(III)/thiocyanate complex disappears as a result of the significant decrease in the iron(III) concentration at the equivalence point.

19F POTENTIOMETRIC END POINTS

We can observe end points for many oxidation/reduction titrations by making the solution of the analyte part of the cell

reference electrode ‖ analyte solution | Pt

By measuring the potential of this cell during a titration, data for curves analogous to those shown in Figures 19-3 and 19-5 can be generated. End points are easily estimated from such curves. Potentiometric end points are discussed in detail in Chapter 21.

WEB WORKS

Most professions have associated organizations, such as the American Chemical Society, whose objectives for scientific societies range from the promulgation of scientific information to social programs catering to members of the profession. Subdisciplines such as electrochemistry also foster similar organizations. Browse to the Electrochemical Society (ECS) website at **http://www.electrochem.org/**. Explore the site and determine the goals and objectives of the ECS. What publications are produced under the auspices of The Society? Briefly describe the nature of each publication. Using the search blank on the ECS home page, enter the title "The Next Frontier: Electrodeposition for Solar Cell Fabrication" and click on the Go button. The article should appear in your search results. In what publication did the article appear? At the time of publication of the article, what was the optimum efficiency of then state-of-the-art crystalline solar cells? Why is this issue important, according to the authors?

Now, use a search engine to locate the website of a second organization called the Society for Electroanalytical Chemistry (SEAC) and perform a similar analysis of the information that you find. Compare and contrast the missions of ECS and SEAC.

QUESTIONS AND PROBLEMS

*19-1. Briefly define the electrode potential of a system that contains two or more redox couples.

19-2. For an oxidation/reduction titration, briefly distinguish between
   *(a) equilibrium and equivalence.
   (b) a true oxidation/reduction indicator and a specific indicator.

19-3. What is unique about the condition of equilibrium in an oxidation/reduction reaction?

*19-4. How is an oxidation/reduction titration curve generated through the use of standard electrode potentials for the analyte species and the volumetric titrant?

19-5. How does calculation of the electrode potential of the system at the equivalence point differ from that for any other point of an oxidation/reduction titration?

*19-6. Under what circumstance is the curve for an oxidation/reduction titration asymmetric about the equivalence point?

19-7. Calculate the theoretical potential of the following cells. Indicate whether the reaction will proceed spontaneously in the direction considered (oxidation on the left, reduction on the right) or whether an external voltage source is needed to force this reaction to occur.
   (a) $Pb|Pb^{2+}(0.120\ M)\|Cd^{2+}(0.0500)|Cd$
   (b) $Zn|Zn^{2+}(0.0420\ M)\|Tl^{3+}(9.06 \times 10^{-2}\ M),$ $Tl^{+}(0.0400\ M)|Pt$
   (c) $Pt, H_2(757\ torr)|HCl(2.00 \times 10^{-4}\ M)\|$ $Ni^{2+}(0.0400\ M)|Ni$
   (d) $Pb|PbI_2(sat'd), I^{-}(0.0220\ M)\|Hg^{2+}$ $(2.60 \times 10^{-3}\ M)|Hg$
   (e) $Pt, H_2(1.00\ atm)|NH_3(0.400\ M), NH_4^{+}$ $(0.200\ M)\|SHE$
   (f) $Pt|TiO^{2+}(0.0450\ M), Ti^{3+}(0.00320\ M),$ $H^{+}(3.00 \times 10^{-2}\ M)\|VO^{2+}(0.1600\ M),$ $V^{3+}(0.0800\ M), H^{+}(0.0100\ M)|Pt$

*19-8. Calculate the theoretical cell potential of the following cells. If the cell is short-circuited, indicate the direction of the spontaneous cell reaction.

(a) $Zn|Zn^{2+}(0.1000 \text{ M})||Co^{2+}(5.87 \times 10^{-4} \text{ M})|Co$

(b) $Pt|Fe^{3+}(0.1600 \text{ M}), Fe^{2+}(0.0700 \text{ M})||$
$Hg^{2+}(0.0350 \text{ M})|Hg$

(c) $Ag|Ag^{+}(0.0575 \text{ M})|H^{+}(0.0333 \text{ M})|O_2(1.12 \text{ atm}), Pt$

(d) $Cu|Cu^{2+}(0.0420 \text{ M})||I^{-}(0.1220 \text{ M}),$
$AgI(sat'd)|Ag$

(e) $SHE||HCOOH(0.1400 \text{ M}), HCOO^{-}(0.0700 \text{ M})|$
$H_2(1.00 \text{ atm}), Pt$

(f) $Pt|UO_2^{2+}(8.00 \times 10^{-3} \text{ M}), U^{4+}(4.00 \times 10^{-2} \text{ M}),$
$H^{+}(1.00 \times 10^{-3} \text{ M})||Fe^{3+}(0.003876 \text{ M}),$
$Fe^{2+}(0.1134 \text{ M})|Pt$

19-9. Calculate the potential of the following two half-cells that are connected by a salt bridge:

*(a) a galvanic cell consisting of a lead electrode (right electrode) immersed in 0.0220 M Pb^{2+} and a zinc electrode in contact with 0.1200 M Zn^{2+}.

(b) a galvanic cell with two platinum electrodes, the one on the left immersed in a solution that is 0.0445 M in Fe^{3+} and 0.0890 M in Fe^{2+}, the one on the right in a solution that is 0.00300 M in $Fe(CN)_6^{4-}$ and 0.1564 M in $Fe(CN)_6^{3-}$.

*(c) a galvanic cell consisting of a standard hydrogen electrode on the left and a platinum electrode immersed in a solution that is 3.50×10^{-3} M in TiO^{2+}, 0.07000 M in Ti^{3+}, and buffered to a pH of 3.00.

19-10. Use the shorthand notation (page 450) to describe the cells in Problem 19-9. Each cell is supplied with a salt bridge to provide electrical contact between the solutions in the two cell compartments.

19-11. Generate equilibrium constant expressions for the following reactions. Calculate numerical values for K_{eq}.

*(a) $Fe^{3+} + V^{2+} \rightleftharpoons Fe^{2+} + V^{3+}$

(b) $Fe(CN)_6^{3-} + Cr^{2+} \rightleftharpoons Fe(CN)_6^{4-} + Cr^{3+}$

*(c) $2V(OH)_4^{+} + U^{4+} \rightleftharpoons 2VO^{2+} + UO_2^{2+} + 4H_2O$

(d) $Tl^{3+} + 2Fe^{2+} \rightleftharpoons Tl^{+} + 2Fe^{3+}$

*(e) $2Ce^{4+} + H_3AsO_3 + H_2O \rightleftharpoons 2Ce^{3+} + H_3AsO_4 + 2H^{+}$ (1 M $HClO_4$)

(f) $2V(OH)_4^{+} + H_2SO_3 \rightleftharpoons SO_4^{2-} + 2VO^{2+} + 5H_2O$

*(g) $VO^{2+} + V^{2+} + 2H^{+} \rightleftharpoons 2V^{3+} + H_2O$

(h) $TiO^{2+} + Ti^{2+} + 2H^{+} \rightleftharpoons 2Ti^{3+} + H_2O$

19-12. Calculate the electrode potential of the system at the equivalence point for each of the reactions in Problem 19-11. Use 0.100 M where a value for $[H^{+}]$ is needed and is not otherwise specified.

19-13. If you start with 0.1000 M solutions and the first-named species is the titrant, what will be the concentration of each reactant and product at the equivalence point of titrations (a), (c), (f), and (g) in Problem 19-11? Assume that there is no change in $[H^{+}]$ during the titration.

*19-14. Select an indicator from Table 19-3 that might be suitable for each of the titrations in Problem 19-11. Write NONE if no indicator listed in Table 19-3 is suitable.

19-15. Use a spreadsheet and construct curves for the following titrations. Calculate potentials after the addition of 10.00, 25.00, 49.00, 49.90, 50.00, 50.10, 51.00, and 60.00 mL of the reagent. Where necessary, assume that $[H^{+}] = 1.00$ throughout.

(a) 50.00 mL of 0.1000 M V^{2+} with 0.05000 M Sn^{4+}.

(b) 50.00 mL of 0.1000 M $Fe(CN)_6^{3-}$ with 0.1000 M Cr^{2+}.

(c) 50.00 mL of 0.1000 M $Fe(CN)_6^{4-}$ with 0.05000 M Tl^{3+}.

(d) 50.00 mL of 0.1000 M Fe^{3+} with 0.05000 M Sn^{2+}.

(e) 50.00 mL of 0.05000 M U^{4+} with 0.02000 M MnO_4^{-}.

19-16. **Challenge Problem:** As a part of a study to measure the dissociation constant of acetic acid, Harned and Ehlers[3] needed to measure E^0 for the following cell:

$$Pt, H_2(1 \text{ atm})|HCl(m), AgCl(sat'd)|Ag$$

(a) Write an expression for the potential of the cell.

(b) Show that the expression can be written as

$$E = E^0 - \frac{RT}{F} \ln (\gamma_{H_3O^+})(\gamma_{Cl^-})m_{H_3O^+}m_{Cl^-}$$

where $\gamma_{H_3O^+}$ and γ_{Cl^-} are the activity coefficients of hydronium ion and chloride ion, respectively, and $m_{H_3O^+}$ and m_{Cl^-} are their respective molal (mole solute/kg solvent) concentrations.

(c) Under what circumstances is this expression valid?

(d) Show that the expression in (b) may be written $E + 2k \log m = E^0 - 2k \log \gamma$, where $k = \ln 10 RT/F$. What are m and γ?

(e) A considerably-simplified version of the Debye-Hückel expression that is valid for very dilute solutions is $\log \gamma = -0.5\sqrt{m} + bm$, where b is a constant. Show that the expression for the cell potential in (d) may be written as $E + 2k \log m - k\sqrt{m} = E^0 - 2kcm$.

(f) The previous expression is a "limiting law" that becomes linear as the concentration of the electrolyte approaches zero. The equation assumes the form $y = ax + b$, where $y = E + 2k \log m - k\sqrt{m}$, $x = m$, the slope $a = -2kc$, and the y-intercept $b = E^0$. Harned and Ehlers very accurately measured the potential of the cell without liquid junction presented at the beginning of the problem as a function of concentration of HCl (molal) and temperature and obtained the data in the table below. For example, they measured the potential of the cell at 25°C with an HCl concentration of 0.01 m and obtained a value of 0.46419 volts.

[3]H. S. Harned, R. W. Ehlers, *J. Am. Chem. Soc.*, **1932**, *54*(4), 1350–57, **DOI:** 10.1021/ja01343a013.

Potential Measurements of Cell Pt,H$_2$(1 atm)|HCl(m), AgCl(sat'd)|Ag without Liquid Junction as a Function of Concentration (Molal) and Temperature (°C)

m, Molal	E_0	E_5	E_{10}	E_{15}	E_{20}	E_{25}	E_{30}	E_{35}
				E_T, volts				
0.005	0.48916	0.49138	0.49338	0.49521	0.44690	0.49844	0.49983	0.50109
0.006	0.48089	0.48295	0.48480	0.48647	0.48800	0.48940	0.49065	0.49176
0.007	0.4739	0.47584	0.47756	0.47910	0.48050	0.48178	0.48289	0.48389
0.008	0.46785	0.46968	0.47128	0.47270	0.47399	0.47518	0.47617	0.47704
0.009	0.46254	0.46426	0.46576	0.46708	0.46828	0.46937	0.47026	0.47103
0.01	0.4578	0.45943	0.46084	0.46207	0.46319	0.46419	0.46499	0.46565
0.02	0.42669	0.42776	0.42802	0.42925	0.42978	0.43022	0.43049	0.43058
0.03	0.40859	0.40931	0.40993	0.41021	0.41041	0.41056	0.41050	0.41028
0.04	0.39577	0.39624	0.39668	0.39673	0.39673	0.39666	0.39638	0.39595
0.05	0.38586	0.38616	0.38641	0.38631	0.38614	0.38589	0.38543	0.38484
0.06	0.37777	0.37793	0.37802	0.37780	0.37749	0.37709	0.37648	0.37578
0.07	0.37093	0.37098	0.37092	0.37061	0.37017	0.36965	0.36890	0.36808
0.08	0.36497	0.36495	0.36479	0.36438	0.36382	0.36320	0.36285	0.36143
0.09	0.35976	0.35963	0.35937	0.35888	0.35823	0.35751	0.35658	0.35556
0.1	0.35507	0.35487	0.33451	0.35394	0.35321	0.35240	0.35140	0.35031
E^0	0.23627	0.23386	0.23126	0.22847	0.22550	0.22239	0.21918	0.21591

Construct a plot of $E + 2k \log m - k\sqrt{m}$ versus m and note that the plot is quite linear at low concentration. Extrapolate the line to the y-intercept and estimate a value for E^0. Compare your value with the value determined by Harned and Ehlers and explain any difference. Also compare the value to the one shown in Table 18-1. The simplest way to carry out this exercise is to place the data in a spreadsheet and use the Excel function INTERCEPT(known_y's, known_x's) to determine the extrapolated value for E^0. Use only the data from 0.005 to 0.01 m to find the intercept.

(g) If you have used a spreadsheet to carry out the data analysis in (f), enter the data for all temperatures into the spreadsheet and determine values for E^0 at all temperatures from 5°C to 35°C. Alternatively, you may download an Excel spreadsheet containing the entire data table from our companion website.

(h) There are two typographical errors in the table above that appeared in the original published paper. Find the errors, and correct them. How can you justify these corrections? What statistical criteria can you apply to justify your action? In your judgment, is it likely that these errors have been detected previously? Explain your answer.

(i) Why do you think that these workers used molality in their studies rather than molarity or weight molarity? Explain whether it matters which of these concentration units are used.

19-17. Challenge Problem: As we saw in Problem 19-16, as a preliminary experiment in their effort to measure the dissociation constant of acetic acid, Harned and Ehlers[5] measured E^0 for the cell without liquid junction shown. To complete the study and determine the dissociation constant, these workers also measured the potential of the following cell:

Pt, H$_2$(1 atm)|HOAc(m_1), NaOAc(m_2),

NaCl(m_3), AgCl(sat'd)|Ag

(a) Show that the potential of this cell is given by

$$E = E^0 - \frac{RT}{F} \ln (\gamma_{H_3O^+})(\gamma_{Cl^-}) m_{H_3O^+} m_{Cl^-}$$

where $\gamma_{H_3O^+}$ and γ_{Cl^-} are the activity coefficients of hydronium ion and chloride ion, respectively, and $m_{H_3O^+}$ and m_{Cl^-} are their respective molal (mole solute/kg solvent) concentrations.

(b) The dissociation constant for acetic acid is given by

$$K = \frac{(\gamma_{H_3O^+})(\gamma_{OAc^-})}{\gamma_{HOAc}} \frac{m_{H_3O^+} m_{OAc^-}}{m_{HOAc}}$$

where γ_{OAc^-} and γ_{HOAc} are the activity coefficients of acetate ion and acetic acid, respectively, and m_{OAc^-} and m_{HOAc} are their respective equilibrium molal (mole solute/kg solvent) concentrations. Show that the potential of the cell in part (a) is given by

[5]H. S. Harned, R. W. Ehlers, *J. Am. Chem. Soc.*, **1932**, *54*(4), 1350–57, **DOI:** 10.1021/ja01343a013.

$$E = E^0 + \frac{RT}{F} \ln \frac{m_{HOAc}m_{Cl^-}}{m_{OAc^-}}$$

$$= -\frac{RT}{F} \ln \frac{(\gamma_{H_3O^+})(\gamma_{Cl^-})(\gamma_{HOAc})}{(\gamma_{H_3O^+})(\gamma_{OAc^-})} - \frac{RT}{F} \ln K$$

(c) As the ionic strength of the solution approaches zero, what happens to the right-hand side of the equation in (b)?

(d) As a result of answer to part (c), we can write the right-hand-side of the equation as $-(RT/F)\ln K'$. Show that

$$K' = \exp\left[-\frac{(E - E^0)F}{RT} \ln\left(\frac{m_{HOAc}m_{Cl^-}}{m_{OAc^-}}\right)\right]$$

(e) The ionic strength of the solution in the cell without liquid junction calculated by Harned and Ehlers is

$$\mu = m_2 + m_3 + m_{H^+}$$

Show that this expression is correct.

(f) These workers prepared solutions of various molal analytical concentrations of acetic acid, sodium acetate, and sodium chloride and measured the potential of the cell presented at the beginning of this problem. Their results are shown in the following table.

Potential Measurements of Cell Pt,H₂(1 atm)|HOAc(c_{HOAc}),NaOAc(c_{NaOAc}),NaCl(c_{NaCl}),AgCl(sat'd)|Ag without Liquid Junction as a Function of Ionic Strength (Molality) and Temperature (°C)

c_{HOAc}, m	c_{NaOAc}, m	c_{NaCl}, m	E_0	E_5	E_{10}	E_{15}	E_{20}	E_{25}	E_{30}	E_{35}
0.004779	0.004599	0.004896	0.61995	0.62392	0.62789	0.63183	0.63580	0.63959	0.64335	0.64722
0.012035	0.011582	0.012326	0.59826	0.60183	0.60538	0.60890	0.61241	0.61583	0.61922	0.62264
0.021006	0.020216	0.021516	0.58528	0.58855	0.59186	0.59508	0.59840	0.60154	0.60470	0.60792
0.04922	0.04737	0.05042	0.56546	0.56833	0.57128	0.57413	0.57699	0.57977	0.58257	0.58529
0.08101	0.07796	0.08297	0.55388	0.55667	0.55928	0.56189	0.56456	0.56712	0.56964	0.57213
0.09056	0.08716	0.09276	0.55128	0.55397	0.55661	0.55912	0.56171	0.56423	0.56672	0.56917

The notation for molal concentration up to this point in our discussion of the Harned and Ehlers paper has been in terms of the variables m_x, where x is the species of interest. Do these symbols represent molal analytical concentrations, molal equilibrium concentrations, or both? Explain. Note that the symbols for concentration in the table adhere to the convention that we have used throughout this book, not the notation of Harned and Ehlers.

(g) Calculate the ionic strength of each of the solutions using the expression for the K_a of acetic acid to calculate $[H_3O^+]$, $[OAc^-]$, and $[HOAc]$ with the usual suitable approximations and a provisional value of $K_a = 1.8 \times 10^{-5}$. Use the potentials in the table at 25°C to calculate values

for K' with the expression in part (d). Construct a plot of K' versus μ and extrapolate the graph to infinite dilution ($\mu = 0$) to find a value for K_a at 25°C. Compare the extrapolated value to the provisional value used to calculate μ. What effect does the provisional value of K_a have on the extrapolated value of K_a? You can perform these calculations most easily using a spreadsheet.

(h) If you have made these computations using a spreadsheet, determine the dissociation constant for acetic acid at all other temperatures for which data are available. How does K_a vary with temperature? At what temperature does the maximum in K_a occur?

Applications of Oxidation/Reduction Titrations

Linus Pauling (1901–1994) was one of the most influential and famous chemists of the twentieth century. His work in chemical bonding, X-ray crystallography, and related areas had a tremendous impact on chemistry, physics, and biology; spanned eight decades; and led to nearly every award available to chemists. He is the only person to receive two unshared Nobel prizes: for chemistry (1954) and, for his efforts to ban nuclear weapons, the peace prize (1962). In his last years, Pauling devoted his immense intellect and energy to the study of various diseases and their cures. He became convinced that vitamin C, or ascorbic acid, was a panacea. His many books and articles on the subject fueled the popularity of alternative therapies and especially the wide use of vitamin C for preventative maintenance of health. This photo of Pauling tossing an orange into the air is symbolic of this work and the importance of being able to determine concentrations of ascorbic acid at all levels in fruits, vegetables, and commercial vitamin preparations. Redox titrations with iodine are widely used to determine ascorbic acid.

© Roger Ressmeyer/CORBIS

In this chapter, we describe the preparation of standard solutions of oxidants and reductants and their applications in analytical chemistry. In addition, auxiliary reagents that convert an analyte to a single oxidation state are discussed.[1]

20A AUXILIARY OXIDIZING AND REDUCING REAGENTS

The analyte in an oxidation/reduction titration must be in a single oxidation state at the outset. Often, however, the steps that precede the titration, such as dissolving the sample and separating interferences, convert the analyte to a mixture of oxidation states. For example, when a sample containing iron is dissolved, the resulting solution usually contains a mixture of iron(II) and iron(III) ions. If we choose to use a standard oxidant for determining iron, we must first treat the sample solution with an auxiliary reducing agent to convert all of the iron to iron(II). On the other hand, if we plan to titrate with a standard reductant, pretreatment with an auxiliary oxidizing reagent is needed.[2]

[1]For further reading on redox titrimetry, see J. A. Dean, *Analytical Chemistry Handbook*, Section 3, pp. 3.65–3.75, New York: McGraw-Hill, 1995.
[2]For a brief summary of auxiliary reagents, see J. A. Goldman and V. A. Stenger, in *Treatise on Analytical Chemistry*, I. M. Kolthoff and P. J. Elving, eds., Part I, Vol. 11, pp. 7204–6, New York: Wiley, 1975.

Figure 20-1 A Jones reductor.

To be useful as a preoxidant or a prereductant, a reagent must react quantitatively with the analyte. In addition, any reagent excess must be easily removable because the excess reagent usually interferes with the titration by reacting with the standard solution.

20A-1 Auxiliary Reducing Reagents

A number of metals are good reducing agents and have been used for the prereduction of analytes. Included among these reductants are zinc, aluminum, cadmium, lead, nickel, copper, and silver (in the presence of chloride ion). Sticks or coils of the metal can be immersed directly in the analyte solution. After reduction is judged complete, the solid is removed manually and rinsed with water. The analyte solution must be filtered to remove granular or powdered forms of the metal. An alternative to filtration is the use of a **reductor**, such as that shown in **Figure 20-1**.[3] In the reductor, the finely divided metal is held in a vertical glass tube through which the solution is drawn under a mild vacuum. The metal in a reductor is normally sufficient for hundreds of reductions.

A typical **Jones reductor** has a diameter of about 2 cm and holds a 40- to 50-cm column of amalgamated zinc. Amalgamation is accomplished by allowing zinc granules to stand briefly in a solution of mercury(II) chloride, where the following reaction occurs:

$$2Zn(s) + Hg^{2+} \rightarrow Zn^{2+} + Zn(Hg)(s)$$

Zinc amalgam is nearly as effective for reductions as the pure metal and has the important virtue of inhibiting the reduction of hydrogen ions by zinc. This side reaction needlessly uses up the reducing agent and also contaminates the sample solution with a large amount of zinc(II) ions. Solutions that are quite acidic can be passed through a Jones reductor without significant hydrogen formation.

Table 20-1 lists the principal applications of the Jones reductor. Also listed in this table are reductions that can be accomplished with a **Walden reductor**, in which

TABLE 20-1

Uses of the Walden Reductor and the Jones Reductor*	
Walden $Ag(s) + Cl^- \rightarrow AgCl(s) + e^-$	**Jones** $Zn(Hg)(s) \rightarrow Zn^{2+} + Hg + 2e^-$
$Fe^{3+} + e^- \rightarrow Fe^{2+}$	$Fe^{3+} + e^- \rightleftharpoons Fe^{2+}$
$Cu^{2+} + e^- \rightarrow Cu^+$	$Cu^{2+} + 2e^- \rightleftharpoons Cu(s)$
$H_2MoO_4 + 2H^+ + e^- \rightarrow MoO_2^+ + 2H_2O$	$H_2MoO_4 + 6H^+ + 3e^- \rightleftharpoons Mo^{3+} + 3H_2O$
$UO_2^{2+} + 4H^+ + 2e^- \rightarrow U^{4+} + 2H_2O$	$UO_2^{2+} + 4H^+ + 2e^- \rightleftharpoons U^{4+} + 2H_2O$
	$UO_2^{2+} + 4H^+ + 3e^- \rightleftharpoons U^{3+} + 2H_2O^\dagger$
$V(OH)_4^+ + 2H^+ + e^- \rightarrow VO^{2+} + 3H_2O$	$V(OH)_4^+ + 4H^+ + 3e^- \rightleftharpoons V^{2+} + 4H_2O$
TiO^{2+} not reduced	$TiO^{2+} + 2H^+ + e^- \rightleftharpoons Ti^{3+} + H_2O$
Cr^{3+} not reduced	$Cr^{3+} + e^- \rightleftharpoons Cr^{2+}$

*I.M. Kolthoff and R. Belcher, *Volumetric Analysis,* Vol. 3, p. 12. New York: Interscience, 1957. John Wiley & Sons, Inc. Reproduced with permission of John Wiley & Sons Inc.
†A mixture of oxidation states is obtained. The Jones reductor may still be used for the determination of uranium, however, because any U^{2+} formed can be converted to U^{4+} by shaking the solution with air for a few minutes.

[3]For a discussion of reductors, see F. Hecht, in *Treatise on Analytical Chemistry,* I. M. Kolthoff and P. J. Elving, eds., Part I. Vol. 11, pp. 6703–7, New York: Wiley, 1975.

granular metallic silver held in a narrow glass column is the reductant. Silver is not a good reducing agent unless chloride or some other ion that forms a silver salt of low solubility is present. For this reason, prereductions with a Walden reductor are generally carried out from hydrochloric acid solutions of the analyte. The coating of silver chloride produced on the metal is removed periodically by dipping a zinc rod into the solution that covers the packing. Table 20-1 suggests that the Walden reductor is somewhat more selective in its action than is the Jones reductor.

20A-2 Auxiliary Oxidizing Reagents

Sodium Bismuthate

Sodium bismuthate is a powerful oxidizing agent capable, for example, of converting manganese(II) quantitatively to permanganate ion. This bismuth salt is a sparingly soluble solid with a formula that is usually written as $NaBiO_3$, although its exact composition is somewhat uncertain. Oxidations are performed by suspending the bismuthate in the analyte solution and boiling for a brief period. The unused reagent is then removed by filtration. The half-reaction for the reduction of sodium bismuthate can be written as

$$NaBiO_3(s) + 4H^+ + 2e^- \rightleftharpoons BiO^+ + Na^+ + 2H_2O$$

Ammonium Peroxydisulfate

Ammonium peroxydisulfate, $(NH_4)_2S_2O_8$, is also a powerful oxidizing agent. In acidic solution, it converts chromium(III) to dichromate, cerium(III) to cerium(IV), and manganese(II) to permanganate. The half-reaction is

$$S_2O_8^{2-} + 2e^- \rightleftharpoons 2SO_4^{2-}$$

The oxidations are catalyzed by traces of silver ion. The excess reagent is easily decomposed by a brief period of boiling:

$$2S_2O_8^{2-} + 2H_2O \rightarrow 4SO_4^{2-} + O_2(g) + 4H^+$$

Sodium Peroxide and Hydrogen Peroxide

Peroxide is a convenient oxidizing agent either as the solid sodium salt or as a dilute solution of the acid. The half-reaction for hydrogen peroxide in acidic solution is

$$H_2O_2 + 2H^+ + 2e^- \rightleftharpoons 2H_2O \qquad E^0 = 1.78 \text{ V}$$

After oxidation is complete, the solution is freed of excess reagent by boiling:

$$2H_2O_2 \rightarrow 2H_2O + O_2(g)$$

20B APPLYING STANDARD REDUCING AGENTS

Standard solutions of most reductants tend to react with atmospheric oxygen. For this reason, reductants are seldom used for the direct titration of oxidizing analytes; indirect methods are used instead. The two most common reductants, iron(II) and thiosulfate ions, are discussed in the paragraphs that follow.

Molecular model of thiosulfate ion. Sodium thiosulfate, formerly called sodium hyposulfite or **hypo**, is used to "fix" photographic images and to extract silver from ore, as well as an antidote in cyanide poisoning, as a mordant in the dye industry, as a bleaching agent in a variety of applications, as the solute in the supersaturated solution of hot packs, and of course, as an analytical reducing agent. The action of thiosulfate as a photographic fixer is based on its capacity to form complexes with silver and thus dissolve unexposed silver bromide from the surface of photographic film and paper. Thiosulfate is often used as a dechlorinating agent to make aquarium water safe for fish and other aquatic life.

> In its reaction with iodine, each thiosulfate ion loses one electron.

> Sodium thiosulfate is one of the few reducing agents that is not oxidized by air.

20B-1 Iron(II) Solutions

Solutions of iron(II) are easily prepared from iron(II) ammonium sulfate, $Fe(NH_4)_2(SO_4)_2 \cdot 6H_2O$ (Mohr's salt), or from the closely related iron(II) ethylenediamine sulfate, $FeC_2H_4(NH_3)_2(SO_4)_2 \cdot 4H_2O$ (Oesper's salt). Air oxidation of iron(II) takes place rapidly in neutral solutions but is inhibited in the presence of acids, with the most stable preparations being about 0.5 M in H_2SO_4. Such solutions are stable for no more than one day, if that long. Numerous oxidizing agents are conveniently determined by treatment of the analyte solution with a measured excess of standard iron(II) followed by immediate titration of the excess with a standard solution of potassium dichromate or cerium(IV) (see Sections 20C-1 and 20C-2). Just before or just after the analyte is titrated, the volumetric ratio between the standard oxidant and the iron(II) solution is established by titrating two or three aliquots of iron(II) with the oxidant. This procedure has been applied to the determination of organic peroxides; hydroxylamine; chromium(VI); cerium(IV); molybdenum(VI); nitrate, chlorate, and perchlorate ions; and numerous other oxidants (see for example, Problems 20-20 and 20-21).

20B-2 Sodium Thiosulfate

Thiosulfate ion ($S_2O_3^{2-}$) is a moderately strong reducing agent that has been widely used to determine oxidizing agents by an indirect procedure in which iodine is an intermediate. With iodine, thiosulfate ion is oxidized quantitatively to tetrathionate ion ($S_4O_6^{2-}$) according to the half-reaction

$$2S_2O_3^{2-} \rightleftharpoons S_4O_6^{2-} + 2e^-$$

The quantitative reaction with iodine is unique. Other oxidants can oxidize the tetrathionate ion to sulfate ion.

The scheme used to determine oxidizing agents involves adding an unmeasured excess of potassium iodide to a slightly acidic solution of the analyte. Reduction of the analyte produces a stoichiometrically equivalent amount of iodine. The liberated iodine is then titrated with a standard solution of sodium thiosulfate, $Na_2S_2O_3$, one of the few reducing agents that is stable toward air oxidation. An example of this procedure is the determination of sodium hypochlorite in bleaches. The reactions are

$$OCl^- + 2I^- + 2H^+ \rightarrow Cl^- + I_2 + H_2O \qquad \text{(unmeasured excess KI)}$$
$$I_2 + 2S_2O_3^{2-} \rightarrow 2I^- + S_4O_6^{2-} \qquad \text{(20-1)}$$

The quantitative conversion of thiosulfate ion to tetrathionate ion shown in Equation 20-1 requires a pH smaller than 7. If strongly acidic solutions must be titrated, air oxidation of the excess iodide must be prevented by blanketing the solution with an inert gas, such as carbon dioxide or nitrogen.

Detecting End Points in Iodine/Thiosulfate Titrations

A solution that is about 5×10^{-6} M in I_2 has a discernible color, which corresponds to less than one drop of a 0.05 M iodine solution in 100 mL. Thus, provided the analyte solution is colorless, the disappearance of the iodine color can serve as the indicator in titrations with sodium thiosulfate.

Most often, iodine titrations are performed with a suspension of starch as an indicator. The deep blue color that develops in the presence of iodine is believed to arise from the absorption of iodine into the helical chain of β-amylose (see **Figure 20-2**), a macromolecular component of most starches. The closely related α-amylose forms a red adduct with iodine. This reaction is not easily reversible and is thus undesirable. In commercially available *soluble starch*, the alpha fraction has been removed to leave principally β-amylose. Indicator solutions are easily prepared from this product.

Aqueous starch suspensions decompose within a few days, primarily because of bacterial action. The decomposition products tend to interfere with the indicator properties of the preparation and may also be oxidized by iodine. The rate of decomposition can be inhibited by preparing and storing the indicator under sterile conditions and by adding mercury(II) iodide or chloroform as a bacteriostat. Perhaps the simplest alternative is to prepare a fresh suspension of the indicator, which requires only a few minutes, on the day it is to be used.

Starch decomposes irreversibly in solutions containing large concentrations of iodine. Therefore, in titrating solutions of iodine with thiosulfate ion, as in the indirect determination of oxidants, addition of the indicator is delayed until the color of the solution changes from red-brown to yellow; at this point, the titration is nearly complete. The indicator can be introduced at the outset when thiosulfate solutions are being titrated directly with iodine.

Stability of Sodium Thiosulfate Solutions

Although sodium thiosulfate solutions are resistant to air oxidation, they do tend to decompose to give sulfur and hydrogen sulfite ion:

$$S_2O_3^{2-} + H^+ \rightleftharpoons HSO_3^- + S(s)$$

Variables that influence the rate of this reaction include pH, the presence of microorganisms, the concentration of the solution, the presence of copper(II) ions, and exposure to sunlight. These variables may cause the concentration of a thiosulfate solution to change by several percent over a period of a few weeks. Proper attention to detail will produce solutions that need only occasional restandardization. The rate of the decomposition reaction increases markedly as the solution becomes acidic.

The most important single cause for the instability of neutral or slightly basic thiosulfate solutions is bacteria that metabolize thiosulfate ion to sulfite and sulfate ions as well as to elemental sulfur. To minimize this problem, standard solutions of the reagent are prepared under reasonably sterile conditions. Bacterial activity appears to be at a minimum at a pH between 9 and 10, which accounts, at least in part, for the reagent's greater stability in slightly basic solutions. The presence of a bactericide, such as chloroform, sodium benzoate, or mercury(II) iodide, also slows decomposition.

Standardizing Thiosulfate Solutions

Potassium iodate is an excellent primary standard for thiosulfate solutions. In this application, weighed amounts of primary-standard-grade reagent are dissolved in water containing an excess of potassium iodide. When this mixture is acidified with a strong acid, the reaction

$$IO_3^- + 5I^- + 6H^+ \rightleftharpoons 3I_2 + 3H_2O$$

> Starch undergoes decomposition in solutions with high I_2 concentrations. In titrations of excess I_2 with $Na_2S_2O_3$, addition of the indicator must be deferred until most of the I_2 has been reduced.

> When sodium thiosulfate is added to a strongly acidic medium, a cloudiness develops almost immediately as a consequence of the precipitation of elemental sulfur. Even in neutral solution, this reaction proceeds at such a rate that standard sodium thiosulfate must be restandardized periodically.

$n > 1000$

(a)

(b)

Figure 20-2 Thousands of glucose molecules polymerize to form huge molecules of β-amylose, as shown schematically in (a). Molecules of β-amylose tend to assume a helical structure. The iodine species I_5^-, as shown in (b), is incorporated into the amylose helix. (Reprinted (adapted) with permission from R. C. Teitelbaum, S. L. Ruby, and T. J. Marks, *J. Amer. Chem. Soc.,* **1980,** *102,* 3322. Copyright 1980 American Chemical Society.)

occurs instantaneously. The liberated iodine is then titrated with the thiosulfate solution. The stoichiometry of the reactions is

$$1 \text{ mol } IO_3^- = 3 \text{ mol } I_2 = 6 \text{ mol } S_2O_3^{2-}$$

EXAMPLE 20-1

A solution of sodium thiosulfate was standardized by dissolving 0.1210 g KIO_3 (214.00 g/mol) in water, adding a large excess of KI, and acidifying with HCl. The liberated iodine required 41.64 mL of the thiosulfate solution to decolorize the blue starch/iodine complex. Calculate the molar concentration of the $Na_2S_2O_3$.

Solution

$$\text{amount } Na_2S_2O_3 = 0.1210 \text{ g } \cancel{KIO_3} \times \frac{1 \text{ mmol } \cancel{KIO_3}}{0.21400 \text{ g } \cancel{KIO_3}} \times \frac{6 \text{ mmol } Na_2S_2O_3}{\text{mmol } \cancel{KIO_3}}$$

$$= 3.3925 \text{ mmol } Na_2S_2O_3$$

$$c_{Na_2S_2O_3} = \frac{3.3925 \text{ mmol } Na_2S_2O_3}{41.64 \text{ mL } Na_2S_2O_3} = 0.08147 \text{ M}$$

Other primary standards for sodium thiosulfate are potassium dichromate, potassium bromate, potassium hydrogen iodate, potassium hexacyanoferrate(III), and metallic copper. All these compounds liberate stoichiometric amounts of iodine when treated with excess potassium iodide.

Applications of Sodium Thiosulfate Solutions

Numerous substances can be determined by the indirect method involving titration with sodium thiosulfate. Typical applications are summarized in **Table 20-2**.

TABLE 20-2

Some Applications of Sodium Thiosulfate as a Reductant

Analyte	Half-Reaction	Special Conditions
IO_4^-	$IO_4^- + 8H^+ + 7e^- \rightleftharpoons \frac{1}{2}I_2 + 4H_2O$	Acidic solution
	$IO_4^- + 2H^+ + 2e^- \rightleftharpoons IO_3^- + H_2O$	Neutral solution
IO_3^-	$IO_3^- + 6H^+ + 5e^- \rightleftharpoons \frac{1}{2}I_2 + 3H_2O$	Strong acid
BrO_3^-, ClO_3^-	$XO_3^- + 6H^+ + 6e^- \rightleftharpoons X^- + 3H_2O$	Strong acid
Br_2, Cl_2	$X_2 + 2I^- \rightleftharpoons I_2 + 2X^-$	
NO_2^-	$HNO_2 + H^+ + e^- \rightleftharpoons NO(g) + H_2O$	
Cu^{2+}	$Cu^{2+} + I^- + e^- \rightleftharpoons CuI(s)$	
O_2	$O_2 + 4Mn(OH)_2(s) + 2H_2O \rightleftharpoons 4Mn(OH)_3(s)$	Basic solution
	$Mn(OH)_3(s) + 3H^+ + e^- \rightleftharpoons Mn^{2+} + 3H_2O$	Acidic solution
O_3	$O_3(g) + 2H^+ + 2e^- \rightleftharpoons O_2(g) + H_2O$	
Organic peroxide	$ROOH + 2H^+ + 2e^- \rightleftharpoons ROH + H_2O$	

20C APPLYING STANDARD OXIDIZING AGENTS

Table 20-3 summarizes the properties of five of the most widely used volumetric oxidizing reagents. Note that the standard potentials for these reagents vary from 0.5 to 1.5 V. The choice among them depends on the strength of the analyte as a reducing agent, the rate of reaction between oxidant and analyte, the stability of the standard oxidant solutions, the cost, and the availability of a satisfactory indicator.

20C-1 The Strong Oxidants: Potassium Permanganate and Cerium(IV)

Solutions of permanganate ion and cerium(IV) ion are strong oxidizing reagents whose applications closely parallel one another. Half-reactions for the two are

$$MnO_4^- + 8H^+ + 5e^- \rightleftharpoons Mn^{2+} + 4H_2O \qquad E^0 = 1.51 \text{ V}$$

$$Ce^{4+} + e^- \rightleftharpoons Ce^{3+} \qquad E^{0\prime} = 1.44 \text{ V}(1 \text{ M } H_2SO_4)$$

The formal potential shown for the reduction of cerium(IV) is for solutions that are 1 M in sulfuric acid. In 1 M perchloric acid and 1 M nitric acid, the potentials are 1.70 V and 1.61 V, respectively. Solutions of cerium(IV) in the latter two acids are not very stable and thus find limited application.

The half-reaction shown for permanganate ion occurs only in solutions that are 0.1 M or greater in strong acid. In less acidic media, the product may be Mn(III), Mn(IV), or Mn(VI), depending on conditions.

Comparing the Two Reagents

For all practical purposes, the oxidizing strengths of permanganate and cerium(IV) solutions are comparable. Solutions of cerium(IV) in sulfuric acid,

Molecular model of permanganate ion, MnO_4^-. In addition to its use as an analytical reagent, usually in the form of its potassium salt, permanganate is very useful as an oxidizing agent in synthetic organic chemistry. It is used as a bleaching agent with fats, oils, cotton, silk, and other fibers. It has also been used as an antiseptic and anti-infective and as a component in outdoor survival kits, as well as for destroying organic matter in fish ponds, in manufacturing printed wiring boards, for neutralizing the effects of the pesticide rotenone, and for scrubbing flue gases in the determination of mercury. Solid potassium permanganate reacts violently with organic matter, and this effect is often used as a demonstration in general chemistry courses. To further explore these and other uses of permanganate, use a browser and search *permanganate uses*.

TABLE 20-3

Some Common Oxidants Used as Standard Solutions

Reagent and Formula	Reduction Product	Standard Potential, V	Standardized With	Indicator*	Stability†
Potassium permanganate, $KMnO_4$	Mn^{2+}	$1.51^{\ddagger}$	$Na_2C_2O_4$, Fe, As_2O_3	MnO_4^-	(b)
Potassium bromate, $KBrO_3$	Br^-	$1.44^{\ddagger}$	$KBrO_3$	(1)	(a)
Cerium(IV), Ce^{4+}	Ce^{3+}	$1.44^{\ddagger}$	$Na_2C_2O_4$, Fe, As_2O_3	(2)	(a)
Potassium dichromate, $K_2Cr_2O_7$	Cr^{3+}	$1.33^{\ddagger}$	$K_2Cr_2O_7$, Fe	(3)	(a)
Iodine, I_2	I^-	$0.536^{\ddagger}$	$BaS_2O_3, \cdot H_2O$, $Na_2S_2O_3$	starch	(c)

*(1) α-Napthoflavone; (2) 1,10-phenanthroline iron(II) complex (ferroin); and (3) diphenylamine sulfonic acid.

†(a) indefinitely stable; (b) moderately stable, requires periodic standardization; and (c) somewhat unstable, requires frequent standardization.

‡$E^{0\prime}$ in 1 M H_2SO_4.

however, are stable indefinitely, but permanganate solutions decompose slowly and thus require occasional restandardization. Furthermore, cerium(IV) solutions in sulfuric acid do not oxidize chloride ion and can be used to titrate hydrochloric acid solutions of analytes. In contrast, permanganate ion cannot be used with hydrochloric acid solutions unless special precautions are taken to prevent the slow oxidation of chloride ion that leads to overconsumption of the standard reagent. A further advantage of cerium(IV) is that a primary-standard-grade salt of the reagent is available, thus making possible the direct preparation of standard solutions.

Despite these advantages of cerium solutions over permanganate solutions, the latter are more widely used. One reason is the color of permanganate solutions, which is intense enough to serve as an indicator in titrations. A second reason for the popularity of permanganate solutions is their modest cost. The cost of 1 L of 0.02 M $KMnO_4$ solution is about one-tenth the cost 1 L of a comparable strength Ce(IV) solution ($1/100\times$ if primary-standard-grade Ce(IV) reagent is used). Another disadvantage of cerium(IV) solutions is their tendency to form precipitates of basic salts in solutions that are less than 0.1 M in strong acid.

Detecting the End Points

A useful property of a potassium permanganate solution is its intense purple color, which is sufficient to serve as an indicator for most titrations. If you add as little as 0.01 to 0.02 mL of a 0.02 M solution of permanganate to 100 mL of water, you can perceive the purple color of the resulting solution. If the solution is very dilute, diphenylamine sulfonic acid or the 1,10-phenanthroline complex of iron(II) (see Table 19-3) provides a sharper end point.

The permanganate end point is not permanent because excess permanganate ions react slowly with the relatively large concentration of manganese(II) ions present at the end point, according to the reaction

$$2MnO_4^- + 3Mn^{2+} + 2H_2O \rightleftharpoons 5MnO_2(s) + 4H^+$$

The equilibrium constant for this reaction is about 10^{47}, indicating that the equilibrium concentration of permanganate ion is incredibly small even in highly acidic media. Fortunately, the rate at which this equilibrium is approached is so slow that the end point fades only gradually over a period of perhaps 30 seconds.

Solutions of cerium(IV) are yellow-orange, but the color is not intense enough to act as an indicator in titrations. Several oxidation/reduction indicators are available for titrations with standard solutions of cerium(IV). The most widely used of these is the iron(II) complex of 1,10-phenanthroline or one of its substituted derivatives (see Table 19-3).

The Preparation and Stability of Standard Solutions

Aqueous solutions of permanganate are not entirely stable because of water oxidation:

$$4MnO_4^- + 2H_2O \rightarrow 4MnO_2(s) + 3O_2(g) + 4OH^-$$

Although the equilibrium constant for this reaction indicates that the products are favored, permanganate solutions, when properly prepared, are reasonably stable because the decomposition reaction is slow. It is catalyzed by light, heat, acids, bases, manganese(II), and manganese dioxide.

Moderately stable solutions of permanganate ion can be prepared if the effects of these catalysts, particularly manganese dioxide, are minimized. Manganese dioxide is a contaminant in even the best grade of solid potassium permanganate. Furthermore, this compound forms in freshly prepared solutions of the reagent as a consequence of the reaction of permanganate ion with organic matter and dust present in the water used to prepare the solution. Removal of manganese dioxide by filtration before standardization markedly improves the stability of standard permanganate solutions. Before filtration, the reagent solution is allowed to stand for about 24 hours or is heated for a brief period to hasten oxidation of the organic species generally present in small amounts in distilled and deionized water. Paper cannot be used for filtering because permanganate ion reacts with it to form additional manganese dioxide.

Standardized permanganate solutions should be stored in the dark. Filtration and restandardization are required if any solid is detected in the solution or on the walls of the storage bottle. In any event, restandardization every one or two weeks is a good precautionary measure.

Solutions containing excess standard permanganate should never be heated because they decompose by oxidizing water. This decomposition cannot be compensated for with a blank. It is possible to titrate hot, acidic solutions of reductants with permanganate without error if the reagent is added slowly enough so that large excesses do not accumulate.

> Permanganate solutions are moderately stable provided they are free of manganese dioxide and stored in a dark container.

FEATURE 20-1

Determination of Chromium Species in Water Samples

Chromium is an important metal to monitor in environmental samples. Not only is the total amount of chromium of interest, but the oxidation state in which the chromium is found is quite important. In water, chromium can exist as the Cr(III) or as Cr(VI) species. Chromium(III) is an essential nutrient and nontoxic. Chromium(VI), however, is a known carcinogen. Hence, the determination of the amount of chromium in each of these oxidation states is often of more interest than the total amount of chromium. There are several good methods available for determining Cr(VI) selectively. One of the most popular utilizes the oxidation of the reagent 1,5-diphenylcarbohydrazide (diphenylcarbazide) by Cr(VI) in acid solution. The reaction produces a red-purple chelate of Cr(III) and diphenylcarbazide that can be monitored colorimetrically. The direct reaction of Cr(III) itself and the reagent is so slow that essentially only the Cr(VI) is measured. To determine Cr(III), the sample is oxidized with excess permanganate in alkaline solution to convert all the Cr(III) to Cr(VI). The excess oxidant is destroyed with sodium azide. A new colorimetric measurement is made that now determines total chromium (the original Cr(VI) plus that formed by oxidation of Cr(III)). The amount of Cr(III) present is then obtained by subtracting the amount of Cr(VI) obtained in the original measurement from the amount of total chromium obtained after permanganate oxidation. Note that in this instance permanganate is being used as an auxiliary oxidizing agent.

Chromium has long been prized for its beauty as a polished coating on metals (see photo) and for its anticorrosive properties in stainless steel and other alloys. In trace amounts, chromimum(III) is an essential nutrient. Chromium(VI) in the form of sodium dichromate is widely used in aqueous solution as a corrosion inhibitor in large-scale industrial processes. See margin note on page 523 for more details on chromium.

© Don Conrard/CORBIS

EXAMPLE 20-2

Describe how you would prepare 2.0 L of an approximately 0.010 M solution of KMnO4 (158.03 g/mol).

Solution

$$\text{mass KMnO}_4 \text{ needed} = 2.0\ \text{L} \times 0.010\ \frac{\text{mol } \cancel{\text{KMnO}_4^-}}{\text{L}} \times 158.03\ \frac{\text{g KMnO}_4}{\text{mol } \cancel{\text{KMnO}_4^-}}$$

$$= 3.16\ \text{g KMnO}_4$$

Dissolve about 3.2 g of $KMnO_4$ in a little water. After solution is complete, add water to bring the volume to about 2.0 L. Heat the solution to boiling for a brief period and let stand until it is cool. Filter through a glass filtering crucible and store in a clean dark bottle.

The most widely used compounds for the preparation of solutions of cerium(IV) are listed in **Table 20-4**. Primary-standard-grade cerium ammonium nitrate is available commercially and can be used to prepare standard solutions of the cation directly by mass. More commonly, less expensive reagent-grade cerium(IV) ammonium nitrate

TABLE 20-4

Analytically Useful Cerium(IV) Compounds

Name	Formula	Molar Mass
Cerium(IV) ammonium nitrate	$Ce(NO_3)_4 \cdot 2NH_4NO_3$	548.2
Cerium(IV) ammonium sulfate	$Ce(SO_4)_2 \cdot 2(NH_4)_2SO_4 \cdot 2H_2O$	632.6
Cerium(IV) hydroxide	$Ce(OH)_4$	208.1
Ce(IV) hydrogen sulfate	$Ce(HSO_4)_4$	528.4

or ceric hydroxide is used to prepare solutions that are subsequently standardized. In either case, the reagent is dissolved in a solution that is at least 0.1 M in sulfuric acid to prevent the precipitation of basic salts. Sulfuric acid solutions of cerium(IV) are remarkably stable and can be stored for months or heated at 100°C for prolonged periods without change in concentration.

Standardizing Permanganate and Ce(IV) Solutions

Sodium oxalate is a widely used primary standard. In acidic solutions, the oxalate ion is converted to the undissociated acid. Thus, its reaction with permanganate can be described by

$$2MnO_4^- + 5H_2C_2O_4 + 6H^+ \rightarrow 2Mn^{2+} + 10CO_2(g) + 8H_2O$$

The reaction between permanganate ion and oxalic acid is complex and proceeds slowly even at elevated temperature unless manganese(II) is present as a catalyst. Therefore, when the first few milliliters of standard permanganate are added to a hot solution of oxalic acid, several seconds are required before the color of the permanganate ion disappears. As the concentration of manganese(II) builds up, however, the reaction proceeds more and more rapidly as a result of **autocatalysis**.

It has been found that, when solutions of sodium oxalate are titrated at 60°C to 90°C, the consumption of permanganate is from 0.1 to 0.4% less than theoretical, probably due to the air oxidation of a fraction of the oxalic acid. This small error can be avoided by adding 90 to 95% of the required permanganate to a cool solution of the oxalate. After the added permanganate is completely consumed (as indicated by the disappearance of color), the solution is heated to about 60°C and titrated to a pink color that persists for about 30 s. The disadvantage of this procedure is that it requires prior knowledge of the approximate concentration of the permanganate solution so that a proper initial volume can be added. For most purposes, direct titration of the hot oxalic acid solution is adequate (usually 0.2 to 0.3% high). If greater accuracy is required, a direct titration of the hot solution of one portion of the primary standard can be followed by titration of two or three portions in which the solution is not heated until the end.

Sodium oxalate is also widely used to standardize Ce(IV) solutions. The reaction between Ce^{4+} and $H_2C_2O_4$ is

$$2Ce^{4+} + H_2C_2O_4 \rightarrow 2Ce^{3+} + 2CO_2(g) + 2H^+$$

Cerium(IV) standardizations against sodium oxalate are usually performed at 50°C in a hydrochloric acid solution containing iodine monochloride as a catalyst.

> **Autocatalysis** is a type of catalysis in which the product of a reaction catalyses the reaction. This phenomenon causes the rate of the reaction to increase as the reaction proceeds.

> ❮ Solutions of $KMnO_4$ and Ce^{4+} can also be standardized with electrolytic iron wire or with potassium iodide.

EXAMPLE 20-3

You wish to standardize the solution in Example 20-2 against primary $Na_2C_2O_4$ (134.00 g/mol). If you want to use between 30 and 45 mL of the reagent for the standardization, what range of masses of the primary standard should you weigh out?

(continued)

Solution

For a 30-mL titration,

$$\text{amount KMnO}_4 = 30 \text{ mL } \cancel{\text{KMnO}_4} \times 0.010 \frac{\text{mmol KMnO}_4}{\text{mL } \cancel{\text{KMnO}_4}}$$

$$= 0.30 \text{ mmol KMnO}_4$$

$$\text{mass Na}_2\text{C}_2\text{O}_4 = 0.30 \text{ mmol } \cancel{\text{KMnO}_4} \times \frac{5 \text{ mmol } \cancel{\text{Na}_2\text{C}_2\text{O}_4}}{2 \text{ mmol } \cancel{\text{KMnO}_4}}$$

$$\times 0.134 \frac{\text{g Na}_2\text{C}_2\text{O}_4}{\text{mmol } \cancel{\text{Na}_2\text{C}_2\text{O}_4}}$$

$$= 0.101 \text{ g Na}_2\text{C}_2\text{O}_4$$

Proceeding in the same way, we find for a 45-mL titration,

$$\text{mass Na}_2\text{C}_2\text{O}_4 = 45 \times 0.010 \times \frac{5}{2} \times 0.134 = 0.151 \text{ g Na}_2\text{C}_2\text{O}_4$$

Thus, you should weigh between 0.10 and 0.15 g samples of the primary standard.

EXAMPLE 20-4

A 0.1278-g sample of primary-standard $Na_2C_2O_4$ required exactly 33.31 mL of the permanganate solution in Example 20-2 to reach the end point. What was the molar concentration of the $KMnO_4$ reagent?

Solution

$$\text{amount Na}_2\text{C}_2\text{O}_4 = 0.1278 \text{ g } \cancel{\text{Na}_2\text{C}_2\text{O}_4} \times \frac{1 \text{ mmol Na}_2\text{C}_2\text{O}_4}{0.13400 \text{ g } \cancel{\text{Na}_2\text{C}_2\text{O}_4}}$$

$$= 0.95373 \text{ mmol Na}_2\text{C}_2\text{O}_4$$

$$c_{\text{KMnO}_4} = 0.95373 \text{ mmol } \cancel{\text{Na}_2\text{C}_2\text{O}_4} \times \frac{2 \text{ mmol KMnO}_4}{5 \text{ mmol } \cancel{\text{Na}_2\text{C}_2\text{O}_4}} \times \frac{1}{33.31 \text{ mL KMnO}_4}$$

$$= 0.01145 \text{ M}$$

Using Potassium Permanganate and Cerium(IV) Solutions

Table 20-5 lists some of the many applications of permanganate and cerium(IV) solutions to the volumetric determination of inorganic species. Both reagents have also been applied to the determination of organic compounds with oxidizable functional groups.

TABLE 20-5

Some Applications of Potassium Permanganate and Cerium(IV) Solutions

Substance Sought	Half-Reaction	Conditions
Sn	$Sn^{2+} \rightleftharpoons Sn^{4+} + 2e^-$	Prereduction with Zn
H_2O_2	$H_2O_2 \rightleftharpoons O_2(g) + 2H^+ + 2e^-$	
Fe	$Fe^{2+} \rightleftharpoons Fe^{3+} + e^-$	Prereduction with $SnCl_2$ or with Jones or Walden reductor
$Fe(CN)_6^{4-}$	$Fe(CN)_6^{4-} \rightleftharpoons Fe(CN)_6^{3-} + e^-$	
V	$VO^{2+} + 3H_2O \rightleftharpoons V(OH)_4^+ + 2H^+ + e^-$	Prereduction with Bi amalgam or SO_2
Mo	$Mo^{3+} + 4H_2O \rightleftharpoons MoO_4^{2-} + 8H^+ + 3e^-$	Prereduction with Jones reductor
W	$W^{3+} + 4H_2O \rightleftharpoons WO_4^{2-} + 8H^+ + 3e^-$	Prereduction with Zn or Cd
U	$U^{4+} + 2H_2O \rightleftharpoons UO_2^{2+} + 4H^+ + 2e^-$	Prereduction with Jones reductor
Ti	$Ti^{3+} + H_2O \rightleftharpoons TiO^{2+} + 2H^+ + e^-$	Prereduction with Jones reductor
$H_2C_2O_4$	$H_2C_2O_4 \rightleftharpoons 2CO_2 + 2H^+ + 2e^-$	
Mg, Ca, Zn, Co, Pb, Ag	$H_2C_2O_4 \rightleftharpoons 2CO_2 + 2H^+ + 2e^-$	Sparingly soluble metal oxalates filtered, washed, and dissolved in acid; liberated oxalic acid titrated
HNO_2	$HNO_2 + H_2O \rightleftharpoons NO_3^- + 3H^+ + 2e^-$	15-min reaction time; excess $KMnO_4$ back-titrated
K	$K_2NaCo(NO_2)_6 + 6H_2O \rightleftharpoons Co^{2+} + 6NO_3^- + 12H^+ + 2K^+ + Na^+ + 11e^-$	Precipitated as $K_2NaCo(NO_2)_6$; filtered and dissolved in $KMnO_4$; excess $KMnO_4$ back-titrated
Na	$U^{4+} + 2H_2O \rightleftharpoons UO_2^{2+} + 4H^+ + 2e^-$	Precipitated as $NaZn(UO_2)_3(OAc)_9$; filtered, washed, dissolved; U determined as above

EXAMPLE 20-5

Aqueous solutions containing approximately 3% (w/w) H_2O_2 are sold in drug stores as a disinfectant. Propose a method for determining the peroxide content of such a preparation using the standard solution described in Examples 20-3 and 20-4. Assume that you wish to use between 30 and 45 mL of the reagent for a titration. The reaction is

$$5H_2O_2 + 2MnO_4^- + 6H^+ \rightarrow 5O_2 + 2Mn^{2+} + 8H_2O$$

Solution

The amount of $KMnO_4$ in 35 to 45 mL of the reagent is between

$$\text{amount } KMnO_4 = 35 \text{ mL } KMnO_4 \times 0.01145 \frac{\text{mmol } KMnO_4}{\text{mL } KMnO_4}$$

$$= 0.401 \text{ mmol } KMnO_4$$

and

$$\text{amount } KMnO_4 = 45 \times 0.01145 = 0.515 \text{ mmol } KMnO_4$$

The amount of H_2O_2 consumed by 0.401 mmol of $KMnO_4$ is

$$\text{amount } H_2O_2 = 0.401 \text{ mmol } KMnO_4 \times \frac{5 \text{ mmol } H_2O_2}{2 \text{ mmol } KMnO_4} = 1.00 \text{ mmol } H_2O_2$$

and

$$\text{amount } H_2O_2 = 0.515 \times \frac{5}{2} = 1.29 \text{ mmol } H_2O_2$$

(continued)

We, therefore, need to take samples that contain from 1.00 to 1.29 mmol H_2O_2.

$$\text{mass sample} = 1.00 \text{ mmol } H_2O_2 \times 0.03401 \frac{\text{g } H_2O_2}{\text{mmol } H_2O_2} \times \frac{100 \text{ g sample}}{3 \text{ g } H_2O_2}$$

$$= 1.1 \text{ g sample}$$

to

$$\text{mass sample} = 1.29 \times 0.03401 \times \frac{100}{3} = 1.5 \text{ g sample}$$

Thus, our samples should weigh between 1.1 and 1.5 g. These should be diluted to perhaps 75 to 100 mL with water and made slightly acidic with dilute H_2SO_4 before titration.

FEATURE 20-2

Antioxidants[4]

Oxidation can have deleterious effects on the cells and tissues of the human body. There is a considerable body of evidence that reactive oxygen and nitrogen species, such as superoxide ion O_2^-, hydroxyl radical $OH\cdot$, peroxyl radicals $RO_2\cdot$, alkoxyl radicals $RO\cdot$, nitric oxide $NO\cdot$, and nitrogen dioxide $NO_2\cdot$, damage cells and other body components. A group of compounds known as antioxidants can help counteract the influence of reactive oxygen and nitrogen species. Antioxidants are reducing agents that are so easily oxidized that they can protect other compounds in the body from oxidation. Typical antioxidants include Vitamins A, C, and E; minerals, such as selenium; and herbs, such as ginkgo, rosemary, and milk thistle.

Several mechanisms for antioxidant action have been proposed. The presence of antioxidants may result in the decreased formation of the reactive oxygen and nitrogen species in the first place. Antioxidants may also scavenge the reactive species or their precursors. Vitamin E is an example of this latter behavior in its inhibition of lipid oxidation by its reaction with radical intermediates generated from polyunsaturated fatty acids. Some antioxidants can bind the metal ions needed to catalyze the formation of the reactive oxidants. Other antioxidants can repair oxidative damage to biomolecules or can influence enzymes that catalyze repair mechanisms.

Vitamin E, or α-tocopherol, is thought to deter atherosclerosis, accelerate wound healing, and protect lung tissue from inhaled pollutants. It may also reduce the risk for heart disease and prevent premature skin aging. Researchers suspect that Vitamin E has several other beneficial effects ranging from alleviating rheumatoid arthritis to preventing cataracts. Most of us get enough Vitamin E through our diet and do not require supplements. Dark-green leafy vegetables, nuts, vegetable oils, seafood, eggs, and avocados are food sources rich in Vitamin E.

Molecular model of Vitamin E.

[4]B. Halliwell, *Nutr. Rev.*, **1997**, *55*(1), S44, **DOI:** 10.1111/j.1753-4887.1997.tb06100.x.

Selenium has antioxidant effects that complement those of Vitamin E. It is a required constituent of several enzymes that remove reactive oxidants. The metal may support the immune function and neutralize some heavy metal poisons. It may also aid in deterring heart disease and some cancers. Good sources of selenium in the diet are whole grains, asparagus, garlic, eggs, mushrooms, lean meats, and seafood. Usually diet alone provides sufficient selenium for good health. Supplements should be taken only if prescribed by a doctor because high doses can be toxic.

Molecular model of dichromate ion. For many years, dichromate in the form of its ammonium, potassium, or sodium salts was used in nearly all areas of chemistry as a powerful oxidizing agent. In addition to its use as a primary standard in analytical chemistry, it has been used as an oxidizing agent in synthetic organic chemistry; as a pigment in the paint, dye, and photographic industries; as a bleaching agent; and as a corrosion inhibitor. Chromic acid solution made from sodium dichromate and sulfuric acid was once the reagent of choice for thorough cleaning of glassware. Dichromate has been used as the analytical reagent in the alcohol Breathalyzer®, but in recent years, these devices have largely been replaced by analyzers based on the absorption of infrared radiation. Early color photography utilized the colors produced by chromium compounds in the so-called gum bichromate process, but this process has been replaced by silver bromide-based processes. The use of chromium compounds in general and dichromate in particular has decreased over the last decade because of the discovery that chromium compounds are carcinogenic. In spite of this danger, many millions of pounds of chromium compounds are manufactured and consumed by industry each year. Before using dichromate in laboratory work, read the MSDS for potassium dichromate (see the Web Works for this chapter) and explore its chemical, toxicological, and carcinogenic properties. Observe all precautions in handling this useful but potentially hazardous chemical either in the solid form or in solution.

20C-2 Potassium Dichromate

In its analytical applications, dichromate ion is reduced to green chromium(III) ion:

$$Cr_2O_7^{2-} + 14H^+ + 6e^- \rightleftharpoons 2Cr^{3+} + 7H_2O \qquad E^0 = 1.33 \text{ V}$$

Dichromate titrations are generally carried out in solutions that are about 1 M in hydrochloric or sulfuric acid. In these media, the formal potential for the half-reaction is 1.0 to 1.1 V.

Potassium dichromate solutions are indefinitely stable, can be boiled without decomposition, and do not react with hydrochloric acid. Moreover, primary-standard reagent is available commercially and at a modest cost. The disadvantages of potassium dichromate compared with cerium(IV) and permanganate ion are its lower electrode potential and the slowness of its reaction with certain reducing agents.

Preparing Dichromate Solutions

For most purposes, reagent-grade potassium dichromate is sufficiently pure to permit the direct preparation of standard solutions, the solid simply being dried at 150°C to 200°C before being weighed.

The orange color of a dichromate solution is not intense enough for use in end-point detection. However, diphenylamine sulfonic acid (see Table 19-3) is an excellent indicator for titrations with this reagent. The oxidized form of the indicator is violet, and its reduced form is essentially colorless; thus, the color change observed in a direct titration is from the green of chromium(III) to violet.

Applying Potassium Dichromate Solutions

The principal use of dichromate is for the volumetric titration of iron(II) based on the reaction

$$Cr_2O_7^{2-} + 6Fe^{2+} + 14H^+ \rightarrow 2Cr^{3+} + 6Fe^{3+} + 7H_2O$$

Often, this titration is performed in the presence of moderate concentrations of hydrochloric acid.

The reaction of dichromate with iron(II) has been widely used for the indirect determination of a variety of oxidizing agents. In these applications, a measured excess of an iron(II) solution is added to an acidic solution of the analyte. The excess

◄ Standard solutions of $K_2Cr_2O_7$ have the great advantage that they are indefinitely stable and do not oxidize HCl. Furthermore, primary-standard grade is inexpensive and commercially available.

iron(II) is then back-titrated with standard potassium dichromate (see Section 20B-1). Standardization of the iron(II) solution by titration with the dichromate is performed concurrently with the determination because solutions of iron(II) tend to be air oxidized. This method has been applied to the determination of nitrate, chlorate, permanganate, and dichromate ions as well as organic peroxides and several other oxidizing agents.

EXAMPLE 20-6

A 5.00-mL sample of brandy was diluted to 1.000 L in a volumetric flask. The ethanol (C_2H_5OH) in a 25.00-mL aliquot of the diluted solution was distilled into 50.00 mL of 0.02000 M $K_2Cr_2O_7$ and oxidized to acetic acid with heating:

$$3C_2H_5OH + 2Cr_2O_7^{2-} + 16H^+ \rightarrow 4Cr^{3+} + 3CH_3COOH + 11H_2O$$

After cooling, 20.00 mL of 0.1253 M Fe^{2+} was pipetted into the flask. The excess Fe^{2+} was then titrated with 7.46 mL of the standard $K_2Cr_2O_7$ to a diphenylamine sulfonic acid end point. Calculate the percent (w/v) C_2H_5OH (46.07 g/mol) in the brandy.

Solution

total amount $K_2Cr_2O_7$

$$= (50.00 + 7.46) \text{ mL } K_2Cr_2O_7 \times 0.02000 \frac{\text{mmol } K_2Cr_2O_7}{\text{mL } K_2Cr_2O_7}$$

$$= 1.1492 \text{ mmol } K_2Cr_2O_7$$

amount $K_2Cr_2O_7$ consumed by Fe^{2+}

$$= 20.00 \text{ mL } Fe^{2+} \times 0.1253 \frac{\text{mmol } Fe^{2+}}{\text{mL } Fe^{2+}} \times \frac{1 \text{ mmol } K_2Cr_2O_7}{6 \text{ mmol } Fe^{2+}}$$

$$= 0.41767 \text{ mmol } K_2Cr_2O_7$$

amount $K_2Cr_2O_7$ consumed by $C_2H_5OH = (1.1492 - 0.41767) \text{ mmol } K_2Cr_2O_7$

$$= 0.73153 \text{ mmol } K_2Cr_2O_7$$

mass C_2H_5OH

$$= 0.73153 \text{ mmol } K_2Cr_2O_7 \times \frac{3 \text{ mmol } C_2H_5OH}{2 \text{ mmol } K_2Cr_2O_7} \times 0.04607 \frac{\text{g } C_2H_5OH}{\text{mmol } C_2H_5OH}$$

$$= 0.050552 \text{ g } C_2H_5OH$$

$$\text{percent } C_2H_5OH = \frac{0.050552 \text{ g } C_2H_5OH}{5.00 \text{ mL sample} \times 25.00 \text{ mL}/1000 \text{ mL}} \times 100\%$$

$$= 40.4\% \text{ } C_2H_5OH$$

20C-3 Iodine

Iodine is a weak oxidizing agent used primarily for the determination of strong reductants. The most accurate description of the half-reaction for iodine in these applications is

$$I_3^- + 2e^- \rightleftharpoons 3I^- \qquad\qquad E^0 = 0.536 \text{ V}$$

where I_3^- is the triiodide ion.

Standard iodine solutions have relatively limited application compared with the other oxidants we have described because of their significantly smaller electrode potential. Occasionally, however, this low potential is advantageous because it imparts a degree of selectivity that makes possible the determination of strong reducing agents in the presence of weak ones. An important advantage of iodine is the availability of a sensitive and reversible indicator for the titrations. On the other hand, iodine solutions lack stability and must be restandardized regularly.

Properties of Iodine Solutions

Iodine is not very soluble in water (0.001 M). To prepare solutions having analytically useful concentrations of the element, iodine is usually dissolved in moderately concentrated solutions of potassium iodide. In this medium, iodine is reasonably soluble as a consequence of the reaction

$$I_2(s) + I^- \rightleftharpoons I_3^- \qquad K = 7.1 \times 10^2$$

Iodine dissolves only very slowly in solutions of potassium iodide, particularly if the iodide concentration is low. To ensure complete solution, the iodine is always dissolved in a small volume of concentrated potassium iodide, care being taken to avoid dilution of the concentrated solution until the last trace of solid iodine has disappeared. Otherwise, the concentration of the diluted solution gradually increases with time. This problem can be avoided by filtering the solution through a sintered glass crucible before standardization.

Iodine solutions lack stability for several reasons, one being the volatility of the solute. Losses of iodine from an open vessel occur in a relatively short time even in the presence of an excess of iodide ion. In addition, iodine slowly attacks most organic materials. Therefore, cork or rubber stoppers are never used to close containers of the reagent, and precautions must be taken to protect standard solutions from contact with organic dusts and fumes.

Air oxidation of iodide ion also causes changes in the concentration of an iodine solution:

$$4I^- + O_2(g) + 4H^+ \rightarrow 2I_2 + 2H_2O$$

In contrast to the other effects, this reaction causes the concentration of the iodine to increase. Air oxidation is promoted by acids, heat, and light.

Standardizing and Using Iodine Solutions

Iodine solutions can be standardized against anhydrous sodium thiosulfate or barium thiosulfate monohydrate, both of which are available commercially. The reaction between iodine and sodium thiosulfate is discussed in detail in Section 20B-2. Often, solutions of iodine are standardized against solutions of sodium thiosulfate that have in turn been standardized against potassium iodate or potassium dichromate (see Section 20B-2). **Table 20-6** summarizes methods that use iodine as an oxidizing agent.

> Solutions prepared by dissolving iodine in a concentrated solution of potassium iodide are properly called *triiodide solutions*. In practice, however, they are often termed *iodine solutions* because this terminology accounts for the stoichiometric behavior of these solutions ($I_2 + 2e^- \rightarrow 2I^-$)

TABLE 20-6

Some Applications of Iodine Solutions

Substance Determined	Half-Reaction
As	$H_3AsO_3 + H_2O \rightleftharpoons H_3AsO_4 + 2H^+ + 2e^-$
Sb	$H_3SbO_3 + H_2O \rightleftharpoons H_3SbO_4 + 2H^+ + 2e^-$
Sn	$Sn^{2+} \rightleftharpoons Sn^{4+} + 2e^-$
H_2S	$H_2S \rightleftharpoons S(s) + 2H^+ + 2e^-$
SO_2	$SO_3^{2-} + H_2O \rightleftharpoons SO_4^{2-} + 2H^+ + 2e^-$
$S_2O_3^{2-}$	$2S_2O_3^{2-} \rightleftharpoons S_4O_6^{2-} + 2e^-$
N_2H_4	$N_2H_4 \rightleftharpoons N_2(g) + 4H^+ + 4e^-$
Ascorbic acid	$C_6H_8O_6 \rightleftharpoons C_6H_6O_6 + 2H^+ + 2e^-$

20C-4 Potassium Bromate as a Source of Bromine

Primary-standard potassium bromate is available from commercial sources and can be used directly to prepare standard solutions that are stable indefinitely. Direct titrations with potassium bromate are relatively few. Instead, the reagent is a convenient and widely used stable source of bromine.[5] In this application, an unmeasured excess of potassium bromide is added to an acidic solution of the analyte. When a measured volume of standard potassium bromate is introduced, a stoichiometric quantity of bromine is produced.

1 mol $KBrO_3 = 3$ mol Br_2

$$\underset{\substack{\text{standard} \\ \text{solution}}}{BrO_3^-} + \underset{\text{excess}}{5Br^-} + 6H^+ \rightarrow 3Br_2 + 3H_2O$$

This indirect generation circumvents the problems associated with the use of standard bromine solutions, which lack stability.

The primary use of standard potassium bromate is for the determination of organic compounds that react with bromine. Few of these reactions are rapid enough to make direct titration feasible. Instead, a measured excess of standard bromate is added to the solution that contains the sample plus an excess of potassium bromide. After acidification, the mixture is allowed to stand in a glass-stoppered vessel until the bromine/analyte reaction is judged complete. To determine the excess bromine, an excess of potassium iodide is introduced so that the following reaction occurs:

$$2I^- + Br_2 \rightarrow I_2 + 2Br^-$$

The liberated iodine is then titrated with standard sodium thiosulfate (Equation 20-1).

Substitution Reactions

Bromine is incorporated into an organic molecule either by substitution or by addition. In halogen substitution, a hydrogen in an aromatic ring is replaced by a halogen. Substitution methods have been successfully applied to the determination of aromatic compounds that contain strong ortho-para-directing groups, particularly amines and phenols.

[5]For a discussion of bromate solutions and their applications, see M. R. F. Ashworth, *Titrimetric Organic Analysis*, Part I, pp. 118–30, New York: Interscience, 1964.

EXAMPLE 20-7

A 0.2981-g sample of an antibiotic powder was dissolved in HCl and the solution diluted to 100.0 mL. A 20.00-mL aliquot was transferred to a flask and followed by 25.00 mL of 0.01767 M $KBrO_3$. An excess of KBr was added to form Br_2, and the flask was stoppered. After 10 min, during which time the Br_2 brominated the sulfanilamide, an excess of KI was added. The liberated iodine titrated with 12.92 mL of 0.1215 M sodium thiosulfate. The reactions are

$$BrO_3^- + 5Br^- + 6H^+ \rightarrow 3Br_2 + 3H_2O$$

Molecular model of sulfanilamide. In the 1930s, sulfanilamide was found to be an effective antibacterial agent. In an effort to provide a solution of the drug that could be conveniently administered to patients, drug companies distributed sulfanilamide elixir containing a high concentration of ethylene glycol, which is toxic to the kidneys. Unfortunately, over one hundred people died from the effects of the solvent. This event led to the rapid passage of the 1938 Federal Food, Drug, and Cosmetic Act, which required toxicity testing prior to marketing and listing of active ingredients on product labels. For more information on the history of drug laws, see the U.S. Food and Drug Administration website.

sulfanilamide

$$Br_2 + 2I^- \rightarrow 2Br^- + I_2 \quad \text{(excess KI)}$$

$$I_2 + 2S_2O_3^{2-} \rightarrow 2S_4O_6^{2-} + 2I^-$$

Calculate the percent sulfanilamide ($NH_2C_6H_4SO_2NH_2$, 172.21 g/mol) in the powder.

Solution

$$\text{total amount Br}_2 = 25.00 \text{ mL KBrO}_3^- \times 0.01767 \frac{\text{mmol KBrO}_3^-}{\text{mL KBrO}_3^-} \times \frac{3 \text{ mmol Br}_2}{\text{mmol KBrO}_3^-}$$

$$= 1.32525 \text{ mmol Br}_2$$

We next calculate how much Br_2 was in excess over that required to brominate the analyte:

$$\text{amount excess Br}_2 = \text{amount I}_2$$

$$= 12.92 \text{ mL Na}_2\text{S}_2\text{O}_3^- \times 0.1215 \frac{\text{mmol Na}_2\text{S}_2\text{O}_3}{\text{mL Na}_2\text{S}_2\text{O}_3} \times \frac{1 \text{ mmol I}_2}{2 \text{ mmol Na}_2\text{S}_2\text{O}_3}$$

$$= 0.78489 \text{ mmol Br}_2$$

The amount of Br_2 consumed by the sample is given by

$$\text{amount Br}_2 = 1.32525 - 0.78489 = 0.54036 \text{ mmol Br}_2$$

$$\text{mass analyte} = 0.54036 \text{ mmol Br}_2 \times \frac{1 \text{ mmol analyte}}{2 \text{ mmol Br}_2} \times 0.17221 \frac{\text{g analyte}}{\text{mmol analyte}}$$

$$= 0.046528 \text{ g analyte}$$

(continued)

$$\text{percent analyte} = \frac{0.046528 \text{ g analyte}}{0.2891 \text{ g sample} \times 20.00 \text{ mL}/100 \text{ mL}} \times 100\%$$

$$= 80.47\% \text{ sulfanilamide}$$

Molecular model of 8-hydroxyquinoline.

An important example of the use of a bromine substitution reaction is the determination of 8-hydroxyquinoline:

$$\text{(structure)} + 2Br_2 \longrightarrow \text{(structure)} + 2HBr$$

In contrast to most bromine substitutions, this reaction takes place rapidly enough in hydrochloric acid solution to make direct titration feasible. The titration of 8-hydroxyquinoline with bromine is particularly significant because the former is an excellent precipitating reagent for cations (see Section 12C-3). For example, aluminum can be determined according to the sequence

$$Al^{3+} + 3HOC_9H_6N \xrightarrow{\text{pH 4-9}} Al(OC_9H_6N)_3(s) + 3H^+$$

$$Al(OC_9H_6N)_3(s) \xrightarrow{\text{hot 4 M HCl}} 3HOC_9H_6N + Al^{3+}$$

$$3HOC_9H_6N + 6Br_2 \longrightarrow 3HOC_9H_4NBr_2 + 6HBr$$

The stoichiometric relationships in this case are

$$1 \text{ mol } Al^{3+} = 3 \text{ mol } HOC_9H_6N = 6 \text{ mol } Br_2 = 2 \text{ mol } KBrO_3$$

Addition Reactions

In addition reactions, olefinic double bonds are opened. For example, 1 mole of ethylene reacts with 1 mole of bromine in the reaction

$$\text{H—C=C—H} + Br_2 \longrightarrow \text{H—C—C—H}$$

The literature contains numerous references to the use of bromine for the estimation of olefinic unsaturation in fats, oils, and petroleum products. A method for the determination of ascorbic acid in vitamin C tablets is given in Section 38I-3.

20C-5 Determining Water with the Karl Fischer Reagent

In industry and commerce, one of the most widely used analytical methods is the Karl Fischer titration procedure for the determination of water in various types of solids and organic liquids. This important titrimetric method is based on an oxidation/reduction reaction that is relatively specific for water.[6]

Describing the Reaction Stoichiometry

The Karl Fischer reaction is based on the oxidation of sulfur dioxide by iodine. In a solvent that is neither acidic nor basic—an aprotic solvent—the reaction can be summarized by

$$I_2 + SO_2 + 2H_2O \rightarrow 2HI + H_2SO_4$$

In this reaction, two moles of water are consumed for each mole of iodine. The stoichiometry, however, can vary from 2:1 to 1:1 depending on the presence of acids and bases in the solution.

Classical Chemistry. In order to stabilize the stoichiometry and shift the equilibrium further to the right, Fischer added pyridine (C_5H_5N) and used anhydrous methanol as the solvent. A large excess of pyridine was used to complex the I_2 and SO_2. The classic reaction has been shown to occur in two steps. In the first step, I_2 and SO_2 react in the presence of pyridine and water to form pyridinium sulfite and pyridinium iodide.

$$C_5H_5N \cdot I_2 + C_5H_5N \cdot SO_2 + C_5H_5N + H_2O \rightarrow$$
$$2C_5H_5N \cdot HI + C_5H_5N \cdot SO_3 \quad \text{(20-2)}$$

$$C_5H_5N^+ \cdot SO_3^- + CH_3OH \rightarrow C_5H_5N(H)SO_4CH_3 \quad \text{(20-3)}$$

where I_2, SO_2, and SO_3 are shown as complexed by the pyridine. This second step is important because the pyridinium sulfite can also consume water:

$$C_5H_5N^+ \cdot SO_3^- + H_2O \rightarrow C_5H_5NH^+SO_4H^- \quad \text{(20-4)}$$

This last reaction is undesirable because it is not as specific for water. It can be prevented completely by having a large excess of methanol present. Note that the stoichiometry is 1 mole of I_2 per mole of H_2O present.

For volumetric analysis, the classical Karl Fischer reagent consists of I_2, SO_2, pyridine, and anhydrous methanol or another suitable solvent. The reagent decomposes on standing and must be standardized often. Stabilized Karl Fischer reagents are available commercially from several suppliers. For ketones and aldehydes, specially formulated reagents are available from commercial sources. For coulometric methods (see Chapter 22), the Karl Fischer reagent contains KI instead of I_2 since, as we will see, the I_2 is generated electrochemically.

---

[6]For a review of the composition and uses of the Karl Fischer reagent see S. K. MacLeod, *Anal. Chem.*, **1991**, *63*, 557A, **DOI**: 10.1021/ac00010a720; J. D. Mitchell, Jr. and D.M. Smith, *Aquametry*, 2nd ed., Vol. 3. New York: Wiley, 1977.

Pyridine-Free Chemistry. In recent years, pyridine, and its objectionable odor, have been replaced in the Karl Fischer reagent by other amines, particularly imidazole, shown in the margin. These pyridine-free reagents are available commercially for both volumetric and coulometric Karl Fischer procedures. More detailed studies of the reaction have been reported.[7] The reaction is now thought to occur as follows:

pyridine   imidazole

1. Solvolysis $2ROH + SO_2 \rightleftharpoons RSO_3^- + ROH_2^+$
2. Buffering $B + RSO_3^- + ROH_2^+ \rightleftharpoons BH^+SO_3R^- + ROH$
3. Redox $B \cdot I_2 + BH^+SO_3R^- + B + H_2O \rightleftharpoons BH^+SO_4R^- + 2BH^+I^-$

Note that the stoichiometry is again one mole of I_2 consumed for each mole of H_2O present in the sample.

Interfering Reactions. There are several reactions that can occur that cause interferences in the Karl Fischer titration. These undesired reactions can cause results to be too high, too low, or just imprecise. Oxidation of iodide in the coulometric reagent by oxidizing agents such as Cu(II), Fe(III), nitrite, Br_2, Cl_2, or quinones produces I_2, which can react with H_2O and cause low results since not as much generated I_2 is needed. The carbonyl groups on aldehydes and ketones can react with SO_2 and H_2O to form bisulfite complexes. Since this reaction consumes water, the titration results are again too low. Substitution of a weaker base like pyridine for imidazole can lessen the problem.

The iodine generated coulometrically or present in the reagent can be reduced by oxidizable species such as ascorbic acid, ammonia, thiols, Tl^+, Sn^{2+}, In^+, hydroxyl amines, and thiosulfite. This reduction results in consumption of I_2 and water determinations that are too high. Phenolic derivatives and bicarbonates also cause reduction of I_2.

Some interfering compounds react to produce water, which causes the water results to be too high. Carboxylic acids can react with alcohols to produce an ester and water. To minimize this problem, the alcohol can be eliminated in the reagent, or an alcohol that reacts at a slower rate than methanol can be used. The pH of the reagent can be increased because the formation of esters is usually acid catalyzed. Ketones and aldehydes can react with alcoholic solvents to form ketals and acetals with the production of water according to

$$R_2C{=}O + 2CH_3OH \rightarrow R_2C(OCH_3)_2 + H_2O$$

Aromatic ketones are less reactive than aliphatic ketones. Aldehydes are much more reactive than ketones. Some commercial reagent preparations have been formulated to minimize this problem by using alcohols that react slowly and a higher pH.

Silanols and cyclic siloxanes also can react with alcohols to produce ethers and water. Some metal oxides, hydroxides, and carbonates can react with HI to produce water. All of these reactions increase the amount of I_2 consumed and produce results that are too high.

[7]E. Scholz, *Karl Fischer Titration*, (Berlin: Springer-Verlag, 1984).

Detecting the End Point

An end point in a Karl Fischer titration can be observed visually based on the brown color of the excess reagent. More commonly, however, end points are obtained by electroanalytical measurements. Several instrument manufacturers offer automatic or semiautomatic instruments for performing Karl Fischer titrations. All of these instruments are based on electrometric end point detection.

Reagent Properties

Karl Fischer reagent decomposes on standing. Because decomposition is particularly rapid immediately after preparation, it is common practice to prepare the reagent a day or two before it is to be used. Its strength must be established at least daily against a standard solution of water in methanol. A proprietary commercial Karl Fischer reagent reported to require only occasional restandardization is now available.

It is obvious that great care must be exercised to keep atmospheric moisture from contaminating the Karl Fischer reagent and the sample. All glassware must be carefully dried before use, and the standard solution must be stored out of contact with air. It is also necessary to minimize contact between the atmosphere and the solution during the titration.

Applications

Karl Fischer reagent has been applied to the determination of water in numerous types of samples. There are several variations of the basic technique depending on the solubility of the material, the state in which the water is retained, and the physical state of the sample. If the sample can be dissolved completely in methanol, a direct and rapid titration is usually feasible. This method has been applied to the determination of water in many organic acids, alcohols, esters, ethers, anhydrides, and halides. The hydrated salts of most organic acids, as well as the hydrates of a number of inorganic salts that are soluble in methanol, can also be determined by direct titration.

Direct titration of samples that are only partially dissolved in the reagent usually leads to incomplete recovery of the water. Satisfactory results with this type of sample are often obtained, however, by the addition of excess reagent and back-titration with a standard solution of water in methanol after a suitable reaction time. An effective alternative is to extract the water from the sample by refluxing with anhydrous methanol or other organic solvents. The resulting solution is then titrated directly with the Karl Fischer solution.

WEB WORKS

Use a search engine to locate one of the many MSDS websites. Find and read the MSDS for potassium dichromate and explore its chemical, toxicological, and carcinogenic properties. Locate a second MSDS website and again explore the properties of potassium dichromate. What differences did you find in the two documents? Which site provided the most detailed information, particularly regarding health effects? Did either site tend to emphasize certain properties over others? What do you conclude from this exercise?

QUESTIONS AND PROBLEMS

20-1. Write balanced net ionic equations to describe
   *(a) the oxidation of Mn^{2+} to MnO_4^- by ammonium peroxydisulfate.
   (b) the oxidation of Ce^{3+} to Ce^{4+} by sodium bismuthate.
   *(c) the oxidation of U^{4+} to UO_2^{2+} by H_2O_2.
   (d) the reaction of $V(OH)_4^+$ in a Walden reductor.
   *(e) the titration of H_2O_2 with $KMnO_4$.
   (f) the reaction between KI and ClO_3^- in acidic solution.

*20-2.** Why is a Walden reductor always used with solutions that contain appreciable concentrations of HCl?

20-3. Write a balanced net ionic equation for the reduction of UO_2^{2+} in a Walden reductor.

*20-4.** Why are standard solutions of reductants less often used for titrations than standard solutions of oxidants?

20-5. Why are Ce^{4+} solutions never used for the titration of reductants in basic solutions?

*20-6.** Why are $KMnO_4$ solutions filtered before they are standardized?

20-7. Why are solutions of $KMnO_4$ and $Na_2S_2O_3$ generally stored in dark reagent bottles?

*20-8.** What is the primary use of standard $K_2Cr_2O_7$ solutions?

20-9. A standard solution of I_2 increased in concentration with standing. Write a balanced net ionic equation that accounts for the increase.

*20-10.** Suggest a way in which a solution of KIO_3 could be used as a source of known quantities of I_2.

20-11. Write balanced equations showing how $K_2Cr_2O_7$ could be used as a primary standard for solutions of $Na_2S_2O_3$.

*20-12.** In the titration of I_2 solutions with $Na_2S_2O_3$, the starch indicator is never added until just before chemical equivalence. Why?

20-13. A solution prepared by dissolving a 0.2541-g sample of electrolytic iron wire in acid was passed through a Jones reductor. The iron(II) in the resulting solution required a 36.76-mL titration. Calculate the molar oxidant concentration if the titrant used was
   *(a) Ce^{4+} (product: Ce^{3+}).
   (b) $Cr_2O_7^{2-}$ (product: Cr^{3+}).
   *(c) MnO_4^- (product: Mn^{2+}).
   (d) $V(OH)_4^+$ (product: VO^{2+}).
   *(e) IO_3^- (product: ICl_2^-).

*20-14.** How would you prepare 1.000 L of 0.05000 M $KBrO_3$?

20-15. How would you prepare 2.5 L of approximately 0.06 M I_3^- solution? Calculate the molar concentration of $KMnO_4$ in this solution.

*20-16.** A 0.2219-g sample of pure iron wire was dissolved in acid, reduced to the +2 state, and titrated with 34.65 mL of cerium(IV). Calculate the molar concentration of the Ce^{4+} solution.

20-17. A 0.1298-g sample of $KBrO_3$ was dissolved in dilute HCl and treated with an unmeasured excess of KI. The liberated iodine required 41.32 mL of a sodium thiosulfate solution. Calculate the molar concentration of the $Na_2S_2O_3$.

*20-18.** Calculate the percentage of MnO_2 in a mineral specimen if the I_2 liberated by a 0.1267-g sample in the net reaction

$$MnO_2(s) + 4H^+ + 2I^- \rightarrow Mn^{2+} + I_2 + 2H_2O$$

required 29.62 mL of 0.08041 M $Na_2S_2O_3$.

20-19. A 0.7120-g specimen of iron ore was dissolved and passed through a Jones reductor. Titration of the Fe(II) produced required 41.63 mL of 0.01926 M $KMnO_4$. Express the results of this analysis in terms of (a) percent Fe and (b) percent Fe_2O_3.

*20-20.** Treatment of hydroxylamine (H_2NOH) with an excess of Fe(III) results in the formation of N_2O and an equivalent amount of Fe(II):

$$2H_2NOH + 4Fe^{3+} \rightarrow N_2O(g) + 4Fe^{2+} + 4H^+ + H_2O$$

Calculate the molar concentration of an H_2NOH solution if the Fe(II) produced by treatment of a 25.00-mL aliquot required 14.48 mL of 0.01528 M $K_2Cr_2O_7$.

20-21. The $KClO_3$ in a 0.1862-g sample of an explosive was determined by reaction with 50.00 mL of 0.01162 M Fe^{2+}

$$ClO_3^- + 6Fe^{2+} + 6H^+ \rightarrow Cl^- + 3H_2O + 6Fe^{3+}$$

When the reaction was complete, the excess Fe^{2+} was back-titrated with 13.26 mL of 0.07654 M Ce^{4+}. Calculate the percentage of $KClO_3$ in the sample.

*20-22.** An 8.13-g sample of an ant-control preparation was decomposed by wet-ashing with H_2SO_4 and HNO_3. The As in the residue was reduced to the trivalent state with hydrazine. After removal of the excess reducing agent, the As(III) required a 31.46-mL titration with 0.03142 M I_2 in a faintly alkaline medium. Express the results of this analysis in terms of percentage of As_2O_3 in the original sample.

20-23. The ethyl mercaptan concentration in a mixture was determined by shaking a 2.043-g sample with 50.00 mL of 0.01204 M I_2 in a tightly stoppered flask:

$$2C_2H_5SH + I_2 \rightarrow C_2H_5SSC_2H_5 + 2I^- + 2H^+$$

The excess I_2 was back-titrated with 18.23 mL of 0.01437 M $Na_2S_2O_3$. Calculate the percentage of C_2H_5SH (62.13 g/mol).

*__20-24.__ A sensitive method for I^- in the presence of Cl^- and Br^- entails oxidation of the I^- to IO_3^- with Br_2. The excess Br_2 is then removed by boiling or by reduction with formate ion. The IO_3^- produced is determined by addition of excess I^- and titration of the resulting I_2. A 1.307-g sample of mixed halides was dissolved and analyzed by the foregoing procedure. A volume of 19.72 mL of 0.04926 M thiosulfate was required for the titration. Calculate the percentage of KI in the sample.

20-25. A 2.667-g sample containing both Fe and V was dissolved under conditions that converted the elements to Fe(III) and V(V). The solution was diluted to 500.0 mL, and a 50.00-mL aliquot was passed through a Walden reductor and titrated with 18.31 mL of 0.1000 M Ce^{4+}. A second 50.00-mL aliquot was passed through a Jones reductor and required 42.41 mL of the same Ce^{4+} solution to reach an end point. Calculate the percentage of Fe_2O_3 and V_2O_5 in the sample.

*__20-26.__ A gas mixture was passed at the rate of 2.50 L/min through a solution of sodium hydroxide for a total of 59.00 min. The SO_2 in the mixture was retained as sulfite ion:

$$SO_2(g) + 2OH^- \rightarrow SO_3^{2-} + H_2O$$

After acidification with HCl, the sulfite was titrated with 5.15 mL of 0.002997 M KIO_3:

$$IO_3^- + 2H_2SO_3 + 2Cl^- \rightarrow ICl_2^- + 2SO_4^{2-} + 2H^+ + H_2O$$

Use 1.20 g/L for the density of the mixture and calculate the concentration of SO_2 in ppm.

20-27. A 25.00-L air sample was passed through an absorption tower containing a solution of Cd^{2+}, where H_2S was retained as CdS. The mixture was acidified and treated with 25.00 mL of 0.00432 M I_2. After the reaction

$$S^{2-} + I_2 \rightarrow S(s) + 2I^-$$

was complete, the excess iodine was titrated with 15.62 mL of 0.01143 M thiosulfate. Calculate the concentration of H_2S in ppm; use 1.20 g/L for the density of the gas stream.

*__20-28.__ The Winkler method for dissolved oxygen in water is based on the rapid oxidation of solid $Mn(OH)_2$ to $Mn(OH)_3$ in alkaline medium. When acidified, the Mn(III) readily releases iodine from iodide. A 250-mL water sample, in a stoppered vessel, was treated with 1.00 mL of a concentrated solution of NaI and NaOH and 1.00 mL of a manganese(II) solution. Oxidation of the $Mn(OH)_2$ was complete in about 1 min. The precipitates were then dissolved by addition of 2.00 mL of concentrated H_2SO_4, whereupon an amount of iodine equivalent to the $Mn(OH)_3$ (and hence to the dissolved O_2) was liberated. A 25.0-mL aliquot (of the 254 mL) was titrated with 14.6 mL of 0.00897 M thiosulfate. Calculate the mass in milligrams of O_2 per milliliter sample. Assume that the concentrated reagents are O_2 free and take their dilutions of the sample into account.

20-29. Use a spreadsheet to do the calculations and plot the titration curves for the following titrations. Calculate potentials after the addition of titrant corresponding to 10%, 20%, 30%, 40%, 50%, 60%, 70%, 80%, 90%, 95%, 99%, 99.9%, 100%, 101%, 105%, 110%, and 120% of the equivalence-point volume.

(a) 20.00 mL of 0.0500 M $SnCl_2$ with 0.100 M $FeCl_3$.

(b) 25.00 mL of 0.08467 M $Na_2S_2O_3$ with 0.10235 M I_2.

(c) 0.1250 g of primary-standard-grade $Na_2C_2O_4$ with 0.01035 M $KMnO_4$. Assume $[H^+] = 1.00$ M and $p_{CO_2} = 1$ atm.

(d) 20.00 mL of 0.1034 M Fe^{2+} with 0.01500 M $K_2Cr_2O_7$. Assume $[H^+] = 1.00$ M.

20-30. **Challenge Problem:** Verdini and Lagier[8] developed an iodimetric titration procedure for determining ascorbic acid in vegetables and fruits. They compared the results of their titration experiments with similar results from an HPLC method (see Chapter 33). The results of their comparison are shown in the following table.

Method Comparison*		
Sample	HPLC, mg/100 g	Voltammetry, mg/100 g
1	138.6	140.0
2	126.6	120.6
3	138.3	140.9
4	126.2	123.7

*Ascorbic acid content determined in kiwi fruit samples by means of HPLC with UV detection and voltammetric titration.

(a) Find the mean and standard deviation of each set of data.

(b) Determine whether there is a difference in the variances of the two data sets at the 95% level.

[8]R. A. Verdini, and C. M. Lagier, *J. Agric. Food Chem.*, **2000**, *48*, 2812. **DOI:** 10.1021/jf990987s

(c) Determine whether the difference in the means is significant at the 95% level.

These workers also carried out a recovery study in which they determined ascorbic acid in samples, then spiked the samples with additional ascorbic acid and redetermined the mass of the analyte. Their results are shown in the following table.

Recovery Study

Sample	1	2	3	4
		Kiwi fruit		
Amounts				
Initial, mg	9.32	7.29	7.66	7.00
Added, mg	6.88	7.78	8.56	6.68
Found, mg	15.66	14.77	15.84	13.79
		Spinach		
Initial, mg	6.45	7.72	5.58	5.21
Added, mg	4.07	4.32	4.28	4.40
Found, mg	10.20	11.96	9.54	9.36

(d) Calculate the percent recovery for total ascorbic acid in each sample.

(e) Find the mean and standard deviation of the percent recovery, first for the kiwi fruit and then for the spinach.

(f) Determine whether the variances of the percent recovery between the kiwi fruit and the spinach are different at the 95% confidence level.

(g) Determine whether the difference in the percent recovery of ascorbic acid is significant at the 95% confidence level.

(h) Discuss how you would apply the iodimetric method for the determination of ascorbic acid to various samples of fruits and vegetables. In particular, comment on how you would apply the results of your analysis of the data to the analysis of new samples.

(i) References to several papers are listed[9–15] on determining ascorbic acid using different analytical techniques. If the papers are available in your library, examine them and briefly describe the methods used in each.

(j) Comment on how each of the methods in (i) might be used and under what circumstances they might be chosen rather than iodimetry. For each method, including iodimetry, compare such factors as speed, convenience, cost of analysis, and quality of the resulting data.

[9]A. Campiglio, *Analyst*, **1993**, *118*, 545, **DOI**: 10.1039/AN9931800545.

[10]L. Cassella, M. Gulloti, A. Marchesini, and M. Petrarulo, *J. Food Sci.*, **1989**, *54*, 374, **DOI**: 10.1111/j.1365-2621.1989.tb03084.x.

[11]Z. Gao, A. Ivaska, T. Zha, G. Wang, P. Li, and Z. Zhao, *Talanta*, **1993**, *40*, 399, **DOI**: 10.1016/0039-9140(93)80251-L.

[12]O. W. Lau, K. K. Shiu, and S. T. Chang, *J. Sci. Food Agric.*, **1985**, *36*, 733, **DOI**: 10.1002/jsfa.2740360814.

[13]A. Marchesini, F. Montuori, D. Muffato, and D. Maestri, *J. Food Sci.*, **1974**, *39*, 568, **DOI**: 10.1111/j.1365-2621.1974.tb02950.x.

[14]T. Moeslinger, M. Brunner, I. Volf, and P. G. Spieckermann, *Clin. Chem.*, **1995**, *41*, 1177.

[15]L. A. Pachla and P. T. Kissinger, *Anal. Chem.*, **1976**, *48*, 364, **DOI**: 10.1021/ac60366a045.

Potentiometry

CHAPTER 21

The research vessel *Meteor*, shown in the photo, is owned by the Federal Republic of Germany through the Ministry of Research and Technology and is operated by the German Research Foundation. It is used by a multinational group of chemical oceanographers to collect data in an effort to better understand the changing chemical composition of the earth's atmosphere and oceans. For example, during April, 2012, a group from the Uni Bjerknes Centre and the Bjerknes Centre for Climate Research in Bergen, Norway, were aboard *Meteor* in the North Atlantic Ocean west of Norway performing measurements related to the oceanic cycling of carbon as well as measurements estimating the flux of oxygen directly involved in biological activity. An important observation in these experiments is the total alkalinity of sea water, which is determined by potentiometric titration, a method that is discussed in this chapter.

© DANIEL†BOCKWOLDT/epa/Corbis

Potentiometric methods of analysis are based on measuring the potential of electrochemical cells without drawing appreciable current. For nearly a century, potentiometric techniques have been used for locating end points in titrations. In more recent methods, ion concentrations are measured directly from the potential of ion-selective membrane electrodes. These electrodes are relatively free from interferences and provide a rapid, convenient, and nondestructive means for quantitatively determining numerous important anions and cations.[1]

Analysts make more potentiometric measurements than perhaps any other type of chemical instrumental measurement. The number of potentiometric measurements made on a daily basis is staggering. Manufacturers measure the pH of many consumer products, clinical laboratories determine blood gases as important indicators of disease states, industrial and municipal effluents are monitored continuously to determine pH and concentrations of pollutants, and oceanographers determine carbon dioxide and other related variables in seawater. Potentiometric measurements are also used in fundamental studies to determine thermodynamic equilibrium constants, such as K_a, K_b, and K_{sp}. These examples are but a few of the many thousands of applications of potentiometric measurements.

The equipment for potentiometric methods is simple and inexpensive and includes a reference electrode, an indicator electrode, and a potential-measuring device. The principles of operation and design of each of these components are described in the initial sections of this chapter. Following these discussions, we investigate analytical applications of potentiometric measurements.

---

[1]R. S. Hutchins and L. G. Bachas, in *Handbook of Instrumental Techniques for Analytical Chemistry*, F. A. Settle, ed., Ch. 38, pp. 727–48, Upper Saddle River, NJ: Prentice-Hall, 1997.

21A GENERAL PRINCIPLES

In Feature 18-3, we showed that absolute values for individual half-cell potentials cannot be determined in the laboratory, that is, only relative cell potentials can be measured experimentally. **Figure 21-1** shows a typical cell for potentiometric analysis. This cell can be represented as

$$\text{reference electrode} | \text{salt bridge} | \text{analyte solution} | \text{indicator electrode}$$

$$\underbrace{\qquad\qquad}_{E_{ref}} \qquad \underbrace{\qquad}_{E_j} \qquad\qquad \underbrace{\qquad\qquad}_{E_{ind}}$$

A **reference electrode** is a half-cell having a known electrode potential that remains constant at constant temperature and is independent of the composition of the analyte solution.

An **indicator electrode** has a potential that varies in a known way with variations in the concentration of an analyte.

> As shown in Figure 21-1, reference electrodes are *always* treated as the left-hand electrode. This practice, which we adopt throughout this text, is consistent with the International Union of Pure and Applied Chemistry (IUPAC) convention for electrode potentials, discussed in Section 18C-4, in which the reference is the standard hydrogen electrode and is the electrode on the left in a cell diagram.

> A hydrogen electrode is seldom used as a reference electrode for day-to-day potentiometric measurements because it is inconvenient to use and maintain and is also a fire hazard.

The **reference electrode** in this diagram is a half-cell with an accurately known electrode potential, E_{ref}, that is independent of the concentration of the analyte or any other ions in the solution under study. It can be a standard hydrogen electrode but seldom is because a standard hydrogen electrode is somewhat troublesome to maintain and use. By convention, the reference electrode is always treated as the left-hand electrode in potentiometric measurements. The **indicator electrode**, which is immersed in a solution of the analyte, develops a potential, E_{ind}, that depends on the activity of the analyte. Most indicator electrodes used in potentiometry are selective in their responses. The third component of a potentiometric cell is a salt bridge that prevents the components of the analyte solution from mixing with those of the reference electrode. As noted in Chapter 18, a potential develops across the liquid junctions at each end of the salt bridge. These two potentials tend to cancel one another if the mobilities of the cation and the anion in the bridge solution are approximately the same. Potassium chloride is a nearly ideal electrolyte for the salt bridge because the mobilities of the K^+ ion and the Cl^- ion are nearly equal. The net potential across the salt bridge, E_j, is thereby reduced to a few millivolts or less. For most electroanalytical methods, the junction potential is small enough to be neglected. In the potentiometric methods discussed in this chapter, however, the junction potential and its uncertainty can be factors that limit the measurement accuracy and precision.

Digital meter

84.2 mV

Reference electrode, E_{ref}

Metallic indicator electrode, E_{ind}

Salt bridge, E_j

Analyte solution

Porous membrane

$$E_{cell} = E_{ind} - E_{ref} + E_j$$

Figure 21-1   A cell for potentiometric determinations.

The potential of the cell we have just considered is given by the equation

$$E_{cell} = E_{ind} - E_{ref} + E_j \qquad (21\text{-}1)$$

The first term in this equation, E_{ind}, contains the information that we are looking for—the concentration of the analyte. To make a potentiometric determination of an analyte then, we must measure a cell potential, correct this potential for the reference and junction potentials, and compute the analyte concentration from the indicator electrode potential. Strictly, the potential of a galvanic cell is related to the activity of the analyte. Only through proper calibration of the electrode system with solutions of known concentration can we determine the concentration of the analyte.

In the sections that follow, we discuss the nature and origin of the three potentials shown on the right side of Equation 21-1.

21B REFERENCE ELECTRODES

The ideal reference electrode has a potential that is accurately known, constant, and completely insensitive to the composition of the analyte solution. In addition, this electrode should be rugged, easy to assemble, and should maintain a constant potential while passing minimal currents.

21B-1 Calomel Reference Electrodes

Calomel reference electrodes consist of mercury in contact with a solution that is saturated with mercury(I) chloride (calomel) and that also contains a known concentration of potassium chloride. Calomel half-cells can be represented as follows:

$$Hg \mid Hg_2Cl_2(\text{sat'd}), KCl(x\ M)\|$$

where x represents the molar concentration of potassium chloride in the solution. The electrode potential for this half-cell is determined by the reaction

$$Hg_2Cl_2(s) + 2e^- \rightleftharpoons 2Hg(l) + 2Cl^-(aq)$$

and depends on the chloride concentration. Thus, the KCl concentration must be specified in describing the electrode.

> The "saturated" in a saturated calomel electrode refers to the KCl concentration and not the calomel concentration. All calomel electrodes are saturated with Hg_2Cl_2 (calomel).

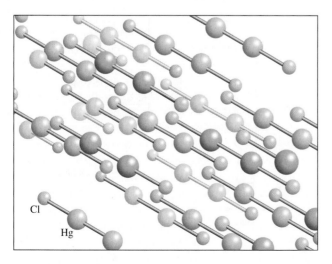

Cl

Hg

The crystal structure of calomel, Hg_2Cl_2, which has a limited solubility in water ($K_{sp} = 1.8 \times 10^{-18}$ at 25°C). Notice the Hg—Hg bond in the structure. There is considerable evidence that a similar type of bonding occurs in aqueous solution, and so mercury(I) is represented as Hg_2^{2+}.

TABLE 21-1

Formal Electrode Potentials for Reference Electrodes as a Function of Composition and Temperature

Temperature, °C	Potential versus SHE, V				
	0.1 M Calomel*	3.5 M Calomel†	Sat'd Calomel*	3.5 M Ag/AgCl†	Sat'd Ag/AgCl†
15	0.3362	0.254	0.2511	0.212	0.209
20	0.3359	0.252	0.2479	0.208	0.204
25	0.3356	0.250	0.2444	0.205	0.199
30	0.3351	0.248	0.2411	0.201	0.194
35	0.3344	0.246	0.2376	0.197	0.189

*From R. G. Bates, in *Treatise on Analytical Chemistry*, 2nd ed., I. M. Kolthoff and P. J. Elving, eds., Part I, Vol. 1, p. 793, New York: Wiley, 1978.
†From D. T. Sawyer, A. Sobkowiak, and J. L. Roberts, Jr., *Electrochemistry for Chemists*, New York: Wiley, 1995, p. 192.

A salt bridge is easily constructed by filling a U-tube with a conducting gel prepared by heating about 5 g of agar in 100 mL of an aqueous solution containing about 35 g of potassium chloride. When the liquid cools, it sets up into a gel that is a good conductor but prevents the two solutions at the ends of the tube from mixing. If either of the ions in potassium chloride interfere with the measurement process, ammonium nitrate may be used as the electrolyte in salt bridges.

Agar, which is available as translucent flakes, is a heteropolysaccharide that is extracted from certain East Indian seaweed. Solutions of agar in hot water set to a gel when they are cooled.

Table 21-1 lists the compositions and formal electrode potentials for the three most common calomel electrodes. Note that each solution is saturated with mercury(I) chloride (calomel) and that the cells differ only with respect to the potassium chloride concentration. Several convenient calomel electrodes, such as the electrode illustrated in **Figure 21-2**, are available commercially. The H-shape body of the electrode is made of glass of dimensions shown in the diagram. The right arm of the electrode contains a platinum electrical contact, a small quantity of mercury/mercury(I) chloride paste in saturated potassium chloride, and a few crystals of KCl. The tube is filled with saturated KCl to act as a salt bridge (see Section 18B-2) through a piece of porous Vycor ("thirsty glass") sealed in the end of the left arm. This type of junction has a relatively high resistance (2000 to 3000 Ω) and a limited current-carrying capacity, but contamination of the analyte solution due to leakage of potassium chloride is minimal. Other configurations of SCEs are available with much lower resistance and better electrical contact to the analyte solution, but they tend to leak small amounts of saturated potassium chloride into the sample. Because of concerns with mercury contamination, SCEs are less common than they once were, but for some applications, they are superior to Ag-AgCl reference electrodes, which are described next.

Figure 21-2 Diagram of a typical commercial saturated calomel electrode. (Reprinted with permission of Bioanalytical Systems, W. Lafayette, IN.)

(a)

21B-2 Silver/Silver Chloride Reference Electrodes

The most widely marketed reference electrode system consists of a silver electrode immersed in a solution of potassium chloride that has been saturated with silver chloride:

$$Ag|AgCl(sat'd), KCl(sat'd)\|$$

The electrode potential is determined by the half-reaction

$$AgCl(s) + e^- \rightleftharpoons Ag(s) + Cl^-$$

Normally, this electrode is prepared with either a saturated or a 3.5 M potassium chloride solution; potentials for these electrodes are given in Table 21-1. **Figure 21-3** shows a commercial model of this electrode, which is little more than a piece of glass tubing that has a narrow opening at the bottom connected to a Vycor plug for making contact with the analyte solution. The tube contains a silver wire coated with a layer of silver chloride that is immersed in a potassium chloride solution saturated with silver chloride.

Silver–silver chloride electrodes have the advantage that they can be used at temperatures greater than 60°C, while calomel electrodes cannot. On the other hand, mercury(II) ions react with fewer sample components than do silver ions (which can react with proteins, for example). Such reactions can lead to plugging of the junction between the electrode and the analyte solution.

> At 25°C, the potential of the saturated calomel electrode versus the standard hydrogen electrode is 0.244 V. For the saturated silver/silver chloride electrode, it is 0.199 V.

21C LIQUID-JUNCTION POTENTIALS

When two electrolyte solutions of different composition are in contact with one another, there is a potential difference across the interface. This junction potential is the result of an unequal distribution of cations and anions across the boundary due

(b)

Figure 21-3 Diagram of a silver/silver chloride electrode showing the parts of the electrode that produce the reference electrode potential, E_{ref}, and the junction potential, E_j. (Reprinted with permission of Bioanalytical Systems, W. Lafayette, IN.)

Figure 21-4 Schematic representation of a liquid junction, showing the source of the junction potential, E_j. The lengths of the arrows correspond to the relative mobilities of the ions.

> The junction potential across a typical KCl salt bridge is a few millivolts.

> The results of potentiometric determinations are the activities of analytes in contrast to most analytical methods that give the concentrations of analytes. Recall that the activity of a species a_X is related to the molar concentration of X by Equation 10-2
>
> $$a_X = \gamma_X[X]$$
>
> where γ_X is the activity coefficient of X, a parameter that varies with the ionic strength of the solution. Because potentiometric data are dependent on activities, it will not be necessary in most cases to make the usual approximation that $a_X \approx [X]$ in this chapter.

to differences in the rates at which these species diffuse. **Figure 21-4** shows a very simple liquid junction consisting of a 1 M hydrochloric acid solution that is in contact with a solution that is 0.01 M in that acid. An inert porous barrier, such as a fritted glass plate, prevents the two solutions from mixing. The liquid junction may be represented as

$$HCl(1\ M)|HCl(0.01\ M)$$

Both hydrogen ions and chloride ions tend to diffuse across this boundary from the more concentrated to the more dilute solution, that is, left to right. The driving force for each ion is proportional to the activity difference between the two solutions. In the present example, hydrogen ions are substantially more mobile than chloride ions. Thus, hydrogen ions diffuse more rapidly than chloride ions, and as shown in the Figure 21-4, a separation of charge results. The more dilute side of the boundary becomes positively charged because of the more rapid diffusion of hydrogen ions. The concentrated side, therefore, acquires a negative charge from the excess of slower-moving chloride ions. The charge developed tends to counteract the differences in diffusion rates of the two ions so that a condition of equilibrium is attained rapidly. The potential difference resulting from this charge separation may be several hundredths of a volt.

The magnitude of the liquid-junction potential can be minimized by placing a salt bridge between the two solutions. The salt bridge is most effective if the mobilities of the negative and positive ions in the bridge are nearly equal and if their concentrations are large. A saturated solution of potassium chloride is good from both standpoints. The junction potential with such a bridge is typically a few millivolts.

21D INDICATOR ELECTRODES

An ideal indicator electrode responds rapidly and reproducibly to changes in the concentration of an analyte ion (or group of analyte ions). Although no indicator electrode is absolutely specific in its response, a few are now available that are remarkably selective. Indicator electrodes are of three types: metallic, membrane, and ion-sensitive field effect transistors.

21D-1 Metallic Indicator Electrodes

It is convenient to classify metallic indicator electrodes as **electrodes of the first kind**, **electrodes of the second kind**, and **inert redox electrodes**.

Electrodes of the First Kind

An electrode of the first kind is a pure metal electrode that is in direct equilibrium with its cation in the solution. A single reaction is involved. For example, the equilibrium between a copper and its cation Cu^{2+} is

$$Cu^{2+}(aq) + 2e^- \rightleftharpoons Cu(s)$$

for which

$$E_{ind} = E^0_{Cu} - \frac{0.0592}{2}\log\frac{1}{a_{Cu^{2+}}} = E^0_{Cu} + \frac{0.0592}{2}\log a_{Cu^{2+}} \qquad (21\text{-}2)$$

where E_{ind} is the electrode potential of the metal electrode and $a_{Cu^{2+}}$ is the activity of the ion (or in dilute solution, approximately its molar concentration, $[Cu^{2+}]$).

We often express the electrode potential of the indicator electrode in terms of the p-function of the cation ($pX = -\log a_{Cu^{2+}}$). Thus, substituting this definition of pCu into Equation 21-2 gives

$$E_{ind} = E^0_{Cu} + \frac{0.0592}{2}\log a_{Cu^{2+}} = E^0_{Cu} - \frac{0.0592}{2}pCu$$

A general expression for any metal and its cation is

$$E_{ind} = E^0_{X^{n+}/X} + \frac{0.0592}{n}\log a_{X^{n+}} = E^0_{X^{n+}/X} - \frac{0.0592}{n}pX \qquad (21\text{-}3)$$

This function is plotted in **Figure 21-5**.

Electrode systems of the first kind are not widely used for potentiometric determinations for several reasons. For one, metallic indicator electrodes are not very selective and respond not only to their own cations but also to other more easily reduced cations. For example, a copper electrode cannot be used for the determination of copper(II) ions in the presence of silver(I) ions because the electrode potential is also a function of the Ag^+ concentration. In addition, many metal electrodes, such as zinc and cadmium, can only be used in neutral or basic solutions because they dissolve in the presence of acids. Third, other metals are so easily oxidized that they can be used only when analyte solutions are deaerated to remove oxygen. Finally, certain harder metals, such as iron, chromium, cobalt, and nickel, do not provide reproducible potentials. For these electrodes, plots of E_{ind} versus pX yield slopes that differ significantly and irregularly from the theoretical ($-0.0592/n$). For these reasons, the only electrode systems of the first kind that have been used in potentiometry are Ag/Ag^+ and Hg/Hg^{2+} in neutral solutions and Cu/Cu^{2+}, Zn/Zn^{2+}, Cd/Cd^{2+}, Bi/Bi^{3+}, Tl/Tl^+, and Pb/Pb^{2+} in deaerated solutions.

Electrodes of the Second Kind

Metals not only serve as indicator electrodes for their own cations but also respond to the activities of anions that form sparingly soluble precipitates or stable complexes with such cations. The potential of a silver electrode, for example, correlates reproducibly with the activity of chloride ion in a solution saturated with silver chloride. In this situation, the electrode reaction can be written as

$$AgCl(s) + e^- \rightleftharpoons Ag(s) + Cl^-(aq) \qquad E^0_{AgCl/Ag} = 0.222 \text{ V}$$

The Nernst expression for this process at 25°C is

$$E_{ind} = E^0_{AgCl/Ag} - 0.0592\log a_{Cl^-} = E^0_{AgCl/Ag} + 0.0592 \text{ pCl} \qquad (21\text{-}4)$$

Equation 21-4 shows that the potential of a silver electrode is proportional to pCl, the negative logarithm of the chloride ion activity. Thus, in a solution saturated with silver chloride, a silver electrode can serve as an indicator electrode of the second kind for chloride ion. Note that the sign of the log term for an electrode of this type is opposite that for an electrode of the first kind (see Equation 21-3). A plot of the potential of the silver electrode versus pCl is shown in **Figure 21-6**.

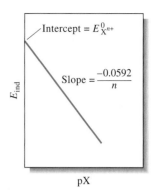

Figure 21-5 A plot of Equation 21-3 for an electrode of the first kind.

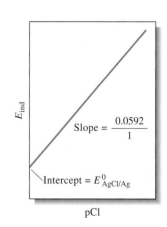

Figure 21-6 A plot of Equation 21-4 for an electrode of the second kind for Cl^-.

Inert Metallic Electrodes for Redox Systems

As noted in Chapter 18, several relatively inert conductors respond to redox systems. Such materials as platinum, gold, palladium, and carbon can be used to monitor redox systems. For example, the potential of a platinum electrode immersed in a solution containing cerium(III) and cerium(IV) is

$$E_{\text{ind}} = E^0_{\text{Ce}^{4+}/\text{Ce}^{3+}} - 0.0592 \log \frac{a_{\text{Ce}^{3+}}}{a_{\text{Ce}^{4+}}}$$

A platinum electrode is a convenient indicator electrode for titrations involving standard cerium(IV) solutions.

21D-2 Membrane Indicator Electrodes[2]

For nearly a century, the most convenient method for determining pH has involved measurement of the potential that appears across a thin glass membrane that separates two solutions with different hydrogen ion concentrations. The phenomenon on which the measurement is based was first reported in 1906 and by now has been extensively studied by many investigators. As a result, the sensitivity and selectivity of glass membranes toward hydrogen ions are reasonably well understood. Furthermore, this understanding has led to the development of other types of membranes that respond selectively to many other ions.

Membrane electrodes are sometimes called **p-ion electrodes** because the data obtained from them are usually presented as p-functions, such as pH, pCa, or pNO_3. In this section, we consider several types of p-ion membranes.

It is important to note at the outset of this discussion that membrane electrodes are fundamentally different from metal electrodes both in design and in principle. We shall use the glass electrode for pH measurements to illustrate these differences.

21D-3 The Glass Electrode for Measuring pH

Figure 21-7a shows a typical *cell* for measuring pH. The cell consists of a glass indicator electrode and a saturated calomel reference electrode immersed in the solution of unknown pH. The indicator electrode consists of a thin pH-sensitive glass membrane sealed onto one end of a heavy-walled glass or plastic tube. A small volume of dilute hydrochloric acid saturated with silver chloride is contained in the tube. The inner solution in some electrodes is a buffer containing chloride ion. A silver wire in this solution forms a silver/silver chloride reference electrode, which is connected to one of the terminals of a potential-measuring device. The calomel electrode is connected to the other terminal.

Figure 21-7a and the representation of this cell in **Figure 21-8** show that a glass-electrode system contains two reference electrodes: the external calomel electrode and the internal silver/silver chloride electrode. While the internal reference electrode is a part of the glass electrode, it is not the pH-sensing element. *It is the thin glass membrane bulb at the tip of the electrode that responds to pH.* At first, it may seem unusual that an insulator like glass (see margin note) can be used to detect ions, but keep in mind that whenever there is a charge imbalance across any material, there is an

> The membrane of a typical glass electrode (with a thickness of 0.03 to 0.1 mm) has an electrical resistance of 50 to 500 MΩ.

---

[2]Some suggested sources for additional information on this topic are R. S. Hutchins and L. G. Bachas, in *Handbook of Instrumental Techniques for Analytical Chemistry,* F. A. Settle, ed., Upper Saddle River, NJ: Prentice-Hall, 1997; A. Evans, *Potentiometry and Ion-Selective Electrodes*, New York: Wiley, 1987; J. Koryta, *Ions, Electrodes, and Membranes*, 2nd ed., New York: Wiley, 1991.

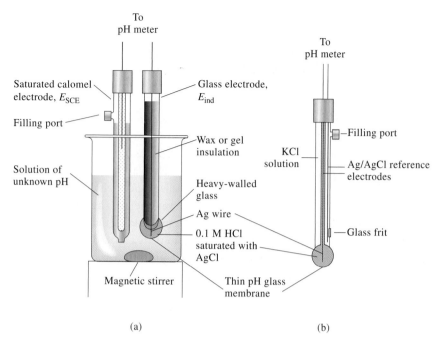

(a)   (b)

Figure 21-7 Typical electrode system for measuring pH. (a) Glass electrode (indicator) and SCE (reference) immersed in a solution of unknown pH. (b) Combination probe consisting of both an indicator glass electrode and a silver/silver chloride reference. A second silver/silver chloride electrode serves as the internal reference for the glass electrode. The two electrodes are arranged concentrically with the internal reference in the center and the external reference outside. The reference makes contact with the analyte solution through the glass frit or other suitable porous medium. Combination probes are the most common configuration of glass electrode and reference for measuring pH.

Figure 21-8 Diagram of glass/calomel cell for the measurement of pH. E_{SCE} is the potential of the reference electrode, E_j is the junction potential, a_1 is the activity of hydronium ions in the analyte solution, E_1 and E_2 are the potentials on either side of the glass membrane, E_b is the boundary potential, and a_2 is the activity of hydronium ion in the internal reference solution.

electrical potential difference across the material. In the case of the glass electrode, the concentration (and the activity) of protons inside the membrane is constant. The concentration outside the membrane is determined by the activity of hydrogen ions in the analyte solution. This concentration difference produces the potential difference that we measure with a pH meter. Notice that the internal and external reference electrodes are just the means of making electrical contact with the two sides of the glass membrane and that their potentials are essentially constant except for the junction potential, which depends to a small extent on the composition of the analyte solution. The potentials of the two reference electrodes depend on the electrochemical characteristics of their respective redox couples, but the potential across the glass membrane depends on the physicochemical characteristics of the glass and its response to ionic concentrations on both sides of the membrane. To understand how the glass electrode works, we must explore the mechanism of the creation of the charge differential across the membrane that produces the membrane potential. In the next few sections, we investigate this mechanism and the important characteristics of these membranes.

In Figure 21-7b, we see the most common configuration for measuring pH with a glass electrode. In this arrangement, the glass electrode and its Ag/AgCl internal reference electrode are positioned in the center of a cylindrical probe. Surrounding the glass electrode is the external reference electrode, which is most often of the Ag/AgCl type. The presence of the external reference electrode is not as obvious as in the dual-probe arrangement of Figure 21-7a, but the single-probe, or combination, variety is much more convenient and can be made much smaller than the dual system. The pH-sensitive glass membrane is attached to the tip of the electrode. These glass pH electrodes are manufactured in many different physical shapes and sizes (5 cm to 5 μm) to suit a broad range of laboratory and industrial applications.

The Composition and Structure of Glass Membranes

Much research has been devoted to the effects of glass composition on the sensitivity of membranes to protons and other cations, and a number of formulations are now used for the manufacture of electrodes. Corning 015 glass, which has been widely used for membranes, consists of approximately 22% Na_2O, 6% CaO, and 72% SiO_2. Membranes made from this glass exhibit excellent specificity to hydrogen ions up to a pH of about 9. At higher pH values, however, the glass becomes somewhat responsive to sodium as well as to other singly charged cations. Other glass formulations are now in use in which sodium and calcium ions are replaced to various degree by barium and lithium ions. These membranes have superior selectivity and lifetime.

As shown in **Figure 21-9**, a silicate glass used for membranes consists of an infinite three-dimensional network of groups in which each silicon atom is bonded to four oxygen atoms and each oxygen atom is shared by two silicon atoms. Within the empty spaces (interstices) inside this structure are enough cations to balance the negative charge of the silicate groups. Singly charged cations, such as sodium and lithium, can move around in the lattice and are responsible for electrical conduction within the membrane.

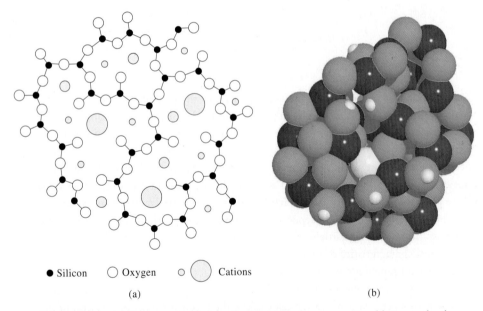

● Silicon   ◯ Oxygen   ○ ◯ Cations

(a)   (b)

Figure 21-9   (a) Cross-sectional view of a silicate glass structure. In addition to the three Si—O bond shown, each silicon is bonded to an additional oxygen atom, either above or below the plane of the paper. (Reprinted (adapted) with permission from G. A. Perley, *Anal. Chem.*, **1949**, *21*, 395, **DOI**: 10.1021/ac60027a013. Copyright 1949 American Chemical Society.) (b) Model showing three-dimensional structure of amorphous silica with Na^+ ion (large dark green) and several H^+ ions (small dark green) incorporated. Note that the Na^+ ion is surrounded by a cage of oxygen atoms and that each proton in the amorphous lattice is attached to an oxygen. The cavities in the structure, the small size, and the high mobility of the proton ensure that protons can migrate deep into the surface of the silica. Other cations and water molecules may be incorporated into the interstices of the structure as well.

The two surfaces of a glass membrane must be hydrated before it will function as a pH electrode. Nonhygroscopic glasses show no pH function. Even hygroscopic glasses lose their pH sensitivity after dehydration by storage over a desiccant. The effect is reversible, however, and the response of a glass electrode can be restored by soaking it in water.

The hydration of a pH-sensitive glass membrane involves an ion-exchange reaction between singly charged cations in the interstices of the glass lattice and hydrogen ions from the solution. The process involves $+1$ cations exclusively because $+2$ and $+3$ cations are too strongly held within the silicate structure to exchange with ions in the solution. The ion-exchange reaction can then be written as

$$\underset{\text{soln}}{H^+} + \underset{\text{glass}}{Na^+Gl^-} \rightleftharpoons \underset{\text{soln}}{Na^+} + \underset{\text{glass}}{H^+Gl^-} \qquad (21\text{-}5)$$

Oxygen atoms attached to only one silicon atom are the negatively charged Gl^- sites shown in this equation. The equilibrium constant for this process is so large that the surfaces of a hydrated glass membrane normally consist entirely of silicic acid (H^+Gl^-). There is an exception to this situation in highly alkaline media, where the hydrogen ion concentration is extremely small and the sodium ion concentration is large. Under this condition, a significant fraction of the sites are occupied by sodium ions.

Membrane Potentials

The lower part of Figure 21-8 shows four potentials that develop in a cell when pH is being determined with a glass electrode. Two of these potentials, $E_{Ag,AgCl}$ and E_{SCE}, are reference electrode potentials that are constant. There is a third potential, the junction potential, E_j, across the salt bridge that separates the calomel electrode from the analyte solution. This junction and its associated potential are found in all cells used to make potentiometric measurements of ion concentration. The fourth, and most important, potential shown in Figure 21-8 is the **boundary potential**, E_b, *which varies with the pH of the analyte solution.* The two reference electrodes simply provide electrical contacts with the solutions so that changes in the boundary potential can be measured.

The Boundary Potential

Figure 21-8 shows that the boundary potential is determined by potentials, E_1 and E_2, which appear at the two *surfaces* of the glass membrane. The source of these two potentials is the charge that accumulates as a consequence of the reactions

$$\underset{\text{glass}_1}{H^+Gl^-(s)} \rightleftharpoons \underset{\text{soln}_1}{H^+(aq)} + \underset{\text{glass}_1}{Gl^-(s)} \qquad (21\text{-}6)$$

$$\underset{\text{glass}_2}{H^+Gl^-(s)} \rightleftharpoons \underset{\text{soln}_2}{H^+(aq)} + \underset{\text{glass}_2}{Gl^-(s)} \qquad (21\text{-}7)$$

where subscript 1 refers to the interface between the exterior of the glass and the analyte solution and subscript 2 refers to the interface between the internal solution and the interior of the glass. These two reactions cause the two glass surfaces to be negatively charged with respect to the solutions with which they are in contact. These negative charges at the surfaces produce the two potentials E_1 and E_2 shown in Figure 21-8. The hydrogen ion concentrations in the solutions on the two sides of the membrane control the positions of the equilibria of Equations 21-7 and 21-8 that in turn determine E_1 and E_2. When the positions of the two equilibria differ, the surface where the greater dissociation has occurred is negative with respect to the

Glasses that absorb water are said to be **hygroscopic**.

other surface. The resulting difference in potential between the two surfaces of the glass is the boundary potential, which is related to the activities of hydrogen ions in each of the solutions by the Nernst-like equation

$$E_b = E_1 - E_2 = 0.0592 \log \frac{a_1}{a_2} \qquad (21\text{-}8)$$

where a_1 is the activity of the analyte solution and a_2 is that of the internal solution. For a glass pH electrode, the hydrogen ion activity of the internal solution, a_2, is held constant so that Equation 21-8 simplifies to

$$E_b = L' + 0.0592 \log a_1 = L' - 0.0592 \, pH \qquad (21\text{-}9)$$

where

$$L' = -0.0592 \log a_2$$

The boundary potential is then a measure of the hydrogen ion activity (pH) of the external solution.

The significance of the potentials and the potential differences shown in Equation 21-8 is illustrated by the potential profiles shown in **Figure 21-10**. The profiles are plotted across the membrane from the analyte solution on the left

Figure 21-10   Potential profile across a glass membrane from the analyte solution to the internal reference solution. The reference electrode potentials are not shown.

through the membrane to the internal solution on the right. The important thing to note about these profiles is that regardless of the absolute potential inside the hygroscopic layers or the glass, the boundary potential is determined by the *difference* in potential on either side of the glass membrane that is in turn determined by the proton activity on each side of the membrane.

The Asymmetry Potential

When identical solutions and reference electrodes are placed on the two sides of a glass membrane, the boundary potential should in principle be zero. Frequently, however, we find a small asymmetry potential that changes gradually with time.

The sources of the asymmetry potential are obscure but undoubtedly include such causes as differences in strain on the two surfaces of the membrane created during manufacture, mechanical abrasion on the outer surface during use, and chemical etching of the outer surface. To eliminate the bias caused by the asymmetry potential, all membrane electrodes must be calibrated against one or more standard analyte solutions. Calibrations should be carried out at least daily and more often when the electrode is heavily used.

The Glass Electrode Potential

The potential of a glass indicator electrode, E_{ind}, has three components: (1) the boundary potential, given by Equation 21-8; (2) the potential of the internal Ag/AgCl reference electrode; and (3) the small asymmetry potential, E_{asy}, which changes slowly with time. In equation form, we may write

$$E_{ind} = E_b + E_{Ag/AgCl} + E_{asy}$$

Substitution of Equation 21-9 for E_b gives

$$E_{ind} = L' + 0.0592 \log a_1 + E_{Ag/AgCl} + E_{asy}$$

or

$$E_{ind} = L + 0.0592 \log a_1 = L - 0.0592 \text{ pH} \qquad (21\text{-}10)$$

where L is a combination of the three constant terms. Compare Equations 21-10 and 21-3. Although these two equations are similar in form and both potentials are produced by separation of charge, remember that *the mechanisms of charge separation that result in these expressions are considerably different.*

The Alkaline Error

In basic solutions, glass electrodes respond to the concentration of both hydrogen ion and alkali metal ions. The magnitude of the resulting alkaline error for four different glass membranes is shown in **Figure 21-11** (curves *C* to *F*). These curves refer to solutions in which the sodium ion concentration was held constant at 1 M while the pH was varied. Note that the error ($pH_{read} - pH_{true}$) is negative (that is, the measured pH values are lower than the true values), suggesting that the electrode is responding to sodium ions as well as to protons. This observation is confirmed by data obtained for solutions containing different sodium ion concentrations. Thus, at pH 12, the electrode with a Corning 015 membrane (curve *C* in Figure 21-11) registered a pH of 11.3 when immersed in a solution having a sodium ion concentration of 1 M but 11.7 in a solution that was 0.1 M in this ion. All singly charged cations induce an alkaline error whose magnitude depends on both the cation in question and the composition of the glass membrane.

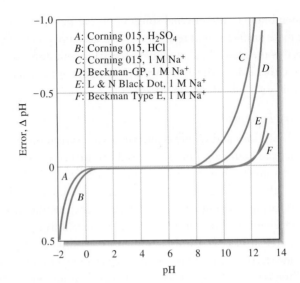

Figure 21-11 Acid and alkaline errors for selected glass electrodes at 25°C. (R.G. Bates, *Determination of pH*, 2nd ed., p.265. New York: Wiley, 1973. Reprinted by permission of the author's estate.)

The alkaline error can be satisfactorily explained by assuming an exchange equilibrium between the hydrogen ions on the glass surface and the cations in solution. This process is simply the reverse of that shown in Equation 21-5:

$$\underset{\text{glass}}{H^+Gl^-} + \underset{\text{soln}}{B^+} \rightleftharpoons \underset{\text{glass}}{B^+Gl^-} + \underset{\text{soln}}{H^+}$$

where B^+ represents some singly charged cation, such as sodium ion.

The equilibrium constant for this reaction is

> In Equation 21-11, b_1 represents the activity of some singly charged cation such as Na^+ or K^+.

$$K_{ex} = \frac{a_1 b_1'}{a_1' b_1} \tag{21-11}$$

where a_1 and b_1 represent the activities of H^+ and B^+ in solution and a_1' and b_1' are the activities of these ions on the glass surface. Equation 21-11 can be rearranged to give ratio of the activities B^+ to H^+ on the glass surface:

$$\frac{b_1'}{a_1'} = K_{ex} \frac{b_1}{a_1}$$

For the glasses used for pH electrodes, K_{ex} is typically so small that the activity ratio b_1'/a_1' is minuscule. The situation differs in strongly alkaline media, however. For example, b_1'/a_1' for an electrode immersed in a pH 11 solution that is 1 M in sodium ions (see Figure 21-11) is $10^{11} \times K_{ex}$. Under these conditions, the activity of the sodium ions relative to that of the hydrogen ions becomes so large that the electrode responds to both species.

Describing Selectivity

The effect of an alkali metal ion on the potential across a membrane can be accounted for by inserting an additional term in Equation 21-9 to give

$$E_b = L' + 0.0592 \log (a_1 + k_{H,B}b_1) \tag{21-12}$$

> The **selectivity coefficient** is a measure of the response of an ion-selective electrode to other ions.

where $k_{H,B}$ is the **selectivity coefficient** for the electrode. Equation 21-12 applies not only to glass indicator electrodes for hydrogen ion but also to all other types of membrane electrodes. Selectivity coefficients range from zero (no interference) to values greater than unity. Thus, if an electrode for ion A responds 20 times more strongly to ion B than to ion A, $k_{H,B}$ has a value of 20. If the response of the electrode to ion C is 0.001 of its response to A (a much more desirable situation), $k_{H,B}$ is 0.001.[3]

The product $k_{H,B}b_1$ for a glass pH electrode is usually small relative to a_1 provided that the pH is less than 9; under these conditions, Equation 21-12 simplifies to Equation 21-9. At high pH values and at high concentrations of a singly charged ion, however, the second term in Equation 21-12 assumes a more important role in determining E_b, and an alkaline error is encountered. For electrodes specifically designed for work in highly alkaline media (curve E in Figure 21-11), the magnitude of $k_{H,B}b_1$ is appreciably smaller than for ordinary glass electrodes.

The Acid Error

As shown in Figure 21-11, the typical glass electrode exhibits an error, opposite in sign to the alkaline error, in solution of pH less than about 0.5. The negative error ($pH_{read} - pH_{true}$) indicates that pH readings tend to be too high in this region. The magnitude of the error depends on a variety of factors and is generally not very reproducible. All the causes of the acid error are not well understood, but one source is a saturation effect that occurs when all the surface sites on the glass are occupied with H^+ ions. Under these conditions, the electrode no longer responds to further increases in the H^+ concentration, and the pH readings are too high.

21D-4 Glass Electrodes for Other Cations

The alkaline error in early glass electrodes led to investigations concerning the effect of glass composition on the magnitude of this error. One consequence has been the development of glasses for which the alkaline error is negligible below about pH 12 (see curves E and F, Figure 21-11). Other studies have discovered glass compositions that permit the determination of cations other than hydrogen. Incorporation of Al_2O_3 or B_2O_3 in the glass has the desired effect. Glass electrodes that permit the direct potentiometric measurement of such singly charged species as Na^+, K^+, NH_4^+, Rb^+, Cs^+, Li^+, and Ag^+ have been developed. Some of these glasses are reasonably selective toward particular singly charged cations. Glass electrodes for Na^+, Li^+, NH_4^+, and total concentration of univalent cations are now available from commercial sources.

21D-5 Liquid-Membrane Electrodes

The potential of liquid-membrane electrodes develops across the interface between the solution containing the analyte and a liquid-ion exchanger that selectively bonds with the analyte ion. These electrodes have been developed for the direct potentiometric measurement of numerous polyvalent cations as well as certain anions.

Figure 21-12 is a schematic of a liquid-membrane electrode for calcium. It consists of a conducting membrane that selectively binds calcium ions, an internal solution containing a fixed concentration of calcium chloride, and a silver electrode that is coated with silver chloride to form an internal reference electrode. Notice the similarities between the liquid-membrane electrode and the glass electrode, as shown in

---

[3]For tables of selectivity coefficients for a variety of membranes and ionic species, see Y. Umezawa, *CRC Handbook of Ion Selective Electrodes: Selectivity Coefficients*, Boca Raton, FL: CRC Press, 1990.

Figure 21-12 Diagram of a liquid-membrane electrode for Ca^{2+}.

Figure 21-13 Comparison of a liquid-membrane calcium ion electrode with a glass pH electrode. (Courtesy of Thermo Orion, Beverly, MA.)

Hydrophobia means fear of water. The hydrophobic disk is porous toward organic liquids but repels water.

Figure 21-13. The active membrane ingredient is an ion exchanger that consists of a calcium dialkyl phosphate that is nearly insoluble in water. In the electrode shown in Figures 21-12 and 21-13, the ion exchanger is dissolved in an immiscible organic liquid that is forced by gravity into the pores of a hydrophobic porous disk. This disk then serves as the membrane that separates the internal solution from the analyte solution. In a more recent design, the ion exchanger is immobilized in a tough polyvinyl chloride gel attached to the end of a tube that holds the internal solution and reference electrode (see Figure 21-13, right). In either design, a dissociation equilibrium develops at each membrane interface that is analogous to Equations 21-6 and 21-7:

$$[(RO)_2POO]_2Ca \rightleftharpoons 2(RO)_2POO^- + Ca^{2+}$$
$$\text{organic} \qquad \text{organic} \qquad \text{aqueous}$$

where R is a high-molecular-mass aliphatic group. As with the glass electrode, a potential develops across the membrane when the extent of dissociation of the ion exchanger dissociation at one surface differs from that at the other surface.

This potential is a result of differences in the calcium ion activity of the internal and external solutions. The relationship between the membrane potential and the calcium ion activities is given by an equation that is similar to Equation 21-8:

$$E_b = E_1 - E_2 = \frac{0.0592}{2} \log \frac{a_1}{a_2} \tag{21-13}$$

where a_1 and a_2 are the activities of calcium ion in the external analyte and internal standard solutions, respectively. Since the calcium ion activity of the internal solution is constant,

$$E_b = N + \frac{0.0592}{2} \log a_1 = N - \frac{0.0592}{2} \text{pCa} \tag{21-14}$$

where N is a constant (compare Equations 21-14 and 21-9). Note that, because calcium is divalent, the value of n in the denominator of the coefficient of the logarithmic term is 2.

The sensitivity of the liquid-membrane electrode for calcium ion is reported to be 50 times greater than for magnesium ion and 1000 times greater than for sodium or potassium ions. Calcium ion activities as low as 5×10^{-7} M can be measured. Performance of the electrode is independent of pH in the range between 5.5 and 11. At lower pH levels, hydrogen ions undoubtedly replace some of the calcium ions on the exchanger; the electrode then becomes sensitive to pH as well as to pCa.

The calcium ion liquid-membrane electrode is a valuable tool for physiological investigations because this ion plays important roles in such processes as nerve conduction, bone formation, muscle contraction, cardiac expansion and contraction, renal tubular function, and perhaps hypertension. Most of these processes are more influenced by the activity than the concentration of the calcium ion; activity, of course, is the parameter measured by the membrane electrode. Therefore, the calcium ion electrode as well as the potassium ion electrode and others are important tools in studying physiological processes.

A liquid-membrane electrode specific for potassium ion is also of great value for physiologists because the transport of neural signals appears to involve movement of this ion across nerve membranes. Investigation of this process requires an electrode that can detect small concentrations of potassium ion in media that contain much larger concentrations of sodium ion. Several liquid-membrane electrodes show promise in meeting this requirement. One is based on the antibiotic valinomycin, a cyclic ether that has a strong affinity for potassium ion. Of equal importance is the observation that a liquid membrane consisting of valinomycin in diphenyl ether is about 10^4 times as responsive to potassium ion as to sodium ion.[4] **Figure 21-14** is a photomicrograph of a tiny electrode used for determining the potassium content of a single cell.

Table 21-2 lists some liquid-membrane electrodes available from commercial sources. The anion-sensitive electrodes listed make use of a solution containing an anion-exchange resin in an organic solvent. Liquid-membrane electrodes in which the exchange liquid is held in a polyvinyl chloride gel have been developed for Ca^{2+}, K^+, NO_3^-, and BF_4^-. These have the appearance of crystalline electrodes, which are considered in the following section. A homemade liquid-membrane ion-selective electrode is described in Feature 21-1.

> Ion-selective microelectrodes can be used to make measurements of ion activities within a living organism.

Figure 21-14 Photograph of a potassium liquid ion exchanger microelectrode with 125 μm of ion exchanger inside the tip. The magnification of the original photo was 400×. (Reprinted with permission from *Anal. Chem.*, March **1971**, *43*(3), 89A-93A. Copyright 1971 American Chemical Society.)

---

[4]M. S. Frant and J. W. Ross, Jr., *Science*, **1970**, *167*, 987, **DOI**: 10.1126/science.167.3920.987.

TABLE 21-2

Characteristics of Liquid-Membrane Electrodes*

Analyte Ion	Concentration Range, M[†]	Major Interferences[‡]
NH_4^+	10^0 to 5×10^{-7}	<1 H$^+$, 5×10^{-1} Li$^+$, 8×10^{-2}, Na$^+$, 6×10^{-4} K$^+$, 5×10^{-2} Cs$^+$, >1 Mg^{2+}, >1 Ca^{2+}, >1 Sr^{2+}, >0.5 Sr^{2+}, 1×10^{-2} Zn^{2+}
Cd^{2+}	10^0 to 5×10^{-7}	Hg^{2+} and Ag$^+$ (poisons electrode at $>10^{-7}$ M), Fe^{3+} (at >0.1[Cd^{2+}]), Pb^{2+} (at $>$[Cd^{2+}]), Cu^{2+} (possible)
Ca^{2+}	10^0 to 5×10^{-7}	10^{-5} Pb^{2+}; 4×10^{-3} Hg^{2+}, H$^+$, 6×10^{-3} Sr^{2+}; 2×10^{-2} Fe^{2+}; 4×10^{-2} Cu^{2+}; 5×10^{-2} Ni^{2+}; 0.2 NH$_3$; 0.2 Na$^+$; 0.3 Tris$^+$; 0.3 Li$^+$; 0.4 K$^+$; 0.7 Ba^{2+}; 1.0 Zn^{2+}; 1.0 Mg^{2+}
Cl^-	10^0 to 5×10^{-6}	Maximum allowable ratio of interferent to [Cl$^-$]: OH$^-$ 80, Br$^-$ 3×10^{-3}; I$^-$ 5×10^{-7}, S^{2-} 10^{-6}, CN$^-$ 2×10^{-7}, NH$_3$ 0.12, S$_2$O$_3^{2-}$ 0.01
BF_4^-	10^0 to 7×10^{-6}	5×10^{-7} ClO$_4^-$; 5×10^{-6} I$^-$; 5×10^{-5} ClO$_3^-$; 5×10^{-4} CN$^-$; 10^{-3} Br$^-$; 10^{-3} NO$_2^-$; 5×10^{-3} NO$_3^-$; 3×10^{-3}HCO$_3^-$, 5×10^{-2} Cl$^-$; 8×10^{-2} H$_2$PO$_4^-$, HPO$_4^{2-}$, PO$_4^{3-}$; 0.2 OAc$^-$; 0.6 F$^-$; 1.0 SO$_4^{2-}$
NO_3^-	10^0 to 7×10^{-6}	10^{-7}ClO$_4^-$; 5×10^{-6}; I$^-$; 5×10^{-5} ClO$_3^-$; 10^{-4} CN$^-$; 7×10^{-4} Br$^-$; 10^{-3} HS$^-$; 10^{-2} HCO$_3^-$, 2×10^{-2} CO$_3^{2-}$; 3×10^{-2} Cl$^-$; 5×10^{-2} H$_2$PO$_4^-$, HPO$_4^{2-}$, PO$_4^{3-}$; 0.2 OAc$^-$; 0.6 F$^-$; 1.0 SO$_4^{2-}$
NO_2^-	1.4×10^{-6} to 3.6×10^{-6}	7×10^{-1} salicylate, 2×10^{-3} I$^-$, 10^{-1} Br$^-$, 3×10^{-1} ClO$_3^-$, 2×10^{-1} acetate, 2×10^{-1} HCO$_3^-$, 2×10^{-1} NO$_3^-$, 2×10^{-1} SO$_4^{2-}$, 1×10^{-1} Cl$^-$, 1×10^{-1} ClO$_4^-$, 1×10^{-1} F$^-$
ClO_4^-	10^0 to 7×10^{-6}	2×10^{-3} I$^-$; 2×10^{-2} ClO$_3^-$; 4×10^{-2} CN$^-$, Br$^-$; 5×10^{-2} NO$_2^-$, NO$_3^-$; 2 HCO$_3^-$, CO$_3^{2-}$; Cl$^-$, H$_2$PO$_4^-$, HPO$_4^{2-}$, PO$_4^{3-}$, OAc$^-$, F$^-$, SO$_4^{2-}$
K^+	10^0 to 1×10^{-6}	3×10^{-4} Cs$^+$; 6×10^{-3} NH$_4^+$, Tl$^+$; 10^{-2} H$^+$; 1.0 Ag$^+$, Tris$^+$; 2.0 Li$^+$, Na$^+$
Water hardness ($Ca^{2+} + Mg^{2+}$)	10^{-3} to 6×10^{-6}	3×10^{-5} Cu^{2+}, Zn^{2+}; 10^{-4} Ni^{2+}; 4×10^{-4} Sr^{2+}; 6×10^{-5} Fe^{2+}; 6×10^{-4} Ba^{2+}; 3×10^{-2} Na$^+$; 0.1 K$^+$

All electrodes are the plastic-membrane type. All values are selectivity coefficients unless otherwise noted.
[†]From product catalog, Boston, MA: Thermo Orion, 2006.
[‡]From product instruction manuals, Boston, MA: Thermo Orion, 2003.

FEATURE 21-1

An Easily Constructed Liquid-Membrane Ion-Selective Electrode

You can make a liquid-membrane ion-selective electrode with glassware and chemicals available in most laboratories.[5] All you need are a pH meter, a pair of reference electrodes, a fritted-glass filter crucible or tube, trimethylchlorosilane, and a liquid ion exchanger.

First, cut the filter crucible (or alternatively, a fritted tube), as shown in **Figure 21F-1.** Carefully clean and dry the crucible and then draw a small amount of trimethylchlorosilane into the frit. This coating makes the glass in the frit hydrophobic. Rinse the frit with water, dry, and apply a commercial liquid ion exchanger to it. After a minute, remove the excess exchanger. Add a few milliliters of a 10^{-2} M solution of the ion of interest to the crucible, insert a reference electrode into the solution, and voilá, you have a very nice ion-selective electrode. The exact details of washing, drying, and preparing the electrode are provided in the original article.

Connect the ion-selective electrode and the second reference electrode to the pH meter, as shown in Figure 21F-1. Prepare a series of standard solutions of the ion of interest, measure the cell potential for each concentration, plot a working curve of

---

[5]See T. K. Christopoulus and E. P. Diamandis, *J. Chem. Educ.*, **1988**, *65*, 648, **DOI**: 10.1021/ed065p648.

Figure 21F-1 A homemade liquid-membrane electrode.

E_{cell} versus log c, and perform a least-squares analysis on the data (see Chapter 8). Compare the slope of the line with the theoretical slope of (0.0592 V)/n. Measure the potential for an unknown solution of the ion and calculate the concentration from the least-squares parameters.

21D-6 Crystalline-Membrane Electrodes

Considerable work has been devoted to the development of solid membranes that are selective toward anions in the same way that some glasses respond to cations. We have seen that anionic sites on a glass surface account for the selectivity of a membrane toward certain cations. By analogy, a membrane with cationic sites might be expected to respond selectively toward anions.

Membranes prepared from cast pellets of silver halides have been used successfully in electrodes for the selective determination of chloride, bromide, and iodide ions. In addition, an electrode based on a polycrystalline Ag_2S membrane is offered by one manufacturer for the determination of sulfide ion. In both types of membranes, silver ions are sufficiently mobile to conduct electricity through the solid medium. Mixtures of PbS, CdS, and CuS with Ag_2S provide membranes that are selective for Pb^{2+}, Cd^{2+}, and Cu^{2+}, respectively. Silver ion must be present in these membranes to conduct electricity because divalent ions are immobile in crystals. The potential that develops across crystalline solid-state electrodes is described by a relationship similar to Equation 21-9.

A crystalline electrode for fluoride ion is available from commercial sources. The membrane consists of a slice of a single crystal of lanthanum fluoride that has been doped with europium(II) fluoride to improve its conductivity. The membrane, supported between a reference solution and the solution to be measured, shows a theoretical response to changes in fluoride ion activity from 10^0 to 10^{-6} M. The electrode is selective for fluoride ion over other common anions by several orders of magnitude; only hydroxide ion appears to offer serious interference.

Some solid-state electrodes available from commercial sources are listed in **Table 21-3**.

TABLE 21-3

Characteristics of Solid-State Crystalline Electrodes*		
Analyte Ion	**Concentration Range, M**	**Major Interferences**
Br^-	10^0 to 5×10^{-6}	CN^-, I^-, S^{2-}
Cd^{2+}	10^{-1} to 1×10^{-7}	$Fe^{2+}, Pb^{2+}, Hg^{2+}, Ag^+, Cu^{2+}$
Cl^-	10^0 to 5×10^{-5}	$CN^-, I^-, Br^-, S^{2-}, OH^-, NH_3$
Cu^{2+}	10^{-1} to 1×10^{-8}	Hg^{2+}, Ag^+, Cd^{2+}
CN^-	10^{-2} to 1×10^{-6}	S^{2-}, I^-
F^-	Sat'd to 1×10^{-6}	OH^-
I^-	10^0 to 5×10^{-8}	CN^-
Pb^{2+}	10^{-1} to 1×10^{-6}	Hg^{2+}, Ag^+, Cu^{2+}
Ag^+/S^{2-}	$Ag^+: 10^0$ to 1×10^{-7}	Hg^{2+}
	$S^{2-}: 10^0$ to 1×10^{-7}	
SCN^-	10^0 to 5×10^{-6}	I^-, Br^-, CN^-, S^{2-}

*From *Orion Guide to Ion Analysis,* Boston, MA: Thermo Orion, 1992.

21D-7 Ion-Sensitive Field Effect Transistors (ISFETs)

The **field effect transistor**, or the **metal oxide field effect transistor (MOSFET)**, is a tiny solid-state semiconductor device that is widely used in computers and other electronic circuits as a switch to control current flow in circuits. One of the problems in using this type of device in electronic circuits has been its pronounced sensitivity to ionic surface impurities, and a great deal of money and effort has been expended by the electronic industry in minimizing or eliminating this sensitivity in order to produce stable transistors.

Scientists have exploited the sensitivities of MOSFETs to surface ionic impurities for the selective potentiometric determination of various ions. These studies have led to the development of a number of different **ion-sensitive field effect transistors** termed **ISFETs**. The theory of their selective ion sensitivity is well understood and is described in Feature 21-2.[6]

ISFETs offer a number of significant advantages over membrane electrodes including ruggedness, small size, inertness toward harsh environments, rapid response, and low electrical impedance. In contrast to membrane electrodes,

ISFETs stands for ion-sensitive field effect transistors.

FEATURE 21-2

The Structure and Performance of Ion-Sensitive Field Effect Transistors

The metal oxide field effect transistor (MOSFET) is a solid-state semiconductor device that is used widely for switching signals in computers and many other types of electronic circuits. **Figure 21F-2** shows a cross-sectional diagram (a) and a circuit symbol (b) for an *n*-channel enhancement mode MOSFET. Modern semiconductor fabrication techniques are used to construct the MOSFET on the surface of a piece of *p*-type semiconductor called the substrate. For a discussion of the characteristics of *p*-type and *n*-type semiconductors, refer to the paragraphs on silicon photodiodes in Section 25A-4. As shown in Figure 21F-2a, two islands of *n*-type semiconductors are formed on the surface of the *p*-type substrate, and the surface is then covered by insulating SiO_2. The last step in the fabrication process is the deposition of metallic

---

[6]For a detailed explanation of the theory of ISFETs, see J. Janata, *Principles of Chemical Sensors,* 2nd ed., New York: Plenum, 2009, pp. 156–167.

conductors that are used to connect the MOSFET to external circuits. There are a total of four such connections to the drain, the gate, the source, and the substrate as shown in the figure.

The area on the surface of the *p*-type material between the drain and source is called the channel (see the dark shaded area in Figure 21F-2a). Note that the channel is separated from the gate connection by an insulating layer of SiO_2. When an electrical potential is applied between the gate and the source, the electrical conductivity of the channel is enhanced by a factor that is related to the size of the applied potential.

Figure 21F-2   A metal oxide field effect transistor (MOSFET). (a) Cross-sectional diagram. (b) Circuit symbol.

The **ion-sensitive field effect transistor**, or **ISFET**, is very similar in construction and function to an *n*-channel enhancement mode MOSFET. The ISFET differs only in that variation in the concentration of the ions of interest provides the variable gate voltage to control the conductivity of the channel. As shown in **Figure 21F-3**, instead of the usual metallic contact, the face of the ISFET is covered with an insulating layer of silicon nitride. The analytical solution, containing hydronium ions in this example, is in contact with this insulating layer and with a reference electrode. The surface of the gate insulator functions very much like the surface of a glass electrode. Protons from the hydronium ions in the test solution are absorbed by available microscopic sites on the silicon nitride. Any change in the hydronium ion concentration (or activity) of the solution results in a change in the concentration of adsorbed protons. The change in concentration of adsorbed protons then gives rise to a changing electrochemical potential between the gate and the source that in turn changes the conductivity of the channel of the ISFET. The conductivity of the channel can be monitored electronically to provide a signal that is proportional to the logarithm of the activity of hydronium ion in the solution. Note that the entire ISFET except the gate insulator is coated with a polymeric encapsulant to insulate all electrical connections from the analyte solution.

The ion-sensitive surface of the ISFET is naturally sensitive to pH changes, but the device may be modified so that it becomes sensitive to other species by coating the silicon nitride gate insulator with a polymer containing molecules that tend to form complexes with species other than hydronium ion. Furthermore, several ISFETs

(continued)

Figure 21F-3   An ion-sensitive field effect transistor (ISFET) for measuring pH.

may be fabricated on the same substrate so that multiple measurements may be made at the same time. All of the ISFETs may detect the same species to enhance accuracy and reliability, or each ISFET may be coated with a different polymer so that measurements of several different species may be made. Their small size (about 1 to 2 mm²), rapid response time relative to glass electrodes, and ruggedness suggest that ISFETs may be the ion detectors of the future for many applications.

ISFETs do not require hydration before use and can be stored indefinitely in the dry state. Despite these many advantages, no ISFET-specific ion electrodes appeared on the market until the early 1990s, over 20 years after their invention. The reason for this delay is that manufacturers were unable to develop the technology of encapsulating the devices to give a product that did not exhibit drift and instability. Several companies now produce ISFETs for the determination of pH, but as of the writing of this text, these electrodes are certainly not as routinely used as the glass pH electrode.

21D-8 Gas-Sensing Probes

Figure 21-15 illustrates the essential features of a potentiometric gas-sensing probe, which consists of a tube containing a reference electrode, a specific ion electrode, and an electrolyte solution. A thin, replaceable, gas-permeable membrane attached to one end of the tube serves as a barrier between the internal and analyte solutions. As can be seen from Figure 21-15, this device is a complete electrochemical cell and is more properly referred to as a probe rather than an electrode, a term that is frequently encountered in advertisements by instrument manufacturers. Gas-sensing probes are used widely for determining dissolved gases in water and other solvents.

Membrane Composition

A *microporous membrane* is fabricated from a hydrophobic polymer. As the name implies, the membrane is highly porous (the average pore size is less than 1 μm) and allows the free passage of gases; at the same time, the water-repellent polymer prevents

A **gas-sensing probe** is a galvanic *cell* whose potential is related to the concentration of a gas in a solution. In instrument brochures, these devices are often called gas-sensing electrodes, which is a misnomer as discussed later in this section.

Reference electrode

Indicator electrode

Internal solution

Gas-permeable membrane

Figure 21-15   Diagram of a gas-sensing probe.

water and solute ions from entering the pores. The thickness of the membrane is about 0.1 mm.

The Mechanism of Response

Using carbon dioxide as an example, we can represent the transfer of gas to the internal solution in Figure 21-15 by the following set of equations:

$$\underset{\text{analyte solution}}{CO_2(aq)} \rightleftharpoons \underset{\text{membrane pores}}{CO_2(g)}$$

$$\underset{\text{membrane pores}}{CO_2(g)} \rightleftharpoons \underset{\text{internal solution}}{CO_2(aq)}$$

$$\underset{\text{internal solution}}{CO_2(aq) + 2H_2O} \rightleftharpoons \underset{\text{internal solution}}{HCO_3^- + H_3O^+}$$

The last equilibrium causes the pH of the internal surface film to change. This change is then detected by the internal glass/calomel electrode system. A description of the overall process is obtained by adding the equations for the three equilibria to give

$$\underset{\text{analyte solution}}{CO_2(aq) + 2H_2O} \rightleftharpoons \underset{\text{internal solution}}{HCO_3^- + H_3O^+}$$

The thermodynamic equilibrium constant K for this overall reaction is

$$K = \frac{(a_{H_3O^+})_{int}(a_{HCO_3^-})_{int}}{(a_{CO_2})_{ext}}$$

For a neutral species such as CO_2, $a_{CO_2} = [CO_2(aq)]$ so that

$$K = \frac{(a_{H_3O^+})_{int}(a_{HCO_3^-})_{int}}{[CO_2(aq)]_{ext}}$$

where $[CO_2(aq)]_{ext}$ is the molar concentration of the gas in the analyte solution. For the measured cell potential to vary linearly with the logarithm of the carbon dioxide concentration of the external solution, the hydrogen carbonate activity of the internal solution must be sufficiently large that it is not altered significantly by the carbon dioxide entering from the external solution. Assuming then that $(a_{HCO_3^-})_{int}$ is constant, we can rearrange the previous equations to

$$\frac{(a_{H_3O^+})_{int}}{[CO_2(aq)]_{ext}} = \frac{K}{(a_{HCO_3^-})_{int}} = K_g$$

If we allow a_1 to be the hydrogen ion activity of the internal solution, we rearrange this equation to give

$$(a_{H_3O^+})_{int} = a_1 = K_g[CO_2(aq)]_{ext} \qquad (21\text{-}15)$$

By substituting Equation 21-15 into Equation 21-10, we find

$$E_{ind} = L + 0.0592 \log a_1 = L + 0.0592 \log K_g[CO_2(aq)]_{ext}$$
$$= L + 0.0592 \log K_g + 0.0592 \log [CO_2(aq)]_{ext}$$

Combining the two constant terms to give a new constant L' leads to

$$E_{ind} = L' + 0.0592 \log [CO_2(aq)]_{ext} \qquad (21\text{-}16)$$

Finally, since

$$E_{cell} = E_{ind} - E_{ref}$$

then

$$E_{cell} = L' + 0.0592 \log [CO_2(aq)]_{ext} - E_{ref} \qquad (21\text{-}17)$$

or

$$E_{cell} = L'' + 0.0592 \log [CO_2(aq)]_{ext}$$

where

$$L'' = L + 0.0592 \log K_g - E_{ref}$$

Thus, the potential between the glass electrode and the reference electrode in the internal solution is determined by the CO_2 concentration in the external solution. Note that no electrode comes in direct contact with the analyte solution. Therefore, these devices are gas-sensing cells, or probes, rather than gas-sensing electrodes. Nevertheless, they continue to be called electrodes is some literature and many advertising brochures.

> Although sold as gas-sensing electrodes, these devices are complete electrochemical cells and should be called gas-sensing probes.

The only species that interfere are other dissolved gases that permeate the membrane and then affect the pH of the internal solution. The specificity of gas probes depends only on the permeability of the gas membrane. Gas-sensing cells for CO_2, NO_2, H_2S, SO_2, HF, HCN, and NH_3 are now available from commercial sources.

FEATURE 21-3

Point-of-Care Testing: Blood Gases, and Blood Electrolytes with Portable Instrumentation

Modern medicine relies heavily on analytical measurements for diagnosis and treatment in emergency rooms, operating rooms, and intensive care units. Prompt reporting of blood gas values, blood electrolyte concentrations, and other variables is especially important to physicians in these areas. In critical life-and-death situations, there is seldom sufficient time to transport blood samples to the clinical laboratory, perform required analyses, and transmit the results back to the bedside. In this feature, we describe an automated blood gas and electrolyte monitor, designed specifically to analyze blood samples at the bedside.[7] The iSTAT® Portable Clinical Analyzer, shown in **Figure 21F-4**, is a handheld device that can measure a variety of important clinical analytes such as potassium, sodium, pH, pCO_2, pO_2, and hematocrit (see margin note). In addition, the computer-based analyzer calculates bicarbonate, total carbon dioxide, base excess, O_2 saturation, and hemoglobin in whole blood. In a study of the performance of the iSTAT system in a neonatal and pediatric intensive care unit, the results shown in the following table were obtained.[8] The results were judged to be sufficiently reliable and cost effective to substitute for similar measurements made in a traditional remote clinical laboratory.

Most of the analytes (pCO_2, Na^+, K^+, Ca^{2+}, and pH) are determined by potentiometric measurements using membrane-based ion-selective electrode technology. The hematocrit is measured by electrolytic conductivity detection and pO_2 is determined with a Clark voltammetric sensor (see Section 23C-4). Other results are calculated from these data.

The central component of the monitor is the single-use disposable electrochemical i-STAT sensor array, depicted in **Figure 21F-5**. The individual microfabricated sensor electrodes are located on chips along a narrow flow channel, as shown in the figure. Each new sensor array is automatically calibrated prior to the measurement step.

> Hematocrit (Hct) is the ratio of the volume of red blood cells to the total volume of a blood sample expressed as a percent.

Analyte	Range	Precision, %RSD	Resolution
pO_2	5–800 mm Hg	3.5	1 mm Hg
pCO_2	5–130 mm Hg	1.5	0.1 mm Hg
Na^+	100–180 mmol/L	0.4	1 mmol/L
K^+	2.0–9.0 mmol/L	1.2	0.1 mmol/L
Ca^{2+}	0.25–2.50 mmol/L	1.1	0.01 mmol/L
pH	6.5–8.0	0.07	0.001

Figure 21F-4   Photo of iSTAT 1 portable clinical analyzer. (Courtesy of Abbott Point of Care, Inc., Princeton, NJ.)

(continued)

Figure 21F-5   Exploded view of iSTAT sensor array cartridge. (Abbott Point of Care, Prinston, NJ. Reprinted by permission.)

Labels for Figure 21F-5: Cartridge label; Sample entry well gasket; Fluid channel; Cartridge cover; Sample entry well; Tape gasket; Biosensor chips; Calibrant pouch; Puncturing barb; Cartridge base; Air bladder

---

[7]Abbott Point of Care, Inc., Princeton, NJ 08540.

[8]J. N. Murthy, J. M. Hicks, and S. J. Soldin, *Clin. Biochem.,* **1997,** *30,* 385.

A blood sample withdrawn from the patient is deposited into the sample entry well, and the cartridge is inserted into the iSTAT analyzer. The calibrant pouch, which contains a standard buffered solution of the analytes, is punctured by the iSTAT analyzer and compressed to force the calibrant through the flow channel across the surface of the sensor array. When the calibration step is complete, the analyzer compresses the air bladder, which forces the blood sample through the flow channel to expel the calibrant solution to waste and bring the blood into contact with the sensor array. Electrochemical measurements are then made, results are calculated, and the data are presented on the liquid crystal display of the analyzer. The results are stored in the memory of the analyzer and may be transmitted wirelessly to the hospital laboratory data management system for permanent storage and retrieval.

This feature shows how modern ion-selective electrode technology coupled with computer control of the measurement process and data reporting can be used to provide rapid, essential measurements of analyte concentrations in whole blood at a patient's bedside.

21E INSTRUMENTS FOR MEASURING CELL POTENTIAL

Most cells containing a membrane electrode have very high electrical resistance (as much as 10^8 ohms or more). In order to measure potentials of such high-resistance circuits accurately, it is necessary that the voltmeter have an electrical resistance that is several orders of magnitude greater than the resistance of the cell being measured. If the meter resistance is too low, current is drawn from the cell, which has the effect of lowering its output potential, thus creating a negative *loading error*. When the meter and the cell have the same resistance a relative error of -50% results. When this ratio is 10, the error is about -9%. When it is 1000, the error is less than 0.1% relative.

FEATURE 21-4

The Loading Error in Potential Measurements

When we measure voltages in electrical circuits, the meter becomes a part of the circuit, perturbs the measurement process, and produces a **loading error** in the measurement. This situation is not unique to potential measurements. In fact, it is a basic example of a general limitation to any physical measurement. In other words, the process of measurement inevitably disturbs the system of interest so that the quantity actually measured differs from its value prior to the measurement. This type of error can never be completely eliminated, but it can often be reduced to an insignificant level.

The size of the loading error in potential measurements depends on the ratio of the internal resistance of the meter to the resistance of the circuit being studied. The percent relative loading error, E_r, associated with the measured potential, V_M, in **Figure 21F-6** is given by

$$E_r = \frac{V_M - V_x}{V_x} \times 100\%$$

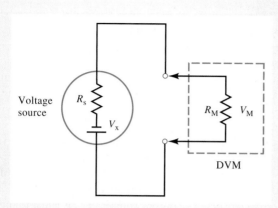

Figure 21F-6 Measurement of output V_x from a potential source with a digital voltmeter.

where V_x is the true voltage of the power source. The voltage drop across the resistance of the meter is given by

$$V_M = V_x \frac{R_M}{R_M + R_s}$$

Substituting this equation into the previous one and rearranging gives

$$E_r = \frac{-R_s}{R_M + R_s} \times 100\%$$

Note in this equation that the relative loading error becomes smaller as the meter resistance, R_M, becomes larger relative to the source resistance R_s. **Table 21F-1** illustrates this effect. Digital voltmeters offer the great advantage of having huge internal resistances (10^{11} to 10^{12} ohms), thus avoiding loading errors except in circuits having load resistances greater than about 10^9 ohms.

TABLE 21F-1

Effect of Meter Resistance on the Accuracy of Potential Measurements

Meter Resistance R_M, Ω	Resistance of Source R_s, Ω	R_M/R_s	Relative Error, %
10	20	0.50	−67
50	20	2.5	−29
500	20	25	−3.8
1.0×10^3	20	50	−2.0
1.0×10^4	20	500	−0.2

Numerous high-resistance, direct-reading digital voltmeters with internal resistances of $> 10^{11}$ ohms are now on the market. These meters are commonly called **pH meters** but could more properly be referred to **as pIon meters** or **ion meters** since they are frequently used for the measurement of concentrations of other ions as well. A photo of a typical pH meter is shown in **Figure 21-16**.

Figure 21-16 Photo of a typical benchtop pH meter. (Courtesy of Mettler Toledo, Inc., Columbus, OH.)

FEATURE 21-5

Operational Amplifier Voltage Measurements

One of the most important developments in chemical instrumentation over the last three decades has been the advent of compact, inexpensive, versatile integrated-circuit amplifiers (op amps).[9] These devices allow us to make potential measurements on high-resistance cells, such as those that contain a glass electrode, without drawing appreciable current. Even a small current (10^{-7}–10^{-10} A) in a glass electrode produces a large error in the measured voltage due to loading (see Feature 21-4) and electrode polarization (see Chapter 22). One of the most important uses for op amps is to isolate voltage sources from their measurement circuits. The basic **voltage follower**, which permits this type of measurement, is shown in **Figure 21F-7a**. This circuit has two important characteristics. The output voltage, E_{out}, is equal to the input voltage, E_{in}, and the input current, I_i, is essentially zero (10^{-7}–10^{-10} A).

A practical application of this circuit is in measuring cell potentials. We simply connect the cell to the op amp input, as shown in **Figure 21F-7b**, and we connect the output of the op amp to a digital voltmeter to measure the voltage. Modern op amps are nearly ideal voltage-measurement devices and are incorporated in most ion meters and pH meters to monitor high-resistance indicator electrodes with minimal error.

Modern ion meters are digital, and some are capable of a precision on the order of 0.001 to 0.005 pH unit. Seldom is it possible to measure pH with a comparable degree of *accuracy*. Inaccuracies of ±0.02 to ±0.03 pH unit are typical.

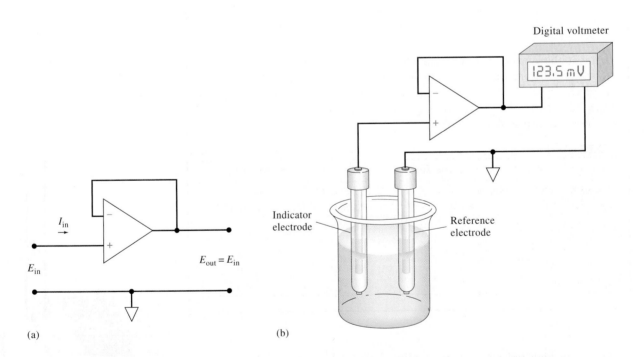

(a)   (b)

Figure 21F-7   (a) A voltage-follower operational amplifier. (b) Typical arrangement for potentiometric measurements with a membrane electrode.

---

[9]For a detailed description of op amp circuits, see H. V. Malmstadt, C. G. Enke, and S. R. Crouch, *Microcomputers and Electronic Instrumentation: Making the Right Connections*, Ch. 5, Washington, DC: American Chemical Society, 1994.

21F DIRECT POTENTIOMETRY

Direct potentiometric measurements provide a rapid and convenient method for determining the activity of a variety of cations and anions. The technique requires only a comparison of the potential developed in a cell containing the indicator electrode in the analyte solution with its potential when immersed in one or more standard solutions of known analyte concentration. If the response of the electrode is specific for the analyte, as it often is, no preliminary separation steps are required. Direct potentiometric measurements are also readily adapted to applications requiring continuous and automatic recording of analytical data.

21F-1 Equations Governing Direct Potentiometry

The sign convention for potentiometry is consistent with the convention described in Chapter 18 for standard electrode potential. In this convention, the indicator electrode is always treated as the right-hand electrode and the reference electrode as the left-hand electrode. For direct potentiometric measurements, the potential of a cell can then be expressed in terms of the potentials developed by the indicator electrode, the reference electrode, and a junction potential, as described in Section 21A:

$$E_{cell} = E_{ind} - E_{ref} + E_j \tag{21-18}$$

In Section 21D, we described the response of various types of indicator electrodes to analyte activities. For the cation X^{n+} at 25°C, the electrode response takes the general *Nernstian* form

$$E_{ind} = L - \frac{0.0592}{n} \, pX = L + \frac{0.0592}{n} \log a_X \tag{21-19}$$

where L is a constant and a_X is the activity of the cation. For metallic indicator electrodes, L is usually the standard electrode potential; for membrane electrodes, L is the summation of several constants, including the time-dependent asymmetry potential of uncertain magnitude.

Substitution of Equation 21-19 into Equation 21-18 yields with rearrangement

$$pX = -\log a_X = -\left[\frac{E_{cell} - (E_j - E_{ref} + L)}{0.0592/n} \right] \tag{21-20}$$

The constant terms in parentheses can be combined to give a new constant K.

$$pX = -\log a_X = -\frac{(E_{cell} - K)}{0.0592/n} = -\frac{n(E_{cell} - K)}{0.0592} \tag{21-21}$$

For an anion A^{n-}, the sign of Equation 21-21 is reversed:

$$pA = \frac{(E_{cell} - K)}{0.0592/n} = \frac{n(E_{cell} - K)}{0.0592} \tag{21-22}$$

All direct potentiometric methods are based on Equation 21-21 or 21-22. The difference in sign in the two equations has a subtle but important consequence in the way that ion-selective electrodes are connected to pH meters and pIon meters. When the two equations are solved for E_{cell}, we find that for cations

$$E_{cell} = K - \frac{0.0592}{n} pX \qquad (21\text{-}23)$$

and for anions

$$E_{cell} = K + \frac{0.0592}{n} pA \qquad (21\text{-}24)$$

Equation 21-23 shows that, for a cation-selective electrode, an increase in pX results in a *decrease* in E_{cell}. Thus, when a high-resistance voltmeter is connected to the cell in the usual way, with the indicator electrode attached to the positive terminal, the meter reading decreases as pX increases. Another way of saying this is that, as the concentration (and activity) of the cation X increases, pX = −log [X] decreases, and E_{cell} increases. Notice that the sense of these changes is exactly the opposite of our sense of how pH meter readings change with increasing hydronium ion concentration. To eliminate this reversal from our sense of the pH scale, instrument manufacturers generally reverse the leads so that cation-sensitive electrodes such as glass electrodes are connected to the negative terminal of the voltage measuring device. Meter readings then increase with increases of pX, and as a result, they decrease with increasing concentration of the cation.

Anion-selective electrodes, on the other hand, are connected to the positive terminal of the meter so that increases in pA also yield larger readings. This sign-reversal conundrum is often confusing so that it is always a good idea to look carefully at the consequences of Equations 21-23 and 21-24 rationalize the output of the instrument with changes in concentration of the analyte anion or cation and corresponding changes in pX or pA.

21F-2 The Electrode-Calibration Method

> The electrode-calibration method is also referred to as the method of external standards, which is described in some detail in Section 8D-2.

As we have seen from our discussions in Section 21D, the constant K in Equations 21-21 and 21-22 is made up of several constants, at least one of which, the junction potential, cannot be measured directly or calculated from theory without assumptions. Thus, before these equations can be used for the determination of pX or pA, K must be evaluated experimentally with a standard solution of the analyte.

In the electrode-calibration method, K in Equations 21-21 and 21-22 is determined by measuring E_{cell} for one or more standard solutions of known pX or pA. The assumption is then made that K is unchanged when the standard is replaced by the analyte solution. The calibration is normally performed at the time pX or pA for the unknown is determined. With membrane electrodes, recalibration may be required if measurements extend over several hours because of slow changes in the asymmetry potential.

The electrode-calibration method offers the advantages of simplicity, speed, and applicability to the continuous monitoring of pX or pA. It suffers, however, from a somewhat limited accuracy because of uncertainties in junction potentials.

Inherent Error in the Electrode-Calibration Procedure

A serious disadvantage of the electrode-calibration method is the inherent error that results from the assumption that K in Equations 21-21 and 21-22 remains constant after calibration. This assumption can seldom, if ever, be exactly true because the electrolyte composition of the unknown almost inevitably differs from that of the solution used for calibration. The junction potential term contained in K varies slightly as a consequence, even when a salt bridge is used. This error is frequently on the order of 1 mV or more. Unfortunately, because of the nature of the potential/activity relationship, such an uncertainty has an amplified effect on the inherent accuracy of the analysis.

The magnitude of the error in analyte concentration can be estimated by differentiating Equation 21-21 while assuming E_{cell} constant.

$$-\log_{10} e \frac{da_x}{a_x} = -0.434 \frac{da_x}{a_x} = -\frac{dK}{0.0592/n}$$

$$\frac{da_x}{a_x} = \frac{ndK}{0.0257} = 38.9 \, ndK$$

When we replace da_x and dK with finite increments and multiply both sides of the equation by 100%, we obtain

$$\text{percent relative error} = \frac{\Delta a_x}{a_x} \times 100\% = 38.9n\Delta K \times 100\%$$

$$= 3.89 \times 10^3 n\Delta K\% \approx 4000n\Delta K\%$$

The quantity $\Delta a_x/a_x$ is the relative error in a_x associated with an absolute uncertainty ΔK in K. If, for example, ΔK is ± 0.001 V, a relative error in activity of about $\pm 4n\%$ can be expected. *It is important to appreciate that this error is characteristic of all measurements involving cells that contain a salt bridge and that this error cannot be eliminated by even the most careful measurements of cell potentials or the most sensitive and precise measuring devices.*

Activity versus Concentration

Electrode response is related to analyte activity rather than analyte concentration. We are usually interested in concentration, however, and the determination of this quantity from a potentiometric measurement requires activity coefficient data. Activity coefficients are seldom available because the ionic strength of the solution either is unknown or else is so large that the Debye-Hückel equation is not applicable.

The difference between activity and concentration is illustrated by **Figure 21-17** in which the response of a calcium ion electrode is plotted against a logarithmic

Figure 21-17 Response of a liquid-membrane electrode to variations in the concentration and activity of calcium ion. (Courtesy of Thermo Electron Corp., Beverly, MA.)

function of calcium chloride concentration. The nonlinearity is due to the increase in ionic strength—and the consequent decrease in the activity of calcium ion—with increasing electrolyte concentration. The upper curve is obtained when these concentrations are converted to activities. This straight line has the theoretical slope of 0.0296 (0.0592/2).

Activity coefficients for singly charged species are less affected by changes in ionic strength than are the coefficients for ions with multiple charges. Thus, the effect shown in Figure 21-17 is less pronounced for electrodes that respond to H^+, Na^+, and other univalent ions.

In potentiometric pH measurements, the pH of the standard buffer used for calibration is generally based on the activity of hydrogen ions. Therefore, the results are also on an activity scale. If the unknown sample has a high ionic strength, the hydrogen ion *concentration* will differ appreciably from the activity measured.

An obvious way to convert potentiometric measurements from activity to concentration is to make use of an empirical calibration curve, such as the lower plot in Figure 21-17. For this approach to be successful, it is necessary to make the ionic composition of the standards essentially the same as that of the analyte solution. Matching the ionic strength of standards to that of samples is often difficult, particularly for samples that are chemically complex.

Where electrolyte concentrations are not too great, it is often useful to swamp both samples and standards with a measured excess of an inert electrolyte. The added effect of the electrolyte from the sample matrix becomes negligible under these circumstances, and the empirical calibration curve yields results in terms of concentration. This approach has been used, for example, in the potentiometric determination of fluoride ion in drinking water. Both samples and standards are diluted with a solution that contains sodium chloride, an acetate buffer, and a citrate buffer; the diluent is sufficiently concentrated so that the samples and standards have essentially identical ionic strengths. This method provides a rapid means for measuring fluoride concentrations in the part-per-million range with an accuracy of about 5% relative.

> Many chemical reactions of physiological importance depend on the activity of metal ions rather than their concentration.

A total ionic strength adjustment buffer (TISAB) is used to control the ionic strength and the pH of samples and standards, in ion-selective electrode measurements.

21F-3 The Standard Addition Method

The standard addition method (see Section 8D-3) involves determining the potential of the electrode system before and after a measured volume of a standard has been added to a known volume of the analyte solution. Multiple additions can also be made. Often, an excess of an electrolyte is introduced into the analyte solution to prevent any major shift in ionic strength that might accompany the addition of standard. It is also necessary to assume that the junction potential remains constant during the two measurements.

EXAMPLE 21-1

A cell consisting of a saturated calomel electrode and a lead ion electrode developed a potential of -0.4706 V when immersed in 50.00 mL of a sample. A 5.00-mL addition of standard 0.02000 M lead solution caused the potential to shift to -0.4490 V. Calculate the molar concentration of lead in the sample.

Solution

We shall assume that the activity of Pb^{2+} is approximately equal to $[Pb^{2+}]$ and apply Equation 21-21. Thus,

$$pPb = -\log[Pb^{2+}] = -\frac{E'_{cell} - K}{0.0592/2}$$

where E'_{cell} is the initial measured potential (-0.4706 V).

After the standard solution is added, the potential becomes E''_{cell} (-0.4490 V), and

$$-\log\frac{50.00 \times [Pb^{2+}] + 5.00 \times 0.0200}{50.00 + 5.00} = -\frac{E''_{cell} - K}{0.0592/2}$$

$$-\log(0.9091[Pb^{2+}] + 1.818 \times 10^{-3}) = -\frac{E''_{cell} - K}{0.0592/2}$$

Subtracting this equation from the first leads to

$$-\log\frac{[Pb^{2+}]}{0.09091[Pb^{2+}] + 1.818 \times 10^{-3}} = \frac{2(E''_{cell} - E'_{cell})}{0.0592}$$

$$= \frac{2[-0.4490 - (-0.4706)]}{0.0592}$$

$$= 0.7297$$

$$\frac{[Pb^{2+}]}{0.09091[Pb^{2+}] + 1.818 \times 10^{-3}} = \text{antilog}(-0.7297) = 0.1863$$

$$[Pb^{2+}] = 3.45 \times 10^{-4}\,M$$

21F-4 Potentiometric pH Measurement with the Glass Electrode[10]

The glass electrode is unquestionably the most important indicator electrode for hydrogen ion. It is convenient to use and subject to few of the interferences that affect other pH-sensing electrodes.

The glass/calomel electrode system is a remarkably versatile tool for the measurement of pH under many conditions. It can be used without interference in solutions containing strong oxidants, strong reductants, proteins, and gases; the pH of viscous or even semisolid fluids can be determined. Electrodes for special applications are available. Included among these electrodes are small ones for pH measurements in one drop (or less) of solution, in a tooth cavity, or in the sweat on the skin; microelectrodes that permit the measurement of pH inside a living cell; rugged electrodes for insertion in a flowing liquid stream to provide a continuous monitoring of pH; and small electrodes that can be swallowed to measure the acidity of the stomach contents (the calomel electrode is kept in the mouth).

Errors Affecting pH Measurements

The ubiquity of the pH meter and the general applicability of the glass electrode tend to lull the chemist into the attitude that any measurement obtained with such

---

[10]For a detailed discussion of potentiometric pH measurements, see R. G. Bates, *Determination of pH*, 2nd ed., New York: Wiley, 1973.

equipment is surely correct. The reader must be alert to the fact that there are distinct limitations to the electrode, some of which were discussed in earlier sections:

1. *The alkaline error.* The ordinary glass electrode becomes somewhat sensitive to alkali metal ions and gives low readings at pH values greater than 9.
2. *The acid error.* Values registered by the glass electrode tend to be somewhat high when the pH is less than about 0.5.
3. *Dehydration.* Dehydration may cause erratic electrode performance.
4. *Errors in low ionic strength solutions.* It has been found that significant errors (as much as 1 or 2 pH units) may occur when the pH of samples of low ionic strength, such as lake or stream water, is measured with a glass/calomel electrode system.[11] The prime source of such errors has been shown to be nonreproducible junction potentials, which apparently result from partial clogging of the fritted plug or porous fiber that is used to restrict the flow of liquid from the salt bridge into the analyte solution. To overcome this problem, free diffusion junctions of various types have been designed, one of which is produced commercially.
5. *Variation in junction potential.* A fundamental source of uncertainty for which a correction cannot be applied is the junction-potential variation resulting from differences in the composition of the standard and the unknown solution.
6. *Error in the pH of the standard buffer.* Any inaccuracies in the preparation of the buffer used for calibration or any changes in its composition during storage cause an error in subsequent pH measurements. The action of bacteria on organic buffer components is a common cause for deterioration.

> Particular care must be taken in measuring the pH of approximately neutral unbuffered solutions, such as samples from lakes and streams.

The Operational Definition of pH

> Perhaps the most common analytical instrumental technique is the measurement of pH.

The utility of pH as a measure of the acidity and alkalinity of aqueous media, the wide availability of commercial glass electrodes, and the relatively recent proliferation of inexpensive solid-state pH meters have made the potentiometric measurement of pH perhaps the most common analytical technique in all of science. It is thus extremely important that pH be defined in a manner that is easily duplicated at various times and in various laboratories throughout the world. To meet this requirement, it is necessary to define pH in operational terms, that is, by the way the measurement is made. Only then will the pH measured by one worker be the same as that by another.

> By definition, pH is what you measure with a glass electrode and a pH meter. It is approximately equal to the theoretical definition of pH = $-\log a_{H^+}$.

The operational definition of pH is endorsed by the National Institute of Standards and Technology (NIST), similar organizations in other countries, and the IUPAC. It is based on the direct calibration of the meter with carefully prescribed standard buffers followed by potentiometric determination of the pH of unknown solutions.

Consider, for example, one of the glass/reference electrode pairs of Figure 21-7. When these electrodes are immersed in a standard buffer, Equation 21-21 applies, and we can write

$$pH_S = \frac{E_S - K}{0.0592}$$

[11]See W. Davison and C. Woof, *Anal. Chem.*, **1985**, *57*, 2567, **DOI:** 10.1021/ac00290a031;
T. R. Harbinson and W. Davison, *Anal. Chem.*, **1987**, *59*, 2450, **DOI:** 10.1021/ac00147a002.

where E_S is the cell potential when the electrodes are immersed in the buffer. Similarly, if the cell potential is E_U when the electrodes are immersed in a solution of unknown pH, we have

$$pH_U = -\frac{E_U - K}{0.0592}$$

By subtracting the first equation from the second and solving for pH_U, we find

$$pH_U = pH_S - \frac{(E_U - E_S)}{0.0592} \tag{21-25}$$

Equation 21-25 has been adopted throughout the world as the *operational definition of pH*.

Workers at the NIST and elsewhere have used cells without liquid junctions to study primary-standard buffers extensively. Some of the properties of these buffers are discussed in detail elsewhere.[12] Note that the NIST buffers are described by their molal concentrations (mol solute/kg solvent) for accuracy and precision of preparation. For general use, the buffers can be prepared from relatively inexpensive laboratory reagents; for careful work, however, certified buffers can be purchased from the NIST.

It should be emphasized that the strength of the operational definition of pH is that it provides a coherent scale for the determination of acidity or alkalinity. However, measured pH values cannot be expected to yield a detailed picture of solution composition that is entirely consistent with solution theory. This uncertainty stems from our fundamental inability to measure single ion activities, that is, the operational definition of pH does not yield the exact pH as defined by the equation

$$pH = -\log \gamma_{H^+}[H^+]$$

> An operational definition of a quantity defines the quantity in terms of how it is measured.

21G POTENTIOMETRIC TITRATIONS

In a **potentiometric titration**, we measure the potential of a suitable indicator electrode as a function of titrant volume. The information provided by a potentiometric titration is different from the data obtained in a direct potentiometric measurement. For example, the direct measurement of 0.100 M solutions of hydrochloric and acetic acids yields two substantially different hydrogen ion concentrations because the weak acid is only partially dissociated. In contrast, the potentiometric titration of equal volumes of the two acids would require the same amount of standard base because both solutes have the same number of titratable protons.

Potentiometric titrations provide data that are more reliable than data from titrations that use chemical indicators and are particularly useful with colored or turbid solutions and for detecting the presence of unsuspected species. Potentiometric titrations have been automated in a variety of different ways, and commercial titrators are available from a number of manufacturers. Manual potentiometric titrations, on the other hand, suffer from the disadvantage of being more time consuming than those involving indicators.

---

[12]R. G. Bates, *Determination of pH*, 2nd ed., Ch. 4., New York: Wiley, 1973.

Automatic *titrators* for carrying out potentiometric titrations are available from several manufacturers. The operator of the instrument simply adds the sample to the titration vessel and pushes a button to initiate, the titration. The instrument adds titrant, records the potential versus volume data, and analyzes the data to determine the concentration of the unknown solution. A photo of such a device is shown on the opening page of Chapter 14.

Potentiometric titrations offer additional advantages over direct potentiometry. Because the measurement is based on the titrant volume that causes a rapid *change* in potential near the equivalence point, potentiometric titrations are not dependent on measuring absolute values of E_{cell}. This characteristic makes the titration relatively free from junction potential uncertainties because the junction potential remains approximately constant during the titration. Titration results, instead, depend most heavily on having a titrant of accurately known concentration. The potentiometric instrument merely signals the end point and thus behaves in an identical fashion to a chemical indicator. Problems with electrodes fouling or not displaying Nernstian response are not nearly as serious when the electrode system is used to monitor a titration. Likewise, the reference electrode potential does not need to be known accurately in a potentiometric titration. Another advantage of a titration is that the result is analyte concentration even though the electrode responds to activity. For this reason, ionic strength effects are not important in the titration procedure.

Figure 21-18 illustrate a typical apparatus for performing a manual potentiometric titration. The operator measures and records the cell potential (in units of millivolts or pH, as appropriate) after each addition of reagent. The titrant is added in large increments early in the titration and in smaller and smaller increments as the end point is approached (as indicated by larger changes in cell potential per unit volume).

21G-1 Detecting the End Point

Several methods can be used to determine the end point of a potentiometric titration. In the most straightforward approach, a direct plot or other recording is made of cell potential as a function of reagent volume. In **Figure 21-19a**, we plot the data of **Table 21-4** and visually estimate the inflection point in the steeply rising portion of the curve and take it as the end point.

Buret

pH meter with millivolt scale

Combination pH electrode

Magnetic stirrer

Figure 21-18   Apparatus for a potentiometric titration.

TABLE 21-4

Potentiometric Titration Data for 2.433 mmol of Chloride with 0.1000 M Silver Nitrate

Volume AgNO$_3$, mL	E vs. SCE, V	$\Delta E/\Delta V$, V/mL	$\Delta^2 E/\Delta V^2$, V^2/mL^2
5.00	0.062		
15.00	0.085	0.002	
20.00	0.107	0.004	
22.00	0.123	0.008	
23.00	0.138	0.015	
23.50	0.146	0.016	
23.80	0.161	0.050	
24.00	0.174	0.065	
24.10	0.183	0.09	
24.20	0.194	0.11	2.8
24.30	0.233	0.39	4.4
24.40	0.316	0.83	−5.9
24.50	0.340	0.24	−1.3
24.60	0.351	0.11	−0.4
24.70	0.358	0.07	
25.00	0.373	0.050	
25.50	0.385	0.024	
26.00	0.396	0.022	
28.00	0.426	0.015	

A second approach to end-point detection is to calculate the change in potential per unit volume of titrant ($\Delta E/\Delta V$), that is, we estimate the numerical first derivative of the titration curve. A plot of the first derivative data (see Table 21-4, column 3) as a function of the average volume V produces a curve with a maximum that corresponds to the point of inflection, as shown in **Figure 21-19b**. Alternatively, this ratio can be evaluated during the titration and recorded rather than the potential. From the plot, it can be seen that the maximum occurs at a titrant volume of about 24.30 mL. If the titration curve is symmetrical, the point of maximum slope coincides with the equivalence point. For the asymmetrical titration curves that are observed when the titrant and analyte half-reactions involve different numbers of electrons, a small titration error occurs if the point of maximum slope is used.

Figure 21-19c shows that the second derivative for the data changes sign at the point of inflection. This change is used as the analytical signal in some automatic titrators. The point at which the second derivative crosses zero is the inflection point, which is taken as the end point of the titration, and this point can be located quite precisely.

All of the methods of end-point detection discussed in the previous paragraphs are based on the assumption that the titration curve is symmetric about the equivalence point and that the inflection in the curve corresponds to this point. This assumption is valid if the titrant and analyte react in a 1:1 ratio and if the electrode reaction is reversible. Many oxidation/reduction reactions, such as the reaction of iron(II) with permanganate, do not occur in equimolar fashion. Even so, such titration curves are often so steep at the end point that very little error is introduced by assuming that the curves are symmetrical.

21G-2 Neutralization Titrations

Experimental neutralization curves closely approximate the theoretical curves described in Chapters 14 and 15. Usually, the experimental curves are somewhat displaced from the theoretical curves along the pH axis because concentrations rather than activities are used in their derivation. This displacement has little effect on determining end points, and so potentiometric neutralization titrations are quite useful for analyzing mixtures of acids or polyprotic acids. The same is true of bases.

Determining Dissociation Constants

An approximate numerical value for the dissociation constant of a weak acid or base can be estimated from potentiometric titration curves. This quantity can be computed from the pH at any point along the curve, but a very convenient point is the half-titration point. At this point on the curve,

$$[HA] \approx [A^-]$$

Therefore,

$$K_a = \frac{[H_3O^+][A^-]}{[HA]} = [H_3O^+]$$

$$pK_a = pH$$

(a)

(b)

(c)

Figure 21-19 Titration of 2.433 mmol of chloride ion with 0.1000 M silver nitrate. (a) Titration curve. (b) First-derivative curve. (c) Second-derivative curve.

It is important to note the use of concentrations instead of activities may cause the value for K_a to differ from its published value by a factor of 2 or more. A more correct form of the dissociation constant for HA is

$$K_a = \frac{a_{H_3O^+}a_{A^-}}{a_{HA}} = \frac{a_{H_3O^+}\gamma_{A^-}[A^-]}{\gamma_{HA}[HA]} \tag{21-26}$$

$$K_a = \frac{a_{H_3O^+}\gamma_{A^-}}{\gamma_{HA}}$$

Since the glass electrode provides a good approximation of $a_{H_3O^+}$, the measured value of K_a differs from the thermodynamic value by the ratio of the two activity coefficients. The activity coefficient in the denominator of Equation 21-26 doesn't change significantly as ionic strength increases because HA is a neutral species. The activity coefficient for A^-, on the other hand, decreases as the electrolyte concentration increases. This decrease means that the observed hydrogen ion activity must be numerically larger than the thermodynamic dissociation constant.

EXAMPLE 21-2

In order to determine K_1 and K_2 for H_3PO_4 from titration data, careful pH measurements are made after 0.5 and 1.5 mol of base is added for each mole of acid. It is then assumed that the hydrogen ion activities computed from these data are identical to the desired dissociation constants. Calculate the relative error incurred by the assumption if the ionic strength is 0.1 at the time of each measurement. (From Appendix 3, K_1 and K_2 for H_3PO_4 are 7.11×10^{-3} and 6.34×10^{-8}, respectively.)

Solution

If we rearrange Equation 21-26, we find that

$$K_a(\text{exptl}) = a_{H_3O^+} = K\left(\frac{\gamma_{HA}}{\gamma_{A^-}}\right)$$

The activity coefficient for H_3PO_4 is approximately equal to 1 since the free acid has no charge. In Table 10-2, we find that the activity coefficient for $H_2PO_4^-$ is 0.77 and that for HPO_4^{2-} is 0.35. When we substitute these values into the equations for K_1 and K_2, we find that

$$K_1(\text{exptl}) = 7.11 \times 10^{-3}\left(\frac{1.00}{0.77}\right) = 9.23 \times 10^{-3}$$

$$\text{error} = \frac{9.23 \times 10^{-3} - 7.11 \times 10^{-3}}{7.11 \times 10^{-3}} \times 100\% = 30\%$$

$$K_2(\text{exptl}) = 6.34 \times 10^{-8}\left(\frac{0.77}{0.35}\right) = 1.395 \times 10^{-7}$$

$$\text{error} = \frac{1.395 \times 10^{-7} - 6.34 \times 10^{-8}}{6.34 \times 10^{-8}} \times 100\% = 120\%$$

It is possible to identify an unknown pure acid by performing a single titration to determine its equivalent mass (molar mass if the acid is monoprotic) and its dissociation constant.

21G-3 Oxidation/Reduction Titrations

An inert indicator electrode constructed of platinum is usually used to detect end points in oxidation/reduction titrations. Occasionally, other inert metals, such a silver, palladium, gold, and mercury, are used instead. Titration curves similar to those constructed in Section 19D are usually obtained, although they may be displaced along the potential (vertical) axis as a consequence of the high ionic strengths. End points are determined by the methods described earlier in this chapter.

21H POTENTIOMETRIC DETERMINATION OF EQUILIBRIUM CONSTANTS

Numerical values for solubility-product constants, dissociation constants, and formation constants are conveniently evaluated through the measurement of cell potentials. One important virtue of this technique is that the measurement can be made without appreciably affecting any equilibria that may be present in the solution. For example, the potential of a silver electrode in a solution containing silver ion, cyanide ion, and the complex formed between them depends on the activities of the three species. It is possible to measure this potential with negligible current. Since the activities of the participants are not altered during the measurement, the position of the equilibrium

$$Ag^+ + 2CN^- \rightleftharpoons Ag(CN)_2^-$$

is likewise undisturbed.

EXAMPLE 21-3

Calculate the formation constant K_f for $Ag(CN)_2^-$:

$$Ag^+ + 2CN^- \rightleftharpoons Ag(CN)_2^-$$

if the cell

$$SCE \,\|\, Ag(CN)_2^-\,(7.50 \times 10^{-3}\,M),\ CN^-\,(0.0250\,M) \,|\, Ag$$

develops a potential of -0.625 V.

Solution

Proceeding as in the earlier examples, we have

$$Ag^+ + e^- \rightleftharpoons Ag(s) \qquad E^0 = +0.799\,V$$

$$-0.625 = E_{right} - E_{left} = E_{Ag^+} - 0.244$$

$$E_{Ag^+} = -0.625 + 0.244 = -0.381\,V$$

(continued)

We then apply the Nernst equation for the silver electrode to find that

$$-0.381 = 0.799 - \frac{0.0592}{1} \log \frac{1}{[Ag^+]}$$

$$\log [Ag^+] = \frac{-0.381 - 0.799}{0.0592} = -19.93$$

$$[Ag^+] = 1.2 \times 10^{-20}$$

$$K_f = \frac{[Ag(CN)_2^-]}{[Ag^+][CN^-]^2} = \frac{7.50 \times 10^{-3}}{(1.2 \times 10^{-20})(2.5 \times 10^{-2})^2}$$

$$= 1.0 \times 10^{21} \approx 1 \times 10^{21}$$

In theory, any electrode system in which hydrogen ions are participants can be used to evaluate dissociation constants for acids and bases.

EXAMPLE 21-4

Calculate the dissociation constant K_{HP} for the weak acid HP if the cell

$$SCE \parallel HP(0.010\ M),\ NaP(0.040\ M) \mid Pt,\ H_2\ (1.00\ atm)$$

develops a potential of -0.591 V.

Solution

The diagram for this cell indicates that the saturated calomel electrode is the left-hand electrode. Thus,

$$E_{cell} = E_{right} - E_{left} = E_{right} - 0.244 = -0.591\ V$$

$$E_{right} = -0.591 + 0.244 = -0.347\ V$$

We then apply the Nernst equation for the hydrogen electrode to find that

$$-0.347 = 0.000 - \frac{0.0592}{2} \log \frac{1.00}{[H_3O^+]^2}$$

$$= 0.000 + \frac{2 \times 0.0592}{2} \log [H_3O^+]$$

$$\log [H_3O^+] = \frac{-0.347 - 0.000}{0.0592} = -5.86$$

$$[H_3O^+] = 1.38 \times 10^{-6}$$

By substituting this value of the hydronium ion concentration as well as the concentrations of the weak acid and its conjugate base into the dissociation constant expression, we obtain

$$K_{HP} = \frac{[H_3O^+][P^-]}{HP} = \frac{(1.38 \times 10^{-6})(0.040)}{0.010} = 5.5 \times 10^{-6}$$

Use a Web search engine, such as Google, to find sites dealing with potentiometric titrators. This search should turn up such companies as Spectralab, Analyticon, Fox Scientific, Metrohm, Mettler-Toledo, and Thermo Orion. Set your browser to one or two of these and explore the types of titrators that are commercially available. At the sites of two different manufacturers, find application notes or bulletins for determining two analytes by potentiometric titration. For each, list the analyte, the instruments and the reagents that are necessary for the determination, and the expected accuracy and precision of the results. Describe the detailed chemistry behind each determination and the experimental procedure.

QUESTIONS AND PROBLEMS

21-1. Briefly describe or define
  *(a) indicator electrode.
  (b) reference electrode.
  *(c) electrode of the first kind.
  (d) electrode of the second kind.

21-2. Briefly describe or define
  *(a) liquid-junction potential.
  (b) boundary potential.
  *(c) asymmetry potential.
  (d) combination electrode.

*21-3. You need to choose between determining an analyte by measuring an electrode potential or by performing a titration. Explain which you would choose if you needed to know
  (a) the absolute amount of the analyte to a few parts per thousand.
  (b) the activity of the analyte.

21-4. What is meant by Nernstian behavior in an indicator electrode?

*21-5. Describe the source of pH dependence in a glass membrane electrode.

21-6. Why is it necessary for the glass in the membrane of a pH-sensitive electrode to be appreciably hygroscopic?

*21-7. List several sources of uncertainty in pH measurements with a glass/calomel electrode system.

21-8. What experimental factor places a limit on the number of significant figures in the response of a membrane electrode?

*21-9. Describe the alkaline error in the measurement of pH. Under what circumstances is this error appreciable? How are pH data affected by alkaline error?

21-10. How does a gas-sensing probe differ from other membrane electrodes?

21-11. What is the source of
  (a) the asymmetry potential in a membrane electrode?
  *(b) the boundary potential in a membrane electrode?
  (c) a junction potential in a glass/calomel electrode system?
  *(d) the potential of a crystalline membrane electrode used to determine the concentration of F^-?

*21-12. How does information supplied by a direct potentiometric measurements of pH differ from that obtained from a potentiometric acid/base titration?

21-13. Give several advantages of a potentiometric titration over a direct potentiometric measurement.

21-14. What is the "operational definition of pH"? Why is it used?

*21-15. (a) Calculate E^0 for the process

$$AgIO_3(s) + e^- \rightleftharpoons Ag(s) + IO_3^-$$

  (b) Use the shorthand notation to describe a cell consisting of a saturated calomel reference electrode and a silver indicator electrode that could be used to measure pIO_3.
  (c) Develop an equation that relates the potential of the cell in (b) to pIO_3.
  (d) Calculate pIO_3 if the cell in (b) has a potential of 0.306 V.

21-16. (a) Calculate E^0 for the process

$$PbI_2(s) + e^- \rightleftharpoons Pb(s) + 2I^-$$

  (b) Use the shorthand notation to describe a cell consisting of a saturated calomel reference electrode and a lead indicator electrode that could be used for the measurement of pI.
  (c) Generate an equation that relates the potential of this cell to pI.
  (d) Calculate pI if this cell has a potential of –0.402 V.

21-17. Use the shorthand notation to describe a cell consisting of a saturated calomel reference electrode and a silver indicator electrode for the measurement of
  *(a) pI.
  (b) pSCN.
  *(c) pPO_4.
  (d) pSO_3.

21-18. Generate an equation that relates pAnion to E_{cell} for each of the cells in Problem 21-17. (For Ag_2SO_3, $K_{sp} = 1.5 \times 10^{-14}$; for Ag_3PO_4, $K_{sp} = 1.3 \times 10^{-20}$.)

21-19. Calculate
*(a) pI if the cell in Problem 21-17(a) has a potential of −196 mV.
(b) pSCN if the cell in Problem 21-17(b) has a potential of 0.137 V.
*(c) pPO$_4$ if the cell in Problem 21-17(c) has a potential of 0.211 V.
(d) pSO$_3$ if the cell in Problem 21-17(d) has a potential of 285 mV.

*21-20. The cell

$$SCE\|Ag_2CrO_4(sat'd), (x\ M)|Ag$$

is used for the determination of pCrO$_4$. Calculate pCrO$_4$ when the cell potential is 0.389 V.

*21-21. The cell

$$SCE\|H^+(a = x)|glass\ electrode$$

has a potential of 0.2106 V when the solution in the right-hand compartment is a buffer of pH 4.006. The following potentials are obtained when the buffer is replaced with unknowns: (a) −0.2902 V and (b) +0.1241 V. Calculate the pH and the hydrogen ion activity of each unknown. (c) Assuming an uncertainty of 0.002 V in the junction potential, what is the range of hydrogen ion activities within which the true value might be expected to lie?

*21-22. A 0.4021-g sample of a purified organic acid was dissolved in water and titrated potentiometrically. A plot of the data revealed a single end point after 18.62 mL of 0.1243 M NaOH had been introduced. Calculate the molecular mass of the acid.

21-23. Calculate the potential of a silver indicator electrode versus the standard calomel electrode after the addition of 5.00, 15.00, 25.00, 30.00, 35.00, 39.00, 39.50, 36.60, 39.70, 39.80, 39.90, 39.95, 39.99, 40.00, 40.01, 40.05, 40.10, 40.20, 40.30, 40.40, 40.50, 41.00, 45.00, 50.00, 55.00, and 70.00 mL of 0.1000 M AgNO$_3$ to 50.00 mL of 0.0800 M KSeCN. Construct a titration curve and a first and second derivative plot from these data. (K_{sp} for AgSeCN = 4.20×10^{-16}.)

21-24. A 40.00-mL aliquot of 0.05000 M HNO$_2$ is diluted to 75.00 mL and titrated with 0.0800 M Ce^{4+}. The pH of the solution is maintained at 1.00 throughout the titration; the formal potential of the cerium system is 1.44 V.
(a) Calculate the potential of the indicator electrode with respect to a saturated calomel reference electrode after the addition of 5.00, 10.00, 15.00, 25.00, 40.00, 49.00, 49.50, 49.60, 49.70, 49.80, 49.90, 49.95, 49.99, 50.00, 50.01, 50.05, 50.10, 50.20, 50.30, 50.40, 50.50, 51.00, 60.00, 75.00, and 90.00 mL of cerium(IV).

(b) Draw a titration curve for these data.
(c) Generate a first and second derivative curve for these data. Does the volume at which the second derivative curve crosses zero correspond to the theoretical equivalence point? Why or why not?

21-25. The titration of Fe(II) with permanganate yields a particularly asymmetrical titration curve because of the different number of electrons involved in the two half-reactions. Consider the titration of 25.00 mL of 0.1 M Fe(II) with 0.1 M MnO$_4^-$. The H$^+$ concentration is maintained at 1.0 M throughout the titration. Use a spreadsheet to generate a theoretical titration curve and a first and second derivative plot. Do the inflection points obtained from the maximum of the first derivative plot or the zero crossing of the second derivative plot correspond to the equivalence point? Explain why or why not.

*21-26. The Na$^+$ concentration of a solution was determined by measurements with a sodium ion-selective electrode. The electrode system developed a potential of −0.2462 V when immersed in 10.0 mL of the solution of unknown concentration. After addition of 1.00 mL of 2.00×10^{-2} M NaCl, the potential changed to −0.1994 V. Calculate the Na$^+$ concentration of the original solution.

21-27. The F$^-$ concentration of a solution was determined by measurements with a liquid-membrane electrode. The electrode system developed a potential of 0.5021 V when immersed in 25.00 mL of the sample, and 0.4213 V after the addition of 2.00 mL of 5.45×10^{-2} M NaF. Calculate pF for the sample.

21-28. A lithium ion-selective electrode gave the potentials given below for the following standard solutions of LiCl and two samples of unknown concentration:

Solution (a_{Li^+})	Potential vs. SCE, mV
0.100 M	+1.0
0.050 M	−30.0
0.010 M	−60.0
0.001 M	−138.0
Unknown 1	−48.5
Unknown 2	−75.3

(a) Construct a calibration curve of potential versus log a_{Li^+} and determine if the electrode follows the Nernst equation.
(b) Use a linear least-squares procedure to determine the concentrations of the two unknowns.

21-29. A fluoride electrode was used to determine the amount of fluoride in drinking water samples. The results given in the table below were obtained for four standards and two unknowns. Constant ionic strength and pH conditions were used.

Solution Containing F^-	Potential vs. SCE, mV
5.00×10^{-4} M	0.02
1.00×10^{-4} M	41.4
5.00×10^{-5} M	61.5
1.00×10^{-5} M	100.2
Unknown 1	38.9
Unknown 2	55.3

(a) Plot a calibration curve of potential versus $\log[F^-]$. Determine whether the electrode system shows Nernstian response.

(b) Determine the concentration of F^- in the two unknown samples by a linear least-squares procedure.

21-30. **Challenge Problem:** Ceresa, Pretsch, and Bakker[13] investigated three ion-selective electrodes for determining calcium concentrations. All three electrodes used the same membrane but differed in the composition of the inner solution. Electrode 1 was a conventional ISE with an inner solution of 1.00×10^{-3} M $CaCl_2$ and 0.10 M NaCl. Electrode 2 (low activity of Ca^{2+}) had an inner solution containing the same analytical concentration of $CaCl_2$, but with 5.0×10^{-2} M EDTA adjusted to a pH of 9.0 with 6.0×10^{-2} M NaOH. Electrode 3 (high Ca^{2+} activity) had an inner solution of 1.00 M $Ca(NO_3)_2$.

(a) Determine the Ca^{2+} concentration in the inner solution of Electrode 2.

(b) Determine the ionic strength of the solution in Electrode 2.

(c) Use the Debye-Hückel equation and determine the activity of Ca^{2+} in Electrode 2. Use 0.6 nm for the α_X value for Ca^{2+}.

(d) Electrode 1 was used in a cell with a calomel reference electrode to measure standard calcium solutions with activities ranging from 0.001 M to 1.00×10^{-9} M. The following data were obtained:

Activity of Ca^{2+}, M	Cell Potential, mV
1.0×10^{-3}	93
1.0×10^{-4}	73
1.0×10^{-5}	37
1.0×10^{-6}	2
1.0×10^{-7}	-23
1.0×10^{-8}	-51
1.0×10^{-9}	-55

Plot the cell potential versus the pCa and determine the pCa value where the plot deviates

significantly from linearity. For the linear portion, determine the slope and intercept of the plot. Does the plot obey the expected Equation 21-23?

(e) For Electrode 2, the following results were obtained:

Activity of Ca^{2+}	Cell Potential, V
1.0×10^{-3}	228
1.0×10^{-4}	190
1.0×10^{-5}	165
1.0×10^{-6}	139
5.6×10^{-7}	105
3.2×10^{-7}	63
1.8×10^{-7}	36
1.0×10^{-7}	23
1.0×10^{-8}	18
1.0×10^{-9}	17

Again, plot cell potential versus pCa and determine the range of linearity for Electrode 2. Determine the slope and intercept for the linear portion. Does this electrode obey Equation 21-23 for the higher Ca^{2+} activities?

(f) Electrode 2 is said to be super-Nernstian for concentrations from 10^{-7} M to 10^{-6} M. Why is this term used? If you have access to a library that subscribes to *Analytical Chemistry* or has Web access to the journal, read the article. This electrode is said to have Ca^{2+} uptake. What does this mean and how might it explain the response?

(g) Electrode 3 gave the following results:

Activity of Ca^{2+}, M	Cell Potential, mV
1.0×10^{-3}	175
1.0×10^{-4}	150
1.0×10^{-5}	123
1.0×10^{-6}	88
1.0×10^{-7}	75
1.0×10^{-8}	72
1.0×10^{-9}	71

Plot the cell potential versus pCa and determine the range of linearity. Again, determine the slope and intercept. Does this electrode obey Equation 21-23?

(h) Electrode 3 is said to have Ca^{2+} release. Explain this term from the article and describe how it might explain the response.

(i) Does the article give any alternative explanations for the experimental results? If so, describe these alternatives.

[13]A. Ceresa, E. Pretsch, and E. Bakker, *Anal. Chem.,* **2000**, *72*, 2054, **DOI:** 10.1021/ac991092h.

CHAPTER 22

Bulk Electrolysis: Electrogravimetry and Coulometry

Sue Ogrocki/Reuters/Corbis

Electrolysis is widely used commercially to provide attractive metal coverings to objects such as car bumpers, which are chromium plated; silverware, which is often silver-plated; and jewelry, which may be electroplated with various precious metals. Another example of an electroplated object is the Oscar (shown in the photograph) which is given to recipients of Academy Awards. Each Oscar stands 13.5 inches tall, not including the base, and weighs 8.5 pounds. The statuette is hand cast in brittanium, an alloy of tin, copper, and antimony, in a steel mold. The cast is then electroplated with copper. Nickel electroplating is applied to seal the pores of the metal. The statuette is then washed in a silver plate, which adheres well to gold. Finally, after polishing, the statuette is electroplated with 24-karat gold and finished in a baked lacquer. The amount of gold deposited on an Oscar could be determined by weighing the statuette before and after the final electrolysis step. This technique, called electrogravimetry, is one of the subjects of this chapter. Alternatively, the current during the electroplating process could be integrated to find the total amount of charge required to electroplate Oscar. The number of moles of electrons needed could then be used to calculate the mass of gold deposited. This method, known as coulometry, is also a subject of this chapter.

I n this chapter, we describe two bulk electroanalytical methods: electrogravimetry, and coulometry.[1] Unlike the potentiometric methods described in Chapters 18 through 21, these methods are electrolytic methods in which there is a net current and a net cell reaction. Electrogravimetry and coulometry are related methods that are based on electrolysis carried out long enough to ensure complete oxidation or reduction of the analyte to a product of known composition. In electrogravimetry, the goal is to determine the amount of analyte present by converting it to a product that is weighed as a deposit on one of the electrodes. In coulometric procedures, we determine the amount of analyte by measuring the quantity of electrical charge needed for complete conversion to a product.

Electrogravimetry and coulometry are moderately sensitive and among the most accurate and precise techniques available. Like the gravimetric techniques discussed in Chapter 12, electrogravimetry requires no preliminary calibration against chemical standards because

> Electrogravimetry and coulometry can often exhibit accuracies of a few parts per thousand.

---

[1]For further information concerning the methods in this chapter, see A. J. Bard and L. R. Faulkner, *Electrochemical Methods*, 2nd ed., Ch. 11, New York: Wiley, 2001; J. A. Dean, *Analytical Chemistry Handbook*, Section 14, pp. 14.93–14.133, New York: McGraw-Hill, 1995.

the functional relationship between the quantity measured and the analyte concentration can be derived from theory and atomic mass data.

We have not previously considered what happens when current is present in an electrochemical cell. Thus, we first discuss the effect of current in a cell. Then, we describe the bulk electrolysis methods in detail. The voltammetric methods described in Chapter 23 also require a net current in the cell, but they use such small electrode areas that no appreciable changes in bulk concentrations occur.

22A THE EFFECT OF CURRENT ON CELL POTENTIAL

When there is a net current in an electrochemical cell, the measured potential across the two electrodes is no longer simply the difference between the two electrode potentials as calculated from the Nernst equation. Two additional phenomena, **IR drop** and **polarization**, must be considered when current is present. Because of these phenomena, potentials larger than the thermodynamic potential are needed to operate an electrolytic cell. When present in a galvanic cell, IR drop and polarization result in the development of potentials smaller than predicted.

Let us now examine these two phenomena in detail. As an example, we will consider the following electrolytic cell for the determination of cadmium(II) in hydrochloric acid solutions by electrogravimetry or coulometry:

$$Ag|AgCl(s), Cl^-(0.200\ M), Cd^{2+}(0.00500\ M)|Cd$$

Similar cells can be used to determine Cu(II) and Zn(II) in acid solution. In this cell, the right-hand electrode is a metal electrode that has been coated with a layer of cadmium. Because this is the electrode at which the reduction of Cd^{2+} ions occurs, this **working electrode** operates as a cathode. The left-hand electrode is a silver/silver chloride electrode whose electrode potential remains nearly constant during the analysis. The left-hand electrode is thus the **reference electrode**. Note that this is an example of a cell without liquid junction. As shown in Example 22-1, this cell, as written, has a thermodynamic potential of -0.734 V. The negative sign for the cell potential indicates that the spontaneous reaction is *not* the reduction of Cd^{2+} on the right and the oxidation of Ag on the left. In order to reduce Cd^{2+} to Cd we must construct an electrolytic cell and *apply* a potential somewhat more negative than -0.734 V. Such a cell is shown in **Figure 22-1**. Applying a potential more negative than the thermodynamic potential, we force the Cd electrode to become the cathode and cause the net reaction shown in Equation 22-1 to occur in the left-to-right direction.

$$Cd^{2+} + 2Ag(s) + 2Cl^- \rightarrow Cd(s) + 2AgCl(s) \qquad (22\text{-}1)$$

Note that this cell is reversible so that in the absence of the external voltage source shown in the figure, the spontaneous cell reaction is in the right-to-left direction (oxidation of $Cd(s)$ to Cd^{2+}). If the spontaneous reaction were allowed to occur by short-circuiting the galvanic cell, the Cd electrode would be the anode.

22A-1 Ohmic Potential: IR Drop

Electrochemical cells, like metallic conductors, resist the flow of charge. Ohm's law describes the effect of this resistance on the magnitude of the current in the cell. The product of the resistance R of a cell in ohms (Ω) and the current I in amperes (A) is called the ohmic potential or the *IR* drop of the cell.

André Marie Ampère (1775–1836), French mathematician and physicist, was the first to apply mathematics to the study of electrical current. Consistent with Benjamin Franklin's definitions of positive and negative charge, Ampère defined a positive current to be the direction of flow of positive charge. Although we now know that negative electrons carry current in metals, Ampère's definition has survived to the present. The unit of current, the ampere, is named in his honor.

Current is the rate of charge flow in a circuit or solution. One ampere of current is a charge flow rate of one coulomb per second (1 A = 1 C/s). **Voltage**, the electrical potential difference, is the potential energy that results from the separation of charges. One volt of electrical potential results when one joule of potential energy is required to separate one coulomb of charge (1 V = 1 J/C).

◀ Ohm's law: $E = IR$, or $I = E/R$. The units of resistance are ohms (Ω). One ohm equals one volt per ampere. Thus, the product IR has the units of amperes × volts/ampere = volts.

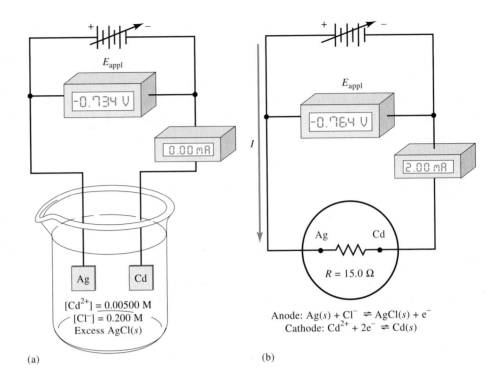

Figure 22-1   An electrolytic cell for determining Cd^{2+}. (a) Current = 0.00 mA. (b) Schematic of cell in (a) with internal resistance of cell represented by a 15.0-Ω resistor and $E_{applied}$ increased to give a current of 2.00 mA.

Direct current (dc) is current that is always in one direction, that is, unidirectional. The direction of **alternating current (ac)** reverses periodically. We can also speak of voltage sources that are unidirectional (dc) or of alternating polarity (ac). The terms ac and dc are also used to describe power supplies, circuits, and components designed for alternating or unipolar operation, respectively. A dc voltage source is often given the battery symbol with + and − polarities indicated like those shown in Figure 22-1. An arrow through the battery indicates that the source voltage can be changed to another dc value.

In Figure 22-1b, we have used a resistor R to represent the cell resistance in Figure 22-1a. In order to generate a current of I amperes in this cell, we must apply a potential that is IR volts more negative than the thermodynamic cell potential, $E_{cell} = E_{right} - E_{left}$, that is,

$$E_{applied} = E_{cell} - IR \qquad (22\text{-}2)$$

Usually we try to minimize the IR drop in the cell by having a very small cell resistance (high ionic strength) or by using a special **three-electrode cell** (see Section 22C-2) in which the current passes between the working electrode and an **auxiliary**, or **counter**, **electrode**. With this arrangement, only a very small current passes between the working electrode and the reference electrode, therefore minimizing the IR drop.

EXAMPLE 22-1

The following cell has been used for the determination of cadmium in the presence of chloride ions by both electrogravimetry and coulometry:

$$Ag\,|\,AgCl(s),\ Cl^-(0.200\ M),\ Cd^{2+}(0.00500\ M)\,|\,Cd$$

Calculate the potential that (a) must be applied to prevent a current from developing in the cell when the two electrodes are connected and (b) that must be applied to cause an electrolytic current of 2.00 mA to develop. Assume that the internal resistance of the cell is 15.0 Ω.

Solution

(a) In Appendix 5, we find the following standard reduction potentials:

$$Cd^{2+} + 2e^- \rightleftharpoons Cd(s) \qquad E^0 = -0.403 \text{ V}$$

$$AgCl(s) + e^- \rightleftharpoons Ag(s) + Cl^- \qquad E^0 = 0.222 \text{ V}$$

The potential of the cadmium electrode is

$$E_{\text{right}} = -0.403 - \frac{0.0592}{2} \log \frac{1}{0.00500} = -0.471 \text{ V}$$

and that of the silver electrode is

$$E_{\text{left}} = 0.222 - 0.0592 \log (0.200) = 0.263 \text{ V}$$

Since the current is to be 0.00 mA, we find from Equation 22-2,

$$E_{\text{applied}} = E_{\text{cell}} = E_{\text{right}} - E_{\text{left}}$$

$$= -0.471 - 0.263 = -0.734 \text{ V}$$

To prevent the passage of current in this cell, we would need to apply a voltage of -0.734 V, as shown in Figure 22-1a. Note that in order to obtain a current of 0.00 mA, the applied voltage must exactly match the galvanic cell potential. This requirement is the basis for a very precise null comparison measurement of the galvanic cell potential. We use a variable, standard voltage source as the applied voltage and adjust its output until a current of 0.00 mA. At this *null point*, the standard voltage is read on a voltmeter to obtain the value of E_{cell}. Since there is no current, this type of voltage measurement prevents the loading error discussed in Section 21E.

(b) In order to calculate the applied potential needed to develop a current of 2.00 mA, or 2.00×10^{-3} A, we substitute into Equation 22-2 to give

$$E_{\text{applied}} = E_{\text{cell}} - IR$$

$$= -0.734 - 2.00 \times 10^{-3} \text{ A} \times 15 \text{ } \Omega$$

$$= -0.734 - 0.030 = -0.764 \text{ V}$$

We see that to obtain a 2.00 mA current as in Figure 22-1b, an applied potential of -0.764 V is required.

22A-2 Polarization Effects

If we solve Equation 22-2 for the current I, we obtain

$$I = \frac{E_{\text{cell}} - E_{\text{applied}}}{R} = -\frac{E_{\text{applied}}}{R} + \frac{E_{\text{cell}}}{R} \qquad (22\text{-}3)$$

Note that a plot of current in an electrolyte cell versus applied potential should be a straight line with a slope equal to the negative reciprocal of the resistance, $-1/R$, and an intercept equal to E_{cell}/R. As can be seen in **Figure 22-2**, the plot is indeed linear for small currents. In this experiment, the measurements were made in a brief

Figure 22-2 Experimental current/voltage curve for operation of the cell shown in Figure 22-1. Dashed line is the theoretical curve assuming no polarization. Overvoltage Π is the potential difference between the theoretical curve and the experimental.

Polarization is the departure of the electrode potential from its theoretical Nernst equation value on the passage of current. **Overvoltage** is the potential difference between the theoretical cell potential from Equation 22-2 and the actual cell potential at a given level of current.

> Factors that influence polarization are (1) electrode size, shape, and composition; (2) composition of the electrolyte solution; (3) temperature and stirring rate; (4) current level; and (5) physical state of the species participating in the cell reaction.

Mass transfer is the movement of material, such as ions, from one location to another.

enough time so that neither electrode potential is changed significantly as a consequence of the electrolytic reaction. As the applied voltage increases, the current ultimately begins to deviate from linearity.

The term **polarization** refers to the deviation of the electrode potential from the value predicted by the Nernst equation on the passage of current. Cells that exhibit nonlinear behavior at higher currents exhibit polarization, and the degree of polarization is given by an **overvoltage**, or **overpotential**, which is symbolized by Π in the figure. Note that polarization requires the application of a potential greater than the theoretical value to give a current of the expected magnitude. Thus, the overpotential required to achieve a current of 7.00 mA in the electrolytic cell in Figure 22-2 is about −0.23 V. For an electrolytic cell affected by overvoltage, Equation 22-2 then becomes

$$E_{applied} = E_{cell} - IR - \Pi \qquad (22\text{-}4)$$

Polarization is an electrode phenomenon that may affect either or both of the electrodes in a cell. The degree of polarization of an electrode varies widely. In some instances, it approaches zero, while in others, it can be so large that the current in the cell becomes independent of potential. Under this circumstance, polarization is said to be complete. Polarization phenomena can be divided into two categories: **concentration polarization** and **kinetic polarization**.

Concentration Polarization

Concentration polarization occurs because of the finite rate of mass transfer from the solution to the electrode surface. Electron transfer between a reactive species in a solution and an electrode can take place only from the interfacial region located immediately adjacent to the surface of the electrode. This region is only a fraction of a nanometer thick and contains a limited number of reactive ions or molecules. In order for there to be a steady current in a cell, the interfacial region must be continuously replenished with reactant from the bulk of the solution. In other words, as reactant ions or molecules are consumed by the electrochemical reaction, more must be transported into the surface layer at a rate that is sufficient to maintain the current. For example, in order to have a current of 2.0 mA in the cell described in Figure 22-1b, it is necessary to transport cadmium ions to the cathode surface at a rate of about 1×10^{-8} mol/s or 6×10^{15} cadmium ions per second. Similarly, silver ions must be removed from the anode surface at a rate of 2×10^{-8} mol/s.[2]

Concentration polarization occurs when reactant species do not arrive at the surface of the electrode or product species do not leave the surface of the electrode fast enough to maintain the desired current. When these events happen, the current is limited to values less than that predicted by Equation 22-2.

Reactants are transported to the surface of an electrode by three mechanisms: **diffusion**, **migration**, and **convection**. Products are removed from electrode surfaces in the same ways.

Diffusion. When there is a concentration difference between two regions of a solution, ions or molecules move from the more concentrated region to the more dilute. This process is called **diffusion** and ultimately leads to a disappearance of the concentration gradient. The rate of diffusion is directly proportional to the concentration difference. For example, when cadmium ions are deposited at a cadmium electrode, as illustrated in **Figure 22-3a**, the concentration of Cd^{2+} at the electrode surface, $[Cd^{2+}]_0$, becomes lower than the bulk concentration. The difference between the concentration at the surface and the concentration in the bulk solution, $[Cd^{2+}]$, creates a concentration *gradient* that causes cadmium ions to diffuse from the bulk of the solution to the surface layer near the electrode (see **Figure 22-3b**).

> Reactants are transported to and products away from an electrode by diffusion, migration, and convection.

> **Diffusion** is the movement of a species under the influence of a concentration gradient. It is the process that causes ions or molecules to move from a more concentrated part of a solution to a more dilute.

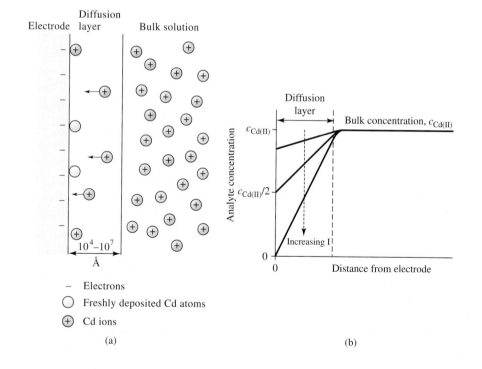

(a)

(b)

Figure 22-3 Pictorial diagram (a) and concentration versus distance plot (b) showing concentration changes at the surface of a cadmium electrode. As Cd^{2+} ions are reduced to Cd atoms at the electrode surface, the concentration of Cd^{2+} at the surface becomes smaller than the bulk concentration. Ions then diffuse from the bulk of the solution to the surface as a result of the concentration gradient. The higher the current, the larger the concentration gradient until the surface concentration falls to zero, its lowest possible value. At this point, the maximum possible current, called the limiting current, is obtained.

---

[2]For more details, see D. A. Skoog, F. J. Holler, and S. R. Crouch, *Principles of Instrumental Analysis*, 6th ed., Belmont, CA: Brooks/Cole, 2007, pp. 647–52.

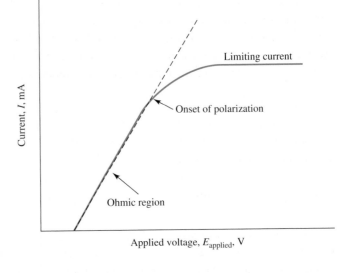

Figure 22-4 Current-potential curve for electrolysis showing the linear or ohmic region, the onset of polarization, and the limiting current plateau. In the limiting current region, the electrode is said to be completely polarized since its potential can be changed widely without affecting the current.

The rate of diffusion is given by

$$\text{rate of diffusion to cathode surface} = k'([Cd^{2+}] - [Cd^{2+}]_0) \qquad (22\text{-}5)$$

where $[Cd^{2+}]$ is the reactant concentration in the bulk of the solution, $[Cd^{2+}]_0$ is its equilibrium concentration at the electrode surface, and k' is a proportionality or rate constant. The value of $[Cd^{2+}]_0$ at any instant is fixed by the potential of the electrode and can be calculated from the Nernst equation. In the present example, we find the surface cadmium ion concentration from the relationship

$$E_{\text{cathode}} = E^0_{Cd^{2+}/Cd} - \frac{0.0592}{2} \log \frac{1}{[Cd^{2+}]_0}$$

Migration is the movement of ions through a solution as a result of electrostatic attraction between the electrodes and the ions.

where E_{cathode} is the potential applied to the cathode. As the applied potential becomes more and more negative, $[Cd^{2+}]_0$ becomes smaller and smaller. The result is that the rate of diffusion and the current become correspondingly larger until the surface concentration falls to zero, and the maximum or **limiting current** is reached as illustrated in **Figure 22-4**.

Migration.   The electrostatic process by which ions move under the influence of an electric field is called **migration**. This process, shown schematically in **Figure 22-5**, is the primary cause of mass transfer in the bulk of the solution in a cell. The rate at which ions migrate to or away from an electrode surface generally increases as the electrode potential increases. This charge movement constitutes a current, which also increases with potential. Migration causes anions to be attracted to the positive electrode and cations to the negative electrode. Migration of analyte species is undesirable in most types of electrochemistry. We want to reduce anions as well as cations at an electrode of negative polarity and oxidize cations as well as anions at a positive electrode. Migration of analyte species can be minimized by having a high concentration of an inert electrolyte, called a **supporting electrolyte**, present in the cell. The current in the cell is then primarily due to charges carried by ions from the supporting electrolyte. The supporting electrolyte also serves to reduce the resistance of the cell, decreasing the *IR* drop.

Figure 22-5 The motion of ions through a solution because of the electrostatic attraction between the ions and the electrodes is called migration.

Convection.   Reactants can also be transferred to or from an electrode by mechanical means. **Forced convection**, such as stirring or agitation, tends to decrease the thickness of the diffusion layer at the surface of an electrode and thus decrease concentration polarization. **Natural convection** resulting from temperature or density differences also contributes to the transport of molecules and ions to and from an electrode.

The Importance of Concentration Polarization.   As noted previously, concentration polarization sets in when the effects of diffusion, migration, and convection are insufficient to transport a reactant to or from an electrode surface at a rate that produces a current of the magnitude given by Equation 22-2. Concentration polarization requires applied potentials that are larger than calculated from Equation 22-2 to maintain a given current in an electrolytic cell (see Figure 22-2). Similarly, the phenomenon causes a galvanic cell potential to be smaller than the value predicted on the basis of the theoretical potential and the *IR* drop.

Kinetic Polarization

In kinetic polarization, the magnitude of the current is limited by the rate of one or both of the electrode reactions, that is, the rate of electron transfer between the reactants and the electrodes. In order to offset kinetic polarization, an additional potential, or overvoltage, is required to overcome the activation energy of the half-reaction.

Kinetic polarization is most pronounced for electrode processes that yield gaseous products because the kinetics of the gas evolution process are complicated and often slow. Kinetic polarization can be negligible for deposition or dissolution of such metals as Cu, Ag, Zn, Cd, and Hg. Kinetic polarization can be significant, however, for reactions involving transition metals, such as Fe, Cr, Ni, and Co. Kinetic effects usually decrease with increasing temperature and decreasing current density. These effects also depend on the composition of the electrode and are most pronounced with softer metals, such as lead, zinc, and particularly mercury. The magnitude of overvoltage effects cannot be predicted from present theory and can only be estimated from empirical information in the literature.[3] Just as with *IR* drop, overvoltage effects require the application of voltages larger than calculated to operate an electrolytic cell at a desired current. Kinetic polarization also causes the potential of a galvanic cell to be smaller than calculated from the Nernst equation and the *IR* drop (see Equation 22-2).

The overvoltages associated with the formation of hydrogen and oxygen are often 1 V or more and are quite important because these molecules are frequently produced by electrochemical reactions. For example, the influence of hydrogen overvoltage on the lead/acid storage battery used in automobiles is discussed in Feature 22-1. The high overvoltage of hydrogen on such metals as copper, zinc, lead, and mercury is particularly interesting for analytical purposes. These metals and several others can, therefore, be deposited without interference from hydrogen evolution. In theory, it is not possible to deposit zinc from a neutral aqueous solution because hydrogen forms at a potential that is considerably less than that required for zinc deposition. In fact, zinc can be deposited on a copper electrode with no significant hydrogen formation

> **Convection** is the transport of ions or molecules through a solution as a result of stirring, vibration, or temperature gradients.

> ◀ The experimental variables that influence the degree of concentration polarization are (1) reactant concentration, (2) total electrolyte concentration, (3) mechanical agitation, and (4) electrode size.

> ◀ The current in a kinetically polarized cell is governed by the rate of electron transfer rather than the rate of mass transfer.

> **Current density** is the current per unit surface area of the electrode (A/cm^2).

> **Kinetic polarization** is most commonly encountered when the reactant or product in an electrochemical cell is gas.

---

[3]Overvoltage data for various gaseous species on different electrode surfaces have been compiled in J. A. Dean, *Analytical Chemistry Handbook*, Section 14, pp. 14.96–14.97, New York: McGraw-Hill, 1995.

FEATURE 22-1

Overvoltage and the Lead/Acid Battery

If it were not for the high overvoltage of hydrogen on lead and lead oxide electrodes, the lead/acid storage batteries found in automobiles and trucks (see **Figure 22F-1**) would not operate because of hydrogen formation at the cathode both during charging and use. Certain trace metals in the system lower this overvoltage and eventually lead to gassing, or hydrogen formation, therefore limiting the lifetime of the battery. The basic difference between a battery with a 48-month warranty and a 72-month warranty is the concentration of these trace metals in the system. The overall cell reaction when the cell is discharging is

$$Pb(s) + PbO_2(s) + 2HSO_4^- + 2H^+ \rightarrow 2PbSO_4(s) + 2H_2O$$

The lead/acid storage battery behaves as a galvanic cell during discharge and as an electrolytic cell when it is being charged. Such batteries acting as galvanic cells were once used as voltage sources for electrolysis. They have been supplanted for this purpose by modern line-operated power supplies.

Anode

Cathode

Sulfuric
acid
electrolyte

Lead grills
filled with
spongy lead

Lead grills filled
with PbO_2

Figure 22F-1   The lead/acid storage battery.

---

because the rate at which the gas forms on both zinc and copper is negligible, as shown by the high hydrogen overvoltage associated with these metals.

22B THE SELECTIVITY OF ELECTROLYTIC METHODS

In principle, electrolytic methods are reasonably selective for separating and determining a number of ions. The feasibility of a separation and the theoretical conditions for accomplishing it can be derived from the standard electrode potentials of the species of interest as illustrated in Example 22-2.

EXAMPLE 22-2

Is a quantitative separation of Cu^{2+} and Pb^{2+} by electrolytic deposition feasible in principle? If so, what range of cathode potentials versus the saturated calomel electrode (SCE) can be used? Assume that the sample solution is initially 0.1000 M in each ion and that quantitative removal of an ion is realized when only 1 part in 10,000 remains undeposited.

Solution

In Appendix 5, we find

$$Cu^{2+} + 2e^- \rightleftharpoons Cu(s) \qquad E^0 = 0.337 \text{ V}$$

$$Pb^{2+} + 2e^- \rightleftharpoons Pb(s) \qquad E^0 = -0.126 \text{ V}$$

Note that based on the standard potentials, copper will begin to deposit at more positive applied voltages than lead. Let us first calculate the potential required to reduce the Cu^{2+} concentration to 10^{-4} of its original concentration (that is, from 0.1000 M to 1.00×10^{-5} M). Substituting into the Nernst equation, we obtain

$$E = 0.337 - \frac{0.0592}{2} \log \frac{1}{1.00 \times 10^{-5}} = 0.189 \text{ V}$$

Similarly, we can derive the potential at which lead begins to deposit:

$$E = -0.126 - \frac{0.0592}{2} \log \frac{1}{0.1000} = -0.156 \text{ V}$$

Therefore, if the cathode potential is maintained between 0.189 V and -0.156 V (versus the standard hydrogen electrode [SHE]), we should get a quantitative separation. Now, we can convert these to potentials versus the SCE by subtracting E_{SCE}:

$$E_{cell} = E_{cathode} - E_{SCE} = 0.189 - 0.244 = -0.055 \text{ V} \qquad \text{for depositing Cu}$$

and

$$E_{cell} = E_{cathode} - E_{SCE} = -0.156 - 0.244 = -0.400 \text{ V} \qquad \text{for depositing Pb}$$

These calculations indicate that the cathode potential should be kept between -0.055 V and -0.400 V versus the SCE to deposit Cu without depositing any appreciable amounts of Pb.

Calculations such as those in Example 22-2 permit us to find the differences in standard electrode potentials that are theoretically needed to determine one ion without interference from another. These differences range from about 0.04 V for triply charged ions to approximately 0.24 V for singly charged species.

These theoretical separation limits can be approached only by maintaining the potential of the working electrode (usually the cathode) at which a metal deposits at the level required. The potential of this electrode can be controlled only by varying the potential applied to the cell, however. Equation 22-4 indicates that variations in $E_{applied}$ affect not only the cathode potential but also the anode potential, the IR drop, and the overpotential. Because of these effects, the only practical way of achieving separation of species whose electrode potentials differ by a few tenths of a volt is to measure the cathode potential continuously against a reference electrode whose potential is known. The applied cell potential can then be adjusted to maintain the cathode potential at the desired level. An analysis performed in this way is called a **controlled-potential electrolysis**. Controlled-potential methods are discussed in Sections 22C-2 and 22D-4.

22C ELECTROGRAVIMETRIC METHODS

Electrolytic deposition has been used for over a century for the gravimetric determination of metals. In most applications, the metal is deposited on a weighed platinum cathode, and the increase in mass is determined. Some methods use anodic deposition such as the determination of lead as lead dioxide on platinum and of chloride as silver chloride on silver.

There are two general types of electrogravimetric methods. In one, the potential of the working electrode is uncontrolled, and the applied cell potential is held at a more or less constant level that provides a large enough current to complete the electrolysis in a reasonable length of time. The second type of electrogravimetric method is the controlled-potential or **potentiostatic method**.

In a **potentiostatic method,** the potential of the working electrode is maintained at a constant level versus a reference electrode, such as a SCE.

A **working electrode** is the electrode at which the analytical reaction occurs.

22C-1 Electrogravimetry without Potential Control

Electrolytic procedures in which the potential of the working electrode is uncontrolled use simple and inexpensive equipment and require little operator attention. In these procedures, the potential applied across the entire cell is maintained at a more-or-less constant level throughout the electrolysis.

Instrumentation

As shown in **Figure 22-6**, the apparatus for an analytical electrodeposition without cathode-potential control consists of a suitable cell and a 6- to 12-V direct-current

Figure 22-6 Apparatus for electrodeposition of metals without cathode-potential control. Note that this is a two-electrode cell.

power supply. The voltage applied to the cell is controlled by the variable resistor, R. An ammeter and a voltmeter indicate the approximate current and applied voltage. To perform an analytical electrolysis with this apparatus, the applied voltage is adjusted with potentiometer R to give a current of several tenths of an ampere. The voltage is then maintained at about the initial level until the deposition is judged to be complete.

Electrolysis Cells

Figure 22-6 shows a typical cell for the deposition of a metal on a solid electrode. Often, the working electrode is a large surface area platinum gauze cylinder 2 or 3 cm in diameter and perhaps 6 cm in length. Copper gauze cathodes and various alloys have also been used. Frequently, as shown, the anode takes the form of a solid platinum stirring paddle that is located inside and connected to the cathode through the external circuit.

Physical Properties of Electrolytic Precipitates

Ideally, an electrolytically deposited metal should be strongly adherent, dense, and smooth so that it can be washed, dried, and weighed without mechanical loss or reaction with the atmosphere. Good metallic deposits are fine grained and have a metallic luster. Spongy, powdery, or flaky precipitates are usually less pure and less adherent than fine-grained deposits.

The principal factors that influence the physical characteristics of deposits are current density, temperature, and the presence of complexing agents. The best deposits are usually formed at low current densities, typically less than 0.1 A/cm^2. Gentle stirring usually improves the quality of a deposit. The effects of temperature are unpredictable and must be determined empirically.

Often, when metals are deposited from solutions of metal complexes, they form smoother and more adherent films than when deposited from the simple ions. Cyanide and ammonia complexes often provide the best deposits.

Applications of Electrogravimetric Methods

In practice, electrolysis at a constant cell potential is limited to the separation of easily reduced cations from those that are more difficult to reduce than hydrogen ion or nitrate ion. The reason for this limitation is illustrated in **Figure 22-7**, which shows the changes of current, IR drop, and cathode potential during an electrolysis in the cell in Figure 22-6. The analyte is copper(II) ions in a solution containing an excess of sulfuric or nitric acid. Initially, R is adjusted so that the potential applied to the cell is about -2.5 V, which as shown in Figure 22-7a, leads to a current of about 1.5 A. The electrolytic deposition of copper is then completed at this applied potential.

As shown in Figure 22-7b, the IR drop decreases continually as the reaction proceeds. The reason for this decrease is primarily concentration polarization at the cathode, which limits the rate at which copper ions are brought to the electrode surface and thus the current. From Equation 22-4, it is can be seen that the decrease in IR must be offset by an increase in the cathode potential since the applied cell potential is constant.

Ultimately, the decrease in current and the increase in cathode potential are slowed at point B by the reduction of hydrogen ions. Because the solution contains a large excess of acid, the current is no longer limited by concentration polarization, and codeposition of copper and hydrogen occurs simultaneously until the remainder

(a)

Figure 22-7 (a) Current. (b) *IR* drop and cathode potential change during electrolytic deposition of copper at a constant applied cell potential. The current (a) and *IR* drop (b) decrease steadily with time. The cathode potential shifts negative to offset the decrease in *IR* drop (b). At point *B*, the cathode becomes depolarized by the reduction of hydrogen ions. Metals that deposit at points *A* or *D* interfere with copper because of codeposition. A metal that deposits at point *C* does not interfere.

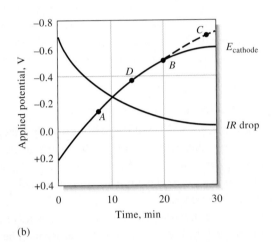

(b)

A **depolarizer** is a species that is easily reduced (or oxidized). It helps maintain the potential of the working electrode at a relatively small constant value and prevents reactions that would occur under more reducing or oxidizing conditions.

of the copper ions are deposited. Under these conditions, the cathode is said to be depolarized by hydrogen ions.

Consider the fate of some metal ion, such as lead(II), which begins to deposit at point *A* on the cathode potential curve. Lead(II) would codeposit well before copper deposition was complete and would, therefore, interfere with the determination of copper. In contrast, a metal ion, such as cobalt(II), that reacts at a cathode potential corresponding to point *C* on the curve would not interfere because depolarization by hydrogen gas formation prevents the cathode from reaching this potential.

Codeposition of hydrogen during electrolysis often leads to formation of deposits that do not adhere well. These are usually unsatisfactory for analytical purposes. This problem can be resolved by introducing another species that is reduced at a less negative potential than hydrogen ion and does not adversely affect the physical properties of the deposit. One such cathode **depolarizer** is nitrate ion. Hydrazine and hydroxylamine are also commonly used.

Electrolytic methods performed without electrode-potential control, while limited by their lack of selectivity, do have several applications of practical importance. **Table 22-1** lists the common elements that are often determined by this procedure.

22C-2 Controlled-Potential Electrogravimetry

In the discussion that follows, we assume that the working electrode is a cathode where the analyte is deposited as a metal. The principles can be extended, however, to an anodic working electrode where nonmetallic deposits are formed.

TABLE 22-1

Some Applications of Electrogravimetry without Potential Control

Analyte	Weighed As	Cathode	Anode	Conditions
Ag^+	Ag	Pt	Pt	Alkaline CN^- solution
Br^-	AgBr (on anode)	Pt	Ag	
Cd^{2+}	Cd	Cu on Pt	Pt	Alkaline CN^- solution
Cu^{2+}	Cu	Pt	Pt	H_2SO_4/HNO_3 solution
Mn^{2+}	MnO_2 (on anode)	Pt	Pt dish	HCOOH/HCOONa solution
Ni^{2+}	Ni	Cu on Pt	Pt	Ammoniacal solution
Pb^{2+}	PbO_2 (on anode)	Pt	Pt	HNO_3 solution
Zn^{2+}	Zn	Cu on Pt	Pt	Acidic citrate solution

The determination of Br^- by forming AgBr and of Mn^{2+} by forming MnO_2 are examples of anodic depositions.

Instrumentation

To separate species with electrode potentials that differ by only a few tenths of a volt, we must use a more sophisticated approach than the one just described. Otherwise, concentration polarization at the cathode causes the potential of that electrode to become so negative that codeposition of the other species present begins before the analyte is completely deposited (see Figure 22-7). A large negative drift in the cathode potential can be avoided by employing the three-electrode system shown in **Figure 22-8** instead of the two-electrode system of Figure 22-5.

The controlled-potential apparatus shown in Figure 22-8 is made up of two independent electrical circuits that share a common electrode, the working electrode

Figure 22-8 Apparatus for controlled-potential, electrolysis. The digital voltmeter monitors the potential between the working and the reference electrode. The voltage applied between the working and the counter electrode is varied by adjusting contact *C* on the potentiometer to maintain the working electrode (cathode in this example) at a constant potential versus a reference electrode. The current in the reference-electrode is essentially zero at all times. Modern potentiostats are fully automatic and often computer controlled. The electrode symbols shown (—○ Working, → Reference, —| Counter) are the currently accepted notation.

Figure 22-9 Changes in cell potential (A) and current (B) during a controlled-potential deposition of copper. The cathode is maintained at −0.36 V (versus SCE) throughout the experiment. (Data from J. J. Lingane, *Anal. Chim. Acta*, **1948**, *2*, 584, **DOI**: 10.1016/s0003-2670(01)93842-5.)

The electrolysis current passes between the working electrode and a **counter electrode**. The counter electrode has no effect on the reaction at the working electrode.

A **potentiostat** maintains the working electrode potential at a constant value relative to a reference electrode.

CHALLENGE: Would Pb^{2+} be expected to interfere with the electrolysis shown in Figure 22-9? Why or why not?

where the analyte is deposited. The electrolysis circuit consists of a dc source, a potentiometer (ACB) that permits the voltage applied between the working electrode and a counter electrode to be continuously varied, and a current meter. The control circuit is made up of a reference electrode (often a SCE), a high-resistance digital voltmeter, and the working electrode. The electrical resistance of the control circuit is so large that the electrolysis circuit supplies essentially all of the current for the electrolysis. The control circuit monitors continuously the voltage between the working electrode and the reference electrode and maintains it at a controlled value.

The current and the cell potential changes that occur in a typical constant-potential electrolysis are illustrated in **Figure 22-9**. Note that the applied cell potential has to be decreased continuously throughout the electrolysis, which is tedious and time consuming when done manually. Modern controlled-potential electrolyses are performed with instruments called **potentiostats**, which automatically maintain the working electrode potential at a controlled value versus the reference electrode.

Electrolysis Cells

Electrolysis cells are similar to those shown in Figure 22-6. Tall-form beakers are often used, and solutions are usually mechanically stirred to minimize concentration polarization. The anode is often rotated to act as a mechanical stirrer.

The working electrode is usually a metallic gauze cylinder, as shown in Figure 22-6. Electrodes are often constructed of platinum, although copper, brass, and other metals find occasional use. Some metals, such as bismuth, zinc, and gallium, cannot be deposited directly onto platinum without causing permanent damage to the electrode. Because of this incompatibility, a protective coating of copper is deposited on the platinum electrode before electrolyzing these metals.

The Mercury Cathode

A mercury cathode, such as that shown in **Figure 22-10**, is particularly useful for removing easily reduced elements as a preliminary step in an analysis. For example, copper, nickel, cobalt, silver, and cadmium are separated at this electrode from such

Wire
anode

+

Wire
cathode

Solution

−

Hg
cathode

Figure 22-10   A mercury cathode
for the electrolytic removal of metal
ions from solution.

ions as aluminum, titanium, the alkali metals, sulfates, and phosphates. The deposited metals dissolve in the mercury with little hydrogen evolution because of the high
overvoltage of hydrogen on mercury. The metals dissolve in the mercury to form
amalgams that are important in several forms of voltammetry (see Section 23B-1).
The deposited metals are not normally determined after electrolysis but are merely
removed from the analyte solution.

Applications of Controlled-Potential Electrogravimetry

Controlled-potential electrolysis is a potent tool for separating and determining
metallic species having standard potentials that differ by only a few tenths of a
volt. For example, copper, bismuth, lead, cadmium, zinc, and tin can be determined in mixtures by successive deposition of the metals on a weighed platinum
cathode. The first three elements are deposited from a nearly neutral solution containing tartrate ion to complex the tin(IV) and prevent its deposition.
Copper is first reduced quantitatively by maintaining the cathode potential at
-0.2 V with respect to the SCE. After being weighed, the copper-plated cathode
is returned to the solution, and bismuth is removed at a potential of -0.4 V. Lead
is then deposited quantitatively by increasing the cathode potential to -0.6 V.
When lead deposition is complete, the solution is made strongly ammoniacal, and
cadmium and zinc are deposited successively at -1.2 and -1.5 V. Finally, the
solution is acidified in order to decompose the tin/tartrate complex by the formation of undissociated tartaric acid. Tin is then deposited at a cathode potential of
-0.65 V. A fresh cathode must be used here because the zinc redissolves under
these conditions. A procedure such as this is particularly attractive for use with
computer-controlled potentiostats because little operator time is required for the
complete analysis.

Table 22-2 lists some other separations performed by controlled-potential electrolysis. Because of limited sensitivity and the time required for washing, drying, and
weighing the electrodes, many electrogravimetric methods have been replaced by the
coulometric methods discussed in the next section.

TABLE 22-2

Some Applications of Controlled-potential Electrolysis*

Metal	Potential versus SCE	Electrolyte	Other Elements That Can Be Present
Ag	+0.10	Acetic acid/acetate buffer	Cu and heavy metals
Cu	−0.30	Tartrate + hydrazine + Cl^-	Bi, Sb, Pb, Sn, Ni, Cd, Zn
Bi	−0.40	Tartrate + hydrazine + Cl^-	Pb, Zn, Sb, Cd, Sn
Sb	−0.35	HCl + hydrazine at 70°C	Pb, Sn
Sn	−0.60	HCl + hydroxylamine	Cd, Zn, Mn, Fe
Pb	−0.60	Tartrate + hydrazine	Cd, Sn, Ni, Zn, Mn, Al, Fe
Cd	−0.80	HCl + hydroxylamine	Zn
Ni	−1.10	Ammoniacal tartrate + sodium sulfite	Zn, Al, Fe

*Sources: H. Diehl, *Electrochemical Analysis with Graded Cathode Potential Control*, G. F. Smith Chemical Co., Columbus, OH, 1948; H. J. S. Sand, *Electrochemistry and Electrochemical Analysis*, Blackie and Sons, Ltd, London, Vol. II, 1940; J. J. Lingane and S. L. Jones, *Anal. Chem.*, **1951**, *23*, 1798, **DOI**: 10.1021/ac60060a023; J. J. Lingane, *Anal. Chim. Acta*, **1948**, *2*, 584, **DOI**: 10.1016/s0003-2670(01)93842-5.

22D COULOMETRIC METHODS

In coulometric methods, the quantity of electrical charge required to convert a sample of an analyte quantitatively to a different oxidation state is measured. Coulometric and gravimetric methods share the common advantage that the proportionality constant between the quantity measured and the analyte mass is calculated from accurately known physical constants, which can eliminate the need for calibration with chemical standards. In contrast to gravimetric methods, coulometric procedures are usually rapid and do not require that the product of the electrochemical reaction be a weighable solid. Coulometric methods are as accurate as conventional gravimetric and volumetric procedures and in addition are easily automated.[4]

22D-1 Determining the Electrical Charge

Electrical charge is the basis of the other electrical quantities, current, voltage, and power. The charge on an electron (and proton) is defined as 1.6022×10^{-19} coulombs (C). A rate of charge flow equal to one coulomb per second is the definition of one ampere (A) of current. Thus, a coulomb can be considered as that charge carried by a constant current of one ampere for one second. The charge Q that results from a constant current of I amperes operated for t seconds is

$$Q = It \tag{22-6}$$

For a variable current i, the charge is given by the integral

$$Q = \int_0^t i\, dt \tag{22-7}$$

The **coulomb (C)** is the amount of charge required to produce 0.00111800 g of silver metal from silver ions. One coulomb = 1 ampere × 1 s = 1 A s.

In describing electrical current, it is common to use the upper case symbol I for a static or direct current (dc). A variable or alternating current (ac) is commonly given the lower case symbol i. Likewise, dc and ac voltages are given the symbols E and e respectively.

---

[4]For additional information about coulometric methods, see J. A. Dean, *Analytical Chemistry Handbook*, Section 14, pp. 14.118–14.133, New York: McGraw-Hill, 1995; D. J. Curran, *in Laboratory Technique in Electroanalytical Chemistry*, 2nd ed., pp. 739–68, P. T. Kissinger and W. R. Heinemann, eds., New York: Marcel Dekker, 1996; J. A. Plambeck, *Electroanalytical Chemistry*, Ch. 12, New York: Wiley, 1982.

The faraday (F) is the quantity of charge that corresponds to one mole or 6.022×10^{23} electrons. Since each electron has a charge of 1.6022×10^{-19} C, the faraday also equals 96,485 C.

Faraday's law relates the number of moles of the analyte n_A to the charge Q

$$n_A = \frac{Q}{nF} \tag{22-8}$$

where n is the number of moles of electrons in the analyte half-reaction. As shown in Example 22-3, we can use these definitions to calculate the mass of a chemical species that is formed at an electrode by a current of known magnitude.

The full constants for fundamental quantities are available from the National Institute of Standards and Technology on their web site at http://physics.nist.gov/cuu/Constants/index.html. The 2010 value for the Faraday is 96485.3365 C mol^{-1} with a standard uncertainty of 0.0021C mol^{-1}. The value for the electron charge is $1.602176565 \times 10^{-19}$ C with a standard uncertainty of $0.000\ 000\ 035 \times 10^{-19}$ C. A detailed description of the data and the analysis that led to the values can be found in the preprint: http://physics.nist.gov/cuu/Constants/Preprints/lsa2010.pdf.

EXAMPLE 22-3

A constant current of 0.800 A is used to deposit copper at the cathode and oxygen at the anode of an electrolytic cell. Calculate the number of grams of each product formed in 15.2 min, assuming no other redox reaction occurs.

Solution

The two half-reactions are

$$Cu^{2+} + 2e^- \rightarrow Cu(s)$$

$$2H_2O \rightarrow 4e^- + O_2(g) + 4H^+$$

Thus, 1 mol of copper is equivalent to 2 mol of electrons, and 1 mol of oxygen corresponds to 4 mol of electrons.

Substituting into Equation 22-6 yields

$$Q = 0.800\ A \times 15.2\ min \times 60\ s/min = 729.6\ A{\cdot}s = 729.6\ C$$

We can find the number of moles of Cu and O_2 from Equation 22-8

$$n_{Cu} = \frac{729.6\ \cancel{C}}{2\ \cancel{mol\ e^-}/mol\ Cu \times 96,485\ \cancel{C}/mol\ \cancel{e^-}} = 3.781 \times 10^{-3} = mol\ Cu$$

$$n_{O_2} = \frac{729.6\ \cancel{C}}{4\ \cancel{mol\ e^-}/mol\ O_2 \times 96,485\ \cancel{C}/mol\ \cancel{e^-}} = 1.890 \times 10^{-3}\ mol\ O_2$$

The masses of Cu and O_2 are given by

$$mass\ Cu = 3.781 \times 10^{-3}\ \cancel{mol} \times \frac{63.55\ g\ Cu}{\cancel{mol}} = 0.240\ g\ Cu$$

$$mass\ O_2 = 1.890 \times 10^{-3}\ \cancel{mol} \times \frac{32.00\ g\ O_2}{\cancel{mol}} = 0.0605\ g\ O_2$$

Michael Faraday (1791–1867) was one of the foremost chemists and physicists of his time. Among his most important discoveries were Faraday's law of electrolysis. Faraday, a simple man who lacked mathematical sophistication, was a superb experimentalist and an inspiring teacher and lecturer. The quantity of charge equal to a mole of electrons is named in his honor.

22D-2 Characterizing Coulometric Methods

Two methods have been developed that are based on measuring the quantity of charge: **controlled-potential (potentiostatic) coulometry** and **controlled-current coulometry**, often called **coulometric titrimetry**. Potentiostatic methods are performed in much the same way as controlled-potential gravimetric methods, with the

Constant-current coulometry is also called coulometric titrimetry.

potential of the working electrode being maintained at a constant value relative to a reference electrode throughout the electrolysis. However, in controlled-potential coulometry, the electrolysis current is recorded as a function of time to give a curve similar to curve B in Figure 22-9. The analysis is then completed by integrating the current-time curve (see Equation 22-7) to obtain the charge and from Faraday's law the amount of analyte (see Equation 22-8).

Coulometric titrations are similar to other titrimetric methods in that analyses are based on measuring the combining capacity of the analyte with a standard reagent. In the coulometric procedure, the reagent consists of electrons, and the standard solution is a constant current of known magnitude. Electrons are added to the analyte (via the direct current) or to some species that immediately reacts with the analyte until an end point is reached. At that point, the electrolysis is discontinued. The amount of analyte is determined from the magnitude of the current and the time required to complete the titration. The magnitude of the current in amperes is analogous to the molar concentration of a standard solution, and the time measurement is analogous to the volume measurement in conventional titrimetry.

> **Electrons are the reagent in a coulometric titration.**

22D-3 Current Efficiency Requirements

A fundamental requirement for all coulometric methods is 100% current efficiency, that is, each faraday of electricity must bring about chemical change in the analyte equivalent to one mole of electrons. Note that 100% current efficiency can be achieved without direct participation of the analyte in electron transfer at an electrode. For example, chloride ion can be determined using potentiostatic coulometry or using coulometric titrations with silver ion at a silver anode. Silver ion then reacts with chloride to form a precipitate or deposit of silver chloride. The quantity of electricity required to complete the silver chloride formation serves as the analytical variable. In this instance, 100% current efficiency is realized because the number of moles of electrons is equal to the number of moles of chloride ion in the sample despite the fact that these ions do not react directly at the electrode surface.

One equivalent of chemical change is the change brought about by 1 mol of electrons. Thus, for the two half-reactions in Example 22-3, one equivalent of chemical change produces 1/2 mole of Cu or 1/4 mole of O_2.

22D-4 Controlled-Potential Coulometry

In controlled-potential coulometry, the potential of the working electrode is maintained at a constant level such that only the analyte is responsible for conducting charge across the electrode/solution interface. The charge required to convert the analyte to its reaction product is then determined by recording and integrating the current-versus-time curve during the electrolysis.

Instrumentation

The instrumentation for potentiostatic coulometry consists of an electrolysis cell, a potentiostat, and a device for determining the charge consumed by the analyte.

Cells.   **Figure 22-11** illustrates two types of cells that are used for potentiostatic coulometry. The first, Figure 22-11a, consists of a platinum-gauze working electrode, a platinum-wire counter electrode, and a saturated calomel reference electrode. The counter electrode is separated from the analyte solution by a salt bridge that usually contains the same electrolyte as the solution being analyzed. The salt bridge is needed to prevent the reaction products formed at the counter electrode from diffusing into the analyte solution and interfering. For example, hydrogen gas is a common product at a cathodic counter electrode. Unless hydrogen is physically isolated from the

Figure 22-11 Electrolysis cells for potentiostatic coulometry. Working electrode: (a) platinum-gauze, (b) mercury-pool. (Reprinted with permission from J. E. Harrar and C. L. Pomernacki, *Anal. Chem.*, **1973**, *45*, 57, **DOI**: 10.1021/ac60323a003. Copyright 1973 by American Chemical Society.)

analyte solution by the bridge, it will react directly with many of the analytes that are determined by oxidation at the working anode.

The second type of cell, shown in Figure 22-11b, is a mercury-pool type. A mercury cathode is particularly useful for separating easily reduced elements as a preliminary step in an analysis. In addition, however, it has found considerable use for the coulometric determination of several metallic cations that form metals soluble in mercury. In these applications, little or no hydrogen evolution occurs even at high applied potentials because of the large overvoltage of hydrogen on mercury. A coulometric cell such as that shown in Figure 22-11b is also useful for the coulometric determination of certain types of organic compounds.

Potentiostats and Coulometers. For controlled-potential coulometry, we use a potentiostat similar in design to that shown in Figure 22-8. Generally, however, the potentiostat is automated and equipped with a computer or an electronic current integrator that gives the charge in coulombs necessary to complete the reaction, as shown in **Figure 22-12**.

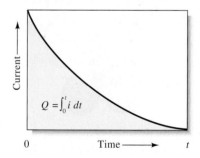

$$Q = \int_0^t i\, dt$$

Figure 22-12 For a current that varies with time, the quantity of charge Q in a time t is the shaded area under the curve, obtained by integration of the current-time curve.

EXAMPLE 22-4

The Fe(III) in a 0.8202-g sample was determined by coulometric reduction to Fe(II) at a platinum cathode. Calculate the percentage of $Fe_2(SO_4)_3$ ($\mathcal{M} = 399.88$ g/mol) in the sample if 103.2775 C were required for the reduction.

(continued)

Solution

Since 1 mol of $Fe_2(SO_4)_3$ consumes 2 mol of electrons, we may write from Equation 22-8

$$n_{Fe_2(SO_4)_3} = \frac{103.2775 \ \mathcal{C}}{2 \ \text{mol e}^-/\text{mol } Fe_2(SO_4)_3 \ \times \ 96,485 \ \mathcal{C}/\text{mol e}^-}$$

$$= 5.3520 \times 10^{-4} \ \text{mol } Fe_2(SO_4)_3$$

$$\text{mass } Fe_2(SO_4)_3 = 5.3520 \times 10^{-4} \ \text{mol } Fe_2(SO_4)_3 \times \frac{399.88 \ \text{g } Fe_2(SO_4)_3}{\text{mol } Fe_2(SO_4)_3}$$

$$= 0.21401 \ \text{g } Fe_2(SO_4)_3$$

$$\text{Percentage } Fe_2(SO_4)_3 = \frac{0.21401 \ \text{g } Fe_2(SO_4)_3}{0.8202 \ \text{g sample}} \times 100\% = 26.09\%$$

Applications of Controlled-Potential Coulometry

Controlled-potential coulometric methods have been used to determine more than 55 elements in inorganic compounds.[5] Methods have been described for the deposition of more than two dozen metals at a mercury cathode. The method has been used in the nuclear energy field for the relatively interference-free determination of uranium and plutonium.

Controlled-potential coulometry also offers possibilities for the electrolytic determination (and synthesis) of organic compounds. For example, trichloroacetic acid and picric acid are quantitatively reduced at a mercury cathode whose potential is suitably controlled:

$$Cl_3CCOO^- + H^+ + 2e^- \rightarrow Cl_2HCCOO^- + Cl^-$$

Picric acid

Coulometric measurements permit the determination of these compounds with a relative error of a few tenths of a percent.

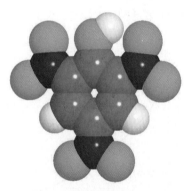

Molecular model of picric acid. Picric acid (2,4,6-trinitrophenol) is a close relative of trinitrotoluene (TNT). It is an explosive compound and has military applications. Picric acid has also been used as a yellow dye and staining agent and as an antiseptic.

> **Spreadsheet Summary** In the first experiment in Chapter 11 of *Applications of Microsoft® Excel in Analytical Chemistry*, 2nd ed., numerical integration methods are investigated. These methods are used for determining the charge required to electrolyze a reagent in a controlled-potential coulometric determination. A trapezoidal method and a Simpson's rule method are studied. From the charge, Faraday's law is used to determine the amount of analyte.

---

[5]For a summary of applications, see J. A. Dean, *Analytical Chemistry Handbook*, Section 14, pp. 14.119–14.123, New York: McGraw-Hill, 1995; A. J. Bard and L. R. Faulkner, *Electrochemical Methods*, 2nd ed., New York: Wiley, 2001, pp. 427–31.

22D-5 Coulometric Titration[6]

Coulometric titrations are performed with a constant-current source, sometimes called a **galvanostat**, which senses decreases in current in a cell and responds by increasing the potential applied to the cell until the current is restored to its original level. Because of the effects of concentration polarization, 100% current efficiency with respect to the analyte can be maintained only by having in large excess an auxiliary reagent that is oxidized or reduced at the electrode to give a product that reacts with the analyte. As an example, consider the coulometric titration of iron(II) at a platinum anode. At the beginning of the titration, the primary anodic reaction directly consumes Fe^{2+} and is

$$Fe^{2+} \rightarrow Fe^{3+} + e^-$$

As the concentration of iron(II) decreases, however, the requirement of a constant current results in an increase in the applied cell potential. Because of concentration polarization, this increase in potential causes the anode potential to increase to the point where the decomposition of water becomes a competing process:

$$2H_2O \rightarrow O_2(g) + 4H^+ + 4e^-$$

The quantity of electricity required to complete the oxidation of iron(II) then exceeds that demanded by theory, and the current efficiency is less than 100%. The lowered current efficiency is prevented, however, by introducing at the outset an unmeasured excess of cerium(III), which is oxidized at a lower potential than is water:

$$Ce^{3+} \rightarrow Ce^{4+} + e^-$$

With stirring, the cerium(IV) produced is rapidly transported from the surface of the electrode to the bulk of the solution where it oxidizes an equivalent amount of iron(II):

$$Ce^{4+} + Fe^{2+} \rightarrow Ce^{3+} + Fe^{3+}$$

The net effect is an electrochemical oxidation of iron(II) with 100% current efficiency, even though only a fraction of that species is *directly* oxidized at the electrode surface.

Detecting the End Point

Coulometric titrations, like their volumetric counterparts, require a means for determining when the reaction between analyte and reagent is complete. Generally, the end points described in the chapters on volumetric methods are applicable to coulometric titrations as well. Thus, for the titration of iron(II) just described, an oxidation/reduction indicator, such as 1,10-phenanthroline, can be used. As an alternative, the end point can be determined potentiometrically. Potentiometric or

Constant-current generators are sometimes called **galvanostats**.

◀ Auxiliary reagents are essential in coulometric titrations.

---

[6]For further details on this technique, see D. J. Curran, in *Laboratory Techniques in Electroanalytical Chemistry*, 2nd ed., pp. 750–68, P. T. Kissinger and W. R. Heineman, eds., New York: Marcel Dekker, 1996.

amperometric (see Section 23C-4) end points are used in Karl Fischer titrators. Some coulometric titrations utilize a photometric end point (see Section 26A-4).

Instrumentation

As shown in **Figure 22-13**, the equipment required for a coulometric titration includes a source of constant current from one to several hundred milliamperes, a titration cell, a switch, a timer, and a device for monitoring current. Moving the switch to position 1 simultaneously starts the timer and initiates a current in the titration cell. When the switch is moved to position 2, the electrolysis and the timing are discontinued. With the switch in this position, however, current continues to be drawn from the source and passes through a dummy resistor, R_D, that has about the same electrical resistance as the cell. This arrangement ensures continuous operation of the source, thus aiding in maintaining a constant current.

Current Sources.   The constant-current source for a coulometric titration is an electronic device capable of maintaining a current of 200 mA or more that is constant to a few hundredths percent. Such constant-current sources are available from several instrument manufacturers. The electrolysis time can be measured very accurately with a digital timer or a computer-based timing system.

Cells for Coulometric Titrations.   Figure 22-14 shows a typical coulometric titration cell consisting of a working electrode at which the reagent is produced and a counter (auxiliary) electrode to complete the circuit. The working electrode used to generate reactants in situ is often referred to as the generator electrode. It is usually a platinum rectangle, a coil of wire, or a gauze cylinder with a relatively large surface area to minimize polarization effects. The counter electrode is usually isolated from the reaction medium by a sintered disk or other porous medium to prevent interference by the reaction products from this electrode. For example, hydrogen is sometimes evolved at this electrode. Since hydrogen is an oxidizing agent, a positive systematic error can occur unless the gas is produced in a separate compartment.

Figure 22-13   Conceptual diagram of a coulometric titration apparatus. Commercial coulometric titrators are totally electronic and usually computer controlled.

To constant-current source

Mercury

Electrolyte solution

Generator electrode

Counter electrode

Stirring bar

Sintered glass disk

Magnetic stirrer

Figure 22-14 A typical coulometric titration cell.

An alternative to isolation of the counter electrode is to generate the reagent externally with a device similar to that shown in **Figure 22-15**. The external generation cell is arranged so that electrolyte flow continues briefly after the current is switched off, thereby flushing the residual reagent into the titration vessel. Note that the generation device shown in Figure 22-15 provides either hydrogen or hydroxide ions, depending on which arm is used. External generation cells have also been used for the generation of other reagents such as iodine.

Comparing Coulometric and Conventional Titrations

The various components of the titrator in Figure 22-13 have their counterparts in the reagents and apparatus required for a volumetric titration. The constant-current source of known magnitude serves the same function as the standard solution in a volumetric method. The digital timer and switch correspond to the buret and stopcock, respectively. Electricity is passed through the cell for relatively long periods of time at the outset of a coulometric titration, but the time intervals are made smaller and smaller as chemical equivalence is approached. Note that these steps are analogous to the operation of a buret in a conventional titration.

A coulometric titration offers several significant advantages over a conventional volumetric procedure. Coulometric titrations eliminate the problems associated with

❮ Coulometric methods are as accurate and precise as comparable volumetric methods. When not limited by end-point detection, they can be more accurate and precise particularly for small quantities.

Electrolyte solution from reservoir

Cathode reaction
$2e^- + 2H_2O \longrightarrow H_2 + 2OH^-$

Anode reaction
$H_2O \longrightarrow \frac{1}{2}O_2 + 2H^+ + 2e^-$

Cathode
−

Anode
+

Glass wool

Source of OH^-

Source of H^+

Figure 22-15 A cell for the external coulometric generation of acid and base.

the preparation, standardization, and storage of standard solutions. This advantage is particularly significant with reagents such as chlorine, bromine, and titanium(III) ion, which are sufficiently unstable in aqueous solution to seriously limit their value as volumetric reagents. Their use in a coulometric determination is, however, straightforward because they are consumed as soon as they are generated.

Coulometric methods also excel when small amounts of analyte have to be titrated because tiny quantities of reagent are generated with ease and accuracy through the proper choice of current. With conventional titrations, it is inconvenient and often inaccurate to use very dilute solutions and small volumes.

A further advantage of the coulometric procedure is that a single constant-current source provides reagents for precipitation, complex formation, neutralization, or oxidation/reduction titrations. Finally, coulometric titrations are more readily automated since it is easier to control electrical current than liquid flow.

The current-time measurements required for a coulometric titration are inherently as accurate as or more accurate than the comparable volume/concentration measurements of a conventional volumetric method, particularly in situations where only small quantities of reagent are required. When the accuracy of a titration is limited by the sensitivity of the end point, the two titration methods have comparable accuracies.

Applications of Coulometric Titrations

Coulometric titrations have been developed for all types of volumetric reactions.[7] Selected applications are described in this section.

Neutralization Titrations.   Hydroxide ion can be generated at the surface of a platinum cathode immersed in a solution containing the analyte acid:

$$2H_2O + 2e^- \rightarrow 2OH^- + H_2(g)$$

The platinum anode must be isolated by a diaphragm to eliminate potential interference from the hydrogen ions produced by anodic oxidation of water. As a convenient alternative, a silver wire can be substituted for the platinum anode, provided chloride or bromide ions are added to the analyte solution. The anode reaction then becomes

$$Ag(s) + Br^- \rightarrow AgBr(s) + e^-$$

Silver bromide does not interfere with the neutralization reaction.

Coulometric titrations of acids are much less susceptible to the carbonate error encountered in volumetric methods (see Section 16A-3). The error can be avoided if carbon dioxide is removed from the solvent by boiling it or by bubbling an inert gas, such as nitrogen, through the solution for a brief period.

Hydrogen ions generated at the surface of a platinum anode can be used for the coulometric titration of strong as well as weak bases:

$$2H_2O \rightarrow O_2 + 4H^+ + 4e^-$$

In this case, the cathode must be isolated from the analyte solution to prevent interference from hydroxide ion.

---

[7]For a summary of applications, see J. A. Dean, *Analytical Chemistry Handbook*, Section 14, pp. 14.127–14.133, New York: McGraw-Hill, 1995.

TABLE 22-3

Summary of Coulometric Titrations Involving Neutralization, Precipitation, and Complex-formation Reactions

Species Determined	Generator Electrode Reaction	Secondary Analytical Reaction
Acids	$2H_2O + 2e^- \rightleftharpoons 2OH^- + H_2$	$OH^- + H^+ \rightleftharpoons H_2O$
Bases	$H_2O \rightleftharpoons 2H^+ + \frac{1}{2}O_2 + 2e^-$	$H^+ + OH^- \rightleftharpoons H_2O$
Cl^-, Br^-, I^-	$Ag \rightleftharpoons Ag^+ + e^-$	$Ag^+ + X^- \rightleftharpoons AgX(s)$
Mercaptans (RSH)	$Ag \rightleftharpoons Ag^+ + e^-$	$Ag^+ + RSH \rightleftharpoons AgSR(s) + H^+$
Cl^-, Br^-, I^-	$2Hg \rightleftharpoons Hg_2^{2+} + 2e^-$	$Hg_2^{2+} + 2X^- \rightleftharpoons Hg_2X_2(s)$
Zn^{2+}	$Fe(CN)_6^{3-} + e^- \rightleftharpoons Fe(CN)_6^{4-}$	$3Zn^{2+} + 2K^+ + Fe(CN)_6^{4-} \rightleftharpoons K_2Zn_3[Fe(CN)_6]_2(s)$
$Ca^{2+}, Cu^{2+}, Zn^{2+}, Pb^{2+}$	See Equation 22-9	$HY^{3-} + Ca^{2+} \rightleftharpoons CaY^{2-} + H^+$, etc.

Precipitation and Complex-Formation Reactions. Coulometric titrations with EDTA are carried out by reduction of the ammine mercury(II) EDTA chelate at a mercury cathode:

$$HgNH_3Y^{2-} + NH_4^+ + 2e^- \rightarrow Hg(l) + 2NH_3 + HY^{3-} \qquad (22\text{-}9)$$

Because the mercury chelate is more stable than the corresponding complexes of calcium, zinc, lead, or copper, complexation of these ions occurs only after the ligand has been freed by the electrode process.

As shown in **Table 22-3**, several precipitating reagents can be generated coulometrically. The most widely used of these is silver ion, which is generated at a silver anode, as discussed in Feature 22-2.

Oxidation/Reduction Titrations. Coulometric titrations have been developed for many, but not all, redox titrations. **Table 22-4** reveals that a variety of redox

---

FEATURE 22-2

Coulometric Titration of Chloride in Biological Fluids

The accepted reference method for determining chloride in blood serum, plasma, urine, sweat, and other body fluids is the coulometric titration procedure.[8] In this technique, silver ions are generated coulometrically. The silver ions then react with chloride ions to form insoluble silver chloride. The end point is usually detected by amperometry (see Section 23C-4) when a sudden increase in current occurs on the generation of a slight excess of Ag^+. In principle, the absolute amount of Ag^+ needed to react quantitatively with Cl^- can be obtained from Faraday's law. In practice, calibration is used. The time t_s required to titrate a chloride standard solution with a known number of moles of chloride, $(n_{Cl^-})_s$, using a constant current I is first measured. The same constant current is next used in the titration of the unknown solution, and the time t_u is measured. The number of moles of chloride in the unknown, $(n_{Cl^-})_u$, is then obtained as follows:

$$(n_{Cl^-})_u = \frac{t_u}{t_s} \times (n_{Cl^-})_s$$

(continued)

❮ CHALLENGE: Derive the equation shown in Feature 22-2 for the number of moles of chloride ion in the unknown. Begin with Faraday's law.

---

[8]L. A Kaplan and A. J. Pesce, *Clinical Chemistry: Theory, Analysis, and Correlation,* St. Louis: Mosby, 1984, p. 1060.

Figure 22F-2   A commercial digital chloridometer. This coulometric titrator is used to determine chloride ion in clinical samples and in food and environmental samples. The end point is detected by a pair of electrodes measuring solution conductance (Courtesy of Wescor, GMBH, Berlin).

If the volumes of the standard solution and the unknown solution are the same, concentrations can be substituted for number of moles in the above equation. A commercial coulometric titrator called a chloridometer is shown in **Figure 22F-2**.

Other popular methods for chloride determination are ion-selective electrodes (see Section 21D), photometric titrations (see Section 26A-4), and isotope dilution mass spectrometry.

reagents can be generated coulometrically. For example, the coulometric generation of bromine forms the basis for a large number of coulometric methods. Of interest as well are reagents, such as silver(II), manganese(III), and the chloride complex of copper(I), which are too unstable to be used in conventional volumetric analysis.

TABLE 22-4

Summary of Coulometric Titrations Involving Oxidation/Reduction Reactions

Reagent	Generator Electrode Reaction	Substance Determined
Br_2	$2Br^- \rightleftharpoons Br_2 + 2e^-$	As(III), Sb(III), U(IV), Ti(I), I^-, SCN^-, NH_3, N_2H_4, NH_2OH, phenol, aniline, mustard gas, mercaptans, 8-hydroxyquinoline, oleins
Cl_2	$2Cl^- \rightleftharpoons Cl_2 + 2e^-$	As(III), I^-, styrene, fatty acids
I_2	$2I^- \rightleftharpoons I_2 + 2e^-$	As(III), Sb(III), $S_2O_3^{2-}$, H_2S, ascorbic acid
Ce^{4+}	$Ce^{3+} \rightleftharpoons Ce^{4+} + e^-$	Fe(II), Ti(III), U(IV), As(III), I^-, $Fe(CN)_6^{4-}$
Mn^{3+}	$Mn^{2+} \rightleftharpoons Mn^{3+} + e^-$	$H_2C_2O_4$, Fe(II), As(III)
Ag^{2+}	$Ag^+ \rightleftharpoons Ag^{2+} + e^-$	Ce(III), V(IV), $H_2C_2O_4$, As(III)
Fe^{2+}	$Fe^{3+} + e^- \rightleftharpoons Fe^{2+}$	Cr(VI), Mn(VII), V(V), Ce(IV)
Ti^{3+}	$TiO^{2+} + 2H^+ + e^- \rightleftharpoons Ti^{3+} + H_2O$	Fe(III), V(V), Ce(IV), U(VI)
$CuCl_3^{2-}$	$Cu^{2+} + 3Cl^- + e^- \rightleftharpoons CuCl_3^{2-}$	V(V), Cr(VI), IO_3^-
U^{4+}	$UO_2^{2+} + 4H^+ + 2e^- \rightleftharpoons U^{4+} + 2H_2O$	Cr(VI), Ce(IV)

Automatic Coulometric Titrations

Several instrument manufacturers offer automatic coulometric titrators, most of which employ a potentiometric end point. Some of these instruments are multipurpose and can be used for the determination of a variety of species. Others are designed for a single type of analysis. Examples of the latter are chloride titrators in which silver ion is generated coulometrically; sulfur dioxide monitors in which anodically generated bromine oxidizes the analyte to sulfate ions; carbon dioxide monitors in which the gas, absorbed in monoethanolamine, is titrated with coulometrically generated base; and water titrators in which Karl Fischer reagent (see Section 20C-5) is generated electrolytically.

 Spreadsheet Summary   In the second experiment in Chapter 11 of *Applications of Microsoft® Excel in Analytical Chemistry*, 2nd ed., a spreadsheet is developed to plot a coulometric titration curve. The end point is located by first- and second-derivative methods.

WEB WORKS   Go to **www.cengage.com/chemistry/skoog/fac9** and choose Chapter 22. Click on the link to Bioanalytical Systems. Investigate the electrochemical instruments produced by this instrument company. In particular, determine if this company makes a cell for bulk electrolysis. If so, describe the cell and list its features and specifications. Use a search engine and search for companies that make coulometers. Compare the features of two instruments from two different instrument companies.

QUESTIONS AND PROBLEMS

22-1. Briefly distinguish between
   *(a) concentration polarization and kinetic polarization.
   (b) a coulomb and an ampere.
   *(c) diffusion and migration.
   (d) a working electrode and a reference electrode.
   *(e) the electrolysis circuit and the control circuit for controlled-potential methods.

22-2. Briefly define
   *(a) ohmic potential.
   (b) overvoltage.
   *(c) controlled-potential electrolysis.
   (d) coulometric titration.
   *(e) current efficiency.
   (f) galvanostat.

*22-3. Describe three mechanisms responsible for the transport of dissolved species to and from an electrode surface.

22-4. How does a current in an electrochemical cell affect its potential?

*22-5. What experimental variables affect concentration polarization in an electrochemical cell?

22-6. How do concentration polarization and kinetic polarization resemble one another? How do they differ?

*22-7. Describe conditions that favor kinetic polarization in an electrochemical cell.

22-8. What is a supporting electrolyte and what is its role in electrochemistry?

*22-9. How do electrogravimetric and coulometric methods differ from potentiometric methods? Consider currents, voltages, and instrumentation in your answer.

22-10. What is the purpose of a depolarizer?

*22-11. Why is the working electrode normally isolated from the counter electrode in a controlled-potential coulometric analysis?

22-12. Why is an auxiliary reagent always required in a coulometric titration?

22-13. Determine the number of ions undergoing electron transfer at the surface of an electrode during each second that an electrochemical cell is operated at 0.0175 A at 100% current efficiency and the participating ions are
(a) univalent.
*(b) divalent.
(c) trivalent.

22-14. Calculate the theoretical potential at 25°C needed to initiate the deposition of
*(a) copper from a solution that is 0.250 M in Cu^{2+} and buffered to a pH of 3.00. Oxygen is evolved at the anode at 1.00 atm.
(b) tin from a solution that is 0.220 M in Sn^{2+} and buffered to a pH of 4.00. Oxygen is evolved at the anode at 770 torr.
*(c) silver bromide on a silver anode from a solution that is 0.0964 M in Br^- and buffered to a pH of 3.70. Hydrogen is evolved at the cathode at 765 torr.
(d) Tl_2O_3 from a solution that is 5.00×10^{-3} M in Tl^+ and buffered to a pH of 7.50. The solution is also made 0.010 M in Cu^{2+}, which acts as a cathode depolarizer for the process

$$Tl_2O_3 + 3H_2O + 4e^- \rightleftharpoons 2Tl^+ + 6OH^-$$
$$E^0 = 0.020 \text{ V}$$

*22-15.** Calculate the initial potential needed for a current of 0.065 A in the cell

$$Co|Co^{2+}(5.90 \times 10^{-3} \text{ M})||Zn^{2+}(2.95 \times 10^{-3} \text{ M})|Zn$$

if this cell has a resistance of 4.50 Ω.

22-16. The cell

$$Sn|Sn^{2+}(7.83 \times 10^{-4} \text{ M})||Cd^{2+}(6.59 \times 10^{-2} \text{ M})|Cd$$

has a resistance of 4.95 Ω. Calculate the initial potential that will be needed for a current of 0.062 A in this cell.

*22-17.** Copper is to be deposited from a solution that is 0.250 M in Cu(II) and is buffered to a pH of 4.00. Oxygen is evolved from the anode at a partial pressure of 730 torr. The cell has a resistance of 3.60 Ω, and the temperature is 25°C. Calculate
(a) the theoretical potential needed to initiate deposition of copper from this solution.
(b) the *IR* drop associated with a current of 0.15 A in this cell.
(c) the initial potential, given that the overvoltage of oxygen is 0.50 V under these conditions.
(d) the potential of the cell when $[Cu^{2+}]$ is 7.00×10^{-6}, assuming that *IR* drop and O_2 overvoltage remain unchanged.

22-18. Nickel is to be deposited on a platinum cathode (area = 120 cm^2) from a solution that is 0.150 M in Ni^{2+} and buffered to a pH of 2.00. Oxygen is evolved at a partial pressure of 1.00 atm at a platinum anode with an area of 80 cm^2. The cell has a resistance of 3.55 Ω, and the temperature is 25°C. Calculate
(a) the thermodynamic potential needed to initiate the deposition of nickel.
(b) the *IR* drop for a current of 1.00 A.
(c) the current density at the anode and the cathode.
(d) the initial applied potential, given that the overvoltage of oxygen on platinum is approximately 0.52 V under these conditions.
(e) the applied potential when the nickel concentration has decreased to 1.00×10^{-4} M (assume that all variables other than $[Ni^{2+}]$ remain constant).

*22-19.** A solution is 0.200 M in Co^{2+} and 0.0650 M in Cd^{2+}. Calculate
(a) the Co^{2+} concentration in the solution as the first cadmium starts to deposit.
(b) the cathode potential needed to lower the Co^{2+} concentration to 1.00×10^{-5} M.
(c) based on (a) and (b) above, can Co^{2+} be quantitatively separated from Cd^{2+}?

22-20. A solution is 0.0450 M in BiO^+ and 0.0350 M in Co^{2+} and has a pH of 2.50.
(a) What is the concentration of the more easily reduced cation at the onset of deposition of the less reducible one?
(b) What is the potential of the cathode when the concentration of the more easily reduced species is 1.00×10^{-6} M?
(c) Can we achieve a quantitative separation based on your results in (a) and (b) above?

*22-21.** Electrogravimetric analysis with control of the cathode potential is proposed as a means for separating Bi^{3+} and Sn^{2+} in a solution that is 0.250 M in each ion and buffered to pH 1.95.
(a) Calculate the theoretical cathode potential at the start of deposition of the more easily reduced ion.
(b) Calculate the residual concentration of the more readily reduced species at the outset of the deposition of the less easily reduced species.
(c) Propose a range (versus SCE), if such exists, within which the cathode potential should be maintained. Consider a residual concentration less than 10^{-6} M as constituting quantitative removal.

22-22. A solution is 0.200 M in each of two reducible cations, A and B. Removal of the more reducible species (A) is deemed complete when [A] has been decreased to 1.00×10^{-5} M. What minimum difference in standard electrode potentials will permit the isolation of A without interference from B when

A is	B is
*(a) univalent	univalent
(b) divalent	univalent
*(c) trivalent	univalent
(d) univalent	divalent
*(e) divalent	divalent
(f) trivalent	divalent
*(g) univalent	trivalent
(h) divalent	trivalent
*(i) trivalent	trivalent

*22-23. Calculate the time needed for a constant current of 0.8510 A to deposit 0.250 g of Co(II) as
(a) elemental cobalt on the surface of a cathode.
(b) Co_3O_4 on an anode.
Assume 100% current efficiency for both gases.

22-24. Calculate the time needed for a constant current of 1.00 A to deposit 0.450 g of
(a) Tl(III) as the element on a cathode.
(b) Tl(I) as Tl_2O_3 on an anode.
(c) Tl(I) as the element on a cathode.

*22-25. A 0.1330-g sample of a purified organic acid was neutralized by the hydroxide ion produced in 5 min and 24 s by a constant current of 300 mA. Calculate the equivalent mass of the acid in grams.

22-26. The CN^- concentration of 10.0 mL of a plating solution was determined by titration with electrogenerated hydrogen ion to a methyl orange end point. A color change occurred after 4 min and 11 s with a current of 55.6 mA. Calculate the number of grams of NaCN per liter of solution. Also calculate the number of ppm of NaCN in the solution.

*22-27. An excess of $HgNH_3Y^{2-}$ was introduced to 25.00 mL of well water. Express the hardness of the water in terms of ppm $CaCO_3$ if the EDTA needed for the titration was generated at a mercury cathode (Equation 22-9) in 3.52 min by a constant current of 39.4 mA. Assume 100% current efficiency.

22-28. Electrolytically generated I_2 was used to determine the amount of H_2S in 100.0 mL of brackish water. Following addition of excess KI, a titration at a constant current of 46.3 mA required 11.05 min. The reaction was

$$H_2S + I_2 \rightarrow S(s) + 2H^+ + 2I^-$$

Express the results of the analysis in terms of ppm H_2S.

*22-29. The nitrobenzene in 300 mg of an organic mixture was reduced to phenylhydroxylamine at a constant potential of −0.96 V (versus SCE) applied to a mercury cathode:

$$C_6H_5NO_2 + 4H^+ + 4e^- \rightarrow C_6H_5NHOH + H_2O$$

The sample was dissolved in 100 mL of methanol. After electrolysis for 30 min, the reaction was judged complete. An electronic coulometer in series with the cell indicated that the reduction required 33.47 C. Calculate the percentage of $C_6H_5NO_2$ in the sample.

22-30. The phenol content of water downstream from a coking furnace was determined by coulometric analysis. A 100-mL sample was rendered slightly acidic, and an excess of KBr was introduced. To produce Br_2 for the reaction

$$C_6H_5OH + 3Br_2 \rightarrow BrC_5H_5OH(s) + 2HBr$$

a steady current of 0.0503 A for 6 min and 22 s was required. Express the results of this analysis in terms of parts of C_6H_5OH per million parts of water. (Assume that the density of water is 1.00 g/mL.)

22-31. At a potential of −1.0 V (versus SCE), CCl_4 in methanol is reduced to $CHCl_3$ at a mercury cathode:

$$2CCl_4 + 2H^+ + 2e^- + 2Hg(l) \rightarrow 2CHCl_3 + Hg_2Cl_2(s)$$

At −1.80 V, the $CHCl_3$ further reacts to give CH_4:

$$2CHCl_3 + 6H^+ + 6e^- + 6Hg(l) \rightarrow 2CH_4 + 3Hg_2Cl_2(s)$$

Several different 0.750-g samples containing CCl_4, $CHCl_3$, and inert organic species were dissolved in methanol and electrolyzed at −1.0 V until the current approached zero. A coulometer indicated the charge required to complete the reaction as given in the middle column of the table below. The potential of the cathode was then adjusted to −1.8 V. The additional charge given in the last column of the table was required at this potential.

Sample No.	Charge required at −1.0 V, C	Charge required at −1.8 V, C
1	11.63	68.60
2	21.52	85.33
3	6.22	45.98
4	12.92	55.31

Calculate the percentage of CCl_4 and $CHCl_3$ in each mixture.

22-32. A single mixture containing only $CHCl_3$ and CH_2Cl_2 was divided into five parts to obtain samples for replicate determinations. Each sample was dissolved in methanol and electrolyzed in a cell containing a mercury cathode. The potential of the cathode was held constant at −1.80 V (versus SCE). Both compounds were reduced to CH_4 (see Problem 22-31 for the reaction). Calculate the mean value of the percentages of $CHCl_3$ and CH_2Cl_2 in the mixture. Find the standard deviations and the relative standard deviations.

Sample	Mass of sample, g	Charge required, C
1	0.1309	306.72
2	0.1522	356.64
3	0.1001	234.54
4	0.0755	176.91
5	0.0922	216.05

22-33. Construct a coulometric titration curve of 100.0 mL of a 1 M H_2SO_4 solution containing Fe(II) titrated with Ce(IV) generated from 0.075 M Ce(III). The titration is monitored by potentiometry. The initial amount of Fe(II) present is 0.05182 mmol. A constant current of 20.0 mA is used. Find the time corresponding to the equivalence point. Then, for about ten values of time prior to the equivalence point, use the stoichiometry of the reaction to calculate the amount of Fe^{3+} produced and the amount of Fe^{2+} remaining. Use the Nernst equation to find the system potential. Find the equivalence-point potential in the usual manner for a redox titration. For about ten times after the equivalence point, calculate the amount of Ce^{4+} produced from the electrolysis and the amount of Ce^{3+} remaining. Plot the curve of system potential versus electrolysis time.

*22-34. Traces of aniline, $C_6H_5NH_2$, in drinking water can be determined by reaction with an excess of electrolytically generated Br_2:

aniline

The polarity of the working electrode is then reversed, and the excess Br_2 is determined by a coulometric titration involving the generation of Cu(I):

$$Br_2 + 2Cu^+ \rightarrow 2Br^- + 2Cu^{2+}$$

Suitable quantities of KBr and $CuSO_4$ were added to a 25.0-mL sample containing aniline. Calculate the number of micrograms of $C_6H_5NH_2$ in the sample from the data:

Working Electrode Functioning As	Generation Time with Constant Current of 1.51 mA, min
Anode	3.76
Cathode	0.270

*22-35. Quinone can be reduced to hydroquinone with an excess of electrolytically generated Sn(II):

quinone   hydroquinone

The polarity of the working electrode is then reversed, and the excess Sn(II) is oxidized with Br_2 generated in a coulometric titration:

$$Sn^{2+} + Br_2 \rightarrow Sn^{4+} + 2\ Br^-$$

Appropriate quantities of $SnCl_4$ and KBr were added to a 50.0-mL sample. Calculate the mass of $C_6H_4O_2$ in the sample from the data:

Working Electrode Functioning As	Generation Time with Constant Current of 1.062 mA, min
Cathode	8.34
Anode	0.691

22-36. Challenge Problem: Sulfide ion (S^{2-}) is formed in wastewater by the action of anaerobic bacteria on organic matter. Sulfide can be readily protonated to form volatile, toxic H_2S. In addition to the toxicity and noxious odor, sulfide and H_2S cause corrosion problems because they can be easily converted to sulfuric acid when conditions change to aerobic. One common method to determine sulfide is by coulometric titration with generated silver ion. At the generator electrode, the reaction is $Ag \rightarrow Ag^+ + e^-$. The titration reaction is $S^{2-} + 2Ag^+ \rightarrow Ag_2S(s)$.

(a) A digital chloridometer was used to determine the mass of sulfide in a wastewater sample. The chloridometer reads out directly in ng Cl^-. In chloride determinations, the same generator reaction is used, but the titration reaction is $Cl^- + Ag^+ \rightarrow AgCl(s)$. Derive an equation that relates the desired quantity, ng S^{2-}, to the chloridometer readout in ng Cl^-.

(b) A particular wastewater standard gave a reading of 1689.6 ng Cl^-. What total charge in coulombs was required to generate the Ag^+ needed to precipitate the sulfide in this standard?

(c) The following results were obtained on 20.00-mL samples containing known amounts of sulfide (D. T. Pierce, M. S. Applebee, C. Lacher, and

J. Bessie, *Environ. Sci. Technol.*, **1998**, *32*, 1734, **DOI**: 10.1021/es970924v). Each standard was analyzed in triplicate and the mass of chloride recorded. Convert each of the chloride results to ng S^{2-}.

Known mass S^{2-}, ng	Mass Cl^- determined, ng		
6365	10447.0	10918.1	10654.9
4773	8416.9	8366.0	8416.9
3580	6528.3	6320.4	6638.9
1989	3779.4	3763.9	3936.4
796	1682.9	1713.9	1669.7
699	1127.9	1180.9	1174.3
466	705.5	736.4	707.7
373	506.4	521.9	508.6
233	278.6	278.6	247.7
0	−22.1	−19.9	−17.7

(d) Determine the average mass of S^{2-} in ng, the standard deviation, and the % RSD of each standard.

(e) Prepare a plot of the average mass of S^{2-} determined (ng) versus the actual mass (ng). Determine the slope, the intercept, the standard error, and the R^2 value. Comment on the fit of the data to a linear model.

(f) Determine the detection limit (DL) in ng and in ppm using a k factor of 2 (see Equation 8-22).

(g) An unknown wastewater sample gave an average reading of 893.2 ng Cl^-. What is the mass of sulfide in ng? If 20.00 mL of the wastewater sample was introduced into the titration vessel, what is the concentration of S^{2-} in ppm?

CHAPTER 23

Voltammetry

Courtesy of Bioanalytical Systems, Inc.

Voltammetric methods are based on measuring current as a function of the potential applied to a small electrode.

Polarography is voltammetry at the dropping mercury electrode.

Lead poisoning in children can cause anorexia, vomiting, convulsions, and permanent brain damage. Lead can enter drinking water by being leached from the solder used to join copper pipes and tubes. Anodic stripping voltammetry, discussed in this chapter, is one of the most sensitive analytical methods for determining heavy metals like lead. Shown in the photo is a three-electrode cell used for anodic stripping voltammetry. The working electrode is a glassy carbon electrode on which a thin mercury film has been deposited. An electrolysis step is used to deposit lead into the mercury film as an amalgam. After the electrolysis step, the potential is scanned anodically toward positive values to oxidize (strip) the metal from the film. Levels as low as a few parts per billion can be detected.

The term **voltammetry** refers to a group of electroanalytical methods in which we acquire information about the analyte by measuring current in an electrochemical cell as a function of applied potential. We obtain this information under conditions that promote polarization of a small indicator, or working, electrode. When current proportional to analyte concentration is monitored at a fixed potential, the technique is called **amperometry**. To enhance polarization, working electrodes in voltammetry and amperometry have surface areas of a few square millimeters at the most and in some applications, a few square micrometers or less. Voltammetry is widely used by inorganic, physical, and biological chemists for fundamental studies of oxidation and reduction processes in various media, adsorption processes on surfaces, and electron transfer mechanisms at chemically modified electrode surfaces.

In voltammetry, the current that develops in an electrochemical cell is measured under conditions of complete concentration polarization. Recall from Section 22A-2 that a polarized electrode is one to which we have applied a voltage in excess of that predicted by the Nernst equation to cause oxidation or reduction to occur. In contrast, potentiometric measurements are made at currents that approach zero and where polarization is absent. Voltammetry differs from coulometry in that, with coulometry, measures are taken to minimize or compensate for the effects of concentration polarization. Furthermore, in voltammetry, there is minimal consumption of analyte, while in coulometry essentially all of the analyte is converted to another state.

Historically, the field of voltammetry developed from **polarography**, which is a particular type of voltammetry that was invented by the Czechoslovakian chemist Jaroslav Heyrovsky in the early 1920s.[1] Polarography differs from other types of voltammetry in that the working

[1]J. Heyrovsky, *Chem. Listy,* **1922**, *16,* 256. Heyrovsky was awarded the 1959 Nobel Prize in chemistry for his discovery and development of polarography.

electrode is the unique **dropping mercury electrode**. At one time, polarography was an important tool used by chemists for the determination of inorganic ions and certain organic species in aqueous solutions. In recent years, the number of applications of polarography in the analytical laboratory has declined dramatically. This decline has been largely a result of concerns about the use of mercury in the laboratory and possible contamination of the environment, the somewhat cumbersome nature of the apparatus, and the broad availability of faster and more convenient (mainly spectroscopic) methods. Because both working and teaching laboratories still perform polarography experiments, we include an abbreviated discussion of it in Section 23D.

While polarography has declined in importance, voltammetry and amperometry at working electrodes other than the dropping mercury electrode have grown at an astonishing pace. Furthermore, voltammetry and amperometry coupled with liquid chromatography have become powerful tools for the analysis of complex mixtures. Modern voltammetry also continues to be an excellent tool in diverse areas of chemistry, biochemistry, materials science and engineering, and the environmental sciences for studying oxidation, reduction, and adsorption processes.[2]

Jaroslav Heyrovsky was born in Prague in 1890. He was awarded the 1959 Nobel Prize in chemistry for his discovery and development of polarography. His invention of the polarographic method dates from 1922, and he concentrated the remainder of his career to the development of this new branch of electrochemistry. He died in 1967.

23A EXCITATION SIGNALS IN VOLTAMMETRY

In voltammetry, a variable potential excitation signal is impressed on a working electrode in an electrochemical cell. This excitation signal produces a characteristic current response, which is the measurable quantity. The waveforms of four of the most common excitation signals used in voltammetry are shown in **Figure 23-1**. The classical voltammetric excitation signal is the linear scan shown in Figure 23-1a in which the voltage applied to the cell increases linearly (usually over a 2- to 3-V range) as a function of time. The current in the cell is then recorded as a function of time and thus as a function of the applied voltage. In amperometry, current is recorded at fixed applied voltage.

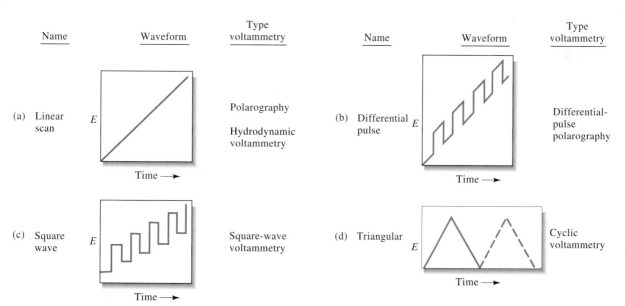

Figure 23-1 Voltage versus time excitation signals used in voltammetry.

[2]Some general references on voltammetry include A. J. Bard and L. R. Faulkner, *Electrochemical Methods*, 2nd ed., New York: Wiley, 2001; S. P. Kounaves, in *Handbook of Instrumental Techniques for Analytical Chemistry*, Frank A. Settle, ed., Upper Saddle River, NJ: Prentice-Hall, 1997, pp. 711–28; *Laboratory Techniques in Electroanalytical Chemistry*, 2nd ed., P. T. Kissinger and W. R. Heineman, eds., New York: Marcel Dekker, 1996; M. R. Smyth and F. G. Vos, eds., *Analytical Voltammetry*, New York: Elsevier, 1992.

Two pulse excitation signals are shown in Figures 23-1b and 1c. Currents are measured at various times during the lifetime of these pulses. With the triangular waveform shown in Figure 23-1d, the potential is cycled between two values, first increasing linearly to a maximum and then decreasing linearly with the same slope to its original value. This process may be repeated numerous times as the current is recorded as a function of time. A complete cycle may take 100 or more seconds or be completed in less than one second.

To the right of each of the waveforms of Figure 23-1 is listed the types of voltammetry that use the various excitation signals. We discuss these techniques in the sections that follow.

23B VOLTAMMETRIC INSTRUMENTATION

Figure 23-2 shows the components of a simple apparatus for carrying out linear-sweep voltammetric measurements. The cell is made up of three electrodes immersed in a solution containing the analyte and also an excess of a nonreactive electrolyte called a **supporting electrolyte**. (Note the similarity of this cell to the one for controlled-potential electrolysis shown in Figure 22-8.) One of the three electrodes is the **working electrode** (WE), whose potential versus a reference electrode is varied linearly with time. The dimensions of the working electrode are kept small to enhance its tendency to become polarized. The reference electrode (RE) has a potential that remains constant throughout the experiment. The third electrode is a **counter electrode**(CE), which is often a coil of platinum wire or a pool of mercury. The current in the cell passes between the working electrode and the counter electrode.[3]

A **supporting electrolyte** is a salt added in excess to the analyte solution. Most commonly, it is an alkali metal salt that does not react at the working electrode at the potentials being used. The salt reduces the effects of migration and lowers the resistance of the solution.

The **working electrode** is the electrode at which the analyte is oxidized or reduced. The potential between the working electrode and the **reference electrode** is controlled. Electrolysis current passes between the working electrode and a **counter electrode**.

Figure 23-2 A manual potentiostat for voltammetry.

---

[3]Early voltammetry was performed with a two-electrode system rather than the three-electrode system shown in Figure 23-2. With a two-electrode system, the second electrode is either a large metal electrode or a reference electrode large enough to prevent its polarization during an experiment. This second electrode combines the functions of the reference electrode and the counter electrode in Figure 23-2. In the two-electrode system, we assume that the potential of this second electrode is constant throughout a scan so that the microelectrode potential is simply the difference between the applied potential and the potential of the second electrode. With solutions of high electrical resistance, however, this assumption is not valid because the *IR* drop is significant and increases as the current increases. Distorted voltammograms are the result. Almost all voltammetry is now performed with three-electrode systems.

The signal source is a variable dc voltage source consisting of a battery in series with a variable resistor R. The desired excitation potential is selected by moving the contact C to the proper position on the resistor. The digital voltmeter has such a high electrical resistance ($>10^{11}\Omega$) that there is essentially no current in the circuit containing the meter and the reference electrode. Thus, virtually all the current from the source passes between the counter electrode and the working electrode. A voltammogram is recorded by moving the contact C in Figure 23-2 and recording the resulting current as a function of the potential between the working electrode and the reference electrode.

In principle, the manual potentiostat of Figure 23-2 could be used to generate a linear-sweep voltammogram. In such an experiment, contact C is moved at a constant rate from A to B to produce the excitation signal shown in Figure 23-1a. The current and voltage are then recorded at consecutive equal time intervals during the voltage (or time) scan. In modern voltammetric instruments, however, the excitation signals shown in Figure 23-1 are generated electronically. These instruments vary the potential in a systematic way with respect to the reference electrode and record the resulting current. The independent variable in this experiment is the potential of the working electrode versus the reference electrode and not the potential between the working electrode and the counter electrode. A potentiostat that is designed for linear-sweep voltammetry is described in Feature 23-1. Figure 23F-2 is a schematic showing the components of a modern operational amplifier potentiostat (see Section 22C-2) for carrying out linear-scan voltammetric measurements.

FEATURE 23-1

Voltammetric Instruments Based on Operational Amplifiers

In Feature 21-5, we described the use of operational amplifiers (op amps) to measure the potential of electrochemical cells. "Op amps" also can be used to measure currents and accomplish a variety of other control and measurement tasks. Consider the measurement of current, as illustrated in **Figure 23F-1**.

Figure 23F-1 An op amp circuit for measuring voltammetric current.

In this circuit, a voltage source E is attached to one electrode of an electrochemical cell, which produces a current I in the cell. Because of the very high input resistance of the op amp, essentially all the current passes through the resistor R to the output of the op amp. The voltage at the output of the op amp is given by $E_{out} = -IR$. The minus

(continued)

sign arises because the amplifier output voltage E_{out} must be opposite in sign to the voltage drop across resistance R for the potential difference between the op amp inputs to be close to zero volts. By solving this equation for I, we have

$$I = \frac{-E_{out}}{R}$$

In other words, the current in the electrochemical cell is proportional to the voltage output of the op amp. The value of the current can then be calculated from the measured values of E_{out} and the resistance R. The circuit is called a **current-to-voltage converter**.

Op amps can be used to construct an automatic three-electrode potentiostat, as illustrated in **Figure 23F-2**. Notice that the current-measuring circuit of Figure 23F-1 is connected to the working electrode of the cell (op amp C). The reference electrode is attached to a voltage follower (op amp B). As discussed in Feature 21-5, the voltage follower monitors the potential of the reference electrode without drawing any current from the cell. The output of op amp B, which is the reference electrode potential, feeds back to the input of op amp A to complete the circuit. The functions of op amp A are (1) to provide the current in the electrochemical cell between the counter electrode and the working electrode and (2) to maintain the potential difference between the reference electrode and the working electrode at a value provided by the linear-sweep voltage generator. In operation, the voltage generator sweeps the potential between the reference and working electrodes, as shown in Figure 23-1a, and the current in the cell is monitored by op amp C. The output voltage of op amp C, which is proportional to the current I in the cell, is recorded or acquired by a computer for data analysis and presentation.[4]

Figure 23F-2 An op amp potentiostat. The three-electrode cell has a working electrode (WE), reference electrode (RE), and a counterelectrode (CE).

The electrical resistance of the control circuit containing the reference electrode is so large ($>10^{11}\Omega$) that it draws essentially no current. Therefore, the entire current from the source is carried from the counter electrode to the working electrode. Furthermore, the control circuit adjusts this current so that the potential between the working electrode and the reference electrode is identical to the output potential from the linear voltage generator. The resulting current, which is directly proportional to

---

[4]For a complete discussion of op amp three-electrode potentiostats, see P. T. Kissinger, in *Laboratory Techniques in Electroanalytical Chemistry*, P. T. Kissinger and W. R. Heineman, eds., New York: Marcel Dekker, 1996, pp. 165–94.

the potential between the working/reference-electrode pair, is then converted to a voltage and recorded as a function of time by the data acquisition system. It is important to emphasize that the independent variable in this experiment is the potential of the working electrode versus the reference electrode and not the potential between the working electrode and the counter electrode. The working electrode is held very close to ground potential (virtual common) throughout the course of the experiment by op amp *C*.

23B-1 Working Electrodes[5]

The working electrodes used in voltammetry take a variety of shapes and forms. Often, they are small flat disks of a conductor that are press fitted into a rod of an inert material, such as Teflon or Kel-F that has imbedded in it a wire contact (see **Figure 23-3a**). The conductor may be anoble metal, such as platinum or gold; carbon paste, carbon fiber, pyrolytic graphite, glassy carbon, diamond, or carbon nanotubes; a semiconductor, such as tin or indium oxide; or a metal coated with a film of mercury. As shown in **Figure 23-4**, the range of potentials that can be used with these electrodes in aqueous solutions varies and depends not only on electrode material but also on the composition of the solution in which it is immersed. Generally, the positive potential limitations are caused by the large currents that develop due to oxidation of the water to give molecular oxygen. The negative limits arise from the reduction of water to produce hydrogen. Note that relatively large negative potentials can be tolerated with mercury electrodes because of the high overvoltage of hydrogen on this metal.

> Large negative potentials can be used with mercury electrodes.

Mercury working electrodes have been widely used in voltammetry for several reasons. One is the relatively large negative potential range just described. An additional advantage of mercury electrodes is that many metal ions are reversibly reduced to amalgams at the surface of a mercury electrode, simplifying the chemistry. Mercury electrodes take several forms. The simplest is a mercury film electrode formed by electrodeposition of the metal onto a disk electrode, such as that shown in Figure 23-3a. **Figure 23-3b** illustrates a hanging mercury drop electrode (HMDE). This electrode, which is available from commercial sources, consists of a very fine capillary tube connected to a mercury-containing reservoir. The metal is forced out of the capillary by a piston arrangement driven by a micrometer screw. The micrometer permits formation of drops having surface areas that are reproducible to 5 percent or better.

> Metals which are soluble in mercury form liquid alloys known as amalgams.

Figure 23-3c shows a typical commercial microelectrode. Such electrodes consist of small diameter metal wires or fibers (5 to 100 μm) sealed within tempered glass bodies. The flattened end of the microelectrode is polished to a mirror finish, which can be maintained using alumina and/or diamond polish. The electrical connection is a 0.060″ gold-plated pin. Microelectrodes are available in a variety of materials including carbon fiber, platinum, gold, and silver. Other materials can be incorporated into microelectrodes if they are available as a wire or a fiber and form a good seal with epoxy. The electrode shown is approximately 7.5 cm long and 4 mm outside diameter.

> Historically, working electrodes with surface areas smaller than a few square millimeters were called **microelectrodes**. Recently, this term has come to signify electrodes with areas on the micrometer scale. In the older literature, micrometer-sized electrodes were sometimes called **ultramicroelectrodes**.

---

[5]Many of the working electrodes that we describe in this chapter have dimensions in the millimeter range. There is now intense interest in studies with electrodes having dimensions in the micrometer range and smaller. We will term such electrodes *microelectrodes*. Such electrodes have several advantages over classical working electrodes. We shall describe some of the unique characteristics of microelectrodes in Section 23I.

Figure 23-3 Some common types of commercial voltammetric electrodes. (a) Disk electrode. (b) Hanging mercury drop electrode (HMDE). (c) Microelectrode. (d) Sandwich-type flow electrode. (e) Dropping mercury electrode (DME). (Reprinted by permission of Bioanalytical Systems, Inc., West Lafayette, IN.)

Figure 23-4 Potential ranges for three types of electrodes in various supporting electrolytes. (A. J. Bard and L. R. Faulkner, *Electrochemical Methods,* 2nd ed., New York: Wiley, 2001, back cover. Reprinted by permission of John Wiley & Sons, Inc.)

Figure 23-3d shows a commercially available sandwich-type working electrode for voltammetry (or amperometry) in flowing streams. The block is made of polyetheretherketone (PEEK) and is available in several formats with different size electrodes (3 mm and 6 mm; see the green area in the figure) and various arrays (dual 3 mm and quad 2 mm). See Figure 23-15 and 23-16 for a diagram showing how the electrodes are used in flowing streams. The working electrodes can be made of glassy carbon, carbon paste, gold, copper, nickel, platinum, or other suitable custom materials.

Figure 23-3e shows a typical dropping mercury electrode (DME), which was used in nearly all early polarographic experiments. It consists of roughly 10 cm of a fine capillary tubing (inside diameter = 0.05 mm) through which mercury is forced by a mercury head of perhaps 50 cm. The diameter of the capillary is such that a new drop forms and breaks every 2 to 6 s. The diameter of the drop is 0.5 to 1 mm and is highly reproducible. In some applications, the drop time is controlled by a mechanical knocker that dislodges the drop at a fixed time after it begins to form. Furthermore, a fresh metallic surface is formed by simply producing a new drop. The fresh reproducible surface is important because the currents measured in voltammetry are quite sensitive to cleanliness and freedom from irregularities.

> ◀ The DME has a high overvoltage for the reduction of H^+ and a renewable metal surface with each droplet. Reproducible currents are attained very rapidly with the DME.

23B-2 Modified Electrodes[6]

An active area of research in electrochemistry is the development of electrodes that are produced by chemical modification of various conductive substrates. Such electrodes have been tailored to accomplish a broad range of functions. Modifications include applying irreversibly adsorbing substances with desired functionalities, covalent bonding of components to the surface, and coating the electrode with polymer films or films of other substances.

Modified electrodes have many potential applications. A primary interest has been in the area of electrocatalysis. In this application, electrodes capable of reducing oxygen to water have been sought for use in fuel cells and batteries. Another application is in the production of electrochromic devices that change color on oxidation and reduction. Such devices are used in displays or *smart windows* and *mirrors*. Electrochemical devices that could serve as molecular electronic devices, such as diodes and transistors, are also under intense study. Finally, the most important analytical use for such electrodes is as analytical sensors that are prepared to be selective for a particular species or functional group.

23B-3 Voltammograms

Figure 23-5 illustrates the appearance of a typical linear-scan voltammogram for an electrolysis involving the reduction of an analyte species A to give a product P at a mercury film electrode. In this example, the working electrode is assumed to be connected to the negative terminal of the linear-scan generator so that the applied potentials are given a negative sign as shown. By convention, cathodic currents are always taken to be positive whereas anodic currents are given a negative sign. In this hypothetical experiment, the solution is assumed to be about 10^{-4} M in A, 0.0 M in P,

> ◀ The American sign convention for voltammetry considers cathodic currents to be positive and anodic currents to be negative. Voltammograms are plotted with positive current in the top hemisphere and negative currents in the bottom. For mostly historical reasons, the potential axis is arranged such that potentials become less positive (more negative) going from left to right.

---

[6]For more information, see R. W. Murray, "Molecular Design of Electrode Surfaces," in *Techniques in Chemistry*, Vol. 22, W. Weissberger, founding ed., New York: Wiley, 1992; A. J. Bard, *Integrated Chemical Systems*, New York: Wiley, 1994.

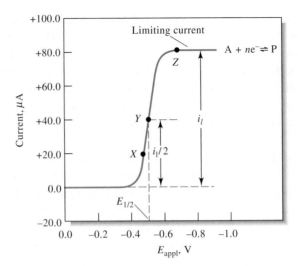

Figure 23-5 Linear-sweep voltammogram for the reduction of a hypothetical species A to give a product P. The limiting current i_l is proportional to the analyte concentration and is used for quantitative analysis. The half-wave potential $E_{1/2}$ is related to the standard potential for the half-reaction and is often used for qualitative identification of species. The half-wave potential is the applied potential at which the current i is $i_{1/2}$.

 A **voltammetric wave** is an ∫-shaped wave that appears in current-voltage plots in voltammetry.

The **limiting current** in voltammetry is the current plateau that is observed at the top of the voltammetric wave. It occurs because the surface concentration of the analyte falls to zero. At this point, the mass-transfer rate is its maximum value. The limiting current plateau is an example of complete concentration polarization.

The **half-wave potential** occurs when the current is equal to one half of the limiting value.

Hydrodynamic voltammetry is a type of voltammetry in which the analyte solution is kept in continuous motion.

and 0.1 M in KCl, which serves as the supporting electrolyte. The half-reaction at the working electrode is the reversible reaction

$$A + ne^- \rightleftharpoons P \qquad E^0 = -0.26\,V \qquad (23\text{-}1)$$

For convenience, we have neglected the charges on A and P and also have assumed that the standard potential for the half-reaction is –0.26 V.

Linear-scan voltammograms generally have a sigmoidal shape and are called **voltammetric waves**. The constant current beyond the steep rise is called the **limiting current**, i_l, because the rate at which the reactant can be brought to the surface of the electrode by mass-transport processes limits the current. Limiting currents are usually directly proportional to reactant concentration. Thus, we may write

$$i_l = kc_A \qquad (23\text{-}2)$$

where c_A is the analyte concentration and k is a constant. Quantitative linear-scan voltammetry relies on this relationship.

The potential at which the current is equal to one half the limiting current is called the **half-wave potential** and given the symbol $E_{1/2}$. After correction for the reference electrode potential (0.242 V with a saturated calomel electrode), the half-wave potential is closely related to the standard potential for the half-reaction but is usually not identical to it. Half-wave potentials are sometimes useful for identification of the components of a solution.

Reproducible limiting currents can be achieved rapidly when either the analyte solution or the working electrode is in continuous and reproducible motion. Linear-scan voltammetry in which the solution or the electrode is in constant motion is called **hydrodynamic voltammetry**. In this chapter, we will focus much of our attention on hydrodynamic voltammetry.

23C HYDRODYNAMIC VOLTAMMETRY

Hydrodynamic voltammetry is performed in several ways. In one method the solution is stirred vigorously while it is in contact with a fixed working electrode. A typical cell for hydrodynamic voltammetry is pictured in **Figure 23-6**. In this cell, stirring is accomplished with an ordinary magnetic stirrer. Another approach is to rotate the working electrode at a constant high speed in the solution to provide the stirring action (see Figure 23-19). Still another way of performing hydrodynamic voltammetry is to pass an analyte solution through a tube fitted with a working electrode (see Figures 23-15 and 23-16). The last technique is widely used for detecting oxidizable or reducible analytes as they exit from a liquid chromatographic column (see Section 33A-5).

As described in Section 22A-2, during an electrolysis, reactant is carried to the surface of an electrode by three mechanisms: migration under the influence of an electric field, convection resulting from stirring or vibration, and diffusion due to concentration differences between the film of liquid at the electrode surface and the bulk of the solution. In voltammetry, we try to minimize the effect of migration by introducing an excess of an inactive supporting electrolyte. When the concentration of supporting electrolyte exceeds that of the analyte by 50- to 100-fold, the fraction of the total current carried by the analyte approaches zero. As a result, the rate of migration of the analyte toward the electrode of opposite charge becomes essentially independent of applied potential.

23C-1 Concentration Profiles at Electrode Surfaces

Throughout this discussion we will consider that the electrode reaction shown in Equation 23-1 takes place at an electrode in a solution of A that also contains an excess of a supporting electrolyte. We assume that the initial concentration of A is c_A while that of the product P is zero. We also assume that the reduction reaction is rapid and reversible so that the concentrations of A and P in the film of solution immediately adjacent to the electrode is given at any instant by the Nernst equation:

$$E_{appl} = E_A^0 - \frac{0.0592}{n} \log \frac{c_P^0}{c_A^0} - E_{ref} \qquad (23\text{-}3)$$

where E_{appl} is the potential between the working electrode and the reference electrode and c_P^0 and c_A^0 are the molar concentrations of P and A in a thin layer of solution at the electrode surface only. We also assume that because the electrode is so very small, the electrolysis, over short periods of time, does not alter the bulk concentration of the solution appreciably. As a result, the concentration of A in the bulk of the solution c_A is unchanged by the electrolysis, and the concentration of P in the bulk of the solution c_P continues to be, for all practical purposes, zero ($c_P \approx 0$).

Profiles for Planar Electrodes in Unstirred Solutions

Before describing the behavior of an electrode in this solution under hydrodynamic conditions, it is instructive to consider what occurs when a potential is applied to a planar electrode, such as that shown in Figure 23-3a, in the absence of convection— that is, in an unstirred solution—and migration. Under these conditions mass transport of the analyte to the electrode surface occurs by diffusion alone.

Let us assume that a pulsed excitation potential E_{appl} is applied to the working electrode for a period of t s, as shown in **Figure 23-7a**. Let us further assume that E_{appl} is large enough so that the ratio c_P^0/c_A^0 in Equation 23-3 is 1000 or greater. Under this condition, the concentration of A at the electrode surface is, for all practical purposes, immediately reduced to zero ($c_A^0 \rightarrow 0$). The current response to this step-excitation signal is shown in Figure 23-7b. Initially, the current rises to a peak value that is required to convert essentially all of A in the surface layer of solution to P. Diffusion from the bulk of the solution then brings more A into this surface layer where further reduction occurs. The current required to keep the concentration of A at the level required by Equation 23-3 decreases rapidly with time, however, because A must travel greater and greater distances to reach the surface layer where it can be reduced. Thus, as seen in Figure 23-7b, the current drops off rapidly after its initial surge.

> Mass-transport processes include diffusion, migration, and convection.

Figure 23-6 A three-electrode cell for hydrodynamic voltammetry.

> Electrolysis at a small voltammetric electrode does not significantly change the bulk concentration of the analyte solution during the course of a voltammetric experiment.

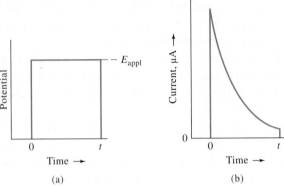

Figure 23-7 Current response to a stepped potential for a planar electrode in an unstirred solution. (a) Excitation potential. (b) Current response.

Figure 23-8 shows concentration profiles for A and P after 0, 1, 5, and 10 ms of electrolysis in the system under discussion. In this example, the concentration of A (solid black lines) and P (solid green lines) are plotted as a function of distance from the electrode surface. The graph on the left shows that the solution is homogeneous before application of the stepped potential with the concentration of A being c_A at the electrode surface and in the bulk of the solution as well; the concentration of P is zero in both of these regions. One millisecond after application of the potential, the profiles have changed dramatically. At the surface of the electrode, the concentration of A has been reduced to essentially zero while the concentration of P has increased and become equal to the original concentration of A, that is, $c_P^0 = c_A$. Moving away from the surface, the concentration of A increases linearly with distance and approaches c_A at about 0.01 mm from the surface. A linear decrease in the concentration of P occurs in this same region. As shown in the figure, with time, these concentration gradients extend farther and farther into the solution. The current i required to produce these gradients is proportional to the slopes of the straight line portions of the solid lines in Figure 23-8b, that is,

$$i = nFAD_A\left(\frac{\partial c_A}{\partial x}\right) \tag{23-4}$$

where i is the current in amperes, n is the number of moles of electrons per mole of analyte, F is the faraday, A is the electrode surface area in cm^2, D_A is the diffusion coefficient for A in $cm^2 s^{-1}$, and c_A is the concentration of A in mol cm^{-3}. As shown

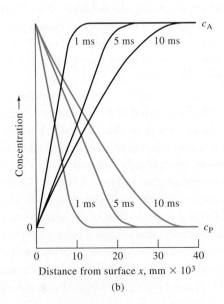

Figure 23-8 Concentration distance profiles during the diffusion-controlled reduction of A to give P at a planar electrode. (a) $E_{appl} = 0$ V. (b) $E_{appl} = $ point Z in Figure 23-5. Elapsed time: 1, 5, and 10 ms.

in the figure, these slopes ($\partial c_A/\partial x$) become smaller with time as does the current. The product $D_A(\partial c_A/\partial x)$ is called the *flux,* which is the number of moles of A per unit time per unit area diffusing to the electrode.

It is not practical to obtain limiting currents with planar electrodes in unstirred solutions because the currents continually decrease with time as the slopes of the concentration profiles become smaller.

Profiles for Electrodes in Stirred Solutions

Let us now consider concentration/distance profiles when the reduction described in the previous section is performed at an electrode immersed in a solution that is stirred vigorously. To understand the effect of stirring, we must develop a picture of liquid flow patterns in a stirred solution containing a small planar electrode. We can identify two types of flow depending on the average flow velocity, as shown in **Figure 23-9**. *Laminar flow* occurs at low flow velocities and has smooth and regular motion, as depicted on the left in the figure. *Turbulent flow*, on the other hand, happens at high velocities and has irregular, fluctuating motion, as shown on the right. In a stirred electrochemical cell, we have a region of turbulent flow in the bulk of solution far from the electrode and a region of laminar flow close to the electrode. These regions are illustrated in **Figure 23-10**. In the laminar flow region, the layers of liquid slide by one another in a direction parallel to the electrode surface. Very near the

Figure 23-9 Visualization of flow patterns in a flowing stream. Turbulent flow, shown on the right, becomes laminar flow as the average velocity decreases to the left. In turbulent flow, the molecules move in an irregular, zigzag fashion, and there are swirls and eddies in the movement. In laminar flow, the streamlines become steady as layers of liquid slide by each other in a regular manner. (From *An Album of Fluid Motion,* assembled by Milton Van Dyke, No. 152, photograph by Thomas Corke and Hassan Nagib, Stanford, CA:Parabolic Press, 1982.)

Figure 23-10 Flow patterns and regions of interest near the working electrode in hydrodynamic voltammetry.

electrode, at a distance δ cm from the surface, frictional forces give rise to a region where the flow velocity is essentially zero. The thin layer of solution in this region is a stagnant layer, called the *Nernst diffusion layer*. It is only within the stagnant Nernst diffusion layer that the concentrations of reactant and product vary as a function of distance from the electrode surface and that there are concentration gradients. In other words, throughout the laminar flow and turbulent flow regions, convection maintains the concentration of A at its original value and the concentration of P at a very small level.

Figure 23-11 shows two sets of concentration profiles for A and P at three potentials shown as X, Y, and Z in Figure 23-5. In Figure 23-11a, the solution is divided into two regions. One makes up the bulk of the solution, and consists of both the turbulent and laminar flow regions shown in Figure 23-10, where mass transport takes place by mechanical convection brought about by the stirrer. The concentration of A throughout this region is c_A, whereas c_P is essentially zero. The second region is the Nernst diffusion layer, which is immediately adjacent to the electrode surface and has a thickness of δ cm. Typically, δ ranges from 10^{-2} to 10^{-3} cm, depending on the efficiency of the stirring and the viscosity of the liquid. Within the static diffusion layer, mass transport takes place by diffusion alone, just

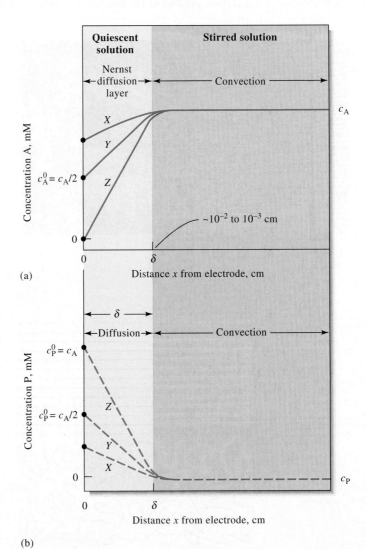

Figure 23-11 Concentration profiles at an electrode/solution interface during the electrolysis $A + ne^- \rightarrow P$ from a stirred solution of A. See Figure 23-5 for potentials corresponding to curves X, Y, and Z.

$$H_2O + I_3^-(aq) + H_3AsO_3(aq) \rightarrow \qquad \leftarrow 3I^-(aq) + H_3AsO_4(aq) + H^+(aq)$$
$$\text{(a)} \qquad\qquad\qquad\qquad \text{(b)}$$

Color Plate 1 Chemical Equilibrium 1: Reaction between iodine and arsenic(III) at pH 1. (a) One mmol I_3^- added to one mmol H_3AsO_3. (b) Three mmol I^- added to one mmol H_3AsO_4. Both combinations of solutions produce the same final equilibrium state (see Section 9B-1, page 202).

$$H_2O + I_3^-(aq) + H_3AsO_3(aq) \rightarrow \qquad \leftarrow 3I^-(aq) + H_3AsO_4(aq) + H^+(aq)$$
$$\text{(a)} \qquad\qquad\qquad\qquad \text{(b)}$$

Color Plate 2 Chemical Equilibrium 2: The same reaction as in color plate 1 carried out at pH 7, producing a different equilibrium state than that in Color Plate 1, and although similar to the situation in Color Plate 1, the same state is produced from either the forward (a) or the reverse (b) direction (see Section 9B-1, page 202).

$$I_3^-(aq) + 2Fe(CN)_6^{4-}(aq) \rightarrow \qquad \leftarrow 3I^-(aq) + 2Fe(CN)_6^{3-}(aq)$$
$$\text{(a)} \qquad\qquad\qquad\qquad\qquad \text{(b)}$$

Color Plate 3 Chemical Equilibrium 3: Reaction between iodine and ferrocyanide. (a) One mmol I_3^- added to two mmol $Fe(CN)_6^{4-}$. (b) Three mmol I^- added to two mmol $Fe(CN)_6^{3-}$. Both combinations of solutions produce the same final equilibrium state. (see Section 9B-1, page 202).

Charles D. Winters

Color Plate 5   Crystallization of sodium acetate from a supersaturated solution (see Section 12A-2, page 280). A tiny seed crystal is dropped into the center of a petri dish containing a supersaturated solution of the compound. The time sequence of photos taken about once per second shows the growth of the beautiful crystals of sodium acetate.

Charles D. Winters

Color Plate 4   The common-ion effect. The test tube on the left contains a saturated solution of silver acetate, AgOAc. The following equilibrium is established in the test tube:

$$AgOAc(s) \rightleftharpoons Ag^+(aq) + OAc^-(aq)$$

When $AgNO_3$ is added to the test tube, the equilibrium shifts to the left to form more AgOAc, as shown in the test tube on the right (see Section 9B-5, page 209).

Color Plate 6 The Tyndall effect. The photo shows two cuvettes: the one on the left contains only water while the one on the right contains a solution of starch. As red and green laser beams pass through the water in the left cuvette, they are invisible. Colloidal particles in the starch solution in the right cuvette scatter the light from the two lasers, so the beams become visible (see Section 12A-2, margin note, page 280).

Color Plate 7 When dimethylglyoxime is added to a slightly basic solution of $Ni^{2+}(aq)$, shown on the left, a bright red precipitate of $Ni(C_4H_7N_2O_2)_2$ is formed as seen in the beaker on the right (see Section 12C-3, page 294).

Methyl violet (0.0–1.6)

Chlorophenol red (4.8–6.7)

Bromophenol blue (3.0–4.6)

m-Cresol purple (7.4–9.0)

Methyl orange (3.2–4.4)

Thymol blue (8.0–9.2)

Bromocresol green (3.8–5.4)

Phenolphthalein (8.0–10.0)

Alizarin red (4.6–6.0)

o-Cresolphthalein (8.2–9.8)

Bromothymol (6.0–7.0)

Thymolphthalein (8.8–10.5)

Charles D. Winters

Color Plate 8 Acid/base indicators and their transition pH ranges (see Section 14A-2, page 323).

Color Plate 9   End point in an acid/base titration with phenolphthalein as indicator. The end point is achieved when the barely perceptible pink color of phenolphthalein persists. The flask on the left shows the titration less than half a drop prior to the end point; the middle flask shows the end point. The flask on the right shows what happens when a slight excess of base is added to the titration mixture. The solution turns a deep pink color, and the end point has been exceeded (see Section 13A-1, page 304).

Color Plate 10   Reduction of silver(I) by direct reaction with copper, forming the "silver tree" (see Section 18A-2, page 445).

$$2Fe^{3+} + 3I^- \rightleftharpoons 2Fe^{2+} + I_3^-$$

Color Plate 12   Reaction between Iron(III) and iodide. The species in each beaker are indicated by the colors of the solutions. Iron (III) is pale yellow, iodide is colorless, and triiodide is intense red-orange (see margin note, Section 18C-6, page 464).

Color Plate 11   A modern version of the Daniell Cell (see Feature 18-2, page 450).

Color Plate 13   The time dependence of the reaction between permanganate and oxalate (see Section 20C-1, page 515).

(a)

Simon Tulloch

(b)

(c)

Simon Tulloch

Color Plate 14 (a) Typical linear CCD arrays for spectrophotometers. The array on the right has 4096 pixels, and the array on the left has 2048 pixels. In both arrays, each pixel has the dimensions of 14 μm $\times$ 14 μm. These devices have a spectral range of 200-1000 nm, a dynamic range of 2500:1 (see Section 8E-2), and are available with low-cost glass or UV-enhanced fused silica windows. In addition to the sizes shown, the arrays are available in lengths of 512 and 1024 pixels. (b) Photomicrograph of a section of a two-dimensional CCD array that is used for imaging and spectroscopy. Light falling on the millions of pixels in the upper left of the photo creates charge that is transferred to the vertical channels at the bottom of the photo and shifted from left to right along the string of channels until it reaches the output amplifier section shown in (c). The amplifier provides a voltage proportional to the charge accumulated on each pixel, which is in turn proportional to the intensity of light falling on the pixel (see Section 25A-4, page 705, for a discussion of charge-transfer devices).

Charles D. Winters

Color Plate 15 Series of standards (left) and two unknowns (right) for the spectrophotometric determination of Fe(II) using 1,10-phenanthroline as reagent (see Section 26A-3 and Problem 26-26, page 757). The color is due to the complex Fe(phen)$_3^{2+}$. The absorbance of the standards is measured, and a working curve is analyzed using linear least-squares (see Section 8C-2, page 172). The equation for the line is then used to determine the concentrations of the unknown solutions from their measured absorbances.

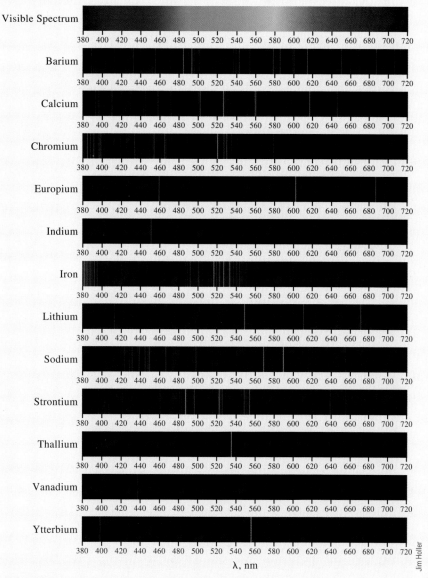

Color Plate 16 Spectrum of white light and emission spectra of selected elements (see Chapter 28).

(a)

(a)

Color Plate 17 The solar spectrum. (a) Expanded color version of the solar spectrum shown in black and white in Feature 24-1 (see Figure 24F-1, page 657). The huge number of dark absorption lines are produced by all of the elements in the sun. See if you can spot some prominent lines like the famous sodium doublet. (b) Compact version of the solar spectrum in (a) compared to the emission spectra of hydrogen, helium, and iron. It is relatively easy to spot lines in the emission spectra of hydrogen and iron that correspond to absorption lines in the solar spectrum, but the lines of helium are quite obscure. In spite of this problem, helium was discovered when these lines were observed in the solar spectrum (see Section 28D). (Images created by Dr. Donald Mickey, University of Hawaii Institute for Astronomy from National Solar Observatory spectral data/NSO/Kitt Peak FTS data by NSF/NOAO.)

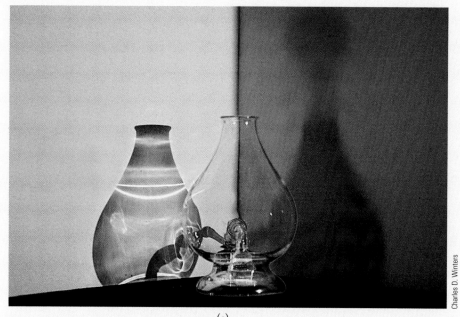

(a)

Color Plate 18 (a) Demonstration of atomic absorption by mercury vapor. (b) White light from the source on the right passes through the mercury vapor above the flask and no shadow appears on the fluorescent screen on the left. Light from the mercury lamp on the left containing the characteristic UV lines of the element is absorbed by the vapor in and above the flask, which casts a shadow on the screen on the right of the plume of mercury vapor (see Section 28D-4, page 797).

Charles D. Winters

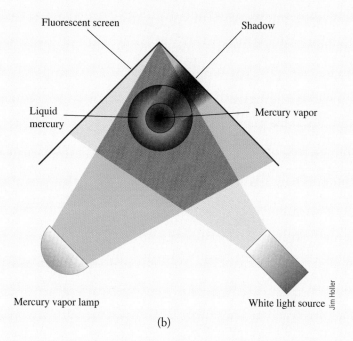

Fluorescent screen   Shadow

Liquid
mercury   Mercury vapor

Mercury vapor lamp   White light source

Jim Holler

(b)

(a)

(b)

(c)

(d)

(e)

Color Plate 19 Weighing by difference the old-fashioned way. (a) Zero the balance. (b) Place a weighing bottle containing the solute on the balance pan. (c) Read the mass (33.2015 g). (d) Transfer the desired amount of solute to a flask. (e) Replace the weighing bottle on the pan and read the mass (33.0832 g). Finally, calculate the mass of the solute transferred to the flask: 33.2015 g − 33.0832 g = 0.1131 g (see Section 2E-4, page 27). (Electronic balance provided by Mettler-Toledo, Inc.)

Charles D. Winters

(a)

(b)

(c)

(d)

Charles D. Winters

Color Plate 20 Weighing by difference the modern way. Place a weighing bottle containing the solute on the balance pan and (a) depress the tare or zero button. The balance should then read 0.0000 g, as shown in (b). (c) Transfer the desired amount of solute to a flask. Replace the weighing bottle on the pan, and (d) the balance reads the decrease in mass directly as −0.1070 g (see Section 2E-4, page 27). Many modern balances have built-in computers with programs to perform a variety of weighing tasks. For example, it is possible to dispense many consecutive quantities of a substance and automatically read out the loss in mass following each dispensing. Many balances also have computer intterfaces so that reading may be logged directly to programs running on the computer. (Electronic balance provided by Mettler-Toledo, Inc.)

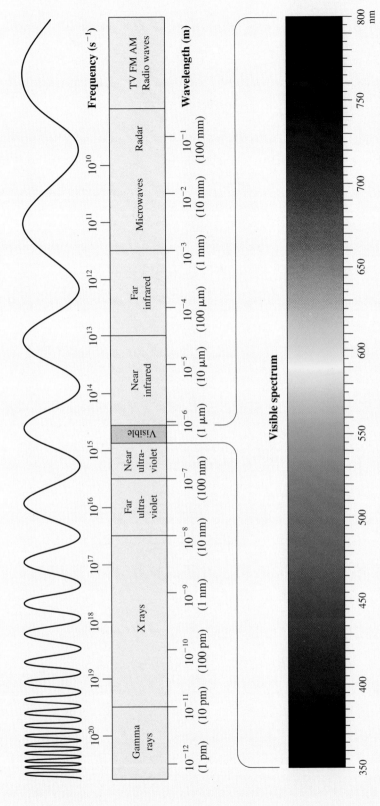

Frequency (s⁻¹) — $\text{Frequency (s}^{-1})$

| Gamma rays | X rays | Far ultra-violet | Near ultra-violet | Visible | Near infrared | Far infrared | Microwaves | Radar | TV FM AM Radio waves |

Wavelength (m)

10^{20} 10^{19} 10^{18} 10^{17} 10^{16} 10^{15} 10^{14} 10^{13} 10^{12} 10^{11} 10^{10}

10^{-12} (1 pm) 10^{-11} (10 pm) 10^{-10} (100 pm) 10^{-9} (1 nm) 10^{-8} (10 nm) 10^{-7} (100 nm) 10^{-6} (1 μm) 10^{-5} (10 μm) 10^{-4} (100 μm) 10^{-3} (1 mm) 10^{-2} (10 mm) 10^{-1} (100 mm)

Visible spectrum

350 400 450 500 550 600 650 700 750 800 nm

Color Plate 21 Electromagnetic spectrum. The spectrum extends from high-energy (frequency) gamma rays to low-energy (frequency) radio waves (see Section 24B-1, page 654). Note that the visible region is only a tiny fraction of the spectrum. The visible region, broken out in the lower portion, extends from the violet ($\approx$380 nm) to the red region ($\approx$800 nm). (Courtesy of Ebbing and Gammon, *General Chemistry*, 10th ed.)

as was the case with the unstirred solution. With the stirred solution, however, diffusion is limited to a narrow layer of liquid, which even with time, cannot extend out indefinitely into the solution. As a result, steady, diffusion-controlled currents appear shortly after applying a voltage.

As is shown in Figure 23-11, at potential X, the equilibrium concentration of A at the electrode surface has been reduced to about 80% of its original value while the equilibrium concentration P has increased by an equivalent amount, that is, $c_P^0 = c_A - c_A^0$. At potential Y, which is the half-wave potential, the equilibrium concentrations of the two species at the surface are approximately the same and equal to $c_A/2$. Finally, at potential Z and beyond, the surface concentration of A approaches zero, while that of P approaches the original concentration of A, c_A. Thus, at potentials more negative than Z, essentially all A ions entering the surface layer are instantaneously reduced to P. As is shown in Figure 23-11b, at potentials greater than Z, the concentration of P in the surface layer remains constant at $c_P^0 = c_A$ because of diffusion of P back into the stirred region.

23C-2 Voltammetric Currents

The current at any point in the electrolysis we have just discussed is determined by the rate of transport of A from the outer edge of the diffusion layer to the electrode surface. Because the product of the electrolysis P diffuses away from the surface and is ultimately swept away by convection, a continuous current is required to maintain the surface concentrations demanded by the Nernst equation. Convection, however, maintains a constant supply of A at the outer edge of the diffusion layer. Therefore, a steady-state current results that is determined by the applied potential. This current is a quantitative measure of how fast A is being brought to the surface of the electrode, and this rate is given by $\partial c_A/\partial x$, where x is the distance in centimeters from the electrode surface.

Note that $\partial c_A/\partial x$ in Equation 23-4 is the slope of the initial part of the concentration profiles shown in Figure 23-11a and that these slopes can be approximated by $(c_A - c_A^0)/\delta$. When this approximation is valid, Equation 23-4 reduces to

$$i = \frac{nFAD_A}{\delta}(c_A - c_A^0) = k_A(c_A - c_A^0) \tag{23-5}$$

where the constant k_A is equal to $nFAD_A/\delta$.

Equation 23-5 shows that as c_A^0 becomes smaller as a result of a larger negative applied potential, the current increases until the surface concentration approaches zero at which point the current becomes constant and independent of the applied potential. Thus, when $c_A^0 \to 0$, the current becomes the limiting current i_l, and Equation 23-5 reduces to

$$i_l = \frac{nFAD_A}{\delta}c_A = k_A c_A \tag{23-6}$$

The derivation leading to Equation 23-6 is based on an oversimplified picture of the diffusion layer in that the interface between the moving and stationary layers is viewed as a sharply defined edge where transport by convection ceases and transport by diffusion begins. Nevertheless, this simplified model does provide a reasonable approximation of the relationship between the current and the variables that affect the current.

CHALLENGE: Show that the units of Equation 23-6 are amperes if the units of the quantities in the equation are as follows:

Quantity	Units
n	mol electrons/mol analyte
F	coulomb/mol electrons
A	cm^2
D_A	cm^2 s^{-1}
c_A	mol analyte/cm^3
δ	cm

Current/Voltage Relationships for Reversible Reactions

To develop an equation for the sigmoidal curve shown in Figure 23-5, we substitute Equation 23-6 into Equation 23-5 and rearrange, giving

$$c_A^0 = \frac{i_l - i}{k_A} \tag{23-7}$$

Although our model is oversimplified, it provides a reasonably accurate picture of the processes occurring at the electrode/solution interface.

The surface concentration of P can also be expressed in terms of the current by using a relationship similar to Equation 23-5, that is,

$$i = -\frac{nFAD_P}{\delta}(c_P - c_P^0) \tag{23-8}$$

where the minus sign results from the negative slope of the concentration profile for P. Note that D_P is now the diffusion coefficient of P. But we have said earlier that throughout the electrolysis the concentration of P approaches zero in the bulk of the solution and, therefore, when $c_P \approx 0$,

$$i = \frac{-nFAD_P c_P^0}{\delta} = k_P c_P^0 \tag{23-9}$$

where $k_P = -nAD_P/\delta$. Rearranging gives

$$c_P^0 = i/k_P \tag{23-10}$$

Substituting Equations 23-7 and 23-10 into Equation 23-3 yields, after rearrangement,

$$E_{appl} = E_A^0 - \frac{0.0592}{n} \log \frac{k_A}{k_P} - \frac{0.0592}{n} \log \frac{i}{i_l - i} - E_{ref} \tag{23-11}$$

When $i = i_l/2$, the third term on the right side of this equation is equal to zero, and by definition, E_{appl} is the **half-wave potential**, that is,

The **half-wave potential** is an identifier for the redox couple and is closely related to the standard reduction potential.

$$E_{appl} = E_{1/2} = E_A^0 - \frac{0.0592}{n} \log \frac{k_A}{k_P} - E_{ref} \tag{23-12}$$

Substituting this expression into Equation 23-11 gives an expression for the voltammogram in Figure 23-5:

$$E_{appl} = E_{1/2} - \frac{0.0592}{n} \log \frac{i}{i_l - i} \tag{23-13}$$

An electrochemical process such as $A + ne^- \rightleftharpoons P$ is said to be **reversible** if it obeys the Nernst equation under the conditions of the experiment. In a **totally irreversible system**, either the forward or the reverse reaction is so slow as to be completely negligible. In a **partially reversible system**, the reaction in one direction is much slower than the other, although not totally insignificant. A process that appears reversible when the potential is changed slowly may show signs of irreversibility when a faster rate of change of potential is applied.

Often, the ratio k_A/k_P in Equation 23-11 and in Equation 23-12 is nearly unity so that we may write for the species A

$$E_{1/2} \approx E_A^0 - E_{ref} \tag{23-14}$$

Current/Voltage Relationships for Irreversible Reactions

Many voltammetric electrode processes, particularly those associated with organic systems, are irreversible, leading to drawn-out and less well-defined waves. To describe these waves quantitatively requires an additional term in Equation 23-12 involving the activation energy of the reaction to account for the kinetics of the electrode process. Although half-wave potentials for irreversible reactions ordinarily show some dependence on concentration, diffusion currents remain linearly related to concentration. Irreversible processes are, therefore, easily adapted to quantitative analysis if suitable calibration standards are available.

Voltammograms for Mixtures of Reactants

The reactants of a mixture generally behave independently of one another at a working electrode. Thus, a voltammogram for a mixture is just the sum of the waves for the individual components. **Figure 23-12** shows the voltammograms for a pair of two-component mixtures. The half-wave potentials of the two reactants differ by about 0.1 V in curve A and by about 0.2 V in curve B. Note that a single voltammogram may permit the quantitative determination of two or more species provided there is sufficient difference between succeeding half-wave potentials to permit evaluation of individual diffusion currents. Generally, a difference of 0.1 to 0.2 V is required if the more easily reducible species undergoes a two-electron reduction; a minimum of about 0.3 V is needed if the first reduction is a one-electron process.

Anodic and Mixed Anodic/Cathodic Voltammograms

Anodic waves as well as cathodic waves are encountered in voltammetry. An example of an anodic wave is illustrated in curve A of **Figure 23-13**, where the electrode reaction is the oxidation of iron(II) to iron(III) in the presence of citrate ion. A limiting current is observed at about +0.1 V (versus SCE), which is due to the half-reaction

$$Fe^{2+} \rightleftharpoons Fe^{3+} + e^{-}$$

As the potential is made more negative, a decrease in the anodic current occurs; at about −0.02 V, the current becomes zero because the oxidation of iron(II) ion has ceased.

Curve *C* represents the voltammogram for a solution of iron(III) in the same medium. For *C*, a cathodic wave results from reduction of iron(III) to iron(II). The half-wave potential is identical with that for the anodic wave, indicating that the oxidation and reduction of the two iron species are perfectly reversible at the working electrode.

Curve *B* is the voltammogram of an equimolar mixture of iron(II) and iron(III). The portion of the curve below the zero-current line corresponds to the oxidation of the iron(II); this reaction ceases at an applied potential equal to the half-wave potential. The upper portion of the curve is due to the reduction of iron(III).

Figure 23-12 Voltammograms for two-component mixtures. Half-wave potentials differ by 0.1 V in curve *A* and by 0.2 V in curve *B*.

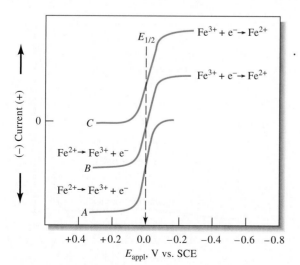

Figure 23-13 Voltammetric behavior of iron(II) and iron(III) in a citrate medium. Curve *A*: anodic wave for a solution in which $c_{Fe^{2+}} = 1 \times 10^{-4}$ M. Curve *B*: anodic/cathodic wave for a solution in which $c_{Fe^{2+}} = c_{Fe^{3+}} = 0.5 \times 10^{-4}$ M. Curve *C*: cathodic wave for a solution in which $c_{Fe^{3+}} = 1 \times 10^{-4}$ M.

23C-3 Oxygen Waves

Dissolved oxygen is easily reduced at a working electrode. Thus, as shown in **Figure 23-14**, an aqueous solution saturated with air exhibits two distinct oxygen waves. The first results from the reduction of oxygen to hydrogen peroxide:

$$O_2(g) + 2H^+ + 2e^- \rightleftharpoons H_2O_2$$

In Figure 23-14 the second wave shows the overall reduction of oxygen to water.

At a more negative potential, the hydrogen peroxide can be further reduced:

$$H_2O_2 + 2H^+ + 2e^- \rightleftharpoons 2H_2O$$

Because both reactions are two-electron reductions, the two waves are of equal height.

Voltammetric measurements offer a convenient and widely used method for determining dissolved oxygen in solutions. However, the presence of oxygen often interferes with the accurate determination of other species. Therefore, oxygen removal is usually the first step in amperometric procedures. Oxygen can be removed by passing an inert gas through the analyte solution for several minutes (**sparging**). A stream of the same gas, usually nitrogen, is passed over the surface of the solution during analysis to prevent reabsorption of oxygen. The lower curve in Figure 23-14 is a voltammogram of an oxygen-free solution.

Sparging is a process in which dissolved gases are swept out of a solvent by bubbling an inert gas, such as nitrogen, argon, or helium, through the solution.

23C-4 Applications of Hydrodynamic Voltammetry

The most important uses of hydrodynamic voltammetry include (1) detection and determination of chemical species as they exit from chromatographic columns or flow-injection apparatus; (2) routine determination of oxygen and certain species of biochemical interest, such as glucose, lactose, and sucrose; (3) detection of end points in coulometric and volumetric titrations; and (4) fundamental studies of electrochemical processes.

Voltammetric Detectors in Chromatography and Flow-Injection Analysis

Hydrodynamic voltammetry is widely used for detection and determination of oxidizable or reducible compounds or ions that have been separated by liquid chromatography or that are produced by flow-injection methods. A thin-layer cell, such

Figure 23-14 Voltammogram for the reduction of oxygen in an air-saturated 0.1 M KCl solution. The lower curve is for a 0.1 M KCl solution in which the oxygen is removed by bubbling nitrogen through the solution.

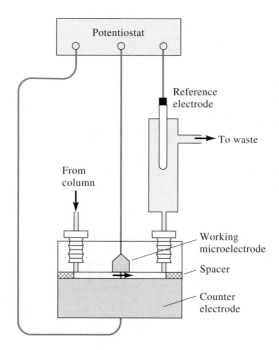

Figure 23-15   A schematic of a voltammetric system for detecting electroactive species as they elute from a column. The cell volume is determined by the thickness of the gasket.

as the one shown schematically in **Figure 23-15**, is used in these applications. The working electrode in these cells is usually imbedded in the wall of an insulating block that is separated from a counter electrode by a thin spacer as shown. The volume of such a cell is typically 0.1 to 1 µL. A voltage corresponding to the limiting current region for analytes is applied between the working electrode and a silver/silver chloride reference electrode that is located downstream from the detector. We present an exploded view of a commercial flow cell in **Figure 23-16a**, which shows clearly how the sandwiched cell is assembled and held in place by the quick release mechanism. A locking collar in the counter electrode block, which is electrically connected to the potentiostat, retains the reference electrode. Five different configurations of working electrode are shown in **Figure 23-16b**. These configurations permit optimization of detector sensitivity under a variety of experimental conditions. Working electrode blocks and electrode materials are described in Section 23B-1. This type of application of voltammetry (or amperometry) has detection limits as low as 10^{-9} to 10^{-10} M. We discuss voltammetric detection for liquid chromatography in more detail in Section 33A-5.

Voltammetric and Amperometric Sensors[7]

In Section 21D, we described how the specificity of potentiometric sensors could be enhanced by applying molecular recognition layers to the electrode surfaces. There has been much research in recent years to apply the same concepts to voltammetric electrodes. A number of voltammetric systems are available commercially for the determination of specific species in industrial, biomedical, environmental, and research applications. These devices are sometimes called electrodes or detectors but are, in fact, complete voltammetric cells and are better referred to as sensors. In the sections that follow, we describe two commercially available sensors.

---

[7]For a review of electrochemical sensors, see E. Bakker and Yu Qin, *Anal. Chem.*, **2006**, *78*, 3965, **DOI**: 10.1021/ac060637m.

Figure 23-16 (a) Detail of a commercial flow cell assembly. (b) Configurations of working electrode blocks. Arrows show the direction of flow in the cell. (Reprinted by permission of Bioanalytical Systems, Inc., West Lafayette, IN.)

Oxygen Sensors. The determination of dissolved oxygen in a variety of aqueous environments, such as seawater, blood, sewage, effluents from chemical plants, and soils, is of tremendous importance to industry, biomedical and environmental research, and clinical medicine. One of the most common and convenient methods for making such measurements is with the **Clark oxygen sensor**, which was patented by L. C. Clark, Jr., in 1956.[8] A schematic of the Clark oxygen sensor is shown in **Figure 23-17**. The cell consists of a cathodic platinum-disk working electrode embedded in a centrally located cylindrical insulator. Surrounding the lower end of this insulator is a ring-shaped silver anode. The tubular insulator and electrodes are mounted inside a second cylinder that contains a buffered solution of potassium chloride. A thin (≈ 20-μm) replaceable, oxygen permeable membrane of Teflon or polyethylene is held in place at the bottom end of the tube by an O-ring. The thickness of the electrolyte solution between the cathode and the membrane is approximately 10 μm.

> The Clark oxygen sensor is widely used in clinical laboratories for the determination of dissolved O_2 in blood and other body fluids.

---

[8]For a detailed discussion of the Clark oxygen sensor, see M. L. Hitchman, *Measurement of Dissolved Oxygen*, Chs. 3–5, New York: Wiley, 1978

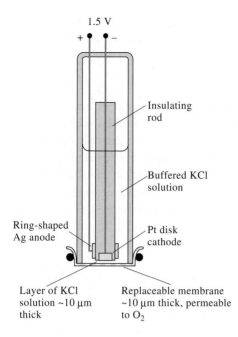

1.5 V

+ −

Insulating rod

Buffered KCl solution

Ring-shaped Ag anode

Pt disk cathode

Layer of KCl solution ~10 μm thick

Replaceable membrane ~10 μm thick, permeable to O_2

Figure 23-17 The Clark voltammetric oxygen sensor. Cathodic reaction: $O_2 + 4H^+ + 4e^- \rightleftharpoons 2H_2O$. Anodic reaction: $Ag + Cl^- \rightleftharpoons AgCl(s) + e^-$.

When the oxygen sensor is immersed in a flowing or stirred solution of the analyte, oxygen diffuses through the membrane into the thin layer of electrolyte immediately adjacent to the disk cathode, where it diffuses to the electrode and is immediately reduced to water. In contrast with a normal hydrodynamic electrode, two diffusion processes are involved: one through the membrane and the other through the solution between the membrane and the electrode surface. In order for a steady-state condition to be reached in a reasonable period (10 to 20 s), the thickness of the membrane and the electrolyte film must be 20 μm or less. Under these conditions, it is the rate of equilibration of the transfer of oxygen across the membrane that determines the steady-state current that is reached.

Enzyme-based Sensors. A number of enzyme-based voltammetric sensors are available commercially. An example is a glucose sensor that is widely used in clinical laboratories for the routine determination of glucose in blood serums. This device is similar in construction to the oxygen sensor shown in Figure 23-17. The membrane in this case is more complex and consists of three layers. The outer layer is a polycarbonate film that is permeable to glucose but impermeable to proteins and other constituents of blood. The middle layer is an immobilized enzyme, glucose oxidase in this example. The inner layer is a cellulose acetate membrane, which is permeable to small molecules, such as hydrogen peroxide. When this device is immersed in a glucose-containing solution, glucose diffuses through the outer membrane into the immobilized enzyme, where the following catalytic reaction occurs:

$$\text{glucose} + O_2 \xrightarrow{\text{glucose oxidase}} H_2O_2 + \text{gluconic acid}$$

The hydrogen peroxide then diffuses through the inner layer of membrane and to the electrode surface, where it is oxidized to give oxygen, that is,

$$H_2O_2 + OH^- \rightarrow O_2 + H_2O + 2e^-$$

The resulting current is directly proportional to the glucose concentration of the analyte solution.

> Enzyme-based sensors can be based on detecting hydrogen peroxide, oxygen, or H^+, depending on the analyte and enzyme. Voltammetric sensors are used for H_2O_2 and O_2, while a potentiometric pH electrode is used for H^+.

Molecular model of hydrogen peroxide. Hydrogen peroxide is a strong oxidizing agent that plays an important role in biological and environmental processes. Hydrogen peroxide is produced in enzyme reactions involving the oxidation of sugar molecules. Peroxide radicals can damage cells and body tissues (see Feature 20-2). They occur in smog and can attack unburned fuel molecules in the environment.

(a)

(b)

(c)

Figure 23-18 Typical amperometric titration curves. (a) Analyte is reduced; reagent is not. (b) Reagent is reduced; analyte is not. (c) Both reagent and analyte are reduced.

A variation on this type of sensor is often found in the home glucose monitors that are now widely used by diabetic patients. This device is one of the largest-selling chemical instruments in the world.

Amperometric Titrations

Hydrodynamic voltammetry can be used to estimate the equivalence point of titrations if at least one of the participants or products of the reaction involved is oxidized or reduced at a working electrode. In this case, the current at some fixed potential in the limiting current region is measured as a function of the reagent volume or of time if the reagent is generated by a constant-current coulometric process. Plots of the data on either side of the equivalence point are straight lines with different slopes; the end point is established by extrapolation to the intersection of the lines.[9]

Amperometric titration curves typically take one of the forms shown in **Figure 23-18**. Figure 23-18a represents a titration in which the analyte reacts at the working electrode while the reagent does not. Figure 23-18b is typical of a titration in which the reagent reacts at the working electrode and the analyte does not. Figure 23-18c corresponds to a titration in which both the analyte and the titrant react at the working electrode.

There are two types of amperometric electrode systems. One uses a single polarizable electrode coupled to a reference, while the other uses a pair of identical solid-state electrodes immersed in a stirred solution. For the first, the working electrode is often a rotating platinum electrode constructed by sealing a platinum wire into the side of a glass tube that is connected to a stirring motor.

Amperometric titrations with one indicator electrode have, with one notable exception, been confined to titrations in which a precipitate or a stable complex is the product. Precipitating reagents include silver nitrate for halide ions, lead nitrate for sulfate ion, and several organic reagents, such as 8-hydroxyquinoline, dimethylglyoxime, and cupferron, for various metallic ions that are reducible at working electrodes. Several metal ions have also been determined by titration with standard solutions of EDTA. The exception just noted involves titrations of organic compounds, such as certain phenols, aromatic amines, and olefins; hydrazine; and arsenic(III) and antimony(III) with bromine. The bromine is often generated coulometrically. It has also been formed by adding a standard solution of potassium bromate to an acidic solution of the analyte that also contains an excess of potassium bromide. Bromine is formed in the acidic medium by the reaction

$$BrO_3^- + 5Br^- + 6H^+ \rightarrow 3Br_2 + 3H_2O$$

This type of titration has been carried out with a rotating platinum electrode or twin platinum electrodes. There is no current prior to the equivalence point; after the equivalence point, there is a rapid increase in current because of the electrochemical reduction of the excess bromine.

There are two advantages in using a pair of identical metallic electrodes to establish the equivalence point in amperometric titrations: simplicity of equipment and not having to purchase or prepare and maintain a reference electrode. This type of system has been incorporated in instruments designed for routine automatic determination of a single species, usually with a coulometrically generated reagent. An instrument of this type is often used for the automatic determination of chloride in samples of serum, sweat, tissue extracts, pesticides, and food products.

---

[9]S. R. Crouch and F. J. Holler, *Applications of Microsoft Excel® in Analytical Chemistry*, 2nd ed., Belmont, CA: Brooks/Cole, 2014, Ch. 11.

The reagent in this system is silver ion coulometrically generated from a silver anode. A voltage of about 0.1 V is applied between a pair of twin silver electrodes that serve as the indicator system. Short of the equivalence point in the titration of chloride ion, there is essentially no current because no electroactive species is present in the solution. Because of this, there is no electron transfer at the cathode, and the electrode is completely polarized. Note that the anode is not polarized because the reaction

$$Ag \rightleftharpoons Ag^+ + e^-$$

occurs in the presence of a suitable cathodic reactant or depolarizer.

Past the equivalence point, the cathode becomes depolarized because silver ions are present. These ions react to give silver:

$$Ag^+ + e^- \rightleftharpoons Ag$$

This half-reaction and the corresponding oxidation of silver at the anode produce a current whose magnitude is, as in other amperometric methods, directly proportional to the concentration of the excess reagent. Thus, the titration curve is similar to that shown in Figure 23-18b. In the automatic titrator just mentioned, an electronic circuit senses the amperometric detection current signal and shuts off the coulometric generator current. The chloride concentration is then computed from the magnitude of the titration current and the generation time. The instrument has a range of 1 to 999.9 mM Cl^- per liter, a precision of 0.1%, and an accuracy of 0.5%. Typical titration times are about 20 s.

The most common end-point detection method for the Karl Fischer titration for determining water (see Section 20C-5) is the amperometric method with dual polarized electrodes. Several manufacturers offer fully automated instruments for use in performing these titrations. A closely related end-point detection method for Karl Fischer titrations measures the potential difference between two identical electrodes through which a small constant current is passed.

Rotating Electrodes

To carry out theoretical studies of oxidation/reduction reactions, it is often of interest to know how k_A in Equation 23-6 is affected by the hydrodynamics of the system. A common method for obtaining a rigorous description of the hydrodynamic flow of stirred solution is based on measurements made with a rotating disk electrode (RDE), such as the one illustrated in **Figures 23-19a** and **23-19b**. When the disk electrode is rotated rapidly, the flow pattern shown by the arrows in the figure is set up. At the surface of the disk, the liquid moves out horizontally from the center of the device, producing an upward axial flow to replenish the displaced liquid. A rigorous treatment of the hydrodynamics is possible in this case[10] and leads to the *Levich equation*[11]

$$i_l = 0.620nFAD\omega^{1/2}\nu^{-1/6}c_A \tag{23-15}$$

The terms *n, F, A,* and *D* in this equation have the same meaning as in Equation 23-5, ω is the angular velocity of the disk in radians per second, and ν is the *kinematic*

[10] A. J. Bard and L. R. Faulkner, *Electrochemical Methods*, 2nd ed., New York: Wiley, 2001, pp. 335–39.
[11] V. G. Levich, *Acta Physicochimica URSS*, **1942**, *17*, 257.

Figure 23-19 (a) Side view of a rotating disk electrode showing solution flow pattern. (b) Bottom view of a disk electrode. (c) Photo of a commercial RDE. (Photo courtesy of Bioanalytical Systems, Inc., W. Lafayette, IN.) (d) Bottom view of a ring-disk electrode.

viscosity in centimeters squared per second, which is the ratio of the viscosity of the solution to its density. Voltammograms for reversible systems generally have the ideal shape shown in Figure 23-5. Numerous studies of the kinetics and the mechanisms of electrochemical reactions have been performed with rotating disk electrodes. A common experiment with the RDE is to study the dependence of i_l on $\omega^{1/2}$. A plot of i_l versus $\omega^{1/2}$ is known as a *Levich plot*, and deviations from the linear relationship often indicate kinetic limitations on the electron transfer process. For example, if i_l becomes independent of ω at large values of $\omega^{1/2}$, the current is not limited by mass transport of the electroactive species to the electrode surface, but instead, the rate of the reaction is the limiting factor. RDEs, such as the versatile commercial model shown in **Figure 23-19c**, have attracted renewed interest in recent years for both fundamental and quantitative analytical studies as enthusiasm for the dropping mercury electrode (polarography) has faded. RDE detection with a mercury-film electrode is sometimes referred to as *pseudopolarography*.

The *rotating ring-disk electrode* is a modified rotating disk electrode that is useful for studying electrode reactions; it has little use in analysis. **Figure 23-19d** shows that a ring-disk electrode contains a second ring-shaped electrode that is electrically isolated from the center disk. After an electroactive species is generated at the disk, it is then swept passed the ring where it undergoes a second electrochemical reaction. **Figure 23-20** shows voltammograms from a typical ring-disk experiment. Figure 23-20a depicts the voltammogram for the reduction of oxygen to hydrogen peroxide at the disk electrode. Figures 23-20b shows the *anodic* voltammogram for the oxidation of the hydrogen peroxide as it flows past the ring electrode. Note that, when the potential of the disk electrode becomes sufficiently negative that the reduction product is hydroxide rather than hydrogen peroxide, the current in the ring electrode decreases to zero. Studies of this type provide much useful information about mechanisms and intermediates in electrochemical reactions.

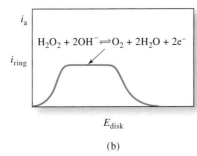

Figure 23-20 Disk (a) and ring (b) current for reduction of oxygen at the rotating ring-disk electrode. (From P. T. Kissinger and W. R. Heineman, eds., *Laboratory Techniques in Electroanalytical Chemistry*, 2nd ed., New York: Marcel Dekker, 1996, p. 117. Laboratory techniques in electroanalytical chemistry by KISSINGER, PETER T., ; HEINEMAN, WILLIAM R. Copyright 1996 Reproduced with permission of TAYLOR & FRANCIS GROUP LLC - BOOKS in the format Textbook via Copyright Clearance Center.)

Spreadsheet Summary Amperometric titrations are the subject of the final exercise in Chapter 11 of *Applications of Microsoft® Excel in Analytical Chemistry*, 2nd ed. An amperometric titration to determine gold in an ore sample is used as an example. Titration curves consisting of two linear segments are extrapolated to find the end point.

23D POLAROGRAPHY

Linear-scan polarography was the first type of voltammetry to be discovered and used. It differs from hydrodynamic voltammetry in two significant ways. First, there is essentially no convection or migration, and second, a dropping mercury electrode (DME), such as that shown in Figure 23-3e, is used as the working electrode. Because there is no convection, diffusion alone controls polarographic limiting currents. Compared with hydrodynamic voltammetry, however, polarographic limiting currents are an order of magnitude or more smaller since convection is absent in polarography.[12]

> Polarographic currents are controlled by diffusion alone, not by convection.

Polarographic Currents

The current in a cell containing a dropping mercury electrode undergoes periodic fluctuations corresponding in frequency to the drop rate. As a drop dislodges from the capillary, the current falls toward zero, as shown in **Figure 23-21**. As the surface area of a new drop increases, so does the current. The diffusion current is usually taken at the maximum of the current fluctuations. In the older literature, the *average current* was measured because instruments responded slowly and damped the oscillations. As shown by the straight lines of Figure 23-21, some modern polarographs have electronic filtering that allows either the maximum or the average current to be determined if the drop rate t is reproducible. Note the effect of irregular drops in the upper part of the curve, probably caused by vibration of the apparatus.

Polarograms

Figure 23-21 shows a polarogram for a solution that is 1.0 M in KCl and 3×10^{-4} M in lead ion. The polarographic wave arises from $Pb^{2+} + 2e^- + Hg \rightleftharpoons Pb(Hg)$, where Pb(Hg) represents elemental lead dissolved in mercury to form an amalgam. The sharp increase in current at about -1.2 V in the polarogram is caused by the reduction of hydrogen ions to give hydrogen. If we examine the polarogram to the left of the wave, we find that there is a small current, called the **residual current**, in the

> The **residual current** in polarography is the small current observed in the absence of an electroactive species.

---

[12]References dealing with polarography include A. J. Bard and L. R. Faulkner, *Electrochemical Methods*, 2nd ed., Ch. 7, pp. 261–304, New York: Wiley, 2001; *Laboratory Techniques in Electroanalytical Chemistry*, 2nd ed., P. T. Kissinger and W. R. Heineman, eds., New York: Marcel Dekker, 1996, pp. 444–61.

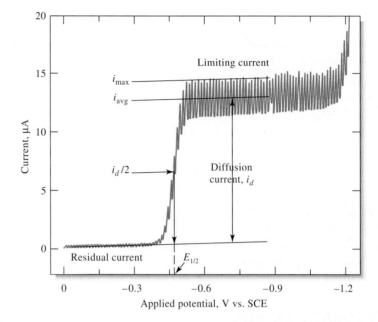

Figure 23-21 Polarogram for 1 M solution of KCl that is 3×10^{-4} M in Pb^{2+}.

Diffusion current is the limiting current observed in polarography when the current is limited only by the rate of diffusion to the dropping mercury electrode surface.

The diffusion current in polarography is proportional to the concentration of analyte.

In polarography, currents are usually recorded in microamperes. The constant 708 in Equation 23-16 carries units such that, when (i_d) is in microamperes, D is in cm²/s, m is in mg/s, t is in s, and the concentration c is in millimoles per liter.

cell even when lead ions are not being reduced. Note the straight line drawn through the residual current and extrapolated to the right below the polarographic wave. This extrapolation permits the determination of the **diffusion current**, as shown in the figure and discussed in the next paragraph.

As in hydrodynamic voltammetry, limiting currents are observed when the magnitude of the current is limited by the rate at which analyte can be brought up to the electrode surface. In polarography, however, the only mechanism of mass transport is diffusion. For this reason, polarographic limiting currents are usually termed diffusion currents and given the symbol i_d. As shown in Figure 23-21, the diffusion current is the difference between the maximum (or average) limiting current and the residual current. The diffusion current is directly proportional to analyte concentration in the bulk of solution, as shown next.

Diffusion Current at the Dropping Mercury Electrode

To derive an equation for polarographic diffusion currents, we must take into account the rate of growth of the spherical electrode, which is related to the drop time in seconds t and the rate of flow of mercury through the capillary m in mg/s and the diffusion coefficient of the analyte D in cm²/s. These variables are taken into account in the Ilkovic equation:

$$(i_d)_{max} = 708 \, nD^{1/2}m^{2/3}t^{1/6}c \tag{23-16}[13]$$

where $(i_d)_{max}$ is the maximum diffusion current in μA and c is the analyte concentration in mM.

Residual Currents

Figure 23-22 shows a residual current curve (obtained at high sensitivity) for a 0.1 M solution of HCl. This current has two sources. The first is the reduction of trace impurities that are inevitably present in the blank solution. The contributors include small amounts of dissolved oxygen, heavy metal ions from the distilled water, and impurities present in the salt used as the supporting electrolyte.

---

[13]If the average diffusion current is measured instead of the maximum, the constant 708 in the Ilkovic equation becomes 607 because $(i_d)_{avg} = 6/7 \, (i_d)_{max}$.

Figure 23-22 Residual current for a 0.1 M solution of HCl.

The second component of the residual current is the so-called **charging**, or **capacitive, current** resulting from a flow of electrons that charge the mercury droplets with respect to the solution; this current may be either negative or positive. At potentials more negative than about −0.4 V, an excess of electrons from the dc source provides the surface of each droplet with a negative charge. These excess electrons are carried down with the drop as it breaks. Since each new drop is charged as it forms, a small but continuous current results. At applied potentials less negative than about −0.4 V, the mercury tends to be positive with respect to the solution. Thus, as each drop is formed, electrons are repelled from the surface toward the bulk of mercury, and a negative current is the result. At about −0.4 V, the mercury surface is uncharged, and the charging current is zero. This potential is called the **potential of zero charge**. The charging current is a type of **nonfaradaic current** in the sense that charge is carried across an electrode/solution interface without an accompanying oxidation/reduction process.

Ultimately, the accuracy and sensitivity of the polarographic method depend on the magnitude of the nonfaradaic residual current and the accuracy with which a correction for its effect can be determined. For these reasons and others mentioned earlier, polarography has declined in importance, while voltammetry and amperometry at working electrodes other than the dropping mercury electrode have grown at an astonishing pace over the past three decades.

A **faradaic current** in an electrochemical cell is the current that results from an oxidation/reduction process. A **nonfaradaic current** is a charging current that results because the mercury drop is expanding and must be charged to the electrode potential. The charging of the double layer is similar to charging a capacitor.

> **Spreadsheet Summary** Polarography is a subject of the voltammetry exercise in Chapter 11 of *Applications of Microsoft® Excel in Analytical Chemistry*, 2nd ed. A polarographic calibration curve is constructed first. Then an accurate determination of half-wave potential is made. Finally, the formation constant and formula of a complex are determined from polarographic data.

23E CYCLIC VOLTAMMETRY[14]

In *cyclic voltammetry* (CV), the current response of a small stationary electrode in an unstirred solution is excited by a triangular voltage waveform, such as that shown in **Figure 23-23**. In this example, the potential is first varied linearly from +0.8 V

---

[14]For brief reviews, see P. T. Kissinger and W. R. Heineman, *J. Chem. Educ.*, 1983, 60, 702, **DOI**: 10.1021/ed060p702; D. H. Evans, K. M. O'Connell, T. A. Petersen, and M. J. Kelly, *J. Chem. Educ.*, **1983**, *60*, 290, **DOI**: 10.1021/ed060p290.

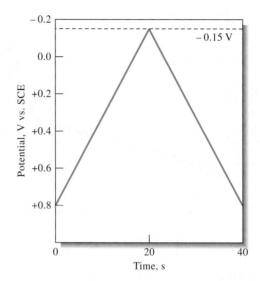

Figure 23-23 Cyclic voltammetric excitation signal.

to −0.15 V versus a saturated calomel electrode. When the extreme of −0.15 V is reached, the scan direction is reversed, and the potential is returned to its original value of +0.8 V. The scan rate in either direction is 50 mV/s. This excitation cycle is often repeated several times. The voltage extrema at which reversal takes place (in this case, −0.15 and +0.8 V) are called *switching potentials*. The range of switching potentials chosen for a given experiment is one in which a diffusion-controlled oxidation or reduction of one or more analytes occurs. The direction of the initial scan may be either negative, as shown, or positive, depending on the composition of the sample (a scan in the direction of more negative potentials is termed a *forward scan*, while one in the opposite direction is called a *reverse scan*). Generally, cycle times range from 1 ms or less to 100 s or more. In this example, the cycle time is 40 s.

Figure 23-24b shows the current response when a solution that is 6 mM in $K_3Fe(CN)_6$ and 1 M in KNO_3 is subjected to the cyclic excitation signal shown in Figures 23-23 and 23-24a. The working electrode was a carefully polished stationary platinum electrode, and the reference electrode was a saturated calomel electrode. At the initial potential of +0.8 V, a tiny anodic current is observed, which immediately decreases to zero as the scan is continued. This initial negative current arises from the oxidation of water to give oxygen (at more positive potentials, this current rapidly increases and becomes quite large at about +0.9 V). No current is observed between a potential of +0.7 and +0.4 V because no reducible or oxidizable species is present in this potential range. When the potential becomes less positive than approximately +0.4 V, a cathodic current begins to develop (point *B*) because of the reduction of the hexacyanoferrate(III) ion to hexacyanoferrate(II) ion. The reaction at the cathode is then

$$Fe(CN)_6^{3-} + e^- \rightleftharpoons Fe(CN)_6^{4-}$$

A rapid increase in the current occurs in the region of *B* to *D* as the surface concentration of $Fe(CN)_6^{3-}$ becomes smaller and smaller. The current at the peak is made up of two components. One is the initial current surge required to adjust the surface concentration of the reactant to its equilibrium concentration as given by the Nernst equation. The second is the normal diffusion-controlled current. The first current then decays rapidly (points *D* to *F*) as the diffusion layer is extended farther and farther away from the electrode surface (see also Figure 23-8b). At point

Figure 23-24 (a) Potential versus time waveform. (b) Cyclic voltammogram for a solution that is 6.0 mM in $K_3Fe(CN)_6$ and 1.0 M in KNO_3. (Reprinted (adapted) with permission from P. T. Kissinger and W. H. Heineman, *J. Chem. Educ.*, **1983**, *60*, 702, **DOI**: 10.1021/ed060p702. Copyright © 1983; Division of Chemical Education, Inc. Copyright 1983 American Chemical Society.)

$F(-0.15 \text{ V})$, the scan direction is switched. The current, however, continues to be cathodic even though the scan is toward more positive potentials because the potentials are still negative enough to cause reduction of $Fe(CN)_6^{3-}$. As the potential sweeps in the positive direction, eventually reduction of $Fe(CN)_6^{3-}$ no longer occurs, and the current goes to zero and then becomes anodic. The anodic current results from the reoxidation of $Fe(CN)_6^{4-}$ that has accumulated near the surface during the forward scan. This anodic current peaks and then decreases as the accumulated $Fe(CN)_6^{4-}$ is used up by the anodic reaction.

Important variables in a cyclic voltammogram are the cathodic peak potential E_{pc}, the anodic peak potential E_{pa}, the cathodic peak current i_{pc}, and the anodic peak current i_{pa}. The definitions and measurements of these parameters are illustrated in Figure 23-24. For a reversible electrode reaction, anodic and cathodic peak currents are approximately equal in absolute value but opposite in sign. For a reversible electrode reaction at 25°C, the difference in peak potentials, ΔE_p, is expected to be

$$\Delta E_p = |E_{pa} - E_{pc}| = 0.0592/n \tag{23-17}$$

where n is the number of electrons involved in the half-reaction. Irreversibility because of slow electron transfer kinetics results in ΔE_p exceeding the expected value. While an electron transfer reaction may appear reversible at a slow sweep rate, increasing the sweep rate may lead to increasing values of ΔE_p, a sure sign of irreversibility. Hence, to detect slow electron transfer kinetics and to obtain rate constants, ΔE_p is measured for different sweep rates.

Quantitative information is obtained from the Randles-Sevcik equation, which at 25°C is

$$i_p = 2.686 \times 10^5 n^{3/2} A c D^{1/2} v^{1/2} \tag{23-18}$$

where i_p is the peak current in amperes, A is the electrode area in cm^2, D is the diffusion coefficient in cm^2/s, c is the concentration in mol/cm^3, and v is the scan rate in V/s. CV offers a way of determining diffusion coefficients if the concentration, electrode area, and scan rate are known.

Fundamental Studies

The primary use of CV is as a tool for fundamental and diagnostic studies that provides qualitative information about electrochemical processes under various conditions. As an example, consider the cyclic voltammogram for the agricultural insecticide parathion that is shown in **Figure 23-25**.[15] In this example, the switching potentials were about −1.2 V and +0.3 V. The initial forward scan was, however,

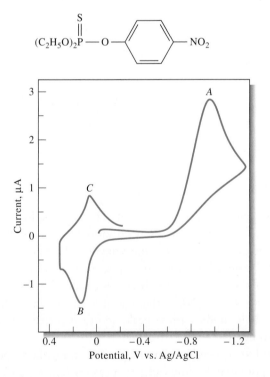

Figure 23-25 Cyclic voltammogram of the insecticide parathion in 0.5 M pH 5 sodium acetate buffer in 50% ethanol. Hanging mercury drop electrode. Scan rate: 200 m V/s. (W.R. Heineman and P.T. Kissinger, *Amer. Lab.*, **1982** (11) 34, Copyright 1982. Reprinted with permission from CompareNetworks, Inc.)

[15]This discussion and the voltammogram are from W. R. Heineman and P. T. Kissinger, *Amer. Lab.,* **1982** (11), 29.

started at 0.0 V and not +0.3 V. Three peaks are observed. The first cathodic peak (*A*) results from a four-electron reduction of the parathion to give a hydroxylamine derivative

$$\phi NO_2 + 4e^- + 4H^+ \rightarrow \phi NHOH + H_2O \qquad (23\text{-}19)$$

The anodic peak at *B* arises from the oxidation of the hydroxylamine to a nitroso derivative during the reverse scan. The electrode reaction is

$$\phi NHOH \rightarrow \phi NO + 2H^+ + 2e^- \qquad (23\text{-}20)$$

The cathodic peak at *C* results from the oxidation of the nitroso compound to the hydroxylamine, as shown by the equation

$$\phi NO + 2e^- + 2H^+ \rightarrow \phi NHOH \qquad (23\text{-}21)$$

Cyclic voltammograms for authentic samples of the two intermediates confirmed the identities of the compounds responsible for peaks *B* and *C*.

CV is widely used as an investigative tool in organic and inorganic chemistry. It is often the first technique selected for exploring systems likely to contain electroactive species. For example, CV is often used to investigate the behavior of modified electrodes and new materials that are suspected to be electroactive. Often cyclic voltammograms reveal the presence of intermediates in oxidation/reduction reactions (for example, see Figure 23-25). Platinum electrodes are often used in CV. For negative potentials, mercury film electrodes can be used. Other popular working electrode materials include glassy carbon, carbon paste, graphite, gold, diamond, and recently, carbon nanotubes.

Peak currents in CV are directly proportional to analyte concentration. Although it is not common to use CV peak currents in routine analytical work, occasionally such applications do appear in the literature, and they are appearing with increasing frequency.

23F PULSE VOLTAMMETRY

By the 1960s, linear-scan voltammetry ceased to be an important analytical tool in most laboratories. The reason for the decline in use of this once popular technique was not only the appearance of several more convenient spectroscopic methods but also the inherent disadvantages of the method including slowness, inconvenient apparatus, and particularly, poor detection limits. Many of these limitations were overcome by the development of pulse methods. We will discuss the two most important pulse techniques, **differential-pulse voltammetry** and **square-wave voltammetry**. The idea behind all pulse-voltammetric methods is to measure the current at a time when the difference between the desired faradaic curve and the interfering charging current is large.

> ◀ The detection limit for classical polarography is about 10^{-5} M. Routine determinations usually involve concentrations in the mM range.

23F-1 Differential-Pulse Voltammetry

Figure 23-26 shows the two most common excitation signals that are used in commercial instruments for differential-pulse voltammetry. The first (see Figure 23-26a), which is usually used in analog instruments, is obtained by superimposing a periodic

Figure 23-26 Excitation signals for differential-pulse voltammetry.

Derivative voltammograms yield peaks that are convenient for qualitative identification of analytes based on the peak potential, E_{peak}.

The detection limits for differential-pulse polarography are two to three orders of magnitude lower than for classical polarography.

Figure 23-27 Voltammogram for a differential-pulse voltammetry experiment. In this example, $\Delta i = i_{S_2} - i_{S_1}$ (see Figure 23-26). The peak potential, E_{peak}, is closely related to the polarographic half-wave potential.

pulse on a linear scan. The second waveform (see Figure 23-26b), which is typically used in digital instruments, is the sum of a pulse and a staircase signal. In either case, a small pulse, typically 50 mV, is applied during the last 50 ms of the lifetime of the period of the excitation signal.

As shown in Figure 23-26, two current measurements are made alternately: one (at S_1), which is 16.7 ms prior to the dc pulse, and one for 16.7 ms (at S_2) at the end of the pulse. The difference in current per pulse (Δi) is recorded as a function of the linearly increasing excitation voltage. A differential curve results, consisting of a peak (see **Figure 23-27**) the height of which is directly proportional to concentration. For a reversible reaction, the peak potential is approximately equal to the standard potential for the half-reaction.

One advantage of the derivative-type voltammogram is that individual peak maxima can be observed for substances with half-wave potentials differing by as little as 0.04 to 0.05 V; in contrast, classical and normal-pulse voltammetry require a potential difference of about 0.2 V for resolving waves. More important, however, differential-pulse voltammetry increases the sensitivity of voltammetry. Typically, differential-pulse voltammetry provides well-defined peaks at a concentration level that is 2×10^{-3} that for the classic voltammetric wave. Note also that the current scale for Δi is in nanoamperes. Generally, detection limits with differential-pulse voltammetry are two to three orders of magnitude lower than those for classical voltammetry and lie in the range of 10^{-7} to 10^{-8} M.

The greater sensitivity of differential-pulse voltammetry can be attributed to two sources. The first is an enhancement of the faradaic current, and the second is a decrease in the nonfaradaic charging current. To account for the enhancement, let us consider the events that must occur in the surface layer around an electrode as the potential is suddenly increased by 50 mV. If an electroactive species is present in this layer, there will be a surge of current that lowers the reactant concentration to that demanded by the new potential (see Figure 23-7b). As the equilibrium concentration for that potential is approached, however, the current decays to a level just sufficient to counteract diffusion, that is, to the diffusion-controlled current. In classical voltammetry, the initial surge of current is not observed because the time scale of the measurement is long relative to the lifetime of the momentary current. On the other hand, in pulse voltammetry, the current measurement is made before the surge has completely decayed. Thus, the current measured contains both a diffusion-controlled component and a component that has to do with reducing the surface layer to the concentration demanded by the Nernst expression; the total current is typically several times larger than the diffusion current. Note under hydrodynamic conditions, the solution becomes homogeneous with respect to the analyte by the time that the next pulse sequence occurs. Thus, at any given applied voltage, an identical current surge accompanies each voltage pulse.

When the potential pulse is first applied to the electrode, a surge in the nonfaradaic current also occurs as the charge increases. This current, however, decays exponentially with time and approaches zero with time. Therefore, by measuring currents at this time only, the nonfaradaic residual current is greatly reduced, and the signal-to-noise ratio is larger. Enhanced sensitivity results.

Reliable instruments for differential-pulse voltammetry are now available commercially at reasonable cost. The method has thus become one of the most widely used analytical voltammetric procedures and is especially useful for determining trace concentrations of heavy metal ions.

23F-2 Square-Wave Voltammetry[16]

Square-wave voltammetry is a type of pulse voltammetry that offers the advantage of great speed and high sensitivity. An entire voltammogram is obtained in less than 10 ms. Square-wave voltammetry has been used with hanging mercury drop electrodes and with other electrodes (see Figure 23-3) and sensors.

Figure 23-28c shows the excitation signal in square-wave voltammetry that is obtained by superimposing the pulse train shown in 23-28b onto the staircase signal in 23-28a. The length of each step of the staircase and the period τ of the pulses are identical and usually about 5 ms. The potential step of the staircase ΔE_s is typically 10 mV. The magnitude of the pulse $2E_{sw}$ is often 50 mV. Operating under these conditions, corresponding to a pulse frequency of 200 Hz, a 1-V scan requires 0.5 s. For a reversible reduction reaction, the size of a pulse is great enough so that oxidation of the product formed on the forward pulse occurs during the reverse pulse. Thus, as shown in **Figure 23-29**, the forward pulse produces a cathodic current i_1, and the

> Multiple scans from multiple drops can be summed to improve the signal-to-noise ratio of a square-wave voltammogram.

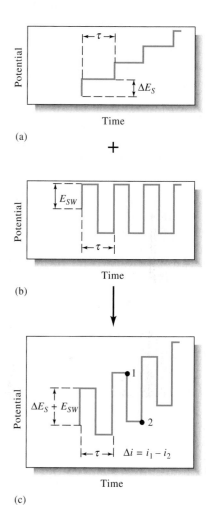

(a)

+

(b)

(c)

Figure 23-28 Generation of a square-wave voltammetry excitation signal. The staircase signal in (a) is added to the pulse train in (b) to give the square-wave excitation signal in (c). The current response, Δi, is equal to the current at potential 1 minus that at potential 2.

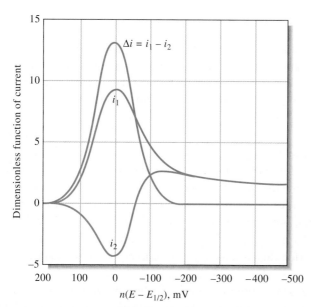

Figure 23-29 Current response for a reversible reaction to excitation signal in Figure 23-28c. This theoretical response plots a dimensionless function of current versus a function of potential, $n(E - E_{1/2})$ in mV. In this example, i_1 = forward current; i_2 = reverse current; and $i_1 - i_2$ = current difference. (From J. J. O'Dea, J. Osteryoung, and R. A. Osteryoung, *Anal. Chem.*, **1981**, *53*, 695, **DOI**: 10.1021/ac00227a028. With permission. Copyright 1981 by the American Chemical Society.)

---

[16]For further information on square-wave voltammetry, see A. J. Bard and L. R. Faulkner, *Electrochemical Methods*, 2nd ed., Ch. 7, pp. 293–99, New York: Wiley, 2001; J. G. Osteryoung and R. A. Osteryoung, *Anal. Chem.*, **1985**, *57*, 101A, **DOI**: 10.1021/ac00279a004.

reverse pulse gives an anodic current i_2. Usually the difference in these currents, Δi, is plotted to give voltammograms. This difference is directly proportional to concentration; the potential of the peak corresponds to the voltammetric half-wave potential. Because of the speed of the measurement, it is possible and practical to increase the precision of analyses by signal-averaging data from several voltammetric scans. Detection limits for square-wave voltammetry are reported to be 10^{-7} to 10^{-8} M.

Commercial instruments for square-wave voltammetry are available from several manufacturers, and as a consequence, this technique is being used routinely for determining inorganic and organic species. Square-wave voltammetry is also being used in detectors for liquid chromatography.

23G APPLICATIONS OF VOLTAMMETRY

In the past, linear-scan voltammetry was used for the quantitative determination of a wide variety of inorganic and organic species, including molecules of biological and biochemical interest. Pulse methods have largely replaced classical voltammetry because of their greater sensitivity, convenience, and selectivity. Generally, quantitative applications are based on calibration curves in which peak heights are plotted as a function of analyte concentration. In some instances the standard-addition method is used in lieu of calibration curves. In either case, it is essential that the composition of standards resemble as closely as possible the composition of the sample, both as to electrolyte concentrations and pH. When this matching is done, relative precisions and accuracies in the 1 to 3% range can often be achieved.

23G-1 Inorganic Applications

Voltammetry is applicable to the analysis of many inorganic substances. Most metallic cations, for example, are reduced at common working electrodes. Even the alkali and alkaline-earth metals are reducible, provided the supporting electrolyte does not react at the high potentials required; in this instance, the tetraalkyl ammonium halides are useful electrolytes because of their high reduction potentials.

The successful voltammetric determination of cations frequently depends on the supporting electrolyte that is used. To aid in this selection, tabular compilations of half-wave potential data are available.[17] The judicious choice of anion often enhances the selectivity of the method. For example, with potassium chloride as a supporting electrolyte, the waves for iron(III) and copper(II) interfere with one another. In a fluoride medium, however, the half-wave potential of iron(III) is shifted by about -0.5 V, while that for copper(II) is altered by only a few hundredths of a volt. The presence of fluoride thus results in the appearance of well-separated waves for the two ions.

Voltammetry is also applicable to the analysis of such inorganic anions as bromate, iodate, dichromate, vanadate, selenite, and nitrite. In general, voltammograms for these substances are affected by the pH of the solution because the hydrogen ion is a participant in their reduction. As a consequence, strong buffering to some fixed pH is necessary to obtain reproducible data (see next section).

---

[17]For example, see J. A. Dean, *Analytical Chemistry Handbook*, Section 14, pp. 14.66–14.70, New York: McGraw-Hill, 1995; D. T. Sawyer, A. Sobkowiak, and J. L. Roberts, *Experimental Electrochemistry for Chemists*, 2nd ed., New York: Wiley, 1995, pp. 102–30.

23G-2 Organic Voltammetric Analysis

Almost from its inception, voltammetry has been used for the study and determination of organic compounds with many papers being devoted to this subject. Several organic functional groups are reduced at common working electrodes, thus making possible the determination of a wide variety of organic compounds.[18] Oxidizable organic functional groups can be studied voltammetrically with platinum, gold, carbon, or various modified electrodes. The number of functional groups that can be oxidized at mercury electrodes is relatively limited because mercury is oxidized at anodic potentials greater than +0.4 V (versus SCE), however.

Solvents for Organic Voltammetry

Solubility considerations frequently dictate the use of solvents other than pure water for organic voltammetry. Aqueous mixtures containing varying amounts of such miscible solvents as glycols, dioxane, acetonitrile, alcohols, Cellosolve, or acetic acid have been used. Anhydrous media such as acetic acid, formamide, diethylamine, and ethylene glycol have also been investigated. Supporting electrolytes are often lithium or tetraalkyl ammonium salts.

More information on applications of voltammetry can be found elsewhere.[19]

23H STRIPPING METHODS

Stripping methods encompass a variety of electrochemical procedures having a common characteristic initial step.[20] In all of these procedures, the analyte is first deposited on a working electrode, usually from a stirred solution. After an accurately measured period, the electrolysis is discontinued, the stirring is stopped, and the deposited analyte is determined by one of the voltammetric procedures that have been described in the previous section. During this second step in the analysis, the analyte is redissolved or stripped from the working electrode; hence the name attached to these methods. In **anodic stripping methods**, the working electrode behaves as a cathode during the deposition step and as an anode during the stripping step, with the analyte being oxidized back to its original form. In a **cathodic stripping method**, the working electrode behaves as an anode during the deposition step and as a cathode during stripping. The deposition step amounts to an electrochemical preconcentration of the analyte, that is, the concentration of the analyte in the surface of the working electrode is far greater than it is in the bulk solution. As a result of the preconcentration step, stripping methods yield the lowest detection limits of all voltammetric procedures. For example, anodic stripping with pulse voltammetry can reach nanomolar detection limits for environmentally important species, such as Pb^{2+}, Ca^{2+}, and Tl^+.

Figure 23-30a illustrates the voltage excitation program that is followed in an anodic stripping method for determining cadmium and copper in an aqueous solution of these ions. A linear-scan method is often used to complete the analysis. Initially, a constant cathodic potential of about −1 V is applied to the working electrode, causing

> The following organic functional groups produce voltammetric waves:
>
> 1. Carbonyl groups
> 2. Certain carboxylic acids
> 3. Most peroxides and epoxides
> 4. Nitro, nitroso, amine oxide, and azo groups
> 5. Most organic halogen groups
> 6. Carbon/carbon double bonds
> 7. Hydroquinones and mercaptans.

In **anodic stripping methods**, the analyte is deposited by reduction and then analyzed by oxidation from the small volume mercury film or drop.

In **cathodic stripping methods**, the analyte is electrolyzed into a small volume of mercury by oxidation and then stripped by reduction.

---

[18]For a detailed discussion of organic electrochemistry, see A. J. Bard, M. Stratmann, and H. J. Schäfer, eds., *Encyclopedia of Electrochemistry*, Vol. 8, *Organic Electrochemistry*, New York: Wiley, 2002; H. Lund and O. Hammerich, eds., *Organic Electrochemistry*, 4th ed., New York: Marcel Dekker, 2001.

[19]D. A. Skoog, F. J. Holler, and S. R. Crouch, *Principles of Instrumental Analysis*, 6th ed., Section 25G, p. 746, Belmont, CA: Brooks/Cole, 2007.

[20]For detailed discussions of stripping methods, see H. D. Dewald, in *Modern Techniques in Electroanalysis*, P. Vanysek, ed., Ch. 4, p. 151, New York: Wiley-Interscience, 1996; J. Wang, *Stripping Analysis*, Deerfield Beach, FL: VCH Publishers, 1985.

Figure 23-30 (a) Excitation signal for stripping determination of Cd^{2+} and Cu^{2+}. (b) Stripping voltammogram.

both cadmium and copper ions to be reduced and deposited as metals. The electrode is maintained at this potential for several minutes until a significant amount of the two metals has accumulated at the electrode. The stirring is then stopped for 30 s or so while the electrode is maintained at −1 V. The potential of the electrode is then decreased linearly to less negative values while the current in the cell is recorded as a function of time, or potential. **Figure 23-30b** shows the resulting differential-pulse voltammogram. At a potential somewhat more negative than −0.6 V, cadmium starts to be oxidized, causing a sharp increase in the current. As the deposited cadmium is consumed, the current peaks and then decreases to its original level. A second peak for oxidation of the copper is then observed when the potential has decreased to approximately −0.1 V. The heights of the two peaks are proportional to the weights of deposited metal.

Stripping methods are important in trace work because the preconcentration step permits the determination of minute amounts of an analyte with reasonable accuracy. Thus, the analysis of solutions in the 10^{-6} to 10^{-9} M range becomes feasible by methods that are both simple and rapid.

23H-1 Electrodeposition Step

> A major advantage of stripping analysis is the capability for electrochemically preconcentrating the analyte prior to the measurement step.

Only a fraction of the analyte is usually deposited during the electrodeposition step. Hence, quantitative results depend not only on control of electrode potential but also on such factors as electrode size, time of deposition, and stirring rate for both the sample and standard solutions used for calibration.

Working electrodes for stripping methods have been formed from a variety of materials, including mercury, gold, silver, platinum, and carbon in various forms. The most popular electrode is the hanging mercury drop electrode (HMDE), which consists of a single drop of mercury in contact with a platinum wire. Hanging drop electrodes are available from several commercial sources. These electrodes often consist of a microsyringe with a micrometer for exact control of drop size. The drop is then formed at the tip of a capillary by displacement of the mercury in the syringe-controlled delivery system (see Figure 23-3b). Rotating disk electrodes may also be used in stripping analysis.

To carry out the determination of a metal ion by anodic stripping, a fresh hanging drop is formed, stirring is begun, and a potential is applied that is a few tenths of a volt more negative than the half-wave potential for the ion of interest. Deposition is allowed to occur for a carefully measured period that can vary from a minute or less for 10^{-7} M solutions to 30 min or longer for 10^{-9} M solutions. We should reemphasize that these times seldom result in complete removal of the ion. The electrolysis period is determined by the sensitivity of the method ultimately used for completion of the analysis.

23H-2 Voltammetric Completion of the Analysis

The analyte collected in the working electrode can be determined by any of several voltammetric procedures. For example, in a linear anodic scan procedure, as described at the beginning of this section, stirring is discontinued for 30 s or so after stopping the deposition. The voltage is then decreased at a linear fixed rate from its original cathodic value, and the resulting anodic current is recorded as a function of the applied voltage. This linear scan produces a curve of the type shown in Figure 23-30b. Analyses of this type are generally based on calibration with standard solutions of the cations of interest. With reasonable care, analytical precisions of about 2% relative can be obtained.

Most of the other voltammetric procedures described in the previous section have also been applied in the stripping step. The most widely used of these appears to be an anodic differential-pulse technique. Often, narrower peaks are produced by this procedure, which is desirable when mixtures are analyzed. Another method of obtaining narrower peaks is to use a mercury film electrode. A thin mercury film is electrodeposited on an inert electrode such as glassy carbon. Usually, the mercury deposition is carried out simultaneously with the analyte deposition. Because the average diffusion path length from the film to the solution interface is much shorter than that in a drop of mercury, escape of the analyte is hastened. The consequence is narrower and larger voltammetric peaks, leading to greater sensitivity and better resolution of mixtures. On the other hand, the hanging drop electrode appears to give more reproducible results, especially at higher analyte concentrations. Thus, for most applications, the hanging drop electrode is used. **Figure 23-31** is a differential-pulse anodic stripping voltammogram for five cations in a sample of mineralized honey, which had been spiked with 1×10^{-5} M $GaCl_3$. The voltammogram demonstrates good resolution and adequate sensitivity for many purposes.

Many other variations of the stripping technique have been developed. For example, a number of cations have been determined by electrodeposition on a platinum cathode. The quantity of electricity required to remove the deposit is then measured coulometrically. Once again, the method is particularly advantageous for trace analyses. Cathodic stripping methods for the halides have also been developed. In these methods, the halide ions are first deposited as mercury(I) salts on a mercury anode. Stripping is then performed by a cathodic current.

23I VOLTAMMETRY WITH MICROELECTRODES

Over the last two decades, a number of voltammetric studies have been carried out with microelectrodes that have dimensions that are smaller by an order of magnitude or more than the electrodes we have described so far. The electrochemical behavior of these tiny electrodes is significantly different from classical electrodes and appears to

Figure 23-31 Differential-pulse anodic stripping voltammogram in the analysis of a mineralized honey sample spiked with $GaCl_3$ (final concentration in the analysis solution: 1×10^{-5} M). Deposition potential: -1.20V. Deposition time: 1200 s in unstirred solution. Pulse height: 50 mV. Anodic potential scan rate: 5 mVs^{-1}. (Reprinted (adapted) from G. Sannaa et al., *Anal. Chim. Acta,* **2000**, *415*, 165, **DOI**: 10.1016/ S0003-2670(00)00864-3, with permission from Elsevier.)

offer advantages in certain analytical applications.[21] Such electrodes are often called microscopic electrodes, or **microelectrodes**, to distinguish them from classical electrodes. The dimensions of such electrodes are typically smaller than about 20 μm and may be as small as a 30 nm in diameter and 2 μm in length ($A \approx 0.2$ μm^2).

Microelectrodes assume a number of useful forms. The most common is a planar electrode formed by sealing a carbon fiber with a radius of 5 μm or a gold or platinum wire having dimensions from 0.3 to 20 μm into a fine capillary tube. Many other shapes and sizes down to 20 Å have been used in a range of applications. Mercury microelectrodes are formed by electrodeposition of the metal onto carbon or metal electrodes. There are several other forms of these electrodes.

Generally, the instrumentation used with microelectrodes is simpler than that shown in Figure 23-2 or Figure 23F-2 because there is no need to employ a three-electrode system. The reason that the reference electrode can be eliminated is that the currents are so small (in the picoampere to nanoampere range) that the *IR* drop does not distort the voltammetric waves the way microampere currents do.

One of the reasons for the early interest in microscopic microelectrodes was the desire to study chemical processes in single cells (see **Figure 23-32**) or processes inside organs of living species, such as in mammalian brains. One approach to this problem is to use electrodes that are small enough not to cause significant alteration in the function of the organ. It was also realized that microelectrodes have certain advantages that justify their application to other kinds of analytical problems. Among these advantages are the very small *IR* drops, which make them applicable to solvents having low dielectric constants, such as toluene. Second, capacitive charging currents, which often limit detection with ordinary voltammetric electrodes, are reduced to insignificant proportions as the electrode size is diminished. Third, the rate of mass transport to and from an electrode increases as the size

[21]See R. M. Wightman, *Science,* **1988**, *240*, 415, **DOI**:10.1126/science.240.4851.415; R. M. Wightman, *Anal. Chem.,* **1981**, *53*, 1125A, **DOI**: 10.1021/ac00232a004; S. Pons and M. Fleischmann, *Anal. Chem.,* **1987**, *59*, **DOI**: 10.1021/ac00151a001; J. Heinze, *Angew. Chem., Int. Ed.,* **1993**, *32*, 1268; R. M. Wightman and D. O. Wipf, in *Electroanalytical Chemistry,* A. J. Bard, ed., Vol. 15, New York: Marcel Dekker, 1989; A. C. Michael and R. M. Wightman, in *Laboratory Techniques in Electroanalytical Chemistry,* 2nd ed., P. T. Kissinger and W. R. Heineman, eds., Ch. 12, New York: Marcel Dekker, 1996; C. G. Zoski, in *Modern Techniques in Electroanalysis,* P. Vanysek, ed., Ch. 6, New York: Wiley, 1996.

Figure 23-32 Optical image using brightfield microscopy showing a carbon fiber microelectrode adjacent to a bovine chromaffin cell from the adrenal medulla. The extracellular solution was 10 mM TRIS buffer containing 150 mM NaCl, 2 mM CaCl$_2$, 1.2 mM MgCl$_2$, and 5 mM glucose. The black scale bar is 50 μm. (From L. Buhler and R. M. Wightman, unpublished work. With permission.)

of an electrode decreases. As a result, steady-state currents are established in unstirred solutions in less than a microsecond rather than in a millisecond or more, as is the case with classical electrodes. Such high-speed measurements permit the study of intermediates in rapid electrochemical reactions. In light of the tremendous present-day interest in nanomaterials and biosensors for determining analytes in minuscule volumes of solution, it is likely that research and development in this fertile area will continue for some time.

WEB WORKS

Use your favorite search engine to find companies that make anodic stripping voltammetry (ASV) instruments. In your search, you should discover links to companies such as ESA, Inc.; Cypress Systems, Inc.; and Bioanalytical Systems. For two instrument manufacturers, compare the working electrodes used for anodic stripping voltammetry. Consider the types of electrodes (thin-film, hanging mercury drop, and so on), whether they are rotating electrodes, and whether they pose any health risks. Also, compare the specifications of two instruments from two different manufacturers. Consider in your comparison, the deposition potential ranges, the available deposition times, the scanning potential ranges, the scanning sweep rates, and the prices.

QUESTIONS AND PROBLEMS

23-1. Distinguish between
　*(a) voltammetry and amperometry.
　(b) linear-scan voltammetry and pulse voltammetry.
　*(c) differential-pulse voltammetry and square-wave voltammetry.
　(d) a rotating disk electrode and a ring-disk electrode.
　*(e) a limiting current and a diffusion current.
　(f) laminar flow and turbulent flow.
　*(g) the standard electrode potential and the half-wave potential for a reversible reaction at a working electrode.
　(h) stripping methods and standard voltammetry.

23-2. Define
　(a) voltammograms.
　(b) hydrodynamic voltammetry.
　(c) Nernst diffusion layer.
　(d) mercury film electrode.

　(e) half-wave potential.
　(f) diffusion current.

*23-3. Why is a high supporting electrolyte concentration used in most electroanalytical procedures?

23-4. Why is the reference electrode placed near the working electrode in a three-electrode cell?

*23-5. Why is it necessary to buffer solutions in organic voltammetry?

23-6. Why are stripping methods more sensitive than other voltammetric procedures?

*23-7. What is the purpose of the electrodeposition step in stripping analysis?

23-8. List the advantages and disadvantages of the hanging mercury drop electrode compared with platinum or carbon electrodes.

*23-9. Suggest how Equation 23-13 could be used to determine the number of electrons n involved in a reversible reaction at an electrode.

23-10. Quinone undergoes a reversible reduction at a voltammetric working electrode. The reaction is

Quinone   Hydroquinone

(a) Assume that the diffusion coefficient for quinone and hydroquinone are approximately the same and calculate the approximate half-wave potential (versus SCE) for the reduction of hydroquinone at a rotating disk electrode from a solution buffered to a pH of 7.0.

(b) Repeat the calculation in (a) for a solution buffered to a pH of 5.0.

23-11. Sulfate ion can be determined by an amperometric titration procedure using Pb^{2+} as the titrant. If the potential of a rotating mercury film electrode is adjusted to -1.00 V versus SCE, the current can be used to monitor the Pb^{2+} concentration during the titration. In a calibration experiment, the limiting current, after correction for background and residual currents, was found to be related to the Pb^{2+} concentration by $i_1 = 10\ c_{Pb^{2+}}$, where i_1 is the limiting current in mA and $c_{Pb^{2+}}$ is the Pb^{2+} concentration in mM. The titration reaction is

$$SO_4^{2-} + Pb^{2+} \rightleftharpoons PbSO_4(s) \qquad K_{sp} = 1.6 \times 10^{-8}$$

If 25 mL of 0.025 M Na_2SO_4 is titrated with 0.040 M $Pb(NO_3)_2$, develop the titration curve in spreadsheet format and plot the limiting current versus the volume of titrant.

*23-12. It has been suggested that many polarograms can be obtained on a solution without depleting the electroactive analyte. Suppose that in a polarographic experiment we monitor the limiting current for 45 minutes in 60 mL of 0.08 M Cu^{2+}. If the average current during the time of the experiment is 6.0 μA, what fraction of the copper is removed from the solution?

*23-13. An unknown cadmium(II) solution was analyzed polarographically by the method of standard additions. A 25.00-mL sample of the unknown solution produced a diffusion current of 1.86 μA. Following addition of a 5.00-mL aliquot of 2.12×10^{-3} M Cd^{2+} standard solution to the unknown solution, a diffusion current of 5.27 μA was produced. Calculate the concentration of Cd^{2+} in the unknown solution.

23-14. (a) What are the advantages of performing voltammetry with microelectrodes? (b) Is it possible for an electrode to be too small? Explain your answer.

23-15. Challenge Problem: A method for determining ultrasmall (nL) volumes by anodic stripping voltammetry has been proposed (W. R. Vandaveer and I. Fritsch, *Anal. Chem.*, **2002**, *74*, 3575, **DOI**: 10.1021/ac011036s). In this method, a metal is exhaustively deposited from the small volume to be measured onto an electrode from which it is later stripped. The solution volume V_s is related to the total charge Q required to strip the metal by

$$V_s = \frac{Q}{nFC}$$

where n is the number of moles of electrons per mole of analyte, F is the faraday, and C is the molar concentration of the metal ion before electrolysis.

(a) Beginning with Faraday's law (see Equation 22-8), derive the above equation for V_s.

(b) In one experiment, the metal deposited was Ag(s) from a solution that was 8.00 mM in $AgNO_3$. The solution was electrolyzed for 30 min at a potential of -0.700 V versus a gold top layer as a pseudoreference. A tubular nanoband electrode was used. The silver was then anodically stripped off the electrode using a linear-sweep rate of 0.10 V/s. The following table represents idealized anodic stripping results. By integration, determine the total charge required to strip the silver from the tubular electrode. You can do a manual Simpson's rule integration or refer to *Applications of Microsoft® Excel in Analytical Chemistry*, 2nd ed., Chapter 11, to do the integration with Excel. From the charge, determine the volume of the solution from which the silver was deposited

Potential, V	Current, nA	Potential, V	Current, nA
−0.50	0.000	−0.123	−1.10
−0.45	−0.02	−0.10	−0.80
−0.40	−0.001	−0.115	−1.00
−0.30	−0.10	−0.09	−0.65
−0.25	−0.20	−0.08	−0.52
−0.22	−0.30	−0.065	−0.37
−0.20	−0.44	−0.05	−0.22
−0.18	−0.67	−0.025	−0.12
−0.175	−0.80	0.00	−0.05
−0.168	−1.00	0.05	−0.03
−0.16	−1.18	0.10	−0.02
−0.15	−1.34	0.15	−0.005
−0.135	−1.28		

(c) Suggest experiments to show whether all the Ag^+ was reduced to Ag(s) in the deposition step.

(d) Would it matter if the droplet were not a hemispherical shape? Why or why not?

(e) Describe an alternative method against which you might test the proposed method.

Spectrochemical Analysis

PART V

Introduction to Spectrochemical Methods

CHAPTER 24

Size of Earth

M. Sigwarth, J. Elrod, K.S. Balasubramaniam,
S. Fletcher/NSO/AURA/NSF

> Methods that use or produce UV, visible, or IR radiation are often called optical spectroscopic methods. Other useful methods include those that use the γ-ray, X-ray, microwave, and RF spectral regions.

This composite image is a sunspot group collected with the Dunn Solar Telescope at the Sacramento Peak Observatory in New Mexico on March 29, 2001. The lower portion consisting of four frames was collected at a wavelength of 393.4 nm, and the upper portion was collected at 430.4 nm. The lower image represents calcium ion concentration, with the intensity of the radiation proportional to the amount of the ion in the sunspot. The upper image shows the presence of the CH molecule. Using data like these, it is possible to determine the location and abundance of virtually any chemical species in the universe. Note that the Earth could fit in the large black core sunspot at the upper left of each of the composite images.

Measurements based on light and other forms of electromagnetic radiation are widely used throughout analytical chemistry. The interactions of radiation and matter are the subject of the science called **spectroscopy**. Spectroscopic analytical methods are based on measuring the amount of radiation produced or absorbed by molecular or atomic species of interest.[1] We can classify spectroscopic methods according to the region of the electromagnetic spectrum used or produced in the measurement. The γ-ray, X-ray, ultraviolet (UV), visible, infrared (IR), microwave, and radio-frequency (RF) regions have been used. Indeed, current usage extends the meaning of spectroscopy yet further to include techniques such as acoustic, mass, and electron spectroscopy in which electromagnetic radiation is not a part of the measurement.

Spectroscopy has played a vital role in the development of modern atomic theory. In addition, **spectrochemical methods** have provided perhaps the most widely used tools for the elucidation of molecular structure as well as the quantitative and qualitative determination of both inorganic and organic compounds.

In this chapter, we discuss the basic principles that are necessary to understand measurements made with electromagnetic radiation, particularly those dealing with the absorption of UV, visible, and IR radiation. The nature of electromagnetic radiation and its interactions with matter are stressed. The next five chapters are devoted to spectroscopic instruments (Chapter 25), molecular absorption spectroscopy (Chapter 26), molecular fluorescence spectroscopy (Chapter 27), atomic spectroscopy (Chapter 28), and mass spectrometry (Chapter 29).

---

[1]For further study, see D. A. Skoog, F. J. Holler, and S. R. Crouch, *Principles of Instrumental Analysis,* 6th ed., Sections 2–3, Belmont, CA: Brooks/Cole, 2007; F. Settle, ed., *Handbook of Instrumental Techniques for Analytical Chemistry,* Sections III–IV, Upper Saddle River, NJ: Prentice-Hall, 1997; J. D. Ingle, Jr., and S. R. Crouch, *Spectrochemical Analysis,* Upper Saddle River, NJ: Prentice-Hall, 1988; E. J. Meehan, in *Treatise on Analytical Chemistry,* 2nd ed., P. J. Elving, E. J. Meehan, and I. M. Kolthoff, eds., Part I, Vol. 7, Chs. 1–3, New York: Wiley, 1981.

24A PROPERTIES OF ELECTROMAGNETIC RADIATION

Electromagnetic radiation is a form of energy that is transmitted through space at enormous velocities. We will call electromagnetic radiation in the UV/visible and sometimes in the IR region, **light**, although strictly speaking the term refers only to visible radiation. Electromagnetic radiation can be described as a wave with properties of wavelength, frequency, velocity, and amplitude. In contrast to sound waves, light requires no transmitting medium; thus, it can travel readily through a vacuum. Light also travels nearly a million times faster than sound.

The wave model fails to account for phenomena associated with the absorption and emission of radiant energy. For these processes, electromagnetic radiation can be treated as discrete packets of energy or particles called **photons** or **quanta**. These dual views of radiation as particles and waves are not mutually exclusive but complementary. In fact, the energy of a photon is directly proportional to its frequency as we shall see. Similarly, this duality applies to streams of electrons, protons, and other elementary particles, which can produce interference and diffraction effects that are typically associated with wave behavior.

Richard P. Feynman (1918–1988) was one of the most renowned scientists of the twentieth century. He was awarded the Nobel Prize in Physics in 1965 for his role in the development of quantum electrodynamics. In addition to his many and varied scientific contributions, he was a skilled teacher, and his lectures and books had a major influence on physics education and science education in general.

24A-1 Wave Properties

In dealing with phenomena such as reflection, refraction, interference, and diffraction, electromagnetic radiation is conveniently modeled as waves consisting of perpendicularly oscillating electric and magnetic fields, as shown in **Figure 24-1a**. The electric field for a single frequency wave oscillates sinusoidally in space and time, as shown in **Figure 24-1b**. The electric field is represented as a vector whose length is proportional to the field strength. The x axis in this plot is either time as the radiation passes a fixed point in space or distance at a fixed time. Note that the direction in which the field oscillates is perpendicular to the direction in which the radiation propagates.

Now we know how the electrons and photons behave. But what can I call it? If I say they behave like particles I give the wrong impression; also if I say they behave like waves. They behave in their own inimitable way, which technically could be called a quantum mechanical way. They behave in a way that is like nothing that you have ever seen before. — R. P. Feynman[2]

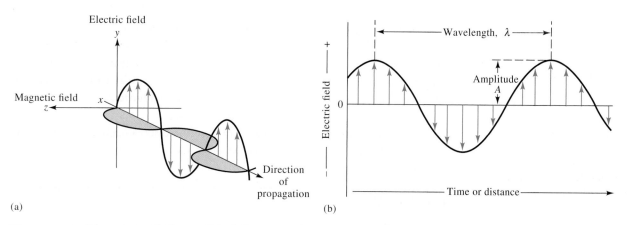

Figure 24-1 Wave nature of a beam of single frequency electromagnetic radiation. In (a), a plane-polarized wave is shown propagating along the x axis. The electric field oscillates in a plane perpendicular to the magnetic field. If the radiation were unpolarized, a component of the electric field would be seen in all planes. In (b), only the electric field oscillations are shown. The amplitude of the wave is the length of the electric field vector at the wave maximum, while the wavelength is the distance between successive maxima.

[2]R. P. Feynman, *The Character of Physical Law*, New York: Random House, 1994, p. 122.

The **amplitude** of an electromagnetic wave is a vector quantity that provides a measure of the electric or magnetic field strength at a maximum in the wave.

The **period** of an electromagnetic wave is the time in seconds for successive maxima or minima to pass a point in space.

The **frequency** of an electromagnetic wave is the number of oscillations that occur in one second.

The unit of frequency is the **hertz** (Hz), which corresponds to one cycle per second, that is, $1 \text{ Hz} = 1 \text{ s}^{-1}$. The frequency of a beam of electromagnetic radiation does not change as it passes through different media.

❯ Radiation velocity and wavelength both decrease as the radiation passes from a vacuum or from air to a denser medium. Frequency remains constant.

❯ Note in Equation 24-1, v (distance/time) = ν waves/time) × λ(distance/wave)

❯ To three significant figures, Equation 24-2 is equally applicable in air or vacuum.

The **refractive index**, η, of a medium measures the extent of interaction between electromagnetic radiation and the medium through which it passes. It is defined by $\eta = c/v$. For example, the refractive index of water at room temperature is 1.33, which means that radiation passes through water at a rate of $c/1.33$ or 2.26×10^{10} cm s^{-1}. In other words, light travels 1.33 times slower in water than it does in vacuum. The velocity and wavelength of radiation become proportionally smaller as the radiation passes from a vacuum or from air to a denser medium while the frequency remains constant.

TABLE 24-1

Wavelength Units for Various Spectral Regions

Region	Unit	Definition
X-ray	Angstrom unit, Å	10^{-10} m
Ultraviolet/visible	Nanometer, nm	10^{-9} m
Infrared	Micrometer, μm	10^{-6} m

Wave Characteristics

In Figure 24-1b, the **amplitude** of the sine wave is shown, and the wavelength is defined. The time in seconds required for the passage of successive maxima or minima through a fixed point in space is called the **period**, p, of the radiation. The **frequency**, ν, is the number of oscillations of the electric field vector per unit time and is equal to $1/p$.

The frequency of a light wave or any wave of electromagnetic radiation is determined by the source that emits it and remains constant regardless of the medium traversed. In contrast, the **velocity**, v, of the wave front through a medium depends on both the medium and the frequency. The **wavelength**, λ, is the linear distance between successive maxima or minima of a wave, as shown in Figure 24-1b. The product of the frequency in waves per unit time and the wavelength in distance per wave is the velocity v of the wave in distance per unit time (cm s^{-1} or m s^{-1}), as shown in Equation 24-1. Note that both the velocity and the wavelength depend on the medium.

$$v = \nu\lambda \tag{24-1}$$

Table 24-1 gives the units used to express wavelengths in various regions of the spectrum.

The Speed of Light

In a vacuum, light travels at its maximum velocity. This velocity, which is given the special symbol c, is 2.99792×10^8 m s^{-1}. The velocity of light in air is only about 0.03 % less than its velocity in vacuum. Thus, for a vacuum, or for air, Equation 24-1 can be written to three significant figures as

$$c = \nu\lambda = 3.00 \times 10^8 \text{ m s}^{-1} = 3.00 \times 10^{10} \text{ cm s}^{-1} \tag{24-2}$$

In a medium containing matter, light travels with a velocity less than c because of interaction between the electromagnetic field and electrons in the atoms or molecules of the medium. Since the frequency of the radiation is constant, the wavelength must decrease as the light passes from a vacuum to a medium containing matter (see Equation 24-1). This effect is illustrated in **Figure 24-2** for a beam of visible radiation. Note that the effect can be quite large.

The **wavenumber**, $\bar{\nu}$, is another way to describe electromagnetic radiation. It is defined as the number of waves per centimeter and is equal to $1/\lambda$. By definition, $\bar{\nu}$ has the units of cm^{-1}.

EXAMPLE 24-1

Calculate the wavenumber of a beam of infrared radiation with a wavelength of 5.00 μm.

Solution

$$\bar{\nu} = \frac{1}{\lambda} = \frac{1}{5.00 \text{ μm} \times 10^{-4} \text{ cm/μm}} = 2000 \text{ cm}^{-1}$$

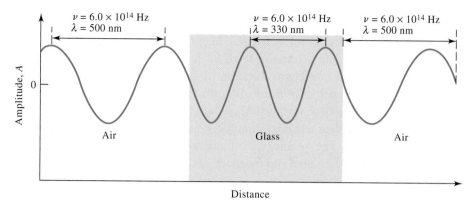

Figure 24-2 Change in wavelength as radiation passes from air into a dense glass and back to air. Note that the wavelength shortens by nearly 200 nm, or more than 30%, as it passes into glass; a reverse change occurs as the radiation again enters air.

Radiant Power and Intensity

The **radiant power**, P, in watts (W) is the energy of a beam that reaches a given area per unit time. The **intensity** is the radiant power-per-unit solid angle.[3] Both quantities are proportional to the square of the amplitude of the electric field (see Figure 24-1b). Although not strictly correct, radiant power and intensity are frequently used interchangeably.

24A-2 The Particle Nature of Light: Photons

In many radiation/matter interactions, it is useful to emphasize the particle nature of light as a stream of photons or quanta. We relate the energy of a single photon to its wavelength, frequency, and wavenumber by

$$E = h\nu = \frac{hc}{\lambda} = hc\bar{\nu} \tag{24-3}$$

where h is Planck's constant (6.63×10^{-34} J·s). Note that the wavenumber and frequency in contrast to the wavelength are directly proportional to the photon energy. Wavelength is inversely proportional to energy. The radiant power of a beam of radiation is directly proportional to the number of photons per second.

> **EXAMPLE 24-2**
>
> Calculate the energy in joules of one photon of radiation with the wavelength given in Example 24-1.
>
> **Solution**
>
> Applying Equation 24-3, we can write
>
> $$E = hc\bar{\nu} = 6.63 \times 10^{-34} \text{ J} \cdot s \times 3.00 \times 10^{10} \frac{\text{cm}}{s} \times 2000 \text{ cm}^{-1}$$
> $$= 3.98 \times 10^{-20} \text{ J}$$

The **wavenumber** $\bar{\nu}$ in cm^{-1} (Kayser) is most often used to describe radiation in the infrared region. The most useful part of the infrared spectrum for the detection and determination of organic species is from 2.5 to 15 µm, which corresponds to a wavenumber range of 4000 to 667 cm^{-1}. As shown below, the wavenumber of a beam of electromagnetic radiation is directly proportional to its energy and thus its frequency.

A **photon** is a particle of electromagnetic radiation having zero mass and an energy of $h\nu$.

❮ Equation 24-3 gives the energy of radiation in SI units of **joules**, where one joule (J) is the work done by a force of one newton (N) acting over a distance of one meter.

❮ Both frequency and wavenumber are proportional to the energy of a photon.

❮ We sometimes speak of "a mole of photons", meaning 6.022×10^{23} packets of radiation of a given wavelength. The energy of one mole of photons with a wavelength of 5.00 µm is 6.022×10^{23} photons/mol × 1 mol × 3.98×10^{-20} J/photon = 2.40×10^4 J = 24.0 kJ.

---

[3]Solid angle is the three dimensional spread at the vertex of a cone measured as the area intercepted by the cone on a unit sphere whose center is at the vertex. The angle is measured in stereradians (sr).

24B INTERACTION OF RADIATION AND MATTER

The most interesting and useful interactions in spectroscopy are those in which transitions occur between different energy levels of chemical species. Other interactions, such as reflection, refraction, elastic scattering, interference, and diffraction, are often related to the bulk properties of materials rather than to the unique energy levels of specific molecules or atoms. Although these bulk interactions are also of interest in spectroscopy, we will limit our discussion here to those interactions in which energy level transitions occur. The specific types of interactions observed depend strongly on the energy of the radiation used and the mode of detection.

24B-1 The Electromagnetic Spectrum

The electromagnetic spectrum covers an enormous range of energies (frequencies) and thus wavelengths (see **Table 24-2**). Useful frequencies vary from $>10^{19}$ Hz (γ-ray) to 10^3 Hz (radio waves). An X-ray photon ($\nu \approx 3 \times 10^{18}$ Hz, $\lambda \approx 10^{-10}$ m), for example, is 10,000 times as energetic as a photon emitted by an ordinary light bulb ($\nu \approx 3 \times 10^{14}$ Hz, $\lambda \approx 10^{-6}$ m) and 10^{15} times as energetic as a radio-frequency photon ($\nu \approx 3 \times 10^3$ Hz $\lambda \approx 10^5$ m).

The major divisions of the spectrum are shown in color in Color Plate 21. Note that the visible region, to which our eyes respond, is only a tiny fraction of the entire spectrum. Different types of radiation such as gamma (γ) rays or radio waves differ from visible light only in the energy (frequency) of their photons.

Figure 24-3 shows the regions of electromagnetic spectrum that are used for spectroscopic analyses. Also shown are the types of atomic and molecular transitions that result from interactions of the radiation with a sample. Note that the

TABLE 24-2

Regions of the UV, Visible, and IR Spectrum

Region	Wavelength Range
UV	180–380 nm
Visible	380–780 nm
Near-IR	0.78–2.5 μm
Mid-IR	2.5–50 μm

One easy way to recall the order of the colors in the spectrum is by the mnemonic **ROY G BIV**, which is short for **R**ed, **O**range, **Y**ellow, **G**reen, **B**lue, **I**ndigo, and **V**iolet.

The **visible region** of the spectrum extends from about 400 nm to almost 800 nm (see Table 24-2).

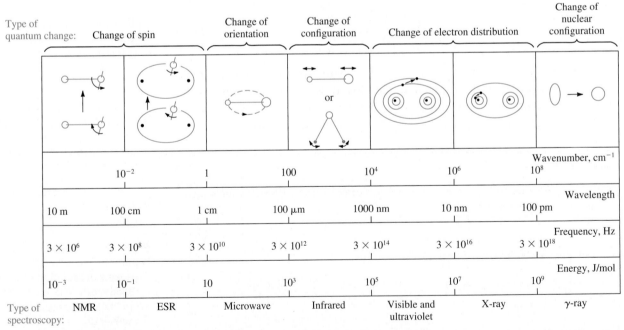

Figure 24-3 The regions of the electromagnetic spectrum. Interaction of an analyte with electromagnetic radiation can result in the types of changes shown. Note that changes in electron distributions occur in the UV/visible region. The wavenumber, wavelength, frequency, and energy are characteristics that describe electromagnetic radiation. (From C. N. Banwell, *Fundamentals of Molecular Spectroscopy,* 3rd ed., New York; McGraw-Hill, 1983, p. 7.)

low-energy radiation used in nuclear magnetic resonance (NMR) and electron spin resonance (ESR) spectroscopy causes subtle changes, such as changes in spin; the high-energy radiation used in γ-ray spectroscopy can cause much more dramatic changes, such as nuclear configuration changes.

Spectrochemical methods that use not only visible but also ultraviolet and infrared radiation are often called **optical methods** in spite of the fact that the human eye is not sensitive to UV or IR radiation. This terminology arises from the many common features of instruments for the three spectral regions and the similarities in the way we view the interactions of the three types of radiation with matter.

Optical methods are spectroscopic methods based on ultraviolet, visible, and infrared radiation.

24B-2 Spectroscopic Measurements

Spectroscopists use the interactions of radiation with matter to obtain information about a sample. Several of the chemical elements were discovered by spectroscopy (see Feature 24-1). The sample is usually stimulated in some way by applying energy in the form of heat, electrical energy, light, particles, or a chemical reaction. Prior to applying the stimulus, the analyte is predominately in its lowest-energy or **ground state**. The stimulus then causes some of the analyte species to undergo a transition to a higher-energy or **excited state**. We acquire information about the analyte by measuring the electromagnetic radiation emitted as it returns to the ground state or by measuring the amount of electromagnetic radiation absorbed as a result of excitation.

Figure 24-4 illustrates the processes that occur in emission and chemiluminescence spectroscopy. The analyte is stimulated by applying heat or electrical energy or by a chemical reaction. The term **emission spectroscopy** usually refers to methods in which the stimulus is heat or electrical energy, while **chemiluminescence spectroscopy** refers to excitation of the analyte by a chemical reaction. In both cases, measurement of the radiant power emitted as the analyte returns to the ground state can give information about its identity and concentration. The results of such a measurement are often expressed graphically by a **spectrum**, which is a plot of the emitted radiation as a function of frequency or wavelength.

A familiar example of **chemiluminescence** is found in the light emitted by a firefly. In the firefly reaction, an enzyme luciferase catalyzes the oxidative phosphorylation reaction of luciferin with adenosine triphosphate (ATP) to produce oxyluciferin, carbon dioxide, adenosine monophosphate (AMP), and light. Chemiluminescence involving a biological or enzyme reaction is often termed **bioluminescence**. The popular light stick is another familiar example of chemiluminescence.

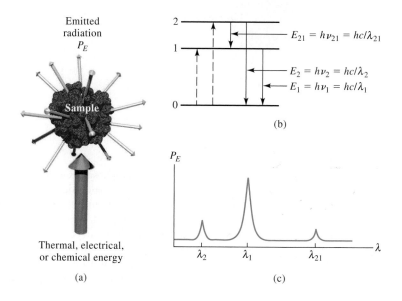

Figure 24-4 Emission or chemiluminescence processes. In (a), the sample is excited by applying thermal, electrical, or chemical energy. No radiant energy is used to produce excited states, and so, these are called non-radiative processes. In the energy level diagram (b), the dashed lines with upward pointing arrows symbolize these nonradiative excitation processes, while the solid lines with downward pointing arrows indicate that the analyte loses its energy by emission of a photon. In (c), the resulting spectrum is shown as a measurement of the radiant power emitted, P_E, as a function of wavelength, λ.

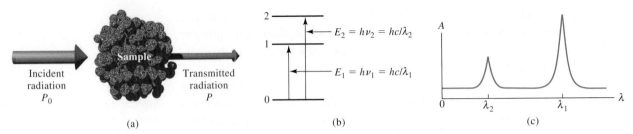

Figure 24-5 Absorption methods. In (a), radiation of incident radiant power P_0 can be absorbed by the analyte, resulting in a transmitted beam of lower radiant power P. For absorption to occur the energy of the incident beam must correspond to one of the energy differences shown in (b). The resulting absorption spectrum is shown in (c).

When the sample is stimulated by applying an external electromagnetic radiation source, several processes are possible. For example, the radiation can be scattered or reflected. What is important to us is that some of the incident radiation can be absorbed and promote some of the analyte species to an excited state, as shown in **Figure 24-5**. In **absorption spectroscopy**, we measure the amount of light absorbed as a function of wavelength. Absorption measurements can give both qualitative and quantitative information about the sample. In **photoluminescence spectroscopy** (see **Figure 24-6**), the emission of photons is measured following absorption. The most important forms of photoluminescence for analytical purposes are **fluorescence** and **phosphorescence spectroscopy**.

We focus here on absorption spectroscopy in the UV/visible region of the spectrum because it is so widely used in chemistry, biology, forensic science, engineering, agriculture, clinical chemistry, and many other fields. Note that the processes shown in Figures 24-4 through 24-6 can occur in any region of the electromagnetic spectrum; the different energy levels can be nuclear levels, electronic levels, vibrational levels, or spin levels.

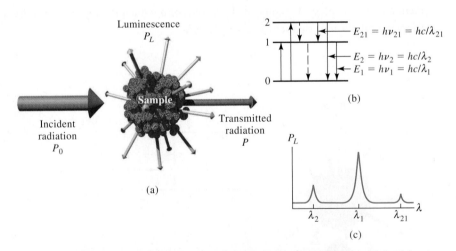

Figure 24-6 Photoluminescence methods (fluorescence and phosphorescence). Fluorescence and phosphorescence result from absorption of electromagnetic radiation and then dissipation of the energy by emission of radiation, as shown in (a). In (b), the absorption can cause excitation of the analyte to state 1 or state 2. Once excited, the excess energy can be lost by emission of a photon (luminescence shown as solid lines) or by nonradiative processes (dashed lines). The emission occurs over all angles, and the wavelengths emitted (c) correspond to energy differences between levels. The major distinction between fluorescence and phosphorescence is the time scale of emission with fluorescence being prompt and phosphorescence being delayed.

Spectroscopy and the Discovery of Elements

The modern era of spectroscopy began with the observation of the spectrum of the sun by Sir Isaac Newton in 1672. In his experiment, Newton passed rays from the sun through a small opening into a dark room where they struck a prism and dispersed into the colors of the spectrum. The first description of spectral features beyond the simple observation of colors was in 1802 by Wollaston, who noticed dark lines on a photographic image of the solar spectrum. These lines along with more than 500 others, which are shown in the solar spectrum of **Figure 24F-1**, were later described in detail by Fraunhofer. Based on his observations, which began in 1817, Fraunhofer gave the prominent lines letters starting with "A" at the red end of the spectrum. The solar spectrum is shown in color plate 17.

It remained, however, for Gustav Kirchhoff and Robert Wilhelm Bunsen in 1859 and 1860 to explain the origin of the Fraunhofer lines. Bunsen had invented his famous burner (see **Figure 24F-2**) a few years earlier, which made possible spectral observations of emission and absorption phenomena in a nearly transparent flame. Kirchhoff concluded that the Fraunhofer "D" lines were due to sodium in the sun's atmosphere and the "A" and "B" lines were due to potassium. To this day, we call the emission lines of sodium the sodium "D" lines. These lines are responsible for the familiar yellow color seen in flames containing sodium or in sodium vapor lamps. The absence of lithium in the sun's spectrum led Kirchhoff to conclude that there was little lithium present in the sun. During these studies, Kirchhoff also developed his famous laws relating the absorption and emission of light from bodies and at interfaces. Together with Bunsen, Kirchhoff observed that different elements could impart different colors to flames and produce spectra exhibiting differently colored bands or lines. Kirchhoff and Bunsen are thus credited with discovering the use of spectroscopy for chemical analysis. Emission spectra of several elements are shown in color plate 16. The method was soon put to many practical uses, including the discovery of new elements. In 1860, the elements cesium and rubidium were discovered with spectroscopy, followed in 1861 by thallium and in 1864 by indium. The age of spectroscopic analysis had begun.

Figure 24F-1 The solar spectrum. The dark vertical lines are the Fraunhofer lines. See color plate 17 for a full-color version of the spectrum. Data for the image were collated by Dr. Donald Mickey, University of Hawaii Institute for Astronomy, from National Solar Observatory spectral data. NSO/Kitt Peak FTS data used here were produced by NSF/NOAO.

Figure 24F-2 Bunsen burner of the type used in early spectroscopic studies with a prism spectroscope of type used by Kirchhoff. (From H. Kayser, *Handbuch der Spectroscopie,* Stuttgart, Germany: S. Hirzel Verlag GmbH, 1900.)

24C ABSORPTION OF RADIATION

Every molecular species is capable of absorbing its own characteristic frequencies of electromagnetic radiation, as described in Figure 24-5. This process transfers energy to the molecule and results in a decrease in the intensity of the incident electromagnetic radiation. Absorption of the radiation thus **attenuates** the beam in accordance with the absorption law as described in Section 24C-1.

24C-1 The Absorption Process

The absorption law, also known as the **Beer-Lambert law** or just **Beer's law**, tells us quantitatively how the amount of attenuation depends on the concentration of the absorbing molecules and the path length over which absorption occurs. As light traverses a medium containing an absorbing analyte, the intensity decreases as the analyte becomes excited. For an analyte solution of a given concentration, the longer the length of the medium through which the light passes (path length of light), the more absorbers are in the path, and the greater the attenuation. Similarly, for a given path length of light, the higher the concentration of absorbers, the stronger the attenuation.

Figure 24-7 depicts the attenuation of a parallel beam of **monochromatic radiation** as it passes through an absorbing solution of thickness b cm and concentration c moles per liter. Because of interactions between the photons and absorbing particles (recall Figure 24-5), the radiant power of the beam decreases from P_0 to P. The **transmittance** T of the solution is the fraction of incident radiation transmitted by the solution, as shown in Equation 24-4. Transmittance is often expressed as a percentage and called the **percent transmittance**.

$$T = P/P_0 \tag{24-4}$$

Absorbance

The **absorbance**, A, of a solution is related to the transmittance in a logarithmic manner as shown in Equation 24-5. Notice that as the absorbance of a solution increases, the transmittance decreases. The relationship between transmittance and absorbance is illustrated by the conversion spreadsheet shown in **Figure 24-8**. The scales on earlier instruments were linear in transmittance or sometimes in absorbance. In modern instruments, a computer calculates absorbance from measured quantities.

$$A = -\log T = -\log \frac{P}{P_0} = \log \frac{P_0}{P} \tag{24-5}$$

> In spectroscopy, **attenuate** means to decrease the energy per unit area of a beam of radiation. In terms of the photon model, attenuate means to decrease the number of photons per second in the beam.

> The term **monochromatic radiation** refers to radiation of a single color; that is, a single wavelength or frequency. In practice, it is virtually impossible to produce a single color of light. We discuss the practical problems associated with producing monochromatic radiation in Chapter 25.

> Percent transmittance = %T
> $$= \frac{P}{P_0} \times 100\%.$$

> Absorbance can be calculated from percent transmittance as follows:
> $$T = \frac{\%T}{100\%}$$
> $$A = -\log T$$
> $$= -\log \%T + \log 100$$
> $$= 2 - \log \%T$$

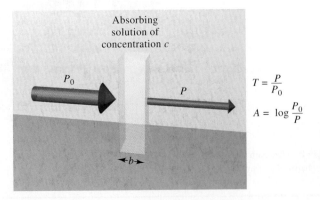

Figure 24-7 Attenuation of a beam of radiation by an absorbing solution. The larger arrow on the incident beam signifies a higher radiant power P_0 than that transmitted by the solution P. The path length of the absorbing solution is b and the concentration is c.

	A	B	C	D
1	Calculation of Absorbance from Transmittance			
2	T	%T	A = -log T	A = 2-log %T
3	0.001	0.1	3.000	3.000
4	0.010	1.0	2.000	2.000
5	0.050	5.0	1.301	1.301
6	0.075	7.5	1.125	1.125
7	0.100	10.0	1.000	1.000
8	0.200	20.0	0.699	0.699
9	0.300	30.0	0.523	0.523
10	0.400	40.0	0.398	0.398
11	0.500	50.0	0.301	0.301
12	0.600	60.0	0.222	0.222
13	0.700	70.0	0.155	0.155
14	0.800	80.0	0.097	0.097
15	0.900	90.0	0.046	0.046
16	1.000	100.0	0.000	0.000
17				
18	Spreadsheet Documentation			
19	Cell B3=A3*100			
20	Cell C3=-LOG10(A3)			
21	Cell D3=2-LOG10(B3)			

Figure 24-8 Conversion spreadsheet relating transmittance T, percent transmittance $\%T$, and absorbance A. The transmittance data to be converted are entered in cells A3 through A16. The percent transmittance is calculated in cells B3 by the formula shown in the documentation section, cell A19. This formula is copied into cells B4 through B16. The absorbance is calculated from $-\log T$ in cells C3 through C16 and from $2-\log \%T$ in cells D3 through D16. The formulas for the first cell in the C and D columns are shown in cells A20 and A21.

Measuring Transmittance and Absorbance

Transmittance and absorbance, as defined by Equations 24-4 and 24-5 and depicted in Figure 24-7, usually cannot be measured as shown because the solution to be studied must be held in a container (cell or cuvette). Reflection and scattering losses can occur at the cell walls, as shown in **Figure 24-9**. These losses can be substantial. For example, about 8.5% of a beam of yellow light is lost by reflection when it passes through a glass cell. Light can also be scattered in all directions from the surface of large molecules or particles, such as dust, in the solvent, and this scattering can cause further attenuation of the beam as it passes through the solution.

To compensate for these effects, the power of the beam transmitted through a cell containing the analyte solution is compared with one that traverses an identical cell containing only the solvent, or a reagent blank. An experimental absorbance that closely approximates the true absorbance for the solution is thus obtained, that is,

$$A = \log \frac{P_0}{P} \approx \log \frac{P_{\text{solvent}}}{P_{\text{solution}}} \qquad (24\text{-}6)$$

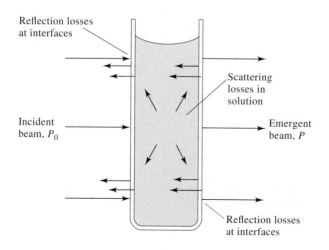

Reflection losses at interfaces

Scattering losses in solution

Incident beam, P_0

Emergent beam, P

Reflection losses at interfaces

Figure 24-9 Reflection and scattering losses with a solution contained in a typical glass cell. Losses by reflection can occur at all the boundaries that separate the different materials. In this example, the light passes through the following boundaries, called interfaces: air-glass, glass-solution, solution-glass, and glass-air.

Because of this close approximation, the terms P_0 and P will henceforth refer to the power of a beam that has passed through cells containing the solvent (or blank) and the analyte solution, respectively.

Beer's Law

According to Beer's law, absorbance is directly proportional to the concentration of the absorbing species, c, and to the path length, b, of the absorbing medium as expressed by Equation 24-7.

$$A = \log(P_0/P) = abc \qquad (24\text{-}7)$$

In Equation 24-7, a is a proportionality constant called the **absorptivity**. Because absorbance is a unitless quantity, the absorptivity must have units that cancel the units of b and c. If, for example, c has the units of g L^{-1} and b has the units of cm, absorptivity has the units of L g^{-1} cm^{-1}.

When we express the concentration in Equation 24-7 in moles per liter and b in cm, the proportionality constant is called the **molar absorptivity** and is given the symbol ε. Thus,

$$A = \varepsilon bc \qquad (24\text{-}8)$$

where ε has the units of L mol^{-1} cm^{-1}.

> The molar absorptivity of a species at an absorption maximum is characteristic of that species. Peak molar absorptivities for many organic compounds range from 10 or less to 10,000 or more. Some transition metal complexes have molar absorptivities of 10,000 to 50,000. High molar absorptivities are desirable for quantitative analysis because they lead to high analytical sensitivity.

FEATURE 24-2

Deriving Beer's Law

To derive Beer's law, we consider the block of absorbing matter (solid, liquid, or gas) shown in **Figure 24F-3**. A beam of parallel monochromatic radiation with power P_0 strikes the block perpendicular to a surface; after passing through a length b of the material, which contains n absorbing particles (atoms, ions, or molecules), its power is decreased to P as a result of absorption. Consider now a cross section of the block having an area S and an infinitesimal thickness dx. Within this section, there are dn absorbing particles. Associated with each particle, we can imagine a surface at which photon capture will occur, that is, if a photon reaches one of these

Figure 24F-3 Attenuation of a beam of electromagnetic radiation with initial power P_0 by a solution containing c mol/L of absorbing solute and a path length of b cm. The transmitted beam has a radiant power P ($P < P_0$).

areas by chance, absorption will follow immediately. The total projected area of these capture surfaces within the section is designated as dS; the ratio of the capture area to the total area then is dS/S. On a statistical average, this ratio represents the probability for the capture of photons within the section. The power of the beam entering the section, P_x, is proportional to the number of photons per square centimeter per second, and dP_x represents the quantity removed per second within the section. The fraction absorbed is then $-dP_x/P_x$, and this ratio also equals the average probability for capture. The term is given a minus sign to indicate that *the* radiant power undergoes a decrease. Thus,

$$-\frac{dP_x}{P_x} = \frac{dS}{S} \tag{24-9}$$

Recall now that dS is the sum of the capture areas for particles within the section. It must, therefore, be proportional to the number of particles, or

$$dS = a \times dn \tag{24-10}$$

where dn is the number of particles and a is a proportionality constant, which is called the *capture cross section*. By combining Equations 24-9 and 24-10 and integrating over the interval between 0 and n, we obtain

$$-\int_{P_x}^{P}\frac{dP_x}{P_x} = \int_{0}^{n}\frac{a \times dn}{S}$$

which, when integrated, gives

$$-\ln\frac{P}{P_0} = \frac{an}{S}$$

We then convert to base 10 logarithms, invert the fraction to change the sign, and obtain

$$\log\frac{P_0}{P} = \frac{an}{2.303\ S} \tag{24-11}$$

where n is the total number of particles within the block shown in Figure 24F-3. The cross-sectional area S can be expressed in terms of the volume of the block V in cm^3 and its length b in cm. Thus,

$$S = \frac{V}{b}\ cm^2$$

By substituting this quantity into Equation 24-11, we find

$$\log\frac{P_0}{P} = \frac{anb}{2.303\ V} \tag{24-12}$$

Notice that n/V has the units of concentration (that is, number of particles per cubic centimeter). To convert n/V to moles per liter, we find the number of moles by

$$\text{number mol} = \frac{n\ \text{particles}}{6.022 \times 10^{23}\ \text{particles/mol}}$$

(continued)

The concentration c in mol/L is then

$$c = \frac{n}{6.022 \times 10^{23}} \text{ mol} \times \frac{1000 \text{ cm}^3/\text{L}}{V \text{ cm}^3}$$

$$= \frac{1000\,n}{6.022 \times 10^{23}} \text{ mol/L}$$

By combining this relationship with Equation 24-12, we have

$$\log \frac{P_0}{P} = \frac{6.022 \times 10^3 \, abc}{2.303 \times 1000}$$

Finally, the constants in this equation can be collected into a single term ε to give

$$A = \log \frac{P_0}{P} = \varepsilon bc \qquad (24\text{-}13)$$

which is Beer's law.

Terms Used in Absorption Spectrometry

In addition to the terms we have introduced to describe absorption of radiant energy, you may encounter other terms in the literature or with older instruments. The terms, symbols, and definitions given in **Table 24-3** are recommended by the Society for Applied Spectroscopy and the American Chemical Society. The third column contains the older names and symbols. Because a standard nomenclature is highly desirable to avoid ambiguities, we urge you to learn and use the recommended terms and symbols and to avoid the older terms.

TABLE 24-3

Important Terms and Symbols Employed in Absorption Measurements

Term and Symbol*	Definition	Alternative Name and Symbol
Incident radiant power, P_0	Radiant power in watts incident on sample	Incident intensity, I_0
Transmitted radiant power, P	Radiant power transmitted by sample	Transmitted intensity, I
Absorbance, A	$\log(P_0/P)$	Optical density, D; extinction, E
Transmittance, T	P/P_0	Transmission, T
Path length of sample, b	Length over which attenuation occurs	l, d
Absorptivity†, a	$A/(bc)$	α, k
Molar absorptivity‡, ε	$A/(bc)$	Molar absorption coefficient

*Compilation of terminology recommended by the American Chemical Society and the Society for Applied Spectroscopy (*Appl. Spectrosc.,* **2012,** *66,* 132).
† c may be expressed in g L^{-1} or in other specified concentration units; b may be expressed in cm or other units of length.
‡ c is expressed in mol L^{-1}; b is expressed in cm.

Using Beer's Law

Beer's law, as expressed in Equations 24-6 and 24-8, can be used in several different ways. We can calculate molar absorptivities of species if the concentration is known, as shown in Example 24-3. We can use the measured value of absorbance to obtain concentration if absorptivity and path length are known. Absorptivities, however, are functions of such variables as solvent, solution composition, and temperature. Because of variations in absorptivity with conditions, it is never a good idea to depend on literature values for quantitative work. Hence, a standard solution of the analyte in the same solvent and at a similar temperature is used to obtain the absorptivity at the time of the analysis. Most often, we use a series of standard solutions of the analyte to construct a calibration curve, or working curve, of A versus c or to obtain a linear regression equation (for the method of external standards and linear regression, see Section 8D-2). It may also be necessary to duplicate closely the overall composition of the analyte solution in order to compensate for matrix effects. Alternatively, the method of standard additions (see Section 8D-3 and Section 26A-3) is used for the same purpose.

EXAMPLE 24-3

A 7.25×10^{-5} M solution of potassium permanganate has a transmittance of 44.1% when measured in a 2.10-cm cell at a wavelength of 525 nm. Calculate (a) the absorbance of this solution and (b) the molar absorptivity of $KMnO_4$.

Solution

(a) $A = -\log T = -\log 0.441 = -(-0.356) = 0.356$
(b) From Equation 24-8,

$$\varepsilon = A/bc = 0.356/(2.10 \text{ cm} \times 7.25 \times 10^{-5} \text{mol L}^{-1})$$

$$= 2.34 \times 10^{3} \text{ L mol}^{-1} \text{ cm}^{-1}$$

Spreadsheet Summary In the first exercise in Chapter 12 of *Applications of Microsoft® Excel in Analytical Chemistry,* 2nd ed., a spreadsheet is developed to calculate the molar absorptivity of permanganate ion. A plot of absorbance versus permanganate concentration is constructed, and least-squares analysis of the linear plot is carried out. The data are analyzed statistically to determine the uncertainty of the molar absorptivity. In addition, other spreadsheets are presented for calibration in quantitative spectrophotometric experiments and for calculating concentrations of unknown solutions.

Applying Beer's Law to Mixtures

Beer's law also applies to solutions containing more than one kind of absorbing substance. Provided that there is no interaction among the various species, the total absorbance for a multicomponent system at a single wavelength is the sum of the individual absorbances. In other words,

> Absorbances are additive if the absorbing species do not interact.

$$A_{total} = A_1 + A_2 + \cdots + A_n = \varepsilon_1 b c_1 + \varepsilon_2 b c_2 + \cdots + \varepsilon_n b c_n \qquad (24\text{-}14)$$

where the subscripts refer to absorbing components $1, 2, \ldots, n$.

24C-2 Absorption Spectra

> A bit of Latin. One plot of absorbance versus wavelength is called a spect**rum**; two or more plots are called spect**ra**.

An **absorption spectrum** is a plot of absorbance versus wavelength, as illustrated in **Figure 24-10**. Absorbance could also be plotted against wavenumber or frequency. Modern scanning spectrophotometers produce such an absorption spectrum directly. Older instruments sometimes displayed transmittance and produced plots of T or $\%T$ versus wavelength. Occasionally plots with log A as the ordinate are used. The logarithmic axis leads to a loss of spectral detail, but it is convenient for comparing solutions of widely different concentrations. A plot of molar absorptivity ε as a function of wavelength is independent of concentration. This type of spectral plot is characteristic for a given molecule and is sometimes used to aid in identifying or confirming the identity of a particular species. The color of a solution is related to its absorption spectrum (see Feature 24-3).

Figure 24-10   Typical absorption spectra of potassium permanganate at five different concentrations. The numbers adjacent to the curves indicate concentration of manganese in ppm, and the absorbing species is permanganate ion, MnO_4^-. The cell path length b is 1.00 cm. A plot of absorbance at the peak wavelength at 525 nm versus concentration of permanganate is linear and thus the absorber obeys Beer's law.

FEATURE 24-3

Why Is a Red Solution Red?

An aqueous solution of the complex $Fe(SCN)^{2+}$ is not red because the complex adds red radiation to the solvent. Instead, it absorbs green from the incoming white radiation and transmits the red component (see **Figure 24F-4**). Thus, in a colorimetric determination of iron based on its thiocyanate complex, the maximum change in absorbance with concentration occurs with green radiation; the absorbance change with red radiation is negligible. In general, then, the radiation used for a colorimetric analysis should be the complementary color of the analyte solution. The following table shows this relationship for various parts of the visible spectrum.

The Visible Spectrum

Wavelength Region Absorbed, nm	Color of Light Absorbed	Complementary Color Transmitted
400–435	Violet	Yellow-green
435–480	Blue	Yellow
480–490	Blue-green	Orange
490–500	Green-blue	Red
500–560	Green	Purple
560–580	Yellow-green	Violet
580–595	Yellow	Blue
595–650	Orange	Blue-green
650–750	Red	Green-blue

White light   Lens $Fe(SCN)^{2+}$ Solution   Absorbs at 460–500 nm   Human eye detector

Figure 24F-4 Color of a solution. White light from a lamp or the sun strikes an aqueous solution of $Fe(SCN)^{2+}$. The fairly broad absorption spectrum shows a maximum absorbance in the 460 to 500 nm range (see Figure 26-4a). The complementary red color is transmitted.

Atomic Absorption

When a beam of polychromatic ultraviolet or visible radiation passes through a medium containing gaseous atoms, only a few frequencies are attenuated by absorption, and when recorded on a very high resolution spectrometer, the spectrum consists of a number of very narrow absorption lines.

Figure 24-11 Partial energy level diagram for sodium, showing the transitions resulting from absorption at 590, 330, and 285 nm.

The **electron volt** (eV) is a unit of energy. When an electron with charge $q = 1.60 \times 10^{-19}$ coulombs is moved through a potential difference of 1 volt = 1 joule/coulomb, the energy expended (or released) is then equal to $E = qV = (1.60 \times 10^{-19}$ coulombs) (1 joule/coulomb) = 1.60×10^{-19} joule = 1 eV.

$$1 \text{ eV} = 1.60 \times 10^{-19} \text{ J}$$
$$= 3.83 \times 10^{-20} \text{ calories}$$
$$= 1.58 \times 10^{-21} \text{ L atm}$$

In an **electronic transition,** an electron moves from one orbital to another. Transitions occur between atomic orbitals in atoms and between molecular orbitals in molecules.

Vibrational and rotational transitions occur with polyatomic species because only this type of species has vibrational and rotational states with different energies.

The **ground state** of an atom or a molecular species is the minimum energy state of the species. At room temperature, most atoms and molecules are in their ground state.

Figure 24-11 is a partial energy level diagram for sodium that shows the major atomic absorption transitions. The transitions, shown as colored arrows between levels, occur when the single outer electron of sodium is excited from its room temperature or ground state 3s orbital to the 3p, 4p, and 5p orbitals. These excitations are brought on by absorption of photons of radiation whose energies exactly match the differences in energies between the excited states and the 3s ground state. Transitions between two different orbitals are termed **electronic transitions**. Atomic absorption spectra are not usually recorded because of instrumental difficulties. Instead, atomic absorption is measured at a single wavelength using a very narrow, nearly monochromatic source (see Section 28D).

EXAMPLE 24-4

The energy difference between the 3p and the 3s orbitals in Figure 24-11b is 2.107 eV. Calculate the wavelength of radiation that would be absorbed in exciting the 3s electron to the 3p state (1 eV = 1.60 × 10⁻¹⁹ J).

Solution

Rearranging Equation 24-3 gives

$$\lambda = \frac{hc}{E}$$

$$= \frac{6.63 \times 10^{-34} \text{ J·s} \times 3.00 \times 10^{10} \text{ cm/s} \times 10^7 \text{ nm/cm}}{2.107 \text{ eV} \times 1.60 \times 10^{-19} \text{ J/eV}}$$

$$= 590 \text{ nm}$$

Molecular Absorption

Molecules undergo three types of quantized transitions when excited by ultraviolet, visible, and infrared radiation. For ultraviolet and visible radiation, excitation occurs when an electron residing in a low-energy molecular or atomic orbital is promoted to a higher-energy orbital. We mentioned previously that the energy $h\nu$ of the photon must be exactly the same as the energy difference between the two orbital energies.

In addition to electronic transitions, molecules exhibit two other types of radiation-induced transitions: **vibrational transitions** and **rotational transitions**. Vibrational transitions occur because a molecule has a multitude of quantized energy levels, or vibrational states, associated with the bonds that hold the molecule together.

Figure 24-12 is a partial energy level diagram that depicts some of the processes that occur when a polyatomic species absorbs infrared, visible, and ultraviolet radiation. The energies E_1 and E_2, two of the several electronically excited states of a molecule, are shown relative to the energy of the ground state E_0. In addition, the relative energies of a few of the many vibrational states associated with each electronic state are indicated by the lighter horizontal lines.

You can get an idea of the nature of vibrational states by picturing a bond in a molecule as a vibrating spring with atoms attached to both ends. In **Figure 24-13a**, two types of stretching vibration are shown. With each vibration, atoms first approach and then move away from one another. The potential energy of such a system at any instant depends on the extent to which the spring is stretched or compressed. For a real-world macroscopic spring, the energy of the system varies continuously and reaches a maximum when the spring is fully stretched or fully compressed. In contrast, the energy of a spring system of atomic dimensions (a chemical bond) can have only certain discrete energies called vibrational energy levels.

Figure 24-13b shows four other types of molecular vibrations. The energies associated with these vibrational states usually differ from one another and from the energies associated with stretching vibrations. Some of the vibrational energy levels associated with each of the electronic states of a molecule are depicted by the lines labeled 1, 2, 3, and 4 in Figure 24-12 (the lowest vibrational levels are labeled 0). Note that the differences in energy among the vibrational states are significantly smaller than among energy levels of the electronic states (typically an order of magnitude smaller). Although they are not shown, a molecule has many quantized rotational states that are associated with the rotational motion of a molecule around its center of gravity. These rotational energy states are superimposed on each of the vibrational states shown in the energy diagram. The energy differences among these states are smaller than those among vibrational states by an order of magnitude and so are not shown in the diagram. The total energy E associated with a molecule is then given by

$$E = E_{\text{electronic}} + E_{\text{vibrational}} + E_{\text{rotational}} \tag{24-15}$$

where $E_{\text{electronic}}$ is the energy associated with the electrons in the various outer orbitals of the molecule, $E_{\text{vibrational}}$ is the energy of the molecule as a whole due to interatomic vibrations, and $E_{\text{rotational}}$ accounts for the energy associated with rotation of the molecule about its center of gravity.

Infrared Absorption. Infrared radiation generally is not energetic enough to cause electronic transitions, but it can induce transitions in the vibrational and rotational states associated with the ground electronic state of the molecule. Four of these

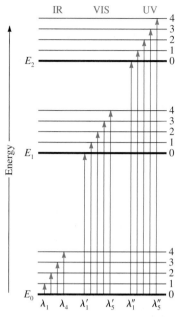

Figure 24-12 Energy level diagram showing some of the energy changes that occur during absorption of infrared (IR), visible (VIS), and ultraviolet (UV) radiation by a molecular species. Note that with some molecules a transition from E_0 to E_1 may require UV radiation instead of visible radiation. With other molecules, the transition from E_0 to E_2 may occur with visible radiation instead of UV radiation. Only a few vibrational levels (0–4) are shown. The rotational levels associated with each vibrational level are not shown because they are too closely spaced.

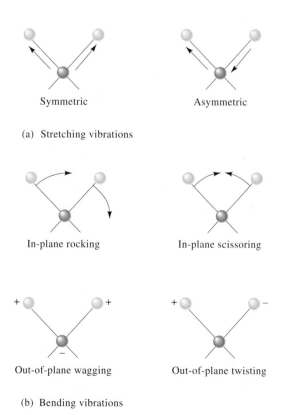

(a) Stretching vibrations

In-plane rocking

In-plane scissoring

Out-of-plane wagging

Out-of-plane twisting

(b) Bending vibrations

Figure 24-13 Types of molecular vibrations. The plus sign indicates motion out of the page; the minus sign indicates motion into the page.

transitions are depicted in the lower left part of Figure 24-12 (λ_1 to λ_4). For absorption to occur, the radiation source has to emit frequencies corresponding exactly to the energies indicated by the lengths of the four arrows.

Absorption of Ultraviolet and Visible Radiation.   The center arrows in Figure 24-12 suggest that the molecules under consideration absorb visible radiation of five wavelengths (λ_1' to λ_5'), thereby promoting electrons to the five vibrational levels of the excited electronic level E_1. Ultraviolet photons that are more energetic are required to produce the absorption indicated by the five arrows to the right.

Figure 24-12 suggests that molecular absorption in the ultraviolet and visible regions produces **absorption bands** made up of closely spaced lines. A real molecule has many more energy levels than can be shown in the diagram. Thus, a typical absorption band consists of a large number of lines. In a solution, the absorbing species are surrounded by solvent molecules, and the band nature of molecular absorption often becomes blurred because collisions tend to spread the energies of the quantum states, giving smooth and continuous absorption peaks.

Figure 21-14 shows visible spectra for 1,2,4,5-tetrazine that were obtained under three different conditions: gas phase, nonpolar solvent, and polar solvent (aqueous solution). Notice that in the gas phase (see Figure 24-14a), the individual tetrazine molecules are sufficiently separated from one another to vibrate and rotate freely so that many individual absorption peaks resulting from transitions among the various vibrational and rotational states appear in the spectrum. In the liquid state and in nonpolar solvents (see Figure 24-14b), however, tetrazine molecules are unable to rotate freely so that we see no fine structure in the spectrum. Furthermore, in a polar solvent such as water (see Figure 24-14c), frequent collisions and interactions between tetrazine and water molecules cause the vibrational levels to be modified

Figure 24-14   Typical visible absorption spectra. The compound is 1,2,4,5-tetrazine. In (a), the spectrum is shown in the gas phase where many lines due to electronic, vibrational, and rotational transitions are seen. In a nonpolar solvent (b), the electronic transitions can be observed, but the vibrational and rotational structure has been lost. In a polar solvent (c), the strong intermolecular forces have caused the electronic peaks to blend together to give only a single smooth absorption peak. (Reproduced from S. F. Mason, *J. Chem. Soc.*, **1959**, 1263, **DOI**: 10.1039/JR9590001263, with permission of The Royal Society of Chemistry.)

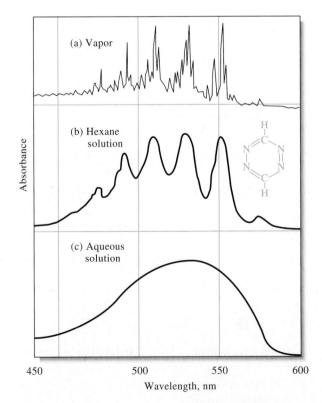

energetically in an irregular way. Hence, the spectrum appears as a single broad peak. The trends shown in the spectra of tetrazine in this figure are typical of UV-visible spectra of other molecules recorded under similar conditions.

24C-3 Limits to Beer's Law

There are few exceptions to the linear relationship between absorbance and path length at a fixed concentration. We frequently observe deviations from the direct proportionality between absorbance and concentration, however, when the path length b is a constant. Some of these deviations, called **real deviations**, are fundamental and represent real limitations to the law. Others are a result of the method that we use to measure absorbance (**instrumental deviations**) or from chemical changes that occur when the concentration changes (**chemical deviations**).

Real Limitations to Beer's Law

Beer's law describes the absorption behavior only of dilute solutions and in this sense is a **limiting law**. At concentrations exceeding about 0.01 M, the average distances between ions or molecules of the absorbing species are diminished to the point where each particle affects the charge distribution and thus the extent of absorption of its neighbors. Because the extent of interaction depends on concentration, the occurrence of this phenomenon causes deviations from the linear relationship between absorbance and concentration. A similar effect sometimes occurs in dilute solutions of absorbers that contain high concentrations of other species, particularly electrolytes. When ions are very close to one another, the molar absorptivity of the analyte can be altered because of electrostatic interactions which can lead to departures from Beer's law.

> Limiting laws in science are those that hold under limiting conditions such as dilute solutions. In addition to Beer's law, the Debye-Hückel law (see Chapter 10) and the law of independent migration that describes the conductance of electricity by ions are limiting laws.

Chemical Deviations

As shown in Example 24-5, deviations from Beer's law appear when the absorbing species undergoes association, dissociation, or reaction with the solvent to give products that absorb differently from the analyte. The extent of such departures can be predicted from the molar absorptivities of the absorbing species and the equilibrium constants for these equilibria. Unfortunately, since we are usually unaware that such processes are affecting the analyte, there is often no opportunity to correct the measurement. Typical equilibria that give rise to this effect include monomer-dimer equilibria, metal complexation equilibria where more than one complex is present, acid/base equilibria, and solvent-analyte association equilibria.

EXAMPLE 24-5

Solutions containing various concentrations of the acidic indicator HIn with $K_a = 1.42 \times 10^{-5}$ were prepared in 0.1 M HCl and 0.1 M NaOH. In both media, plots of absorbance at either 430 nm or 570 nm versus the total indicator concentration are nonlinear. However, in both media, the individual species HIn or In$^-$ obey Beer's law at 430 nm and 570 nm. Hence, if we knew the equilibrium concentrations of HIn and In$^-$, we could compensate for the fact that dissociation of HIn occurs. Usually, though, the individual concentrations are unknown, and only the total concentration $c_{total} = [HIn] + [In^-]$ is known. Let us now calculate the absorbance for a solution with $c_{total} = 2.00 \times 10^{-5}$ M. The magnitude of the

(continued)

acid dissociation constant suggests that, for all practical purposes, the indicator is entirely in the undissociated form (HIn) in the HCl solution and completely dissociated as In^- in NaOH. The molar absorptivities at the two wavelengths were found to be

	ε_{430}	ε_{570}
HIn (HCl solution)	6.30×10^2	7.12×10^3
In^- (NaOH solution)	2.06×10^4	9.60×10^2

We would now like to find the absorbances (1.00-cm cell) of unbuffered solutions of the indicator ranging in concentration from 2.00×10^{-5} M to 16.00×10^{-5} M. Let us first find the concentration of HIn and In^- in the unbuffered 2×10^{-5} M solution. From the equation for the dissociation reaction, we know that $[H^+] = [In^-]$. Furthermore, the mass-balance expression for the indicator tells us that $[In^-] + [HIn] = 2.00 \times 10^{-5}$ M. By substituting these relationships into the K_a expression, we find that

$$\frac{[In^-]^2}{2.00 \times 10^{-5} - [In^-]} = 1.42 \times 10^{-5}$$

This equation can be solved to give $[In^-] = 1.12 \times 10^{-5}$ M and $[HIn] = 0.88 \times 10^{-5}$ M. The absorbances at the two wavelengths are found by substituting the values for ε, b, and c into Equation 24-13 (Beer's Law). The result is that $A_{430} = 0.236$ and $A_{570} = 0.073$. We could similarly calculate A for several other values of c_{total}. Additional data, obtained in the same way, are shown in **Table 24-4**. **Figure 24-15** shows plots at the two wavelengths that were constructed from data obtained in a similar manner.

CHALLENGE: Perform calculations to confirm that $A_{430} = 0.596$ and $A_{570} = 0.401$ for a solution in which the analytical concentration of HIn is 8.00×10^{-5} M.

The plots of Figure 24-15 illustrate the kinds of departures from Beer's law that occur when the absorbing system undergoes dissociation or association. Notice that the direction of curvature is opposite at the two wavelengths.

TABLE 24-4

Absorbance Data for Several Concentrations of the Indicator in Example 24-5

c_{HIn}, M	[HIn]	$[In^-]$	A_{430}	A_{570}
2.00×10^{-5}	0.88×10^{-5}	1.12×10^{-5}	0.236	0.073
4.00×10^{-5}	2.22×10^{-5}	1.78×10^{-5}	0.381	0.175
8.00×10^{-5}	5.27×10^{-5}	2.73×10^{-5}	0.596	0.401
12.0×10^{-5}	8.52×10^{-5}	3.48×10^{-5}	0.771	0.640
16.0×10^{-5}	11.9×10^{-5}	4.11×10^{-5}	0.922	0.887

❮ Deviations from Beer's law often occur when polychromatic radiation is used to measure absorbance.

Instrumental Deviations: Polychromatic Radiation

Beer's law strictly applies only when measurements are made with monochromatic source radiation. In practice, polychromatic sources that have a continuous distribution of wavelengths are used in conjunction with a grating or a filter to isolate a nearly symmetric band of wavelengths surrounding the wavelength to be employed (see Chapter 25, Section 25A-3).

The following derivation shows the effect of polychromatic radiation on Beer's law. Consider a beam of radiation consisting of just two wavelengths λ' and λ''. Assuming that Beer's law applies strictly for each wavelength, we may write for λ'

$$A' = \log \frac{P_0'}{P'} = \varepsilon'bc$$

or

$$\frac{P_0'}{P'} = 10^{\varepsilon'bc}$$

where P_0' is the incident power and P' is the resultant power at λ'. The symbols b and c are the path length and concentration of the absorber, and ε' is the molar absorptivity at λ'. Then,

$$P' = P_0'10^{-\varepsilon'bc}$$

Similarly, for λ'',

$$P'' = P_0''10^{-\varepsilon''bc}$$

When an absorbance measurement is made with radiation composed of both wavelengths, the power of the beam emerging from the solution is the sum of the powers emerging at the two wavelengths $P' + P''$. Likewise, the total incident power is the sum $P_0' + P_0''$. Therefore, the measured absorbance A_m is

$$A_m = \log\left(\frac{P_0' + P_0''}{P' + P''}\right)$$

We then substitute for P' and P'' and find that

$$A_m = \log\left(\frac{P_0' + P_0''}{P_0' \, 10^{-\varepsilon' bc} + P_0'' \, 10^{-\varepsilon'' bc}}\right)$$

or

$$A_m = \log(P_0' + P_0'') - \log(P_0' \, 10^{-\varepsilon' bc} + P_0'' \, 10^{-\varepsilon'' bc})$$

We see that, when $\varepsilon' = \varepsilon''$, this equation simplifies to

$$A_m = \log(P_0' + P_0'') - \log[(P_0' + P_0'')(10^{-\varepsilon' bc})]$$
$$= \log(P_0' + P_0'') - \log(P_0' + P_0'') - \log(10^{-\varepsilon' bc})$$
$$= \varepsilon' bc = \varepsilon'' bc$$

High-quality spectrophotometers produce narrow bands of radiation and are less likely to suffer deviations from Beer's law due to polychromatic radiation than low-quality instruments.

Polychromatic light, literally multicolored light, is light of many wavelengths, such as that from a tungsten light bulb. Light that is essentially monochromatic can be produced by filtering, diffracting, or refracting polychromatic light, as discussed in Chapter 25, Section 25A-3.

and Beer's law is followed. As shown in **Figure 24-16**, however, the relationship between A_m and concentration is no longer linear when the molar absorptivities differ. In addition, as the difference between ε' and ε'' increases, the deviation from linearity increases. When this derivation is expanded to include additional wavelengths, the effect remains the same.

If the band of wavelengths selected for spectrophotometric measurements corresponds to a region of the absorption spectrum in which the molar absorptivity of the analyte is essentially constant, departures from Beer's law will be minimal. Many molecular bands in the UV/visible region of the spectrum fit this description. For these bands, Beer's law is obeyed, as demonstrated by Band A in **Figure 24-17**. On the other hand, some absorption bands in the UV-visible region and many in the IR region are very narrow, and departures from Beer's law are common, as illustrated for Band B in Figure 24-17. To avoid such deviations, it is best to select a wavelength

Figure 24-16 Deviations from Beer's law with polychromatic radiation. The absorber has the indicated molar absorptivities at the two wavelengths λ' and λ''.

band near the wavelength of maximum absorption where the analyte absorptivity changes little with wavelength. Atomic absorption lines are so narrow that they require special sources to obtain adherence to Beer's law as discussed in Chapter 25, Section 25A-2.

Instrumental Deviations: Stray Light

Stray radiation, commonly called **stray light**, is defined as radiation from the instrument that is outside the nominal wavelength band chosen for the determination. This stray radiation often is the result of scattering and reflection off the surfaces of gratings, lenses or mirrors, filters, and windows. When measurements are made in the presence of stray light, the observed absorbance A' is given by

$$A' = \log\left(\frac{P_0 + P_s}{P + P_s}\right)$$

where P_s is the radiant power of the stray light. **Figure 24-18** shows a plot of the apparent absorbance A' versus concentration for various levels of P_s, relative to P_0. Stray light always causes the apparent absorbance to be lower than the true absorbance. The deviations due to stray light are most significant at high absorbance values. Because stray radiation levels can be as high as 0.5% in modern instruments, absorbance levels above 2.0 are rarely measured unless special precautions are taken or special instruments with extremely low stray light levels are used. Some inexpensive filter instruments show deviations from Beer's law at absorbances as low as 1.0 because of high stray light levels and/or the presence of polychromatic light.

Mismatched Cells

Another almost trivial, but important, deviation from adherence to Beer's law is caused by mismatched cells. If the cells holding the analyte and blank solutions are not of equal path length and equivalent in optical characteristics, an intercept will occur in the calibration curve, and $A = \varepsilon bc + k$ will be the equation for the

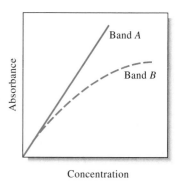

Figure 24-17   The effect of polychromatic radiation on Beer's law. In the absorption spectrum at the top, the absorptivity of the analyte is seen to be nearly constant over Band *A* from the source. Note in the Beer's law plot at the bottom that using Band *A* gives a linear relationship. In the spectrum, band *B* coincides with a region of the spectrum over which the absorptivity of the analyte changes. Note the dramatic deviation from Beer's law that results in the lower plot.

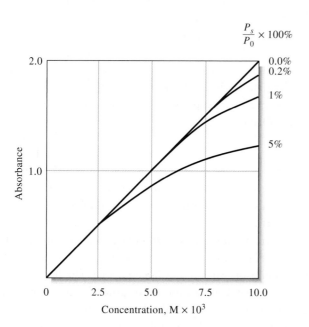

Figure 24-18   Deviation from Beer's law caused by various levels of stray light. Note that absorbance begins to level off with concentration at high stray light levels. Stray light always limits the maximum absorbance that can be obtained because, when the absorbance is high, the radiant power transmitted through the sample can become comparable to or lower than the stray light level.

curve instead of Equation 24-8. This error can be avoided either by using carefully matched cells or by using a linear regression procedure to calculate both the slope and intercept of the calibration curve. In most cases, linear regression is the best strategy because an intercept can also occur if the blank solution does not totally compensate for interferences. Another way to avoid the mismatched-cell problem with single beam instruments is to use only one cell and keep it in the same position for both blank and analyte measurements. After obtaining the blank reading, the cell is emptied by aspiration, washed, and filled with analyte solution.

> **Spreadsheet Summary**   In Chapter 12 of *Applications of Microsoft® Excel in Analytical Chemistry,* 2nd ed., spreadsheets are presented for modeling the effects of chemical equilibria and stray light on absorption measurements. Chemical and physical variables may be changed to observe their effects on instrument readouts.

24D EMISSION OF ELECTROMAGNETIC RADIATION

> Chemical species can be caused to emit light by (1) bombardment with electrons; (2) heating in a plasma, flame, or an electric arc; or (3) irradiation with a beam of light.

Atoms, ions, and molecules can be excited to one or more higher energy levels by any of several processes, including bombardment with electrons or other elementary particles; exposure to a high-temperature plasma, flame, or electric arc; or exposure to a source of electromagnetic radiation. The lifetime of an excited species is generally transitory (10^{-9} to 10^{-6} s), and relaxation to a lower energy level or the ground state takes place with a release of the excess energy in the form of electromagnetic radiation, heat, or perhaps both.

24D-1 Emission Spectra

Radiation from a source is conveniently characterized by means of an emission spectrum, which usually takes the form of a plot of the relative power of the emitted radiation as a function of wavelength or frequency. **Figure 24-19** illustrates a typical emission spectrum, which was obtained by aspirating a brine solution into an oxyhydrogen flame. Three types of spectra are superimposed in the figure: a **line spectrum**, a **band spectrum**, and a **continuum spectrum**. The line spectrum, marked lines in Figure 24-19, consists of a series of sharp, well-defined spectral lines caused by excitation of individual atoms. The band spectrum, marked bands, is comprised of several groups of lines so closely spaced that they are not completely resolved. The source of the bands is small molecules or radicals in the source flame. Finally, the continuum spectrum, shown as a green dashed line in the figure, is responsible for the increase in the background that appears above about 350 nm. The line and band spectra are superimposed on this continuum. The source of the continuum is described on page 677.

> The line widths of atoms in a medium such as a flame or plasma are about 0.1–0.01 Å. The wavelengths of atomic lines are unique for each element and are often used for qualitative analysis.

Line Spectra

Line spectra occur when the radiating species are individual atoms or ions that are well separated, as in a gas. The individual particles in a gaseous medium behave independently of one another, and the spectrum in most media consists of a series of sharp lines with widths of $10^{-1} - 10^{-2}$ Å ($10^{-2} - 10^{-3}$ nm). In Figure 24-19, lines for sodium, potassium, strontium, calcium, and magnesium are identified.

The energy level diagram in **Figure 24-20** shows the source of three of the lines that appear in the emission spectrum of Figure 24-19. The horizontal line labeled $3s$ in Figure 24-20 corresponds to the lowest, or ground state, energy of the atom E_0. The horizontal lines labeled $3p$, $4p$, and $4d$ are three higher-energy electronic levels

Figure 24-19 Emission spectrum of a brine sample obtained with an oxyhydrogen flame. The spectrum consists of the superimposed line, band, and continuum spectra of the constituents of the sample and flame. The characteristic wavelengths of the species contributing to the spectrum are listed beside each feature. (R. Hermann and C. T. J. Alkemade, *Chemical Analysis by Flame Photometry*, 2nd ed., New York: Interscience, 1979, p. 484.)

of sodium. Note that each of the *p* and *d* states are split into two closely spaced energy levels as a result of electron spin. The single outer shell electron in the ground state 3*s* orbital of a sodium atom can be excited into either of these levels by absorption of thermal, electrical, or radiant energy. Energy levels E_{3p} and E'_{3p} then represent the energies of the atom when this electron has been promoted to the two 3*p* states by absorption. The promotion to these states is depicted by the green line between the 3*s* and the two 3*p* levels in Figure 24-20. A few nanoseconds after excitation, the electron returns from the 3*p* state to the ground state, emitting a photon whose wavelength is given by Equation 24-3.

$$\lambda_1 = \frac{hc}{(E_{3p} - E_0)} = 589.6 \text{ nm}$$

In a similar way, relaxation from the 3*p*′ state to the ground state yields a photon with $\lambda_2 = 589.0$ nm. This emission process is once again shown by the green line between the 3*s* and 3*p* levels in Figure 24-20. The result is that the emission process from the two closely spaced 3*p* levels produces two corresponding closely spaced lines

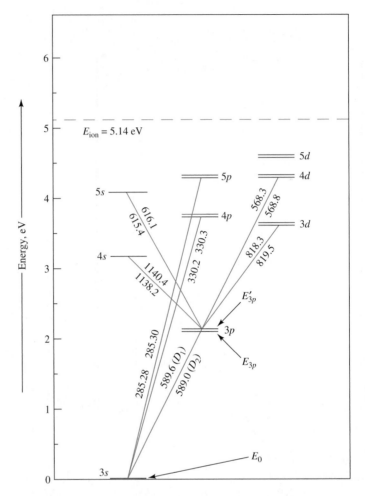

Figure 24-20 Energy level diagram for sodium in which the horizontal lines represent the atomic orbitals, which are identified with their respective labels. The vertical scale is orbital energy in electron volts (eV), and the energies of excited states relative to the ground state $3s$ orbital can be read from the vertical axis. The lines in color show the allowed transitions resulting in emission of various wavelengths (in nm), indicated adjacent to the lines. The horizontal dashed line represents the ionization energy of sodium. (INGLE, JAMES D., CROUCH, STANLEY R., *SPECTROCHEMICAL ANALYSIS*, 1st Edition, © 1988, p.206. Reprinted by permission of Pearson Education, Inc., Upper Saddle River, NJ.)

in the emission spectrum called a **doublet**. These lines, indicated by the transitions labeled D_1 and D_2 in Figure 24-20, are the famous Fraunhofer "D" lines discussed in Feature 24-1. They are so intense that they are completely off scale in the upper right corner of the emission spectrum of Figure 24-19.

The transition from the more energetic $4p$ state to the ground state (see Figure 24-20) produces a second doublet at a shorter wavelength. The line appearing at about 330 nm in Figure 24-19 results from these transitions. The $4d$-to-$3p$ transition provides a third doublet at about 568 nm. Notice that all three of these doublets appear in the emission spectrum of Figure 24-19 as just single lines. This is a result of the limited resolution of the spectrometer used to produce the spectrum, as discussed in Sections 25A-3 and 28A-4. It is important to note that the emitted wavelengths are identical to the wavelengths of the absorption peaks for sodium (see Figure 24-11) because the transitions are between the same pairs of states.

At first glance, it may appear that radiation could be absorbed and emitted by atoms between any pair of the states shown in Figure 24-20, but in fact, only certain transitions are allowed, while others are forbidden. The transitions that are allowed and forbidden to produce lines in the atomic spectra of the elements are determined by the laws of quantum mechanics in what are called **selection rules**. These rules are beyond the scope of our discussion.[4]

---

[4]See J. D. Ingle, Jr., and S. R. Crouch, *Spectrochemical Analysis,* Upper Saddle River, NJ: Prentice-Hall, 1988, p. 205.

Band Spectra

Band spectra are often produced in spectral sources because of the presence of gaseous radicals or small molecules. For example, in Figure 24-19, bands for OH, MgOH, and MgO are labeled and consist of a series of closely spaced lines that are not fully resolved by the instrument used to obtain the spectrum. Bands arise from the numerous quantized vibrational levels that are superimposed on the ground state electronic energy level of a molecule. For further discussion of band spectra, see Section 28B-3.

> ❮ An emission band spectrum is made up of many closely spaced lines that are difficult to resolve.

Continuum Spectrum

As shown in **Figure 24-21**, a spectral continuum of radiation is produced when solids such as carbon and tungsten are heated to incandescence. Thermal radiation of this kind, which is called **blackbody radiation**, is more characteristic of the temperature of the emitting surface than of its surface material. Blackbody radiation is produced by the innumerable atomic and molecular oscillations excited in the condensed solid by the thermal energy. Note that the energy peaks in Figure 24-21 shift to shorter wavelengths with increasing temperature. As the figure shows, very high temperatures are required to cause a thermally excited source to emit a substantial fraction of its energy as ultraviolet radiation.

> ❮ A spectral continuum has no line character and is generally produced by heating solids to a high temperature.

Part of the continuum background radiation in the flame spectrum shown in Figure 24-19 is probably thermal emission from incandescent particles in the flame. Note that this background decreases rapidly as the wavelength approaches the ultraviolet region of the spectrum.

Heated solids are important sources of infrared, visible, and longer-wavelength ultraviolet radiation for analytical instruments, as we will see in Chapter 25.

Effect of Concentration on Line and Band Spectra

The radiant power P of a line or a band depends directly on the number of excited atoms or molecules, which in turn is proportional to the total concentration c of the species present in the source. Thus, we can write

$$P = kc \qquad (24\text{-}16)$$

where k is a proportionality constant. This relationship is the basis of quantitative emission spectroscopy, which is described in some detail in Section 28C.

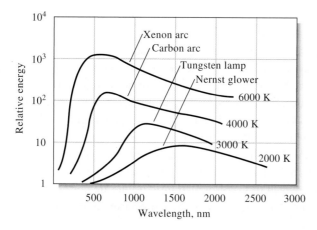

Figure 24-21 Blackbody radiation curves for various light sources. Note the shift in the wavelengths of maximum emission as the temperature of the sources changes.

© Bettmann/CORBIS

In 1900, Max Planck (1858–1947) discovered a formula (now often called the Planck radiation law) that modeled curves like those shown in Figure 24-21 nearly perfectly. He followed this discovery by developing a theory that made two bold assumptions regarding the oscillating atoms or molecules in blackbody radiators. He assumed (1) that these species could have only discrete energies and (2) that they could absorb or emit energy in discrete units, or quanta. These assumptions, which are implicit in Equation 24-3, laid the foundation for the development of quantum theory.

Resonance fluorescence is radiation that is identical in wavelength to the radiation that excited the fluorescence.

24D-2 Emission by Fluorescence and Phosphorescence

Fluorescence and phosphorescence are analytically important emission processes in which atoms or molecules are excited by the absorption of a beam of electromagnetic radiation. The excited species then relax to the ground state, giving up their excess energy as photons. Fluorescence takes place much more rapidly than phosphorescence and is generally complete in 10^{-5} s or less from the time of excitation. Phosphorescence emission may extend for minutes or even hours after irradiation has ceased. Fluorescence is considerably more important than phosphorescence in analytical chemistry, so our discussions focus primarily on fluorescence.

Atomic Fluorescence

Gaseous atoms fluoresce when they are exposed to radiation that has a wavelength that exactly matches that of one of the absorption (or emission) lines of the element in question. For example, gaseous sodium atoms are promoted to the excited energy state, E_{3p}, shown in Figure 24-20 through absorption of 589-nm radiation. Relaxation may then take place by reemission of radiation of the identical wavelength. When excitation and emission wavelengths are the same, the resulting emission is called **resonance fluorescence**. Sodium atoms could also exhibit resonance fluorescence when exposed to 330-nm or 285-nm radiation. In addition, however, the element could also produce nonresonance fluorescence by first relaxing from E_{5p} or E_{4p} to energy level E_{3p} through a series of nonradiative collisions with other species in the medium. Further relaxation to the ground state can then take place either by the emission of a 589-nm photon or by further collisional deactivation.

Molecular Fluorescence

Fluorescence is a photoluminescence process in which atoms or molecules are excited by absorption of electromagnetic radiation, as shown in **Figure 24-22a**. The excited species then relax back to the ground state, giving up their excess energy as photons. As we have noted, the lifetime of an excited species is brief because there are several mechanisms for an excited atom or molecule to give up its excess energy and relax to its ground state. Two of the most important of these mechanisms, **nonradiative relaxation** and **fluorescence emission**, are illustrated in **Figures 24-22b** and **c**.

Nonradiative Relaxation.   Two types of nonradiative relaxation are shown in Figure 24-22b. **Vibrational deactivation**, or **relaxation**, depicted by the short wavy arrows between vibrational energy levels, takes place during collisions between excited molecules and molecules of the solvent. During the collisions, the excess vibrational energy is transferred to solvent molecules in a series of steps as indicated in the figure. The gain in vibrational energy of the solvent is reflected in a tiny increase in the temperature of the medium. Vibrational relaxation is such an efficient process that the average lifetime of an excited vibrational state is only about 10^{-15} s. Nonradiative relaxation between the lowest vibrational level of an excited electronic state and the upper vibrational level of another electronic state can also occur. This type of relaxation, which is called **internal conversion**, depicted by the two longer wavy arrows in Figure 24-22b, is much less efficient than vibrational relaxation so that the average lifetime of an electronic excited state is between 10^{-9} and 10^{-6} s. The mechanisms by which this type of relaxation occurs are not fully understood, but the net effect is again a very small rise in the temperature of the medium.

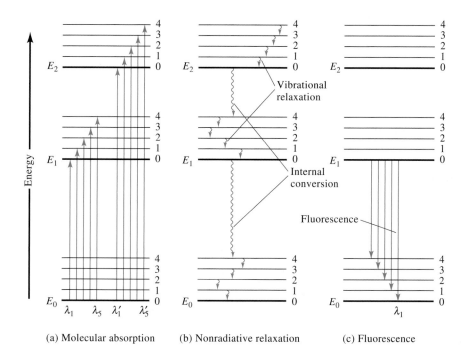

(a) Molecular absorption (b) Nonradiative relaxation (c) Fluorescence

Figure 24-22 Energy level diagram showing some of the energy changes that occur during absorption, nonradiative relaxation, and fluorescence by a molecular species.

Fluorescence. The relative number of molecules that fluoresce is small because fluorescence requires structural features that slow the rate of the nonradiative relaxation processes illustrated in Figure 24-22b and enhance the rate of fluorescence emission shown in Figure 24-22c. Most molecules lack these features and undergo nonradiative relaxation at a rate that is significantly greater than the radiative relaxation rate, and so fluorescence does not occur. As shown in Figure 24-22c, bands of radiation are produced when molecules relax from the lowest-lying vibrational state of an excited state, E_1, to the many vibrational levels of the ground state, E_0. Like molecular absorption bands, molecular fluorescence bands are made up of a large number of closely spaced lines that are usually difficult to resolve. Notice that the transition from E_1 to the lowest-lying vibrational state of the ground state (λ_1) has the highest energy of all of the transitions in the band. As a result, all of the other lines that terminate in higher vibrational levels of the ground state are lower in energy and produce fluorescence emission at longer wavelengths than λ_1. In other words, molecular fluorescence bands consist largely of lines that are longer in wavelength than the band of absorbed radiation responsible for their excitation. This shift in wavelength is called the **Stokes shift**. A more detailed discussion of molecular fluorescence is given in Chapter 27.

The **Stokes shift** refers to fluorescence radiation that occurs at wavelengths that are longer than the wavelength of radiation used to excite the fluorescence.

WEB WORKS

To learn more about Beer's law, use a search engine to find the IUPAC "Glossary of Terms Used in Photochemistry." Find how the molar absorptivity (the IUPAC "Glossary" uses **molar absorption coefficient**) of a compound (ε) relates to the absorption cross section (σ). Multiply the absorption cross section by Avogadro's number and note the result. How would the result change if absorbance were expressed as $A = -\ln(P/P_0)$ rather than the usual definition in terms of base 10 logarithms? What are the units of σ? Which of the quantities ε or σ is a macroscopic quantity? Which of the terms, molar absorptivity or molar absorption coefficient, is most descriptive? Explain and justify your answer.

QUESTIONS AND PROBLEMS

*24-1. In a solution of pH 5.3, the indicator bromocresol purple exhibits a yellow color, but when the pH is 6.0, the indicator solution changes to purple. Discuss why these colors are observed in terms of the wavelength regions and colors absorbed and transmitted.

24-2. What is the relationship between
*(a) absorbance and transmittance?
(b) absorptivity a and molar absorptivity ε?

*24-3. Identify factors that cause the Beer's law relationship to be nonlinear.

24-4. Describe the differences between "real" deviations from Beer's law and those due to instrumental or chemical factors.

24-5. How does an electronic transition resemble a vibrational transition? How do they differ?

24-6. Calculate the frequency in hertz of
*(a) an X-ray beam with a wavelength of 2.65 Å.
(b) an emission line for copper at 211.0 nm.
*(c) the line at 694.3 nm produced by a ruby laser.
(d) the output of a CO_2 laser at 10.6 μm.
*(e) an infrared absorption peak at 19.6 μm.
(f) a microwave beam at 1.86 cm.

24-7. Calculate the wavelength in centimeters of
*(a) an airport tower transmitting at 118.6 MHz.
(b) a VOR (radio navigation aid) transmitting at 114.10 kHz.
*(c) an NMR signal at 105 MHz.
(d) an infrared absorption peak with a wavenumber of 1210 cm^{-1}.

24-8. A sophisticated ultraviolet/visible/near-IR instrument has a wavelength range of 185 to 3000 nm. What are its wavenumber and frequency ranges?

*24-9. A typical simple infrared spectrophotometer covers a wavelength range from 3 to 15 μm. Express its range (a) in wavenumbers and (b) in hertz.

24-10. Calculate the frequency in hertz and the energy in joules of an X-ray photon with a wavelength of 2.70 Å.

*24-11. Calculate the wavelength and the energy in joules associated with a signal at 220 MHz.

24-12. Calculate the wavelength of
*(a) the sodium line at 589 nm in an aqueous solution with a refractive index of 1.35.
(b) the output of a ruby laser at 694.3 nm when it is passing through a piece of quartz that has a refractive index of 1.55.

24-13. What are the units for absorptivity when the path length is given in centimeters and the concentration is expressed in
*(a) parts per million?
(b) micrograms per liter?
*(c) mass-volume percent?
(d) grams per liter?

24-14. Express the following absorbances in terms of percent transmittance
*(a) 0.0356
(b) 0.895
*(c) 0.379
(d) 0.167
*(e) 0.485
(f) 0.753

24-15. Convert the accompanying transmittance data to absorbances.
*(a) 27.2%
(b) 0.579
*(c) 30.6%
(d) 3.98%
*(e) 0.093
(f) 63.7%

24-16. Calculate the percent transmittance of solutions that have twice the absorbance of the solutions in Problem 24-14.

24-17. Calculate the absorbances of solutions with half the transmittance of those in Problem 24-15.

24-18. Evaluate the missing quantities in the accompanying table. Where needed, use 200 for the molar mass of the analyte.

	A	$\%T$	ε L mol^{-1} cm^{-1}	a cm^{-1} ppm^{-1}	b cm	c M	c ppm
*(a)	0.172		4.23×10^3		1.00		
(b)		44.9		0.0258		1.35×10^{-4}	
*(c)	0.520		7.95×10^3		1.00		
(d)		39.6		0.0912			1.76
*(e)			3.73×10^3		0.100	1.71×10^{-3}	
(f)		83.6			1.00	8.07×10^{-6}	
*(g)	0.798				1.50		33.6
(h)		11.1	1.35×10^4			7.07×10^{-5}	
*(i)		5.23	9.78×10^3				5.24
(j)	0.179				1.00	7.19×10^{-5}	

24-19. A solution containing 4.48 ppm KMnO$_4$ exhibits 85.9 %T in a 1.00-cm cell at 520 nm. Calculate the molar absorptivity of KMnO$_4$ at this wavelength.

24-20. Beryllium(II) forms a complex with acetylacetone (166.2 g/mol). Calculate the molar absorptivity of the complex, given that a 2.25 ppm solution has a transmittance of 37.5% when measured in a 1.00-cm cell at 295 nm, the wavelength of maximum absorption.

*24-21. At 580 nm, the wavelength of its maximum absorption, the complex Fe(SCN)$^{2+}$ has a molar absorptivity of 7.00×10^3 L cm^{-1} mol^{-1}. Calculate
(a) the absorbance of a 3.40×10^{-5} M solution of the complex at 580 nm in a 1.00-cm cell.
(b) the absorbance of a solution in which the concentration of the complex is twice that in (a).
(c) the transmittance of the solutions described in (a) and (b).
(d) the absorbance of a solution that has half the transmittance of that described in (a).

24-22. A 2.50-mL aliquot of a solution that contains 4.33 ppm iron(III) is treated with an appropriate excess of KSCN and diluted to 50.0 mL. What is the absorbance of the resulting solution at 580 nm in a 2.50-cm cell? See Problem 24-21 for absorptivity data.

*24-23. A solution containing the complex formed between Bi(III) and thiourea has a molar absorptivity of 9.32×10^3 L cm^{-1} mol^{-1} at 470 nm.
(a) What is the absorbance of a 5.67×10^{-5} M solution of the complex at 470 nm in a 1.00-cm cell?
(b) What is the percent transmittance of the solution described in (a)?
(c) What is the molar concentration of the complex in a solution that has the absorbance described in (a) when measured at 470 nm in a 2.50-cm cell?

24-24. The complex formed between Cu(I) and 1,10-phenanthroline has a molar absorptivity of 7000 L cm^{-1} mol^{-1} at 435 nm, the wavelength of maximum absorption. Calculate
(a) the absorbance of a 6.17×10^{-5} M solution of the complex when measured in a 1.00-cm cell at 435 nm.
(b) the percent transmittance of the solution in (a).
(c) the concentration of a solution that in a 5.00-cm cell has the same absorbance as the solution in (a).
(d) the path length through a 3.13×10^{-5} M solution of the complex that is needed for an absorbance that is the same as the solution in (a).

*24-25. A solution with a "true" absorbance [$A = -\log(P_0/P)$] of 2.10 was placed in a spectrophotometer with a stray light percentage (P_s/P_0) of 0.75. What absorbance A' would be measured? What percentage error would result?

24-26. A compound X is to be determined by UV/visible spectrophotometry. A calibration curve is constructed from standard solutions of X with the following results: 0.50 ppm, $A = 0.24$; 1.5 ppm, $A = 0.36$; 2.5 ppm, $A = 0.44$; 3.5 ppm, $A = 0.59$; and 4.5 ppm, $A = 0.70$. Find the slope and intercept of the calibration curve, the standard error in Y, the concentration of the solution of unknown X concentration, and the standard deviation in the concentration of X. Construct a plot of the calibration curve and determine the unknown concentration by hand from the plot.

24-27. One common way to determine phosphorus in urine is to treat the sample after removing the protein with molybdenum (VI) and then reducing the resulting 12-molybdophosphate complex with ascorbic acid to give an intense blue-colored species called molybdenum blue. The absorbance of molybdenum blue can be measured at 650 nm. A 24-hour urine sample was collected, and the patient produced 1122 mL in 24 hours. A 1.00 mL aliquot of the sample was treated with Mo(VI) and ascorbic acid and diluted to a volume of 50.00 mL. A calibration curve was prepared by treating 1.00 mL aliquots of phosphate standard solutions in the same manner as the urine sample. The absorbances of the standards and the urine sample were obtained at 650 nm and the following results obtained:

Solution	Absorbance at 650 nm
1.00 ppm P	0.230
2.00 ppm P	0.436
3.00 ppm P	0.638
4.00 ppm P	0.848
Urine sample	0.518

(a) Find the slope, intercept, and standard error in y of the calibration curve. Construct a calibration curve. Determine the concentration number of phosphorus in ppm in the urine sample and its standard deviation from the least-squares equation of the line. Compare the unknown concentration to that obtained manually from a calibration curve.
(b) What mass in grams of phosphorus was eliminated per day by the patient?
(c) What is the phosphate concentration in urine in mM?

24-28. Nitrite is commonly determined by a colorimetric procedure using a reaction called the Griess reaction. In this reaction, the sample containing nitrite is reacted with sulfanilimide and N-(1-Napthyl) ethylenediamine to form a colored species that absorbs at 550 nm. Using an automated flow analysis instrument, the following results were obtained for standard solutions of nitrite and for a sample containing an unknown amount:

Solution	Absorbance at 550 nm
2.00 μM	0.065
6.00 μM	0.205
10.00 μM	0.338
14.00 μM	0.474
18.00 μM	0.598
Unknown	0.402

(a) Find the slope, intercept, and standard deviation of the calibration curve.

(b) Construct the calibration curve.

(c) Determine the concentration of nitrite in the sample and its standard deviation.

24-29. The equilibrium constant for the reaction

$$2CrO_4^{2-} + 2H^+ \rightleftharpoons Cr_2O_7^{2-} + H_2O$$

is 4.2×10^{14}. The molar absorptivities for the two principal species in a solution of K_2CrO_7 are

λ, nm	ε_1 (CrO_4^{2-})	ε_2 ($Cr_2O_7^{2-}$)
345	1.84×10^3	10.7×10^2
370	4.81×10^3	7.28×10^2
400	1.88×10^3	1.89×10^2

Four solutions were prepared by dissolving 4.00×10^{-4}, 3.00×10^{-4}, 2.00×10^{-4}, and 1.00×10^{-4} moles of $K_2Cr_2O_7$ in water and diluting to 1.00 L with a pH 5.60 buffer. Calculate theoretical absorbance values (1.00-cm cells) for each solution and plot the data for (a) 345 nm; (b) 370 nm; and (c) 400 nm.

24-30. **Challenge Problem**: NIST maintains a database of the spectra of the elements at *http://www.nist.gov/pml/data/asd_contents.cfm*. The following energy levels for neutral lithium were obtained from this database:

Electronic Configuration	Level, eV
$1s^2 2s^1$	0.00000
$1s^2 2p^1$	1.847818
	1.847860
$1s^2 3s^1$	3.373129
$1s^2 3p^1$	3.834258
	3.834258
$1s^2 3d^1$	3.878607
	3.878612
$1s^2 4s^1$	4.340942
$1s^2 4p^1$	4.521648
	4.521648
$1s^2 4d^1$	4.540720
	4.540723

(a) Construct a partial energy level diagram similar to the one in Figure 24-20. Label each energy level with its corresponding orbital.

(b) Browse to the NIST website and click on the Physical Reference Data link. Locate and click on the link for the Atomic Spectral Database and click on the Lines icon. Use the form to retrieve the spectral lines for Li I between 300 nm and 700 nm, including energy level information. Note that the retrieved table contains wavelength, relative intensity, and changes in electron configuration for the transitions that give rise to each line. Add connecting lines to the partial energy level diagram from (a) to illustrate the transitions and label each line with the wavelength of the emission. Which of the transitions in your diagram are doublets?

(c) Use the intensity versus wavelength data that you retrieved in (b) to sketch an emission spectrum for lithium. If you placed a sample of $LiCO_3$ in a flame, what color would the flame be?

(d) Describe how the flame spectrum of an ionic lithium compound, such as $LiCO_3$, displays the spectrum of neutral lithium atoms.

(e) There appear to be no emission lines for lithium between 544 nm and 610 nm. Why is this?

(f) Describe how the information obtained in this problem could be used to detect the presence of lithium in urine. How would you determine the amount of lithium quantitatively?

Instruments for Optical Spectrometry

CHAPTER 25

The bright star in the middle of the photograph is Supernova 1987a, which was the first supernova visible to the naked eye to appear in over 400 years. The black dots over the star's image were produced by superimposing a negative of a photo taken two years before the supernova appeared. Nearly coincident with the supernova was an unusual burst of neutrinos, which was detected by a facility beneath Lake Erie and by a similar facility in Japan. The newly refurbished Irvine-Michigan-Brookhaven underground detector in Ohio consists of a 6800-cubic meter volume of water surrounded by 2048 high-sensitivity, large-area photomultiplier tubes and housed in a salt mine under Lake Erie. When at least 20 of the photomultipliers detect a pulse of blue Cherenkov radiation from the impact of neutrinos with water molecules in the detector within a time window of 55 ns, a neutrino event is judged to have occurred. The Lake Erie detector and others like it were built in an effort to detect the spontaneous decay of protons in the water molecules. These experiments are very long term, and data from the Lake Erie detector are recorded continuously. As a result, the detector was poised to monitor the neutrino burst from Supernova 1987a. The photomultiplier is one of the radiation detectors described in this chapter.

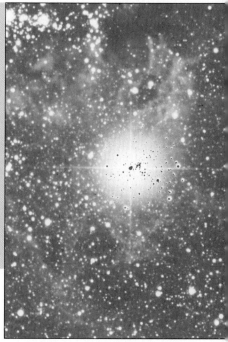

© Australian Astronomical Observatory/Photograph by David Malin from AAT plates

The basic components of analytical instruments for absorption, as well as for emission and fluorescence spectroscopy, are remarkably similar in function and in general performance requirements regardless of whether the instruments are designed for UV, visible, or IR radiation. Because of these similarities, such instruments are frequently referred to as **optical instruments** even though the human eye is only sensitive to the visible region. In this chapter, we first examine the characteristics of the components common to optical instruments. We then consider the characteristics of typical instruments designed for UV, visible, and IR absorption spectroscopy.

25A INSTRUMENT COMPONENTS

Most spectroscopic instruments in the UV/visible and IR regions are made up of five components: (1) a stable source of radiant energy; (2) a wavelength selector to isolate a limited region of the spectrum for measurement; (3) one or more sample containers; (4) a radiation detector, to convert radiant energy to a measurable electrical signal; and (5) a signal-processing and readout unit consisting of electronic hardware and in modern instruments a computer. **Figure 25-1** illustrates the three ways these components are configured for making optical spectroscopic measurements. The figure shows that components (3), (4), and (5) have similar configurations for each type of measurement.

We often call the UV/visible and IR regions of the spectrum the optical region. Even though the human eye is responsive only to visible radiation, the other regions are included because the lenses, mirrors, prisms, and gratings used are similar and function in a comparable manner. Spectroscopy in the UV/visible and IR regions is, therefore, often called **optical spectroscopy**.

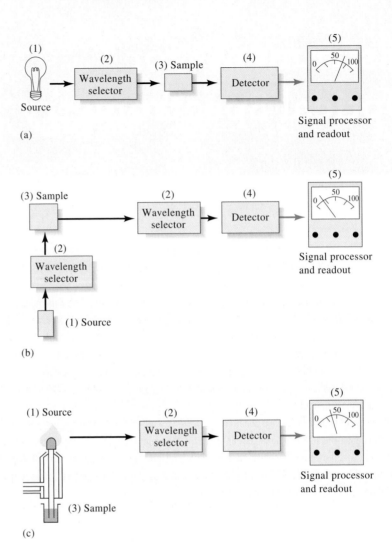

Figure 25-1 Components of various instruments used in optical spectroscopy. In (a), the arrangement for absorption measurements is shown. Note that source radiation of the selected wavelength is sent through the sample, and the transmitted radiation is measured by the detector/signal-processing/readout unit. With some instruments, the position of the sample and wavelength selector is reversed. In (b), the configuration for fluorescence measurements is shown. For this measurement, two wavelength selectors are needed to select the excitation and the emission wavelengths. The selected source radiation is incident on the sample and the radiation emitted is measured, usually at right angles to avoid detecting the source radiation and to minimize scattering. In (c), the configuration for emission spectroscopy is shown. In this instrument, a source of thermal energy, such as a flame, produces an analyte vapor that emits radiation isolated by the wavelength selector and converted to an electrical signal by the detector.

The first two designs, for absorption and fluorescence, require an external source of radiation. In absorption measurements (see Figure 25-1a), the attenuation of the source radiation at the selected wavelength is measured. In fluorescence measurements (see Figure 25-1b), the source excites the analyte and causes the emission of characteristic radiation, which is usually measured perpendicular to the incident source beam. In emission spectroscopy (see Figure 25-1c), the sample itself is the emitter and no external radiation source is needed. In emission methods, the sample is usually introduced into a plasma or a flame that provides enough thermal energy to cause the analyte to emit characteristic radiation. Fluorescence and emission methods are described in more detail in Chapters 27 and 28, respectively.

25A-1 Optical Materials

The cells, windows, lenses, mirrors, and wavelength-selecting elements in an optical spectroscopic instrument must transmit radiation in the wavelength region being investigated. **Figure 25-2** shows the functional wavelength ranges for several optical materials that are used in the UV, visible, and IR regions of the spectrum. Ordinary

Figure 25-2 Transmittance ranges for various optical materials. Simple glasses are fine in the visible region, while fused silica or quartz is necessary in the UV region (<380 nm). Halide salts (KBr, NaCl, and AgCl) are often used in the IR, although they have the disadvantages of being expensive and somewhat water soluble.

silicate glass is satisfactory for the visible region and has the considerable advantage of low cost. In the UV region, at wavelengths shorter than about 380 nm, glass begins to absorb and fused silica or quartz must be substituted. Also, in the IR region, glass, quartz, and fused silica all absorb at wavelengths longer than about 2.5 μm. Hence, optical elements for IR spectrometry are typically made from halide salts or in some cases polymeric materials.

25A-2 Spectroscopic Sources

To be suitable for spectroscopic studies, a source must generate a beam of radiation that is sufficiently powerful for easy detection and measurement. In addition, its output power should be stable for reasonable periods of time. Typically, for good stability, the power supply for the source must be well regulated. Spectroscopic sources are of two types: **continuum sources**, which emit radiation that changes in intensity only slowly as a function of wavelength, and **line sources**, which emit a limited number of spectral lines, each of which spans a very narrow wavelength range. The distinction between these sources is illustrated in **Figure 25-3**. Sources can also be classified as **continuous sources**, which refer to the fact that they emit radiation continuously with time, or **pulsed sources**, which emit radiation in bursts.

A continuum source provides a broad distribution of wavelengths within a particular spectral range. This distribution is known as a **spectral continuum**. A line source emits a limited number of narrow spectral lines.

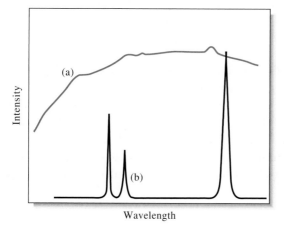

Figure 25-3 Spectra of two different spectral sources. The spectrum of a continuum source (a) is much broader than that of a line source (b).

(a)

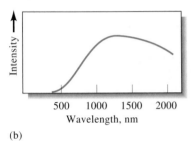

(b)

Figure 25-4 (a) A tungsten lamp of the type used in spectroscopy and its spectrum (b). Intensity of the tungsten source is usually quite low at wavelengths shorter than about 350 nm. Note that the intensity reaches a maximum in the near-IR region of the spectrum ($\approx$1200 nm in this case).

TABLE 25-1

Continuum Sources for Optical Spectroscopy

Source	Wavelength Region, nm	Type of Spectroscopy
Xenon arc lamp	250–600	Molecular fluorescence
H_2 and D_2 lamps	160–380	UV molecular absorption
Tungsten/halogen lamp	240–2500	UV/visible/near-IR molecular absorption
Tungsten lamp	350–2200	Visible/near-IR molecular absorption
Nernst glower	400–20,000	IR molecular absorption
Nichrome wire	750–20,000	IR molecular absorption
Globar	1200–40,000	IR molecular absorption

Continuum Sources in the Ultraviolet/Visible Region

The most widely used continuum sources are listed in **Table 25-1**. An ordinary tungsten filament lamp provides a broad distribution of wavelengths from 320 to 2500 nm (see **Figure 25-4**). Generally these lamps are operated at a temperature of around 2900 K, thereby producing useful radiation from about 350 to 2500 nm.

Tungsten/halogen lamps, also called quartz/halogen lamps, contain a small amount of iodine within the quartz envelope that houses the filament. Quartz allows the filament to be operated at a temperature of about 3500 K, leading to higher intensities and extending the range of the lamp well into the UV. The lifetime of a tungsten/halogen lamp is more than double that of an ordinary tungsten lamp, which is limited by sublimation of tungsten from the filament. In the presence of iodine, the sublimed tungsten reacts to give gaseous WI_2 molecules. These molecules then diffuse back to the hot filament where they decompose, redeposit W atoms on the filament, and release iodine. Tungsten/halogen lamps are finding ever-increasing use in spectroscopic instruments because of their extended wavelength range, greater intensity, and longer life.

Deuterium (and also hydrogen) lamps are most often used to provide continuum radiation in the UV region. A deuterium lamp consists of a cylindrical tube containing deuterium at low pressure with a quartz window from which the radiation exits, as shown in **Figure 25-5**. The lamp emits continuum radiation when deuterium (or hydrogen) is stimulated by electrical energy to produce excited molecule of D_2^* (or H_2^*). The excited-state species then dissociates to give two hydrogen or deuterium atoms plus an ultraviolet photon. The reactions for hydrogen are

$$H_2 + E_e \rightarrow H_2^* \rightarrow H' + H'' + h\nu$$

where E_e is the electrical energy absorbed by the molecule. The energy for the overall process is

$$E_e = E_{H_2^*} = E_{H'} + E_{H''} + h\nu$$

where $E_{H_2^*}$ is the fixed quantized energy of H_2^* and $E_{H'}$ and $E_{H''}$ are the kinetic energies of the two hydrogen atoms. The sum of the latter two energies can vary from zero to $E_{H_2^*}$. Thus, the energy and the frequency of the photon can also vary within this range of energies. That is, when the two kinetic energies are by chance small, $h\nu$ is large, and when the two energies are large, $h\nu$ is small. As a result, hydrogen lamps produce a true spectral continuum from about 160 nm to the beginning of the visible region. Today, most lamps for generating ultraviolet radiation contain deuterium and are of a low voltage type in which an arc is formed between a heated, oxide-coated filament and a metal electrode (see Figure 25-5a). The heated filament provides electrons to maintain a direct current at a potential of about 40 V; a regulated power supply is required for constant intensities. Both deuterium and hydrogen lamps provide a useful spectral continuum in the region from 160 to 375 nm, as shown in Figure 25-5b. The deuterium lamp is more widely used than the hydrogen lamp, however, because the deuterium lamp is more intense. At longer wavelengths (>360 nm), the lamps generate emission lines that are superimposed on the continuum. For many applications, these lines are a nuisance, but they are useful for wavelength calibration of absorption instruments.

Other Ultraviolet/Visible Sources

In addition to the continuum sources just discussed, line sources are also important for use in the UV/visible region. Low-pressure mercury arc lamps are common sources for use in liquid chromatography detectors. The dominant line emitted by these sources is the 253.7-nm Hg line. Hollow cathode lamps are also common line sources that are specifically used for atomic absorption spectroscopy, as discussed in Chapter 28. Lasers (see Feature 25-1) are also used in many spectroscopic applications, both for single wavelength and for scanning purposes.

(a)

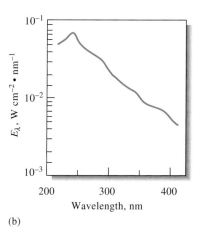

(b)

Figure 25-5 (a) A deuterium lamp of the type used in spectrophotometers and (b) its spectrum. Note the maximum intensity, proportional to the irradiance E_λ, occurs at ≈ 225 nm. Typically, instruments switch from deuterium to tungsten at ≈ 350 nm.

FEATURE 25-1

Laser Sources: The Light Fantastic

Lasers have become widely used as sources in certain types of analytical spectroscopy. To help us understand how a laser works, consider an assembly of atoms or molecules interacting with an electromagnetic wave. For simplicity, we will consider the atoms or molecules to have two energy levels: an upper level 2 with energy E_2 and a lower level 1 with energy E_1. If the electromagnetic wave is of a frequency corresponding to the energy difference between the two levels, excited species in level 2 can be stimulated to emit radiation of the same frequency and phase as the original electromagnetic wave. Each **stimulated emission** generates a photon while each absorption removes a photon. The number of photons per second, called the **radiant flux Φ**, changes with distance as the radiation interacts with the assembly of atoms or molecules. The change in flux, $d\Phi$, is proportional to the flux itself, to the difference in the populations of the levels, $n_2 - n_1$, and to the path length of the interaction, dz, according to

$$d\Phi = k\Phi(n_2 - n_1)dz$$

where k is a proportionality constant related to the absorptivity of the absorbing species. If the upper-level population can be made to exceed that of the lower level, there will be a net gain in flux, and the system will behave as an amplifier. If $n_2 > n_1$, the atomic or molecular system is said to be an **active medium** and to have undergone **population inversion**. The resulting amplifier is called a **laser**, which stands for light amplification by stimulated emission of radiation.

The optical amplifier can be converted into an oscillator by placing the active medium inside a resonant cavity made from two mirrors as shown in **Figure 25F-1**.

(continued)

Figure 25F-1   Laser cavity. The electromagnetic wave travels back and forth between the mirrors, and the wave is amplified with each pass. The output mirror is partially transparent to allow only a fraction of the beam to pass out of the cavity.

When the gain of the active medium equals the losses in the system, laser oscillation begins.

Population inversion is often achieved by a multilevel atomic or molecular system in which the excitation process, called **pumping**, is accomplished by electrical means, by optical methods, or by chemical reactions. In some cases, the population inversion can be sustained to produce a **continuous wave (CW)** output beam that is continuous with respect to time. In other cases, the lasing action is **self-terminating** so that the laser is operated in a pulsed mode to produce a repetitive pulse train or a single shot action.[1]

There are many types of lasers available. The first operating lasers were **solid-state lasers** in which the active medium was a ruby crystal. In addition to the ruby laser, there are many other solid-state lasers. A widely used material contains a small concentration of Nd^{3+} embedded in a yttrium-aluminum-garnet (YAG) host. The active material is shaped into a rod and pumped optically by a flashlamp, as illustrated in **Figure 25F-2a**. The pump and laser transitions are shown in **Figure 25F-2b**. The Nd:YAG laser generates nanosecond pulses with a very high output power at a wavelength of 1.06 μm. The Nd:YAG laser is popular as a pumping source for tunable dye lasers.

Several other rare earth elements, such as ytterbium, holmium, and erbium, are also used as dopants in solid-state lasers. Titanium-doped sapphire (Ti:sapphire) is used to produce a tunable infrared laser. Some versions generate ultrashort pulses of very high output power.

The very common helium-neon (He-Ne) laser is a **gas laser** that operates in a CW mode. The He-Ne laser is widely used as an optical alignment aid and as a source for some types of spectroscopy. The nitrogen laser lases on a transition of the nitrogen molecule at 337.1 nm. It is a self-terminating pulsed laser that requires a very short electrical pulse for pumping the appropriate transitions. The N_2 laser is also used for pumping tunable dye lasers, as discussed later. **Excimer** (excited dimer or trimer) **lasers** are among the newest gas lasers. Rare-gas halide excimer lasers were first demonstrated in 1975. In one popular type, a gas mixture of Ar, F_2, and He produces ArF excimers when subjected to an electrical discharge. The excimer laser is an important UV source for photochemical studies, for fluorescence applications, and for pumping tunable dye lasers.

Dye lasers are liquid lasers containing a fluorescent dye such as one of the rhodamines, a coumarin, or a fluorescein. These have been made to lase at wavelengths from the IR to the UV. Lasing typically occurs between the first excited **singlet state** and the ground state. The lasers can be pumped by flashlamps or by another laser such as those discussed previously. Lasing can be sustained over

A **singlet state** is an electronic state of a molecule in which all electron spins are paired.

---

[1]For additional information, see J. D. Ingle and S. R. Crouch, *Spectrochemical Analysis*, Upper Saddle River, NJ: Prentice-Hall, 1988.

(a)

(b)

Figure 25F-2 Schematic of a Nd:YAG laser (a) and energy levels (b). The pump transitions are in the red region of the spectrum, and the laser output is in the near infrared. The laser is flashlamp pumped. The region between the two mirrors is the laser cavity.

a continuous range of wavelengths on the order of 40 to 50 nm. The broad band over which lasing occurs makes the dye laser suitable for tuning by inserting a grating, a filter, a prism, or an interferometric element into the laser cavity. Dye lasers are very useful for molecular fluorescence spectroscopy and many other applications.

Semiconductor lasers, also known as **diode lasers**, obtain population inversion between the conduction band and the **valence** band of a *pn*-junction diode. Various compositions of the semiconductor material can be used to give different output wavelengths. Diode lasers can be tuned over small wavelength intervals and can produce outputs in the IR region of the spectrum. They have become extremely useful in CD and DVD players, in CD-ROM drives, in laser printers, and in spectroscopic applications, such as Raman spectroscopy.

Laser radiation is highly directional, spectrally pure, coherent,[2] and highly intense. These properties have made possible many unique research applications that cannot easily be achieved with conventional sources. Despite the many advances in laser science and technology, only recently have lasers become routinely useful in analytical instruments. Even today, many high-powered or ultrafast lasers can be somewhat difficult to align, maintain, and use.

---

[2]Coherent radiation is radiation in which the waves are in phase with one another.

Continuum Sources in the Infrared Region

The continuum sources for IR radiation are normally heated inert solids. A **Globar** source consists of a silicon carbide rod. Infrared radiation is emitted when the Globar is heated to about 1500°C by passing electricity through it. Table 25-1 gives the wavelength range of these sources.

A **Nernst glower** is a cylinder of zirconium and yttrium oxides that emits IR radiation when heated to a high temperature by an electric current. Electrically heated spirals of nichrome wire also serve as inexpensive IR sources.

25A-3 Wavelength Selectors

Spectroscopic instruments in the UV and visible regions are usually equipped with one or more devices to restrict the radiation being measured to a narrow band that is absorbed or emitted by the analyte. Such devices greatly enhance both the selectivity and the sensitivity of an instrument. In addition, for absorption measurements, as we saw in Section 24C-3, narrow bands of radiation greatly diminish the chance for Beer's law deviations due to polychromatic radiation. Many instruments use a **monochromator** or a **filter** to isolate the desired wavelength band so that only the band of interest is detected and measured. Others use a **spectrograph** to spread out, or disperse, the wavelengths so that they can be detected with a multichannel detector.

Monochromators and Polychromators

Monochromators generally have a diffraction grating (see Feature 25-3) to disperse the radiation into its component wavelengths as shown in **Figure 25-6a**. Older

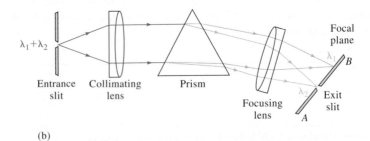

Figure 25-6 Types of monochromators: (a) grating monochromator; (b) prism monochromator. The monochromator design in (a) is a Czerny-Turner design, while the prism monochromator in (b) is a Bunsen design. In both cases, $\lambda_1 > \lambda_2$.

instruments used prisms for this purpose as seen in **Figure 25-6b**. By rotating the grating, different wavelengths can be made to pass through an exit slit. The output wavelength of a monochromator is thus continuously variable over a considerable spectral range. The wavelength range passed by a monochromator, called the **spectral bandpass** or **effective bandwidth**, can be less than 1 nm for moderately expensive instruments to greater than 20 nm for inexpensive systems. Because of the ease with which the wavelength can be changed with a monochromator-based instrument, these systems are widely used for spectral scanning applications as well as applications requiring a fixed wavelength. With an instrument containing a **spectrograph**, the sample and wavelength selector are reversed from the configuration shown in Figure 25-1a. Like the monochromator, the spectrograph contains a diffraction grating to disperse the spectrum. However, the spectrograph has no exit slit, so the dispersed spectrum impinges on a multiwavelength detector. Still other instruments used for emission spectroscopy contain a device called a **polychromator**, which contains multiple exit slits and multiple detectors. This arrangement allows many discrete wavelengths to be measured simultaneously.

Figure 25-6a shows the design of a typical grating monochromator. Radiation from a source enters the monochromator via a narrow rectangular opening or slit. The radiation is then collimated by a concave mirror, which produces a parallel beam that strikes the surface of a reflection grating. Angular dispersion results from diffraction, which occurs at the reflective surface. To illustrate, the radiation entering the monochromator is shown as consisting of just two wavelengths, λ_1 and λ_2, where λ_1 is longer than λ_2. The pathway of the longer wavelength radiation after it is reflected from the grating is shown by the dashed lines; the solid lines show the path of the shorter wavelength. Note that the shorter wavelength radiation λ_2 is reflected off the grating at a sharper angle than is λ_1. That is, **angular dispersion** of the radiation takes place at the grating surface. The two wavelengths are focused by another concave mirror onto the **focal plane** of the monochromator, where they appear as two images of the entrance slit, one for λ_1 and the other as λ_2. By rotating the grating either one of these images can be focused on the exit slit. If a detector is located at the exit slit of the monochromator in Figure 25-6a and the grating is rotated so that one of the lines shown (say, λ_1) is scanned across the slit from $\lambda_1 - \delta\lambda$ to $\lambda_1 + \delta\lambda$ (where $\delta\lambda$ is a small wavelength difference), the output of the detector takes the shape shown in **Figure 25-7**.[3] The effective bandwidth of the monochromator, which is defined in the figure, depends on the size and quality of the dispersing element, the slit widths, and the focal length of the monochromator. A high-quality monochromator will exhibit an effective bandwidth of a few tenths of a nanometer or less in the ultraviolet/visible region. The effective bandwidth of a monochromator that is satisfactory for most quantitative applications is from about 1 to 20 nm.

Many monochromators are equipped with adjustable slits to permit some control over the bandwidth. A narrow slit decreases the **effective bandwidth** but also diminishes the power of the emergent beam. Therefore, the minimum practical bandwidth may be limited by the sensitivity of the detector. For qualitative analysis, narrow slits and minimum effective bandwidths are required if a spectrum is made up of narrow peaks. For quantitative work, on the other hand, wider slits permit operation of the detector system at lower amplification, which in turn provides greater reproducibility of response.

A **spectrograph** is a device that uses a grating to disperse a spectrum. It contains an entrance slit to define the area of the source to be viewed. A large opening at its exit allows a range of wavelengths to strike a multiwavelength detector. A **monochromator** is a device that contains an entrance slit and an exit slit. The exit slit is used to isolate a small band of wavelengths. One band at a time is isolated and different bands can be transmitted sequentially by rotating the grating. A **polychromator** contains multiple exit slits so that several wavelength bands can be isolated simultaneously.

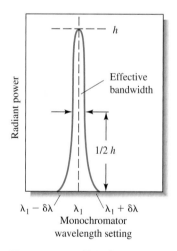

Figure 25-7 Exit slit output as monochromator is scanned from $\lambda_1 - \Delta\lambda$ to $\lambda_1 + \Delta\lambda$.

The **effective bandwidth** of a wavelength selector is the width of the band of radiation in wavelength units at half-peak height.

---

[3]The slit function itself is approximately triangular. Various instrumental factors combine to produce the shape shown in Figure 25-7.

Figure 25-8 Mechanism of diffraction from an echellette-type grating. Angle *i* from the grating normal is the angle of the incident beam; angle *r* is the angle of the reflected beam. The distance between successive rulings is *d*.

Gratings

Most gratings in modern monochromators are replica gratings, which are obtained by making castings of a master grating. The latter consists of a hard, optically flat, polished surface on which have been ruled with a suitably shaped diamond tool a large number of parallel and closely spaced grooves. A magnified cross-sectional view of a few typical grooves is shown in **Figure 25-8**. A grating for the ultraviolet and visible region typically has from 50 to 6000 grooves/mm, with 1200 to 2400 being most common. The construction of good master grating is tedious, time consuming, and expensive because the grooves must be identical in size, exactly parallel, and equally spaced over the length of the grating (3 to 10 cm). Replica gratings are formed from a master grating by a liquid resin casting process that closely preserves the optical accuracy of the original master grating on a clear resin surface. This surface is usually coated with aluminum or sometimes gold or platinum so that it reflects electromagnetic radiation.

The Echellette Grating.   One of the most common types of reflection gratings is the echellette grating. Figure 25-8 shows a schematic representation of this type of grating, which is grooved or **blazed** such that it has relatively broad faces where reflection occurs and narrow unused faces.[4] This geometry provides highly efficient diffraction of radiation. In Figure 25-8, a parallel beam of monochromatic radiation approaches the grating surface at an angle *i* relative to the grating normal. The incident beam is depicted as consisting of three parallel beams that make up a wave front labeled 1, 2, 3. The diffracted beam is reflected at the angle *r*, which depends on the wavelength of the radiation. In Feature 25-2, we show that the angle of reflection *r* is related to the wavelength of the incoming radiation by the equation

$$\mathbf{n}\lambda = d(\sin i + \sin r) \tag{25-1}$$

---

[4]The echellette grating is blazed for use in relatively low orders, but an **echelle grating** is used in high orders (>10). The echelle grating is often used with a second dispersive element, such as a prism, to sort out overlapping orders and to provide cross dispersion. For more on echelle gratings and how they are used, see D. A. Skoog, F. J. Holler, and S. R. Crouch, *Principles of Instrumental Analysis,* 6th ed., Section 10A-3, Belmont, CA: Brooks/Cole, 2007; J. D. Ingle, Jr., and S. R. Crouch, *Spectrochemical Analysis,* Section 3-5, Englewood Cliffs, NJ: Prentice-Hall, 1988.

Equation 25-1 suggests that there are several values of λ for a given diffraction angle r. Thus, if a first-order line ($\mathbf{n} = 1$) of 900 nm is found at r, second-order (450 nm) and third-order (300 nm) lines also appear at this angle. Usually, the first-order line is the most intense, and it is possible to design gratings that concentrate as much as 90% of the incident intensity in this order. The higher-order lines can generally be removed by filters or a prism. For example, glass, which absorbs radiation below 350 nm, eliminates the high-order spectra associated with first-order radiation in most of the visible region.

FEATURE 25-2

Origin of Equation 25-1

In Figure 25-8, parallel beams of monochromatic radiation labeled 1 and 2 are shown striking two of the broad faces at an incident angle i relative to the grating normal. Maximum constructive interference occurs at the reflected angle r. Beam 2 travels a greater distance than beam 1, and this difference is equal to $\overline{CB} + \overline{BD}$. For constructive interference to occur, this difference must equal $\mathbf{n}\lambda$:

$$\mathbf{n}\lambda = \overline{CB} + \overline{BD}$$

where the small integer $\mathbf{n}$ is called the **diffraction order**. Note, however, that angle CAB is equal to angle i and that angle DAB is identical to angle r. Therefore, from trigonometry,

$$\overline{CB} = d \sin i$$

where d is the spacing between the reflecting surfaces. We also see that

$$\overline{BD} = d \sin r$$

Substituting the last two expressions into the first gives Equation 25-1. That is,

$$\mathbf{n}\lambda = d(\sin i + \sin r)$$

Note that, when diffraction occurs to the left of the grating normal, values of $\mathbf{n}$ are positive, and when diffraction occurs to the right of the grating normal, $\mathbf{n}$ is negative. Thus, $\mathbf{n} = \pm 1, \pm 2, \pm 3$, etc.

A major advantage of a grating monochromator is that, in contrast to a prism monochromator, the dispersion along the focal plane is for all practical purposes linear. Figure 25-9 demonstrates this property which greatly simplifies the design of monochromators.

Concave Gratings. Gratings can be formed on a concave surface in much the same way as on a plane surface. A concave grating permits the design of a monochromator without auxiliary collimating and focusing mirrors or lenses because the concave surface both disperses the radiation and focuses it on the exit slit. Monochromators containing a concave grating are cost efficient, and the reduction in the number of optical surfaces increases their energy throughput.

EXAMPLE 25-1

An echellette grating containing 1450 blazes per millimeter was irradiated with a polychromatic beam at an incident angle 48 degrees to the grating normal. Calculate the wavelengths of radiation that would appear at angles of reflection of +20, +10, and 0 deg (angle r, Figure 25-8).

Solution

To obtain d in Equation 25-1, we write

$$d = \frac{1 \text{ mm}}{1450 \text{ blazes}} \times 10^6 \frac{\text{nm}}{\text{mm}} = 689.7 \frac{\text{nm}}{\text{blaze}}$$

When r in Figure 25-8 equals +20 deg, λ can be obtained by substituting into Equation 25-1. Therefore,

$$\lambda = \frac{689.7 \text{ nm}}{n}(\sin 48 + \sin 20) = \frac{748.4}{n} \text{ nm}$$

and the wavelengths for the first-, second-, and third-order reflections are 748, 374, and 249 nm, respectively. Similar calculations, shown in the table that follows, reveal that the wavelength in the second order is one half that in the first order, the wavelength in the third order is one-third that in the first order, and so forth.

	Wavelength (nm) for		
r, deg	$n = 1$	$n = 2$	$n = 3$
20	748	374	249
10	632	316	211
0	513	256	171

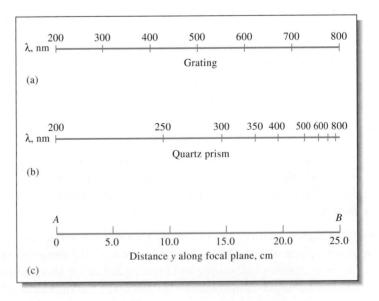

Figure 25-9 Dispersion of radiation along the focal plane *AB* of a typical (a) grating and (b) quartz prism. The positions of *A* and *B* in the scale in (c) are shown in Figure 25-6.

Holographic Gratings.[5]   One of the products that emerged from laser technology is an optical (rather than mechanical) technique for forming gratings on plane or concave glass surfaces. Holographic gratings produced in this way appear in ever-increasing numbers in modern optical instruments, even some of the less expensive ones. Because holographic gratings are not subject to the mechanical errors of the ruling engine, they exhibit superior groove shape and flatness and thus produce spectra that are freer from stray radiation and ghosts (double images). Replica holographic gratings are essentially indistinguishable from the master grating.[6] Feature 25-3 describes the ruling process for both mechanically ruled and holographically ruled gratings.

FEATURE 25-3

Producing Ruled and Holographic Gratings

Dispersion of UV/visible radiation can be brought about by directing a polychromatic beam through a **transmission grating** or onto the surface of a **reflection grating**. The reflection grating is by far the more common. **Replica gratings**, which are used in many monochromators, are manufactured from a **master grating**. The master grating consists of a large number of parallel and closely spaced grooves ruled on a hard, polished surface with a suitably shaped diamond tool. For the UV/visible region, a grating will contain from 50 to 6000 grooves mm^{-1}, with 1200 to 2400 being most common. Master gratings are ruled by a diamond ruling tool that is operated by a ruling engine. The construction of a good master grating is

tedious, time consuming, and expensive, because the grooves must be identical in size, exactly parallel, and equally spaced over the typical 3 to 10 cm length of the grating. Because of the difficulty in construction, few master gratings are produced.

The modern era of gratings dates back to the 1880s when Henry Rowland constructed an engine capable of ruling gratings of up to 6 inches in width with over 100,000 grooves. A simplified drawing of the Rowland engine is shown in Figure 25F-3. With this machine, a high-precision screw moves the grating carriage, while a diamond stylus cuts the tiny parallel grooves. Imagine manually ruling a grating with 100,000 grooves in a 6-inch width! The engine required around

NYPL/Science Source/Getty Images

Henry A. Rowland (1848–1901) was an American physicist and the first president of the American Physical Society. He was also the first chairman of the Physics Department at Johns Hopkins University. Although he did outstanding work in several areas of electricity and magnetism, he is best known for developing methods to produce high-quality diffraction gratings.

(continued)

---

[5]See J. Flamand, A. Grillo, and G. Hayat, *Amer. Lab.,* **1975**, *7*(5), 47; J. M. Lerner et al., *Proc. Photo-Opt. Instrum. Eng.,* **1980**, *240*, 72, 82.
[6]I. R. Altelmose, *J. Chem. Educ.,* **1986**, *63*, A216, **DOI**: 10.1021/ed063pA216.

Figure 25F-3 Simplified diagram of the Rowland ruling engine. A single high-precision screw moves the grating carriage. A diamond point then travels over the grating, which is ruled on a concave mirror surface. Machines of this type were the models for many of the ruling engines constructed after Rowland's time. Ruling engines are among the most sensitive and precise macroscopic mechanical devices ever made. The resulting gratings have played an integral role in many of the most important advances in science over the past century.

5 hours just to warm up to a nearly uniform temperature. After this warm-up period, nearly 15 hours more were needed to obtain a uniform layer of lubricant on the surface. Only after this time was the diamond lowered to begin the ruling process. Large gratings required almost a week to produce.

Two important improvements were made by Strong in the 1930s. The most significant was the vacuum deposition of aluminum onto glass blanks as a medium. The thin layer of aluminum gave a much smoother surface and reduced wear on the diamond tool. Strong's second improvement was to move the grating blank instead of the diamond tool.

Today, ruling engines use interferometric (see Feature 25-7) control over the ruling process. Fewer than fifty ruling engines are in use around the world. Even if all these engines were operated 24 hours a day, they could not begin to meet the demand for gratings. Fortunately, modern coating and resin technology has made it possible to produce replica gratings of very high quality. Replica gratings are formed from the master grating by vacuum deposition of aluminum onto a ruled master grating. The aluminum layer is subsequently coated by an epoxy-type material. The material is then polymerized, and the replica is separated from the master. The replica gratings of today are superior to the master gratings produced in the past.

Another way that gratings are made is a result of laser technology. These **holographic gratings** are made by coating a flat glass plate with a material that is photosensitive (photoresist). Beams from a pair of identical lasers then strike the coated glass surface. The resulting interference fringes (see Feature 25-7) from the two beams sensitize the photoresist, producing areas that can be dissolved away, leaving a grooved structure. Aluminum is then vacuum deposited on the surface to produce a reflection grating. The spacing of the grooves can be changed by changing the angle of the two laser beams with respect to one another. Nearly perfect gratings with as many as 6000 lines per mm can be manufactured in this way at a relatively low cost. Holographic gratings are not quite as efficient in terms of their light output as ruled gratings; however, they can eliminate false lines, called **grating ghosts**, and reduce scattered light that results from ruling errors.

Radiation Filters

Filters operate by blocking or absorbing all but a restricted band of radiation. As shown in **Figure 25-10**, two types of filters are used in spectroscopy: **interference filters** and **absorption filters**. Interference filters are typically used for absorption measurements. These filters generally transmit a much greater fraction of radiation at their nominal wavelengths than do absorption filters.

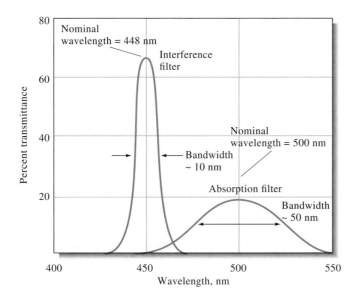

Figure 25-10 Bandwidths for two types of filters.

Interference Filters. Interference filters are used with ultraviolet and visible radiation, as well as with wavelengths as long as about 14 μm in the infrared region. As the name implies, an interference filter relies on optical interference to provide a relatively narrow band of radiation, typically 5 to 20 nm in width. As shown in **Figure 25-11a**, an interference filter consists of a very thin layer of a transparent **dielectric** material (frequently calcium fluoride or magnesium fluoride) coated on both sides with a film of metal that is thin enough to transmit approximately half of the radiation striking it and to reflect the other half. This array is sandwiched

A **dielectric** is a nonconducting substance or insulator. Such materials are usually optically transparent.

(a)

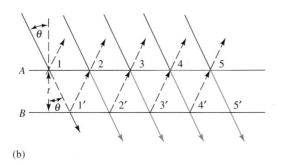

(b)

Figure 25-11 (a) Schematic cross section of an interference filter. Note that the drawing is not to scale, and the three central bands are much narrower than shown. (b) Schematic to show the conditions for constructive interference.

between two glass plates that protect it from the atmosphere. When radiation strikes the central array at a 90-degree angle, approximately half is transmitted by the first metallic layer and the other half reflected. The transmitted radiation undergoes a similar partition when it reaches the second layer of metal. If the reflected portion from the second layer is of the proper wavelength, it is partially reflected from the inner portion of the first layer in phase with the incoming light of the same wavelength. The result is constructive interference of the radiation of this wavelength and destructive removal of most other wavelengths. As shown in Feature 25-4, the nominal wavelength for an interference filter λ_{max} is given by the equation

$$\lambda_{max} = \frac{2t\eta}{\mathbf{n}}$$

(25-2)

where t is the thickness of the central fluoride layer, η is its refractive index, and $\mathbf{n}$ is an integer called the interference order. The glass layers of the filter are often selected to absorb all but one of the wavelengths transmitted by the central layer, thus restricting the transmission of the filter to a single order.

FEATURE 25-4

Basis of Equation 25-2

The relationship between the thickness of the dielectric layer t and the transmitted wavelength λ can be found with the aid of **Figure 25-11b**. For purposes of clarity, the incident beam is shown arriving at an angle θ from the perpendicular. At point 1, the radiation is partially reflected and partially transmitted to point 1′ where partial reflection and transmission again takes place. The same process occurs at 2, 2′, and so forth. For reinforcement to occur at point 2, the distance traveled by the beam reflected at 1′ must be some multiple of its wavelength in the medium λ'. Since the path length between surfaces can be expressed as $t/\cos\theta$, the condition for reinforcement is that $\mathbf{n}\lambda' = 2t/\cos\theta$, where $\mathbf{n}$ is a small whole number.

In ordinary use, θ approaches zero, and $\cos\theta$ approaches unity so that the equation derived from Figure 25-11 simplifies to

$$\mathbf{n}\lambda' = 2t$$

where λ' is the wavelength of radiation *in the dielectric* and t is the thickness of the dielectric. The corresponding wavelength in air is given by

$$\lambda = \lambda'\eta$$

where η is the refractive index of the dielectric medium. Thus, the wavelengths of radiation transmitted by the filter are

$$\lambda = \frac{2t\eta}{\mathbf{n}}$$

Figure 25-10 illustrates the performance characteristics of a typical interference filter. Most filters of this type have bandwidths of less than 1.5% of the nominal wavelength, although this figure is lowered to 0.15% in some narrow-band filters. Narrow-band filters have a maximum transmittance of about 10%.

Absorption Filters.   Absorption filters, which are generally less expensive and more rugged than interference filters, are limited in use to the visible region. This type of filter usually consists of a colored glass plate that absorbs part of the incident radiation and transmits the desired band of wavelengths. Absorption filters have effective bandwidths that range from perhaps 30 to 250 nm. Filters that provide the narrowest bandwidths also absorb a significant fraction of the desired radiation and may have a transmittance of 1% or less at their band peaks. Figure 25-10 contrasts the performance characteristics of a typical absorption filter with its interference counterpart. Glass filters with transmittance maxima throughout the entire visible region are available from commercial sources. While their performance characteristics are distinctly inferior to those of interference filters, their cost is appreciably less, and they can be adequate for many routine applications.

Filters have the advantages of simplicity, ruggedness, and low cost. However, since one filter can only isolate a single band of wavelengths, a new filter must be used for a different wavelength band. Therefore, filter instruments are used only when measurements are made at a fixed wavelength or when the wavelength is changed infrequently.

In the IR region of the spectrum, most modern instruments do not disperse the spectrum at all, although this practice was common with older instruments. Instead, an **interferometer** is used, and the constructive and destructive interference of electromagnetic waves are used to obtain spectral information through a technique called Fourier transformation. These IR instruments are further discussed in Feature 25-7 and in Section 26C-2.

25A-4 Detecting and Measuring Radiant Energy

To obtain spectroscopic information, the radiant power transmitted, fluoresced or emitted, must be detected in some manner and converted into a measurable quantity. A **detector** is a device that identifies, records, or indicates a change in one of the variables in its environment such as pressure, temperature, or electromagnetic radiation. Familiar examples of detectors include photographic film for indicating the presence of electromagnetic or radioactive radiation, the pointer of a balance for indicating mass differences, and the mercury level in a thermometer for indicating temperature. The human eye is also a detector; it converts visible radiation into an electrical signal that is passed to the brain via a chain of neurons in the optic nerve and produces vision.

Invariably in modern instruments, the information of interest is encoded and processed as an electrical signal. A **transducer** converts nonelectrical quantities, such as light intensity, pH, mass, and temperature, into **electrical signals** that can be subsequently amplified, manipulated, and finally converted into numbers proportional to the magnitude of the original quantity. We discuss only radiation transducers in this section.

> A **transducer** converts various types of chemical and physical quantities into electrical signals, such as electrical charge, current, or voltage.

Properties of Radiation Transducers

The ideal transducer for electromagnetic radiation responds rapidly to low levels of radiant energy over a broad wavelength range. In addition, it produces an electrical signal that is easily amplified and has a low electrical noise level (see Feature 25-5).

> Common noise sources include vibration, pickup from 60-Hz lines, temperature variations, and frequency or voltage fluctuations in the power.

TABLE 25-2

Common Detectors for Absorption Spectroscopy

Type	Wavelength Range, nm
Photon Detectors	
Phototubes	150–1000
Photomultiplier tubes	150–1000
Silicon photodiodes	350–1100
Photoconductive cells	1000–50,000
Thermal Detectors	
Thermocouples	600–20,000
Bolometers	600–20,000
Pneumatic cells	600–40,000
Pyroelectric devices	1000–20,000

Finally, the electrical signal produced by the transducer should be linearly related to the radiant power P of the beam, as shown in Equation 25-3:

$$G = KP + K'$$ (25-3)

where G is the electrical response of the detector in units of current, voltage, or charge. The proportionality constant K measures the sensitivity of the detector in terms of electrical response per unit of radiant power input.

Many transducers exhibit a small constant response K', known as a **dark current**, even when no radiation strikes their surfaces. Instruments with transducers that have a significant dark-current response are usually equipped with an electronic circuit or computer program to automatically subtract the dark current. Thus, under ordinary circumstances, we can simplify Equation 25-3 to

$$G = KP$$ (25-4)

Types of Transducers

As shown in **Table 25-2**, there are two general types of transducers: one type responds to photons, the other to heat. All photon detectors are based on the interaction of radiation with a reactive surface either to produce electrons (**photoemission**) or to promote electrons to energy states in which they can conduct electricity (**photoconduction**). Only UV, visible, and near-IR radiation possess enough energy to cause photoemission to occur; therefore, photoemissive detectors are limited to wavelengths shorter than about 2 μm (2000 nm). Photoconductors can be used in the near-, mid-, and far-IR regions of the spectrum.

> **Dark current** is a current produced by a radiation transducer when no light strikes the device.

> Generally, the signals produced by analytical instruments fluctuate in a random way because a large number of variables are not controlled. These fluctuations, which limit the sensitivity of an instrument, are called **noise**. The terminology originated in radio engineering where the presence of unwanted signal fluctuations is audible as static, or noise.

FEATURE 25-5

Signals, Noise, and the Signal-to-Noise Ratio

The output of an analytical instrument fluctuates in a random way. These fluctuations limit the precision of the instrument and are the net result of a large number of uncontrolled random variables in the instrument and in the chemical system under study. An example of such a variable is the random arrival of photons at the photocathode of a photomultiplier tube. The term *noise* is used to describe these fluctuations, and each uncontrolled variable is a noise source. The term comes from audio

Figure 25F-4 Absorption spectra of hemoglobin with identical signal levels but different amounts of noise. Note that the curves have been offset on the absorbance axis for clarity.

and electronic engineering where undesirable signal fluctuations appear to the ear as static, or noise. The average value of the output of an electronic device is called the *signal*, and the standard deviation of the signal is a measure of the noise.

An important figure of merit for analytical instruments, stereos, compact-disk players, and many other types of electronic devices is the signal-to-noise ratio (*S/N*). The **signal-to-noise ratio** is usually defined as the ratio of the average value of the output signal to its standard deviation. The signal-to-noise behavior of an absorption spectrophotometer is illustrated in the spectra of hemoglobin shown in **Figure 25F-4**. The spectrum at the bottom of the figure has *S/N* = 100, and you can easily pick out the absorption maxima at 540 nm and 580 nm. As the *S/N* degrades to about two in the second spectrum from the top of the figure, the peaks are barely visible. Somewhere between *S/N* = 2 and *S/N* = 1, the peaks disappear altogether into the noise and are impossible to identify. As modern instruments have become computerized and controlled by sophisticated electronic circuits, various methods have been developed to increase the signal-to-noise ratio of instrument outputs. These methods include analog filtering, lock-in amplification, boxcar averaging, smoothing, and Fourier transformation.[7]

---

[7]D. A. Skoog, F. J. Holler, and S. R. Crouch, *Principles of Instrumental Analysis,* 6th ed., Ch. 5, Belmont, CA: Brooks/Cole, 2007.

Generally, we detect IR radiation by measuring the temperature rise of a thermally sensitive material located in the path of the beam or by measuring the increase in electrical conductivity of a photoconducting material when it absorbs IR radiation. Because the temperature changes resulting from the absorption of the IR energy are tiny, ambient temperature must be controlled carefully to avoid large errors. The detector system often limits the sensitivity and precision of an IR instrument.

Photon Detectors

Widely used types of photon detectors include phototubes, photomultiplier tubes, silicon photodiodes, photodiode arrays, and charge-transfer devices such as charge-coupled and charge-injection devices.

Phototubes and Photomultiplier Tubes.   The response of a phototube or a photomultiplier tube results from the photoelectric effect. As shown in **Figure 25-12**, a phototube consists of a semicylindrical photocathode and a wire anode sealed inside an evacuated transparent glass or quartz envelope. The concave surface of the cathode supports a layer of photoemissive material, such as an alkali metal or metal oxide that emits electrons when irradiated with light of the appropriate energy. When a voltage is applied across the electrodes, the emitted **photoelectrons** are attracted to the positively charged wire anode. These electrons produce a **photocurrent** in the circuit shown in Figure 25-12. This current can then be amplified and measured. The number of photoelectrons ejected from the photocathode per unit time is directly proportional to the radiant power of the beam striking the surface. With an applied voltage of about 90 V or more, all these photoelectrons are collected at the anode to produce a photocurrent that is also proportional to the radiant power of the beam.

The **photomultiplier tube** (PMT) is similar in construction to the phototube but is significantly more sensitive. The photocathode is similar to that of the phototube with electrons being emitted on exposure to radiation. However, in place of a single wire anode, the PMT has a series of electrodes called **dynodes**, as shown in **Figure 25-13**. The electrons emitted from the cathode are accelerated toward the first dynode that is maintained 90–100 V positive with respect to the cathode. Each accelerated photoelectron that strikes the dynode surface produces several electrons, called secondary electrons, that are then accelerated to dynode 2, which is

Photoelectrons are electrons that are ejected from a photosensitive surface by electromagnetic radiation. A photocurrent is the current in an external circuit that is limited by the rate of ejection of photoelectrons.

Photomultiplier tubes are among the most widely used types of transducers for detecting ultraviolet/visible radiation.

A major advantage of photomultipliers is internal amplification. About 10^6 to 10^7 electrons are produced at the anode for each photon that strikes the photocathode of a photomultiplier tube (PMT).

Figure 25-12   A phototube and accompanying circuit. The photocurrent induced by the radiation causes a voltage ($V = IR$) across the measuring resistor; this voltage is then amplified and measured.

(a)

(b)

(c)

Figure 25-13 Diagram of a photomultiplier tube. (a) photograph; (b) cross-sectional view; and (c) electrical diagram illustrating dynode polarization and photocurrent measurement. Radiation striking the photosensitive cathode (b) gives rise to photoelectrons by the photoelectric effect. Dynode D1 is held at a positive voltage with respect to the photocathode. Electrons emitted by the cathode are attracted to the first dynode and accelerated in the field. Each electron striking dynode D1 thus gives rise to two to four secondary electrons. These are attracted to dynode D2, which is again positive with respect to dynode D1. The resulting amplification at the anode can be 10^6 or greater. The exact amplification factor depends on the number of dynodes and the voltage difference between each dynode. This automatic internal amplification is one of the major advantages of photomultiplier tubes. With modern instrumentation, the arrival of individual photocurrent pulses can be detected and counted instead of being measured as an average current. This technique, called *photon counting*, is advantageous at very low light levels.

> With modern electronic instrumentation, it is possible to detect the electron pulses resulting from the arrival of individual photons at the photocathode of a PMT. The pulses are counted, and the accumulated count is a measure of the intensity of the electromagnetic radiation impinging on the PMT. **Photon counting** is advantageous when the light intensity, or the frequency of arrival of photons at the photocathode, is low.

held 90–100 V more positive than dynode 1. Again, electron amplification (gain) occurs. By the time this process has been repeated at each of the dynodes, 10^5 to 10^7 electrons have been produced for each incident photon. This cascade of electrons is finally collected at the anode to provide an average current that is further amplified electronically and measured.

Photoconductive Cells. Photoconductive transducers consist of a thin film of a semiconductor material, such as lead sulfide, mercury cadmium telluride (MCT), or indium antimonide, deposited often on a nonconducting glass surface and sealed in an evacuated envelope. Absorption of radiation by these materials promotes non-conducting valence electrons to a higher energy state, which decreases the electrical resistance of the semiconductor. Typically, a photoconductor is placed in series with a voltage source and load resistor, and the voltage drop across the load resistor serves as a measure of the radiant power of the beam of radiation. The PbS and InSb detectors are quite popular in the near-IR region of the spectrum. The MCT detector is useful in the mid- and far-IR regions when cooled with liquid N_2 to minimize thermal noise. This application is important in infrared Fourier transform spectrometers.

Silicon Photodiodes and Photodiode Arrays. Crystalline silicon is a semiconductor, a material whose electrical conductivity is less than that of a metal but greater than that of an electrical insulator. Silicon is a Group IV element and thus has four valence electrons. In a silicon crystal, each of these electrons is combined with electrons from four other silicon atoms to form four covalent bonds. At room temperature, sufficient thermal agitation occurs in this structure to liberate an occasional electron from its bonded state, leaving it free to move throughout the crystal. Thermal excitation of an electron leaves behind a positively charged region termed a hole, which, like the electron, is also mobile. The mechanism of hole movement is stepwise, with a bound electron from a neighboring silicon atom jumping into the electron-deficient region (the hole) and thereby creating another positive hole in its wake. The motion of electrons and holes in opposite directions in semiconductors is the source of conduction in these devices.

The conductivity of silicon can be greatly enhanced by doping, a process in which a tiny controlled amount (approximately 1 ppm) of a Group V or Group III element is distributed homogeneously throughout a silicon crystal. For example, when a crystal is doped with a Group V element, such as arsenic, four out of five of the valence electrons of the dopant form covalent bonds with four silicon atoms leaving one electron free to conduct (see **Figure 25-14**). When the silicon is doped with a Group III element, such as gallium, which has but three valence electrons, an excess of holes develops, also enhancing conductivity (see **Figure 25-15**). A semiconductor containing unbonded electrons (negative charges) is termed an *n*-type semiconductor, and one containing an excess of holes (positive charges) is a *p*-type. In an *n*-type semiconductor, electrons are the majority carrier; in a *p*-type, holes are the majority carrier.

Present silicon technology makes it possible to fabricate what is called a *pn* junction or a *pn* diode, which is conductive in one direction and not in the other. **Figure 25-16a** is a schematic of a silicon diode. The *pn* junction is shown as a dashed line through the middle of the crystal. Electrical wires are attached to both ends of the device. **Figure 25-16b** shows the junction in its conduction mode, wherein the positive terminal of a dc source is connected to the *p* region and the negative terminal to the *n* region. The diode is said to be **forward biased** under these conditions. The excess electrons in the *n* region and the positive holes in the *p* region move toward the junction, where they combine and annihilate each other. The negative terminal of the source injects new electrons into the *n* region, which can continue the conduction process. The positive terminal extracts electrons from the *p* region, thus creating new holes that are free to migrate toward the *pn* junction.

Photodiodes are semiconductor *pn*-junction devices that respond to incident light by forming electron–hole pairs. When a voltage is applied to the *pn* diode such

A **semiconductor** is a substance having a conductivity that lies between that of a metal and that of a dielectric (an insulator).

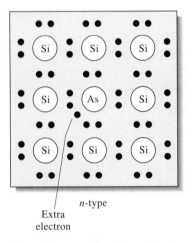

n-type

Extra electron

Figure 25-14 Two-dimensional representation of *n*-type silicon showing "impurity" atom.

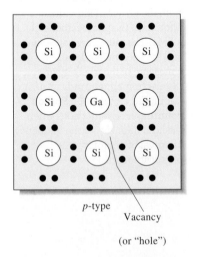

p-type

Vacancy

(or "hole")

Figure 25-15 Two-dimensional representation of *p*-type silicon showing "impurity" atom.

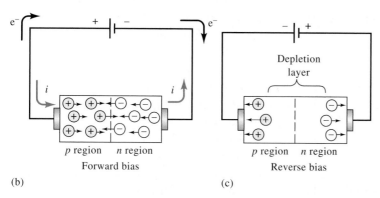

Figure 25-16 (a) Schematic of a silicon diode. (b) Flow of electricity under forward bias. (c) Formation of depletion layer, which prevents flow of electricity under reverse bias.

that the p-type semiconductor is negative with respect to the n-type semiconductor, the diode is said to be **reversed biased**. **Figure 25-16c** illustrates the behavior of a silicon diode under reverse biasing. The majority carriers are drawn away from the junction, leaving a nonconductive **depletion layer**. The conductance under reverse bias is only about 10^{-6} to 10^{-8} of that under forward biasing. In other words, a silicon diode conducts in one direction but not in the other and is said to be a current rectifier.

A reverse-biased silicon diode can serve as a radiation transducer because ultraviolet and visible photons are sufficiently energetic to create additional electrons and holes when they strike the depletion layer of a pn junction. The resulting increase in current can be measured and is directly proportional to radiant power. A silicon-diode detector is more sensitive than a simple vacuum phototube but less sensitive than a photomultiplier tube.

Diode-Array Detectors. Silicon photodiodes have become important recently because 1000 or more can be fabricated side by side on a single small silicon chip with the width of individual diodes being only about 0.02 mm. With one or two of the diode-array detectors placed along the length of the focal plane of a monochromator, all wavelengths passed can be monitored simultaneously, thus making high-speed spectroscopy possible. If the number of light-induced charges per unit time is large compared to thermally produced charge carriers, the current in an external circuit, under reverse-bias conditions, is directly related to the incident radiant power. Silicon photodiode detectors respond extremely rapidly, usually in nanoseconds. Diode arrays can also be obtained commercially with front-end devices called **image intensifiers** to provide gain and allow the detection of low light levels.

Charge-Transfer Devices. Photodiode arrays cannot match the performance of photomultiplier tubes in terms of sensitivity, dynamic range, and signal-to-noise ratio. Thus, their use has been limited to situations where the multichannel advantage outweighs their other shortcomings. In contrast, performance characteristics

In electronics, a **bias** is a dc voltage, sometimes called a polarizing voltage, applied to a circuit element to establish a reference level for operation.

❮ A silicon photodiode is a reverse-biased silicon diode that is used for measuring radiant power.

❮ Photodiode arrays are used not only in spectroscopic instruments but also in optical scanners and bar-code readers.

of **charge-transfer device** (CTD) detectors approach or sometimes surpass those of photomultiplier tubes in addition to having the multichannel advantage. As a result, this type of detector is now appearing in ever-increasing numbers in modern spectroscopic instruments.[8] A further advantage of charge-transfer detectors is that they are two dimensional in the sense that individual detector elements are arranged in rows and columns. For example, one detector that we describe in the next section consists of 244 rows of detector elements, each row being made up of 388 detector elements, therefore giving a two-dimensional array of 94,672 individual detectors, or pixels, contained on a silicon chip having dimensions of 6.5 mm by 8.7 mm. With this device, it becomes possible to record an entire two-dimensional spectrum.

> Silica is silicon dioxide, SiO_2, which is an electrical insulator.

Charge-transfer detectors operate much like photographic film in the sense that they integrate signal information as radiation strikes them. **Figure 25-17** is a cross-sectional depiction of one of the pixels making up a charge-transfer array. In this case, the pixel consists of two conductive electrodes overlying an insulating layer of silica (SiO_2). (A pixel in some charge-transfer devices is made up of more than two electrodes.) This silica layer separates the electrodes from a region of n-doped silicon. This assembly constitutes a metal oxide semiconductor capacitor that stores the charges formed when radiation strikes the doped silicon. When, as shown, a negative charge is applied to the electrodes, a charge inversion region is created under the electrodes, which is energetically favorable for the storage of positive holes. The mobile holes created by the absorption of photons by the silicon then migrate and collect in this region. Typically, this region, which is called a potential well, is capable of holding as many as 10^5 to 10^6 charges before overflowing into an adjacent pixel. In Figure 25-17, one electrode is shown as being more negative than the other, making the accumulation of charge under this electrode more favorable. The amount of charge generated during exposure to radiation is measured in either of two ways. In a **charge-injection device** (CID) detector, the voltage change arising from movement of the charge from the region under one electrode to the region under the other is measured. In a **charge-coupled device** (CCD) detector, the charge is moved to a charge-sensing amplifier for measurement.

Charge-coupled devices are also available with front-end image intensifiers to provide gain. Such intensified CCDs (ICCDs) can be gated on and off at selected intervals to provide time resolution for lifetime studies or for chemical kinetics experiments or to discriminate against undesirable signals. A recent development

Figure 25-17 Cross section of one of the pixels of a charge-transfer device. The positive hole produced by the photon hv is collected under the negative electrode.

[8]For details on charge-transfer devices, see J. V. Sweedler, K. L. Ratzlaff, and M. B. Denton, eds., *Charge-Transfer Devices in Spectroscopy*, New York: VCH, 1994; J. V. Sweedler, Crit. *Rev. Anal. Chem.,* **1993**, *24*, 59, **DOI**:10.1080/10408349308048819; J. V. Sweedler, R. B. Bilhorn, P. M. Epperson, G. R. Sims, and M. B. Denton, *Anal. Chem.,* **1988**, *60*, 282A, **DOI**: 10.1021/ac00155a002; P. M. Epperson, J. V. Sweedler, R. B. Bilhorn, G. R. Sims, and M. B. Denton, *Anal. Chem.,* **1988**, *60*, 327A, **DOI**: 10.1021/ac00156a001.

in CCD cameras is the electron-multiplying CCD (EMCCD) in which a gain register is inserted prior to the output amplifier. Both ICCDs and EMCCDs are capable of single photon detection. Because of the image intensifier, ICCDs are more expensive than EMCCDs. However, the EMCCD must be cooled to low temperatures ($\approx 170K$), which leads to additional expense and often condensation problems.

CCDs and CIDs are appearing in ever-increasing numbers in modern spectroscopic instruments. In spectroscopic applications, charge-transfer devices are used in conjunction with multichannel instruments as discussed in Section 25B-3. In addition to spectroscopic applications, charge-transfer devices find widespread applications in digital cameras, in solid-state television cameras, in microscopy, and in astronomical applications, such as the Hubble Space Telescope.

Thermal Detectors

The convenient photon detectors discussed in the previous section cannot be used to measure infrared radiation because photons in this region lack sufficient energy to cause photoemission of electrons. Historically, thermal detectors, such as thermocouples, bolometers, and pneumatic devices, were used to detect all but the shortest IR wavelengths. These detectors are still found in older dispersive IR spectrometers. The characteristics of most thermal detectors are, however, much inferior to those of the photon detectors used in the UV/visible region. Most Fourier transform IR spectrometers use a pyroelectric transducer or the MCT photoconductive detector discussed earlier.

A thermal detector consists of a tiny blackened surface that absorbs infrared radiation and increases in temperature as a result. The temperature rise is converted to an electrical signal that is amplified and measured. Under the best of circumstances, the temperature changes involved are minuscule, amounting to a few thousandths of a degree Celsius. The difficulty of measurement is compounded by thermal radiation from the surroundings, which is always a potential source of uncertainty. To minimize the effects of this background radiation, or noise, thermal detectors are housed in a vacuum and are carefully shielded from their surroundings. To further minimize the effects of this external noise, the beam from the source is made to alternate between maximum intensity and zero intensity by a rotating slotted disk, called a chopper, which is inserted between source and detector.[9] The transducer converts this periodic radiation signal to an alternating electrical current that can be amplified and separated from the dc signal resulting from the background radiation. Despite all these measures, infrared measurements are significantly less precise than measurements of ultraviolet and visible radiation.

As shown in Table 25-2, four types of thermal transducers are used for infrared spectroscopy.[10] The most widely used is a tiny thermocouple or a group of thermocouples called a **thermopile**. The **bolometer** consists of a conducting element whose electrical resistance changes as a function of temperature. A **pneumatic detector** consists of a small cylindrical chamber that is filled with xenon and contains a blackened membrane to absorb infrared radiation and heat the gas. **Pyroelectric detectors** are manufactured from crystals of a pyroelectric material, such as barium titanate or deuterated triglycine sulfate. A crystal of either of these compounds sandwiched between a pair of electrodes produces a temperature-dependent voltage when

---

[9]See D. A. Skoog, F. J. Holler, and S. R. Crouch, *Principles of Instrumental Analysis,* 6th ed., Belmont, CA: Brooks/Cole, 2007, pp. 115–16.

[10]Ibid., pp. 200–202.

exposed to infrared radiation. Pyroelectric transducers are used in IR spectrometers, particularly the Fourier transform instruments described in Section 25C-2.

25A-5 Signal Processors and Readout Devices

A signal processor is an electronic device that may amplify the electrical signal from the detector (see Feature 25-6). In addition, the signal processor may convert the signal from dc to ac (or the reverse), change the phase of the signal, and filter it to remove unwanted components. The signal processor may also perform such mathematical operations on the signal as differentiation, integration, or conversion to logarithms. Several types of readout devices are found in modern instruments. Digital meters and computer monitors are two examples. Computers are often used to control various instrumental parameters, to process and store data, to print results and spectra, to compare results with various databases, and to communicate with other computers and network devices.

FEATURE 25-6

Measuring Photocurrents with Operational Amplifiers

The current produced by a reverse-biased silicon photodiode is typically 0.1 μA to 100 μA. These currents, as well as those generated by photomultipliers and phototubes, are so small that they must be converted to a voltage that is large enough to be measured with a digital voltmeter or other voltage-measuring device. We can perform such a conversion with the operational amplifier (op amp) circuit shown in **Figure 25F-5**. Light striking the reverse-biased photodiode causes a current I in the circuit. Because the op amp has a very large input resistance, essentially no current enters the op amp input designated by the minus sign. Thus, current in the photodiode must pass through the resistor R. The current is conveniently calculated from Ohm's law: $E_{out} = -IR$. Since the current is proportional to the radiant power (P) of the light striking the photodiode, $I = kP$, where k is a constant, and therefore, $E_{out} = -IR = -kPR = k'P$. A voltmeter is connected to the output of the op amp to give a direct readout that is proportional to the radiant power of the light falling on the phoodiode. This same circuit can also be used with vacuum photodiodes or photomultipliers.[11]

Figure 25F-5 An operational amplifier current to voltage converter used to monitor the current in a solid-state photodiode.

25A-6 Sample Containers

Sample containers, which are usually called **cells** or **cuvettes**, must have windows that are transparent in the spectral region of interest. The various transmittance ranges for optical materials are shown in Figure 25-2. As can be seen, quartz or fused silica is

---

[11]For additional information on operational amplifiers, see D. A. Skoog, F. J. Holler, and S. R. Crouch, *Principles of Instrumental Analysis,* 6th ed., Ch. 3, Belmont, CA: Brooks/Cole, 2007.

required for the UV region (wavelengths less than 350 nm) and may be used in the visible region and out to about 3000 nm (3 μm) in the IR. Silicate glass is usually used for the 375–2000 nm region because of its low cost compared to quartz. Plastic cells are also used in the visible. The most common window material for IR studies is crystalline sodium chloride, which is soluble in water and in some other solvents.

The best cells have windows that are perpendicular to the direction of the beam in order to minimize reflection losses. The most common cell path length for studies in the UV and visible regions is 1 cm; matched, calibrated cells of this size are available from several commercial sources. Many other shorter and longer path length cells can also be purchased. Some typical UV/visible cells are shown in **Figure 25-18**.

For reasons of economy, cylindrical cells are sometimes used. Particular care must be taken to duplicate the position of such cells with respect to the beam. Otherwise, variations in path length and reflection losses at the curved surfaces can cause significant error, as discussed in Section 24C-3.

The quality of spectroscopic data is critically dependent on the way that cells are used and maintained. Fingerprints, grease, or other deposits on the walls may alter significantly the transmission characteristics of a cell. Thus, it is imperative to thoroughly clean cells both before and after use, and the windows must not be touched after cleaning is complete. Matched cells should never be dried by heating in an oven or over a flame because this may cause physical damage or may change the path length. Matched cells should be calibrated against each other regularly with an absorbing solution.

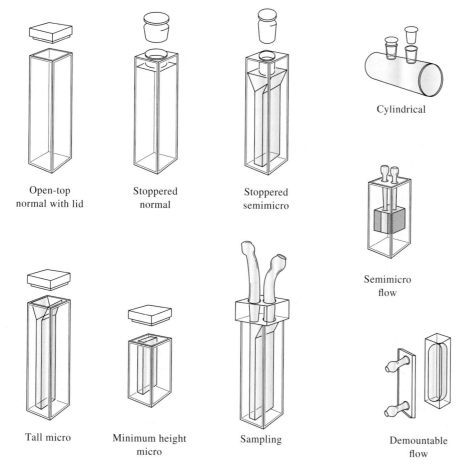

Open-top
normal with lid

Stoppered
normal

Stoppered
semimicro

Cylindrical

Semimicro
flow

Tall micro

Minimum height
micro

Sampling

Demountable
flow

Figure 25-18 Typical examples of commercially available cells for the UV/visible region.

25B ULTRAVIOLET/VISIBLE PHOTOMETERS AND SPECTROPHOTOMETERS

The optical components described in Figure 25-1 have been combined in various ways to produce two types of instruments for absorption measurements. Several common terms are used to describe complete instruments. A **spectrometer** is a spectroscopic instrument that uses a monochromator or polychromator in conjunction with a transducer to convert the radiant intensities into electrical signals. **Spectrophotometers** are spectrometers that allow measurement of the ratio of the radiant powers of two beams, a requirement to measure absorbance (recall from Chapter 24, Equation 24-6 on page 659 that $A = \log P_0/P \approx \log P_{solvent}/P_{solution}$). **Photometers** use a filter for wavelength selection in conjunction with a suitable radiation transducer. Spectrophotometers offer the considerable advantage that the wavelength used can be varied continuously, making it possible to record absorption spectra. Photometers have the advantages of simplicity, ruggedness, and low cost. Several dozen models of spectrophotometers are available commercially. Most spectrophotometers cover the UV/visible and occasionally the near-infrared region, while photometers are most often used for the visible region. Photometers find considerable use as detectors for chromatography, electrophoresis, immunoassays, or continuous flow analysis. Both photometers and spectrophotometers can be obtained in single- and double-beam varieties.

25B-1 Single-Beam Instruments

Figure 25-19 shows the design of a simple and inexpensive spectrophotometer, the Spectronic 20, which is designed for the visible region of the spectrum. This instrument first appeared on the market in the mid-1950s, and the modified version shown in the figure is still being manufactured and widely sold. More of these instruments are currently in use throughout the world than any other single spectrophotometer model.

The Spectronic 20 reads out in transmittance or in absorbance on a **liquid-crystal display (LCD)**; older analog instruments read out in transmittance on a meter. The instrument is equipped with an **occluder**, which is a vane that automatically falls between the beam and the detector whenever the cylindrical cell is removed from its holder. The light control device consists of a V-shaped aperture that is moved in and out of the beam to control the amount of light reaching the exit slit.

To obtain a percent transmittance reading, the digital readout is first zeroed with the sample compartment empty so that the occluder blocks the beam and no radiation reaches the detector. This process is called the **0% T calibration**, or **adjustment**. A cell containing the blank (often the solvent) is then inserted into the cell holder, and the pointer is brought to the 100% T mark by adjusting the position of the light control aperture and thus the amount of light reaching the detector. This adjustment is called the **100% T calibration**, or **adjustment**. Finally, the sample is placed in the cell compartment, and the percent transmittance or the absorbance is read directly from the LCD display.

The spectral range of the Spectronic 20 is 400 to 900 nm. Other specifications include a spectral bandpass of 20 nm, a wavelength accuracy of ± 2.5 nm, and a photometric accuracy of $\pm 4\%$ T. An option allows the instrument to be interfaced to a computer for data storage and analysis. A newer version, the Spectronic 200, has a < 4 nm bandpass and can scan spectra.

The 0% T and 100% T adjustments should be made immediately before each transmittance or absorbance measurement. To obtain reproducible transmittance measurements, the radiant power of the source must remain constant during the time that the 100% T adjustment is made and the % T is displayed.

(a)

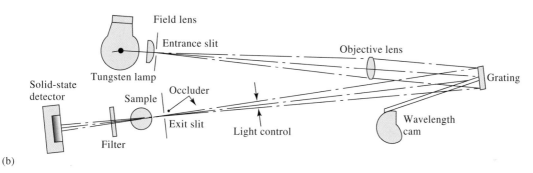

(b)

Figure 25-19   The Spectronic 20 spectrophotometer. A photograph is shown in (a), while the optical diagram is seen in (b). Radiation from the tungsten filament source passes through an entrance slit into the monochromator. A reflection grating diffracts the radiation, and the selected wavelength band passes through the exit slit into the sample chamber. A solid-state detector converts the light intensity into a related electrical signal which is amplified and displayed on a digital readout. The newer Spectronic 200 has a reverse optical path. (Courtesy of Thermo Fisher Scientific, Inc. Madison, WI.)

Single-beam instruments of the type just described are well suited for quantitative absorption measurements at a single wavelength. With these instruments, simplicity of instrumentation, low cost, and ease of maintenance offer distinct advantages. Several instrument manufacturers offer single-beam spectrophotometers and photometers of the single-wavelength type. Prices for these instruments are in the range of a thousand to several thousand dollars. In addition, basic single-beam multichannel instruments based on array detectors are widely available as discussed in Section 25B-3.

25B-2 Double-Beam Instruments

Many modern photometers and spectrophotometers are based on a double-beam design. **Figure 25-20** shows two double-beam designs (b and c) compared with a

(a)

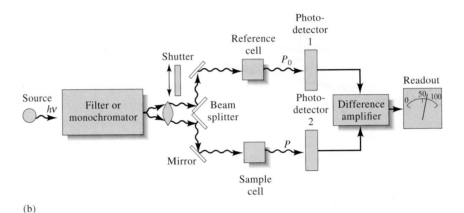

(b)

Figure 25-20 Instrumental designs for UV/visible photometers or spectrophotometers. In (a), a single-beam instrument is shown. Radiation from the filter or monochromator passes through either the reference cell or the sample cell before striking the photodetector. In (b), a double-beam-in-space instrument is shown. In this instrument, radiation from the filter or monochromator is split into two beams that simultaneously pass through the reference and sample cells before striking two matched photodetectors. In the double-beam-in-time instrument (c), the beam is alternately sent through reference and sample cells before striking a single photodetector. A period of only a few milliseconds separates the beams as they pass through the two cells.

(c)

single-beam system (a). **Figure 25-20b** illustrates a double-beam-in-space instrument in which two beams are formed by a V-shaped mirror called a **beam-splitter**. One beam passes through the reference solution to a photodetector, and the second simultaneously passes through the sample to a second, matched photodetector. The two outputs are amplified, and their ratio, or the logarithm of their ratio, is obtained electronically or computed and displayed on the output device.

Figure 25-20c illustrates a double-beam-in-time spectrophotometer. In this design, the beams are separated in time by a rotating sector mirror that directs the entire beam through the reference cell and then through the sample cell. The pulses of radiation are then recombined by another mirror that transmits the reference beam and reflects the sample beam to the detector. The double-beam-in-time approach is generally preferred over the double-beam-in-space because of the difficulty in matching two detectors.

Double-beam instruments offer the advantage that they compensate for all but the most rapid fluctuations in the radiant output of the source. They also compensate for wide variations of source intensity with wavelength. Furthermore, the double-beam design is well suited for continuous recording of absorption spectra.

25B-3 Multichannel Instruments

Photodiode arrays and charge-transfer devices, discussed in Section 25A-4, are the basis of multichannel instruments for UV/visible absorption. These instruments are usually of the single-beam design illustrated in **Figure 25-21**. With multichannel systems, the dispersive system is a grating spectrograph placed after the sample or reference cell. The photodiode array or CCD array is placed in the focal plane of the spectrograph. These detectors allow the measurement of an entire spectrum in less than 1 s. A computer is required for obtaining spectra. With single-beam designs, the array dark current is acquired and stored in computer memory. Next, the spectrum of the source is obtained and stored in memory after dark current subtraction. Finally, the raw spectrum of the sample is obtained, and after dark current subtraction, the sample values are divided by the source values at each wavelength to produce the absorption spectrum. Multichannel instruments can also be configured as double-beam-in-time spectrophotometers.

The spectrophotometer shown in Figure 25-21 can be controlled by most personal computers. The instrument (without the computer) can be purchased for around $10,000. Several instrument companies combine array detector systems with fiber optic probes that transport the light to and from the sample. These instruments allow measurements in convenient locations that are remote to the spectrometer.

25C INFRARED SPECTROPHOTOMETERS

Two types of spectrometers are used in IR spectroscopy: the dispersive type and the Fourier transform variety.

25C-1 Dispersive Infrared Instruments

The older IR instruments were almost always dispersive double-beam designs. These were often of the double-beam-in-time variety shown in Figure 25-20c except that the location of the cell compartment with respect to the monochromator was reversed. In most UV/visible instruments the cell is located between the monochromator and the detector in order to avoid photodecomposition of the sample, which may occur if samples are exposed to the full power of the source. Note that photodiode

Figure 25-21 Diagram of a multichannel spectrometer based on a grating spectrograph with a photodiode array detector.

Albert Abraham Michelson (1852–1931) was one of the most gifted and inventive experimentalists of all time. He was a graduate of the United States Naval Academy and eventually became professor of physics at The University of Chicago. He studied the properties of light and performed several experiments that laid the foundation for our modern view of the universe. He invented the interferometer described in Feature 25-7 to determine the effect of the Earth's motion on the velocity of light. For his many inventions and their application to the study of light, Michelson won the 1907 Nobel Prize in Physics. At the time of his death, Michelson and his collaborators were attempting to measure the speed of light in a mile-long vacuum tube located in what is now Irvine, California.

> Fourier transform spectrometers detect all the IR wavelengths all the time. They have greater light-gathering power than dispersive instruments and consequently better precision. Although the calculations of the Fourier transform are computationally intense, they are easily accomplished with high-speed personal computers and appropriate software.

array instruments avoid this problem because of the short exposure time of the sample to the beam. Infrared radiation, in contrast, is not sufficiently energetic to bring about photodecomposition. Also, most samples are good emitters of IR radiation. Because of these factors, the cell compartment is usually located between the source and the monochromator in an IR instrument.

As discussed earlier in this section, the components of IR instruments differ significantly from those in UV/visible instruments. Thus, IR sources are heated solids, and IR detectors respond to heat rather than to photons. Furthermore, the optical components of IR instruments are constructed from polished salts, such as sodium chloride or potassium bromide.

25C-2 Fourier Transform Instruments

When Fourier transform infrared (FTIR) spectrometers first appeared on the market in the early 1970s, they were bulky, expensive (more than $100,000), and required frequent mechanical adjustments. For these reasons, their use was limited to special applications in which their unique characteristics (great speed, high resolution, high sensitivity, and excellent wavelength precision and accuracy) were essential. Since the 1990s, however, FTIR spectrometers have been reduced to benchtop size and have become very reliable and easy to maintain. Furthermore, the simple models are now similarly priced to dispersive spectrometers. Hence, FTIR spectrometers have largely displaced dispersive instruments in most laboratories.

Fourier transform IR instruments contain no dispersing element, and all wavelengths are detected and measured simultaneously. Instead of a monochromator, an interferometer is used to produce interference patterns that contain the infrared spectral information. The same types of sources used in dispersive instruments are used in FTIR spectrometers. Transducers are typically triglycine sulfate, a pyroelectric transducer, or mercury cadmium telluride, a photoconductive transducer. In order to obtain radiant power as a function of wavelength, the interferometer modulates the source signal in such a way that it can be decoded by the mathematical technique of Fourier transformation. This operation requires a high-speed computer to do the necessary calculations. The theory of Fourier transform measurements is discussed in Feature 25-7.[12]

Most benchtop FTIR spectrometers are of the single-beam type. In order to collect the spectrum of a sample, the background spectrum is first obtained by Fourier transformation of the interferogram from the background (solvent, ambient water, and carbon dioxide). Next, the sample spectrum is acquired. Finally, the ratio of the single-beam sample spectrum to that of the background spectrum is calculated and absorbance or transmittance versus wavelength or wavenumber is plotted. Often benchtop instruments purge the spectrometer with an inert gas or dry, CO_2-free air to reduce the background absorption from water vapor and CO_2.

The major advantages of FTIR instruments over dispersive spectrometers include better speed and sensitivity, better light-gathering power, more accurate wavelength calibration, simpler mechanical design, and the virtual elimination of any contribution from stray light and IR emission. Because of these advantages, nearly all new IR instruments are FTIR systems.

---

[12]See also, J. D. Ingle, Jr., and S. R. Crouch, *Spectrochemical Analysis*, Englewood Cliffs, NJ: Prentice-Hall, 1988; D. A. Skoog, F. J. Holler, and S. R. Crouch, *Principles of Instrumental Analysis*, 6th ed., Belmont, CA: Brooks/Cole, 2007.

How Does a Fourier Transform Infrared Spectrometer Work?

Fourier transform infrared (FTIR) spectrometers utilize an ingenious device called a **Michelson interferometer**, which was developed many years ago by A. A. Michelson for making precise measurements of the wavelengths of electromagnetic radiation and for making incredibly accurate distance measurements. The principles of interferometry are utilized in many areas of science including chemistry, physics, astronomy, and metrology and are applicable in many regions of the electromagnetic spectrum.

A diagram of a Michelson interferometer is shown in **Figure 25F-6**. It consists of a collimated light source, shown on the left of the diagram, a stationary mirror at the top, a moveable mirror at the right, a beam-splitter, and a detector. The light source may be a continuum source as in FTIR spectroscopy, or it may be a monochromatic source such as a laser or a sodium arc lamp for other uses such as, for example, measuring distances. The mirrors are precision-polished ultraflat glass with a reflective coating vapor deposited on their front surfaces. The moveable mirror is usually mounted on a very precise linear bearing that allows it to move along the direction

of the light beam while remaining perpendicular to it as shown in the diagram.

The key to the operation of the interferometer is the *beam-splitter*, which is usually a partially silvered mirror similar to the "two-way" mirrors often seen in retail stores and police interrogation rooms. The beam-splitter allows a fraction of the light falling on it to pass through the mirror, and another fraction is reflected. This device works in both directions, so that light falling on either side of the beam-splitter is partially reflected and partially transmitted.

For simplicity, we will use as our light source the blue line of an argon-ion laser. Beam *A* from the source impinges on the beam-splitter, which is tilted at 45° to the incoming beam. Our beam-splitter is coated on the right side, so Beam *A* enters the glass and is partially reflected off the back side of the coating. It emerges from the beam-splitter as Beam *A'* and moves up toward the stationary mirror where it is reflected back down toward the beam-splitter. Part of the beam is then transmitted down through the beam-splitter toward the detector. Although the beam loses some intensity with each interaction with the

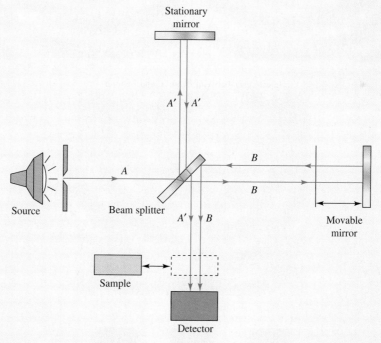

Figure 25F-6 Diagram of a Michelson interferometer. A beam from the light source at left is split into two beams by the beam-splitter. The two beams travel two separate paths and converge on the detector. The two beams *A'* and *B* converge in the same region of space and form an interference pattern. As the movable mirror on the right is moved, the interference pattern shifts across the detector and modulates the optical signal. The resulting reference interferogram is recorded and used as a measure of the power of the incident beam at all wavelengths. An absorbing sample is then inserted into the beam, and a sample interferogram is recorded. The two interferograms are then used to compute the absorption spectrum of the sample.

(continued)

stationary mirror and the beam-splitter, the net effect is that a fraction (Beam A') of incident Beam A ends up at the detector.

In its first interaction with the beam-splitter, the fraction of Beam A that is transmitted emerges to the right toward the moveable mirror as Beam B. It then is reflected back to the left to the beam-splitter where it is reflected down toward the detector. With careful alignment, both Beam A' and Beam B (shown separately in the diagram for clarity), are collinear and impinge on the detector at the same spot.

The overall purpose of the interferometer optics is to split the incident beam into two beams that move through space along separate paths and then recombine at the detector. It is in this region that the two beams, or wavefronts, interact to form

an **interference pattern**. The origin of the interference pattern is illustrated in **Figure 25F-7**, which is a two-dimensional representation of the interaction of the two spherical wavefronts. Beam A' and Beam B converge and interact as two point sources of light represented in the upper portion of the figure. When the two beams interfere, they form a pattern similar to the one shown. In regions where the waves interfere constructively, bright bands appear, and where destructive interference occurs, dark bands form. The alternating light and dark bands are called **interference fringes**. These fringes appear at the detector as the output image shown at the bottom of the figure. In the earliest versions of the Michelson interferometer, the detector was the human eye aided by a

Beam A'

Beam B

Interference between Beam A' and Beam B

Image at output

Figure 25F-7 A two-dimensional representation of the interference of two monochromatic wavefronts of the same frequency. Beam A' and Beam B at the top form the interference pattern in the middle and the two wavefronts constructively and destructively interfere. The image shown at the bottom would appear at the output of the Michelson interferometer perpendicular to the plane of the two-dimensional interference pattern.

telescope. The fringes could be counted or measured through the telescope.

When the moveable mirror is moved to the left at constant velocity, the interference pattern gradually sweeps past the detector as the path that Beam *B* follows is gradually shortened. The form of the interference pattern remains the same, but the positions of constructive and destructive interference are shifted as the path difference changes. For example, if the wavelength of our laser source is λ, as we move the mirror a distance of $\lambda/4$, the path difference between the two beams changes by $\lambda/2$, and where we had constructive interference, we now have destructive interference. If we move the mirror another $\lambda/4$, the path difference changes again by $\lambda/2$, and we again return to constructive interference. As the mirror moves, the two wavefronts are shifted

in space relative to one another, and alternate light and dark fringes sweep across the detector, as illustrated in **Figure 25F-8a**. At the detector, we find the sinusoidal intensity profile shown in **Figure 25F-8b**. This profile is called an **interferogram**. The net effect of the constant uniform motion of the mirror is that the light intensity at the output of the interferometer is **modulated**, or systematically varied, in a precisely controlled way as shown in the figure. In practice, it turns out not to be very easy to move the interferometer mirror at a constant, precisely controlled velocity. There is a better and much more precise way to monitor the mirror motion using a second parallel interferometer.[13] In this example, we will just assume that we can measure and/or monitor the progress of the mirror and compensate for any nonuniform motion computationally.

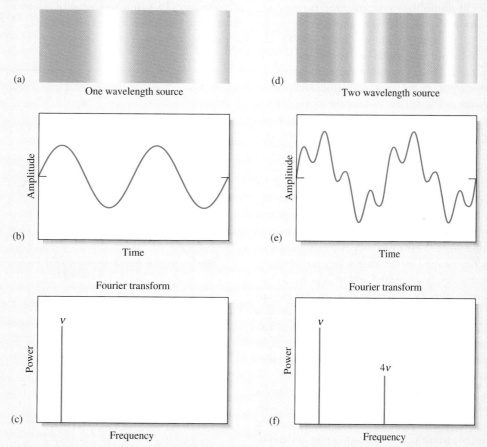

Figure 25F-8 Formation of interferograms at the output of the Michelson interferometer. (a) Interference pattern at the output of the interferometer resulting from a monochromatic source. (b) Sinusoidally varying signal produced at the detector by the pattern in (a). (c) Frequency spectrum of the monochromatic light source resulting from the Fourier transformation of the signal in (b). (d) Interference pattern at the output of the interferometer resulting from a two-color source. (e) Complex signal produced by the interference patter of (d) as it falls on the detector. (f) Frequency spectrum of the two-color source.

(continued)

[13]D. A. Skoog, F. J. Holler, and S. R. Crouch, *Principles of Instrumental Analysis,* 6th ed., Chs. 5 and 16, p. 440, Belmont, CA: Brooks/Cole, 2007.

We have established that a Michelson interferometer with a monochromatic light source produces a sinusoidally varying signal at the detector when the mirror is moved at constant velocity. Now, we must investigate what happens to the signal once it is recorded. Although the characteristics of Michelson interferometers have been well known for over a century and the mathematical apparatus for dealing with the data has been in place for nearly two centuries, the device could not be used routinely for spectroscopy until two developments occurred. First, high-speed, inexpensive computers had to become available. Second, appropriate computational methods had to be invented to handle the huge number of rather routine calculations that must be applied to the raw data acquired in interferometric experiments. Briefly, the principles of Fourier synthesis and analysis tell us that any waveform can be represented as a series of sinusoidal waveforms, and correspondingly, any combination of sinusoidal waveforms can be broken down into a series of sinusoids of known frequency. We can apply this idea to the sinusoidal signal detected at the output of the Michelson interferometer shown in Figure 25F-8b.

If we subject the signal in the figure to Fourier analysis via a computer algorithm called the fast Fourier transform (FFT), we obtain the frequency spectrum illustrated in **Figure 25F-8c**. Notice that the original waveform in Figure 25F-8b is a time-dependent signal; the resultant output from the FFT is a frequency-dependent signal. In other words, the FFT takes amplitude signals in the **time domain** and converts them to power in the **frequency domain**. Since the output of the interferometer is a sine wave of a single frequency, the frequency spectrum shows a single spike of frequency v, the frequency of the original sine wave. This frequency is proportional to the optical frequency emitted by the laser source but of much lower value so that it can be measured and manipulated electronically. We now modify the interferometer so that we can obtain a second sine wave at the output. One way to do this is to simply add a second wavelength to our light source. Experimentally, a second laser or another monochromatic light source at the input of the interferometer gives us a beam that contains just two wavelengths.

For example, let us assume that the second wavelength is one quarter of our first one so that our second frequency is $4v$. We further assume that its intensity is one half the intensity of our original source. As a result, the signal appearing at the output of the interferometer exhibits a pattern somewhat more complex than in the single-wavelength example, as shown in **Figure 25F-8d**. The detector signal plot appears as the sum of two sine waves as depicted in **Figure 25F-8e**. We then apply the FFT to the complex sinusoidal signal to produce the frequency spectrum of **Figure 25F-8f**. This spectrum reveals just two frequencies at v and $4v$, and the relative magnitudes

of the two frequency spikes are proportional to the amplitudes of the two sine waves composing the original signal. The two frequencies correspond to the two wavelengths in our interferometer light source, and the FFT has revealed the intensities of the source at those two wavelengths.

To illustrate how the Michelson interferometer is used in practical experiments, we place a continuum infrared light source (see **Figure 25F-9a**) containing a huge number of wavelengths at the input of the interferometer. As the mirror moves along its path, all wavelengths are modulated simultaneously, producing the very interesting interferogram shown in **Figure 25F-9b**. This interferogram contains all of the information that we require in a spectroscopy experiment regarding the intensity of the light source at all of its component wavelengths.

As suggested in the previous section, there are a number of advantages to acquiring intensity information in this way rather than using a scanning spectrometer.[14] First, there is the advantage of speed. The mirror can be moved in a matter of

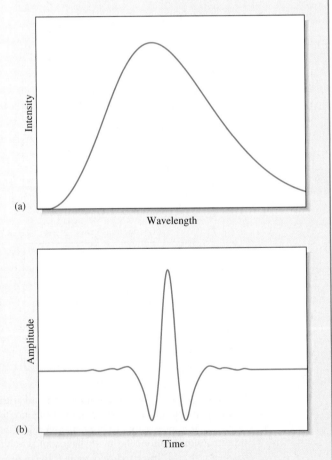

(a)

(b)

Figure 25F-9   (a) Spectrum of a continuum light source. (b) Interferogram of the light source in (a) produced at the output of the Michelson interferometer.

---

[14]See J. D. Ingle, Jr., and S. R. Crouch, *Spectrochemical Analysis*, Englewood Cliffs, NJ: Prentice-Hall, 1988, pp. 425–26.

seconds, and a computer attached to the detector can collect all necessary data during the course of the mirror scan. In just a few more seconds, the computer can perform the FFT and produce the frequency spectrum containing all of the intensity information. Next is **Fellgett's advantage**, which suggests that Michelson interferometers are capable of producing higher signal-to-noise ratios in shorter time than equivalent dispersive spectrometers. Finally, we have the throughput, or **Jacquinot's advantage**, which permits 10 to 200 times more radiation to pass through a sample compared to standard dispersive spectrometers. These advantages are often partially offset by the lower sensitivity of detectors that are used in FTIR spectrometers. Under these circumstances, the speed of the measurement process and the simplicity and reliability of FTIR spectrometers become primary considerations. We discuss some of these issues further in Chapter 26.

Up to this point in our discussion of the FTIR spectrometer, we have shown how the Michelson interferometer can provide intensity information for a light source as a function of wavelength. To collect the IR spectrum of a sample, we must first obtain a reference interferogram of the source with no sample in the light path, as shown in Figure 25F-6. Then, the sample is placed in the path as indicated by the arrow and dashed box in the figure, and once again, we scan the mirror and acquire a second interferogram. In FTIR spectrometry, the sample absorbs infrared radiation, which attenuates the beams in the interferometer. The difference between the second (sample) interferogram and the reference interferogram is then computed. Since the difference interferogram depends only on the absorption of radiation by the sample, the FFT is performed on the resulting data, which produces the IR spectrum of the sample. We discuss a specific example of this process in Chapter 26. Finally, we should note that the FFT can be accomplished using the most basic modern personal computer equipped with the appropriate software. Many software packages such as Mathcad, Mathematica, Matlab, and even the Data Analysis Toolpak of Microsoft® Excel have Fourier analysis functions built in. These tools are widely used in science and engineering for a broad range of signal-processing tasks.[15]

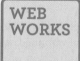

WEB WORKS Use a search engine to find companies that manufacture monochromators. Navigate to several websites of these companies and find a UV/visible monochromator of the Czerny-Turner design that has better than 0.1 nm resolution. List several other important specifications of monochromators and describe what they mean and how they affect the quality of analytical spectroscopic measurements. From the specifications and, if available, prices, determine the factors that have the most significant effect on the cost of the monochromators.

QUESTIONS AND PROBLEMS

25-1. Describe the differences between the following pairs of terms and list any particular advantages possessed by one over the other:

*(a) solid-state photodiodes and phototubes as detectors for electromagnetic radiation.

(b) phototubes and photomultiplier tubes.

*(c) filters and monochromators as wavelength selectors.

(d) conventional and diode-array spectrophotometers.

25-2. Define the term *effective bandwidth of a monochromator*.

*25-3. Why do quantitative and qualitative analyses often require different monochromator slit widths?

25-4. Why are photomultiplier tubes unsuited for the detection of infrared radiation?

*25-5. Why is iodine sometimes introduced into a tungsten lamp?

25-6. Describe the differences between the following pairs of terms and list any particular advantages of one over the other:

*(a) spectrophotometers and photometers.

(b) spectrographs and polychromators.

*(c) monochromators and polychromators.

(d) single-beam and double-beam instruments for absorbance measurements.

[15]See also D. A. Skoog, F. J. Holler, and S. R. Crouch, *Principles of Instrumental Analysis*, 6th ed., Belmont, CA: Brooks/Cole, 2007, pp. 98–103.

25-7. The Wien displacement law states that the wavelength maximum in micrometers for blackbody radiation is

$$\lambda_{max} T = 2.90 \times 10^3$$

where T is the temperature in kelvins. Calculate the wavelength maximum for a blackbody that has been heated to *(a) 4000 K, (b) 3000 K, *(c) 2000 K, and (d) 1000 K.

25-8. Stefan's law states that the total energy emitted by a blackbody per unit time and per unit area is

$$E_t = \alpha T^4$$

where α is 5.69×10^{-8} W/m^{2}K^4. Calculate the total energy output in W/m^2 for the blackbodies described in Problem 25-7.

*25-9. The relationships described in Problems 25-7 and 25-8 may be of help in solving the following.

(a) Calculate the wavelength of maximum emission of a tungsten-filament bulb operated at 2870 K and at 3000 K.

(b) Calculate the total energy output of the bulb in W/cm^2.

25-10. What minimum requirement is needed to obtain reproducible results with a single-beam spectrophotometer?

*25-11. What is the purpose of (a) the 0% T adjustment and (b) the 100% T adjustment of a spectrophotometer?

25-12. What experimental variables must be controlled to assure reproducible absorbance data?

*25-13. What are the major advantages of Fourier transform IR instruments over dispersive IR instruments?

25-14. A photometer with a linear response to radiation gave a reading of 625 mV with a blank in the light path and 149 mV when the blank was replaced by an absorbing solution. Calculate

*(a) the percent transmittance and absorbance of the absorbing solution.

(b) the expected percent transmittance if the concentration of absorber is one half that of the original solution.

*(c) the percent transmittance to be expected if the light path through the original solution is doubled.

25-15. A portable photometer with a linear response to radiation registered a photocurrent of 75.9 μA with a blank solution in the light path. Replacement of the blank with an absorbing solution yielded a response of 23.5 μA. Calculate

(a) the percent transmittance of the sample solution.

*(b) the absorbance of the sample solution.

(c) the transmittance to be expected for a solution in which the concentration of the absorber is one third that of the original sample solution.

*(d) the transmittance to be expected for a solution that has twice the concentration of the sample solution.

25-16. Why does a deuterium lamp produce a continuum rather than a line spectrum in the ultraviolet?

*25-17. What are the differences between a photon detector and a thermal detector?

25-18. Describe the basic design difference between a spectrometer for absorption measurements and one for emission studies.

*25-19. Describe how an absorption photometer and a fluorescence photometer differ from each other.

25-20. What data are needed to describe the performance characteristics of an interference filter?

25-21. Define

*(a) transducer.

(b) dark current.

*(c) n-type semiconductor.

(d) majority carrier.

*(e) depletion layer.

(f) scattered radiation (in a monochromator).

25-22. An interference filter is to be constructed for isolation of the CS$_2$ absorption band at 4.54 μm.

(a) If the determination is to be based on first-order interference, how thick should the dielectric layer be (refractive index 1.34)?

(b) What other wavelengths will be transmitted?

25-23. The following data were taken from a diode array spectrophotometer in an experiment to measure the spectrum of the Co(II)-EDTA complex. The column labeled $P_{solution}$ is the relative signal obtained with sample solution in the cell after subtraction of the dark signal. The column labeled $P_{solvent}$ is the reference signal obtained with only solvent in the cell after subtraction of the dark signal. Find the transmittance at each wavelength and the absorbance at each wavelength. Plot the spectrum of the compound.

Wavelength, nm	$P_{solvent}$	$P_{solution}$
350	0.002689	0.002560
375	0.006326	0.005995
400	0.016975	0.015143
425	0.035517	0.031648
450	0.062425	0.024978
475	0.095374	0.019073
500	0.140567	0.023275
525	0.188984	0.037448
550	0.263103	0.088537
575	0.318361	0.200872
600	0.394600	0.278072
625	0.477018	0.363525
650	0.564295	0.468281
675	0.655066	0.611062
700	0.739180	0.704126
725	0.813694	0.777466
750	0.885979	0.863224
775	0.945083	0.921446
800	1.000000	0.977237

25-24. Challenge Problem: Horlick has described the mathematical principles of the Fourier transform, interpreted them graphically, and described how they may be used in analytical spectroscopy.[16] Read the article, and answer the following questions.

(a) Define *time domain* and *frequency domain*.

(b) Write the equations for the Fourier integral and its transformation and define each of the terms in the equations.

(c) The paper shows the time-domain signals for a 32-cycle cosine wave, a 21-cycle cosine wave, and a 10-cycle cosine wave as well as the Fourier transforms of these signals. How does the shape of the frequency-domain signal change as the number of cycles in the original waveform changes?

(d) The author describes the phenomenon of *damping*. What effect does damping have on the original cosine waves? What effect does it have on the resulting Fourier transformations?

(e) What is a resolution function?

(f) What is the process of convolution?

(g) Discuss how the choice of the resolution function can affect the appearance of a spectrum.

(h) Convolution may be used to decrease the amount of noise in a noisy spectrum. Consider the following plots of time-domain and frequency-domain signals. Label the axes of the five plots. For example, (b) should be labeled as amplitude versus time. Characterize each plot as either a time-domain or a frequency-domain signal.

(i) Describe the mathematical relationships among the plots. For example, how could you arrive at (a) from (d) and (e)?

(j) Discuss the practical importance of being able to reduce noise in spectroscopic signals.

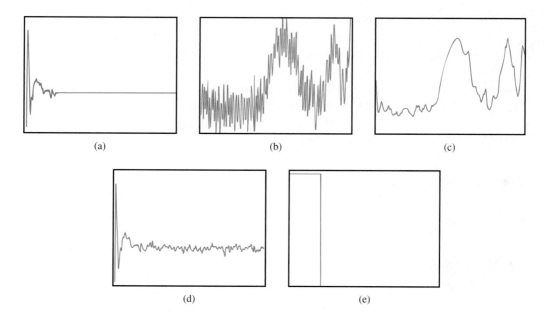

(a) (b) (c)

(d) (e)

[16]G. Horlick, *Anal. Chem.*, **1971**, *43*(8), 61A–66A, **DOI**: 10.1021/ac60303a029.

CHAPTER **26**

Molecular Absorption Spectrometry

© Thomas A. Heinz

Glassmaking is among the oldest technologies, dating from the Neolithic period nearly 10,000 years ago. Ordinary glass is transparent because valence electrons in the silicate structure do not receive sufficient energy from visible light to be excited from their ground states in the valence band of the silicate structure to the conduction band. Beginning with the Egyptians in the second millennium B.C.E., glassmakers learned to add various compounds to glasses to produce colored glass. These additives often contain transition metals to provide accessible energy levels so that absorption of light occurs, and the resulting glass is colored. Colored glass is used widely in art and architecture as, for example, in the stained glass window shown here. Optical spectroscopy is used to characterize colored glasses by recording their absorption spectra. This information is used in several different fields. For example, in art history absorption spectra are used to characterize, identify, and trace the origin and development of works of art, in archeology spectra are used to explore the origins of humankind, and in forensics they are used to correlate evidence in crime investigations.

The absorption of ultraviolet, visible, and infrared radiation is widely used to identify and determine many inorganic, organic, and biochemical species.[1] Ultraviolet and visible molecular absorption spectroscopy is used primarily for quantitative analysis and is probably applied more extensively in chemical and clinical laboratories than any other single technique. Infrared absorption spectroscopy is a very powerful tool for determining the identity and structure of both inorganic and organic compounds. In addition, it now plays an important role in quantitative analysis, particularly in the area of environmental pollution.

26A ULTRAVIOLET AND VISIBLE MOLECULAR ABSORPTION SPECTROSCOPY

Several types of molecular species absorb ultraviolet and visible radiation. Molecular absorption by these species can be used for qualitative and quantitative analyses. UV-visible absorption is also used to monitor titrations and to study the composition of

---

[1]For more detailed treatment of absorption spectroscopy, see E. J. Meehan, in *Treatise on Analytical Chemistry,* 2nd ed., P. J. Elving, E. J. Meehan, and I. M. Kolthoff, eds., Part I, Vol. 7, Ch. 2, New York: Wiley, 1981; C. Burgess and A. Knowles, eds., *Techniques in Visible and Ultraviolet Spectrometry,* Vol. 1, New York: Chapman and Hall, 1981; J. D. Ingle, Jr., and S. R. Crouch, *Spectrochemical Analysis,* Chs. 12–14, Englewood Cliffs, NJ: Prentice-Hall, 1988; D. A. Skoog, F. J. Holler, and S. R. Crouch, *Principles of Instrumental Analysis,* 6th ed., Chs. 13, 14, 16, 17, Belmont, CA: Brooks/Cole, 2007.

complex ions. The use of absorption spectrometry to follow the kinetics of chemical reactions for quantitative purposes is described in Chapter 30.

26A-1 Absorbing Species

As noted in Section 24C-2, absorption of ultraviolet and visible radiation by molecules generally occurs in one or more electronic absorption bands, each of which is made up of many closely packed but discrete lines. Each line arises from the transition of an electron from the ground state to one of the many vibrational and rotational energy states associated with each excited electronic energy state. Because there are so many of these vibrational and rotational states and because their energies differ only slightly, the number of lines contained in the typical band is quite large and their separation from one another is very small.

As we saw previously in Figure 24-14a, the visible absorption spectrum for 1,2,3,4-tetrazine vapor shows the fine structure that is due to the numerous rotational and vibrational levels associated with the excited electronic states of this aromatic molecule. In the gaseous state, the individual tetrazine molecules are sufficiently separated from one another to vibrate and rotate freely, and many individual absorption lines appear as a result of the large number of vibrational and rotational energy states. As a pure liquid or in solution, however, the tetrazine molecules have little freedom to rotate, so lines due to differences in rotational energy levels disappear. Furthermore, when solvent molecules surround the tetrazine molecules, energies of the various vibrational levels are modified in a nonuniform way, and the energy of a given state in a sample of solute molecules appears as a single broad peak. This effect is more pronounced in polar solvents, such as water, than in nonpolar hydrocarbon media. This solvent effect is illustrated in Figures 24-14b and 24-14c.

Absorption by Organic Compounds

Absorption of radiation by organic molecules in the wavelength region between 180 and 780 nm results from interactions between photons and electrons that either participate directly in bond formation (and are thus associated with more than one atom) or that are localized about such atoms as oxygen, sulfur, nitrogen, and the halogens.

The wavelength of absorption of an organic molecule depends on how tightly its electrons are bound. The shared electrons in carbon-carbon or carbon-hydrogen single bonds are so firmly held that their excitation requires energies corresponding to wavelengths in the vacuum ultraviolet region below 180 nm. Single-bond spectra have not been widely exploited for analytical purposes because of the experimental difficulties of working in this region. These difficulties occur because both quartz and atmospheric components absorb in this region, which requires that evacuated spectrophotometers with lithium fluoride optics be used.

Electrons in double and triple bonds of organic molecules are not as strongly held and are therefore more easily excited by electromagnetic radiation. Thus, species with unsaturated bonds generally exhibit useful absorption bands. Unsaturated organic functional groups that absorb in the ultraviolet or visible regions are known as **chromophores**. Table 26-1 lists common chromophores and the approximate wavelengths at which they absorb. The wavelength and peak intensity data are only rough guides since both are influenced by solvent effects as well as structural details of the molecule. In addition, conjugation between two or more chromophores

❮ A band consists of a large number of closely spaced vibrational and rotational lines. The energies associated with these lines differ little from one another.

Chromophores are unsaturated organic functional groups that absorb in the ultraviolet or visible region.

Figure 26-1   Absorption spectra for typical organic compounds.

TABLE 26-1

Absorption Characteristics of Some Common Organic Chromophores

Chromophore	Example	Solvent	λ_{max}, nm	ε_{max}
Alkene	$C_6H_{13}CH{=}CH_2$	*n*-Heptane	177	13,000
Conjugated alkene	$CH_2{=}CHCH{=}CH_2$	*n*-Heptane	217	21,000
Alkyne	$C_5H_{11}C{\equiv}C{-}CH_3$	*n*-Heptane	178	10,000
			196	2,000
			225	160
Carbonyl	CH_3CCH_3 (C=O)	*n*-Hexane	186	1,000
			280	16
	CH_3CH (C=O)	*n*-Hexane	180	Large
			293	12
Carboxyl	CH_3COH (C=O)	Ethanol	204	41
Amido	CH_3CNH_2 (C=O)	Water	214	60
Azo	$CH_3N{=}NCH_3$	Ethanol	339	5
Nitro	CH_3NO_2	Isooctane	280	22
Nitroso	C_4H_9NO	Ethyl ether	300	100
			665	20
Nitrate	$C_2H_5ONO_2$	Dioxane	270	12
Aromatic	Benzene	*n*-Hexane	204	7,900
			256	200

tends to cause shifts in absorption maxima to longer wavelengths. Finally, it is often difficult to determine precisely an absorption maximum because vibrational effects broaden absorption bands in the ultraviolet and visible regions. Typical spectra for organic compounds are shown in **Figure 26-1**.

Saturated organic compounds containing such heteroatoms as oxygen, nitrogen, sulfur, or halogens have nonbonding electrons that can be excited by radiation in the 170- to 250-nm range. **Table 26-2** lists a few examples of such compounds. Some of these compounds, such as alcohols and ethers, are common solvents. Their absorption in this region prevents measuring absorption of analytes dissolved in these solvents at wavelengths shorter than 180 to 200 nm. Occasionally, absorption in this region is used for determining halogen and sulfur-bearing compounds.

Absorption by Inorganic Species

In general, the ions and complexes of elements in the first two transition series absorb broad bands of visible radiation in at least one of their oxidation states. As a result, these compounds are colored (see, for example, **Figure 26-2**). Absorption occurs when electrons make transitions between filled and unfilled *d*-orbitals with energies that depend on the ligands bonded to the metal ions. The energy differences between these *d*-orbitals (and thus the position of the corresponding absorption maxima) depend on the position of the element in the periodic table, its oxidation state, and the nature of the ligand bonded to it.

Absorption spectra of ions of the lanthanide and actinide series differ substantially from those shown in Figure 26-2. The electrons responsible for absorption by these elements (4*f* and 5*f*, respectively) are shielded from external influences by electrons

TABLE 26-2

Absorption by Organic Compounds Containing Unsaturated Heteroatoms

Compound	λ_{max}, nm	ε_{max}
CH_3OH	167	1480
$(CH_3)_2O$	184	2520
CH_3Cl	173	200
CH_3I	258	365
$(CH_3)_2S$	229	140
$(CH_3)NH_2$	215	600
$(CH_3)_3N$	227	900

that occupy orbitals with larger principal quantum numbers. As a result, the bands tend to be narrow and relatively unaffected by the species bonded by the outer electrons, as shown in **Figure 26-3**.

Charge-Transfer Absorption

Charge-transfer absorption is particularly important for quantitative analysis because molar absorptivities are unusually large ($\varepsilon > 10,000$ L mol^{-1} cm^{-1}), which leads to high sensitivity. Many inorganic and organic complexes exhibit this type of absorption and are therefore called charge-transfer complexes.

A **charge-transfer complex** consists of an electron-donor group bonded to an electron acceptor. When this product absorbs radiation, an electron from the donor is transferred to an orbital that is largely associated with the acceptor. The excited state is thus the product of a kind of internal oxidation/reduction process. This behavior differs from that of an organic chromophore in which the excited electron is in a molecular orbital that is shared by two or more atoms.

Familiar examples of charge-transfer complexes include the phenolic complex of iron(III), the 1,10-phenanthroline complex of iron(II), the iodide complex of molecular iodine, and the ferro/ferricyanide complex responsible for the color of Prussian blue. The red color of the iron(III)/thiocyanate complex is yet another example of charge-transfer absorption. Absorption of a photon results in the transfer of an electron from the thiocyanate ion to an orbital that is largely associated with the iron(III) ion. The product is an excited species involving predominantly iron(II) and the thiocyanate radical SCN. As with other types of electronic excitation, the electron in this complex normally returns to its original state after a brief period. Occasionally, however, an excited complex may dissociate and produce photochemical oxidation/reduction products. Three spectra of charge-transfer complexes are shown in **Figure 26-4**.

In most charge-transfer complexes containing a metal ion, the metal serves as the electron acceptor. Exceptions are the 1,10-phenanthroline complexes of iron(II) (see Section 38N-2) and copper(I), where the ligand is the acceptor and the metal ion the donor. A few additional examples of this type of complex are known.

26A-2 Qualitative Applications of Ultraviolet/ Visible Spectroscopy

Spectrophotometric measurements with ultraviolet radiation are useful for detecting chromophoric groups, such as those shown in Table 26-1.[2] Because large parts of even the most complex organic molecules are transparent to radiation longer than 180 nm, the appearance of one or more absorption bands in the region from 200 to 400 nm is clear indication of the presence of unsaturated groups or of atoms such as sulfur or halogens. Often, you can get an idea as to the identity of the absorbing groups by comparing the spectrum of an analyte with those of simple molecules containing various chromophoric groups.[3] Usually, however, ultraviolet spectra do

---

[2] For a detailed discussion of ultraviolet absorption spectroscopy in the identification of organic functional groups, see R. M. Silverstein and F. X. Webster, *Spectrometric Identification of Organic Compounds,* 6th ed., Ch. 7, New York: Wiley, 1998.

[3] H. H. Perkampus, *UV-VIS Atlas of Organic Compounds,* 2nd ed., Weinheim, Germany: Wiley-VCH, 1992. In addition, in the past, several organizations have published catalogs of spectra that may still be useful, including American Petroleum Institute, Ultraviolet Spectral Data, A.P.I. Research Project 44, Pittsburgh: Carnegie Institute of Technology; *Sadtler Handbook of Ultraviolet Spectra.* Philadelphia: Sadtler Research Laboratories, 1979; American Society for Testing Materials, Committee E-13, Philadelphia.

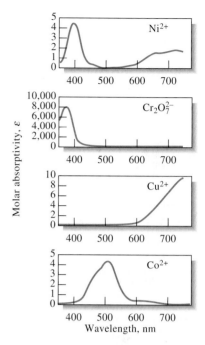

Figure 26-2 Absorption spectra of aqueous solutions of transition metal ions.

A **charge-transfer complex** is a strongly absorbing species that is made up of an electron-donating species that is bonded to an electron-accepting species.

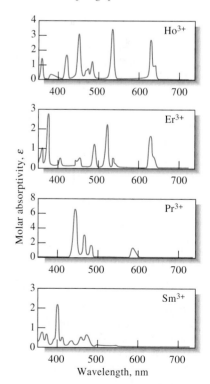

Figure 26-3 Absorption spectra of aqueous solutions of rare earth ions.

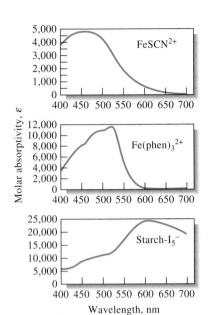

Figure 26-4 Absorption spectra of aqueous charge-transfer complexes.

Use small slit widths for qualitative studies to preserve maximum spectral detail.

Figure 26-5 Spectra for reduced cytochrome *c* obtained with four spectral bandwidths: (1) 1 nm, (2) 5 nm, (3) 10 nm, and (4) 20 nm. At bandwidths < 1 nm, the noise on the absorption bands becomes pronounced. (Courtesy of Varian Instrument Division, Palo Alto, CA.)

not have sufficient fine structure to permit an analyte to be identified unambiguously. Thus, ultraviolet qualitative data must be supplemented with other physical or chemical evidence such as infrared, nuclear magnetic resonance, and mass spectra as well as solubility and melting- and boiling-point information.

Solvents

Ultraviolet spectra for qualitative analysis are usually measured using dilute solutions of the analyte. For volatile compounds, however, gas-phase spectra are often more useful than liquid-phase or solution spectra (for example, compare Figure 24-14a and 24-14b). Gas-phase spectra can often be obtained by allowing a drop or two of the pure liquid to evaporate and equilibrate with the atmosphere in a stoppered cuvette.

A solvent for ultraviolet/visible spectroscopy must be transparent in the region of the spectrum where the solute absorbs. The analyte must be sufficiently soluble in the solvent to give a well-defined analyte. In addition, we must consider possible interactions of the solvent with the absorbing species. For example, polar solvents, such as water, alcohols, esters, and ketones, tend to obliterate vibrational fine structure and should thus be avoided to preserve spectral detail. Spectra in nonpolar solvents, such as cyclohexane, often more closely approach gas-phase spectra (compare, for example, the three spectra in Figure 24-14). In addition, solvent polarity often influences the position of absorption maxima. For qualitative analysis, analyte spectra should thus be compared to spectra of known compounds taken in the same solvent.

Table 26-3 lists common solvents for studies in the ultraviolet and visible regions and their approximate lower wavelength limits. These limits strongly depend on the purity of the solvent. For example, ethanol and the hydrocarbon solvents are frequently contaminated with benzene, which absorbs below 280 nm.[4]

The Effect of Slit Width

The effect of variation in slit width, and hence effective bandwidth, is illustrated by the spectra in **Figure 26-5**. The four traces show that peak heights and peak separation are distorted at wider bandwidths. To avoid this type of distortion, spectra for qualitative applications should be measured with the smallest slit widths that provide adequate signal-to-noise ratios.

TABLE 26-3

Solvents for the Ultraviolet and Visible Regions			
Solvent	**Lower Wavelength Limit, nm**	**Solvent**	**Lower Wavelength Limit, nm**
Water	180	Carbon tetrachloride	260
Ethanol	220	Diethyl ether	210
Hexane	200	Acetone	330
Cyclohexane	200	Dioxane	320
		Cellosolve	320

[4]Most major suppliers of reagent chemicals in the United States offer spectrochemical grades of solvents. Spectral-grade solvents have been treated so as to remove absorbing impurities and meet or exceed the requirements set forth in *Reagent Chemicals, American Chemical Society Specifications,* 10th ed., Washington, DC: American Chemical Society, 2005, available online or in hard bound forms.

The Effect of Stray Radiation at the Wavelength Extremes of a Spectrophotometer

Previously, we demonstrated that stray radiation may lead to instrumental deviations from Beer's law (see Section 24C-3). Another undesirable effect of this type of radiation is that it occasionally causes false peaks to appear when a spectrophotometer is being operated at its wavelength extremes. **Figure 26-6** shows an example of such behavior. Curve *B* is the true spectrum for a solution of cerium(IV) produced with a research-quality spectrophotometer responsive down to 200 nm or less. Curve *A* was obtained for the same solution with an inexpensive instrument operated with a tungsten source designed for work in the visible region only. The false peak at about 360 nm is directly attributable to stray radiation, which was not absorbed because it was made up of wavelengths longer than 400 nm. Under most circumstances, such stray radiation has a negligible effect because its power is only a tiny fraction of the total power of the beam exiting from the monochromator. At wavelength settings below 380 nm, however, radiation from the monochromator is greatly attenuated as a result of absorption by glass optical components and cuvettes. In addition, both the output of the source and the transducer sensitivity fall off dramatically below 380 nm. These factors combine to cause a substantial fraction of the measured absorbance to be due to the stray radiation of wavelengths to which cerium(IV) is transparent. A false absorption maximum results. This same effect is sometimes observed with ultraviolet/visible instruments when attempts are made to measure absorbances at wavelengths lower than about 190 nm.

Figure 26-6 Spectra of cerium(IV) obtained with a spectrophotometer having glass optics (*A*) and quartz optics (*B*). The apparent absorption band in *A* occurs when stray radiation is transmitted at long wavelengths.

26A-3 Quantitative Applications

Ultraviolet and visible molecular absorption spectroscopy is one of the most useful tools available for quantitative analysis. The important characteristics of spectrophotometric and photometric methods are

- *Wide applicability.* A majority of inorganic, organic, and biochemical species absorb ultraviolet or visible radiation and are thus amenable to direct quantitative determination. Many nonabsorbing species can also be determined after chemical conversion to absorbing derivatives. Of the determinations performed in clinical laboratories, a large majority is based on ultraviolet and visible absorption spectroscopy.
- *High sensitivity.* Typical detection limits for absorption spectroscopy range from 10^{-4} to 10^{-5} M. This range can often be extended to 10^{-6} or even 10^{-7} M with procedural modifications.
- *Moderate to high selectivity.* Often a wavelength can be found at which the analyte alone absorbs. Furthermore, where overlapping absorption bands do occur, corrections based on additional measurements at other wavelengths sometimes eliminate the need for a separation step. When separations are required, spectrophotometry often provides the means for detecting the separated species (see Section 33A-5).
- *Good accuracy.* The relative errors in concentration encountered with a typical spectrophotometric or photometric procedure lie in the range from 1% to 5%. Such errors can often be decreased to a few tenths of a percent with special precautions.
- *Ease and convenience.* Spectrophotometric and photometric measurements are easily and rapidly performed with modern instruments. In addition, the methods lend themselves to automation quite nicely.

Scope

The applications of molecular absorption measurements are not only numerous but also touch on every area in which quantitative information is sought. You can get an idea of the scope of spectrophotometry by consulting specialized monographs on the subject.[5]

Applications to Absorbing Species. Table 26-1 (page 724) lists many common organic chromophores. Spectrophotometric determination of organic compounds containing one or more of these groups is thus potentially feasible. Many such applications can be found in the literature.

A number of inorganic species also absorb. We have noted that many ions of the transition metals are colored in solution and can thus be determined by spectrophotometric measurement. In addition, a number of other species show characteristic absorption bands, including nitrite, nitrate, and chromate ions, the oxides of nitrogen, the elemental halogens, and ozone.

Applications to Nonabsorbing Species. Many nonabsorbing analytes can be determined photometrically by causing them to react with chromophoric reagents to produce products that absorb strongly in the ultraviolet and visible regions. The successful application of these color-forming reagents usually requires that their reaction with the analyte be forced to near completion unless methods such as kinetic methods (see Chapter 30) are used.

Typical inorganic reagents include the following: thiocyanate ion for iron, cobalt, and molybdenum; hydrogen peroxide for titanium, vanadium, and chromium; and iodide ion for bismuth, palladium, and tellurium. Organic chelating reagents that form stable colored complexes with cations are even more important. Common examples include diethyldithiocarbamate for the determination of copper, diphenylthiocarbazone for lead, 1,10-phenanthroline for iron (see color plate 15), and dimethylglyoxime for nickel; **Figure 26-7** shows the color-forming reaction for the first two of these reagents. The structure of the 1,10-phenanthroline complex of iron(II) is shown on page 503, and the reaction of nickel with dimethylglyoxime to form a red precipitate is described on page 294 (see also color plate 7). In the application of the dimethylglyoxime reaction to the photometric determination of nickel, an aqueous solution of the cation is extracted with a solution of the chelating agent in an immiscible organic liquid. The absorbance of the resulting bright red organic layer serves as a measure of the concentration of the metal.

Other reagents are available that react with organic functional groups to produce colors that are useful for quantitative analysis. For example, the red color of the 1:1 complexes that form between low-molecular-mass aliphatic alcohols and cerium(IV) can be used for the quantitative estimation of these alcohols.

[5]M. L. Bishop, E. P Fody, and L. E. Schoeff, *Clinical Chemistry: Techniques, Principles, Correlations,* Part I, Ch. 5, Part II, Philadelphia: Lippincott, Williams, and Wilkins, 2009; O. Thomas, *UV-Visible Spectrophotometry of Water and Wasterwater,* Vol. 27, *Techniques and Instrumentation in Analytical Chemistry,* Amsterdam: Elsevier, 2007; S. Görög, *Ultraviolet-Visible Spectrophotometry in Pharmaceutical Analysis,* Boca Rotan, FL: CRC Press, 1995; H. Onishi, *Photometric Determination of Traces of Metals,* 4th ed., Parts IIA and IIB, New York: Wiley, 1986, 1989; *Colorimetric Determination of Nonmetals,* 2nd ed., D. F. Boltz, ed., New York: Interscience, 1978.

(a)

(b)

Figure 26-7 Typical chelating reagents for absorption spectrophotometry. (a) Diethyldithiocarbamate. (b) Diphenylthiocarbazone.

Procedural Details

A first step in any photometric or spectrophotometric analysis is the development of conditions that yield a reproducible relationship (preferably linear) between absorbance and analyte concentration.

Wavelength Selection. In order to realize maximum sensitivity, spectrophotometric absorbance measurements are usually made at the wavelength of maximum absorption because the change in absorbance per unit of concentration is greatest at this point. In addition, the absorption curve is often flat at a maximum, leading to good adherence to Beer's law (see Figure 24-17) and less uncertainty from failure to reproduce precisely the wavelength setting of the instrument.

Variables That Influence Absorption. Common variables that influence the absorption spectrum of a substance include the nature of the solvent, the pH of the solution, the temperature, high electrolyte concentrations, and the presence of interfering substances. The effects of these variables must be known and conditions for the determination chosen such that the absorbance will not be materially affected by small, uncontrolled variations.

The Relationship between Absorbance and Concentration. The calibration standards for a photometric or a spectrophotometric method should approximate as closely as possible the overall composition of the actual samples and should encompass a reasonable range of analyte concentrations. A calibration curve of absorbance versus the concentrations of several standards is usually obtained to evaluate the relationship. It is seldom, if ever, safe to assume that Beer's law holds and to use only a single standard to determine the molar absorptivity. Unless there is no other choice, it is never a good idea to base the results of a determination solely on a literature value for the molar absorptivity. In cases where matrix effects are a problem, the standard addition method may improve results by providing compensation for some of these effects.

The Standard Addition Method. Ideally, the composition of calibration standards should approximate the composition of the samples to be analyzed. This is true not only for the analyte concentration but for the concentrations of the other species in the sample matrix. Approximating the sample composition should minimize the effects of various components of the sample on the measured absorbance. For example, the absorbance of many colored complexes of metal ions is decreased in the presence of sulfate and phosphate ions because of the tendency of these anions to

Molecular model of diphenylthiocarbazone.

❮ Absorption spectra are affected by such variables as temperature, pH, electrolyte concentration, and the presence of interferences.

form colorless complexes with metal ions. As a result, the color formation reaction is often less complete, and the sample absorbance is lowered. The matrix effect of sulfate and phosphate can often be counteracted by introducing into the standards amounts of the two species that approximate the amounts found in the samples. Unfortunately, when complex materials such as soils, minerals, and plant ash are being analyzed, preparing standards that match the samples is often impossible or extremely difficult. When this is the case, the standard addition method can be helpful in counteracting matrix effects.

The standard addition method can take several forms as discussed in Section 8D-3; the single-point method was described in Example 8-8.[6] The multiple-additions method is often chosen for photometric or spectrophotometric analyses, and this method is described here. In the multiple additions technique, several increments of a standard solution are added to sample aliquots of the same size. Each solution is then diluted to a fixed volume before measuring its absorbance. When the amount of sample is limited, standard additions can be carried out by successive addition of increments of the standard to a single measured aliquot of the unknown. The measurements are made on the original solution and after each addition of standard analyte.

Assume that several identical aliquots V_x of the unknown solution with a concentration c_x are transferred to volumetric flasks having a volume V_t. To each of these flasks is added a variable volume V_s mL of a standard solution of the analyte having a known concentration c_s. The color development reagents are then added, and each solution is diluted to volume. If the chemical system follows Beer's law, the absorbance of the solutions is described by

$$A_s = \frac{\varepsilon b V_s c_s}{V_t} + \frac{\varepsilon b V_x c_x}{V_t}$$

$$= k V_s c_s + k V_x c_x \tag{26-1}$$

where k is a constant equal to $\varepsilon b / V_t$. A plot of A_s as a function of V_s should yield a straight line of the form

$$A_s = m V_s + b$$

where the slope m and the intercept b are given by

$$m = k c_s$$

and

$$b = k V_x c_x$$

Least-squares analysis (see Section 8D-2) of the data can be used to determine m and b. The unknown concentration c_x can then be calculated from the ratio of these two quantities and the known values of V_x and V_s. Thus,

$$\frac{m}{b} = \frac{k c_s}{k V_x c_x}$$

which rearranges to

$$c_x = \frac{b c_s}{m V_x} \tag{26-2}$$

If we assume that the uncertainties in c_s, V_s, and V_t are negligible with respect to those in m and b, the standard deviation in c_x can be estimated. It follows then

[6]See M. Bader, *J. Chem. Educ.*, **1980**, *57*, 703, **DOI:** 10.1021/ed057p703.

that the relative variance of the result $(s_c/c_x)^2$ is the sum of the relative variances of m and b, that is,

$$\left(\frac{s_c}{c_x}\right)^2 = \left(\frac{s_m}{m}\right)^2 + \left(\frac{s_b}{b}\right)^2$$

where s_m and s_b are the standard deviations of the slope and intercept, respectively. By taking the square root of this equation, we can solve for the standard deviation in concentration, s_c:

$$s_c = c_x\sqrt{\left(\frac{s_m}{m}\right)^2 + \left(\frac{s_b}{b}\right)^2} \qquad (26\text{-}3)$$

EXAMPLE 26-1

Ten-millimeter aliquots of a natural water sample were pipetted into 50.00 mL volumetric flasks. Exactly 0.00, 5.00, 10.00, 15.00, and 20.00 mL of a standard solution containing 11.1 ppm of Fe^{3+} were added to each, followed by an excess of thiocyanate ion to give the red complex $Fe(SCN)^{2+}$. After dilution to volume, absorbances for the five solutions, measured with a photometer equipped with a green filter, were found to be 0.240, 0.437, 0.621, 0.809, and 1.009, respectively (0.982-cm cells). (a) What was the concentration of Fe^{3+} in the water sample? (b) Calculate the standard deviation of the slope, the intercept, and the concentration of Fe.

Solution

(a) In this problem, $c_s = 11.1$ ppm, $V_x = 10.00$ mL, and $V_t = 50.00$ mL. A plot of the data, shown in **Figure 26-8**, demonstrates that Beer's law is obeyed. To obtain the equation for the line in Figure 26-8, the procedure illustrated in Example 8-4 (pages 174–175) is followed. The result is $m = 0.03820$, and $b = 0.2412$. Thus,

$$A_s = 0.03820V_s + 0.2412$$

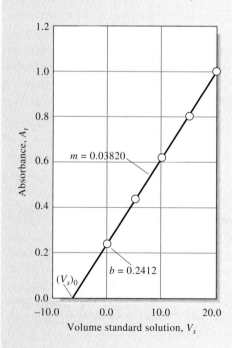

Figure 26-8 Plot of data for standard addition determination of Fe^{3+} as the $Fe(SCN)^{2+}$ complex.

(continued)

Substituting into Equation 26-2 gives

$$c_x = \frac{(0.2412)(11.1 \text{ ppm } Fe^{3+})}{(0.03820 \text{ mL}^{-1})(10.00 \text{ mL})} = 7.01 \text{ ppm } Fe^{3+}$$

(b) Equations 8-16 and 8-17 give the standard deviation of the slope and the intercept. That is, $s_m = 3.07 \times 10^{-4}$, and $s_b = 3.76 \times 10^{-3}$.
Substituting into Equation 26-3 gives

$$s_c = 7.01 \text{ ppm } Fe^{3+} \sqrt{\left(\frac{3.07 \times 10^{-4}}{0.03820}\right)^2 + \left(\frac{3.76 \times 10^{-3}}{0.2412}\right)^2}$$

$$= 0.12 \text{ ppm } Fe^{3+}$$

In the interest of saving time or sample, it is possible to perform a standard addition analysis using only two increments of sample. In that case, a single addition of V_s mL of standard is added to one of the two samples, and we can write

$$A_1 = \varepsilon b c_x$$

$$A_2 = \frac{\varepsilon b V_x c_x}{V_t} + \frac{\varepsilon b V_s c_s}{V_t}$$

where A_1 and A_2 are absorbances of the sample and the sample plus standard, respectively, and V_t is $V_x + V_s$. If we solve the first equation for εb, substitute into the second equation, and solve for c_x, we find

$$c_x = \frac{A_1 c_s V_s}{A_2 V_t - A_1 V_x} \tag{26-4}$$

Single-point standard additions methods are inherently more risky than multiple-point methods. There is no check on linearity with single-point methods, and results depend strongly on the reliability of one measurement.

 Spreadsheet Summary   In Chapter 12 of *Applications of Microsoft® Excel in Analytical Chemistry*, 2nd ed., we investigate the multiple standard additions method for determining solution concentration. Conventional and weighted linear regression methods are also used to determine concentrations and standard deviations.

EXAMPLE 26-2

The single-point standard addition method was used in the determination of phosphate by the molybdenum blue method. A 2.00-mL urine specimen was treated with molybdenum blue reagents to produce a species absorbing at 820 nm, after which the sample was diluted to 100 mL. A 25.00-mL aliquot of this solution gave an absorbance of 0.428 (solution 1). Addition of 1.00 mL of a solution containing 0.0500 mg of phosphate to a second 25.0-mL aliquot gave an absorbance of 0.517 (solution 2). Use these data to calculate the mass of phosphate in milligrams per millimeter of the specimen.

Solution

We substitute into Equation 26-4 and obtain

$$c_x = \frac{A_1 c_s V_s}{A_2 V_t - A_1 V_x} = \frac{(0.428)(0.0500 \text{ mg PO}_4^{3-}/\text{mL})(1.00 \text{ mL})}{(0.517)(26.00 \text{ mL}) - (0.428)(25.00 \text{ mL})}$$

$$= 0.0780 \text{ mg PO}_4^{3-}/\text{mL}$$

This is the concentration of the diluted sample. To obtain the concentration of the original urine sample, we need to multiply by 100.00/2.00. Thus,

$$\text{concentration of phosphate} = 0.0780 \frac{\text{mg}}{\text{mL}} \times \frac{100.00 \text{ mL}}{2.00 \text{ mL}}$$

$$= 0.390 \text{ mg/mL}$$

Analysis of Mixtures. The total absorbance of a solution at any given wavelength is equal to the sum of the absorbances of the individual components in the solution (Equation 24-14). This relationship makes it possible in principle to determine the concentrations of the individual components of a mixture even if their spectra overlap completely. For example, **Figure 26-9** shows the spectrum of a solution containing a mixture of species M and species N as well as absorption spectra for the individual components. It is apparent that there is no wavelength where the absorbance is due to just one of these components. To analyze the mixture, molar absorptivities for M and N are first determined at wavelengths λ_1 and λ_2. The concentrations of the standard solutions of M and N should be such that Beer's law is obeyed over an absorbance range that encompasses the absorbance of the sample. As shown in Figure 26-9, wavelengths should be selected so that the molar absorptivities of the two components differ significantly. Thus, at λ_1, the molar absorptivity of component M is much larger than that for component N. The reverse is true for λ_2. To complete the analysis, the absorbance of the mixture is determined at the same two wavelengths. From the known molar absorptivities and path length, the following equations hold

$$A_1 = \varepsilon_{M_1} b c_M + \varepsilon_{N_1} b c_N \qquad (26\text{-}5)$$

$$A_2 = \varepsilon_{M_2} b c_M + \varepsilon_{N_2} b c_N \qquad (26\text{-}6)$$

where the subscript 1 indicates measurement at λ_1, and the subscript 2 indicates measurement a λ_2. With the known values of ε and b, Equations 26-5 and 26-6 are two equations in two unknowns (c_M and c_N) and can be solved as demonstrated in Example 26-3.

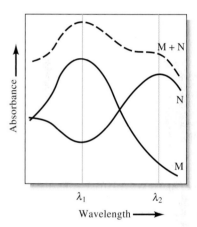

Figure 26-9 Absorption spectrum of a two-component mixture (M + N), with spectra of the individual components M and N.

EXAMPLE 26-3

Palladium(II) and gold(III) can be determined simultaneously by reaction with methiomeprazine ($C_{19}H_{24}N_2S_2$). The absorption maximum for the Pd complex occurs at 480 nm, while that for the Au complex is at 635 nm. Molar absorptivity data at these wavelengths are as follows:

(continued)

	ε, L mol^{-1} cm^{-1}	
	480 nm	**635 nm**
Pd complex	3.55×10^3	5.64×10^2
Au complex	2.96×10^3	1.45×10^4

A 25.0-mL sample was treated with an excess of methiomeprazine and subsequently diluted to 50.0 mL. Calculate the molar concentrations of Pd(II), c_{Pd}, and Au(III), c_{Au}, in the sample if the diluted solution had an absorbance of 0.533 at 480 nm and 0.590 at 635 nm when measured in a 1.00-cm cell.

Solution

At 480 nm from Equation 26-5,

$$A_{480} = \varepsilon_{Pd(480)}bc_{Pd} + \varepsilon_{Au(480)}bc_{Au}$$

$$0.533 = (3.55 \times 10^3 \text{ M}^{-1} \text{ cm}^{-1})(1.00 \text{ cm})c_{Pd}$$
$$+ (2.96 \times 10^3 \text{ M}^{-1}\text{cm}^{-1})(1.00 \text{ cm})c_{Au}$$

or

$$c_{Pd} = \frac{0.533 - 2.96 \times 10^3 \text{ M}^{-1}c_{Au}}{3.55 \times 10^3 \text{ M}^{-1}}$$

At 635 nm from Equation 26-6,

$$A_{635} = \varepsilon_{Pd(635)}bc_{Pd} + \varepsilon_{Au(635)}bc_{Au}$$

$$0.590 = (5.64 \times 10^2 \text{ M}^{-1} \text{ cm}^{-1})(1.00 \text{ cm})c_{Pd}$$
$$+ (1.45 \times 10^4 \text{ M}^{-1} \text{ cm}^{-1})(1.00 \text{ cm})c_{Au}$$

Substitution for c_{Pd} in this expression gives

$$0.590 = \frac{(5.64 \times 10^2 \text{ M}^{-1})(0.533 - 2.96 \times 10^3 \text{ M}^{-1}c_{Au})}{3.55 \times 10^3 \text{ M}^{-1}}$$
$$+ (1.45 \times 10^4 \text{ M}^{-1})c_{Au}$$

$$= 0.0847 - (4.70 \times 10^2 \text{ M}^{-1})c_{Au} + (1.45 \times 10^4 \text{ M}^{-1})c_{Au}$$

$$c_{Au} = \frac{(0.590 - 0.0847)}{(1.45 \times 10^4 \text{ M}^{-1} - 4.70 \times 10^2 \text{ M}^{-1})} = 3.60 \times 10^{-5} \text{ M}$$

and

$$c_{Pd} = \frac{0.533 - (2.96 \times 10^3 \text{ M}^{-1})(3.60 \times 10^{-5} \text{ M})}{3.55 \times 10^3 \text{ M}^{-1}} = 1.20 \times 10^{-4} \text{ M}$$

Since solutions were diluted twofold, the concentrations of Pd(II) and Au(III) in the original sample were 7.20×10^{-5} M and 2.40×10^{-4} M, respectively.

Mixtures containing more than two absorbing species can be analyzed, in principle at least, if one additional absorbance measurement is made for each extra component. The uncertainties in the resulting data become greater, however, as the number of measurements increases. Some computerized spectrophotometers are capable of minimizing these uncertainties by overdetermining the system. These instruments use many more data points than unknowns and effectively match the entire spectrum of the unknown as closely as possible by calculating synthetic spectra for various concentrations of the components. The calculated spectra are then added, and the sum is compared with the spectrum of the analyte solution until a close match is found. The spectra for standard solutions of each component of the mixture are acquired and stored in computer memory prior to making measurements on the analyte mixture.

 Spreadsheet Summary In Chapter 12 of *Applications of Microsoft® Excel in Analytical Chemistry*, 2nd ed., we use spreadsheet methods to determine concentrations of mixtures of analytes. Solutions to sets of simultaneous equations are evaluated using iterative techniques, the method of determinants, and matrix manipulations.

The Effect of Instrumental Uncertainties[7]

The accuracy and precision of spectrophotometric analyses are often limited by the indeterminate error, or noise, associated with the instrument. As pointed out in Chapter 25, a spectrophotometric absorbance measurement entails three steps: setting or measuring 0% T, setting or measuring 100% T, and measuring the % T of the sample. The random errors associated with each of these steps combine to give a net random error for the final value obtained for T. The relationship between the noise encountered in the measurement of T and the resulting *concentration uncertainty* can be derived by writing Beer's law in the form

$$c = -\frac{1}{\varepsilon b} \log T = \frac{-0.434}{\varepsilon b} \ln T$$

Taking the partial derivative of this equation while holding εb constant leads to the expression

$$\partial c = \frac{-0.434}{\varepsilon b T} \partial T$$

where ∂c can be interpreted as the uncertainty in c that results from the noise (or uncertainty) in T. Dividing this equation by the previous one gives

$$\frac{\partial c}{c} = \frac{0.434}{\log T}\left(\frac{\partial T}{T}\right) \tag{26-7}$$

where $\partial T/T$ is the relative random error in T attributable to the noise in the three measurement steps, and $\partial c/c$ is the resulting relative random concentration error.

The best and most useful measure of the random error ∂T is the standard deviation σ_T, which may be measured conveniently for a given instrument by making 20 or more replicate transmittance measurements of an absorbing solution. Substituting σ_T and σ_c for the corresponding differential quantities in Equation 26-7 leads to

$$\frac{\sigma_c}{c} = \frac{0.434}{\log T}\left(\frac{\sigma_T}{T}\right) \tag{26-8}$$

where σ_T/T is the relative standard deviation in transmittance and σ_c/c is the resulting relative standard deviation in concentration.

Equation 26-8 shows that the uncertainty in a photometric concentration measurement varies in a complex way with the magnitude of the transmittance. The situation is even more complicated than suggested by the equation, however, because

In the context of this discussion, **noise** refers to random variations in the instrument output due to electrical fluctuations and also variables such as the temperature of the solution, the position of the cell in the light beam, and the output of the source. With older instruments, the way the operator reads the meter can also result in a random variation.

[7]For further reading, see J. D. Ingle, Jr., and S. R. Crouch, *Spectrochemical Analysis*, Ch. 5, Englewood Cliffs, NJ: Prentice Hall, 1988; J. Galbán, S. de Marcos, I. Sanz, C. Ubide, and J.Zuriarrain. *Anal. Chem.*, **2007**, *79*, 4763, **DOI:** 10.1021/ac071933h.

TABLE 26-4

Categories of Instrumental Indeterminate Errors in Transmittance Measurements

Category	Sources	Effect of T on Relative Standard Deviation in Concentration	
$\sigma_T = k_1$	Readout resolution, thermal detector noise, dark current, and amplifier noise	$\dfrac{\sigma_c}{c} = \dfrac{0.434}{\log T}\left(\dfrac{k_1}{T}\right)$	(26-9)
$\sigma_T = k_2\sqrt{T^2 + T}$	Photon detector shot noise	$\dfrac{\sigma_c}{c} = \dfrac{0.434}{\log T} \times k_2\sqrt{1 + \dfrac{1}{T}}$	(26-10)
$\sigma_T = k_3 T$	Cell positioning uncertainty, fluctuations in source intensity	$\dfrac{\sigma_c}{c} = \dfrac{0.434}{\log T} \times k_3$	(26-11)

Note: σ_T is the standard deviation of the transmittance, σ_c/c is the relative standard deviation in concentration, T is transmittance, and k_1, k_2, and k_3 are constants for a given instrument.

> Uncertainties in spectrophotometric concentration measurements depend on the magnitude of the transmittance (absorbance) in a complex way. The uncertainties can be independent of T, proportional to $\sqrt{T^2 + T}$, or proportional to T.

the uncertainty σ_T is, under many circumstances, also dependent on T. In a detailed theoretical and experimental study, Rothman, Crouch, and Ingle[8] described several sources of instrumental random errors and showed the net effect of these errors on the precision of concentration measurements. The errors fall into three categories: those for which the magnitude of σ_T is (1) independent of T, (2) proportional to $\sqrt{T^2 + T}$, and (3) proportional to T. **Table 26-4** summarizes information about these sources of uncertainty. When the three relationships for σ_T in the first column are substituted into Equation 26-8, we obtain three equations for the relative standard deviation in concentration σ_c/c. These derived equations are shown in the third column of Table 26-4.

Concentration Errors When $\sigma_T = k_1$. For many photometers and spectrophotometers, the standard deviation in the measurement of T is constant and independent of the magnitude of T. We often see this type of random error in direct-reading instruments with analog meter readouts, which have somewhat limited resolution. The size of a typical scale is such that a reading cannot be reproduced to better than a few tenths of a percent of the full-scale reading, and the magnitude of this uncertainty is the same from one end of the scale to the other. For typical inexpensive instruments, we find standard deviations of about 0.003 ($\sigma_T = \pm 0.003$).

EXAMPLE 26-4

A spectrophotometric analysis was performed with an instrument that exhibited an absolute standard deviation of ±0.003 throughout its transmittance range. Find the relative standard deviation in concentration if the analyte solution has an absorbance of (a) 1.000 and (b) 2.000.

Solution

(a) To convert absorbance to transmittance, we write

$$\log T = -A = -1.000$$

$$T = \text{antilog}(-1.000) = 0.100$$

---

[8]L. D. Rothman, S. R. Crouch, and J. D. Ingle, Jr., *Anal. Chem.*, **1975**, *47*, 1226, **DOI:** 10.1021/ac60358a029.

For this instrument, $\sigma_T = k_1 = \pm 0.003$ (see first entry in Table 26-4). Substituting this value and $T = 0.100$ into Equation 26-8 yields

$$\frac{\sigma_c}{c} = \frac{0.434}{\log 0.100}\left(\frac{\pm 0.003}{0.100}\right) = \pm 0.013 \quad (1.3\%)$$

(b) At $A = 2.000$, $T = $ antilog $(-2.000) = 0.010$

$$\frac{\sigma_c}{c} = \frac{0.434}{\log 0.010}\left(\frac{\pm 0.003}{0.010}\right) = \pm 0.065 \quad (6.5\%)$$

The data plotted as curve A in **Figure 26-10** were obtained from calculations similar to those in Example 26-4. Note that the relative standard deviation in concentration passes through a minimum at an absorbance of about 0.5 and rises rapidly when the absorbance is less than about 0.1 or greater than approximately 1.5.

Figure 26-11a is a plot of the relative standard deviation for experimentally determined concentrations as a function of absorbance. It was obtained with a spectrophotometer similar to the one shown in Figure 25-19 but with an old-fashioned analog panel meter rather than a digital readout. The striking similarity between this curve and curve A in Figure 26-10 indicates that the instrument studied is affected by an absolute indeterminate error in transmittance of about ±0.003 and that this error is independent of transmittance. The source of this uncertainty is probably the limited resolution of the manual transmittance scale. A digital readout with sufficient resolution, such as that shown in Figure 25-19, is less susceptible to this type of error.

Many infrared spectrophotometers also exhibit an indeterminate error that is independent of transmittance. The source of the error in these instruments lies in the thermal detector. Fluctuations in the output of this type of transducer are independent of the output; indeed, fluctuations are observed even in the absence of radiation. An experimental plot of data from an infrared spectrophotometer is similar in appearance to Figure 26-11a. The curve is displaced upward, however, because of the greater standard deviation characteristic of infrared measurements.

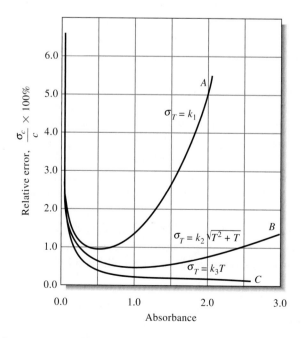

Figure 26-10   Error curves for various categories of instrumental uncertainties.

Figure 26-11 Experimental curves relating relative concentration uncertainties to absorbance for two spectrophotometers. Data obtained with (a) a Spectronic 20, a low-cost instrument (see Figure 25-19), and (b) a Cary 118, a research-quality instrument. (W. E. Harris and B. Kratochvil, An *Introduction to Chemical Analysis*, p. 384. Philadelphia: Saunders College Publishing, 1981. Reprinted by permission of the authors.)

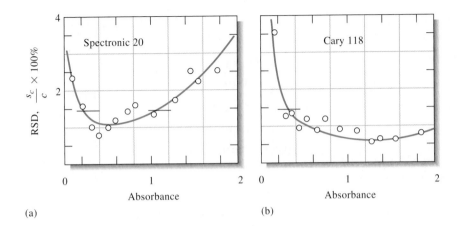

Concentration Errors When $\sigma_T = k_2\sqrt{T^2 + T}$. This type of random uncertainty is characteristic of the highest-quality spectrophotometers. It has its origin in the shot noise that causes the output of photomultipliers and phototubes to fluctuate randomly about a mean value. Equation 26-10 in Table 26-4 describes the effect of shot noise on the relative standard deviation of concentration measurements. A plot of this relationship appears as curve *B* in Figure 26-10. We calculated these data assuming that $k_2 = \pm 0.003$, a value that is typical for high-quality spectrophotometers.

Figure 26-11b shows an analogous plot of experimental data obtained with a high-quality research-type ultraviolet/visible spectrophotometer. Note that, in contrast to the less expensive instrument, absorbances of 2.0 or greater can be measured here without serious deterioration in the quality of the data.

Concentration Errors When $\sigma_T = k_3 T$. Substituting $\sigma_T = k_3 T$ into Equation 26-8 reveals that the relative standard deviation in concentration from this type of uncertainty is inversely proportional to the logarithm of the transmittance (Equation 26-11 in Table 26-4). Curve *C* in Figure 26-10, which is a plot of Equation 25-11, reveals that this type of uncertainty is important at low absorbances (high transmittances) but approaches zero at high absorbances.

At low absorbances, the precision obtained with high-quality double-beam instruments is often described by Equation 26-11. The source of this behavior is failure to position cells reproducibly with respect to the beam during replicate measurements. This position dependence is probably the result of small imperfections in the cell windows, which cause reflective losses and transparency to differ from one area of the window to another.

It is possible to evaluate Equation 26-11 by comparing the precision of absorbance measurements made in the usual way with measurements in which the cells are left undisturbed at all times with replicate solutions being introduced with a syringe. Experiments of this kind with a high-quality spectrophotometer yielded a value of 0.013 for k_3.[9] Curve *C* in Figure 26-10 was obtained by substituting this numerical value into Equation 26-11. Cell positioning errors affect all types of spectrophotometric measurements in which cells are repositioned between measurements.

Fluctuations in source intensity also yield standard deviations that are described by Equation 26-11. This type of behavior sometimes occurs in inexpensive single-beam instruments that have unstable power supplies and in infrared instruments.

[9]L. D. Rothman, S. R. Crouch, and J. D. Ingle, Jr., *Anal. Chem.,* **1975**, *47*, 1226,
DOI: 10.1021/ac60358a029.

Spreadsheet Summary   In Chapter 12 of *Applications of Microsoft®* *Excel in Analytical Chemistry*, 2nd ed., we explore errors in spectrophotometric measurements by simulating error curves such as those shown in Figures 26-10 and 26-11.

26A-4 Photometric and Spectrophotometric Titrations

Photometric and spectrophotometric measurements are useful for locating the equivalence points of titrations.[10] This application of absorption measurements requires that one or more of the reactants or products absorb radiation or that an absorbing indicator be added to the analyte solution.

Titration Curves

A photometric titration curve is a plot of absorbance (corrected for volume change) as a function of titrant volume. If conditions are chosen properly, the curve consists of two straight-line regions with different slopes, one occurring prior to the equivalence point of the titration and the other located well beyond the equivalence-point region. The end point is taken as the intersection of extrapolated linear portions of the two lines.

Figure 26-12 shows typical photometric titration curves. Figure 26-12a is the curve for the titration of a nonabsorbing species with an absorbing titrant that reacts with the titrant to form a nonabsorbing product. An example is the titration of thiosulfate ion with triiodide ion. The titration curve for the formation of an absorbing product from nonabsorbing reactants is shown in Figure 26-12b. An example is the titration of iodide ion with a standard solution of iodate ion to form triiodide. The remaining figures illustrate the curves obtained with various combinations of absorbing analytes, titrants, and products.

To obtain titration curves with linear portions that can be extrapolated, the absorbing system(s) must obey Beer's law. In addition, absorbances must be corrected for volume changes by multiplying the observed absorbance by $(V + v)/V$, where V is the original volume of the solution and v is the volume of added titrant. In some cases, adequate end points can be obtained even for systems in which Beer's law is not strictly obeyed. An abrupt change in the slope of the titration curve signals the location of the end-point volume.

Instrumentation

Photometric titrations are usually performed with a spectrophotometer or a photometer that has been modified so that the titration vessel is held stationary in the light path. After the instrument is set to a suitable wavelength or an appropriate filter is inserted, the 0%T adjustment is made in the usual way. With radiation passing through the analyte solution to the detector, the instrument is then adjusted to a convenient absorbance reading by varying the source intensity or the detector sensitivity. It is not usually necessary to measure the true absorbance since relative values are adequate for end-point detection. Titration data are then collected without changing the instrument settings. The power of the radiation source and the response of the detector must remain constant during a photometric titration. Cylindrical containers are often used in photometric titrations, and it is important to avoid moving the cell

---

[10] For further information, see J. B. Headridge, *Photometric Titrations,* New York: Pergamon Press, 1961.

Figure 26-12 Typical photometric titration curves. Molar absorptivities of the substance titrated, the product, and the titrant are ε_s, ε_p, and ε_t, respectively.

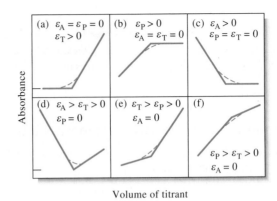

so that the path length remains constant. Both filter photometers and spectrophotometers have been used for photometric titrations.

Applications of Photometric Titrations

Photometric titrations often provide more accurate results than a direct photometric determination because the data from several measurements are used to determine the end point.

Furthermore, the presence of other absorbing species may not interfere since only a change in absorbance is being measured.

An advantage of end points determined from linear-segment photometric titration curves is that the experimental data are collected well away from the equivalence-point region where the absorbance changes gradually. Consequently, the equilibrium constant for the reaction need not be as large as that required for a sigmoidal titration curve that depends on observations near the equivalence point (for example, potentiometric or indicator end points). For the same reason, more dilute solutions may be titrated using photometric detection.

The photometric end point has been applied to many types of reactions. For example, most standard oxidizing agents have characteristic absorption spectra and thus produce photometrically detectable end points. Although standard acids or bases do not absorb, the introduction of acid/base indicators permits photometric neutralization titrations. The photometric end point has also been used to great advantage in titrations with EDTA and other complexing agents. **Figure 26-13** illustrates the application of this technique to the successive titration of bismuth(III) and copper(II). At 745 nm, the cations, the reagent, and the bismuth complex formed do not absorb but the copper complex does. Thus, during the first segment of the titration when the bismuth-EDTA complex is being formed ($K_f = 6.3 \times 10^{22}$), the solution exhibits no absorbance until essentially all the bismuth has been titrated. With the first formation of the copper complex ($K_f = 6.3 \times 10^{18}$), an increase in absorbance occurs. The increase continues until the copper equivalence point is reached. Further additions of titrant cause no additional absorbance change. Two well-defined end points result as shown in Figure 26-13.

The photometric end point has also been adapted to precipitation titrations. The suspended solid product causes a decrease in the radiant power of the light source by scattering from the particles of the precipitate. The equivalence point occurs when the precipitate stops forming, and the amount of light reaching the detector becomes constant. This type of end-point detection is called **turbidimetry** because the amount of light reaching the detector is a measure of the **turbidity** of the solution.

> Photometric titrations are often more accurate than direct photometric determinations.

Figure 26-13 Photometric titration curve at 745 nm for 100 mL of a solution that was 2.0×10^{-3} M in Bi^{3+} and Cu^{2+}. (Reprinted with permission from A. L. Underwood, *Anal. Chem.*, **1954**, 26, 1322, **DOI**: 10.1021/ac60092a017. Copyright 1954 by the American Chemical Society.)

Spreadsheet Summary   In Chapter 12 of *Applications of Microsoft® Excel in Analytical Chemistry*, 2nd ed., methods for treating data from spectrophotometric titrations are explored. We analyze titration data using least-squares procedures and use the resulting parameters to compute the concentration of the analyte.

26A-5 Spectrophotometric Studies of Complex Ions

Spectrophotometry is a valuable tool for determining the composition of complex ions in solution and for determining their formation constants. The power of the technique lies in the fact that quantitative absorption measurements can be performed without disturbing the equilibria under consideration. Although in many spectrophotometric studies of systems of complexes, a reactant or a product absorbs, nonabsorbing systems can also be investigated successfully. For example, the composition and formation constant for a complex of iron(II) and a nonabsorbing ligand may often be determined by measuring the absorbance decreases that occur when solutions of the absorbing iron(II) complex of 1,10-phenanthroline are mixed with various amounts of the nonabsorbing ligand. The success of this approach depends on the well-known values of the formation constant ($K_f = 2 \times 10^{21}$) and the composition of the 1,10-phenanthroline (3:1) complex of iron(II).

The three most common techniques used for complex-ion studies are (1) the method of continuous variations, (2) the mole-ratio method, and (3) the slope-ratio method. We illustrate these methods for metal ion-ligand complexes, but the principles apply to other types.

The Method of Continuous Variations

In the method of continuous variations, cation and ligand solutions with identical analytical concentrations are mixed in such a way that the total volume and the total moles of reactants in each mixture are constant but the mole ratio of reactants varies systematically (for example, 1:9, 8:2, 7:3, and so forth). The absorbance of each solution is then measured at a suitable wavelength and corrected for any absorbance the mixture might exhibit if no reaction had occurred. The corrected absorbance is plotted against the volume fraction of one reactant, that is, $V_M / (V_M + V_L)$, where V_M is the volume of the cation solution and V_L is the volume of the ligand solution. A typical continuous-variations plot is shown in **Figure 26-14**. A maximum (or minimum if the complex absorbs less than the reactants) occurs at a volume ratio V_M/V_L, corresponding to the combining ratio of metal ion and ligand in the complex. In Figure 26-14,

❮ The composition of a complex in solution can be determined without actually isolating the complex as a pure compound.

Figure 26-14 Continuous-variation plot for the 1:2 complex ML_2.

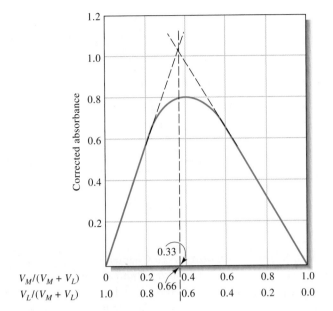

$V_M/(V_M + V_L)$ is 0.33, and $V_L/(V_M + V_L)$ is 0.66; thus, V_M/V_L is 0.33/0.66, suggesting that the complex has the formula ML_2.

The curvature of the experimental lines in Figure 26-14 is the result of incompleteness of the complex-formation reaction. A formation constant for the complex can be evaluated from measurements of the deviations from the theoretical straight lines, which represent the curve that would result if the reaction between the ligand and the metal proceeded to completion.

The Mole-Ratio Method

In the mole-ratio method, a series of solutions is prepared in which the analytical concentration of one reactant (usually the metal ion) is held constant while that of the other is varied. A plot of absorbance versus mole ratio of the reactants is then prepared. If the formation constant is reasonably favorable, two straight lines of different slopes that intersect at a mole ratio that corresponds to the combining ratio in the complex are obtained. Typical mole-ratio plots are shown in **Figure 26-15**. Notice that the ligand of the 1:2 complex absorbs at the wavelength selected so that the slope beyond the equivalence point is greater than zero. We deduce that the uncomplexed cation of the 1:1 complex absorbs because the initial point has an absorbance greater than zero.

Formation constants can be evaluated from the data in the curved portion of mole-ratio plots where the reaction is least complete.

Figure 26-15 Mole-ratio plots for a 1:1 and a 1:2 complex. The 1:2 complex is the more stable of the two complexes as indicated by closeness of the experimental curve to the extrapolated lines. The closer the curve is to the extrapolated lines, the larger the formation constant of the complex; the larger the deviation from the straight lines, the smaller the formation constant of the complex.

EXAMPLE 26-5

Derive equations to calculate the equilibrium concentrations of all the species in the 1:2 complex-formation reaction illustrated in Figure 26-15.

Derivation

Two mass-balance expressions based on the preparatory data can be written. Thus, for the reaction

$$M + 2L \rightleftharpoons ML_2$$

we can write

$$c_M = [M] + [ML_2]$$

$$c_L = [L] + 2[ML_2]$$

where c_M and c_L are the molar concentrations of M and L before reaction occurs. For 1-cm cells, the absorbance of the solution is

$$A = \varepsilon_M[M] + \varepsilon_L[L] + \varepsilon_{ML_2}(ML_2)$$

From the mole-ratio plot, we see that $\varepsilon_M = 0$. Values for ε_{ML} and ε_{ML_2} can be obtained from the two straight-line portions of the curve. With one or more measurements of A in the curved region of the plot, sufficient data are available to calculate the three equilibrium concentrations and thus the formation constant.

A mole-ratio plot may reveal the stepwise formation of two or more complexes as successive slope changes if the complexes have different molar absorptivities and their formation constants are sufficiently different from each other.

The Slope-Ratio Method

The slope-ratio approach is particularly useful for weak complexes but is applicable only to systems in which a single complex is formed. The method assumes (1) that the complex-formation reaction can be forced to completion by a large excess of either reactant, (2) that Beer's law is followed under these circumstances, and (3) that only the complex absorbs at the wavelength chosen.

Consider the reaction in which the complex M_xL_y is formed by the reaction of x moles of the cation M with y moles of a ligand L:

$$x\text{M} + y\text{L} \rightleftharpoons \text{M}_x\text{L}_y$$

Mass-balance expressions for this system are

$$c_M = [M] + x[M_xL_y]$$

$$c_L = [L] + y[M_xL_y]$$

where c_M and c_L are the molar analytical concentrations of the two reactants. We now assume that, at very high analytical concentrations of L, the equilibrium is shifted far to the right and $[M] \ll x[M_xL_y]$. Under this condition, the first mass-balance expression simplifies to

$$c_M = x[M_xL_y]$$

If the system obeys Beer's law,

$$A_1 = \varepsilon b[M_xL_y] = \varepsilon b c_M/x$$

where ε is the molar absorptivity of M_xL_y and b is the path length. A plot of absorbance as a function of c_M is linear when there is sufficient L present to justify the assumption that $[M] \ll x[M_xL_y]$. The slope of this plot is $\varepsilon b/x$.

When c_M is made very large, we assume that $[L] \ll y[M_xL_y]$, and the second mass-balance equation reduces to

$$c_L = y[M_xL_y]$$

and

$$A_2 = \varepsilon b[M_xL_y] = \varepsilon b c_L/y$$

Again, if our assumptions are valid, we find that a plot of A versus c_L is linear at high concentrations of M. The slope of this line is $\varepsilon b/y$.

The ratio of the slopes of the two straight lines gives the combining ratio between M and L:

$$\frac{\varepsilon b/x}{\varepsilon b/y} = \frac{y}{x}$$

 Spreadsheet Summary In Chapter 12 of *Applications of Microsoft® Excel in Analytical Chemistry*, 2nd ed., we investigate the method of continuous variations using the slope and intercept functions and learn how to produce inset plots.

26B AUTOMATED PHOTOMETRIC AND SPECTROPHOTOMETRIC METHODS

The first fully automated instrument for chemical analysis (the Technicon AutoAnalyzer®) appeared on the market in 1957. This instrument was designed to fulfill the needs of clinical laboratories where blood and urine samples are routinely analyzed for a dozen or more chemical species. The number of such analyses demanded by modern medicine is enormous, so it is necessary to keep their cost at a reasonable level. These two considerations motivated the development of analytical systems that perform several analyses simultaneously with a minimum input of human labor. The use of automatic instruments has spread from clinical laboratories to laboratories for the control of industrial processes and the routine determination of a wide spectrum of species in air, water, soils, and pharmaceutical and agricultural products. In the majority of these applications, the measurement step in the analyses is accomplished by photometry, spectrophotometry, or fluorometry.

In Section 8C, we described various automated sample handling techniques including discrete and continuous flow methods. In this section, we explore the instrumentation and two applications of flow-injection analysis (FIA) with photometric detection.

26B-1 Instrumentation

Figure 26-16a is a flow diagram of the simplest of all flow-injection systems. In this example, a colorimetric reagent for chloride ion is pumped by a peristaltic pump directly into a valve that permits injection of samples into the flowing stream. The sample and reagent then pass through a 50-cm reactor coil where the reagent mixes with the sample plug and produces a colored product by the sequence of reaction

$$Hg(SCN)_2(aq) + 2Cl^- \rightleftharpoons HgCl_2(aq) + 2SCN^-$$

$$Fe^{3+} + SCN^- \rightleftharpoons Fe(SCN)^{2+}$$
$$\text{red}$$

From the reactor coil, the solution passes into a flow-through photometer equipped with a 480-nm interference filter for absorbance measurement.

The signal output from this system for a series of standards containing from 5 to 75 ppm of chloride is shown in Figure 26-16b. Notice that four injections of each standard were made to demonstrate the reproducibility of the system. The two

Figure 26-16 Flow-injection determination of chloride: (a) Flow diagram. (b) Recorder readout for quadruplicate runs on standards containing 5 to 75 ppm of chloride ion. (c) Fast scan of two of the standards to demonstrate the low analyte carryover (less than 1%) from run to run. Notice that the point marked 1% corresponds to where the response would just begin for a sample injected at time S_2. (Reprinted with permission from E. H. Hansen and J. Ruzicka, *J. Chem. Educ.*, **1979**, 56, 677, **DOI:** 10.1021/ ed056p677. Copyright by the American Chemical Society.)

curves in Figure 26-16c are high-speed recorder scans of one of the samples containing 30 ppm (R_{30}) and another containing 75 ppm (R_{75}) chloride. These curves demonstrate that cross-contamination between successive samples is minimal in this unsegmented stream. Thus, less than 1% of the first analyte is present in the flow cell after 28 s, the time of the next injection (S_2). This system has been successfully used for the routine determination of chloride ion in brackish and waste waters as well as in serum samples.

Sample and Reagent Transport System

Normally, the solution in a flow-injection analysis is pumped through flexible tubing in the system by a peristaltic pump, a device in which a fluid (liquid or gas) is squeezed through plastic tubing by rollers. **Figure 26-17** illustrates the operating principle of the peristaltic pump. The spring-loaded cam, or band, pinches the tubing against two or more of the rollers at all times, thus forcing a continuous flow of fluid through the tubing. These pumps generally have 8 to 10 rollers, arranged in a circular configuration so that half are squeezing the tube at any instant. This design leads to a flow that is relatively pulse free. The flow rate is controlled by the speed of the motor, which should be greater than 30 rpm, and the inside diameter of the tube. A wide variety of tube sizes (i.d. = 0.25 to 4 mm) are available commercially that permit flow rates as small as 0.0005 mL/min and as great as 40 mL/min. The rollers of typical commercial peristaltic pumps are long enough so that several reagent and sample streams can be pumped simultaneously. Syringe pumps and electroosmosis are also used to induce flow in flow-injection systems. Flow-injection systems have been miniaturized through the use of fused silica capillaries (i.d. 25-100 μm) or through **lab-on-a-chip** technology (see Feature 8-1).

Figure 26-17 Diagram showing one channel of a peristaltic pump. Several additional tubes may be located under the one shown (below the plane of the diagram) to carry multiple channels of reagent or sample. (Reprinted from B. Karl- berg and G. E. Pacey, *Flow Injection Analysis. A Practical Guide*, New York: Elsevier, 1989, p. 34 with permission from Elsevier.)

Sample Injectors and Detectors

Sample sizes for flow-injection analysis range from 5 to 200 μL, with 10 to 30 μL being typical for most applications. For a successful determination, it is important to inject the sample solution rapidly as a plug, or pulse, of liquid; in addition, the injections must not disturb the flow of the carrier stream. The most useful and convenient injector systems are based on sampling loops similar to those used in chromatography (see, for example, Figure 33-6). The method of operation of a sampling loop is illustrated in Figure 26-16a. With the valve of the loop in the position shown, reagents flow through the bypass. When a sample has been injected into the loop and the valve turned 90 deg, the sample enters the flow as a single, well-defined zone. For all practical purposes, flow through the bypass ceases with the valve in this position because the diameter of the sample loop is significantly greater than that of the bypass tubing.

The most common detectors in flow-injection analysis are spectrophotometers, photometers, and fluorometers. Electrochemical systems, refractometers, atomic emission, and atomic absorption spectrometers have also been used.

Advanced Flow-Injection Techniques[11]

Flow-injection methods have been used to accomplish separations, titrations, and kinetic methods. In addition, several variations of flow injection have been shown to be useful. These include flow reversal FIA, sequential injection FIA, and lab-on-a-valve technology.

Separations by dialysis, by liquid/liquid extraction, and by gaseous diffusion can be accomplished automatically with flow-injection systems.

26B-2 A Typical Application of Flow-Injection Analysis

Figure 26-18 illustrates a flow-injection system designed for the automatic spectrophotometric determination of caffeine in acetyl salicylic acid drug preparations after extraction of the caffeine into chloroform. The chloroform solvent, after cooling in an ice bath to minimize evaporation, is mixed with the alkaline sample stream in a T-tube (see lower insert). After passing through the 2-m extraction coil, the mixture enters a T-tube separator, which is differentially pumped so that about 35% of the organic phase containing the caffeine passes into the flow cell, the other 65% accompanying the aqueous solution containing the rest of the sample to waste. In order to avoid contaminating the flow cell with water, Teflon fibers, which are not wetted by water, are twisted into a thread and inserted in the inlet to the T-tube in such a way as to form a smooth downward bend. The chloroform flow then follows this bend to the photometer cell where the caffeine concentration is determined based on its absorption peak at 275 nm. The output of the photometer is similar in appearance to that shown in Figure 26-16b.

26C INFRARED ABSORPTION SPECTROSCOPY

Infrared spectroscopy is a powerful tool for identifying pure organic and inorganic compounds because, with the exception of a few homonuclear molecules such as O_2, N_2, and Cl_2, all molecular species absorb infrared radiation. In addition, with the exception of chiral molecules in the crystalline state, every molecular compound has a unique infrared absorption spectrum. Therefore, an exact match between the

> Flow-injection analyzers can be fairly simple, consisting of a pump, an injection valve, plastic tubing, and a detector. Filter photometers and spectrophotometers are the most common detectors.

---

[11] For more information on FIA methods, see D. A. Skoog, F. J. Holler, and S. R. Crouch, *Principles of Instrumental Analysis,* 6th ed., Belmont, CA: Brooks/Cole, 2007, pp. 933–41.

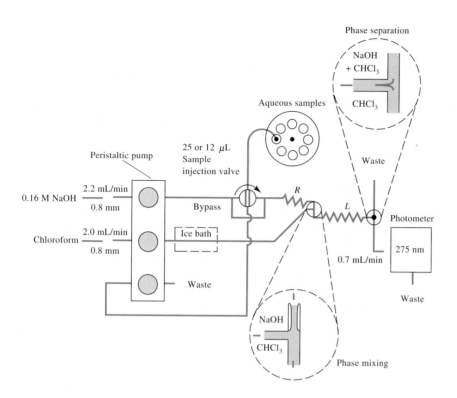

Figure 26-18 Flow-injection apparatus for the determination of caffeine in acetylsalicylic acid preparations. With the valve rotated at 90 deg, the flow in the bypass is essentially zero because of its small diameter. *R* and *L* are Teflon coils with 0.8-mm inside diameters; *L* has a length of 2 m, and the distance from the injection point through *R* to the mixing point is 0.15 m. (Reprinted from B. Karlberg and S. Thelander, *Anal. Chim. Acta*, **1978**, 98, 2, **DOI**: 10.1016/S0003-2670(01)83231-1 with permission from Elsevier.)

spectrum of a compound of known structure and the spectrum of an analyte unambiguously identifies the analyte.

Infrared spectroscopy is a less satisfactory tool for quantitative analyses than its ultraviolet and visible counterparts because of lower sensitivity and frequent deviations from Beer's law. Additionally, infrared absorbance measurements are considerably less precise. Nevertheless, in instances where modest precision is adequate, the unique nature of infrared spectra provides a degree of selectivity in a quantitative measurement that may offset these undesirable characteristics.[12]

26C-1 Infrared Absorption Spectra

The energy of infrared radiation can excite vibrational and rotational transitions, but it is insufficient to excite electronic transitions. As shown in **Figure 26-19**, infrared spectra exhibit narrow, closely spaced absorption peaks resulting from transitions among the various vibrational quantum levels. Variations in rotational levels may also give rise to a series of peaks for each vibrational state. With liquid or solid samples, however, rotation is often hindered or prevented, and the effects of these small energy differences are not detected. Thus, a typical infrared spectrum for a liquid, such as that in Figure 26-19, consists of a series of vibrational bands.

The number of ways a molecule can vibrate is related to the number of atoms, and thus the number of bonds, it contains. For even a simple molecule, the number of possible vibrations is large. For example, *n*-butanal ($CH_3CH_2CH_2CHO$) has 33 vibrational modes, most differing from each other in energy. Not all of these vibrations produce infrared peaks, but, as shown in Figure 26-19, the spectrum for *n*-butanal is relatively complex.

---

[12] For a detailed discussion of infrared spectroscopy, see N. B. Colthup, L. H. Daly, and S. E. Wiberley, *Introduction to Infrared and Raman Spectroscopy*, 3rd ed., New York: Academic Press, 1990.

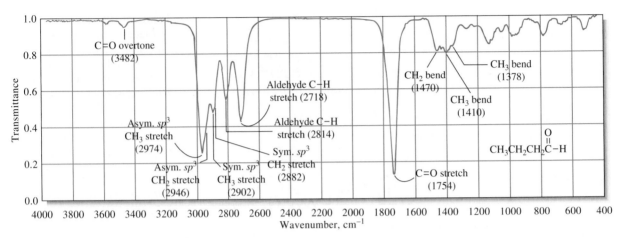

Figure 26-19 Infrared spectrum for *n*-butanal (*n*-butyraldehyde). The vertical scale is plotted as transmittance, as has been common practice in the past. The horizontal scale is linear in wavenumbers, which is proportional to frequency and thus energy. Most modern IR spectrometers are capable of providing data plotted as either transmittance or absorbance on the vertical axis and wavenumber or wavelength on the horizontal axis. IR spectra are usually plotted with frequency increasing from right to left, which is a historical artifact. Early IR spectrometers produced spectra with wavelength increasing from left to right, which led to an auxiliary frequency scale from right to left. Note that several of the bands have been labeled with assignments of the vibrations that produce the bands. Data from NIST Mass Spec Data Center, S. E. Stein, director, "Infrared Spectra," in NIST Chemistry WebBook, NIST Standard Reference Database Num- ber 69, P. J. Linstrom and W .G. Mallard, eds., Gaithersburg MD: National Institute of Standards and Technology, March 2003 (http:// webbook.nist.gov).

Molecular model of *n*-butanal.

Infrared absorption occurs not only with organic molecules but also with covalently bonded metal complexes, which are generally active in the longer-wavelength infrared region. Infrared studies have provided important information about complex metal ions.

26C-2 Instruments for Infrared Spectrometry

Three types of infrared instruments are found in modern laboratories; dispersive spectrometers (spectrophotometers), Fourier transform spectrometers, and filter photometers. The first two are used for obtaining complete spectra for qualitative identification, while filter photometers are designed for quantitative work. Fourier transform and filter instruments are nondispersive in the sense that neither uses a grating or prism to disperse radiation into its component wavelengths.[13]

Dispersive Instruments

With one difference, dispersive infrared instruments are similar in general design to the double-beam (in time) spectrophotometers shown in Figures 25-20c. The difference lies in the location of the cell compartment with respect to the monochromator. In ultraviolet/visible instruments, cells are always located between the monochromator and the detector in order to avoid photochemical decomposition, which may occur if samples are exposed to the full power of an ultraviolet or visible source. Infrared radiation, in contrast, is not sufficiently energetic to bring about photodecomposition; thus, the cell compartment can be located between the source and the monochromator. This arrangement is advantageous because any scattered radiation generated in the cell compartment is largely removed by the monochromator.

---

[13] For a discussion of the principles of Fourier transform spectroscopy, see D. A. Skoog, F. J. Holler, and S. R. Crouch, *Principles of Instrumental Analysis*, 6th ed., Belmont, CA: Brooks/Cole, 2007, pp. 439–47.

As shown in Section 25A, the components of infrared instruments differ considerably in detail from those in ultraviolet and visible instruments. Therefore, infrared sources are heated solids rather than deuterium or tungsten lamps, infrared gratings are much coarser than those required for ultraviolet/visible radiation, and infrared detectors respond to heat rather than photons. In addition, the optical components of infrared instruments are constructed from polished solids, such as sodium chloride or potassium bromide.

Fourier Transform Spectrometers

Fourier transform infrared (FTIR) spectrometers offer the advantages of high sensitivity, resolution, and speed of data acquisition (data for an entire spectrum can be obtained in 1 s or less). In the early days of FTIR, instruments were large, intricate, costly devices controlled by expensive laboratory computers. Since the 1980s, the instrumentation has evolved, and the price of spectrometers and computers have dropped dramatically. Today, FTIR spectrometers are commonplace, having replaced older, dispersive instruments in most laboratories.

Fourier transform instruments contain no dispersing element, and all wavelengths are detected and measured simultaneously using a Michelson interferometer as we described in Feature 25-7. In order to separate wavelengths, it is necessary to modulate the source signal and pass it through the sample in such a way that it can be recorded as an **interferogram**. The interferogram is subsequently decoded by Fourier transformation, a mathematical operation that is conveniently carried out by the computer, which is now an integral part of all spectrometers. Although the detailed mathematical theory of Fourier transform measurements is beyond the scope of this book, the qualitative treatment presented in Feature 25-7 and in Feature 26-1 should give you an idea of how the IR signal is collected and how spectra are extracted from the data.

Figure 26-20 is a photo of a typical benchtop FTIR spectrometer. A personal computer is needed for data acquisition, analysis, and presentation. The instrument is relatively inexpensive (ca. $10,000), has a resolution better than 0.8 cm^{-1}, and achieves a signal-to-noise ratio of 8000 for a five-second measurement. The measured spectrum appears on the computer screen where the included software allows many different display options (%T, A, zoom, peak height, and peak area). Various processing tools, such as baseline correction, spectral subtraction, and spectral interpretation, are featured in the software package. A number of different sampling accessories allow gaseous, liquid, and solid samples to be measured and techniques such as attenuated total reflectance (ATR) to be implemented. Some benchtop FTIR spectrometers are self-contained with a built-in computer for data acquisition, analysis, and presentation. These instruments typically are less flexible in terms of software, display modes, and data storage than units with a separate computer.

A research-quality instrument may cost more than $50,000. It can have a resolution of 0.10 cm^{-1} or better and can exhibit a signal-to-noise ratio of 50,000 or greater for a one-minute measurement period. Research-grade spectrometers typically have multiple scanning ranges (27000 to 15 cm^{-1}) and a variety of scanning velocities. They have excellent wavenumber precision (0.01 cm^{-1}). Research-grade instruments can accommodate numerous sampling modes (solids, gases, liquids, polymers, attenuated total reflectance, diffuse reflectance, and microscope attachments, among others). Typically, a research-grade instrument will be connected to a separate computer, providing a number of advantages. Software and databases of spectra can be installed and used to process spectral data and to match measured spectra to known spectra in the database. In addition, a personal computer provides considerable flexibility for archiving data on CDs or DVDs, and if the computer is connected to a local area network, spectra may

◀ The FTIR spectrometer is the most common type of IR spectrometer. The great majority of infrared instruments sold today are FTIR systems.

An **interferogram** is a recording of the signal produced by a Michelson interferometer. The signal is processed by a mathematical process known as the Fourier transform to produce an IR spectrum.

Figure 26-20 Photo of a basic student-grade benchtop FTIR spectrometer. A separate laptop or desktop computer is required. Spectra are recorded in a few seconds and displayed on the computer screen for viewing and interpretation. (Courtesy of Thermo Fisher Scientific Inc)

be transmitted to colleagues or coworkers, and software or firmware upgrades may be conveniently downloaded and installed on the computer or in the spectrometer.

Filter Photometers

Infrared photometers designed to monitor the concentration of air pollutants, such as carbon monoxide, nitrobenzene, vinyl chloride, hydrogen cyanide, and pyridine, are often used to ensure compliance with regulations established by the Occupational Safety and Health Administration (OSHA). Interference filters, each designed for the determination of a specific pollutant, are available. These transmit narrow bands of radiation in the range of 3 to 14 μm. There are also nondispersive spectrometers for monitoring gas streams for a single component.[14]

26C-3 Qualitative Applications of Infrared Spectrometry

An infrared absorption spectrum, even one for a relatively simple compound, often contains a bewildering array of sharp peaks and minima. Peaks useful for the identification of functional groups are located in the shorter-wavelength region of the infrared (from about 2.5 to 8.5 μm), where the positions of the maxima are only slightly affected by the carbon skeleton of the molecule. This region of the spectrum thus abounds with information regarding the overall constitution of the molecule under investigation. **Table 26-5** gives the positions of characteristic maxima for some common functional groups.[15]

Identifying functional groups in a molecule is seldom sufficient to positively identify the compound, and the entire spectrum from 2.5 to 15 μm must be compared with that of known compounds. Collections of spectra are available for this purpose.[16]

---

[14] For more information, see D. A. Skoog, F. J. Holler, and S. R. Crouch, *Principles of Instrumental Analysis*, 6th ed., Belmont, CA: Brooks/Cole, 2007, pp. 447–48.

[15] For more detailed information, see R. M. Silverstein, F. X. Webster, and D. Kiemle, *Spectrometric Identification of Organic Compounds*, 7th ed., Ch. 2, New York: Wiley, 2005.

[16] See *Sadtler Standard Spectra*, Informatics/Sadtler Group, Bio-Rad Laboratories, Philadelphia, PA; C. J. Pouchert, *The Aldrich Library of Infrared Spectra*, 3rd ed., Milwaukee, WI: Aldrich Chemical, 1981; *NIST Chemistry WebBook*, NIST Standard Reference Database Number 69, Gaithersburg, MD: National Institute of Standards and Technology, 2008 (http://webbook.nist.gov).

Producing Spectra with an FTIR Spectrometer

In Feature 25-7, we described the basic operating principles of the Michelson interferometer and the function of the Fourier transform to produce a frequency spectrum from a measured interferogram. **Figure 26F-1** shows an optical diagram for a Michelson interferometer similar to the one in the spectrometer depicted in Figure 26-20. The interferometer is actually two parallel interferometers, one to modulate the IR radiation from the source before it passes through the sample and a second to modulate the red light from the He-Ne laser to provide a reference signal for acquiring data from the IR detector. The output of the detector is digitized and stored in the memory of the instrument computer.

Figure 26F-1   Instrument diagram for a basic FTIR spectrometer. Radiation of all frequencies from the IR source are reflected into the interferometer where it is modulated by the moving mirror on the left. The modulated radiation is then reflected from the two mirrors on the right through the sample in the compartment at the bottom. After passing through the sample, the radiation falls on the detector. A data acquisition system attached to the detector records the signal and stores it in the memory of a computer as an interferogram. (Reprinted by permission of Thermo Fisher Scientific.)

The first step in producing an IR spectrum is to collect and store a reference interferogram with no sample in the sample cell. Then, the sample is placed in the cell, and a second interferogram is collected. **Figure 26F-2a** shows an interferogram collected using an FTIR spectrometer with methylene chloride, CH_2Cl_2, in the sample cell. The Fourier transform is then applied to the two interferograms to compute the IR spectra of the reference and the sample. The ratio of the two spectra can then be computed to produce an IR spectrum of the analyte such as the one illustrated in Figure 26F-2b.

(continued)

Notice that the methylene chloride IR spectrum exhibits little noise. Since a single interferogram can be scanned in only a second or two, many interferograms can be scanned in a relatively short time and summed in the memory of the computer. This process, which is often called **signal averaging**, reduces the noise on the resulting signal and improves the signal-to-noise ratio of the spectrum, as described in Feature 25-5 and illustrated in Figure 25F-4. This capability of noise reduction and speed coupled with Fellgett's advantage and Jacquinot's advantage (see Feature 25-7) makes the FTIR spectrometer a marvelous tool for a broad range of qualitative and quantitative analyses.

Figure 26F-2 (a) Interferogram obtained from a typical FTIR spectrometer for methylene chloride. The plot shows detector signal output as a function of time, or displacement of the moving mirror of the interferometer. (b) IR spectrum of methylene chloride produced by the Fourier transformation of the data in (a). Note that the Fourier transform takes signal intensity collected as a function of time and produces transmittance as a function of frequency after subtraction of a background interferogram and proper scaling.

26C-4 Quantitative Infrared Spectrometry

Quantitative infrared absorption methods differ somewhat from their ultraviolet and visible counterparts because of the greater complexity of the spectra, the narrowness of the absorption bands, and the capabilities of the instruments available for measurements in this spectral region.[17]

---

[17]For an extensive discussion of quantitative infrared analysis, see A. L. Smith, in *Treatise on Analytical Chemistry*, 2nd ed., P. J. Elving, E. J. Meehan, and I. M. Kolthoff, eds., Part I, Vol. 7, pp. 415–56, New York: Wiley, 1981.

TABLE 26-5

Some Characteristic Infrared Absorption Peaks

| | | Absorption Peaks | |
	Functional Group	Wavenumber, cm^{-1}	Wavelength, μm
O—H	Aliphatic and aromatic	3600–3000	2.8–3.3
NH_2	Also secondary and tertiary	3600–3100	2.8–3.2
C—H	Aromatic	3150–3000	3.2–3.3
C—H	Aliphatic	3000–2850	3.3–3.5
C≡N	Nitrile	2400–2200	4.2–4.6
C≡C—	Alkyne	2260–2100	4.4–4.8
COOR	Ester	1750–1700	5.7–5.9
COOH	Carboxylic acid	1740–1670	5.7–6.0
C=O	Aldehydes and ketones	1740–1660	5.7–6.0
$CONH_2$	Amides	1720–1640	5.8–6.1
C=C—	Alkene	1670–1610	6.0–6.2
ϕ—O—R	Aromatic	1300–1180	7.7–8.5
R—O—R	Aliphatic	1160–1060	8.6–9.4

Absorbance Measurements

Using matched cuvettes for solvent and analyte is seldom practical for infrared measurements because it is difficult to obtain cells with identical transmission characteristics. Part of this difficulty results from degradation of the transparency of infrared cell windows (typically polished sodium chloride) with use due to attack by traces of moisture in the atmosphere and in samples. In addition, path lengths are hard to reproduce because infrared cells are often less than 1 mm thick. Such narrow cells are required to permit the transmission of measurable intensities of infrared radiation through pure samples or through very concentrated solutions of the analyte. Measurements on dilute analyte solutions, as is done in ultraviolet or visible spectroscopy, are usually difficult because there are few good solvents that transmit over appreciable regions of the IR spectrum.

For these reasons, a reference absorber is often dispensed with entirely in qualitative infrared work, and the intensity of the radiation passing through the sample is simply compared with that of the unobstructed beam; alternatively, a salt plate may be used as a reference. Either way, the resulting transmittance is often less than 100%, even in regions of the spectrum where the sample is totally transparent.

Applications of Quantitative Infrared Spectroscopy

Infrared spectrophotometry offers the potential for determining an unusually large number of substances because nearly all molecular species absorb in the IR region. Moreover, the uniqueness of an IR spectrum provides a degree of specificity that is matched or exceeded by relatively few other analytical methods. This specificity has particular application to the analysis of mixtures of closely related organic compounds.

The recent proliferation of government regulations on atmospheric contaminants has demanded the development of sensitive, rapid, and highly specific methods for a variety of chemical compounds. IR absorption procedures appear to meet this need better than any other single analytical tool.

Table 26-6 illustrates the variety of atmospheric pollutants that can be determined with a simple, portable filter photometer equipped with a separate interference filter for each analyte species. Of the more than 400 chemicals for which maximum tolerable limits have been set by OSHA, half or more have absorption characteristics that make

TABLE 26-6

Examples of Infrared Vapor Analysis for OSHA Compliance*

Compound	Allowable Exposure, ppm†	Wavelength, μm	Minimum Detectable Concentration, ppm‡
Carbon disulfide	4	4.54	0.5
Chloroprene	10	11.4	4
Diborane	0.1	3.9	0.05
Ethylenediamine	10	13.0	0.4
Hydrogen cyanide	4.7§	3.04	0.4
Methyl mercaptan	0.5	3.38	0.4
Nitrobenzene	1	11.8	0.2
Pyridine	5	14.2	0.2
Sulfur dioxide	2	8.6	0.5
Vinyl chloride	1	10.9	0.3

*Courtesy of The Foxboro Company, Foxboro, MA 02035.
†1992–1993 OSHA exposure limits for 8-hr weighted average.
‡For 20.25-m cell.
§Short-term exposure limit:15-min time-weighted average that shall not be exceeded at any time during the work day.

them amenable to determination by infrared photometry or spectrophotometry. With so many compounds absorbing, overlapping peaks are quite common. In spite of this potential disadvantage, the method provides a moderately high degree of selectivity.

WEB WORKS

Locate the *NIST Chemistry WebBook* on the web, and perform a search for 1,3-dimethyl benzene. What data are available for this compound on the NIST site? Click on the link to the IR spectrum and notice that there are several versions of the spectrum. How are they alike, and how do they differ? Where did the spectra originate? Select the gas-phase 2-cm^{-1} resolution spectrum. Click on View Image of Digitized Spectrum and print a copy of the spectrum. Now, return to the IR spectrum and its links. Under gas phase, choose the highest resolution spectrum with boxcar apodization. Click on the desired resolution to load the spectrum. Note that this spectrum presents molar absorptivity versus wavenumber while the previous lower resolution spectrum shows transmittance versus wavenumber. What are the major spectral differences noted? Does the additional resolution give any extra information? How might the molar absorptivity spectrum be used for quantitative analysis? Try a few other compounds and compare the low resolution vapor-phase spectra to the high-resolution quantitative spectra.

QUESTIONS AND PROBLEMS

26-1. Describe the differences between the following pairs of terms and list any particular advantages of one over the other:
   *(a) spectrophotometers and photometers.
   (b) single-beam and double-beam instruments for absorbance measurements.
   *(c) conventional and diode-array spectrophotometers.

26-2. What minimum requirement is needed to obtain reproducible results with a single-beam spectrophotometer?

*26-3. What experimental variables must be controlled to assure reproducible absorbance data?

26-4. What is(are) advantage(s) of the multiple standard addition method over the single-point standard addition method?

*26-5. The molar absorptivity for the complex formed between bismuth(III) and thiourea is 9.32×10^3 L cm^{-1} mol^{-1} at 470 nm. Calculate the range of permissible concentrations for the complex if the absorbance is to be no less than 0.10 nor greater than 0.90 when the measurements are made in 1.00-cm cells.

26-6. The molar absorptivity for aqueous solutions of phenol at 211 nm is 6.17×10^3 L cm^{-1} mol^{-1}. Calculate the permissible range of phenol concentrations if the transmittance is to be less than 85% and greater than 7% when the measurements are made in 1.00-cm cells.

*26-7. The logarithm of the molar absorptivity for acetone in ethanol is 2.75 at 366 nm. Calculate the range of acetone concentrations that can be used if the absorbance is to be greater than 0.100 and less than 2.000 with a 1.50-cm cell.

26-8. The logarithm of the molar absorptivity of phenol in aqueous solution is 3.812 at 211 nm. Calculate the range of phenol concentrations that can be used if the absorbance is to be greater than 0.150 and less than 1.500 with a 1.25-cm cell.

26-9. A photometer with a linear response to radiation gave a reading of 690 mV with a blank in the light path and 169 mV when the blank was replaced by an absorbing solution. Calculate

  *(a) the transmittance and absorbance of the absorbing solution.

  (b) the expected transmittance if the concentration of absorber is one half that of the original solution.

  *(c) the transmittance to be expected if the light path through the original solution is doubled.

26-10. A portable photometer with a linear response to radiation registered 75.5 μA with a blank solution in the light path. Replacement of the blank with an absorbing solution yielded a response of 23.7 μA. Calculate

  (a) the percent transmittance of the sample solution.

  *(b) the absorbance of the sample solution.

  (c) the transmittance to be expected for a solution in which the concentration of the absorber is one third that of the original sample solution.

  *(d) the transmittance to be expected for a solution that has twice the concentration of the sample solution.

26-11. Sketch a photometric titration curve for the titration of Sn^{2+} with MnO$_4^-$. What color radiation should be used for this titration? Explain.

26-12. Iron(III) reacts with thiocyanate ion (SCN) to form the red complex, Fe(SCN)$^{2+}$. Sketch a photometric titration curve for Fe(III) with thiocyanate ion when a photometer with a green filter is used to collect data. Why is a green filter used?

*26-13. Ethylenediaminetetraacetic acid displaces bismuth(III) from its thiourea complex:

$$Bi(tu)_6^{3+} + H_2Y^{2-} \rightarrow BiY^- + 6tu + 2H^+$$

where tu is the thiourea molecule, $(NH_2)_2CS$. Predict the shape of a photometric titration curve based on this process, given that the Bi(III)/thiourea complex is the only species in the system that absorbs at 465 nm, the wavelength selected for the titration.

26-14. The accompanying data (1.00-cm cells) were obtained for the spectrophotometric titration of 10.00 mL of Pd(II) with M Nitroso R (O. W. Rollins and M. M. Oldham, *Anal. Chem.,* **1971,** *43,* 262, **DOI:** 10.1021/ac60297a026):

Volume of Nitroso R, mL	A500
0	0
1.00	0.147
2.00	0.271
3.00	0.375
4.00	0.371
5.00	0.347
6.00	0.325
7.00	0.306
8.00	0.289

Calculate the concentration of the Pd(II) solution, given that the ligand-to-cation ratio in the colored product is 2:1.

26-15. A 4.97-g petroleum specimen was decomposed by wet ashing and subsequently diluted to 500 mL in a volumetric flask. Cobalt was determined by treating 25.00-mL aliquots of this diluted solution as follows:

	Reagent Volume		
Co(II), 3.00 ppm	Ligand	H$_2$O	Absorbance
0.00	20.00	5.00	0.398
5.00	20.00	0.00	0.510

Assume that the Co(II)/ligand chelate obeys Beer's law and calculate the percentage of cobalt in the original sample.

*26-16. Iron(III) forms a complex with thiocyanate ion that has the formula Fe(SCN)$^{2+}$. The complex has an absorption maximum at 580 nm. A specimen of well water was assayed according to the scheme below. Calculate the concentration of iron in parts per million.

			Volumes, mL			
Sample	Sample Volume	Oxidizing Reagent	Fe(II) 2.75 ppm	KSCN 0.050 M	H$_2$O	Absorbance, 580 nm (1.00-cm cells)
1	50.00	5.00	5.00	20.00	20.00	0.549
2	50.00	5.00	0.00	20.00	25.00	0.231

26-17. A. J. Mukhedkar and N. V. Deshpande (*Anal. Chem.*, **1963**, *35*, 47, **DOI:** 10.1021/ac60194a014) report on a simultaneous determination for cobalt and nickel based on absorption by their 8-quinolinol complexes. Molar absorptivities (L mol^{-1} cm^{-1}) are $\varepsilon_{Co} = 3529$ and $\varepsilon_{Ni} = 3228$ at 365 nm and $\varepsilon_{Co} = 428.9$ and $\varepsilon_{Ni} = 0$ at 700 nm. Calculate the concentration of nickel and cobalt in each of the following solutions (1.00-cm cells):

Solution	A_{365}	A_{700}
1	0.617	0.0235
2	0.755	0.0714
3	0.920	0.0945
4	0.592	0.0147
5	0.685	0.0540

*26-18. Molar absorptivity data for the cobalt and nickel complexes with 2,3-quinoxalinedithiol are $\varepsilon_{Co} = 36,400$ and $\varepsilon_{Ni} = 5520$ at 510 nm and $\varepsilon_{Co} = 1240$ and $\varepsilon_{Ni} = 17,500$ at 656 nm. A 0.425-g sample was dissolved and diluted to 50.0 mL. A 25.0-mL aliquot was treated to eliminate interferences; after addition of 2,3-quinoxalinedithiol, the volume was adjusted to 50.0 mL. This solution had an absorbance of 0.446 at 510 nm and 0.326 at 656 nm in a 1.00-cm cell. Calculate the concentration in parts per million of cobalt and nickel in the sample.

26-19. The indicator HIn has an acid dissociation constant of 4.80×10^{-6} at ordinary temperatures. The accompanying absorbance data are for 8.00×10^{-5} M solutions of the indicator measured in 1.00-cm cells in strongly acidic and strongly alkaline media:

	Absorbance	
λ, nm	pH 1.00	pH 13.00
420	0.535	0.050
445	0.657	0.068
450	0.658	0.076
455	0.656	0.085
470	0.614	0.116
510	0.353	0.223
550	0.119	0.324
570	0.068	0.352
585	0.044	0.360
595	0.032	0.361
610	0.019	0.355
650	0.014	0.284

Estimate the wavelength at which absorption by the indicator becomes independent of pH (that is, the isosbestic point).

26-20. Calculate the absorbance (1.00-cm cells) at 450 nm of a solution in which the total molar concentration of the indicator described in Problem 26-19 is 8.00×10^{-5} M and the pH is *(a) 4.92, (b) 5.46, *(c) 5.93, and (d) 6.16.

*26-21. What is the absorbance at 595 nm (1.00-cm cells) of a solution that is 1.25×10^{-4} M in the indicator of Problem 26-19 and has a pH of (a) 5.30, (b) 5.70, and (c) 6.10?

26-22. Several buffer solutions were made 1.00×10^{-4} M in the indicator of Problem 26-19. Absorbance data (1.00-cm cells) are

Solution	A_{450}	A_{595}
*A	0.344	0.310
B	0.508	0.212
*C	0.653	0.136
D	0.220	0.380

Calculate the pH of each solution.

26-23. Construct an absorption spectrum for an 7.00×10^{-5} M solution of the indicator of Problem 26-19 when measurements are made with 1.00-cm cells and

(a) $\dfrac{[\text{HIn}]}{[\text{In}^-]} = 3$

(b) $\dfrac{[\text{HIn}]}{[\text{In}^-]} = 1$

(c) $\dfrac{[\text{HIn}]}{[\text{In}^-]} = \dfrac{1}{3}$

26-24. Solutions of P and Q individually obey Beer's law over a large concentration range. Spectral data for these species in 1.00-cm cells are

	Absorbance	
λ, nm	8.55×10^{-5} M P	2.37×10^{-4} M Q
400	0.078	0.500
420	0.087	0.592
440	0.096	0.599
460	0.102	0.590
480	0.106	0.564
500	0.110	0.515
520	0.113	0.433
540	0.116	0.343
580	0.170	0.170
600	0.264	0.100
620	0.326	0.055
640	0.359	0.030
660	0.373	0.030
680	0.370	0.035
700	0.346	0.063

(a) Plot an absorption spectrum for a solution that is 6.45×10^{-5} M in P and 3.21×10^{-4} M in Q.

(b) Calculate the absorbance (1.00-cm cells) at 440 nm of a solution that is 3.86×10^{-5} M in P and 5.37×10^{-4} M in Q.

(c) Calculate the absorbance (1.00-cm cells) at 620 nm of a solution that is 1.89×10^{-4} M in P and 6.84×10^{-4} M in Q.

26-25. Use the data in Problem 26-24 to calculate the molar concentration of P and Q in each of the following solutions:

	A_{440}	A_{620}
*(a)	0.357	0.803
(b)	0.830	0.448
*(c)	0.248	0.333
(d)	0.910	0.338
*(e)	0.480	0.825
(f)	0.194	0.315

26-26. A standard solution was put through appropriate dilutions to give the concentrations of iron shown in the accompanying table. The iron(II)-1,10,phenanthroline complex was then formed in 25.0-mL aliquots of these solutions, following which each was diluted to 50.0 mL (see color plate 15). The absorbances in the table (1.00-cm cells) were recorded at 510 nm.

Fe(II) Concentration in Original Solution, ppm	A_{510}
4.00	0.160
10.0	0.390
16.0	0.630
24.0	0.950
32.0	1.260
40.0	1.580

(a) Plot a calibration curve from these data.
*(b) Use the method of least squares to find an equation relating absorbance and the concentration of iron(II).
*(c) Calculate the standard deviation of the slope and intercept.

26-27. The method developed in Problem 26-26 was used for the routine determination of iron in 25.0-mL aliquots of ground water. Express the concentration (as ppm Fe) in samples that yielded the accompanying absorbance data (1.00-cm cell). Calculate the relative standard deviation of the result. Repeat the calculation assuming the absorbance data are means of three measurements.
(a) 0.143
(b) 0.675
(c) 0.068
(d) 1.009
(e) 1.512
(f) 0.546

*26-28. The sodium salt of 2-quinizarinsulfonic acid (NaQ) forms a complex with Al^{3+} that absorbs strongly at 560 nm.[18] The data collected on this system are shown in the accompanying table. (a) Find the formula of the complex from the data. In all solutions,

$c_{Al} = 3.7 \times 10^{-5}$ M, and all measurements were made in 1.00-cm cells. (b) Find the molar absorptivity of the complex.

c_Q, M	A_{560}
1.00×10^{-5}	0.131
2.00×10^{-5}	0.265
3.00×10^{-5}	0.396
4.00×10^{-5}	0.468
5.00×10^{-5}	0.487
6.00×10^{-5}	0.498
8.00×10^{-5}	0.499
1.00×10^{-4}	0.500

26-29. The accompanying data were obtained in a slope-ratio investigation of the complex formed between Ni^{2+} and 1-cyclopentene-1-dithiocarboxylic acid (CDA). The measurements were made at 530 nm in 1.00-cm cells.

$c_{CDA} = 1.00 \times 10^{-3}$ M		$c_{Ni} = 1.00 \times 10^{-3}$ M	
c_{Ni}, M	A_{530}	c_{CDA}, M	A_{530}
5.00×10^{-6}	0.051	9.00×10^{-6}	0.031
1.20×10^{-5}	0.123	1.50×10^{-5}	0.051
3.50×10^{-5}	0.359	2.70×10^{-5}	0.092
5.00×10^{-5}	0.514	4.00×10^{-5}	0.137
6.00×10^{-5}	0.616	6.00×10^{-5}	0.205
7.00×10^{-5}	0.719	7.00×10^{-5}	0.240

(a) Determine the formula of the complex. Use linear least squares to analyze the data.
(b) Find the molar absorptivity of the complex and its uncertainty.

*26-30. The accompanying absorption data were recorded at 390 nm in 1.00-cm cells for a continuous-variation study of the colored product formed between Cd^{2+} and the complexing reagent R.

	Reagent Volumes, mL		
Solution	$c_{Cd} = 1.25 \times 10^{-4}$ M	$c_R = 1.25 \times 10^{-4}$ M	A_{390}
0	10.00	0.00	0.000
1	9.00	1.00	0.174
2	8.00	2.00	0.353
3	7.00	3.00	0.530
4	6.00	4.00	0.672
5	5.00	5.00	0.723
6	4.00	6.00	0.673
7	3.00	7.00	0.537
8	2.00	8.00	0.358
9	1.00	9.00	0.180
10	0.00	10.00	0.000

(a) Find the ligand-to-metal ratio in the product.

[18]E. G. Owens and J. H. Yoe, *Anal. Chem.*, **1959**, *31*, 384, **DOI:** 10.1021/ac60147a016.

(b) Calculate an average value for the molar absorptivity of the complex and its uncertainty. Assume that in the linear portions of the plot the metal is completely complexed.

(c) Calculate K_f for the complex using the stoichiometric ratio determined in (a) and the absorption data at the point of intersection of the two extrapolated lines.

26-31. Palladium(II) forms an intensely-colored complex at pH 3.5 with arsenazo III at 660 nm.[19] A meteorite was pulverized in a ball mill, and the resulting powder was digested with various strong mineral acids. The resulting solution was evaporated to dryness, dissolved in dilute hydrochloric acid, and separated from interferents by ion-exchange chromatography (see Section 33D). The resulting solution containing an unknown amount of Pd(II) was then diluted to 50.00 mL with pH 3.5 buffer. Ten-milliliter aliquots of this analyte solution were then transferred to six 50-mL volumetric flasks. A standard solution was then prepared that was 1.00×10^{-5} M in Pd(II). Volumes of the standard solution shown in the table were then pipetted into the volumetric flasks along with 10.00 mL of 0.01 M arsenazo III. Each solution was then diluted to 50.00 mL, and the absorbance of each solution was measured at 660 nm in 1.00-cm cells.

Volume Standard Solution, mL	A_{660}
0.00	0.209
5.00	0.329
10.00	0.455
15.00	0.581
20.00	0.707
25.00	0.833

(a) Enter the data into a spreadsheet and construct a standard additions plot.

(b) Determine the slope and intercept of the line.

(c) Determine the standard deviation of the slope and of the intercept.

(d) Calculate the concentration of Pd(II) in the analyte solution.

(e) Find the standard deviation of the measured concentration.

26-32. Mercury(II) forms a 1:1 complex with triphenyltetrazolium chloride (TTC) that exhibits an absorption maximum at 255 nm.[20] The mercury(II) in a soil sample was extracted into an organic solvent containing an excess of TTC, and the resulting solution was diluted to 100.0 mL in a volumetric flask. Five-milliliter aliquots of the analyte solution were then transferred to

six 25-mL volumetric flasks. A standard solution was then prepared that was 5.00×10^{-6} M in Hg(II). Volumes of the standard solution shown in the table were then pipetted into the volumetric flasks, and each solution was then diluted to 25.00 mL. The absorbance of each solution was measured at 255 nm in 1.00-cm quartz cells.

Volume Standard Solution, mL	A_{255}
0.00	0.582
2.00	0.689
4.00	0.767
6.00	0.869
8.00	1.009
10.00	1.127

(a) Enter the data into a spreadsheet, and construct a standard-additions plot.

(b) Determine the slope and intercept of the line.

(c) Determine the standard deviation of the slope and of the intercept.

(d) Calculate the concentration of Hg(II) in the analyte solution.

(e) Find the standard deviation of the measured concentration.

*26-33. Estimate the frequencies of the absorption maxima in the IR spectrum of methylene chloride shown in Figure 26F-2. From these frequencies, assign molecular vibrations of methylene chloride to each of the bands. Notice that some of the group frequencies that you will need are not listed in Table 26-5, so you will have to look elsewhere.

26-34. Challenge Problem: (a) Show that the overall formation constant for the complex ML_n is

$$K_f = \frac{\left(\dfrac{A}{A_{extr}}\right)c}{\left[c_M - \left(\dfrac{A}{A_{extr}}\right)c\right]\left[c_L - n\left(\dfrac{A}{A_{extr}}\right)c\right]^n}$$

where A is the experimental absorbance at a given value on the x-axis in a continuous variations plot, A_{extr} is the absorbance determined from the extrapolated lines corresponding to the same point on the x-axis, c_M is the molar analytical concentration of the ligand, c_L is the molar analytical concentration of the metal, and n is the ligand-to-metal ratio in the complex.[21]

(b) Under what assumptions is the equation valid?

(c) What is c?

(d) Discuss the implications of the occurrence of the maximum in a continuous variations plot at a value of less than 0.5.

[19]J. G. Sen Gupta, *Anal. Chem.*, **1967**, *39*, 18, **DOI:** 10.1021/ac60245a029.
[20]M. Kamburova, *Talanta.*, **1993**, *40*(5), 719, **DOI:** 10.1016/0039-9140(93)80285-y

[21] J. Inczédy, *Analytical Applications of Complex Equilibria*, New York: Wiley, 1976.

(e) Using the method of continuous variations, Calabrese and Khan[22] characterized the complex formed between I_2 and I^-. They combined 2.60×10^{-4} M solutions of I_2 and I^- in the usual way to obtain the following data set. Use the data to find the composition of the I_2/I^- complex.

$V(I_2$ soln), mL	A_{350}
0.00	0.002
1.00	0.121
2.00	0.214
3.00	0.279
4.00	0.312
5.00	0.325
6.00	0.301
7.00	0.258
8.00	0.188
9.00	0.100
10.00	0.001

[22] V. T. Calabrese and A. Khan, *J. Phys. Chem. A*, **2000**, *104*, 1287, **DOI:** 10.1021/jp992847r.

(f) The continuous variations plot appears to be asymmetrical. Consult the paper by Calabrese and Khan and explain this asymmetry.

(g) Use the equation in part (a) to determine the formation constant of the complex for each of the three central points on the continuous variations plot.

(h) Explain any trend in the three values of the formation constant in terms of the asymmetry of the plot.

(i) Find the uncertainty in the formation constant determined by this method.

(j) What effect, if any, does the formation constant have on the ability to determine the composition of the complex using the method of continuous variations?

(k) Discuss the various advantages and potential pitfalls of using the method of continuous variations as a general method for determining the composition and formation constant of a complex compound.

Molecular Fluorescence Spectroscopy

CHAPTER 27

Dr. Gopal Murti/Science Photo Library/Photo Researchers, Inc.

The photograph is an immunofluorescent light micrograph of HeLa cancer cells. The cell in the center of the photo is in the prophase stage of mitotic cell division. The chromosomes have condensed before dividing to form two nuclei. The cells are stained to reveal actin microfilaments and microtubules of the cytoskeleton, which appear as the filamentary structures surrounding the cell nuclei. The nuclei of the cells are visualized by exposing the cells to structure-specific fluorescent antibodies, prepared by covalently attaching ordinary antibodies to fluorescent molecules. The antibodies collect in the nuclei so that when they are exposed to UV radiation, they glow as shown in the photo. Similar chemistry is used in the fluorescence immunoassay described in Feature 11-2.

Fluorescence is a photoluminescence process in which atoms or molecules are excited by absorption of electromagnetic radiation (recall Figure 24-6). The excited species then relax to the ground state, giving up their excess energy as photons. One of the most attractive features of molecular fluorescence is its inherent sensitivity, which is often one to three orders of magnitude better than absorption spectroscopy. In fact, single molecules of selected species have been detected by fluorescence spectroscopy under controlled conditions. Another advantage is the large linear concentration ranges of fluorescence methods, which are significantly broader than linear concentration ranges in absorption spectroscopy. Fluorescence methods are, however, less widely applicable than absorption methods because of the smaller number of chemical systems that show appreciable fluorescence. Fluorescence is also subject to many more environmental interference effects than absorption methods. We consider here some of the most important aspects of molecular fluorescence methods.[1]

27A THEORY OF MOLECULAR FLUORESCENCE

Molecular fluorescence is measured by exciting the sample at an absorption wavelength, also called the excitation wavelength, and measuring the emission at a longer wavelength called the emission or fluorescence wavelength. For example, the reduced form of the coenzyme nicontinamide adenine dinucleotide (NADH) absorbs radiation at 340 nm, and the molecule emits photoluminescence radiation with an emission maximum at 465 nm. Usually photoluminescence emission is measured at right angles to the incident beam to avoid measuring the incident radiation (recall Figure 25-1b). The short-lived emission that occurs is called **fluorescence**, while luminescence that is much longer lasting is called **phosphorescence**.

> Fluorescence emission occurs in 10^{-5} s or less. In contrast, phosphorescence may last for several minutes or even hours. Fluorescence is much more widely used for chemical analysis than phosphorescence.

---

[1]For more information on molecular fluorescence spectroscopy, see J. R.. Lakowicz, *Principles of Fluorescence Spectroscopy*, New York: Springer, 2006.

27A-1 Relaxation Processes

Figure 27-1 shows a partial energy level diagram for a hypothetical molecular species. Three electronic energy states are shown, E_0, E_1, and E_2; the ground state is E_0, and the excited states are E_1 and E_2. Each of the electronic states is shown as having four excited vibrational levels. When this species is irradiated with a band of wavelengths λ_1 to λ_5 (see Figure 27-1a), the five vibrational levels of the first excited electronic state, E_1, are momentarily populated. Similarly, when the molecules are irradiated with a more energetic band made up of shorter wavelengths λ_1' to λ_5', the five vibrational levels of the higher energy electronic state E_2 become populated briefly.

Once the molecule is excited to E_1 or E_2, several processes can occur that cause the molecule to lose its excess energy. Two of the most important of these processes, **nonradiative relaxation** and **fluorescence emission**, are illustrated in Figure 27-1b and c.

The two most important nonradiative relaxation methods that compete with fluorescence are illustrated in Figure 27-1b. **Vibrational relaxation**, depicted by the short wavy arrows between vibrational energy levels, takes place during collisions between excited molecules and molecules of the solvent. Nonradiative relaxation between the lower vibrational levels of an excited electronic state and the higher vibrational levels of another electronic state can also occur. This type of relaxation, sometimes called **internal conversion**, is depicted by the two longer wavy arrows in Figure 27-1b. Internal conversion is much less efficient than vibrational relaxation so that the average lifetime of an electronic excited state is between 10^{-9} and 10^{-6} s. The exact mechanism by which these two relaxational processes occur is currently under study, but the net result is a tiny increase in the temperature of the medium.

Figure 27-1c illustrates the relaxation process that is desired: the fluorescence process. Fluorescence is almost always observed from the lowest-lying excited electronic state E_1 to the ground state E_0. Also, the fluorescence usually occurs only from the

Vibrational relaxation involves transfer of the excess energy of a vibrationally excited species to molecules of the solvent. This process takes place in less than 10^{-15} s and leaves the molecules in the lowest vibrational state of an electronic excited state.

Internal conversion is a type of relaxation that involves transfer of the excess energy of a species in the lowest vibrational level of an excited electronic state to solvent molecules and conversion of the excited species to a lower electronic state.

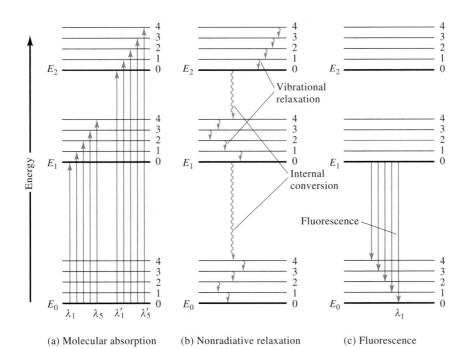

Figure 27-1 Energy level diagram shows some of the processes that occur during (a) absorption of incident radiation, (b) nonradiative relaxation, and (c) fluorescence emission by a molecular species. Absorption typically occurs in 10^{-15} s, while vibrational relaxation occurs in 10^{-11} to 10^{-10} s. Internal conversion between different electronic states is also very rapid (10^{-12} s), while fluorescence lifetimes are typically 10^{-10} to 10^{-5} s.

(a) Molecular absorption   (b) Nonradiative relaxation   (c) Fluorescence

> Fluorescence bands consist of a large number of closely spaced lines.

lowest vibrational level of E_1 to various vibrational levels of E_0, because the internal conversion and vibrational relaxation processes are very rapid compared to fluorescence. Hence, a fluorescence spectrum usually consists of only one band with many closely spaced lines that represent transitions from the lowest vibrational level of E_1 to the many different vibrational levels of E_0.

The line in Figure 27-1c that terminates the fluorescence band on the short-wavelength or high-energy side (λ_1) is identical in energy to the line labeled λ_1 in the absorption diagram in Figure 27-1a. Since fluorescence lines in this band originate in the lowest vibrational state of E_1, all of the other lines in the band are of lower energy or longer wavelength than the line corresponding to λ_1. Molecular fluorescence bands are mostly made up of lines that are longer in wavelength, higher in frequency, and thus lower in energy than the band of absorbed radiation responsible for their excitation. This shift to longer wavelength is called the **Stokes shift**.

> **Stokes-shifted fluorescence** is longer in wavelength than the radiation that caused the excitation.

Relationship between Excitation Spectra and Fluorescence Spectra

Because the energy differences between vibrational states is about the same for both ground and excited states, the absorption, or **excitation spectrum**, and the fluorescence spectrum for a compound often appear as approximate mirror images of one another with overlap occurring near the origin transition (0 vibrational level of E_1 to 0 vibrational level of E_0). This effect is demonstrated by the spectra for anthracene shown in **Figure 27-2**. There are many exceptions to this mirror-image rule, particularly when the excited and ground states have different molecular geometries or when different fluorescence bands originate from different parts of the molecule.

Molecular model of anthracene.

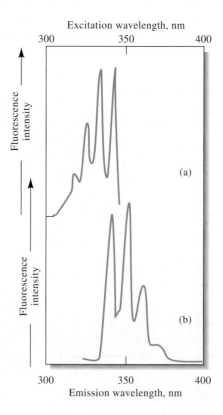

Figure 27-2 Fluorescence spectra for 1 ppm anthracene in alcohol: (a) excitation spectrum; (b) emission spectrum.

27A-2 Fluorescent Species

As shown in Figure 27-1, fluorescence is one of several mechanisms by which a molecule returns to the ground state after it has been excited by absorption of radiation. All absorbing molecules have the potential to fluoresce, but most compounds do not because their structure allows radiationless pathways for relaxation to occur *at a greater rate* than fluorescence emission. The **quantum yield** of molecular fluorescence is simply the ratio of the number of molecules that fluoresce to the total number of excited molecules, or the ratio of photons emitted to photons absorbed. Highly fluorescent molecules, such as fluorescein, have quantum efficiencies that approach unity under some conditions. Species that do not fluoresce or that show very weak fluorescence have quantum efficiencies that are essentially zero.

Fluorescence and Structure

Compounds containing aromatic rings give the most intense and most useful molecular fluorescence emission. While certain aliphatic and alicyclic carbonyl compounds as well as highly conjugate double-bonded structures also fluoresce, there are very few of these compared to the number of fluorescent compounds containing aromatic systems.

Most unsubstituted aromatic hydrocarbons fluoresce in solution, with the quantum efficiency increasing with the number of rings and their degree of condensation. The simplest heterocyclics, such as pyridine, furan, thiophene, and pyrrole, do not exhibit molecular fluorescence (see **Figure 27-3**), but fused-ring structures containing these rings often do (see **Figure 27-4**). Substitution on an aromatic ring causes shifts in the wavelength of absorption maxima and corresponding changes in the fluorescence bands. In addition, substitution frequently affects the fluorescence efficiency. These effects are demonstrated by the data in the **Table 27-1**.

The Effect of Structural Rigidity

Experiments show that fluorescence is particularly favored in rigid molecules. For example, under similar measurement conditions, the quantum efficiency of fluorene is nearly 1.0 while that of biphenyl is about 0.2 (see **Figure 27-5**). The difference in behavior is a result of the increased rigidity provided by the bridging methylene group in fluorene. This rigidity lowers the rate of nonradiative relaxation to the point where relaxation by fluorescence has time to occur. There are many similar examples of this type of behavior. In addition, enhanced emission frequently results when fluorescing dyes are adsorbed on a solid surface. Once again, the added rigidity provided by the solid may account for the observed effect.

The influence of rigidity also explains the increase in fluorescence of certain organic chelating agents when they are complexed with a metal ion. For example, the fluorescence intensity of 8-hydroxyquinoline is much less than that of the zinc complex (see **Figure 27-6**).

Temperature and Solvent Effects

In most molecules, the quantum efficiency of fluorescence decreases with increasing temperature because the increased frequency of collision at elevated temperatures increases the probability of collisional relaxation. A decrease in solvent viscosity leads to the same result.

Quantum efficiency is described by the **quantum yield of fluorescence**, Φ_F.

$$\Phi_F = \frac{k_F}{k_F + k_{nr}}$$

where k_F is the first-order rate constant for fluorescence relaxation and k_{nr} is the rate constant for radiationless relaxation. See Chapter 30 for a discussion of rate constants.

❮ Many unsubstituted aromatic compounds fluoresce.

pyridine   furan

thiophene   pyrrole

Figure 27-3 Typical aromatic molecules that do not fluoresce.

❮ Rigid molecules or complexes are often fluorescent.

quinoline   isoquinoline

indole

Figure 27-4 Typical aromatic compounds that fluoresce.

TABLE 27-1

Effect of Substitution on the Fluorescence of Benzene Derivatives*

Compound	Relative Intensity of Fluorescence
Benzene	10
Toluene	17
Propylbenzene	17
Fluorobenzene	10
Chlorobenzene	7
Bromobenzene	5
Iodobenzene	0
Phenol	18
Phenolate ion	10
Anisole	20
Aniline	20
Anilinium ion	0
Benzoic acid	3
Benzonitrile	20
Nitrobenzene	0

*In ethanol solution. From W. West, *Chemical Applications of Spetroscopy, Techniques of Organic Chemistry*, Vol. IX, p. 730. New York: Interscience, 1956. Reprinted by permission of John Wiley & Sons.

fluorene
$\Phi \rightarrow 1$

biphenyl
$\Phi \rightarrow 0.2$

Figure 27-5 Effect of molecular rigidity on quantum yield. The fluorine molecule is held rigid by the central ring, so fluorescence is enhanced. The planes of the two benzene rings in biphenyl can rotate relative to one another, so fluorescence is suppressed.

nonfluorescing   fluorescing

Figure 27-6 Effect of rigidity on quantum yield in complexes. Free 8-hydroxyquinoline molecules in solution are easily deactivated through collisions with solvent molecules and do not fluoresce. The rigidity of the Zn-8-hydroxyquinoline complex enhances fluorescence.

27B EFFECT OF CONCENTRATION ON FLUORESCENCE INTENSITY

The radiant power of fluorescence emitted F is proportional to the radiant power of the excitation beam absorbed by the system:

$$F = K'(P_0 - P) \tag{27-1}$$

where P_0 is the radiant power of the beam incident on the solution and P is its power after it passes through a length b of the medium. The constant K' depends on the quantum efficiency of the fluorescence. In order to relate F to the concentration c of the fluorescing particle, we write Beer's law in the form

$$\frac{P}{P_0} = 10^{-\varepsilon bc} \tag{27-2}$$

where ε is the molar absorptivity of the fluorescing species and εbc is the absorbance. By substituting Equation 27-2 into Equation 27-1, we obtain

$$F = K'P_0(1 - 10^{-\varepsilon bc}) \tag{27-3}$$

Expansion of the exponential term in Equation 27-3 leads to

$$F = K'P_0\left[2.3\varepsilon bc - \frac{(-2.3\varepsilon bc)^2}{2!} - \frac{(-2.3\varepsilon bc)^3}{3!} - \cdots\right] \tag{27-4}$$

When $\varepsilon bc = A < 0.05$, the first term inside the brackets, $2.3\varepsilon bc$, is much larger than subsequent terms, and we can write

$$F = 2.3K'\varepsilon bcP_0 \tag{27-5}$$

or when the incident power P_0 is constant,

$$F = Kc \tag{27-6}$$

Thus, a plot of the fluorescence power emitted versus the concentration of the emitting species should be linear at low concentrations. When c becomes large enough that the absorbance exceeds about 0.05 (or the transmittance is smaller than about 0.9), the relationship represented by Equation 27-6 become nonlinear, and F lies below an extrapolation of the linear plot. This effect is a result of **primary absorption** in which the incident beam is absorbed so strongly that fluorescence is no longer proportional to concentration as shown in the more complete Equation 27-4. At very high concentrations, F reaches a maximum and may even begin to decrease with increasing concentration because of **secondary absorption**. This phenomenon occurs because of absorption of the emitted radiation by other analyte molecules. A typical plot of F versus concentration is shown in **Figure 27-7**. Note that primary and secondary absorption effects, sometimes called **inner-filter effects**, can also occur because of absorption by nonanalyte molecules in the sample matrix.

27C FLUORESCENCE INSTRUMENTATION

There are several different types of fluorescence instruments. All follow the general block diagram of Figure 25-1b. Optical diagrams of typical instruments are shown in **Figure 27-8**. If the two wavelength selectors are both filters, the instrument is called a **fluorometer**. If both wavelength selectors are monochromators, the instrument is a **spectrofluorometer**. Some instruments are hybrids and use an excitation filter along with an emission monochromator. Fluorescence instruments can incorporate a double-beam design in order to compensate for changes in the source radiant power with time and wavelength. Instruments that correct for the source spectral distribution are called **corrected spectrofluorometers**.

Radiation sources for fluorescence are usually more powerful than typical absorption sources. In fluorescence, the radiant power emitted is directly proportional to the source intensity (Equation 27-5), but absorbance is essentially independent

Figure 27-7 Calibration curve for the spectrofluorometric determination of tryptophan in soluble proteins from the lens of a mammalian eye.

Figure 27-8 Typical fluorescence instruments. A filter fluorometer is shown in (a). Note that the emission is measured at right angles to the mercury arc lamp source. Fluorescence radiation is emitted in all directions, and the 90-deg geometry avoids the detector viewing the source. The spectrofluorometer (b) uses two grating monochromators and also views the emission at right angles. The two monochromators allow the scanning of excitation spectra (excitation wavelength scanned at a fixed emission wavelength), emission spectra (emission wavelength scanned at a fixed excitation wavelength), or synchronous spectra (both wavelengths scanned with a fixed wavelength offset between the two monochromators).

of source intensity because it is related to the ratio of radiant powers as shown in Equation 27-7.

$$c = kA = k \log \left(\frac{P_0}{P} \right) \tag{27-7}$$

As a result of these different dependencies on source intensity, fluorescence methods are generally one to three orders of magnitude more sensitive than methods based on absorption. Mercury arc lamps, xenon arc lamps, xenon-mercury arc lamps, and lasers are typical fluorescence sources. Monochromators and transducers are typically similar to those used in absorption spectrophotometers. Photomultipliers are still widely used in high-sensitivity spectrofluorometers, but CCDs and photodiode arrays have become popular in recent years. The sophistication, performance characteristics, and cost of fluorometers and spectrofluorometers vary widely as with absorption spectrophotometers. Generally, fluorescence instruments are more expensive than absorption instruments of corresponding quality.

> Fluorescence methods are 10 to 1000 times more sensitive than absorption methods.

27D APPLICATIONS OF FLUORESCENCE METHODS

Fluorescence spectroscopy is not a major structural or qualitative analysis tool because molecules with subtle structural differences often have similar fluorescence spectra. Also, fluorescence bands in solution are relatively broad at room temperature. However, fluorescence has proved to be a valuable tool in oil spill identification. The source of an oil spill can often be identified by comparing the fluorescence emission spectrum of the spill sample to that of a suspected source. The vibrational structure of polycyclic hydrocarbons present in the oil makes this type of identification possible.

Fluorescence methods are used to study chemical equilibria and kinetics in much the same way as absorption spectrophotometry. Often, it is possible to study chemical reactions at lower concentrations because of the higher sensitivity of fluorescence methods. In many cases where fluorescence monitoring is not feasible, fluorescent probes or tags can be bound covalently to specific sites in molecules such as proteins, thus making them detectable via fluorescence. These tags can be used to provide information about energy transfer processes, the polarity of the protein, and the distances between reactive sites (see for example Feature 27-1).

Quantitative fluorescence methods have been developed for inorganic, organic, and biochemical species. Inorganic fluorescence methods can be divided into two classes: direct methods and indirect methods. Direct methods are based on the reaction of the analyte with a complexing agent to form a fluorescent complex. Indirect methods depend on the decrease in fluorescence, also called **quenching**, as a result of interaction of the analyte with a fluorescent reagent. Quenching methods are primarily used for the determination of anions and dissolved oxygen. Some fluorescence reagents for cations are shown in **Figure 27-9**.

Nonradiative relaxation of transition-metal chelates is so efficient that these species seldom fluoresce. It is worth noting that most transition metals absorb in the UV or visible region, while nontransition-metal ions do not. For this reason, fluorescence is often considered complementary to absorption for the determination of cations.

The number of applications of fluorescence methods to organic and biochemical problems is impressive. Among the compound types that can be determined by fluorescence are amino acids, proteins, coenzymes, vitamins, nucleic acids, alkaloids,

Figure 27-9 Some fluorometric chelating agents for metal cations. Alizarin garnet R can detect Al^{3+} at levels as low as 0.007 μg/mL. Detection of F^- with alizarin garnet R is based on quenching of the fluorescence of the Al^{3+} complex. Flavanol can detect Sn^{4+} at the 0.1-μg/mL level.

8-hydroxyquinoline
(reagent for Al, Be, and other metal ions)

alizarin garnet R
(reagent for Al, F^-)

flavanol
(reagent for Zr and Sn)

benzoin
(reagent for B, Zn, Ge, and Si)

Use of Fluorescence Probes in Neurobiology: Probing the Enlightened

Fluorescent indicators have been widely used to probe biological events in individual cells. A particularly interesting probe is the so-called ion probe that changes its excitation or emission spectrum when it binds to specific ions such as Ca^{2+} or Na^+. These indicators can be used to record events that take place in different parts of individual neurons or to monitor simultaneously the activity of a collection of neurons. In neurobiology, for example, the dye Fura-2 has been used to monitor the free intracellular calcium concentration following some pharmacological or electrical stimulation. By following the fluorescence changes with time at specific sites in the neuron, researchers can determine when and where a calcium-dependent electrical event took place.

One cell that has been studied is the Purkinje neuron in the cerebellum, which is one of the largest in the central nervous system. When this cell is loaded with the Fura-2 fluorescent indicator, sharp changes in fluorescence can be measured that correspond to individual calcium action potentials. The changes are correlated to specific sites in the cell by means of fluorescence imaging techniques. **Figure 27F-1** shows the fluorescent image on the right along with fluorescence transients, recorded as the change in fluorescence relative to the steady fluorescence $\Delta F/F$, correlated with sodium action potential spikes. The interpretation of these kinds of patterns can have important implications in understanding the details of synaptic activity.

Figure 27F-1 Calcium transients in a cerebellar Purkinje cell. The image on the right is of the cell filled with a fluorescent dye that responds to the calcium concentration. Fluorescent transients are shown on the top left recorded at areas d, p, and s in the cell. The transients in region d correspond to the dendrite region of the cell. Specific calcium signals can be correlated to the action potentials shown on the bottom left. (From V. Lev-Ram, H. Mikayawa, N. Lasser-Ross, and W. N. Ross, *J. Neurophysiol.,* **1992**, *68*, 1167. With permission of the American Physiological Society.)

Some typical polycyclic aromatic hydrocarbons found in oil spills are chrysene, perylene, pyrene, fluorene, and 1,2-benzofluorene. Most of these compounds are carcinogenic.

Molecular model of pyrene.

porphryins, steroids, flavonoids, and many metabolites.[2] Because of its sensitivity, fluorescence is widely used as a detection technique for liquid chromatographic methods (see Chapter 33), for flow analysis methods, and for electrophoresis. In addition to methods that are based on measurements of fluorescence intensity, there are many methods involving measurements of fluorescence lifetimes. Several instruments have been developed that provide microscopic images of specific species based on fluorescence lifetimes.[3]

27D-1 Methods for Inorganic Species

The most successful fluorometric reagents for the determination of cations are aromatic compounds having two or more donor functional groups that form chelates with the metal ion. A typical example is 8-hydroxyquinoline, the structure of which is given in Section 12C-3. A few other fluorometric reagents and their applications are found in **Table 27-2**. With most of these reagents, the cation is extracted into a solution of the reagent in an immiscible organic solvent, such as chloroform. The fluorescence of the organic solution is then measured. For a more complete summary of fluorometric methods for inorganic substances, see the handbook by Dean.[4]

27D-2 Methods for Organic and Biochemical Species

The number of applications of fluorometric methods to organic problems is impressive. Dean summarizes the most important of these methods in a table.[5] There are more than 200 entries under the heading "Fluorescence Spectroscopy of Some Organic Compounds," including such diverse compounds as adenine, anthranilic acid, aromatic polycyclic hydrocarbons, cysteine, guanine, isoniazid, naphthols, nerve gases sarin and tabun, proteins, salicylic acid, skatole, trypotophan, uric acid, and warfarin (Coumadin). Many medicinal agents that can be determined fluorometrically are listed, including adrenaline, morphine, penicillin, phenobarbital, procaine, reserpine, and lysergic acid diethylamide (LSD). The most important applications of

TABLE 27-2

Selected Fluorometric Methods for Inorganic Species*

Ion	Reagent	Wavelength, nm		Sensitivity, μg/mL	Interference
		Absorption	Fluorescence		
Al^{3+}	Alizarin garnet R	470	500	0.007	Be, Co, Cr, Cu, F^-, NO_3^-, Ni, PO_4^{3-}, Th, Zr
F^-	Al complex of Alizarin garnet R (quenching)	470	500	0.001	Be, Co, Cr, Cu, Fe, Ni, PO_4^{3-}, Th, Zr
$B_4O_7^{2-}$	Benzoin	370	450	0.04	Be, Sb
Cd^{2+}	2-(o-Hydroxyphenyl)-benzoxazole	365	Blue	2	NH_3
Li^+	8-Hydroxyquinoline	370	580	0.2	Mg
Sn^{4+}	Flavanol	400	470	0.1	F^-, PO_4^{3-}, Zr
Zn^{2+}	Benzoin	—	Green	10	B, Be, Sb, colored ions

*From L. Meites, ed., *Handbook of Analytical Chemistry*, New York: McGraw-Hill, 1963, pp. 6-178-6-181.

[2]See O. S. Wolfbeis, in *Molecular Luminescence Spectroscopy: Methods and Applications*, S. G. Schulman, ed., Part I, Ch. 3, New York: Wiley-Interscience, 1985.

[3]See J. R. Lakowicz, H. Szmacinski, K. Nowacyzk, K. Berndt, and M. L. Johnson, in *Fluorescence Spectroscopy: New Methods and Applications*, O. S. Wolfbeis, ed., Ch. 10, Berlin: Springer-Verlag, 1993.

[4]J. A. Dean, *Analytical Chemistry Handbook*, New York: McGraw-Hill, 1995, pp. 5.60–5.62.

[5]*Ibid.*, pp. 5.63–5.69.

fluorometry include the analysis of food products, pharmaceuticals, clinical samples, and natural products. The sensitivity and selectivity of molecular fluorescence make it a particularly valuable tool in these fields.

27E MOLECULAR PHOSPHORESCENCE SPECTROSCOPY

Phosphorescence is a photoluminescence phenomenon that is quite similar to fluorescence. To understand the difference between these two phenomena, we must consider electron spin and the difference between a **singlet state** and a **triplet state**. Ordinary molecules that are not free radicals exist in the ground state with their electron spins paired. A molecular electronic state in which all electron spins are paired is said to be a **singlet state**. The ground state of a free radical, on the other hand, is a **doublet state**, because the odd electron can assume two orientations in a magnetic field.

When one of a pair of electrons in a molecule is excited to a higher energy level, a singlet or a triplet state can be produced. In the excited singlet state, the spin of the promoted electron is still opposite that of the remaining electron. In the triplet state, however, the spins of the two electrons become unpaired and are thus parallel. These states can be represented as illustrated in **Figure 27-10**. The excited triplet state is less energetic than the corresponding excited singlet state.

Transitions from an excited singlet state to the ground singlet state produce molecular fluorescence. This singlet-singlet transition is highly probable, and thus, the lifetime of an excited singlet state is very short (10^{-5} s or less). On the other hand, transitions from an excited triplet state to the ground singlet state produce molecular phosphorescence. Because the triplet-singlet transition produces a change in electron spin, it is much less probable. As a result, the triplet state has a much longer lifetime (typically 10^{-4} to 10^4 s).

The long lifetime of phosphorescence is also one of its drawbacks. Because the excited state is relatively long-lived, nonradiational processes have time to compete with phosphorescence for deactivation. Therefore, the efficiency of the phosphorescence process, as well as the corresponding phosphorescence intensity, is relatively low. To increase the efficiency, phosphorescence is commonly observed at low temperatures in rigid media, such as glasses. Another approach is to adsorb the analyte on a solid surface or enclose it in a molecular cavity (micelle or cyclodextrin cavity), which protects the fragile triplet state. This technique is known as **room temperature phosphorescence**.

Phosphorescence materials and pigments, called **phosphors**, find many uses, including the markings of safety-related signs, such as highway exit and stop signs. Luminous watches contain a phosphor consisting of alkaline earth metal aluminates doped with rare earth elements, such as europium. Cathode-ray tubes, used in some oscilloscopes, computer monitors, and older television sets have solid-state phosphors coated on the screen, allowing actions of the electron beam to be visualized.

In room temperature phosphorescence, the triplet state of the analyte can be protected by being incorporated into a surfactant aggregate called a micelle. In aqueous solutions the aggregate has a nonpolar core due to repulsion of the polar head groups. The opposite occurs in nonpolar solvents. Cyclodextrin cavities are also used.

Micelle in aqueous solvent

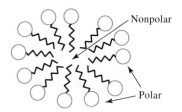

Micelle in nonaqueous solvent

Structure of micelles.

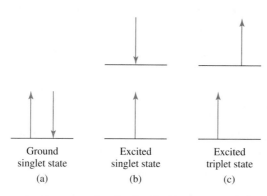

Figure 27-10 Electronic spin states of molecules. In (a), the ground electronic state is shown. In the lowest energy or ground state, the spins are always paired, and the state is a singlet state. In (b) and (c), excited electronic states are shown. If the spins remain paired in the excited state, the molecule is in an excited singlet state (b). If the spins become unpaired, the molecule is in an excited triplet state (c).

Because of its weak intensity, phosphorescence is much less widely applicable than fluorescence. However, molecular phosphorescence has been used for the determination of a variety of organic and biochemical species, including nucleic acids, amino acids, pyrine and pyrimidine, enzymes, polycyclic hydrocarbons, and pesticides. Many pharmaceutical compounds exhibit measurable phosphorescence signals. The instrumentation for phosphorescence is also somewhat more complex than that for fluorescence. Phosphorescence instruments usually discriminate phosphorescence from fluorescence by delaying the phosphorescence measurement until the fluorescence has decayed to nearly zero. Many fluorescence instruments have attachments, called **phosphoroscopes**, that allow the same instrument to be used for phosphorescence measurements.

27F CHEMILUMINESCENCE METHODS

> Fireflies produce light by the phenomenon of bioluminescence. Different firefly species flash with different on-off cycle times. Fireflies mate only with their own species. The familiar bioluminescence reaction occurs when the firefly is looking for a mate.

Chemiluminescence is produced when a chemical reaction yields an electronically excited molecule, which emits light as it returns to the ground state. Chemiluminescence reactions occur in a number of biological systems, where the process is often termed, **bioluminescence**. Examples of species exhibiting bioluminescence include the firefly, the sea pansy, certain jellyfish, bacteria, protozoa, and crustacea.[6]

One attractive feature of chemiluminescence for analytical uses is the very simple instrumentation required. Since no external source of radiation is needed for excitation, the instrument may consist of only a reaction vessel and a photomultiplier tube. Generally, no wavelength selection device is needed because the only source of radiation is the emission caused by the chemical reaction.

> Several commercial analyzers for the determination of gases are based on chemiluminescence. Nitrous oxide (NO) can be determined by reaction with ozone (O_3). The reaction converts the NO to excited NO_2 with the subsequent emission of light.

Chemiluminescence methods are known for their high sensitivities. Typical detection limits range from parts per million to parts per billion or lower. Applications include the determination of gases, such as oxides of nitrogen, ozone, and sulfur compounds; determination of inorganic species, such as hydrogen peroxide and some metal ions; immunoassay techniques; DNA probe assays; and polymerase chain reaction methods.[7]

WEB WORKS

One of the problems plaguing quantitative fluorescence measurements has been that of excessive absorption effects, often called inner-filter effects. Use a browser to find this interesting article, Q. Gu and J. E. Kenny, *Anal. Chem.*, **2009**, *81*, 420–26, **DOI:** 10.1021/ac801676j, which describes an approach for correcting fluorescence measurements for excessive absorption. (If your school does not subscribe to this online journal, find a hard copy in the school library.) Their correction method extends the range of linearity for fluorescence measurements to systems in which the absorbance is quite high. Discuss the model used by Gu and Kenny for their correction scheme. How does it differ from previous schemes that allowed corrections for absorbances up to $A \approx 2.0$? What were the major restrictions of these prior correction methods? What is the cell-shift method, and how can it be used for correcting fluorescence values? How does the Gu and Kenny approach allow corrections with common instrument geometries? How large can absorbances be in the Gu and Kenny scheme for linear fluorescence results to be obtained in the case of only a primary inner-filter effect and for both a primary- and a secondary-inner-filter effect?

---

[6]For more information on chemiluminescence and bioluminescence see O. Shimomura, *Bioluminescence: Chemical Principles and Methods,* Singapore: World Scientific Publishing, 2006; A. Roda, ed., *Chemiluminescence and Bioluminescence: Past, Present and Future,* London: Royal Society of Chemistry, 2010.

[7]See T. A. Nieman, in *Handbook of Instrumental Techniques for Analytical Chemistry,* F. A. Settle, ed., Ch. 27, Upper Saddle River, NJ: Prentice Hall, 1997.

QUESTIONS AND PROBLEMS

27-1. Briefly describe or define
*(a) fluorescence.
 (b) vibrational relaxation.
*(c) internal conversion.
 (d) phosphorescence.
*(e) Stokes shift.
 (f) quantum yield.
*(g) inner-filter effect.
 (h) excitation spectrum.

27-2. Why is spectrofluorometry potentially more sensitive than spectrophotometry?

27-3. Which compound in each of the pairs below would you expect to have a greater fluorescence quantum yield? Explain.

*(a)

phenolphthalein

fluorescein

(b)

o,o'-dihydroxyazobenzene

bis(*o*-hydroxyphenyl) hydrazine

27-4. Why do some absorbing compounds fluoresce while others do not?

*27-5. Describe the characteristics of organic compounds that fluoresce.

27-6. Explain why molecular fluorescence often occurs at a longer wavelength than the exciting radiation.

*27-7. Describe the components of a filter fluorometer and a spectrofluorometer.

27-8. Why are most fluorescence instruments double beam in design?

*27-9. Why are fluorometers often more useful than spectrofluorometers for quantitative analysis?

27-10. The reduced form of nicotinamide adenine dinucleotide (NADH) is an important and highly fluorescent coenzyme. It has an absorption maximum of 340 nm and an emission maximum at 465 nm. Standard solutions of NADH gave the following fluorescence intensities:

Concn NADH, μmol/L	Relative Intensity
0.100	2.24
0.200	4.52
0.300	6.63
0.400	9.01
0.500	10.94
0.600	13.71
0.700	15.49
0.800	17.91

(a) Construct a spreadsheet and use it to draw a calibration curve for NADH.
*(b) Find the least-squares slope and intercept for the plot in (a).
(c) Calculate the standard deviation of the slope and the standard deviation about regression for the curve.
*(d) An unknown exhibits a relative fluorescence intensity of 11.34. Use the spreadsheet to calculate the concentration of NADH.
*(e) Calculate the relative standard deviation for the result in part (d).
(f) Calculate the relative standard deviation for the result in part (d) if the reading of 12.16 was the mean of three measurements.

27-11. The volumes of a 1.10 ppm standard solution of Zn^{2+} shown in the table below were pipetted into separatory funnels each containing 5.00 mL of an unknown zinc solution. Each was extracted with three 5-mL aliquots of CCl_4 containing an excess of 8-hydroxyquinoline. The extracts were then diluted to 25.0 mL, and their fluorescence measured with a fluorometer. The results were:

Volume Std Zn^{2+}, mL	Fluorometer Reading
0.000	6.12
4.00	11.16
8.00	15.68
12.00	20.64

(a) Construct a working curve from the data.
(b) Calculate a linear least-squares equation for the data.

(c) Calculate the standard deviation of the slope and the standard deviation about regression.

(d) Calculate the concentration of zinc in the sample.

(e) Calculate a standard deviation for the result in part (d).

*27-12. Quinine in a 1.664-g antimalarial tablet was dissolved in sufficient 0.10 M HCl to give 500 mL of solution. A 15.00-mL aliquot was then diluted to 100.0 mL with the acid. The fluorescence intensity for the diluted sample at 347.5 nm provided a reading of 288 on an arbitrary scale. A standard 100 ppm quinine solution registered 180 when measured under conditions identical to those for the diluted sample. Calculate the mass of quinine in milligrams in the tablet.

27-13. The determination in Problem 27-12 was modified to use the standard addition method. In this case, a 2.196-g tablet was dissolved in sufficient 0.10 M HCl to give 1.000 L. Dilution of a 20.00-mL aliquot to 100 mL produced a solution that gave a reading of 540 at 347.5 nm. A second 20.00-mL aliquot was mixed with 10.0 mL of 50 ppm quinine solution before dilution to 100 mL. The fluorescence intensity of this solution was 600. Calculate the concentration in parts per million of quinine in the tablet.

27-14. **Challenge Problem:** The following volumes of a standard 10.0 ppb F^- solution were added to four 10.00-mL aliquots of a water sample: 0.00, 1.00, 2.00, and 3.00 mL. Precisely 5.00 mL

of a solution containing an excess of the strongly absorbing Al-acid Alizarin Garnet R complex was added to each of the four solutions, and they were each diluted to 50.0 mL. The fluorescence intensity of the four solutions were as follows:

V_s, mL	Meter reading
0.00	68.2
1.00	55.3
2.00	41.3
3.00	28.8

(a) Explain the chemistry of the analytical method.

(b) Construct a plot of the data.

(c) Use the fact that the fluorescence decreases with increasing amounts of the F^- standard to derive a relationship like Equation 26-1 for multiple standard additions. Then use this relationship to obtain an equation for the unknown concentration c_x in terms of the slope and intercept of the standard additions plot, similar to Equation 26-2.

(d) Use linear least squares to find the equation for the line representing the relationship between the decrease in fluorescence and the volume of standard fluoride V_s.

(e) Calculate the standard deviation of the slope and intercept.

(f) Calculate the concentration of F^- in the sample in ppb.

(g) Calculate the standard deviation of the result in (e).

Atomic Spectroscopy

CHAPTER 28

Water pollution remains a serious problem in the United States and in other industrial countries. The photo here shows land left over from strip mining in Belmont County, Ohio. The various water pools shown are contaminated with waste chemicals. The large pool to the right of center contains sulfuric acid, and the smaller pools contain manganese and cadmium. Trace metals in contaminated water samples are often determined by a multielement technique such as inductively coupled plasma atomic emission spectroscopy. Single-element techniques such as atomic absorption spectrometry are also used. Atomic emission and atomic absorption methods are described in this chapter.

© Charles E. Rotkin/CORBIS

Atomic spectroscopic methods are used for the qualitative and quantitative determination of more than 70 elements. Typically, these methods can detect parts-per-million to parts-per-billion amounts, and in some cases, even smaller concentrations. Atomic spectroscopic methods are also rapid, convenient, and usually of high selectivity. These methods can be divided into two groups; **optical atomic spectrometry**[1] and **atomic mass spectrometry**. We discuss optical methods in this chapter and mass spectrometry in Chapter 29.

Spectroscopic determination of atomic species can only be performed on a gaseous medium in which the individual atoms or elementary ions, such as Fe^+, Mg^+, or Al^+, are well separated from one another. Consequently, the first step in all atomic spectroscopic procedures is **atomization**, a process in which a sample is volatilized and decomposed in such a way as to produce gas-phase atoms and ions. The efficiency and reproducibility of the atomization step can have a large influence on the sensitivity, precision, and accuracy of the method. In short, atomization is a critical step in atomic spectroscopy.

Table 28-1 lists several methods that are used to atomize samples for atomic spectroscopy. Inductively coupled plasmas, flames, and electrothermal atomizers are the most widely used atomization methods. We consider these three atomization methods and the direct current plasma in this chapter. Flames and electrothermal atomizers are found in atomic absorption (AA) spectrometry, while the inductively coupled plasma is used in optical emission and in atomic mass spectrometry.

Atomization is a process in which a sample is converted into gas-phase atoms or elementary ions.

---

[1]References that deal with the theory and applications of optical atomic spectroscopy include Jose A. C. Broekaert, *Analytical Atomic Spectrometry with Flames and Plasma,* Weinheim, Germany: Wiley-VCH, 2002; L. H. J. Lajunen and P. Peramaki, *Spectrochemical Analysis by Atomic Absorption and Emission,* 2nd ed, Cambridge: Royal Society of Chemistry, 2004; J. D. Ingle and S. R. Crouch, *Spectrochemical Analysis,* Chs. 7–11, Upper Saddle River, NJ: Prentice-Hall, 1988.

TABLE 28-1

Classification of Atomic Spectroscopic Methods

Atomization Method	Typical Atomization Temperature, °C	Types of Spectroscopy	Common Name and Abbreviation
Inductively coupled plasma	6000–8000	Emission	Inductively coupled plasma atomic emission spectroscopy, ICPAES
		Mass	Inductively coupled plasma mass spectrometry, ICP-MS (see Chapter 29)
Flame	1700–3150	Absorption	Atomic absorption spectroscopy, AAS
		Emission	Atomic emission spectroscopy, AES
		Fluorescence	Atomic fluorescence spectroscopy, AFS
Electrothermal	1200–3000	Absorption	Electrothermal AAS
		Fluorescence	Electrothermal AFS
Direct-current plasma	5000–10,000	Emission	DC plasma spectroscopy, DCP
Electric arc	3000–8000	Emission	Arc-source emission spectroscopy
Electric spark	Varies with time and position	Emission	Spark-source emission spectroscopy
		Mass	Spark-source mass spectroscopy

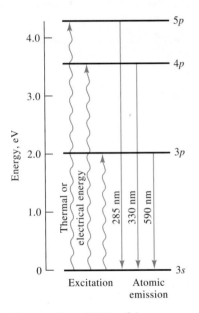

Figure 28-1   Origin of three sodium emission lines.

Atomic p orbitals are in fact split into two energy levels that differ only slightly in energy. The energy difference between the two levels is so small that the emission appears to be a single line, as suggested by Figure 28-1. With a very high-resolution spectrometer, each of the lines appears as two closely spaced lines known as a **doublet**.

28A   ORIGINS OF ATOMIC SPECTRA

Once the sample has been converted into gaseous atoms or ions, various types of spectroscopy can be performed. We consider here only optical spectrometric methods. With gas-phase atoms or ions, there are no vibrational or rotational energy states. This absence means that only electronic transitions occur. Thus, atomic emission, absorption, and fluorescence spectra are made up of a limited number of narrow **spectral lines**.

28A-1 Emission Spectra

In atomic emission spectroscopy, analyte atoms are excited by heat or electrical energy, as illustrated in Figure 24-4 (see color plate 16 for emission spectra of several elements). The energy typically is supplied by a plasma, a flame, a low-pressure discharge, or by a high-powered laser. **Figure 28-1** is a partial energy level diagram for atomic sodium showing the source of three of the most prominent emission lines. Before the external energy source is applied, the sodium atoms are usually in their lowest-energy or **ground state**. The applied energy then causes the atoms to be momentarily in a higher-energy or **excited state**. With sodium atoms, for example, in the ground state, the single valence electrons are in the $3s$ orbital. External energy promotes the outer electrons from their ground state $3s$ orbitals to $3p$, $4p$, or $5p$ excited-state orbitals. After a few nanoseconds, the excited atoms relax to the ground state giving up their energy as photons of visible or ultraviolet radiation. As shown in Figure 28-1, the wavelength of the emitted radiation is 590, 330, and 285 nm. A transition to or from the ground state is called a **resonance transition**, and the resulting spectral line is called a **resonance line**.

28A-2 Absorption Spectra

In atomic absorption spectroscopy, an external source of radiation impinges on the analyte vapor, as illustrated in Figure 24-5. If the source radiation is of the appropriate frequency (wavelength), it can be absorbed by the analyte atoms and promote them to excited states. **Figure 28-2a** shows three of several absorption lines for sodium vapor. The source of these spectral lines is indicated in the partial energy diagram shown in Figure 28-2b. In this instance, absorption of radiation of 285, 330, and 590 nm excites the single outer electron of sodium from its ground state 3s energy level to the excited 3p, 4p, and 5p orbitals, respectively. After a few nanoseconds, the excited atoms relax to their ground state by transferring their excess energy to other atoms or molecules in the medium.

The absorption and emission spectra for sodium are relatively simple and consist of only a few lines. For elements that have several outer electrons that can be excited, absorption and emission spectra may be much more complex.

28A-3 Fluorescence Spectra

In atomic fluorescence spectroscopy, an external source is used just as in atomic absorption, as shown in Figure 24-6. Instead of measuring the attenuated source radiant power, the radiant power of fluorescence, P_F, is measured, usually at right angles to the source beam. In such experiments, we must avoid or discriminate against scattered source radiation. Atomic fluorescence is often measured at the same wavelength as the source radiation and then is called **resonance fluorescence**.

28A-4 Widths of Atomic Spectral Lines

Atomic spectral lines have finite widths. With ordinary spectrometers, the observed line widths are determined not by the atomic system but by the spectrometer properties. With very high resolution spectrometers or with interferometers, the actual widths of spectral lines can be measured. Several factors contribute to atomic spectral line widths.

Natural Broadening

The natural width of an atomic spectral line is determined by the lifetime of the excited state and Heisenberg's uncertainty principle. The shorter the lifetime, the broader the line and vice versa. Typical radiative lifetimes of atoms are on the order of 10^{-8} s, leading to natural line widths on the order of 10^{-5} nm.

Collisional Broadening

Collisions between atoms and molecules in the gas-phase lead to deactivation of the excited state and thus broadening of the spectral line. The amount of broadening increases with the concentrations (partial pressures) of the collision partners. As a result, collisional broadening is sometimes called **pressure broadening**. Pressure broadening increases with increasing temperature. Collisional broadening is highly dependent on the gaseous medium. For Na atoms in flames, such broadening can be as large as 3×10^{-3} nm. In energetic media, such as flames and plasmas, collisional broadening greatly exceeds natural broadening.

Doppler Broadening

Doppler broadening results from the rapid motion of atoms as they emit or absorb radiation. Atoms moving toward the detector emit wavelengths that are slightly shorter than the wavelengths emitted by atoms moving at right angles to the detector. This difference is a manifestation of the well-known Doppler effect shown in **Figure 28-3a**. The effect is reversed for atoms moving away from the detector as can be seen in Figure 28-3b. The net effect is an increase in the width of the emission line. For precisely the same reason, the Doppler effect also causes broadening of absorption lines. This type of broadening becomes more pronounced as the

> Note that the wavelengths of the absorption and emission lines for sodium are identical.

(a)

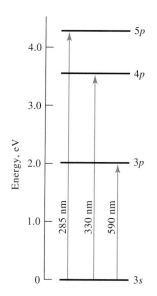

(b)

Figure 28-2   (a) Partial absorption spectrum for sodium vapor. (b) Electronic transitions responsible for the absorption lines in (a).

(a)

(b)

Figure 28-3 Cause of Doppler broadening. (a) When an atom moves toward a photon detector and emits radiation, the detector sees wave crests more often and detects radiation of higher frequency. (b) When an atom moves away from a photon detector and emits radiation, the detector sees crests less frequently and thus detects radiation of lower frequency. The result in an energetic medium is a statistical distribution of frequencies and thus a broadening of the spectral lines.

Both Doppler broadening and pressure broadening are temperature dependent.

flame temperature increases because of the increased velocity of the atoms. Doppler broadening can be a major contributor to overall line widths. For Na, in flames, the Doppler line widths are on the order of 4×10^{-3} to 5×10^{-3} nm.

28B   PRODUCTION OF ATOMS AND IONS

In all atomic spectroscopic techniques, we must atomize the sample, converting it into gas-phase atoms, and ions. Samples usually enter the atomizer in solution form, although we sometimes introduce gases and solids. Hence, the atomization device must normally perform the complex task of converting analyte species in solution into gas-phase free atoms and/or elementary ions.

28B-1 Sample Introduction Systems

Atomization devices fall into two classes: **continuous atomizers** and **discrete atomizers**. With continuous atomizers, such as plasmas and flames, samples are introduced in a steady, continuous stream. With discrete atomizers, individual samples are injected by means of a syringe or autosampler. The most common discrete atomizer is the **electrothermal atomizer**.

The general methods for introducing solution samples into plasma and flames are illustrated in **Figure 28-4**. Direct **nebulization** is most often used. In this case, the **nebulizer** constantly introduces the sample in the form of a fine spray of droplets, called an **aerosol**. When a sample is introduced into a flame or plasma continuously, a steady-state population of atoms, molecules, and ions develops. When flow injection or liquid chromatography is used, a plug of sample is introduced with a concentration that varies with time. This procedure results in a time-dependent vapor population. The complex processes that must occur in order to produce free atoms or elementary ions are illustrated in **Figure 28-5**.

Discrete solution samples are introduced by transferring an aliquot of the sample to the atomizer. The vapor cloud produced with electrothermal atomizers is transient because of the limited amount of sample available and the removal of vapor through diffusion and other processes.

To **nebulize** means to convert a liquid into a fine spray or mist.

An **aerosol** is a suspension of finely divided liquid or solid particles in a gas.

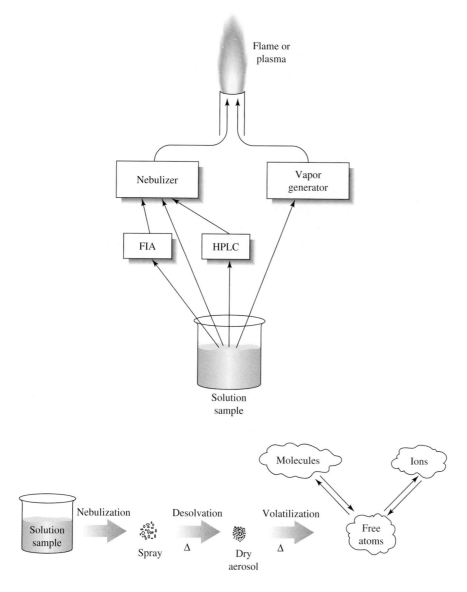

Figure 28-4   Continuous sample introduction methods. Samples are frequently introduced into plasma or flames by means of a nebulizer which produces a mist or spray. Samples can be introduced directly to the nebulizer or by means of a flow injection (FIA) or high-performance liquid chromatography (HPLC). In some cases, samples are separately converted to a vapor by a vapor generator, such as a hydride generator or an electrothermal vaporizer.

Figure 28-5   Processes leading to atoms, molecules, and ions with continuous sample introduction into a plasma or flame. The solution sample is converted into a spray by the nebulizer. The high temperature of the flame or plasma causes the solvent to evaporate leaving dry aerosol particles. Further heating volatilizes the particles producing atomic, molecular, and ionic species. These species are often in equilibrium at least in localized regions.

Solid samples can be introduced into plasmas by vaporizing them with an electrical spark or with a laser beam. Laser volatilization, often called **laser ablation**, has become a popular method for introducing samples into inductively coupled plasmas. In laser ablation, a high-powered laser beam, often a Nd:YAG or excimer laser, is directed onto a portion of the solid sample. The sample is then vaporized by radiative heating. The plume of vapor produced is swept into the plasma by means of a carrier gas.

28B-2 Plasma Sources

Plasma atomizers, which became available commercially in the mid-1970s, offer several advantages for analytical atomic spectrosocopy.[2] Plasma atomization has been used for atomic emission, atomic fluorescence, and atomic mass spectrometry (see Chapter 29).

---

[2]For a detailed discussion of the various plasma sources, see S. J. Hill, *Inductively Coupled Plasma Spectrometry and Its Applications*, 2nd ed., Oxford, UK: Wiley-Blackwell, 2007; *Inductively Coupled Plasmas in Analytical Atomic Spectroscopy*, 2nd ed., A. Montaser and D. W. Golightly, eds., New York: Wiley-VCH Publishers, 1992; *Inductively Coupled Plasma Emission Spectroscopy*, Parts 1 and 2, P. W. J. M. Boumans, ed., New York: Wiley 1987.

A **plasma** is a hot, partially ionized gas. It contains relatively high concentrations of ions and electrons.

By definition, a **plasma** is a conducting gaseous mixture containing a significant concentration of ions and electrons. In the argon plasma used for atomic spectroscopy, argon ions and electrons are the principal conducting species, although cations from the sample also contribute. Once argon ions are formed in a plasma, they are capable of absorbing sufficient power from an external source to maintain the temperature at a level at which further ionization sustains the plasma indefinitely. Temperatures as great as 10,000 K are achieved in this way.

Three power sources have been used in argon plasma spectroscopy. One is a dc arc source capable of maintaining a current of several amperes between electrodes immersed in the argon plasma. The second and third are powerful radio-frequency and microwave-frequency generators through which the argon flows. Of the three, the radio-frequency, or **inductively coupled plasma** (ICP), source offers the greatest advantage in terms of sensitivity and freedom from interference. It is commercially available from a number of instrument companies for use in optical emission and mass spectroscopy. A second source, the **dc plasma source** (DCP), has seen some commercial success and has the virtues of simplicity and lower cost.

Inductively Coupled Plasmas

Figure 28-6 is a schematic drawing of an inductively coupled plasma source. The source consists of three concentric quartz tubes through which streams of argon flow at a total rate of between 11 and 17 L/min. The diameter of the largest tube is about 2.5 cm. Surrounding the top of this tube is a water-cooled induction coil powered by a radio-frequency generator, which radiates 0.5 to 2 kW of power at 27.12 MHz or 40.68 MHz. Ionization of the flowing argon is initiated by a spark from a Tesla coil. The resulting ions and their associated electrons then interact with the fluctuating magnetic field (labeled H in Figure 28-6) produced by the induction coil I. This interaction causes the ions and electrons within the coil to flow in the closed annular

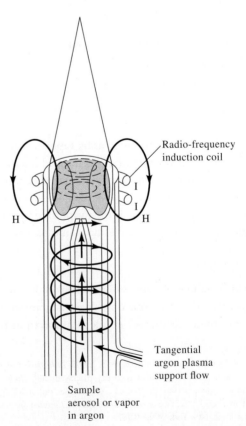

Radio-frequency
induction coil

Tangential
argon plasma
support flow

Sample
aerosol or vapor
in argon

Figure 28-6   Inductively coupled plasma source. (From V. A. Fassel, Science, 1978, 202, 185. Reprinted with permission from AAAS.)

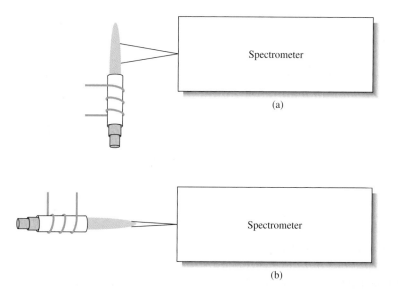

(a)

(b)

Figure 28-7 Viewing geometries for ICP sources. (a) Radial geometry used in many ICP atomic emission spectrometers. (b) Axial geometry used in ICP mass spectrometer and in several ICP atomic emission spectrometers.

paths shown in the figure. The resistance of the ions and electrons to this flow of charge causes ohmic heating of the plasma.

The temperature of the ICP is high enough that it must be thermally isolated from the quartz cylinder. Isolation is achieved by flowing argon tangentially around the walls of the tube, as indicated by the arrows in Figure 28-6. The tangential flow cools the inside walls of the central tube and centers the plasma radially.

Viewing the plasma at right angles as shown in **Figure 28-7a** is called the **radial viewing geometry**. Recent ICP instruments have incorporated the **axial viewing geometry**, as shown in Figure 28-7b, in which the torch is turned 90°. The axial geometry was originally made popular for torches that were used as ionization sources for mass spectrometry (see Chapter 29). More recently, axial torches have become available for emission spectrometry. Several companies, in fact, manufacture torches that can be switched from axial to radial viewing geometry in atomic emission spectrometry. The radial geometry provides better stability and precision, while the axial geometry is used for achieving lower detection limits.

During the 1980s, low-flow, low-power torches appeared on the market. Typically, these torches require a total argon flow of lower than 10 L/min and require less than 800 W of radio-frequency power.

Sample Introduction.   Samples can be introduced into the ICP by argon flowing at about 1 L/min through the central quartz tube. The sample can be an aerosol, a thermally generated vapor, or a fine powder. The most common sample introduction is by means of the concentric glass nebulizer shown in **Figure 28-8**. The sample is transported to the tip by the **Bernoulli effect**. This transport process is called **aspiration**. The high-velocity gas breaks the liquid up into fine droplets of various sizes, which are then carried into the plasma.

Another popular type of nebulizer is the cross-flow design. In this nebulizer, a high-velocity gas flows across a capillary tip at right angles causing the same Bernoulli effect. Often, in this type of nebulizer, the liquid is pumped through the capillary with a peristaltic pump. Many other types of nebulizers are available for higher efficiency, for samples with high solids content, and for producing ultrafine mists.

Plasma Appearance and Spectra.   The typical plasma has a very intense, brilliant white, opaque core topped by a flamelike tail. The core, which extends a few millimeters above the tube, produces a spectral continuum with the atomic spectrum for argon superimposed. The continuum is typical of ion-electron recombination

Figure 28-8   The Meinhard nebulizer. The nebulizing gas flows through an opening that concentrically surrounds the capillary. This arrangement causes a reduced pressure at the tip and aspiration of the sample. The high-velocity gas at the tip breaks up the solution into a mist or spray of various sized droplets. (Courtesy of Meinhard-Elemental Scientific.)

reactions and **bremsstrahlung**, which is continuum radiation produced when charged particles are slowed or stopped.

In the region 10 to 30 mm above the core, the continuum fades, and the plasma becomes slightly transparent. Spectral observations are generally made 15 to 20 mm above the induction coil where the temperatures can be as high as 5000 to 6000 K. In this region, the background radiation consists primarily of Ar lines, OH band emission, and some other molecular bands. Many of the most sensitive analyte lines in this region of the plasma are from ions such as Ca^+, Cd^+, Cr^+, and Mn^+. Above this second region is the "tail flame" where temperatures are similar to those in an ordinary flame ($\approx$ 3000 K). This lower temperature region can be used to determine easily excited elements such as alkali metals.

Analyte Atomization and Ionization.   By the time the analyte atoms and ions reach the observation point in the plasma, they have spent about 2 ms in the plasma at temperatures ranging from 6000 to 8000 K. The residence times are two to three times longer, and the temperatures are substantially higher than those attainable in the hottest combustion flames (acetylene/nitrous oxide). As a consequence, desolvation and vaporization are essentially complete, and the atomization efficiency is quite high. Therefore, there are fewer chemical interferences in ICPs than in combustion flames. Surprisingly, ionization interference effects are small or nonexistent because the large concentration of electrons from the ionization of argon maintains a more or less constant electron concentration in the plasma.

Several other advantages are associated with the ICP when compared with flames and other plasma sources. Atomization occurs in a chemically inert environment in contrast to flames where the environment is violent and highly reactive. In addition, the temperature cross section of the plasma is relatively uniform. The plasma also has a rather thin optical path length that minimizes self-absorption (see Section 28C-2). As a result, calibration curves are usually linear over several orders of magnitude of concentration. Ionization of analyte elements can be significant in typical ICPs. This characteristic has led to the use of the ICP as an ionization source for mass spectrometry, as discussed in Chapter 29. One significant disadvantage of the ICP is that it is not very tolerant of organic solvents. Carbon deposits tend to build up on the quartz tube and can lead to cross-contamination and clogging.

DC and Other Plasma Sources

Direct-current plasma jets were first described in the 1920s and have been systematically investigated as sources for emission spectroscopy. In the early 1970s, the first commercial direct-current plasma (DCP) was introduced. The source was quite popular, particularly among soil scientists and geochemists for multielement analysis.

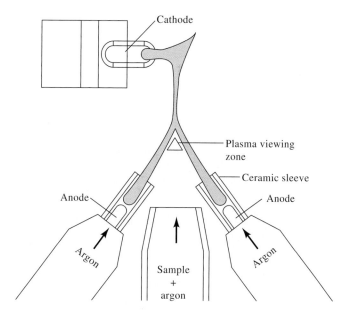

Figure 28-9   Diagram of a three-electrode dc plasma jet. Two separate dc plasmas have a single common cathode. The overall plasma burns in the form of an upside down Y. Sample can be introduced as an aerosol from between the two graphite anodes. Observation of emission in the region beneath the strongly emitting plasma core avoids much of the plasma background emission.

Figure 28-9 is a diagram of a dc plasma source for the excitation of emission spectra. This plasma-jet source consists of three electrodes arranged in an inverted Y configuration. A graphite anode is located in each arm of the Y, and a tungsten cathode is located at the inverted base. Argon flows from the two anode blocks toward the cathode. The plasma jet is formed when the cathode is momentarily brought into contact with the anodes. Ionization of the argon occurs, and the current that develops (≈ 14 A) generates additional ions to sustain itself indefinitely. The temperature is more than 8000 K in the arc core and about 5000 K in the viewing region. The sample is aspirated into the area between the two arms of the Y, where it is atomized, excited, and its spectrum viewed.

Spectra produced by the DCP tend to have fewer lines than those produced by the ICP, and the lines formed in the DCP are largely from atoms rather than ions. Sensitivities achieved with the DCP appear to range from an order of magnitude lower to about the same as those obtainable with the ICP. The reproducibilities of the two systems are similar. Significantly less argon is required for the dc plasma, and the auxiliary power supply is simpler and less expensive. Also, the DCP is able to handle organic solutions and aqueous solutions with high solids content better than the ICP. Sample volatilization is often incomplete with the DCP, however, because of the short residence times in the high-temperature region. Also, the optimum viewing region with the DCP is quite small so that optics have to be carefully aligned to magnify the source image. In addition, the graphite electrodes must be replaced every few hours, whereas the ICP requires little maintenance.

28B-3 Flame Atomizers

A flame atomizer consists of a pneumatic nebulizer, which converts the sample solution into a mist, or aerosol, that is then introduced into a burner. The same types of nebulizers that are used with ICPs are used with flame atomizers. The concentric nebulizer is the most popular. In most atomizers, the high-pressure gas is the oxidant, with the aerosol-containing oxidant being mixed subsequently with the fuel.

The burners used in flame spectroscopy are most often premixed, laminar flow burners. **Figure 28-10** is a diagram of a typical commercial laminar flow burner for atomic absorption spectroscopy that employs a concentric tube nebulizer. The aerosol flows into a **spray chamber** where it encounters a series of baffles that remove all but the finest droplets. As a result, most of the sample collects in the bottom of the

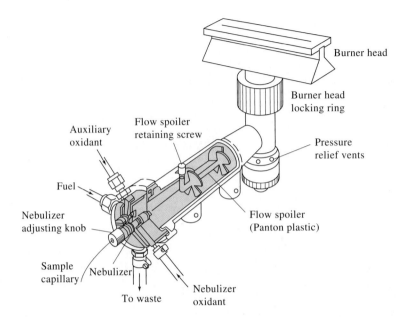

Figure 28-10 A laminar flow burner used in flame atomic absorption spectroscopy. (Reprinted by permission of PerkinElmer Corporation, Waltham, MA.)

Modern flame atomic absorption instruments use laminar flow burners almost exclusively. ❯

spray chamber, where it is drained to a waste container. Typical solution flow rates are 2 to 5 mL/min. The sample spray is also mixed with fuel and oxidant gas in the spray chamber. The aerosol, oxidant, and fuel are then burned in a slotted burner, which provides a flame that is usually 5 or 10 cm in length.

Laminar flow burners of the type shown in Figure 28-10 provide a relatively quiet flame and a long path length. These properties tend to enhance sensitivity for atomic absorption and reproducibility. The mixing chamber in this type of burner contains a potentially explosive mixture, which can be ignited by flashback if the flow rates are not sufficient. Note that, for this reason, the burner in Figure 28-10 is equipped with pressure relief vents.

Properties of Flames

When a nebulized sample is carried into a flame, the droplets are desolvated in the **primary combustion zone**, which is located just above the tip of the burner, as shown in **Figure 28-11**. The resulting finely divided solid particles are carried to a region in the center of the flame called the **inner cone**. Here, in this hottest part of

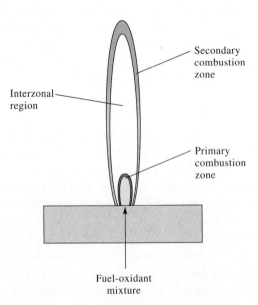

Figure 28-11 Regions of a flame.

the flame, the particles are vaporized and converted to gaseous atoms, elementary ions, and molecular species (see Figure 28-5). Excitation of atomic emission spectra also takes place in this region. Finally, the atoms, molecules, and ions are carried to the outer edge, or **outer cone**, where oxidation may occur before the atomization products disperse into the atmosphere. Because the velocity of the fuel/oxidant mixture through the flame is high, only a fraction of the sample undergoes all these processes. Unfortunately, a flame is not a very efficient atomizer.

Types of Flames Used in Atomic Spectroscopy

Table 28-2 lists the common fuels and oxidants found in flame spectroscopy and the approximate range of temperatures achieved with each of these mixtures. Note that, when the oxidant is air, temperatures are in the range of 1700 to 2400°C. At these temperatures, only easily excitable species, such as the alkali and alkaline earth metals, produce usable emission spectra. For heavy-metal species, which are not so easily excited, oxygen or nitrous oxide must be used as the oxidant. These oxidants produce temperatures of 2500 to 3100°C with common fuels.

Effects of Flame Temperature

Both emission and absorption spectra are affected in a complex way by variations in flame temperature. In both cases, higher temperatures increase the total atom population of the flame and thus the sensitivity. With certain elements, such as the alkali metals, however, this increase in atom population is more than offset by the loss of atoms by ionization.

Flame temperature determines to a large extent the efficiency of atomization, which is the fraction of the analyte that is desolvated, vaporized, and converted to free atoms and/or ions. The flame temperature also determines the relative number of excited and unexcited atoms in a flame. In an air/acetylene flame, for example, calculations show that the ratio of excited to unexcited magnesium atoms is about 10^{-8}, whereas in an oxygen/acetylene flame, which is about 700°C hotter, this ratio is about 10^{-6}. Hence, control of temperature is very important in flame emission methods. For example, with a 2500°C flame, a temperature increase of 10°C causes the number of sodium atoms in the excited $3p$ state to increase by about 3%. In contrast, the corresponding *decrease* in the much larger number of ground state atoms is only about 0.002%. Therefore, at first glance, emission methods, based as they are on the population of *excited atoms*, require much closer control of flame temperature than do absorption procedures in which the analytical signal depends upon the number of *unexcited atoms*. However, in practice because of the temperature dependence of the atomization step, both methods show similar dependencies.

The number of unexcited atoms in a typical flame exceeds the number of excited atoms by a factor of 10^3 to 10^{10} or more. This fact suggests that absorption methods should show lower detection limits (DLs) than emission methods. In fact, however, several other variables also influence detection limits, and the two methods tend to complement each other in this regard. **Table 28-3** illustrates this point.

Absorption and Emission Spectra in Flames

Both atomic and molecular emission and absorption can be measured when a sample is atomized in a flame. A typical flame-emission spectrum was shown in Figure 24-19. Atomic emissions in this spectrum are made up of narrow lines, such as that for sodium at about 330 nm, potassium at approximately 404 nm, and calcium at 423 nm. Atomic spectra are thus called **line spectra**. Also present are emission bands that result from excitation of molecular species such as MgOH, MgO,

TABLE 28-2

Flames Used in Atomic Spectroscopy

Fuel and Oxidant	Temperature, °C
*Gas/Air	1700–1900
*Gas/O$_2$	2700–2800
H$_2$/air	2000–2100
H$_2$/O$_2$	2500–2700
†C$_2$H$_2$/air	2100–2400
†C$_2$H$_2$/O$_2$	3050–3150
†C$_2$H$_2$/N$_2$O	2600–2800

*Propane or natural gas
†Acetylene

The width of atomic emission lines in flames is on the order of 10^{-3} nm. The width can be measured with an interferometer.

TABLE 28-3

Comparison of Detection Limits for Various Elements by Flame Atomic Absorption and Flame Atomic Emission Methods*

Flame Emission Shows Lower DLs	DLs about the Same	AA Shows Lower DLs
Al, Ba, Ca, Eu, Ga, Ho, In, K, La, Li, Lu, Na, Nd, Pr, Rb, Re, Ru, Sm, Sr, Tb, Tl, Tm, W, Yb	Cr, Cu, Dy, Er, Gd, Ge, Mn, Mo, Nb, Pd, Rh, Sc, Ta, Ti, V, Y, Zr	Ag, As, Au, B, Be, Bi, Cd, Co, Fe, Hg, Ir, Mg, Ni, Pb, Pt, Sb, Se, Si, Sn, Te, Zn

*Adapted with permission from E. E. Pickett and S. R. Koirtyohann, *Anal. Chem.*, **1969**, 41, 28A–42A. **DOI**: 10.1021/ac50159a003. Copyright 1969 American Chemical Society.

CaOH, and OH. These bands form when vibrational transitions are superimposed on electronic transitions to produce many closely spaced lines that are not completely resolved by the spectrometer. Because of this, molecular spectra are often referred to as **band spectra**.

Atomic absorption spectra are seldom recorded because a high-resolution spectrometer or an interferometer would be required. A high-resolution absorption spectrum would have much the same general appearance as Figure 24-19 and would contain both atomic and molecular absorption components. The vertical axis in this case would be absorbance rather than relative power.

Ionization in Flames

Because all elements ionize to some degree in a flame, the hot medium contains a mixture of atoms, ions, and electrons. For example, when a sample containing barium is atomized, the equilibrium

$$Ba \rightleftharpoons Ba^+ + e^-$$

> Ionization of an atomic species in a flame is an equilibrium process that can be described by the usual mathematics of chemical equilibria.

> The spectrum of an atom is entirely different from that of its ion.

is established in the inner cone of the flame. The position of this equilibrium depends on the temperature of the flame and the total concentration of barium as well as on the concentration of the electrons produced from the ionization of *all elements* present in the sample. At the temperatures of the hottest flames (> 3000 K), nearly half of the barium is present in ionic form. Because the emission and absorption spectra of Ba and Ba^+ are totally different, two spectra for barium appear, one for the atom and one for its ion. Flame temperature again plays an important role in determining the fraction of the analyte ionized.

28B-4 Electrothermal Atomizers

Electrothermal atomizers, which first appeared on the market about 1970, generally provide enhanced sensitivity because the entire sample is atomized in a short period and because the average residence time of the atoms in the optical path is a second or more.[3] Also, samples are introduced into a confined-volume furnace, and so, they are not diluted nearly as much as they would be in a plasma or flame. Electrothermal atomizers are used for atomic absorption and atomic fluorescence measurements, but they have not been generally applied for emission work. They are, however, also used for vaporizing samples in inductively coupled plasma emission spectroscopy.

With electrothermal atomizers, a few microliters of sample are deposited in the furnace by syringe or autosampler. Next, a programmed series of heating events occurs: **drying**, **ashing**, and **atomization**. During the drying step, the sample is evaporated at a

---

[3]For detailed discussions of electrothermal atomizers, see L. H. J. Lajunen and P. Peramaki, *Spectrochemical Analysis by Atomic Absorption and Emission,* 2nd ed., Ch. 3, Cambridge, Royal Society of Chemistry, 2004; B. E. Erickson, *Anal. Chem.,* **2000**, *72,* 543A; *Electrothermal Atomization for Analytical Atomic Spectrometry,* K. W. Jackson, ed., New York: Wiley, 1999; D. J. Butcher and J. Sneddon, *A Practical Guide to Graphite Furnace Atomic Absorption Spectrometry,* New York: Wiley, 1998; C. W. Fuller, *Electrothermal Atomization for Atomic Absorption Spectroscopy,* London: Chemical Society, 1977.

relatively low temperature, usually 110°C. The temperature is then increased to 300 to 1200°C, and the organic matter is ashed or converted to H_2O and CO_2. After ashing, the temperature is rapidly increased to perhaps 2000 to 3000°C, causing the sample to vaporize and atomize. Atomization of the sample occurs in a period of a few milliseconds to seconds. The absorption or fluorescence of the vapor is then measured in the region immediately above the heated surface before the vapor can escape the furnace.

Atomizer Designs

Commercial electrothermal atomizers are small, electrically heated tubular furnaces. **Figure 28-12a** is a cross-sectional view of a commercial electrothermal atomizer. Atomization occurs in a cylindrical graphite tube that is open at both ends and has a central hole for introduction of sample. The tube is about 5 cm long and has an internal diameter of somewhat less than 1 cm. The interchangeable graphite tube fits snugly into a pair of cylindrical graphite electrical contacts located at the two ends of the tube. These contacts are held in a water-cooled metal housing. An external stream of inert gas bathes the tube and prevents it from being incinerated in air. A second internal stream flows into the two ends of the tube and out the central sample port. This stream not only excludes air but also serves to carry away vapors generated from the sample matrix during the first two heating stages.

Figure 28-12b shows the L'vov platform, which is often used in graphite furnaces. The platform is also graphite and is located beneath the sample entrance port. The sample is evaporated and ashed on this platform in the usual way. When the tube temperature is raised rapidly, however, atomization is delayed since the sample is no longer directly on the furnace wall. As a consequence, atomization occurs in an environment in which the temperature is not changing as rapidly as in other atomizers. The resulting signals are more reproducible than those from conventional systems.

Several other designs of electrothermal atomizers are available commercially.

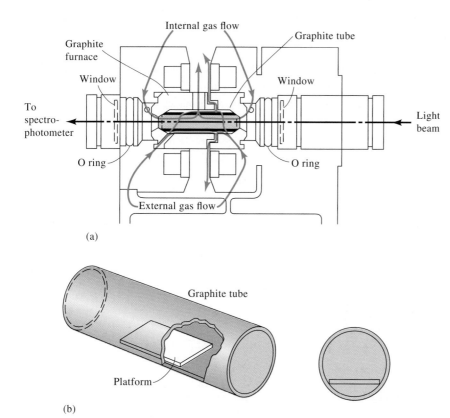

(a)

(b)

Unless otherwise noted, all content on this page is © Cengage Learning.

Figure 28-12 (a) Cross-sectional view of a graphite furnace atomizer. (b) The L'vov platform and its position in the graphite furnace. a: Reprinted by permission of PerkinElmer Corporation, Waltham, MA; b: Reprinted with permission from W. Slavin, *Anal. Chem.,* **1982**, 54, 685A, **DOI**: 10.1021/ac00243a001. Copyright 1982 American Chemical Society.

Output Signals

The output signals in electrothermal AA are transient, not the steady-state signals seen with flame atomization. The atomization step produces a pulse of atomic vapor that lasts only a few seconds, and the vapor is lost from the furnace by diffusion and other processes. The transient absorption signal produced by the pulse of vapor must be acquired and recorded rapidly by an appropriate data acquisition system.

28B-5 Other Atomizers

Many other types of atomization devices have been used in atomic spectroscopy. Gas discharges operated at reduced pressure have been investigated as sources of atomic emission. The **glow discharge** is generated between two planar electrodes in a cylindrical glass tube filled with gas to a pressure of a few torr. High-powered lasers have been employed to vaporize samples and to cause **laser-induced breakdown**. In the latter technique, dielectric breakdown of a gas occurs at the laser focal point. A laser-induced breakdown spectrometer (LIBS) is part of the Mars Science Laboratory aboard the rover Curiosity which arrived on Mars in August 2012.

In the early days of atomic spectroscopy, dc and ac arcs and high-voltage sparks were popular sources for exciting atomic emission. Such sources have almost entirely been replaced by the ICP.

> A **dielectric** is a material that does not conduct electricity. By applying high voltages or radiation from a high-powered laser, a gas can be made to break down into ions and electrons, a phenomenon known as **dielectric breakdown**.

28C   ATOMIC EMISSION SPECTROMETRY

Atomic emission spectrometry is widely used in elemental analysis. The ICP is now the most popular source for emission spectrometry, although the DCP and flames are still used in some situations.

28C-1 Instrumentation

The block diagram of a typical ICP emission spectrometer is shown in **Figure 28-13**. Atomic or ionic emission from the plasma is separated into its constituent wavelengths by the wavelength isolation device. This separation can take place in a **monochromator**, a **polychromator**, or a **spectrograph**. The monochromator isolates one wavelength at a time at a single exit slit, while a polychromator isolates several wavelengths simultaneously at multiple exit slits. The spectrograph provides a large aperture at its output

Figure 28-13 Block diagram of a typical ICP atomic emission spectrometer.

to allow a range of wavelengths to exit. The isolated radiation is then converted into electrical signals by a single transducer, multiple transducers, or an array detector. The electrical signals are then processed and provided as input to the computer system.

Flame emission spectrometers and DCP emission spectrometers follow the same block diagram except that a flame or DCP is substituted for the ICP of Figure 28-13. Flame spectrometers most often isolate a single wavelength, while DCP spectrometers may isolate multiple wavelengths with a polychromator.

Wavelength Isolation

Emission spectrometry is often used for multielement determinations. There are two types of instruments generally available for this purpose. The **sequential spectrometer** uses a monochromator and scans to different emission lines in sequence. Usually the wavelengths to be used are set by the user in a computer program, and the monochromator rapidly slews from one wavelength to the next. Alternatively, monochromators can scan a range of wavelengths. True **simultaneous spectrometers** use polychromators or spectrographs. The **direct reading spectrometer** uses a polychromator with as many as 64 detectors located at exit slits in the focal plane. Some spectrometers use spectrographs and one or more array detectors to monitor multiple wavelengths simultaneously. Some can even combine a scanning function with a spectrographic function to present different wavelength regions to an array detector. The dispersive devices in these spectrometers can be gratings, grating/prism combinations, and echelle gratings. Simultaneous instruments are usually more expensive than sequential systems.

For routine flame emission determinations of alkali metals and alkaline earth elements, simple filter photometers often suffice. A low-temperature flame is employed to prevent excitation of more difficult to excite metals. As a consequence, the spectra are simple, and interference filters can be used to isolate the desired emission lines. Flame emission was once widely used in the clinical laboratory for the determination of sodium and potassium. These methods have largely been replaced by methods using ion-selective electrodes (see Section 21D).

Radiation Transducers

Single-wavelength instruments most often use photomultiplier transducers as do direct reading spectrometers. The charge-coupled device (CCD) has now become very popular as an array detector for simultaneous and some sequential spectrometers. Such devices are available with over 1 million pixels to allow a fairly wide wavelength coverage. One commercial instrument uses a segmented-array, charge-coupled device detector to allow more than one wavelength region to be monitored simultaneously.

Computer Systems and Software

Commercial spectrometers are now equipped with powerful computers and software. Most of the newer ICP emission systems provide software that can assist in wavelength selection, calibration, background correction, interelement correction, spectral deconvolution, standard additions calibration, quality control charts, and report generation.

28C-2 Sources of Nonlinearity in Atomic Emission Spectrometry

Quantitative results in atomic emission spectrometry are usually based on the method of external standards (see Section 8D-2). For many reasons, we prefer calibration curves that are linear or at least follow a predicted relationship. At high concentrations, the major cause of nonlinearity when resonance transitions are used is **self-absorption**. Even at high concentrations, most of the analyte atoms are in the ground state with only a small fraction being excited. When the excited analyte atoms emit radiation, the resulting photons can be absorbed by ground state analyte atoms since these atoms have precisely the same energy levels for absorption. In media where the temperature is not homogeneous, resonance lines can be severely broadened and even have a dip in the center due to a phenomenon known as **self-reversal**. In flame emission, self-absorption is usually seen at solution concentrations between 10 and 100 μg/mL. In plasmas, self-absorption is often not seen until concentrations are higher because the optical path length is shorter for absorption in the plasma than in the flame.

At low concentrations, ionization of the analyte can cause nonlinearity in calibration curves when atomic lines are used. With ICP and DCP sources, the high electron concentrations in the plasma tend to act as a buffer against changes in the extent of ionization of the analyte with concentration. When ionic emission lines are used with the ICP, nonlinearities due to further ionization are few since removing a second electron is more difficult than removing the first electron. Changes in atomizer characteristics, such as flow rate, temperature, and efficiency, with analyte concentration can also be a cause of nonlinearity.

Flame emission calibration curves are often linear over two or three decades in concentration. ICP and DCP sources can manifest very broad linear ranges, often four to five decades in concentration.

28C-3 Interferences in Plasma and Flame Atomic Emission Spectrometry

Many of the interference effects caused by concomitants are similar in plasma and flame atomic emission. Some techniques, however, may be prone to certain interferences and may exhibit freedom from others. The interference effects are conveniently divided into blank interferences and analyte interferences.

Blank Interferences

A **blank,** or **additive**, **interference** produces an effect that is independent of the analyte concentration. These effects could be reduced or eliminated if a perfect blank could be prepared and analyzed under the same conditions. A **spectral interference** is an example. In emission spectroscopy, any element other than the analyte that emits radiation within the bandpass of the wavelength selection device or that causes stray light to appear within the bandpass causes a blank interference.

> Spectral interferences are examples of blank interferences. They produce an interference effect that is independent of the analyte concentration.

An example of a blank interference is the effect of Na emission at 285.28 nm on the determination of Mg at 285.21 nm. With a moderate-resolution spectrometer, any sodium in the sample will cause high readings for magnesium unless a blank with the correct amount of sodium is subtracted. Such line interferences can, in principle, be reduced by improving the resolution of the spectrometer. In practice, however, the user rarely has the opportunity to change the spectrometer resolution. In multielement spectrometers, measurements at multiple wavelengths can be used at times to determine correction factors to apply for an interfering species. Such interelement corrections are commonplace with modern computer-controlled ICP spectrometers.

Molecular band emission can also cause a blank interference. This interference is particularly troublesome in flame spectrometry where the lower temperature and reactive atmosphere are more likely to produce molecular species. As an example, a high concentration of Ca in a sample can produce band emission from CaOH, which can cause a blank interference if it occurs at the analyte wavelength. Usually improving the resolution of the spectrometer will not reduce band emission since the narrow analyte lines are superimposed on a broad molecular emission band. Flame or plasma background radiation is usually well compensated by measurements on a blank solution.

Analyte Interferences

Analyte, or **multiplicative**, **interferences** change the magnitude of the analyte signal itself. Such interferences are usually not spectral in nature but rather are physical or chemical effects.

Physical interferences can alter the aspiration, nebulization, desolvation, or volatilization processes. Substances in the sample that change the solution viscosity, for example, can alter the flow rate and the efficiency of the nebulization process. Combustible constituents, such as organic solvents, can change the atomizer temperature and thus affect the atomization efficiency indirectly.

Chemical interferences are usually specific to particular analytes. They occur in the conversion of the solid or molten particle after desolvation into free atoms or elementary ions. Constituents that influence the volatilization of analyte particles cause this type of interference and are often called **solute volatilization interferences**. For example, in some flames, the presence of phosphate in the sample can alter the atomic concentration of calcium in the flame due to the formation of relatively nonvolatile complexes. Such effects can sometimes be eliminated or moderated by the use of higher temperatures. Alternatively, **releasing agents**, which are species that react preferentially with the inteferent and prevent its interaction with the analyte, can be used. For example, the addition of excess Sr or La minimizes the phosphate interference on calcium because these cations form more stable phosphate compounds than Ca and release the analyte.

Protective agents prevent interference by preferentially forming stable but *volatile* species with the analyte. Three common reagents for this purpose are EDTA, 8-hydroxyquinoline, and APDC (the ammonium salt of 1-pyrrolidine-carbodithioc acid). For example, the presence of EDTA has been shown to minimize or eliminate interferences by silicon, phosphate, and sulfate in the determination of calcium.

Substances that alter the ionization of the analyte also cause **ionization interferences**. The presence of an easily ionized element, such as K, can alter the extent of ionization of a less easily ionized element, such as Ca. In flames, relatively large effects can occur unless an easily ionized element is purposely added to the sample in relatively large amounts. These **ionization suppressants** contain elements such as K, Na, Li, Cs, or Rb. When ionized in the flame, these elements produce electrons which then shift the ionization equilibrium of the analyte to favor neutral atoms.

Chemical, physical, and ionization interferences are examples of **analyte interferences**. These interferences influence the magnitude of the analyte signal itself.

Releasing agents are cations that react selectively with anions and prevent their interfering in the determination of a cationic analyte.

An **ionization suppressant** is an easily ionized species that produces a high concentration of electrons in a flame and represses ionization of the analyte.

28C-4 Applications

The ICP has become the most widely used source for emission spectroscopy. Its success stems from its high stability, low noise, low background, and freedom from many interferences. The ICP is, however, relatively expensive to purchase and to operate. Additionally, users require extensive training to manage and maintain these instruments. Still, modern computerized systems with their sophisticated software have eased the burden substantially.

The ICP is widely used in determining trace metals in environmental samples, such as drinking water, waste water, and ground water supplies. It is also used for determining trace metals in petroleum products, in foodstuffs, in geological samples, and in biological materials. The ICP has proven especially useful in industrial quality control. The DCP has found a significant niche in trace metal determinations in soil and geological samples. Flame emission is still used in some clinical laboratories for determining Na and K.

Simultaneous, multielement determinations using plasma sources have gained in popularity. Such determinations make it possible to identify correlations and to draw conclusions that were impossible with single-element determinations. For example, multielement trace metal determinations can aid in determining the origins of petroleum products found in oil spills or in identifying sources of pollution.

28D   ATOMIC ABSORPTION SPECTROMETRY

Flame atomic absorption spectroscopy (AAS) is currently the most widely used of all the atomic methods listed in Table 28-1 because of its simplicity, effectiveness, and relatively low cost. The technique was introduced in 1955 by Walsh in Australia and by Alkemade and Milatz in Holland.[4] The first commercial atomic absorption (AA) spectrometer was introduced in 1959, and use of the technique grew explosively after that. The reason that atomic absorption methods were not widely used until that time was directly related to problems created by the very narrow widths of atomic absorption lines, as discussed in Section 28A-4 (see color plate 17 for the solar spectrum and some atomic absorption lines).

28D-1   Line-Width Effects in Atomic Absorption

> The widths of atomic absorption lines are much less than the effective bandwidths of most monochromators.

No ordinary monochromator is capable of yielding a band of radiation as narrow as the width of an atomic absorption line (0.002 to 0.005 nm). As a result, the use of radiation that has been isolated from a continuum source by a monochromator inevitably causes instrumental departures from Beer's law (see the discussion of instrument deviations from Beer's law in Section 24C-3). In addition, since the fraction of radiation absorbed from such a beam is small, the transducer receives a signal that is less attenuated (that is, $P \rightarrow P_0$), and the sensitivity of the measurement is reduced. This effect is illustrated by the lower curve in Figure 24-17 (page 673).

The problem created by narrow absorption lines was surmounted by using radiation from a source that emits not only a *line of the same wavelength* as the one selected for absorption measurements but also one that is *narrower*. For example, a mercury vapor lamp is selected as the external radiation source for the determination of mercury. Gaseous mercury atoms that are electrically excited in such a lamp return to the ground state by *emitting* radiation with wavelengths that are identical to the wavelengths *absorbed* by the analyte mercury atoms in the flame. Since the lamp is operated at a temperature lower than that of the flame, the Doppler and pressure broadening of the mercury emission lines from the lamp is less than the corresponding broadening of the analyte absorption lines in the hot flame that holds the sample. The effective bandwidths of the lines emitted by the lamp are, therefore, significantly less than the corresponding bandwidths of the absorption lines for the analyte in the flame.

---

[4]A. Walsh, *Spectrochim. Acta*, **1955**, *7*, 108, **DOI:** 10.1016/0371-1951(55)80013-6;
C. Th. J. Alkemade and J. M. W. Milatz, *J. Opt. Soc. Am.*, **1955**, *45*, 583.

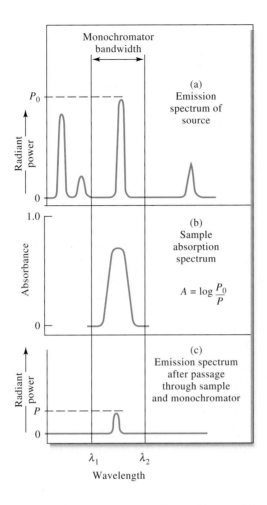

Figure 28-14 Atomic absorption of a narrow emission line from a source. The source lines in (a) are very narrow. One line is isolated by a monochromator. The line is absorbed by the broader absorption line of the analyte in the flame (b), resulting in attenuation (c) of the source radiation. Since most of the source radiation occurs at the peak of the absorption line, Beer's law is obeyed.

Figure 28-14 illustrates the strategy generally used in measuring absorbances in atomic absorption methods. Figure 28-14a shows four narrow *emission* lines from a typical atomic absorption source. Also shown is how one of these lines is isolated by a filter or monochromator. Figure 28-14b shows the flame *absorption spectrum* for the analyte between the wavelengths λ_1 and λ_2. Note that the width of the absorption line in the flame is significantly greater than the width of the emission line from the lamp. As shown in Figure 28-14c, the intensity of the incident beam P_0 has been decreased to P after passing through the sample. Since the bandwidth of the emission line from the lamp is significantly less than the bandwidth of the absorption line in the flame, $\log P_0/P$ is expected to be linearly related to concentration.

28D-2 Instrumentation

The instrumentation for AA can be fairly simple, as shown in **Figure 28-15** for a single-beam AA spectrometer.

Line Sources

The most useful radiation source for atomic absorption spectroscopy is the **hollow-cathode lamp**, shown schematically in **Figure 28-16**. It consists of a tungsten anode and a cylindrical cathode sealed in a glass tube containing an inert gas, such as argon, at a pressure of 1 to 5 torr. The cathode either is fabricated from the analyte metal or else serves as a support for a coating of that metal.

If a potential difference of about 300 V is applied across the electrodes, the argon ionizes, and as the argon cations and electrons migrate to the two electrodes, a

Figure 28-15 Block diagram of a single-beam atomic absorption spectrometer. Radiation from a line source is focused on the atomic vapor in a flame or electrothermal atomizer. The attenuated source radiation then enters a monochromator that isolates the line of interest. Next, the radiant power from the source, attenuated by absorption, is converted into an electrical signal by the photomultiplier tube (PMT). The signal is then processed and directed to a computer system for output.

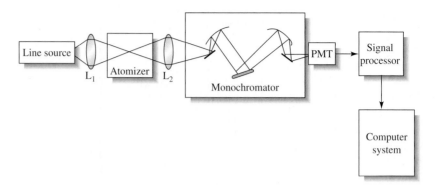

Sputtering is a process in which atoms or ions are ejected from a surface by a beam of charged particles.

current of 5 to 10 mA is generated. If the potential is large enough, the cations strike the cathode with sufficient energy to dislodge some of the metal atoms and to produce an atomic cloud. This process is called **sputtering**. Some of the sputtered metal atoms are in an excited state and emit their characteristic wavelengths as they return to the ground state. Recall that the atoms producing emission lines in the lamp are at a significantly lower temperature and pressure than the analyte atoms in the flame. As a result, the emission lines from the lamp are narrower than the absorption lines in the flame. The sputtered metal atoms eventually diffuse back to the cathode surface or to the walls of the lamp and are deposited.

Hollow-cathode lamps for about 70 elements are available from commercial sources. For some elements, high-intensity lamps are available that provide about an order of magnitude higher intensity than normal lamps. Some hollow cathode lamps have a cathode containing more than one element and thus provide spectral lines for the determination of several species. The development of the hollow-cathode lamp is widely regarded as the single most important event in the evolution of atomic absorption spectroscopy.

> Hollow-cathode lamps made atomic absorption spectroscopy practical.

In addition to hollow-cathode lamps, **electrodeless-discharge lamps** are useful sources of atomic line spectra. These lamps are often one to two orders of magnitude more intense than their hollow-cathode counterparts. A typical electrodeless-discharge lamp is constructed from a sealed quartz tube containing an inert gas, such as argon, at a pressure of a few torr and a small quantity of the analyte metal (or its salt). The lamp contains no electrodes, but instead it is energized by an intense field of radio-frequency or microwave radiation. The argon ionizes in this field, and the ions are accelerated by the high-frequency component of the field until they gain sufficient energy to excite (by collision) the atoms of the analyte metal.

Electrodeless-discharge lamps are available commercially for several elements. They are particularly useful for elements, such as As, Se, and Te, where hollow-cathode lamp intensities are low.

Figure 28-16 Diagram of a hollow cathode lamp.

Source Modulation

In an atomic absorption measurement, it is necessary to discriminate between radiation from the hollow-cathode or electrodeless-discharge lamp and radiation from the atomizer. Much of the atomizer radiation is eliminated by the monochromator, which is always located between the atomizer and the detector. The thermal excitation of a fraction of the analyte atoms in a flame, however, produces radiation of the wavelength at which the monochromator is set. Because such radiation is not removed, it acts as a potential source of interference.

The effect of analyte emission is overcome by **modulating** the output from the hollow-cathode lamp so that its intensity fluctuates at a constant frequency. The transducer thus receives an alternating signal from the hollow-cathode lamp and a continuous signal from the flame and converts these signals into the corresponding types of electric current. An electronic system then eliminates the unmodulated dc signal produced by the flame and passes the ac signal from the source to an amplifier and finally to the readout device.

Modulation can be accomplished by placing a motor-driven circular chopper *between the source and the flame* as shown in **Figure 28-17**. Segments of the metal chopper have been removed so that radiation passes through the device half of the time and is blocked the other half. By rotating the chopper at a constant speed, the beam reaching the flame varies periodically from zero intensity to some maximum intensity and then back to zero. Alternatively, the power supply for the source can be designed to pulse the hollow-cathode lamps in an alternating manner.

Complete AA Instrument

An atomic absorption instrument contains the same basic components as an instrument designed for molecular absorption measurements, as shown in Figure 28-15 for a single-beam system. Both single- and double-beam instruments are offered by numerous manufacturers. The range of sophistication and the cost (upward from a few thousand dollars) are both substantial.

Photometers.   At a minimum, an instrument for atomic absorption spectroscopy must be capable of providing a sufficiently narrow bandwidth to isolate the line chosen for a measurement from other lines that may interfere with or diminish the sensitivity of the method. A photometer equipped with a hollow-cathode source and filters is satisfactory for measuring concentrations of the alkali metals, which have only a few widely spaced resonance lines in the visible region. A more versatile photometer is sold with readily interchangeable interference filters and lamps. A separate filter and lamp are used for each element. Satisfactory results for the determination of 22 metals are claimed.

Spectrophotometers.   Most measurements in AAS are made with instruments equipped with an ultraviolet/visible grating monochromator. Figure 28-17 is a schematic of a typical double-beam instrument. Radiation from the

> **Modulation** is defined as changing some property of a waveform, called the **carrier**, by the desired signal such that the carrier conveys information about the desired signal. Properties that are typically altered are frequency, amplitude, or wavelength. In AAS, the source radiation is amplitude modulated, but the background and analyte emission are not and are observed as dc signals.

> ❮ Modulation of the source is often accomplished by a beam chopper or by pulsing the source electronically.

Figure 28-17 Optical paths in a double-beam atomic absorption spectrophotometer. The chopper converts the hollow-cathode radiation into an alternating signal at the detector, while the flame emission is a continuous dc signal.

hollow-cathode lamp is chopped and mechanically split into two beams, one of which passes through the flame and the other around the flame. A half-silvered mirror returns both beams to a single path by which they pass alternately through the monochromator and to the detector. The signal processor then separates the ac signal generated by the chopped light source from the dc signal produced by the flame. The logarithm of the ratio of the reference and sample components of the ac signal is then computed and sent to a computer or readout device for display as absorbance.

Background Correction

Absorption by the flame atomizer itself as well as by concomitants introduced into the flame or electrothermal atomizer can cause serious problems in atomic absorption. Because hollow-cathode lines are so narrow, interferences by absorption of the analyte line by other atoms are rare. On the other hand, molecular species can absorb the radiation and cause errors in AA measurements.

The total measured absorbance, A_T, in AA is the sum of the analyte absorbance, A_A, plus the background absorbance, A_B:

$$A_T = A_A + A_B \tag{28-1}$$

Background correction schemes attempt to measure A_B in addition to A_T. The true absorbance $A_A = A_T - A_B$ is then calculated.

Continuum Source Background Correction.   A popular background correction scheme in commercial AA spectrometers is the continuum lamp technique. In this scheme, a deuterium lamp and the analyte hollow cathode are directed through the atomizer at different times. The hollow-cathode lamp measures the total absorbance, A_T, while the deuterium lamp provides an estimate of the background absorbance, A_B. The computer system or processing electronics calculates the difference and reports the background-corrected absorbance. This method has limitations for elements with lines in the visible because the D_2 lamp intensity becomes quite low in this region.

Pulsed Hollow-Cathode Lamp Background Correction.   In this technique, often called **Smith-Hieftje background correction**, the analyte hollow cathode is pulsed at a low current (5 to 20 mA) for typically 10 ms and then at a high current (100 to 500 mA) for 0.3 ms. During the low current pulse, the analyte absorbance plus the background absorbance is measured (A_T). During the high-current pulse, the hollow-cathode emission line becomes broadened. The center of the line can be strongly self-absorbed so that much of the line at the analyte wavelength is missing. Hence, during the high-current pulse, a good estimate of the background absorbance, A_B, is obtained. The instrument computer then calculates the difference which is an estimate of A_A, the true analyte absorption.

Zeeman Effect Background Correction.   Background correction with electrothermal atomizers can be done by means of the Zeeman effect. In Zeeman background correction, a magnetic field splits spectral lines that are normally of the same energy (degenerate) into components with different polarization characteristics. Analyte and background absorption can be separated because of their different magnetic and polarization behaviors.[5]

Continuum source background correction uses a deuterium lamp to obtain an estimate of the background absorbance. A hollow-cathode lamp obtains the total absorbance. The corrected absorbance is then obtained calculating the difference between the two.

Smith-Hieftje background correction uses a single hollow-cathode lamp pulsed with first a low current and then with a high current. The low-current mode obtains the total absorbance, while the background is estimated during the high-current pulse.

[5]For more information, see D. A. Skoog, F. J. Holler, and S. R. Crouch, *Principles of Instrumental Analysis*, 6th ed., Belmont, CA: Brooks/Cole, 2007, pp. 242–43.

28D-3 Flame Atomic Absorption

Flame AA provides a sensitive means for determining some 60 to 70 elements. The method is well suited for routine measurements by relative inexperienced operators. Because a different hollow-cathode lamp is required for each element, only a single element can be determined at a time, and this is the major drawback of AA.

Region of the Flame for Quantitative Measurements

Figure 28-18 shows the absorbance of three elements as a function of distance above the burner head. For magnesium and silver, the initial rise in absorbance is a consequence of the longer exposure to the high temperature of the flame, leading to a greater concentration of atoms in the radiation path. The absorbance for magnesium, however, reaches a maximum near the center of the flame and then falls off as oxidation of the magnesium to magnesium oxide takes place. Silver does not suffer this effect because it is much more resistant to oxidation. For chromium, which forms very stable oxides, maximum absorbance is found immediately above the burner. Chromium oxide formation begins as soon as chromium atoms are formed.

Figure 28-18 shows that the optimum region of a flame used in a determination must change from element to element and that the position of the flame with respect to the source must be reproduced closely during calibration and measurement. Generally, the flame position is adjusted to give a maximum absorbance reading for the element being determined.

Quantitative Analysis

Quantitative analyses are frequently based on external standard calibration (see Section 8D-2). In atomic absorption, departures from linearity occur more often than in molecular absorption. Thus, analyses should *never* be based on the measurement of a single standard with the assumption that Beer's law is being followed. In addition, the production of an atomic vapor involves so many uncontrollable variables that the absorbance of at least one standard solution should be measured each time an analysis is performed. Often, two standards are used whose absorbances fall above and below (bracket) the absorbance of the unknown. Any deviation of the standard from its original calibration value can then be applied as a correction.

Standard addition methods, discussed in Section 8D-3, are also used extensively in AAS in an attempt to compensate for differences between the composition of the standards and the unknowns.

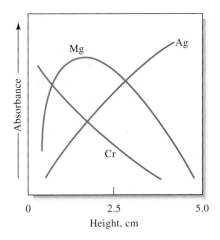

Figure 28-18 Height profiles for three elements in AAS. The plot shows absorbance versus height above the burner for Mg, Ag, and Cr.

TABLE 28-4

Detection Limits (ng/mL) for Some Elements by Atomic Spectroscopy*

Element	Flame AA	Electrothermal AA[†]	Flame Emission	ICP Emission	ICPMS
Ag	3	0.02	20	0.2	0.003
Al	30	0.2	5	0.2	0.06
Ba	20	0.5	2	0.01	0.002
Ca	1	0.5	0.1	0.0001	2
Cd	1	0.02	2000	0.07	0.003
Cr	4	0.06	5	0.08	0.02
Cu	2	0.1	10	0.04	0.003
Fe	6	0.5	50	0.09	0.45
K	2	0.1	3	75	1
Mg	0.2	0.004	5	0.003	0.15
Mn	2	0.02	15	0.01	0.6
Mo	5	1	100	0.2	0.003
Na	0.2	0.04	0.1	0.1	0.05
Ni	3	1	600	0.2	0.005
Pb	5	0.2	200	1	0.007
Sn	15	10	300	1	0.02
V	25	2	200	8	0.005
Zn	1	0.01	200	0.1	0.008

*Values taken from V. A. Fassel and R. N. Knisely, *Anal. Chem.*, **1974**, 46, 1110A, **DOI**: 10.1021/ac60349a023; J. D. Ingle, Jr., and S. R. Crouch, *Spectrochemical Analysis*, Englewood Cliffs, NJ: Prentice-Hall, 1988; C. W. Fuller, *Electrothermal Atomization for Atomic Absorption Spectroscopy*, London: Chemical Society, 1977; *Ultrapure Water Specifications, Quantitative ICP-MS Detection Limits*, Fremont, CA, Balazs Analytical Services, 1993.

[†]Based on a 10 μL sample.

Detection Limits and Accuracy

Column 2 of **Table 28-4** shows detection limits for a number of common elements determined by flame atomic absorption and compares them with results from other atomic spectroscopic methods. Under usual conditions, the relative error of flame absorption analysis is of the order of 1 to 2%. With special precautions, this figure can be lowered to a few tenths of one percent. Note that flame AA detection limits are generally better than flame AE detection limits except for the easily excited alkali metals.

28D-4 Atomic Absorption with Electrothermal Atomization

Electrothermal atomizers offer the advantage of unusually high sensitivity for small volumes of sample. Typically, sample volumes are between 0.5 and 10 μL. Under these circumstances, absolute detection limits often lie in the picogram range. In general, electrothermal AA detection limits are best for the more volatile elements. Detection limits for electrothermal AA vary considerably from one manufacturer to the next because they depend on atomizer design and atomization conditions.

The relative precision of electrothermal methods is generally in the range of 5 to 10% compared with the 1% or better that can be expected for flame or plasma atomization. Furthermore, furnace methods are slow and typically require several minutes per element. Still another disadvantage is that chemical interference effects are often more severe with electrothermal atomization than with flame atomization. A final disadvantage is that the analytical range is low, usually less than two orders of magnitude. Because of these disadvantages, electrothermal atomization is typically applied only when flame or plasma atomization provides inadequate detection limits or when sample sizes are extremely limited.

Another AA method applicable to volatile elements and compounds is the cold-vapor technique. Mercury is a volatile metal and can be determined by the method described in Feature 28-1 (see color plate 18 for mercury absorption). Other metals form volatile metal hydrides that can also be determined by the cold-vapor technique.

FEATURE 28-1

Determining Mercury by Cold-Vapor Atomic Absorption Spectroscopy

Our fascination with mercury began when prehistoric cave dwellers discovered the mineral cinnabar (HgS) and used it as a red pigment. Our first written record of the element came from Aristotle who described it as "liquid silver" in the fourth century B.C. Today, there are thousands of uses of mercury and its compounds in medicine, metallurgy, electronics, agriculture, and many other fields. Because it is a liquid metal at room temperature, mercury is used to make flexible and efficient electrical contacts in scientific, industrial, and household applications. Thermostats, silent light switches, and fluorescent light bulbs are but a few examples of electrical applications.

A useful property of metallic mercury is that it forms amalgams with other metals, which have a host of uses. For example, metallic sodium is produced as an amalgam by electrolysis of molten sodium chloride. Dentists use a 50% amalgam with an alloy of silver for fillings.

The toxicological effects of mercury have been known for many years. The bizarre behavior of the Mad Hatter in Lewis Carroll's *Alice in Wonderland* (see **Figure 28F-1**) was a result of the effects of mercury and mercury compounds on the Hatter's brain. Mercury that has been absorbed through the skin and lungs destroys brain cells, which are not regenerated. Hatters of the nineteenth century used mercury compounds in processing fur to make felt hats. These workers and workers in other industries have suffered the debilitating symptoms of mercurialism such as loosening of teeth, tremors, muscle spasms, personality changes, depression, irritability, and nervousness.

The toxicity of mercury is complicated by its tendency to form both inorganic and organic compounds. Inorganic mercury is relatively insoluble in body tissues and fluids, so it is expelled from the body about ten times faster than organic mercury. Organic mercury, usually in the form of alkyl compounds such as methyl mercury, is somewhat soluble in fatty tissues such as the liver. Methyl mercury accumulates to toxic levels and is expelled from the body quite slowly. Even experienced scientists must take extreme precautions in handling organo-mercury compounds. In 1997, Dr. Karen Wetterhahn of Dartmouth College died as a result of mercury poisoning despite being one of the world's leading experts in handling methyl mercury.

Mercury concentrates in the environment, as illustrated in **Figure 28F-2**. Inorganic mercury is converted to organic mercury by anaerobic bacteria in sludge deposited at the bottom of lakes, streams, and other bodies of water. Small aquatic animals consume the organic mercury and are in turn eaten by larger life forms. As the element moves up the food chain from microbes to shrimp to fish and ultimately to larger animals such as swordfish, the mercury becomes ever more concentrated. Some sea creatures such as oysters may concentrate mercury by a factor of 100,000. At the top of the food chain, the concentration of mercury reaches levels as high as 20 ppm. The Food and Drug Administration has set a legal limit of 1 ppm in fish for human consumption. As a result, mercury levels in some areas threaten local fishing industries. The Environmental Protection Agency has set a limit of 2 ppb of

Figure 28F-1   Mad Hatter from Alice in Wonderland

(continued)

Figure 28F-2 Biological concentration of mercury in the environment.

mercury in drinking water, and the Occupational Safety and Health Administration has set a limit of 0.1 mg/m^3 in air.

Analytical methods for the determination of mercury play an important role in monitoring the safety of food and water supplies. One of the most useful methods is based on the atomic absorption by mercury of 253.7 nm radiation. Color plate 18 shows the striking absorption of UV light by mercury vapor that forms over the metal at room temperature. **Figure 28F-3** shows an apparatus that is used to determine mercury by atomic absorption at room temperature.[6]

A sample suspected of containing mercury is decomposed in a hot mixture of nitric acid and sulfuric acid, which converts the mercury to the +2 state. The resulting Hg^{2+} and any remaining compounds are reduced to the metal with a mixture of hydroxylamine sulfate, and tin(II) sulfate. Air is then pumped through the solution to carry the resulting mercury-containing vapor through the drying tube and into the observation cell.

Water vapor is trapped by Drierite in the drying tube so that only mercury vapor and air pass through the cell. The monochromator of the atomic absorption spectrophotometer is tuned to a band around 254 nm. Radiation from the 253.7 nm line of the mercury hollow-cathode lamp passes through the quartz windows of the observation cell, which is placed in the light path of the instrument. The absorbance is directly proportional to the concentration of mercury in the cell, which is in turn proportional to the concentration of mercury in the sample. Solutions of known mercury concentration are treated in a similar way to calibrate the apparatus. The method depends on the low solubility of mercury in the reaction mixture and its appreciable vapor pressure, which is 2×10^{-3} torr at 25°C. The sensitivity of the method is about 1 ppb, and it is used to determine mercury in foods, metals, ores, and environmental samples. The method has the advantages of sensitivity, simplicity, and room temperature operation.

Figure 28F-3 Apparatus for cold-vapor atomic absorption determination of mercury.

[6]W. R. Hatch and W. L. Ott, *Anal. Chem.* **1968,** *40,* 2085, **DOI:** 10.1021/ac50158a025.

28D-5 Interferences in Atomic Absorption

Flame atomic absorption is subject to many of the same chemical and physical interferences as flame atomic emission (see Section 28C-2). Spectral interferences by elements that absorb at the analyte wavelength are rare in AA. Molecular constituents and radiation scattering can cause interferences, however. These are often corrected by the background correction schemes discussed in Section 28D-2. In some cases, if the source of interference is known, an excess of the interferent can be added to both the sample and the standards. The added substance is sometimes called a **radiation buffer**.

> A **radiation buffer** is a substance that is added in large excess to both samples and standards to swamp out the effect of matrix species and thus to minimize interference.

28E ATOMIC FLUORESCENCE SPECTROMETRY

Atomic fluorescence spectrometry (AFS) is the newest of the optical atomic spectroscopic methods. Like atomic absorption, an external source is used to excite the element of interest. However, instead of measuring the attenuation of the source, the radiation emitted as a result of absorption is measured, often at right angles to avoid measuring the source radiation.

For most elements, atomic fluorescence with conventional hollow-cathode or electrodeless-discharge sources has no significant advantages over atomic absorption or atomic emission. As a result, the commercial development of atomic fluorescence instrumentation has been quite slow. Sensitivity advantages have been shown, however, for elements such as Hg, Sb, As, Se, and Te.

Laser-excited atomic fluorescence spectrometry is capable of extremely low detection limits, particularly when combined with electrothermal atomization. Detection limits in the femtogram (10^{-15} g) to attogram (10^{-18} g) range have been shown for many elements. Commercial instrumentation has not been developed for laser-based AFS probably because of its expense and the nonroutine nature of high-powered lasers. Atomic fluorescence has the disadvantage of being a single-element method unless tunable lasers with their inherent complexities are used.

> ◀ Despite its potential advantages of high sensitivity and selectivity, atomic fluorescence spectrometry has never been commercially successful. Difficulties can be attributed partly to the lack of reproducibility of the high-intensity sources required and to the single-element nature of AFS.

WEB WORKS   Use a search engine to find the Laboratory for Spectrochemistry at Indiana University. Locate the list of research projects dealing with fundamental plasma studies. Find a project on mechanisms of matrix effects in the ICP and describe the project in detail. Include the purpose of the project, the instrumentation used, and results obtained. Click on the list of publications for the laboratory. Find a paper entitled, "Algorithm to determine matrix-effect crossover points for overcoming interferences in inductively coupled plasma-atomic emission spectrometry." Describe four features incorporated into the algorithm.

QUESTIONS AND PROBLEMS

*28-1. Describe the basic differences among atomic emission, atomic absorption, and atomic fluorescence spectroscopy.

28-2. Define
    *(a) atomization.
    (b) collisional broadening.
    *(c) Doppler broadening.
    (d) nebulization.
    *(e) plasma.
    (f) laminar flow burner

*(g) hollow-cathode lamp.

(h) sputtering.

*(i) additive interference.

(j) spectral interference.

*(k) chemical interference.

(l) radiation buffer.

*(m) protective agent.

(n) ionization suppressor.

*28-3. Why is atomic emission more sensitive to flame instability than atomic absorption?

28-4. Why are ionization interferences usually not as severe in the ICP as they are in flames?

*28-5. Why is source modulation used in atomic absorption spectroscopy?

28-6. Why are higher resolution monochromators found in ICP atomic emission spectrometers than in flame atomic absorption spectrometers?

*28-7. Why are the lines from a hollow-cathode lamp generally narrower than the lines emitted by atoms in a flame?

28-8. In flame AA with a hydrogen/oxygen flame, the absorbance for iron decreased in the presence of large concentrations of sulfate ion.

(a) Suggest an explanation for this observation.

(b) Suggest three possible methods for overcoming the potential interference of sulfate in a quantitative determination of iron.

*28-9. Name four characteristics of inductively coupled plasmas that make them suitable for atomic emission spectrometry.

28-10. Why is the ICP rarely used for atomic absorption measurements?

*28-11. Discuss the differences that result in ICP atomic emission when the plasma is viewed axially rather than radially.

28-12. In the atomic absorption determination of uranium, there is a linear relationship between the absorbance at 351.5 nm and concentration from 500 to 2000 ppm of U. At concentrations much lower than 500 ppm, the relationship becomes nonlinear unless about 2000 ppm of an alkali metal salt is introduced. Explain.

*28-13. A 5.00-mL sample of blood was treated with trichloroacetic acid to precipitate proteins. After centrifugation, the resulting solution was brought to pH 3 and extracted with two 5-mL portions of methyl isobutyl ketone containing the lead-complexing agent APCD. The extract was aspirated directly into an air/acetylene flame and yielded an absorbance of 0.502 at 283.3 nm. Five-milliliter aliquots of standard solutions containing 0.400 and 0.600 ppm of lead were treated in the same way and yielded absorbances of 0.396 and 0.599. Find the concentration of lead in the sample in ppm assuming that Beer's law is followed.

28-14. The chromium in a series of steel samples was determined by ICP emission spectroscopy. The spectrometer was calibrated with a series of standards containing 0, 2.0, 4.0, 6.0, and 8.0 μg $K_2Cr_2O_7$ per milliliter. The instrument readings for these solutions were 3.1, 21.5, 40.9, 57.1, and 77.3 in arbitrary units.

(a) Plot the data.

(b) Find the equation for the regression line.

(c) Calculate standard deviations for the slope and the intercept of the line in (b).

(d) The following data were obtained for replicate 1.00-g samples of cement dissolved in HCl and diluted to 100.0 mL after neutralization:

| | Emission Readings | | |
	Blank	Sample A	Sample B	Sample C
Replicate 1	5.1	28.6	40.7	73.1
Replicate 2	4.8	28.2	41.2	72.1
Replicate 3	4.9	28.9	40.2	spilled

Calculate the percent Cr_2O_3 in each sample. What are the absolute and relative standard deviations for the average of each determination?

28-15. The copper in an aqueous sample was determined by atomic absorption flame spectrometry. First, 10.0 mL of the unknown were pipetted into each of five 50.0-mL volumetric flasks. Various volumes of a standard containing 12.2 ppm Cu were added to the flasks, and the solutions were then diluted to volume.

Unknown, mL	Standard, mL	Absorbance
10.0	0.0	0.201
10.0	10.0	0.292
10.0	20.0	0.378
10.0	30.0	0.467
10.0	40.0	0.554

(a) Plot absorbance as a function of volume of standard.

*(b) Derive an expression relating absorbance to the concentrations of standard and unknown (c_s and c_x) and the volumes of the standards and unknown (V_s and V_x) as well as the volume to which the solutions were diluted (V_t).

*(c) Derive expressions for the slope and the intercept of the straight line obtained in (a) in terms of the variables listed in (b).

(d) Show that the concentration of the analyte is given by the relationship $c_x = bc_s/mV_x$, where m and b are the slope and the intercept of the straight line in (a).

*(e) Determine values for m and b by the method of least squares.

(f) Calculate the standard deviation for the slope and the intercept in (e).

*(g) Calculate the copper concentration in ppm Cu in the sample using the relationship given in (d).

28-16. Challenge Problem: Seawater samples were examined by ICP-atomic emission spectrometry (ICP-AES) in a multielement study. Vanadium was one of the elements determined. Standard solutions in a synthetic seawater matrix were prepared and determined by ICP-AES. The following results were obtained:

Concentration, pg/mL	Intensity, arbitrary units
0.0	2.1
2.0	5.0
4.0	9.2
6.0	12.5
8.0	17.4
10.0	20.9
12.0	24.7

(a) Determine the least-squares regression line.

(b) Determine the standard deviations of the slope and intercept.

(c) Test the hypothesis that the slope is equal to 2.00.

(d) Test the hypothesis that the intercept is equal to 2.00.

(e) Three seawater solutions gave readings for vandadium of 3.5, 10.7, and 15.9. Determine their concentrations and the standard deviation of their concentrations.

(f) Determine the 95% confidence limits for the three unknowns in part (e).

(g) Estimate the limit of detection for determining vanadium in seawater from the data (see Section 8D-1). Use a k value of 3 in your DL estimate.

(h) The second seawater sample with a reading of 10.7 units was a certified reference standard with a known vanadium concentration of 5.0 pg/mL. What was the absolute and percent error in its determination?

(i) Test the hypothesis that the value determined in part (e) for the second seawater sample (reading of 10.7) is identical to the certified concentration of 5.0 pg/mL.

CHAPTER 29

Mass Spectrometry

Mass spectrometry has rapidly become one of the most important of all analytical techniques. The photo shows NASA's flying mass spectrometer laboratory aboard a DC-8 jet airplane. The mass spectrometer is being used to study the impact of air pollution on remote areas of the planet such as the Arctic region. The amounts and types of airborne particulates are measured by the mass spectrometer to study the influence of pollution on climate change. Mass spectrometry is widely used in chemistry and biology to determine the structures of complex molecules and to identify the molecules present in many different samples. It has also become very important in geology, in paleontology, in forensic science, and in clinical chemistry.

David McNew/Getty Images

Mass spectrometry (MS) is a powerful and versatile analytical tool for obtaining information about the identity of an unknown compound, its molecular mass, its elemental composition, and in many cases, its chemical structure. Mass spectrometry can be conveniently divided into atomic, or elemental, mass spectrometry and molecular mass spectrometry. Atomic mass spectrometry is a quantitative tool that can determine nearly all the elements in the periodic table. Detection limits are often several orders of magnitude better than optical methods. On the other hand, molecular mass spectrometry is capable of providing information about the structures of inorganic, organic, and biological molecules and about the qualitative and quantitative composition of complex mixtures. We first discuss the principles that are common to all forms of mass spectrometry and the components that constitute a mass spectrometer.

29A PRINCIPLES OF MASS SPECTROMETRY

In the mass spectrometer, analyte molecules are converted to ions by applying energy to them. The ions formed are separated on the basis of their mass-to-charge ratio (m/z) and directed to a transducer that converts the number of ions (abundance) into an electrical signal. The ions of different mass-to-charge ratios are directed to the transducer sequentially by scanning or made to strike a multichannel transducer simultaneously. The ion abundance plotted against mass-to-charge ratio is called a **mass spectrum**. Often, singly charged ions are produced in the ionization source, and the mass-to-charge ratio is shortened to just mass so that the spectrum is plotted as number of ions versus mass, as shown in **Figure 29-1** for an elemental mass spectrum of a geological sample. This convenient simplification is only applicable, however, to singly charged ions.

A **mass spectrum** is a plot of ion abundance versus mass-to-charge ratio (see Section 29A-2) or just mass for singly charged ions.

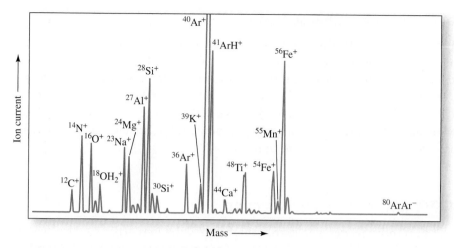

Figure 29-1 The mass spectrum of a geological sample obtained by laser ablation/ICP-MS. Ion current on the *y* axis is proportional to the number of ions (ion abundance). Mass on the *x* axis is proportional to the mass-to-charge ratio for singly charged ions. Major components (%): Na, 1.80; Mg, 3.62; Al, 4.82.1; Si, 26.61; K, 0.37; Ti, 0.65; Fe, 9.53; Mn 0.15. (Reproduced (adapted) from A. L. Gray, *Analyst*, **1985**, *110*, 55, **DOI**:10:1039/AN9851000551, with permission of The Royal Society of Chemistry.)

29A-1 Atomic Masses

Atomic and molecular masses are usually expressed in terms of **the atomic mass scale**, based on a specific isotope of carbon. One **unified atomic mass unit** on this scale is equal to 1/12 the mass of a neutral $^{12}_{6}C$ atom. The unified atomic mass is given the symbol u. One unified mass unit is commonly termed one dalton (Da), which has become the accepted term even though it is not an official SI unit. The older term, atomic mass unit (amu), is to be discouraged since it was based on the most abundant stable isotope of oxygen ^{16}O.

> The $^{12}_{6}C$ isotope is assigned a value of exactly 12 unified atomic mass units, or commonly 12 daltons.

In mass spectrometry, in contrast to most types of chemistry, we are often interested in the exact mass *m* of particular isotopes of an element or the exact mass of compounds containing a particular set of isotopes. Thus, we may need to distinguish between the masses of compounds such as

$$^{12}C^1H_4 \qquad m = 12.0000 \times 1 + 1.008 \times 4$$
$$= 16.03200 \text{ Da}$$
$$^{13}C^1H_4 \qquad m = 13.0000 \times 1 + 1.008 \times 4$$
$$= 17.0320 \text{ Da}$$
$$^{12}C^1H_3{}^2H_1 \qquad m = 12.0000 \times 1 + 1.008 \times 3 + 2.0160 \times 1$$
$$= 17.0400 \text{ Da}$$

The isotopic masses in the calculations above are shown with four digits to the right of the decimal point. We normally quote exact masses to three or four figures to the right of the decimal point because typical high-resolution mass spectrometers make measurements at this level of precision.

The **chemical atomic mass**, or the **average atomic mass**, of an element in nature is given by summing the exact masses of each isotope weighted by its fractional abundance in nature. The chemical atomic mass is the type of mass of interest to chemists for most purposes. The average or chemical molecular mass of a compound is then the sum of the chemical atomic masses for the atoms appearing in the formula of the compound. Thus, the chemical molecular mass of CH_4 is $12.011 + 4 \times 1.008 = 16.043$ Da. The atomic or molecular mass expressed without units is the **mass number**.

> The **mass number** is the atomic or molecular mass expressed without units.

29A-2 Mass-to-Charge Ratio

The **mass-to-charge ratio**, m/z, of an ion is the quantity of most interest because the mass spectrometer separates ions according to this ratio. The mass-to-charge ratio of an ion is the unitless ratio of its mass number to the number of fundamental charges z on the ion. Thus, for $^{12}C^1H_4^+$, $m/z = 16.032/1 = 16.032$. For $^{13}C^1H_4^{2+}$, $m/z = 17.032/2 = 8.516$. Strictly speaking, referring to the mass-to-charge ratio as the mass of an ion is only correct for singly charged ions, but this terminology is commonly used in the mass spectrometry literature.

29B   MASS SPECTROMETERS

The **mass spectrometer** is an instrument that produces ions, separates them according to their m/z values, detects them, and plots the mass spectrum. Such instruments vary widely in size, resolution, flexibility, and cost. Their components, however, are remarkably similar.

29B-1 Components of Mass Spectrometer

Figure 29-2 illustrates the principal components of all types of mass spectrometers. In molecular mass spectrometry, samples enter the evacuated region of the mass spectrometer through the inlet system. Solids, liquids, and gases may be introduced depending on the nature of the ionization source. The purpose of the inlet system is to introduce a micro amount of sample into the ion source where the components of the sample are converted into gaseous ions by bombardment with electrons, photons, ions, or molecules. In atomic mass spectrometry, the ionization source is outside the evacuated region and also serves as the inlet. In atomic mass spectrometers, ionization is accomplished by applying thermal or electrical energy. The output of the ion source is a stream of positive (most common) or negative gaseous ions. These ions are accelerated into the mass analyzer, which then separates them according to their mass-to-charge ratios. The ions of particular m/z values are then collected and converted into an electrical signal by the ion transducer. The data handling system processes the results to produce the mass spectrum. The processing may also include comparison to known spectra, tabulation of results, and data storage.

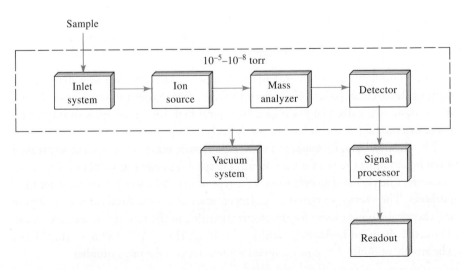

Figure 29-2 Components of a mass spectrometer.

TABLE 29-1

Common Mass Analyzers for Mass Spectrometry

Basic Type	Analysis Principle
Magnetic sector	Deflection of ions in a magnetic field. Ion trajectories depend on *m/z* value.
Double-focusing	Electrostatic focusing followed by magnetic field deflection. Trajectories depend on *m/z* values.
Quadrupole	Ion motion in dc and radio-frequency fields. Only certain *m/z* values are passed.
Ion trap	Storage of ions in space defined by ring and end cap electrodes. Electric field sequentially ejects ions of increasing *m/z* values.
Ion cyclotron resonance	Trapping of ions in cubic cell under influence of trapping voltage and magnetic field. Orbital frequency related inversely to *m/z* value.
Time-of-flight	Equal kinetic energy ions enter drift tube. Drift velocity and thus arrival time at detector depend on mass.

Mass spectrometers require an elaborate vacuum system to maintain a low pressure in all of the components except the signal processor and display. Low pressure ensures a relatively low collision frequency between various species in the mass spectrometer that is vital for the production and maintenance of free ions and electrons.

In the sections that follow, we first describe the mass analyzers that are used in mass spectrometers. Then, we consider the various transducer systems that are used in both molecular and elemental mass spectrometry. Section 29C-1 contains material on the nature and operation of common ion sources for atomic mass spectrometers, while section 29D-2 describes ionization sources for molecules.

> Mass spectrometers are operated at low pressure so that free ions and electrons can be maintained.

29B-2 Mass Analyzers

Ideally, the mass analyzer should distinguish minute mass differences and simultaneously permit the passage of a sufficient number of ions to yield measurable ion currents. Because these two properties are not entirely compatible, design compromises have resulted in many different types of mass analyzers. **Table 29-1** lists six of the most common analyzers. We describe in detail magnetic and electric sector analyzers, quadrupole mass analyzers, and time-of-flight systems. Several other analyzer types are used in mass spectrometry including ion traps and Fourier transform ion cyclotron resonance spectrometers.[1]

Resolution of Mass Spectrometers

The capability of a mass spectrometer to differentiate between masses is usually stated in terms of its *resolution*, *R*, which is defined as

$$R = \frac{m}{\Delta m} \tag{29-1}$$

> A resolution of 100 means that unit mass (1 Da) can be distinguished at a nominal mass of 100.

where Δm is the mass difference between two adjacent peaks that are just resolved and *m* is the nominal mass of the first peak (the mean mass of the two peaks is sometimes used instead).

The resolution required in a mass spectrometer depends greatly on its intended use. For example, to detect differences in mass among ions of the same nominal mass, such as $C_2H_4^+$, CH_2N^+, N_2^+, and CO^+ (all ions of nominal mass 28 Da but

---

[1]For information on ion traps and ion cyclotron resonance spectrometers, see D. A. Skoog, F. J. Holler, and S. R. Crouch, *Principles of Instrumental Analysis*, 6th ed., Belmont, CA: Brooks/Cole, 2007, pp. 369–73.

Figure 29-3   Schematic of a magnetic sector spectrometer. The kinetic energy, KE, of an ion of mass m and charge z exiting slit B is KE $= zeV = \frac{1}{2}mv^2$. If all ions have the same kinetic energy, heavier ions travel at lower velocities than lighter ions. The balancing of centripetal force and magnetic force results in ions of different mass traveling different paths as shown.

exact masses of 28.054, 28.034, 28.014, and 28.010 Da, respectively), requires an instrument with a resolution of several thousand. On the other hand, low-molecular-mass ions differing by a unit of mass or more, such, as NH_3^+ ($m = 17$) and CH_4^+ ($m = 16$), can be distinguished with an instrument having a resolution smaller than 50. Commercial spectrometers are available with resolutions ranging from about 500 to 500,000.

Sector Analyzers[2]

In the magnetic sector analyzer, shown in **Figure 29-3**, separation is based on the deflection of ions in a magnetic field. The trajectories that ions take depend on their m/z values. Typically, the magnetic field is slowly changed to bring ions of different m/z value to a detector. In the double-focusing mass spectrometer, an electric sector precedes the magnetic sector. The electrostatic field serves to focus a beam of ions having only a narrow range of kinetic energies onto a slit that leads to the magnetic sector. Such instruments are capable of very high resolution.

Quadrupole Mass Analyzers

The quadrupole mass analyzer consists of four cylindrical rods, as illustrated in **Figure 29-4**. Quadrupole analyzers are mass filters that only allow ions of a certain mass-to-charge ratio to pass. Ion motion in electric fields is the basis of separation. Rods opposite each other are connected to dc and radio-frequency (RF) voltages. With proper adjustment of the voltages, a stable path is created for ions of a certain m/z ratio to pass through the analyzer to the transducer. The mass spectrum is obtained by scanning the voltages applied to the rods. Quadrupole analyzers have relatively high throughput but relatively low resolution. Unit mass (1 Da) is the typical resolution of a quadrupole analyzer. This resolution may be sufficient in many forms of elemental mass spectrometry or in cases where a mass spectrometer serves as a detector for molecules separated by gas or liquid chromatography.

---

[2]For information on mass analyzers, see D. A. Skoog, F. J. Holler, and S. R. Crouch, *Principles of Instrumental Analysis,* 6th ed., Belmont, CA: Brooks/Cole, 2007, pp. 366–73.

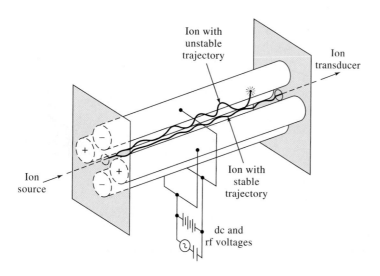

Figure 29-4 A quadrupole
mass analyzer.

Time-of-Flight Mass Analyzers

The time-of-flight (TOF) mass spectrometer represents another approach to mass analysis. In a TOF analyzer, a packet of ions with nearly identical kinetic energies is rapidly sampled, and the ions enter a field-free region. Since the kinetic energy, KE, is $\frac{1}{2}mv^2$, the ion velocity v varies inversely with its mass, as shown by Equation 29-2:

$$v = \sqrt{\frac{2KE}{m}} \tag{29-2}$$

The time required for the ions to travel a fixed distance to the detector is thus inversely related to the ion mass. In other words, ions with low m/z arrive at the detector more rapidly than those with high m/z. Each m/z value is then detected in sequence. Flight times are quite brief, leading to analysis times that are typically on the order of microseconds.

Time-of-flight instruments are relatively simple and rugged and have nearly unlimited mass range. The TOF analyzer suffers, however, from limited resolution and sensitivity. As a result, TOF analyzers are less widely used than magnetic sector and quadrupole analyzers.

29B-3 Transducers for Mass Spectrometry

Several types of ion transducers are available for mass spectrometry.[3] The most common transducer is the electron multiplier, illustrated in **Figure 29-5**. The discrete-dynode electron multiplier operates much like the photomultiplier transducer for UV/visible radiation, discussed in Section 25A-4. When energetic ions or electrons strike a Cu-Be cathode, secondary electrons are emitted. These electrons are attracted to dynodes that are each held at a successively higher positive voltage. Electron multipliers with up to 20 dynodes are available. These devices can multiply the signal strength by a factor of up to 10^7.

Continuous-dynode electron multipliers are also popular. These multipliers are trumpet-shaped devices made of glass heavily doped with lead. A potential of 1.8 to 2 kV is imposed across the length of the device. Ions that strike the surface eject electrons that skip along the inner surface ejecting more electrons with each impact.

---

[3]For information on ion transducers, see D. A. Skoog, F. J. Holler, and S. R. Crouch, *Principles of Instrumental Analysis,* 6th ed., Belmont, CA: Brooks/Cole, 2007, pp. 284–87.

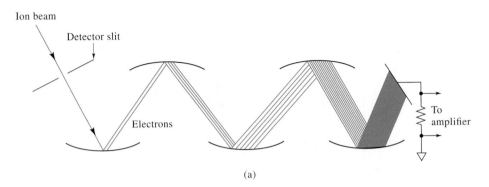

Ion beam

Detector slit

Electrons

To
amplifier

(a)

Figure 29-5   Discrete-dynode electron multiplier. Dynodes are kept at successively higher voltages by means of a multistage voltage divider.

In addition to electron multiplier transducers, Faraday cup transducers and array transducers have become available for mass spectrometry. As in optical spectrometry, array transducers allow the simultaneous detection of multiple resolution elements. Microchannel plate arrays and micro-Faraday arrays have been used.

29C | ATOMIC MASS SPECTROMETRY

Atomic mass spectrometry has been around for many years, but the introduction of the inductively coupled plasma (ICP) in the 1970s and its subsequent development for mass spectrometry[4] led to the successful commercialization of ICPMS by several instrument companies. Today, ICPMS is a widely used technique for the simultaneous determination of over 70 elements in a few minutes. The ion source is the major difference between atomic and molecular mass spectrometry. For atomic mass spectrometry, the ion source must be very energetic to convert the sample into simple gas phase ions and atoms. In molecular mass spectrometry, the ion source is much less energetic and converts the sample into molecular ions and fragment ions.

29C-1 Sources for Atomic Mass Spectrometry

Several different ionization sources have been proposed for atomic mass spectrometry. **Table 29-2** lists the most common ion sources and the typical mass analyzers used with each.

The Inductively Coupled Plasma

The inductively coupled plasma is described extensively in Section 28B-2 in connection with its use in atomic emission spectrometry. The axial geometry shown in Figure 28-7 is most often used in ICPMS. In MS applications, the ICP serves as both an

TABLE 29-2

Common Ionization Sources for Atomic Mass Spectrometry

Name	Acronym	Atomic Ion Sources	Typical Mass Analyzer
Inductively coupled plasma	ICPMS	High-temperature argon plasma	Quadrupole
Direct current plasma	DCPMS	High-temperature argon plasma	Quadrupole
Microwave-induced plasma	MIPMS	High-temperature argon plasma	Quadrupole
Spark source	SSMS	Radio-frequency electric spark	Double-focusing
Glow-discharge	GDMS	Glow-discharge plasma	Double-focusing

[4]R. S. Houk, V. A. Fassel, G. D. Flesch, H. J. Svec, A. L. Gray, and C. E. Taylor, *Anal. Chem.*, **1980**, *52*, 2283, **DOI**: 10.1021/ac50064a012.

atomizer and an ionizer. Solution samples may be introduced by a conventional or an ultrasonic nebulizer. Solid samples can be dissolved in solution or volatized by means of a high-voltage spark or high-powered laser prior to introduction into the ICP. Ions formed in the plasma are then introduced into the mass analyzer, often a quadrupole, where they are sorted according to mass-to-charge ratio and detected.

Extracting ions from the plasma can present a major technical problem in ICPMS. While an ICP operates at atmospheric pressure, a mass spectrometer operates at high vacuum, typically less than 10^{-6} torr. The interface region between the ICP and the mass spectrometer is thus critical to ensure that a substantial fraction of the ions produced are transported into the mass analyzer. The interface usually consists of two metal cones, called the **sampler** and the **skimmer**. Each cone has a small orifice (≈ 1 mm) to allow the ions to pass through to ion optics that then guide them into the mass analyzer.[5] The beam introduced into the mass spectrometer has about the same ionic composition as the plasma region from which the ions are extracted. **Figure 29-6** shows that ICPMS spectra are often remarkably simple compared with conventional ICP atomic emission spectra. The ICPMS spectrum shown in the figure consists of a simple series of isotope peaks for each element present along with some background ionic peaks. Background ions include Ar^+, ArO^+, ArH^+, H_2O^+, O^+, O_2^+, and Ar_2^+, as well as argon adducts with metals. In addition, some polyatomic ions from constituents in the sample are also found in ICP mass spectra. Such background ions can interfere with the determination of analytes as described in Section 29C-2.

Commercial instruments for ICPMS have been on the market since 1983. ICPMS spectra are used for identifying the elements present in the sample and for determining these elements quantitatively. Usually, quantitative analyses are based on calibration curves in which the ratio of the ion signal for the analyte to that for an internal standard is plotted as a function of concentration.

Other Ionization Sources for Atomic Mass Spectrometry

Of the sources listed in Table 29-2, the spark source and the glow discharge have received the most attention. Spark source atomic mass spectrometry (SSMS) was first introduced in the 1930s as a general tool for multielement and isotope trace analyses. It was not until 1958, however, that the first commercial spark source mass spectrometer appeared on the market. After a period of rapid development in the 1960s, the use of this technique leveled off and then declined with the appearance of ICPMS. Currently, spark source mass spectrometry is still applied to solid samples that are not easily dissolved and analyzed by ICP. In addition, spark sources are used in conjunction with ICP sources to volatilize and atomize solid samples before introduction to the plasma.

As discussed in Section 28B-5, the glow-discharge source is a useful device for various types of atomic spectroscopy. In addition to atomizing samples, it also produces a cloud of positive analyte ions from solid samples. This device consists of a simple two-electrode closed system containing argon at a pressure of 0.1 to 10 torr. A voltage of 5 to 15 kV from a pulsed dc power supply is applied between the electrodes, causing the formation of positive argon ions, which are then accelerated toward the cathode. The cathode is fabricated from the sample, or the sample is deposited on an inert metal cathode. Just as in the hollow-cathode lamp (see Section 28D-2), atoms of the sample are sputtered from the cathode into the region between the two electrodes, where they are converted to positive ions by collision with electrons or positive argon ions. Analyte ions are then drawn into the mass spectrometer by **differential pumping**.

> In a vacuum system, two chambers are said to be differentially pumped if they are connected by a small orifice and evacuated by two separate vacuum pumps. The pumps are connected to the chambers through large conduits. Such an arrangement allows gas to enter one chamber with little change in pressure in the second chamber.

[5]For more information, see R. S. Houk, *Acc. Chem. Res.*, **1994**, *27*, 333, **DOI**: 10.1021/ar00047a003.

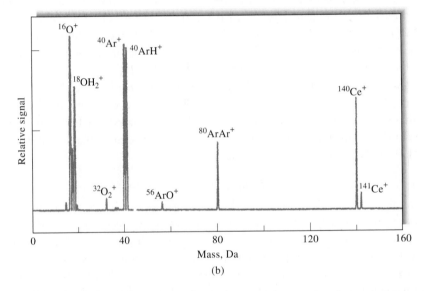

Figure 29-6 Comparison of ICP atomic emission spectrum for 100 ppm cerium (a) with ICP mass spectrum for 10 ppm cerium (b). (Adapted from M. Selby and G. M. Hieftje, *Amer. Lab.*, **1987**, *19*, 16.)

The ions are then filtered in a quadrupole analyzer or dispersed with a magnetic sector analyzer for detection and determination. Glow-discharge sources, like spark sources, are often used with ICP torches. The glow discharge serves as the atomizer, and the ICP torch is the ionizer.

29C-2 Atomic Mass Spectra and Interferences

High-resolution mass analyzers, such as double-focusing analyzers, can reduce or eliminate many spectral interferences in ICPMS.

Because the ICP source predominates in atomic mass spectrometry, we focus our discussion on ICPMS. The simplicity of ICPMS spectra, such as the cerium spectrum shown in Figure 29-6b, led early workers in the field to have hopes of an "interference-free method." Unfortunately, this hope was not realized in further studies, and serious interference problems are sometimes encountered in atomic mass spectrometry, just as in optical atomic spectroscopy. Interference effects in atomic mass spectroscopy fall into two broad categories: spectroscopic interferences and matrix interferences. Spectroscopic

[6]For additional discussion of interferences in ICPMS, see K. E. Jarvis, A. L. Gray, and R. S. Houk, *Handbook of Inductively Coupled Plasma Mass Spectrometry*, Ch. 5, New York: Blackie, 1992; G. Horlick and Y. Shao, in *Inductively Coupled Plasmas in Analytical Atomic Spectrometry*, 2nd ed., A. Montaser and D. W. Golightly, eds., New York: VCH-Wiley, 1992, pp. 571–96.

interferences occur when an ionic species in the plasma has the same m/z value as an analyte ion. Most of these interferences are from polyatomic ions, elements having isotopes with essentially the same mass, doubly charged ions, and refractory oxide ions.[6] High-resolution spectrometers can reduce or eliminate many of these interferences.

Matrix effects become noticeable when the concentrations of matrix species exceed about 500 to 1000 µg/mL. Usually, these effects cause a reduction in the analyte signal, although enhancements are sometimes observed. Generally, such effects can be minimized by diluting the sample, by altering the introduction procedure, or by separating the interfering species. The effects can also be minimized by the use of an appropriate internal standard, an element that has about the same mass and ionization potential as the analyte (see Section 8D-3).

29C-3 Applications of Atomic Mass Spectrometry

ICPMS is well suited for multielement analysis and for determinations such as isotope ratios. The technique has a wide dynamic range, typically four orders of magnitude, and produces spectra that are, in general, simpler and easier to interpret than optical emission spectra. ICPMS is finding widespread use in the semiconductor and electronics industry, in geochemistry, in environmental analyses, in biological and medical research, and in many other areas.

Detection limits for ICPMS are listed in Table 28-4, where they are compared to those from several other atomic spectrometric methods. Most elements can be detected well below the part per billion level. Quadrupole instruments typically allow ppb detection for their entire mass range. High-resolution instruments can routinely achieve sub-part-per-trillion detection limits because the background levels in these instruments are extremely low.

> Detection limits for quadrupole ICPMS instruments are often less than 1 ppb.

Quantitative analysis is normally performed by preparing calibration curves using external standards. To compensate for instrument drifts, instabilities, and matrix effects, an internal standard can be added to the standards and to the sample. Multiple internal standards are sometimes used to optimize matching the characteristics of the standard to that of various analytes.

For simple solutions where the composition is known or the matrix can be matched well between samples and standards, accuracies can be better than 2% for analytes at concentrations 50 times the detection limit. For solutions of unknown composition, accuracies of 5% are typical.

29D MOLECULAR MASS SPECTROMETRY

Molecular mass spectrometry was first used for routine chemical analysis in the early 1940s when the petroleum industry adopted the technique for quantitative analysis of hydrocarbon mixtures produced in catalytic crackers. Beginning in the 1950s, commercial instruments began to be adapted by chemists for the identification and structural elucidation of a wide variety of organic compounds. This use of the mass spectrometer combined with the invention of nuclear magnetic resonance and the development of infrared spectrometry revolutionized the way organic chemists identify and determine the structure of molecules. This application of mass spectrometry is still extremely important.

Applications of molecular mass spectrometry dramatically changed in the decade of the 1980s as a result of the development of new methods for producing ions from nonvolatile or thermally unstable molecules, such as those frequently encountered in the biological sciences. Since about 1990, there has been an explosive growth in the

Figure 29-7   Mass spectrum of ethyl benzene.

area of biological mass spectrometry brought about by these new ionization methods. Currently, mass spectrometry is being applied to the determination of the structure of polypeptides, proteins, and other high-molecular-mass biopolymers.

We consider here the nature of molecular mass spectra and the types of information that can be obtained. The ionization sources that are commonly used are described along with mass spectrometric instrumentation. Finally, we describe several current applications.[7]

29D-1 Molecular Mass Spectra

Figure 29-7 illustrates the way in which mass spectral data are usually presented. The analyte is ethyl benzene, which has a nominal molecular mass of 106 daltons (Da). To obtain this spectrum, ethyl benzene vapor was bombarded with a stream of electrons that led to the loss of an electron by the analyte and formation of the molecular ion M^+, as shown by the reaction

$$C_6H_5CH_2CH_3 + e^- \rightarrow C_6H_5CH_2H_3^{\cdot +} + 2e^- \qquad (29\text{-}3)$$

The charged species $C_6H_5CH_2H_3^{\cdot +}$ is the **molecular ion**. As indicated by the dot, the molecular ion is a radical ion that has the same molecular mass as the molecule.

> Fragment ions peaks may dominate molecular mass spectra.

The collision between energetic electrons and analyte molecules usually imparts enough energy to the molecules to leave them in an excited state. Relaxation then often occurs by fragmentation of part of the molecular ions to produce ions of lower masses. For example, a major product in the case of ethyl benzene is $C_6H_5CH_2^+$, which results from the loss of a CH_3 group. Other smaller positively charged fragments are also formed in lesser amounts.

The positive ions produced on electron impact are attracted through the slit of a mass spectrometer, where they are sorted according to their mass-to-charge ratios and displayed in the bar graph form of a mass spectrum. Note in Figure 29-7, that the largest peak at $m/z = 91$, termed the **base peak**, has been arbitrarily assigned a value of 100. The heights of the remaining peaks are then computed as a percentage of the base-peak height.

---

[7]For a detailed discussion of mass spectrometry, see D. M. Desiderio and N. M. Nibbering, eds., *Mass Spectrometry: Instrumentation, Interpretation, and Applications*, Hoboken, NJ: Wiley, 2009; J. T. Watson and O. D. Sparkman, *Introduction to Mass Spectrometry: Instrumentation, Applications and Strategies for Data Interpretation*, 4th ed., Chichester, UK: Wiley, 2007; R. M. Smith, *Understanding Mass Spectra: A Basic Approach*, 2nd ed., New York: Wiley, 2004.

29D-2 Ion Sources

The starting point for a mass spectrometric analysis is the formation of gaseous analyte ions, and the scope and the utility of a mass spectrometric method is dictated by the ionization process. The appearance of mass spectra for a given molecular species is highly dependent on the method used for ion formation. Table 29-3 lists many of the ion sources that have been used in molecular mass spectrometry.[8] Note that these methods fall into two major categories: **gas-phase sources** and **desorption sources**. With a gas-phase source, the sample is first vaporized and then ionized. With a desorption source, the sample in a solid or liquid state is converted directly into gaseous ions. An advantage of desorption sources is that they are applicable to nonvolatile and thermally unstable samples. Currently, commercial mass spectrometers are equipped with accessories that permit use of several of these sources interchangeably.

The most widely used source is the electron impact (EI) source. In this source, molecules are bombarded with a high-energy beam of electrons. This produces positive ions, negative ions, and neutral species. The positive ions are directed toward the analyzer by electrostatic repulsion.

In EI, the electron beam is so energetic that many fragments are produced. These fragments, however, are very useful in identifying the molecular species entering the mass spectrometer. Mass spectra for many libraries of MS data have been collected using EI sources.

There has been a good deal of activity in the area of ambient sampling and ionization sources for mass spectrometry.[9] These sources make use of many of the established ionization methods, such as ESI, CI, and plasmas, but in an open-air, direct-ionization environment. Such an environment allows ionization with minimal sample pretreatment on samples of unusual sizes and shapes that are not easily examined under high-vacuum conditions. A wide assortment of ambient MS techniques exists, but desorption electrospray ionization (DESI) and direct analysis in real time (DART) are the leading techniques. In addition, low-temperature plasma probe ionization (LTP), easy ambient sonic-spray ionization (EASI), and laser ablation electrospray ionization (LAESI) have shown promise.

> Most ion sources for molecular mass spectrometry are either gas-phase sources or desorption sources.

> The majority of mass spectral libraries contain mass spectra collected using electron impact ionization.

TABLE 29-3

Common Ion Sources for Molecular Mass Spectrometry			
Basic Type	**Name and Acronym**	**Method of Ionization**	**Type of Spectra**
Gas-phase	Electron impact (EI)	Energetic electrons	Fragmentation patterns
	Chemical ionization (CI)	Reagent gaseous ions	Proton adducts, few fragments
Desorption	Fast atom bombardment (FAB)	Energetic atomic beam	Molecular ions and fragments
	Matrix assisted laser desorption/ionization (MALDI)	High-energy photons	Molecular ions, multiply charged ions
	Electrospray ionization (ESI)	Electric field produces charged spray which desolvates	Multiply charged molecular ions

[8]For more information about modern ion sources, see D. A. Skoog, F. J. Holler, and S. R. Crouch, *Principles of Instrumental Analysis,* 6th ed., Belmont, CA: Brooks/Cole, 2007, pp. 551–63; J. T. Watson and O. D. Sparkman, *Introduction to Mass Spectrometry: Instrumentation, Applications and Strategies for Data Interpretation,* 4th ed., Chichester, UK: Wiley, 2007.
[9]G. A. Harris, A. S. Galhena, and F. M. Fernandez, *Anal. Chem.,* **2011**, *83*, 4508, **DOI**: 10.1021/ac200918u.

29D-3 Molecular Mass Spectrometric Instrumentation

Molecular mass spectrometers follow the basic block diagram of Figure 29-2. We concentrate here on components of molecular mass spectrometers that differ from the atomic mass spectrometers described in Section 29C.

Inlet Systems[10]

The purpose of the inlet system is to introduce a representative sample to the ion source with minimal loss of vacuum. Most modern mass spectrometers are equipped with several types of inlets to accommodate various kinds of samples. The major types of inlets can be classified as **batch inlets**, **direct probe inlets**, **chromatographic inlets**, and **electrophoretic inlets.**

The conventional (and simplest) inlet system is the batch type in which the sample is volatilized externally and then allowed to leak into the evacuated ionization region. Liquids and gases can be introduced in this way.

Solids can be placed on the tip of a probe, inserted into the vacuum chamber, and evaporated or sublimed by heating. Nonvolatile liquids can be introduced through special controlled-flow inlets, or they can be desorbed from a surface on which they are coated as a thin film. In general, samples for molecular mass spectrometry must be pure because the fragmentation that occurs causes the mass spectrum of mixtures to be difficult to interpret. Gas chromatography (see Chapter 32) is an ideal way to introduce mixtures because the components are separated from the mixture by the chromatograph prior to introduction to the mass spectrometer. The combination of gas chromatography and mass spectrometry is often called GC/MS. **Figure 29-8** shows the schematic of a typical GC/MS instrument. High-performance liquid chromatography and capillary electrophoresis can also be coupled with a mass spectrometer through the use of specialized interfaces.

> The batch inlet is the most common for introducing liquids and gases.

Mass Analyzers

All of the mass analyzers listed in Table 29-1 are used in molecular mass spectrometry. The quadrupole mass analyzer is commonly used with GC/MS systems. Higher-resolution spectrometers (magnetic sector, double-focusing, time-of-flight, Fourier transform) are often used when fragmentation patterns are to be analyzed for structural or identification purposes.

Tandem mass spectrometry, also called **mass spectrometry-mass spectrometry** (MS/MS), is a technique that allows the mass spectrum of a preselected or fragmented

Figure 29-8 Schematic of a typical capillary GC/MS instrument. The effluent from the GC is passed into the inlet of the mass spectrometer, where the molecules in the gas are ionized and fragmented, analyzed, and detected.

---

[10]For additional information on inlet systems, see D. A. Skoog, F. J. Holler, and S. R. Crouch, *Principles of Instrumental Analysis,* 6th ed., Belmont, CA: Brooks/Cole, 2007, pp. 564–66.

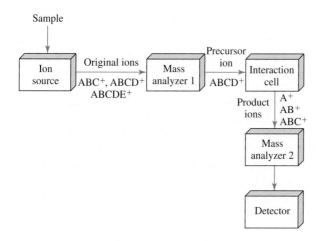

Figure 29-9 Block diagram of a tandem mass spectrometer.

ion to be obtained. **Figure 29-9** illustrates the basic concept. With a tandem mass spectrometer, an ionization source produces molecular ions and fragment ions. These are then the input to the first mass analyzer, which selects a particular ion (the **precursor ion**) and sends it to the interaction cell. In the interaction cell, the precursor ion can decompose spontaneously, react with a collision gas, or interact with an intense laser beam to produce fragments, or **product ions**. These ions are then mass analyzed by the second mass analyzer and detected by the ion detector.

Tandem mass spectrometers can produce a variety of different spectra. **Production-ion spectra** are obtained by scanning mass analyzer 2 while mass analyzer 1 is held constant to act as a mass selector for the precursor ion. A **precursor-ion spectrum** can be obtained by scanning mass analyzer 1 and selecting a given product ion with mass analyzer 2. If both mass analyzers are scanned with a small offset in mass between them, a **neutral loss spectrum** can be obtained. A neutral loss spectrum might be used, for example, to identify the m/z values of all ions losing a common molecule, such as water. Finally, a complete **three-dimensional MS/MS spectrum** can be acquired by recording a product ion spectrum for each selected precursor ion, that is, by scanning mass analyzer 2 for various settings of mass analyzer 1.

Tandem mass spectrometry can produce an enormous amount of information and has proven useful in structural elucidation as well as in the analysis of mixtures. Conventional mass spectrometry of mixtures usually requires chromatographic or electrophoretic separation to present a single compound at a time to the mass spectrometer.

> Several different types of spectra can be produced with a tandem mass spectrometer.

29D-4 Applications of Molecular Mass Spectrometry

The applications of molecular mass spectrometry are so numerous and widespread that describing them adequately in a brief space is not possible. **Table 29-4** lists several of the most important applications to provide some idea of the capabilities of mass spectrometry. We describe a few of these applications in this section.

Identification of Pure Compounds

The mass spectrum of a pure compound provides several kinds of data that are useful for its identification. The first is the molecular mass of the compound, and the second is its molecular formula. In addition, study of fragmentation patterns revealed by the mass spectrum often provides information about the presence or absence of various functional groups. Finally, the actual identity of a compound can often be

TABLE 29-4

Applications of Molecular Mass Spectrometry
Elucidation of the structure of organic and biological molecules
Determination of the molecular mass of peptides, proteins, and oligonucleotides
Identification of components in thin-layer and paper chromatograms
Determination of amino acid sequences in sample of polypeptides and proteins
Detection and identification of species separated by chromatography and capillary electrophoresis
Identification of drugs of abuse and metabolites of drugs of abuse in blood, urine, and saliva
Monitoring gases in patient's breath during surgery
Testing for the presence of drugs in blood in thoroughbred race horses and in Olympic athletes
Dating archaeological specimens
Analysis of aerosol particles
Determination of pesticide residues in food
Monitoring volatile organic species in water supplies

established by comparing its mass spectrum with those of known compounds until a close match is found.

Analysis of Mixtures

While ordinary mass spectrometry is a powerful tool for the identification of pure compounds, its usefulness for analysis of all but the simplest mixtures is limited because of the immense number of fragments of differing m/z values produced. It is often impossible to interpret the resulting complex spectrum. For this reason, chemists have developed methods in which mass spectrometers are coupled with various efficient separation devices. When two or more analytical techniques or instruments are combined to form a new, more efficient device, the resulting methodology is often termed a **hyphenated method**.

Gas chromatography/mass spectrometry has become one of the most powerful tools available for the analysis of complex organic and biochemical mixtures. In this application, spectra are collected for compounds as they exit from a chromatographic column. These spectra are then stored in a computer for subsequent processing. Mass spectrometry has also been coupled with liquid chromatography (LC/MS) for the analysis of samples that contain nonvolatile constituents.

Tandem mass spectrometry offers some of the same advantages as GC/MS and LC/MS and is significantly faster. While separations on a chromatographic column are achieved in a time scale of a few minutes to hours, equally satisfactory separations in tandem mass spectrometers are complete in milliseconds. In addition, the chromatographic techniques require dilution of the sample with large excesses of a mobile phase and subsequent removal of the mobile phase, greatly enhancing the probability of introducing interferences. As a result, tandem mass spectrometry is potentially more sensitive than either of the hyphenated chromatographic techniques because the chemical noise associated with its use is generally smaller. A current disadvantage of tandem mass spectrometry with respect to the other two chromatographic procedures is the greater cost of the required equipment; this gap appears to be narrowing as tandem mass spectrometers gain wider use.

For some complex mixtures the combination of GC or LC and MS does not provide enough resolution. In recent years, it has become feasible to couple chromatographic methods with tandem mass spectrometers to form GC/MS/MS and LC/MS/MS systems.

Quantitative Determinations

Applications of mass spectrometry for quantitative analyses fall into two categories. The first is the quantitative determination of molecular species or types of molecular species in organic, biological, and occasionally inorganic samples. Several of these applications are listed in Table 29-4. The second category is the determination of the concentration of elements in inorganic and, less commonly, organic and biological samples as discussed in Section 29C-3.

WEB WORKS Use a search engine to find "distance-of-flight mass spectrometry" (DOF). Locate a patent issued on the DOF approach. To whom was the patent issued? Describe this technique and how it differs from the time-of-flight (TOF) approach. What are its advantages and disadvantages? Can the DOF approach be used in a tandem MS arrangement? Describe how a DOF spectrometer might be coupled with a TOF analyzer to achieve a full two-dimensional precursor/product ion spectrum.

QUESTIONS AND PROBLEMS

29-1. Define
  *(a) Dalton.
  (b) quadrupole mass filter.
  *(c) mass number.
  (d) sector analyzer.
  *(e) time-of-flight analyzer.
  (f) electron multiplier.

29-2. Name three characteristics of inductively coupled plasmas that make them suitable for atomic mass spectrometry.

*29-3. What function does the ICP torch serve in mass spectrometry?

29-4. What are the ordinate and the abscissa of an ordinary mass spectrum?

*29-5. What types of interferences are encountered in ICPMS?

29-6. What is the purpose of an internal standard in ICPMS?

*29-7. Why are detection limits for ICPMS often lower with double-focusing mass spectrometers than with quadrupole mass spectrometers?

29-8. How do gaseous and desorption ionization sources differ? What are the advantages of each?

*29-9. Why are fragments often produced with electron impact ionization?

29-10. Discuss why it is much easier to couple a gas chromatograph with a mass spectrometer than it is to couple a liquid chromatograph with a mass spectrometer.

*29-11.What is the difference between a precursor ion and a product ion in tandem mass spectrometry?

29-12. Some ionization sources, known as soft ionization sources, do not produce as many fragments as an electron impact source, which is a hard ionization source. Which type of ionization source (hard or soft) is more useful for structure elucidation? Which for molecular mass determination? Which for compound identification? Give your reasoning with your answer.

29-13. **Challenge Problem:**
  (a) The kinetic energy KE imparted to an ion of mass m bearing a charge z in a TOF analyzer is KE $= zeV = \frac{1}{2} mv^2$, where e is the electronic charge, V is the electric field voltage, and v is the ion velocity. If the field-free drift tube has a length L, show that the flight time t_F is given by

$$t_F = L\sqrt{\frac{m}{2zeV}}$$

  (b) An ion M^+ has a mass of 286.1930 Da. What is the mass of the ion in kg?
  (c) Show that a kinetic energy of 1 eV is equal to 1.6×10^{-19} kg m^2 s^{-2}.
  (d) If the ion receives a kinetic energy of 3000 eV prior to introduction to a flight tube, what is its velocity in m/s?
  (e) If the flight tube has a length of 1.5 m, how long will it take the ion to reach the detector at the end of the flight tube?
  (f) What would be the flight time for an impurity ion of mass 285.0410 Da?
  (g) What resolution is required to separate M^+ from the impurity?

PART VI

Kinetics and Separations

Kinetic Methods of Analysis

CHAPTER 30

A modern automobile is equipped with a three-way catalytic converter to lower to acceptable levels the emissions of nitrogen oxides, unburned hydrocarbons, and carbon monoxide. The converter must oxidize CO and unburned hydrocarbons to CO_2 and H_2O, and it must reduce nitrogen oxides to N_2 gas. Hence, two different catalysts are used, an oxidation catalyst and a reduction catalyst. Three different converter styles are shown in the photograph. Many cars use the honeycomb catalyst structure, shown on the lower right, to maximize the exposure of the catalysts to the exhaust stream. The catalysts are usually metals such as platinum, rhodium, or palladium.

The amount of catalyst can be determined by measuring how much the rate of a chemical reaction is affected. Catalytic methods, which are among the most sensitive of all analytical methods, are used for trace analysis of metals in the environment, organics in a variety of samples, and enzymes in biological systems.

© Raymond Reuter/Sygma/Corbis

Kinetic methods of analysis differ in a fundamental way from the equilibrium, or thermodynamic, methods we have dealt with in previous chapters. In **kinetic methods**, measurements are made under dynamic conditions in which the concentrations of reactants and products are changing as a function of time. In contrast, thermodynamic methods are performed on systems that have come to equilibrium or steady state so that concentrations are static.

The distinction between the two types of methods is illustrated in **Figure 30-1**, which shows the progress over time of the reaction

$$A + R \rightleftharpoons P \tag{30-1}$$

where A represents the analyte, R the reagent, and P the product. Thermodynamic methods operate in the region beyond time t_e, when the bulk concentrations of reactants and product have become constant and the chemical system is at equilibrium. In contrast, kinetic methods are carried out during the time interval from 0 to t_e when reactant and product concentrations are changing continuously.

Selectivity in kinetic methods is achieved by choosing reagents and conditions that produce differences in the rates at which the analyte and potential interferences react. Selectivity in thermodynamic methods is realized by choosing reagents and conditions that create differences in equilibrium constants.

Kinetic methods significantly extend the number of chemical reactions that can be used for analytical purposes because they permit the use of reactions that are too slow or too incomplete for thermodynamic-based procedures. Kinetic methods can be based on complexation

In **kinetic methods**, measurements are made while net changes in the extent of the reaction are still occurring. In **equilibrium methods**, measurements are made under conditions of equilibrium or steady state.

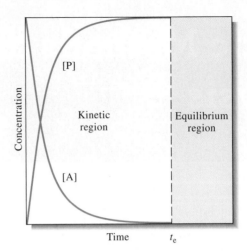

Figure 30-1 Change in concentration of analyte [A] and product [P] as a function of time. Until time t_e, the analyte and product concentrations are continuously changing. This is the kinetic regime. After t_e, the analyte and product concentrations are static.

reactions, acid/base reactions, redox reactions, and many others. Many kinetic methods are based on catalyzed reactions. In one type of catalytic method, the analyte is the catalyst and is determined from its effect on an indicator reaction with reactants or products that are determined conveniently. Such methods are among the most sensitive analytical reactions. In another catalyzed reaction, the catalyst is introduced to accelerate the reaction between analyte and reagent. This approach is often highly selective, or even specific, particularly when the catalyst is an enzyme. Undoubtedly, the most widespread use of kinetic methods is in biochemical and clinical laboratories, where the number of analyses based on kinetics exceeds those based on thermodynamics.[1]

30A RATES OF CHEMICAL REACTIONS

We provide here a brief introduction to chemical kinetics, which is needed to understand the basis for kinetic methods of analysis.

30A-1 Reaction Mechanisms and Rate Laws

The **mechanism** by which a chemical reaction proceeds consists of a series of chemical equations describing the individual elementary steps that lead to products being formed from reactants. Much of what chemists know about mechanisms has been gained from studies in which the rate at which reactants are consumed or products formed is measured as a function of such variables as reactant and product concentration, temperature, pressure, pH, and ionic strength. Such studies lead to an empirical **rate law** that relates the reaction rate to the concentrations of reactants, products, and intermediates at any instant. Mechanisms are discovered by postulating a series of elementary steps that are chemically reasonable and consistent with the empirical rate law. Often such mechanisms are further tested by doing studies designed to discover or monitor any transient intermediate species predicted by the mechanism.

The **rate law** for a reaction is an experimentally determined relationship between the rate of a reaction and the concentration of reactants, products, and other species such as catalysts, activators, and inhibitors.

---

[1]H. O. Mottola, *Kinetic Aspects of Analytical Chemistry*, New York: Wiley, 1988.

Concentration Terms in Rate Laws

Rate laws are algebraic expressions consisting of concentration terms and constants, which often look somewhat like an equilibrium-constant expression (see Equation 30-2). You should realize, however, that the square-bracketed terms in a rate expression represent molar concentrations *at a particular instant* rather than equilibrium molar concentrations (as in equilibrium-constant expressions). This meaning is frequently emphasized by adding a subscript to show the time to which the concentration refers. Thus, $[A]_t$, $[A]_0$, and $[A]_\infty$ indicate the concentration of A at time t, time zero, and infinite time, respectively. Infinite time is regarded as any time greater than required for equilibrium to be achieved, that is, $t_\infty > t_e$ in Figure 30-1.

> In the context of chemical kinetics, molar concentrations, symbolized with square brackets, change with time.

Reaction Order

Let us assume that the empirical rate law for the general reaction shown as Equation 30-1 is found by experiment to take the form

$$\text{rate} = -\frac{d[A]}{dt} = -\frac{d[R]}{dt} = \frac{d[P]}{dt} = k[A]^m[R]^n \qquad (30\text{-}2)$$

where the rate is the derivative of the concentration of A, R, or P with respect to time. Note that the first two rates carry a negative sign because the concentrations of A and R decrease as the reaction proceeds. In this rate expression, k is the **rate constant**, m is the **order of the reaction with respect to A**, and n is the **order of the reaction with respect to R**. The **overall order** of the reaction is $p = m + n$. Thus, if $m = 1$ and $n = 2$, the reaction is said to be first order in A, second order in R, and third order overall.

> Because A and R are being depleted, the rates of change of [A] and [R] with respect to time are negative.

Units for Rate Constants

Since reaction rates are always expressed in terms of concentration per unit time, the units of the rate constant are determined by the overall order p of the reaction according to the relation

$$\frac{\text{concentration}}{\text{time}} = (\text{units of } k)(\text{concentration})^p$$

where $p = m + n$. Rearranging leads to

$$\text{units of } k = (\text{concentration})^{1-p} \times \text{time}^{-1}$$

Thus, the units for a first-order rate constant are s^{-1}, and the units for a second-order rate constant are $M^{-1}s^{-1}$.

> The units of the rate constant k depend on the overall order of the reaction. For a first-order reaction, the units are s^{-1}.

30A-2 The Rate Law for First-Order Reactions

The simplest case in the mathematical analysis of reaction kinetics is that of a spontaneous irreversible decomposition of a species A:

$$A \xrightarrow{\ k\ } P \qquad (30\text{-}3)$$

> Radioactive decay is an example of a spontaneous decomposition.

The reaction is first order in A, and the rate is

$$\text{rate} = -\frac{d[A]}{dt} = k[A] \tag{30-4}$$

Pseudo-First-Order Reactions

A first-order decomposition reaction per se is generally of no use in analytical chemistry because a determination is usually based on reactions involving at least two species, an analyte and a reagent.[2] Usually, however, the rate law for a reaction involving two species is sufficiently complex that simplifications are needed for analytical purposes. In fact, the majority of useful kinetic methods are performed under conditions that permit the chemist to simplify complex rate laws to a form analogous to Equation 30-4. A higher-order reaction that is executed so that such a simplification is feasible is termed a **pseudo-first-order reaction**. Methods for converting higher-order reactions to pseudo-first-order are dealt with in later sections.

Mathematics Describing First-Order Behavior

Because the vast majority of kinetic determinations are performed under pseudo-first-order conditions, it is worthwhile to examine in detail some of the characteristics of reactions having rate laws that approximate Equation 30-4.

By rearranging Equation 30-4, we obtain

$$\frac{d[A]}{[A]} = -kdt \tag{30-5}$$

The integral of this equation from time zero, when $[A] = [A]_0$, to time t, when $[A] = [A]_t$, is

$$\int_{[A]_0}^{[A]_t} \frac{d[A]}{[A]} = -k\int_0^t dt$$

Evaluation of the integrals gives

$$\ln\frac{[A]_t}{[A]_0} = -kt \tag{30-6}$$

Finally, by taking the exponential of both sides of Equation 30-6, we obtain

$$\frac{[A]_t}{[A]_0} = e^{-kt} \qquad \text{or } [A]_t = [A]_0 e^{-kt} \tag{30-7}$$

This integrated form of the rate law gives the concentration of A as a function of the initial concentration $[A]_0$, the rate constant k, and the time t. A plot of this relationship is depicted in Figure 30-1. Example 30-1 illustrates the use of this equation in finding a reactant concentration at a particular time.

---

[2] Radioactive decay is an exception to this statement. The teachnique of neutron activation analysis is based on the measurement of the spontaneous decay of radionuclides created by irradiation of a sample in a nuclear reactor.

EXAMPLE 30-1

A reaction is first order with $k = 0.0370 \text{ s}^{-1}$. Calculate the concentration of reactant remaining 18.2 s after initiation of the reaction if its initial concentration is 0.0100 M.

Solution

Substituting into Equation 30-7 gives

$$[A]_{18.2} = (0.0100 \text{ M})e^{-(0.0370 \text{ s}^{-1}) \times (18.2 \text{ s})} = 0.00510 \text{ M}$$

When the rate of a reaction is being followed by monitoring the rate of appearance of a product P rather than the rate of disappearance of analyte A, it is useful to modify Equation 30-7 to relate the concentration of P at time t to the initial analyte concentration $[A]_0$. The concentration of A at any time is equal to its original concentration minus the concentration of product (when 1 mol of product forms for 1 mol of analyte). Thus,

$$[A]_t = [A]_0 - [P]_t \tag{30-8}$$

Substituting this expression for $[A]_t$ into Equation 30-7 and rearranging gives

$$[P]_t = [A]_0(1 - e^{-kt}) \tag{30-9}$$

A plot of this relationship is also shown in Figure 30-1.

The form of Equations 30-7 and 30-9 is that of a pure exponential, which appears widely in science and engineering. A pure exponential in this case has the useful characteristic that equal elapsed times give equal fractional decreases in reactant concentration or increases in product concentration. For example, let us see what happens over the time interval $t = \tau = 1/k$. When we substitute this time into Equation 30-7, we find

> The fraction of reactant used (or product formed) in a first-order reaction is the same for any given period of time.

$$[A]_\tau = [A]_0 e^{-k\tau} = [A]_0 e^{-k/k} = (1/e)[A]_0$$

In a similar way, for a period $t = 2\tau = 2/k$, we have

$$[A]_{2\tau} = (1/e)^2[A]_0$$

and so on for successive periods, as shown in **Figure 30-2**.

The period $\tau = 1/k$ is sometimes referred to as the **natural lifetime** of species A. During time τ, the concentration of A decreases to $1/e$ of its original value. A second period, from $t = \tau$ to $t = 2\tau$, produces an equivalent fractional decrease in concentration to $1/e$ of the value at the beginning of the second interval, which is $(1/e)^2$ of $[A]_0$. A more familiar example of this property of exponentials is found in the half-life $t_{1/2}$ of radionuclides. During a period $t_{1/2}$, half of the atoms in a sample of a radioactive element decay to products. A second period of $t_{1/2}$ reduces the amount of the element to one quarter of its original number, and so on for succeeding periods. Regardless of the time interval chosen, equal elapsed times produce equal fractional decreases in reactant concentration for a first-order process.

> CHALLENGE: Derive an expression for $t_{1/2}$ in terms of τ.

Figure 30-2 Progress curve for a first-order reaction showing that equal elapsed times produce equal fractional decreases in analyte concentration.

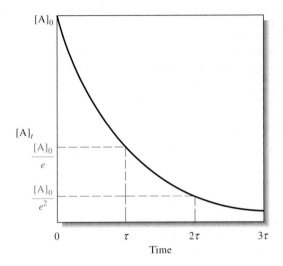

EXAMPLE 30-2

Calculate the time required for a first-order reaction with $k = 0.0500 \text{ s}^{-1}$ to proceed to 99.0% completion.

Solution

For 99.0% completion, $[A]_t/[A]_0 = (100 - 99)/100 = 0.010$. Substitution into Equation 30-6 then gives

$$\ln 0.010 = -kt = -(0.0500 \text{ s}^{-1})t$$

$$t = -\frac{\ln 0.010}{0.0500 \text{ s}^{-1}} = 92 \text{ s}$$

30A-3 Rate Laws for Second-Order and Pseudo-First-Order Reactions

Consider a typical analytical reaction in which 1 mol of analyte A reacts with 1 mol of reagent B to give a single product P. For now, we assume the reaction is irreversible and write

$$A + R \xrightarrow{k} P \tag{30-10}$$

If the reaction occurs in a single elementary step, the rate is proportional to the concentration of each of the reactants, and the rate law is

$$-\frac{d[A]}{dt} = k[A][R] \tag{30-11}$$

The reaction is first order in each of the reactants and second order overall. If the concentration of R is chosen such that $[R] \gg [A]$, the concentration of R changes very little during the course of the reaction, and we can write $k[R] = \text{constant} = k'$. Equation 30-11 is then rewritten as

$$-\frac{d[A]}{dt} = k'[A] \qquad (30\text{-}12)$$

which is identical in form to the first-order case of Equation 30-4. Hence, the reaction is said to be **pseudo-first-order** in A (see Example 30-3).

> Second-order or other higher-order reactions can usually be made pseudo-first-order through control of experimental conditions.

EXAMPLE 30-3

For a pseudo-first-order reaction in which the reagent is present in 100-fold excess, find the relative error resulting from the assumption that $k[R]$ is constant when the reaction is 40% complete.

Solution

The initial concentration of the reagent can be expressed as

$$[R]_0 = 100[A]_0$$

At 40% reaction, 60% of A remains. Thus,

$$[A]_{40\%} = 0.60[A]_0$$
$$[R]_{40\%} = [R]_0 - 0.40[A]_0 = 100[A]_0 - 0.40[A]_0 = 99.6[A]_0$$

Assuming pseudo-first-order behavior, the rate at 40% reaction is

$$-\frac{d[A]_{40\%}}{dt} = k[R]_0[A]_{40\%}$$

The true rate at 40% reaction is $k(99.6[A]_0)(0.60[A]_0)$. Therefore, the relative error is

$$\frac{k(100[A]_0)(0.60[A]_0) - k(99.6[A]_0)(0.60[A]_0)}{k(99.6[A]_0)(0.60[A]_0)} = 0.004 \quad (\text{or } 0.4\%)$$

As Example 30-3 shows, the error associated with the determination of the rate of a pseudo-first-order reaction with a 100-fold excess of reagent is quite small. A 50-fold reagent excess leads to a 1% error, which is usually deemed acceptable in kinetic methods. Moreover, the error is even less significant at times when the reaction is less than 40% complete.

Reactions are seldom completely irreversible, and a rigorous description of the kinetics of a second-order-reaction that occurs in a single step must take into account the reverse reaction. The rate of the reaction is the difference between the forward rate and the reverse rate:

$$-\frac{d[A]}{dt} = k_1[A][R] - k_{-1}[P]$$

where k_1 is the second-order rate constant for the forward reaction and k_{-1} is the first-order rate constant for the reverse reaction. In deriving this equation, we have assumed for simplicity that a single product is formed, but more complex cases can be described as well.[3] As long as conditions are maintained such that k_{-1} and/or [P] are relatively small, the rate of the reverse reaction is negligible, and little error is introduced by the assumption of pseudo-first-order behavior.

 Spreadsheet Summary   In Chapter 13 of *Applications of Microsoft® Excel in Analytical Chemistry*, 2nd ed., the first exercise explores the properties of first- and second-order reactions. The time behavior of both types of reactions is considered, and linear plotting methods are studied. Conditions needed for obtaining pseudo-first-order behavior are also investigated.

30A-4 Catalyzed Reactions

> Enzymes are high-molecular-mass molecules that catalyze reactions in biological systems. They can serve as highly selective analytical reagents.

Catalyzed reactions, particularly those in which enzymes serve as catalysts, are widely used for the determination of many biological and biochemical species as well as a number of inorganic cations and anions. We shall, therefore, use enzyme-catalyzed reactions to illustrate catalytic rate laws and to show how these rate laws can be reduced to relatively simple algebraic relationships, such as the pseudo-first-order equation shown as Equation 30-12. These simplified relationships can then be used for analytical purposes.

Enzyme-Catalyzed Reactions

The species acted on by an enzyme is called a **substrate**. Species that enhance the rate of a reaction but do not take part in the stoichiometric reaction are called **activators**. Species that do not participate in the stoichiometric reaction but decrease the reaction rate are called **inhibitors**.

Enzymes are high-molecular-mass protein molecules that catalyze reactions of importance in biology and biomedicine. Feature 30-1 discusses the basic features of enzymes. Enzymes are particularly useful as analytical reagents because many are quite selective catalysts for reactions with molecules known as **substrates**. For example, the enzyme glucose oxidase quite selectively catalyzes the reaction of its substrate β-D-glucose with oxygen to form a gluconolactone. In addition to the determination of substrates, enzyme-catalyzed reactions are used for the determination of activators, inhibitors, and of course, enzymes themselves.[4]

The behavior of many enzymes is consistent with the general mechanism

$$E + S \underset{k_{-1}}{\overset{k_1}{\rightleftharpoons}} ES \xrightarrow{k_2} P + E \tag{30-13}$$

In this **Michaelis-Menten mechanism**, the enzyme E reacts reversibly with the substrate S to form an enzyme-substrate complex ES. This complex then decomposes irreversibly to form the product(s) and the regenerated enzyme. The rate law for this mechanism assumes one of two forms, depending on the relative rates of the two steps. In the most general case, the rates of the two steps are fairly comparable in magnitude. In this steady-state case, ES decomposes as rapidly as it is

[3]See J. H. Espenson, *Chemical Kinetics and Reaction Mechanisms,* 2nd ed., New York: McGraw Hill, 1995, pp. 49–52.

[4]For a review of catalyzed reactions for kinetic methods, see S. R. Crouch, A. Scheeline, and E.W. Kirkor, *Anal. Chem.,* **2000**, *72*, 53R, **DOI**: 10.1021/a1000004b.

FEATURE 30-1

Enzymes

Enzymes are proteins that catalyze reactions necessary to sustain life. Like other proteins, enzymes consist of chains of amino acids. The structural formulas of a few important amino acids are shown in **Figure 30F-1**. Molecules formed by linking two or more amino acids are called **peptides**. Each amino acid in a peptide is called a **residue**. Molecules with many amino acid linkages are **polypeptides**, and those with long polypeptide chains are **proteins**. Enzymes differ from other proteins in that a specific area of the structure, called the active site, assists in the catalysis. As a result, enzyme catalysis is often quite specific favoring a particular substrate over other closely related compounds.

The protein structure is very important to its function. The **primary structure** is the sequence of amino acids in the protein. The **secondary structure** is the shape that the polypeptide chain assumes. Two types of secondary structures exist, the α-helix and the β-pleated sheet. The α-helix depicted in **Figure 30F-2** is the most common shape adopted by animal proteins. In this structure, the helical shape is maintained by hydrogen bonds between neighboring residues. The β-pleated sheet structure is shown in

Figure 30F-3. In this structure, the peptide chain is nearly fully extended, and the hydrogen bonding is between parallel sections of peptide chains rather than between close neighbors as in the α-helix. The β-pleated sheet structure is found in fibers like silk.

The **tertiary structure** is the overall three-dimensional shape into which the α-helix or β-pleated sheet folds as a result of interactions between residues far apart in the primary structure. Proteins may also have a **quaternary structure**, which describes how polypeptide chains stack together in a multichain protein.

The effectiveness of an enzyme as a catalyst is called the **enzyme activity**. The activity is closely linked to the three-dimensional shape of the protein, particularly to its active site. In general, the active site is a part of the protein that binds the substrate. The specificity of the enzyme depends to a large extent on the structure in the active site region. One explanation of the role of the active site is the "lock and key" model. The precise stereochemical fit of the substrate to the active site is deemed responsible for the specificity of the catalysis. Several more complex models, such as the induced-fit model, have been proposed.

Figure 30F-1 Some important amino acids. There are 20 different amino acids found in nature.

(continued)

Figure 30F-2 The α-helix. In the model on the left, the hydrogen bonding between neighboring amino acid residues that leads to the helical structure is shown. In the model on the right, only the atoms in the polypeptide chain are shown to reveal more clearly the helical structure. (Adapted from D. L. Reger, S. R. Goode, and D. W. Ball, *Chemistry: Principles and Practice,* 3rd ed, Belmont, CA: Brooks/Cole, 2010.)

A huge number of enzymes have been discovered. Only a fraction of these have been isolated and purified. The commercial availability of some of the most useful enzymes has spurred a good deal of interest in their analytical use. Enzymes have been covalently bonded to solid supports or encapsulated in gels and membranes to make them reusable and to decrease the analysis cost.

Figure 30F-3 The β-pleated sheet. Note that hydrogen bonding is between different sections of a polypeptide chain or between different chains, leading to a more extended structure. (Adapted from D. L. Reger, S. R. Goode, and D. W. Ball, *Chemistry: Principles and Practice,* 3rd ed. Belmont, CA: Brooks/Cole, 2010.)

formed, and its concentration can be assumed to be small and relatively constant throughout much of the reaction. If the second step is considerably slower than the first, the reactants and ES are always essentially at equilibrium. This equilibrium case can be derived from the general case. In the sections that follow, we show that in both cases, the reaction conditions can be arranged to yield simple relationships between rate and analyte concentration.

Steady-State Case

In the most general treatment, we derive the rate law corresponding to the mechanism of Equation 30-13 by using the **steady-state approximation**. In this approximation, the concentration of ES is presumed to be small and relatively constant throughout the reaction. The enzyme-substrate complex forms in the first step, with rate constant k_1. It decomposes by two paths; the reverse of the first step (rate constant k_{-1}) and the second step to form product (rate constant k_2). Assuming that [ES] stays constant throughout is the same as assuming that the rate of change of [ES], $d[ES]/dt$, is zero. Thus, mathematically, the steady-state assumption is written

$$\frac{d[ES]}{dt} = k_1[E][S] - k_{-1}[ES] - k_2[ES] = 0 \tag{30-14}$$

In Equation 30-14, the concentrations of enzyme [E] and substrate refer to the free concentrations at any time t. Usually we want to express the rate law in terms of the total concentration of enzyme that is known or measurable. By mass balance, the total (initial) enzyme concentration $[E]_0$ is given by

$$[E]_0 = [E] + [ES] \tag{30-15}$$

The rate of formation of product is given by

$$\frac{d[P]}{dt} = k_2[ES] \tag{30-16}$$

If we solve Equation 30-14 for [ES], we obtain

$$[ES] = \frac{k_1[E][S]}{k_{-1} + k_2} \tag{30-17}$$

If we now substitute for [E] the expression given in equation 30-15 and re-solve for [ES], we find

$$[ES] = \frac{k_1[E]_0[S]}{k_{-1} + k_2 + k_1[S]} \tag{30-18}$$

Substituting this value for [ES] into Equation 30-16 and rearranging leads to the rate law

$$\frac{d[P]}{dt} = \frac{k_2[E]_0[S]}{\dfrac{k_{-1} + k_2}{k_1} + [S]} = \frac{k_2[E]_0[S]}{K_m + [S]} \tag{30-19}$$

where the term $K_m = (k_{-1} + k_2)/k_1$ is known as the **Michaelis constant**. Equation 30-19 is often called the **Michaelis-Menten equation**. From Equation 30-17, it can be seen that the Michaelis constant K_m is given by

$$K_m = \frac{k_{-1} + k_2}{k_1} = \frac{[E][S]}{[ES]} \qquad (30\text{-}20)$$

The Michaelis constant is quite similar to the equilibrium constant for the dissociation of the enzyme-substrate complex. It is sometimes referred to as a **pseudo-equilibrium constant** since the k_2 in the numerator prevents it from being a "true" equilibrium constant. The Michaelis constant is usually expressed in units of millimoles/liter (mM) and ranges from 0.01 mM to 100 mM for many enzymes as can be seen in **Table 30-1**.

The rate equation given in Equation 30-19 can be simplified so that the reaction rate is proportional to either enzyme or substrate concentration. For example, if the concentration of substrate is made large enough that it greatly exceeds the Michaelis constant, $[S] \gg K_m$, Equation 30-19 reduces to

> To determine enzymes, the substrate concentration should be large compared to the Michaelis constant, $[S] \gg K_m$.

$$\frac{d[P]}{dt} = k_2[E]_0 \qquad (30\text{-}21)$$

Under these conditions, when the rate is independent of substrate concentration, the reaction is said to be **pseudo-zero-order** in substrate, and the rate is directly proportional to the concentration of enzyme. The enzyme is said to be **saturated** with substrate.

When conditions are such that the concentration of S is small or when K_m is relatively large, then $[S] \ll K_m$, and Equation 30-19 simplifies to

$$\frac{d[P]}{dt} = \frac{k_2}{K_m}[E]_0[S] = k'[S]$$

TABLE 30-1

Michaelis Constants for Some Enzymes		
Enzyme	**Substrate**	**K_m, mM**
Alkaline phosphatase	p-Nitrophenylphosphate	0.1
Catalase	H_2O_2	25
Hexokinase	Glucose	0.15
	Fructose	1.5
Creatine phosphokinase	Creatine	19
Carbonic anhydrase	HCO_3^-	9.0
Chymotrypsin	n-Benzoyltyrosinamide	2.5
	n-Formyltyrosinamide	12.0
	n-Acetyltyrosinamide	32
	Glycyltyrosinamide	122
Glucose oxidase	Glucose saturated with O_2	0.013
Lactate dehydrogenase	Lactate	8.0
	Pyruvate	0.125
L-amino acid oxidase	L-leucine	1.0
Urease	Urea	2.0
Uricase	Uric acid saturated with O_2	0.0175

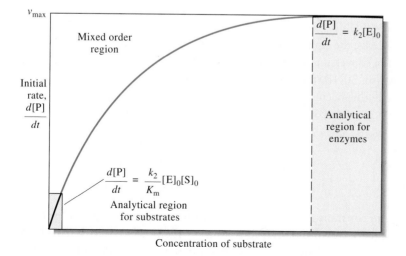

D-glucose   D-fructose

Molecular models of glucose and fructose. Glucose and fructose are important monosaccharides. Glucose is a polyhydroxyaldehyde, while fructose is a poly-hydroxyketone. Glucose is the primary fuel for biological cells. Fructose is the major sugar in fruits and vegetables. Both sugars are substrates for one or more enzymes.

where $k' = k_2[E]_0/K_m$. Hence, the kinetics are first order in substrate. In order to use this equation for determining analyte concentrations, it is necessary to measure $d[P]/dt$ at the beginning of the reaction, where $[S] \approx [S]_0$, so that

$$\frac{d[P]}{dt} \approx k'[S]_0 \qquad (30\text{-}22)$$

To determine substrates, conditions should be arranged such that the substrate concentration is small compared to the Michaelis constant, $[S] \ll K_m$.

The regions where Equations 30-21 and 30-22 are applicable are illustrated in **Figure 30-3** in which the initial rate of an enzyme-catalyzed reaction is plotted as a function of substrate concentration. When the substrate concentration is low, Equation 30-22, which is linear in substrate concentration, governs the shape of the curve, and it is this region that is used to determine the amount of substrate present.

Figure 30-3 Plot of initial rate of product formation as a function of substrate concentration that shows the parts of the curve useful for the determination of substrate and enzyme.

If it is desired to determine the amount of enzyme, the region of high substrate concentration is employed, where Equation 30-21 applies, and the rate is independent of substrate concentration. The limiting rate of the reaction at large values of [S] is the maximum rate that can be achieved at a given enzyme concentration, v_{max}, as indicated in Figure 30-3. It can be shown that the value of the substrate concentration at exactly $v_{max}/2$ is equal to the Michaelis constant, K_m. Example 30-4 illustrates the use of the Michaelis-Menten equation.

EXAMPLE 30-4

The enzyme urease, which catalyzes the hydrolysis of urea, is widely used for determining urea in blood. Details of this application are given in Feature 30-3 on page 842. The Michaelis constant for urease at room temperature is 2.0 mM, and $k_2 = 2.5 \times 10^4$ s^{-1} at pH 7.5. (a) Calculate the initial rate of the reaction when the urea concentration is 0.030 mM and the urease concentration is 5.0 μM and (b) find v_{max}.

Solution

(a) From Equation 30-19,

$$\frac{d[P]}{dt} = \frac{k_2[E]_0[S]}{K_m + [S]}$$

At the beginning of the reaction, $[S] = [S]_0$, and

$$\frac{d[P]}{dt} = \frac{(2.5 \times 10^4\ s^{-1})(5.0 \times 10^{-6}M)(0.030 \times 10^{-3}M)}{2.0 \times 10^{-3}M + 0.030 \times 10^{-3}M}$$

$$= 1.8 \times 10^{-3}M\ s^{-1}$$

(b) Figure 30-3 reveals that $d[P]/dt = v_{max}$ when the concentration of substrate is large, and therefore, Equation 30-21 applies. Thus,

$$d[P]/dt = v_{max} = k_2[E]_0 = (2.5 \times 10^4\ s^{-1})(5.0 \times 10^{-6}\ M) = 0.125\ M\ s^{-1}$$

The Equilibrium Case

We can derive the equilibrium case from the general, steady-state case just discussed. When the conversion of ES to products is slow compared to the reversible first step of Equation 30-13, the first step is essentially at equilibrium throughout. Mathematically this occurs when k_2 is much smaller than k_{-1}. Under these conditions, Equation 30-19 becomes

$$\frac{d[P]}{dt} = \frac{k_2[E]_0[S]}{\dfrac{k_{-1}}{k_1} + [S]} = \frac{k_2[E]_0[S]}{K + [S]} \tag{30-23}$$

where the constant K is now a true equilibrium constant given by $K = k_{-1}/k_1$. Note that the form of Equation 30-23 is identical to the Michaelis-Menten equation, 30-19. There is only a subtle difference in the definition of K_m and K.

Hence, substrate and enzyme concentrations can be determined in the same manner as for the steady-state case for enzyme reactions in which k_2 is small and the equilibrium assumption holds. Enzyme concentrations are determined under conditions where the substrate concentration is large, whereas substrate concentrations are determined when $[S] \ll K$.

There are many more complex mechanisms for enzyme reactions involving reversible reactions, multiple substrates, activators, and inhibitors. Techniques for modeling and analyzing these systems are available.[5]

Although our discussion thus far has been concerned with enzymatic methods, an analogous treatment for ordinary catalysis gives rate laws that are similar in form to those for enzymes. These expressions often reduce to the first-order case for ease of data treatment, and many examples of kinetic-catalytic methods are found in the literature.[6]

 Spreadsheet Summary   The second exercise in Chapter 13 of *Applications of Microsoft® Excel in Analytical Chemistry*, 2nd ed., concerns enzyme catalysis. A linear transformation is made so that the Michaelis constant, K_m, and the maximum velocity, v_{max}, can be determined from a least-squares procedure. The nonlinear regression method is used with Excel's Solver to find these parameters by fitting the nonlinear Michaelis-Menten equation.

30B DETERMINING REACTION RATES

Several methods are used for the determination of reaction rates. In this section, we describe some of these methods and how they are used.

30B-1 Experimental Methods

The way rates are measured depends on whether the reaction of interest is fast or slow. A reaction is generally regarded as fast if it proceeds to 50% of completion in 10 s or less. Analytical methods that use fast reactions generally require special equipment that permits rapid mixing of reagents and fast recording of data, as discussed in Feature 30-2.

A **fast reaction** is 50% complete in 10 s or less.

FEATURE 30-2

Fast Reactions and Stopped-Flow Mixing

One of the most popular and most reliable methods for carrying out rapid reactions is stopped-flow mixing. In this technique, streams of reagent and sample are mixed rapidly, and the flow of mixed solution is stopped suddenly. The reaction progress is then monitored at a position slightly downstream from the mixing point. A stopped-flow mixing system is shown in **Figure 30F-4**.

To illustrate the operation of the instrument, we begin with the drive syringes filled with reagent and sample and with valves A, B, and C closed. The stop syringe is empty. The drive mechanism is then activated to move the drive syringe plungers forward rapidly. The reagent and sample pass into the mixer where they are mixed and pass immediately into the observation cell, as indicated by the green arrows.

(continued)

---

[5]For example, see Heino Prinz, *Numerical Methods for the Life Scientist*, Heidelberg: Springer-Verlag, 2011; P. F. Cook and W. W. Cleland, *Enzyme Kinetics and Mechanisms*, New York: Garland Science, 2007.

[6]See D. Perez-Bendito and M. Silva, *Kinetic Methods in Analytical Chemistry*, New York: Halsted Press-Wiley, 1988; H. A. Mottola, *Kinetic Aspects of Analytical Chemistry*, New York: Wiley, 1988.

Figure 30F-4 Stopped-flow mixing apparatus.

The reaction mixture then passes into the stop syringe. When the stop syringe fills, the stop syringe plunger strikes the stop block. This event causes the flow to cease almost instantly with a recently mixed plug of solution in the observation cell. In this example, the observation cell is transparent so that a light beam can be passed through to make absorption measurements. In this way, the progress of the reaction can be monitored. All that is required is that the dead time, or the time between the mixing of reagents and the arrival of the sample in the observation cell, be short relative to the time required for the reaction to proceed to completion. For well-designed systems in which the turbulent flow of the mixer provides very rapid and efficient mixing, the dead time is on the order of 2 to 4 ms. Thus, first-order or pseudo-first-order reactions with $\tau \approx 25$ ms ($k \approx 40$ s^{-1}) can be examined using the stopped-flow technique.

When the reaction is complete, valve C is opened, and the stop syringe plunger is then pushed down to purge the stop syringe of its contents (gray arrow). Valve C is then closed, valves A and B are opened, and the drive mechanism is moved down to fill the drive syringes with solution (black arrows). At this point, the apparatus is ready for another rapid mixing experiment. The entire apparatus can be placed under the control of a computer, which can also collect and analyze the data. Stopped-flow mixing has been used for fundamental studies of rapid reactions and for routine kinetic determinations of analytes involved in fast reactions. The principles of fluid dynamics that make stopped-flow mixing possible as well as the solution-handling capabilities of this and other similar devices are used in many contexts for automatically mixing solutions and measuring analyte concentrations in numerous industrial and clinical laboratories.

If a reaction is sufficiently slow, conventional methods of analysis can be used to determine the concentration of a reactant or product as a function of time. Often, however, the reaction of interest is too rapid for many static measurement techniques, that is, concentrations change appreciably during the measurement process. Under these circumstances, either the reaction must be stopped while the measurement is made or an instrumental technique that records concentrations continuously as the reaction proceeds must be employed. In the former case, an aliquot is removed from the reaction mixture and rapidly quenched by mixing it with a reagent that combines with one of the reactants to stop the reaction. Alternatively, quenching is accomplished by lowering the temperature rapidly to slow the reaction to an acceptable level for the measurement step. Unfortunately, quenching techniques tend to be laborious and often time consuming and are thus not widely used for analytical purposes.

The most convenient approach for obtaining kinetic data is to monitor the progress of the reaction continuously by spectrophotometry, conductometry, potentiometry, amperometry, or some other instrumental technique. With the advent of inexpensive computers, instrumental readings proportional to concentrations of reactants and/or products are often recorded directly as a function of time, stored in the computer's memory, and retrieved later for data processing. Stopped-flow principles can also be employed with flow injection analyzers (see Section 8C) by turning off the pump or otherwise stopping the flow while the reaction mixture is in the observation chamber.[7] Although not a technique for fast reactions like conventional stopped-flow mixing, stopped-flow, flow-injection has been used successfully in several enzyme-based determinations.

In the following sections, we explore some strategies used in kinetic methods to permit analyte concentrations to be determined from reaction progress plots.

30B-2 Types of Kinetic Methods

Kinetic methods are classified according to the type of relationship that exists between the measured variable and the analyte concentration.

The Differential Method

In the **differential method**, concentrations are computed from reaction rates by means of a differential form of a rate expression. Rates are determined by measuring the slope of a curve relating analyte or product concentration to reaction time. To illustrate, let us substitute $[A]_t$ from Equation 30-7 for $[A]$ in Equation 30-4:

$$\text{rate} = -\left(\frac{d[A]}{dt}\right) = k[A]_t = k[A]_0 e^{-kt} \tag{30-24}$$

As an alternative, the rate can be expressed in terms of the product concentration, that is,

$$\text{rate} = \left(\frac{d[P]}{dt}\right) = k[A]_0 e^{-kt} \tag{30-25}$$

[7]J. Ruzicka and E. H. Hansen, *Anal. Chim. Acta*, **1978**, *99*, 37; J. Ruzicka and E. H. Hansen, *Anal. Chim. Acta*, **1979**, *106*, 207.

Equations 30-24 and 30-25 show the dependence of the rate on k, t, and, most important, $[A]_0$, the initial concentration of the analyte. At any fixed time t, the factor ke^{-kt} is a constant, and the rate is directly proportional to the initial analyte concentration. Example 30-5 illustrates the use of the differential method to calculate the initial analyte concentration.

EXAMPLE 30-5

The rate constant for a pseudo-first-order reaction is 0.156 s^{-1}. Find the initial concentration of the reactant if its rate of disappearance 10.00 s after the initiation of the reaction is 2.79×10^{-4} M s^{-1}.

Solution

The proportionality constant ke^{-kt} is

$$ke^{-kt} = (0.156\ s^{-1})e^{-(0.156\ s^{-1})(10.00\ s)} = 3.28 \times 10^{-2}\ s^{-1}$$

Rearranging Equation 30-24 and substituting numerical values, we have

$$[A]_0 = rate/ke^{-kt}$$

$$= (2.79 \times 10^{-4}\ M\ s^{-1})/(3.28 \times 10^{-2}\ s^{-1})$$

$$= 8.51 \times 10^{-3}\ M$$

The choice of the time at which a reaction rate is measured is often based on such factors as convenience, the existence of interfering side reactions, and the inherent precision of making the measurement at a particular time. It is often advantageous to make the measurement near $t = 0$ because this portion of the exponential curve is nearly linear (for example, see the initial parts of the curves in Figure 30-1), and we can estimate the slope from the tangent to the curve. Also, if the reaction is pseudo-first-order, such a small amount of excess reagent is consumed that there is no error from changes in k resulting from changes in reagent concentration. Finally, the *relative error* in determining the slope is minimal at the beginning of the reaction because the slope is a maximum in this region.

Figure 30-4 illustrates how the differential method is used to determine the concentration of an analyte $[A]_0$ from experimental rate measurements for the reaction shown as Equation 30-1. The solid curves in Figure 30-4a are plots of the experimentally measured product concentration [P] as a function of reaction time for four

Figure 30-4 A plot of data for the determination of A by the differential method. (a) Solid lines are the experimental plots of product concentration as a function of time for four initial concentrations of A. Dashed lines are tangents to the curve at $t \rightarrow 0$. (b) A plot of the slopes obtained from the tangents in (a) as a function of the analyte concentration.

(a)

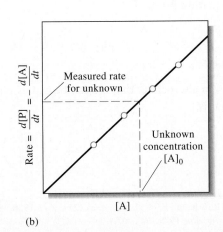

(b)

standard solutions of A. These curves are then used to prepare the calibration plot shown in Figure 30-4b. To obtain the rates, tangents are drawn to each of the curves in 30-4a at a time near zero (dashed lines). The slopes of the tangents are then plotted as a function of [A], giving the straight line shown in 30-4b. Unknowns are treated in the same way, and analyte concentrations are determined from the calibration curve.

It is not necessary to record the entire rate curve, as has been done in Figure 30-4a, since only a small portion of the plot is used for measuring the slope. As long as enough data points are collected to determine the initial slope precisely, time is saved, and the entire procedure is simplified. More sophisticated data-handling procedures and numerical analysis of the data make possible high-precision rate measurements at later times as well. Under certain circumstances such measurements are more accurate and precise than those made near $t = 0$.

Integral Methods

In contrast to the differential method, **integral methods** take advantage of integrated forms of rate laws, such as those shown by Equations 30-6, 30-7, and 30-9.

Graphical Methods. Equation 30-6 may be rearranged to give

$$\ln [A]_t = -kt + \ln [A]_0 \tag{30-26}$$

A plot of the natural logarithm of experimentally measured concentrations of A (or P) as a function of time should yield a straight line with a slope of $-k$ and a y intercept of $\ln [A]_0$. Use of this procedure for the determination of nitromethane is illustrated in Example 30-6.

EXAMPLE 30-6

The data in the first two columns of Table 30-2 were recorded for the pseudo-first-order decomposition of nitromethane in the presence of excess base. Find the initial concentration of nitromethane and the pseudo-first-order rate constant for the reaction.

Solution

Computed values for the natural logarithms of nitromethane concentrations are shown in the third column of **Table 30-2**. The data are plotted in **Figure 30-5**. A least-squares analysis of the data (Section 8D-2) leads to an intercept b of

$$b = \ln[CH_3NO_2]_0 = -5.129$$

which after exponentiation gives

$$[CH_3NO_2]_0 = 5.92 \times 10^{-3} \, M$$

The least-squares analysis also gives the slope of the line m, which in this case is

$$m = -1.62 = -k$$

and thus,

$$k = 1.62 \, s^{-1}$$

TABLE 30-2

Data for the Decomposition of Nitromethane

Time, s	$[CH_3NO_2]$, M	$\ln[CH_3NO_2]$
0.25	3.86×10^{-3}	-5.557
0.50	2.59×10^{-3}	-5.956
0.75	1.84×10^{-3}	-6.298
1.00	1.21×10^{-3}	-6.717
1.25	0.742×10^{-3}	-7.206

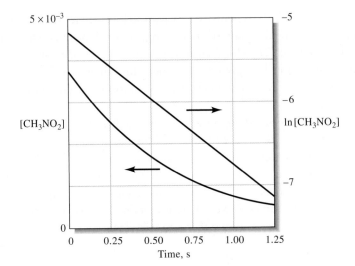

Figure 30-5 Plots of nitromethane concentration and the natural logarithm of nitromethane concentration as a function of time. The data are from Example 30-6.

Fixed-Time Methods. Fixed-time methods are based on Equation 30-7 or 30-9. The former can be rearranged to

$$[A]_0 = \frac{[A]_t}{e^{-kt}} \tag{30-27}$$

The simplest way of using this relationship is to perform a calibration experiment with a standard solution that has a known concentration $[A]_0$. After a carefully measured reaction time t, $[A]_t$ is determined and used to evaluate the constant e^{-kt} by means of Equation 30-27. Unknowns are then analyzed by measuring $[A]_t$ after exactly the same reaction time and employing the calculated value for e^{-kt} to compute the analyte concentrations.

Equation 30-27 can be modified for the situation where $[P]$ is measured experimentally rather than $[A]$. Equation 30-9 can be rearranged to solve for $[A]_0$, that is,

$$[A]_0 = \frac{[P]_t}{1-e^{-kt}} \tag{30-28}$$

A more desirable approach is to measure $[A]$ or $[P]$ at two times t_1 and t_2. For example, if the product concentration is determined, we can write

$$[P]_{t_1} = [A]_0(1 - e^{-kt_1})$$
$$[P]_{t_2} = [A]_0(1 - e^{-kt_2})$$

Subtracting the first equation from the second and rearranging yields

$$[A]_0 = \frac{[P]_{t_2} - [P]_{t_1}}{e^{-kt_1} - e^{-kt_2}} = C([P]_{t_2} - [P]_{t_1}) \tag{30-29}$$

The reciprocal of the denominator is constant for constant t_1 and t_2 and is assigned the symbol C.

The use of Equation 30-29 has the fundamental advantage common to most kinetic methods in that the absolute determination of concentration or of a variable proportional to concentration is unnecessary. It is the difference between two concentrations that is proportional to the initial concentration of the analyte.

An important example of an uncatalyzed method is the fixed-time method for the determination of thiocyanate ion based on spectrophotometric measurements of the red iron(III) thiocyanate complex. The reaction in this application is

$$Fe^{3+} + SCN^- \underset{k_{-1}}{\overset{k_1}{\rightleftharpoons}} Fe(SCN)^{2+}$$

Under conditions of excess Fe^{3+}, the reaction is pseudo-first-order in SCN^-. The curves in **Figure 30-6a** show the increase in absorbance due to the appearance of $Fe(SCN)^{2+}$ versus time following the rapid mixing of 0.100 M Fe^{3+} with various concentrations of SCN^- at pH 2. Since the concentration of $Fe(SCN)^{2+}$ is related to the absorbance by Beer's law, the experimental data can be used directly without conversion to concentration. Thus, the change in absorbance, ΔA, between times t_1 and t_2 is computed and plotted versus $[SCN^-]_0$ as in **Figure 30-6b**. Unknown concentrations are then determined by evaluating ΔA under the same experimental conditions and finding the concentration of thiocyanate ion from the calibration curve or the least-squares equation.

Fixed-time methods are advantageous because the measured quantity is directly proportional to the analyte concentration and because measurements can be made *at any time* during the progress of first-order reactions. When instrumental methods are used for monitoring reactions by fixed-time procedures, the precision of the analytical results approaches the precision of the instrument used.

Curve-Fitting Methods. With computers attached to instruments, the fitting of a mathematical model to the concentration or signal versus time curve is straightforward. Curve-fitting techniques compute values of the model parameters, including the initial concentration of analyte, that "best fit" the data. The most sophisticated of these methods use the parameters of the model to estimate the value of the equilibrium or steady-state response. These methods can provide error

> A major advantage of kinetic methods is their immunity to errors resulting from long-term drift of the measurement system.

(a)

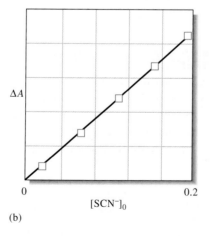

(b)

Figure 30-6 (a) Absorbance due to the formation of $Fe(SCN)^{2+}$ as a function of time for five concentrations of SCN^-. (b) A plot of the difference in absorbance ΔA at times t_2 and t_1 as a function of SCN^- concentration.

Figure 30-7 The predictive approach in kinetic methods. A mathematical model, shown as the blue squares, is used to fit the response, shown as the solid line, during the kinetic regime of a reaction. The model is then used to predict the equilibrium value of the signal, A_e, which is related to the analyte concentration. In the example shown, the absorbance is plotted versus time, and the early-time data used to predict A_e, the equilibrium value, is shown as the blue circle. Reprinted (adapted) with permission from G. L. Mieling and H. L. Pardue, *Anal. Chem.*, **1978**, *50*, 1611, **DOI**: 10.1021/ac50034a011. Copyright 1978 American Chemical Society.

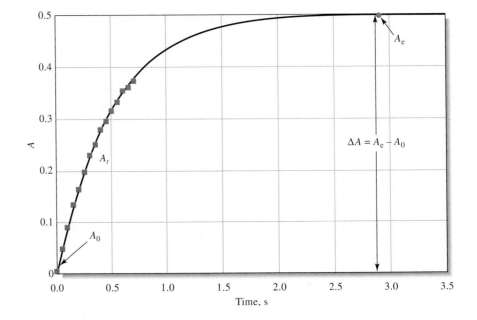

compensation because the equilibrium position is less sensitive to such experimental variables as temperature, pH, and reagent concentrations. **Figure 30-7** illustrates the use of this approach to predict the equilibrium absorbance from data obtained during the kinetic regime of the response curve. The equilibrium absorbance is then related to the analyte concentration in the usual way.

The computer is enabling many innovative techniques for kinetic methods. Some recent error-compensation methods do not require prior knowledge of the reaction order for the system employed, but instead, they use a generalized model. Still other methods calculate the model parameters as the data are collected instead of employing batch processing methods.

> **Spreadsheet Summary** In the final exercise of Chapter 13 of *Applications of Microsoft® Excel in Analytical Chemistry*, 2nd ed., the initial rate method is explored for determining the concentration of an analyte. Initial rates are determined from a linear least-squares analysis and used to establish a calibration curve and equation. An unknown concentration is determined.

30C APPLICATIONS OF KINETIC METHODS

The reactions used in kinetic methods fall into two categories, **catalyzed** and **uncatalyzed**. As noted previously, catalyzed reactions are the most widely used because of their superior sensitivity and selectivity. Uncatalyzed reactions are used to advantage when high-speed, automated measurements are required, however, or when the sensitivity of the detection method is great.[8]

---

[8]For reviews of applications of kinetic methods, see H. O. Mottola, *Kinetic Aspects of Analytical Chemistry*, New York: Wiley, 1988, pp. 88–121.; D. Perez-Bendito and M. Silva, *Kinetic Methods in Analytical Chemistry*, New York: Halsted Press-Wiley, 1988, pp. 31–189.

30C-1 Catalytic Methods

Catalytic methods have been used to determine both inorganic and organic compounds.

Determination of Inorganic Species

Many inorganic cations and anions catalyze indicator reactions, that is, reactions with rates that can be measured by instrumental methods, such as absorption spectrophotometry, fluorescence spectrometry, or electrochemistry. Conditions are then employed such that the rate is proportional to the concentration of catalyst, and from the rate data, the concentration of catalyst is determined. Such catalytic methods often allow extremely sensitive detection of the catalyst concentration. Kinetic methods based on catalysis by inorganic analytes are widely applicable. For example, the literature in the area lists more than 40 cations and 15 anions that have been determined by a variety of indicator reactions.[9] Table 30-3 gives catalytic methods for several inorganic species along with the indicator reactions used, the method of detection, and the detection limit.

Determination of Organic Species

Enzymes are used as catalysts in the most important applications of catalyzed reactions to organic analyses. These methods have been used for the determination of both enzymes and substrates and serve as the basis for many of the routine and automated screening tests performed by the thousands in clinical laboratories throughout the world.

Many different enzyme substrates have been determined with enzyme-catalyzed reactions. Table 30-4 lists some of the substrates that are determined in various applications. One important application is for determining the quantity of urea in blood and is called the blood urea nitrogen (BUN) test. A description of this determination is given in Feature 30-3.

Molecular model of sucrose. Sucrose is a disaccharide, consisting of two monosaccharide units linked together. One of the units in sucrose is a glucose ring (6-member), and the other is a fructose ring (5-member). Sucrose is common table sugar.

Molecular model of uric acid. Uric acid is essential to the digestive process. However, if the body produces too much uric acid or if not enough is excreted, high levels of this substance in the blood can lead to crystals of sodium urate being concentrated in the joints and tendons. These cause the inflammation, the pressure, and the severe pain associated with gouty arthritis, or gout.

TABLE 30-3

Catalytic Methods for Inorganic Species

Analyte	Indicator reaction	Detection method	Detection limit, ng/mL
Cobalt	Catechol + H_2O_2	Spectrophotometry	3
Copper	Hydroquinone + H_2O_2	Spectrophotometry	0.2
Iron	$H_2O_2 + I^-$	Potentiometry	50
Mercury	$Fe(CN)_6^{4-} + C_6H_5NO$	Spectrophotometry	60
Molybdenum	$H_2O_2 + I^-$	Spectrophotometry	10
Bromide	Decomposition of BrO_3^-	Spectrophotometry	3
Chloride	$Fe^{2+} + ClO_3^-$	Spectrophotometry	100
Cyanide	Reduction of o-dinitrobenzene	Spectrophotometry	100
Iodide	Ce(IV) + As(III)	Potentiometry	0.2
Oxalate	Rhodamine B + $Cr_2O_7^{2-}$	Spectrophotometry	20

---

[9]M. Kopanica and V. Stara, in *Comprehensive Analytical Chemistry*, G. Svehla, ed., Vol. 18, pp. 11–227, New York: Elsevier, 1983.

TABLE 30-4

Some Important Substrates[10]

Substrate	Enzyme	Application
Ethanol	Alcohol dehydrogenase	Law enforcement, alcoholism
Galactose	Galactose oxidase	Diagnosis of galactosemia
Glucose	Glucose oxidase	Diagnosis of diabetes
Lactose	Lactase	Food products
Maltose	α-Glucosidase	Food products
Penicillin	Penicillinase	Pharmaceutical preparations
Phenol	Tyrosinase	Water and wastewaters
Sucrose	Invertase	Food products
Urea	Urease	Diagnosis of liver and kidney disease
Uric acid	Uricase	Diagnosis of gout, leukemia, lymphoma

FEATURE 30-3

The Enzymatic Determination of Urea

The determination of urea in blood and urine is frequently carried out by measuring the rate of hydrolysis of urea in the presence of the enzyme urease. The equation for the reaction is

$$CO(NH_2)_2 \ + \ 2H_2O \ \xrightarrow{\text{urease}} \ NH_4^+ \ + \ HCO_3^-$$

As suggested in Example 30-4, urea can be determined by measuring the initial rate of production of the products of this reaction. The high selectivity of the enzyme permits the use of nonselective detection methods, such as electrical conductance, for initial rate measurements. There are commercial instruments that operate on this principle. The sample is mixed with a small amount of an enzyme-buffer solution in a conductivity cell. The maximum rate of increase in conductance is measured within 10 s of mixing, and the concentration of urea is determined from a calibration curve consisting of a plot of maximum initial rate as a function of urea concentration. The precision of the instrument is on the order of 2 to 5% for concentrations in the physiological range of 2 to 10 mM.

Another method for following the rate of urea hydrolysis is based on a specific-ion electrode for ammonium ions (Section 21D). With this method, the production of NH_4^+ is monitored potentiometrically and used to obtain the reaction rate. In yet another approach, the urease can be immobilized on the surface of a pH electrode, and the rate of change of pH monitored. Many enzymes have now been immobilized onto supports such as gels, membranes, tubing walls, glass beads, polymers, and thin films. **Immobilized enzymes** often show enhanced stability over their soluble counterparts. In addition, they can be reused often for hundreds or thousands of analyses.

O
‖
$H_2N — C — NH_2$

Molecular model of urea. Urea is the diamide of carbonic acid. It is excreted by mammals as a waste product from the metabolism of proteins.

Enzymes can be immobilized by entrapment in a gel, by adsorption to a solid support, or by covalent bonding to a solid.

A number of inorganic species can also be determined by enzyme-catalyzed reactions, including ammonia, hydrogen peroxide, carbon dioxide, and hydroxylamine, as well as nitrate, phosphate, and pyrophosphate ions.

---

[10]For more information, see G. G. Guilbault, *Analytical Uses of Immobilized Enzymes*, New York: Dekker, 1984; P. W. Carr and L. D. Bowers, *Immobilized Enzymes in Analytical and Clinical Chemistry*, New York: Wiley, 1980.

Kinetic methods have been described for the quantitative determination of several hundred enzymes. Some of the enzymes that are important in diagnosing liver diseases are glutamic-oxaloacetic transaminase (GOT), glutamatic-pyruvic transaminase (GPT), and lactate dehydrogenase (LDH). Elevated levels of GOT, GPT, and LDH can also occur after heart attacks. These enzymes and creatine phosphokinase are often diagnostic for myocardial infarction. Other enzymes of interest in diagnosis include such hydrolases as amylase, lipase and alkaline phosphatase, phosphohexose isomerase, and aldolase.

In addition, some two dozen inorganic cations and anions are known to decrease the rates of certain enzyme-catalyzed indicator reactions. These **inhibitors** can thus be determined from the decrease in rate brought about by their presence.

Enzyme activators are substances, often inorganic ions, that are required for certain enzymes to become active as catalysts. Activators can be determined by their effect on the rates of enzyme-catalyzed reactions. For example, it has been reported that magnesium at concentrations as low as 10 ppb can be determined in blood plasma based on activation of the enzyme isocitric dehydrogenase by the magnesium ion.

> ◀ Kinetic methods are necessary in determining enzyme activities since the enzyme is a catalyst and affects only the reaction rate.

> ◀ Enzymes can be used for the determination of activators and inhibitors. Activators increase the reaction rate, while inhibitors decrease the rate.

30C-2 Uncatalyzed Reactions

As noted earlier, kinetic methods based on uncatalyzed reactions are not nearly as widely used as catalyzed reactions. We have already described two of these methods (pages 000 and 000).

Generally, uncatalyzed reactions are useful when selective reagents are used with sensitive detection methods. For example, the selectivity of complexing agents can be controlled by adjusting the pH of the medium in the determination of metal ions, as discussed in Section 17D-8. Sensitivity can be achieved through the use of spectrophotometric detection to monitor reagents that form complexes with large molar absorptivities. The determination of Cu^{2+} presented in Problem 30-13 is an example. A highly sensitive alternative is to select complexes that fluoresce so that the rate of change of fluorescence is used as a measure of analyte concentration (see Problem 30-14).

The precision of both noncatalytic and catalytic kinetic methods depends on such experimental conditions as pH, ionic strength, and temperature. With careful control of these variables, relative standard deviations of 1 to 10% are typical. Automation of kinetic methods and computerized data analysis can often improve the relative precision to 1% or less.

30C-3 Kinetic Determination of Components in Mixtures

An important application of kinetic methods is in the determination of closely related species in mixtures, such as alkaline earth cations or organic compounds with the same functional groups. For example, suppose two species A and B react with a common excess reagent to form products under pseudo-first-order conditions:

$$A + R \xrightarrow{k_A} P$$

$$B + R \xrightarrow{k_B} P'$$

Generally, k_A and k_B differ from each other. Thus, if $k_A > k_B$, A is depleted before B. It is possible to show that, if the ratio k_A/k_B is greater than about 500, the consumption of A is approximately 99% complete before 1% of B is used up.

Therefore, a differential determination of A with no significant interference from B is possible provided the rate is measured shortly after mixing.

When the ratio of the two rate constants is small, determination of both species is still possible by more complex methods of data treatment. Many of these methods use chemometric, multivariate calibration techniques similar to those described in Feature 8-3. The details of **multicomponent kinetic methods** are beyond the scope of this text.[11]

WEB WORKS

Use a search engine to find instrument makers that produce glucose analyzers based on enzymatic reactions. Find one company that makes a spectrophotometric analyzer and one that makes an electrochemical analyzer. Compare and contrast the features of the two instruments including accuracy, precision, dynamic range, and cost.

QUESTIONS AND PROBLEMS

30-1. Define the following terms as they are used in kinetic methods of analysis.
*(a) order of a reaction   *(e) Michaelis constant
(b) pseudo-first-order   (f) differential method
*(c) enzyme   *(g) integral method
(d) substrate   (h) indicator reaction

30-2. The analysis of a multicomponent mixture by kinetic methods is sometimes referred to as a "kinetic separation." Explain the meaning of this term.

*30-3. List three advantages of kinetic methods. Can you think of two possible limitations of kinetic methods when compared to equilibrium methods?

30-4. Explain why pseudo-first-order conditions are utilized in many kinetic methods.

*30-5. Derive an expression for the half-life of the reactant in a first-order process in terms of the rate constant k.

30-6. Find the natural lifetime in seconds for first-order reactions corresponding to
*(a) $k = 0.497 \text{ s}^{-1}$.
(b) $k = 6.62 \text{ h}^{-1}$.
*(c) $[A]_0 = 3.16 \text{ M}$, and $[A]_t = 0.496 \text{ M}$ at $t = 3876 \text{ s}$.
(d) $[P]_\infty = 0.176 \text{ M}$, and $[P]_t = 0.0423 \text{ M}$ at $t = 9.54 \text{ s}$ (assume 1 mol of product is formed for each mol of analyte reacted).
*(e) half-life, $t_{1/2} = 26.5$ years.
(f) $t_{1/2} = 0.583 \text{ s}$.

30-7. Find the first-order rate constant for a reaction that is 75.0% complete in
*(a) 0.0100 s.   *(c) 1.00 s.   *(e) 26.8 μs.
(b) 0.100 s.   (d) 5280 s.   (f) 8.86 ns.

30-8. Find the number of half-lives required to reach the following levels of completion:
*(a) 10%.   *(c) 90%.   *(e) 99.9%.
(b) 50%.   (d) 99%.   (f) 99.99%.

30-9. Calculate the number of lifetimes τ required for a pseudo-first-order reaction to achieve the levels of completion listed in Problem 30-8.

30-10. Find the relative error associated with the assumption that k' is invariant during the course of a pseudo-first-order reaction under the following conditions:

	Extent of Reaction, %	Excess of Reagent
*(a)	1	5×
(b)	1	10×
*(c)	1	50×
(d)	1	100×
*(e)	5	5×
(f)	5	10×
*(g)	5	100×
(h)	63.2	5×
*(i)	63.2	10×
(j)	63.2	50×
*(k)	63.2	100×

---

[11]For some applications of kinetic methods to multicomponent mixtures, see, H. O. Mottola, *Kinetic Aspects of Analytical Chemistry*, New York: Wiley, 1988, pp. 122–148; D. Perez-Bendito and M. Silva, *Kinetic Methods in Analytical Chemistry*, New York: Halsted Press-Wiley, 1988, pp. 172–189.

30-11. Show that, for an enzyme reaction obeying Equation 30-19, the substrate concentration for which the rate equals $v_{max}/2$ is equal to K_m.

*30-12. Equation 30-19 can be rearranged to produce the equation

$$\frac{1}{d[P]/dt} = \frac{K_m}{v_{max}[S]} + \frac{1}{v_{max}}$$

where $v_{max} = k_2[E]_0$, the maximum velocity when [S] is large.

(a) Suggest a way to use this equation in the construction of a calibration (working) curve for the enzymatic determination of substrate.

(b) Describe how the resulting working curve can be used to find K_m and v_{max}.

*30-13. Copper(II) forms a 1:1 complex with the organic complexing agent R in acidic medium. The formation of the complex can be monitored by spectrophotometry at 480 nm. Use the following data collected under pseudo-first-order conditions to construct a calibration curve of rate versus concentration of R. Find the concentration of copper(II) in an unknown whose rate under the same conditions was $6.2 \times 10^{-3} \, A \, s^{-1}$. Also find the standard deviation of the concentration.

$c_{Cu^{2+}}$, ppm	Rate, $A \, s^{-1}$
3.0	3.6×10^{-3}
5.0	5.4×10^{-3}
7.0	7.9×10^{-3}
9.0	1.03×10^{-2}

30-14. Aluminum forms a 1:1 complex with 2-hydroxy-1-naphthaldehyde p-methoxybenzoylhydraxonal that exhibits fluorescence emission at 475 nm. Under pseudo-first-order conditions, a plot of the initial rate of the reaction (emission units per second) versus the concentration of aluminum (in µM) yields a straight line described by the equation

$$rate = 1.74c_{Al} - 0.225$$

Find the concentration of aluminum in a solution that exhibits a rate of 0.76 emission units per second under the same experimental conditions.

*30-15. The enzyme monoamine oxidase catalyzes the oxidation of amines to aldehydes. For tryptamine, K_m for the enzyme is 4.0×10^{-4} M, and $v_{max} = k_2[E]_0 = 1.6 \times 10^{-3}$ µM/min at pH 8. Find the concentration of a solution of tryptamine that reacts at a rate of 0.18 µM/min in the presence of monoamine oxidase under the above conditions. Assume that $[tryptamine] \ll K_m$.

30-16. The following data represent the product concentrations versus time during the initial stages of pseudo-first-order reactions with different initial concentrations of analyte $[A]_0$:

t, s	[P], M				
0	0.00000	0.00000	0.00000	0.00000	0.00000
10	0.00004	0.00018	0.00027	0.00037	0.00014
20	0.00007	0.00037	0.00055	0.00073	0.00029
50	0.00018	0.00091	0.00137	0.00183	0.00072
100	0.00036	0.00181	0.00272	0.00362	0.00144
$[A]_0$, M	0.01000	0.05000	0.07500	0.10000	unknown

For each concentration of analyte, find the average initial reaction rate for the five time slots given. Plot the initial rate versus the concentration of analyte. Obtain the least-squares slope and intercept of the plot and determine the unknown concentration. Hint: A good way to calculate the initial rate for a given analyte concentration is to find $\Delta[P]/\Delta t$ for the 0 to 10 s interval, the 10 to 20 s interval, the 20 to 50 s interval, and the 50 to 100 s interval; then average the 4 values obtained. Alternatively, the least-squares slope of a plot of [P] versus t for the 0-to-100-s interval can be used.

*30-17. Calculate the product concentrations versus time for a pseudo-first-order reaction with $k' = 0.015 \, s^{-1}$ and $[A]_0 = 0.005$ M. Use times of 0.000 s, 0.001 s, 0.01 s, 0.1 s, 0.2 s 0.5 s, 1.0 s, 2.0 s, 5.0 s, 10.0 s, 20.0 s, 50.0 s, 100.0 s, 200.0 s, 500.0 s, and 1000.0 s. From the two earliest time values, find the "true" initial rate of the reaction. Determine approximately what percentage completion of the reaction occurs before the initial rate drops to (a) 99 % and (b) 95% of the true value.

30-18. Challenge Problem: The hydrolysis of N-glutaryl-L-phenylalanine-p-nitroanilide (GPNA) by the enzyme α-chymotrypsin (CT) to form p-nitroaniline and N-glutaryl-L-phenylalanine follows the Michaelis-Menten mechanism in its early stages.

(a) Show that Equation 30-19 can be manipulated to give the following transformation:

$$\frac{1}{v_i} = \frac{K_m}{v_{max}[S]_0} + \frac{1}{v_{max}}$$

where v_i is the initial rate, $(d[P]/dt)_i$, v_{max} is $k_2[E]_0$, and $[S]_0$ is the initial GPNA concentration. The above equation is often called the Lineweaver-Burke equation. A plot of $1/v_i$ versus $1/[S]_0$ is called a Lineweaver-Burke plot.

(b) For $[CT] = 4.0 \times 10^{-6}$ M, use the following results and the Lineweaver-Burke plot to determine K_m, v_{max}, and k_2.

$[GPNA]_0$, mM	v_i, µM s^{-1}
0.250	0.037
0.500	0.063
10.0	0.098
15.0	0.118

(c) Show that the Michaelis-Menten equation for the initial rate can be transformed to give the Hanes-Woolf equation below:

$$\frac{[S]}{v_i} = \frac{[S]_0}{v_{max}} + \frac{K_m}{v_{max}}$$

Use a Hanes-Woolf plot of the data in part (b) to determine K_m, v_{max}, and k_2.

(d) Show that the Michaelis-Menten equation for the initial rate can be transformed to give the Eadie-Hofster equation

$$v_i = -\frac{K_m v_i}{[S]_0} + v_{max}$$

Use a Eadie-Hofster plot of the data in part (b) to determine K_m, v_{max}, and k_2.

(e) Comment on which of the plots made above should be most accurate for determining K_m and v_{max} under the circumstances given. Justify your answer!

(f) The substrate GPNA is to be determined in a biological sample using the data in part (b) to construct a calibration curve. Three samples were analyzed under the same conditions as part (b) and gave initial rates of 0.069, 0.102, and 0.049 μM s^{-1}. What were the GNPA concentrations in these samples? What are the standard deviations of the concentrations?

Introduction to Analytical Separations

CHAPTER 31

Separations are extremely important in synthesis, in industrial chemistry, in the biomedical sciences, and in chemical analyses. Shown in the photograph is a petroleum refinery. The first step in the refining process is to separate petroleum into fractions on the basis of boiling point in large distillation towers. The petroleum is fed into a large still, and the mixture is heated. The materials with the lowest boiling points vaporize first. The vapor moves up the tall distillation column or tower where it recondenses into a much purer liquid. By regulating the temperatures of the still and the column, the boiling point range of the fraction condensed can be controlled.

Analytical separations occur on a much smaller laboratory scale than the industrial-scale distillation shown in the photograph. The separation methods introduced in this chapter include precipitation, distillation, extraction, ion exchange, and various chromatographic techniques.

© Charles E. Rotkin/CORBIS

Few, if any, measurement techniques used for chemical analysis are specific for a single chemical species. Because of this, for most analyses, we must consider how to treat foreign species that attenuate the signal from the analyte or produce a signal that is indistinguishable from that of the analyte. A substance that affects an analytical signal or the background is called an **interference** or an **interferent**.

Several methods can be used to deal with interferences in analytical procedures, as discussed in Section 8D-3. **Separations** isolate the analyte from potentially interfering constituents. In addition, techniques such as matrix modification, masking, dilution, and saturation are often used to offset the effects of interferents. The internal standard and standard addition methods can sometimes be used to compensate for or to reduce interference effects. In this chapter, we focus on separation methods that are the most powerful and widely used methods for treating interferences.

The basic principles of a separation are depicted in **Figure 31-1**.[1] As shown, separations can be complete or partial. In the separation process, material is transported while its components are spatially redistributed. We should note that a separation always requires energy because the reverse process, *mixing* at constant volume, is spontaneous, being accompanied by an increase in entropy. Separations can be *preparative* or *analytical*. We focus on analytical separations although many of the same principles are at play in preparative separations.

> An **interferent** is a chemical species that causes a systematic error in an analysis by enhancing or attenuating the analytical signal or the background.

---

[1]See J. C. Giddings, *Unified Separation Science*, New York: Wiley, 1991, pp. 1–7.

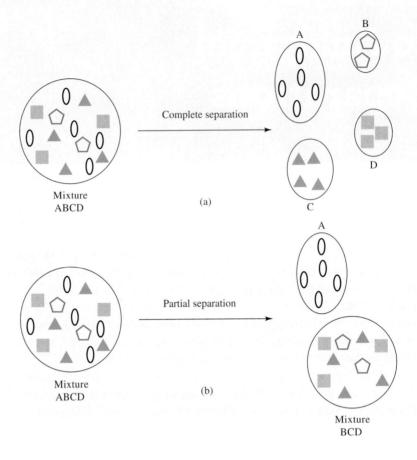

Figure 31-1 Separation principles. In (a), a mixture of four components is completely separated so that each component occupies a different spatial region. In (b), a partial separation is shown. In the partial separation, species A is isolated from the remaining mixture of B, C, and D. The reverse of the separation process shown is mixing at constant volume.

The goals of an analytical separation are usually to eliminate or reduce interferences so that quantitative analytical information can be obtained from complex mixtures. Separations can also allow identification of the separated constituents if appropriate correlations are made or a structurally sensitive measurement technique, such as mass spectrometry, is used. With techniques such as chromatography, quantitative information is obtained nearly simultaneously with the separation. In other procedures, the separation step is distinct and quite independent of the measurement step that follows.

Table 31-1 lists several separation methods that are in common use including (1) chemical or electrolytic precipitation, (2) distillation, (3) solvent extraction, (4) ion exchange, (5) chromatography, (6) electrophoresis, and (7) field-flow fractionation. The first four of these are discussed in Sections 31A through 31D of this chapter. An introduction to chromatography is presented in Section 31E. Chapters 32 and 33 treat gas and liquid chromatography, respectively, while Chapter 34 deals with electrophoresis, field-flow fractionation, and other separation methods.

31A SEPARATION BY PRECIPITATION

Separations by precipitation require large solubility differences between the analyte and potential interferents. The theoretical feasibility of this type of separation can be determined by solubility calculations such as those shown in Section 11C. Unfortunately, several other factors may preclude the use of precipitation to achieve

TABLE 31-1

Separation Methods

Method	Basis of Method
1. Mechanical phase separation	
a. Precipitation and filtration	Difference in solubility of compounds formed
b. Distillation	Difference in volatility of compounds
c. Extraction	Difference in solubility in two immiscible liquids
d. Ion exchange	Difference in interaction of reactants with ion-exchange resin
2. Chromatography	Difference in rate of movement of a solute through a stationary phase
3. Electrophoresis	Difference in migration rate of charged species in an electric field
4. Field-flow fractionation	Difference in interaction with a field or gradient applied perpendicular to transport direction

a separation. For example, the various coprecipitation phenomena described in Section 12A-5 may cause extensive contamination of a precipitate by an unwanted component even though the solubility product of the contaminant has not been exceeded. Likewise, the rate of an otherwise feasible precipitation may be too slow to be useful for a separation. Finally, when precipitates form as colloidal suspensions, coagulation may be difficult and slow, particularly when the isolation of a small quantity of a solid phase is attempted.

Many precipitating agents have been used for quantitative inorganic separations. Some of the most generally useful are descried in the sections that follow.

31A-1 Separations Based on Control of Acidity

There are enormous differences among the solubilities of the hydroxides, hydrous oxides, and acids of various elements. Moreover, the concentration of hydrogen or hydroxide ions in a solution can be varied by a factor of 10^{15} or more and can be easily controlled by the use of buffers. As a result, many separations based on pH control are in theory possible. In practice, these separations can be grouped in three categories: (1) those made in relatively concentrated solutions of strong acids, (2) those made in buffered solutions at intermediate pH values, and (3) those made in concentrated solutions of sodium or potassium hydroxide. **Table 31-2** lists common separations that can be achieved by control of acidity.

TABLE 31-2

Separations Based on Control of Acidity

Reagent	Species Forming Precipitates	Species Not Precipitated
Hot concd HNO_3	Oxides of W(VI), Ta(V), Nb(V), Si(IV), Sn(IV), Sb(V)	Most other metal ions
NH_3/NH_4Cl buffer	Fe(III), Cr(III), Al(III)	Alkali and alkaline earths, Mn(II), Cu(II), Zn(II), Ni(II), Co(II)
$HOAc/NH_4OAc$ buffer	Fe(III), Cr(III), Al(III)	Cd(II), Co(II), Cu(II), Fe(II) Mg(II), Sn(II), Zn(II)
$NaOH/Na_2O_2$	Fe(III), most +2 ions, rare earths	Zn(II), Al(III), Cr(VI), V(V), U(VI)

Recall from Equation 11-42 that

$$[S^{2-}] = \frac{1.2 \times 10^{-22}}{[H_3O^+]^2}$$

31A-2 Sulfide Separations

With the exception of the alkali metals and alkaline-earth metals, most cations form sparingly soluble sulfides whose solubilities differ greatly from one another. Because it is relatively easy to control the sulfide ion concentration of an aqueous solution of H_2S by adjustment of pH (see Section 11C-2), separations based on the formation of sulfides have found extensive use. Sulfides can be conveniently precipitated from homogeneous solution, with the anion being generated by the hydrolysis of thioacetamide (see Table 12-1).

The ionic equilibria influencing the solubility of sulfide precipitates were considered in Section 11C-2. These treatments, however, may not always produce realistic conclusions about the feasibility of separations because of coprecipitation and the slow rates at which some sulfides form. For these reasons, we often rely on previous results or empirical observations to indicate whether a given separation is likely to be successful.

Table 31-3 shows some common separations that can be accomplished with hydrogen sulfide through control of pH.

31A-3 Separations by Other Inorganic Precipitants

No other inorganic ions are as generally useful for separations as hydroxide and sulfide ions. Phosphate, carbonate, and oxalate ions are often used as precipitants for cations, but they are not selective. Because of this drawback, separations are usually performed prior to precipitation.

Chloride and sulfate are useful because of their highly selective behavior. Chloride can separate silver from most other metals, and sulfate can isolate a group of metals that includes lead, barium, and strontium.

31A-4 Separations by Organic Precipitants

Selected organic reagents for the isolation of various inorganic ions were discussed in Section 12C-3. Some of these organic precipitants, such as dimethylglyoxime, are useful because of their remarkable selectivity in forming precipitates with only a few ions. Other reagents, such as 8-hydroxyquinoline, yield slightly soluble compounds with many different cations. The selectivity of this sort of reagent is due to the wide range of solubility among its reaction products and also to the fact that the precipitating reagent is usually an anion that is the conjugate base of a weak acid. Thus, separations based on pH control can be realized just as with hydrogen sulfide.

TABLE 31-3

Precipitation of Sulfides		
Elements	**Conditions of Precipitation***	**Conditions for No Precipitation***
Hg(II), Cu(II), Ag(I)	1, 2, 3, 4	
As(V), As(III), Sb(V), Sb(III)	1, 2, 3	4
Bi(III), Cd(II), Pb(II), Sn(II)	2, 3, 4	1
Sn(IV)	2, 3	1, 4
Zn(II), Co(II), Ni(II)	3, 4	1, 2
Fe(II), Mn(II)	4	1, 2, 3

*1 = 3 M HCl; 2 = 0.3 M HCl; 3 = buffered to pH 6 with acetate; 4 = buffered to pH 9 with $NH_3/(NH_4)_2S$.

31A-5 Separation of Species Present in Trace Amounts by Precipitation

A problem often encountered in trace analysis is that of isolating from the major components of the sample the species of interest, which may be present in microgram quantities. Although such a separation is sometimes based on a precipitation, the techniques required differ from those used when the analyte is present in large amounts.

Several problems can accompany the quantitative separation of a trace element by precipitation even when solubility losses are not important. Supersaturation often delays formation of the precipitate, and coagulation of small amounts of a colloidally dispersed substance is often difficult. In addition, it is common to lose an appreciable fraction of the solid during transfer and filtration. To minimize these difficulties, a quantity of some other ion that also forms a precipitate with the reagent is often added to the solution. The precipitate from the added ion is called a **collector** and carries the desired minor species out of solution. For example, in isolating manganese as the sparingly soluble manganese dioxide, a small amount of iron(III) is frequently added to the analyte solution before the introduction of ammonia as the precipitating reagent. The basic iron(III) oxide brings down even the smallest traces of the manganese dioxide. Other examples include basic aluminum oxide as a collector of trace amounts of titanium and copper sulfide for collection of traces of zinc and lead. Many other collectors are described by Sandell and Onishi.[2]

A collector may entrain a trace constituent as a result of similarities in their solubilities. Other collectors function by coprecipitation in which the minor component is adsorbed on or incorporated into the collector precipitate as the result of mixed-crystal formation. We must be sure that the collector does not interfere with the method selected for determining the trace component.

> A **collector** is used to remove trace constituents from solution.

31A-6 Separation by Electrolytic Precipitation

Electrolytic precipitation is a highly useful method for accomplishing separations. In this process, the more easily reduced species, either the wanted or the unwanted component of the sample, is isolated as a separate phase. The method becomes particularly effective when the potential of the working electrode is controlled at a predetermined level (see Section 22B).

The mercury cathode (page 593) has found wide application in the removal of many metal ions prior to the analysis of the residual solution. In general, metals more easily reduced than zinc are conveniently deposited in the mercury, leaving such ions as aluminum, beryllium, the alkaline earths, and the alkali metals in solution. The potential required to decrease the concentration of a metal ion to any desired level can be calculated from voltammetric data. Stripping methods (see Section 23H) use an electrodeposition step for separation followed by voltammetry for completion of the analysis.

31A-7 Salt-Induced Precipitation of Proteins

A common way to separate proteins is by adding a high concentration of salt. This procedure is termed **salting out** the protein. The solubility of protein molecules shows a complex dependence on pH, temperature, ionic strength, the nature of the protein, and the concentration of the salt used. At low salt concentrations, solubility is usually increased with increasing salt concentration. This **salting in effect** is explained by the Debye-Hückel theory. The counter ions of the salt surround the protein, and the screening

---

[2]E. B. Sandell and H. Onishi, *Colorimetric Determination of Traces of Metals*, 4th ed., New York: Interscience, 1978, pp. 709–21.

results in decreasing the electrostatic attraction of protein molecules for each other. This decrease, in turn, leads to increasing solubility with increasing ionic strength.

At high concentrations of salt, however, the repulsive effect of like charges is reduced as are the forces leading to solvation of the protein. When these forces are reduced enough, the protein precipitates and salting out is observed. Ammonium sulfate is an inexpensive salt and is widely used because of its effectiveness and high inherent solubility.

At high concentrations, protein solubility, S is given by the following empirical equation:

$$\log S = C - K\mu \tag{31-1}$$

where C is a constant that is a function of pH, temperature, and the protein; K is the salting out constant that is a function of the protein and the salt used; and μ is the ionic strength.

Proteins are commonly least soluble at their isoelectric points. Hence, a combination of high salt concentration and pH control is used to achieve salting out. Protein mixtures can be separated by a stepwise increase in the ionic strength. Care must be taken with some proteins because ammonium sulfate can denature the protein. Alcoholic solvents are sometimes used in place of salts. They reduce the dielectric constant and subsequently reduce solubility by lowering protein-solvent interactions.

31B SEPARATION OF SPECIES BY DISTILLATION

Distillation is widely used to separate volatile analytes from nonvolatile interferents. Distillation is based on differences in the boiling points of the materials in a mixture. A common example is the separation of nitrogen analytes from many other species by converting the nitrogen to ammonia, which is then distilled from basic solution. Other examples include separating carbon as carbon dioxide and sulfur as sulfur dioxide. Distillation is widely used in organic chemistry to separate components in mixtures for purification purposes.

There are many types of distillation. **Vacuum distillation** is used for compounds that have very high boiling points. Lowering the pressure to the vapor pressure of the compound of interest causes boiling and is often more effective for high boilers than raising the temperature. **Molecular distillation** occurs at very low pressure (<0.01 torr) such that the lowest possible temperature is used with the least damage to the distillate. **Pervaporation** is a method for separating mixtures by partial volatilization through a nonporous membrane. **Flash evaporation** is a process in which a liquid is heated and then sent through a reduced pressure chamber. The reduction in pressure causes partial vaporization of the liquid.

31C SEPARATION BY EXTRACTION

The extent to which solutes, both inorganic and organic, distribute themselves between two immiscible liquids differs enormously, and these differences have been used for decades to separate chemical species. This section considers applications of the distribution phenomenon to analytical separations.

31C-1 Principles

The partition of a solute between two immiscible phases is an equilibrium process that is governed by the **distribution law**. If the solute species A is allowed to distribute itself between water and an organic phase, the resulting equilibrium may be written as

$$A_{aq} \rightleftharpoons A_{org}$$

where the subscripts refer to the aqueous and the organic phases, respectively. Ideally, the ratio of activities for A in the two phases will be constant and independent of the total quantity of A so that, at any given temperature,

$$K = \frac{(a_A)_{org}}{(a_A)_{aq}} \approx \frac{[A]_{org}}{[A]_{aq}}$$ (31-2)

where $(a_A)_{org}$ and $(a_A)_{aq}$ are the activities of A in each of the phases and the bracketed terms are molar concentrations of A. As with many other equilibria, under many conditions, molar concentrations can be substituted for activities without serious error. The equilibrium constant K is known as the **distribution constant**. Generally, the numerical value for K approximates the ratio of the solubility of A in each solvent.

Distribution constants are useful because they permit us to calculate the concentration of an analyte remaining in a solution after a certain number of extractions. They also provide guidance as to the most efficient way to perform an extractive separation. Thus, we can show (see Feature 31-1) that for the simple system described by Equation 31-2, the concentration of A remaining in an aqueous solution after i extractions with an organic solvent ($[A]_i$) is given by the equation

$$[A]_i = \left(\frac{V_{aq}}{V_{org}K + V_{aq}} \right)^i [A]_0$$ (31-3)

where $[A]_i$ is the concentration of A remaining in the aqueous solution after extracting V_{aq} mL of the solution with an original concentration of $[A]_0$ with i portions of the organic solvent, each with a volume of V_{org}. Example 31-1 illustrates how this equation can be used to decide on the most efficient way to perform an extraction.

EXAMPLE 31-1

The distribution constant for iodine between an organic solvent and H_2O is 85. Find the concentration of I_2 remaining in the aqueous layer after extraction of 50.0 mL of 1.00×10^{-3} M I_2 with the following quantities of the organic solvent: (a) 50.0 mL; (b) two 25.0-mL portions; (c) five 10.0-mL portions.

Solution

Substitution into Equation 31-3 gives

(a) $[I_2]_1 = \left(\dfrac{50.0}{50.0 \times 85 + 50.0} \right)^1 \times 1.00 \times 10^{-3} = 1.16 \times 10^{-5}$ M

(b) $[I_2]_2 = \left(\dfrac{50.0}{25.0 \times 85 + 50.0} \right)^2 \times 1.00 \times 10^{-3} = 5.28 \times 10^{-7}$ M

(c) $[I_2]_5 = \left(\dfrac{50.0}{10.0 \times 85 + 50.0} \right)^5 \times 1.00 \times 10^{-3} = 5.29 \times 10^{-10}$ M

❮ It is always better to use several small portions of solvent to extract a sample than to extract with one large portion.

Note the increased extraction efficiencies that result from dividing the original 50 mL of solvent into two 25-mL or five 10-mL portions.

Figure 31-2 Plot of Equation 31-3 assuming that $K = 2$ and $V_{aq} = 100$ mL. The total volume of the organic solvent was assumed to be 100 mL so that $V_{org} = 100/n_i$.

Figure 31-2 shows that the improved efficiency of multiple extractions falls off rapidly as a total fixed volume is subdivided into smaller and smaller portions. We see that there is little to be gained by dividing the extracting solvent into more than five or six portions.

FEATURE 31-1

Derivation of Equation 31-3

Consider a simple system that is described by Equation 31-2. Suppose n_0 mmol of the solute A in V_{aq} mL of aqueous solution is extracted with V_{org} mL of an immiscible organic solvent. At equilibrium, n_1 mmol of A will remain in the aqueous layer, and $(n_0 - n_1)$ mmol will have been transferred to the organic layer. The concentrations of A in the two layers will then be

$$[A]_1 = \frac{n_1}{V_{aq}}$$

and

$$[A]_{org} = \frac{(n_0 - n_1)}{V_{org}}$$

Substitution of these quantities into Equation 31-2 and rearrangement gives

$$n_1 = \left(\frac{V_{aq}}{V_{org}K + V_{aq}} \right) n_0$$

Similarly, the number of millimoles, n_2, remaining after a second extraction with the same volume of solvent will be

$$n_2 = \left(\frac{V_{aq}}{V_{org}K + V_{aq}} \right) n_1$$

Substitution of the previous equation into this expression gives

$$n_2 = \left(\frac{V_{aq}}{V_{org}K + V_{aq}} \right)^2 n_0$$

By the same argument, the number of millimoles, n_i, that remain after i extractions is given by the expression

$$n_i = \left(\frac{V_{aq}}{V_{org}K + V_{aq}} \right)^i n_0$$

Finally, this equation can be written in terms of the initial and final concentrations of A in the aqueous layer by substituting the relationships

$$n_i = [A]_i V_{aq} \quad \text{and} \quad n_0 = [A]_0 V_{aq}$$

Therefore,

$$[A]_i = \left(\frac{V_{aq}}{V_{org}K + V_{aq}} \right)^i [A]_0$$

which is Equation 31-3.

31C-2 Extracting Inorganic Species

An extraction is often more attractive than a precipitation method for separating inorganic species. The processes of equilibration and separation of phases in a separatory funnel are less tedious and time consuming than conventional precipitation, filtration, and washing.

Separating Metal Ions as Chelates

Many organic chelating agents are weak acids that react with metal ions to give uncharged complexes that are highly soluble in organic solvents such as ethers, hydrocarbons, ketones, and chlorinated species (including chloroform and carbon tetrachloride).[3] Most uncharged metal chelates, on the other hand, are nearly insoluble in water. Similarly, the chelating agents themselves are often quite soluble in organic solvents but of limited solubility in water.

Figure 31-3 shows the equilibria that develop when an aqueous solution of a divalent cation, such as zinc(II), is extracted with an organic solution containing a large excess of 8-hydroxyquinoline (see Section 12C-3 for the structure and reactions of this chelating agent). Four equilibria are shown. In the first, 8-hydroxyquinoline, HQ, is distributed between the organic and aqueous layers. The second is the acid dissociation of the HQ to give H^+ and Q^- ions in the aqueous layer. The third equilibrium is the complex-formation reaction giving MQ_2. Fourth is distribution of the chelate between the two solvents. If it were not for the fourth equilibrium, MQ_2

Figure 31-3 Equilibria in the extraction of an aqueous cation M^{2+} into an immiscible organic solvent containing 8-hydroxyquinoline.

---

[3]The use of chlorinated solvents is decreasing because of concerns about their health effects and their possible role in the ozone layer depletion.

would precipitate out of the aqueous solution. The overall equilibrium is the sum of these four reactions or

$$2HQ(org) + M^{2+}(aq) \rightleftharpoons MQ_2(org) + 2H^+(aq)$$

The equilibrium constant for this reaction is

$$K' = \frac{[MQ_2]_{org}[H^+]^2_{aq}}{[HQ]^2_{org}[M^{2+}]_{aq}}$$

Usually, HQ is present in the organic layer in large excess with respect to M^{2+} in the aqueous phase so that $[HQ]_{org}$ remains essentially constant during the extraction. The equilibrium-constant expression can then be simplified to

$$K'[HQ]^2_{org} = K = \frac{[MQ_2]_{org}[H^+]^2_{aq}}{[M^{2+}]_{aq}}$$

or

$$\frac{[MQ_2]_{org}}{[M^{2+}]_{aq}} = \frac{K}{[H^+]^2_{aq}}$$

Thus, we see that the ratio of concentration of the metal species in the two layers is inversely proportional to the square of the hydrogen ion concentration of the aqueous layer. Equilibrium constants K vary widely from metal ion to metal ion, and these differences often make it possible to selectively extract one cation from another by buffering the aqueous solution at a level where one is extracted nearly completely and the second remains largely in the aqueous phase.

Several useful extractive separations with 8-hydroxyquinoline have been developed. There are also several other chelating agents that behave in a similar way and are described in the literature.[4] As a result, pH-controlled extractions can be powerful tools for separating metallic ions.

Extracting Metal Chlorides and Nitrates

A number of inorganic species can be separated by extraction with suitable solvents. For example, a single ether extraction of a 6 M hydrochloric acid solution will cause better than 50% of several ions to be transferred to the organic phase, including iron(III), antimony(V), titanium(III), gold(III), molybdenum(VI), and tin(IV). Other ions, such as aluminum(III) and the divalent cations of cobalt, lead, manganese, and nickel, are not extracted.

Uranium(VI) can be separated from such elements as lead and thorium by ether extraction of a solution that is 1.5 M in nitric acid and saturated with ammonium nitrate. Bismuth and iron(III) are also extracted to some extent from this medium.

31C-3 Solid-Phase Extraction

Liquid-liquid extractions have several limitations. With extractions from aqueous solutions, the solvents that can be used must be immiscible with water and must not form emulsions. A second difficulty is that liquid-liquid extractions use

---

[4]For example, see J. A. Dean, *Analytical Chemistry Handbook*, New York: McGraw-Hill, 1995, p. 2.24.

relatively large volumes of solvent, which can cause a problem with waste disposal. Also, most extractions are performed manually, which makes them somewhat slow and tedious.

Solid-phase extraction, or liquid-solid extraction, can overcome several of these problems.[5] Solid-phase extraction techniques use membranes or small disposable syringe-barrel columns or cartridges. A hydrophobic organic compound is coated or chemically bonded to powdered silica to form the solid extracting phase. The compounds can be nonpolar, moderately polar, or polar. For example, an octadecyl (C_{18}) bonded silica (ODS) is a common packing. The functional groups bonded to the packing attract hydrophobic compounds in the sample by van der Waals interactions and extract them from the aqueous solution.

A typical cartridge system for solid-phase extractions is shown in **Figure 31-4**. The sample is placed in the cartridge and pressure is applied by the syringe or from an air or nitrogen line. Alternatively, a vacuum can be used to pull the sample through the extractant. Organic molecules are then extracted from the sample and concentrated in the solid phase. They can later be displaced from the solid phase by a solvent such as methanol. By extracting the desired components from a large volume of water and then flushing them out with a small volume of solvent, the components can be concentrated. Preconcentration methods are often necessary for trace analytical methods. For example, solid-phase extractions are used in determining organic constituents in drinking water by methods approved by the Environmental Protection Agency. In some solid-phase extraction procedures, impurities are extracted into the solid phase while compounds of interest pass through unretained.

In addition to packed cartridges, solid-phase extraction can be accomplished by using small membranes or extraction disks. These have the advantages of reducing extraction time and lowering solvent use. Solid-phase extraction can also be done in continuous flow systems, which can automate the preconcentration process.

A related technique, called **solid-phase microextraction**, uses a fused silica fiber coated with a nonvolatile polymer to extract organic analytes directly from aqueous samples or from the headspace above the samples.[6] The analyte partitions between the fiber and the liquid phase. The analytes are then desorbed thermally in the heated injector of a gas chromatograph (see Chapter 32). The extracting fiber is mounted in a holder that is much like an ordinary syringe. This technique combines sampling and sample preconcentration in a single step.

Figure 31-4 Solid-phase extraction performed in a small cartridge. The sample is placed in the cartridge and pressure is applied via a syringe plunger. Alternatively, a vacuum can be used to pull the sample through the extracting agent.

31D SEPARATING IONS BY ION EXCHANGE

Ion exchange is a process by which ions held on a porous, essentially insoluble solid are exchanged for ions in a solution that is brought in contact with the solid. The ion-exchange properties of clays and zeolites have been recognized and studied for more

In the ion-exchange process, ions held on an ion-exchange resin are exchanged for ions in a solution brought into contact with the resin.

---

[5]For more information, see N. J. K. Simpson, ed., *Solid-Phase Extraction: Principles, Techniques and Applications*, New York: Dekker, 2000; M. J. Telepchak, T. F. August, and G. Chaney, *Forensic and Clinical Applications of Solid Phase Extraction*, Totowa, NJ: Human Press, 2004; J. S. Fritz, *Analytical Solid-Phase Extraction*, New York: Wiley, 1999; E. M. Thurman and M. S. Mills, *Solid-Phase Extraction: Principles and Practice*, New York: Wiley, 1998.

[6]For more information, see S. A. S. Wercinski, ed., *Solid-Phase Microextraction: A Practical Guide*, New York: Dekker, 1999; J. Pawliszyn, ed., *Applications of Solid Phase Microextraction*, London: Royal Society of Chemistry, 1999.

Figure 31-5 Structure of a cross-linked polystyrene ion-exchange resin. Similar resins are used in which the —SO₃H⁺ group is replaced by —COO⁻H⁺, —NH₃⁺OH, and —N(CH₃)₃⁺OH groups.

than a century. Synthetic ion-exchange resins were first produced in the mid-1930s and have since found widespread application in water softening, water deionization, solution purification, and ion separation.

31D-1 Ion-Exchange Resins

Synthetic ion-exchange resins are high-molecular-mass polymers that contain large numbers of an ionic functional group per molecule. Cation-exchange resins contain acidic groups, while anion-exchange resins have basic groups. Strong-acid-type exchangers have sulfonic acid groups ($-SO_3^-H^+$) attached to the polymeric matrix (see **Figure 31-5**) and have wider application than weak-acid-type exchangers, which owe their action to carboxylic acid ($-COOH$) groups. Similarly, strong-base anion exchangers contain quaternary amine [$-N(CH_3)_3^+OH^-$] groups, while weak-base types contain secondary or tertiary amines.

Cation exchange is illustrated by the equilibrium

$$x\text{RSO}_3^-\text{H}^+ \ + \ \text{M}^{x+} \ \rightleftharpoons \ (\text{RSO}_3^-)_x\text{M}^{x+} \ + \ x\text{H}^+$$
$$\text{solid} \qquad \text{soln} \qquad \text{solid} \qquad \text{soln}$$

where M^{x+} represents a cation and R represents *that part of a resin molecule that contains one sulfonic acid group*. The analogous equilibrium involving a strong-base anion exchanger and an anion A^{x-} is

$$x\text{RN}(\text{CH}_3)_3^+\text{OH}^- \ + \ \text{A}^{x-} \ \rightleftharpoons \ [\text{RN}(\text{CH}_3)_3^+]_x\text{A}^{x-} \ + \ x\text{OH}^-$$
$$\text{solid} \qquad \text{soln} \qquad \text{solid} \qquad \text{soln}$$

31D-2 Ion-Exchange Equilibria

The law of mass action can be used to treat ion-exchange equilibria. For example, when a dilute solution containing calcium ions is passed through a column packed with a sulfonic acid resin, the following equilibrium is established:

$$\text{Ca}^{2+}(aq) \ + \ 2\text{H}^+(res) \ \rightleftharpoons \ \text{Ca}^{2+}(res) \ + \ 2\text{H}^+(aq)$$

for which the equilibrium constant K' is given by

$$K' = \frac{[\text{Ca}^{2+}]_{\text{res}}[\text{H}^+]^2_{\text{aq}}}{[\text{Ca}^{2+}]_{\text{aq}}[\text{H}^+]^2_{\text{res}}} \tag{31-4}$$

As usual, the bracketed terms are molar concentrations (strictly speaking, activities) of the species in the two phases. Note that $[\text{Ca}^{2+}]_{\text{res}}$ and $[\text{H}^+]_{\text{res}}$ are molar concentrations of the two ions *in the solid phase*. In contrast to most solids, however, these concentrations can vary from zero to some maximum value when all of the negative sites on the resin are occupied by one species only.

Ion-exchange separations are usually performed under conditions in which one ion predominates in *both* phases. Thus, in the removal of calcium ions from a dilute and somewhat acidic solution, the calcium ion concentration will be much smaller than that of hydrogen ion in both the aqueous and resin phases, that is,

$$[\text{Ca}^{2+}]_{\text{res}} \ll [\text{H}^+]_{\text{res}}$$

and

$$[Ca^{2+}]_{aq} \ll [H^+]_{aq}$$

As a result, the hydrogen ion concentration is essentially constant in both phases, and Equation 31-4 can be rearranged to

$$\frac{[Ca^{2+}]_{res}}{[Ca^{2+}]_{aq}} = K' \frac{[H^+]^2_{res}}{[H^+]^2_{aq}} = K \qquad (31\text{-}5)$$

where K is a distribution constant analogous to the constant that governs an extraction equilibrium (Equation 31-2). Note that K in Equation 31-5 represents the affinity of the resin for calcium ion relative to another ion (here, H^+). In general, where K for an ion is large, there is a strong tendency for the resin phase to retain that ion. With a small value of K, there is only a small tendency for retention of the ion by the resin phase. Selection of a common reference ion (such as H^+) permits a comparison of distribution constants for various ions on a given type of resin. Such experiments reveal that polyvalent ions are much more strongly retained than singly charged species. Within a given charge group, the differences among values for K appear to be related to the size of the hydrated ion as well as other properties. Therefore, for a typical sulfonated cation-exchange resin, values of K for univalent ions decrease in the order $Ag^+ > Cs^+ > Rb^+ > K^+ > NH_4^+ > Na^+ > H^+ > Li^+$. For divalent cations, the order is $Ba^{2+} > Pb^{2+} > Sr^{2+} > Ca^{2+} > Ni^{2+} > Cd^{2+} > Cu^{2+} > Co^{2+} > Zn^{2+} > Mg^{2+} > UO_2^{2+}$.

31D-3 Applications of Ion-Exchange Methods

There are many uses for ion-exchange resins. They are used in many cases to eliminate ions that would otherwise interfere with an analysis. For example, iron(III), aluminum(III), and many other cations tend to coprecipitate with barium sulfate during the determination of sulfate ion. Passing the solution that contains sulfate through a cation-exchange resin results in the retention of these interfering cations and the release of an equivalent number of hydrogen ions. Sulfate ions pass freely through the column and can be precipitated as barium sulfate from the effluent.

Another valuable application of ion-exchange resins is to concentrate ions from a dilute solution. Thus, traces of metallic elements in large volumes of natural waters can be collected on a cation-exchange column and subsequently liberated from the resin by treatment with a small volume of an acidic solution. The result is a considerably more concentrated solution for analysis by atomic absorption or ICP emission spectrometry (see Chapter 28).

The total salt content of a sample can be determined by titrating the hydrogen ion released as an aliquot of sample passes through a cation exchanger in the acidic form. Similarly, a standard hydrochloric acid solution can be prepared by diluting to known volume the effluent resulting from treatment of a cation-exchange resin with a known mass of sodium chloride. Substitution of an anion-exchange resin in its hydroxide form will permit the preparation of a standard base solution. Ion-exchange resins are also widely used in household water softeners as discussed in Feature 31-2. As shown in Section 33D, ion-exchange resins are particularly useful for the chromatographic separation of both inorganic and organic ionic species.

FEATURE 31-2

Home Water Softeners

Hard water is water that is rich in the salts of calcium, magnesium, and iron. The cations of hard water combine with fatty acid anions from soap to form insoluble salts known as **curd** or **soap curd**. In areas with particularly hard water, these precipitates can be seen as gray rings around bathtubs and sinks.

One method of solving the problem of hard water in homes is to exchange the calcium, magnesium, and iron cations for sodium ions, which form soluble fatty acid salts. A commercial water softener consists of a tank containing an ion-exchange resin, a storage reservoir for sodium chloride, and various valves and regulators for controlling the flow of water, as shown in **Figure 31F-1**. During the charging, or regeneration cycle, concentrated salt water from the reservoir is directed through the ion-exchange resin where the resin sites are occupied by Na^+ ions.

$$(RSO_3^-)_xM^{x+} + xNa^+ \rightleftharpoons xRSO_3^-Na^+ + M^{x+} \text{ (regeneration)}$$
$$\text{solid} \qquad \text{water} \qquad \text{solid} \qquad \text{water}$$

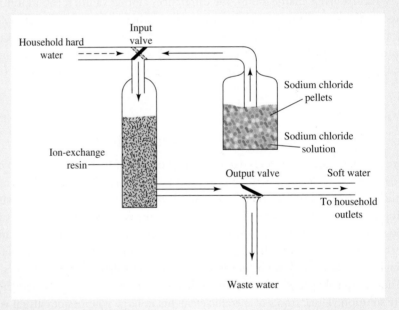

Figure 31F-1   Schematic of a home water softener. During the charging cycle the valves are in the positions shown. Salt water from the storage reservoir passes through the ion-exchange resin to waste. Sodium ions from the salt water exchange with ions on the resin to leave the resin in the sodium form. During water use, the valves switch, and hard water passes through the resin where the calcium, magnesium, and iron cations replace the sodium ions attached to the resin.

The M^{x+} cations (calcium, magnesium, or iron) released are sent to waste during this cycle.

After the regeneration cycle, the valves controlling the inlet to the ion-exchange resin and the outlet from the resin change so that water from the household supply passes through the resin and out to the household faucets. When the hard water passes through the resin, the M^{x+} cations are exchanged for Na^+ ions, and the water is softened.

$$xRSO_3^- Na^+ + M^{x+} \rightleftharpoons (RSO_3^-)_x M^{x+} + xNa^+ \text{ (household use)}$$

solid   water   solid   water

With use, the ion-exchange resin gradually accumulates the cations from the hard water. Hence, the softener periodically must be recharged by passing salt water through it and venting the hard water ions to waste. After softening, soaps are much more effective because they remain dispersed in the water and do not form soap curds. Potassium chloride is also used instead of sodium chloride and is particularly advantageous for people on a restricted sodium diet. Potassium chloride is, however, more expensive to use than sodium chloride.

31E CHROMATOGRAPHIC SEPARATIONS

Chromatography is a widely used method for the separation, identification, and determination of the chemical components in complex mixtures. No other separation method is as powerful and generally applicable as is chromatography.[7] The remainder of this chapter is devoted to the general principles that apply to all types of chromatography. Chapters 32 through 34 deal with some of the applications of chromatography and related methods for analytical separations.

31E-1 General Description of Chromatography

The term **chromatography** is difficult to define rigorously because the name has been applied to several systems and techniques. All of these methods, however, have in common the use of a **stationary phase** and a **mobile phase**. Components of a mixture are carried through the stationary phase by the flow of a mobile phase, and separations are based on differences in migration rates among the mobile-phase components.

31E-2 Classification of Chromatographic Methods

Chromatographic methods are of two basic types. In **column chromatography**, the stationary phase is held in a narrow tube, and the mobile phase is forced through the tube under pressure or by gravity. In **planar chromatography**, the stationary phase is supported on a flat plate or in the pores of a paper, and the mobile phase moves through the stationary phase by capillary action or under the influence of gravity. We consider here only column chromatography. Planar chromatography is discussed in Section 34B.

As shown in the first column of **Table 31-4**, chromatographic methods fall into three categories based on the nature of the mobile phase: liquid, gas, and supercritical fluid. The second column of the table reveals that there are five types of liquid

Chromatography is a technique in which the components of a mixture are separated based on differences in the rates at which they are carried through a fixed or **stationary phase** by a gaseous or liquid **mobile phase**.

The **stationary phase** in chromatography is a phase that is fixed in place either in a column or on a planar surface.

The **mobile phase** in chromatography is a phase that moves over or through the stationary phase carrying with it the analyte mixture. The mobile phase may be a gas, a liquid, or a supercritical fluid.

Planar and **column chromatography** are based on the same types of equilibria.

❮ Gas chromatography and supercritical fluid chromatography require the use of a column. Only liquid mobile phases can be used on planar surfaces.

[7]Some general references on chromatography include J. M. Miller, *Chromatography: Concepts and Contrasts*, 2nd ed., New York: Wiley, 2005; R. L Wixom and C. W. Gehrke, eds., *Chromatography: A Science of Discovery*, Hoboken, NJ: Wiley, 2010; E. F. Heftman, ed., *Chromatography: Fundamentals of Chromatography and Related Differential Migration Methods*, Amsterdam: Elsevier, 2004; C. F. Poole, *The Essence of Chromatography*, Amsterdam: Elsevier, 2003; J. Cazes and R. P. W. Scott, *Chromatography Theory*, New York: Dekker, 2002; A. Braithwaite and F. J. Smith, *Chromatographic Methods*, 5th ed., London: Blackie, 1996; R. P. W. Scott, *Techniques and Practice of Chromatography*, New York: Dekker, 1995; J. C. Giddings, *Unified Separation Science*, New York: Wiley, 1991.

TABLE 31-4

Classification of Column Chromatographic Methods

General Classification	Specific Method	Stationary Phase	Type of Equilibrium
1. Gas chromatography (GC)	a. Gas-liquid (GLC)	Liquid adsorbed or bonded to a solid surface	Partition between gas and liquid
	b. Gas-solid	Solid	Adsorption
2. Liquid Chromatography (LC)	a. Liquid-liquid, or partition	Liquid adsorbed or bonded to a solid surface	Partition between immiscible liquids
	b. Liquid-solid, or adsorption	Solid	Adsorption
	c. Ion exchange	Ion-exchange resin	Ion exchange
	d. Size exclusion	Liquid in interstices of a polymeric solid	Partition/sieving
	e. Affinity	Group specific liquid bonded to a solid surface	Partition between surface liquid and mobile liquid
3. Supercritical fluid chromatography (SFC) (mobile phase: supercritical fluid)		Organic species bonded to a solid surface	Partition between supercritical fluid and bonded surface

chromatography and two types of gas chromatography that differ in the nature of the stationary phase and the types of equilibria between phases.

31E-3 Elution in Column Chromatography

Figure 31-6a shows how two components A and B of a sample are resolved on a packed column by **elution**. The column consists of narrow-bore tubing that is packed with a finely divided inert solid that holds the stationary phase on its surface. The mobile phase occupies the open spaces between the particles of the packing. Initially, a solution of the sample containing a mixture of A and B in the mobile phase is introduced at the head of the column as a narrow plug as shown in Figure 31-6a at time t_0. The two components distribute themselves between the mobile phase and the stationary phase. Elution then occurs by forcing the sample components through the column by continuously adding fresh mobile phase.

With the first introduction of fresh mobile phase, the **eluent**, the portion of the sample contained in the mobile phase moves down the column, where further partitioning between the mobile phase and the stationary phase occurs (time t_1). Partitioning between the fresh mobile phase and the stationary phase takes place simultaneously at the site of the original sample.

Further additions of solvent carry solute molecules down the column in a continuous series of transfers between the two phases. Because solute movement can occur only in the mobile phase, the average *rate* at which a solute migrates *depends on the fraction of time it spends in that phase*. This fraction is small for solutes that are strongly retained by the stationary phase (component B in **Figure 31-6**, for example) and large where retention in the mobile phase is more likely (component A). Ideally, the resulting differences in rates cause the components in a mixture to separate into **bands**, or **zones,** along the length of the column (see **Figure 31-7**). Isolation of the separated species is then accomplished by passing a sufficient quantity of mobile phase through the column to cause the individual bands to pass out the end (to be **eluted** from the column), where they can be collected or detected (times t_3 and t_4 in Figure 31-6a).

Elution is a process in which solutes are washed through a stationary phase by the movement of a mobile phase. The mobile phase that exits the column is termed the **eluate**.

An **eluent** is a solvent used to carry the components of a mixture through a stationary phase.

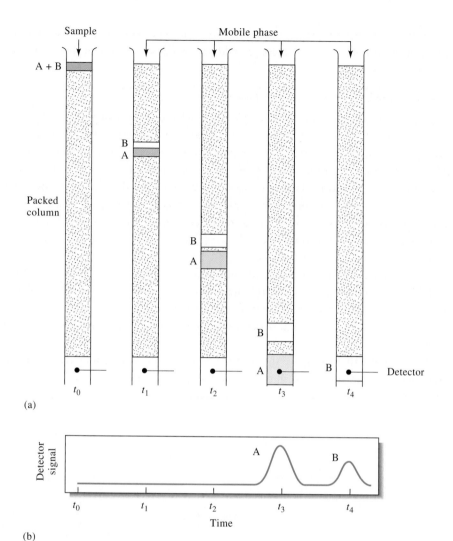

(a)

(b)

Figure 31-6 (a) Diagram showing the separation of a mixture of components A and B by column elution chromatography. (b) The detector signal at the various stages of elution shown in (a).

Chromatograms

If a detector that responds to solute concentration is placed at the end of the column during elution and its signal is plotted as a function of time (or of volume of added mobile phase), a series of peaks is obtained, as shown in Figure 31-6b. Such a plot, called a **chromatogram**, is useful for both qualitative and quantitative analysis. The

A **chromatogram** is a plot of some function of solute concentration versus elution time or elution volume.

Figure 31-7 Concentration profiles of solute bands A and B at two different times in their migration down the column in Figure 31-6. The times t_1 and t_2 are indicated in Figure 31-6.

The Russian botanist Mikhail Tswett (1872–1919) invented chromatography shortly after the turn of the twentieth century. He used the technique to separate various plant pigments, such as chlorophylls and xanthophylls, by passing solutions of these species through glass columns packed with finely divided calcium carbonate. The separated species appeared as colored bands on the column, which accounts for the name he chose for the method (Greek *chroma* meaning "color" and *graphein* meaning "to write").

positions of the peak maxima on the time axis can be used to identify the components of the sample. The peak areas provide a quantitative measure of the amount of each species.

Methods for Improving Column Performance

Figure 31-7 shows concentration profiles for the bands containing solutes A and B on the column in Figure 31-6a at time t_1 and at a later time t_2.[8] Because B is more strongly retained by the stationary phase than is A, B lags during the migration. We see that the distance between the two increases as they move down the column. At the same time, however, broadening of both bands takes place, lowering the efficiency of the column as a separating device. While band broadening is inevitable, conditions can often be found where it occurs more slowly than band separation. Thus, as shown in Figure 31-7, a clean separation of species is possible provided the column is sufficiently long.

Several chemical and physical variables influence the rates of band separation and band broadening. As a result, improved separations can often be realized by the control of variables that either (1) increase the rate of band separation or (2) decrease the rate of band spreading. These alternatives are illustrated in **Figure 31-8**.

The variables that influence the relative rates at which solutes migrate through a stationary phase are described in the next section. Following this discussion, we turn to those factors that play a part in zone broadening.

Figure 31-8 Two-component chromatogram illustrating two methods for improving separation. (a) Original chromatogram with overlapping peaks. (b) Improvement brought about by an increase in band separation. (c) Improvement brought about by a decrease in band widths.

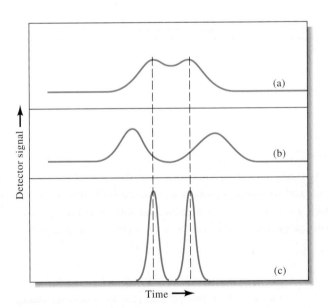

[8]Note that the relative positions of the bands for A and B in the concentration profile in Figure 31-7 appear to be reversed from their positions in Figure 31-6b. The difference is that the abscissa is distance along the column in Figure 31-7., but it is time in Figure 31-6b. Thus, in Figure 31-6b, the *front* of a peak lies to the left and the *tail* to the right; in Figure 31-7, the reverse is true.

31E-4 Migration Rates of Solutes

The effectiveness of a chromatographic column in separating two solutes depends in part on the relative rates at which the two species are eluted. These rates in turn are determined by the ratios of the solute concentrations in each of the two phases.

Distribution Constants

All chromatographic separations are based on differences in the extent to which solutes are distributed between the mobile and the stationary phase. For the solute species A, the equilibrium is described by the equation

$$A(\textit{mobile}) \rightleftharpoons A(\textit{stationary}) \tag{31-6}$$

The equilibrium constant K_c for this reaction is called a **distribution constant**, which is defined as

$$K_c = \frac{(a_A)_S}{(a_A)_M} \tag{31-7}$$

where $(a_A)_S$ is the activity of solute A in the stationary phase and $(a_A)_M$ is the activity in the mobile phase. We often substitute c_S, the molar analytical concentrations of the solute in the stationary phase, for $(a_A)_S$ and c_M, the molar analytical concentration in the mobile phase, for $(a_A)_M$. Hence, we often write equation 31-7 as

$$K_c = \frac{c_S}{c_M} \tag{31-8}$$

Ideally, the distribution constant is constant over a wide range of solute concentrations, that is, c_S is directly proportional to c_M.

> The **distribution constant** for a solute in chromatography is equal to the ratio of its molar concentration in the stationary phase to its molar concentration in the mobile phase.

Retention Times

Figure 31-9 is a simple chromatogram of a two-component mixture. The small peak on the left is for a species that is *not* retained by the stationary phase. The time t_M after sample injection for this peak to appear is sometimes called the **dead** or **void time**. The dead time provides a measure of the average rate of migration of the mobile phase and is an important parameter in identifying analyte peaks. All components spend at least time t_M in the mobile phase. To aid in measuring t_M, an unretained species can be added if one is not already present in the sample or the mobile phase. The larger peak on the right in Figure 31-9 is that of an analyte species. The time required for this zone to reach the detector after sample injection is called

> The **dead time** (void time), t_M, is the time it takes for an unretained species to pass through a chromatographic column. All components spend at least this amount of time in the mobile phase. Separations are based on the different times, t_S, that components spend in the stationary phase.

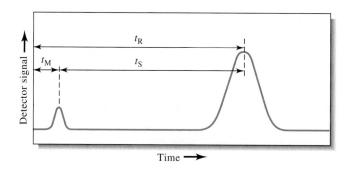

Figure 31-9 A typical chromatogram for a two-component mixture. The small peak on the left represents a solute that is not retained on the column and so reaches the detector almost immediately after elution is begun. Thus, its retention time, t_M, is approximately equal to the time required for a molecule of the mobile phase to pass through the column.

The **retention time**, t_R, is the time between injection of a sample and the appearance of a solute peak at the detector of a chromatographic column.

the **retention time** and is given the symbol t_R. The analyte has been retained because it spends a time t_S in the stationary phase. The retention time is then

$$t_R = t_S + t_M \tag{31-9}$$

The average linear rate of solute migration, $\bar{v}$ (usually cm/s), is

$$\bar{v} = \frac{L}{t_R} \tag{31-10}$$

where L is the length of the column packing. Similarly, the average linear velocity, u, of the mobile phase molecules is

$$u = \frac{L}{t_M} \tag{31-11}$$

Volumetric Flow Rate and Linear Flow Velocity

Experimentally in chromatography the mobile phase flow is usually characterized by the volumetric flow rate, F (cm³/min), at the column outlet. For an open tubular column, F is related to the linear velocity at the column outlet u_o by

$$F = u_o A = u_o \times \pi r^2 \tag{31-12}$$

where A is the cross-sectional area of the tube (πr^2). For a packed column, the entire column volume is not available to the liquid, and so, Equation 31-12 must be modified to

$$F = \pi r^2 u_o \varepsilon \tag{31-13}$$

where ε is the fraction of the total column volume available to the liquid (column porosity).

Migration Rates and Distribution Constants

To relate the rate of migration of a solute to its distribution constant, we express the rate as a fraction of the velocity of the mobile phase:

$$\bar{v} = u \times \text{fraction of time solute spends in mobile phase}$$

This fraction, however, equals the average number of moles of solute in the mobile phase at any instant divided by the total number of moles of solute in the column:

$$\bar{v} = u \times \frac{\text{no. of moles of solute in mobile phase}}{\text{total no. of moles of solute}}$$

The total number of moles of solute in the mobile phase is equal to the molar concentration, c_M, of the solute in that phase multiplied by its volume, V_M. Similarly, the number of moles of solute in the stationary phase is given by the product of c_S, the concentration of the solute in the stationary phase, and its volume, V_S. Therefore,

$$\bar{v} = u \times \frac{c_M V_M}{c_M V_M + c_S V_S} = u \times \frac{1}{1 + c_S V_S / c_M V_M}$$

Substituting Equation 31-8 into this equation gives an expression for the rate of solute migration as a function of its distribution constant as well as a function of the volumes of the stationary and mobile phases:

$$\bar{v} = u \times \frac{1}{1 + K_c V_S / V_M} \tag{31-14}$$

The two volumes can be estimated from the method by which the column is prepared.

The Retention Factor, k

The retention factor is an important experimental parameter that is widely used to compare the migration rates of solutes on columns.[9] For solute A, the retention factor k_A is defined as

$$k_A = \frac{K_A V_S}{V_M} \tag{31-15}$$

where K_A is the distribution constant for solute A. Substituting Equation 31-15 into 31-14 yields

$$\bar{v} = u \times \frac{1}{1 + k_A} \tag{31-16}$$

To show how k_A can be calculated from a chromatogram, we substitute Equations 31-10 and 31-11 into Equation 31-16:

$$\frac{L}{t_R} = \frac{L}{t_M} \times \frac{1}{1 + k_A} \tag{31-17}$$

We rearrange this equation to

$$k_A = \frac{t_R - t_M}{t_M} = \frac{t_S}{t_M} \tag{31-18}$$

The **retention factor**, k_A, for solute A is related to the rate at which A migrates through a column. It is the amount of time a solute spends in the stationary phase relative to the time it spends in mobile phase.

As shown in Figure 31-9, t_R and t_M are easily obtained from a chromatogram. A retention factor much less than unity means that the solute emerges from the column at a time near that of the void time. When retention factors are larger than perhaps 20 to 30, elution times become inordinately long. Ideally, separations are performed under conditions in which the retention factors for the solutes of interest in a mixture lie in the range between 1 and 5.

❮ Ideally, the **retention factors** for analytes in a sample are between 1 and 5.

---

[9]In the older literature, this constant was called the capacity factor and symbolized by k'. In 1993, however, the IUPAC Committee on Analytical Nomenclature recommended that this constant be termed the *retention factor* and symbolized by k.

In gas chromatography, retention factors can be varied by changing the temperature and the column packing, as discussed in Chapter 32. In liquid chromatography, retention factors can often be manipulated to give better separations by varying the composition of the mobile phase and the stationary phase, as illustrated in Chapter 33.

The Selectivity Factor

The **selectivity factor**, α, of a column for the two solutes A and B is defined as

> The **selectivity factor**, α, for solutes A and B is defined as the ratio of the distribution constant of the more strongly retained solute (B) to the distribution constant for the less strongly held solute (A).

$$\alpha = \frac{K_B}{K_A} \tag{31-19}$$

where K_B is the distribution constant for the more strongly retained species B and K_A is the constant for the less strongly held or more rapidly eluted species A. According to this definition, α *is always greater than unity*.

If we substitute Equation 31-15 and the analogous equation for solute B into Equation 31-19, we obtain the relationship between the selectivity factor for two solutes and their retention factors:

> The selectivity factor for two analytes in a column provides a measure of how well the column will separate the two.

$$\alpha = \frac{k_B}{k_A} \tag{31-20}$$

where k_B and k_A are the retention factors for B and A, respectively. Substituting Equation 31-18 for the two solutes into Equation 31-20, we obtain an expression that permits the determination of α from an experimental chromatogram:

$$\alpha = \frac{(t_R)_B - t_M}{(t_R)_A - t_M} \tag{31-21}$$

In Section 31E-7, we show how retention and selectivity factors influence column resolution.

31E-5 Band Broadening and Column Efficiency

The amount of band broadening that occurs as a solute passes through a chromatographic column strongly affects the column efficiency. Before defining column efficiency in more quantitative terms, let us examine the reasons that bands become broader as they move down a column.

Rate Theory of Chromatography

The **rate theory** of chromatography describes the shapes and breadths of elution bands in quantitative terms based on a random-walk mechanism for the migration of molecules through a column. A detailed discussion of the rate theory is beyond the scope of this text. We can, however, give a qualitative picture of why bands broaden and what variables improve column efficiency.[10]

---

[10]For more information see J. C. Giddings, *Unified Separation Science,* New York: Wiley, 1991, pp. 94–96.

If you examine the chromatograms shown in this and the next chapter, you will see that the elution peaks look very much like the Gaussian or normal error curves discussed in Chapters 6 and 7. As shown in Section 6A-2, normal error curves are rationalized by assuming that the uncertainty associated with any single measurement is the summation of a much larger number of small, individually undetectable, and random uncertainties, each of which has an equal probability of being positive or negative. In a similar way, the typical Gaussian shape of a chromatographic band can be attributed to the additive combination of the random motions of the various molecules as they move through the column. We assume in the following discussion that a narrow zone has been introduced so that the injection width is not the limiting factor determining the overall width of the band that elutes. It is important to realize that the widths of eluting bands can never be *narrower* than the width of the injection zone.

Consider a single solute molecule as it undergoes many thousands of transfers between the stationary and the mobile phases during elution. Residence time in either phase is highly irregular. Transfer from one phase to the other requires energy, and the molecule must acquire this energy from its surroundings. Therefore, the residence time in a given phase may be very short after some transfers and relatively long after others. Recall that movement through the column can occur *only while the molecule is in the mobile phase*. As a result, certain particles travel rapidly by virtue of their accidental inclusion in the mobile phase for a majority of the time while others lag because they happen to be incorporated in the stationary phase for a greater-than-average length of time. The result of these random individual processes is a symmetric spread of velocities around the mean value, which represents the behavior of the average analyte molecule.

As shown in **Figure 31-10**, some chromatographic peaks are nonideal and exhibit **tailing** or **fronting**. In the former case, the tail of the peak, appearing to the right on the chromatogram, is drawn out while the front is steepened. With fronting, the reverse is the case. A common cause of tailing and fronting is a distribution constant that varies with concentration. Fronting also arises when the amount of sample introduced onto a column is too large. Distortions of this kind are undesirable because they lead to poorer separations and less reproducible elution times. In the discussion that follows, tailing and fronting are assumed to be minimal.

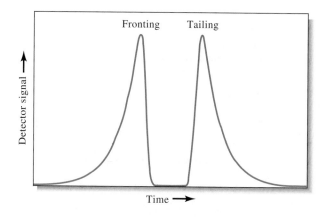

Figure 31-10 Illustration of fronting and tailing in chromatographic peaks.

A Quantitative Description of Column Efficiency

Two related terms are widely used as quantitative measures of chromatographic column efficiency: (1) **plate height**, H, and (2) **plate count** or **number of theoretical plates**, N. The two are related by the equation

$$N = \frac{L}{H} \tag{31-22}$$

where L is the length (usually in centimeters) of the column packing. The efficiency of chromatographic columns increases as the plate count N becomes greater and as the plate height H becomes smaller. Enormous differences in efficiencies are encountered in columns as a result of differences in column type and in mobile and stationary phases. Efficiencies in terms of plate numbers can vary from a few hundred to several hundred thousand, while plate heights ranging from a few tenths to one thousandth of a centimeter or smaller are not uncommon.

In Section 6B-2, we pointed out that the breadth of a Gaussian curve is described by the standard deviation σ and the variance σ^2. Because chromatographic bands are often Gaussian and because the efficiency of a column is reflected in the breadth of chromatographic peaks, the variance per unit length of column is used by chromatographers as a measure of column efficiency. That is, the column efficiency H is defined as

$$H = \frac{\sigma^2}{L} \tag{31-23}$$

This definition of column efficiency is illustrated in **Figure 31-11**, which shows a column having a packing L cm in length (Figure 31-11a) and a plot (Figure 31-11b) showing the distribution of molecules along the length of the column at the moment the analyte peak reaches the end of the packing (that is, at the retention time). The curve is Gaussian, and the locations of $L + 1\sigma$ and $L - 1\sigma$ are indicated as broken vertical lines. Note that L carries units of centimeters and σ^2 units of centimeters squared. Thus H represents a linear distance in centimeters (Equation 31-23). In fact, the plate height can be thought of as the length of column that contains a fraction of the analyte that lies between L and $L - \sigma$. Because the area under a normal error curve bounded by $\pm\sigma$ is about 68% of the total area (page 101), the plate height, as defined, contains 34% of the analyte.

Figure 31-11   Definition of plate height $H = \sigma^2/L$. In (a), the column length is shown as the distance from the sample entrance point to the detector. In (b), the Gaussian distribution of sample molecules is shown.

FEATURE 31-3

What Is the Source of the Terms *Plate* and *Plate Height*?

The 1952 Nobel Prize was awarded to two Englishmen, A. J. P. Martin and R. L. M. Synge, for their work in the development of modern chromatography. In their theoretical studies, they adapted a model that was first developed in the early 1920s to describe separations on fractional distillation columns. Fractionating columns, which were first used in the petroleum industry for separating closely related hydrocarbons, consist of numerous interconnected bubble-cap plates (see **Figure 31F-2**) at which vapor-liquid equilibria is established when the column is operated under reflux conditions.

Martin and Synge treated a chromatographic column as if it were made up of a series of contiguous bubble-cap-like plates within which equilibrium conditions always prevail. This plate model successfully accounts for the Gaussian shape of chromatographic peaks as well as for factors that influence differences in solute-migration rates. The plate model does not adequately account for zone broadening, however, because of its basic assumption that equilibrium conditions prevail throughout a column during elution. This assumption can never be valid in the dynamic state of a chromatographic column, where phases are moving past one another fast enough that there is not adequate time for equilibration.

Because the plate model is not a very good representation of a chromatographic column, we strongly advise you (1) to avoid attaching any special significance to the terms plate and plate height and (2) to view these terms as designators of column efficiency that are retained for historic reasons only and not because they have physical significance. Unfortunately, these terms are so well entrenched in the chromatographic literature that their replacement by more appropriate designations seems unlikely, at least in the near future.

Figure 31F-2 Plates in a fractionating column.

Determining the Number of Plates in a Column

The number of theoretical plates, N, and the plate height, H, are widely used in the literature and by instrument manufacturers as measures of column performance. **Figure 31-12** shows how N can be determined from a chromatogram. In the figure, the retention time of a peak t_R and the width of the peak at its base W (in units of time) are measured. It can be shown (see Feature 31-4) that the number of plates can then be computed by the relationship

$$N = 16\left(\frac{t_R}{W}\right)^2 \qquad (31\text{-}24)^{11}$$

Figure 31-12 Determination of the number of plates, $N = 16\left(\dfrac{t_R}{W}\right)^2$.

---

[11]Many chromatographic data systems report the width at half-height, $W_{1/2}$, in which case $N = 5.54(t_R/W_{1/2})^2$.

FEATURE 31-4

Derivation of Equation 31-24

The variance of the peak shown in Figure 31-12 has units of seconds squared because the x-axis is time in seconds (or sometimes in minutes). This time-based variance is usually designated as τ^2 to distinguish it from σ^2, which has units of centimeters squared. The two standard deviations τ and σ are related by

$$\tau = \frac{\sigma}{L/t_R} \tag{31-25}$$

where L/t_R is the average linear velocity of the solute in centimeters per second.

Figure 31-12 illustrates one method for approximating τ from an experimental chromatogram. Tangents at the inflection points on the two sides of the chromatographic peak are extended to form a triangle with the base line. The area of this triangle can be shown to be approximately 96% of the total area under the peak. In Section 6B-2, it was shown that about 96% of the area under a Gaussian peak is included within plus or minus two standard deviations ($\pm 2\sigma$) of its maximum. Thus, the intercepts shown in Figure 31-12 occur at approximately $\pm 2\tau$ from the maximum, and $W = 4\tau$, where W is the magnitude of the base of the triangle. Substituting this relationship into Equation 31-25 and rearranging yields

$$\sigma = \frac{LW}{4t_R}$$

When σ from this equation is substituted into Equation 3-23, we obtain

$$H = \frac{LW^2}{16t_R^2} \tag{31-26}$$

To obtain N, we substitute into Equation 31-22 and rearrange to get

$$N = 16\left(\frac{t_R}{W}\right)^2$$

Thus, N can be calculated from two time measurements, t_R and W. To obtain H, the length of the column packing L must also be known.

31E-6 Variables Affecting Column Efficiency

Band broadening reflects a loss of column efficiency. The slower the rate of mass-transfer processes occurring while a solute migrates through a column, the broader the band at the column exit. Some of the variables that affect mass-transfer rates are controllable and can be exploited to improve separations. **Table 31-5** lists the most important of these variables.

Effect of Mobile-Phase Flow Rate

The extent of band broadening depends on the length of time the mobile phase is in contact with the stationary phase, which in turn depends on the flow rate of the mobile phase. For this reason, efficiency studies generally have been carried

TABLE 31-5

Variables That Influence Column Efficiency

Variable	Symbol	Usual Units
Linear velocity of mobile phase	u	cm s^{-1}
Diffusion coefficient in mobile phase*	D_M	cm^2 s^{-1}
Diffusion coefficient in stationary phase*	D_S	cm^2 s^{-1}
Retention factor (Equation 31-18)	k	unitless
Diameter of packing particles	d_p	cm
Thickness of liquid coating on stationary phase	d_f	cm

*Increases as temperature increases and viscosity decreases.

(a) Liquid chromatography

(b) Gas chromatography

Figure 31-13 Effect of mobile-phase flow rate on plate height for (a) liquid chromatography and (b) gas chromatography.

out by determining H (by means of Equation 31-26) as a function of mobile-phase velocity. The plots for liquid chromatography and for gas chromatography shown in **Figure 31-13** are typical of the data obtained from such studies. While both show a minimum in H (or a maximum in efficiency) at low linear flow rates, the minimum for liquid chromatography usually occurs at flow rates that are well below those for gas chromatography. Often these flow rates are so low that the minimum H is not observed for liquid chromatography under normal operating conditions.

Generally, liquid chromatograms are obtained at lower linear flow rates than gas chromatograms. Also, as shown in Figure 31-13, plate heights for liquid

Linear flow rate and **volumetric flow rate** are two different but related quantities. Recall that the linear flow rate is related to the volumetric flow rate by the cross-sectional area and porosity (packed column) of the column (Equations 31-12 and 31-13).

chromatographic columns are an order of magnitude or more smaller than those encountered with gas chromatographic columns. Offsetting this advantage is the fact that it is impractical to use liquid chromatographic columns that are longer than about 25 to 50 cm because of high pressure drops. In contrast, gas chromatographic columns may be 50 m or more in length. As a result, the total number of plates, and thus overall column efficiency, are usually superior with gas chromatographic columns.

Theory of Band Broadening

Researchers have devoted an enormous amount of theoretical and experimental effort to develop quantitative relationships describing the effects of the experimental variables listed in Table 31-5 on plate heights for various types of columns. Perhaps a dozen or more expressions for calculating plate height have been put forward and applied with various degrees of success. None of these models is entirely adequate to explain the complex physical interactions and effects that lead to zone broadening and thus lower column efficiencies. Some of the equations, though imperfect, have been very useful, however, in pointing the way toward improved column performance. One of these is presented here.

The efficiency of capillary chromatographic columns and packed chromatographic columns at low flow velocities can be approximated by the expression

$$H = \frac{B}{u} + C_S u + C_M u \tag{31-27}$$

where H is the plate height in centimeters and u is the linear velocity of the mobile phase in centimeters per second.[12] The quantity B is the **longitudinal diffusion coefficient**, while C_S and C_M are mass-transfer coefficients for the stationary and mobile phases, respectively.

At high flow velocities in packed columns where flow effects dominate diffusion, the efficiency can be approximated by

$$H = A + \frac{B}{u} + C_S u \tag{31-28}$$

The Longitudinal Diffusion Term, B/u.   Diffusion is a process in which species migrate from a more concentrated part of a medium to a more dilute region. The rate of migration is proportional to the concentration difference between the regions and to the **diffusion coefficient** D_M of the species. The latter, which is a measure of the mobility of a substance in a given medium, is a constant for a given species equal to the velocity of migration under a unit concentration gradient.

In chromatography, longitudinal diffusion results in the migration of a solute from the concentrated center of a band to the more dilute regions on either side (that is, toward and opposed to the direction of flow). Longitudinal diffusion is a common source of band broadening in gas chromatography where the rate at which molecules diffuse is high. The phenomenon is of little significance in liquid chromatography where diffusion rates are much smaller. The magnitude of the B term in Equation 31-27 is largely determined by the diffusion coefficient D_M of the analyte in the mobile phase and is directly proportional to this constant.

Theoretical studies of zone broadening in the 1950s by Dutch chemical engineers led to the **van Deemter equation**, which can be written in the form

$$H = A + B/u + Cu$$

where the constants A, B, and C are coefficients of multiple path effects, longitudinal diffusion, and mass transfer, respectively. Today, we consider the van Deemter equation to be appropriate only for packed columns at high flow velocities. For other cases, Equation 31-27 is usually a better description.

[12]S. J. Hawkes, *J. Chem. Educ.*, **1983**, *60*, 393, **DOI**: 10.1021/ed060p393.

As shown by Equation 31-27, the contribution of longitudinal diffusion to plate height is inversely proportional to the linear velocity of the eluent. Such a relationship is not surprising inasmuch as the analyte is in the column for a briefer period when the flow rate is high. Thus, diffusion from the center of the band to the two edges has less time to occur.

The initial decreases in H shown in both curves in Figure 31-13 are a direct result of longitudinal diffusion. Note that the effect is much less pronounced in liquid chromatography because of the much lower diffusion rates in the liquid mobile phase. The striking difference in plate heights shown by the two curves in Figure 31-13 can also be explained by considering the relative rates of longitudinal diffusion in the two mobile phases. In other words, diffusion coefficients in gaseous media are orders of magnitude larger than in liquids. Therefore, band broadening occurs to a much greater extent in gas chromatography than in liquid chromatography.

> Diffusion coefficients in gases are usually about 1000 times larger than diffusion coefficients in liquids.

The Stationary Phase Mass-Transfer Term, $C_S u$. When the stationary phase is an immobilized liquid, the mass-transfer coefficient is directly proportional to the square of the thickness of the film on the support particles, d_f^2, and inversely proportional to the diffusion coefficient, D_S, of the solute in the film. These effects can be understood by realizing that both of these quantities reduce the average frequency at which analyte molecules reach the interface where transfer to the mobile phase can occur. That is, with thick films, molecules must on the average travel farther to reach the surface, and with smaller diffusion coefficients, they travel slower. The result is a slower rate of mass transfer and an increase in plate height.

When the stationary phase is a solid surface, the mass-transfer coefficient C_S is directly proportional to the time required for a species to be adsorbed or desorbed, which in turn is inversely proportional to the first-order rate constant for the processes.

The Mobile Phase Mass-Transfer Term, $C_M u$. The mass-transfer processes that occur in the mobile phase are sufficiently complex that we do not yet have a complete quantitative description. On the other hand, we have a good qualitative understanding of the variables affecting zone broadening from this cause, and this understanding has led to vast improvements in all types of chromatographic columns.

The mobile-phase mass-transfer coefficient C_M is known to be inversely proportional to the diffusion coefficient of the analyte in the mobile phase D_M. For packed columns, C_M is proportional to the square of the particle diameter of the packing material, d_p^2. For capillary columns, C_M is proportional to the square of the column diameter, d_c^2, and a function of the flow rate.

The contribution of mobile-phase mass transfer to plate height is the product of the mass-transfer coefficient C_M (which is a function of solvent velocity) as well as the velocity of the solvent itself. Thus, the net contribution to plate height is not linear in u (see the curve labeled $C_M u$ in Figure 31-15) but bears a complex dependency on solvent velocity.

Zone broadening in the mobile phase is due in part to the multitude of pathways by which a molecule (or ion) makes its way through a packed column. As shown in **Figure 31-14**, the lengths of these pathways can differ significantly. This difference means that the residence times in the column for molecules of the same species vary. Solute molecules then reach the end of the column over a range of times, leading to a broadened band. This multiple path effect, which is sometimes

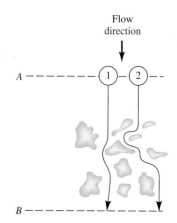

Figure 31-14 Typical pathways of two molecules during elution. Note that the distance traveled by molecule 2 is greater than that traveled by molecule 1. Therefore, molecule 2 will arrive at B later than molecule 1.

called **eddy diffusion**, would be independent of solvent velocity if it were not partially offset by ordinary diffusion, which results in molecules being transferred from a stream following one pathway to a stream following another. If the velocity of flow is very low, a large number of these transfers will occur, and each molecule in its movement down the column will sample numerous flow paths, spending a brief time in each. As a result, the rate at which each molecule moves down the column tends to approach that of the average. Thus, at low mobile-phase velocities, the molecules are not significantly dispersed by the multiple path effect. At moderate or high velocities, however, sufficient time is not available for diffusion averaging to occur, and band broadening due to the different path lengths is observed. At sufficiently high velocities, the effect of eddy diffusion becomes independent of flow rate.

> Pathways for the mobile phase through the column are numerous and have different lengths.

Superimposed on the eddy diffusion effect is one that arises from stagnant pools of the mobile phase retained in the stationary phase. Thus, when a solid serves as the stationary phase, its pores are filled with static volumes of mobile phase. Solute molecules must then diffuse through these stagnant pools before transfer can occur between the moving mobile phase and the stationary phase. This situation applies not only to solid stationary phases but also to liquid stationary phases immobilized on porous solids because the immobilized liquid does not usually fully fill the pores.

> Stagnant pools of solvent contribute to increases in H.

The presence of stagnant pools of mobile phase slows the exchange process and results in a contribution to the plate height that is directly proportional to the mobile-phase velocity and inversely proportional to the diffusion coefficient for the solute in the mobile phase. An increase in internal volume then accompanies increases in particle size.

Effect of Mobile-Phase Velocity on Terms in Equation 31-27.   **Figure 31-15** shows the variation of the three terms in Equation 31-27 as a function of mobile-phase velocity. The top curve is the summation of these various effects. Note that there is an optimum flow rate at which the plate height is a minimum and the separation efficiency is a maximum.

Summary of Methods for Reducing Band Broadening.   For packed columns, one variable that affects column efficiency is the diameter of the particles making up the packing. For capillary columns, the diameter of the column itself is an important variable. The effect of particle diameter is demonstrated by the data shown in **Figure 31-16** for gas chromatography.

> For packed columns, band broadening is minimized by small particle diameters. For capillary columns, small column diameters reduce band broadening.

Figure 31-15   Contribution of various mass-transfer terms to plate height. $C_S u$ arises from the rate of mass transfer to and from the stationary phase, $C_M u$ comes from a limitation in the rate of mass transfer in the mobile phase, and B/u is associated with longitudinal diffusion.

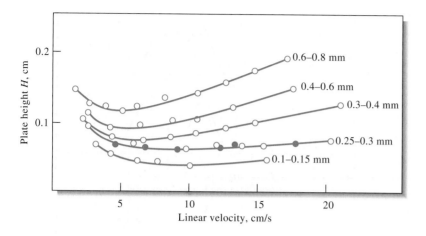

Figure 31-16 Effect of particle size on plate height for a packed gas chromatography column. The numbers to the right of each curve are particle diameters. (From J. Boheman and J. H. Purnell, in *Gas Chromatography 1958*, D. H. Desty, ed., New York: Academic Press, 1958.)

A similar plot for liquid chromatography is shown in Figure 33-1. To take advantage of the effect of column diameter, narrower and narrower columns have been used in recent years.

With gaseous mobile phases, the rate of longitudinal diffusion can be reduced appreciably by lowering the temperature and thus the diffusion coefficient. The result is significantly smaller plate heights at lower temperatures. This effect is usually not noticeable in liquid chromatography because diffusion is slow enough that the longitudinal diffusion term has little effect on overall plate height. With liquid stationary phases, the thickness of the layer of adsorbed liquid should be minimized since C_S in Equation 31-27 is proportional to the square of this variable.

> The diffusion coefficient D_M has a greater effect in gas chromatography than in liquid chromatography.

31E-7 Column Resolution

The **resolution**, R_s, of a column tells us how far apart two bands are relative to their widths. The resolution provides a quantitative measure of the ability of the column to separate two analytes. The significance of this term is illustrated in Figure 31-17, which consists of chromatograms for species A and B on three columns with different resolving powers. The resolution of each column is defined as

> The **resolution** of a chromatographic column is a quantitative measure of its ability to separate analytes A and B.

$$R_s = \frac{\Delta Z}{\dfrac{W_A}{2} + \dfrac{W_B}{2}} = \frac{2\Delta Z}{W_A + W_B} = \frac{2[(t_R)_B - (t_R)_A]}{W_A + W_B}$$

(31-29)

where all of the terms on the right side are as defined in the figure.

It is evident from **Figure 31-17** that a resolution of 1.5 gives an essentially complete separation of A and B, but a resolution of 0.75 does not. At a resolution of 1.0, zone A contains about 4% B, and zone B contains about 4% A. At a resolution of 1.5, the overlap is about 0.3%. The resolution for a given stationary phase can be improved by lengthening the column, thus increasing the number of plates. The added plates, however, result in an increase in the time required for separating the components.

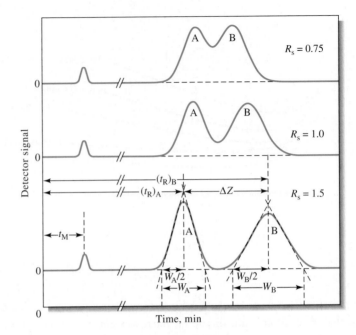

Figure 31-17 Separation at three resolution values: $R_s = 2\Delta Z/(W_A + W_B)$.

Effect of Retention Factor and Selectivity Factor on Resolution

We can derive a very useful equation that relates the resolution of a column to the number of plates it contains as well as to the retention and selectivity factors of a pair of solutes on the column. Thus, it can be shown[13] that for the two solutes A and B in Figure 31-17, the resolution is given by the equation

$$R_s = \frac{\sqrt{N}}{4}\left(\frac{\alpha - 1}{\alpha}\right)\left(\frac{k_B}{1 + k_B}\right) \tag{31-30}$$

where k_B is the retention factor of the slower-moving species and α is the selectivity factor. This equation can be rearranged to give the number of plates needed to realize a given resolution:

$$N = 16R_s^2\left(\frac{\alpha}{\alpha - 1}\right)^2\left(\frac{1 + k_B}{k_B}\right)^2 \tag{31-31}$$

Effect of Resolution on Retention Time

As mentioned previously, the goal in chromatography is the highest possible resolution in the shortest possible elapsed time. Unfortunately, these goals tend to be incompatible, and a compromise between them is usually necessary. The time $(t_R)_B$ required to elute the two species in Figure 31-17 with a resolution of R_s is given by

$$(t_R)_B = \frac{16R_s^2 H}{u}\left(\frac{\alpha}{\alpha - 1}\right)^2\frac{(1 + k_B)^3}{(k_B)^2} \tag{31-32}$$

where u is the linear velocity of the mobile phase.

---

[13]See D. A. Skoog, F. J. Holler, and S. R. Crouch, *Principles of Instrumental Analysis*, 6th ed., Belmont, CA: Brooks/Cole, 2007, pp. 776–777.

EXAMPLE 31-2

Substances A and B have retention times of 16.40 and 17.63 min, respectively, on a 30.0-cm column. An unretained species passes through the column in 1.30 min. The peak widths (at base) for A and B are 1.11 and 1.21 min, respectively. Calculate (a) the column resolution, (b) the average number of plates in the column, (c) the plate height, (d) the length of column required to achieve a resolution of 1.5, and (e) the time required to elute substance B on the column that gives an R_s value of 1.5.

Solution

(a) Using Equation 31-29, we find

$$R_s = \frac{2(17.63 - 16.40)}{1.11 + 1.21} = 1.06$$

(b) Equation 31-24 permits computation of N:

$$N = 16\left(\frac{16.40}{1.11}\right)^2 = 3493 \quad \text{and} \quad N = 16\left(\frac{17.63}{1.21}\right)^2 = 3397$$

$$N_{avg} = \frac{3493 + 3397}{2} = 3445$$

(c) $H = \dfrac{L}{N} = \dfrac{30.0}{3445} = 8.7 \times 10^{-3}$ cm

(d) The quantities k and α do not change greatly with increasing N and L. Thus, substituting N_1 and N_2 into Equation 31-30 and dividing one of the resulting equations by the other yield

$$\frac{(R_s)_1}{(R_s)_2} = \frac{\sqrt{N_1}}{\sqrt{N_2}}$$

where the subscripts 1 and 2 refer to the original and longer columns, respectively. Substituting the appropriate values for N_1, $(R_s)_1$, and $(R_s)_2$ gives

$$\frac{1.06}{1.5} = \frac{\sqrt{3445}}{\sqrt{N_2}}$$

$$N_2 = 3445\left(\frac{1.5}{1.06}\right)^2 = 6.9 \times 10^3$$

But

$$L = NH = 6.9 \times 10^3 \times 8.7 \times 10^{-3} = 60 \text{ cm}$$

(e) Substituting $(R_s)_1$, and $(R_s)_2$ into Equation 31-32 and dividing yield

$$\frac{(t_R)_1}{(t_R)_2} = \frac{(R_s)_1^2}{(R_s)_2^2} = \frac{17.63}{(t_R)_2} = \frac{(1.06)^2}{(1.5)^2}$$

$$(t_R)_2 = 35 \text{ min}$$

So, to obtain the improved resolution, the column length and thus the separation time must be doubled.

Optimization Techniques

Equation 31-30 and 31-32 serve as guides for choosing conditions that lead to a desired degree of resolution with a minimum expenditure of time. Each equation is made up of three parts. The first describes the efficiency of the column in terms of $\sqrt{N}$ or H. The second, which is the quotient containing α, is a selectivity term that depends on the properties of the two solutes. The third component is the retention factor term, which is the quotient containing k_B, the term that depends on the properties of both the solute and the column.

Variation in Plate Height.   As shown by Equation 31-30, the resolution of a column improves as the square root of the number of plates increases. Example 31-2e reveals, however, that increasing the number of plates is expensive in terms of time unless the increase is achieved by reducing the plate height and not by increasing column length.

Methods for minimizing plate height, discussed in Section 31E-6, include reducing the particle size of the packing material, the diameter of the column, and the thickness of the liquid film. Optimizing the flow rate of the mobile phase is also helpful.

Variation in the Retention Factor.   Often, a separation can be improved significantly by manipulation of the retention factor k_B. Increases in k_B generally enhance resolution (but at the expense of elution time). To determine the optimum range for k_B, it is convenient to write Equation 31-30 in the form

$$R_s = Q\left(\frac{k_B}{1 + k_B}\right)$$

and Equation 31-32 as

$$(t_R)_B = Q'\left(\frac{(1 + k_B)^3}{(k_B)^2}\right)$$

where Q and Q' contain the rest of the terms in the two equations. **Figure 31-18** is a plot of R_s/Q and $(t_R)_B/Q'$ as a function of k_B, assuming Q and Q' remain approximately constant. It is clear that values of k_B greater than about 10 should be avoided because they provide little increase in resolution but markedly increase the time required for separations. The minimum in the elution-time curve occurs at $k_B \approx 2$. Often, then, the optimal value of k_B lies in the range from 1 to 5.

Usually, the easiest way to improve resolution is by optimizing k. For gaseous mobile phases, k can often be improved by temperature changes. For liquid

Figure 31-18   Effect of retention factor k_B on resolution R_s and elution time $(t_R)_B$. It is assumed that Q and Q' remain constant with variations in k_B.

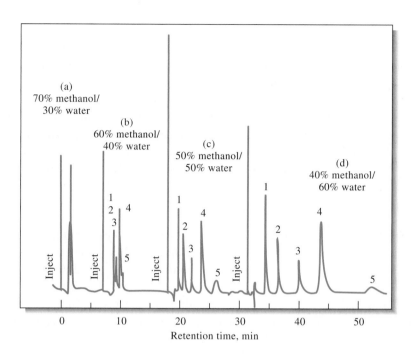

Figure 31-19 Effect of solvent variation on chromatograms. Analytes are (1) 9,10-anthraquinone; (2) 2-methyl-9,10-anthraquinone; (3) 2-ethyl-9,10-anthraquinone; (4) 1,4-dimethyl-9,10-anthraquinone; and (5) 2-t-butyl-9,10-anthraquinone.

mobile phases, changes in the solvent composition often permit manipulation of k to yield better separations. An example of the dramatic effect that relatively simple solvent changes can bring about is demonstrated in **Figure 31-19**. In the figure, modest variations in the methanol/water ratio convert unsatisfactory chromatograms (a and b) to chromatograms with well-separated peaks for each component (c and d). For most purposes, the chromatogram shown in (c) is best since it shows adequate resolution in minimum time. The retention factor is also influenced by the stationary phase film thickness.

Variation in the Selectivity Factor.   Optimizing k and increasing N are not sufficient to give a satisfactory separation of two solutes in a reasonable time when α approaches unity. A means must be sought to increase α while maintaining k in the range of 1 to 10. At least four options are available. These options in decreasing order of their desirability as determined by potential and convenience are (1) changing the composition of the mobile phase, (2) changing the column temperature, (3) changing the composition of the stationary phase, and (4) using special chemical effects.

An example of the use of option 1 has been reported for the separation of anisole ($C_6H_5OCH_3$) and benzene.[14] With a mobile phase that was a 50% mixture of water and methanol, k was 4.5 for anisole and 4.7 for benzene while α was only 1.04. Substitution of an aqueous mobile phase containing 37% tetrahydrofuran gave k values of 3.9 and 4.7 and an α value of 1.20. Peak overlap was significant with the first solvent system and negligible with the second.

A less convenient but often highly effective method for improving α while maintaining values for k in their optimal range is to alter the chemical composition of the stationary phase. To take advantage of this option, most laboratories that frequently use chromatography maintain several columns that can be interchanged with a minimum of effort.

Molecular model of 9,10-anthraquinone.

---

[14]L. R. Snyder and J. J. Kirkland, *Introduction to Modern Liquid Chromatography*, 2nd ed., New York: Wiley, 1979, p. 75.

Increases in temperature usually cause increases in k but have little effect on α values in liquid-liquid and liquid-solid chromatography. In contrast, with ion-exchange chromatography, temperature effects can be large enough to make exploration of this option worthwhile before resorting to a change in column packing material.

A final method to enhance resolution is to incorporate into the stationary phase a species that complexes or otherwise interacts with one or more components of the sample. A well-known example occurs when an adsorbent impregnated with a silver salt is used to improve the separation of olefins. The improvement is a result of the formation of complexes between the silver ions and unsaturated organic compounds.

The General Elution Problem

Figure 31-20 shows hypothetical chromatograms for a six-component mixture made up of three pairs of components with widely different distribution constants and thus widely different retention factors. In chromatogram (a), conditions have been adjusted so that the retention factors for components 1 and 2 (k_1 and k_2) are in the optimal range of 1 to 5. The factors for the other components are far larger than the optimum, however. Thus, the bands corresponding to components 5 and 6 appear only after an inordinate length of time has passed; furthermore, the bands are so broad that they may be difficult to identify unambiguously.

As shown in chromatogram (b), changing conditions to optimize the separation of components 5 and 6 bunches the peaks for the first four components to the point where their resolution is unsatisfactory. In this case, however, the total elution time is ideal.

The phenomenon illustrated in Figure 31-20 is encountered often enough to be given a name: the **general elution problem**. A common solution to this problem is to change conditions that determine the values of k as the separation proceeds. These changes can be performed in a stepwise manner or continuously. Therefore, for the mixture shown in Figure 31-20, conditions at the outset could be those producing chromatogram (a). Immediately after the elution of components 1 and 2, conditions could be changed to those that are optimal for separating components 3 and 4 (as in chromatogram c). With the appearance of peaks for these components, the

Figure 31-20 The general elution problem in chromatography.

elution could be completed under the conditions used for producing chromatogram (b). Often such a procedure leads to satisfactory separation of all the components of a mixture in minimal time.

For liquid chromatography, variations in k are brought about by varying the composition of the mobile phase during elution. Such a procedure is called **gradient elution** or **solvent programming**. Elution under conditions of constant mobile-phase composition is called **isocratic elution.** For gas chromatography, the temperature can be changed in a known fashion to bring about changes in k. This **temperature-programming** mode can help achieve optimal conditions for many separations.

31E-8 Applications of Chromatography

Chromatography is a powerful and versatile tool for separating closely related chemical species. In addition, it can be used for the qualitative identification and quantitative determination of separated species. Examples of the applications of the various types of chromatography are given in Chapters 32 through 34.

> **Spreadsheet Summary** In Chapter 14 of *Applications of Microsoft®* *Excel in Analytical Chemistry*, 2nd ed., several exercises involving chroma-tography are suggested. In the first, a chromatogram of a three-component mixture is simulated. The resolution, number of theoretical plates, and retention times are varied, and their effect on the chromatograms noted. The number of theoretical plates needed to achieve a given resolution is the subject of another exercise. A spreadsheet is constructed to find N for various retention factors of a two-component mixture. An exponentially modified Gaussian is investigated as a function of the time constant of the exponential. The optimization of chromatographic methods is illustrated by plotting the van Deemter equation for various flow veloci-ties, longitudinal diffusion, and mass-transfer coefficient values. Solver is then used to find best-fit values of the van Deemter coefficients.

WEB WORKS Use a search engine to locate websites that deal with peak tailing in reverse-phase liquid chromatography. Describe the phenomenon and discuss ways in which tailing can be mini-mized. Also, perform a search on temperature effects in liquid chromatography. Describe how temperature influences liquid chromatographic separations. Based on what you learn, would temperature programming be a valuable aid to separation in liquid chromatography? Why or why not?

QUESTIONS AND PROBLEMS

*31-1. What is a collector ion and how is it used?

31-2. What does the term salting out a protein mean? What is the salting in effect?

*31-3. What two events accompany the separation process?

31-4. Name three methods based on mechanical phase separation.

31-5. Define
  *(a) elution.
  (b) mobile phase.
  *(c) stationary phase.
  (d) distribution constant.
  *(e) retention time.
  (f) retention factor.
  *(g) selectivity factor.
  (h) plate height.

31-6. How do strong- and weak-acid synthetic ion-exchange resins differ in structure?

*31-7. List the variables that lead to band broadening in chromatography.

31-8. What are the major differences between gas-liquid and liquid-liquid chromatography?

*31-9. Describe a method for determining the number of plates in a column.

31-10. Describe two general methods for improving the resolution of two substances on a chromatographic column.

*31-11. The distribution constant for X between n-hexane and water is 8.9. Calculate the concentration of X remaining in the aqueous phase after 50.0 mL of 0.200 M X is treated by extraction with the following quantities of n-hexane:
- (a) one 40.0-mL portion.
- (b) two 20.0-mL portions.
- (c) four 10.0-mL portions.
- (d) eight 5.00-mL portions.

31-12. The distribution coefficient for Z between n-hexane and water is 5.85. Calculate the percent of Z remaining in 25.0 mL of water that was originally 0.0550 M in Z after extraction with the following volumes of n-hexane:
- (a) one 25.0-mL portion.
- (b) two 12.5-mL portions.
- (c) five 5.00-mL portions.
- (d) ten 2.50-mL portions.

*31-13. What volume of n-hexane is required to decrease the concentration of X in Problem 31-11 to 1.00×10^{-4} M if 25.0 mL of 0.0500 M X is extracted with
- (a) 25.0-mL portions?
- (b) 10.0-mL portions?
- (c) 2.0-mL portions?

31-14. What volume of n-hexane is required to decrease the concentration of Z in Problem 31-12 to 1.00×10^{-5} M if 40.0 mL of 0.0200 M Z is extracted with
- (a) 50.0-mL portions of n-hexane?
- (b) 25.0-mL portions?
- (c) 10.0-mL portions?

*31-15. What is the minimum distribution coefficient that permits removal of 99% of a solute from 50.0 mL of water with
- (a) two 25.0-mL extractions with toluene?
- (b) five 10.0-mL extractions with toluene?

31-16. If 30.0 mL of water that is 0.0500 M in Q is to be extracted with four 10.0-mL portions of an immiscible organic solvent, what is the minimum distribution coefficient that allows transfer of all but the following percentages of the solute to the organic layer:
- *(a) 1.00×10^{-4}
- (b) 1.00×10^{-3}
- (c) 1.00×10^{-2}

*31-17. A 0.150 M aqueous solution of the weak organic acid HA was prepared from the pure compound, and three 50.0-mL aliquots were transferred to 100.0-mL volumetric flasks. Solution 1 was diluted to 100.0 mL with 1.0 M $HClO_4$, solution 2 was diluted to the mark with 1.0 M NaOH, and solution 3 was diluted to the mark with water. A 25.0-mL aliquot of each was extracted with 25.0 mL of n-hexane. The extract from solution 2 contained no detectable trace of A-containing species, indicating that A^- is not soluble in the organic solvent. The extract from solution 1 contained no ClO_4^- or $HClO_4$ but was found to be 0.0454 M in HA (by extraction with standard NaOH and back-titration with standard HCl). The extract from solution 3 was found to be 0.0225 M in HA. Assume that HA does not associate or dissociate in the organic solvent, and calculate
- (a) the distribution ratio for HA between the two solvents.
- (b) the concentration of the *species* HA and A^- in aqueous solution 3 after extraction.
- (c) the dissociation constant of HA in water.

31-18. To determine the equilibrium constant for the reaction

$$I_2 + 2SCN^- \rightleftharpoons I(SCN)_2^- + I^-$$

25.0 mL of a 0.0100 M aqueous solution of I_2 was extracted with 10.0 mL of $CHCl_3$. After extraction, spectrophotometric measurements revealed that the I_2 concentration *of the aqueous layer* was 1.12×10^{-4} M. An aqueous solution that was 0.0100 M in I_2 and 0.100 M in KSCN was then prepared. After extraction of 25.0 mL of this solution with 10.0 mL of $CHCl_3$, the concentration of I_2 *in the $CHCl_3$ layer* was found from spectrophotometric measurement to be 1.02×10^{-3} M.
- (a) What is the distribution constant for I_2 between $CHCl_3$ and H_2O?
- (b) What is the formation constant for $I(SCN)_2^-$?

*31-19. The total cation content of natural water is often determined by exchanging the cations for hydrogen ions on a strong-acid ion-exchange resin. A 25.0-mL sample of a natural water was diluted to 100 mL with distilled water, and 2.0 g of a cation-exchange resin was added. After stirring, the mixture was filtered and the solid remaining on the filter paper was washed with three 15.0-mL portions of water. The filtrate and washings required 15.3 mL of 0.0202 M NaOH to give a bromocresol green end point.
- (a) Calculate the number of millimoles of cation present in exactly 1.00 L of sample.
- (b) Report the results in terms of milligrams of $CaCO_3$ per liter.

31-20. Describe the preparation of exactly 2.00 L of 0.1500 M HCl from primary-standard-grade NaCl using a cation-exchange resin.

*31-21. An aqueous solution containing $MgCl_2$ and HCl was analyzed by first titrating a 25.00-mL aliquot to a bromocresol green end point with 17.53 mL of 0.02932 M NaOH. A 10.00-mL aliquot was then diluted to 50.00 mL with distilled water and passed through a strong-acid ion-exchange resin. The eluate and washings required 35.94 mL of the NaOH solution to reach the same end point. Report the molar concentrations of HCl and $MgCl_2$ in the sample.

31-22. An open tubular column used for gas chromatography had an inside diameter of 0.25 mm. A volumetric flow rate of 0.95 mL/min was used. Find the linear flow velocity in cm/s at the column outlet.

*31-23. A packed column in gas chromatography had an inside diameter of 5.0 mm. The measured volumetric flow rate at the column outlet was 48.0 mL/min. If the column porosity was 0.43, what was the linear flow velocity in cm/s?

31-24. The following data are for a liquid chromatographic column:

Length of Packing	24.7 cm
Flow rate	0.313 mL/min
V_M	1.37 mL
V_S	0.164 mL

A chromatogram of a mixture of species A. B, C, and D provided the following data:

	Retention Time, min	Width of Peak Base (W), min
Nonretained	3.1	—
A	5.4	0.41
B	13.3	1.07
C	14.1	1.16
D	21.6	1.72

Calculate
(a) the number of plates from each peak.
(b) the mean and the standard deviation for N.
(c) the plate height for the column.

*31-25. From the data in Problem 31-24, calculate for A, B, C, and D
(a) the retention factor.
(b) the distribution constant.

31-26. From the data in Problem 31-24, calculate for species B and C
(a) the resolution.
(b) the selectivity factor.
(c) the length of column necessary to separate the two species with a resolution of 1.5.
(d) the time required to separate the two species on the column in part (c).

31-27. From the data in Problem 31-24, calculate for species C and D
(a) the resolution.
(b) the length of column necessary to separate the two species with a resolution of 1.5.

31-28. The following data were obtained by gas-liquid chromatography on a 40-cm packed column:

Compound	t_R, min	W, min
Air	1.9	—
Methylcyclohexane	10.0	0.76
Methylcyclohexene	10.9	0.82
Toluene	13.4	1.06

Calculate
(a) an average number of plates from the data.
(b) the standard deviation for the average in (a).
(c) an average plate height for the column.

31-29. Referring to Problem 31-28, calculate the resolution for
(a) methylcyclohexene and methylcyclohexane.
(b) methylcyclohexene and toluene.
(c) methylcyclohexane and toluene.

*31-30. If a resolution of 1.75 is desired in separating methylcyclohexane and methylcyclohexene in Problem 31-28,
(a) how many plates are required?
(b) how long must the column be if the same packing is used?
(c) what is the retention time for methylcyclohexene on the column of part b?

31-31. If V_S and V_M for the column in Problem 31-28 are 19.6 and 62.6 mL, respectively, and a nonretained air peak appears after 1.9 min, calculate
(a) the retention factor for each compound.
(b) the distribution constant for each compound.
(c) the selectivity factor for methylcyclohexane and methylcyclohexene.

*31-32. From distribution studies, species M and N are known to have water/hexane distribution constants of 5.99 and 6.16 ($K = [X]_{H_2O}/[X]_{hex}$), where X = M or N. The two species are to be separated by elution with hexane in a column packed with silica gel containing adsorbed water. The ratio V_S/V_M for the packing is 0.425.
(a) Calculate the retention factor for each solute.
(b) Calculate the selectivity factor.
(c) How many plates are needed to provide a resolution of 1.5?
(d) How long a column is needed if the plate height of the packing is 1.5×10^{-3} cm?
(e) If the flow rate is 6.75 cm/min, how long will it take to elute the two species?

31-33. Repeat the calculations in Problem 31-32 assuming $K_M = 5.81$ and $K_N = 6.20$.

31-34. Challenge Problem: A chromatogram of a two-component mixture on a 25-cm packed liquid chromatography column is shown in the figure below. The flow rate was 0.40 mL/min.

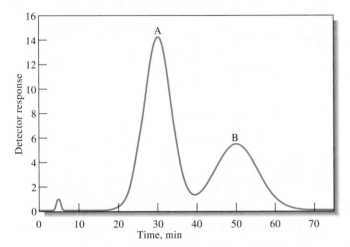

A chromatogram of a two-component mixture

(a) Find the times that components A and B spend in the stationary phase.
(b) Find the retention times for A and B.
(c) Determine the retention factors for the two components.
(d) Find the full widths of each peak and the full width at half maximum values.
(e) Find the resolution of the two peaks.
(f) Find the average number of plates for the column.
(g) Find the average plate height.
(h) What column length would be needed to achieve a resolution of 1.75?
(i) What time would be required to achieve the resolution in part h.
(j) Assume that the column length is fixed at 25 cm and the packing material is fixed. What measures could you take to increase the resolution to achieve baseline separation?
(k) Are there any measures you could use to achieve a better separation in a shorter time with the same column as in part j?

Gas Chromatography

CHAPTER 32

Gas chromatography is one of most widely used techniques for qualitative and quantitative analysis. Shown in the photo is a benchtop gas chromatograph/mass spectrometer system that can provide high-resolution separations and identification of the compounds separated. Such systems are invaluable in industrial, biomedical, and forensic laboratories.

This chapter considers gas chromatography in detail, including the columns and stationary phases that are most widely used. Various detection systems, including mass spectrometry, are described. Although the chapter is primarily concerned with gas-liquid chromatography, there is a brief discussion of gas-solid chromatography.

Shimadzu Corp

In gas chromatography, the components of a vaporized sample are separated by being distributed between a mobile gaseous phase and a liquid or a solid stationary phase held in a column.[1] In performing a gas chromatographic separation, the sample is vaporized and injected onto the head of a chromatographic column. Elution is brought about by the flow of an inert gaseous mobile phase. In contrast to most other types of chromatography, the mobile phase does not interact with molecules of the analyte. The only function of the mobile phase is to transport the analyte through the column.

Two types of gas chromatography are encountered: **gas-liquid chromatography** (GLC) and **gas-solid chromatography** (GSC). Gas-liquid chromatography finds widespread use in all fields of science where its name is usually shortened to **gas chromatography** (GC). Gas-solid chromatography is based on a solid stationary phase in which retention of analytes occurs because of physical adsorption. Gas-solid chromatography has limited application because of semipermanent retention of active or polar molecules and severe tailing of elution peaks. The tailing is due to the nonlinear character of adsorption process. Thus, this technique has not found wide application except for the separation of certain low-molecular-mass gaseous species; we discuss the method briefly in Section 32D.

Gas-liquid chromatography is based on partitioning of the analyte between a gaseous mobile phase and a liquid phase immobilized on the surface of an inert solid packing or on the walls of capillary tubing. The concept of gas-liquid chromatography was first enunciated in 1941 by Martin

In **gas-liquid chromatography**, the mobile phase is a gas, and the stationary phase is a liquid that is retained on the surface of an inert solid by adsorption or chemical bonding.

In **gas-solid chromatography**, the mobile phase is a gas, and the stationary phase is a solid that retains the analytes by physical adsorption. Gas-solid chromatography permits the separation and determination of low-molecular-mass gases, such as air components, hydrogen sulfide, carbon monoxide, and nitrogen oxides.

---

[1]For detailed treatment of GC, see C. Poole, ed., *Gas Chromatography*, Amsterdam: Elsevier, 2012; H. M. McNair and J. M Miller, *Basic Gas Chromatography*, 2nd ed., Hoboken, NJ: Wiley, 2009; R. L. Grob and E. F. Barry, ed. *Modern Practice of Gas Chromatography*, 4th ed., Hoboken, NJ: Wiley-Interscience, 2004; R. P. W. Scott, *Introduction to Analytical Gas Chromatography*, 2nd ed., New York: Marcel Dekker, 1997.

and Synge, who were also responsible for the development of liquid-liquid partition chromatography. More than a decade was to elapse, however, before the value of gas-liquid chromatography was demonstrated experimentally and this technique began to be used as a routine laboratory tool. In 1955, the first commercial apparatus for gas-liquid chromatography appeared on the market. Since that time, the growth in applications of this technique has been phenomenal. Currently, several hundred thousand gas chromatographs are in use throughout the world.

32A INSTRUMENTS FOR GAS-LIQUID CHROMATOGRAPHY

Many changes and improvements in gas chromatographic instruments have appeared in the marketplace since their commercial introduction. In the 1970s, electronic integrators and computer-based data-processing equipment became common. The 1980s saw computers being used for automatic control of such instrument parameters as column temperature, flow rates, and sample injection. This same decade also saw the development of very high-performance instruments at moderate costs and, perhaps most important, the introduction of open tubular columns that are capable of separating components of complex mixtures in relatively short times. Today, some 50 instrument manufacturers offer about 150 different models of gas chromatographic equipment at costs that vary from $1000 to over $50,000. The basic components of a typical instrument for performing gas chromatography are shown in **Figure 32-1** and are described briefly in this section.

32A-1 Carrier Gas System

The mobile phase gas in gas chromatography is called the **carrier gas** and must be chemically inert. Helium is the most common mobile phase, although argon, nitrogen, and hydrogen are also used. These gases are available in pressurized tanks. Pressure regulators, gauges, and flow meters are required to control the flow rate of the gas.

Classically, flow rates in gas chromatographs were regulated by controlling the gas inlet pressure. A two-stage pressure regulator at the gas cylinder and some sort of pressure regulator or flow regulator mounted in the chromatograph were used. Inlet pressures usually range from 10 to 50 psi (lb/in^2) above room pressure, yielding flow rates of 25 to 150 mL/min with packed columns and 1 to 25 mL/min for open tubular capillary columns. With pressure-controlled devices, it is assumed that flow rates are constant if the inlet pressure remains constant. Newer chromatographs use electronic pressure controllers both for packed and for capillary columns.

Figure 32-1   Block diagram of a typical gas chromatograph.

With any chromatograph, it is desirable to measure the flow through the column. The classical soap-bubble meter shown in **Figure 32-2** is still widely used. A soap film is formed in the path of the gas when a rubber bulb containing an aqueous solution of soap or detergent is squeezed; the time required for this film to move between two graduations on the buret is measured and converted to volumetric flow rate (see Figure 32-2). Note that volumetric flow rates and linear flow velocities are related by Equation 31-12 or 31-13. Bubble flow meters are now available with digital readouts that eliminate some human reading errors. Usually, the flow meter is located at the end of the column, as shown in Figure 32-1. The use of electronic flow meters has become increasingly common. Digital flow meters are available that measure mass flow, volume flow, or both. Volumetric flow measurements are independent of the gas composition. Mass flow meters are calibrated for specific gas compositions, but, unlike volumetric meters, they are independent of temperature and pressure.

32A-2 Sample Injection System

For high column efficiency, a suitably sized sample should be introduced as a "plug" of vapor. Slow injection or oversized samples cause band spreading and poor resolution. Calibrated microsyringes, such as those shown in **Figure 32-3**, are used to inject liquid samples through a rubber or silicone diaphragm, or septum, into a heated sample port located at the head of the column. The sample port (see **Figure 32-4**) is usually kept at about 50°C greater than the boiling point of the least volatile component of the sample. For ordinary packed analytical columns, sample sizes range from a few tenths of a microliter to 20 μL. Capillary columns require samples that are smaller by a factor of 100 or more. For these columns, a sample splitter is often needed to deliver a small known fraction (1:100 to 1:500) of the injected sample, with the remainder going to waste. Commercial gas chromatographs intended for use with capillary columns incorporate such splitters, and they also allow for splitless injection when packed columns are used.

For the most reproducible sample injection, newer gas chromatographs use autoinjectors and autosamplers, such as the system shown in **Figure 32-5**. With such autoinjectors, syringes are filled, and the sample injected into the chromatograph

Figure 32-2 A soap-bubble flow meter.

Figure 32-3 A set of microsyringes for sample injection.

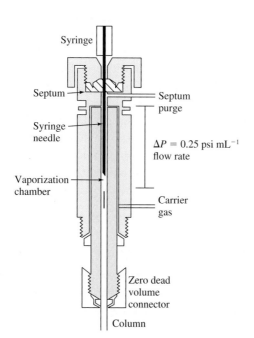

Syringe

Septum

Septum purge

Syringe needle

$\Delta P = 0.25$ psi mL^{-1} flow rate

Vaporization chamber

Carrier gas

Zero dead volume connector

Column

Figure 32-4 Cross-sectional view of a microflash vaporizer direct injector.

Figure 32-5 An autoinjection system with autosampler for gas chromatography.

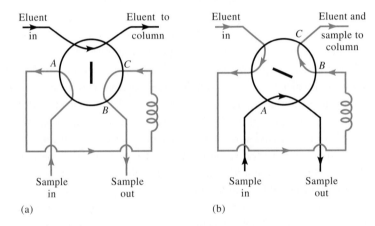

Figure 32-6 A rotary sample valve. Valve position (a) is for filling the sample loop *ACB*; position (b) is for introduction of sample into the column.

automatically. In the autosampler, samples are contained in vials on a sample turntable. The autoinjector syringe picks up the sample through a septum on the vial and injects the sample through a septum on the chromatograph. With the unit shown, up to 150 sample vials can be placed on the turntable. Injection volumes can vary from 0.1 μL with a 10-μL syringe to 200 μL with a 200-μL syringe. Standard deviations as low as 0.3% are common with autoinjection systems.

For introducing gases, a sample valve, such as that shown in **Figure 32-6**, is often used instead of a syringe. With such devices, sample sizes can be reproduced to better than 0.5% relative. Liquid samples can also be introduced through a sampling valve. Solid samples are introduced as solutions or alternatively are sealed into thin-walled vials that can be inserted at the head of the column and punctured or crushed from the outside.

32A-3 Column Configurations and Column Ovens

The columns in gas chromatography are of two general types: **packed columns** or **capillary columns**. In the past, the vast majority of gas chromatographic analyses used packed columns. For most current applications, packed columns have been replaced by more efficient and faster capillary columns.

Chromatographic columns vary in length from less than 2 m to 60 m or more. They are constructed of stainless steel, glass, fused silica, or Teflon. In order to fit into an oven for thermostating, they are usually formed as coils having diameters of 10 to 30 cm (see **Figure 32-7**). A detailed discussion of columns, column packings, and stationary phases is found in Section 32-B.

Column temperature is an important variable that must be controlled to a few tenths of a degree for precise work. Thus, the column is normally housed in a thermostated oven. The optimum column temperature depends on the boiling point of the sample and the degree of separation required. Roughly, a temperature equal to or slightly above the average boiling point of a sample results in a reasonable elution time (2 to 30 min). For samples with a broad boiling range, it is often desirable to use **temperature programming** whereby the column temperature is increased either continuously or in steps as the separation proceeds. **Figure 32-8** shows the improvement in a chromatogram brought about by temperature programming.

In general, optimum resolution is associated with minimal temperature. The cost of lowered temperature, however, is an increase in elution time and, therefore, the time required to complete an analysis. Figures 32-8a and 32-8b illustrate this principle.

Analytes of limited volatility can sometimes be determined by forming derivatives that are more volatile. Likewise, derivatization is used at times to enhance detection or improve chromatographic performance.

Figure 32-7 Fused-silica capillary columns.

Temperature programming in gas chromatography is achieved by increasing the column temperature continuously or in steps during elution.

Figure 32-8 Effect of temperature on gas chromatograms. (a) Isothermal at 45°C. (b) Isothermal at 145°C. (c) Programmed at 30° to 180°C. (From W. E. Harris and H. W. Habgood, *Programmed Temperature Gas Chromatography*, New York: Wiley, 1966, p. 10. Reprinted with permission of the author.)

32A-4 Chromatographic Detectors

Dozens of detectors have been investigated and used with gas chromatographic separations.[2] We first describe the characteristics that are most desirable in a gas chromatographic detector and then discuss the most widely used devices.

Characteristics of the Ideal Detector

The ideal detector for gas chromatography has the following characteristics:

1. Adequate sensitivity. In general, the sensitivities of present-day detectors lie in the range of 10^{-8} to 10^{-15} g solute/s.
2. Good stability and reproducibility.
3. A linear response to solutes that extends over several orders of magnitude.
4. A temperature range from room temperature to at least 400°C.
5. A short response time that is independent of flow rate.
6. High reliability and ease of use. To the greatest extent possible, the detector should be foolproof in the hands of inexperienced operators.
7. Similarity in response toward all solutes or, alternatively, a highly predictable and selective response toward one or more classes of solutes.
8. Nondestructive of sample.

Needless to say, no current detector exhibits all of these characteristics. Some of the more common detectors are listed in **Table 32-1**. Four of the most widely used detectors are described in the paragraphs that follow.

Flame Ionization Detectors

The **flame ionization detector** (FID) is the most widely used and generally applicable detector for gas chromatography. With a FID, such as that shown in **Figure 32-9**, effluent from the column is directed into a small air/hydrogen flame. Most organic compounds produce ions and electrons when pyrolyzed at the temperature of an air/hydrogen flame. These compounds are detected by monitoring the current produced by collecting the ions and electrons. A few hundred volts applied between the burner tip and a collector electrode located above the flame serves to collect the ions and electrons. The resulting current ($\sim 10^{-12}$ A) is then measured with a sensitive picoammeter.

TABLE 32-1

Gas Chromatographic Detectors		
Type	**Applicable Samples**	**Typical Detection Limit**
Flame ionization	Hydrocarbons	1 pg/s
Thermal conductivity	Universal detector	500 pg/mL
Electron capture	Halogenated compounds	5 fg/s
Mass spectrometer (MS)	Tunable for any species	0.25 to 100 pg
Thermionic	Nitrogen and phosphorous compounds	0.1 pg/s (P) 1 pg/s (N)
Electrolytic conductivity (Hall)	Compounds containing halogens, sulfur, or nitrogen	0.5 pg Cl/s 2 pg S/s 4 pg N/s
Photoionization	Compounds ionized by UV radiation	2 pg C/s
Fourier transform IR (FTIR)	Organic compounds	0.2 to 40 ng

[2]See L. A. Colon and L. J. Baird, in *Modern Practice of Gas Chromatography*, R. L. Grob and E. F. Barry, eds., 4th ed., Ch. 6, Hoboken, NJ: Wiley-Interscience, 2004.

Figure 32-9 A typical flame ionization detector. (Courtesy of Agilent Technologies.)

The ionization of carbon compounds in a flame is a poorly understood process, although it is observed that the number of ions produced is roughly proportional to the number of *reduced* carbon atoms in the flame. Because the flame ionization detector responds to the number of carbon atoms entering the detector per unit of time, it is a *mass-sensitive* rather than a *concentration-sensitive* device. As such, this detector has the advantage that changes in flow rate of the mobile phase have little effect on detector response.

Functional groups, such as carbonyl, alcohol, halogen, and amine, yield fewer ions or none at all in a flame. In addition, the detector is insensitive toward noncombustible gases, such as H_2O, CO_2, SO_2, and NO_x. These properties make the flame ionization detector a most useful general detector for the analysis of most organic samples including those that are contaminated with water and the oxides of nitrogen and sulfur.

The FID exhibits a high sensitivity ($\sim 10^{-13}$ g/s), large linear response range ($\sim 10^7$), and low noise. It is generally rugged and easy to use. Disadvantages of the flame ionization detector are that it destroys the sample during the combustion step and requires additional gases and controllers.

Thermal Conductivity Detectors

The **thermal conductivity detector** (TCD), which was one of the earliest detectors for gas chromatography, still finds wide application. This device consists of an electrically heated source whose temperature at constant electric power depends on the thermal conductivity of the surrounding gas. The heated element may be a fine platinum, gold, or tungsten wire or, alternatively, a small thermistor. The electrical resistance of this element depends on the thermal conductivity of the gas. **Figure 32-10a** shows a cross-sectional view of one of the temperature-sensitive elements in a TCD.

Four thermally sensitive resistive elements are often used. A *reference pair* is located ahead of the sample injection chamber and a *sample pair* immediately beyond the column. Alternatively, the gas stream can be split. The detectors are incorporated in two arms of a simple bridge circuit, as shown in **Figure 32-10b**, such that the thermal conductivity of the carrier gas is canceled. In addition, the effects of variations in temperature, pressure, and electric power are minimized. Modulated single-filament TCDs are also available.

(a)

(b)

Figure 32-10 Schematic of (a) a thermal conductivity detector cell and (b) an arrangement of two sample detector cells (R_2 and R_3) and two reference detector cells (R_1 and R_4). (Reprinted from F. Rastrelloa, P. Placidi, A. Scorzonia, E. Cozzanib, M. Messinab, I. Elmib, S. Zampollib, and G. C. Cardinali, Sensors and Actuators A, **2012**, *178*, 49, **DOI**:10.1016/j.sna.2012.02.008. Copyright (2012), with permission from Elsevier.)

The thermal conductivities of helium and hydrogen are roughly six to ten times greater than those of most organic compounds. Thus, even small amounts of organic species cause relatively large decreases in the thermal conductivity of the column effluent, resulting in a marked rise in the temperature of the detector. Detection by thermal conductivity is less satisfactory with carrier gases whose conductivities closely resemble those of most sample components.

The advantages of the TCD are its simplicity, its large linear dynamic range (about five orders of magnitude), its general response to both organic and inorganic species, and its nondestructive character, which permits collection of solutes after detection. The chief limitation of this detector is its relatively low sensitivity ($\sim 10^8$ g/s solute/mL carrier gas). Other detectors exceed this sensitivity by factors of 10^4 to 10^7. The low sensitivities of TCDs often precludes their use with capillary columns where sample amounts are very small.

Electron Capture Detectors

The **electron capture detector** (ECD) has become one of the most widely used detectors for environmental samples because this detector selectively responds to halogen-containing organic compounds, such as pesticides and polychlorinated biphenyls. In this detector, the sample eluate from a column is passed over a radioactive β emitter, usually nickel-63. An electron from the emitter causes ionization of the carrier gas (often nitrogen) and the production of a burst of electrons. In the absence of organic species, a constant standing current between a pair of electrodes results from this ionization process. The current decreases markedly, however, in the presence of organic molecules containing electronegative functional groups that tend to capture electrons. Compounds, such as halogens, peroxides, quinones, and nitro groups, are detected with high sensitivity. The detector is insensitive to functional groups such as amines, alcohols, and hydrocarbons.

Electron capture detectors are highly sensitive and have the advantage of not altering the sample significantly (in contrast to the flame ionization detector, which consumes the sample). The linear response of the detector, however, is limited to about two orders of magnitude.

Mass Spectrometry Detectors

One of the most powerful detectors for GC is the **mass spectrometer**. The combination of gas chromatography and mass spectrometry is known as **GC/MS**.[3] As discussed in Chapter 29, a mass spectrometer measures the mass-to-charge ratio (m/z) of ions that have been produced from the sample. Most of the ions produced are singly charged ($z = 1$) so that mass spectrometrists often speak of measuring the mass of ions when mass-to-charge ratio is actually measured.

Currently, some 50 instrument companies offer GC/MS equipment. The flow rate from capillary columns is usually low enough that the column output can be fed directly into the ionization chamber of the mass spectrometer. The schematic of a typical GC/MS system was shown previously in Figure 29-8. Prior to the advent of capillary columns, when packed columns were used, it was necessary to minimize the large volume of carrier gas eluting from the GC. Various jet, membrane, and effusion separators were used for this purpose. Presently, capillary columns are invariably used in GC/MS instruments, and such separators are no longer needed.

The most common ion sources for GC/MS are electron impact and chemical ionization. The most common mass analyzers are quadrupole and ion-trap analyzers. Sources and analyzers for mass spectrometry are also described in Chapter 29.

In GC/MS, the mass spectrometer scans the masses repetitively during a chromatographic experiment. If a chromatographic run is 10 minutes, for example, and a scan is taken each second, 600 mass spectra are recorded. A computer data system is needed to process the large amount of data obtained. The data can be analyzed in several ways. First, the ion abundance in each spectrum can be summed and plotted as a function of time to give a **total-ion chromatogram**. This plot is similar to a conventional chromatogram. Second, one can also display the mass spectrum at a particular time during the chromatogram to identify the species eluting at that time. Finally, a single mass-to-charge (m/z) value can be selected and monitored throughout the chromatographic experiment, a technique known as **selected-ion monitoring**. Mass spectra of selected ions during a chromatographic experiment are known as **mass chromatograms**.

GC/MS instruments have been used for the identification of thousands of components that are present in natural and biological systems. An example of one application of GC/MS is shown in **Figure 32-11**. The total-ion chromatogram of a methanol extract from a termite sample is shown in part (a). The selected-ion chromatogram in part (b) is that of the ion at a mass-to-charge ratio of 168. To complete the identification, the complete mass spectrum of the species eluting at 10.46 min was taken and shown in (c) allowing the compound to be identified as β-carboline norharmane, an alkaloid.

Mass spectrometry can also be used to acquire information about incompletely separated components. For example, the mass spectrum of the front edge of a GC peak may be different from that of the trailing edge if multiple components are eluting at

---

[3]See O. D. Sparkman, Z. E. Penton, and F. G. Kitson *Gas Chromatography and Mass Spectrometry,* 2nd ed., Amsterdam: Elsevier, 2011; M. C. McMaster, *GC/MS: A Practical User's Guide,* 2nd ed., New York: Wiley, 2008.

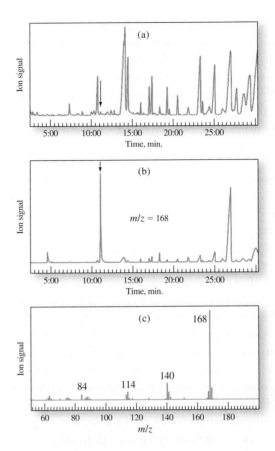

Figure 32-11 Typical outputs for a GC/MS system. In (a) the total ion chromatogram of an extract from a termite sample is shown. In (b) the ion at $m/z = 168$ was monitored during the chromatogram. In (c), the complete mass spectrum of the compound eluting at $t = 10.46$ minutes is presented, allowing it to be identified as β-carboline norharmane, an important alkaloid. From S. Ikatura, S. Kawabata, H. Tanaka, and A. Enoki, *J. Insect Sci.*, **2008**, 8: 13.

the same time. With mass spectrometry, we can not only determine that a peak is due to more than one component, but we can also identify the various unresolved species. GC has also been coupled with tandem mass spectrometers and with Fourier transform mass spectrometers to give GC/MS/MS or GC/MS^n systems, which are very powerful tools for identifying components in mixtures.

Other GC Detectors

Other important GC detectors include the thermionic detector, the electrolytic conductivity or Hall detector, and the photoionization detector. The thermionic detector is similar in construction to the FID. With the thermionic detector, nitrogen- and phosphorous-containing compounds produce increased currents in a flame in which an alkali metal salt is vaporized. The thermionic detector is widely used for organophosphorous pesticides and pharmaceutical compounds.

With the electrolytic conductivity detector, compounds containing halogens, sulfur, or nitrogen are mixed with a reaction gas in a small reactor tube. The products are then dissolved in a liquid that produces a conductive solution. The change in conductivity as a result of the presence of the active compound is then measured. In the photoionization detector, molecules are photoionized by UV radiation. The ions and electrons produced are then collected with a pair of biased electrodes, and the resulting current is measured. The detector is often used for aromatic and other molecules that are easily photoionized.

Gas chromatography is often coupled with the selective techniques of spectroscopy and electrochemistry. We have discussed GC/MS, but gas chromatography can be combined with several other techniques like infrared spectroscopy and NMR spectroscopy to provide the chemist with powerful tools for identifying the components of complex mixtures. These combined techniques are sometimes called **hyphenated methods**.[4]

In early hyphenated methods, the eluates from the chromatographic column were collected as separate fractions in a cold trap, with a nondestructive, nonselective detector used to indicate their appearance. The composition of each fraction was then investigated by nuclear magnetic resonance, infrared, or mass spectrometry or by electroanalytical measurements. A serious limitation to this approach was the very small (usually micromolar) quantities of solute contained in a fraction.

Most modern hyphenated methods monitor the effluent from the chromatographic column continuously by spectroscopic methods. The combination of two techniques based on different principles can achieve tremendous selectivity.

Hyphenated methods couple the separation capabilities of chromatography with the qualitative and quantitative detection capabilities of spectral methods.

32B GAS CHROMATOGRAPHIC COLUMNS AND STATIONARY PHASES

The pioneering gas-liquid chromatographic studies in the early 1950s were carried out on packed columns in which the stationary phase was a thin film of liquid retained by adsorption on the surface of a finely divided, inert solid support. From theoretical studies made during this early period, it became apparent that unpacked columns having inside diameters of a few tenths of a millimeter could provide separations that were superior to packed columns in both speed and column efficiency.[5] In such **capillary columns**, the stationary phase was a film of liquid a few tenths of a micrometer thick that uniformly coated the interior of a capillary tubing. In the late 1950s, such **open tubular columns** were constructed, and the predicted performance characteristics were experimentally confirmed in several laboratories, with open tubular columns having 300,000 plates or more being described.[6]

Despite such spectacular performance characteristics, capillary columns did not gain widespread use until more than two decades after their invention. The reasons for the delay were several, including small sample capacities, fragility of columns, mechanical problems associated with sample introduction and connection of the column to the detector, difficulties in coating the column reproducibly, short lifetimes of poorly prepared columns, tendencies of columns to clog, and patents, which limited commercial development to a single manufacturer (the original patent expired in 1977). The most significant development in capillary GC occurred in 1979 when fused-silica capillaries were introduced. Since then an impressive list of commercially available capillary columns for various applications has appeared. As a result, the majority of applications that have appeared in the past few years use capillary columns.[7]

---

[4]For reviews on hyphenated methods, see C. L. Wilkins, *Science*, **1983**, *222*, 291, **DOI**:10.1126/science.6353577; C. L. Wilkins, *Anal. Chem.*, **1989**, *59*, 571A, **DOI**: 10.1021/ac00135a001.

[5]For a thorough discussion of packed and capillary column technology, see E. F. Barry and R. L. Grob, *Columns for Gas Chromatography*, Hoboken, NJ: Wiley-Interscience, 2007.

[6]In 1987, a world record for length of an open tubular column and number of theoretical plates was set by Chrompack International Corporation of the Netherlands, as attested in the *Guinness Book of Records*. The column was a fused-silica column drawn in one piece and having an internal diameter of 0.32 mm and a length of 2100 m or 1.3 miles. The column was coated with a 0.1-m film of polydimethyl siloxane. A 1300-m section of this column contained over 2 million plates.

[7]For more information on capillary columns, see E. F. Barry, in *Modern Practice of Gas Chromatography*, R. L. Grob and E. F. Barry, eds., 4th ed., Ch. 3, New York: Wiley-Interscience, 2004.

32B-1 Capillary Columns

Capillary columns are also called open tubular columns because of the open flow path through them. They are of two basic types: **wall-coated open tubular** (WCOT) and **support-coated open tubular** (SCOT).[8] Wall-coated columns are capillary tubes coated with a thin layer of the liquid stationary phase. In support-coated open tubular columns, the inner surface of the capillary is lined with a thin film ($\sim$30 μm) of a solid support material, such as diatomaceous earth, on which the liquid stationary phase is adsorbed. This type of column holds several times as much stationary phase as does a wall-coated column and thus has a greater sample capacity. Generally, the efficiency of a SCOT column is less than that of a WCOT column but significantly greater than that of a packed column.

Early WCOT columns were constructed of stainless steel, aluminum, copper, or plastic. Subsequently, glass was used. Often, an alkali or borosilicate glass was leached with gaseous hydrochloric acid, strong aqueous hydrochloric acid, or potassium hydrogen fluoride to give an inert surface. Subsequent etching roughened the surface, which bonded the stationary phase more tightly.

Fused-silica open tubular (FSOT) columns are currently the most widely used GC columns.

Fused-silica capillaries are drawn from specially purified silica that contain minimal amounts of metal oxides. These capillaries have much thinner walls than their glass counterparts. They are given added strength by an outside protective polyimide coating, which is applied as the capillary tubing is being drawn. The resulting columns are quite flexible and can be bent into coils with diameters of a few inches. Figure 32-7 shows a picture of fused-silica capillary columns. Commercial fused-silica columns offer several important advantages over glass columns, such as physical strength, much lower reactivity toward sample components, and flexibility. For most applications, they have replaced the older type WCOT glass columns.

Fused-silica columns with inside diameters of 0.32 and 0.25 mm are very popular. Higher-resolution columns are also sold with diameters of 0.20 and 0.15 mm. Such columns are more troublesome to use and are more demanding on the injection and detection systems. Thus, a sample splitter must be used to reduce the size of the sample injected onto the column, and a more sensitive detector system with a rapid response time is required.

Capillary columns with 530 μm inside diameters, sometimes called **megabore columns**, are also available commercially. These columns will tolerate sample sizes that are similar to those for packed columns. The performance characteristics of megabore capillary columns are not as good as those of smaller diameter columns but significantly better than those of packed columns.

Table 32-2 compares the performance characteristics of fused-silica capillary columns with other types of wall-coated columns as well as with support-coated and packed columns.

32B-2 Packed Columns

Modern packed columns are fabricated from glass or metal tubing. They are typically 2 to 3 m long and have inside diameters of 2 to 4 mm. These tubes are densely packed with a uniform, finely divided packing material, or solid support, that is coated with a thin layer (0.05 to 1 μm) of the stationary liquid phase. The tubes are usually formed as coils with diameters of roughly 15 cm so that they can be conveniently placed in a temperature-controlled oven.

---

[8]For a detailed description of open tubular columns, see M. L. Lee, F. J. Yang, and K. D. Bartle, *Open Tubular Column Gas Chromatography: Theory and Practice*, New York: Wiley, 1984.

TABLE 32-2

Properties and Characteristics of Typical GC Columns

	Type of Column			
	FSOT*	WCOT†	SCOT‡	Packed
Length, m	10–100	10–100	10–100	1–6
Inside Diameter, mm	0.1–0.3	0.25–0.75	0.5	2–4
Efficiency, plates/m	2000–4000	1000–4000	600–1200	500–1000
Sample Size, ng	10–75	10–1000	10–1000	10–10^6
Relative Pressure	Low	Low	Low	High
Relative Speed	Fast	Fast	Fast	Slow
Flexible?	Yes	No	No	No
Chemical Inertness	Best	→		Poorest

*Fused-silica open tubular column.
†Wall-coated open tubular column
‡Support-coated open tubular column (also called porous layer open tubular, or PLOT)

Solid Support Materials

The packing, or solid support, in a packed column serves to hold the liquid stationary phase in place so that as large a surface area as possible is exposed to the mobile phase. The ideal support consists of small, uniform, spherical particles with good mechanical strength and a specific surface area of at least 1 m²/g. In addition, the material should be inert at elevated temperatures and be uniformly wetted by the liquid phase. No substance that meets all of these criteria perfectly is yet available.

The earliest, and still the most widely used, packings for gas chromatography were prepared from naturally occurring diatomaceous earth, which consists of the skeletons of thousands of species of single-celled plants that inhabited ancient lakes and seas (see **Figure 32-12**, an enlarged photo of a diatom obtained with a scanning electron microscope). These support materials are often treated chemically with dimethylchlorosilane, which gives a surface layer of methyl groups. This treatment reduces the tendency of the packing to adsorb polar molecules.

Figure 32-12 A photomicrograph of a diatom. Magnification 5000×.

Particle Size of Supports

As shown in Figure 31-16 (page 877), the efficiency of a gas chromatographic column increases rapidly with decreasing particle diameter of the packing. The pressure difference required to maintain an acceptable flow rate of carrier gas, however, varies inversely as the square of the particle diameter. The latter relationship has placed lower limits on the size of particles used in gas chromatography because it is not convenient to use pressure differences that are greater than about 50 psi. As a result, the usual support particles are 60 to 80 mesh (250 to 170 μm) or 80 to 100 mesh (170 to 149 μm).

32B-3 Liquid Stationary Phases

Desirable properties for the immobilized liquid phase in a gas-liquid chromatographic column include (1) *low volatility* (ideally, the boiling point of the liquid should be at least 100°C higher than the maximum operating temperature for the column), (2) *thermal stability*, (3) *chemical inertness*, and (4) *solvent characteristics* such that k and α (see Section 31E-4) values for the solutes to be resolved fall within a suitable range.

Many liquids have been proposed as stationary phases in the development of gas-liquid chromatography. Currently, fewer than a dozen are commonly used.

The proper choice of stationary phase is often crucial to the success of a separation. Qualitative guidelines for stationary phase selection can be based on a literature review, an Internet search, or recommendations from vendors of chromatographic equipment and supplies.

The retention time for an analyte on a column depends on its distribution constant, which in turn is related to the chemical nature of the liquid stationary phase. To separate various sample components, their distribution constants must be sufficiently different to accomplish a clean separation. At the same time, these constants must not be extremely large or extremely small because large distribution constants lead to prohibitively long retention times and small constants produce such short retention times that separations are incomplete.

To have a reasonable residence time on the column, an analyte must show some degree of compatibility (solubility) with the stationary phase. The principle of "like dissolves like" applies, where "like" refers to the polarities of the analyte and the immobilized liquid. The polarity of a molecule, as indicated by its dipole moment, is a measure of the electric field produced by separation of charge within the molecule. Polar stationary phases contain functional groups such as —CN, —CO, and —OH. Hydrocarbon-type stationary phases and dialkyl siloxanes are nonpolar, whereas polyester phases are highly polar. Polar analytes include alcohols, acids, and amines; solutes of medium polarity include ethers, ketones, and aldehydes. Saturated hydrocarbons are nonpolar. Generally, the polarity of the stationary phase should match that of the sample components. When the match is good, the order of elution is determined by the boiling point of the eluents.

> The polarities of common organic functional groups in increasing order are as follows: aliphatic hydrocarbons < olefins < aromatic hydrocarbons < halides < sulfides < ethers < nitro compounds < esters, aldehydes, ketones < alcohols, amines < sulfones < sulfoxides < amides < carboxylic acids < water.

Some Widely Used Stationary Phases

Table 32-3 lists the most widely used stationary phases for both packed and open tubular column gas chromatography in order of increasing polarity. These six liquids can probably provide satisfactory separations for 90% or more of samples.

Five of the liquids listed in Table 32-3 are polydimethyl siloxanes that have the general structure

$$R-\underset{\underset{R}{|}}{\overset{\overset{R}{|}}{Si}}-O-\left[\underset{\underset{R}{|}}{\overset{\overset{R}{|}}{Si}}-O\right]_n\underset{\underset{R}{|}}{\overset{\overset{R}{|}}{Si}}-R$$

TABLE 32-3

Some Common Liquid Stationary Phases for Gas-Liquid Chromatography

Stationary Phase	Common Trade Name	Maximum Temperature, °C	Common Applications
Polydimethyl siloxane	OV-1, SE-30	350	General-purpose nonpolar phase, hydrocarbons, polynuclear aromatics, steroids, PCBs
5% Phenyl-polydimethyl siloxane	OV-3, SE-52	350	Fatty acid methyl esters, alkaloids, drugs, halogenated compounds
50% Phenyl-polydimethyl siloxane	OV-17	250	Drugs, steroids, pesticides, glycols
50% Trifluoropropyl-polydimethyl siloxane	OV-210	200	Chlorinated aromatics, nitroaromatics, alkyl-substituted benzenes
Polyethylene glycol	Carbowax 20M	250	Free acids, alcohols, ethers, essential oils, glycols
50% Cyanopropyl- polydimethyl siloxane	OV-275	240	Polyunsaturated fatty acids, rosin acids, free acids, alcohols

In the first of these, polydimethyl siloxane, the —R groups are all —CH₃ giving a liquid that is relatively nonpolar. In the other polysiloxanes shown in the table, a fraction of the methyl groups are replaced by functional groups such as phenyl (—C_6H_5), cyanopropyl (—C_3H_6CN), and trifluoropropyl (—$C_3H_6CF_3$). The percentages listed before some of the stationary phases in Table 32-3 give the amount of substitution of the named group for methyl groups on the polysiloxane backbone. Thus, for example, 5% phenyl polydimethyl siloxane has a phenyl ring bonded to 5% (by number) of the silicon atoms in the polymer. These substitutions increase the polarity of the liquids to various degrees.

The fifth entry in Table 32-3 is a polyethylene glycol with the structure

$$-HO-CH_2-CH_2-(O-CH_2-CH_2)_n-OH$$

It finds widespread use for separating polar species.

Bonded and Cross-Linked Stationary Phases

Commercial columns are advertised as having bonded and/or cross-linked stationary phases. The purpose of bonding and cross-linking is to provide a longer lasting stationary phase that can be rinsed with a solvent when the film becomes contaminated. With use, untreated columns slowly lose their stationary phase due to "bleeding" in which a small amount of immobilized liquid is carried out of the column during the elution process. Bleeding is exacerbated when a column must be rinsed with a solvent to remove contaminants. Chemical bonding and cross-linking inhibit bleeding.

Bonding consists of attaching a monomolecular layer of the stationary phase to the silica surface of the column by a chemical reaction. For commercial columns, the nature of the reaction is usually proprietary.

Cross-linking is accomplished in situ after a column is coated with one of the polymers listed in Table 32-3. One way of cross-linking is to incorporate a peroxide into the original liquid. When the film is heated, reaction between the methyl groups in the polymer chains is initiated by a free radical mechanism. The polymer molecules are then cross-linked through carbon-to-carbon bonds. The resulting films are less extractable and have considerably greater thermal stability than do untreated films. Cross-linking has also been initiated by exposing the coated columns to gamma radiation.

Film Thickness

Commercial columns are available having stationary phases that vary in thickness from 0.1 to 5 μm. Film thickness primarily affects the retentive character and the capacity of a column as discussed in Section 31E-6. Thick films are used with highly volatile analytes because such films retain solutes for a longer time, thus providing a greater time for separation to take place. Thin films are useful for separating species of low volatility in a reasonable length of time. For most applications with 0.25 or 0.32 mm columns, a film thickness of 0.25 μm is recommended. With megabore columns, 1 to 1.5 μm films are often used. Today, columns with 8-μm films are marketed.

32C APPLICATIONS OF GAS-LIQUID CHROMATOGRAPHY

Gas-liquid chromatography is applicable to species that are appreciably volatile and thermally stable at temperatures up to a few hundred degrees Celsius. A large number of important compounds have these qualities. As a result, gas chromatography has been widely applied to the separation and determination of the components in a variety of sample types. **Figure 32-13** shows chromatograms for a few such applications.

Figure 32-13 Typical chromatograms from open tubular columns coated with (a) polydimethyl siloxane, (b) 5% (phenylmethyldimethyl) siloxane, (c) 50% (phenylmethyldimethyl) siloxane, (d) 50% poly(trifluoropropyldimethyl) siloxane, (e) polyethylene glycol, (f) 50% poly(cyanopropyldimethyl) siloxane. (Courtesy of J & W Scientific.)

32C-1 Qualitative Analysis

Gas chromatography is widely used to establish the purity of organic compounds. Contaminants, if present, are revealed by the appearance of additional peaks in the chromatogram. The areas under these extraneous peaks provide rough estimates of the extent of contamination. The technique is also useful for evaluating the effectiveness of purification procedures.

In theory, GC retention times should be useful for identifying components in mixtures. In fact, however, the applicability of such data is limited by the number of variables that must be controlled in order to obtain reproducible results. Nevertheless, gas chromatography provides an excellent means of confirming the presence or absence of a suspected compound in a mixture, provided that an authentic sample of the substance is available. If we add a small amount of the suspected compound to the mixture, no new peaks in the chromatogram of the mixture should appear, and enhancement of an existing peak should be observed. The evidence is particularly convincing if the effect can be duplicated on different columns and at different temperatures. On the other hand, because a chromatogram provides but a single piece of information about each species in a mixture (the retention time), the application of the technique to the qualitative analysis of complex samples of unknown composition is limited. This limitation has been largely overcome by linking chromatographic columns directly with ultraviolet, infrared, and mass spectrometers to produce hyphenated instruments (see Section 32A-4). An example of the use of mass spectroscopy combined with gas chromatography for the identification of constituents in blood is given in Feature 32-1.

Although a chromatogram may not lead to positive identification of the species in a sample, it often provides sure evidence of the *absence* of species. Thus, failure of a sample to produce a peak at the same retention time as a standard obtained under identical conditions is strong evidence that the compound in question is absent (or present at a concentration below the detection limit of the procedure).

Use of GC/MS to Identify a Drug Metabolite in Blood[9]

A comatose patient was suspected of taking an overdose of a prescription drug glutethimide (Doriden™) because an empty prescription bottle had been found near where the patient was discovered. A gas chromatogram was obtained of a blood plasma extract, and two peaks were found as shown in **Figure 32F-1**. The retention time for peak 1 corresponded to the retention time of glutethimide, but the compound responsible for peak 2 was not known. The possibility that the patient had taken another drug was considered. However, the retention time for peak 2 under the conditions used did not correspond to any other drug available to the patient or to a known drug of abuse. Hence, gas chromatography/mass spectrometry was called on to establish the identity of peak 2 and to confirm the identity of peak 1 before treating the patient.

The plasma extract was subjected to GC/MS analysis, and the mass spectrum depicted in **Figure 32F-2a** confirmed that peak 1 was due to glutethimide. A peak in the mass spectrum at a mass-to-charge ratio of 217 is the correct ratio for the glutethimide molecular ion, and the mass spectrum was identical to that from a known sample of glutethimide. The mass spectrum of peak 2, however, showed a molecular-ion peak at a mass-to-charge ratio of 233, as shown in **Figure 32F-2b**. This number differs from the molecular ion of glutethimide by 16 mass units. Several other peaks in the mass spectrum from GC peak 2 differed from those of glutethimide by 16 mass units indicating incorporation of oxygen into the glutethimide molecule. This finding led the investigators to believe that peak 2 was due to a 4-hydroxy metabolite of the parent drug.

Structure and molecular model of glutethimide.

Time

Figure 32F-1 Gas chromatogram of a blood plasma extract from a drug overdose victim. Peak 1 was at the appropriate retention time to be glutethimide, but the compound responsible for peak 2 was unknown until GC/MS was done.

(continued)

[9]From J. T. Watson and O. D. Sparkman, *Introduction to Mass Spectrometry*, 4th ed., New York: Wiley, 2007, pp. 29–32.

Figure 32F-2 (a) Mass spectrum obtained during elution of peak 1 of the gas chromatogram in Figure 32F-1. This mass spectrum is identical with that of glutethimide. (b) Mass spectrum obtained during elution of peak 2 of the gas chromatogram in Figure 32F-1. In both cases, electron-impact ionization was used in the mass spectrometer. Different ions, produced by fragmentation of the two compounds, aids in their identification. Peak A at $m/z = 217$ in spectrum (a) corresponds to the molar mass of glutethimide and is thus due to the molecular ion. The mass spectrum conclusively identifies peak 1 in the chromatogram as glutethimide. Peak B in mass spectrum (b) appears at $m/z = 233$, exactly 16 mass units more than glutethimide. Other peaks in spectrum (b) also appear 16 mass units higher than in the glutethimide spectrum. This evidence suggests the presence of an extra oxygen atom in the molecule, corresponding to the 4-hydroxy metabolite shown below.

Structure and molecular model of 4-hydroxy metabolite.

FOR THE CHEMIST ON THE GO: LAPTOP GAS CHROMATOGRAPH / MASS SPECTROMETER

ScienceCartoonsPlus.com

An acetic anhydride derivative of the peak 2 material was then prepared and found to be identical to the acetate derivative of 4-hydroxy-2-ethyl-2-phenylglutarimide, the metabolite shown in the molecular model on the previous page. This metabolite was known to exhibit toxicity in animals. The patient was then subjected to hemodialysis which removed the polar metabolite more rapidly than the less polar parent drug. Soon thereafter, the patient regained consciousness.

32C-2 Quantitative Analysis

Gas chromatography owes its enormous growth in part to its speed, simplicity, relatively low cost, and wide applicability to separations. It is doubtful, however, that GC would have become so widely used were it not able to provide quantitative information about separated species as well.

Quantitative GC is based on comparison of either the height or the area of an analyte peak with that of one or more standards. If conditions are properly controlled, both of these parameters vary linearly with concentration. Peak area is independent of the broadening effects discussed earlier. From this standpoint, therefore, area is a more satisfactory analytical variable than peak height. Peak heights are more easily measured than areas, however, and for narrow peaks, they may be determined more accurately. Most modern chromatographic instruments are equipped with computers that provide measurements of relative peak areas. If such equipment is not available, a manual estimate must be made. A simple method that works well for symmetric peaks of reasonable widths is to multiply peak height by the width at one-half peak height.

Calibration with Standards

In the most straightforward method for quantitative gas-chromatographic analyses, a series of standard solutions that approximate the composition of the unknown is prepared (see Section 8D-2 for general information on the external standard method). Chromatograms for the standards are then obtained, and peak heights or areas are plotted as a function of concentration to obtain a working curve. A plot of the data should yield a straight line passing through the origin; quantitative analyses are based on this plot. Frequent standardization is necessary for the highest accuracy.

The Internal Standard Method

The highest precision for quantitative GC is obtained using internal standards because the uncertainties introduced by sample injection, flow rate, and variations in column conditions are minimized. In this procedure, a carefully measured quantity of an internal standard is introduced into each standard and sample (see Section 8D-3), and the ratio of analyte peak area (or height) to internal standard peak area (or height) is used as the analytical parameter (see Example 32-1). For this method to be successful, it is necessary that the internal standard peak be well separated from the peaks of all other components in the sample. However, it must appear close to the analyte peak. Of course, the internal standard should be absent in the sample to be analyzed. With a suitable internal standard, precisions of 0.5% to 1% relative are reported.

EXAMPLE 32-1

Gas chromatographic peaks can be influenced by a variety of instrumental factors. We can often compensate for variations in these factors by using the internal standard method. With this method, we add the same amount of internal standard to mixtures containing known amounts of the analyte and to the samples of unknown analyte concentration. We then calculate the ratio of peak height (or area) for the analyte to that of the internal standard.

(continued)

The data shown in the table were obtained for the determination of a C_7 hydrocarbon with a closely related compound added to each standard and to the unknown as an internal standard.

Percent Analyte	Peak Height Analyte	Peak Height, Internal Standard
0.05	18.8	50.0
0.10	48.1	64.1
0.15	63.4	55.1
0.20	63.2	42.7
0.25	93.6	53.8
Unknown	58.9	49.4

Construct a spreadsheet to determine the peak height ratio of the analyte to internal standard and plot this ratio versus the analyte concentration. Determine the concentration of the unknown and its standard deviation.

Solution

The spreadsheet is shown in **Figure 32-14**. The data are entered into columns A through C, as shown. In cells D4 through D9, the peak height ratio is calculated by the formula shown in documentation cell A22. A plot of the calibration curve is also shown in the figure. The linear regression statistics are calculated in cells B11 through B20 using the same approach as described in Section 8D-2. The statistics are calculated by the formulas in documentation cells A23 through A31. The percentage of the analyte in the unknown is found to be 0.163 ± 0.008.

	A	B	C	D	E	F	G
1	**Quantitative GC using an internal standard method**						
2		Peak height	Peak height,	Peak height ratio,			
3	Percent analyte	analyte	internal standard	analyte/internal std.			
4	0.050	18.8	50.0	0.38			
5	0.100	48.1	64.1	0.75			
6	0.150	63.4	55.1	1.15			
7	0.200	63.2	42.7	1.48			
8	0.250	93.6	53.8	1.74			
9	Unknown	58.9	49.4	1.19			
10	**Regression equation**						
11	Slope	6.914515					
12	Intercept	0.062202					
13	Concentration of unknown	0.163440					
14	**Error Analysis**						
15	Standard error in Y	0.049960					
16	N	5					
17	S_{xx}	0.025					
18	y bar (average ratio)	1.1					
19	M	1					
20	Standard deviation in c	0.007939					
21	**Spreadsheet Documentation**						
22	Cell D4=B4/C4						
23	Cell B11=SLOPE(D4:D8,A4:A8)						
24	Cell B12=INTERCEPT(D4:D8,A4:A8)						
25	Cell B13=(D9-B12)/B11						
26	Cell B15=STEYX(D4:D8,A4:A8)						
27	Cell B16=COUNT(A4:A8)						
28	Cell B17=B16*VARP(A4:A8)						
29	Cell B18=AVERAGE(D4:D8)						
30	Cell B19=enter no. of replicates						
31	Cell B20=B15/B11*SQRT(1/B19+1/B16+((D9-B18)^2)/((B11^2)*B17))						

Figure 32-14   Spreadsheet to illustrate the internal standard method for the GC determination of a C_7 hydrocarbon.

32C-3 Advances in GC

Although GC is a very mature technique, there have been many developments in recent years in theory, instrumentation, columns, and practical applications. Some developments in high-speed GC, miniaturization, and multidimensional GC are briefly described here.

High-Speed Gas Chromatography[10]

Researchers in GC have often focused on achieving ever-higher resolution in order to separate more and more complex mixtures. In most separations, conditions are varied in order to separate the most difficult-to-separate pair of components, the so-called *critical pair*. Many of the components of interest under these conditions are highly over-separated. The basic idea of high-speed GC is that, for many separations of interest, higher speed can be achieved, albeit at the expense of some selectivity and resolution.

The principles of high-speed separations can be demonstrated by substituting Equation 31-11 into Equation 31-17

$$\frac{L}{t_R} = u \times \frac{1}{1 + k_n} \tag{32-1}$$

where k_n is the retention factor for the last component of interest in the chromatogram. If we rearrange Equation 32-1 and solve for the retention time of the last component of interest, we obtain

$$t_R = \frac{L}{u} \times (1 + k_n) \tag{32-2}$$

Equation 32-2 tells us that we can achieve faster separations by using short columns, higher-than-usual carrier gas velocities, and small retention factors. The price to be paid is reduced resolving power caused by increased band broadening and reduced peak capacity (the number of peaks that will fit in the chromatogram).

Research workers in the field have been designing instrumentation and chromatographic conditions to optimize separation speed at the lowest cost in terms of resolution and peak capacity.[11] They have designed systems to achieve tunable columns and high-speed temperature programming. A tunable column is a series combination of a polar and a nonpolar column. **Figure 32-15** shows the separation of 12 compounds prior to initiating a programmed temperature ramp and 19 compounds after the temperature program was begun. The total time required was 140 s. These workers have also been using high-speed GC with mass spectrometry detection including time-of-flight detection.[12]

Miniaturized GC Systems

For many years, there has been a desire to miniaturize GC systems to the microchip level. Miniature GC systems are useful in space exploration, in portable instruments for field use, and in environmental monitoring.

Most of the research in this area has concentrated on miniaturizing individual components of the chromatographic systems such as columns and detectors. Microfabricated columns were designed using substrates of silicon, metals, and polymers.[13]

[10]For more information, see R. D. Sacks, in *Modern Practice of Gas Chromatography*, R. L. Grob and E. F. Barry, eds., 4th ed., Ch. 5, New York: Wiley-Interscience, 2004.

[11]H. Smith and R. D. Sacks, *Anal. Chem.*, **1998**, *70*, 4960, **DOI:** 10.1021/ac980463b.

[12]C. Leonard and R. Sacks, *Anal. Chem.*, **1999**, *71*, 5177, **DOI:** 10.1021/ac990631f.

[13]G. Lambertus et al., *Anal. Chem.*, **2004**, *76*, 2629, **DOI:** 10.1021/ac030367x.

Figure 32-15 High-speed chromatogram obtained with isothermal operation (30°C) for 37 s followed by a 35°C/min temperature ramp to 90°C. (Reprinted (adapted) with permission from H. Smith and R. D. Sacks, *Anal. Chem.*, **1998**, *70*, 4960. Copyright 1998 by the American Chemical Society.)

Relatively deep, narrow channels are etched into the substrate. These channels have low dead volume to reduce band broadening and high surface area to increase stationary phase volume. Recent reports have described complete microfabricated ensembles with interconnected injectors, columns, and detectors.[14] One instrument was specifically designed for measurement of trichloroethylene vapors due to the migration of volatile organic compounds from contaminated soil or groundwater. The miniature GC could be deployed in the field and was capable of sub ppb detection of the vapors.

Multidimensional Gas Chromatography

In multidimensional GC, two or more capillary columns of differing selectivities are connected in series. Therefore, with two columns, one might contain a nonpolar stationary phase, while the second might have a polar stationary phase. Subjecting a sample to separation in one dimension followed by separations in one or more additional dimensions can give rise to extremely high selectivity and resolution.

Multidimensional GC can take several forms. In one implementation, called **heart cutting**, a portion of the eluent from the first column containing the species of interest is switched to a second column for further separation.[15] This approach has been successfully implemented in commercial instrumentation.

In another methodology, known as comprehensive two-dimensional GC or GC × GC, the effluent from the first column is continuously switched to a second short column.[16] Although the resolving power of the second column is necessarily limited, the fact that a column precedes it produces high-resolution separations. This approach has also been developed into commercial instrumentation.

The multidimensional GC techniques have also been combined with mass spectrometry, resulting in separations that are not only of high resolution but that are also able to identify minor components, distinguish closely related compounds, and unravel coeluting species.[17]

[14]S. Zampolli et al., *Sens. Actuators, B*, **2009**, 141, 322, **DOI**:10.1016/j.snb.2009.06.021; S. K. Kim, H. Chang, and E. T. Zellers, *Anal. Chem.*, **2011**, *83*, 7198, **DOI**: 10.1021/ac201788q.

[15]P. Q. Tranchida, D. Sciaronne, P. Dugo, and L. Mondello, *Anal. Chim. Acta*, **2012**, *716*, 66, **DOI**: 10.1016/j.aca.2011.12.015.

[16]M. Adahchour, J. Beens, and U. A. Th. Brinkman, *J. Chromatogr. A*, **2008**, *1186*, 67, **DOI**: 10.1016/j.chroma.2008.01.002.

[17]T. Veriotti and R. Sacks, *Anal. Chem.*, **2003**, *75*, 4211, **DOI**: 10.1021/ac020522s.

32D GAS-SOLID CHROMATOGRAPHY

Gas-solid chromatography is based on adsorption of gaseous substances on solid surfaces. Distribution coefficients are generally much larger than those for gas-liquid chromatography. This property renders gas-solid chromatography useful for separating species that are not retained by gas-liquid columns, such as the components of air, hydrogen sulfide, carbon disulfide, nitrogen oxides, carbon monoxide, carbon dioxide, and the rare gases.

Gas-solid chromatography is performed with both packed and open tubular columns. For the latter, a thin layer of the adsorbent is affixed to the inner walls of the capillary. Such columns are sometimes called **porous layer open tubular columns**, or PLOT columns. **Figure 32-16** shows a typical application of a PLOT column.

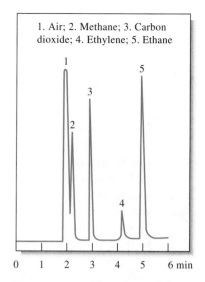

Figure 32-16 Typical gas-solid chromatogram on a PLOT column.

WEB WORKS Do an Internet search and find several makers of gas chromatographic instruments. Find an instrument company that makes both a premium GC instrument and a routine GC instrument. Investigate the features of both types of GC systems. Compare and contrast these features. Pay close attention in your comparison to the size of the oven, the uncertainty in oven temperature, the ability to use temperature programming, the type of detectors available, and the type and sophistication of the data analysis system. Find an instrument company that makes a multidimensional GC and discuss whether it is a capable of comprehensive two-dimensional GC, heart-cutting multidimensional GC, or both. Can the system be conveniently interfaced to a mass spectrometer?

QUESTIONS AND PROBLEMS

*32-1. How do gas-liquid and gas-solid chromatography differ?

32-2. Why is gas-solid chromatography not used as extensively as gas-liquid chromatography?

*32-3. What kind of mixtures are separated by gas-solid chromatography?

32-4. What types of flow meters are used in GC?

*32-5. Describe a chromatogram and explain what type of information it contains.

32-6. What is meant by temperature programming in gas chromatography?

*32-7. Describe the physical differences between capillary and packed columns. What are the advantages and disadvantages of each?

32-8. What variables must be controlled if satisfactory qualitative data are to be obtained from chromatograms?

*32-9. What variables must be controlled if satisfactory quantitative data are to be obtained from chromatograms?

32-10. Describe the principle on which each of the following GC detectors are based: (a) thermal conductivity, (b) flame ionization, (c) electron capture, (d) thermionic, and (e) photoionization.

*32-11. What are the principal advantages and the principal limitations of each of the detectors listed in Problem 32-10?

32-12. What are *hyphenated* gas-chromatographic methods? Briefly describe three hyphenated methods.

*32-13. What are megabore open tubular columns? Why are they used?

32-14. How do the following capillary columns differ?
(a) PLOT columns (b) WCOT columns
(c) SCOT columns

*32-15. Why are gas chromatographic stationary phases often bonded and cross-linked? What do these terms mean?

32-16. What properties should the stationary phase liquid for gas chromatography possess?

*32-17. What are the advantages of fused-silica capillary columns compared with glass or metal columns?

32-18. What is the effect of stationary phase film thickness on gas chromatograms?

*32-19. List the variables that lead to (a) band broadening and (b) band separation in gas-liquid chromatography.

32-20. One method for quantitative determination of the concentration of constituents in a sample analyzed by gas chromatography is the area normalization method. In this method, complete elution of all of the sample constituents is necessary. The area of each peak is then measured and corrected for differences in detector response to the different eluates. This correction is accomplished by dividing the area by an empirically determined correction factor. The concentration of the analyte is found from the ratio of its corrected area to the total corrected area of all peaks. For a chromatogram containing three peaks, the relative areas were found to be 16.4, 45.2, and 30.2 in the order of increasing retention time. Calculate the percentage of each compound if the relative detector responses were 0.60, 0.78, and 0.88, respectively.

*32-21. Peak areas and relative detector responses are to be used to determine the concentration of the five species in a sample. The area normalization method described in Problem 32-20 is to be used. The relative areas for the five gas chromatographic peaks are given in the accompanying table. Also shown are the relative responses of the detector. Calculate the percentage of each component in the mixture.

Compound	Relative Peak Area	Relative Detector Response
A	32.5	0.70
B	20.7	0.72
C	60.1	0.75
D	30.2	0.73
E	18.3	0.78

32-22. For the data given in Example 32-1, compare the method of external standards to the internal standard method. Plot the analyte peak height versus percent analyte and determine the unknown without using the internal standard results. Are your results any more precise using the internal standard method? If so, give some possible reasons.

32-23. **Challenge Problem:** Cinnamaldehyde is the component responsible for cinnamon flavor. It is also a potent antimicrobial compound present in essential oils

(see M. Friedman, N. Kozukue, and L. A. Harden, *J. Agric. Food Chem.*, **2000**, *48*, 5702, **DOI:** 10.1021/jf000585g). The GC response of an artificial mixture containing six essential oil components and methyl benzoate as an internal standard is shown in the figure.

Chromatogram. (From M. Friedman, N. Kozukue, and L.A. Harden, *J. Agric. Food Chem.*, **2000**, *48*, 5702. Copyright 2000 American Chemical Society.)

(a) The figure that follows is an idealized enlargement of the region near the cinnamaldehyde peak.

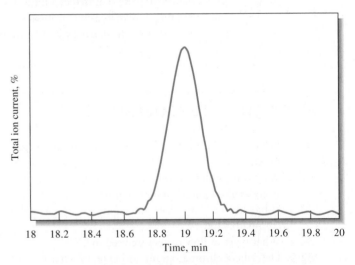

Enlarged chromatogram.

Determine the retention time for cinnamaldehyde.

(b) From the figure in part (a), determine the number of theoretical plates for the column.

(c) The fused-silica column was 0.25 mm × 30 cm with a 0.25-μm film. Determine the height equivalent to a theoretical plate from the data in parts (a) and (b).

(d) Quantitative data were obtained by using methyl benzoate as the internal standard. The following results were obtained for calibration curves of cinnamaldehyde, eugenol, and thymol. The values under each component represent the peak area of the component divided by the peak area of the internal standard.

Concentration, mg sample/ 200 μL	Cinnamaldehyde	Eugenol	Thymol
0.50		0.4	
0.65			1.8
0.75	1.0	0.8	
1.10		1.2	
1.25	2.0		
1.30			3.0
1.50		1.5	
1.90	3.1	2.0	4.6
2.50	4.0		5.8

Determine the calibration curve equations for each component. Include the R^2 values.

(e) From the data in part (d), determine which of the components has the highest calibration curve sensitivity? Which has the lowest?

(f) A sample containing the three essential oils in part (d) gave the peak areas relative to the internal standard area: cinnamaldehyde, 2.6; eugenol, 0.9; and thymol, 3.8. Determine the concentrations of each of the oils in the sample and the standard deviations in concentration.

(g) A study was made of the decomposition of cinnamaldehyde in cinnamon oil. The oil was heated for various times at different temperature. The data given below were obtained:

Temp, °C	Time, min	% Cinnamaldehyde
25, initial		90.9
40	20	87.7
	40	88.2
	60	87.9
60	20	72.2
	40	63.1
	60	69.1
100	20	66.1
	40	57.6
	60	63.1
140	20	64.4
	40	53.7
	60	57.1
180	20	62.3
	40	63.1
	60	52.2
200	20	63.1
	40	645
	60	63.3
210	20	74.9
	40	73.4
	60	77.4

Use ANOVA to determine whether there is an effect of temperature on the decomposition of cinnamaldehyde. In the same way, determine if there is an effect of time of heating.

(h) With the data in part (g), assume that decomposition begins at 60°C. Test the hypothesis that there is no effect of temperature or time.

High-Performance Liquid Chromatography

CHAPTER 33

High-performance liquid chromatography has become an indispensable analytical tool. The crime labs in television forensic and police dramas, such as *NCIS, NCIS: Los Angeles, CSI, CSI: New York, CSI: Miami,* and *Law and Order*, often use HPLC in the processing of evidence. The photo shows *NCIS* laboratory technician Abby Sciuto (Pauley Perrette) explaining the results of an HPLC analysis to NCIS Special Agent Leroy Jethro Gibbs (Mark Harmon).

This chapter considers the theory and practice of HPLC, including partition, adsorption, ion-exchange, size-exclusion, affinity, and chiral chromatography. HPLC has applications not only in forensics but also in biochemistry, environmental science, food science, pharmaceutical chemistry, and toxicology.

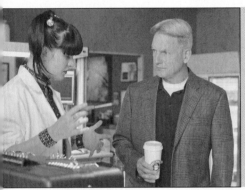

Sonja Flemming/CBS via Getty Images

High-performance liquid chromatography (HPLC) is the most versatile and widely used type of elution chromatography. The technique is used by scientists for separating and determining species in a variety of organic, inorganic, and biological materials. In liquid chromatography, the mobile phase is a liquid solvent containing the sample as a mixture of solutes. The types of high-performance liquid chromatography are often classified by the separation mechanism or by the type of stationary phase. These include (1) **partition**, or **liquid-liquid, chromatography**; (2) **adsorption**, or **liquid-solid, chromatography**; (3) **ion-exchange**, or **ion, chromatography**; (4) **size-exclusion chromatography**; (5) **affinity chromatography**; and (6) **chiral chromatography.**

Early liquid chromatography was performed in glass columns having inside diameters of perhaps 10 to 50 mm. The columns were packed with 50- to 500-cm lengths of solid particles coated with an adsorbed liquid that formed the stationary phase. To ensure reasonable flow rates through this type of stationary phase, the particle size of the solid was kept larger than 150 to 200 μm. Even with these particles, flow rates were a few tenths of a milliliter per minute at best. Attempts to speed up this classic procedure by application of vacuum or pressure were not effective because increases in flow rates were accompanied by increases in plate heights and accompanying decreases in column efficiency.

Early in the development of the theory of liquid chromatography, it was recognized that large decreases in plate heights would be realized if the particle size of packings were reduced. This effect is shown by the data in **Figure 33-1**. Note that the minimum shown in Figure 31-13a (page 873) is not reached in any of these plots. The reason for this difference is that diffusion in liquids is much slower than in gases, and therefore, its effect on plate heights is observed only at extremely low flow rates.

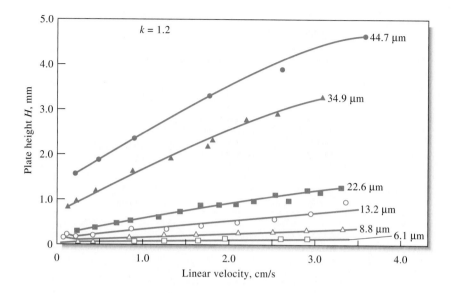

Figure 33-1   Effect of particles size of packing and flow rate on plate height in liquid chromatography. (From R. E. Majors, *J. Chromatrogr. Sci,* **1973**, Vol. 11, (2), 1973: 88–95, Fig 5. Reprinted by permission of Oxford University Press.)

Not until the late 1960s, was the technology developed for producing and using packings with particle diameters as small as 3 to 10 μm. This technology required instruments capable of much higher pumping pressures than the simple devices that preceded them. Simultaneously, detectors were developed for continuous monitoring of column effluents. The name **high-performance liquid chromatography** (HPLC) is often used to distinguish this technology from the simple column chromatographic procedures that preceded them.[1] Simple column chromatography, however, still finds considerable use for preparative purposes.

Applications of the most widely used types of HPLC for various analyte species are shown in **Figure 33-2**. Note that the various types of liquid chromatography tend to be complementary in their application. For example, for analytes having molecular masses greater than 10,000, one of the two size-exclusion methods is often used: gel permeation for nonpolar species and gel filtration for polar or ionic compounds. For ionic species, ion-exchange chromatography is often the method of choice. In most cases for nonionic small molecules, reversed-phase methods are suitable.

> **High-performance liquid chromatography, HPLC,** is a type of chromatography that combines a liquid mobile phase and a very finely divided stationary phase. In order to obtain satisfactory flow rates, the liquid must be pressurized to several hundred or more pounds per square inch.

33A INSTRUMENTATION

Pumping pressures of several hundred atmospheres are required to achieve reasonable flow rates with packings in the 3- to 10-μm size range, which are common in modern liquid chromatography. Because of these high pressures, the equipment for high-performance liquid chromatography tends to be considerably more elaborate and expensive than that encountered in other types of chromatography. **Figure 33-3** is a diagram showing the important components of a typical HPLC instrument.

---

[1]For a detailed discussion of HPLC systems, see L. R. Snyder, J. J. Kirkland, and J. W. Dolan, *Introduction to Modern Liquid Chromatography,* 4th ed., Hoboken, NJ: Wiley, 2010; V. Meyer, *Practical High-Performance Liquid Chromatography,* 5th ed., Chichester, UK: Wiley 2010.

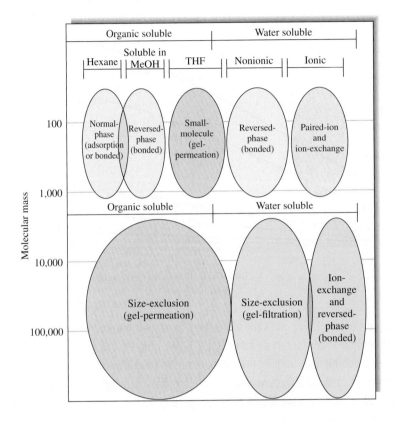

Figure 33-2 Applications of liquid chromatography. Methods can be chosen based on solubility and molecular mass. In many cases, for small molecules, reversed-phase methods are appropriate. Techniques toward the bottom of the diagram are best suited for high molecular mass ($M > 2000$). (*High Performance Liquid Chromatography*, 2nd ed., S. Lindsay and H. Barnes, eds. Copyright 1987, 1992, Thames Polytechnic, London, UK. New York: Wiley, 1992.)

Figure 33-3 Block diagram showing components of a typical apparatus for HPLC. (Courtesy of PerkinElmer, Inc., Waltham, MA.)

33A-1 Mobile-Phase Reservoirs and Solvent Treatment Systems

A modern HPLC instrument is equipped with one or more glass reservoirs, each of which contains 500 mL or more of a solvent. Provisions are often included to remove dissolved gases and dust from the liquids. Dissolved gases can lead to irreproducible flow rates and band spreading. In addition, both bubbles and dust interfere with the performance of most detectors. Degassers may consist of a vacuum pumping system, a distillation system, a device for heating and stirring, or, as shown in Figure 33-3, a system for **sparging** in which the dissolved gases are swept out of solution by fine bubbles of an inert gas that is not soluble in the mobile phase.

An elution with a single solvent or solvent mixture of constant composition is termed an **isocratic elution**. In **gradient elution**, two (and sometimes more) solvent systems that differ significantly in polarity are used and varied in composition during the separation. The ratio of the two solvents is varied in a preprogrammed way, sometimes continuously and sometimes in a series of steps. As shown in **Figure 33-4**, gradient elution frequently improves separation efficiency, just as temperature programming helps in gas chromatography. Modern HPLC instruments are

Sparging is a process in which dissolved gases are swept out of a solvent by bubbles of an inert, insoluble gas.

An **isocratic elution** in HPLC is one in which the solvent composition remains constant.

A **gradient elution** in HPLC is one in which the composition of the solvent is changed continuously or in a series of steps.

(a) Gradient elution

Peak identity

1. Benzene
2. Monochlorobenzene
3. Orthodichlorobenzene
4. 1,2,3-trichlorobenzene
5. 1,3,5-trichlorobenzene
6. 1,2,4-trichlorobenzene
7. 1,2,3,4-tetrachlorobenzene
8. 1,2,4,5-tetrachlorobenzene
9. Pentachlorobenzene
10. Hexachlorobenzene

(b) Isocratic elution

Retention time, min

Figure 33-4 Improvement in separation effectiveness by using gradient elution. (J. J. Kirkland, *Modern Practice of Liquid Chromatography*, p. 88. New York: Interscience, 1971. Reprinted by permission of the Chromatography Forum of the Delaware Valley.)

often equipped with proportioning valves that introduce liquids from two or more reservoirs at ratios that can be varied continuously (see Figure 33-3).

33A-2 Pumping Systems

The requirements for liquid chromatographic pumps include (1) the generation of pressures of up to 6000 psi (lb/in^2), (2) pulse-free output, (3) flow rates ranging from 0.1 to 10 mL/min, (4) flow reproducibilities of 0.5% relative or better, and (5) resistance to corrosion by a variety of solvents. The high pressures generated by liquid chromatographic pumps are not an explosion hazard because liquids are not very compressible. Thus, rupture of a component results only in solvent leakage. Such leakage may constitute a fire or environmental hazard with some solvents, however.

Two major types of pumps are used in HPLC instruments: the screw-driven syringe type and the reciprocating pump. Reciprocating types are used in almost all commercial instruments. Syringe-type pumps produce a pulse-free delivery whose flow rate is easily controlled. They suffer, however, from relatively low capacity (~250 mL) and are inconvenient when solvents must be changed. **Figure 33-5** illustrates the operating principles of the reciprocating pump. This device consists of a small cylindrical chamber that is filled and then emptied by the back-and-forth motion of a piston. The pumping motion produces a pulsed flow that must be subsequently damped because the pulses appear as baseline noise on the chromatogram. Modern HPLC instruments use dual pump heads or elliptical cams to minimize such pulsations. Advantages of reciprocating pumps include small internal volume (35 to 400 μL), high output pressure (up to 10,000 psi), ready adaptability to gradient elution, and constant flow rates, which are largely independent of column back-pressure and solvent viscosity.

As part of their pumping systems, many commercial instruments are equipped with computer-controlled devices for measuring the flow rate by determining the pressure drop across a restrictor located at the pump outlet. Any difference in signal from a preset value is then used to increase or decrease the speed of the pump motor. Most instruments also have a means for varying the composition of the solvent either continuously or in a stepwise fashion. For example, the instrument shown in Figure 33-3 contains a proportioning valve that permits mixing of up to four solvents in a preprogrammed and continuously variable way.

33A-3 Sample Injection Systems

The most widely used method of sample introduction in liquid chromatography is based on a sampling loop, such as that shown in **Figure 33-6**. These devices are often

Figure 33-5   A reciprocating pump for HPLC.

an integral part of liquid chromatography equipment and have interchangeable loops capable of providing a choice of sample sizes ranging from 1 to 100 μL or more. The reproducibility of injections with a typical sampling loop is a few tenths of a percent relative. Many HPLC instruments incorporate an autosampler with an automatic injector. These injectors can introduce continuously variable volumes from containers on the autosampler.

33A-4 Columns for HPLC

Liquid chromatographic columns are usually constructed from stainless steel tubing, although glass and polymer tubing, such as polyetheretherketone (PEEK), are sometimes used. In addition, stainless steel columns lined with glass or PEEK are also available. Hundreds of packed columns differing in size and packing can be purchased from HPLC suppliers. The cost of standard-sized, nonspecialty columns ranges from $200 to more than $500. Specialized columns, such as chiral columns, can cost more than $1000.

Analytical Columns

Most columns range in length from 5 to 25 cm and have inside diameters of 3 to 5 mm. Straight columns are invariably used. The most common particle size of packings is 3 or 5 μm. Commonly used columns are 10 or 15 cm long, 4.6 mm in inside diameter, and packed with 5-μm particles. Columns of this type provide 40,000 to 70,000 plates/m.

In the 1980s, microcolumns became available with inside diameters of 1 to 4.6 mm and lengths of 3 to 7.5 cm. These columns, which are packed with 3- or 5-μm particles, contain as many as 100,000 plates/m and have the advantage of speed and minimal solvent consumption. This latter property is of considerable importance because the high-purity solvents required for liquid chromatography are expensive to purchase and to dispose of after use. **Figure 33-7** illustrates the speed with which a separation can be performed on a microbore column. In this example, MS/MS was used to monitor the separation of rosuvastatin from human plasma components on a column that was 5 cm in length with an inside diameter of 1.0 mm. The column was packed with 3-μm particles. Less than 3 minutes were required for the separation.

Precolumns

Two types of precolumns are used. A precolumn between the mobile phase reservoir and the injector is used for mobile-phase conditioning and is termed a **scavenger column**. The solvent partially dissolves the silica packing and ensures that the mobile

Figure 33-6 A sampling loop for liquid chromatography. (Courtesy of Beckman Coulter, Fullerton, CA.)

A **scavenger column** between the mobile-phase container and the injector is used to condition the mobile phase.

Figure 33-7 High-speed gradient elution separation of rosuvastatin from human plasma-related components. Column: 5 cm × 1.0 mm i.d. Luna C18.3 μm. Monitored by MS/MS at m/z = 488.2 and 264.2. (Reprinted from K. A. Oudhoff, T. Sangster, E. Thomas, I. D. Wilson, *J. Chromatogr. B*, **2006**, *832*, 191. Copyright 2006, with permission from Elsevier.)

Molecular model of *p*-xylene. There are three xylene isomers, ortho, meta, and para. Para-xylene is used for the production of artificial fibers. Xylol is a mixture of the three isomers and is used as a solvent.

A **guard column** between the injector and the column removes particulates and other solvent impurities.

phase is saturated with silicic acid prior to entering the analytical column. This saturation minimizes losses of the stationary phase from the analytical column.

A second type of precolumn is a **guard column**, positioned between the injector and the analytical column. A guard column is a short column packed with a similar stationary phase as the analytical column. The purpose of the guard column is to prevent impurities, such as highly retained compounds and particulate matter, from reaching and contaminating the analytical column. The guard column is replaced regularly and serves to increase the lifetime of the analytical column.

Column Temperature Control

For some applications, close control of column temperature is not necessary, and columns are operated at room temperature. Often, however, better, more reproducible chromatograms are obtained by maintaining constant column temperature. Most modern commercial instruments are equipped with heaters that control column temperatures to a few tenths of a degree from near room temperature to 150°C. Columns can also be fitted with water jackets fed from a constant-temperature bath to give precise temperature control. Many chromatographers consider temperature control to be essential for reproducible separations.

Column Packings

Two types of packings are used in HPLC, *pellicular* and *porous particle*. The original pellicular particles were spherical, nonporous, glass or polymer beads with typical diameters of 30 to 40 μm. A thin, porous layer of silica, alumina, a polystyrene-divinyl benzene synthetic resin, or an ion-exchange resin was deposited on the surface of these beads. Small porous microparticles have completely replaced these large pellicular particles. In recent years, small (≈ 5 μm) pellicular packings have been reintroduced for separation of proteins and large biomolecules.

The typical porous particle packing for liquid chromatography consists of porous microparticles having diameters ranging from 3 to 10 μm; for a given size particle, a very narrow particle size distribution is desirable. The particles are composed of silica, alumina, the synthetic resin polystyrene-divinyl benzene, or an ion-exchange resin. Silica is by far the most common packing in liquid chromatography. Silica particles are often coated with thin organic films, which are chemically or physically bonded to the surface. Column packings for specific chromatographic modes are discussed in later sections of this chapter.

33A-5 HPLC Detectors

The ideal detector for HPLC should have all the characteristics of the ideal GC detector listed in Section 32A-4 except that it need not have as great a temperature range. In addition, an HPLC detector must have low internal volume (dead volume) to minimize extra-column band broadening. The detector should be small and compatible with liquid flow. Unfortunately, no highly sensitive, universal detector system is available for high-performance liquid chromatography. Thus, the detector used will depend on the nature of the sample. Table 33-1 lists some of the common detectors and their properties.[2]

The most widely used detectors for liquid chromatography are based on absorption of ultraviolet or visible radiation (see Figure 33-8). Both photometers and spectrophotometers, specifically designed for use with chromatographic columns, are available from commercial sources. Photometers often make use of the 254- and 280-nm lines from a mercury source because many organic functional groups absorb in the region. Deuterium sources or tungsten-filament sources with interference filters also provide a simple means of detecting absorbing species. Some modern instruments are equipped with filter wheels that contain several interference filters, which can be rapidly switched into place. Spectrophotometric detectors are considerably more versatile than photometers and are also widely used in high-performance instruments. Modern instruments use diode-array detectors that can display an entire spectrum as an analyte exits the column.

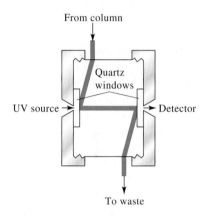

Figure 33-8 A UV-visible absorption detector for HPLC.

TABLE 33-1

Performances of HPLC Detectors*

HPLC Detector	Commercially Available	Mass LOD[†] (Typical)	Linear Range[‡] (Decades)
Absorbance	Yes	10 pg	3–4
Fluorescence	Yes	10 fg	5
Electrochemical	Yes	100 pg	4–5
Refractive index	Yes	1 ng	3
Conductivity	Yes	100 pg–1 ng	5
Mass spectrometry	Yes	< 1 pg	5
FTIR	Yes	1 μg	3
Light scattering	Yes	1 μg	5
Optical activity	No	1 ng	4
Element selective	No	1 ng	4–5
Photoionization	No	< 1 pg	4

*From manufacturer's literature; F. Settle, ed., *Handbook of Instrumental Techniques for Analytical Chemistry*, Upper Saddle River, NJ: Prentice-Hall, 1997; E. S. Yeung and R. E. Synovec, *Anal. Chem.*, **1986**, *58*, 1237A, **DOI**: 10.1021/ac00125a002.

[†]Mass LODs (limits of detection) are dependent on compound, instrument, and HPLC conditions, but those given are typical values with commercial systems when available.

[‡]Typical values from the sources.

---

[2]For a more extensive discussion of HPLC detectors, see D. A. Skoog, F. J. Holler, and S. R. Crouch, *Principles of Instrumental Analysis*, 6th ed., Belmont, CA: Brooks/Cole, 2007, pp. 823–28.

The combination of HPLC with a mass spectrometry detector produces a very powerful analytical tool as shown in Figure 33-7. Such LC/MS systems can identify the analytes exiting from the HPLC column, as discussed in Feature 33-1.[3]

Another detector, which has found considerable application, is based on the changes in the refractive index of the solvent that is caused by analyte molecules. In contrast to most of the other detectors listed in Table 33-1, the refractive index detector is general rather than selective and responds to the presence of all solutes.

LC/MS and LC/MS/MS

The combination of liquid chromatography and mass spectrometry would seem to be an ideal merger of separation and detection. Just as in gas chromatography, a mass spectrometer could identify species as they elute from the chromatographic column. There are major problems though in the coupling of these two techniques. A gas-phase sample is needed for mass spectrometry, while the output of the LC column is a solute dissolved in a solvent. As a first step, the solvent must be vaporized. When vaporized, however, the LC solvent produces a gas volume that is 10 to 1000 times greater than the carrier gas in GC. Hence, most of the solvent must also be removed. There have been several devices developed to solve the problems of solvent removal and LC column interfacing. Today, the most popular approaches are to use a low flow-rate atmospheric pressure ionization technique. The block diagram of a typical LC/MS system is shown in **Figure 33F-1**. The HPLC system is typically a nanoscale capillary LC system with flow rates in the μL/min range. Alternatively, some interfaces allow flow rates as high as 1 to 2 mL/min, which is typical of conventional HPLC conditions. The most common ionization sources are electrospray ionization and atmospheric pressure chemical ionization (see Section 29D-2). The combination of HPLC and mass spectrometry gives high selectivity since unresolved peaks can be isolated by monitoring only a selected mass. The LC/MS technique can provide fingerprinting of a particular eluate instead of relying on retention time as in conventional HPLC. The combination also can give molecular mass and structural information as well as accurate quantitative analysis.[4]

For some complex mixtures, the combination of LC and MS does not provide enough resolution. In recent years, it has become feasible to couple two or more mass analyzers

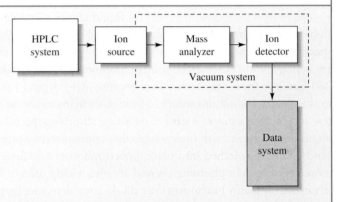

Figure 33F-1 Block diagram of an LC/MS system. The effluent from the LC column is introduced to an atmospheric pressure ionization source, such as an electrospray or a chemical ionization source. The ions produced are sorted by the mass analyzer and detected by the ion detector.

together in a technique known as tandem mass spectrometry (see Section 29D-3). When combined with LC, the tandem mass spectrometry system is called an LC/MS/MS instrument. Tandem mass spectrometers are usually triple quadrupole systems or quadrupole ion trap spectrometers.

To attain higher resolution than can be achieved with a quadrupole, the final mass analyzer in a tandem MS system can be a time-of-flight mass spectrometer. Sector mass spectrometers can also be combined to give tandem systems. Ion cyclotron resonance and ion trap mass spectrometers can be operated in such a way as to provide not only two stages of mass analysis but n stages. Such MS^n systems provide the analysis steps sequentially within a single mass analyzer. These spectrometers have been combined with LC systems in LC/MS^n instruments.

---

[3]See W. M. A. Niessen, *Liquid Chromatography-Mass Spectrometry*, 3rd ed., Boca Raton: CRC Press, 2006; R. E. Ardrey, *Liquid Chromatography-Mass Spectrometry: An Introduction*, Chichester, UK: Wiley, 2003.

[4]For a review of commercial LC/MS systems, see B. E. Erickson, *Anal. Chem.*, **2000**, *72*, 711A, **DOI**: 10.1021/ac0029758.

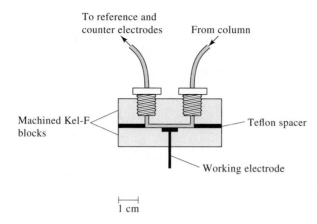

To reference and counter electrodes

From column

Machined Kel-F blocks

Teflon spacer

Working electrode

1 cm

The disadvantage of this detector is its somewhat limited sensitivity. Several electrochemical detectors that are based on potentiometric, conductometric, and voltammetric measurements have also been introduced. An example of an amperometric detector is shown in **Figure 33-9**.

33B PARTITION CHROMATOGRAPHY

The most widely used type of HPLC is **partition chromatography** in which the stationary phase is a second liquid that is immiscible with the liquid mobile phase. Partition chromatography can be subdivided into **liquid-liquid** and **liquid-bonded-phase** chromatography. The difference between the two lies in the way that the stationary phase is held on the support particles of the packing. The liquid is held in place by physical adsorption in liquid-liquid chromatography, while it is attached by chemical bonding in bonded-phase chromatography. Early partition chromatography was exclusively liquid-liquid; now, however, bonded-phase methods predominate because of their greater stability and compatibility with gradient elution. Liquid-liquid packings are today relegated to certain special applications. We restrict our discussion in this section to bonded-phase partition chromatography.[5]

In **liquid-liquid partition chromatography**, the stationary phase is a solvent held in place by adsorption of the surface of the packing particles.

In **liquid-bonded-phase chromatography**, the stationary phase is an organic species that is attached to the surface of the packing particles by chemical bonds.

33B-1 Bonded-Phase Packings

Most bonded-phase packings are prepared by reaction of an organochlorosilane with the —OH groups formed on the surface of silica particles by hydrolysis in hot dilute hydrochloric acid. The product is an organosiloxane. The reaction for one such SiOH site on the surface of a particle can be written as

$$
-\overset{|}{\underset{|}{Si}}-OH + Cl-\overset{CH_3}{\underset{CH_3}{\overset{|}{\underset{|}{Si}}}}-R \longrightarrow -\overset{|}{\underset{|}{Si}}-O-\overset{CH_3}{\underset{CH_3}{\overset{|}{\underset{|}{Si}}}}-R
$$

---

[5]For a report on retention mechanisms in bonded-phase chromatography, see J. G. Dorsey and W. T. Cooper, *Anal. Chem.*, **1994**, *66*, 857A, **DOI**: 10.1021/ac00089a002.

where R is often a straight chain octyl- or octyldecyl-group. Other organic functional groups that have been bonded to silica surfaces include aliphatic amines, ethers, and nitriles as well as aromatic hydrocarbons. Thus, many different polarities for the bonded stationary phase are available.

33B-2 Normal- and Reversed-Phase Packings

Two types of partition chromatography are distinguishable based on the relative polarities of the mobile and stationary phases. Early work in liquid chromatography was based on highly polar stationary phases such as triethylene glycol or water; a relatively nonpolar solvent such as hexane or *i*-propyl ether then served as the mobile phase. For historic reasons, this type of chromatography is now called **normal-phase chromatography**. In **reversed-phase chromatography**, the stationary phase is nonpolar, often a hydrocarbon, and the mobile phase is a relatively polar solvent (such as water, methanol, acetonitrile, or tetrahydrofuran).[6]

In normal-phase chromatography, the *least* polar component is eluted first; *increasing* the polarity of the mobile phase then *decreases* the elution time. In contrast, with reversed-phase chromatography, the *most* polar component elutes first, and *increasing* the mobile phase polarity *increases* the elution time.

It has been estimated that more than three-quarters of all HPLC separations are currently performed with reversed-phase, bonded, octyl- or octyldecyl siloxane packings. With such preparations, the long-chain hydrocarbon groups are aligned parallel to one another and perpendicular to the surface of the particle, giving a brushlike, nonpolar hydrocarbon surface. The mobile phase used with these packings is often an aqueous solution containing various concentrations of such solvents as methanol, acetonitrile, or tetrahydrofuran.

Ion-pair chromatography is a subset of reversed-phase chromatography in which easily ionizable species are separated on reversed-phase columns. In this type of chromatography, an organic salt containing a large organic counterion, such as a quarternary ammonium ion or alkyl sulfonate, is added to the mobile phase as an ion-pairing reagent. Two mechanisms for separation are postulated. In the first, the counterion forms an uncharged ion pair with a solute ion of opposite charge in the mobile phase. This ion pair then partitions into the nonpolar stationary phase giving differential retention of solutes based on the affinity of the ion pair for the two phases. Alternatively, the counterion is retained strongly by the normally neutral stationary phase and imparts a charge to this phase. Separation of organic solute ions of the opposite charge then occurs by formation of reversible ion-pair complexes with the more strongly retained solutes forming the strongest complexes with the stationary phase. Some unique separations of both ionic and nonionic compounds in the same sample can be accomplished by this form of partition chromatography. **Figure 33-10** illustrates the separation of ionic and nonionic compounds using alkyl sulfonates of various chain lengths as ion-pairing agents. Note that a mixture of C_5- and C_7-alkyl sulfonates gives the best separation results.

33B-3 Choice of Mobile and Stationary Phases

Successful partition chromatography requires a proper balance of intermolecular forces among the three participants in the separation process: the analyte, the mobile

In **normal-phase partition chromatography**, the stationary phase is polar and the mobile phase nonpolar. In **reversed-phase partition chromatography**, the polarity of these phases is reversed.

> In normal-phase chromatography, the least polar analyte is eluted first. In reversed-phase chromatography, the least polar analyte is eluted last.

Molecular model of octyldecyl-siloxane.

---

[6]For a detailed discussion of reversed-phase HPLC, see L. R. Snyder, J. J. Kirkland, and J. W. Dolan, *Introduction to Modern Liquid Chromatography*, 3rd ed., Chs. 6–7, Hoboken, NJ: Wiley, 2010.

Column: μ-Bondapak C$_{18}$
(4 mm × 30 cm)
Solvent: MeOH/H$_2$O with
C$_7$-alkyl sulfonate

(a)

Column: μ-Bondapak C$_{18}$
(4 mm × 30 cm)
Solvent: MeOH/H$_2$O with
C$_5$-alkyl sulfonate

(b)

Column: μ-Bondapak C$_{18}$
(4 mm × 30 cm)
Solvent: MeOH/H$_2$O with
(50/50) mixture
C-5/C-7-alkyl sulfonates

(c)

Figure 33-10 Chromatograms illustrating separations of mixtures of ionic and nonionic compounds by ion-pair chromatography. Compounds: (1) niacinamide, (2) pyridoxine, (3) riboflavin, and (4) thiamine. At pH 3.5, niacinamide is strongly ionized, while riboflavin is nonionic. Pyridoxine and thiamine are weakly ionized. Column: μ-Bondapak, C$_{18}$, 4 mm × 30 cm. Mobile phase: (a) MeOH/H$_2$O with C$_7$-alkyl sulfonate, (b) MeOH/H$_2$O with C$_5$-alkyl sulfonate, and (c) MeOH/H$_2$O with 1:1 mixture of C$_5$- and C$_7$-alkyl sulfonate. (Courtesy of Waters Corp., Milford, MA.)

phase, and the stationary phase. These intermolecular forces are described qualitatively in terms of the relative polarity possessed by each of the three components. In general, the polarities of common organic functional groups in increasing order are hydrocarbons < ethers < ketones < aldehydes < amides < alcohols. Water is more polar than compounds containing any of the preceding functional groups.

Often, in choosing a column and mobile phase, the polarity of the stationary phase is matched roughly with that of the analytes; a mobile phase of considerably different polarity is then used for elution. This procedure is generally more successful than one in which the polarities of the analyte and the mobile phase are matched but are different from that of the stationary phase. In this latter case, the stationary phase cannot compete successfully for the sample components; retention times then become too short for practical application. At the other extreme is the situation where the polarities of the analyte and stationary phase are too much alike; then, retention times become inordinately long.

33B-4 Applications

Figure 33-11 illustrates typical applications of bonded-phase partition chromatography for separating soft drink additives and organophosphate insecticides. **Table 33-2** further illustrates the variety of samples to which the technique is applicable.

> The order of polarities of common mobile phase solvents are water > acetonitrile > methanol > ethanol > tetrahydrofuran > propanol > cyclohexane > hexane.

Molecular model of acetonitrile. Acetonitrile (CH$_3$C≡N) is a widely used organic solvent. Its use as an LC mobile phase stems from its being more polar than methanol but less polar than water.

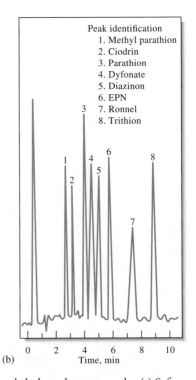

Figure 33-11 Typical applications of bonded-phase chromatography. (a) Soft drink additives. Column: 4.6 × 250 mm packed with polar (nitrile) bonded-phase packing. Isocratic elution with 6% HOAc/94% H_2O. Flow rate: 1.0 mL/min. (Courtesy of BTR Separations, a DuPont ConAgra affiliate.) (b) Organophosphate insecticides. Column: 4.5 × 250 mm packed with 5 μm C_8 bonded-phase particles. Gradient elution: 67% CH_3OH/33% H_2O to 80% CH_3OH/20% H_2O. Flow rate: 2 mL/min. Both used 254-nm UV detectors.

TABLE 33-2

Typical Applications of High-Performance Partition Chromatography

Field	Typical Mixtures Separated
Pharmaceuticals	Antibiotics, sedatives, steroids, analgesics
Biochemical	Amino acids, proteins, carbohydrates, lipids
Food products	Artificial sweeteners, antioxidants, aflatoxins, additives
Industrial chemicals	Condensed aromatics, surfactants, propellants, dyes
Pollutants	Pesticides, herbicides, phenols, polychlorinated biphenyls (PCBs)
Forensic science	Drugs, poisons, blood alcohol, narcotics
Clinical chemistry	Bile acids, drug metabolites, urine extracts, estrogens

33C ADSORPTION CHROMATOGRAPHY

Adsorption, or liquid-solid, chromatography is the classic form of liquid chromatography first introduced by Tswett at the beginning of the twentieth century. Because of the strong overlap between normal-phase partition chromatography and adsorption chromatography, many of the principles and techniques used for the former apply to adsorption chromatography. In fact, in many normal-phase separations, adsorption/displacement processes govern retention.

Finely divided silica and alumina are the only stationary phases that find use for adsorption chromatography. Silica is preferred for most applications because of its higher sample capacity. The adsorption characteristics of the two substances parallel one another. For both, retention times become longer as the polarity of the analyte increases.

Because of the versatility and ready availability of bonded stationary phases, traditional adsorption chromatography with solid stationary phases has seen decreasing use in recent years in favor of normal-phase chromatography.

33D ION CHROMATOGRAPHY

In Section 31D, we described some of the applications of ion-exchange resins to analytical separations. In addition, these materials are useful as stationary phases for liquid chromatography where they are used to separate charged species. Ion chromatography as it is practiced today was first developed in the mid-1970s when it was shown that anion or cation mixtures can be resolved on HPLC columns packed with anion-exchange or cation-exchange resins. At that time, detection was generally performed with conductivity measurements, which were not ideal because of high electrolyte concentrations in the mobile phase. The development of low-exchange-capacity columns allowed the use of low-ionic-strength mobile phases that could be further deionized (ionization suppressed) to allow high sensitivity conductivity detection. Currently, several other detector types are available for ion chromatography, including spectrophotometric and electrochemical.[7]

Two types of ion chromatography are currently in use: **suppressor-based** and **single-column**. They differ in the method used to prevent the conductivity of the eluting electrolyte from interfering with the measurement of analyte conductivities.

33D-1 Ion Chromatography Based on Suppressors

Conductivity detectors have many of the properties of the ideal detector. They can be highly sensitive, they are universal for charged species, and as a general rule, they respond in a predictable way to concentration changes. Furthermore, such detectors are simple to operate, inexpensive to construct and maintain, easy to miniaturize, and usually give prolonged, trouble-free service. The only limitation to the use of conductivity detectors, which delayed their general application to ion chromatography until the mid-1970s, was due to the high electrolyte concentrations required to elute most analyte ions in a reasonable time. As a result, the conductivity from the mobile-phase components tends to swamp that from the analyte ions, greatly reducing the detector sensitivity.

In 1975, the problem created by the high conductance of eluents was solved by the introduction of an **eluent suppressor column** immediately following the ion-exchange column.[8] The suppressor column is packed with a second ion-exchange resin that effectively converts the ions of the eluting solvent to a molecular species of limited ionization without affecting the conductivity due to analyte ions. For example, when cations are being separated and determined, hydrochloric acid is chosen as the eluting reagent, and the suppressor column is an anion-exchange resin in the hydroxide form. The product of the reaction in the suppressor is water, that is

$$H^+(aq) + Cl^-(aq) + resin^+OH^-(s) \rightarrow resin^+Cl^-(s) + H_2O$$

The analyte cations are not retained by this second column.

> The conductivity detector is well suited for ion chromatography.

---

[7]For brief reviews of ion chromatography, see J. S. Fritz, *Anal. Chem.*, **1987**, *59*, 335A, **DOI**: 10.1021/ac00131a002; P. R. Haddad, *Anal. Chem.*, **2001**, *73*, 266A, **DOI**: 10.1021/ac012440u. For a detailed description of the method, see H. Small, *Ion Chromatography*, New York: Plenum Press, 1989; J. S. Fritz and D. T. Gjerde, *Ion Chromatography*, 4th ed., Weinheim, Germany: Wiley-VCH, 2009.

[8]H. Small, T. S. Stevens, and W. C. Bauman, *Anal. Chem.*, **1975**, *47*, 1801, **DOI**: 10.1021/ac60361a017.

For anion separations, the suppressor packing is the acid form of a cation-exchange resin, and sodium bicarbonate or carbonate is the eluting agent. The reaction in the suppressor is

$$Na^+(aq) + HCO_3^-(aq) + resin^-H^+(s) \rightarrow resin^-Na^+(s) + H_2CO_3(aq)$$

The largely undissociated carbonic acid does not contribute significantly to the conductivity.

An inconvenience associated with the original suppressor columns was the need to regenerate them periodically (typically every 8 to 10 hr) in order to convert the packing back to the original acid or base form. In the 1980s, however, micromembrane suppressors that operate continuously became available.[9] For example, where sodium carbonate or bicarbonate is to be removed, the eluent is passed over a series of ultrathin cation-exchange membranes that separate it from a stream of acidic regenerating solution that flows continuously in the opposite direction. The sodium ions from the eluent exchange with hydrogen ions on the inner surface of the exchanger membrane and then migrate to the other surface for exchange with hydrogen ions from the regenerating reagent. Hydrogen ions from the regeneration solution migrate in the reverse direction, thus preserving electrical neutrality. Micromembrane separators are capable of removing essentially all the sodium ions from a 0.1M NaOH solution with an eluent flow rate of 2 mL/min.

Figure 33-12 shows two applications of ion chromatography based on a suppressor column and conductometric detection. In each, the ions were present in

> In **suppressor-based ion chromatography**, the ion-exchange column is followed by a **suppressor column**, or a **suppressor membrane**, that converts an ionic eluent into a nonionic species that does not interfere with the conductometric detection of analyte ions.

Figure 33-12 Typical applications of ion chromatography. (a) Separation of anions on an anion-exchange column. Eluent: 0.0028 M NaHCO₃/0.0023 M Na₂CO₃. Sample size: 50 μL. (b) Separation of alkaline earth ions on a cation-exchange column. Eluent: 0.025 M phenylenediamine dihydrochloride/0.0025 M HCl. Sample size: 100 μL. (Courtesy of Dionex, Inc., Sunnyvale, CA.)

---

[9]J. S. Fritz and D. T. Gjerde, *Ion Chromatography*, 4th ed., Chs 6–7, Weinheim, Germany: Wiley-VCH, 2009.

the parts-per-million range; the sample size was 50 μL in one case and 100 μL in the other. The method is particularly important for anion analysis because there is no other rapid and convenient method for handling mixtures of this type.

33D-2 Single-Column Ion Chromatography

Commercial ion chromatography instrumentation that requires no suppressor column is also available. This approach depends on the small differences in conductivity between sample ions and the prevailing eluent ions. To amplify these differences, low-capacity exchangers are used that permit elution with solutions with low electrolyte concentrations. Furthermore, eluents of low conductivity are chosen.

Single-column ion chromatography offers the advantage of not requiring special equipment for suppression. However, it is a somewhat less sensitive method for determining anions than suppressor-column methods.

In **single-column ion-exchange chromatography**, analyte ions are separated on a low-capacity ion exchanger by means of a low-ionic strength eluent that does not interfere with the conductometric detection of analyte ions.

33E SIZE-EXCLUSION CHROMATOGRAPHY

Size-exclusion, or gel chromatography, is a powerful technique that is particularly applicable to high-molecular-mass species.[10] Packings for size-exclusion chromatography consist of small (~10 μm) silica or polymer particles containing a network of uniform pores into which solute and solvent molecules can diffuse. While in the pores, molecules are effectively trapped and removed from the flow of the mobile phase. The average residence time of analyte molecules depends on their effective size. Molecules that are significantly larger than the average pore size of the packing are excluded and thus suffer no retention, that is, they travel through the column at the rate of the mobile phase. Molecules that are appreciably smaller than the pores can penetrate throughout the pore maze and are thus entrapped for the greatest time; they are last to elute. Between these two extremes are intermediate-size molecules whose average penetration into the pores of the packing depends on their diameters. The fractionation that occurs within this group is directly related to molecular size and, to some extent, molecular shape. Note that size-exclusion separations differ from the other chromatographic procedures in the respect that there are no chemical or physical interactions between analytes and the stationary phase. Indeed, such interactions are avoided because they lead to lower column efficiencies. Note also that, unlike other forms of chromatography, there is an upper limit to retention time because no analyte species is retained longer than those small molecules that totally permeate the stationary phase.

In **size-exclusion chromatography**, fractionation is based on molecular size.

33E-1 Column Packings

Two types of packing for size-exclusion chromatography are encountered: polymer beads and silica-based particles, both of which have diameters of 5 to 10 μm. Silica particles are more rigid, which leads to easier packing and permits higher pressures to be used. They are also more stable, allowing a great range of solvents to be used and exhibiting more rapid equilibration with new solvents.

---

[10]For monographs on this subject, see A. Striegel, W. W. Yau, J. J. Kirkland, and D. D. Bly, *Modern Size-Exclusion Chromatography: Practice of Gel Permeation and Gel Filtration Chromatography*, 2nd ed., Hoboken, NJ: Wiley, 2009; C. S. Wu, ed., *Handbook of Size Exclusion Chromatography*, 2nd ed., New York: Dekker, 2004; C. S. Wu, ed., *Column Handbook for Size Exclusion Chromatography*, San Diego: Academic Press, 1999.

Figure 33-13 Applications of size-exclusion chromatography. (a) Separation of fatty acids. Column: polystyrene based, 7.5 × 600 mm. Mobile phase: tetrahydrofuran. (b) An analysis of a commercial epoxy resin. (*n* = number of monomeric units in the polymer). Column: porous silica 6.2 × 250 mm. Mobile phase: tetrahydrofuran. (Adapted from BTR Separations a DuPont ConAgra affiliate.)

Gel filtration is a type of size-exclusion chromatography in which the packing is hydrophilic. It is used to separate polar species.

Gel permeation is a type of size-exclusion chromatography in which the packing is hydrophobic. It is used to separate nonpolar species.

Numerous size-exclusion packings are on the market. Some are hydrophilic for use with aqueous mobile phases; others are hydrophobic and are used with nonpolar organic solvents. Chromatography based on the hydrophilic packings is sometimes call **gel filtration**, while that based on hydrophobic packings is termed **gel permeation**. With both types of packings, many pore diameters are available. Generally, a given packing will accommodate a 2- to 2.5-decade range of molecular mass. The average molecular mass suitable for a given packing may be as small as a few hundred or as large as several million.

33E-2 Applications

Figure 33-13 illustrates typical applications of size-exclusion chromatography. Both chromatograms were obtained with hydrophobic packings in which the eluent was tetrahydrofuran. In Figure 33-13a, the separation of fatty acids with molecular mass $\mathcal{M}$ from 116 to 344 is shown. In Figure 33-13b, the sample was a commercial epoxy resin in which each monomer unit had a molecular mass of 280 (*n* = number of monomer units).

Another important application of size-exclusion chromatography is the rapid determination of the molecular mass or the molecular mass distribution of large polymers or natural products. The key to such determinations is an accurate molecular mass calibration. Calibrations can be accomplished by means of standards of known molecular mass (peak position method) or by the universal calibration method. The latter method relies on the principle that the product of the intrinsic molecular viscosity η and molecular mass $\mathcal{M}$ is proportional to hydrodynamic volume (effective volume including solvation sheath). Ideally, molecules are separated in size-exclusion chromatography according to hydrodynamic volume. Hence, a universal calibration curve can be obtained by plotting log $[\eta \mathcal{M}]$ versus the retention volume, V_r, where $V_r = t_r \times F$. Alternatively, absolute calibration can be achieved by using a molar mass-sensitive detector such as a low-angle, light-scattering detector.

Feature 33-2 illustrates how size-exclusion chromatography can be used in the separation of fullerenes.

Buckyballs: The Chromatographic Separation of Fullerenes

Our ideas about the nature of matter are often profoundly influenced by chance discoveries. No event in recent memory has captured the imagination of both the scientific community and the public as did the serendipitous discovery in 1985 of the soccer ball-shaped molecule C_{60}. This molecule, illustrated in **Figure 33F-2**, its cousin C_{70}, and other similar molecules discovered since 1985 are called **fullerenes**, or more commonly, **buckyballs**.[11] The compounds are named in honor of the famous architect, R. Buckminster Fuller, who designed many geodesic dome buildings having the same hexagonal/

Figure 33F-2 Buckminster fullerene, C_{60}.

pentagonal structure as buckyballs. Since their discovery, thousands of research groups throughout the world have studied various chemical and physical properties of these highly stable molecules. They represent a third allotropic form of carbon besides graphite and diamond.

The preparation of buckyballs is almost trivial. When an ac arc is established between two carbon electrodes in a flowing helium atmosphere, the soot that is collected is rich in C_{60} and C_{70}. Although the preparation is easy, the separation and purification of more than a few milligrams of C_{60} proved tedious and expensive. Relatively large quantities of buckyballs have been separated using size-exclusion chromatography.[12] Fullerenes are extracted from soot prepared as mentioned above and injected on a 199-mm × 30-cm, 500-Å Ultrastyragel column (Waters Corp., Milford, MA), using toluene as the mobile phase and UV/visible detection following separation. A typical chromatogram is shown in **Figure 33F-3**. The peaks in the chromatogram are labeled with their identities and retention times.

Note that C_{60} elutes before C_{70} and the higher fullerenes. This is contrary to what we expect; the smallest molecule, C_{60}, should be retained more strongly than C_{70} and the higher fullerenes. It has been suggested that the interaction between the solute molecules and the gel is on the surface of the gel rather than in its pores. Since C_{70} and the higher fullerenes have larger surface areas than C_{60}, the higher fullerenes are retained

Figure 33F-3 Separation of fullerenes.

[11] R. F. Curl and R. E. Smalley, *Scientific American*, **1991**, *265* (4), 54.
[12] M. S. Meier and J. P. Selegue, *J. Org. Chem.*, **1992**, *57*, 1924, **DOI**: 10.1021/jo00032a057; A. Gugel and K. Mullen, *J. Chromatogr. A*, **1993**, *628*, 23, **DOI**: 10.1016/0021-9673(93)80328-6.

more strongly on the surface of the gel and thus eluted after C_{60}. With an automated apparatus, this method of separation may be used to prepare several grams of 99.8% pure C_{60} from 5 to 10 g of a C_{60} to C_{70} mixture in a twenty-four hour period. These quantities of C_{60} can then be used to prepare and study the chemistry and physics of derivatives of this interesting and unusual form of carbon.

In addition to size-exclusion, HPLC, with an octadecyl silica (ODS)-bonded stationary phase, has been used to separate fullerenes.[13] Both polymeric and monomeric ODS phases have been used, and they provide a higher selectivity than other phases. **Figure 33F-4** shows the preparative separation of whole soot extract and a higher fullerenes fraction on a polymeric ODS column. These were among the first separations of the individual higher fullerenes. Note the excellent resolution compared to the size exclusion separation of Figure 33F-3.

Figure 33F-4 Chromatograms of whole soot extract (a) and a higher fullerene fraction (b) obtained with a polymeric ODS column and an acetonitrile:toluene mobile phase. (Reprinted (adapted) with permission from F. Diederich and R. L. Whetten, *Acc. Chem. Res.*, **1992**, *25*, 121. **DOI**: 10.1021/ar00015a004. Copyright 1992 American Chemical Society.) See article for fullerene nomenclature.

[13]K. Jinno, H. Ohta, and Y. Sato, in *Separation of Fullerenes by Liquid Chromatography*, K. Jinno, ed., Ch. 3, London: Royal Society of Chemistry, 1999.

33F AFFINITY CHROMATOGRAPHY

In affinity chromatography, a reagent called an **affinity ligand** is covalently bonded to a solid support.[14] Typical affinity ligands are antibodies, enzyme inhibitors, or other molecules that reversibly and selectively bind to analyte molecules in the sample. When the sample passes through the column, only the molecules that selectively bind to the affinity ligand are retained. Molecules that do not bind pass through the column with the mobile phase. After the undesired molecules are removed, the retained analytes can be eluted by changing the mobile phase conditions.

The stationary phase for affinity chromatography is a solid, such as agarose, or a porous glass bead to which the affinity ligand is immobilized. The mobile phase in affinity chromatography has two distinct roles to play. First, it must support the strong binding of the analyte molecules to the ligand. Second, once the undesired species are removed, the mobile phase must weaken or eliminate the analyte-ligand interaction so that the analyte can be eluted. Often, changes in pH or ionic strength are used to change the elution conditions during the two stages of the process.

Affinity chromatography has the major advantage of extraordinary specificity. The primary use is in the rapid isolation of biomolecules during preparative work.

33G CHIRAL CHROMATOGRAPHY

Tremendous advances have been made in separating compounds that are nonsuperimposable mirror images of each other, called **chiral compounds**. Such mirror images are called **enantiomers**. Either chiral mobile-phase additives or chiral stationary phases are required for these separations.[15] Preferential complexation between the chiral resolving agent (additive or stationary phase) and one of the isomers results in a separation of the enantiomers. The **chiral resolving agent** must have chiral character itself in order to recognize the chiral nature of the solute.

Chiral stationary phases have received the most attention.[16] A chiral agent is immobilized on the surface of a solid support. Several different modes of interaction can occur between the chiral resolving agent and the solute.[17] In one type, the interactions are due to attractive forces such as those between π bonds, hydrogen bonds, or dipoles. In another type, the solute can fit into chiral cavities in the stationary phase to form inclusion complexes. No matter what the mode, the ability to separate these very closely related compounds is of extreme importance in many fields. **Figure 33-14** shows the separation of a racemic mixtures of an ester on a chiral stationary phase. Note the excellent resolution of the *R* and *S* enantiomers.

A **chiral resolving agent** is a chiral mobile-phase additive or a chiral stationary phase that preferentially complexes one of the enantiomers.

[14]For details on affinity chromatography, see M. Zachariou, ed., *Affinity Chromatography: Methods and Protocols*, 2nd ed., Totowa, NJ: Humana Press, 2007; D. S. Hage ed., *Handbook of Affinity Chromatography*, 2nd ed., Boca Raton: CRC Press, 2006.

[15]G. Subramanian, *Chiral Separation Techniques: A Practical Approach*, Weinheim, Germany: Wiley-VCH, 2007); S. Ahuja, *Chiral Separations by Chromatography*, New York: Oxford University Press, 2000.

[16]For a review on chiral stationary phases, see D. W. Armstrong and B. Zhang, *Anal. Chem.*, **2001**, *73*, 557A, **DOI**: 10.1021/ac012526n.

[17]For a review on chiral interactions, see M. C. Ringo and C. E. Evans, *Anal. Chem.*, **1998**, *70*, 315A, **DOI**: 10.1021/ac9818428.

Figure 33-14 Chromatogram of a racemic mixture of *N*-(1-Naphthyl) leucine ester 1 on a dinitrobenzene-leucine chiral stationary phase. The *R* and *S* enantiomers are seen to be well separated. Column: 4.6 × 50 mm. Mobile phase: 20% 2-propanol in hexane. Flow rate: 1.2 mL/min; UV detector at 254 nm. (Reprinted (adapted) with permission from L. H. Bluhm, Y. Wang, and T. Li, *Anal. Chem.*, **2000**, *72*, 5201, **DOI**: 10.1021/ac000568q. Copyright 2000 American Chemical Society.)

33H COMPARISON OF HIGH-PERFORMANCE LIQUID CHROMATOGRAPHY AND GAS CHROMATOGRAPHY

Table 33-3 provides a comparison between high-performance liquid chromatography and gas-liquid chromatography. When either is applicable, GC offers the advantage of speed and simplicity of equipment. On the other hand, HPLC is applicable to nonvolatile substances (including inorganic ions) and thermally unstable materials, but GC is not. Often the two methods are complementary.

TABLE 33-3

Comparison of High Performance Liquid Chromatography and Gas-Liquid Chromatography

Characteristics of Both Methods
- Efficient, highly selective, widely applicable
- Only small sample required
- May be nondestructive of sample
- Readily adapted to quantitative analysis

Advantages of HPLC
- Can accommodate nonvolatile and thermally unstable compounds
- Generally applicable to inorganic ions

Advantages of GC
- Simple and inexpensive equipment
- Rapid
- Unparalleled resolution (with capillary columns)
- Easy to interface with mass spectrometry

 Spreadsheet Summary In Chapter 15 of *Applications of Microsoft® Excel in Analytical Chemistry*, 2nd ed., begins with an exercise treating the resolution of overlapped Gaussian peaks. The overlapped chromatogram, the response, is modeled as the sum of Gaussian curves. Initial estimates are made for the model parameters. Excel calculates the residuals, the difference between the response and the model, and the sum of the squares of the residuals. Excel's Solver is then used to minimize the sum of the squares of the residuals while displaying the results of each iteration.

Connect to **www.cengage.com/chemistry/skoog/fac9**. Look under Chapter 33 Web Works and find the link to the *LC-GC* magazine Web site. *LC-GC* is a free magazine that contains interesting and timely articles for chromatographers and other users of chromatography equipment. From the *LC-GC* home page, find the article by J. L. Herman and T. Edge entitled "Theoretical Concepts and Applications of Turbulent Flow Chromatography" (2012). What is the definition of turbulent flow? What is the Reynolds number? Why is a turbulent flow profile more difficult to define mathematically than a laminar flow profile? Can turbulent flow chromatography be described as a two-dimensional technique? What types of molecules can be separated by turbulent flow chromatography? How is turbulent flow chromatography useful in LC-MS systems? Is the technique useful for sample cleanup with biological samples? How do plate numbers compare with conventional HPLC? Why is a two-column approach often used in turbulent flow chromatography?

QUESTIONS AND PROBLEMS

33-1. List the types of substances to which each of the following chromatographic methods is most applicable:
  *(a) gas-liquid.
  (b) liquid partition.
  *(c) ion.
  (d) affinity.
  *(e) gel permeation.
  (f) gel filtration.
  *(g) chiral.

33-2. Define
  *(a) isocratic elution.
  (b) gradient elution.
  *(c) normal-phase packing.
  (d) reversed-phase packing.
  *(e) bonded-phase packing
  (f) chiral chromatography.
  *(g) ion-pair chromatography.
  (h) eluent-suppressor column.
  *(i) gel filtration.
  (j) gel permeation.

33-3. Indicate the order in which the following compounds would be eluted from an HPLC column containing a reversed-phase packing:
  *(a) benzene, diethyl ether, *n*-hexane.
  (b) acetone, dichloroethane, acetamide.

33-4. Indicate the order of elution of the following compounds from a normal-phase packed HPLC column:
  *(a) ethyl acetate, acetic acid, dimethylamine.
  (b) propylene, hexane, benzene, dichlorobenzene.

*33-5.** Describe the fundamental difference between adsorption and partition chromatography.

33-6. Describe the fundamental difference between ion-exchange and size-exclusion chromatography.

*33-7.** Describe the difference between gel-filtration and gel-permeation chromatography.

33-8. What types of species can be separated by HPLC but not by GC?

*33-9.** What is the major difference between isocratic elution and gradient elution? For what types of compounds are these two elution methods most suited?

33-10. Describe two types of pumps used in high-performance liquid chromatography. What are the advantages and disadvantages of each?

*33-11.** Describe the differences between single-column and suppressor-column ion chromatography.

33-12. Mass spectrometry is an extremely versatile detection system for gas chromatography. Describe the major reasons why it is more difficult to combine HPLC with mass spectrometry than it is to combine GC with mass spectrometry.

*33-13.** Which of the GC detectors in Table 32-1 are suitable for HPLC? Why are some of these unsuitable for HPLC?

33-14. The ideal detector for GC is described in Section 32A-4. Which of the eight characteristics of an ideal GC detector are applicable to HPLC detectors? What additional characteristics would be added to describe the ideal HPLC detector?

*33-15.** Although temperature does not have nearly the effect on HPLC separations that it has on GC separations, it nonetheless can play an important role. Discuss how and why temperature might or might not influence the following separations:
  (a) a reversed-phase chromatographic separation of a steroid mixture.
  (b) an adsorption chromatographic separation of a mixture of closely related isomers.

33-16. Two components in an HPLC separation have retention times that differ by 22 s. The first peak elutes in 10.5 min and the peak widths are approximately equal. Use a spreadsheet to find the minimum number of theoretical plates needed to achieve the following resolution, R_s, values: 0.50, 0.75, 0.90, 1.0, 1.10, 1.25, 1.50, 1.75, 2.0, and 2.5. How would the results change if peak 2 were twice as broad as peak 1?

33-17. An HPLC method was developed for the separation and determination of ibuprofen in rat plasma samples as part of a study of the time course of the drug in laboratory animals. Several standards were chromatographed and the results below obtained:

Ibuprofen Concentration, μg/mL	Peak Area
0.5	5.0
1.0	10.1
2.0	17.2
3.0	19.8
6.0	39.7
8.0	57.3
10.0	66.9
15.0	95.3

Next, a 10-mg/kg sample of ibuprofen was administered orally to a laboratory rat. Blood samples were drawn at various times after administration of the drug and subjected to HPLC analysis. The following results were obtained:

Time, hr.	Peak Area
0	0
0.5	91.3
1.0	80.2
1.5	52.1
2.0	38.5
3.0	24.2
4.0	21.2
6.0	18.5
8.0	15.2

Find the concentration of ibuprofen in the blood plasma for each of the times given above and plot the concentration versus time. On a percentage basis, during what half-hour period (1st, 2nd, 3rd, etc.) is most of the ibuprofen lost?

33-18. Challenge Problem: Assume for simplicity that the HPLC plate height, H, can be given by Equation 31-27 as

$$H = \frac{B}{u} + C_S u + C_M u = \frac{B}{u} + Cu$$

where $C = C_S + C_M$.

(a) By using calculus to find the minimum H, show that the optimum velocity u_{opt} can be expressed as

$$u_{opt} = \sqrt{\frac{B}{C}}$$

(b) Show that this relationship leads to a minimum plate height H_{min} given by

$$H_{min} = 2\sqrt{BC}$$

(c) Under some conditions for chromatography, C_S is negligible compared to C_M. For packed LC columns, C_M is given by

$$C_M = \frac{\omega d_p^2}{D_M}$$

where ω is a dimensionless constant, d_p is the particle size of the column packing, and D_M is the diffusion coefficient in the mobile phase. The B coefficient can be expressed as

$$B = 2\gamma D_M$$

where γ is also a dimensionless constant. Express u_{opt} and H_{min} in terms of D_M, d_p, and the dimensionless constants γ and ω.

(d) If the dimensionless constants are on the order of unity, show that u_{opt} and H_{min} can be expressed as

$$u_{opt} \approx \frac{D_M}{d_p} \quad \text{and} \quad H_{min} \approx d_p$$

(e) Under the conditions in part (d), how could the plate height be reduced by 1/3? What would happen to the optimum velocity under these conditions? What would happen to the number of theoretical plates N for the same length column?

(f) For the conditions in part (e), how could you maintain the same number of theoretical plates while reducing the plate height by 1/3?

(g) The preceding discussion assumes that band broadening all occurs within the column. Name two sources of extra-column band broadening that might also contribute to the overall width of LC peaks.

Miscellaneous Separation Methods

CHAPTER 34

Capillary electrophoresis (CE), one of the separation methods discussed in this chapter, is a major technique used in DNA profiling for disease diagnosis. One application is to the early detection of Lyme disease, a bacterial disease transmitted by the bite of a deer tick (black-legged tick, shown in the photographs). Deer ticks become infected by feeding on mice that carry the bacteria. In forested areas, the ticks can move from small mammals to deer and to humans, where they can cause Lyme disease. In its early stages, the disease may cause a red rash accompanied by fever, chills, headaches, swollen lymph nodes, and other symptoms. Early diagnosis is critical because Lyme disease can then be treated effectively with antibiotics before becoming chronic and debillitating. The development of diagnostic tests for Lyme disease has led researchers to apply DNA profiling methodology to identify the bacteria responsible. In addition to disease diagnosis, CE is also used in many forensic applications such as the identification or elimination of suspects in criminal investigations and in paternity and familial testing.

This chapter discusses several separation methods that cannot be classified easily. We describe supercritical fluid chromatography, paper chromatography, capillary electrophoresis, capillary electrochromatography, and field-flow fractionation and their applications.

Top: Photo Researchers/Getty Images;
Bottom: Dr. David M. Phillips/Getty Images

I n this chapter, we discuss several additional methods of separation: supercritical fluid chromatography and extractions, thin-layer and paper chromatography, capillary electrophoresis, capillary electrochromatography, and field-flow fractionation. These methods, while not as established as GC and HPLC, nonetheless are very powerful and can provide separations that are impossible or impractical to achieve by conventional methods.

34A SUPERCRITICAL FLUID SEPARATIONS

Supercritical fluids are an important class of solvents that have unique solvating properties. Such fluids have proven quite useful in chromatography and solvent extraction. In supercritical fluid chromatography (SFC), the supercritical fluid acts as the mobile phase. At first, SFC was considered as a hybrid of gas and liquid chromatography, but now it is thought of as being more similar to HPLC in its operation and instrumentation.[1] Supercritical fluid extraction (SFE) can provide some unique

---

[1]For additional information, see M. Caude and D. Thiebaut, eds., *Practical Supercritical Fluid Chromatography and Extraction*, Amsterdam: Harwood, 2000; K. Anton and C. Berger, eds., *Supercritical Fluid Chromatography with Packed Columns, Techniques and Applications*, New York: Dekker, 1998; L. Taylor, in *Handbook of Instrumental Techniques for Analytical Chemistry*, F. Settle ed., Ch. 11, Upper Saddle River, NJ: Prentice Hall, 1997. For recent reviews, see, L. T. Taylor, *Anal. Chem.*, **2010**, *82*, 4925, **DOI**: 10.1021/ac101194x; L. T. Taylor, *Anal. Chem.*, **2008**, *80, 4285*, **DOI**: 10.1021/ac800482d.

TABLE 34-1

Comparison of Properties of Supercritical Fluids, Liquids, and Gases*			
Property	**Gas (STP)**	**Supercritical Fluid**	**Liquid**
Density, g/cm³	$(0.6–2) \times 10^{-3}$	0.2–0.5	0.6–2
Diffusion coefficient, cm²/s	$(1–4) \times 10^{-1}$	$10^{-3}–10^{-4}$	$(0.2–2) \times 10^{-5}$
Viscosity, g cm⁻¹ s⁻¹	$(1–3) \times 10^{-4}$	$(1–3) \times 10^{-4}$	$(0.2–3) \times 10^{-2}$

*All data order of magnitude only

separation capabilities, particularly for complex materials such as environmental, pharmaceutical, and food samples.[2] We first review the properties of supercritical fluids before discussing the principles and applications of SFC and SFE.

34A-1 Properties of Supercritical Fluids

A **supercritical fluid** is the physical state of a substance held above its critical temperature.

The **critical temperature** is the temperature above which a substance cannot be liquified.

The density of a supercritical fluid is 200 to 400 times that of its gaseous state, and it is nearly as dense as its liquid state.

Supercritical fluids are able to dissolve large nonvolatile molecules.

A **supercritical fluid** is formed whenever a substance is heated above its **critical temperature**. Above the critical temperature, a substance can no longer be condensed to a liquid by simply applying pressure. For example, carbon dioxide is a supercritical fluid at temperatures above 31°C. In this state, the molecules of carbon dioxide act independently of one another just as they do in a gas.

As shown by the data in **Table 34-1**, the physical properties of a supercritical fluid can be remarkably different from the same properties in either the liquid or the gaseous state. For example, the density of a supercritical fluid is typically 200 to 400 times greater than that of the corresponding gas and approaches that of the substance in its liquid state. The properties compared in Table 34-1 are important in chromatography and other separations.

An important property of supercritical fluids and one that is related to their high densities (0.2 to 0.5 g/cm³) is their ability to dissolve large nonvolatile molecules. For example, supercritical carbon dioxide is an excellent solvent for *n*-alkanes containing from 5 to 22 carbon atoms, di-*n*-alkylphthalates in which the alkyl groups contain 4 to 16 carbon atoms, and various polycyclic aromatic hydrocarbons consisting of several rings.[3]

Critical temperatures for fluids used in chromatography vary widely, from about 30°C to above 200°C. Lower critical temperatures are advantageous in chromatography from several standpoints. For this reason, much of the work to date has focused on the supercritical fluids shown in **Table 34-2**. Note that these temperatures, and

TABLE 34-2

Properties of Some Supercritical Fluids*			
Fluid	**Critical Temperature, °C**	**Critical Pressure, atm**	**Critical Point Density, g/mL**	**Density at 400 atm, g/mL**
CO₂	31.3	72.9	0.47	0.96
N₂O	36.5	71.7	0.45	0.94
NH₃	132.5	112.5	0.24	0.40
n-Butane	152.0	37.5	0.23	0.50

*From M. L. Lee and K. E. Markides, *Science*, **1987**, *235*, 1342, **DOI**:10.1126/science.235.4794.1342. Reprinted with permission from AAAS.

[2]See G. Brunner, ed., *Supercritical Fluids As Solvents and Reaction Media*, Ch. 4, Amsterdam: Elsevier, 2004; M. C. Henry and C. R. Yonker, *Anal. Chem.*, **2006**, *78*, 3909, **DOI**: 10.1021/ac0605703.

[3]Certain important industrial processes are based on the high solubility of organic species in supercritical carbon dioxide. For example, this medium has been used in extracting caffeine from coffee beans to give decaffeinated coffee and in extracting nicotine from cigarette tobacco.

the pressures at these temperatures, are well within the operating conditions of ordinary high-performance liquid chromatography.

34A-2 Instrumentation and Operating Variables

Instruments for supercritical fluid chromatography are similar in design to high-performance liquid chromatographs except that in SFC the pumping system must include a chilled pump head to keep the fluid in the liquid state and that there must be provision for controlling and measuring the column pressure. Several manufacturers began to offer apparatus for supercritical fluid chromatography in the mid-1980s, although today only a few companies produce such instrumentation.[4]

Effects of Pressure

The density of a supercritical fluid increases rapidly and nonlinearly with pressure increases. Density increases also change retention factors (k) and thus elution times. For example, the elution time for hexadecane is reported to decrease from 25 to 5 min as the pressure of carbon dioxide is raised from 70 to 90 atm. An effect similar to that of temperature programming in GC and gradient elution in HPLC can be achieved by linearly increasing the column pressure or by regulating the pressure to obtain linear density increases. **Figure 34-1** illustrates the improvement in chromatograms realized by pressure programming. The decompression of fluids as they travel through the column can give rise to temperature changes that can affect separations and thermodynamic measurements. The most common pressure profiles used in SFC are often constant (isobaric) for a given length of time followed by a linear or asymptotic approach to a final pressure. In addition to pressure programming, temperature programming and mobile-phase gradients have been used.

> Gradient elution can be achieved in SFC by systematically changing the column pressure or the density of the supercritical fluid.

Columns

Both packed columns and open tubular columns are used in supercritical fluid chromatography. Packed columns can provide more theoretical plates and handle larger

Sample:
    1. cholesteryl octanoate
    2. cholesteryl decylate
    3. cholesteryl laurate
    4. cholesteryl myristate
    5. cholesteryl palmitate
    6. cholesteryl stearate

Column: DB−1
Mobile phase: CO_2
Temperature: 90C
Detector: FID

Figure 34-1 Effect of pressure programming in supercritical fluid chromatography. Note the shorter time for the pressure gradient chromatogram on the right compared with the isobaric chromatogram on the left. (Courtesy of Brownlee labs, Santa Clara, CA.)

---

[4]L. T. Taylor, *Anal. Chem.*, **2010**, *82*, 4925, **DOI:** 10.1021/ac101194x.

Very long columns can be used in SFC because the viscosity of supercritical fluids is so low.

sample volumes than open tubular columns. Because of the low viscosity of supercritical media, columns can be much longer than those used in liquid chromatography, and column lengths of 10 to 20 m and inside diameters of 50 or 100 μm are common. For difficult separations, columns 60 m in length and longer have been used. Well over 100,000 plates can be achieved with packed columns. Open tubular columns are similar to the fused-silica open tubular (FSOT) columns described on page 898. Packed columns are the most widely accepted columns for SFC. Packed column SFC is very similar to normal phase HPLC.

Many of the column coatings used in liquid chromatography have been applied to supercritical fluid chromatography as well. Typically, these are polysiloxanes (see Section 32B-3) that are chemically bonded to the surface of silica particles or to the inner silica wall of capillary tubing. Film thicknesses are 0.05 to 0.4 μm.

Mobile Phases

The most widely used mobile phase for supercritical fluid chromatography is carbon dioxide. It is an excellent solvent for a variety of nonpolar organic molecules. In addition, it transmits in the ultraviolet and is odorless, nontoxic, widely available, and remarkably inexpensive relative to other chromatographic solvents. Its critical temperature of 31°C and its pressure of 73 atm at the critical temperature permit a wide selection of temperatures and pressures without exceeding the operating limits of modern HPLC equipment. In some applications, polar organic modifiers, such as methanol, are introduced in small concentrations ($\approx$ 1%) to modify alpha values for analytes.

Several other substances have served as mobile phases in supercritical chromatography, including ethane, pentane, dichlorodifluoromethane, diethyl ether, and tetrahydrofuran, but CO_2 remains by far the most popular.

Detectors

A major advantage of supercritical fluid chromatography is that the sensitive and universal detectors of gas chromatography are applicable to this technique as well. For example, the convenient flame ionization detector of GC can be applied by simply allowing the supercritical carrier to expand through a restrictor and into an air-hydrogen flame, where ions formed from the analytes are collected at biased electrodes, giving rise to an electrical current.

Mass spectrometry has become an important detection method for SFC.

Many other detectors have been used including UV-visible absorption and light-scattering detectors. Because of the ease with which solvents such as CO_2 can be volatilized, mass spectrometers are easier to interface with SFC systems than with HPLC systems. For this reason, SFC/MS has become a very useful hyphenated technique. Tandem mass spectrometers have also been successfully interfaced to SFC instruments.

34A-3 Supercritical Fluid Chromatography versus Other Column Methods

The information in Table 34-1 and other data reveal that several physical properties of supercritical fluids are intermediate between the properties of gases and liquids. As a result, this type of chromatography combines some of the characteristics of both gas and liquid chromatography. Thus, like gas chromatography, supercritical fluid chromatography is inherently faster than liquid chromatography because of the lower viscosity and higher diffusion rates in the mobile phase. High diffusivity, however, leads to longitudinal band spreading, which is a significant factor with gas but not with liquid chromatography. The intermediate diffusivities and

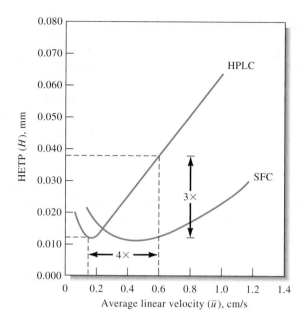

Figure 34-2 Performance characteristics of a 5-μm ODS column when elution is carried out with a conventional mobile phase (HPLC) and supercritical carbon dioxide SFC. (Copyright (2012) Hewlett-Packard Development company, L. P. Reproduced with Permission.)

viscosities of supercritical fluids result in faster separations than are achieved with liquid chromatography accompanied by less zone spreading than is encountered in gas chromatography.

Figure 34-2 shows plots of plate heights H as a function of average linear velocity $\bar{u}$ in cm/s for high-performance liquid chromatography and supercritical fluid chromatography. In both cases, the solute was pyrene, and the stationary phase a reversed-phase octadecyl silane maintained at 40°C. The mobile phase for HPLC was acetonitrile and water, while for SFC, the mobile phase was carbon dioxide. These conditions yielded about the same retention factor (k) for both mobile phases. Note that the minimum in plate height occurred at a flow rate of 0.13 cm/s with the HPLC and 0.40 cm/s for the SFC. The significance of this difference is shown in **Figure 34-3**, where these same conditions are used for the separation of pyrene from biphenyl. Note that the HPLC separation required over twice the time of the SFC separation.

Despite its advantages, SFC has not gained widespread acceptance because of the complexity and cost of the instrumentation and because of the lack of applications for which it provides unique information. However, SFC still fills an important gap in the separations world and provides a significant link between HPLC and GC.

Molecular model of biphenyl, a hazardous aromatic hydrocarbon. It is used as an intermediate in the production of emulsifiers, brighteners, plastics, and many other compounds. Biphenyl has been used as a heat transfer medium in heating fluids, as a dye carrier for textiles and copying paper, and as a solvent in pharmaceutical preparations. Paper impregnated with biphenyl is used in citrus fruit packaging to reduce fruit damage by fungus. Short-term exposure causes eye and skin irritation and toxic effects on the liver, kidneys, and nervous system. Long-term exposure causes kidney damage to laboratory animals and may affect the central nervous system in humans.

34A-4 Applications

Supercritical fluid chromatography appears to have a niche in the spectrum of column chromatographic methods because it can be applied to compounds that are not readily separated by gas or liquid chromatography. These compounds include species that are nonvolatile or thermally unstable and, in addition, contain no chromophoric groups that can be used for photometric detection. Separation of these compounds is possible with supercritical fluid chromatography at temperatures below 100°C; furthermore, detection is readily carried out by means of the highly sensitive flame ionization detector.

Supercritical fluid chromatography is today one of the primary separation methods for chiral compounds, such as those encountered in drug discovery. It has the potential to replace some of the reversed-phase HPLC separations of these compounds.

❮ SFC is widely used for chiral separations in the pharmaceutical industry.

Figure 34-3 Separation of pyrene and biphenyl by (a) HPLC and (b) SFC. (From D. R. Gere, *Science*, **1983**, *222*, 253, **DOI**: 10.1126/science.6414083. Reprinted with permission from AAAS.)

34B PLANAR CHROMATOGRAPHY

Planar chromatographic methods include **thin-layer chromatography** (TLC), **paper chromatography** (PC), and **electrochromatography**.[5] Each makes use of a flat, relatively thin layer of material that is either self-supporting or is coated on a glass, plastic, or metal surface. The mobile phase moves through the stationary phase by capillary action, sometimes assisted by gravity or an electrical potential. Planar chromatography was once called two-dimensional chromatography, although this term has now come to signify the coupling of two chromatographic techniques with different separation mechanisms.

Because most planar chromatography is based on the thin-layer technique, which is faster, has better resolution, and is more sensitive than its paper counterpart, this section is devoted to thin-layer methods. Capillary electrochromatography is described in Section 34D.

34B-1 The Scope of Thin-Layer Chromatography

Thin-layer chromatography (TLC) can be considered a form of liquid-solid chromatography in which the stationary phase is a thin layer on the surface of an appropriate plate. The mobile phase is drawn over the surface by capillary action. Thin-layer and liquid chromatography are quite similar in regards to theory, stationary and mobile phases. In fact, thin-layer plates can be profitably used to develop optimal conditions for separations by column liquid chromatography. The advantages of following this procedure are the speed and low cost of the exploratory thin-layer experiments. The TLC apparatus is much simpler than an HPLC system and much less expensive to operate.

At one time, TLC methods were widely used in the pharmaceutical industry for determining product purity. Today, HPLC techniques have replaced many of these methods. TLC has found widespread use in clinical laboratories and is the backbone of many biochemical and biological studies. It also finds extensive use in industrial laboratories.[6] Because of these many areas of application, TLC remains a very important technique.

34B-2 Principles of Thin-Layer Chromatography

Typical thin-layer separations are performed on a glass plate that is coated with a stationary phase, which consists of a thin and adherent layer of finely divided particles. The particles are similar to those described in the discussion of adsorption, normal- and reversed-phase partition, ion-exchange, and size-exclusion column chromatography. Mobile phases are also similar to those used in high-performance liquid chromatography.

Preparation of Thin-Layer Plates

A thin-layer plate is prepared by spreading an aqueous slurry of the finely ground solid onto the clean surface of a glass or plastic plate or microscope slide. Often a

---

[5]For recent reviews on planar chromatography, see J. Sherma, *Anal. Chem.*, **2010**, *82*, 4895, **DOI**: 10.1021/ac902643v; J. Sherma, **2008**, *80*, 4253, **DOI**: 10.1021/ac7023415.
[6]Monographs devoted to the principles and applications of thin-layer chromatography include B. Spangenberg, C. F. Poole, Ch. Weins, *Quantitative Thin-Layer Chromatography: A Practical Survey*, Berlin: Springer-Verlag, 2011; P. E. Wall, *Thin-Layer Chromatography: A Modern Practical Approach*, London: Royal Society of Chemistry, 2005; J. Sherma and B. Fried, eds., *Handbook of Thin-Layer Chromatography*, 3rd ed., New York: Dekker, 2003.

binder is incorporated into the slurry to enhance adhesion of the solid particles to the glass and to one another. The plate is then allowed to stand until the layer has set and adheres tightly to the surface; for some purposes, it may be heated in an oven for several hours. Several chemical supply houses offer precoated plates of various kinds. Costs are a few dollars per plate. The common plate dimensions in centimeters are 5×20, 10×20, and 20×20.

Commercial plates can be conventional and high-performance plates. Conventional plates have thicker layers (200 to 250 μm) of particles with particles sizes of 20 μm or greater. High-performance plates usually have film thicknesses of 100 μm and particle diameters of 5 μm or less.

Sample Application

Sample application is perhaps the most critical aspect of thin-layer chromatography. Usually the sample is applied as a spot 1 to 2 cm from the edge of the plate. Manual application of samples is performed by touching a capillary tube containing the sample to the plate or by use of a syringe. Mechanical dispensers, which increase the precision and accuracy of sample application, are available commercially.

Plate Development

Plate development is the process by which a sample is carried through the stationary phase by a mobile phase. It is analogous to elution in liquid chromatography. After applying a spot and evaporating the solvent, the plate is placed in a closed container saturated with vapors of the developing solvent. One end of the plate is immersed in the developing solvent, with care being taken to avoid direct contact between the sample and the developer (see **Figure 34-4**). After the developer has traversed one half or two thirds of the length of the plate, the plate is removed from the container and dried. The positions of the components are then determined in any of several ways.

Locating Analytes on the Plate

Several methods are used to locate sample components after separation. Two common methods that can be applied to most organic mixtures involve spraying with a solution of iodine or sulfuric acid. Both of these reagents react with organic compounds to yield dark products. Several specific reagents (such as ninhydrin) are also useful for locating separated species.[7]

Another method of detection is based on incorporating a fluorescent material into the stationary phase. After development, the plate is examined under ultraviolet light. The sample components quench the fluorescence of the material so that all of the plate fluoresces except where the nonfluorescing sample components are located.

The process of locating analytes on a thin-layer plate is often termed **visualization**.

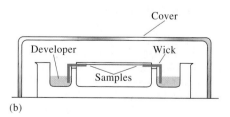

(a) (b)

Figure 34-4 (a) Ascending-flow developing chamber. (b) Horizontal-flow developing chamber in which samples are placed on both ends of the plate and developed toward the middle, thus doubling the number of samples that can be accommodated.

---

[7]For information on the calculations of TLC, see D. A. Skoog, F. J. Holler, and S. R. Crouch, *Principles of Instrumental Analysis*, 6th ed., Belmont, CA: Brooks/Cole, 2007, pp. 850–51.

34B-3 Paper Chromatography

Separations by paper chromatography are performed in the same way as those on thin-layer plates. The papers are manufactured from highly purified cellulose with close control over porosity and thickness. Such papers contain sufficient adsorbed water to make the stationary phase aqueous. Other liquids can be made to displace the water, however, thus providing a different type of stationary phase. For example, paper treated with silicone or paraffin oil permits reversed-phase paper chromatography in which the mobile phase is a polar solvent. Also available commercially are special papers that contain an adsorbent or an ion-exchange resin, thus permitting adsorption and ion-exchange paper chromatography.

34C CAPILLARY ELECTROPHORESIS[8]

Electrophoresis is a separation method based on the differential rates of migration of charged species in an applied dc electric field. This separation technique for macrosize samples was first developed by the Swedish chemist Arne Tiselius in the 1930s for the study of serum proteins; he was awarded the 1948 Nobel Prize for his work.

Electrophoresis on a macro scale has been applied to a variety of difficult analytical separation problems: inorganic anions and cations, amino acids, catecholamines, drugs, vitamins, carbohydrates, peptides, proteins, nucleic acids, nucleotides, polynucleotides, and numerous other species. A particular strength of electrophoresis is its unique ability to separate charged macromolecules of interest to biochemists, biologists, and clinical chemists. For many years, electrophoresis has been the powerhouse method for separating proteins (enzymes, hormones, and antibodies) and nucleic acids (DNA and RNA) for which it offers unparalleled resolution.

Until the appearance of capillary electrophoresis, electrophoretic separations were not carried out in columns but were performed in a flat stabilized medium such as paper or a porous semisolid gel. Remarkable separations were realized in such media, but the technique was slow, tedious, and required a good deal of operator skill. In the early 1980s, scientists began to explore the feasibility of performing these same separations on micro amounts of sample in fused-silica capillary tubes. Their results proved promising in terms of resolution, speed, and potential for automation. As a result, capillary electrophoresis (CE) has developed into an important tool for a wide variety of analytical separation problems and is the only type of electrophoresis that we will consider.[9]

34C-1 Instrumentation for Capillary Electrophoresis

As shown in **Figure 34-5**, the instrumentation for capillary electrophoresis is simple. A buffer-filled fused-silica capillary, which is typically 10 to 100 μm in internal diameter and 40 to 100 cm long, extends between two buffer reservoirs that also

---

[8]For additional discussion of capillary electrophoresis, see M. L. Marina, A. Rios, and M. Valcarcel, eds., *Analysis and Detection by Capillary Electrophoresis,* Vol. 45 of *Comprehensive Analytical Chemistry,* D. Barcelo, ed., Amsterdam: Elsevier, 2005; M. A. Strege and A. L. Lagu, eds., *Capillary Electrophoresis of Proteins and Peptides,* Totowa, NJ: Human Press, 2004; J. R. Petersen and A. A. Mohamad, eds., *Clinical and Forensic Applications of Capillary Electrophoresis,* Totowa, NJ: Human Press, 2001; R. Weinberger, *Practical Capillary Electrophoresis,* 2nd ed., New York: Academic Press, 2000.

[9]For recent reviews, see M. Geiger, A. L. Hogerton, and M. T. Bowser, *Anal. Chem.,* **2012,** *84,* 577, **DOI:** 10.1021/ac203205a; N. W. Frost, M. Jing, and M. T. Bowser, *Anal. Chem.,* **2010,** *82,* 4682, **DOI:** 10.1021/ac101151k.

Figure 34-5 Schematic of a capillary electrophoresis system.

hold platinum electrodes. Sample introduction is performed at one end and detection at the other. A potential of 5 to 30 kV dc is applied between the two electrodes. The positive polarity of the high voltage in Figure 34-5 can be reversed to allow separation of anions.

Sample introduction is accomplished by electrokinetic or pressure injection. In electrokinetic injection, one end of the capillary and its electrode are removed from their buffer compartments and placed in a small sample cup. A voltage is then applied for a measured time, causing the sample to enter the capillary by a combination of ionic migration and electroosmotic flow (see next section). In hydrodynamic injection, the sample introduction end of the capillary is also placed in a small cup containing the sample, but in this case, a pressure difference drives the sample solution into the capillary. The pressure difference can be caused by applying a vacuum at the detector end or by elevating the sample (hydrodynamic injection).

Because the separated analytes move past a common point in most types of capillary electrophoresis, detectors are similar in design and function to those described for HPLC. Table 34-3 lists several of the detection methods that have been reported for capillary electrophoresis. The second column of the table shows representative detection limits for these detectors.

34C-2 Electroosmotic Flow

A unique feature of capillary electrophoresis is **electroosmotic flow**. When a high voltage is applied across a fused-silica capillary tube containing a buffer solution, electroosmotic flow usually occurs in which the solvent migrates toward the cathode.

TABLE 34-3

Detectors for Capillary Electrophoresis*	
Type of Detector	**Representative Detection Limit[‡]** (attomoles detected)
Spectrometry	1–1000
Absorption[†]	1–0.01
Fluorescence	10
Thermal lens[†]	1000
Raman[†]	1–0.0001
Chemiluminescence[†]	1–0.01
Mass spectrometry	
Electrochemical	
Conductivity[†]	100
Potentiometry[†]	1
Amperometry	0.1

*B. Huang, J. J. Li, L. Zhang, and J. K. Cheng, *Anal. Chem.,* **1996**, *68*, 2366, **DOI**: 10.1021/ac9511253; S. C. Beale, *Anal. Chem.,* **1998**, *70*, 279, **DOI**: 10.1021/a19800141; S. N. Krylov and N. J. Dovichi, *Anal. Chem.,* **2000**, *72*, 111, **DOI**: 10.1021/a1000014c; S. Hu and N. J. Dovichi, *Anal. Chem.,* **2002**, *74*, 2833, **DOI**: 10.1021/ac0202379.

[‡]Detection limits quoted have been determined with injection volumes ranging from 18 pL to 10 nL.

[†]Mass detection limit converted from concentration detection limit using a 1-nL injection volume.

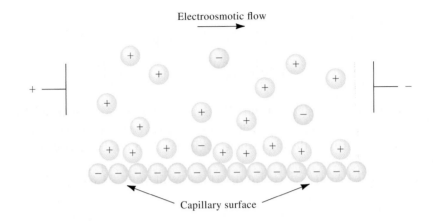

Figure 34-6 Charge distribution at a silica/capillary interface and resulting electroosmotic flow. (Reprinted (adapted) with permission from A. G. Ewing, R. A. Wallingford, and T. M Olefirowicz, *Anal. Chem.*, **1989**, *61*, 292A. **DOI**: 10.1021/ac00179a002. Copyright 1989 American Chemical Society.)

The rate of migration can be substantial. For example, a 50-mM pH 8 buffer has been found to flow through a 50-cm capillary toward the cathode at approximately 5 cm/min with an applied potential of 25 kV.[10]

As shown in **Figure 34-6**, the cause of electroosmotic flow is the electric double layer that develops at the silica/solution interface. At pH values higher than 3, the inside wall of a silica capillary is negatively charged due to ionization of the surface silanol groups (Si—OH). Buffer cations congregate in an electrical double layer adjacent to the negative surface of the silica capillary. The cations in the diffuse outer layer to the double layer are attracted toward the cathode, or negative electrode, and since they are solvated, they drag the bulk solvent along with them. As shown in **Figure 34-7**, electroosmosis leads to bulk solution flow that has a flat profile across the tube because flow originates at the walls of the tubing. This profile is in contrast to the laminar (parabolic) profile that is observed with the pressure-driven flow encountered in HPLC. Because the profile is essentially flat, electroosmotic flow does not contribute significantly to band broadening the way pressure-driven flow does in liquid chromatography.

The rate of electroosmotic flow is generally greater than the electrophoretic migration velocities of the individual ions and effectively becomes the mobile-phase pump of CE. Even through analytes migrate according to their charges within the capillary, the electroosmotic flow rate is usually sufficient to sweep all positive, neutral, and even negative species toward the same end of the capillary so that all can be detected as they pass by a common point (see **Figure 34-8**). The resulting **electropherogram** looks like a chromatogram but with narrower peaks.

Electroosmosis is often desirable in certain types of capillary electrophoresis, but in other types, it is not. It can be minimized by coating the inside capillary

Figure 34-7 Flow profiles for liquids under (a) electroosmotic flow and (b) pressure-induced flow.

[10]J. D. Olechno, J. M. Y. Tso, J. Thayer, and A. Wainright, *Amer. Lab.*, **1990**, *22*(17), 51.

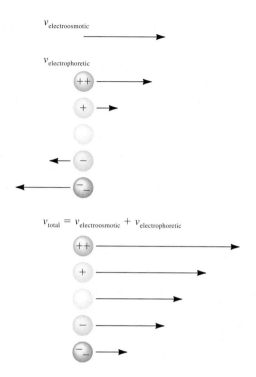

$v_{\text{electroosmotic}}$

$v_{\text{electrophoretic}}$

$v_{\text{total}} = v_{\text{electroosmotic}} + v_{\text{electrophoretic}}$

Figure 34-8 Velocities in the presence of electroosmotic flow. The length of the arrow next to an ion indicates the magnitude of its velocity; the direction of the arrow indicates the direction of motion. The negative electrode would be to the right and the positive electrode to the left of this section of solution.

wall with a reagent like trimethylchlorosilane to eliminate the surface silanol groups.

34C-3 The Basis for Electrophoretic Separations

The migration rate v of an ion in an electric field is given by

$$v = \mu_e E = \mu_e \times \frac{V}{L} \tag{34-1}$$

where E is the electric field strength in volts per centimeter, V is the applied voltage, L is the length of the tube between electrodes, and μ_e is the **electrophoretic mobility**, which is proportional to the charge on the ion and inversely proportional to the frictional retarding force on the ion. The frictional retarding force on an ion is determined by the size and shape of the ion and the viscosity of the medium.

The **electrophoretic mobility** is the ratio of the migration rate of an ion to the applied electric field.

Although CE is not a chromatographic process, separations are often described in a manner similar to chromatography. For example, in electrophoresis, we can calculate the plate count N by

$$N = \frac{\mu_e V}{2D} \tag{34-2}$$

where D is the diffusion coefficient of the solute (cm^2/s). Because resolution increases with plate count, it is desirable to use high applied voltages in order to achieve high-resolution separations. Note that for electrophoresis, contrary to the situation in chromatography, the plate count does not increase with the column length. Typically, capillary electrophoresis plate counts are 100,000 to 200,000 at the usual applied voltages.

34C-4 Applications of Capillary Electrophoresis

Capillary electrophoretic separations are performed in several ways called modes. These modes include **isoelectric focusing**, **isotachophoresis**, and **capillary zone electrophoresis** (CZE). We shall consider only capillary zone electrophoresis in which the buffer composition is constant throughout the region of the separation. The applied field causes each of the different ionic components of the mixture to migrate according to its own mobility and to separate into zones that may be completely resolved or may be partially overlapped. Completely resolved zones have regions of buffer between them. The situation is analogous to elution column chromatography, where regions of mobile phase are located between zones containing separated analytes.

Separation of Small Ions

For most electrophoretic separations of small ions, the smallest analysis time results when the analyte ions move in the same direction as the electroosmotic flow. Thus, for cation separations, the walls of the capillary are untreated, and the electroosmotic flow and the cation movement are toward the cathode. For the separation of anions, on the other hand, the electroosmotic flow is usually reversed by treating the walls of the capillary with an alkyl ammonium salt, such as cetyl trimethylammonium bromide. The positively charged ammonium ions become attached to the negatively charged silica surface and, in turn, create a negatively charged double layer of solution, which is attracted toward the anode, reversing the electroosmotic flow.

In the past, the most common method for analysis of small anions has been ion-exchange chromatography. For cations, the preferred techniques have been atomic absorption spectroscopy and inductively coupled plasma emission or mass spectrometry. In recent years, capillary electrophoretic methods have begun to compete with these traditional methods for small ion analysis. Several major reasons for adoption of electrophoretic methods have been recognized: lower equipment costs, smaller sample size requirements, much greater speed, and better resolution. However, because variations in electroosmotic flow rates make reproducing CE separations difficult, LC methods and atomic spectrometric methods are still widely used for small inorganic ions.

The initial cost of equipment and the expense of maintenance for electrophoresis is generally significantly lower than those for ion chromatographic and atomic spectroscopic instruments. The cost of CE instruments varies significantly depending on the type of detection system desired. Simple CE instruments with UV-visible detection can cost in the $10,000-to-$20,000 range, but instruments with mass spectrometry detection can cost substantially more.

Sample sizes for electrophoresis are in the nanoliter range, but microliter or larger samples are usually needed for other types of small ion analysis. Thus, electrophoretic methods are more sensitive than the other methods on a mass basis (but usually not on a concentration basis).

Figure 34-9 illustrates the unsurpassed quickness and resolution of electrophoretic separations of small anions. Here, 30 anions were separated cleanly in just over three minutes. Typically, an ion-exchange separation of only three or four anions

Figure 34-9 Electropherogram showing the separation of 30 anions. Capillary internal diameter: 50 μm (fused silica). Detection: indirect UV, 254 nm. Peaks: 1 = thiosulfate (4 ppm), 2 = bromide (4 ppm), 3 = chloride (2 ppm), 4 = sulfate (4 ppm), 5 = nitrite (4 ppm), 6 = nitrate (4 ppm), 7 = molybdate (10 ppm), 8 = azide (4 ppm), 9 = tungstate (10 ppm), 10 = monofluorophosphate (4 ppm), 11 = chlorate (4 ppm), 12 = citrate (2 ppm), 13 = fluoride (1 ppm), 14 = formate (2 ppm), 15 = phosphate (4 ppm), 16 = phosphite (4 ppm), 17 = chlorite (4 ppm), 18 = galactarate (5 ppm), 19 = carbonate (4 ppm), 20 = acetate (4 ppm), 21 = ethanesulfonate (4 ppm), 22 = propionate (5 ppm), 23 = propanesulfonate (4 ppm), 24 = butyrate (5 ppm), 25 = butanesulfonate (4 ppm), 26 = valerate (5 ppm), 27 = benzoate (4 ppm), 28 = *l*-glutamate (5 ppm), 29 = pentanesulfonate (4 ppm), 30 = *d*-gluconate (5 ppm). (Reprinted from W. A. Jones and P. Jandik, *J. Chromatogr.*, **1991**, *546*, 445, **DOI**: 10.1016/s0021-9673(01)93043-2, with permission from Elsevier.)

could be accomplished in this brief time period. **Figure 34-10** further illustrates the speed at which separations can be carried out. As can be seen, 19 cations were separated in less than two minutes.

Separation of Molecular Species

A variety of small synthetic herbicides, pesticides, and pharmaceuticals that are ions or can be derivatized to yield ions have been separated and analyzed by CE. **Figure 34-11** is illustrative of this type of application in which anti-inflammatory drugs, which have acidic properties with characteristic pK_as, are separated in less than 15 min.

Figure 34-10 Separation of alkali, alkaline earths, and lanthanides. Capillary: 36.5-cm × 75-μm fused silica, + 30 kV. Injection: hydrostatic, 20 s at 10 cm. Detection: indirect UV, 214 nm. Peaks: 1 = rubidium (2 ppm), 2 = potassium (5 ppm), 3 = calcium (2 ppm), 4 = sodium (1 ppm), 5 = magnesium (1 ppm), 6 = lithium (1 ppm), 7 = lanthanum (5 ppm), 8 = cerium (5 ppm), 9 = praseodymium (5 ppm), 10 = neodymium (5 ppm), 11 = samarium (5 ppm), 12 = europium (5 ppm), 13 = gadolinium (5 ppm), 14 = terbium (5 ppm), 15 = dysprosium (5 ppm), 16 = holmium (5 ppm), 17 = erbium (5 ppm), 18 = thulium (5 ppm), 19 = ytterbium (5 ppm). (Reprinted from A. Weston, P. R. Brown, P. Jandik, W. R. Jones and A. L. Heckenberg, *J. Chromatog. A*, **1992**, *593*, 289, **DOI**: 10.1016/0021-9673(92)80297-8, with permission from Elsevier.)

Figure 34-11 Separation of anti-inflammatory drugs by CZE. Detection UV at 200 nm. Analytes: (1) sulindac, (2) indomethacin, (3) piroxicam, (4) ketoprofen, (5) nimesulide, (6) ibuprofen, (7) naproxen. (From Y. L. Chen and S. M. Wu, *Anal. Bioanal. Chem.,* **2005,** *381,* 907, **DOI:** 10.1007/s00216-004-2970-x. With permission of Springer-Verlag.

Naproxen

Ibuprofen

Tolmetin

Molecular models of anti-inflammatory drugs, naproxen, ibuprofen, and tolmetin. These nonsteroidal anti-inflammatory agents are thought to relieve pain by inhibiting the synthesis of prostaglandins that are produced by the human body in response to the presence of toxins, infectious agents, and tissue fluids that result from the inflammation process. High concentrations of prostaglandins cause fever and pain. Ibuprofen is also known as Motrin, Advil, or Nuprin. Naproxen sodium is Aleve, and tolmetin is tolectin. Each has been used to treat symptoms of arthritis and to relieve the pain caused by gout; bursitis; tendonitis; sprains, strains, and other injuries; and menstrual cramps. Ibuprofen and naproxen are available over-the-counter in the United States.

Proteins, amino acids, and carbohydrates have all been separated in minimum times by CZE. In the case of neutral carbohydrates, the separations are preceded by formation of negatively charged borate complexes. The separation of protein mixtures is illustrated in **Figure 34-12**. Feature 34-1 discusses the use of capillary electrophoresis arrays for DNA sequencing.

Figure 34-12 CZE separation of a model protein mixture. Conditions: pH 2.7 buffer. Absorbance detection: 214 nm; 22 kV, 10 A. Peaks: identified in the following table.

Model Proteins Separated at pH 2.7			
Peak No.	**Proteins**	**Molecular Mass**	**Isoelectric Point, pH**
1	Cytochrome c	12,400	10.7
2	Lysozyme	14,100	11.1
3	Trypsin	24,000	10.1
4	Trypsinogen	23,700	8.7
5	Trypsin inhibitor	20,100	4.5

FEATURE 34-1

Capillary Array Electrophoresis in DNA Sequencing

A major goal of the human genome project was to determine the order of occurrence of the four bases, adenine (A), cytosine (C), guanine (G), and thymine (T), in DNA molecules. The sequence defines an individual's genetic code. The need for sequencing DNA has spawned the development of several new analytical instruments. Among the most attractive of these approaches is capillary array electrophoresis.[11] In this technique, as many as 96 capillaries are operated in parallel. The capillaries are filled with a separation matrix, normally a linear polyacrylamide gel. The capillaries have inner diameters of 35 to 75 μm and are 30 to 60 cm in length.

In sequencing, DNA extracted from cells is fragmented by various approaches. Depending on the terminal base in the fragment, one of four fluorescent dyes is attached to the various fragments. The sample contains many different-sized fragments each with a fluorescent label. Under the influence of the electrophoretic field, lower molecular mass fragments move faster and arrive at the detector sooner than higher molecular mass fragments. The DNA sequence is determined by the dye color sequence of the eluting fragments. Lasers are used to excite the dye fluorescence. Several different techniques have been described for detecting the fluorescence. One method uses a scanning system such that the capillary bundle is moved relative to the excitation laser and the four wavelength detection system. In the detection system illustrated in **Figure 34F-1**, a laser beam is focused onto the capillary array by a lens. The region that is illuminated by the laser is imaged onto a CCD detector (see Section 25A-4). Filters allow wavelength selection to detect the four colors. Simultaneous separation of

11 DNA fragments in 100 capillaries has been reported.[12] Other designs include sheath-flow detector systems and a detector that uses two diode lasers for excitation. Commercial instrumentation is available from several instrument companies. Miniaturized DNA sequencers have been developed using lab-on-a-chip technology. Such miniature systems are becoming more portable, which should allow field use for forensic and other applications.

Figure 34F-1 On-column laser fluorescence detection system for capillary array electrophoresis. A laser is focused as a line onto the array of capillaries at a 45° angle. The fluorescence is filtered and detected by a CCD camera through a wide-angle lens. (Reprinted (adapted) with permission from K. Ueno and E. S. Yeung, *Anal. Chem.*, **1994**, *66*, 1424, **DOI**: 10.1021/ac00081a010. Copyright 1994 American Chemical Society.)

Spreadsheet Summary In Chapter 15 of *Applications of Microsoft® Excel in Analytical Chemistry*, 2nd ed., capillary electrophoresis data are used to determine the mobilities of inorganic ions. Capillary electrophoresis results are also used to determine pK_a values of several weak organic acids.

34D CAPILLARY ELECTROCHROMATOGRAPHY

Capillary electrochromatography (CEC) is a hybrid of HPLC and capillary electrophoresis (CE) that offers some of the best features of the two methods.[13] Like HPLC, it is applicable to the separation of neutral species. Like CE, however, it provides highly efficient separations on microvolumes of sample solution without the

---

[11]For reviews, see I. Kheterpal and R. A. Mathies, *Anal. Chem.*, **1999**, *71*, 31A, **DOI**: 10.1021/ac990099w; M. Geiger, A. L. Hogerton, and M. T. Bowser, *Anal. Chem.*, **2012**, *84*, 577, **DOI**: 10.1021/ac203205a; N. W. Frost, M. Jing, and M. T. Bowser, *Anal. Chem.*, **2010**, *82*, 4682, **DOI**: 10.1021/ac101151k.

[12]K. Ueno and E. S. Yeung, *Anal. Chem.*, **1994**, *66*, 1424, **DOI**: 10.1021/ac00081a010.

[13]For a discussion of this method, see L. A. Colon, Y. Guo, and A. Fermier, *Anal. Chem.*, **1997**, *69*, 461A, **DOI**: 10.1021/ac9717245.

need for the high-pressure pumping system required for HPLC. In CEC, a mobile phase is transported across a stationary phase by electroosmotic flow. As shown in Figure 34-7, electroosmotic pumping leads to a flat plug profile rather than the parabolic profile that results from pressure-induced flow. The flat profile leads to narrow bands and thus high separation efficiencies.

34D-1 Packed Column Electrochromatography

Electrochromatography based on packed columns is the least mature of the various electroseparation techniques. In this method, a polar solvent is usually driven by electroosmotic flow through a capillary that is packed with a reversed-phase HPLC packing. Separations depend on the distribution of the analyte species between the mobile phase and the liquid stationary phase held on the packing. **Figure 34-13** shows a typical electrochromatogram for the separation of 16 polyaromatic hydrocarbons (PAHs) in a 33-cm-long capillary having an inside diameter of 75 μm. The mobile phase consisted of acetonitrile in a 4-mM sodium borate solution. The stationary phase consisted of 3-μm octadecylsilica particles.

34D-2 Micellar Electrokinetic Capillary Chromatography

The capillary electrophoretic methods we have described thus far are not applicable to the separation of uncharged solutes. In 1984, however, Terabe and collaborators[14] described a modification of the method that permitted the separation of low-molecular-mass aromatic phenols and nitro compounds with equipment such as shown in Figure 34-5. In this technique, a surfactant is introduced at a concentration level at which **micelles** form. Micelles form in aqueous solutions when the concentration of an ionic species having a long-chain hydrocarbon tail is increased above a certain level called the **critical micelle concentration** (CMC). At this point, the surfactant begins to form spherical aggregates made up to 40 to 100 ions with their hydrocarbon tails in the interior

Figure 34-13 Electrochromatogram showing the electrochromatographic separation of 16 PAHs (~10^{-6} to 10^{-8} M of each compound). Peaks: (1) naphthalene, (2) acenaphthylene, (3) acenaphthene, (4) fluorene, (5) phenanthrene, (6) anthracene, (7) fluoranthene, (8) pyrene, (9) benz[*a*]anthracene, (10) chrysene, (11) benzo[*b*]fluoranthene, (12) benzo[*k*]fluoranthene, (13) benzo[*a*]pyrene, (14) dibenz[*a,h*] anthracene, (15) benzo[*ghi*]perylene, and (16) indeo[1,2,3-*cd*]pyrene. (Reprinted (adapted) with permission from C. Yan et al., *Anal. Chem.,* **1995**, *67*, 2026, **DOI**: 10.1021/ ac00109a020. Copyright 1995 American Chemical Society.)

[14]S. Terabe et al., *Anal. Chem.,* **1984**, *56*, 111; **DOI**: 10.1021/ac00265a031; S. Terabe, K. Otsuka, and T. Ando, *Anal. Chem.,* **1985**, *57*, 841, **DOI**: 10.1021/ac00281a014; S. Terabe, *Anal. Chem.,* **2004**, 76, 240A, **DOI**: 10.1021/ac0415859.

of the aggregate and their charged ends exposed to water on the outside. Micelles constitute a stable second phase that can incorporate nonpolar compounds in the hydrocarbon interior of the particles, thus *solubilizing* the nonpolar species. Solubilization is commonly encountered when a greasy material or surface is washed with a detergent solution.

Capillary electrophoresis carried out in the presence of micelles is termed **micellar electrokinetic capillary chromatography** and given the acronym MEKC. In this technique, surfactants are added to the operating buffer in amounts that exceed the critical micelle concentration. For most applications to date, the surfactant has been sodium dodecyl sulfate (SDS). The surface of an ionic micelle of this type has a large negative charge, which gives it a large electrophoretic mobility. Most buffers, however, exhibit such a high electroosmotic flow rate toward the negative electrode that the anionic micelles are carried toward that electrode also, although at a much reduced rate. Thus, during an experiment, the buffer mixture consists of a faster-moving aqueous phase and a slower-moving micellar phase. When a sample is introduced into this system, the components distribute themselves between the aqueous phase and the hydrocarbon phase in the interior of the micelles. The positions of the resulting equilibria depend on the polarity of the solutes. With polar solutes, the aqueous solution is favored; with nonpolar compounds, the hydrocarbon environment is preferred.

The phenomena just described are quite similar to what occurs in a liquid partition chromatographic column except that the "stationary phase" is moving along the length of the column, although at a much slower rate than the mobile phase. The mechanism of separations is identical in the two cases and depends on differences in distribution constants for analytes between the mobile aqueous phase and the hydrocarbon **pseudostationary phase**. The process is thus true chromatography; hence, the name micellar electrokinetic capillary *chromatography*. **Figure 34-14** illustrates two typical separations by MEKC.

Molecular model of caffeine. Caffeine stimulates the cerebral cortex by inhibiting an enzyme that inactivates a certain form of adenosine triphosphate, the molecule that supplies energy. Caffeine occurs in coffee, tea, and cola drinks.

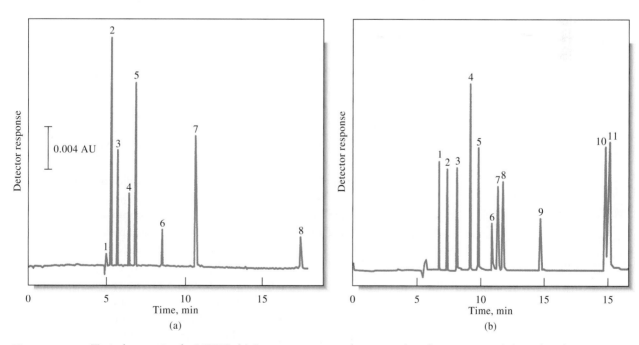

Figure 34-14   Typical separation by MEKC. (a) Some test compounds: 1 = methanol, 2 = resorcinol, 3 = phenol, 4 = *p*-nitroaniline, 5 = nitrobenzene, 6 = toluene, 7 = 2-naphthol, 8 = Sudan III. Capillary: 50-μm inside diameter, 500 mm to the detector. Applied voltage: ca. 15 kV. Detection: UV absorption, 210 nm. (b) Analysis of a cold medicine. Compounds: 1 = acetaminophen, 2 = caffeine, 3 = sulpyrine, 4 = naproxen, 5 = guaiphenesin, 6 = impurity, 7 = phenacetin, 8 = ethenzamide, 9 = 4-isopro-pylantipyrine, 10 = noscapine, 11 = chlorpheniramine and tipepidine. Applied voltage: 20 kV. Capillary: as in (a). Detection: UV absorption, 220 nm. (Reprinted from S. Terabe, *Trends Anal. Chem.*, **1989**, *8*, 129, **DOI**: 10.1016/0165-9936(89)85022-8, with permission from Elsevier.)

Capillary chromatography in the presence of micelles appears to have a promising future. One advantage that this hybrid technique has over HPLC is much higher column efficiencies (100,000 plates or more). In addition, changing the second phase in MEKC is simple, involving only the changing of the micellar composition of the buffer. In contrast, in HPLC, the second phase can only be altered by changing the type of column packing. The MEKC technique seems particularly useful for separating small molecules that are impossible to separate by traditional electrophoresis.

> **Spreadsheet Summary** In the final exercise in Chapter 15 of *Applications of Microsoft® Excel in Analytical Chemistry*, 2nd ed., micellar electrokinetic capillary chromatography is used to determine the critical micelle concentration (CMC) of a surfactant. An equation is developed to relate the retention factor to the CMC. Measured retention times are then used to determine the CMC from a regression analysis.

34E FIELD-FLOW FRACTIONATION

Field-flow fractionation (FFF) describes a group of analytical techniques that are becoming quite useful in the separation and characterization of dissolved or suspended materials such as polymers, large particles and colloids. The FFF concept was first described by Giddings in 1966.[15] However, only recently have practical applications and advantages over other methods been shown.[16]

34E-1 Separation Mechanisms

Separations in FFF occur in a thin ribbonlike flow channel, such as that shown in **Figure 34-15**. The channel is typically 25 to 100 cm in length and 1 to 3 cm in breadth. The thickness of the ribbonlike structure is usually 50 to 500 μm. The channel is usually cut from a thin spacer and sandwiched between two walls. An electrical, thermal or centrifugal field is applied perpendicular to the flow direction. Alternatively, a cross flow perpendicular to the main flow direction can be used.

In practice, the sample is injected at the inlet to the channel. The external field is next applied across the face of the channel, as illustrated in Figure 34-15. In the presence of the field, sample components migrate toward the **accumulation wall** at a velocity determined by the strength of the interaction of the component with the field. Sample components rapidly reach a steady-state concentration distribution near the accumulation wall, as shown in **Figure 34-17**. The mean thickness of the component layer l is related to the diffusion coefficient of the molecule D and to the field-induced velocity U toward the wall. The faster the component moves in the field, the thinner the layer near the wall. The larger the diffusion coefficient, the thicker is the layer. Since the sample components have different values of D and U, the mean layer thickness will vary among components.

[15]J. C. Gidding, *Sep Sci.,* **1966**, *1*, 123, **DOI**: 10.1080/01496396608049439.
[16]For a review of FFF methods, see J. C. Giddings, *Anal. Chem.*, **1995**, *67*, 592A, **DOI**: 10.1021/ac00115a001.

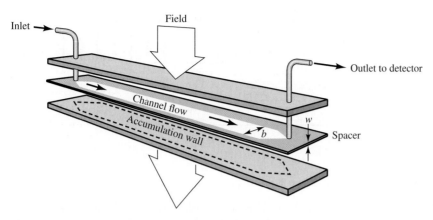

Figure 34-15 Schematic diagram of FFF flow channel sandwiched between two walls. An external field (electrical, thermal, or centrifugal) is applied perpendicular to the flow direction.

Once components have reached their steady-state profiles near the accumulation wall, the channel flow is begun. The flow is laminar, resulting in the parabolic profile shown on the left in **Figure 34-16**. The main carrier flow has its highest velocity in the center of the channel and its lowest velocity near the walls. Components that interact strongly with the field are compressed very near the wall, as shown by component A in Figure 34-17. Here, they are eluted by slow-moving solvent. Components B and C protrude more into the channel and experience a higher solvent velocity. The elution order is thus C, then B, and finally A. Components that are separated by FFF flow through a UV-visible absorption, refractive index, or fluorescence detector located at the end of the flow channel. Detectors used are similar to those used in HPLC separations. The separation results are revealed by a plot of detector response versus time, called a **fractogram**, which is similar to a chromatogram in chromatography.

34E-2 FFF Methods

Different FFF subtechniques result from the application of different types of fields or gradients.[17] The methods that have been used are **sedimentation FFF**, **electrical FFF**, **thermal FFF**, and **flow FFF**.

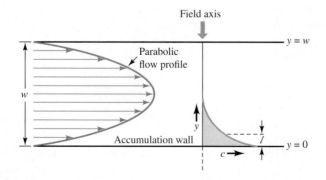

Figure 34-16 When the field is applied in FFF, components migrate to the accumulation wall where there is an exponential concentration profile as seen on the right. Components extend a distance y into the channel. The average thickness of the layer is l, which differs for each component. The main channel flow is then turned on, and the parabolic flow profile of the eluting solvent is shown on the left.

---

[17]For a discussion of the various FFF methods, see J. C. Giddings, *Unified Separation Science*, Ch. 9, New York: Wiley, 1991; M. E. Schimpf, K. Caldwell, and J. C. Giddings, eds., *Field-Flow Fractionation Handbook*, New York: Wiley, 2000.

Figure 34-17 Three components A, B, and C are shown compressed against the accumulation wall in FFF to different degrees because of different interactions with the external field. When the flow is begun, component A experiences the lowest solvent velocity because it is the closest to the wall. Component B protrudes more into the channel where it experiences a higher flow velocity. Component C, which interacts the least with the field, experiences the highest solvent flow velocity and thus is displaced the most rapidly by the flow.

Sedimentation FFF

Sedimentation FFF has been the most widely used form. In this technique, the channel is coiled and made to fit inside a centrifuge basket, as illustrated in **Figure 34-18**. Components with the highest mass and density are driven to the wall by the sedimentation (centrifugation) force and elute last. Low-mass species are eluted first. There is relatively high selectivity between particles of different size in sedimentation FFF. A separation of polystyrene beads of various diameters by sedimentation FFF is shown in **Figure 34-19**.

Because the centrifugation forces are relatively weak for small molecules, sedimentation FFF is most applicable for molecules with molecular masses exceeding 10^6. Such systems as polymers, biological macromolecules, natural and industrial colloids, emulsions, and subcelluar particles appear to be amenable to separation by sedimentation FFF.

Electrical FFF

In electrical FFF, an electric field is applied perpendicular to the flow direction. Retention and separation occur based on electrical charge. Species with the highest charge are driven most effectively toward the accumulation wall. Species of lower charge are not as compacted and protrude more into the higher-flow region. Hence, species of the lowest charge are eluted first, with highly charged species retained the most.

Figure 34-18 Sedimentation FFF apparatus. (Courtesy of Postnova Analytics.)

Figure 34-19 Fractogram illustrating separation of polystyrene beads of various diameters by sedimentation FFF. The channel flow rate was 2 mL/min. (Courtesy of FFFractionation, LLC, Salt Lake City, UT.)

Because electrical fields are quite powerful, even small ions should be amenable to separation by electrical FFF. However, electrolysis effects have limited the applications of this method to the separation of mixtures of proteins and other large molecules.

Thermal FFF

In thermal FFF, a thermal field is applied perpendicular to the flow direction by forming a temperature gradient across the FFF channel. The temperature difference induces thermal diffusion in which the velocity of movement is related to the thermal diffusion coefficient of the species.

Thermal FFF is particularly well suited for the separation of synthetic polymers with molecular masses in the range of 10^3 to 10^7. The technique has significant advantages over size-exclusion chromatography for high-molecular-mass polymers. On the other hand, low-molecular-mass polymers appear to be better separated by size-exclusion methods. In addition to polymers, particles and colloids have been separated by thermal FFF.[18]

Flow FFF

Perhaps the most versatile of all the FFF subtechniques is flow FFF in which the external field is replaced by a slow cross flow of the carrier liquid.[19] The perpendicular flow transports material to the accumulation wall in a nonselective manner. However, steady-state layer thicknesses are different for various components because they depend not only on the transport rate but also on molecular diffusion. Exponential distributions of differing thicknesses are formed as in normal FFF.

[18]P. M. Shiundu, G. Liu, and J. C. Giddings, *Anal. Chem.*, **1995**, *67*, 2705, **DOI:** 10.1021/ac00111a032.

[19]See K. Wahlund and L. Nilsson, in *Field-Flow Fractionation in Biopolymer Analysis*, S. K. R. Williams and K. D. Caldwell, eds., New York: Springer-Verlag, 2012.

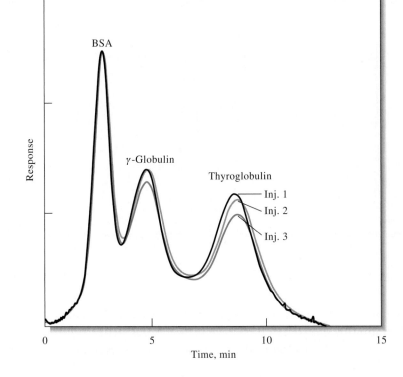

Figure 34-20 Separation of three proteins by flow FFF. Three separate injections are shown. In this experiment, the sample was concentrated at the head of the channel by means of an opposing flow. (Reprinted (adapted) with permission from H. Lee, S. K. R. Williams, and J. C. Giddings, *Anal. Chem.*, **1998**, *70*, 2495, **DOI:** 10.1021/ac9710792. Copyright 1998 American Chemical Society.)

Flow FFF has been applied to the separation of proteins, synthetic polymers, and a variety of colloidal particles. **Figure 34-20** illustrates the separation of three proteins by flow FFF. The reproducibility is illustrated by the fractograms for the three injections.

34E-3 Advantages of FFF over Chromatographic Methods

Field-flow fractionation has several apparent advantages over ordinary chromatographic methods for some applications. First, no packing material or stationary phase is needed for separation to occur. In some chromatographic systems, there may be undesirable interactions between the packing material or stationary phase and the sample constituents. Some solvents or sample materials adsorb or react with the stationary phase or its support. Macromolecules and particles are particularly prone to such adverse interactions.

The geometry and flow profiles of FFF are well characterized. Likewise, the effects of most external fields can be readily modeled. As a result, fairly exact theoretical predictions of retention and plate height can be made in FFF. Chromatographic predictions are still rather inexact in comparison.

Finally, the external field governs FFF retention. With electrical, centrifugal and flow FFF, the perpendicular forces can be varied rapidly and in a time-programmed fashion. This capability imparts to FFF a certain versatility in adapting to different types of samples. Similarly, methods can be easily optimized for resolution and separation speed.

Although field-flow fractionation is a fairly recent addition to the group of analytical separation methods, it has been shown to be highly complementary to chromatography. The FFF methods are best suited at present for macromolecules and particles that are for the most part beyond the molecular mass range of chromatographic methods. On the other hand, chromatographic methods are superior for low-molecular-mass substances.

WEB WORKS Use a search engine to find articles on capillary electrophoresis with mass spectrometry detection. What are the major challenges in coupling a CE capillary to a mass spectrometer? What types of mass spectrometers are most widely used in CE-MS applications? Are tandem mass spectrometers used? What modes of CE are most useful for CE-MS? Are commercial CE-MS systems available? Which instrument companies produce CE-MS systems? Describe the major applications for which CE-MS has been useful. What unique information has been provided by CE-MS?

QUESTIONS AND PROBLEMS

34-1. List the types of substances to which each of the following separation methods is most applicable:
*(a) supercritical fluid chromatography.
(b) thin-layer chromatography.
*(c) capillary zone electrophoresis.
(d) sedimentation FFF.
*(e) flow FFF.
(f) micellar electrokinetic capillary chromatography.

34-2. Define:
*(a) supercritical fluid.
(b) critical point.
*(c) two-dimensional thin-layer chromatography.
(d) electrophoretic mobility.
*(e) critical micelle concentration.
(f) electrical FFF.

*34-3. What properties of a supercritical fluid are important in chromatography?

34-4. Describe the effect of pressure on supercritical fluid chromatography.

*34-5. How do instruments for supercritical fluid chromatography differ from those for (a) HPLC and (b) GC?

34-6. List some of the advantageous properties of supercritical CO_2 as a mobile phase for chromatographic separations.

*34-7. What important property of supercritical fluids is related to their densities?

34-8. Compare supercritical fluid chromatography with other column chromatographic methods.

*34-9. For supercritical carbon dioxide, predict the effect that the following changes will have on the elution time in an SFC experiment.
(a) Increase the flow rate (at constant temperature and pressure).

(b) Increase the pressure (at constant temperature and flow rate).
(c) Increase the temperature (at constant pressure and flow rate).

34-10. What is the effect of pH on the separation of amino acids by electrophoresis? Why?

*34-11. What is electroosmotic flow? Why does it occur?

34-12. How could electroosmotic flow be repressed? Why would one want to repress it?

*34-13. What is the principle of separation by capillary zone electrophoresis?

34-14. A certain inorganic cation has a electrophoretic mobility of 5.13×10^{-4} cm^2 s^{-1} V^{-1}. This same ion has a diffusion coefficient of 9.1×10^{-6} cm^2 s^{-1}. If this ion is separated by capillary zone electrophoresis with a 50-cm capillary, what is the expected plate count, N, at applied voltages of
(a) 5 kV?   (c) 20 kV?
(b) 10 kV?   (d) 30 kV

*34-15. The cationic analyte of Problem 34-14 was separated by capillary zone electrophoresis in a 50-cm capillary at 20 kV. Under the separation conditions, the electroosmotic flow rate was 0.65 mm s^{-1} toward the cathode. If the detector were placed 40 cm from the injection end of the capillary, how long does it take in minutes for the analyte cation to reach the detector after the field is applied?

34-16. What is the principle of micellar electrokinetic capillary chromatography? How does it differ from capillary zone electrophoresis?

*34-17. Describe a major advantage of micellar electrokinetic capillary chromatography over conventional liquid chromatography.

34-18. What determines the elution order in sedimentation FFF?

*34-19. Three large proteins are ionized at the pH at which an electrical FFF separation is carried out. If the ions are designated A^{2+}, B^+, and C^{3+}, predict the order of elution.

34-20. List the major advantages and limitations of FFF compared to chromatographic methods.

34-21. Challenge Problem: Doxorubicin (DOX) is a widely used anthracycline that has been effective in treatments of leukemia and breast cancer in humans (A. B. Anderson, C. M. Ciriaks, K. M. Fuller, and E. A. Ariaga, *Anal. Chem.*, **2003**, *75*, 8, **DOI:** 10.1021/ac020426r). Unfortunately, side effects, such as liver toxicity and drug resistance, have been reported. In their study, Anderson et al. used laser-induced fluorescence (LIF) as a detection mode for capillary electrophoresis (CE) in order to investigate metabolites of DOX in single cells and subcellular fractions. The following are results similar to those obtained by Anderson et al. for quantifying doxorubicin by LIF. The CE peak areas were measured as a function of the DOX concentration in order to construct a calibration curve.

DOX Concentration, nM	Peak Area
0.10	0.10
1.00	0.80
5.00	4.52
10.00	8.32
20.00	15.7
30.00	26.2
50.00	41.5

(a) Find the equation for the calibration curve and the standard deviations of the slope and intercept. Find the R^2 value.

(b) Rearrange the equation found in part (a) to express concentration in terms of the measured area.

(c) The limit of detection (LOD) for DOX was found to be 3×10^{-11} M. If the injection volume was 100 pL, what was the LOD in moles?

(d) Two samples of unknown DOX concentration were injected, and peak areas of 11.3 and 6.97 obtained. What were the concentrations and their standard deviations?

(e) Under certain conditions, the DOX peak required 300 s to reach the LIF detector. What time would be required if the applied voltage were doubled? What time would be required if the capillary length were doubled at the same applied voltage?

(f) The capillary used in part (e) under normal conditions had a plate count of 100,000. What would N be if the capillary length were doubled at the same applied voltage? What would N be if the applied voltage were doubled at the original capillary length?

(g) For a 40.6-cm-long capillary of inside diameter 50 μm, what would the plate height be for a capillary with $N = 100,000$?

(h) For the same capillary as in part (g), what is the variance σ^2 of a typical peak?

Glossary

A

Absolute error   An accuracy measurement equal to the numerical difference between an experimental measurement and its true (or accepted) value.

Absolute standard deviation   A precision estimate based on the deviations between individual members in a set and the mean of that set (see Equation 6-4).

Absorbance, A   The logarithm of the ratio between the initial power of a beam of radiation P_0 and its power after it has traversed an absorbing medium, P. $A = \log(P_0/P) = -\log(P/P_0)$.

Absorption   A process in which a substance is incorporated or assimilated within another. Also, a process in which a beam of electromagnetic radiation is attenuated during passage through a medium.

Absorption of electromagnetic radiation   Processes in which radiation causes transitions in atoms and molecules to excited states. The absorbed energy is lost, usually as heat, as the excited species return to their ground states.

Absorption filter   A colored medium (usually glass) that transmits a band of the visible spectrum.

Absorption spectrum   A plot of absorbance as a function of wavelength.

Absorptivity, a   The proportionality constant in the Beer's law equation, $A = abc$, where b is the path length of radiation (usually in cm) and c is the concentration of the absorbing species (usually in mol/L). Thus, a has the units of length^{-1} concentration^{-1}.

Accuracy   A measure of the agreement between an analytical result and the true or accepted value for the measured quantity. This agreement is measured in terms of error.

Acid dissociation constant, K_a   The equilibrium constant for the dissociation reaction of a weak acid.

Acid error   The tendency of a glass electrode to register anomalously high pH response in highly acidic media.

Acidic flux   A salt that exhibits acidic properties in the molten state. Fluxes are used to convert refractory substances into water-soluble products.

Acid rain   Rainwater that has been rendered acidic from absorption of airborne nitrogen and sulfur oxides produced mainly by humans.

Acids   In the Brønsted-Lowry theory, species that are capable of donating protons to other species that in turn are capable of accepting these protons.

Acid salt   A conjugate base that contains an acidic hydrogen.

Activity, a   The effective concentration of a participant in a chemical equilibrium. The activity of a species is given by the product of the molar equilibrium concentration of the species and its activity coefficient.

Activity coefficient, γ_X   A unitless quantity whose numerical value depends on the ionic strength of a solution. It is the proportionality constant between activity and concentration.

Adsorbed water   Nonessential water that is held on the surface of solids.

Adsorption   A process in which a substance becomes physically bound to the surface of a solid.

Adsorption chromatography   A separation technique in which a solute equilibrates between the eluent and the surface of a finely divided adsorbed solid.

Agar   A polysaccharide that forms a conducting gel with electrolyte solutions; used in salt bridges to provide electric contact between dissimilar solutions without mixing.

Air damper   A device that hastens achievement of equilibrium by the beam of a mechanical analytical balance; also called a *dashpot*.

Aliquot   A volume of liquid that is a known fraction of a larger volume.

Alkaline error   The tendency of many glass electrodes to provide an anomalously low pH response in highly alkaline environments.

Alpha (α) value   The ratio of the molar concentration of a particular species to the molar analytical concentration of the solute from which it is derived.

Alumina   The common name for aluminum oxide. In a finely divided state, used as a stationary phase in adsorption chromatography; also finds application as a support for a liquid stationary phase in HPLC.

Amines   Derivatives of ammonia with one or more organic groups replacing hydrogen.

Amino acids   Weak organic acids that also contain basic amine groups. The amine group is α to the carboxylic acid group in amino acids derived from proteins.

Ammonium-1-pyrrolidinecarbodithiolate (APDC)   A protective agent in atomic spectroscopy that forms volatile species with an analyte.

Amperometric titration   A method based on applying a constant potential to a working electrode in stirred solution and recording the resulting current. A linear segment curve is obtained.

Amperostat   An instrument that maintains a constant current in an electrochemical cell. Can be used for coulometric titrations.

Amphiprotic substances   Species that can either donate protons or accept protons, depending on the chemical environment.

Amylose   A component of starch, the β-form of which is a specific indicator for iodine.

Analysis of Variance (ANOVA)   A collection of statistical procedures for analysis of responses from experiments. Single-factor *ANOVA* allows comparison of more than two means of populations.

Analyte The species in a sample about which analytical information is sought.

Analytical balance An instrument for accurately determinating mass.

Angstrom, Å A unit of length equal to 1×10^{-10} meter.

Angular dispersion, $dr/d\lambda$ A measure of the change in the angle of reflection or refraction of radiation by a prism or grating as a function of wavelength.

Anhydrone® Trade name for magnesium perchlorate, a drying agent.

Anion exchange resins High-molecular-weight polymers to which amine groups are bonded. They permit the exchange of anions in solution for hydroxide ions from the exchanger.

Anode The electrode of an electrochemical cell at which oxidation occurs.

Aqua regia A potent oxidizing solution made by mixing three volumes of concentrated hydrochloric acid and one volume of nitric acid.

Argentometric titration A titration in which the reagent is a solution of a silver salt (usually $AgNO_3$).

Arithmetic mean Synonymous with *mean* or *average*.

Asbestos A fibrous mineral, some varieties of which are carcinogenic. It was once used as a filtering medium in a Gooch crucible but is currently subject to stringent regulation.

Ashing The process whereby an organic material is combusted in air. See also *dry ashing* and *wet ashing*.

Ashless filter paper Paper produced from cellulose fibers that have been treated to eliminate inorganic species, thus leaving no residue when ashed.

Aspiration The process by which a sample solution is drawn by suction in atomic spectroscopy.

Aspirator A device that can be attached to a laboratory faucet to create a vacuum for filtering solutions. Water from the faucet passes through a narrowed channel where the pressure is lowered by the Venturi effect. A hose is connected to the device at the narrowed channel where the vacuum is produced.

Assay The process of determining how much of a given sample is the material indicated by its name.

Asymmetry potential A small potential that results from slight differences between the two surfaces of a glass membrane.

Atomic absorption The process by which unexcited atoms in a flame, furnace, or plasma absorb characteristic radiation from a radiation source and attenuate the radiant power of the source.

Atomic absorption spectroscopy (AAS) An analytical method that is based on absorption of electromagnetic radiation (EMR) in a reservoir of analyte atoms.

Atomic emission The emission of radiation by atoms that have been excited in a plasma, a flame, or an electric arc or spark.

Atomic emission spectroscopy (AES) An analytical method based on emission of electromagnetic radiation from atoms in a reservoir.

Atomic fluorescence Radiant emission from atoms that have been excited by absorption of electromagnetic radiation.

Atomic fluorescence spectroscopy (AFS) An analytical method based on measuring the intensity of EMR from fluorescent atoms in a reservoir.

Atomic mass unit See *unified atomic mass unit*.

Atomization The process of producing an atomic gas by applying energy to a sample.

Atomizer A device such as a plasma, a flame, or a furnace that produces an atomic vapor.

Attenuation In absorption spectroscopy, a decrease in the power of a beam of radiant energy. More generally, any decrease in a measured quantity or signal.

Attenuator A device for diminishing the radiant power in the beam of an optical instrument.

Autocatalysis A condition in which the product of a reaction catalyzes the reaction itself.

Autoprotolysis A process in which a solvent molecule transfers a proton (H^+) to another molecule of solvent, producing a protonated and a deprotonated ion.

Auxiliary balance A generic term for a balance that is less sensitive but more rugged than an analytical balance; synonymous with *laboratory balance*.

Average A number obtained by summing the values in a data set and dividing the sum by the number of data points in the set. Synonymous with *mean* or *arithmetic mean*.

Average current Polarographic current determined by dividing the total charge accumulated by a mercury drop by its lifetime.

Average linear velocity, u The length, L, of a chromatographic column divided by the time, t_M, required for an unretained species to pass through the column.

Azo indicators A group of acid/base indicators that have in common the structure R—N=N—R.

B

Back-titration The titration of an excess of a standard solution that has reacted completely with an analyte.

Ball mill A device for decreasing the particle size of the laboratory sample.

Band Ideally, a Gaussian-shaped distribution of (1) adjacent wavelengths encountered in spectroscopy or (2) the amount of a compound as it exits from a chromatographic or an electrophoretic column.

Band broadening The tendency of zones to spread as they pass through a chromatographic column; caused by various diffusion and mass transfer processes.

Band spectrum A molecular spectrum made up of one or more wavelength regions in which spectral lines are numerous and close together owing to rotational and vibrational transitions.

Bandwidth Usually, the range of wavelengths or frequencies of a spectral absorption or emission peak at half the height of the peak. The range passed by a wavelength isolation device.

Base dissociation constant, K_b The equilibrium constant for the reaction of a weak base with water.

Bases Species that are capable of accepting protons from donors (acids).

Basic flux A substance with basic characteristics in the molten state. Used to solubilize refractory samples, principally silicates.

Beam The principal moving part of a mechanical analytical balance.

Beam arrest A mechanism that lifts the beam from its bearing surface when an analytical balance is not in use or when the load is being changed.

Beam splitter A device for dividing source radiation into two beams.

Beer's law The fundamental relationship for the absorption of radiation by matter, that is, $A = abc$, where a is the absorptivity, b is the path length of the beam of radiation, and c is the concentration of the absorbing species.

Bernoulli effect In atomic spectroscopy, the mechanism by which sample droplets are aspirated into a plasma or flame.

β-amylose That component of starch that serves as a specific indicator for iodine.

Bias The tendency to skew estimates in the direction that favors the anticipated result. Also used to describe the effect of a *systematic error* on a set of measurements. Also a dc voltage that is applied to a circuit element.

Blackbody radiation Continuum radiation produced by a heated solid.

Blank determination The process of performing all steps of an analysis in the absence of sample. Used to detect and compensate for systematic errors in an analysis.

Bolometer A detector for infrared radiation based on changes in resistance with changes in temperature.

Bonded-phase packings In HPLC, a support medium to which a liquid stationary phase is chemically bonded.

Bonded stationary phase A liquid stationary phase that is chemically bonded to the support medium.

Boundary potential, E_b The difference between two potentials that develop at the opposite surfaces of a membrane electrode.

Brønsted-Lowry acids and bases A description of acid-base behavior in which an acid is defined as a proton donor and a base is a proton acceptor. The loss of a proton by an acid results in the formation of a potential proton acceptor, or *conjugate base* of the parent acid.

Buffer capacity The number of moles of strong acid (or strong base) needed to change the pH of 1.00 L of a buffer solution by 1.00 unit.

Buffer solutions Solutions that tend to resist changes in pH as the result of dilution or the addition of small amounts of acids or bases.

Bumping The sudden and often violent boiling of a liquid that results from local overheating.

Buoyancy The displacement of the medium (usually air) by an object, producing an apparent loss of mass. A significant source of error when the densities of the object and the comparison standards (weights) differ.

Buret A graduated tube from which accurately known volumes can be dispensed.

C

Calibration The empirical determination of the relationship between a measured quantity and a known reference or standard value. Used to establish analytical signal versus concentration relationships in a calibration or working curve.

Calomel The compound Hg_2Cl_2.

Calomel electrode A versatile reference electrode based on the half-reaction $Hg_2Cl_2(s) + 2e^- \rightleftharpoons 2Hg(l) + 2Cl^-$

Capillary column A small-diameter chromatographic column for GC or HPLC, fabricated of metal, glass, or fused silica. For GC, the stationary phase is a thin coating of liquid on the interior wall of the tube; for HPLC, capillary columns are often packed.

Capillary electrophoresis High-speed, high-resolution electrophoresis performed in capillary tubes or in microchips.

Carbonate error A systematic error caused by absorption of carbon dioxide by standard solutions of base used in the titration of weak acids.

Carrier gas The mobile phase for gas chromatography.

Catalytic method Analytical method for determining the concentration of a catalyst based on measuring the rate of a catalyzed reaction.

Catalytic reaction A reaction whose progress toward equilibrium is hastened by a substance that is not consumed in the overall process.

Cathode In an electrochemical cell, the electrode at which reduction takes place.

Cathode depolarizer A substance that is more easily reduced than hydrogen ion. Used to prevent evolution of hydrogen during an electrolysis.

Cathodic stripping analysis An electrochemical method in which the analyte is deposited by oxidation into a small-volume electrode and later stripped off by reduction.

Cation-exchange resins High-molecular-weight polymers to which acidic groups are bonded. These resins permit the substitution of cations in solution for hydrogen ions from the exchanger.

Cell (1) In electrochemistry, an array consisting of a pair of electrodes immersed in solutions that are in electrical contact; the electrodes are connected externally by a metallic conductor. (2) In spectroscopy, the container that holds the sample in the light path of an optical instrument. (3) In an electronic balance, a system of constraints that assure alignment of the pan. (4) In a spreadsheet, a location at the intersection of a row and a column where data or formulas can be placed.

Cells without liquid junction Electrochemical cells in which both anode and cathode are immersed in a common electrolyte.

Charge-balance equation An expression relating the concentrations of anions and cations based on charge neutrality in any solution.

Charged-coupled device (CCD) A solid-state two-dimensional detector array used for spectroscopy and imaging.

Charge-injection device (CID) A solid-state photodetector array used in spectroscopy.

Charge-transfer complexes Complexes that are made up of an electron donor group and an electron acceptor group. Absorption of radiation by these complexes involves a transfer of electrons from the donor to the acceptor.

Charging current A positive or a negative nonfaradaic current resulting from an excess or a deficiency of electrons in a mercury droplet at the instant of detachment.

Chelating agents Substances with multiple sites available for coordinate bonding with metal ions. Such bonding typically results in the formation of five- or six-membered rings.

Chelation The reaction between a metal ion and a chelating reagent.

Chemical Abstracts A major hard-copy source of chemical information worldwide. Has been largely supplanted by Scifinder Scholar®, an online database with a rich set of tools for searching for chemical information.

Chemical deviations from Beer's law Deviations from Beer's law that result from association or dissociation of the absorbing species or reaction with the solvent, producing a product that absorbs differently from the analyte. In atomic spectroscopy, chemical interactions of the analyte with interferents that affect the absorption properties of the analyte.

Chemical equilibrium A dynamic state in which the rates of forward and reverse reactions are identical. A system in equilibrium will not spontaneously depart from this condition.

Chemiluminescence The emission of energy as electromagnetic radiation during a chemical reaction.

Chopper A mechanical device that alternately transmits and blocks radiation from a source.

Chromatogram A plot of an analyte signal proportional to concentration or mass as a function of elution time or elution volume.

Chromatograph An instrument for carrying out chromatographic separations.

Chromatographic band The distribution (ideally Gaussian) of the concentration of eluted species about a central value. The result of variations in the time that analyte species reside in the mobile phase.

Chromatographic zone Synonymous with *chromatographic band.*

Chromatography A term for methods of separation based on the interaction of species with a stationary phase while they are being transported by a mobile phase.

Clark oxygen sensor A voltammetric sensor for dissolved oxygen.

Coagulation The process whereby particles with colloidal dimensions are caused to form larger aggregates.

Coefficient of variation (CV) The relative standard deviation, expressed as a percentage.

Colloidal suspension A mixture (commonly of a solid in a liquid) in which the particles are so finely divided that they have no tendency to settle.

Colorimeter A relatively simple optical instrument, often utilizing colored filters, for measuring transmittance or absorbance of electromagnetic radiation in the visible region of the spectrum.

Column chromatography A chromatographic method in which the stationary phase is held within or on the surface of a narrow tube and the mobile phase is forced through the tube, where compound separation occurs. Compare with *planar chromatography.*

Column efficiency A measure of the degree of broadening of a chromatographic band, often expressed in terms of plate height, $H,$ or the number of theoretical plates, $N.$ If the distribution of analyte is Gaussian within the band, the plate height is given by the variance, σ^2, divided by the length, $L,$ of the column.

Column resolution, R Measures the capability of a column to separate two analyte bands.

Common-ion effect The shift in the position of equilibrium caused by the addition of a participating ion.

Complex formation The process whereby a species with one or more unshared electron pairs forms coordinate bonds with metal ions.

Concentration-based equilibrium constant, K' The equilibrium constant based on molar equilibrium concentrations. The numerical value of K' depends on the ionic strength of the medium.

Concentration polarization The deviation of the electrode potential in an electrochemical cell from its equilibrium or Nernstian value on the passage of current as a result of slow transport of species to and from the electrode surface.

Concentration profile The distribution of analyte concentrations with time as they emerge from a chromatographic column; also, the time behavior of reactants or products during a chemical reaction.

Conduction of electricity The movement of charge by ions in solution, by electrochemical reaction at the surfaces of electrodes, or by movement of electrons in metals.

Conductometric detector A detector for charged species that is often used in ion chromatography.

Confidence interval Defines bounds about the experimental mean within which—with a given probability—the true mean should be located.

Confidence limits The values that define the confidence interval.

Conjugate acid/base pairs Species that differ from one another by one proton.

Constant-boiling HCl Solutions of hydrochloric acid with concentrations that depend on the atmospheric pressure.

Constant error A systematic error that is independent of the size of the sample taken for analysis. Its effect on the results of an analysis increases as the sample size decreases.

Constant mass The condition in which the mass of an object is no longer altered by heating or cooling.

Constructive interference Increase in the amplitude of a resultant wave in regions where two or more wave fronts are in phase with one another.

Continuous source A source that emits radiation continuously in time.

Continuum source A source that emits a spectral continuum of wavelengths. Examples include tungsten filament lamps and deuterium lamps used in absorption spectroscopy.

Continuum spectrum Radiation consisting of a band of wavelengths as opposed to discrete lines. Incandescent solids provide continuum output *(blackbody radiation)* in the visible and infrared regions. Deuterium and hydrogen lamps yield continuum spectra in the ultraviolet region.

Control chart A plot that demonstrates statistical control of a product or a service as a function of time.

Control circuit A three-electrode electrochemical apparatus that maintains a constant potential between the working electrode and the reference electrode. See *potentiostat.*

Controlled potential methods Electrochemical methods that use a potentiostat to maintain a constant potential between the working electrode and a reference electrode.

Convection The transport of a species in a liquid or gaseous medium by stirring, mechanical agitation, or temperature gradients.

Coordination compounds Species formed between metal ions and electron-pair donating groups. The product may be anionic, neutral, or cationic.

Coprecipitation The carrying down of otherwise soluble species either within a solid or on the surface of a solid as it precipitates.

Coulomb, C The quantity of charge provided by a constant current of one ampere in one second.

Coulometer A device that measures the quantity of charge consumed during an electrochemical process. Electronic coulometers evaluate the integral of the current/time curve. Chemical coulometers function by measuring the quantity of a reactant consumed or a product formed in a reaction in an auxiliary cell.

Coulometric titration A type of coulometric analysis that involves measuring the time needed for a constant current to produce enough reagent to react completely with an analyte.

Counter electrode The electrode that with the working electrode forms the electrolysis circuit in a three-electrode cell.

Counter-ion layer A layer of solution surrounding a colloidal particle in which there is a quantity of ions sufficient to balance the charge on the surface of the particle. Also, in electrolysis, a layer of electroactive ions of charge opposite of the charge on an electrode. A second layer of ions opposite in charge to the first layer and with the same charge as the electrode is called the counter-ion layer.

Creeping The tendency of some precipitates to spread over a wetted surface.

Critical temperature The temperature above which a substance can no longer exist in the liquid state, regardless of pressure.

Cross-linked stationary phase A polymer stationary phase in a chromatographic column in which covalent bonds link different strands of the polymer, thus creating a more stable phase.

Crystalline membrane electrode Electrode in which the sensing element is a crystalline solid that responds selectively to the activity of an ionic analyte.

Crystalline precipitates Solids that tend to form as large, easily filtered crystals.

Crystalline suspensions Particles with greater-than-colloidal dimensions temporarily dispersed in a liquid.

Current, i The amount of electrical charge that passes through an electrical circuit per unit time. Units of current are amperes, A.

Current density The current per unit area of an electrode in A/m^2.

Current efficiency Measure of the effectiveness of a quantity of electricity in bringing about an equivalent amount of chemical change in an analyte. Coulometric methods require 100% current efficiency.

Current-to-voltage converter A device for converting an electric current into a voltage that is proportional to the current.

Cuvette The container that holds the analyte in the light path in absorption spectroscopy.

D

Dalton Unit of mass. One Dalton is equal to one unified atomic mass unit.

Dark currents Small currents that occur even when no radiation is reaching a photometric transducer.

Dashpot Synonym for *air damper* in an analytical balance.

dc Plasma (DCP) spectroscopy A method that utilizes an electrically induced argon plasma to excite the emission spectra of analyte species.

Dead time In *column chromatography,* the time, t_M, required for an unretained species to traverse the column. Also, in stopped-flow kinetics, the time between the mixing of reactants and the arrival of the mixture at the observation cell.

Debye-Hückel equation An expression that permits calculation of activity coefficients in media with ionic strengths less than 0.1.

Debye-Hückel limiting law A simplified form of the Debye-Hückel equation, applicable to solutions in which the ionic strength is less than 0.01.

Decantation The transfer of supernatant liquid and washings from a container to a filter without disturbing the precipitated solid in the container.

Decrepitation The shattering of a crystalline solid as it is heated that is caused by vaporization of occluded water.

Degrees of freedom The number of members in a statistical sample that provide an independent measure of the precision of the set.

Dehydration The loss of water by a solid.

Dehydrite® Trade name for magnesium perchlorate, a drying agent.

Density The ratio of the mass of an object to its volume, normally measured in units of g/cm^3 for liquids and solids and g/L for gases. The SI unit is kg/m^3.

Depletion layer A nonconductive region in a reverse-biased semiconductor.

Depolarizer An additive that undergoes reaction at an electrode in preference to an otherwise undesirable process. See *cathode depolarizer.*

Derivative titration curve A plot of the change in the quantity measured per unit volume against the volume of titrant added. A derivative curve displays a maximum where there is a point of inflection in a conventional titration curve. See also *second derivative curve.*

Desiccants Drying agents.

Desiccator A container that provides a dry atmosphere for the cooling and storage of samples, crucibles, and precipitates.

Destructive interference A decrease in amplitude of waves resulting from the superposition of two or more wave fronts that are not in phase with one another.

Detection limit The minimum amount of analyte that a system or a method is capable of measuring.

Detector A device that responds to some characteristic of the system under observation and converts that response into a measurable signal.

Determinate error A class of errors that at least, in principle, has a known cause. Synonym for *systematic error.*

Deuterium lamp A source that provides a spectral continuum in the ultraviolet region of the spectrum. Radiation results from applying about 40 V to a pair of electrodes housed in a deuterium atmosphere.

Devarda's alloy An alloy of copper, aluminum, and zinc used to reduce nitrates and nitrites to ammonia in a basic medium.

Deviation The difference between an individual measurement and the mean (or median) value for a set of data.

Diatomaceous earth The siliceous skeletons of unicellular algae used as a solid support in GC.

Differentiating solvents Solvents in which differences in the strengths of solute acids or bases are enhanced. Compare with *leveling solvents.*

Diffraction order, n Integer multiples of a wavelength at which constructive interference occurs.

Diffusion The migration of species from a region of high concentration in a solution to a more dilute region.

Diffusion coefficient (*polarographic, D, chromatographic, D_m*) A measure of the mobility of a species, usually in units of cm^2/s.

Diffusion current, i_d. The limiting current in voltammetry when diffusion is the major form of mass transfer.

Digestion The practice of maintaining an unstirred mixture of freshly formed precipitate and solution from which it was formed at temperatures just below boiling that produces improved purity and particle size.

Dimethylglyoxime A precipitating reagent that is specific for nickel(II). Its formula is $CH_3(C{=}NOH)_2CH_3$.

Diode array detector A silicon chip that usually contains 64 to 4096 photodiodes arranged linearly. The device provides the capability to collect data from entire spectral regions simultaneously.

Diphenylthiocarbazide A chelating reagent, also known as *dithizone.* Its adducts with cations are sparingly soluble in water but are readily extracted with organic solvents.

Dissociation The splitting of molecules of a substance, commonly into two simpler entities.

Distribution constant The equilibrium constant for the distribution of an analyte between two immiscible solvents. It is approximately equal to the ratio of the equilibrium molar concentrations in the two solvents.

Dithizone Common name of *diphenylthiocarbazide.*

Doping The intentional introduction of traces of group III or group V elements to enhance the semiconductor properties of a silicon or germanium crystal.

Doppler broadening Absorption or emission of radiation by a species in rapid motion, resulting in a broadening of spectral lines. Wavelengths that are slightly shorter or longer than nominal are detected, depending on the direction of motion of the species.

Double-beam instrument An optical instrument design that eliminates the need to alternate blank and analyte solutions manually in the light path. A *beam splitter* divides the radiation in a double beam in space spectrometer. A *chopper* directs the beam alternately between blank and analyte in a double beam in time instrument.

Drierite® Trade name for the drying agent.

Dropping mercury electrode An electrode in which mercury is forced through a capillary, producing regular drops.

Dry ashing The elimination of organic matter from a sample by direct heating in air.

Dumas method A method of analysis based on the combustion of nitrogen-containing organic samples by CuO to convert the nitrogen to N_2, which is then measured volumetrically.

Dynode An intermediate electrode in a photomultiplier tube.

E

Echelle grating A grating that is blazed with reflecting surfaces that are larger than the nonreflecting faces.

Eddy diffusion Diffusion of solutes that contributes to broadening of chromatographic bands, the result of differences in the pathways for solutes as they traverse a column.

EDTA An abbreviation of *ethylenediaminetetraacetic acid*, a chelating agent widely used for complex formation titrations. Its formula is $(HOOCCH_2)_2NCH_2CH_2N(CH_2COOH)_2$.

Effective bandwidth The bandwidth of a monochromator or an interference filter at which the transmittance is 50% of that at the nominal wavelength.

Electric double layer Refers to the charge on the surface of a colloidal particle and the counter-ion layer that balances this charge. Also, the two adjacent charged layers on the surface of the working electrode in voltammetry.

Electroanalytical methods A large group of methods that have in common the measurement of an electrical property of the system that is proportional to the amount of analyte in the sample.

Electrochemical cell An array consisting of two electrodes, each of which is in contact with an electrolyte solution. Typically, the two electrolytes are in electrical contact through a *salt bridge*. An external metal conductor connects the two electrodes.

Electrochemical reversibility The ability of some cell processes to reverse themselves when the direction of the current is reversed. In an irreversible cell, reversal of current causes a different reaction at one or both electrodes.

Electrode A conductor at the surface of which electron transfer to or from the surrounding solution takes place.

Electrodeless-discharge lamp A source of atomic line spectra that is powered by radio-frequency or microwave radiation.

Electrode of the first kind A metallic electrode whose potential is proportional to the logarithm of the concentration (strictly, activity) of a cation (or the ratio of cations) derived from the electrode metal.

Electrode of the second kind A metallic electrode whose response is proportional to the logarithm of the concentration (strictly, activity) of an anion that forms either a sparingly soluble species or stable complexes with a cation (or the ratio of cations) derived from the electrode metal.

Electrode potential The potential of an electrochemical cell in which the electrode of interest is the right-hand electrode and the standard hydrogen electrode is the left-hand electrode.

Electrogravimetric analysis A branch of gravimetric analysis that involves measuring the mass of species deposited on an electrode of an electrochemical cell.

Electrolysis circuit In a three-electrode arrangement, a dc source and a voltage divider to permit regulation of the potential between the working electrode and the counter electrode.

Electrolyte effect The dependence of numerical values for equilibrium constants on the ionic strength of the solution.

Electrolytes Solute species whose aqueous solutions conduct electricity.

Electrolytic cell An electrochemical cell that requires an external source of energy to drive the cell reaction. Compare with *galvanic cell*.

Electromagnetic radiation(EMR) A form of energy with properties that can be described in terms of waves or, alternatively, as particulate photons, depending on the method of observation.

Electromagnetic spectrum The power or intensity of electromagnetic radiation plotted as a function of wavelength or frequency.

Electronic balance A balance in which an electromagnetic field supports the pan and its contents. The current needed to restore the loaded pan to its original position is proportional to the mass on the pan.

Electronic transition The promotion of an electron from one electronic state to a second electronic state, and conversely.

Electroosmotic flow The net flow of bulk liquid in an applied electric field through a porous material, capillary tube, membrane, or microchannel.

Electrophoresis A separation method based on the differential rates of migration of charged species in an electric field.

Electrothermal analyzer Any of several devices that form an atomized gas containing an analyte in the light path of an instrument by electrical heating. Used for atomic absorption and atomic fluorescence measurements.

Eluent A mobile phase in chromatography that is used to carry solutes through a stationary phase.

Eluent suppressor column In ion chromatography, a column downstream from the analytical column where ionic eluents are converted to nonconducting species, while analyte ions remain unaffected.

Elution chromatography Describes processes in which analytes are separated from one another on a column owing to differences in the time that they are retained in the column.

Emission spectrum The collection of spectral lines or bands that are observed when species in excited states relax by giving off their excess energy as electromagnetic radiation.

Empirical formula The simplest whole-number combination of atoms in a molecule.

End point An observable change during titration that signals that the amount of titrant added is chemically equivalent to the amount of analyte in the sample.

Enzymatic sensor A membrane electrode that has been coated with an immobilized enzyme. The electrode responds to the amount of analyte in the sample.

Enzyme-substrate complex (ES) The intermediate formed in the process

$$\text{Enzyme (E)} + \text{substrate (S)} \rightleftharpoons \text{ES} \rightarrow \text{product (P)} + \text{E}$$

Eppendorf pipet A type of micropipet that delivers adjustable volumes of liquid.

Equilibrium-constant expression An algebraic expression that describes the equilibrium relationship among the participants in a chemical reaction.

Equilibrium molar concentration The concentration of a solute species (in mol/L or mmol/mL).

Equivalence point That point in a titration at which the amount of standard titrant added is chemically equivalent to the amount of analyte in the sample.

Equivalence-point potential The electrode potential of the system in an oxidation/reduction titration when the amount of titrant that has been added is chemically equivalent to the amount of analyte in the sample.

Equivalent For an oxidation/reduction reaction, that mass of a species that can donate or accept 1 mole of electrons. For an acid/base reaction, that mass of a species that can donate or accept 1 mole of protons.

Equivalent of chemical change The mass of a species that is directly or indirectly equivalent to one faraday (6.02×10^{23} electrons).

Equivalent weight or mass A specialized basis for expressing mass in chemical terms similar to, but different from, *molar mass*. As a consequence of definition, one equivalent of an analyte reacts with one equivalent of a reagent, even if the stoichiometry of the reaction is not one to one.

Error The difference between an experimental measurement and its accepted value.

Essential water Water in a solid that exists in a fixed amount, either within the molecular structure *(water of constitution)* or within the crystalline structure *(water of crystallization)*.

Ethylenediaminetetraacetic acid Probably the most versatile reagent for complex formation titrations. It forms chelates with most cations. See *EDTA*.

Excitation The promotion of an atom, an ion, or a molecule to a state that is more energetic than a lower energy state.

Excitation spectrum In fluorescence spectroscopy, a plot of fluorescence intensity as a function of excitation wavelength.

Exhaustive extraction A cycle in which an organic solvent, after percolation through an aqueous phase containing the solute of interest, is distilled, condensed, and again passed through the aqueous phase.

F

Faradaic current An electric current produced by oxidation/reduction processes in an electrochemical cell.

Faraday, F The quantity of electricity associated with 6.022×10^{23} electrons.

Fast reaction Reaction that is half complete in 10 seconds or less.

Ferroin A common name for the 1,10-phenanthroline-iron(II) complex, which is a versatile redox indicator. Its formula is $(C_{12}H_8N_2)_3Fe^{2+}$.

Flame emission spectroscopy Method that uses a flame to cause an atomized analyte to emit its characteristic emission spectrum; also known as flame photometry.

Flame ionization detector (FID) A detector for gas chromatography based on the collection of ions produced during the pyrolysis of organic analytes in a flame.

Fluorescence Radiation produced by an atom or a molecule that has been excited by photons to a singlet excited state.

Fluorescence bands Groups of fluorescence lines that originate from the same excited electronic state.

Fluorescence spectrum A plot of fluorescence intensity versus wavelength in which either the excitation wavelength (emission spectrum) or the emission wavelength (excitation spectrum) is held constant (see Figure 27-8b).

Fluorometer A filter instrument for quantitative fluorescence measurements.

Fluxes Substances that in the molten state have acidic or basic properties. Used to solubilize the analyte in refractory samples.

Focal plane A plane on which dispersed radiation from a prism or a grating is focused.

Formality, F The number of moles of solute contained in each liter of solution. A synonym of *molar analytical concentration.*

Formal potential, $E^{0\prime}$ The electrode potential for a couple when the analytical concentrations of all participants are unity and the concentrations of other species in the solution are defined.

Formula mass The sum of atomic masses in the chemical formula of a substance. A synonym for gram formula weight and *molar mass.*

Fourier transform spectrometer A spectrometer in which an interferometer and Fourier transformation are used to obtain a spectrum.

Frequency, ν, of electromagnetic radiation The number of oscillations per second with units of hertz (Hz), which is one oscillation per second.

Fritted-glass crucible A filtering crucible equipped with a porous glass bottom. Also called a *sintered-glass crucible.*

Fronting Describes a nonideal chromatographic peak in which the early portions tend to be drawn out. Compare with *tailing.*

F-test A statistical method that permits comparison of the variances of two sets of measurements.

Fused-silica open tubular (FSOT) column A wall-coated gas chromatography column that has been fabricated from purified silica.

G

Galvanic cell An electrochemical cell that provides energy during its operation. Synonym for *voltaic cell.*

Galvanostat Synonym for *amperostat.*

Gas chromatography (GC) Separation methods that use a gaseous mobile phase and a liquid or a solid stationary phase.

Gas electrode An electrode that involves the formation or consumption of a gas during its operation.

Gas-sensing probe An indicator/reference electrode system that is isolated from the analyte solution by a hydrophobic membrane. The membrane is permeable to a gas, and the potential is proportional to the gas content of the analyte solution.

Gaussian distribution A theoretical bell-shaped distribution of results obtained for replicate measurements that are affected by random errors.

GC/MS A combined technique in which a mass spectrometer is used as a detector for gas chromatography.

Gel filtration chromatography A type of *size exclusion chromatography* that uses a hydrophilic packing. Used to separate polar species.

Gel permeation chromatography A type of *size exclusion chromatography* that uses a hydrophobic packing. Used to separate nonpolar species.

General elution problem The compromise between elution time and resolution that is addressed through *gradient elution* (for liquid chromatography) or *temperature programming* (for gas chromatography).

General redox indicators Indicators that respond to changes in E_{system}.

Ghosts Double images in the output of a grating, the result of imperfections in the ruling engine used in its preparation.

Glass electrode An electrode in which a potential develops across a thin glass membrane. It provides a measure of the pH of the solution in which the electrode is immersed.

Gooch crucible A porcelain filtering crucible. Filtration is accomplished by means of a glass fiber mat or a layer of asbestos fiber.

Gradient elution In liquid chromatography, the systematic alteration of mobile phase composition to optimize the chromatographic resolution of the components in a mixture. See also solvent programming.

Graphical kinetic methods Methods of determining reaction rates from plots of the concentration of a reactant or a product as a function of time.

Grating A device consisting of closely spaced grooves that is used to disperse polychromatic radiation by diffracting it into its component wavelengths.

Gravimetric analysis A group of analytical methods in which the amount of analyte is established through the measurement of the mass of a pure substance containing the analyte.

Gravimetric factor, GF The stoichiometric mass ratio of the analyte to the solid weighed in a gravimetric analysis.

Gravimetric titrimetry Titrations in which the mass of standard titrant is measured rather than volume. The concentration of titrant is expressed in mmol/g of solution (rather than the more familiar mmol/mL).

Gross error An occasional error, neither random nor systematic, that results in the occurrence of a questionable outlier result.

Gross sample A representative portion of a whole analytical sample, which with further treatment, becomes the laboratory sample.

Ground state The lowest energy state of an atom or a molecule.

Guard column A precolumn located ahead of an HPLC column. The composition of the packing in the guard column is selected to extend the useful lifetime of the analytical column by removing particulate matter and contaminants and by saturating the eluent with the stationary phase.

H

Half-cell potential The potential of an electrochemical half-cell measured with respect to the standard hydrogen electrode.

Half-life, $t_{1/2}$　The time interval during which the amount of reactant has decreased to one half of its original value.

Half-reaction　A method of portraying the oxidation or the reduction of a species. A balanced equation that shows the oxidized and reduced forms of a species, any H_2O or H^+ needed to balance the hydrogen and oxygen atoms in the system, and the number of electrons required to balance the charge.

Half-wave potential, $E_{1/2}$　The potential versus a reference electrode at which the current of a voltammetric wave is one half the limiting current.

Hanging mercury drop electrode (HMDE)　A microelectrode that can concentrate traces of metals by electrolysis into a small volume; the analysis is completed by voltammetric stripping of the metal from the mercury drop.

Heat detector　A device that is sensitive to changes in the temperature of its surroundings; used to monitor infrared radiation.

Height equivalent of a theoretical plate, H (HETP)　A measure of chromatographic column efficiency; equal to the length of a column divided by the number of theoretical plates in the column.

Henderson-Hasselbalch equation　An expression to calculate the pH of a buffer solution; $pH = pK_a + \log (c_{NaA}/c_{HA})$, where pK_a is the negative logarithm of the dissociation constant for the acid and c_{NaA} and c_{HA} are the molar concentrations of the compounds making up the buffer. Popular with biochemists.

High-performance adsorption chromatography　Synonymous with *liquid-solid chromatography*. See also *adsorption chromatography*.

High-performance ion-exchange chromatography　See *ion chromatography*.

High-performance liquid chromatography (HPLC)　Column chromatography in which the mobile phase is a liquid, often forced through a stationary phase by pressure.

High-performance size-exclusion chromatography　See *size-exclusion chromatography*.

Histogram　A bar graph in which replicate results are grouped according to ranges of magnitude along the horizontal axis and by frequency of occurrence on the vertical axis.

Hollow-cathode lamp　A source used in atomic absorption spectroscopy that emits sharp lines for a single element or sometimes for several elements.

Holographic grating　A grating that has been produced by optical interference on a coated glass plate rather than by mechanical ruling.

Homogeneous precipitation　A technique in which a precipitating agent is generated slowly throughout a solution of an analyte to yield a dense and easily filtered precipitate for gravimetric analysis.

Hundred percent T adjustment　Adjustment of an optical absorption instrument to register 100% T with a suitable blank in the light path.

Hydrodynamic voltammetry　Voltammetry performed with the analyte solution in constant motion relative to the electrode surface; produced by pumping the solution past a stationary electrode, by moving the electrode through the solution, or by stirring the solution.

Hydrogen lamp　A continuum source of radiation in the ultraviolet range that is similar in structure to a deuterium lamp.

Hydronium ion　The hydrated proton whose symbol is H_3O^+.

8-Hydroxyquinoline　A versatile chelating reagent; used in gravimetric analysis, in volumetric analysis as a protective reagent in atomic spectroscopy, and as an extracting reagent; also known as *oxine*. Its formula is HOC_9H_6N.

Hygroscopic glass　A glass that absorbs minute amounts of water on its surface; hygroscopicity is an essential property in the membrane of a glass electrode.

Hyphenated methods　Methods involving the combination of two or more types of instrumentation; the product is an instrument with greater capabilities than any one instrument alone.

Hypothesis testing　The process of testing a tentative assertion with various statistical tests. See *t-test, F-test, Q-test,* and *ANOVA*.

I

Ilkovic equation　An equation that relates the diffusion current to the variables that affect it, that is the number of electrons involved *(n)* in the reaction with the analyte, the square root of the diffusion coefficient *($D^{1/2}$)*, the mass flow rate of mercury *($m^{2/3}$)*, and the drop time *($t^{1/6}$)* of the dropping mercury electrode.

Immobilized enzyme reactor　Tubular reactor or detector surface on which an enzyme has been attached by adsorption, covalent bonding, or entrapment.

Indeterminate error　Synonymous with *random error*.

Indicator electrode　An electrode whose potential is related to the logarithm of the activity of one or more species in contact with the electrode.

Indicator reaction, kinetics　A fast reaction involving an indicator species that can be used to monitor the reaction of interest.

Inductively coupled plasma (ICP) spectroscopy　A method that uses an inert gas (usually argon) plasma formed by the absorption of radio-frequency radiation to atomize and excite a sample for atomic emission spectroscopy.

Inert electrode　An electrode that responds to the potential of the system, E_{system}, and is not otherwise involved in the cell reaction.

Infrared radiation　Electromagnetic radiation in the 0.78 to 300 μm range.

Inhibitor, catalytic　A species that decreases the rate of an enzyme-catalyzed reaction.

Initial rate methods　Kinetic methods based on measurements made near the beginning of a reaction.

Inner-filter effect　Phenomenon causing nonlinear fluorescence calibration curves as a result of excessive absorption of the incident beam or the emitted beam.

Instrumental deviations from Beer's law　Departures from linearity between absorbance and concentration that are attributable to the measuring device.

Integral methods　Kinetic methods based on an integrated form of the rate law.

Intensity, I, of electromagnetic radiation　The power per unit solid angle; often used synonymously with radiant power, P.

Intercept, b, of a regression line　The y value in a regression line when the x value is zero; in an analytical calibration curve, the hypothetical value of the analytical signal when the concentration of analyte is zero.

Interference filter　An optical filter that provides narrow bandwidths caused by constructive interference.

Interference order, n　An integer that along with the thickness and the refractive index of the dielectric material determines the wavelength transmitted by an interference filter.

Interferences, or interferents　Species that affect the signal on which an analysis is based.

Interferometer　A nondispersive device that obtains spectral information through constructive and destructive interference; used in Fourier transform infrared instruments.

Internal standard　A known quantity of a species with properties similar to an analyte that is introduced into solutions of the standard and the unknown; the ratio of the signal from the internal standard to the signal from the analyte serves as the basis for the analysis.

International Union of Pure and Applied Chemistry (IUPAC)   An international organization devoted to developing definitions and usages for the worldwide chemical community.

Ion chromatography   An HPLC technique based on the partitioning of ionic species between a liquid mobile phase and a solid polymeric ionic exchanger; also called ion-exchange chromatography.

Ion-exchange resin   A high-molecular-weight polymer to which a large number of acidic or basic functional groups have been bonded. Cationic resins permit the exchange of hydrogen ion for cations in solution; anionic resins substitute hydroxide ion for anions.

Ionic strength, μ   A property of a solution that depends on the total concentration of ions in the solution as well as on the charge carried by each of these ions, that is, $\mu = \frac{1}{2} \sum c_i Z_i^2$, where c_i is the molar concentration of each ion and Z_i is its charge.

Ionization suppressor   In atomic spectroscopy, an easily ionized species, such as potassium, that is introduced to suppress the ionization of the analyte.

***IR* drop**   The potential drop across a cell due to resistance to the movement of charge; also known as the *ohmic potential drop.*

Irreversible cell   An electrochemical cell in which the chemical reaction as a galvanic cell is different from that which occurs when the current is reversed.

Irreversible electrochemical reaction   A reaction that yields a poorly defined voltammogram caused by the irreversibility of electron transfer at the electrode.

Isocratic elution   Elution with a single solvent; compare with *gradient elution.*

Isoelectric point   The pH at which an amino acid has no tendency to migrate under the influence of an electric field.

IUPAC convention   A set of definitions relating to electrochemical cells and their potentials; also known as the *Stockholm convention.*

J

Jones reductor   A column packed with amalgamated zinc; used for the prereduction of analytes.

Joule   A unit of work equal to a newton-meter.

Junction potential   The potential that develops at the interface between solutions with dissimilar composition; synonymous with *liquid junction* potential.

K

Karl Fischer reagent   A reagent for the titrimetric determination of water.

Kilogram   The base unit of mass in the SI system.

Kinetic methods   Analytical methods based on relating the kinetics of a reaction to the analyte concentration.

Kinetic polarization   Nonlinear behavior of an electrochemical cell caused by the slowness of the reaction at the surface of one or both electrodes.

Kjeldahl flask   A long-necked flask used for the digestion of samples with hot, concentrated sulfuric acid.

Kjeldahl method   A titration method for the determination of nitrogen in organic compounds in which the nitrogen is converted to ammonia, which is then distilled and determined by a neutralization titration.

Knife edge   The nearly friction-free contact between the moving components of a mechanical analytical balance.

L

Laboratory balance   Synonymous with *auxiliary balance.*

Laminar flow   Streamline flow in a liquid near and parallel to a solid boundary. In a tube, this flow results in a parabolic flow profile; near an electrode surface, this results in parallel layers of liquid that slide by one another.

Least-squares method   A statistical method of obtaining the parameters of a mathematical model (such as the equation for a straight line) by minimizing the sum of the squares of the differences between the experimental points and the points predicted by the model.

Le Châtelier principle   A statement that the application of a stress to a chemical system at equilibrium will result in a shift in the position of the equilibrium that tends to relieve the stress.

Leveling solvents   Solvents in which the strength of solute acids or bases tend to be the same; compare with *differentiating solvents.*

Levitation   As applied to electronic balances, the suspension of the pan of the balance in air by a magnetic field.

Ligand   A molecule or an ion with at least one pair of unshared electrons available for coordinate bonding with cations.

Limiting current, i_l   Current plateau reached in voltammetry when the electrode reaction rate is limited by the rate of mass transfer.

Linear-scan voltammetry   Electroanalytical methods that involve measurement of the current in a cell as the electrode potential is linearly increased or decreased with time; the basis for *hydrodynamic voltammetry* and *polarography.*

Linear-segment curve   A titration curve in which the end point is obtained by extrapolating linear regions well before and after the equivalence point; useful for reactions that do not strongly favor the formation of products.

Line source   In atomic spectroscopy, a radiation source that emits sharp atomic lines characteristic of the analyte atoms. See *hollow-cathode lamp* and *electrodeless discharge lamp.*

Liquid bonded-phase chromatography   Partition chromatography that uses a stationary phase that is chemically bonded to the column packing.

Liquid junction   The interface between two liquids with different compositions.

Liquid-liquid chromatography   Chromatography in which the mobile and stationary phases are liquids.

Liquid-solid chromatography   Chromatography in which the mobile phase is a liquid and the stationary phase is a polar solid; synonymous with *adsorption chromatography.*

Liter   One cubic decimeter or 1000 cubic centimeters.

Loading error   An error in the measurement of a voltage due to current being drawn by the measuring device; occurs when the measuring device has resistance that is comparable to that of the voltage source being measured.

Longitudinal diffusion coefficient, B   A measure of the tendency for analyte species to migrate from regions of high concentration to regions of lower concentration; contributes to band broadening in chromatography.

Longitudinal diffusion term, B/u   A term in chromatographic band-broadening models that accounts for longitudinal diffusion.

Lower control limit, LCL   The lower boundary that has been set for satisfactory performance of a process or measurement.

Luminescence   Radiation resulting from photoexcitation (photoluminescence), chemical excitation (chemiluminescence), or thermal excitation (thermoluminescence).

L'vov platform   Device for the electrothermal atomization of samples in atomic absorption spectroscopy.

M

Macro analysis   Analysis of samples of masses more than 0.1 g.

Macrobalance   An analytical balance with a capacity of 160 to 200 g and a precision of 0.1 mg.

Major constituent   A constituent whose concentration is between 1% and 100%.

Majority carrier   The species principally responsible for the transport of charge in a semiconductor.

Masking agent   A reagent that combines with and inactivates matrix species that would otherwise interfere with the determination of an analyte.

Mass   An invariant measure of the amount of matter in an object.

Mass-action effect   The shift in the position of equilibrium through the addition or removal of a participant in the equilibrium. See also *Le Châtelier's principle.*

Mass-balance equation   An expression that relates the equilibrium molar concentrations of various species in a solution to one another and to the molar analytical concentration of the various solutes.

Mass-sensitive detector, chromatography   A detector that responds to the mass of analyte. The *flame ionization detector* is an example.

Mass spectrometry   Methods based on forming ions in the gas phase and separating them on the basis of mass-to-charge ratio.

Mass-transfer coefficients, C_S, C_M   Terms that account for mass transfer in the stationary and mobile phases in chromatography; mass transfer effects contribute to *band broadening.*

Mass transport   The movement of species through a solution caused by diffusion, convection, and electrostatic forces.

Matrix   The medium that contains an analyte.

Mean   Synonym for *arithmetic mean* and *average;* used to report what is considered the most representative value for a set of measurements.

Mean activity coefficient, $\gamma_\pm$ An experimentally measured activity coefficient for an ionic compound. It is not possible to resolve the mean activity coefficient into values for the individual ions.

Measuring pipet   A pipet calibrated to deliver any desired volume up to its maximum capacity; compare with *volumetric pipet.*

Mechanical entrapment   The incorporation of impurities within a growing crystal.

Mechanism of reaction   The elementary steps involved in the formation of products from reactants.

Median   The central value in a set of replicate measurements. For an odd number of data points, there are an equal number of points above and below the median; for an even number of data points, the median is the average of the central pair.

Megabore column   An open tubular column that can accommodate samples that are similar to those in an ordinary packed column.

Melt   The fused mass produced by the action of a flux; usually a fused salt.

Membrane electrode   An indicator electrode whose response is due to ion-exchange processes on each side of a thin membrane.

Meniscus   The curved surface displayed by a liquid held in a vessel.

Mercury electrode   A static or dropping electrode used in voltammetry.

Mercury film electrode   An electrode that has been coated with a thin layer of mercury; used in place of a *hanging mercury drop electrode* in anodic stripping analysis.

Metal oxide field effect transistor (MOSFET)   A semiconductor device; when suitably coated can be used as an ion-selective electrode.

Method uncertainty, s_m   The standard deviation associated with a measurement method; a factor, with the sampling standard deviation, s_s, in determining the overall standard deviation, s_o, of an analysis.

Michaelis constant   A collection of constants in the rate equation for enzyme kinetics; a measure of the dissociation of the enzyme/substrate complex.

Micro analysis   Analysis of samples with masses from 0.0001 to 0.01 g.

Microanalytical balance   An analytical balance with a capacity of 1 to 3 g and a precision of 0.0001 mg.

Microelectrode   An electrode with dimensions on the micrometer scale; used in voltammetry.

Microgram, μg 1×10^{-6} g.

Microliter, μL 1×10^{-6} L.

Microporous membrane   A hydrophobic membrane with a pore size that permits the passage of gases and is impermeable to other species; the sensing element of a *gas-sensing probe.*

Migration   In electrochemistry, mass transport due to electrostatic attraction or repulsion; in chromatography, mass transport in the column.

Migration rate, $\bar{v}$ The rate at which an analyte traverses a chromatographic column.

Milligram, mg 1×10^{-3} g or 1×10^{-6} kg.

Milliliter, mL 1×10^{-3} L.

Millimole, mmol 1×10^{-3} mol.

Minor constituent   A constituent whose concentration is between 0.01% (100 ppm) and 1%.

Mixed-crystal formation   A type of coprecipitation encountered in crystalline precipitates in which some of the ions in the analyte crystals are replaced by nonanalyte ions.

Mobile phase   In chromatography, a liquid or a gas that carries analytes through a liquid or solid stationary phase.

Mobile-phase mass-transfer coefficient, $C_M u$ A quantity that affects band broadening and thus plate height; nonlinear in solvent velocity u and influenced by the diffusion coefficient of the analyte, the particle size of the stationary phase, and the inside diameter of the column.

Modulation   Process of superimposing the analytical signal on a carrier wave. In amplitude modulation, the carrier wave magnitude varies according to the variations in the analytical signal; in frequency modulation, the carrier wave frequency varies with the analytical signal.

Mohr's salt   A common name for iron(II) ammonium sulfate hexahydrate.

Molar absorptivity, ε   The proportionality constant in Beer's law; $\varepsilon = A/bc$, where A is the absorbance, b is the path length in centimeters, and c is the concentration in moles per liter; characteristic of the absorbing species.

Molar analytical concentration, c_X   The number of moles of solute, X, that has been dissolved in sufficient solvent to give one liter of solution. Also numerically equal to the number of millimoles of solute per milliliter of solution. Compare with *equilibrium molar concentration.*

Molar concentration, M   The number of moles of a species contained in one liter of solution or the number of millimoles contained in one milliliter.

Molar mass, $\mathcal{M}$ The mass, in grams, of one mole of a chemical substance.

Molar species concentration   The equilibrium concentration of a species expressed in moles per liter and symbolized with square brackets [ ]; synonymous with *molar equilibrium concentration.*

Mole   The amount of substance that is 6.022×10^{23} particles of that substance.

Molecular absorption   The absorption of ultraviolet, visible, and infrared radiation brought about by quantized transitions in molecules.

Molecular fluorescence   The process whereby singlet excited-state electrons in molecules return to a lower quantum state, with the resulting energy being given off as electromagnetic radiation.

Molecular formula   A formula that includes structural information in addition to the number and identity of the atoms in a molecule.

Molecular weight   Obsolete synonym for molecular mass.

Monochromatic radiation   Ideally, electromagnetic radiation that consists of a single wavelength; in practice, a very narrow band of wavelengths.

Monochromator   A device for resolving polychromatic radiation into its component wavelengths.

Mother liquor   The solution that remains following the precipitation of a solid.

Muffle furnace   A heavy-duty oven capable of maintaining temperatures in excess of 1100°C.

N

Nanometer, nm 1×10^{-9} m.

National Institute of Standards and Technology (NIST)   An agency of the U.S. Department of Commerce; formerly the *National Bureau of Standards* (NBS); a major source for primary standards and analyzed standard reference materials.

Natural lifetime, τ   The radiative lifetime of an excited state; the time period during which the concentration of the reactant in a first-order process decreases to $1/e$ of its original value.

Nebulization   The transformation of a liquid into a spray of small droplets.

Nernst diffusion layer, δ   A thin layer of stagnant solution at the surface of an electrode in which mass transport is controlled only by diffusion. Outside the layer, the concentration of electroactive species is maintained constant by convection.

Nernst equation   A mathematical expression that relates the potential of an electrode to the activities of those species in solution that are responsible for the potential.

Nernst glower   A source of infrared radiation that consists of a cylinder of zirconium and yttrium oxides heated to a high temperature by passage of an electrical current.

Nichrome   A nickel/chromium alloy; when heated to incandescence, a source of infrared radiation.

Noise   Random fluctuations of an analytical signal that result from a large number of uncontrolled variables affecting the signal; any signal that interferes with detection of the analyte signal.

Nominal wavelength   The principal wavelength provided by a wavelength selection device.

Nonessential water   Water that is retained in or on a solid by physical, rather than chemical, forces.

Normal error curve   A plot of a Gaussian distribution of the frequency of results from random errors in a measurement.

Normal hydrogen electrode (NHE)   Synonym for *standard hydrogen electrode*.

Normality, c_N   The number of equivalent weights (masses) of a species in one liter of solution.

Normal-phase chromatography   A type of partition chromatography that involves a polar stationary phase and a nonpolar mobile phase; compare with *reversed-phase chromatography*.

Nucleation   A process involving formation of very small aggregates of a solid during precipitation.

Null hypothesis   A claim that a characteristic of a single population is equal to some specified value or that two or more population characteristics are identical; statistical tests are devised to validate or invalidate the null hypothesis with a specified level of probability.

Number of theoretical plates, N   A characteristic of a chromatographic column used to describe its efficiency.

O

Occluded water   Nonessential water that has been entrained in a growing crystal.

Occlusion   The physical entrainment of soluble impurities in a growing crystal.

Occupational Safety and Health Administration (OSHA)   A federal agency charged with assuring safety in the laboratory and the workplace.

Oesper's salt   Common name for iron(II) ethylenediamine sulfate tetrahydrate.

Ohmic potential drop   Synonymous with *IR drop*.

Open tubular column   A capillary column of glass or fused silica used in gas chromatography; the walls of the tube are coated with a thin layer of the stationary phase.

Operational amplifier   A versatile analog electronic amplifier for performing mathematical tasks and for conditioning output signals from instrument transducers.

Optical instruments   A broad term for instruments that measure absorption, emission, or fluorescence by analyte species based on ultraviolet, visible, or infrared radiation.

Optical methods   Synonymous with *spectrochemical methods*.

Optical wedge   A device used in optical spectroscopy whose transmission decreases linearly along its length.

Order of reaction   The exponent associated with the concentration of a species in the rate law for that reaction.

Outlier   A result that appears at odds with the other members in a data set.

Overall reaction order   The sum of the exponents for the concentrations appearing in the rate law for a chemical reaction.

Overall standard deviation, s_o   The square root of the sum of the variance of the measurement process and the variance of the sampling step.

Overpotential, overvoltage, Π   Excess voltage necessary to produce current in a polarized electrochemical cell.

Oxidant   Synonym for *oxidizing agent*.

Oxidation   The loss of electrons by a species in an oxidation/reduction reaction.

Oxidation potential   The potential of an electrode process that is written as an oxidation.

Oxidizing agent   A substance that acquires electrons in an oxidation/reduction reaction.

Oxine   A common name for 8-hydroxyquinoline.

Oxygen wave   At mercury electrodes, oxygen produces two waves, the first due to formation of peroxide, the second due to further reduction to water; this can be an interference in the determination of other species but is used in the determination of dissolved oxygen.

P

Packed columns   Chromatographic columns packed with porous materials to provide a large surface area for interaction with analyte in the mobile phase.

Pan arrest   A device to support the pans of a balance when a load is being placed on them. Designed to avoid damage to the knife edges.

Parallax   Apparent change in position of an object as a result of the movement of the observer; results in systematic errors in reading burets, pipets, and meters with pointers.

Particle growth   A stage in the precipitation of solids.

Particle properties of electromagnetic radiation   Behavior that consistent with radiation acting as small particles or *quanta* of en

Partition chromatography   A type of chromatography based on the distribution of solutes between a liquid mobile phase and a liquid stationary phase retained on the surface of a solid.

Partition coefficient   An equilibrium constant for the distribution of a solute between two immiscible liquid phases. See *distribution constant.*

Parts per million, ppm   A convenient method of expressing the concentration of a solute species that exists in trace amounts; for dilute aqueous solutions, ppm is synonymous with milligrams of solute per liter of solution.

Peak area, peak height   Properties of peak-shaped signals that can be used for quantitative analysis; used in chromatography, electrothermal atomic absorption, and other techniques.

Peptization   A process in which a coagulated colloid returns to its dispersed state.

Period of electromagnetic radiation   The time required for successive peaks of an electromagnetic wave to pass a fixed point in space.

pH   The negative logarithm of the hydrogen-ion activity of a solution.

Phosphorescence   Emission of light from an excited triplet state; phosphorescence is slower than fluorescence and may occur over several minutes.

Phosphorus pentoxide, P_2O_5   A drying agent.

Photoconductive cell   A detector of electromagnetic radiation whose electrical conductivity increases with the intensity of radiation impinging on it.

Photodecomposition   The formation of new species from molecules excited by radiation; one of several ways by which excitation energy is dissipated.

Photodiode   (1) A vacuum tube consisting of a wire anode and a photosensitive cathode, or photocathode, that produces an electron for each photon absorbed on the surface. (2) A reverse-biased silicon semiconductor that produces electrons and holes when irradiated by electromagnetic radiation. The resulting current provides a measure of the number of photons per second striking the device.

Photodiode array   A linear or two-dimensional array of photodiodes that can detect multiple wavelengths simultaneously. See *diode array detector.*

Photoelectric colorimeter   A photometer that responds to visible radiation.

Photoelectron   An electron released by the absorption of a photon striking a photoemissive surface.

Photoionization detector   A chromatographic detector that uses intense ultraviolet radiation to ionize analyte species; the resulting currents, which are amplified and recorded, are proportional to analyte concentration.

Photometer   An instrument for the measurement of absorbance that incorporates a filter for wavelength selection and a photon detector.

Photomultiplier tube   A sensitive detector of electromagnetic radiation; amplification is accomplished by a series of dynodes that produce a cascade of electrons for each photon received by the tube.

Photon detector   A generic term for transducers that convert an optical signal to an electrical signal.

Photons   Energy packets of electromagnetic radiation; also known as *quanta.*

Phototube   See *photodiode.*

Phthalein indicators   Acid/base indicators derived from phthalic anhydride, the most common of which is phenolphthalein.

pIon meter   An instrument that directly measures the concentration (or activity) of an analyte; consists of a specific ion indicator electrode, reference electrode, and a potential-measuring device.

Pipet   A tubular glass or plastic device for transferring known volumes of solution from one container to another.

Pixel   A single detector element on a diode array detector or a charge-transfer detector.

Planar chromatography   The term used to describe chromatographic methods that use a flat stationary phase; the mobile phase migrates across the surface by gravity or capillary action.

Plasma   A conductive gaseous medium containing ions and electrons.

Plate height, H   A quantity describing the efficiency of a chromatographic column. The term comes from the height of a plate, or distillation stage, in a traditional distillation column.

Platinum electrode   Used extensively in electrochemical systems in which an inert metallic electrode is required.

Plattner diamond mortar   A device for crushing small amounts of brittle materials.

Pneumatic detector   A transducer that converts changes in radiant power to changes in the pressure that a gas exerts on a flexible diaphragm. Changes in the volume of the diaphragm produce a change in signal at the output of the transducer.

p-n junction diode   A semiconductor device containing a junction between electron-rich and electron-deficient regions; permits current in one direction only.

Polarization   (1) In an electrochemical cell, a phenomenon in which the magnitude of the current is limited by the low rate of the electrode reactions (kinetic polarization) or the slowness of transport of reactants to the electrode surface (concentration polarization). (2) The process of causing electromagnetic radiation to oscillate in a plane or a circular pattern.

Polarogram   The current/voltage plot obtained from polarographic measurements.

Polarography   Voltammetry with a dropping mercury electrode.

Polychromatic radiation   Electromagnetic radiation consisting of more than one wavelength; compare with *monochromatic radiation.*

Polyfunctional acids and bases   Species that contain more than one acidic or basic functional group.

Population mean, μ   The mean value for a population of data; the true value for a quantity that is free of systematic error.

Population of data   The total number of values (sometimes assumed to be infinite) that a measurement could take; also referred to as a *universe of data.*

Population standard deviation, σ   A measure of precision based on a population of data.

Porous layer open tube (PLOT) column   A capillary column for gas-solid chromatography in which a thin layer of the stationary phase is adsorbed on the walls of the column.

Potentiometric titration   A titrimetric method involving measurement of the potential between a reference electrode and an indicator electrode as a function of titrant volume.

Potentiometry   That branch of electrochemistry concerned with the relationship between the potential of an electrochemical cell and the concentrations (activities) of the contents of the cell.

Potentiostat   An electronic device that alters the applied potential so that the potential between a working electrode and a reference electrode is maintained at a fixed value.

Potentiostatic methods   Electrochemical methods that use a controlled potential between the working electrode and a reference electrode.

Power, P, of electromagnetic radiation   The energy that reaches a given area per second; often used synonymously with intensity, although the two are not precisely the same.

Precipitation from homogeneous solution Synonymous with *homogeneous precipitation.*

Precipitation methods of analysis Gravimetric and titrimetric methods involving the formation (or less frequently, the disappearance) of a precipitate.

Precision A measure of the agreement among individual data in a set of replicate observations.

Premixed burner Burner in which gases are mixed prior to combustion.

Pressure broadening An effect that increases the width of an atomic spectral line; caused by collisions among atoms that result in slight variations in their energy states.

Primary absorption Absorption of the excitation beam in fluorescence or phosphorescence spectroscopy; compare with *secondary absorption.*

Primary adsorption layer Charged layer of ions on the surface of a solid, resulting from the attraction of lattice ions for ions of opposite charge in the solution.

Primary standard A highly pure chemical compound that is used to prepare or determine the concentrations of standard solutions for titrimetry.

Prism A transparent, glass or quartz polyhedron comprising two parallel triangular faces and three square or rectangular faces that disperses polychromatic radiation into its component wavelengths by refraction.

Proportional error An error whose magnitude increases as the sample size increases.

Protective agent In atomic spectroscopy, species that form soluble complexes with the analyte, thereby preventing the formation of compounds that have low volatility.

Pseudo-order reactions Chemical systems in which the concentration of a reactant (or reactants) is large and essentially invariant with respect to that of the component (or components) of interest.

Pulse polarography Voltammetric methods that periodically impose a pulse on the linearly increasing excitation voltage; the difference in measured current, Δi, yields a peak whose height is proportional to the analyte concentration.

p-Value An expression of the concentration of a solute species as its negative logarithm; the use of p-values permits expression of enormous ranges of concentration in terms of relatively small numbers.

Pyroelectric detector A thermal detector based on the temperature-dependent potential that develops between electrodes separated by a pyroelectric material. A Pyroelectric material becomes polarized and produces a potential difference across its surfaces when its temperature is changed.

Q

Q test A statistical test that indicates—with a specified level of probability—whether an outlying measurement in a set of replicate data is a member of a given Gaussian distribution.

Quality assessment A protocol to assure that quality control methods are providing the information needed to evaluate satisfactory performance of a product or a service.

Quality assurance A protocol designed to demonstrate that a product or a service is meeting criteria that have been established for satisfactory performance.

Quantum A microscopic quantity of energy that can have only discrete values. Quanta are absorbed by and emitted from atoms with energies corresponding to differences in the energies of atomic orbitals. Emitted and absorbed quanta are referred to as *photons*, which have frequencies determined by the Planck relationship, $E = h\nu$.

Quantum yield of fluorescence The fraction of absorbed photons that are emitted as fluorescence photons.

Quenching (1) Process by which molecules in an excited state lose energy to other species without fluorescing. (2) An action that brings about the cessation of a chemical reaction.

R

Radiation buffers Potential interferents that are intentionally added in large amounts to samples and standards to swamp out their effects on atomic emission measurements.

Random errors Uncertainties resulting from the operation of small uncontrolled variables that are inevitable as measurement systems are extended to and beyond their limits.

Range, *w*, of data The difference between extreme values in a set of data; synonymous with *spread.*

Rate constant, *k* A proportionality constant in a rate expression.

Rate-determining step The slow step in the sequence of elementary reactions making up a mechanism.

Rate law The empirical relationship describing the rate of a reaction in terms of the concentrations of participating species.

Rate theory A theory that accounts for the shapes of chromatographic peaks.

Reagent-grade chemicals Highly pure chemicals that meet the standards of the Reagent Chemical Committee of the American Chemical Society.

Redox Synonymous with *oxidation/reduction.*

Redox electrode An inert electrode that responds to the electrode potential of a redox system.

Reducing agent The species that supplies electrons in an oxidation/reduction reaction.

Reductant Synonym for *reducing agent.*

Reduction The process whereby a species acquires electrons.

Reduction potential The potential of an electrode process expressed as a reduction; synonymous with *electrode potential.*

Reductor A column packed with a granular metal through which a sample is passed to prereduce an analyte.

Reference electrode An electrode whose potential relative to the standard hydrogen electrode is known and against which potentials of unknown electrodes may be measured; the potential of a reference electrode is completely independent of the analyte concentration.

Reference standards Complex materials that have been extensively analyzed; a prime source for these standards is the National Institute of Standards and Technology (NIST).

Reflection The return of radiation from a surface.

Reflection grating An optical element that disperses polychromatic radiation into its component wavelengths. Consists of lines ruled on a reflecting surface; dispersion is the result of constructive and destructive interference.

Refractive index The ratio of the velocity of electromagnetic radiation in a vacuum to its velocity in some other medium.

Refractory materials Substances that resist attack by ordinary laboratory acids or bases; brought into solution by high-temperature fusion with a flux.

Regression analysis A statistical technique for determining the parameters of a model. See also *least-squares method.*

Relative electrode potential The potential of an electrode with respect to another (ordinarily the standard hydrogen electrode another reference electrode.)

Relative error   The error in a measurement divided by the true (or accepted) value for the measurement; often expressed as a percentage.

Relative humidity   The ratio, often expressed as a percentage, between the ambient vapor pressure of water and its saturated vapor pressure at a given temperature.

Relative standard deviation (RSD)   The standard deviation divided by the mean value for a set of data; when expressed as a percentage, the relative standard deviation is referred to as the *coefficient of variation.*

Relative supersaturation   The difference between the instantaneous *(Q)* and the equilibrium *(S)* concentrations of a solute in a solution, divided by *S;* provides general guidance as to the particle size of a precipitate formed by addition of reagent to an analyte solution.

Relaxation   The return of excited species to a lower energy level. The process is accompanied by the release of excitation energy as heat or luminescence.

Releasing agent   In atomic absorption spectroscopy, species introduced to combine with sample components that would otherwise interfere by forming compounds of low volatility with the analyte.

Replica grating   An impression of a master grating; used as the dispersing element in most grating instruments, owing to the high cost of a master grating.

Replicate samples   Portions of a material, of approximately the same size, that are carried through an analysis at the same time and in precisely the same way.

Reprecipitation   A method of improving the purity of precipitates involving formation and filtration of the solid, followed by redissolution and reformation of the precipitate.

Residual   The difference between the value predicted by a model and the experimental value.

Residual current   Nonfaradaic currents due to impurities and to charging of the electrical double layer.

Resolution, R_s   Measures the ability of a chromatographic column to separate two analytes; defined as the difference between the retention times for the two peaks divided by their average widths.

Resonance fluorescence   Fluorescence emission at a wavelength that is identical with the excitation wavelength.

Resonance line   A spectral line resulting from a resonance transition.

Resonance transition   A transition to or from the ground electronic state.

Retention factor, k   A term used to describe the migration of a species through a chromatographic column. Its numerical value is given by $k = (t_R - t_M)/t_M$, where t_R is the retention time for a peak and t_M is the dead time; also called the *capacity factor.*

Retention time, t_R   In chromatography, the time between sample injection on a chromatographic column and the arrival of an analyte peak at the detector.

Reversed-phase chromatography   A type of liquid-liquid partition chromatography that uses a nonpolar stationary phase and a polar mobile phase; compare with *normal-phase chromatography.*

Reversible cell   An electrochemical cell in which electron transfer is ... in both directions.

...ostat   A variable resistor used to control the current in a circuit. ...nfigured properly, may be used as a voltage divider.

...tional states   Quantized states associated with the rotation of a ...cule about its center of mass.

...tional transition   A change in quantized rotational energy ... in a molecule.

...er policeman   A small length of rubber tubing that has been ...ed on one end; used to dislodge adherent particles of precipitate ... beaker walls.

S

Salt   An ionic compound formed by the reaction of an acid and a base.

Salt bridge   A device in an electrochemical cell that allows flow of charge between the two electrolyte solutions while minimizing mixing of the two.

Salt effect   Influence of ions on the activities of solutes.

Salt-induced precipitation   Technique used to precipitate proteins. At low salt concentration, adding salt increases solubility (salting-in effect), whereas high salt concentrations induce precipitation (salting-out effect).

Sample of data   A finite group of replicate measurements.

Sample matrix   The medium that contains an analyte.

Sample mean, $\bar{x}$ The arithmetic average of a finite set of measurements.

Sample splitter   A device that permits the introduction of small and reproducible portions of sample to a chromatographic column. In capillary gas chromatography, a reproducible fraction of the injected sample is introduced onto the column, while the remaining portion goes to waste.

Sample standard deviation, s   A precision estimate based on deviations of individual data from the mean, $\bar{x}$, of a data sample; also referred to as the *standard deviation.*

Sampling   The process of collecting a small portion of a material whose composition is representative of the bulk of the material from which it was taken.

Sampling loop   A small piece of tubing used in chromatography that has a sampling valve to inject small quantities of sample.

Sampling uncertainty, s_s   The standard deviation associated with the taking of a sample; a factor—with the method uncertainty—in determining the overall standard deviation of an analysis.

Sampling valve   A rotary valve used to inject small portions of a sample onto a chromatographic column; usually used in conjunction with a *sampling loop.*

Saponification   The cleavage of an ester group to regenerate the alcohol and the acid from which the ester was derived.

Saturated calomel electrode (SCE)   A reference electrode that can be formulated as $Hg \mid Hg_2Cl_2(sat), KCl(sat) \parallel$. Its half-reaction is

$$Hg_2Cl_2(s) + 2e^- \rightleftharpoons 2Hg(l) + 2Cl^-$$

Schöniger apparatus   A device for the combustion of samples in an oxygen-rich environment.

Second derivative curve   A plot of $\Delta^2 E/\Delta V^2$ versus volume for a potentiometric titration; the function undergoes a change in sign at the inflection point in a conventional titration curve.

Secondary absorption   Absorption of the emitted radiation in fluorescence or phosphorescence spectrometry; compare with *primary absorption.*

Secondary standard   A substance whose purity has been established and verified by chemical analysis.

Sector mirror   A disk with portions that are partially mirrored and partially nonreflecting; when rotated, directs radiation from the monochromator of a double-beam spectrophotometer alternately through the sample and the reference cells.

Selectivity   The tendency for a reagent or an instrumental method to react with or respond similarly to only a few species.

Selectivity coefficient, $k_{A,B}$ The selectivity coefficient for a specific ion electrode is a measure of the relative response of the electrode to ions A and B.

Selectivity factor, α   In chromatography, $\alpha = K_B/K_A$, where K_B is the distribution constant for a less strongly retained species and K_A is the constant for a more strongly retained species.

Self-absorption A process in which analyte molecules absorb radiation emitted by other analyte molecules.

Semiconductor A material with electrical conductivity that is intermediate between a metal and an insulator.

Semimicro analysis Analysis of samples with masses from 0.01 g to 0.1g.

Semimicroanalytical balance A balance with a capacity of about 30 g and a precision of 0.01 mg.

Servo system A device in which a small error signal is amplified and used to return the system to a null position.

Sigmoid curve An S-shaped curve; typical of the plot of the p-function of an analyte versus the volume of reagent in titrimetry.

Signal-to-noise ratio, _S/N_ The ratio of the mean analyte output signal to the standard deviation of the signal.

Significant figure convention A system of communicating to the reader information concerning the reliability of numerical data in the absence of any statistical data; in general, all digits known with certainty, plus the first uncertain digit, are considered significant.

Silica Common name for silicon dioxide; used in the manufacture of crucibles and the cells for optical analysis and as a chromatographic support medium.

Silicon photodiode A photon detector based on a reverse-biased silicon diode; exposure to radiation creates new holes and electrons, thereby increasing photocurrent. See _photodiode._

Silver-silver chloride electrode A widely used reference electrode, which can be represented as Ag | AgCl(s), KCl(xM) || . The half-reaction for the electrode is

$$AgCl(s) + e^- \rightleftharpoons Ag(s) + Cl^-(xM)$$

Single-beam instruments Photometric instruments that use only one beam; they require the operator to position the sample and the blank alternately in a single light path.

Single-electrode potential Synonymous with _relative electrode potential._

Single-pan balance An unequal-arm balance with the pan and weights on one side of the fulcrum and an air damper on the other; the weighing operation involves removal of standard weights in an amount equal to the mass of the object on the pan.

Sintered-glass crucible Synonymous with _fritted-glass crucible._

SI units An international system of measurement that uses seven base units; all other units are derived from these seven units.

Size-exclusion chromatography A type of chromatography in which the packing is a finely divided solid having a uniform pore size; separation is based on the size of analyte molecules.

Slope, _m_, of a calibration line A parameter of the linear model $y = mx + b$; determined by regression analysis.

Soap-bubble meter A device for measuring gas flow rates in gas chromatography.

Solubility-product constant, K_{sp} A numerical constant that describes the equilibrium between a saturated solution of a sparingly soluble ionic salt and the solid salt that must be present.

Soluble starch β-amylose, an aqueous suspension that is a specific indicator for iodine.

Solvent programming The systematic alteration of mobile-phase composition to optimize migration rates of solutes in a chromatographic column. See also _gradient elution._

Sorbed water Nonessential water that is retained in the interstices of solid materials.

Sparging The removal of an unwanted dissolved gas by purging with an inert gas.

Special-purpose chemicals Reagents that have been specially purified for a particular end use.

Specific gravity, sp gr The ratio of the density of a substance to that of water at a specified temperature (ordinarily 4°C).

Specific indicator A species that reacts with a particular species in an oxidation/reduction titration.

Specific surface area The ratio between the surface area of a solid and its mass.

Specificity Refers to methods or reagents that respond or react with one and only one analyte.

Spectra Plots of absorbance, transmittance, or emission intensity as a function of wavelength, frequency, or wavenumber.

Spectral interference Emission or absorption by species other than the analyte within the band-pass of the wavelength selection device; causes a blank interference.

Spectrochemical methods Synonymous with _spectrometric methods._

Spectrofluorometer A fluorescence instrument that has monochromators for selecting excitation and emission wavelengths; in some cases, hybrid instruments have a filter and a monochromator.

Spectrograph An optical instrument equipped with a dispersing element, such as a grating or a prism, that allows a range of wavelengths to strike a spatially sensitive detector, such as a diode array, charge coupled device, or photographic plate.

Spectrometer An instrument equipped with a monochromator or a polychromator, a photodetector, and an electronic readout that displays a number proportional to the intensity of an isolated spectral band.

Spectrometric methods Methods based on the absorption, emission, or fluorescence of electromagnetic radiation that is related to the amount of analyte in the sample.

Spectrophotometer A spectrometer designed for the measurement of the absorption of ultraviolet, visible, or infrared radiation. The instrument includes a source of radiation, a monochromator, and an electrical means of measuring the ratio of the intensities of the sample and reference beams.

Spectrophotometric titration A titration monitored by ultraviolet/visible spectrometry.

Spectroscope An optical instrument similar to a spectrometer except that spectral lines can be observed visually.

Spectroscopy A general term used to describe techniques based on the measurement of absorption, emission, or luminescence of electromagnetic radiation.

Spread, _w_, of data A precision estimate; synonymous with _range._

Sputtering The process whereby an atomic vapor is produced by collisions with excited ions on a surface such as the cathode in a hollow-cathode lamp.

Square-wave polarography A variety of _pulse polarography._

Standard-addition method A method of determining the concentration of an analyte in a solution. Small measured increments of the analyte are added to the sample solution, and instrument readings are recorded after one or more additions. The method compensates for some matrix interferences.

Standard deviation, σ or _s_ A measure of how closely replicate data cluster around the mean; in a normal distribution, 67% of the data points can be expected to lie within one standard deviation of the mean.

Standard deviation about regression, s_r The standard error of the deviations from a least-square straight line. A synonym of _standard error of the estimate._

Standard electrode potential, E^0 The potential (relative to the standard hydrogen electrode) of a half-reaction written as a reduction when the activities of all reactants and products are unity.

Standard error of the estimate Synonym for standard deviation about regression.

Standard error of the mean, σ_m or s_m The standard deviation divided by the square root of the number of measurements in the set.

Standard hydrogen electrode (SHE) A gas electrode consisting of a platinized platinum electrode immersed in a solution that has a hydrogen ion activity of 1.00 and is kept saturated with hydrogen at a pressure of 1.00 atm. Its potential is assigned a value of 0.000 V at all temperatures.

Standardization Determination of the concentration of a solution by calibration, directly or indirectly, with a primary standard.

Standard reference materials (SRMs) Samples of various materials in which the concentration of one or more species is known with very high certainty.

Standard solution A solution in which the concentration of a solute is known with high reliability.

Stationary phase In chromatography, a solid or an immobilized liquid on which analyte species are partitioned during passage of a mobile phase.

Stationary phase mass-transfer term, $C_S u$ A measure of the rate at which an analyte molecule enters and is released from the stationary phase.

Statistical control The condition in which performance of a product or a service is deemed within bounds that have been set for quality assurance; defined by upper and lower control limits.

Statistical sample A finite set of measurements, drawn from a population of data, often from a hypothetical infinite number of possible measurements.

Steady-state approximation The assumption that the concentration of an intermediate in a multistep reaction remains essentially constant with time.

Stirrup The link between the beam of a mechanical balance and its pan (or pans).

Stockholm convention A set of conventions relating to electrochemical cells and their potentials; also known as the *IUPAC convention.*

Stoichiometry The combining ratios among molar quantities of species in a chemical reaction.

Stokes shifts Differences in wavelengths of incident and emitted or scattered radiation.

Stop-flow injection In flow-injection analysis, turning off the flow to allow kinetic measurements on a static plug of solution.

Stopped-flow mixing A technique in which the reactants are mixed rapidly and the course of the reaction is monitored downstream after the flow has stopped abruptly.

Stray radiation Radiation of a wavelength other than the wavelength selected for optical measurement.

Strong acids and strong bases Acids and bases that are completely dissociated in a particular solvent.

Strong electrolytes Solutes that are completely dissociated into ions in a particular solvent.

Student's *t* test See *t* test.

Substrate (1) A substance acted on, usually by an enzyme. (2) A solid on which surface modifications are made.

Successive approximations A procedure for solving higher order equations through the use of intermediate estimates of the quantity sought.

Sulfide separations The use of sulfide precipitation to separate cations.

Sulfonic acid group —RSO_3H.

Supercritical fluid A substance that is maintained above its critical temperature; its properties are intermediate between those of a liquid and those of a gas.

Supercritical fluid chromatography Chromatography involving a supercritical fluid as the mobile phase.

Supersaturation A condition in which a solution temporarily contains an amount of solute that exceeds its equilibrium solubility.

Support-coated open tubular (SCOT) columns Capillary gas chromatography columns whose interior walls are lined with a solid support.

Supporting electrolyte A salt added to the solution in a voltammetric cell to eliminate migration of the analyte to the electrode surface.

Suppressor-based chromatography A chromatographic technique involving a column or a membrane located between the analytical column and a conductivity detector; its purpose is to convert ions of the eluting solvent into nonconducting species while passing ions of the sample.

Surface adsorption The retention of a normally soluble species on the surface of a solid.

Swamping The introduction of a potential interferent to both calibration standards and the solution of the analyte in order to minimize the effect of the interferent in the sample matrix.

Systematic error Errors that have a known source; they affect measurements in one and only one way and can, in principle, be accounted for. Also called *determinate error* or *bias.*

T

0% T adjustment A calibration step that eliminates dark current and other background signals from the response of a spectrophotometer.

100% T adjustment Adjustment of a spectrophotometer to register 100% transmittance with a blank in the light path.

Tailing A nonideal condition in a chromatographic peak in which the latter portions are drawn out; compare with *fronting.*

Tare A counterweight used on an analytical balance to compensate for the mass of a container; the act of zeroing a balance.

Temperature programming The systematic adjustment of column temperature in gas chromatography to optimize migration rates for solutes.

THAM *tris*-(hydroxymethyl) aminomethane, a primary standard for bases; its formula is $(HOCH_2)_3CNH_2$.

Thermal conductivity detector A detector used in gas chromatography that depends on measuring the thermal conductivity of the column eluent.

Thermal detector An infrared detector that produces heat as a result of absorption of radiation and converts it to a mechanical or electrical signal.

Thermionic detector (TID) A detector for gas chromatography similar to a flame ionization detector; particularly sensitive for analytes that contain nitrogen or phosphorus.

Thermistor A temperature-sensing semiconductor; used in some bolometers. The electrical resistance varies with the temperature.

Thermodynamic equilibrium constant, K The equilibrium constant expressed in terms of the activities of all reactants and products.

TISAB (total ionic strength adjustment buffer) A solution used to provide a large and constant ionic strength and thus swamp the effect of electrolytes on direct potentiometric analyses.

Titration The procedure whereby a standard solution reacts with known stoichiometry with an analyte to the point of chemical equivalence, which is measured experimentally as the end point. The volume or the mass of the standard needed to reach the end point is used to calculate the amount of analyte present.

Titration error The difference between the titrant volume needed to reach an end point in a titration and the theoretical volume required to obtain an equivalence point.

Titrator   An instrument that performs titrations automatically.

Titrimetry   The process of systematically introducing an amount of titrant that is chemically equivalent to the quantity of analyte in a sample.

Trace constituent   A constituent whose concentration is between 1 ppb and 100 ppm.

Transducer   A device that converts a chemical or physical phenomenon into an electrical signal.

Transfer pipet   Synonym for *volumetric pipet*.

Transition pH range   The span of acidities (frequently about 2 pH units) over which an acid/base indicator changes from its pure acid color to that of its conjugate base.

Transition potential   The range in E_{system} over which an oxidation/reduction indicator changes from the color of its reduced form to that of its oxidized form.

Transmittance, T   The ratio of the power, P, of a beam of radiation after it has traversed an absorbing medium to its original power, P_0; often expressed as a percentage:

$$\%T = (P/P_0) \times 100\%.$$

Transverse wave   A wave motion in which the direction of displacement is perpendicular to the direction of propagation.

Triple-beam balance   A rugged, albeit primitive in the age of electronic balances, laboratory balance that is used to weigh approximate amounts.

TRIS   Synonymous with *THAM*.

t-test   A statistical test used to decide whether an experimental value equals a known or theoretical value or whether two or more experimental values are identical with a given level of confidence; used with s and $\bar{x}$ when good estimates of σ and μ are not available.

Tungsten filament lamp   A convenient source of visible and near-infrared radiation.

Tungsten-halogen lamp   A tungsten lamp that contains a small amount of I_2 within a quartz envelope that permits the lamp to operate at a higher temperature; brighter than a conventional tungsten filament lamp.

Turbulent flow   Describes the random motion of liquid in the bulk of a flowing solution; compare with *laminar flow*.

Tyndall effect   The scattering of radiation by particles in a solution or a gas that have colloidal dimensions.

U

Ultramicro analysis   Analysis of samples whose mass is less than 10^{-4} g.

Ultramicroelectrode   Synonymous with *microelectrode*.

Ultratrace constituent   A constituent whose concentration is less than 1 ppb.

Ultraviolet/visible detector, HPLC   Detector for high-performance liquid chromatography that uses ultraviolet/visible absorption to monitor eluted species as they exit a chromatographic column.

Ultraviolet/visible region   The region of the electromagnetic spectrum between 180 and 780 nm; associated with electronic transitions in atoms and molecules.

Unified atomic mass unit   Basic unit of mass equal to 1/12 the mass of the most abundant isotope of carbon, ^{12}C. Equal to 1 Dalton.

Universe of data   Synonymous with a *population of data*.

V

Valinomycin   An antibiotic that has been used in a membrane electrode for potassium.

van Deemter equation   An equation that expresses plate height in terms of eddy diffusion, longitudinal diffusion, and mass transport.

Variance, σ^2 or s^2   A precision estimate consisting of the square of the standard deviation. Also a measure of column performance; given the symbol τ^2 where the abscissa of the chromatogram has units of time.

V-blender   A device that is used to thoroughly mix dry samples.

Velocity of electromagnetic radiation, v   *In vacuo*, 3×10^{10} cm/sec.

Vernier   An aid for making estimates between graduation marks on a scale.

Vibrational relaxation   A very efficient process in which excited molecules relax to the lowest vibrational level of an electronic state.

Vibrational transitions   Transitions between vibrational states of an electronic state that are responsible for infrared absorption.

Visible radiation   That portion of the electromagnetic spectrum (380 to 780 nm) to which the human eye is responsive.

Volatilization   The process of converting a liquid (or a solid) to the vapor state.

Volatilization method of analysis   A variant of the gravimetric method based on mass loss caused by heating or ignition.

Voltage divider   A resistive network that provides a fraction of the input voltage at its output.

Voltaic cell   Synonymous with *galvanic cell*.

Voltammetric wave   An $\int$-shaped curve that is produced in a voltammetric experiment when the voltage sweeps through the half-wave potential of an electroactive species.

Voltammetry   A group of electroanalytical methods that measure current as a function of the voltage applied to a working electrode.

Voltammogram   A plot of current as a function of the potential applied to a working electrode.

Volume percent (v/v)   The ratio of the volume of a liquid to the volume of its solution, multiplied by 100%.

Volumetric flask   A container for preparing precise volumes of solution.

Volumetric methods   Methods of analysis in which the final measurement is a volume of a standard titrant needed to react with the analyte in a known quantity of sample.

Volumetric pipet   A device that will deliver a precise volume from one container to another; also called a *measuring pipet*.

W

Walden reductor   A column packed with finely divided silver granules; used to prereduce analytes.

Wall-coated open tubular (WCOT) column   A capillary column coated with a thin layer of stationary phase.

Water of constitution   Essential water that is derived from the molecular composition of the species.

Water of crystallization   Essential water that is an integral part of the crystal structure of a solid.

Wavelength, of electromagnetic radiation, λ   The distance between successive maxima (or minima) of a wave.

Wavelength selector   A device that limits the range of wavelengths used in an optical measurement (see Section 25A-3).

Wavenumber, $\bar{\nu}$ The reciprocal of wavelength; has units of cm^{-1}.

Wave properties, electromagnetic radiation   Behavior of radiation as an electromagnetic wave.

Weak acid/conjugate base pairs   In the Brønsted-Lowry view, solute pairs that differ from one another by one proton.

Weak acids and weak bases   Acids and bases that are only partially dissociated in a particular solvent.

Weak electrolytes   Solutes that are incompletely dissociated into ions in a particular solvent.

Weighing bottle A lightweight container for the storage and weighing of analytical samples.

Weighing by difference The process of weighing a container plus the sample, followed by weighing the container after the sample has been removed or before it has been placed in the container.

Weighing form In gravimetric analysis, the species collected whose mass is proportional to the amount of analyte in the sample.

Weight The attraction between an object and its surroundings, terrestrially, the Earth.

Weight molar concentration, M_w The molar concentration of titrant expressed as millimoles per gram.

Weight percent (w/w) The ratio of the mass of a solute to the mass of its solution, multiplied by 100%.

Weight titrimetry Synonymous with *gravimetric titrimetry*.

Weight/volume percent (w/v) The ratio of the mass of a solute to the volume of solution in which it is dissolved, multiplied by 100%.

Wet ashing The use of strong liquid oxidizing reagents to decompose the organic matter in a sample.

Windows, of cells Surfaces of cells through which radiation passes.

Z

Zero percent T adjustment A calibration step that compensates for dark current in the response of a spectrophotometer.

Zimmermann-Reinhardt reagent A solution of manganese(II) in concentrated H_2SO_4 and H_3PO_4 that prevents the induced oxidation of chloride ion by permanganate during the titration of iron(II).

Zones, chromatographic Synonymous with *chromatographic bands*.

Zwitterion The species that results from the transfer in solution of a proton from an acidic group to an acceptor site on the same molecule.

Appendix 1

The Literature of Analytical Chemistry

TREATISES

As used here, the term *treatise* means a comprehensive presentation of one or more broad areas of analytical chemistry.

D. Barcelo, series ed., *Comprehensive Analytical Chemistry*, New York: Elsevier, 1959–2010. As of 2012, 58 volumes of this work have appeared.

N. H. Furman and F. J. Welcher, eds., *Standard Methods of Chemical Analysis*, 6th ed., New York: Van Nostrand, 1962–1966. In five parts, this work is largely devoted to specific applications.

I. M. Kolthoff and P. J. Elving, eds., *Treatise on Analytical Chemistry*, New York: Wiley, 1961–1986. Part I, 2nd ed. (14 volumes), is devoted to theory; Part II (17 volumes) deals with analytical methods for inorganic and organic compounds; and Part III (4 volumes) treats industrial analytical chemistry.

R. A. Meyers, ed., *Encyclopedia of Analytical Chemistry: Applications, Theory and Instrumentation*, New York: Wiley, 2000. A 15-volume reference work for all areas of analytical chemistry. The encyclopedia has been published online since 2007.

B. W. Rossitor and R. C. Baetzold, eds., *Physical Methods of Chemistry*, 2nd ed., New York: Wiley, 1986–1993. This series consists of 12 volumes devoted to various types of physical and chemical measurements performed by chemists.

P. Worsfold, A. Townshend, and C. Poole, eds., *Encyclopedia of Analytical Science*, 2nd ed., Amsterdam: Elsevier, 2005. A 10-volume reference work that covers all areas of analytical science. The work is available in print and online.

OFFICIAL METHODS OF ANALYSIS

These publications are often single volumes that provide a useful source of analytical methods for the determination of specific substances in articles of commerce. The methods have been developed by various scientific societies and serve as standards in arbitration as well as in the courts.

Annual Book of ASTM Standards, Philadelphia: American Society for Testing Materials. This work of 80+ volumes is revised annually and contains methods for both physical testing and chemical analysis. Volumes 3.05, *Analytical Chemistry for Metals, Ores and Related Materials*, and 3.06, *Molecular Spectroscopy and Surface Analysis*, are particularly useful sources. The work is available online or on CD-ROM.

L. S. Clesceri, A. E. Greenberg, and A. D. Eaton, eds., *Standard Methods for the Examination of Water and Wastewater*, 20th ed., New York: American Public Health Association, 1998.

Official Methods of Analysis, 18th ed., Washington, DC: Association of Official Analytical Chemists, 2005. This is a very useful source of methods for the analysis of such materials as drugs, food, pesticides, agricultural materials, cosmetics, vitamins, and nutrients. The online edition is a continuous edition with new and revised methods published as soon as they are approved and ready.

C. A. Watson, *Official and Standardized Methods of Analysis*, 3rd ed., London: Royal Society of Chemistry, 1994.

REVIEW SERIALS

The reviews listed below are general reviews in the field. In addition, there are specific review serials devoted to advances in areas such as chromatography, electrochemistry, mass spectrometry, and many others.

Analytical Chemistry: "Fundamental Reviews" and "Application Reviews," Washington, DC: American Chemical Society. Through 2010, in the June 15 issue of *Analytical Chemistry*, "Fundamental Reviews" appeared in even-numbered years, while "Application Reviews" appeared in odd-numbered years. "Fundamental Reviews" covered significant developments in many areas of analytical chemistry. "Application Reviews" were devoted to specific areas, such as water analysis, clinical chemistry, and petroleum products. In 2011, both types of reviews appeared in the June 15 issue. Beginning in 2012, the annual reviews issue appears in January and focuses on topics in contemporary measurement science.

Annual Review of Analytical Chemistry, Palo Alto, CA: Annual Reviews. Authoritative review articles on important aspects of modern analytical chemistry. The annual review has been published each year since 2008.

Critical Reviews in Analytical Chemistry, Boca Rotan, FL: CRC Press. This publication appears quarterly and provides in-depth articles covering the latest developments in the analysis of biochemical substances.

Reviews in Analytical Chemistry, Berlin: De Gruyter GMBH. A journal devoted to reviews in the field. Four volumes per year are published in all branches of modern analytical chemistry.

TABULAR COMPILATIONS

A. J. Bard, R. Parsons, and T. Jordan, eds., *Standard Potentials in Aqueous Solution*, New York: Marcel Dekker, 1985.

J. A. Dean, *Analytical Chemistry Handbook*, New York: McGraw-Hill, 1995.

A. E. Martell and R. M. Smith, *Critical Stability Constants*, 6 vols., New York: Plenum Press, 1974–1989.

G. Milazzo, S. Caroli, and V. K. Sharma, *Tables of Standard Electrode Potential*, New York: Wiley, 1978.

ADVANCED ANALYTICAL AND INSTRUMENTAL TEXTBOOKS

J. N. Butler, *Ionic Equilibrium: A Mathematical Approach*, Reading, MA: Addison-Wesley, 1964.

J. N. Butler, *Ionic Equilibrium: Solubility and pH Calculations*, New York: Wiley, 1998.

G. D. Christian and J. E. O'Reilly, *Instrumental Analysis*, 2nd ed., Boston: Allyn and Bacon, 1986.

W. B. Guenther, *Unified Equilibrium Calculations*, New York: Wiley, 1991.

H. A. Laitinen and W. E. Harris, *Chemical Analysis*, 2nd ed., New York: McGraw-Hill, 1975.

F. A. Settle, ed., *Handbook of Instrumental Techniques for Analytical Chemistry*, Upper Saddle River, NJ: Prentice Hall, 1997.

D. A. Skoog, F. J. Holler, and S. R. Crouch, *Principles of Instrumental Analysis*, 6th ed., Belmont, CA: Brooks/Cole, 2007.

H. Strobel and W. R. Heineman, *Chemical Instrumentation: A Systematic Approach*, 3rd ed., Boston: Addison-Wesley, 1989.

MONOGRAPHS

Hundreds of monographs devoted to specialized areas of analytical chemistry are available. In general, these monographs are authored by experts and are excellent sources of information. Representative monographs in various areas are listed below.

Gravimetric and Titrimetric Methods

M. R. F. Ashworth, *Titrimetric Organic Analysis*, 2 vols., New York: Interscience, 1965.

R. deLevie, *Aqueous Acid-Base Equilibria and Titrations*, Oxford: Oxford University Press, 1999.

L. Erdey, *Gravimetric Analysis*, Oxford: Pergamon, 1965.

J. S. Fritz, *Acid-Base Titration in Nonaqueous Solvents*, Boston: Allyn and Bacon, 1973.

W. F. Hillebrand, G. E. F. Lundell, H. A. Bright, and J. I. Hoffman, *Applied Inorganic Analysis*, 2nd ed., New York: Wiley, 1953, reissued 1980.

I. M. Kolthoff, V. A. Stenger, and R. Belcher, *Volumetric Analysis*, 3 vols., New York: Interscience, 1942–1957.

T. S. Ma and R. C. Ritner, *Modern Organic Elemental Analysis*, New York: Marcel Dekker, 1979.

L. Safarik and Z. Stransky, *Titrimetric Analysis in Organic Solvents*, Amsterdam: Elsevier, 1986.

E. P. Serjeant, *Potentiometry and Potentiometric Titrations*, New York: Wiley, 1984.

W. Wagner and C. J. Hull, *Inorganic Titrimetric Analysis*, New York: Marcel Dekker, 1971.

Organic Analysis

S. Siggia and J. G. Hanna, *Quantitative Organic Analysis via Functional Groups*, 4th ed., New York: Wiley, 1979.

F. T. Weiss, *Determination of Organic Compounds: Methods and Procedures*, New York: Wiley-Interscience, 1970.

Spectrometric Methods

D. F. Boltz and J. A. Howell, *Colorimetric Determination of Nonmetals*, 2nd ed., New York: Wiley-Interscience, 1978.

J. A. C. Broekaert, *Analytical Atomic Spectrometry with Flames and Plasmas*, Weinheim: Cambridge University Press: Wiley-VCH, 2002.

S. J. Hill, *Inductively Coupled Plasma Spectrometry and Its Applications*, Boca Rotan, Fl: CRC Press, 1999.

J. D. Ingle and S. R. Crouch, *Spectrochemical Analysis*, Upper Saddle River, NJ: Prentice-Hall, 1988.

L. H. J. Lajunen and P. Peramaki, *Spectrochemical Analysis by Atomic Absorption and Emission*, 2nd ed., Cambridge: Royal Society of Chemistry, 2004.

J. R. Lakowiz, *Principles of Fluorescence Spectroscopy*, New York: Plenum Press, 1999.

A. Montaser and D. W. Golightly, eds., *Inductively Coupled Plasmas in Analytical Atomic Spectroscopy*, 2nd ed., New York: Wiley-VCH, 1992.

A. Montaser, ed., *Inductively Coupled Plasma Mass Spectrometry*, New York: Wiley, 1998.

E. B. Sandell and H. Onishi, *Colorimetric Determination of Traces of Metals*, 4th ed., New York: Wiley, 1978–1989. Two volumes.

S. G. Schulman, ed., *Molecular Luminescence Spectroscopy*, 2 parts, New York: Wiley, 1985.

F. D. Snell, *Photometric and Fluorometric Methods of Analysis*, 2 vols., New York: Wiley, 1978–1981.

Electroanalytical Methods

A. J. Bard and L. R. Faulkner, *Electrochemical Methods*, 2nd ed., New York: Wiley, 2001.

P. T. Kissinger and W. R. Heinemann, eds., *Laboratory Techniques in Electroanalytical Chemistry*, 2nd ed., New York: Marcel Dekker, 1996.

J. J. Lingane, *Electroanalytical Chemistry*, 2nd ed., New York: Interscience, 1954.

D. T. Sawyer, A. Sobkowiak, and J. L. Roberts, Jr., *Experimental Electrochemistry for Chemists*, 2nd ed., New York: Wiley, 1995.

J. Wang, *Analytical Electrochemistry*, New York: Wiley, 2000.

Analytical Separations

K. Anton and C. Berger, eds., *Supercritical Fluid Chromatography with Packed Columns, Techniques and Applications*, New York: Dekker, 1998.

P. Camilleri, ed., *Capillary Electrophoresis: Theory and Practice*, Boca Raton, FL: CRC Press, 1993.

M. Caude and D. Thiebaut, eds., *Practical Supercritical Fluid Chromatography and Extraction*, Amsterdam: Harwood, 2000.

B. Fried and J. Sherma, *Thin Layer Chromatography*, 4th ed., New York: Dekker, 1999.

J. C. Giddings, *Unified Separation Science*, New York: Wiley, 1991.

E. Katz, *Quantitative Analysis Using Chromatographic Techniques*, New York: Wiley, 1987.

M. McMaster and C. McMaster, *GC/MS: A Practical User's Guide*, New York: Wiley-VCH, 1998.

H. M. McNair and J. M Miller, *Basic Gas Chromatography*, New York: Wiley, 1998.

W. M. A. Niessen, *Liquid Chromatography-Mass Spectrometry*, 2nd ed., New York: Dekker, 1999.

M. E. Schimpf, K. Caldwell, and J. C. Giddings, eds., *Field-Flow Fractionation Handbook*, New York: Wiley, 2000.

R. P. W. Scott, *Introduction to Analytical Gas Chromatography*, 2nd ed., New York: Marcel Dekker, 1997.

R. P. W. Scott, *Liquid Chromatography for the Analyst*, New York: Marcel Dekker, 1995.

R. M. Smith, *Gas and Liquid Chromatography in Analytical Chemistry*, New York: Wiley, 1988.

L. R. Snyder, J. J. Kirkland, and J. W. Dolan, *Introduction to Modern Liquid Chromatography*, 3rd ed., New York: Wiley, 2010.

R. Weinberger, *Practical Capillary Electrophoresis*, New York: Academic Press, 2000.

Miscellaneous

R. G. Bates, *Determination of pH: Theory and Practice*, 2nd ed., New York: Wiley, 1973.

R. Bock, *Decomposition Methods in Analytical Chemistry*, New York: Wiley, 1979.

G. D. Christian and J. B. Callis, *Trace Analysis*, New York: Wiley, 1986.

J. L. Devore, *Probability and Statistics for Engineering and the Sciences*, 8th ed., Boston: Brooks/Cole, 2012.

J. L. Devore and N. R. Farnum, *Applied Statistics for Engineers and Scientists*, Pacific Grove, CA: Duxbury/Brooks/Cole, 1999.

H. A. Mottola, *Kinetic Aspects of Analytical Chemistry*, New York: Wiley, 1988.

D. Perez-Bendito and M. Silva, *Kinetic Methods in Analytical Chemistry*, New York: Halsted Press-Wiley, 1988.

D. D. Perrin, *Masking and Demasking Chemical Reactions*, New York: Wiley, 1970.

W. Rieman and H. F. Walton, *Ion Exchange in Analytical Chemistry*, Oxford: Pergamon, 1970.

J. Ruzicka and E. H. Hansen, *Flow Injection Analysis*, 2nd ed., New York: Wiley, 1988.

J. T. Watson and O. D. Sparkman, *Introduction to Mass Spectrometry*, 4th ed., Chichester: Wiley, 2007.

PERIODICALS

Numerous journals are devoted to analytical chemistry; these are primary sources of information in the field. Some of the best-known and most widely used titles are listed below. The boldface portion of the title is the *Chemical Abstracts* abbreviation for the journal.

***Analyst**, The*
***Anal**ytical and **Bioanal**ytical **Chem**istry*
***Anal**ytical **Biochem**istry*
***Anal**ytical **Chem**istry*

Analytica Chimica Acta
Analytical Letters
Applied Spectroscopy
Clinical Chemistry
Instrumentation Science and Technology
International Journal of Mass Spectrometry
Journal of the American Society for Mass Spectrometry
Journal of the Association of Official Analytical Chemists
Journal of Chromatographic Science
Journal of Chromatography
Journal of Electroanalytical Chemistry
Journal of Liquid Chromatography and Related Techniques
Journal of Microcolumn Separations
Microchemical Journal
Mikrochimica Acta
Separation Science
Spectrochimica Acta
Talanta
TrAC—Trends Analytical Chemistry

Appendix 2

Solubility Product Constants at 25°C

Compound	Formula	K_{sp}	Notes
Aluminum hydroxide	$Al(OH)_3$	3×10^{-34}	
Barium carbonate	$BaCO_3$	5.0×10^{-9}	
Barium chromate	$BaCrO_4$	2.1×10^{-10}	
Barium hydroxide	$Ba(OH)_2 \cdot 8H_2O$	3×10^{-4}	
Barium iodate	$Ba(IO_3)_2$	1.57×10^{-9}	
Barium oxalate	BaC_2O_4	1×10^{-6}	
Barium sulfate	$BaSO_4$	1.1×10^{-10}	
Cadmium carbonate	$CdCO_3$	1.8×10^{-14}	
Cadmium hydroxide	$Cd(OH)_2$	4.5×10^{-15}	
Cadmium oxalate	CdC_2O_4	9×10^{-8}	
Cadmium sulfide	CdS	1×10^{-27}	
Calcium carbonate	$CaCO_3$	4.5×10^{-9}	Calcite
	$CaCO_3$	6.0×10^{-9}	Aragonite
Calcium fluoride	CaF_2	3.9×10^{-11}	
Calcium hydroxide	$Ca(OH)_2$	6.5×10^{-6}	
Calcium oxalate	$CaC_2O_4 \cdot H_2O$	1.7×10^{-9}	
Calcium sulfate	$CaSO_4$	2.4×10^{-5}	
Cobalt(II) carbonate	$CoCO_3$	1.0×10^{-10}	
Cobalt(II) hydroxide	$Co(OH)_2$	1.3×10^{-15}	
Cobalt(II) sulfide	CoS	5×10^{-22}	α
	CoS	3×10^{-26}	β
Copper(I) bromide	$CuBr$	5×10^{-9}	
Copper(I) chloride	$CuCl$	1.9×10^{-7}	
Copper(I) hydroxide*	Cu_2O^*	2×10^{-15}	
Copper(I) iodide	CuI	1×10^{-12}	
Copper(I) thiocyanate	$CuSCN$	4.0×10^{-14}	
Copper(II) hydroxide	$Cu(OH)_2$	4.8×10^{-20}	
Copper(II) sulfide	CuS	8×10^{-37}	
Iron(II) carbonate	$FeCO_3$	2.1×10^{-11}	
Iron(II) hydroxide	$Fe(OH)_2$	4.1×10^{-15}	
Iron(II) sulfide	FeS	8×10^{-19}	
Iron(III) hydroxide	$Fe(OH)_3$	2×10^{-39}	
Lanthanum iodate	$La(IO_3)_3$	1.0×10^{-11}	
Lead carbonate	$PbCO_3$	7.4×10^{-14}	
Lead chloride	$PbCl_2$	1.7×10^{-5}	
Lead chromate	$PbCrO_4$	3×10^{-13}	
Lead hydroxide	$PbO^{\dagger}$	8×10^{-16}	Yellow
	$PbO^{\dagger}$	5×10^{-16}	Red
Lead iodide	PbI_2	7.9×10^{-9}	
Lead oxalate	PbC_2O_4	8.5×10^{-9}	$\mu = 0.05$
Lead sulfate	$PbSO_4$	1.6×10^{-8}	
Lead sulfide	PbS	3×10^{-28}	
Magnesium ammonium phosphate	$MgNH_4PO_4$	3×10^{-13}	
Magnesium carbonate	$MgCO_3$	3.5×10^{-8}	

continues

Compound	Formula	K_{sp}	Notes
Magnesium hydroxide	$Mg(OH)_2$	7.1×10^{-12}	
Manganese carbonate	$MnCO_3$	5.0×10^{-10}	
Manganese hydroxide	$Mn(OH)_2$	2×10^{-13}	
Manganese sulfide	MnS	3×10^{-11}	Pink
	MnS	3×10^{-14}	Green
Mercury(I) bromide	Hg_2Br_2	5.6×10^{-23}	
Mercury(I) carbonate	Hg_2CO_3	8.9×10^{-17}	
Mercury(I) chloride	Hg_2Cl_2	1.2×10^{-18}	
Mercury(I) iodide	Hg_2I_2	4.7×10^{-29}	
Mercury(I) thiocyanate	$Hg_2(SCN)_2$	3.0×10^{-20}	
Mercury(II) hydroxide	$HgO^{\ddagger}$	3.6×10^{-26}	
Mercury(II) sulfide	HgS	2×10^{-53}	Black
	HgS	5×10^{-54}	Red
Nickel carbonate	$NiCO_3$	1.3×10^{-7}	
Nickel hydroxide	$Ni(OH)_2$	6×10^{-16}	
Nickel sulfide	NiS	4×10^{-20}	α
	NiS	1.3×10^{-25}	β
Silver arsenate	Ag_3AsO_4	6×10^{-23}	
Silver bromide	$AgBr$	5.0×10^{-13}	
Silver carbonate	Ag_2CO_3	8.1×10^{-12}	
Silver chloride	$AgCl$	1.82×10^{-10}	
Silver chromate	$AgCrO_4$	1.2×10^{-12}	
Silver cyanide	$AgCN$	2.2×10^{-16}	
Silver iodate	$AgIO_3$	3.1×10^{-8}	
Silver iodide	AgI	8.3×10^{-17}	
Silver oxalate	$Ag_2C_2O_4$	3.5×10^{-11}	
Silver sulfide	Ag_2S	8×10^{-51}	
Silver thiocyanate	$AgSCN$	1.1×10^{-12}	
Strontium carbonate	$SrCO_3$	9.3×10^{-10}	
Strontium oxalate	SrC_2O_4	5×10^{-8}	
Strontium sulfate	$SrSO_4$	3.2×10^{-7}	
Thallium(I) chloride	$TlCl$	1.8×10^{-4}	
Thallium(I) sulfide	Tl_2S	6×10^{-22}	
Zinc carbonate	$ZnCO_3$	1.0×10^{-10}	
Zinc hydroxide	$Zn(OH)_2$	3.0×10^{-16}	Amorphous
Zinc oxalate	ZnC_2O_4	8×10^{-9}	
Zinc sulfide	ZnS	2×10^{-25}	α
	ZnS	3×10^{-23}	β

Most of these data are taken from A. E. Martell and R. M Smith, *Critical Stability Constants*, Vol. 3–6, New York: Plenum, 1976–1989. In most cases, the values are for infinite dilution (ionic strength μ = 0.0) and the temperature 25°C.

*$Cu_2O(s) + H_2O \rightleftharpoons 2Cu^+ + 2OH^-$

†$PbO(s) + H_2O \rightleftharpoons Pb^{2+} + 2OH^-$

‡$HgO(s) + H_2O \rightleftharpoons Hg^{2+} + 2OH^-$

Appendix 3

Acid Dissociation Constants at 25°C

Acid	Formula	K_1	K_2	K_3
Acetic acid	CH_3COOH	1.75×10^{-5}		
Ammonium ion	NH_4^+	5.70×10^{-10}		
Anilinium ion	$C_6H_5NH_3^+$	2.51×10^{-5}		
Arsenic acid	H_3AsO_4	5.8×10^{-3}	1.1×10^{-7}	3.2×10^{-12}
Arsenous acid	H_3AsO_3	5.1×10^{-10}		
Benzoic acid	C_6H_5COOH	6.28×10^{-5}		
Boric acid	H_3BO_3	5.81×10^{-10}		
1-Butanoic acid	$CH_3CH_2CH_2COOH$	1.52×10^{-5}		
Carbonic acid	H_2CO_3	4.45×10^{-7}	4.69×10^{-11}	
	$CO_2(aq)$	4.2×10^{-7}	4.69×10^{-11}	
Chloroacetic acid	$ClCH_2COOH$	1.36×10^{-3}		
Citric acid	$HOOC(OH)C(CH_2COOH)_2$	7.45×10^{-4}	1.73×10^{-5}	4.02×10^{-7}
Dimethyl ammonium ion	$(CH_3)_2NH_2^+$	1.68×10^{-11}		
Ethanol ammonium ion	$HOC_2H_4NH_3^+$	3.18×10^{-10}		
Ethyl ammonium ion	$C_2H_5NH_3^+$	2.31×10^{-11}		
Ethylene diammonium ion	$^+H_3NCH_2CH_2NH_3^+$	1.42×10^{-7}	1.18×10^{-10}	
Formic acid	$HCOOH$	1.80×10^{-4}		
Fumaric acid	trans-$HOOCCH{:}CHCOOH$	8.85×10^{-4}	3.21×10^{-5}	
Glycolic acid	$HOCH_2COOH$	1.47×10^{-4}		
Hydrazinium ion	$H_2NNH_3^+$	1.05×10^{-8}		
Hydrazoic acid	HN_3	2.2×10^{-5}		
Hydrogen cyanide	HCN	6.2×10^{-10}		
Hydrogen fluoride	HF	6.8×10^{-4}		
Hydrogen peroxide	H_2O_2	2.2×10^{-12}		
Hydrogen sulfide	H_2S	9.6×10^{-8}	1.3×10^{-14}	
Hydroxyl ammonium ion	$HONH_3^+$	1.10×10^{-6}		
Hypochlorous acid	$HOCl$	3.0×10^{-8}		
Iodic acid	HIO_3	1.7×10^{-1}		
Lactic acid	$CH_3CHOHCOOH$	1.38×10^{-4}		
Maleic acid	cis-$HOOCCH{:}CHCOOH$	1.3×10^{-2}	5.9×10^{-7}	
Malic acid	$HOOCCHOHCH_2COOH$	3.48×10^{-4}	8.00×10^{-6}	
Malonic acid	$HOOCCH_2COOH$	1.42×10^{-3}	2.01×10^{-6}	
Mandelic acid	$C_6H_5CHOHCOOH$	4.0×10^{-4}		
Methyl ammonium ion	$CH_3NH_3^+$	2.3×10^{-11}		
Nitrous acid	HNO_2	7.1×10^{-4}		
Oxalic acid	$HOOCCOOH$	5.60×10^{-2}	5.42×10^{-5}	
Periodic acid	H_5IO_6	2×10^{-2}	5×10^{-9}	
Phenol	C_6H_5OH	1.00×10^{-10}		
Phosphoric acid	H_3PO_4	7.11×10^{-3}	6.32×10^{-8}	4.5×10^{-13}
Phosphorous acid	H_3PO_3	3×10^{-2}	1.62×10^{-7}	
o-Phthalic acid	$C_6H_4(COOH)_2$	1.12×10^{-3}	3.91×10^{-6}	
Picric acid	$(NO_2)_3C_6H_2OH$	4.3×10^{-1}		
Piperidinium ion	$C_5H_{11}NH^+$	7.50×10^{-12}		
Propanoic acid	CH_3CH_2COOH	1.34×10^{-5}		

continues

Acid	Formula	K_1	K_2	K_3
Pyridinium ion	$C_5H_5NH^+$	5.90×10^{-6}		
Pyruvic acid	$CH_3COCOOH$	3.2×10^{-3}		
Salicylic acid	$C_6H_4(OH)COOH$	1.06×10^{-3}		
Succinic acid	$HOOCCH_2CH_2COOH$	6.21×10^{-5}	2.31×10^{-6}	
Sulfamic acid	H_2NSO_3H	1.03×10^{-1}		
Sulfuric acid	H_2SO_4	Strong	1.02×10^{-2}	
Sulfurous acid	H_2SO_3	1.23×10^{-2}	6.6×10^{-8}	
Tartaric acid	$HOOC(CHOH)_2COOH$	9.20×10^{-4}	4.31×10^{-5}	
Thiocyanic acid	HSCN	0.13		
Thiosulfuric acid	$H_2S_2O_3$	0.3	2.5×10^{-2}	
Trichloroacetic acid	Cl_3CCOOH	3		
Trimethyl ammonium ion	$(CH_3)_3NH^+$	1.58×10^{-10}		

Most data are infinite dilution values ($\mu = 0$). (From A. E. Martell and R. M. Smith, *Critical Stability Constants*, Vol. 1–6, New York Plenum Press, 1974–1989.)

Appendix 4

Formation Constants at 25°C

Ligand	Cation	log K_1	log K_2	log K_3	log K_4	Ionic Strength
Acetate (CH_3COO^-)	Ag^+	0.73	−0.9			0.0
	Ca^{2+}	1.18				0.0
	Cd^{2+}	1.93	1.22			0.0
	Cu^{2+}	2.21	1.42			0.0
	Fe^{3+}	3.38*	3.1*	1.8*		0.1
	Hg^{2+}	log K_1K_2 = 8.45				0.0
	Mg^{2+}	1.27				0.0
	Pb^{2+}	2.68	1.40			0.0
Ammonia (NH_3)	Ag^+	3.31	3.91			0.0
	Cd^{2+}	2.55	2.01	1.34	0.84	0.0
	Co^{2+}	1.99*	1.51	0.93	0.64	0.0
		log K_5 = 0.06	log K_6 = −0.74			0.0
	Cu^{2+}	4.04	3.43	2.80	1.48	0.0
	Hg^{2+}	8.8	8.6	1.0	0.7	0.5
	Ni^{2+}	2.72	2.17	1.66	1.12	0.0
		log K_5 = 0.67	log K_6 = −0.03			0.0
	Zn^{2+}	2.21	2.29	2.36	2.03	0.0
Bromide (Br^-)	Ag^+	$Ag^+ + 2Br^- \rightleftharpoons AgBr_2^-$		log K_1K_2 = 7.5		0.0
	Hg^{2+}	9.00	8.1	2.3	1.6	0.5
	Pb^{2+}	1.77				0.0
Chloride (Cl^-)	Ag^+	$Ag^+ + 2Cl^- \rightleftharpoons AgCl_2^-$		log K_1K_2 = 5.25		0.0
		$AgCl_2^- + Cl^- \rightleftharpoons AgCl_3^{2-}$		log K_3 = 0.37		0.0
	Cu^+	$Cu^+ + 2Cl^- \rightleftharpoons CuCl_2^-$		log = 5.5*		0.0
	Fe^{3+}	1.48	0.65			0.0
	Hg^{2+}	7.30	6.70	1.0	0.6	0.0
	Pb^{2+}	$Pb^{2+} + 3Cl^- \rightleftharpoons PbCl_3^-$		log $K_1K_2K_3$ = 1.8		0.0
	Sn^{2+}	1.51	0.74	−0.3	−0.5	0.0
Cyanide (CN^-)	Ag^+	$Ag^+ + 2CN^- \rightleftharpoons Ag(CN)_2^-$		log K_1K_2 = 20.48		0.0
	Cd^{2+}	6.01	5.11	4.53	2.27	0.0
	Hg^{2+}	17.00	15.75	3.56	2.66	0.0
	Ni^{2+}	$Ni^{2+} + 4CN^- \rightleftharpoons Ni(CN)_4^-$		log $K_1K_2K_3K_4$ = 30.22		0.0
	Zn^{2+}	log K_1K_2 = 11.07		4.98	3.57	0.0
EDTA	See Table 17-4, page 418.					
Fluoride (F^-)	Al^{3+}	7.0	5.6	4.1	2.4	0.0
	Fe^{3+}	5.18	3.89	3.03		0.0
Hydroxide (OH^-)	Al^{3+}	$Al^{3+} + 4OH^- \rightleftharpoons Al(OH)_4^-$		log $K_1K_2K_3K_4$ = 33.4		0.0
	Cd^{2+}	3.9	3.8			0.0
	Cu^{2+}	6.5				0.0
	Fe^{2+}	4.6				0.0
	Fe^{3+}	11.81	11.5			0.0
	Hg^{2+}	10.60	11.2			0.0
	Ni^{2+}	4.1	4.9	3		0.0
	Pb^{2+}	6.4	$Pb^{2+} + 3OH^- \rightleftharpoons Pb(OH)_3^-$		log $K_1K_2K_3$ = 13.9	0.0
	Zn^{2+}	5.0	$Zn^{2+} + 4OH^- \rightleftharpoons Zn(OH)_4^{2-}$		log $K_1K_2K_3K_4$ = 15.5	0.0

continues

Ligand	Cation	log K_1	log K_2	log K_3	log K_4	Ionic Strength
Iodide (I^-)	Cd^{2+}	2.28	1.64	1.0	1.0	0.0
	Cu^+	$Cu^+ + 2I^- \rightleftharpoons CuI_2^-$	log $K_1K_2 = 8.9$			0.0
	Hg^{2+}	12.87	10.95	3.8	2.2	0.5
	Pb^{2+}	$Pb^{2+} + 3I^- \rightleftharpoons PbI_3^-$	log $K_1K_2K_3 = 3.9$			0.0
		$Pb^{2+} + 4I^- \rightleftharpoons PbI_4^{2-}$	log $K_1K_2K_3K_4 = 4.5$			0.0
Oxalate ($C_2O_4^{2-}$)	Al^{3+}	5.97	4.96	5.04		0.1
	Ca^{2+}	3.19				0.0
	Cd^{2+}	2.73	1.4	1.0		1.0
	Fe^{3+}	7.58	6.23	4.8		1.0
	Mg^{2+}	3.42(18°C)				
	Pb^{2+}	4.20	2.11			1.0
Sulfate (SO_4^{2-})	Al^{3+}	3.89				0.0
	Ca^{2+}	2.13				0.0
	Cu^{2+}	2.34				0.0
	Fe^{3+}	4.04	1.34			0.0
	Mg^{2+}	2.23				0.0
Thiocyanate (SCN^-)	Cd^{2+}	1.89	0.89	0.1		0.0
	Cu^+	$Cu^+ + 3SCN^- \rightleftharpoons Cu(SCN)_3^{2-}$	log $K_1K_2K_3 = 11.60$			0.0
	Fe^{3+}	3.02	0.62*			0.0
	Hg^{2+}	log $K_1K_2 = 17.26$		2.7	1.8	0.0
	Ni^{2+}	1.76				0.0
Thiosulfate ($S_2O_3^{2-}$)	Ag^+	8.82*	4.7	0.7		0.0
	Cu^{2+}	log $K_1K_2 = 6.3$				0.0
	Hg^{2+}	log $K_1K_2 = 29.23$		1.4		0.0

Data from A. E. Martell and R. M. Smith, *Critical Stability Constants,* **Vol. 3–6, New York: Plenum Press, 1974–1989.**
*20°C.

Appendix 5

Standard and Formal Electrode Potentials

Half-Reaction	E^0, V*	Formal Potential, V⁺
Aluminum		
$Al^{3+} + 3e^- \rightleftharpoons Al(s)$	-1.662	
Antimony		
$Sb_2O_5(s) + 6H^+ + 4e^- \rightleftharpoons 2SbO^+ + 3H_2O$	$+0.581$	
Arsenic		
$H_3AsO_4 + 2H^+ + 2e^- \rightleftharpoons H_3AsO_3 + H_2O$	$+0.559$	0.577 in 1 M HCl, HClO₄
Barium		
$Ba^{2+} + 2e^- \rightleftharpoons Ba(s)$	-2.906	
Bismuth		
$BiO^+ + 2H^+ + 3e^- \rightleftharpoons Bi(s) + H_2O$	$+0.320$	
$BiCl_4^- + 3e^- \rightleftharpoons Bi(s) + 4Cl^-$	$+0.16$	
Bromine		
$Br_2(l) + 2e^- \rightleftharpoons 2Br^-$	$+1.065$	1.05 in 4 M HCl
$Br_2(aq) + 2e^- \rightleftharpoons 2Br^-$	$+1.087^\ddagger$	
$BrO_3^- + 6H^+ + 5e^- \rightleftharpoons \frac{1}{2}Br_2(l) + 3H_2O$	$+1.52$	
$BrO_3^- + 6H^+ + 6e^- \rightleftharpoons Br^- + 3H_2O$	$+1.44$	
Cadmium		
$Cd^{2+} + 2e^- \rightleftharpoons Cd(s)$	-0.403	
Calcium		
$Ca^{2+} + 2e^- \rightleftharpoons Ca(s)$	-2.866	
Carbon		
$C_6H_4O_2 \text{ (quinone)} + 2H^+ + 2e^- \rightleftharpoons C_6H_4(OH)_2$	$+0.699$	0.696 in 1 M HCl, HClO₄, H₂SO₄
$2CO_2(g) + 2H^+ + 2e^- \rightleftharpoons H_2C_2O_4$	-0.49	
Cerium		
$Ce^{4+} + e^- \rightleftharpoons Ce^{3+}$		+1.70 in 1 M HClO₄; +1.61 in 1 M HNO₃; 1.44 in 1 M H₂SO₄
Chlorine		
$Cl_2(g) + 2e^- \rightleftharpoons 2Cl^-$	$+1.359$	
$HClO + H^+ + e^- \rightleftharpoons \frac{1}{2}Cl_2(g) + H_2O$	$+1.63$	
$ClO_3^- + 6H^+ + 5e^- \rightleftharpoons \frac{1}{2}Cl_2(g) + 3H_2O$	$+1.47$	
Chromium		
$Cr^{3+} + e^- \rightleftharpoons Cr^{2+}$	-0.408	
$Cr^{3+} + 3e^- \rightleftharpoons Cr(s)$	-0.744	
$Cr_2O_7^{2-} + 14H^+ + 6e^- \rightleftharpoons 2Cr^{3+} + 7H_2O$	$+1.33$	
Cobalt		
$Co^{2+} + 2e^- \rightleftharpoons Co(s)$	-0.277	
$Co^{3+} + e^- \rightleftharpoons Co^{2+}$	$+1.808$	
Copper		
$Cu^{2+} + 2e^- \rightleftharpoons Cu(s)$	$+0.337$	
$Cu^{2+} + e^- \rightleftharpoons Cu^+$	$+0.153$	
$Cu^+ + e^- \rightleftharpoons Cu(s)$	$+0.521$	
$Cu^{2+} + I^- + e^- \rightleftharpoons CuI(s)$	$+0.86$	
$CuI(s) + e^- \rightleftharpoons Cu(s) + I^-$	-0.185	

continues

Half-Reaction	E°, V*	Formal Potential, V[†]
Fluorine		
$F_2(g) + 2H^+ + 2e^- \rightleftharpoons 2HF(aq)$	+3.06	
Hydrogen		
$2H^+ + 2e^- \rightleftharpoons H_2(g)$	0.000	−0.005 in 1 M HCl, HClO$_4$
Iodine		
$I_2(s) + 2e^- \rightleftharpoons 2I^-$	+0.5355	
$I_2(aq) + 2e^- \rightleftharpoons 2I^-$	+0.615[‡]	
$I_3^- + 2e^- \rightleftharpoons 3I^-$	+0.536	
$ICl_2^- + e^- \rightleftharpoons \frac{1}{2}I_2(s) + 2Cl^-$	+1.056	
$IO_3^- + 6H^+ + 5e^- \rightleftharpoons \frac{1}{2}I_2(s) + 3H_2O$	+1.196	
$IO_3^- + 6H^+ + 5e^- \rightleftharpoons \frac{1}{2}I_2(aq) + 3H_2O$	+1.178[‡]	
$IO_3^- + 2Cl^- + 6H^+ + 4e^- \rightleftharpoons ICl_2^- + 3H_2O$	+1.24	
$H_5IO_6 + H^+ + 2e^- \rightleftharpoons IO_3^- + 3H_2O$	+1.601	
Iron		
$Fe^{2+} + 2e^- \rightleftharpoons Fe(s)$	−0.440	
$Fe^{3+} + e^- \rightleftharpoons Fe^{2+}$	+0.771	0.700 in 1 M HCl; 0.732 in 1 M HClO$_4$; 0.68 in 1 M H$_2$SO$_4$
$Fe(CN)_6^{3-} + e^- \rightleftharpoons Fe(CN)_6^{4-}$	+0.36	0.71 in 1 M HCl; 0.72 in 1 M HClO$_4$, H$_2$SO$_4$
Lead		
$Pb^{2+} + 2e^- \rightleftharpoons Ps(s)$	−0.126	−0.14 in 1 M HClO$_4$; −0.29 in 1 M H$_2$SO$_4$
$PbO_2(s) + 4H^+ + 2e^- \rightleftharpoons Pb^{2+} + 2H_2O$	+1.455	
$PbSO_4(s) + 2e^- \rightleftharpoons Pb(s) + SO_4^{2-}$	−0.350	
Lithium		
$Li^+ + e^- \rightleftharpoons Li(s)$	−3.045	
Magnesium		
$Mg^{2+} + 2e^- \rightleftharpoons Mg(s)$	−2.363	
Manganese		
$Mn^{2+} + 2e^- \rightleftharpoons Mn(s)$	−1.180	
$Mn^{3+} + e^- \rightleftharpoons Mn^{2+}$		1.51 in 7.5 M H$_2$SO$_4$
$MnO_2(s) + 4H^+ + 2e^- \rightleftharpoons Mn^{2+} + 2H_2O$	+1.23	
$MnO_4^- + 8H^+ + 5e^- \rightleftharpoons Mn^{2+} + 4H_2O$	+1.51	
$MnO_4^- + 4H^+ + 3e^- \rightleftharpoons MnO_2(s) + 2H_2O$	+1.695	
$MnO_4^- + e^- \rightleftharpoons MnO_4^{2-}$	+0.564	
Mercury		
$Hg_2^{2+} + 2e^- \rightleftharpoons 2Hg(l)$	+0.788	0.274 in 1 M HCl; 0.776 in 1 M HClO$_4$; 0.674 in 1 M H$_2$SO$_4$
$2Hg^{2+} + 2e^- \rightleftharpoons Hg_2^{2+}$	+0.920	0.907 in 1 M HClO$_4$
$Hg^{2+} + 2e^- \rightleftharpoons Hg(l)$	+0.854	
$Hg_2Cl_2(s) + 2e^- \rightleftharpoons 2Hg(l) + 2Cl^-$	+0.268	0.244 in sat'd KCl; 0.282 in 1 M KCl; 0.334 in 0.1 M KCl
$Hg_2SO_4(s) + 2e^- \rightleftharpoons 2Hg(l) + SO_4^{2-}$	+0.615	
Nickel		
$Ni^{2+} + 2e^- \rightleftharpoons Ni(s)$	−0.250	
Nitrogen		
$N_2(g) + 5H^+ + 4e^- \rightleftharpoons N_2H_5^+$	−0.23	
$HNO_2 + H^+ + e^- \rightleftharpoons NO(g) + H_2O$	+1.00	
$NO_3^- + 3H^+ + 2e^- \rightleftharpoons HNO_2 + H_2O$	+0.94	0.92 in 1 M HNO$_3$
Oxygen		
$H_2O_2 + 2H^+ + 2e^- \rightleftharpoons 2H_2O$	+1.776	
$HO_2^- + H_2O + 2e^- \rightleftharpoons 3OH^-$	+0.88	
$O_2(g) + 4H^+ + 4e^- \rightleftharpoons 2H_2O$	+1.229	
$O_2(g) + 2H^+ + 2e^- \rightleftharpoons H_2O_2$	+0.682	
$O_3(g) + 2H^+ + 2e^- \rightleftharpoons O_2(g) + H_2O$	+2.07	
Palladium		
$Pd^{2+} + 2e^- \rightleftharpoons Pd(s)$	+0.987	

continues

Half-Reaction	E^0, V*	Formal Potential, V[†]
Platinum		
$PtCl_4^{2-} + 2e^- \rightleftharpoons Pt(s) + 4Cl^-$	$+0.755$	
$PtCl_6^{2-} + 2e^- \rightleftharpoons PtCl_4^{2-} + 2Cl^-$	$+0.68$	
Potassium		
$K^+ + e^- \rightleftharpoons K(s)$	-2.925	
Selenium		
$H_2SeO_3 + 4H^+ + 4e^- \rightleftharpoons Se(s) + 3H_2O$	$+0.740$	
$SeO_4^{2-} + 4H^+ + 2e^- \rightleftharpoons H_2SeO_3 + H_2O$	$+1.15$	
Silver		
$Ag^+ + e^- \rightleftharpoons Ag(s)$	$+0.799$	0.228 in 1 M HCl; 0.792 in 1 M $HClO_4$; 0.77 in 1 M H_2SO_4
$AgBr(s) + e^- \rightleftharpoons Ag(s) + Br^-$	$+0.073$	
$AgCl(s) + e^- \rightleftharpoons Ag(s) + Cl^-$	$+0.222$	0.228 in 1 M KCl
$Ag(CN)_2^- + e^- \rightleftharpoons Ag(s) + 2CN^-$	-0.31	
$Ag_2CrO_4(s) + 2e^- \rightleftharpoons 2Ag(s) + CrO_4^{2-}$	$+0.446$	
$AgI(s) + e^- \rightleftharpoons Ag(s) + I^-$	-0.151	
$Ag(S_2O_3)_2^{3-} + e^- \rightleftharpoons Ag(s) + 2S_2O_3^{2-}$	$+0.017$	
Sodium		
$Na^+ + e^- \rightleftharpoons Na(s)$	-2.714	
Sulfur		
$S(s) + 2H^+ + 2e^- \rightleftharpoons H_2S(g)$	$+0.141$	
$H_2SO_3 + 4H^+ + 4e^- \rightleftharpoons S(s) + 3H_2O$	$+0.450$	
$SO_4^{2-} + 4H^+ + 2e^- \rightleftharpoons H_2SO_3 + H_2O$	$+0.172$	
$S_4O_6^{2-} + 2e^- \rightleftharpoons 2S_2O_3^{2-}$	$+0.08$	
$S_2O_8^{2-} + 2e^- \rightleftharpoons 2SO_4^{2-}$	$+2.01$	
Thallium		
$Tl^+ + e^- \rightleftharpoons Tl(s)$	-0.336	-0.551 in 1 M HCl; -0.33 in 1 M $HClO_4$, H_2SO_4
$Tl^{3+} + 2e^- \rightleftharpoons Tl^+$	$+1.25$	0.77 in 1 M HCl
Tin		
$Sn^{2+} + 2e^- \rightleftharpoons Sn(s)$	-0.136	-0.16 in 1 M $HClO_4$
$Sn^{4+} + 2e^- \rightleftharpoons Sn^{2+}$	$+0.154$	0.14 in 1 M HCl
Titanium		
$Ti^{3+} + e^- \rightleftharpoons Ti^{2+}$	-0.369	
$TiO^{2+} + 2H^+ + e^- \rightleftharpoons Ti^{3+} + H_2O$	$+0.099$	0.04 in 1 M H_2SO_4
Uranium		
$UO_2^{2+} + 4H^+ + 2e^- \rightleftharpoons U^{4+} + 2H_2O$	$+0.334$	
Vanadium		
$V^{3+} + e^- \rightleftharpoons V^{2+}$	-0.255	
$VO^{2+} + 2H^+ + e^- \rightleftharpoons V^{3+} + H_2O$	$+0.337$	
$V(OH)_4^+ + 2H^+ + e^- \rightleftharpoons VO^{2+} + 3H_2O$	$+1.00$	1.02 in 1 M HCl, $HClO_4$
Zinc		
$Zn^{2+} + 2e^- \rightleftharpoons Zn(s)$	-0.763	

*G. Milazzo, S. Caroli, and V. K. Sharma, *Tables of Standard Electrode Potentials*, London: Wiley, 1978.

[†]E. H. Swift and E. A. Butler, *Quantitative Measurements and Chemical Equilibria*, New York: Freeman, 1972.

[‡]These potentials are hypothetical because they correspond to solutions that are 1.00 M in Br_2 or I_2. The solubilities of these two compounds at 25°C are 0.18 M and 0.0020 M, respectively. In saturated solutions containing an excess of $Br_2(l)$ or $I_2(s)$, the standard potentials for the half-reaction $Br_2(l) + 2e^- \rightleftharpoons 2Br^-$ or $I_2(s) + 2e^- \rightleftharpoons 2I^-$ should be used. In contrast, at Br_2 and I_2 concentrations less than saturation, these hypothetical electrode potentials should be used.

Appendix 6

Use of Exponential Numbers and Logarithms

Scientists frequently find it necessary or convenient to use exponential notation to express numerical data. A brief review of this notation follows.

A6A EXPONENTIAL NOTATION

An exponent is used to describe the process of repeated multiplication or division. For example, 3^5 means

$$3 \times 3 \times 3 \times 3 \times 3 = 3^5 = 243$$

The power 5 is the exponent of the number (or base) 3; thus, 3 raised to the fifth power is equal to 243.

A negative exponent represents repeated division. For example, 3^{-5} means

$$\frac{1}{3} \times \frac{1}{3} \times \frac{1}{3} \times \frac{1}{3} \times \frac{1}{3} = \frac{1}{3^5} = 3^{-5} = 0.00412$$

Note that changing the sign of the exponent yields the *reciprocal* of the number, that is,

$$3^{-5} = \frac{1}{3^5} = \frac{1}{243} = 0.00412$$

A number raised to the first power is the number itself, and any number raised to the zero power has a value of 1. For example,

$$4^1 = 4$$
$$4^0 = 1$$
$$67^0 = 1$$

A6A-1 Fractional Exponents

A fractional exponent symbolizes the process of extracting the root of a number. The fifth root of 243 is 3; this process is expressed exponentially as

$$(243)^{1/5} = 3$$

Other examples are

$$25^{1/2} = 5$$

$$25^{-1/2} = \frac{1}{25^{1/2}} = \frac{1}{5}$$

A6A-2 The Combination of Exponential Numbers in Multiplication and Division

Multiplication and division of exponential numbers having the same base are accomplished by adding and subtracting the exponents. For example,

$$3^3 \times 3^2 = (3 \times 3 \times 3)(3 \times 3) = 3^{(3+2)} = 3^5 = 243$$

$$3^4 \times 3^{-2} \times 3^0 = (3 \times 3 \times 3 \times 3)\left(\frac{1}{3} \times \frac{1}{3}\right) \times 1 = 3^{(4-2+0)} = 3^2 = 9$$

$$\frac{5^4}{5^2} = \frac{5 \times 5 \times 5 \times 5}{5 \times 5} = 5^{(4-2)} = 5^2 = 25$$

$$\frac{2^3}{2^{-1}} = \frac{(2 \times 2 \times 2)}{1/2} = 2^4 = 16$$

In the last equation, the exponent, is given by the relationship

$$3 - (-1) = 3 + 1 = 4$$

A6A-3 Extraction of the Root of an Exponential Number

To obtain the root of an exponential number, the exponent is divided by the desired root. Thus,

$$(5^4)^{1/2} = (5 \times 5 \times 5 \times 5)^{1/2} = 5^{(4/2)} = 5^2 = 25$$

$$(10^{-8})^{1/4} = 10^{(-8/4)} = 10^{-2}$$

$$(10^9)^{1/2} = 10^{(9/2)} = 10^{4.5}$$

A6B THE USE OF EXPONENTS IN SCIENTIFIC NOTATION

Scientists and engineers are frequently called upon to use very large or very small numbers for which ordinary decimal notation is either awkward or impossible. For example, to express Avogadro's number in decimal notation would require 21 zeros following the number 602. In scientific notation, the number is written as a multiple of two numbers, the one number in decimal notation and the other expressed as a power of 10. Thus, Avogadro's number is written as 6.02×10^{23}. Other examples are

$$4.32 \times 10^3 = 4.32 \times 10 \times 10 \times 10 = 4320$$

$$4.32 \times 10^{-3} = 4.32 \times \frac{1}{10} \times \frac{1}{10} \times \frac{1}{10} = 0.00432$$

$$0.002002 = 2.002 \times \frac{1}{10} \times \frac{1}{10} \times \frac{1}{10} = 2.002 \times 10^{-3}$$

$$375 = 3.75 \times 10 \times 10 = 3.75 \times 10^2$$

The scientific notation for a number can be expressed in any of several equivalent forms. Thus,

$$4.32 \times 10^3 = 43.2 \times 10^2 = 432 \times 10^1 = 0.432 \times 10^4 = 0.0432 \times 10^5$$

The number in the exponent is equal to the number of places the decimal must be shifted to convert a number from scientific to purely decimal notation. The shift is to the right if the exponent is positive and to the left if it is negative. The process is reversed when decimal numbers are converted to scientific notation.

A6C ARITHMETIC OPERATIONS WITH SCIENTIFIC NOTATION

Scientific notation is helpful in preventing decimal errors in arithmetic calculations. Some examples follow.

A6C-1 Multiplication

In this example, the decimal parts of the numbers are multiplied, and the exponents are added. Thus,

$$420,000 \times 0.0300 = (4.20 \times 10^5)(3.00 \times 10^{-2})$$
$$= 12.60 \times 10^3 = 1.26 \times 10^4$$

$$0.0060 \times 0.000020 = 6.0 \times 10^{-3} \times 2.0 \times 10^{-5}$$
$$= 12 \times 10^{-8} = 1.2 \times 10^{-7}$$

A6C-2 Division

With division, the decimal parts of the numbers are divided; the exponent in the denominator is subtracted from that in the numerator. For example,

$$\frac{0.015}{5000} = \frac{15 \times 10^{-3}}{5.0 \times 10^3} = 3.0 \times 10^{-6}$$

A6C-3 Addition and Subtraction

Addition or subtraction in scientific notation requires that all numbers to be added or subtracted must be expressed to a common power of 10. The decimal parts are then added or subtracted, as appropriate. Thus,

$$2.00 \times 10^{-11} + 4.00 \times 10^{-12} - 3.00 \times 10^{-10}$$
$$= 2.00 \times 10^{-11} + 0.400 \times 10^{-11} - 30.0 \times 10^{-11}$$
$$= -27.6 \times 10^{-11} = -2.76 \times 10^{-10}$$

A6C-4 Raising to a Power a Number Written in Exponential Notation

Each part of the number is raised to the power separately. For example,

$$(2 \times 10^{-3})^4 = (2.0)^4 \times (10^{-3})^4 = 16 \times 10^{-(3 \times 4)}$$
$$= 16 \times 10^{-12} = 1.6 \times 10^{-11}$$

A6C-5 Extraction of the Root of a Number Written in Exponential Notation

The number is written in such a way that the exponent of 10 is evenly divisible by the root. Thus,

$$(4.0 \times 10^{-5})^{1/3} = \sqrt[3]{40 \times 10^{-6}} = \sqrt[3]{40} \times \sqrt[3]{10^{-6}}$$
$$= 3.4 \times 10^{-2}$$

A6D LOGARITHMS

In this discussion, we will assume that you have an electronic calculator for calculating logarithms and antilogarithms of numbers. (The key for the antilogarithm function on most calculators is designated as 10^x.) It is desirable, however, to understand what a logarithm is as well as some of its properties. The discussion that follows provides this information.

A logarithm (or log) of a number is the power to which some base number (usually 10) must be raised in order to give the desired number. Thus, a logarithm is an exponent of the base 10. From the discussion in the previous paragraphs about exponential numbers, we can draw the following conclusions with respect to logs:

1. The logarithm of a product is the sum of the logarithms of the individual numbers in the product.

$$\log (100 \times 1000) = \log 10^2 + \log 10^3 = 2 + 3 = 5$$

2. The logarithm of a quotient is the difference between the logarithms of the individual numbers.

$$\log (100/1000) = \log 10^2 - \log 10^3 = 2 - 3 = -1$$

3. The logarithm of a number raised to some power is the logarithm of the number multiplied by that power.

$$\log (1000)^2 = 2 \times \log 10^3 = 2 \times 3 = 6$$
$$\log (0.01)^6 = 6 \times \log 10^{-2} = 6 \times (-2) = -12$$

4. The logarithm of a root of a number is the logarithm of that number divided by the root.

$$\log (1000)^{1/3} = \frac{1}{3} \times \log 10^3 = \frac{1}{3} \times 3 = 1$$

The following examples illustrate these statements:

$$\log 40 \times 10^{20} = \log 4.0 \times 10^{21} = \log 4.0 + \log 10^{21}$$
$$= 0.60 + 21 = 21.60$$
$$\log 2.0 \times 10^{-6} = \log 2.0 + \log 10^{-6} = 0.30 + (-6) = -5.70$$

For some purposes, it is helpful to dispense with the subtraction step shown in the last example and report the log as a *negative* integer and a *positive* decimal number, that is,

$$\log 2.0 \times 10^{-6} = \log 2.0 + \log 10^{-6} = \bar{6}.30$$

The last two examples demonstrate that the logarithm of a number is the sum of two parts, a *characteristic* located to the left of the decimal point and a *mantissa* that lies to the right. The characteristic is the logarithm of 10 raised to a power and indicates the location of the decimal point in the original number when that number is expressed in decimal notation. The mantissa is the logarithm of a number in the range between 0.00 and 9.99. Note that the mantissa is *always positive*. As a consequence, the characteristic in the last example is −6, and the mantissa is +0.30.

Appendix 7

Volumetric Calculations Using Normality and Equivalent Weight

The **normality** of a solution expresses the number of equivalents of solute contained in 1 L of solution or the number of milliequivalents in 1 mL. The equivalent and milliequivalent, like the mole and millimole, are units for describing the amount of a chemical species. The equivalent and milliequivalent, however, are defined so that we may state that, at the equivalence point in *any* titration,

$$\text{no. meq analyte present} = \text{no. meq standard reagent added} \qquad \text{(A7-1)}$$

or

$$\text{no. eq analyte present} = \text{no. eq standard reagent added} \qquad \text{(A7-2)}$$

As a result of this equivalence, stoichiometric ratios such as those described in Section 13C-3 (page 308) need not be derived every time a volumetric calculation is performed. Instead, the stoichiometry is taken into account by how the equivalent or milliequivalent weight is defined.

A7A THE DEFINITIONS OF EQUIVALENT AND MILLIEQUIVALENT

In contrast to the mole, the amount of a substance contained in one equivalent can vary from reaction to reaction. Consequently, the weight of one equivalent of a compound can never be computed *without reference to a chemical reaction* in which that compound is, directly or indirectly, a participant. Similarly, the normality of a solution can never be specified *without knowledge about how the solution will be used.*

A7A-1 Equivalent Weights in Neutralization Reactions

One equivalent weight of a substance participating in a neutralization reaction is that amount of substance (molecule, ion, or paired ion such as NaOH) that either reacts with or supplies 1 mol of hydrogen ions *in that reaction.*[1] A milliequivalent is 1/1000 of an equivalent.

The relationship between equivalent weight (eqw) and the molar mass ($\mathcal{M}$) is straightforward for strong acids or bases and for other acids or bases that contain a single reaction hydrogen or hydroxide ion. For example, the equivalent weights of potassium hydroxide, hydrochloric acid, and acetic acid are equal to their molar masses because each has but a single reactive hydrogen ion or hydroxide ion. Barium hydroxide, which

> Once again we find ourselves using the term *weight* when we really mean *mass*. The term *equivalent weight* is so firmly engrained in the literature and vocabulary of chemistry that we retain it in this discussion.

---

[1]The IUPAC defines an equivalent entity as corresponding to the transfer of a H^+ ion in a neutralization reaction, to the transfer of an electron in a redox reaction, or to a magnitude of charge number equal to 1 in ions. Examples: $1/2H_2SO_4$, $1/5KMnO_4$, $1/3Fe^{3+}$. **DOI:** 10.1351/goldbook.E02192.

contains two identical hydroxide ions, reacts with two hydrogen ions in any acid/base reaction, and so its equivalent weight is one half its molar mass:

$$\text{eqw Ba(OH)}_2 = \frac{\mathscr{M}_{\text{Ba(OH)}_2}}{2}$$

The situation is more complex for acids or bases that contain two or more reactive hydrogen or hydroxide ions with different tendencies to dissociate. With certain indicators, for example, only the first of the three protons in phosphoric acid is titrated:

$$\text{H}_3\text{PO}_4 + \text{OH}^- \rightarrow \text{H}_2\text{PO}_4^- + \text{H}_2\text{O}$$

With certain other indicators, a color change occurs only after two hydrogen ions have reacted:

$$\text{H}_3\text{PO}_4 + 2\text{OH}^- \rightarrow \text{HPO}_4^{2-} + 2\text{H}_2\text{O}$$

For a titration involving the first reaction, the equivalent weight of phosphoric acid is equal to the molar mass; for the second, the equivalent weight is one half the molar mass. (Because it is not practical to titrate the third proton, an equivalent weight that is one third the molar mass is not generally encountered for H_3PO_4.) If it is not known which of these reactions is involved, an unambiguous definition of the equivalent weight for phosphoric acid *cannot be made.*

A7A-2 Equivalent Weights in Oxidation/Reduction Reactions

The equivalent weight of a participant in an oxidation/reduction reaction is that amount that directly or indirectly produces or consumes 1 mol of electrons. The numerical value for the equivalent weight is conveniently established by dividing the molar mass of the substance of interest by the change in oxidation number associated with its reaction. As an example, consider the oxidation of oxalate ion by permanganate ion:

$$5\text{C}_2\text{O}_4^{2-} + 2\text{MnO}_4^- + 16\text{H}^+ \rightarrow 10\text{CO}_2 + 2\text{Mn}^+ + 8\text{H}_2\text{O} \qquad \text{(A7-3)}$$

In this reaction, the change in oxidation number of manganese is 5 because the element passes from the +7 to the +2 state; the equivalent weights for MnO_4^- and Mn^{2+} are thus one fifth their molar masses. Each carbon atom in the oxalate ion is oxidized from the +3 to the +4 state, leading to the production of two electrons by that species. Therefore, the equivalent weight of sodium oxalate is one half its molar mass. It is also possible to assign an equivalent weight to the carbon dioxide produced by the reaction. Since this molecule contains but a single carbon atom and since that carbon undergoes a change in oxidation number of 1, the molar mass and equivalent weight of the two are identical.

It is important to note that in evaluating the equivalent weight of a substance, *only its change in oxidation number* during the titration is considered. For example, suppose the manganese content of a sample containing Mn_2O_3 is to be determined by a titration based on the reaction given in Equation A7-3. The fact that each manganese in the Mn_2O_3 has an oxidation number of +3 plays no part in determining equivalent weight.

That is, we must assume that by suitable treatment, all the manganese is oxidized to the $+7$ state before the titration is begun. Each manganese from the Mn_2O_3 is then reduced from the $+7$ to the $+2$ state in the titration step. The equivalent weight is thus the molar mass of Mn_2O_3 divided by $2 \times 5 = 10$.

As in neutralization reactions, the equivalent weight for a given oxidizing or reducing agent is not invariant. Potassium permanganate, for example, reacts under some conditions to give MnO_2:

$$MnO_4^- + 3e^- + 2H_2O \rightarrow MnO_2(s) + 4OH^-$$

The change in the oxidation state of manganese in this reaction is from $+7$ to $+4$, and the equivalent weight of potassium permanganate is now equal to its molar mass divided by 3 (instead of 5, as in the earlier example).

A7A-3 Equivalent Weights in Precipitation and Complex Formation Reactions

The equivalent weight of a participant in a precipitation or a complex formation reaction is that weight which reacts with or provides one mole of the *reacting* cation if it is univalent, one-half mole if it is divalent, one-third mole if it is trivalent, and so on. It is important to note that the cation referred to in this definition is always *the cation directly involved in the analytical reaction* and not necessarily the cation contained in the compound whose equivalent weight is being defined.

EXAMPLE A7-1

Define equivalent weights for $AlCl_3$ and $BiOCl$ if the two compounds are determined by a precipitation titration with $AgNO_3$:

$$Ag^+ + Cl^- \rightarrow AgCl(s)$$

Solution

In this instance, the equivalent weight is based on the number of moles of *silver ions* involved in the titration of each compound. Since 1 mol of Ag^+ reacts with 1 mol of Cl^- provided by one-third mole of $AlCl_3$, we can write

$$\text{eqw } AlCl_3 = \frac{\mathcal{M}_{AlCl_3}}{3}$$

Because each mole of $BiOCl$ reacts with only 1 Ag^+ ion,

$$\text{eqw } BiOCl = \frac{\mathcal{M}_{BiOCl}}{1}$$

Note that Bi^{3+} (or Al^{3+}) being trivalent has no bearing because the definition is based *on the cation involved in the titration:* Ag^+.

A7B THE DEFINITION OF NORMALITY

The normality, c_N, of a solution is the number of milliequivalents of solute contained in 1 mL of solution or the number of equivalents contained in 1 L. Thus, a 0.20 N hydrochloric acid solution contains 0.20 meq of HCl in each milliliter of solution or 0.20 eq in each liter.

The normal concentration of a solution is defined by equations analogous to Equation 4-2. Thus, for a solution of the species A, the normality $c_{N(A)}$ is given by the equations

$$c_{N(A)} = \frac{\text{no. meq A}}{\text{no. mL solution}} \tag{A7-4}$$

$$c_{N(A)} = \frac{\text{no. eq A}}{\text{no. L solution}} \tag{A7-5}$$

A7C SOME USEFUL ALGEBRAIC RELATIONSHIPS

Two pairs of algebraic equations, analogous to Equations 13-1 and 13-2 as well as 13-3 and 13-4 in Chapter 13, apply when normal concentrations are used:

$$\text{amount A} = \text{no. meq A} = \frac{\text{mass A (g)}}{\text{meqw A (g/meq)}} \tag{A7-6}$$

$$\text{amount A} = \text{no. eq A} = \frac{\text{mass A (g)}}{\text{eqw A (g/eq)}} \tag{A7-7}$$

$$\text{amount A} = \text{no. meq A} = V\,(\text{mL}) \times c_{N(A)}(\text{meq/mL}) \tag{A7-8}$$

$$\text{amount A} = \text{no. eq A} = V\,(\text{L}) \times c_{N(A)}(\text{eq/L}) \tag{A7-9}$$

A7D CALCULATION OF THE NORMALITY OF STANDARD SOLUTIONS

Example A7-2 shows how the normality of a standard solution is computed from preparatory data.

EXAMPLE A7-2

Describe the preparation of 5.000 L of 0.1000 N Na_2CO_3 (105.99 g/mol) from the primary-standard solid, assuming the solution is to be used for titrations in which the reaction is

$$CO_3^{2-} + 2H^+ \rightarrow H_2O + CO_2$$

Solution

Applying Equation A7-9 gives

$$\text{amount } Na_2CO_3 = V\,\text{soln (L)} \times c_{N(Na_2CO_3)}(\text{eq/L})$$

$$= 5.000\,\text{L} \times 0.1000\,\text{eq/L} = 0.5000\,\text{eq } Na_2CO_3$$

Rearranging Equation A7-7 gives

$$\text{mass } Na_2CO_3 = \text{no. eq } Na_2CO_3 \times \text{eqw } Na_2CO_3$$

But 2 eq of Na_2CO_3 are contained in each mole of the compound; therefore,

$$\text{mass } Na_2CO_3 = 0.5000 \text{ eq } Na_2CO_3 \times \frac{105.99 \text{ g } Na_2CO_3}{2 \text{ eq } Na_2CO_3} = 26.50 \text{ g}$$

Thus, dissolve 26.50 g in water and dilute to 5.000 L.

Note that, when the carbonate ion reacts with two protons, the weight of sodium carbonate required to prepare a 0.10 N solution is just one half that required to prepare a 0.10 M solution.

A7E THE TREATMENT OF TITRATION DATA WITH NORMALITIES

A7E-1 Calculation of Normalities from Titration Data

Examples A7-3 and A7-4 illustrate how normality is computed from standardization data. Note that these examples are similar to Examples 13-4 and 13-5 in Chapter 13.

EXAMPLE A7-3

Exactly 50.00 mL of an HCl solution required 29.71 mL of 0.03926 N $Ba(OH)_2$ to give an end point with bromocresol green indicator. Calculate the normality of the HCl.

Note that the molar concentration of $Ba(OH)_2$ is one half its normality, that is,

$$c_{Ba(OH)_2} = 0.03926 \frac{\text{meq}}{\text{mL}} \times \frac{1 \text{ mmol}}{2 \text{ meq}} = 0.01963 \text{ M}$$

Solution

Because we are basing our calculations on the milliequivalent, we write

$$\text{no. meq HCl} = \text{no. meq } Ba(OH)_2$$

The number of milliequivalents of standard is obtained by substituting into Equation A7-8:

$$\text{amount } Ba(OH)_2 = 29.71 \text{ mL } \cancel{Ba(OH)_2} \times 0.03926 \frac{\text{meq } Ba(OH)_2}{\text{mL } \cancel{Ba(OH)_2}}$$

To obtain the number of milliequivalents of HCl, we write

$$\text{amount HCl} = (29.71 \times 0.03926) \text{ meq } \cancel{Ba(OH)_2} \times \frac{1 \text{ meq HCl}}{1 \text{ meq } \cancel{Ba(OH)_2}}$$

(continued)

Equating this result to Equation A7-8 yields

$$\text{amount HCl} = 50.00 \text{ mL} \times c_{\text{N(HCl)}}$$

$$= (29.71 \times 0.03926 \times 1) \text{ meq HCl}$$

$$c_{\text{N(HCl)}} = \frac{(29.71 \times 0.03926 \times 1) \text{ meq HCl}}{50.00 \text{ mL HCl}} = 0.02333 \text{ N}$$

EXAMPLE A7-4

A 0.2121-g sample of pure $Na_2C_2O_4$ (134.00 g/mol) was titrated with 43.31 mL of $KMnO_4$. What is the normality of the $KMnO_4$ solution? The chemical reaction is

$$2MnO_4^- + 5C_2O_4^{2-} + 16H^+ \rightarrow 2Mn^{2+} + 10CO_2 + 8H_2O$$

Solution

By definition, at the equivalence point in the titration,

$$\text{no. meq } Na_2C_2O_4 = \text{no. meq } KMnO_4$$

Substituting Equations A7-8 and A7-6 into this relationship gives

$$V_{KMnO_4} \times c_{\text{N(KMnO}_4)} = \frac{\text{mass } Na_2C_2O_4 \text{ (g)}}{\text{meqw } Na_2C_2O_4 \text{ (g/meq)}}$$

$$43.31 \text{ mL } KMnO_4 \times c_{\text{N(KMnO}_4)} = \frac{0.2121 \text{ g } Na_2C_2O_4}{0.13400 \text{ g } Na_2C_2O_4/2 \text{ meq}}$$

$$c_{\text{N(KMnO}_4)} = \frac{0.2121 \text{ g } Na_2C_2O_4}{43.31 \text{ mL } KMnO_4 \times 0.1340 \text{ g } Na_2C_2O_4/2 \text{ meq}}$$

$$= 0.073093 \text{ meq/mL } KMnO_4 = 0.07309 \text{ N}$$

Note that the normality found here is five times the molar concentration computed in Example 13-5.

A7E-2 Calculation of the Quantity of Analyte from Titration Data

The examples that follow illustrate how analyte concentrations are computed when normalities are involved. Note that Example A7-5 is similar to Example 13-6 in Chapter 13.

EXAMPLE A7-5

A 0.8040-g sample of an iron ore was dissolved in acid. The iron was then reduced to Fe^{2+} and titrated with 47.22 mL of 0.1121 N (0.02242 M) $KMnO_4$ solution. Calculate the results of this analysis in terms of (a) percent Fe (55.847 g/mol) and (b) percent Fe_3O_4 (231.54 g/mol). The reaction of the analyte with the reagent is described by the equation

$$MnO_4^- + 5Fe^{2+} + 8H^+ \rightarrow Mn^{2+} + 5Fe^{3+} + 4H_2O$$

Solution

(a) At the equivalence point, we know that

$$\text{no. meq } KMnO_4 = \text{no. meq } Fe^{2+} = \text{no. meq } Fe_3O_4$$

Substituting Equations A7-8 and A7-6 leads to

$$V_{KMnO_4}(\text{mL}) \times c_{N(KMnO_4)}(\text{meq/mL}) = \frac{\text{mass } Fe^{2+} (g)}{\text{meqw } Fe^{2+} (g/\text{meq})}$$

Substituting numerical data into this equation gives, after rearranging,

$$\text{mass } Fe^{2+} = 47.22 \text{ mL } KMnO_4^- \times 0.1121 \frac{\text{meq}}{\text{mL } KMnO_4^-} \times \frac{0.055847 \text{ g}}{1 \text{ meq}}$$

Note that the milliequivalent weight of the Fe^{2+} is equal to its millimolar mass. The percentage of iron is

$$\text{percent } Fe^{2+} = \frac{(47.22 \times 0.1121 \times 0.055847) \text{ g } Fe^{2+}}{0.8040 \text{ g sample}} \times 100\%$$

$$= 36.77\%$$

(b) In this instance,

$$\text{no. meq } KMnO_4 = \text{no. meq } Fe_3O_4$$

and

$$V_{KMnO_4}(\text{mL}) \times c_{N(KMnO_4)}(\text{meq/mL}) = \frac{\text{mass } Fe_3O_4 (g)}{\text{meqw } Fe_3O_4 (g/\text{meq})}$$

Substituting numerical data and rearranging give

$$\text{mass } Fe_3O_4 = 47.22 \text{ mL} \times 0.1121 \frac{\text{meq}}{\text{mL}} \times 0.23154 \frac{\text{g } Fe_3O_4}{3 \text{ meq}}$$

Note that the milliequivalent weight of Fe_3O_4 is one third its millimolar mass because each Fe^{2+} undergoes a one-electron change and the compound is converted to $3Fe^{2+}$ before titration. The percentage of Fe_3O_4 is then

$$\text{percent } Fe_3O_4 = \frac{(47.22 \times 0.1121 \times 0.23154/3) \text{ g } Fe_3O_4}{0.8040 \text{ g sample}} \times 100\%$$

$$= 50.81\%$$

Note that the answers to this example are identical to those in Example 13-6.

EXAMPLE A7-6

A 0.4755-g sample containing $(NH_4)_2C_2O_4$ and inert compounds was dissolved in water and made alkaline with KOH. The liberated NH_3 was distilled into 50.00 mL of 0.1007 N (0.05035 M) H_2SO_4. The excess H_2SO_4 was back-titrated with 11.13 mL of 0.1214 N NaOH. Calculate the percentage of N (14.007 g/mol) and of $(NH_4)_2C_2O_4$ (124.10 g/mol) in the sample.

Solution

At the equivalence point, the number of milliequivalents of acid and base are equal. In this titration, however, two bases are involved: NaOH and NH_3. Thus,

$$\text{no. meq } H_2SO_4 = \text{no. meq } NH_3 + \text{no. meq NaOH}$$

After rearranging,

$$\text{no. meq } NH_3 = \text{no. meq N} = \text{no. meq } H_2SO_4 - \text{no. meq NaOH}$$

Substituting Equations A7-6 and A7-8 for the number of milliequivalents of N and H_2SO_4, respectively, yields

$$\frac{\text{mass N(g)}}{\text{meqw N (g/meq)}} = 50.00 \text{ mL } \cancel{H_2SO_4} \times 0.1007 \frac{\text{meq}}{\text{mL } \cancel{H_2SO_4}}$$

$$- 11.13 \text{ mL } \cancel{NaOH} \times 0.1214 \frac{\text{meq}}{\text{mL } \cancel{NaOH}}$$

$$\text{mass N} = (50.00 \times 0.1007 - 11.13 \times 0.1214) \text{ meq} \times 0.014007 \text{ g N/meq}$$

$$\text{percent N} = \frac{(50.00 \times 0.1007 - 11.13 \times 0.1214) \times 0.014007 \text{ g N}}{0.4755 \text{ g sample}} \times 100\%$$

$$= 10.85\%$$

The number of milliequivalents of $(NH_4)_2C_2O_4$ is equal to the number of milliequivalents of NH_3 and N, but the milliequivalent weight of the $(NH_4)_2C_2O_4$ is equal to one half its molar mass. Thus,

$$\text{mass } (NH_4)_2C_2O_4 = (50.00 \times 0.1007 - 11.13 \times 0.1214) \text{ meq}$$

$$\times 0.12410 \text{ g/2 meq}$$

percent $(NH_4)_2C_2O_4$

$$= \frac{(50.00 \times 0.1007 - 11.13 \times 0.1214) \times 0.06205 \text{ g}\cancel{(NH_4)_2C_2O_4}}{0.4755 \text{ g } \cancel{\text{sample}}} \times 100\%$$

$$= 48.07\%$$

Appendix 8

Compounds Recommended for the Preparation of Standard Solutions of Some Common Elements*

Element	Compound	Molar Mass	Solvent[†]	Notes
Aluminum	Al metal	26.9815386	Hot dil HCl	a
Antimony	$KSbOC_4H_4O_6 \cdot \frac{1}{2}H_2O$	333.94	H_2O	c
Arsenic	As_2O_3	197.840	dil HCl	i,b,d
Barium	$BaCO_3$	197.335	dil HCl	
Bismuth	Bi_2O_3	465.958	HNO_3	
Boron	H_3BO_3	61.83	H_2O	d,e
Bromine	KBr	119.002	H_2O	a
Cadmium	CdO	128.410	HNO_3	
Calcium	$CaCO_3$	100.086	dil HCl	i
Cerium	$(NH_4)_2Ce(NO_3)_6$	548.218	H_2SO_4	
Chromium	$K_2Cr_2O_7$	294.185	H_2O	i,d
Cobalt	Co metal	58.933195	HNO_3	a
Copper	Cu metal	63.546	dil HNO_3	a
Fluorine	NaF	41.9881725	H_2O	b
Iodine	KIO_3	214.000	H_2O	i
Iron	Fe metal	55.845	HCl, hot	a
Lanthanum	La_2O_3	325.808	HCl, hot	f
Lead	$Pb(NO_3)_2$	331.2	H_2O	a
Lithium	Li_2CO_3	73.89	HCl	a
Magnesium	MgO	40.304	HCl	
Manganese	$MnSO_4 \cdot H_2O$	169.01	H_2O	g
Mercury	$HgCl_2$	271.49	H_2O	b
Molybdenum	MoO_3	143.96	1 M NaOH	
Nickel	Ni metal	58.6934	HNO_3, hot	a
Phosphorus	KH_2PO_4	136.09	H_2O	
Potassium	KCl	74.55	H_2O	a
	$KHC_8H_4O_4$	204.22	H_2O	i,d
	$K_2Cr_2O_7$	294.182	H_2O	i,d
Silicon	Si metal	28.085	NaOH, concd	
	SiO_2	60.083	HF	j
Silver	$AgNO_3$	169.872	H_2O	a
Sodium	NaCl	58.44	H_2O	i
	$Na_2C_2O_4$	133.998	H_2O	i,d
Strontium	$SrCO_3$	147.63	HCl	a
Sulfur	K_2SO_4	174.25	H_2O	
Tin	Sn metal	118.71	HCl	

continue

Element	Compound	Molar Mass	Solvent[†]	Notes
Titanium	Ti metal	47.867	H_2SO_4; 1 : 1	a
Tungsten	$Na_2WO_4 \cdot 2H_2O$	329.85	H_2O	h
Uranium	U_3O_8	842.079	HNO_3	d
Vanadium	V_2O_5	181.878	HCl, hot	
Zinc	ZnO	81.38	HCl	a

*The data in this table are taken from a more complete list assembled by B. W. Smith and M. L. Parsons, *J. Chem, Educ.,* 1973, *50,* 679, DOI: 10.1021/ed050p679. Unless otherwise specified, compounds should be dried to constant weight at 110°C.

[†]Unless otherwise specified, acids are concentrated analytical grade.

[a]Conforms well to the criteria listed in Section 13A-2 and approaches primary-standard quality.

[b]Highly toxic.

[c]Loses $\frac{1}{2}H_2O$ at 110°C. After drying, molar mass = 324.92. The dried compound should be weighed quickly after removal from the desiccator.

[d]Available as a primary standard from the National Institute of Standards and Technology.

[e]H_3BO_3 should be weighed directly from the bottle. It loses 1 mole H_2O at 100°C and is difficult to dry to constant weight.

[f]Absorbs CO_2 and H_2O. Should be ignited just before use.

[g]May be dried at 110°C without loss of water.

[h]Loses both waters at 110°C. Molar mass = 293.82. Keep in desiccator after drying.

[i]Primary standard.

[j]HF is highly toxic and dissolves glass.

Appendix 9

Derivation of Error Propagation Equations

In this appendix, we derive several equations that permit the calculation of the standard deviation for the results from various types of arithmetical computations.

A9A PROPAGATION OF MEASUREMENT UNCERTAINTIES

The calculated result for a typical analysis ordinarily requires data from several independent experimental measurements, each of which is subject to a random uncertainty and each of which contributes to the net random error of the final result. For the purpose of showing how such random uncertainties affect the outcome of an analysis, let us assume that a result y is dependent on the experimental variables, $a, b, c, \ldots$, each of which fluctuates in a random and independent way. In other words, y is a function of $a, b, c, \ldots$, so we may write

$$y = f(a, b, c, \ldots) \tag{A9-1}$$

The uncertainty dy_i is generally given in terms of the deviation from the mean or $(y_i - \bar{y})$, which will depend on the size and sign of the corresponding uncertainties $da_i, db_i, dc_i, \ldots$. Thus,

$$dy_i = (y_i - \bar{y}) = f(da_i, db_i, dc_i, \ldots)$$

The uncertainty in dy as a function of the uncertainties in $a, b, c, \ldots$ can be derived by taking the total differential of Equation A9-1. Therefore,

$$dy = \left(\frac{\partial y}{\partial a}\right)_{b,c,\ldots} da + \left(\frac{\partial y}{\partial b}\right)_{a,c,\ldots} db + \left(\frac{\partial y}{\partial c}\right)_{a,b,\ldots} dc + \ldots \tag{A9-2}$$

To develop a relationship between the standard deviation of y and the standard deviations of $a, b,$ and c for N replicate measurements, we employ Equation 6-4 (p. 103), which requires that we square Equation A9-2, sum between $i = 0$ and $i = N$, divide by $N - 1$, and take the square root of the result. The square of Equation A9-2 takes the form

$$(dy)^2 = \left[\left(\frac{\partial y}{\partial a}\right)_{b,c,\ldots} da + \left(\frac{\partial y}{\partial b}\right)_{a,c,\ldots} db + \left(\frac{\partial y}{\partial c}\right)_{a,b,\ldots} dc + \ldots\right]^2 \tag{A9-3}$$

This equation must then be summed between the limits of $i = 1$ to $i = N$.

In squaring Equation A9-2, two types of terms emerge from the right-hand side of the equation: (1) square terms and (2) cross terms. Square terms take the form

$$\left(\frac{\partial y}{\partial a}\right)^2 da^2, \left(\frac{\partial y}{\partial b}\right)^2 db^2, \left(\frac{\partial y}{\partial c}\right)^2 dc^2, \ldots$$

Square terms are always positive and can, therefore, *never* cancel when summed. In contrast, cross terms may be either positive or negative in sign. Examples are

$$\left(\frac{\partial y}{\partial a}\right)\left(\frac{\partial y}{\partial b}\right) dadb, \left(\frac{\partial y}{\partial a}\right)\left(\frac{\partial y}{\partial c}\right) dadc, \ldots$$

If *da, db,* and *dc* represent *independent* and *random uncertainties,* some of the cross terms will be negative and others positive. Thus, the *sum of all such terms should approach zero,* particularly when N is large.

As a consequence of the tendency of cross terms to cancel, the sum of Equation A9-3 from $i = 1$ to $i = N$ can be assumed to be made up exclusively of square terms. This sum then takes the form

$$\Sigma (dy_i)^2 = \left(\frac{\partial y}{\partial a}\right)^2 \Sigma (da_i)^2 + \left(\frac{\partial y}{\partial b}\right)^2 \Sigma (db_i)^2 + \left(\frac{\partial y}{\partial c}\right)^2 \Sigma (dc_i)^2 + \ldots \qquad \text{(A9-4)}$$

Dividing through by $N - 1$ gives

$$\frac{\Sigma (dy_i)^2}{N - 1} = \left(\frac{\partial y}{\partial a}\right)^2 \frac{\Sigma (da_i)^2}{N - 1} + \left(\frac{\partial y}{\partial b}\right)^2 \frac{\Sigma (db_i)^2}{N - 1} + \left(\frac{\partial y}{\partial c}\right)^2 \frac{\Sigma (dc_i)^2}{N - 1} + \ldots \qquad \text{(A9-5)}$$

From Equation 6-4, however, we see that

$$\frac{\Sigma (dy_i)^2}{N - 1} = \Sigma\frac{(y_i - \bar{y})^2}{N - 1} = s_y^2$$

where s_y^2 is the variance of *y.* Similarly,

$$\frac{\Sigma (da_i)^2}{N - 1} = \frac{\Sigma(a_i - \bar{a})^2}{N - 1} = s_a^2$$

and so forth. Thus, Equation A9-5 can be written in terms of the variances of the variables, that is,

$$s_y^2 = \left(\frac{\partial y}{\partial a}\right)^2 s_a^2 + \left(\frac{\partial y}{\partial b}\right)^2 s_b^2 + \left(\frac{\partial y}{\partial c}\right)^2 s_c^2 + \ldots \qquad \text{(A9-6)}$$

A9B THE STANDARD DEVIATION OF COMPUTED RESULTS

In this section, we employ Equation A9-6 to derive relationships that permit calculation of standard deviations for the results produced by five types of arithmetic operations.

A9B-1 Addition and Subtraction

Consider the case where we wish to compute the quantity *y* from the three experimental quantities *a, b,* and *c* by means of the equation

$$y = a + b - c$$

We assume that the standard deviations for these quantities are s_y, s_a, s_b, and s_c. Applying Equation A9-6 leads to

$$s_y^2 = \left(\frac{\partial y}{\partial a}\right)_{b,c}^2 s_a^2 + \left(\frac{\partial y}{\partial b}\right)_{a,c}^2 s_b^2 + \left(\frac{\partial y}{\partial c}\right)_{a,b}^2 s_c^2$$

The partial derivatives of y with respect to the three experimental quantities are

$$\left(\frac{\partial y}{\partial a}\right)_{b,c} = 1; \quad \left(\frac{\partial y}{\partial b}\right)_{a,c} = 1; \quad \left(\frac{\partial y}{\partial c}\right)_{a,b} = -1$$

Therefore, the variance of y is given by

$$s_y^2 = (1)^2 s_a^2 + (1)^2 s_b^2 + (-1)^2 s_c^2 = s_a^2 + s_b^2 + s_c^2$$

or the standard deviation of the result is given by

$$s_y = \sqrt{s_a^2 + s_b^2 + s_c^2} \tag{A9-7}$$

Thus, the *absolute* standard deviation of a sum or difference is equal to the square root of the sum of the squares of the *absolute* standard deviation of the numbers making up the sum or difference.

A9B-2 Multiplication and Division

Let us now consider the case where

$$y = \frac{ab}{c}$$

The partial derivatives of y with respect to a, b, and c are

$$\left(\frac{\partial y}{\partial a}\right)_{b,c} = \frac{b}{c}; \quad \left(\frac{\partial y}{\partial b}\right)_{a,c} = \frac{a}{c}; \quad \left(\frac{\partial y}{\partial c}\right) = -\frac{ab}{c^2}$$

Substituting into Equation A9-6 gives

$$s_y^2 = \left(\frac{b}{c}\right)^2 s_a^2 + \left(\frac{a}{c}\right)^2 s_b^2 + \left(\frac{ab}{c^2}\right)^2 s_c^2$$

Dividing this equation by the square of the original equation ($y^2 = a^2 b^2 / c^2$) gives

$$\frac{s_y^2}{y^2} = \frac{s_a^2}{a^2} + \frac{s_b^2}{b^2} + \frac{s_c^2}{c^2}$$

or

$$\frac{s_y}{y} = \sqrt{\left(\frac{s_a}{a}\right)^2 + \left(\frac{s_b}{b}\right)^2 + \left(\frac{s_c}{c}\right)^2} \tag{A9-8}$$

Hence, for products and quotients, the *relative* standard deviation of the result is equal to the sum of the squares of the *relative* standard deviation of the number making up the product or quotient.

A9B-3 Exponential Calculations

Consider the following computation

$$y = a^x$$

In this instance, Equation A9-6 takes the form

$$s_y^2 = \left(\frac{\partial a^x}{\partial y}\right)^2 s_a^2$$

or

$$s_y = \frac{\partial a^x}{\partial y} s_a$$

But

$$\frac{\partial a^x}{\partial y} = xa^{(x-1)}$$

Thus,

$$s_y = xa^{(x-1)} s_a$$

and dividing by the original equation ($y = a^x$) gives

$$\frac{s_y}{y} = \frac{xa^{(x-1)} s_a}{a^x} = x\frac{s_a}{a} \qquad\qquad \text{(A9-9)}$$

Therefore, the relative error of the result is equal to the relative error of numbers to be exponentiated, multiplied by the exponent.

It is important to note that the error propagated in taking a number to a power is different from the error propagated in multiplication. For example, consider the uncertainty in the square of 4.0($\pm$0.2). The relative error in the result (16.0) is given by Equation A9-9

$$s_y/y = 2 \times (0.2/4) = 0.1 \qquad \text{or} \qquad 10\%$$

Consider now the case when y is the product of two *independently measured* numbers that by chance happen to have values of $a = 4.0(\pm0.2)$ and $b = 4.0(\pm0.2)$. In this case, the relative error of the product $ab = 16.0$ is given by Equation A9-8:

$$s_y/y = \sqrt{(0.2/4)^2 + (0.2/4)^2} \quad \text{or} \quad 7\%$$

The reason for this apparent anomaly is that in the second case the sign associated with one error can be the same or different from that of the other. If they happen to be the same, the error is identical to that encountered in the first case, where the signs *must* be the same. In contrast, the possibility exists that one sign could be positive and the other negative in which case the relative errors tend to cancel one another. Thus, the probable error lies between the maximum (10%) and zero.

A9B-4 Calculation of Logarithms

Consider the computation

$$y = \log_{10} a$$

In this case, we can write Equation A9-6 as

$$s_y^2 = \left(\frac{\partial \log_{10} a}{\partial y} \right)^2 s_a^2$$

But

$$\frac{\partial \log_{10} a}{\partial y} = \frac{0.434}{a}$$

and

$$s_y = 0.434 \frac{s_a}{a} \qquad \text{(A9-10)}$$

This equation shows that, the absolute standard deviation of a logarithm is determined by the *relative* standard deviation of the number.

A9B-5 Calculation of Antilogarithms

Consider the relationship

$$y = \text{antilog}_{10} a = 10^a$$

$$\left(\frac{\partial y}{\partial a} \right) = 10^a \log_e 10 = 10^a \ln 10 = 2.303 \times 10^a$$

$$s_y^2 = \left(\frac{\partial y}{\partial a} \right)^2 s_a^2$$

or

$$s_y = \frac{\partial y}{\partial a} s_a = 2.303 \times 10^a s_a$$

Dividing by the original relationship gives

$$\frac{s_y}{y} = 2.303 s_a \qquad \text{(A9-11)}$$

We see that the *relative* standard deviation of the antilog of a number is determined by the absolute standard deviation of the number.

Answers to Selected Questions and Problems

Chapter 3
3-1. **(a)** SQRT returns a positive square root; **(b)** AVERAGE returns the arithmetic mean; **(c)** PI returns pi to 15 digits; **(d)** FACT returns the factorial of a number; **(e)** EXP returns e raised to a power; **(f)** LOG returns the logarithm of a number to a base specified by the user or the base 10 logarithm if no base is identified.

Chapter 4
4-1. **(a)** The *millimole* is an amount of a chemical species, such as an atom, an ion, a molecule or an electron that contains

$$6.02 \times 10^{23} \frac{particles}{mol} \times 10^{-3} \frac{mol}{mmol} = 6.02 \times 10^{20} \frac{particles}{mmol}$$

(c) The *millimolar mass* is the mass in grams of one millimole of a chemical species.

4-3.
$$1\,L = \frac{1000\,mL}{1L} \times \frac{1\,cm^3}{mL} \times \left(\frac{m}{100\,cm}\right)^3 = 10^{-3}\,m^3$$

$$1\,M = \frac{1\,mol}{L} \times \frac{L}{10^{-3}\,m^3} = \frac{1\,mol}{10^{-3}\,m^3}$$

4-4. **(a)** 320 MHz   **(c)** 84.3 mol   **(e)** 8.96 mm

4-5. For O, 15.999 u = 15.999 g/mol. So, 1 u = 1 g/mol, and 1 g = 1 mol u.

4-7. $3.22 \times 10^{22}\ Na^+$ ion

4-9. **(a)** 0.251 mol   **(b)** 3.07 mmol
 (c) 0.0650 mol   **(d)** 5.20 mmol

4-11. **(a)** 111 mmol   **(b)** 2.44 mmol
 (c) 7.30×10^{-2} mmol   **(d)** 103.5 mmol

4-13. **(a)** 2.31×10^4 mg   **(b)** 9.87×10^3 mg
 (c) 1.00×10^6 mg   **(d)** 2.71×10^6 mg

4-15. **(a)** 1.92×10^3 mg   **(b)** 246 mg

4-16. **(a)** 2.25 g   **(b)** 2.60×10^{-3} g

4-17. **(a)** pNa = 0.984; pCl = 1.197, pOH = 1.395
 (c) pH = 0.398; pCl = 0.222; pZn = 1.00
 (e) pK = 5.94; pOH = 6.291; $pFe(CN)_6$ = 6.790

4-18. **(a)** 4.9×10^{-5} M   **(c)** = 0.26 M
 (e) 2.4×10^{-8} M   **(g)** 5.8 M

4-19. **(a)** pNa = pBr = 1.533   **(c)** pBa = 2.26; pOH = 1.96
 (e) pCa = 2.06; pBa = 2.18

4-20. **(a)** 0.0955 M   **(c)** 1.70×10^{-8} M   **(e)** 4.5×10^{-13} M
 (g) 0.733 M

4-21. **(a)** $[Na^+] = 4.79 \times 10^{-2}$ M; $[SO_4^{2-}] = 2.87 \times 10^{-3}$ M
 (b) pNa = 1.320; pSO_4 = 2.543

4-23. **(a)** 1.04×10^{-2} M   **(b)** 1.04×10^{-2} M   **(c)** 3.12×10^{-2} M
 (d) 0.288% (w/v)   **(e)** 0.78 mmol   **(f)** 407 ppm
 (g) 1.983   **(h)** 1.506

4-25. **(a)** 0.281 M   **(b)** 0.843 M   **(c)** 68.0 g

4-27. **(a)** Dissolve 23.8 g ethanol and add enough water to give a final volume of 500 mL.

(b) Mix 23.8 g ethanol with 476.2 g water.
(c) Dilute 23.8 mL ethanol with enough water to give a final volume of 500 mL.

4-29. Dilute 300 ml to 750 mL with water.

4-31. **(a)** Dissolve 6.37 g $AgNO_3$ in enough water to give a final volume of 500 mL.
 (b) Dilute 47.5 mL of the 6.00 M HCl to 1.00 L using water.
 (c) Dissolve 2.98 g $K_4Fe(CN)_6$ in enough water to give a final volume of 400 mL.
 (d) Dilute 216 mL of the 0.400 M $BaCl_2$ solution to 600 mL using water.
 (e) Dilute 20.3 mL of the concentrated reagent to 2.00 L using water.
 (f) Dissolve 1.7 g Na_2SO_4 in enough water to give a final volume of 9.00 L.

4-33. 5.01 g

4-35. **(a)** 9.214×10^{-2} g   **(b)** 3.12×10^{-2} M

4-37. **(a)** 1.5 g   **(b)** 0.064 M

4-39. 2.93 L

Chapter 5
5-1. **(a)** Random error causes data to be scattered around a mean value, while systematic error causes the mean of a data set to differ from the accepted value.
 (c) The absolute error is the difference between the measured value and the true value, while the relative error is the absolute error divided by the true value.

5-2. (1) Meter stick slightly longer or shorter than 1.0 m— systematic error.
 (2) Markings always read at a given angle—systematic error.
 (3) Variability in placement of metal rule to measure full 3-m width—random error.
 (4) Variability in interpolation of finest division of meter stick— random error.

5-4. (1) Balance out of calibration.
 (2) Fingerprints on weighing vial.
 (3) Sample absorbs water from the atmosphere.

5-5. (1) pipet incorrectly calibrated; (2) temperature different from calibration temperature; (3) meniscus read at an angle.

5-7. Both constant and proportional errors.

5-8. **(a)** -0.08% **(c)** -0.27%

5-9. **(a)** 33 g ore   **(c)** 4.2 g ore

5-10. **(a)** 0.060%   **(b)** 0.30%   **(c)** 0.12%

5-11. **(a)** -1.3% **(c)** -0.13%

5-12.

	Mean	Median	Deviation from Mean	Mean Deviation
(a)	0.0106	0.0105	0.0004, 0.0002, 0.0001	0.0002
(c)	190	189	2, 0, 4, 3	2
(e)	39.59	39.65	0.24, 0.02, 0.34, 0.09	0.17

Chapter 6

6-1. (a) The *standard error of the mean* is the standard deviation of the data set divided by the number of measurements.
(c) *Variance* is the standard deviation squared.

6-2. (a) *Parameter* refers to quantities that characterize a population or distribution of data. A *statistic* is an estimate of a parameter made from a sample.
(c) *Random errors* result from uncontrolled variables; *systematic errors* have a specific cause.

6-3. (a) *Sample standard deviation, s*, is that of a sample of data:

$$s = \sqrt{\frac{\sum_{i=1}^{N}(x_i - \bar{x})^2}{N - 1}}$$

Population standard deviation, σ, is for an entire population:

$$\sigma = \sqrt{\frac{\sum_{i=1}^{N}(x_i - \mu)^2}{N}}$$

where μ is the population mean.

6-5. Probability of a result between 0 and $+1\sigma$ is 0.342; between 1σ and 2σ, it is 0.136.

6-7.

	(a) Mean	(b) Median	(c) Spread	(d) Std Dev	(e) CV, %
A	9.1	9.1	1.0	0.37	4.1
C	0.650	0.653	0.108	0.056	8.5
E	20.61	20.64	0.14	0.07	0.32

6-8.

	Absolute Error	Relative Error, ppt
A	0.1	11.1
C	0.0195	31
E	0.03	1.3

6-9.

	s_y	CV, %	y
(a)	0.03	−1.4	$-2.08(\pm 0.03)$
(c)	0.085×10^{-16}	1.42	$5.94(\pm 0.08) \times 10^{-16}$
(e)	0.00520	6.9	$7.6(\pm 0.5) \times 10^{-2}$

6-10.

	s_y	CV, %	y
(a)	2.83×10^{-10}	4.25	$6.7 \pm 0.3 \times 10^{-9}$
(c)	0.1250	12.5	$14(\pm 2)$
(e)	25	50	$50(\pm 25)$

6-11.

	s_y	CV, %	y
(a)	6.51×10^{-3}	0.18	-3.699 ± 0.006
(c)	0.11	0.69	15.8 ± 0.1

6-12. (a) $s_y = 1.565 \times 10^{-12}$; CV = 2.2%; $y = 7.3(\pm 0.2) \times 10^{-11}$
6-13. $s_V = 0.145$; $V = 5.2(\pm 0.1)$ cm^3
6-15. CV = 0.6%

6-17. (a) $c_X = 2.029 \times 10^{-4}$ M **(b)** $S_{c_X} = 2.22 \times 10^{-6}$
(c) CV = 1.1%
6-19. (a) $s_1 = 0.096$, $s_2 = 0.077$, $s_3 = 0.084$, $s_4 = 0.090$, $s_5 = 0.104$, $s_6 = 0.083$
(b) 0.088
6-21. 3.5

Chapter 7

7-1. The distribution of means is narrower than the distribution of single results. The standard error of the mean of five measurements is, therefore, smaller than the standard deviation of a single result.

7-4.

	A	C	E
$\bar{x}$	2.86	70.19	0.824
s	0.24	0.08	0.051
95% CI	2.86 ± 0.30	70.19 ± 0.20	0.824 ± 0.081

The 95% CI is the interval within which the true mean is expected to lie 95% of the time.

7-5. For Set A, CI = 2.86 ± 0.26; for Set C, CI = 70.19 ± 0.079; for Set E, CI = 0.824 ± 0.088

7-7. (a) 99% CI = 18.5 ± 9.3 μg Fe/mL; 95% CI = 18.5 ± 7.1 μg Fe/mL
(b) 99% CI = 18.5 ± 6.6 μg Fe/mL; 95% CI = 18.5 ± 5.0 μg Fe/mL
(c) 99% CI = 18.5 ± 4.6 μg Fe/mL; 95% CI = 18.5 ± 3.5 μg Fe/mL

7-9. For 95% CI, $N \approx 11$; for 99% CI, $N \approx 18$
7-11. (a) 95% CI = 3.22 ± 0.15 meq Ca/L
(b) 95% CI = 3.22 ± 0.06 meq Ca/L
7-13. (a) 11
7-15. For two of the elements, there is a significant difference, but for three, there is not. Thus, the defendant might have grounds for claiming reasonable doubt.
7-17. Cannot reject the value 5.6 at the 95% confidence level.
7-19. H_0: $\mu_{current} = \mu_{previous}$; H_a: $\mu_{current} > \mu_{previous}$. Type I error is we reject the H_0 when it is true and decide the level of the pollutant is > the previous level when it is not. Type II error is that we accept H_0 when it is false and decide there is no change in the level when it is > than before.
7-20. (a) H_0: $\mu_{ISE} = \mu_{EDTA}$, H_a: $\mu_{ISE} \neq \mu_{EDTA}$. Two-tailed test. Type I error is that we decide the methods agree when they do not. Type II error is that we decide the methods do not agree when they do.
(c) H_0: $\sigma_X^2 = \sigma_Y^2$; H_a $\sigma_X^2 < \sigma_Y^2$. One-tailed test. Type I error is that we decide $\sigma_X^2 < \sigma_Y^2$ when it is not. Type II error is that we decide that $\sigma_X^2 = \sigma_Y^2$ when $\sigma_X^2 < \sigma_Y^2$.
7-21. (a) $t < t_{crit}$, so no significant difference at 95% confidence level.
(b) Significant difference at 95% confidence level.
(c) Large sample-to-sample variability causes s_{Top} and s_{Bottom} to be large and masks the differences.
7-23. We can be between 99% and 99.9% confident that the nitrogen prepared in the two ways is different. The probability of this conclusion being in error is 0.16%.
7-25. (a)

Source	SS	df	MS	F
Between juices	$4 \times 7.715 = 30.86$	$5 - 1 = 4$	$0.913 \times 8.45 = 7.715$	8.45
Within juices	$25 \times 0.913 = 22.825$	$30 - 5 = 25$	0.913	
Total	$30.86 + 22.82 = 50.68$	$30 - 1 = 29$		

(b) H_0: $\mu_{brand1} = \mu_{brand2} = \mu_{brand3} = \mu_{brand4} = \mu_{brand5}$; H_a: at least two of the means differ.

(c) average ascorbic acid contents differ at 95% confidence level.

7-27. (a) H_0: $\mu_{Analyst1} = \mu_{Analyst2} = \mu_{Analyst3} = \mu_{Analyst4}$; H_a: at least two of the means differ.

(b) Analysts differ at 95% confidence level.

(c) Significant difference between analyst 2 and analysts 1 and 4 but not analyst 3. Significant difference between analyst 3 and analyst 1 but not analyst 4. Significant difference between analyst 1 and analyst 4.

7-29. (a) H_0: $\mu_{ISE} = \mu_{EDTA} = \mu_{AA}$; H_a: at least two of the means differ.

(b) We conclude that the three methods give different results at the 95% confidence level.

(c) Significant difference between AA method and EDTA titration. No significant difference between EDTA titration method and ISE method, and there is no significant difference between the AA method and ISE method.

7-31. (a) Cannot reject with 95% confidence; **(b)** can reject with 95% confidence.

Chapter 8

8-1. Micro analysis of trace constituent

8-3. Step 1: Identify population. Step 2: Collect gross sample. Step 3: Reduce gross sample to laboratory sample.

8-5. 0.76%

8-7. (a) 1225; **(b)** 3403; **(c)** 10,000; **(d)** 122,500

8-9. (a) 8714 particles; **(b)** 650 g; **(c)** 0.32 mm

8-11. (a) Mean concentrations vary significantly from day to day.

(b) 79.19

(c) Reduce sampling variance.

8-13. 8

8-15. (b) $y = -29.74\,x + 92.86$

(d) $pCa_{Unk} = 2.608$; SD = 0.079; RSD = 0.030

8-17. (a) $m = 0.07014$; $b = 0.008286$

(b) $s_m = 0.00067$; $s_b = 0.004039$; SE = 0.00558

(c) 95% $CI_m = 0.07014 \pm 0.0019$; 95% $CI_b = 0.0083 \pm 0.0112$

(d) $c_{unk} = 5.77$ mM; $s_{unk} = 0.09$; 95% $CI_{unk} = 5.77 \pm 0.24$ mM

8-19. (b) $m = -8.456$; $b = 10.83$; SE = 0.0459

(c) 38.7 ± 1.1 kcal/mol

(d) No reason to doubt that E_A is not 41.00 kcal/mol at 95% confidence level.

8-21. (c) 5.2 ppm

8-23. 6.23×10^{-4} M

8-25. (c) For $k = 2$, DL = 0.14 ng/mL (92.1% confidence level); $k = 3$, DL = 0.21 ng/mL (98.3% confidence level)

8-27. Mean = 96.52; $s_{pooled} = 1.27$; UCL = 98.08; LCL = 94.97; out of control on Day 22.

Chapter 9

9-1. (a) A *weak electrolyte* only partially ionizes when dissolved in water. H_2CO_3 is an example of a weak electrolyte.

(c) The *conjugate acid of a Brønsted-Lowry base* is the species formed when a Brønsted-Lowry base accepts a proton. NH_4^+ is the conjugate acid of the base NH_3.

(e) An *amphiprotic solvent* can act either as an acid or a base. Water is an example.

(g) *Autoprotolysis* is self-ionization of a solvent to produce both a conjugate acid and a conjugate base.

(i) The *Le Châtelier principle* states that the position of an equilibrium always shifts in such a direction to relieve an applied stress.

9-2. (a) An *amphiprotic solute* is a chemical species that can act as either an acid or base. The dihydrogen phosphate ion, $H_2PO_4^-$, is an example.

(c) A *leveling solvent* is one in which a series of acids (or bases) all dissociate completely. Water is an example since HCl and $HClO_4$ dissociate completely.

9-3. For dilute aqueous solutions, the concentration of water is so much larger than other reactants that it can be assumed to be constant. Thus, its concentration is included in the equilibrium constant, but not in the equilibrium constant expression. For a pure solid, the concentration of the chemical species in the solid phase is constant. As long as some solid exists as a second phase, its effect on the equilibrium is constant and is included in the equilibrium constant.

9-4.

	Acid	Conjugate Base
(a)	HOCl	OCl^-
(c)	NH_4^+	NH_3
(e)	$H_2PO_4^-$	HPO_4^{2-}

9-6. (a) $2H_2O \rightleftharpoons H_3O^+ + OH^-$

(c) $2CH_3NH_2 \rightleftharpoons CH_3NH_3^+ + CH_3NH^-$

9-7. (a) $K_b = \dfrac{K_w}{K_a} = \dfrac{1.00 \times 10^{-14}}{2.3 \times 10^{-11}} = \dfrac{[C_2H_5NH_2^+][OH^-]}{[C_2H_5NH_2]} = 4.3 \times 10^{-4}$

(c) $K_a = \dfrac{[CH_3NH_2][H_3O^+]}{[CH_3NH_3^+]} = 2.3 \times 10^{-11}$

(e) $K_{overall} = \dfrac{[H_3O^+]^3[AsO_4^{3-}]}{[H_3AsO_4]} = K_{a1}K_{a2}K_{a3} = 2.0 \times 10^{-21}$

9-8. (a) $K_{sp} = [Cu^+][Br^-]$; **(b)** $K_{sp} = [Hg^{2+}][Cl^-][I^-]$;

(c) $K_{sp} = [Pb^{2+}][Cl^-]^2$

9-10. (b) $K_{sp} = 4.4 \times 10^{-11}$; **(d)** $K_{sp} = 3.5 \times 10^{-10}$

9-13. (a) 7.04×10^{-8} M; **(b)** 1.48 M

9-15. (a) 0.0225 M; **(b)** 1.6×10^{-2} M;

(c) 1.7×10^{-6} M; **(d)** 1.5×10^{-2} M

9-17. (a) $PbI_2 > BiI_3 > CuI > AgI$

(b) $PbI_2 > CuI > AgI > BiI_3$

(c) $PbI_2 > BiI_3 > CuI > AgI$

9-20. (a) $[H_3O^+] = 1.34 \times 10^{-3}$ M; $[OH^-] = 7.5 \times 10^{-12}$ M

(c) $[OH^-] = 6.37 \times 10^{-3}$ M; $[H_3O^+] = 1.57 \times 10^{-12}$ M

(e) $[OH^-] = 5.66 \times 10^{-6}$ M; $[H_3O^+] = 1.77 \times 10^{-9}$ M

(g) $[H_3O^+] = 5.24 \times 10^{-4}$ M; $[OH^-] = 1.91 \times 10^{-11}$ M

9-21. (a) $[H_3O^+] = 1.58 \times 10^{-2}$ M

(b) $[H_3O^+] = 8.26 \times 10^{-9}$ M

(e) $[H_3O^+] = 2.11 \times 10^{-4}$ M

9-23. *Buffer capacity* of a solution is defined as the number of moles of a strong acid (or a strong base) that causes 1.00 L of a buffer to undergo a 1.00-unit change in pH.

9-25. The solutions all are buffers with the same pH, but they differ in buffer capacity with **(a)** having the greatest and **(c)** the least.

9-26. (a) $C_6H_5NH_3^+/C_6H_5NH_2$;

(c) $C_2H_5NH_3^+/C_2H_5NH_2$ or $CH_3NH_3^+/CH_3NH_2$

9-27. 19.6 g

9-29. 387 mL

Chapter 10

10-1. (a) *Activity*, a_A, is the effective concentration of species A in solution. The *activity coefficient*, γ_A, is the numerical factor necessary to convert the molar concentration of species A to activity: $a_A = \gamma_A[A]$.

(b) The *thermodynamic equilibrium constant* refers to an ideal system within which each chemical species is unaffected by any others. A *concentration equilibrium constant* takes into account the influence exerted by solute species upon one another. The thermodynamic equilibrium constant is constant and independent of ionic strength; the concentration equilibrium constant depends on ionic strength.

10-3. (a) The ionic strength should decrease.

(b) The ionic strength should be unchanged.

(c) The ionic strength should increase.

10-5. Water is a neutral molecule, and its activity equals its concentration at all low to moderate ionic strengths. In such instances, activity coefficients of ions decrease with increasing ionic strength because the ionic atmosphere surrounding the ion causes it to lose some of its chemical effectiveness and its activity is less than its concentration.

10-7. Multiply charged ions deviate from ideality more than singly charged ions.

10-9. (a) 0.12　**(c)** 2.4

10-10. (a) 0.22　**(c)** 0.08

10-12. (a) 1.8×10^{-12}　**(c)** 1.1×10^{-10}

10-13. (a) 5.5×10^{-6} M　**(b)** 7.6×10^{-6} M

(c) 2.8×10^{-13} M　**(d)** 1.5×10^{-7} M

10-14. (a) (1) 1.4×10^{-6} M　(2) 1.0×10^{-6} M

(b) (1) 2.1×10^{-3} M　(2) 1.3×10^{-3} M

(c) (1) 2.9×10^{-5} M　(2) 1.0×10^{-5} M

(d) (1) 1.4×10^{-5} M　(2) 2.0×10^{-6} M

10-15. (a) (1) 2.2×10^{-4} M　(2) 1.8×10^{-4} M

(b) (1) 1.7×10^{-4} M　(2) 1.2×10^{-4} M

(c) (1) 3.3×10^{-8} M　(2) 6.6×10^{-9} M

(d) (1) 1.3×10^{-3} M　(2) 7.8×10^{-4} M

10-16. (a) -19%　**(c)** -40%　**(e)** -48%

10-17. (a) -45%

Chapter 11

11-2. In an equation with sums or differences, assuming a concentration is zero leads to an appropriate result. In an equilibrium-constant equation, multiplying or dividing by zero leads to a meaningless result.

11-4. A charge-balance equation relates the concentration of cations and anions such that no. mol/L positive charge = no. mol/L negative charge. For a doubly charged ion, such as Ba^{2+}, the concentration of charge of each mole is twice the molar concentration. For Fe^{3+}, it is three times the molar concentration. Thus, the molar concentration is always multiplied by the charge in a charge-balance equation.

11-5 (a) $0.20 = [HF] + [F^-]$

(c) $0.10 = [H_3PO_4] + [H_2PO_4^-] + [HPO_4^{2-}] + [PO_4^{3-}]$

(e) $0.0500 + 0.100 = [HClO_2] + [ClO_2^-]$

(g) $0.100 = [Na^+] = [OH^-] + 2[Zn(OH)_4^{2-}]$

(i) $[Pb^{2+}] = \frac{1}{2}([F^-] + [HF])$

11-7. (a) 2.3×10^{-4} M　**(c)** 2.2×10^{-4} M

11-8. (a) 1.9×10^{-4} M　**(c)** 3.1×10^{-5} M

11-9. (a) 1.5×10^{-4} M　**(b)** 1.5×10^{-7} M

11-11. (a) 4.7 M

11-12. 5.1×10^{-4} M

11-14. (a) $Cu(OH)_2$ precipitates first.

(b) 9.8×10^{-10} M

(c) 9.6×10^{-9} M

11-16. (a) 8.3×10^{-11} M　**(b)** 1.4×10^{-11} M;　**(c)** 1.3×10^4;

(d) 1.3×10^4

11-18. 3.754 g

11-20. (a) 0.0101 M; 49%　**(b)** 7.14×10^{-3} M; 70%

Chapter 12

12-1. (a) A *colloidal precipitate* consists of solid particles with dimensions that are less than 10^{-4} cm. A *crystalline precipitate* consists of solid particles with dimensions that are at least 10^{-4} cm or greater. Crystalline precipitates settle rapidly, whereas colloidal precipitates remain suspended in solution.

(c) In *precipitation*, a solid phase forms and is carried out of solution when the solubility product of a chemical species is exceeded. In *coprecipitation*, normally soluble compounds are carried out of solution during precipitate formation.

(e) *Occlusion* is a type of coprecipitation in which a compound is trapped within a pocket formed during rapid crystal formation. *Mixed-crystal formation* is also a type of coprecipitation in which a contaminant ion replaces an ion in the crystal lattice.

12-2. (a) *Digestion* is a process in which a precipitate is heated in the presence of the solution from which it was formed (the *mother liquor*). Digestion improves the purity and filterability of the precipitate.

(c) In *reprecipitation*, the filtered solid precipitate is redissolved and reprecipitated. Because the concentration of the impurity in the new solution is lower, the second precipitate contains less coprecipitated impurity.

(e) The *counter-ion layer* is a layer of solution surrounding a charged particle containing a sufficient excess of oppositely charged ions to balance the surface charge on the particle.

(g) *Supersaturation* is an unstable state in which a solution contains higher solute concentration than a saturated solution. Supersaturation is relieved by precipitation of excess solute.

12-3. A *chelating agent* is an organic compound that contains two or more electron-donor groups located in such a configuration that five- or six-membered rings are formed when the donor groups complex a cation.

12-5. (a) positive charge　**(b)** adsorbed Ag^+　**(c)** NO_3^- ions

12-7. In *peptization*, a coagulated colloid returns to its original dispersed state because of a decrease in the electrolyte concentration of the solution contacting the precipitate. Peptization can be avoided by washing the coagulated colloid with an electrolyte solution instead of pure water.

12-9. (a) mass SO_2 = mass $BaSO_4 \times \dfrac{\mathcal{M}_{SO_2}}{\mathcal{M}_{BaSO_4}}$

(c) mass In = mass $In_2O_3 \times \dfrac{2\mathcal{M}_{In}}{\mathcal{M}_{In_2O_3}}$

(e) mass CuO = mass $Cu_2(SCN)_2 \times \dfrac{2\mathcal{M}_{CuO}}{\mathcal{M}_{Cu_2(SCN)_2}}$

(i) mass $Na_2B_4O_7 \cdot 10H_2O$ = mass $B_2O_3 \times \dfrac{\mathcal{M}_{Na_2B_4O_7 \cdot 10H_2O}}{2\mathcal{M}_{B_2O_3}}$

12-10. 60.59%

12-12. 1.076 g

12-14. 0.178 g

12-18. 17.23%

12-20. 44.58%

12-22. 38.74%

12-24. 0.550 g

12-26. **(a)** 0.239 g **(b)** 0.494 g **(c)** 0.406 g

12-28. 4.72% Cl^- and 27.05% I^-

12-30. 0.764 g

12-32. **(a)** 0.369 g **(b)** 0.0149 g

Chapter 13

13-1. **(a)** The *millimole* is the amount of an elementary species, such as an atom, an ion, a molecule, or an electron. A millimole contains 6.02×10^{20} particles.

(c) *Stoichiometric ratio* is the molar ratio of two species in a balanced chemical reaction.

13-3. **(a)** The *equivalence point* in a titration is that point at which sufficient titrant has been added so that stoichiometrically equivalent amounts of analyte and titrant are present. The *end point* is the point at which an observable physical change signals the equivalence point.

13-5. **(a)** $\dfrac{1 \text{ mol } H_2NNH_2}{2 \text{ mol } I_2}$ **(c)** $\dfrac{1 \text{ mole } Na_2B_4O_7 \cdot 10H_2O}{2 \text{ moles } H^+}$

13-7. **(a)** 0.233 **(b)** 11.34 **(c)** 0.820 **(d)** 11.00

13-9. **(a)** 1.51 g **(b)** 0.00302 g
(c) 0.058 g **(d)** 0.0776 g

13-11. 3.03 M

13-13. **(a)** Dissolve 23.70 g $KMnO_4$ in water and dilute to 1.00 L.
(b) Dilute 139 mL of concentrated (9.00 M) reagent to 2.50 L.
(c) Dissolve 2.78 g MgI_2 in water and bring to 400 mL total volume.
(d) Dilute 57.5 mL of the 0.218 M solution to a volume of 200 mL.
(e) Dilute 16.9 mL of the concentrated reagent to 1.50 L.
(f) Dissolve 42.4 mg $K_4Fe(CN)_6$ in water and dilute to 1.50 L.

13-15. 0.1281 M

13-17. 0.2790 M

13-19. 0.1146 M

13-21. 165.6 ppm

13-23. 7.317%

13-25. 0.6718 g

13-27. **(a)** 0.02966 M **(b)** 47.59%

13-29. **(a)** 0.01346 M **(b)** 0.01346 M
(c) 4.038×10^{-2} M **(d)** 0.374%
(e) 1.0095 mmol **(f)** 526 ppm

Chapter 14

14-1. Because of the limited sensitivity of the eye, the color change requires a roughly tenfold excess of one or the other form of the indicator. This color change corresponds to a pH range of the indicator $pK_a \pm 1$ pH unit, a total range of 2 pH units.

14-3. **(a)** The initial pH of the NH_3 solution will be less than that for the solution containing NaOH. With the first addition of titrant, the pH of the NH_3 solution will decrease rapidly and then level off and become nearly constant throughout the middle part of the titration. In contrast, additions of standard acid to the NaOH solution will cause the pH of the NaOH solution to decrease gradually and nearly linearly until the equivalence point is approached. The equivalence point pH for the NH_3 solution will be well below 7, whereas for the NaOH solution it will be exactly 7.
(b) Beyond the equivalence point, the pH is determined by the excess titrant. Thus, the curves become identical in this region.

14-5. Temperature, ionic strength, the presence of organic solvents, and colloidal particles.

14-6. **(a)** NaOCl **(c)** methylamine

14-7. **(a)** iodic acid **(c)** pyruvic acid

14-9. 3.19

14-11. **(b)** 13.26

14-12. **(b)** 11.26

14-13. 0.078

14-15. 7.04

14-17. **(a)** 2.13 **(b)** 1.74 **(c)** 9.22 **(d)** 9.08

14-19. **(a)** 1.30 **(b)** 1.37

14-21. **(a)** 4.26 **(b)** 4.76 **(c)** 5.76

14-23. **(a)** 11.12 **(b)** 10.62 **(c)** 9.53

14-25. **(a)** 12.04 **(b)** 11.48 **(c)** 9.97

14-27. **(a)** 1.98 **(b)** 2.48 **(c)** 3.56

14-29. **(a)** 2.44 **(b)** 8.32 **(c)** 12.52 **(d)** 3.90

14-31. **(a)** 9.02 **(b)** 9.12

14-33. **(a)** 8.77 **(b)** 12.20 **(c)** 10.11 **(d)** 5.66

14-34. **(a)** 0.00 **(c)** −1.000 **(e)** −0.500 **(g)** 0.000

14-35. **(a)** −5.00 **(c)** −0.097 **(e)** −3.369 **(g)** −0.017

14-37. **(b)** −0.141

14-39. Cresol purple (range 7.6 to 9.2, Table 14-1) would be suitable.

14-41.

Vol, mL	(a) pH	(c) pH
0.00	2.09	2.44
5.00	2.38	2.96
15.00	2.82	3.50
25.00	3.17	3.86
40.00	3.76	4.46
45.00	4.11	4.82
49.00	4.85	5.55
50.00	7.92	8.28
51.00	11.00	11.00
55.00	11.68	11.68
60.00	11.96	11.96

14-43.

Vol, mL	(a) pH	(c) pH
0.00	2.51	4.26
5.00	2.62	6.57
15.00	2.84	7.15
25.00	3.09	7.52
40.00	3.60	8.12
45.00	3.94	8.48
49.00	4.66	9.21
50.00	7.28	10.11
51.00	10.00	11.00
55.00	10.68	11.68
60.00	10.96	11.96

14-44. **(a)** $\alpha_0 = 0.215$; $\alpha_1 = 0.785$
(c) $\alpha_0 = 0.769$; $\alpha_1 = 0.231$
(e) $\alpha_0 = 0.917$; $\alpha_1 = 0.083$

14.45. 0.105 M

14-47. Bolded entries are missing data points.

Acid	c_T	pH	[HA]	[A$^-$]	α_0	α_1
Lactic	0.120	**3.61**	**0.0768**	0.0432	0.640	**0.360**
Butanoic	**0.162**	5.00	0.644	0.0979	0.397	**0.604**
Sulfamic	0.250	1.20	**0.095**	0.155	0.380	**0.620**

Chapter 15

15-1. Not only is NaHA a proton donor, it is also the conjugate base of the parent acid H_2A. In order to calculate the pH of solutions of this type, it is necessary to take both the acid and the basic equilibria into account.

15-4. The species HPO_4^{2-} is such a weak acid ($K_{a3} = 4.5 \times 10^{-13}$) that the change in pH in the vicinity of the third equivalence point is too small to be observable.

15-5. (a) neutral (c) neutral (e) basic (g) acidic

15-6. Bromocresol green would be satisfactory.

15-8. H_3PO_4 could be determined with bromocresol green as an indicator. A titration with phenolphthalein indicator gives the number of millimoles of NaH_2PO_4 plus twice the number of millimoles of H_3PO_4. The amount of NaH_2PO_4 is obtained from the difference in volume for the two titrations.

15-9. (a) Cresol purple (c) Cresol purple
(e) Bromocresol green (g) Phenolphthalein

15-10. (a) 1.86 (c) 1.64 (e) 4.21

15-11. (a) 4.71 (c) 4.28 (e) 9.80

15-12. (a) 12.32 (c) 9.70 (e) 12.58

15-14. (a) 2.42 (b) 7.51 (c) 9.43
(d) 3.66 (e) 3.66

15-16. (a) 1.89 (b) 1.54 (c) 12.58 (d) 12.00

15-18. (a) $[H_2S]/[HS^-] = 0.010$
(b) $[BH^+]/[B] = 8.5$
(c) $[H_2AsO_4^-]/[HAsO_4^{2-}] = 9.1 \times 10^{-3}$
(d) $[HCO_3^-]/[CO_3^{2-}] = 21$

15-20. 49.0 g

15-22. (a) 5.47 (b) 2.92

15-24. Mix 366 mL H_3PO_4 with 634 mL NaOH.

15-28. The volume to the first end point would have to be smaller than one half the total volume to the second end point because in the titration from the first to second end points both analytes are titrated, whereas to the first end point only the H_3PO_4 is titrated.

15-32. (a) $\dfrac{[H_3AsO_4][HAsO_4^{2-}]}{[H_2AsO_4^-]^2} = 1.9 \times 10^{-5}$

(b) $\dfrac{[AsO_4^{3-}][H_2AsO_4^-]}{[HAsO_4^{2-}]^2} = 2.9 \times 10^{-5}$

15-34.

	pH	D	α_0	α_1	α_2	α_3
(a)	2.00	1.112×10^{-4}	0.899	0.101	3.94×10^{-5}	
	6.00	5.500×10^{-9}	1.82×10^{-4}	0.204	0.796	
	10.00	4.379×10^{-9}	2.28×10^{-12}	2.56×10^{-5}	1.000	
(c)	2.00	1.075×10^{-6}	0.931	6.93×10^{-2}	1.20×10^{-4}	4.82×10^{-9}
	6.00	1.882×10^{-14}	5.31×10^{-5}	3.96×10^{-2}	0.685	0.275
	10.00	5.182×10^{-15}	1.93×10^{-16}	1.44×10^{-9}	2.49×10^{-4}	1.000
(e)	2.00	4.000×10^{-4}	0.250	0.750	1.22×10^{-5}	
	6.00	3.486×10^{-9}	2.87×10^{-5}	0.861	0.139	
	10.00	4.863×10^{-9}	2.06×10^{-12}	6.17×10^{-4}	0.999	

Chapter 16

16-1. Nitric acid is an oxidizing agent and can react with reducible species in titrations.

16-3. Carbon dioxide is not strongly bonded by water molecules and is volatilized from aqueous solution by brief boiling. When dissolved in water, gaseous HCl molecules are fully dissociated into H_3O^+ and Cl^-, which are nonvolatile.

16-5. First, the higher molecular mass of $KH(IO_3)_2$ means that the relative mass error is less than with benzoic acid. Second, $KH(IO_3)_2$ is a strong acid while benzoic acid is not.

16-7. If the NaOH solution is used for titrations with an acid-range indicator, CO_3^{2-} in the base solution consumes two H_3O^+ ions, the same as the two hydroxides lost forming Na_2CO_3.

16-9. (a) Dissolve 11 g KOH in water and dilute to 2.00 L.
(b) Dissolve 6.3 g $Ba(OH)_2 \cdot 8H_2O$ in water and dilute to 2.00 L.
(c) Dilute 90 mL reagent to 2.00 L.

16-11. (a) 0.1077 M (b) $s = 0.00061$; CV = 0.57%
(c) Reject 1.0862 at 95% confidence level but retain at 99% CL.

16-13. Error = −29%

16-15. (a) 0.01535 M (b) 0.04175 M (c) 0.03452 M

16-17.

mL HCl	SD TRIS	SD Na_2CO_3	SD $Na_2B_4O_7 \cdot H_2O$
20.00	0.00004	0.00009	0.00003
30.00	0.00003	0.00006	0.00002
40.00	0.00002	0.00005	0.00001
50.00	0.00002	0.00004	0.00001

16-19. 0.1214 g/100 mL

16-21. (a) 46.55% (b) 88.23% (c) 32.21% (d) 10.00%

16-23. 23.7%

16-25. 7.216%

16-27. Probably $MgCO_3$ with a molar mass of 84.31.

16-29. 3.35×10^3 ppm

16-31. 6.333%

16-33. 22.08%

16-35. 3.93%

16-37. (a) 10.09% (b) 21.64% (c) 47.61% (d) 35.81%

16-39. 15.23% $(NH_4)_2SO_4$ and 24.39% NH_4NO_3

16-41. 28.56% $NaHCO_3$; 45.85% Na_2CO_3 and 25.59% H_2

16-43. (a) 12.93 mL (b) 16.17 mL
(c) 24.86 mL (d) 22.64 mL

16-45. (a) 4.31 mg/mL NaOH
(b) 7.985 mg/mL $NaHCO_3$ and 4.358 mg/mL Na_2CO_3
(c) 3.455 mg/mL Na_2CO_3 and 4.396 mg/mL NaOH
(d) 8.215 mg/mL Na_2CO_3
(e) 13.462 mg/mL $NaHCO_3$

16-47. (a) 126.066 (b) 63.03

Chapter 17

17-1. (a) A *ligand* is a species that contains one or more electron pair donor groups to form bonds with metal ions.
(c) A *tetradentate chelating agent* contains four pairs of donor electrons located in such positions that they all can bond to a metal ion, thus forming two rings.
(e) *Argentometric titrations* are based on forming precipitates with standard solutions of silver nitrate.

(g) In an *EDTA displacement titration*, an unmeasured excess of a solution containing the magnesium or zinc complex of EDTA is introduced into the solution of an analyte that forms a more stable complex than that of magnesium or zinc. The liberated magnesium or zinc ions are then titrated with a standard solution of EDTA.

17-3. Direct titration (1), back-titration (2), and displacement titration (3). Method (1) is simple, rapid, but it requires one standard reagent. Method (2) is advantageous for those metals that react very slowly with EDTA or with samples that form precipitates. Method (3) is particularly useful where no satisfactory indicators are available for direct titration.

17-4. (a) $Ag^+ + S_2O_3^{2-} \rightleftharpoons Ag(S_2O_3)^-$ $K_1 = \dfrac{[Ag(S_2O_3)^-]}{[Ag^+][S_2O_3^{2-}]}$

$Ag(S_2O_3)^- + S_2O_3^{2-} \rightleftharpoons Ag(S_2O_3)_2^{3-}$

$$K_2 = \dfrac{[Ag(S_2O_3)_2^{3-}]}{[Ag(S_2O_3)^-][S_2O_3^{2-}]}$$

17-5. The overall formation constant, β_n, is equal to the product of the individual stepwise constants.

17-7. The Fajans method involves a direct titration, while a Volhard titration requires two standard solutions and a filtration step.

17-9. In the beginning stages of a precipitation titration, one of the lattice ions is in excess, and its charge determines the sign of the charge of the particles. After the equivalence point, the oppositely charged ion is in excess and determines the sign of the charge.

17-11. (a) $\alpha_1 = \dfrac{K_a}{[H^+] + K_a}$

(b) $\alpha_2 = \dfrac{K_{a1}K_{a2}}{[H^+]^2 + K_{a1}[H^+] + K_{a1}K_{a2}}$

(c) $\alpha_3 = \dfrac{K_{a1}K_{a2}K_{a3}}{[H^+]^3 + K_{a1}[H^+]^2 + K_{a1}K_{a2}[H^+] + K_{a1}K_{a2}K_{a3}}$

17-13. $\beta_3' = (\alpha_2)^3 \beta_3 = \dfrac{[Fe(Ox)_3^{3-}]}{[Fe^{3+}](c_T)^3}$

17-15. $\beta_n = \dfrac{[ML_n]}{[M][L]^n}$

Taking the logarithm of both sides gives $\log \beta_n = \log[ML_n] - \log[M] - n\log[L]$.
Converting the right side to p functions, $\log \beta_n = pM + npL - pML_n$.

17-17. 0.00918 M

17-19. (a) 32.28 mL   **(b)** 14.98 mL   **(c)** 32.28 mL

17-20. (a) 34.84 mL   **(c)** 45.99 mL   **(e)** 32.34 mL

17-21. 3.244%

17-23. (a) 51.78 mL   **(c)** 10.64 mL   **(e)** 46.24 mL

17-25. (a) 44.70 mL   **(c)** 14.87 mL

17-27. 1.216%

17-29. 184.0 ppm Fe^{3+} and 213.1 ppm Fe^{2+}

17-31. 55.16% Pb and 44.86% Cd

17-33. 83.75% ZnO and 0.230% Fe_2O_3

17-34. 31.48% NaBr and 48.57% $NaBrO_3$

17-36. 13.72% Cr, 56.82% Ni, and 27.44% Fe

17-38. (a) 4.7×10^9 **(b)** 1.1×10^{12} **(c)** 7.5×10^{13}

17-42. (a) 570.5 ppm   **(b)** 350.5 ppm   **(c)** 185.3 ppm

Chapter 18

18-1. (a) *Oxidation* is a process in which a species loses one or more electrons.

(c) A *salt bridge* is a device that provides electrical contact but prevents mixing of dissimilar solutions in an electrochemical cell.

(e) The *Nernst equation* relates the potential to the concentrations (strictly, activities) of the participants in an electrochemical half-cell.

18-2. (a) The *electrode potential* is the potential of an electrochemical cell in which a standard hydrogen electrode acts as the reference electrode on the left and the half-cell of interest is on the right.

(c) The *standard electrode potential* is the potential of a cell consisting of the half-reaction of interest on the right and a standard hydrogen electrode on the left. The activities of all the participants in the half-reaction are specified as having a value of unity.

18-3. (a) *Oxidation* is the process whereby a substance loses electrons; an *oxidizing agent* causes the loss of electrons.

(c) The *cathode* is the electrode at which reduction occurs. The *right-hand electrode* is the electrode on the right in the cell diagram.

(e) The *standard electrode potential* is the potential of a cell in which the standard hydrogen electrode acts as the reference electrode on the left and all participants in the right-hand electrode process have unit activity. The *formal potential* differs in that the molar *concentrations* of all the reactants and products are unity and the concentrations of other species in the solution are carefully specified.

18-4. The first standard potential is for a solution saturated with I_2, which has an $I_2(aq)$ activity significantly less than one. The second potential is for a *hypothetical* half-cell in which the $I_2(aq)$ activity is unity.

18-5. To keep the solution saturated with $H_2(g)$. Only then is the hydrogen activity constant and the electrode potential constant and reproducible.

18-7. (a) $2Fe^{3+} + Sn^{2+} \rightarrow 2Fe^{2+} + Sn^{4+}$

(c) $2NO_3^- + Cu(s) + 4H^+ \rightarrow 2NO_2(g) + 2H_2O + Cu^{2+}$

(e) $Ti^{3+} + Fe(CN)_6^{3-} + H_2O \rightarrow TiO^{2+} + Fe(CN)_6^{4-} + 2H^+$

(g) $2Ag(s) + 2I^- + Sn^{4+} \rightarrow 2AgI(s) + Sn^{2+}$

(i) $5HNO_2 + 2MnO_4^- + H^+ \rightarrow 5NO_3^- + 2Mn^{2+} + 3H_2O$

18-8. (a) Oxidizing agent Fe^{3+}; $Fe^{3+} + e^- \rightleftharpoons Fe^{2+}$;
Reducing agent Sn^{2+}; $Sn^{2+} \rightleftharpoons Sn^{4+} + 2e^-$

(c) Oxidizing agent NO_3^-, $NO_3^- + 2H^+ + e^- \rightleftharpoons NO_2(g) + H_2O$; Reducing agent Cu; $Cu(s) \rightleftharpoons Cu^{2+} + 2e^-$

(e) Oxidizing agent $Fe(CN)_6^{3-}$; $Fe(CN)_6^{3-} + e^- \rightleftharpoons Fe(CN)_6^{4-}$; Reducing agent Ti^{3+};
$Ti^{3+} + H_2O \rightleftharpoons TiO^{2+} + 2H^+ + e^-$

(g) Oxidizing agent Sn^{4+}; $Sn^{4+} + 2e^- \rightleftharpoons Sn^{2+}$; Reducing agent Ag;
$Ag(s) + I^- \rightleftharpoons AgI(s) + e^-$

(i) Oxidizing agent MnO_4^-; $MnO_4^- + 8H^+ + 5e^- \rightleftharpoons Mn^{2+} + 4H_2O$
Reducing agent HNO_2; $HNO_2 + H_2O \rightleftharpoons NO_3^- + 3H^+ + 2e^-$

18-9. (a) $MnO_4^- + 5VO^{2+} + 11H_2O \rightarrow Mn^{2+} + 5V(OH)_4^+ + 2H^+$

(c) $Cr_2O_7^{2-} + 3U^{4+} + 2H^+ \rightarrow 2Cr^{3+} + 3UO_2^{2+} + H_2O$

(e) $IO_3^- + 5I^- + 6H^+ \rightarrow 3I_2 + 3H_2O$

(g) $HPO_3^{2-} + 2MnO_4^- + 3OH^- \rightarrow PO_4^{3-} + 2MnO_4^{2-} + 2H_2O$

(i) $V^{2+} + 2V(OH)_4^+ + 2H^+ \rightarrow 3VO^{2+} + 5H_2O$

18-11. (a) $AgBr(s) + e^- \rightleftharpoons Ag(s) + Br^-$ $V^{2+} \rightleftharpoons V^{3+} + e^-$
$Ti^{3+} + 2e^- \rightleftharpoons Ti^+$ $Fe(CN)_6^{4-} \rightleftharpoons Fe(CN)_6^{3-} + e^-$
$V^{3+} + e^- \rightleftharpoons V^{2+}$ $Zn \rightleftharpoons Zn^{2+} + 2e$
$Fe(CN)_6^{3-} + e^- \rightleftharpoons Fe(CN)_6^{4-}$ $Ag(s) + Br^- \rightleftharpoons AgBr(s) + e^-$
$S_2O_8^{2-} + 2e^- \rightleftharpoons 2SO_4^{2-}$ $Ti^+ \rightleftharpoons Ti^{3+} + 2e^-$

(b), (c)	E^0
$S_2O_8^{2-} + 2e^- \rightleftharpoons 2SO_4^{2-}$	2.01
$Ti^{3+} + 2e^- \rightleftharpoons Ti^+$	1.25
$Fe(CN)_6^{3-} + e^- \rightleftharpoons Fe(CN)_6^{4-}$	0.36
$AgBr(s) + e^- \rightleftharpoons Ag(s) + Br^-$	0.073
$V^{3+} + e^- \rightleftharpoons V^{2+}$	−0.256
$Zn^{2+} + 2e^- \rightleftharpoons Zn(s)$	−0.763

18-13. (a) 0.295 V (b) 0.193 V (c) −0.149 V
(d) 0.061 V (e) 0.002 V

18-16. (a) 0.75 V (b) 0.192 V (c) −0.385 V
(d) 0.278 V (e) 0.177 V (f) 0.86 V

18-18. (a) −0.281 V anode (b) −0.089 V anode
(c) 1.016 V cathode (d) 0.165 V cathode
(e) 0.012 V cathode

18-20. 0.390 V

18-22. −0.96 V

18-24. −1.25 V

18-25. 0.13 V

Chapter 19

19-1. The electrode potential of a system that contains two or more redox couples is the electrode potential of all half-cell processes at equilibrium in the system.

19-2. (a) *Equilibrium* is the state that a system assumes after each addition of reagent. *Equivalence* refers to a particular equilibrium state when a stoichiometric amount of titrant has been added.

19-4. Before the equivalence point, potential data are computed from the analyte standard potential and the analytical concentrations of the analyte and product. Post-equivalence-point data are based on the standard potential for the titrant and its analytical concentrations. The equivalence point potential is computed from the two standard potentials and the stoichiometric relation between the analyte and titrant.

19-6. An asymmetric titration curve will be encountered whenever the titrant and the analyte react in a ratio that is not 1:1.

19-8. (a) 0.420 V, oxidation on the left, reduction on the right.
(b) 0.019 V, oxidation on the left, reduction on the right.
(c) 0.416 V, oxidation on the left, reduction on the right.
(d) −0.393 V, reduction on the left, oxidation on the right.
(e) −0.204 V, reduction on the left, oxidation on the right.
(f) 0.726 V, oxidation on the left, reduction on the right.

19-9. (a) 0.615 V (c) −0.333 V

19-11. (a) 2.2×10^{17} (c) 3×10^{22}
(e) 9×10^{37} (g) 2.4×10^{10}

19-14. (a) phenosafranine
(c) indigo tetrasulfonate or methylene blue
(e) erioglaucin A (g) none

Chapter 20

20-1. (a) $2Mn^{2+} + 5S_2O_8^{2-} + 8H_2O \rightarrow 10SO_4^{2-} + 2MnO_4^- + 16H^+$
(c) $H_2O_2 + U^{4+} \rightarrow UO_2^{2+} + 2H^+$
(e) $2MnO_4^- + 5H_2O_2 + 6H^+ \rightarrow 5O_2 + 2Mn^{2+} + 8H_2O$

20-2. Only in the presence of Cl^- ion is Ag a sufficiently good reducing agent to be very useful for prereductions.

20-4. Standard solutions of reductants are susceptible to air oxidation.

20-6. Freshly prepared solutions of permanganate are inevitably contaminated with small amounts of solid manganese dioxide, which catalyzes the further decompositions of permanganate ion.

20-8. Solutions of $K_2Cr_2O_7$ are used extensively for back-titrating solutions of Fe^{2+} when the latter is being used as a standard reductant for the determination of oxidizing agents.

20-10. When a measured volume of a standard solution of KIO_3 is introduced into an acidic solution containing an excess of iodide ion, a known amount of iodine is produced as a result of:

$$IO_3^- + 5I^- + 6H^+ \rightarrow 3I_2 + 3H_2O$$

20-12. Starch decomposes in the presence of high concentrations of iodine to give products that do not behave satisfactorily as indicators. Delaying the addition of the starch until the iodine concentration is very small prevents this reaction.

20-13. (a) 0.1238 M (c) 0.02475 M (e) 0.03094 M

20-14. Dissolve 8.350 g $KBrO_3$ in water and dilute to 1.000 L.

20-16. 0.1147 M

20-18. 81.71%

20-20. 0.0266 M

20-22. 1.199%

20-24. 2.056%

20-26. 11.2 ppm

20-28. 0.0426 mg/mL sample

Chapter 21

21-1. (a) An *indicator electrode* is an electrode used in potentiometry that responds to variations in the activity of an analyte ion or molecule.
(c) An *electrode of the first kind* is a metal electrode that responds to the activity of its cation in solution.

21-2. (a) A *liquid-junction potential* is the potential that develops across the interface between two solutions having different electrolyte compositions.
(c) The *asymmetry potential* is a potential that develops across an ion-sensitive membrane when the concentrations of the ion are the same on either side of the membrane. This potential arises from dissimilarities between the inner and outer surface of the membrane.

21-3. (a) A titration is generally more accurate than measurements of electrode potential. Therefore, if ppt accuracy is needed, a titration should be picked.
(b) Electrode potentials are related to the activity of the analyte. Thus, pick potential measurements if activity is the desired quantity.

21-5. The potential arises from the difference in positions of dissociation equilibria on each of the two surfaces. These equilibria are described by

$$\underset{membrane}{H^+Gl^-} \rightleftharpoons \underset{solution}{H^+} + \underset{membrane}{Gl^-}$$

The surface exposed to the solution having the higher H^+ concentration becomes positive with respect to the other surface. This charge difference, or potential, serves as the analytical parameter when the pH of the solution on one side of the membrane is held constant.

21-7. Uncertainties include (1) the acid error in highly acidic solutions, (2) the alkaline error in strongly basic solutions, (3) the error that arises when the ionic strength of the calibration standards differs from that of the analyte solution, (4) uncertainties in the pH of the standard buffers, (5) nonreproducible junction potentials with solutions of low ionic strength, and (6) dehydration of the working surface.

21-9. The *alkaline error* arises when a glass electrode is employed to measure the pH of solutions having pH values in the 10 to 12 range or greater. In the presence of alkali ions, the glass surface becomes responsive to not only hydrogen ions but also alkali metal ions. Measured pH values are low as a result.

21-11. (b) The *boundary potential* for a membrane electrode is a potential that develops when the membrane separates two solutions that have different concentrations of a cation or an anion that the membrane binds selectively.

(d) The membrane in a solid-state F^- electrode is crystalline LaF_3, which when immersed in aqueous solution, dissociates according to the equation

$$LaF_3(s) \rightleftharpoons La^{3+} + 3F^-$$

A boundary potential develops across this membrane when it separates two solutions of F^- ion concentration.

21-12. The direct potentiometric measurement of pH provides a measure of the equilibrium activity of hydronium ions in the sample. A potentiometric titration provides information on the amount of reactive protons, both ionized and nonionized, in the sample.

21-15. (a) 0.354 V
(b) $SCE \| IO_3^- (x\,M), AgIO_3(sat'd)|Ag$
(c) $(E_{cell} - 0.110)/0.0592$
(d) 3.31

21-17. (a) $SCE \| I^- (x\,M), AgI (sat'd)|Ag$
(c) $SCE \| PO_4^{3-} (x\,M), Ag_3PO_4 (sat'd)|Ag$

21-19. (a) 3.36
(c) 2.43

21-20. 6.32

21-21. (a) 12.47, 3.42×10^{-13} M
(b) 5.47, 3.41×10^{-6} M
(c) For (a), pH should be 12.43 to 12.50 and a_{H^+} in the range of 3.17 to 3.70×10^{-13} M.
For (b), pH in the range 5.43 to 5.50, a_{H^+} in the range 3.16×10^{-6} to 3.69×10^{-6} M.

21-22. 173.7 g/mol
21-26. 3.5×10^{-4} M

Chapter 22

22-1. (a) In *Concentration polarization*, the current in an electrochemical cell is limited by the rate at which reactants are brought to or removed from the surface of one or both electrodes. In *Kinetic polarization*, the current is limited by the rate at which electrons are transferred between the electrode surfaces and the reactant in solution. For either type, the current is no longer linearly related to cell potential.

(c) *Diffusion* is the movement of species under the influence of a concentration gradient. *Migration* is the movement of an ion under the influence of an electrostatic attractive or repulsive force.

(e) The *electrolysis circuit* consists of a working electrode and a counter electrode. The *control circuit* regulates the applied potential such that the potential between the working electrode and a reference electrode in the control circuit is constant and at a desired level.

22-2. (a) The *Ohmic potential*, or IR drop, of a cell is the product of the current in the cell in amperes and the electrical resistance of the cell in ohms.

(c) In *controlled-potential electrolysis*, the potential applied to a cell is continuously adjusted to maintain a constant potential between the working electrode and a reference electrode.

(e) *Current efficiency* is a measure of agreement between the number of faradays of charge and the number of moles of reactant oxidized or reduced at a working electrode.

22-3. *Diffusion* arises from concentration differences between the electrode surface and the bulk of solution. *Migration* results from electrostatic attraction or repulsion. *Convection* results from stirring, vibration, or temperature differences.

22-5. Temperature, stirring, reactant concentrations, presence or absence of other electrolytes, and electrode surface area.

22-7. Gaseous product, particularly when the electrode is a soft metal such as mercury, zinc, or copper; low temperatures; and high current densities.

22-9. Potentiometric methods are carried out under zero current conditions, and the effect of the measurement on analyte concentration is typically undetectable. In contrast, electrogravimetric and coulometric methods depend on the presence of a net current and a net cell reaction. Two additional phenomena, *IR* drop and polarization, must be considered in electrogravimetric and coulometric methods where current is present. Lastly, the final measurement in electrogravimetric and coulometric methods is the mass of the product produced electrolytically, while in potentiometric methods, it is the cell potential.

22-11. The species produced at the counter electrode are potential interferences by reacting with the products at the working electrode.

22-13. (b) 5.5×10^{16}
22-14. (a) -0.732 V **(c)** -0.352 V
22-15. -0.788 V
22-17. (a) -0.673 V **(b)** -0.54 V **(c)** -1.71 V **(d)** -1.85 V
22-19. (a) 3.6×10^{-6} M **(b)** -0.425 V
(c) If the cathode is maintained between -0.425 V and -0.438 V, quantitative separation is possible in theory.

22-21. (a) 0.231 V **(b)** 7.6×10^{-21} M
(c) -0.12 to -0.398 V

22-22. (a) 0.237 V **(c)** 0.0513 V **(e)** 0.118 V
(g) 0.264 V **(i)** 0.0789 V

22-23. (a) 16.0 min **(b)** 5.34 min
22-25. 132.0 g/eq
22-27. 173 ppm
22-29. 3.56%
22-34. 50.9 μg
22-35. 2.73×10^{-4} g

Chapter 23

23-1. (a) *Voltammetry* is an analytical technique that is based on measuring the current that develops at a small electrode as the applied potential is varied. *Amperometry* is a technique in which the limiting current is measured at a constant potential.

(c) *Differential-pulse* and *square-wave voltammetry* differ in the type of pulse sequence used, as shown in Figure 23-1b and c and 23-27.

(e) In voltammetry, a *limiting current* is a current that is independent of applied potential and limited by the rate at which a reactant is brought to the surface of the electrode by migration, convection, and/or diffusion. A *diffusion current* is a limiting current when analyte transport is solely by diffusion.

(g) The *half-wave potential* is closely related to the *standard potential* for a reversible reaction. That is,

$$E_{1/2} = E_A^0 - \frac{0.0592}{n} \log\left(\frac{k_A}{k_B}\right)$$

where k_A and k_B are constants that are proportional to the diffusion coefficients of the analyte and product. When these are approximately the same, the half-wave potential and the standard potential are essentially equal.

23-3. A high supporting electrolyte concentration is used in most electroanalytical procedures to minimize the contribution of migration to concentration polarization. The supporting electrolyte also reduces the cell resistance, which decreases the *IR* drop.

23-5. Most organic electrode processes consume or produce hydrogen ions. Unless buffered solutions are used, marked pH changes can occur at the electrode surface as the reaction proceeds.

23-7. The purpose of the electrodeposition step in stripping analysis is to preconcentrate the analyte on the surface of the working electrode and to separate it from many interfering species.

23-9. A plot of E_{appl} versus $\log \dfrac{i}{i_l - i}$ should yield a straight line having a slope of $\dfrac{-0.0592}{n}$, and n is readily obtained from the slope.

23-12. $1.7 \times 10^{-3}\%$ Cu^{2+} removed.

23-13. 1.77×10^{-4} M

Chapter 24

24-1. The yellow color comes about because the solution absorbs blue light in the wavelength region 435 to 480 nm and transmits its complementary color (yellow). The purple color comes about because green radiation (500 to 560 nm) is absorbed and its complementary color (purple) is transmitted.

24-2. **(a)** Absorbance, A, is the negative logarithm of transmittance, T ($A = -\log T$).

24-3. Deviations from linearity can occur because of polychromatic radiation, unknown chemical changes, stray light, and molecular or ionic interactions at high concentration.

24-6. **(a)** 1.13×10^{18} Hz **(c)** 4.32×10^{14} Hz **(e)** 1.53×10^{13} Hz

24-7. **(a)** 253.0 cm **(c)** 286 cm

24-9. **(a)** 3.33×10^3 cm^{-1} to 667 cm^{-1}
(b) 1.00×10^{14} Hz to 2.00×10^{13} Hz

24-11. $\lambda = 1.36$ m; $E = 1.46 \times 10^{-25}$ J

24-12. **(a)** 436 nm

24-13. **(a)** ppm^{-1} cm^{-1} **(c)** $\%^{-1}$ cm^{-1}

24-14. **(a)** 92.1% **(c)** 41.8% **(e)** 32.7%

24-15. **(a)** 0.565 **(c)** 0.514 **(e)** 1.032

24-18. **(a)** $\%T = 67.3$, $a = 0.0211$ cm^{-1} ppm^{-1}, $c = 4.07 \times 10^{-5}$ M, $c_{\text{ppm}} = 8.13$ ppm
(c) $\%T = 30.2$, $a = 0.0397$ cm^{-1} ppm^{-1}, $c = 6.54 \times 10^{-5}$ M, $c_{\text{ppm}} = 13.1$ ppm
(e) $A = 0.638$, $\%T = 23.0$, $a = 0.0187$ cm^{-1} ppm^{-1}, $c_{\text{ppm}} = 342$ ppm
(g) $\%T = 15.9$, $\varepsilon = 3.17 \times 10^3$ L mol^{-1} cm^{-1}, $a = 0.0158$ cm^{-1} ppm^{-1}, $c = 1.68 \times 10^{-4}$ M
(i) $A = 1.28$, $a = 0.0489$ cm^{-1} ppm^{-1}, $b = 5.00$ cm, $c = 2.62 \times 10^{-5}$ M

24-21. **(a)** 0.238 **(b)** 0.476
(c) 0.578 and 0.334 **(d)** 0.539

24-23. **(a)** 0.528 **(b)** 29.6% **(c)** 2.27×10^{-5} M

24-25. $A' = 1.81$, error $= -13.6\%$

Chapter 25

25-1. **(a)** *Phototubes* consist of a single photoemissive surface (cathode) and an anode in an evacuated envelope. They exhibit low dark current but have no inherent amplification. *Solid-state photodiodes* are semiconductor *pn*-junction devices that respond to incident light by forming electron-hole pairs. They are more sensitive than phototubes but less sensitive than photomultiplier tubes.

(c) *Filters* isolate a single band of wavelengths and provide low-resolution wavelength selection suitable for quantitative work. *Monochromators* produce high resolution for qualitative and quantitative work. With monochromators, the wavelength can be varied continuously, whereas such manipulation is not possible with filters.

25-3. Quantitative analyses can tolerate rather wide slits since measurements are usually carried out at a wavelength maximum where the slope of the spectrum $dA/d\lambda$ is relatively constant. Qualitative analyses require narrow slits so that any fine structure in the spectrum will be resolved.

25-5. The iodine prolongs the life of the lamp and permits it to operate at a higher temperature. The iodine combines with gaseous tungsten that sublimes from the filament and causes the metal to be redeposited, thus adding to the life of the lamp.

25-6. **(a)** *Spectrophotometers* have monochromators for multiple wavelength operation and for procuring spectra while *photometers* utilize filters for fixed wavelength operation. Spectrophotometers are more complex and more expensive than photometers.

(c) Both a *monochromator* and a *polychromator* use a diffraction grating to disperse the spectrum, but a monochromator contains only one exit slit and detector, while a polychromator contains multiple exit slits and detectors. A monochromator can be used to monitor one wavelength at a time, while a polychromator can monitor several discrete wavelengths simultaneously.

25-7. **(a)** 0.73 μm (730 nm) **(c)** 1.45 μm (1450 nm)

25-9. **(a)** 1010 nm for 2870 K and 967 nm for 3000 K.
(b) 386 W/cm² for 2870 K and 461 W/cm² for 3000 K.

25-11. **(a)** The 0% transmittance is measured with no light reaching the detector and is a measure of the dark current.
(b) The 100% transmittance adjustment is made with a blank in the light path and compensates for any absorption or reflection losses in the cell and optics.

25-13. Fourier transform IR spectrometers have the advantages of higher speed and sensitivity, better light-gathering power, more accurate and precise wavelength settings, simpler mechanical design, and elimination of stray light and IR emission.

25-14. **(a)** $\%T = 23.84$, and A = 0.623
(c) $\%T = 5.7$

25-15. **(b)** $A = 0.509$ **(d)** $T = 0.096$

25-17. A *photon detector* produces a current or voltage as a result of the emission of electrons from a photosensitive surface when struck by photons. A *thermal detector* consists of a darkened surface to absorb infrared radiation and produce a temperature increase. The thermal detector produces an electrical signal whose magnitude is related to the temperature and thus the intensity of the infrared radiation.

25-19. An *absorption photometer* and a *fluorescence photometer* consist of the same components. The basic difference is in the location of the detector. The detector in a fluorometer is positioned at an angle of 90° to the direction of the beam from the source so that emission is detected rather than transmission. In addition, a filter is often positioned in front of the detector to remove radiation from the excitation beam that may result from scattering or other nonfluorescence processes. In a transmission photometer, the detector is positioned in a line with the source, the filter, and the detector.

25-21. **(a)** A *transducer* converts quantities, such as light intensity, pH, mass, and temperature, into electrical signals that can be subsequently amplified, manipulated, and finally converted into numbers proportional to the magnitude of the original quantity.

(c) An *n*-type semiconductor contains unbonded electrons (e.g., produced by doping silicon with a Group V element).

(e) A *depletion layer* results when a reverse bias is applied to a *pn*-junction-type device. Majority carriers are drawn away from the junction leaving a nonconductive depletion layer.

Chapter 26

26-1. (a) *Spectrophotometers* use a grating or a prism to provide narrow bands of radiation, while *photometers* use filters. Spectrophotometers have greater versatility and can obtain entire spectra. Photometers are simple, rugged, and low-cost and have higher light throughput.

(c) *Diode-array spectrophotometers* detect the entire spectral range simultaneously and can produce a spectrum in less than a second. *Conventional spectrophotometers* require several minutes to scan the spectrum.

26-3. Electrolyte concentration, pH, temperature, nature of solvent, and interfering substances.

26-5. $c_{min} = 1.1 \times 10^{-5}$ M; $c_{max} = 9.7 \times 10^{-5}$ M

26-7. $c_{min} = 1.2 \times 10^{-4}$ M; $c_{max} = 2.4 \times 10^{-3}$ M

26-9. (a) $A = 0.611$; $T = 0.245$ **(c)** $T = 0.060$

26-10. (b) $A = 0.503$ **(d)** $T = 0.099$

26-13. The absorbance should decrease approximately linearly with titrant volume until the end point. After the end point, the absorbance becomes independent of titrant volume.

26-16. 0.200 ppm Fe

26-18. 132 ppm Co and 248 ppm Ni

26-20. (a) $A = 0.492$ **(c)** $A = 0.190$

26-21. (a) $A = 0.301$ **(b)** $A = 0.413$ **(c)** $A = 0.491$

26-22. For A, pH = 5.60; for C, pH = 4.80

26-25. (a) $[P] = 2.08 \times 10^{-4}$ M; $[Q] = 4.90 \times 10^{-5}$ M

(c) $[P] = 8.36 \times 10^{-5}$ M; $[Q] = 6.10 \times 10^{-5}$ M

(e) $[P] = 2.11 \times 10^{-4}$ M; $[Q] = 9.64 \times 10^{-5}$ M

26-26. (b) $A = 0.03939c_{Fe} - 0.001008$

(c) $s_m = 1.1 \times 10^{-4}$, and $s_b = 2.7 \times 10^{-3}$

26-28. (a) 1:1 complex **(b)** 1.4×10^4 L mol^{-1} cm^{-1}

26-30. (a) 1:1 complex

(b) $\varepsilon = 1400 \pm 200$ L mol^{-1} cm^{-1}

(c) $K_f = 3.78 \times 10^5$

26-33. (1) 740 cm^{-1} C—Cl stretch; (2) 1270 cm^{-1} CH$_2$ wagging; (3) 2900 cm^{-1} Aliphatic C-H stretch.

Chapter 27

27-1. (a) *Fluorescence* is a photoluminescence process in which atoms or molecules are excited by absorption of electromagnetic radiation and then relax to the ground state, giving up their excess energy as photons.

(c) *Internal conversion* is the nonradiative relaxation of a molecule from a low energy vibrational level of an excited electronic state to a high energy vibrational level of a lower electronic state.

(e) The *Stokes shift* is the difference in wavelength between the radiation used to excite fluorescence and the wavelength of the emitted radiation.

(g) An *inner-filter effect* is a result of excessive absorption of the incident beam (primary absorption) or absorption of the emitted beam (secondary absorption).

27-3. (a) Fluorescein because of its greater structural rigidity due to the bridging —O— groups.

27-5. Organic compounds containing aromatic rings often exhibit fluorescence. Rigid molecules or multiple ring systems tend to have large quantum yields of fluorescence, while flexible molecules generally have lower quantum yields.

27-7. A filter fluorometer usually consists of a light source, a filter for selecting the excitation wavelength, a sample container, an emission filter, and a transducer/readout device. A spectrofluorometer has two monochromators that are the wavelength selectors.

27-9. Fluorometers are more sensitive because filters allow more excitation radiation to reach the sample and more emitted radiation to reach the transducer. In addition, fluorometers are substantially less expensive and more rugged than spectrofluorometer, making them particularly well suited for routine quantitation and remote analysis applications.

27-10. (b) $I_{rel} = 22.3c_{NADH} + 0.0004$.

(d) 0.510 μM NADH

(e) 0.016

27-12. 533 mg quinine

Chapter 28

28-1. In *atomic emission spectroscopy*, the radiation source is the sample itself. The energy for excitation of analyte atoms is supplied by a plasma, a flame, an oven, or an electric arc or spark. The signal is the measured intensity of the source at the wavelength of interest. In *atomic absorption spectroscopy*, the radiation source is usually a line source, such as a hollow cathode lamp, and the signal is the absorbance. The latter is calculated from the radiant power of the source and the resulting power after the radiation has passed through the atomized sample. In *atomic fluorescence spectroscopy*, an external radiation source is used, and the fluorescence emitted, usually at right angles to the source, is measured. The signal is the intensity of the fluorescence emitted.

28-2. (a) *Atomization* is a process in which a sample, often in solution, is volatilized and decomposed to form an atomic vapor.

(c) *Doppler broadening* is an increase in the width of the atomic lines caused by the Doppler effect in which atoms moving toward a detector absorb or emit wavelengths that are slightly shorter than those absorbed or emitted by atoms moving at right angles to the detector. The reverse effect is observed for atoms moving away from the detector.

(e) A *plasma* is a conducting gas that contains a large concentration of ions and/or electrons.

(g) A *hollow-cathode lamp* consists of a tungsten wire anode and a cylindrical cathode sealed in a glass tube that contains argon at a pressure of 1 to 5 torr. The cathode is constructed from or supports the element whose emission is desired.

(i) An *additive interference*, also called a blank interference, produces an effect that is independent of the analyte concentration. It could be eliminated with a perfect blank solution.

(k) A *chemical interference* is encountered when a species interacts with the analyte in such a way as to alter the spectral emission or absorption characteristics of the analyte.

(m) A *protective agent* prevents interference by forming a stable, but volatile, compound with the analyte.

28-3. In atomic emission spectroscopy, the analytical signal is produced by the relatively small number of *excited* atoms or ions, whereas in atomic absorption the signal results from absorption by the much larger number of *unexcited* species. Any small change in flame conditions dramatically influences the number of *excited species*, whereas such changes have a much smaller effect on the number of *unexcited species*.

28-5. The source radiation is modulated to create an ac signal at the detector. The detector is made to reject the dc signal from the flame and measure the modulated signal from the source. In this way, background emission from the flame and atomic emission from the analyte is discriminated against and prevented from causing an interference effect.

28-7. The temperature and pressure in a hollow-cathode lamp are much less than those in an ordinary flame. As a result, Doppler and collisional broadening effects are much less, and narrower lines result.

28-9. The temperatures are high, sample residence times are long, and the atoms and ions are formed in a nearly chemically inert environment. The high and relatively constant electron concentration leads to fewer ionization interferences.

28-11. The radial geometry provides better stability and precision, while the axial geometry can achieve lower detection limits.

28-13. 0.504 ppm Pb

28-15. **(b)** $A_s = \dfrac{\varepsilon b V_s c_s}{V_t} + \dfrac{\varepsilon b V_x c_x}{V_t} = kV_s c_s + kV_x c_x$

(c) $m = kc_s$; $b = kV_x c_x$ **(e)** $m = 0.00881$; $b = 0.202$

(g) $28.0 \, (\pm 0.2)$ ppm Cu

Chapter 29

29-1. **(a)** The *Dalton* is one unified atomic mass unit and equal to $1/12$ the mass of a neutral $^{12}_{6}C$ atom.

(c) The *mass number* is the atomic or molecular mass expressed without units.

(e) In a *time-of-flight* analyzer, ions with nearly the same kinetic energy traverse a field-free region. The time required for an ion to reach a detector at the end of the field-free region is inversely proportional to the mass of the ion.

29-3. The ICP torch serves both as an atomizer and ionizer.

29-5. Interferences are spectroscopic and matrix interferences. In a spectroscopic interference, the interfering species has the same mass-to-charge ratio as the analyte. Matrix effects occur at high concentrations where interfering species can interact chemically or physically to change the analyte signal.

29-7. The higher resolution of the double-focusing spectrometer allows the ions of interest to be better separated from background ions than with a relative low-resolution quadrupole spectrometer. The higher signal-to-background ratio of the double-focusing instrument leads to lower detection limits than with the quadrupole instrument.

29-9. The high energy of the beam of electrons used in EI sources is enough to break some chemical bonds and produce fragment ions.

29-11. The ion selected by the first analyzer is called the precursor ion. It then undergoes thermal decomposition, reaction with a collision gas, or photodecomposition to form product ions that are analyzed by a second mass analyzer.

Chapter 30

30-1. **(a)** The *order of a reaction* is the numerical sum of the exponents of the concentration terms in the rate law for the reaction.

(c) *Enzymes* are high molecular mass organic molecules that catalyze reactions of biological importance.

(e) The *Michaelis constant*, K_m, is an equilibrium-like constant for the dissociation of the enzyme-substrate complex. It is defined by the equation $K_m = (k_{-1} + k_2)/k_1$, where k_1 and k_{-1} are the rate constants for the forward and reverse reactions in the formation of the enzyme-substrate complex. The term k_2 is the rate constant for the dissociation of the complex to give products.

(g) *Integral methods* use integrated forms of the rate equations to calculate concentrations from kinetic data.

30-3. Advantages include the following: (1) measurements are made relatively early in the reaction before side reactions can occur; (2) measurements do not depend on the determination of absolute concentration but rather depend upon differences in concentration; and (3) selectivity is often enhanced in reaction-rate methods, particularly in enzyme-based methods. Limitations include (1) lower sensitivity, (2) greater dependence on conditions, and (3) lower precision.

30-5. $t_{1/2} = \ln 2/k = 0.693/k$

30-6. **(a)** 2.01 s **(c)** 2.093×10^3 s **(e)** 1.2×10^9 s

30-7. **(a)** $28.8 \, s^{-1}$ **(c)** $0.288 \, s^{-1}$ **(e)** $1.07 \times 10^4 \, s^{-1}$

30-8. **(a)** 0.152 **(c)** 3.3 **(e)** 10

30-10. **(a)** 0.2% **(c)** 0.02% **(e)** 1.0%

(g) 0.05% **(i)** 6.7% **(k)** 0.64%

30-12. **(a)** Plot $1/$Rate versus $1/[S]$ for known $[S]$ to give a linear calibration curve. Measure rate for unknown $[S]$, calculate $1/$Rate and $1/[S]_{unknown}$ from the working curve and find $[S]_{unknown}$.

(b) The intercept of the calibration curve is $1/v_{max}$, and the slope is K_m/v_{max}. Use the slope and intercept to calculate $K_m =$ slope/intercept and $v_{max} = 1/$intercept.

30-13. 5.5 ± 0.2 ppm

30-15. 0.045 M

30-17. **(a)** $\approx 2\%$ completion.

(b) a little over 9% completion.

Chapter 31

31-1. A *collector ion* is an ion added to a solution that forms a precipitate with the reagent that carries the desired minor species out of solution.

31-3. Transport of material and a spatial redistribution of the components.

31-5. **(a)** *Elution* is a process in which species are washed through a chromatographic column by additions of fresh mobile phase.

(c) The *stationary phase* is a solid or liquid phase that is fixed in place. The mobile phase then passes over or through the stationary phase.

(e) The *retention time* is the time interval between injection onto a column and appearance at the detector.

(g) The *selectivity factor*, α, of a column toward two species is given by the equation $\alpha = K_B/K_A$, where K_B is the distribution constant for the more strongly retained species B and K_A is the constant for the less strongly held or more rapidly eluting species A.

31-7. Large particle diameters for stationary phases; large column diameters; high temperatures (important only in gas chromatography); for liquid stationary phases, thick layers of the immobilized liquid; and very rapid or very slow flow rates.

31-9. Determine the retention time, t_R, for a solute and the width of the peak at its base, W. Calculate the number of plates, N, from $N = 16(t_R/W)^2$.

31-11. **(a)** 0.0246 M **(b)** 9.62×10^{-3} M

(c) 3.35×10^{-3} M **(d)** 1.23×10^{-3} M

31-13. **(a)** 75 mL **(b)** 50 mL **(c)** 24 mL

31-15. **(a)** $K = 18.0$ **(b)** $K = 7.56$

31-16. **(a)** $K = 91.9$

31-17. **(a)** $K = 1.53$

(b) $[HA]_{aq} = 0.0147$ M; $[A^-] = 0.0378$ M

(c) $K_a = 9.7 \times 10^{-2}$

31-19. **(a)** 12.36 mmol cation/L **(b)** 619 mg $CaCO_3$/L

31-21. 0.02056 M in HCl and 0.0424 M in $MgCl_2$

31-23. 9.5 cm/s

31-25. **(a)** $k_A = 0.74$; $k_B = 3.3$; $k_C = 3.5$; $k_D = 6.0$

(b) $K_A = 6.2$; $K_B = 27$; $K_C = 30$; $K_D = 50$

31-30. **(a)** $N = 6400$ **(b)** $L = 94$ cm **(c)** $t_R = 26$ min

31-32. **(a)** $k_M = 2.55$; $k_N = 2.62$ **(b)** $\alpha = 1.03$

(c) 9.03×10^4 **(d)** 135 cm **(e)** $(t_R)_N = 73$ min

Chapter 32

32-1. In *gas-liquid chromatography*, the stationary phase is a liquid that is immobilized on a solid. Retention of sample constituents involves equilibria between a gaseous and a liquid phase. In *gas-solid chromatography*, the stationary phase is a solid surface that retains analytes by physical adsorption. Separation involves adsorption equilibria.

32-3. Gas-solid chromatography is used primarily for separating low molecular mass gaseous species, such as carbon dioxide, carbon monoxide, and oxides of nitrogen.

32-5. A chromatogram is a plot of detector response versus time. The peak position, retention time, can reveal the identity of the compound eluting. The peak area is related to the concentration of the compound.

32-7. In *open tubular or capillary columns,* the stationary phase is held on the inner surface of a capillary, whereas in *packed columns,* the stationary phase is supported on particles that are contained in a glass or metal tube. Open tubular columns contain an enormous number of plates that permit rapid separations of closely related species. They suffer from small sample capacities.

32-9. Sample injection volume, carrier gas flow rate, and column condition are among the parameters that must be controlled for highest precision quantitative GC. The use of an internal standard can minimize the impact of variations in these parameters.

32-11. **(a)** Advantages of thermal conductivity: general applicability, large linear range, simplicity, and nondestructive. Disadvantage: low sensitivity.
(b) Advantages of flame ionization: high sensitivity, large linear range, low noise, ruggedness, ease of use, and response that is largely independent of flow rate. Disadvantage: destructive.
(c) Advantages of electron capture: high sensitivity selectivity toward halogen-containing compounds and several others and nondestructive. Disadvantage: small linear range.
(d) Advantages of thermionic detector: high sensitivity for compounds containing nitrogen and phosphorus and good linear range. Disadvantages: destructive and not applicable for many analytes.
(e) Advantages of photoionization: versatility, nondestructive, and large linear range. Disadvantages: not widely available, expensive.

32-13. Megabore columns are open tubular columns that have a greater inside diameter (530 μm) than typical open tubular columns (150 to 320 μm). Megabore columns can tolerate sample sizes similar to those for packed columns, although with significantly improved performance characteristics.

32-15. Liquid stationary phases are generally bonded and/or cross-linked in order to provide thermal stability and a more permanent stationary phase that will not leach off the column. Bonding involves attaching a monomolecular layer of the stationary phase to the packing surface by means of chemical bonds. Cross-linking involves treating the stationary phase while it is in the column with a chemical reagent that creates cross links between the molecules making up the stationary phase.

32-17. Fused silica columns have greater physical strength and flexibility than glass open tubular columns and are less reactive toward analytes than either glass or metal columns.

32-19. **(a)** Band broadening arises from very high or very low flow rates, large particles making up packing, thick layers of stationary phase, low temperature, and slow injection rates.
(b) Band separation is enhanced by maintaining conditions so that k lies in the range of 1 to 10, using small particles for packing, limiting the amount of stationary phase so that particle coatings are thin, and injecting the sample rapidly.

32-21. A = 21.1%, B = 13.1%, C = 36.4%, D = 18.8%, and E = 10.7%.

Chapter 33

33-1. **(a)** Substances that are somewhat volatile and are thermally stable.
(c) Substances that are ionic.
(e) High molecular mass compounds that are soluble in nonpolar solvents.
(g) Chiral compounds (enantiomers).

33-2. **(a)** In an *isocratic elution,* the solvent composition is held constant throughout the elution.
(c) In a *normal-phase packing,* the stationary phase is quite polar, and the mobile phase is relatively nonpolar.
(e) In a *bonded-phase packing,* the stationary phase liquid is held in place by chemically bonding it to the solid support.
(g) In *ion-pair chromatography,* a large organic counter-ion is added to the mobile phase as an ion-pairing reagent. Separation is achieved either through partitioning of the neutral ion pair or as a result of electrostatic interactions between the ions in solution and charges on the stationary phase resulting from adsorption of the organic counterion.
(i) *Gel filtration* is a type of size-exclusion chromatography in which the packings are hydrophilic and the eluents are aqueous. It is used for separating high molecular mass polar compounds.

33-3. **(a)** diethyl ether, benzene, *n*-hexane.

33-4. **(a)** ethyl acetate, dimethylamine, acetic acid.

33-5. In *adsorption chromatography,* separations are based on adsorption equilibria between the components of the sample and a solid surface. In *partition chromatography,* separations are based on distribution equilibria between two immiscible liquids.

33-7. *Gel filtration* is a type of size-exclusion chromatography in which the packings are hydrophilic and the eluents are aqueous. It is used for separating high molecular mass polar compounds. *Gel-permeation chromatography* is a type of size-exclusion chromatography in which the packings are hydrophobic and the eluents are nonaqueous. It is used for separating high molecular mass nonpolar species.

33-9. In an *isocratic elution,* the solvent composition is held constant throughout the elution. Isocratic elution works well for many types of samples and is simplest to implement. In a *gradient elution,* two or more solvents are employed, and the composition of the eluent is changed continuously or in steps as the separation proceeds. Gradient elution is best used for samples in which there are some compounds separated well and others with inordinately long retention times.

33-11. In *suppressor-column ion chromatography,* the chromatographic column is followed by a column whose purpose is to convert the ions used for elution to molecular species that are largely nonionic and thus do not interfere with conductometric detection of the analyte species. In *single-column ion chromatography,* low capacity ion exchangers are used so that the concentrations of ions in the eluting solution can be kept low.

33-13. Comparison of Table 33-1 with Table 32-1 suggests that the GC detectors that are suitable for HPLC are the mass spectrometer, FTIR, and possibly photoionization. Many of the GC detectors are unsuitable for HPLC because they require the eluting analyte components to be in the gas-phase.

33-15. **(a)** For a reversed phase chromatographic separation of a steroid mixture, selectivity and, as a consequence, separation could be influenced by temperature-dependent changes in distribution coefficients.
(b) For an adsorption chromatographic separation of a mixture of isomers, selectivity and, as a consequence, separation could be influenced by temperature-dependent changes in distribution coefficients.

Chapter 34

34-1. **(a)** Nonvolatile or thermally unstable species that contain no chromophoric groups.

(c) Inorganic anions and cations, amino acids, catecholamines, drugs, vitamins, carbohydrates, peptides, proteins, nucleic acids, nucleotides, and polynucleotides.

(e) Proteins, synthetic polymers, and colloidal particles.

34-2. **(a)** A *supercritical fluid* is a substance that is maintained above its critical temperature so that it cannot be condensed into a liquid no matter how great the pressure.

(c) *Two-dimensional thin-layer chromatography* is a method in which development is carried out with two solvents that are applied successively at right angles to one another.

(e) *Critical micelle concentration* is the level above which surfactant molecules begin to form spherical aggregates made up to 40 to 100 ions.

34-3. Density, viscosity, and the rates at which solutes diffuse.

34-5. **(a)** Instruments for supercritical fluid chromatography have provisions for controlling and measuring the column pressure.

(b) SFC instruments must be capable of operating at much higher mobile phase pressures than are typically encountered in GC.

34-7. Their ability to dissolve large nonvolatile molecules, such as large *n*-alkanes and polycyclic aromatic hydrocarbons.

34-9. **(a)** An increase in flow rate results in a decrease in retention time.

(b) An increase in pressure results in a decrease in retention time.

(c) An increase in temperature results in a decrease in density of supercritical fluids and thus an increase in retention time.

34-11. *Electroosmotic flow* is the migration of the solvent toward the cathode in an electrophoretic separation. This flow is due to the electrical double layer that develops at the silica/solution interface. At pH values higher than 3 the inside wall of the silica capillary becomes negatively charged leading to a build-up of buffer cations in the electrical double layer adjacent to the wall. The cations in this double layer are attracted to the cathode, and since they are solvated, they drag the bulk solvent along with them.

34-13. Under the influence of an electric field, mobile ions in solution are attracted or repelled by the negative potential of one of the electrodes. The rate of movement toward or away from a negative electrode is dependent on the net charge on the analyte and the size and shape of analyte molecules. These properties vary from species to species. Hence, the rate at which molecules migrate under the influence of the electric field vary, and the time it takes them to traverse the capillary varies, making separations possible.

34-15. 2.5 min

34-17. Higher column efficiencies and the ease with which pseudostationary phase can be altered.

34-19. B^+ followed by A^{2+} followed by C^{3+}.

Index

Bold page references indicate online content, t page references indicate tabular entries, s page references indicate spreadsheet exercises, CP page references indicate color plate entries, and A page references indicate appendix entries.

INTERNATIONAL ATOMIC MASSES

Element	Symbol	Atomic Number	Atomic Mass	Element	Symbol	Atomic Number	Atomic Mass
Actinium	Ac	89	(227)	Mendelevium	Md	101	(258)
Aluminum	Al	13	26.9815386	Mercury	Hg	80	200.59
Americium	Am	95	(243)	Molybdenum	Mo	42	95.96
Antimony	Sb	51	121.760	Neodymium	Nd	60	144.242
Argon	Ar	18	39.948	Neon	Ne	10	20.1797
Arsenic	As	33	74.92160	Neptunium	Np	93	(237)
Astatine	At	85	(210)	Nickel	Ni	28	58.6934
Barium	Ba	56	137.327	Niobium	Nb	41	92.90638
Berkelium	Bk	97	(247)	Nitrogen	N	7	14.007
Beryllium	Be	4	9.012182	Nobelium	No	102	(259)
Bismuth	Bi	83	208.98040	Osmium	Os	76	190.23
Bohrium	Bh	107	(270)	Oxygen	O	8	15.999
Boron	B	5	10.81	Palladium	Pd	46	106.42
Bromine	Br	35	79.904	Phosphorus	P	15	30.973762
Cadmium	Cd	48	112.411	Platinum	Pt	78	195.084
Calcium	Ca	20	40.078	Plutonium	Pu	94	(244)
Californium	Cf	98	(251)	Polonium	Po	84	(209)
Carbon	C	6	12.011	Potassium	K	19	39.0983
Cerium	Ce	58	140.116	Praseodymium	Pr	59	140.90765
Cesium	Cs	55	132.90545	Promethium	Pm	61	(145)
Chlorine	Cl	17	35.45	Protactinium	Pa	91	231.03588
Chromium	Cr	24	51.9961	Radium	Ra	88	(226)
Cobalt	Co	27	58.933195	Radon	Rn	86	(222)
Copernicium	Cn	112	(285)	Rhenium	Re	75	186.207
Copper	Cu	29	63.546	Rhodium	Rh	45	102.90550
Curium	Cm	96	(247)	Roentgenium	Rg	111	(280)
Darmstadtium	Ds	110	(281)	Rubidium	Rb	37	85.4678
Dubnium	Db	105	(268)	Ruthenium	Ru	44	101.07
Dysprosium	Dy	66	162.500	Rutherfordium	Rf	104	(265)
Einsteinium	Es	99	(252)	Samarium	Sm	62	150.36
Erbium	Er	68	167.259	Scandium	Sc	21	44.955912
Europium	Eu	63	151.964	Seaborgium	Sg	106	(271)
Fermium	Fm	100	(257)	Selenium	Se	34	78.96
Flerovium	Fl	114	(289)	Silicon	Si	14	28.085
Fluorine	F	9	18.9984032	Silver	Ag	47	107.8682
Francium	Fr	87	(223)	Sodium	Na	11	22.98976928
Gadolinium	Gd	64	157.25	Strontium	Sr	38	87.62
Gallium	Ga	31	69.723	Sulfur	S	16	32.06
Germanium	Ge	32	72.63	Tantalum	Ta	73	180.94788
Gold	Au	79	196.966569	Technetium	Tc	43	(98)
Hafnium	Hf	72	178.49	Tellurium	Te	52	127.60
Hassium	Hs	108	(277)	Terbium	Tb	65	158.92535
Helium	He	2	4.002602	Thallium	Tl	81	204.38
Holmium	Ho	67	164.93032	Thorium	Th	90	232.03806
Hydrogen	H	1	1.008	Thulium	Tm	69	168.93421
Indium	In	49	114.818	Tin	Sn	50	118.710
Iodine	I	53	126.90447	Titanium	Ti	22	47.867
Iridium	Ir	77	192.217	Tungsten	W	74	183.84
Iron	Fe	26	55.845	Ununoctium	Uuo	118	(294)
Krypton	Kr	36	83.798	Ununpentium	Uup	115	(288)
Lanthanum	La	57	138.90547	Ununseptium	Uus	117	(294)
Lawrencium	Lr	103	(262)	Ununtrium	Uut	113	(284)
Lead	Pb	82	207.2	Uranium	U	92	238.02891
Lithium	Li	3	6.94	Vanadium	V	23	50.9415
Livermorium	Lv	116	(293)	Xenon	Xe	54	131.293
Lutetium	Lu	71	174.9668	Ytterbium	Yb	70	173.054
Magnesium	Mg	12	24.3050	Yttrium	Y	39	88.90585
Manganese	Mn	25	54.938045	Zinc	Zn	30	65.38
Meitnerium	Mt	109	(276)	Zirconium	Zr	40	91.224

The values given in parentheses are the atomic mass numbers of the isotopes of the longest known half-life. From M. E. Wieser and T. B. Coplen, *Pure Appl. Chem.*, **2011**, *83*(2), 359–96, **DOI:** 10.1351/PAC-REP-10-09-14.

MOLAR MASSES OF SOME COMPOUNDS

Compound	Molar Mass	Compound	Molar Mass
$AgBr$	187.772	$K_3Fe(CN)_6$	329.248
$AgCl$	143.32	$K_4Fe(CN)_6$	368.346
Ag_2CrO_4	331.729	$KHC_8H_4O_4$ (phthalate)	204.222
AgI	234.7727	$KH(IO_3)_2$	389.909
$AgNO_3$	169.872	K_2HPO_4	174.174
$AgSCN$	165.95	KH_2PO_4	136.084
Al_2O_3	101.960	$KHSO_4$	136.16
$Al_2(SO_4)_3$	342.13	KI	166.0028
As_2O_3	197.840	KIO_3	214.000
B_2O_3	69.62	KIO_4	229.999
$BaCO_3$	197.335	$KMnO_4$	158.032
$BaCl_2 \cdot 2H_2O$	244.26	KNO_3	101.102
$BaCrO_4$	253.319	KOH	56.105
$Ba(IO_3)_2$	487.130	$KSCN$	97.18
$Ba(OH)_2$	171.341	K_2SO_4	174.25
$BaSO_4$	233.38	$La(IO_3)_3$	663.610
Bi_2O_3	465.958	$Mg(C_9H_6NO)_2$	312.611
CO_2	44.009	(8-hydroxyquinolate)	
$CaCO_3$	100.086	$MgCO_3$	84.313
CaC_2O_4	128.096	$MgNH_4PO_4$	137.314
CaF_2	78.075	MgO	40.304
CaO	56.077	$Mg_2P_2O_7$	222.551
$CaSO_4$	136.13	$MgSO_4$	120.36
$Ce(HSO_4)_4$	528.37	MnO_2	86.936
CeO_2	172.114	Mn_2O_3	157.873
$Ce(SO_4)_2$	332.23	Mn_3O_4	228.810
$(NH_4)_2Ce(NO_3)_6$	548.22	$Na_2B_4O_7 \cdot 10H_2O$	381.36
$(NH_4)_4Ce(SO_4)_4 \cdot 2H_2O$	632.53	$NaBr$	102.894
Cr_2O_3	151.989	$NaC_2H_3O_2$	82.034
CuO	79.545	$Na_2C_2O_4$	133.998
Cu_2O	143.091	$NaCl$	58.44
$CuSO_4$	159.60	$NaCN$	49.008
$Fe(NH_4)_2(SO_4)_2 \cdot 6H_2O$	392.13	Na_2CO_3	105.988
FeO	71.844	$NaHCO_3$	84.006
Fe_2O_3	159.687	$Na_2H_2EDTA \cdot 2H_2O$	372.238
Fe_3O_4	231.531	Na_2O_2	77.978
HBr	80.912	$NaOH$	39.997
$HC_2H_3O_2$ (acetic acid)	60.052	$NaSCN$	81.07
$HC_7H_5O_2$ (benzoic acid)	122.123	Na_2SO_4	142.04
$(HOCH_2)_3CNH_2$ (TRIS)	121.135	$Na_2S_2O_3 \cdot 5H_2O$	248.17
HCl	36.46	NH_4Cl	53.49
$HClO_4$	100.45	$(NH_4)_2C_2O_4 \cdot H_2O$	142.111
$H_2C_2O_4 \cdot 2H_2O$	126.064	NH_4NO_3	80.043
H_5IO_6	227.938	$(NH_4)_2SO_4$	132.13
HNO_3	63.012	$(NH_4)_2S_2O_8$	228.19
H_2O	18.015	NH_4VO_3	116.978
H_2O_2	34.014	$Ni(C_4H_7O_2N_2)_2$	288.917
H_3PO_4	97.994	(dimethylglyoximate)	
H_2S	34.08	$PbCrO_4$	323.2
H_2SO_3	82.07	PbO	223.2
H_2SO_4	98.07	PbO_2	239.2
HgO	216.59	$PbSO_4$	303.3
Hg_2Cl_2	472.08	P_2O_5	141.943
$HgCl_2$	271.49	Sb_2S_3	339.70
KBr	119.002	SiO_2	60.083
$KBrO_3$	166.999	$SnCl_2$	189.61
KCl	74.55	SnO_2	150.71
$KClO_3$	122.55	SO_2	64.06
KCN	65.116	SO_3	80.06
K_2CrO_4	194.189	$Zn_2P_2O_7$	304.70
$K_2Cr_2O_7$	294.182		